Geochemistry

Geochemistry

SECOND EDITION

William M. White

Cornell University
Ithaca, NY, US

WILEY Blackwell

Registered Office(s)
John Wiley & Sons, Inc., 111 River Street, Hoboken, NJ 07030, USA
John Wiley & Sons Ltd, The Atrium, Southern Gate, Chichester, West Sussex, PO19 8SQ, UK

Editorial Office
9600 Garsington Road, Oxford, OX4 2DQ, UK

For details of our global editorial offices, customer services, and more information about Wiley products visit us at www.wiley.com.

Wiley also publishes its books in a variety of electronic formats and by print-on-demand. Some content that appears in standard print versions of this book may not be available in other formats.

Library of Congress Cataloging-in-Publication data applied for
ISBN: 9781119438052 [paperback]

Cover Design: Wiley
Cover Image: © Khritthithat Weerasirirut/Shutterstock

Set in 11/12pt SabonLTStd by SPi Global, Chennai, India
Printed and bound by CPI Group (UK) Ltd, Croydon, CR0 4YY

C9781119438052_310523

Contents

Preface

This book is intended to serve both as a textbook and as a professional reference. It is based on a course that I have taught for over 30 years at Cornell University to advanced undergraduates and graduate students. I was, however, never able to cover the entire contents in a single semester. That is even less possible with this expanded second edition. Rather, it is designed to allow the instructor to pick and choose chapters, and parts of chapters, to focus on particular aspects of geochemistry.

The second edition includes three new chapters that cover the critical zone, the oceans, and economic and environmental geochemistry. Additional material in Chapter 4 covers the thermodynamics of aqueous fluids at high temperatures, and there is expanded coverage of stable isotope clumping, mass independent fractionations, and nontraditional isotopes in Chapter 9. There have also been extensive revisions throughout to bring the book up to date with new developments over the last seven years. Not surprisingly, thermodynamics and kinetics have changed little, although I have added additional material to these chapters. There have been significant advances in isotope geochemistry, cosmochemistry, solid earth geochemistry, and organic geochemistry, and these chapters have been significantly revised. The new edition includes examples using MATLAB, THERMOCALC, and PHREEQC. Chapter 1 now includes an expanded overview of geochemistry to hopefully spark the interest of students at the beginning of the course. Each chapter now includes a summary of the most important points covered.

The ultimate basis of this book was my notes from geochemistry courses taught by my dissertation advisor, Jean-Guy Schilling. When I first began teaching geochemistry at Cornell, I pulled out those old notes as the starting point. I am grateful for the strong grounding he gave me in geochemistry, particularly in thermodynamic fundamentals. In the first few years, what was to become this book were merely outlined notes handed out to students, which slowly evolved into book form. Once the internet became available, I posted them online, mainly to save myself the trouble of copying and handing them out. Others began to access them as well. I am very grateful to my students over the years for pointing out errors and asking questions that I could not answer, which in turn forced me to research them and expand the book. I am also grateful to my colleagues at Cornell and around the world who have pointed out errors and critiqued the content. This expanded version has been produced in response to their comments and reviews.

One sad lesson I have learned about writing books is that you can never correct all the errors, no matter how many times you proofread. I apologize in advance for errors still present in the book.

July, 2019
Ithaca, New York

About the companion website

Don't forget to visit the companion website for this book:

www.wiley.com/go/white/geochemistry

There you will find valuable material designed to enhance your learning, including:

- Learning outcomes for all chapters
- Color version of figures
- Exercises for all chapters
- References for all chapters
- Further reading for all chapters

Scan this QR code to visit the companion website.

Chapter 1

Introduction

1.1 INTRODUCTION

Geochemistry is the chemistry of the natural environment and it encompasses the Earth as well as other Solar System bodies, what Antoine Lavoisier described as "the grand laboratory of nature." As the etymology of the word implies, the field of *geochemistry* is somehow a marriage of the fields of *chemistry* and *geology*, or more broadly, *earth science*. But just how are chemistry and geology combined within geochemistry; what is the relationship between them? Perhaps the best explanation would be to state that *in geochemistry, we use the tools of chemistry to understand the Earth and how it works.* The Earth is part of a closely related family of heavenly bodies, our Solar System, that formed more or less simultaneously. To fully understand the Earth and how it became habitable, we need to understand that formation. Hence, the realm of geochemistry extends beyond the Earth to encompass the entire Solar System.

1.2 BEGINNINGS

The term *geochemistry* was first used by the German-Swiss chemist Christian Friedrich Schönbein* in 1838 who wrote, "In a word, a comparative geochemistry ought to be launched, before geognosy can become

geology, and before the mystery of the genesis of our planets and their inorganic matter may be revealed." The goal of geochemistry does indeed address those mysteries, but with one addendum: it addresses organic matter as well because life and its organic products are an integral part of our planet and its evolution. The roots of geochemistry are, however, much older and intimately intertwined with the roots of chemistry, both of which were nurtured in the fertile soil of alchemy and metallurgy (see Morris, 2003). Indeed, Georg Agricola's 1556 treatise on mining and metallurgy, *De re metallica*, might be considered the first text on geochemistry. A principal goal of alchemy was to transform one substance derived from the Earth into another, such as gold. Even as chemists came to recognize that there were certain fundamental substances, elements, that could not be so transformed and sought to identify them, it was in natural substances that they sought them. However, credit for establishing the modern field of geochemistry is generally given to three individuals: Frank Wigglesworth Clarke (1847–1931), Victor Goldschmidt (1888–1947), and Vladimir Vernadsky (1863–1945). These individuals set the stage for what geochemistry was to become and are collectively known as the fathers of geochemistry.

---

* Christian Friedrich Schönbein (1799–1868) was born in Metzingen in Swabia, Germany and served as professor at the University of Basel from 1835 until 1868. He is best known for his discovery of ozone.

Geochemistry, Second Edition. William M. White.
© 2020 John Wiley & Sons Ltd. Published 2020 by John Wiley & Sons Ltd.
Companion website: www.wiley.com/go/white/geochemistry

Frank Clarke received his B.S. in chemistry from Harvard University in 1868 and moved to the newly established Cornell University as an assistant in chemistry in 1869. Though trained as a chemist, his intense interest in geology led him to explore the many gorges and waterfalls in the region and write a guide book about them. Clarke spent most of his career as chief chemist of the US Geological Survey and made a number of remarkable contributions to both chemistry and geochemistry, among other things serving as president of the American Chemical Society in 1901. His classic work *The Data of Geochemistry* (Clarke, 1906), of which there were several editions, summarized current knowledge of the composition of the Earth and its various components. In addition, he published some 136 papers on topics ranging from the composition of stars, the heat of formation of water from hydrogen and hydroxyl ions, chemical denudation, systematic studies of the chemistry of mineral classes such as micas, silicates, a review of atomic theory and papers on the composition of US rivers and lakes, the composition of igneous rocks and the Earth's crust.

Victor Goldschmidt was born to Jewish parents in Zurich in 1888 and moved to Oslo in 1901 where he received his PhD in mineralogy at the University of Christiania (later Oslo) in 1911. Among his diverse research efforts, he used newly invented X-ray spectrograph to show that the even atomic-numbered rare earth elements (REE) have higher abundances than the odd-numbered ones, a key to nuclear structure. Using the newly invented X-ray diffractometer, he determined the ionic radii of many elements, among other things discovering that the REE atomic radii decreased with atomic number, which he called the lanthanide contraction, which turned out to be a key to atomic electronic structure. He described how the relationships between ionic radius, ionic charge, atomic number, and periodic group, governed substitution of elements in crystal lattices (Goldschmidt, 1937; this will be discussed in Chapter 7). His papers on the formation and differentiation of the Earth foreshadowed modern thinking; he recognized that the cosmic abundances of the elements could best be deduced from meteorites. He used glacial clays to estimate the composition of the continental crust, an estimate that compared well with that of Clarke which was based on a far larger data set. In 1929 Goldschmidt accepted a professorship at the University of Göttingen where, among other things, he published papers on the carbon cycle, pointing out that burning of fossil fuels was increasing the CO_2 content of the atmosphere. In 1935, Goldschmidt resigned his professorship in protest of the Nazi anti-Jewish agitation and returned to Oslo, where his work on the abundance of elements led directly to the concept of "magic numbers" of nuclei and the shell theory of nuclear structure (which we will discuss in Chapter 8). He was arrested and sent to a concentration camp by the Nazi occupiers in 1942 but was released soon afterward. He then escaped to Sweden and went on to England in 1943, where he worked with officials involved with the war effort.

Vladimir Vernadsky was born in St. Petersburg to Ukrainian parents. He received his PhD in mineralogy from Moscow University in 1897. Vernadsky was one the first to recognize the significance of radioactive decay as a terrestrial energy source, establishing a Radium Commission in 1909. Vernadsky was active in prerevolution Russian politics and served as a member of the Constitutional Democratic Party in parliament, but he left for Kiev after the Russian Revolution of October 1917 when the Bolsheviks rose to power. There he began the work for which he is best known: biogeochemistry. He argued that life was not independent of its surroundings and was itself a geologic force, both affecting and being affected by the inert world, and introduced the concept of the biosphere (Verndasky, 1926). His viewpoint is very much substantiated by what we have learned since he first espoused it and one critical to our modern view of the Earth. After several years in Paris working with Marie Curie, he returned to Russia to organize the Biogeochemical Laboratory, first in St. Petersburg and later in Moscow. There he led studies of permafrost, mineral resources, and determining the composition of living organisms and the relationship between health and the abundances of elements such as iodine, calcium, and barium in the local environment. He continued his strong interest in radioactivity and isotopes, leading a program for concentration of heavy water and initiating construction of the first cyclotron in the Soviet Union. His efforts led directly to the establishment of a Uranium Commission and indirectly to the Soviet atomic bomb project.

1.3 GEOCHEMISTRY IN THE TWENTY-FIRST CENTURY

These individuals set the stage that allowed geochemistry to flourish in the quantitative approach that grew to dominate earth science in the second half of the twentieth century. This quantitative approach has produced greater advances in the understanding of our planet in the last 60 years than in all of prior human history. The contributions of geochemistry to this advance has been simply enormous. Much of what we know about how the Earth and the Solar System formed has come from research on the chemistry of meteorites. Through geochemistry, we have quantified the geologic time scale. Through geochemistry, we can determine the depths and temperatures of magma chambers. Through geochemistry, we know the temperatures and pressures at which the various metamorphic rock types form and we can use this information, for example, to determine the throw on ancient faults. Through geochemistry, we know how much and how fast mountain belts have risen. Through geochemistry, we are learning how fast they are eroding. Through geochemistry, we are learning how and when the Earth's crust formed. Through geochemistry, we are learning when the Earth's atmosphere formed and how it has evolved. Through geochemistry, we are learning how the mantle convects. Through geochemistry, we have learned how cold the Ice Ages were and what caused them. The evidence of the earliest life, at 3.8 gigayears (billion, or 10^9 years, which we will henceforth abbreviate as Ga*), is not fossilized remains, but chemical traces of life. And through geochemistry, we have learned that the Earth's atmosphere first became oxidizing about 2.3 to 2.4 Ga but another billion and a half years would pass before multi-celled animals would emerge.

Not surprisingly, instruments for chemical analysis have been key part of probes sent to other heavenly bodies, including Venus, Mars, Jupiter, and Titan. Geochemistry lies at the heart of environmental science and environmental concerns. Problems such as acid rain, ozone holes, the greenhouse effect and global warming, water and soil pollution are geochemical problems. Addressing these problems requires knowledge of geochemistry. Similarly, most of our nonrenewable resources, such as metal ores and petroleum, form through geochemical processes. Locating new sources of these resources increasing requires geochemical approaches. In summary, every aspect of earth science has been advanced through geochemistry.

Although we will rarely discuss it in this book, geochemistry, like much of science, is very much driven by technology. Technology has given modern geochemists tools that allow them to study the Earth in ways that pioneers of the field could not have dreamed possible. The electron microprobe allows us to analyze mineral grains on the scale of microns in minutes; the electron microscope allows us to view the same minerals on almost the atomic scale. Techniques such as X-ray diffraction, nuclear magnetic resonance, and Raman and infrared spectroscopy allow us to examine atomic ordering and bonding in natural materials. Mass spectrometers allow us to determine the age of rocks and the temperature of ancient seas. Ion probes allow us to do these things on micron scale samples. Analytical techniques such as X-ray fluorescence, inductively coupled plasma spectrometry, and laser ablation allow us to perform in minutes analyses that would take days using "classical" techniques. All this is done with ever-increasing precision and accuracy. Computers with gigahertz of power and terabytes of memory allow us to perform in seconds thermodynamic calculations that would have taken years or lifetimes half a century ago and the future promises even more computational power. This makes possible *ab initio* computation, that is, from the first principles governing atomic interactions, of, for example, mineral structures at the enormous pressures in the Earth's deep interior and chemical equilibrium everywhere, something not possible half a century ago

---

* We will be using System International Units as much as practicable throughout the book. A list of these units and their abbreviations can be found in the Appendix.

even though we knew those principles. New instruments and analytical techniques now being developed promise even greater sensitivity, speed, accuracy, and precision. Together, these advances will bring us ever closer to our goal of understanding the Earth and its cosmic environment.

Before we begin our study of geochemistry, we will review some "fundamentals." First, we briefly examine the philosophy and approach that is common to all science. Then we review the most fundamental aspects of chemistry: how matter is organized into atoms of different elements, how the properties of the elements vary, and how these atoms interact to form compounds. Finally, we review a few fundamental aspects of the Earth. Following that we will preview what will come in subsequent chapters.

1.4 THE PHILOSOPHY OF SCIENCE

This book will concentrate on communicating to you the body of knowledge we call geochemistry. Geochemistry is just part of a much larger field of human endeavor known as science. Science is certainly among humanity's greatest successes; without it, our current civilization would not be possible. Among other things, it would simply not be possible to feed, clothe, and shelter the 7 billion people living today. This phenomenal success is due in large part to the philosophy of science.

Science consists of two parts: the knowledge it encompasses and the approach or philosophy that achieves that knowledge. The goal of all science is to understand the world around us. The arts and humanities also seek understanding. Science differs from those fields as much by its approach and philosophy as by its body of knowledge.

This approach and philosophy unite the great diversity of fields that we collectively call science. When one compares the methods and tools of a high-energy physicist with those of a behavioral biologist, for example, it might at first seem that they have little in common. Among other things, their vocabularies are sufficiently different that each would have difficulty communicating his or her research

to the other. In spite of this, they share at least two things. The first is a criterion of "understanding." Both the physicist and the behavioral biologist attempt to explain their observations by the application of a set of rules, which, by comparison to the range of phenomena considered, are both few and simple. Both would agree that a phenomenon is understood if and only if the outcome of an experiment related to that phenomenon can be predicted beforehand by applying those rules to measured variables.* The physicist and biologist also share a common method of seeking understanding, often called the *scientific method*.

1.4.1 Building scientific understanding

Science deals in only two quantities: *observations* and *theories*. The most basic of these is the observation. Measurements, data, analyses, and experiments are all observations in the present sense. An observation might be as simple as a measurement of the dip and strike of a rock formation or as complex as the electromagnetic spectrum of a star. Of course, it is possible to measure both the dip of rock strata and a stellar spectrum incorrectly. Before an observation becomes part of the body of scientific knowledge, we would like some reassurance that it is right. How can we tell whether observations are right or not? The most important way to verify an observation is to replicate it independently. In the strictest sense, *independent* means by a separate observer, team of observers, or laboratory, and preferably by a different technique or instrument. It is not practicable to replicate every observation in this manner, but critical observations, those which appear to be inconsistent with existing theories or which test the predictions of newly established ones should be, and generally are, replicated. But even replication does not guarantee that an observation is correct.

Observations form the basis of *theories*. Theories are also called models, hypotheses, or laws. *Scientific understanding is achieved by constructing and modifying theories to explain observations*. Theories are merely the products of the imagination of scientists, so we

---

* Randomness can affect the outcome of any experiment (though the effect might be slight). By definition, the effect of this randomness cannot be predicted. Where the effects of randomness are large, one performs a large collection, or ensemble, of experiments and then considers the average result.

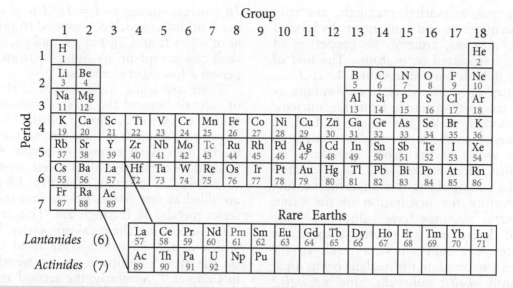

Figure 1.1 The periodic table showing symbols and atomic numbers of naturally occurring elements. Many older periodic tables number the groups as IA-VIIIA and IB-VIIB. This version shows the current International Union of Pure and Applied Chemistry (IUPAC) Convention.

isotopes: ^{3}He and ^{4}He. Both ^{3}He and ^{4}He* have two protons (and a matching number of electrons), but ^{4}He has two neutrons while ^{3}He has only one.

The *atomic weight* of an element depends on both the masses of its various isotopes and on the relative abundances of these isotopes. This bedeviled nineteenth century chemists. William Prout (1785–1850), an English chemist and physiologist, had noted in 1815 that the densities of a number of gases were integer multiples of the density of hydrogen (e.g., 14 for nitrogen, 16 for oxygen). This law appeared to extend to many elemental solids as well, and it seemed reasonable that this might be a universal law. But there were puzzling exceptions. Cl, for example, has an atomic weight of 35.45 times that of hydrogen. The mystery wasn't resolved until Thompson demonstrated the existence of two isotopes of Ne in 1918. The explanation is that while elements such as H, N, O, C, and Si consist almost entirely of a single isotope, and thus have atomic weights very close to the mass number of that isotope, natural Cl consists of about 75% ^{35}Cl and 25% ^{37}Cl†.

1.5.2 Electrons and orbits

We stated above that the atomic number of an element is its most important property. This is true because the number of electrons is determined by atomic number, and it is the electronic structure of an atom that largely dictates its chemical properties. The organization of the elements in the periodic table reflects this electronic structure.

The electronic structure of atoms, and indeed the entire organization of the periodic table, is determined by quantum mechanics and the quantization of energy, angular momentum, magnetic moment, and spin of electrons. Four quantum numbers, called

---

* By convention, the mass number, which is the sum of protons and neutrons in the nucleus, of an isotope is written as a preceding superscript. However, for historical reasons, it is often pronounced "helium–4." Note also that the atomic number or proton number can be readily deduced from the chemical symbol (atomic number of He is 2). The neutron number can be found by subtracting the proton number from the mass number. Thus, the symbol ^{4}He gives a complete description of the nucleus of this atom.

† The actual mass of an atom depends on the number of electrons and the nuclear binding energy as well as the number of protons and neutrons. However, the mass of the electron is over 1000 times less than the mass of the proton and neutron, which have nearly identical masses, and the effect of nuclear binding energy on mass was too small for nineteenth-century chemists to detect.

the principal, azimuthal, magnetic, and spin quantum numbers and conventionally labeled n, l, m, and ms, control the properties of electrons associated with atoms. The first of these, n, which may take values 1, 2, 3, ..., determines most of the electron's energy as well as its mean distance from the nucleus. The second, l, which has values 0, 1, 2, ... $n-1$, determines the total angular momentum and the shape of the orbit. The third, m, which may have values $-l$, ... 0 ... l, determines the z component of angular momentum and therefore the orientation of the orbit. The fourth, ms, may have values of $-\frac{1}{2}$ or $+\frac{1}{2}$ and determines the electron's spin. The first three quantum numbers result in the electrons surrounding the nucleus being organized into *shells*, *subshells*, and *orbitals*.* The *Pauli exclusion principle* requires that no two electrons in an atom may have identical values of all four quantum numbers. Because each orbital corresponds to a unique set of the first three quantum numbers and the spin quantum number has only two possible values, two electrons with opposite spins may occupy a given orbital. In Chapter 8 we will see that the properties of the nucleus are also dictated by quantum mechanics, and that the nucleus may also be thought of as having a shell structure.

Each shell corresponds to a different value of the principal quantum number. The periodic nature of chemical properties reflects the filling of successive shells as additional electrons (and protons) are added. Each shell corresponds to a 'period', or row, in the periodic table. The first shell (the K shell) has one subshell, the *1s*, consisting of a single orbital (with quantum numbers $n = 1$, $l = 0$, $m = 0$. The *1s* orbital accepts up to two electrons. Thus period 1 has two elements: H and He. If another proton and electron are added, the electron is added to the first orbital, *2s*, of the next shell (the L shell). Such a configuration has the chemical properties of lithium, the first element of period 2. The second shell has 2 subshells, *2s* (corresponding to $l = 0$) and

2p (corresponding to $l = 1$). The *p* subshell has 3 orbitals (which correspond to values for m of -1, $+1$, and 0), *px*, *py*, and *pz*, so the L shell can accept up to eight electrons. Thus, period 2 has eight elements.

There are some complexities in the filling of orbitals beyond the M shell, which corresponds to period 3. The *3d* subshell is vacant in period 3 element in their ground states, and in the first two elements of period 4. Only when the *4s* orbital is filled do electrons begin to fill the *3d* orbitals. The five *3d* orbitals are filled as one passes up the first transition series metals, Sc through Zn. This results in some interesting chemical properties, because which of the *3d* orbitals are filled depends on the atom's environment, as we shall see in Chapter 7. Similarly, the second and third transition series metals correspond to filling of the *4d* and *5d* orbitals. The *lanthanide* and *actinide rare earth elements* correspond to the filling of the *4f* and *5f* shells (again resulting in some interesting properties, which we will consider subsequently). The predicted sequence in which orbitals are filled and their energy levels are shown in Figure 1.2. Figure 1.3 shows the electronic configuration of the elements.

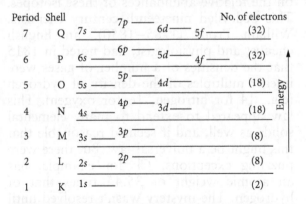

Figure 1.2 The predicted sequence of orbital energies for electrons in atoms. *S* levels can hold 2 electrons, *p*, d, and *f* can hold 6, 10, and 14 respectively.

---

* It is often convenient to think of the electrons orbiting the nucleus much as the planets orbit the Sun. This analogy has its limitations. The electron's position cannot be precisely specified as can a planet's. In quantum mechanics, the Schrödinger wave function, ψ (or more precisely, ψ^2) determines the probability of the electron being located in a given region about the atom. As an example of failure of the classical physical description of the atom, consider an electron in the *1s* orbital. Both quantum number specifying angular momentum, l and m, are equal to 0, and hence the electron has 0 angular momentum, and therefore cannot be in an orbit in the classical sense.

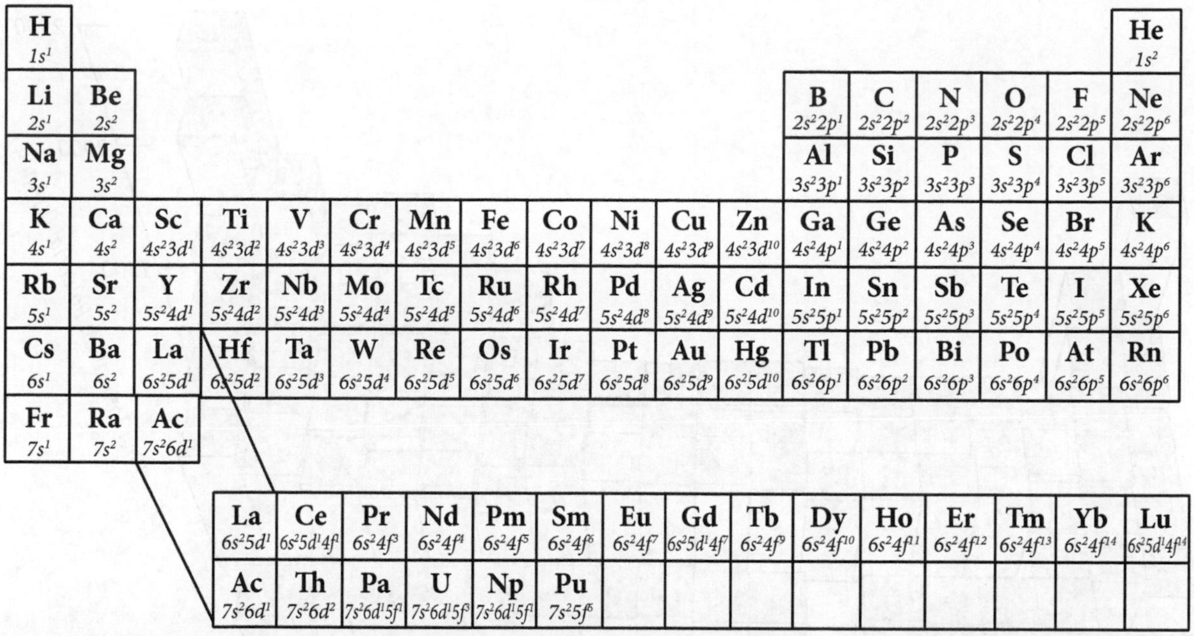

H $1s^1$																	He $1s^2$
Li $2s^1$	Be $2s^2$											B $2s^22p^1$	C $2s^22p^2$	N $2s^22p^3$	O $2s^22p^4$	F $2s^22p^5$	Ne $2s^22p^6$
Na $3s^1$	Mg $3s^2$											Al $3s^23p^1$	Si $3s^23p^2$	P $3s^23p^3$	S $3s^23p^4$	Cl $3s^23p^5$	Ar $3s^23p^6$
K $4s^1$	Ca $4s^2$	Sc $4s^23d^1$	Ti $4s^23d^2$	V $4s^23d^3$	Cr $4s^23d^4$	Mn $4s^23d^5$	Fe $4s^23d^6$	Co $4s^23d^7$	Ni $4s^23d^8$	Cu $4s^23d^9$	Zn $4s^23d^{10}$	Ga $4s^24p^1$	Ge $4s^24p^2$	As $4s^24p^3$	Se $4s^24p^4$	Br $4s^24p^5$	K $4s^24p^6$
Rb $5s^1$	Sr $5s^2$	Y $5s^24d^1$	Zr $5s^24d^2$	Nb $5s^24d^3$	Mo $5s^24d^4$	Tc $5s^24d^5$	Ru $5s^24d^6$	Rh $5s^24d^7$	Pd $5s^24d^8$	Ag $5s^24d^9$	Cd $5s^24d^{10}$	In $5s^25p^1$	Sn $5s^25p^2$	Sb $5s^25p^3$	Te $5s^25p^4$	I $5s^25p^5$	Xe $5s^25p^6$
Cs $6s^1$	Ba $6s^2$	La $6s^25d^1$	Hf $6s^25d^2$	Ta $6s^25d^3$	W $6s^25d^4$	Re $6s^25d^5$	Os $6s^25d^6$	Ir $6s^25d^7$	Pt $6s^25d^8$	Au $6s^25d^9$	Hg $6s^25d^{10}$	Tl $6s^26p^1$	Pb $6s^26p^2$	Bi $6s^26p^3$	Po $6s^26p^4$	At $6s^26p^5$	Rn $6s^26p^6$
Fr $7s^1$	Ra $7s^2$	Ac $7s^26d^1$															

La $6s^25d^1$	Ce $6s^25d^14f^1$	Pr $6s^24f^3$	Nd $6s^24f^4$	Pm $6s^24f^5$	Sm $6s^24f^6$	Eu $6s^24f^7$	Gd $6s^25d^14f^7$	Tb $6s^24f^9$	Dy $6s^24f^{10}$	Ho $6s^24f^{11}$	Er $6s^24f^{12}$	Tm $6s^24f^{13}$	Yb $6s^24f^{14}$	Lu $6s^25d^14f^{14}$
Ac $7s^26d^1$	Th $7s^26d^2$	Pa $7s^26d^15f^1$	U $7s^26d^15f^2$	Np $7s^26d^15f^4$	Pu $7s^25f^6$									

Figure 1.3 The periodic table of naturally occurring elements showing the electronic configuration of the elements. Only the last orbitals filled are shown, thus each element has electrons in the orbitals of all previous Group 18 elements (noble gases) in addition to those shown. Superscripts indicate the number of electrons in each subshell.

1.5.3   Some chemical properties of the elements

It is only the most loosely bound electrons, those in the outermost shells, that participate in chemical bonding, so elements sharing a similar outermost electronic configuration tend to behave similarly. Elements within the same column of the periodic table, or *group*, share outer electronic configurations and hence behave in a similar manner. Thus, the elements of Group 1, the *alkalis*, all have one electron in the outermost *s* orbital, and behave in a similar manner. The Group 18 elements, the *noble*, or *rare*, *gases*, all have a filled *p* subshell, and behave similarly.

Let's now consider several concepts that are useful in describing the behavior of atoms and elements: *ionization potential, electron affinity,* and *electronegativity*. The first ionization potential of an atom is the energy required to remove (i.e., move an infinite distance away) the least tightly bound electron. This is energy gained by the electron in reactions such as:

$$Na \rightarrow Na^+ + e^- \qquad (1.1)$$

The first ionization potential of the elements is illustrated in Figure 1.4. The *Second*

Ionization Potential is the energy required to remove a second electron, etc. The *electron affinity* is the energy given up in reactions such as:

$$F + e^- \rightarrow F^- \qquad (1.2)$$

Electronegativity is another parameter that is often used to characterize the behavior of the elements. It is a relative, unitless quantity determined from the differences in bond energy between an A–B molecule and the mean energies of A–A and B–B molecules. *Electronegativity quantifies the tendency of an element to attract a shared electron when bonded to another element.* For example, F has a higher electronegativity than H (the values are 3.8 and 2.5, respectively), thus the bonding electron in hydrogen fluoride, HF, is more likely to be found in the vicinity of F than of H. It is also useful in characterizing the nature of chemical bonds between elements, as we shall see in a subsequent section. Electronegativities of the elements are shown in Figure 1.5.

In general, first ionization potential, electron affinity, and electronegativities increase from left to right across the periodic table, and to a lesser degree from bottom to top.

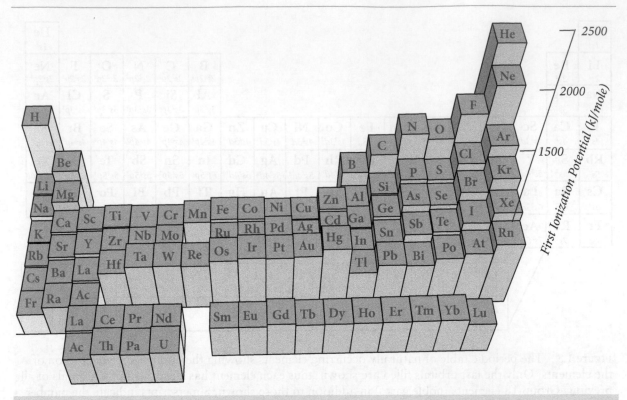

Figure 1.4 First ionization potential of the elements.

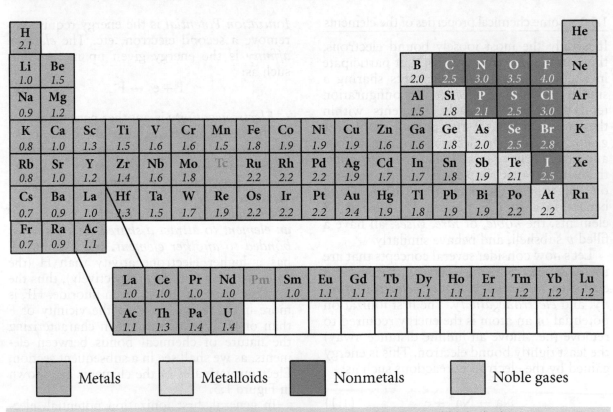

Figure 1.5 Electronegativities of the elements. Nonmetals are characterized by high electronegativity, metals by low electronegativity. Metalloids have intermediate values.

This reflects the shielding of outer electrons, particularly those in *s* orbitals, by inner electrons, particularly those in *p* orbitals, from the charge of the nucleus. Thus the outer *3s* electron of neutral sodium is effectively shielded from the nucleus and is quite easily removed. On the other hand, the *2p* orbitals of oxygen are not very effectively shielded, and it readily accepts two additional electrons. With the addition of these two electrons, the *2p* orbital is filled and the *3s* orbital effectively shielded, so there is no tendency to add a third electron. With the outer *p* (and *s*) orbitals filled, a particularly stable configuration is reached. Thus, Ne and other noble gases have little tendency to either add or give up an electron.

Metallic elements have electronegativities generally ≤ 1.9 and are said to be "electropositive". They tend to form positively charged ions, called *cations*, by giving up electrons. Elements with electronegativities ≥2.5 are nonmetals and tend to form negatively charged ions, called *anions,* by acquiring additional elections. Those with electronegativities in the range of >1.8 and <2.2 are called metalloids or semi-metals and form either type of ion.

The number of electrons that an element will either give up or accept is known as its *valence*. For elements in the wings of the periodic table (i.e., all except the transition metals), valence is easily determined simply by counting how far the element is horizontally displaced from Group 18 in the periodic table. For Group 18, this is 0, so these elements, the noble gases, have 0 valence. For Group 1 it is 1, so these elements have valence of +1;

for Group 17 it is −1, so these elements have valence of −1, etc. Valence of the transition metals is not so simply determined, and these elements can have more than one valence state. Most, however, have valence of 2 or 3, though some, such as U, can have valences as high as 6.

A final characteristic that is important in controlling chemical properties is *ionic radius*. This is deduced from bond length when the atom is bonded to one or more other atoms. Positively charged atoms, or *cations*, have smaller ionic radii than do negatively charged atoms, or *anions*. Also, ionic radius decreases as charge increases for cations. This decrease is due both to loss of outer electrons and to shrinking of the orbits of the remaining electrons. The latter occurs because the charge of the nucleus is shared by fewer electrons and hence has a greater attractive force on each. In addition, ionic radius increases downward in each group in the periodic table, both because of addition of electrons to outer shells and because these outer electrons are increasingly shielded from the nuclear charge by the inner ones. Ionic radius is important in determining geochemical properties such as substitution in solids, solubility, and diffusion rates. Large ions are surrounded, or *coordinated*, by a greater number of oppositely charged ions than do smaller ones. The ionic radii of the elements are illustrated in Figure 1.6.

We can now summarize a few of the more important chemical properties of the various groups in the periodic table. Group 18 does not participate in chemical bonding in nature, hence the term *noble gases*. Group 1 elements,

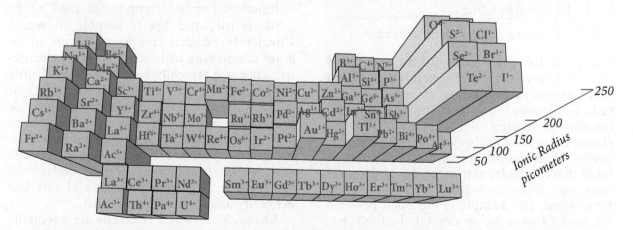

Figure 1.6 Ionic radii of the elements.

the *alkalis*, readily lose an electron and hence are highly reactive. They tend to form *ionic bonds* rather than *covalent* ones and hence weaker bonds to other elements and to be quite soluble in aqueous solutions and can be easily leached from minerals. Because they have only a +1 charge, their ionic radii tend to be larger than those of other cations. Group 2 elements, the *alkaline earths*, have these same characteristics, but somewhat moderated. Group 17 elements, the *halogens*, are highly electronegative and readily accept an electron, are highly reactive, form ionic bonds, and are quite soluble. Their ionic radii tend to be larger than more highly charged anions. Elements of Groups 13–16 tend to form bonds that are predominantly covalent. As a result, they tend to be less reactive and less soluble (except where they form soluble radicals, such as SO_4^{2-}) than Group 1, 2, and 17 elements. Finally, the transition metals are a varied lot. Many form strong bonds (generally with O in nature) and can have quite variable solubility (which, as we shall see in Chapter 6, depends on their valence state). Some, the noble metals (Ru, Rh, Pd, Os, Ir, Pt) in particular, tend to be very unreactive. The rare earths are of interest because all have two electrons in the *6s* outer orbital and vary only by the number of electrons in the *4f* shell. Because they almost always have the same valence (+3), bonding behavior is quite similar. As Goldschmidt found, they vary systematically in ionic radius, which results in a systematic variance in geochemical behavior, as we will see in Chapter 7.

1.5.4   Chemical bonding

1.5.4.1   Covalent, ionic, and metal bonds

Except for the noble gases, atoms rarely exist independently; they are generally bound to other atoms in molecules, crystals, or ionic radicals. Atoms bind to one another through transfer or sharing of electrons, or through electrostatic forces arising from uneven distribution of charge in atoms and molecules. A bond that results from the transfer of electrons from one atom to another is known as an *ionic bond*, an example is the bond between Na and Cl in a halite crystal. In this case, the Na atom (the electropositive element) gives up an electron, becoming positively charged, to the Cl atom (the electronegative element), which becomes negatively charged. Electrostatic forces between the Na^+ and the Cl^- ions hold the ions in place in the crystal. When electrons are shared between atoms, such as in the H_2O or CH_4 molecules or the SO_4^{2-} radical, the bond is known as covalent. In a *covalent bond*, the outer electrons of the atoms involved are in hybrid orbits that encompass both atoms.

Ideal covalent and ionic bonds represent the extremes of a spectrum: most bonds are neither wholly covalent nor wholly ionic. In these intermediate cases, the bonding electrons will spend most, but not all, of their time associated with one atom or another. Electronegativity is useful in describing the degree of ionicity of a bond: a bond is considered ionic when the difference in the electronegativity of the two atoms involved is greater than 2. In Figure 1.5, we see that metals tend to have low electronegativities while the nonmetals have high electronegativities. Thus, bonds between metals and nonmetals (e.g., NaCl) will be ionic while those between nonmetals (e.g., CO_2) will be covalent, as will bonds between two like atoms (e.g., O_2).

Another type of bond occurs in pure metal and metal alloy solids. In the metallic bond, valence electrons are not associated with any single atom or pair of atoms; rather, they are mobile and may be found at any site in the crystal lattice. Since metals rarely occur naturally at the surface of the Earth (they do occur in meteorites and the Earth's core), this type of bond is less important in geochemistry than other bonds.

Ionically bonded compounds tend to be hard, brittle, and highly soluble in water. Covalently bonded compounds tend to be good conductors of heat, but not of electricity. They are typically harder and less brittle than ionic solids but less soluble. In molecular solids, such as ice, atoms within the molecule are covalently bonded. The molecules themselves, which occupy the lattice points of the crystal, are bonded to each other through van der Waals and/or hydrogen bonds. Such solids are comparatively weak and soft and generally have low melting points.

Molecules in which electrons are unequally shared have an asymmetric distribution of charge and are termed *polar*. A good example is the hydrogen chloride molecule. The difference in electronegativity between hydrogen and chlorine is 0.9, so we can predict that the bonding electron will be shared but associated more with the Cl atom than with the H atom in HCl. Thus, the H atom will have a partial positive charge, and the Cl atom a partial negative charge. Such a molecule is said to be a *dipole*. The *dipole moment*, which is the product of one of the charges (the two charges are equal and opposite) times the distance between the charges, is a measure of the asymmetric distribution of charge. Dipole moment is usually expressed in Debye units (1 Debye = 3.3356×10^{-34} coulomb-meters).

1.5.4.2 Van der Waals interactions and hydrogen bonds

Covalent and ionic bonds account for the majority of bonds between atoms in molecules and crystals. There are two other interactions that play a lesser role in interactions between atoms and molecules: *van der Waals* interactions and *hydrogen bonds*. These are much weaker but nevertheless play an important role in chemical interactions, particularly where water and organic substances are involved.

Van der Waals interactions arise from asymmetric distribution of charge in molecules and crystals. There are three sources for van der Waals interactions: *dipole–dipole attraction*, *induction*, and *London dispersion* forces. As we noted above, many molecules, including water, have permanent dipole moments. When two polar molecules encounter each other, they will behave much as two bar magnets: they will tend to orient themselves so that the positive part of one molecule is closest to the negative part of another (Figure 1.7a). This results in a net attractive force between the two molecules. When the distance between molecules is large compared with the distance between charges within molecules, the energy of attraction can be shown to be:

$$U_\infty = -\frac{2}{3}\frac{\mu^4}{r^6}\frac{1}{kT} \qquad (1.3)$$

where U_∞ is the interaction energy, μ is the dipole moment, T is temperature (absolute, or thermodynamic temperature, which we will introduce in the next chapter), k is a constant

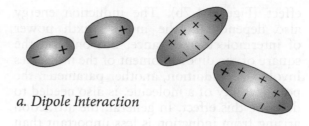

a. Dipole Interaction

b. The Induction Effect

Figure 1.7 Van der Waals interactions arise because of the polar nature of some molecules. Illustrated here are (a) dipole–dipole interactions, which occur when two dipolar molecules orient themselves so oppositely charged sides are closest, and (b) the induction effect, which arises when the electron orbits of one molecule are perturbed by the electromagnetic field of another molecule.

(Boltzmann's constant, which we shall also meet in the next chapter), and r is distance. We do not want to get lost in equations at this point; however, we can infer several important things about dipole–dipole interactions just from a quick glance at it. First, the interaction energy depends inversely on the sixth power of distance. Many important forces, such as electromagnetic and gravitational forces, depend on the inverse square of distance. Thus, we may infer that dipole–dipole forces become weaker with distance very rapidly. Indeed, they are likely to be negligible unless the molecules are very close. Second, the interaction energy depends on the fourth power of the dipole moment, so that small differences in dipole moment will result in large differences in interaction energy. For example, the dipole moment of water (1.84 Debyes) is less than twice that of HCl (1.03 Debyes), yet the dipole interaction energy between two water molecules (716 J/mol) is nearly 10 times as great as that between two HCl molecules (72.24 J/mol) at the same temperature and distance (298 K and 50 pm). Finally, we see that dipole interaction energy will decrease with temperature.

Dipole molecules may also polarize electrons in a neighboring molecule and distort their orbits in such a way that their interaction with the dipole of the first molecule is attractive. This is known as the induction effect (Figure 1.7b). The induction energy also depends on the inverse sixth power of intermolecular distance, but only on the square of the dipole moment of the molecules involved. In addition, another parameter, the polarizability of a molecule, is also needed to describe this effect. In general, the attraction arising from induction is less important than from dipole–dipole interaction. However, because it depends only on the square of dipole moment, the induction attraction can be larger than the dipole–dipole attraction for some weakly dipolar molecules.

Finally, van der Waals forces can also occur as a consequence of fluctuations of charge distribution on molecules that occur on time scales of 10^{-16} seconds. These are known as London dispersion forces. They arise when the instantaneous dipole of one molecule induces a dipole in a neighboring molecule. As was the case in induction, the molecules will orient themselves so that the net forces between them are attractive.

The total energy of all three types of van der Waals interactions between water molecules is about 380 J/mol, assuming an intermolecular distance of 5 pm and a temperature of 298 K (25°C). Though some interaction energies can be much stronger (e.g., CCl_4, 2.8 kJ/mol) or weaker (1 J/mol for He), an energy of a few hundred joules per mole is typical of many substances. By comparison, the hydrogen–oxygen bond energy for each H–O bond in the water molecule is 46.5 kJ/mol. Thus, van der Waals interactions are quite weak compared with typical intramolecular bond energies.

The *hydrogen bond* is similar to van der Waals interactions in that it arises from non-symmetric distribution of charge in molecules. However, it differs from van der Waals inter-actions in a number of ways. First, it occurs exclusively between hydrogen and strongly electronegative atoms, namely oxygen, nitro-gen, and fluorine. Second, it can be several orders of magnitude stronger than van der Waals interactions, though still weak by com-parison with covalent and ionic bonds. In the water molecule, binding between oxygen and hydrogen results in hybridization of s and p orbitals to yield two bonding orbitals between the O and two H atoms, and two nonbinding sp^3 orbitals on the oxygen. The latter are prominent on the opposite side of

the O from the hydrogens. The hydrogen in one water molecule, carrying a net positive charge, is attracted by the nonbinding sp^3 electrons of the oxygen of another water molecule, forming a hydrogen bond with it (Figure 1.8).

Hydrogen bonds typically have energies in the range of 20–40 kJ/mol. These are much higher than expected for electrostatic interactions alone, and indeed approach val-ues similar to intramolecular bond energies. Thus, there is the suspicion that some degree of covalency is also involved in the hydrogen bond. That is to say, the nonbinding electrons of oxygen are to some degree shared with the hydrogen in another molecule. Hydrogen bonds are perhaps most important in water, where they account for some of the extremely usual properties of this compound, such as its high heat of vaporization, but they can also be important in organic molecules and are present in HF and ammonia as well.

1.5.5 Molecules, crystals, and minerals

1.5.5.1 *Molecules*

Molecules, which result from the chemical bonds between atoms discussed above, are a familiar concept. By some definitions, they are electrically neutral; charged species consisting of two or more atoms are known as *radicals*. The properties of molecules are generally quite different from those of their constituent atoms. Carbon dioxide is a good example: the equilibrium form of pure carbon at the room temperature and pressure is a solid, graphite, while that of oxygen is a diatomic gas. The properties of molecules depend on the bond lengths, which depend on the strength of the chemical bond, as well as the geometric arrangement of the atoms. For example, the polar nature of the water molecule, from which many of its unusual properties arise (and which will discuss in Chapter 3), includ-ing the formation of hydrogen bonds, is a consequence of the tendency of valence elec-tron pairs surrounding an atom tend to repel each other which results in an arrangement in which the two hydrogens are separated by an angle of 104.45° (Figure 1.8a) and a partial positive charge on the hydrogens and a partial negative one on the other side of the molecule In contrast, the arrangement

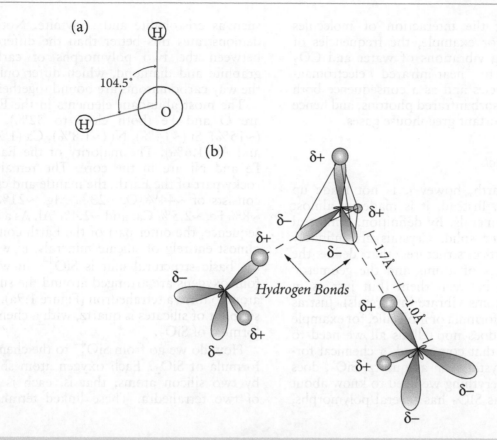

Figure 1.8 (a) Geometry of the water molecule. (b) Hydrogen bonds between water molecules. The $\delta+$ and $\delta-$ indicate partial positive and negative charges, respectively.

of atoms in CO_2 is linear with a bond angle of 180°. CO_2 reacts with water to produce carbonic acid (H_2CO_3), which has a plane trigonal geometry.

Geometry becomes enormously important for organic molecules and life. For example, $C_{12}H_{22}O_{11}$ is the chemical formula for both lactose and sucrose, as well as several other disaccharide carbohydrates, but the atoms are stitched together differently and as a result they have quite different properties. Among other things, all adults (and essentially all animals) can readily digest sucrose, but many adult humans (and most adult mammals) cannot digest lactose. In other molecules, even slight variations in structure, for example, a molecular structure and its mirror image can have quite different properties – a topic we'll explore briefly in Chapter 12.

Molecules are not necessarily static entities. An important feature of some molecules is the ability to dissociate. This is particularly true of both water and carbonic acid, which can give

up hydrogen atoms. Acidity reflects the balance between H^+ (strictly speaking H_3O^+) and OH^- ions; these must be equal in pure water, but a solution of CO_2 in water will have an excess of H^+ and hence be acidic. These hydrogen ions can also reassociate with their parent molecules and do so when H ions become abundant.

The bonds between atoms in molecules are also not static, but rather bond lengths and bond angles continually oscillate about their mean values. For example, the water molecule has three fundamental modes of vibration: two stretching vibrations of the O-H bond (one symmetric, one asymmetric) and a bending vibration in which the bond angle changes. Vibrational frequencies are proportional to bond strength and increase with increasing temperature (in stepwise fashion, as they are quantized), although many molecules remain in the ground state vibrational frequencies over the range of temperatures at the Earth's surface. These vibrations are responsible for some of the interaction of molecules with light. For example, the frequencies of the stretching vibrations of water and CO_2 correspond to near-infrared electromagnetic frequencies and as a consequence both molecules absorb infrared photons, and hence both are important greenhouse gases.

1.5.5.2 Crystals

The solid Earth, however, is not made up of molecules. Instead, it is made up almost entirely of minerals. By definition, a mineral is a crystalline solid. Crystals are infinitely repeating lattice structures that define the fixed positions of atoms and the geometric relationships between them (but just as in molecules, atoms vibrate in crystals). Just as the chemical formula of molecule, for example $C_{12}H_{22}O_{11}$, does not tell us all we need to know about that compound, a chemical formula of a crystal such as quartz, SiO_2, does not tell us everything we need to know about that crystal as SIO_2 has several polymorphs,

such as cristobalite and tridymite. Nothing demonstrates this better than the difference between the two polymorphs of carbon: graphite and diamond, which differ only in the way carbon atoms are bound together.

The most abundant elements in the Earth are O and Fe (both close to 32%), Mg (~15%), Si (~14%), Ni (~1.8%), Ca (1.7%), and Al (1.6%). The majority of the Earth's Fe and Ni are in the core. The remaining rocky part of the Earth, the mantle and crust, consists of ~44% O, ~23% Mg, ~21% Si, ~8% Fe, ~2.5% Ca, and ~2.4% Al. As a consequence, the outer part of the Earth consists almost entirely of silicate minerals, in which the basic structural unit is SiO_4^{4-}, in which four oxygens are arranged around the silicon atom to form a tetrahedron (Figure 1.9a). The simplest of silicates is quartz, with a chemical formula of SiO_2.

How do we go from SiO_4^{4-} to the chemical formula of SiO_2? Each oxygen atom shared by two silicon atoms; that is, each is part of two tetrahedra. These linked tetrahedra

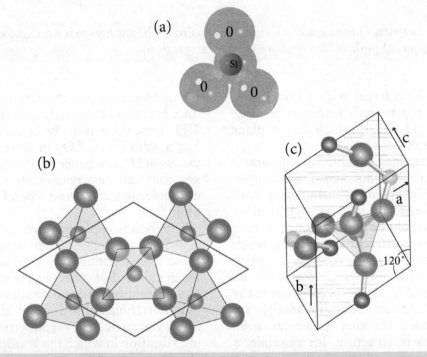

Figure 1.9 (a) The silica tetrahedron, SiO_4^{4-}, consists of a silicon atom surrounded by four oxygens. This is the basic building block of silicate minerals. (b) In quartz, each oxygen (gray) is shared between two tetrahedra to produce a three-dimensional structure. (c) The quartz unit cell consists of a central Si atom whole within the cell and with four Si atoms on the edges that are shared by adjacent unit cells, hence it contains $1 + 4 \times \frac{1}{2}$ Si = 3 Si (faded silicon atoms are outside the cell). Six oxygens are wholly within the cell, so the chemical formula of the cell is Si_3O_6; a, b, and c are the crystallographic axes.

are repeated infinitely to produce a quartz crystal (Figure 1.9b). We can define a *unit cell* that contains all the necessary information to describe both the chemical and structural elements of quartz (Figure 1.9c). The unit cell of lattice is defined by the length of three axes and the angles between them. In the case of quartz, the lattice cell has the chemical formula of Si_3O_6 and axes of lengths: $a = 0.49$ nm, $b = 0.49$ nm, and $c = 0.54$ nm, with the b–c and b–a angles are equal to 90° and the angle between the a and c axes equal to 120°. The quartz crystal is built up by repetition of the unit cell along its principal axes.

Crystals have varying degrees of symmetry that can be divided into seven different systems, which, with decreasing symmetry, are cubic, hexagonal, trigonal, tetragonal, orthorhombic, monoclinic, and triclinic (sometimes trigonal is included with hexagonal to give only six systems). The cubic system has the highest symmetry, with all three axes of equal length and all three angles equal to 90°; triclinic has the lowest, with no axes of equal length and no angles of 90°. Diamond is an example of a cubic mineral. Quartz, with two equal length axes and one 120° angle, is an example of a trigonal mineral. Graphite, in its most common form, is hexagonal. Since all axes are equal in cubic crystals, they transmit light and vibrations, sound, and seismic waves, equally in all directions. Hexagonal, trigonal, and tetragonal have one unique crystallographic axis that transmits light and sound at different velocities than the other two and are said to be uniaxial. The least symmetric classes, orthorhombic, monoclinic, and triclinic as said to be biaxial and transmit light and sound at three different velocities along the three axes.

Minerals are by definition naturally occurring inorganic crystalline solids. Most minerals are "ionic solids" that consist of cations, metals or metalloids, bound to nonmetals or anionic radicals (although the bonds can have a partly covalent character; in quartz for example, bonds between the oxygens and silicon are roughly 50% covalent and quite strong). In the majority of minerals, several metals are present and only a single anion, most often oxygen. For these kinds of crystals, rules elaborated by Linus Pauling in 1929, known as *Pauling's rules*, dictate crystal

structure. The first of these is that the ratio of cation to anion radius determines the number of anions that coordinate the cation and the shape of the coordinated polyhedron of anions. Smaller cations will be surrounded by few anions, larger cations will be coordinated by more anions (Figure 1.10). When the cation/anion radius ratio is <0.22, three anions will coordinate each anion, forming a triangular polyhedron; when the ratio is >0.22 and <0.414, four anions coordinate the cation; the radius Si^{4+} is 40 pm, that of O^{2-} is 126 pm, so the ratio is 0.317. As a result, the four oxygens form a tetrahedron around silicon. For a ratio >0.414 and <0.73, six anions surround the cation in octahedral coordination, as in halite, which has a

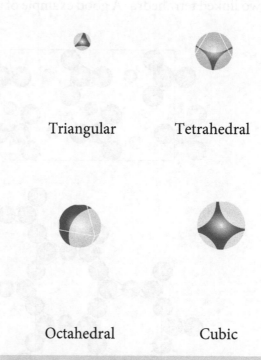

Triangular   Tetrahedral

Octahedral   Cubic

Figure 1.10 Geometric relationships between cations and their coordinating anions. Because they have fewer electrons relative to protons, cations tend to be small, while anions, with excess electrons, tend to be large. The number of anions that immediately surround and bond to a cation depends on the relative radii. Small cations, such as B^{3+} in the borate ion, are coordinated by only three oxygens, Si in quartz by four, and Na in halite by six; if Na^+ is replace by Cs^+, a much larger ion, the coordination number increases to eight.

Na⁺/Cl⁻ radius ratio is 0.56. If the ratio is ≥0.73, eight coordinating anions will form a cube. Pauling's second rule is that the electrostatic bond strength between the cation and each coordinating anion is equal to the cation charge divided by the number of coordinating anions. So highly charged, small cations are more strongly bond than large ones with smaller charge.

Another of Pauling's rules is that in a crystal containing different cations, those of high valency and small coordination number tend not to share polyhedron elements with one another. Most silicate minerals contain more than one cation; like quartz, the Si tetrahedron is the basic structural unit, but unlike quartz, not all oxygens are shared. In orthosilicates, the silica tetrahedra are either completely independent or form dimers – that is, two linked tetrahedra. A good example of a

mineral of this type is olivine, whose structure is illustrated in Figure 1.11a. The chemical formula for olivine is $(Mg,Fe)_2SiO_4$. The notation (Mg,Fe) indicates that either magnesium or iron or both may be present. *Olivine* is an example of a solid solution between the Mg end-member, forsterite (Mg_2SiO_4), and the Fe end-member, fayalite (Fe_2SiO_4). Such solid solutions are quite common among silicates. As the formula indicates, there are two magnesium or iron atoms for each silica tetrahedron. Since each Mg or Fe has a charge of +2, their charge balances the −4 charge of each silica tetrahedron. Olivine constitutes roughly 50% of the Earth's upper mantle and is thus one of the most abundant minerals on Earth.

In chain silicates, the silica tetrahedra are linked together to form infinite chains (Figure 1.11b and c), with two bridging

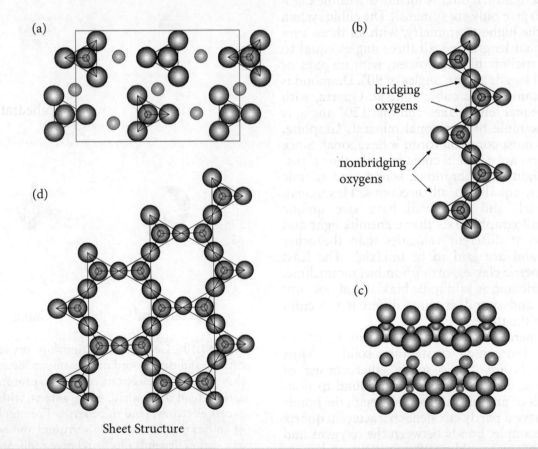

Figure 1.11 Silicate mineral structures. (a) In orthosilicates such as olivine, the tetrahedra are separate and each oxygen is also bound to other metal ions that occupy interstitial sites between the tetrahedra. (b) In pyroxenes, the tetrahedra each share two oxygens and are bound together into chains. (c) Metal ions are located between the chains in pyroxenes. (d) In sheet silicates, such as talc, mica, and clays, the tetrahedra each share three oxygens and are bound together into sheets.

oxygen per tetrahedron. Oxygens shared by two silicons are called *bridging oxygens*. Minerals of this group are known as *pyroxenes* and have the general formula $XSiO_3$ where X is some metal, most commonly Ca, Mg, or Fe, which is located between the chains. Two pyroxenes, orthopyroxene $((Mg,Fe)SiO_3)$ and clinopyroxene $(Ca(Mg,Fe)Si_2O_6)$ are very abundant in the Earth's upper mantle as well as in mafic igneous rocks. The pyroxenes wollastonite $(CaSiO_3)$ and jadeite $(NaAlSi_2O_6)$ are found exclusively in metamorphic rocks.

In double chain silicates, an additional one-half bridging oxygen per tetrahedra joins two chains together. Minerals of this group are known as *amphiboles*, which occur widely in both igneous and metamorphic rocks. Among the important minerals in this group are hornblende $(Ca_2Na(Mg,Fe)_4Al_3Si_8O_{22}(OH)_2)$, tremolite-actinolite $(Ca_2(Mg,Fe)_5Si_8O_{22}(OH)_2)$, and glaucophane $(Ca_2(Mg,Fe)_3Al_3Si_8O_{22}(OH)_2)$. These minerals all contain OH as an essential component (Cl or F sometimes substitutes for OH). They are thus examples of hydrous silicates.

Sharing of a third oxygen links the tetrahedra into sheets, forming the sheet silicates (Figure 1.11d). This group includes *micas* such as biotite $(K(Mg,Fe)_3AlSi_3O_{10}(OH)_2)$ and muscovite $(KAl_3Si_3O_{10}(OH)_2)$, talc $(Mg_3Si_4O_{10}(OH)_2)$, and *clay minerals* such as kaolinite $(Al_2Si_2O_5(OH)_4)$. As in amphiboles, OH is an essential component of sheet silicates. Many of these minerals form through weathering and are thus primary sedimentary minerals. Many of them are found in igneous and metamorphic rocks as well.

When all four oxygens are shared between tetrahedra, the result is a framework. The simplest framework silicate is quartz (SiO_2), which consists solely of linked SiO_4 tetrahedra. The other important group of framework silicates is the *feldspars*, of which there are three end-members: sanidine $(KAlSi_3O_8)$, albite $(NaAlSi_3O_8)$, and anorthite $(CaAl_2Si_2O_8)$. The calcium and sodium feldspars form the plagioclase solid solution, which is stable through a large temperature range. Sodium and potassium feldspars, collectively called alkali feldspar, form more limited solid solutions, as we will find in Chapter 4. Feldspars are the most abundant minerals in the Earth's crust.

Because only a single anion or anionic group is present in the common minerals (the presence of OH^- in many kinds of silicates is an important exception), minerals are classified compositionally by the nature of the anion or anionic group. Silicates, as we noted, are the most abundant compositional class. Other classes of minerals include oxides, in which one or more metals are bound to oxygen, for example, magnetite (Fe_3O_4) and ilmenite $(FeTiO_3)$, Carbonates are a particularly important and abundant class of minerals at the Earth's surface, of which calcite $(CaCO_3)$ is by far the most important. Other important groups include sulfates, such as gypsum $(CaSO_4 \cdot 2H_2O)$, hydroxides such as gibbsite $(Al(OH)_3)$, sulfides such as pyrite FeS_2, halides, such as halite $(NaCl)$ and fluorite (CaF), and phosphates, of which only one, apatite $(Ca_5(PO_4)_3(OH,F,Cl)$ is common. There are others as well, such as "native" minerals, consisting of a single element such as diamond, borates, arsenates, etc.

1.6 A BRIEF LOOK AT THE EARTH

1.6.1 Structure of the Earth

The Earth consists of three principal layers: the core, the mantle, and the crust (Figure 1.12). The core, roughly 3400 km thick and extending about halfway to the surface, consists of Fe–Ni alloy and can be subdivided into an inner and outer core. The outer core is liquid while the inner core is solid. The mantle is nearly 3000 km thick and accounts for about two-thirds of the mass of the Earth, with the core accounting for the other third. The crust is quite thin by comparison, nowhere thicker than 100 km and usually much thinner. Its mass is only about 0.5% of the mass of the Earth. There are two fundamental kinds of crust: oceanic and continental. Ocean crust is thin (about 6 km) and, with the exception of the eastern Mediterranean, is nowhere older than about 180 million years. The continental crust is thicker (about 35–40 km thick on average) and relatively permanent, with an average age of ~2 billion years.

Both the crust and the mantle consist principally of silicates. The mantle is comparatively rich in iron and magnesium, so ferromagnesian silicates, such as olivine and pyroxenes,

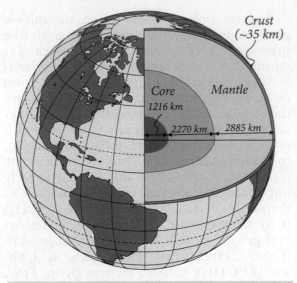

Figure 1.12 The Earth in cross-section. The outer rocky part of the planet, the mantle and crust, consists principally of silicates and is 2885 km thick. The core, divided into a liquid outer core and a solid inner core, consists of iron–nickel alloy and is 3486 km thick.

dominate. Rocks having these compositional characteristics are called *ultramafic*. The continental crust is poor in iron and magnesium, and aluminosilicates such as feldspars dominate. Rocks of this composition are sometimes referred to as *felsic* (or silicic). The oceanic crust is intermediate in composition between the mantle and continental crust and has a *mafic* composition, consisting of a roughly 50:50 mix of ferromagnesian minerals and feldspar. These differences in composition lead to differences in density, which are ultimately responsible for the layering of the Earth: the density of the layers decreases outward. The continental crust is the least of dense of these layers. The fundamental reason why continents stick out above the oceans is that continental crust is less dense than oceanic crust. Not shown in the figure are two additional layers that are even less dense: the oceans or *hydrosphere*, which cover about two-thirds of the surface, and the atmosphere, which extends about 100 km above the surface.

1.6.2 Plate tectonics and the hydrologic cycle

Two sources of energy drive all geologic processes: solar energy and the Earth's internal heat. Solar energy drives atmospheric and oceanic circulation, and with them, the hydrologic cycle. In the hydrologic cycle, water vapor in the atmosphere precipitates on the land as rain or snow, percolates into the soil and, through the action of gravity, makes its way to the oceans. From the oceans, it is evaporated into the atmosphere again and the cycle continues. The hydrologic cycle is responsible for two very important geologic processes: weathering and erosion. Weathering causes rocks to break down into small particles and dissolved components. The particles and dissolved matter are carried by the flow of water (and more rarely by wind and ice) from high elevation to areas of low elevation. Thus, the effect of the hydrologic cycle is to level the surface of the planet. From a geochemical perspective, it releases essential nutrients from rock that make life possible.

The Earth's internal heat is responsible for tectonic processes, which deform the surface of the planet, producing topographic highs and lows. The internal heat has two sources. Some fraction of the heat originated from the gravitational energy released when the Earth formed. The other fraction, 50% or less, of internal heat is produced by the decay of radioactive elements, principally uranium, thorium, and potassium, in the Earth. The Earth's internal heat is being slowly lost over geologic time as it migrates to the surface and is radiated away into space. It is this migration of heat out of the Earth that drives tectonic processes. Heat causes both the outer core and the mantle to convect, as hot regions rise and cold regions sink. Convection within the outer core gives rise to the Earth's magnetic field, and may have other, as yet not understood, geologic consequences. Convection in the mantle is responsible for deformation of the Earth's crust as well as volcanism. Solar energy drives convection in the oceans and atmosphere and is responsible for redistribution heat over the surface of the Earth and wind and ocean currents.

The great revolution in earth science in the 1960s centered on the realization that the outer part of the Earth was divided into a number of "plates" that moved relative to one another. Most tectonic processes, as well as most volcanism, occur at the boundaries between these plates. The outer part of the Earth, roughly the outer 100 km or so, is cool

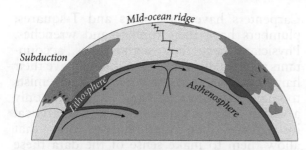

Figure 1.13 Cross-section of the Earth illustrating relationships between lithosphere and asthenosphere and plate tectonic processes. Oceanic crust and lithosphere are created as plates diverge at mid-ocean ridges and are eventually subducted back into the mantle. Continental lithosphere is thicker and lighter than oceanic lithosphere and not easily subducted.

enough (<1000° C) that it is rigid. This rigid outer layer is known as the lithosphere and comprises both the crust and the outermost mantle (Figure 1.13). The mantle below the lithosphere is hot enough (and under sufficient confining pressure) that it flows, albeit extremely slowly, when stressed. This part of the mantle is known as the asthenosphere. Temperature differences in the mantle create buoyancy stresses that produce convective flow. It is this flow that drives the motion of the lithospheric plates. The motion of the plates is extremely slow, a few tens of centimeters per year at most and generally much less. Nevertheless, on geologic time scales they are sufficient to continually reshape the surface of the Earth, creating the Atlantic Ocean, for example, in the last 200 million years and Himalayan Mountains in the last 40 million years.

Rather than thinking of plate motion as being driven by mantle convection, it would be more correct to think of plate motion as part of mantle convection. Where plates move apart, mantle rises to fill the gap. As the mantle does so, it melts; we'll see why this is so in Chapter 7. The melt rises to the surface as magma and creates new oceanic crust at volcanoes along mid-ocean ridges (Figure 1.13). Mid-ocean ridges, such as the East Pacific Rise and the Mid-Atlantic Ridge, thus mark *divergent plate boundaries*. As the oceanic crust moves away from the mid-ocean ridge it cools, along with the mantle

immediately below it. This cooling produces a steadily thickening lithosphere. As this lithosphere cools, it contracts and its density increases. Because of this contraction, the depth of the ocean floor increases away from the mid-ocean ridge. When this lithosphere has cooled sufficiently, after 100 million years or so, it becomes denser than the underlying asthenosphere. The lithosphere may then sink back into the mantle in a process known as *subduction*. As the lithosphere sinks, it creates deep ocean trenches, such as the Peru–Chile Trench, or the Marianas Trench. Chains of volcanoes, known as island arcs, almost invariably occur adjacent to these deep-sea trenches. The volcanism occurs as a result of dehydration of the subducting oceanic crust and lithosphere. Water released from the subducting oceanic crust rises into the overlying mantle, lowering its melting temperature and causing it to melt. The island arcs and deep-sea trenches are collectively called *subduction zones*. Subduction zones thus mark *convergent plate boundaries*. It is primarily the sinking of old, cold lithosphere that drives the motion of plates. Thus the lithosphere does not merely ride upon convecting mantle, its motion is actually part of mantle convection.

The density of the continental crust is always lower than that of the mantle, regardless of how cold the crust becomes. As a result, it cannot be subducted into the mantle. The Indian–Eurasian plate boundary is a good example of what happens when two continental plates converge. Neither plate readily subducts and the resulting compression has produced, and continues to uplift, the Himalayan Mountains and the Tibetan Plateau. This area of continental crust is not only high – it is also deep. The crust beneath this region extends to depths of as much as 100 km, nearly three times the average crustal thickness. Rocks within this thickened crust will experience increased temperatures and pressures, leading to *metamorphism*, a process in which new minerals form in place of the original ones. In the deepest part of the crust, melting may occur, giving rise to granitic magmas, which will then intrude into the upper crust. In such cases of crustal thickening, the lowermost continental crust can become denser than the mantle and can detach and sink, a process called foundering or delamination.

The topographically high Himalayas are subject to extremely high rates of erosion, and the rivers draining the area carry enormous quantities of sediment. These are deposited mainly in the northern Indian Ocean, building the Ganges and Indus Fans outward from the continental margin. As the mountains erode, the mass of crust bearing down on the underlying asthenosphere is reduced. As a result of the decreased downward force, further uplift occurs.

The third kind of plate boundary is known as a transform boundary and occurs where plates slide past one another. A good example of this type of plate boundary is the San Andreas Fault system of California. Here the Pacific Plate is sliding northward past the North American Plate. The passage is not an easy one, however. The two plates occasionally stick together. When they do, stresses steadily build up. Eventually, the stress exceeds the frictional forces holding the plates together, and there is a sudden jump producing an earthquake. Earthquakes are also common in subduction zones and along mid-ocean ridges. They are much rarer in the interior of plates.

Most volcanism and crustal deformation occur along plate boundaries. A few volcanoes, however, are located in plate interiors and appear to be entirely unrelated to plate tectonic processes. Crustal uplift also occurs in association with these volcanoes. Two good examples are Hawaii and Yellowstone. These phenomena are thought to be the result of mantle plumes. Mantle plumes are convective upwellings that are largely independent of the convention driving plate motions. In contrast to the convective upwelling occurring along mid-ocean ridges, which is typically sheet-like, mantle plumes appear to be narrow (~100 km diameter) and approximately cylindrical. Furthermore, it appears that mantle plumes rise from much deeper in the mantle, near the core–mantle boundary, than convection associated with plate motion.

1.7   A LOOK AHEAD

The intent of this book is to introduce you to geochemistry, and through it, paraphrasing Schönbein, reveal the mysteries of our planet. To do this, we must first acquire the tools of the trade. Every trade has a set of tools.

Carpenters have their saws and T-squares; plumbers have their torches and wrenches. Physicians have their stethoscopes, accountants their balance sheets, geologists have their hammers, compasses, and maps. Geochemists too have a set of tools. These include not only a variety of physical tools such as analytical instruments, but interpretative tools that allow them to make sense of the data these instruments produce. The first part of this book is intended to familiarize you with the tools of geochemistry. Once we have a firm grip on these tools, we can use them to dissect the Earth in the second part of the book. There, we begin at the beginning, with the formation of the Solar System and the Earth. We then work our way upward through the solid Earth, from core to mantle and crust, and on to the intersection between geochemistry and life: organic geochemistry, the carbon cycle and climate. We'll then examine the processes at the surface of the Earth, first on land, then in the oceans. Finally, we will briefly consider how geochemistry is applied to practical problems: finding resources and addressing pollution.

In filling our geochemical toolbox, we start with the tools of physical chemistry: thermodynamics and kinetics. Thermodynamics is perhaps the most fundamental tool of geochemistry; most other tools are built around this one. For this reason, Chapters 2, 3, and 4 are devoted to thermodynamics. In Chapter 2, we will introduce the laws of thermodynamics and from them develop a most useful tool: the Gibbs free energy. In Chapters 3 and 4, we'll expand our tool set to deal with solutions. These tools allows us to predict the outcome of chemical reactions under a given set of conditions. In geochemistry, we can, for example, predict the sequence of minerals that will crystallize from a magma under given conditions of temperature and pressure or which should replace them in weathering reactions at the Earth's surface. Thus, thermodynamics provides enormous predictive power for the petrologist. Since geologists and geochemists are more often concerned with understanding the past than with predicting the future, this might seem to be a pointless academic exercise. However, we can also use thermodynamics in the reverse sense: given a suite of minerals in a rock, we can use thermodynamics to determine the temperature and pressure conditions under which the rock formed. We can also use it to determine the temperature and composition of water or magma from which minerals crystallized. This sort of information has been invaluable in reconstructing the past and understanding how the Earth has come to its present condition.

Thermodynamics has an important limitation: it is useful only in equilibrium situations. The rate at which chemical systems achieve equilibrium increases exponentially with temperature. Thermodynamics will be most useful at temperatures relevant to the interior of the Earth, but at temperatures relevant to the surface of the Earth, many geochemical systems will not be in equilibrium and instead be governed by kinetics, the subject of Chapter 5. Kinetics deals with the rates and mechanisms of reactions. Reactions can occur only when reactants are brought together. Unlike gas phase reactions or ones within a solution, this often requires the reactants be transported across an interface. So in this chapter, we will also touch on such topics as diffusion and mineral surfaces.

In Chapter 6, we see how tools of physical chemistry are adapted for use in dealing with natural aqueous solutions. Much of the Earth's surface is covered by water, and water usually is present in pores and fractures to considerable depths even on the continents. This water is not pure but is instead a solution formed by interaction with minerals and atmospheric gases. In Chapter 6, we acquire tools that allow us to deal with the interactions among dissolved species and their interactions with the solids with which they come in contact. These interactions include phenomena such as dissolution and precipitation, complexation, adsorption and ion exchange. The tools of aquatic chemistry are essential to understanding processes such as weathering and precipitation of sedimentary minerals, as well as dealing with environmental problems.

In Chapter 7, we move on to trace element geochemistry. In this chapter we will see that trace elements, which comprise most of the periodic table, have provided remarkable insights into the origin and behavior of magmas. Without question, their value to geochemists far outweighs their abundance. There are several reasons for this. Their concentrations vary much more than do those of the more abundant elements, and their behavior tends often to be simpler and easier to treat than that of major elements (a property we will come to know as Henry's law). Geochemists have developed special tools for dealing with trace elements; the objective of Chapter 7 is to become familiar with them.

Chapters 8 and 9 are devoted to isotope geochemistry. In Chapter 8, we learn that radioactive decay adds the important element of time; radioactivity is nature's clock because the rate at which a radioactive nuclide decays is absolutely constant and independent of all external influences. We can read this clock by measuring the build-up of radiogenic daughter elements, for example ^{206}Pb produced by decay of ^{238}U. In this way we have established the age of the Solar System and the continents, and we have placed firm ages alongside the relative geologic time scale developed in the nineteenth century. Importantly, radiogenic isotope geochemistry has provided some perspective on the rate and manner of evolution of the Earth, and the evolution of our own species by answering questions such as how old are those bones and when were those cave paintings done? We can also use the products of radioactive decay, *radiogenic elements*, as tracers. By following these tracers much as we would dye in a fish tank, we can follow the evolution of a magma, the convection pattern of the mantle, and the circulation of the oceans, and determine from where sediments were derived. Radiogenic isotopes allow us to distinguish magmas produced by melting of the crust from those produced by melting of the mantle and to distinguish a number of distinct chemical reservoirs in the mantle; for example, magmas erupted by oceanic island volcanoes come from different reservoirs than those erupted at mid-ocean ridges.

The isotopes of another set of elements vary not because of radioactive decay, but because of subtle differences in their chemical behavior. These "stable isotopes" are the subject of Chapter 9. The subtle differences in isotopic abundances of elements such as H, C, N, O, and S has provided insights into the evolution of the Earth's atmosphere and reveal the diets of ancient peoples. One of the most significant contributions of stable isotope geochemistry has been to establish temperature changes of the oceans associated with the Pleistocene glacial cycles. Together with radiogenic isotope geochronology, the timing of these cycles and demonstrated that the ultimate driver of them has been small variations in the Earth's orbit and rotation, known as *Milankovitch variations*. Stable isotope geochemistry is the last of our geochemical tools.

With our toolbox full, we are ready to examine the Earth from the geochemical perspective in the second part of the book. Where else to start but at the beginning? The Earth today is the product of its long history, and of all the events in that history, none set the stage more for what Earth would become than its formation.

In Chapter 10 we'll begin by looking at "the big picture": the cosmos and the Solar System. The cosmic beginning was some 13.8 billion years ago. The Big Bang, time's opening act, produced a universe of hydrogen and helium and very little else. Only once stars and galaxies had formed, perhaps half a billion years later, did the universe begin to be seeded with heavier elements. Stars the size of the Sun and larger synthesize the principal elements of life, carbon, nitrogen, and oxygen in their geriatric "red giant" phase and blow them back out into the cosmos in enormous stellar winds. That, however, is not enough to create a planet like Earth, or support life for that matter, both of which require heavier elements as well such as magnesium, silicon, phosphorous, and iron. These are synthesized during the death throes of giant stars and expelled into the cosmos in spectacular explosions called supernovae, which can radiate more energy than an entire galaxy.

Some 9.5 billion years later, part of a vast cloud of gas and dust, not unlike the Great Nebula in Orion visible in the northern hemisphere night sky in winter, began to collapse in on itself, spinning ever more rapidly as it did so like a skater pulling in her arms. The Sun formed in the center of this swirling mass and planets formed in the surrounding disk. The idea that the Solar System formed in this way is an old one: Immanuel Kant postulated it in 1755. But what are the details? We'll find that the details are revealed in leftovers from the process: chondritic meteorites. These meteorites consist of aggregations the dust from which the solar system is formed, although some were metamorphosed in their asteroidal parent bodies. Among other things, they reveal that this nebula, at least in the inner part, was so hot that almost all the dust had evaporated to gas. The first materials to condense, so-called calcium–aluminum inclusions, have been dated with exquisite precision by the decay of U to Pb at 4568.22 ± 0.17 million years. These meteorites also once contained short-lived radioactive nuclides that must have been synthesized within a million years or less of solar system formation, products of nucleosynthesis in our galactic neighborhood. The decay of these radionuclides resulted in the build-up of their daughter products, and we can put our tools of isotope geochemistry to good use to see how this can be used to produce a chronology of events in the young solar system. As samples of the solar system nebular dust, these meteorites provide an inventory of the elements available to build the Earth and in this way place important constraints on the Earth's composition.

The Earth differs in its composition from chondrites, mainly because the region in which it formed was too hot for the more volatile elements to condense. We'll put our thermodynamic tools to good use in understanding the sequence in which the elements condensed from the nebular gas. Other meteorites, the achondrites and irons, come from larger asteroids that broke apart when they collided with each other before they had become full-fledged planets. Remarkably, they had already differentiated into iron cores and silicate mantles within a few million years of the start of the solar system – we know this from radiogenic isotope geochemistry. These meteorites thus provide insights into the process of planetary formation. The chronology established by those short-lived radionuclides reveal that formation of the Earth was a much more drawn-out process that continued for tens of millions of years before a cataclysmic collision between Earth and a Mars-sized body produced the Moon. The abundance of certain trace elements in the Earth's mantle tell us, however, that a bit more material must have accreted to the Earth after that.

In Chapter 11, we turn our attention to solid Earth, its composition and its differentiation into layers: crust, mantle, and core. One question we would like to answer is what is the composition of the Earth? Since the mantle is the largest and most massive of these layers, we begin there. Because we only rarely find mantle material at the surface, geophysical observations such as seismic waves, gravity, and moment of inertia provide particularly critical constraints on the nature of the mantle. The composition of blocks of mantle occasionally thrust to the surface, small pieces of it carried to the surface in volcanic eruptions, and the composition of magma produced by melting of the mantle are also critical in constraining its nature. Finally, chondritic meteorites provide the inventory of materials available to form the Earth; while the Earth certainly differs in its composition from chondrites, we need to relate terrestrial composition to that of chondrites in a way consistent with thermodynamics and what we know of the behavior of the elements. We'll find that most modern estimates converge on an estimate of terrestrial composition within a few percent for the most abundant elements. For trace elements, estimates diverge by 25% or so, but that is nevertheless remarkable, considering how heterogeneous the Earth is.

The heterogeneous nature of the mantle comes into full focus when we examine the trace element and isotopic composition of basalts. Basalts are our most abundant mantle sample, but as partial melts they are not compositionally representative except for isotope ratios. The understanding of trace element behavior in partial melting and fractional crystallization we gain in Chapter 7 nevertheless allows us to constrain mantle trace element compositions. What we find from combining trace elements and isotope ratios is that the mantle consists of identifiable chemical reservoirs whose evolution we can partly reconstruct. Mid-ocean ridge basalts, easily the most voluminous on the planet, come from a shallow reservoir from which melt has previously been extracted to form the crust. Oceanic island and other basalts produced by melting of mantle plumes rising from the deep mantle clearly derive from different reservoirs. Although these too shows evidence of previous melt extraction, they have been reenriched in the elements lost. Furthermore, stable isotope ratios in these basalts demonstrate conclusively that they contain material once at the surface of the Earth. This is truly remarkable: the surface

and deep Earth are connected by a grand geochemical cycle.

The core, as we noted earlier, consists of iron-nickel alloy. You might ask how can we be confident about the composition of something we have never sampled and have no prospect of ever sampling? The answer is again the geophysical constraints, which tell us that the core is very dense, and the composition of chondrites, which tell us that the only elements of sufficient abundance and density to form the core are iron and nickel. That conclusion is reinforced by iron meteorites, most of which are cores of asteroids. There is a problem, however; namely, that any combination of iron and nickel will be denser than the Earth's core at relevant temperatures and pressures. These elements must thus be diluted with perhaps some 5% or so of one or more lighter elements. The meteorite inventory of what was available and the isotopic composition of some of the candidate *light elements*, such as silicon, helps us narrow the possibilities, but we do not yet have a firm answer. Experiments showing how elements partition between silicate and iron liquids together with thermodynamics places important constraints on what is possible. Comparing the composition of the mantle with that of chondritic meteorites show that the mantle is highly depleted in elements, including the most valuable metals such as platinum and gold, that we expect to partition into iron liquid and since this partitioning is temperature and pressure dependent, we can begin to develop scenarios on how the core formed.

Then we turned to the crust, first the oceanic crust, then to the continental crust. The first question is its composition, an easier one to answer than the composition of mantle and core. It is not an easy task, however, given how heterogeneous the crust is and while the surface is easily sampled, the lower crust is not. Nevertheless, we continue to build on the work of Clark and Goldschmidt and refine estimates of crustal composition. Then we turn our attention to how the crust formed. We can certainly establish that the continental crust has formed through partial melting of the mantle, but in what tectonic environment under what circumstances, and when? Did it form early in Earth's history, steadily through time or in pulses, or perhaps only recently? And how permanent is it? We know a lot, but we're still struggling to completely answer these questions.

Life is, of course, ubiquitous at the surface of the Earth and has modified the planet in remarkable ways: life is a geologic force. Organisms produce a vast array of chemicals that find their way into the physical environment. As we noted, modern geochemistry differs from what Schönbein envisioned in that it encompasses organic as well as inorganic matter, and these organic substances are ubiquitous at the surface of the Earth. This is the subject to which we turn in Chapter 12. After briefly exploring the nature and structure of organic compounds and the role they play in life, we'll survey their presence in soils and natural waters. Once outside a cell, organic substances are subject to attack by microbes and begin to degrade almost immediately. Yet some can survive on millennial time scales and longer. An emerging paradigm emphasizes the importance of adsorption of mineral surfaces in resisting degradation. The ability of dissolved organic molecules to adsorb complex inorganic substances is important: it retains nutrients in soil and maintains otherwise insoluble metals in solution. Some of these long surviving molecules, or at least their hydrocarbon skeletons, can be associated with specific biomolecules. Some of these *biomarkers*, or chemical fossils, are restricted to specific groups of organisms and can thus help us reconstruct past environments and biological evolution. Others have proved useful in reconstructing past atmospheric CO_2 levels and paleotemperatures.

Organic substances are an important part of the carbon cycle. Photosynthesis and subsequent sequestration of organic matter in sedimentary rocks transformed the Earth's initial CO_2-rich atmosphere to one containing free oxygen, which first occurred 2.3 billion years ago in the *Great Oxidation Event*. For the next billion and a half years, some atmospheric oxygen was present, but not enough to support metazoans (animals). Then around 600 million years ago, atmospheric oxygen levels began to rise again and just at this time the first animals appear in the fossil record. But as oxygen was produced, atmospheric CO_2 was drawn down. As a greenhouse gas, CO_2 plays a critically important role governing climate and the times oxygen rose in the atmosphere were accompanied by glaciations in the Proterozoic and Paleozoic.

This was not the cause of the Pleistocene glaciations, however. Stable isotope studies demonstrated that glacial-interglacial cycles correlated with small changes in the Earth's orbit and rotation (the Milankovitch variations). These were the pacemaker of the Pleistocene glacial cycles, but it was shuffling of CO_2 between the atmosphere and oceans that actually caused the climate swings.

Burial of organic carbon in sediments has also produced the coal and petroleum that have provided the energy to power the global economy since the Industrial Revolution. We'll examine the processes that transform this buried organic matter into these energy resources. But in burning fossil fuels we are increasing atmospheric CO_2, which, not surprisingly, is warming the planet and initiating a host of other climate changes.

In Chapter 13, we will focus attention on the *Critical Zone,* which is the land surface from the top of vegetation to the bottom of circulating groundwater. It is so called because essentially all terrestrial life lives within it and ultimately all life, including marine life, depends on processes occurring within this zone. It is here that rock comes in contact with water and air, and primary minerals are replaced by new ones. These weathering reactions produce soil and release nutrients that make terrestrial life possible. Some fraction of these nutrients is carried to the oceans by streams and rivers and make marine life possible. Life is an integral part of the weathering and soil development process, as organic acids help to break down rock and movements of metals complexed by organic molecules contribute to the development of distinct soil horizons over time.

Weathering of silicate rocks is another important part of the carbon cycle and consequently influences climate on time scales of tens to hundreds of millions of years. This is because carbonic acid produced by dissolution of CO_2 in water provides most of the acidity necessary to drive weathering reactions. The result is a solution enriched in calcium and bicarbonate, which is then carried to the oceans by streams and rivers to be precipitated as carbonate sediment, thus removing CO_2 from the atmosphere until it is again released by metamorphism or volcanism to the atmosphere as CO_2. Over Earth's history, there has been a net transfer of CO_2 from the atmosphere to sedimentary carbonate, keeping Earth's surface temperature within the habitable range even as the Sun has grown steadily brighter. We'll examine weathering reactions and their rates from the perspective of field studies. We'll find that lithology, climate and hydrology, topography, and the biota all exert important controls on weathering rates. We'll then turn our attention to the composition of streams and rivers and see how these same factors control the composition of streams and rivers. Finally, we look at the composition of saline lakes and see how the process of fractional crystallization leads to a great diversity of their compositions.

In Chapter 14, we follow the rivers to where they lead: the oceans. The oceans are salty and alkaline because, as Anton Lavoisier put it, they are "the rinsings of the Earth," that is, they contain the weathering products of the land surface. Just six components, Na^+, Mg^{2+}, Ca^{2+}, K^+, Cl^-, and SO_4^{2-}, make up 99.3% of the dissolved solids, and these are always present in the same proportions and in the same proportion to the total, the *salinity*, which is about 35 parts per thousand by weight on average. A final component, HCO_3^-, brings the total to 99.7%.

Ultimately, the concentrations of all components in seawater are controlled both by the rates at which they are added from *sources* and the rates at which they are removed by *sinks*. Rivers are the major source of most elements in seawater, but the atmosphere is the major source for dissolved gases as well as a few metals such as Al, Pb, and Th, which reach the ocean in wind-blown dust. Ridge crust hydrothermal activity is an important source of some elements, but it is also an important sink for others. Sediments are the major sink for most elements, and half the ocean floor is covered by the carbonate and siliceous shells of planktonic organisms. Evaporites are the major sink for Na^+, K^+, Cl^-, and SO_4^{2-}, but these form discontinuously through time. The last major evaporite deposit formed in the Mediterranean when tectonics closed the Strait of Gibraltar between 6 and 5.3 million years ago. The Mediterranean dried up nearly entirely, and the resulting drop in base level allowed rivers running into it, such as the Rhone and Nile, to cut channels 1000 m below their present levels, which subsequently filled with sediment when the Gibraltar connection reopened. The vast, thick beds of salt deposited beneath the Mediterranean during this time were enough to decrease global ocean salinity by 5%.

Biological processes exert an extremely important influence on ocean chemistry. Unlike the other major components, the concentration of HCO_3^- varies, mainly due to photosynthesis and respiration, although calcium carbonate precipitation and CO_2 exchange with the atmosphere also contribute to variations. Photosynthesis is restricted by light availability to the upper hundred meters or so, while respiration occurs throughout the ocean. Temperature and salinity establish a strong density gradient in the ocean that has the effect of limiting exchange between this photic zone and the deep ocean. Once it is cooled at high latitudes and sinks into the deep ocean, water remains there on time scales of ~1000 years. Sinking organic remains can fall through the water column and this density barrier and can be remineralized through respiration in the deep water. This transports dissolved CO_2 from the surface to this deep water where it builds up, a phenomenon known as the *biological pump*. Consequently HCO_3^- is present in higher concentration in deep water, which also results in a decrease in pH from ~8.1 in the surface water to ~7.6 in the deep water. Partly as a result of this variation in pH, the ocean becomes undersaturated with respect to calcium carbonate with depth so that carbonate shells formed in the surface water tend to dissolve of depth and do so completely below a depth at ~4500 m. Falling carbonate shells also contribute to the biological pump, and as we noted above, this is also part of the long-term carbon cycle controlling climate. On much shorter time scales, changes in ocean circulation and biological productivity changed the efficiency of this biological pump between glacial and interglacial periods, resulting in a transfer of CO_2 from the atmosphere to the deep ocean, very much amplifying the Milankovitch climate signal.

Unlike the major elements, concentrations of most minor and trace elements are quite variable in the oceans and much of this variation is due to biologic activity that imposes vertical concentration gradients, as these elements are taken up by phytoplankton in the surface water and released by respiration in the deep water. This includes not only nutrients such as P, Si, and Fe, but also nonutilized elements such as Ge because organisms take them up incidentally. A few elements, such as Al and Pb, show the opposite pattern: enrichment in the surface water and depletion in deep water because wind-deposited dust is the primary source of these elements and they are quickly scavenged onto particle surfaces after deposition.

In the final chapter we see how geochemistry can be used to address the needs of society, specifically, its need for mineral resources and environmental protection. The story of civilization is in some respects the story of increasingly sophisticated tools. The Stone Age ended when people learned to produce copper metal from copper sulfide ores around 7000 years ago. Copper tools were subsequently replaced by bronze ones and then by iron ones beginning around 3000 years ago. In a sense, we still live in the Copper and Iron Ages, however, as 21 million tons of copper ore and 2.5 billion tons of iron ore were mined globally in 2018. In the United States, about half the demand for metals is met by recycling, but modern society still need enormous amounts. Furthermore, modern technology requires a great variety of metals, many of which were unknown as recently as two centuries ago. At least 80 different elements are incorporated in smartphones or used in their production, including exotic ones like neodymium, europium, and tantalum. Two other exotic elements, cadmium and tellurium, are used to produce CdTe solar panels, which have the highest efficiency and can be produced in thinner films than other solar cells.

We'll discuss the process of geochemical exploration and consider examples of the formation of a variety of ore deposit types. The first of these is the Bushveld complex of South Africa, which is an example of *orthomagmatic ores*, in which the ore had precipitated directly from magma. The Bushveld, which outcrops over an area the size of Ireland, is a layered mafic intrusion that formed 2 billion years ago and hosts the world's largest reserves of platinum group elements, Cr, and V. Decades of geochemical detective work

have shown that these ores formed as fractional crystallization combined with repeated intrusions of magma and assimilation of surrounding crust periodically saturated the magma in ore-forming minerals, including chromite, magnetite, and sulfides that settled out of the magma chamber to formed distinct bands. In contrast, *hydromagmatic ores* such as porphyry copper deposits, which are the primary source of copper ore, form when a saline aqueous fluid exsolves from a magma and intrudes, often with violent force, into surrounding rock. Laboratory experiments together with analysis of fluid inclusions in these ores have revealed that many metals, including Cu, Zn, Pb, Co, Sn, and Au, form highly soluble chloride and sulfide complexes in these fluids at elevated temperatures and partition into the fluid phase from the magma, then precipitate when the solution cools. These form mainly from subduction-related magmas because they are rich in water and oxidizing; the latter prevents premature precipitation from the magma of the ore metals as sulfides. Many tin deposits form in a similar way but the magmas are produced by melting of Sn-rich sediments within the crust and reducing conditions allow Sn concentrations to build up through fractional crystallization and Sn is often complexed by F rather than Cl.

Hydrothermal ores also precipitate from aqueous solution and chloride complexes are also important in transporting metals in these deposits. The fluid, however, is derived from seawater or formation brines within the crust. These types of deposits include volcanogenic massive sulfides (VMS); mid-ocean ridge hydrothermal systems are actively forming examples of this type of deposit. The ore-forming fluids can be directly sampled and their chemistry determined; study of these systems has provided much insight into how VMS deposits form. Seawater is warmed as it penetrates the hot, young ocean crust and a series of reactions result in the solution becoming acidic and reducing. Under these conditions, metals, most notably Cu, Zn, and Pb, are leached from the rock. When temperatures reach 350–400°C, the fluid rises, eventually mixing with seawater whereupon the metals precipitate as sulfides.

We'll examine two examples of *sedimentary ore deposits*. The first is banded iron formations, which are the principal source of iron ore. Most of these formed around the time the atmosphere first became oxidizing about 2.3–2.4 billion years ago as ferrous iron-rich deep ocean water upwelled to the surface and the iron was oxidized to the insoluble ferric form. Directly or indirectly, the evolution of photosynthetic life appears responsible for them. Brines associated with saline lakes and *salars*, or salt flats, and their associated brines, particularly from the high plateaus of the Andes and Tibet, are becoming the most important source of lithium, which is needed for high performance batteries in everything from cell phones to electric cars. But not all such brines are Li-rich; we learn the conditions under which Li-rich brines form. *Weathering-related ore deposits* include bauxite, the ore of Al, and laterites, which are sources of Ni, Fe, and rare earths. These form through extreme weathering of soils such that little remains but these highly insoluble elements; what we have learned about weathering, soil-forming processes, and the geochemistry of these metals will serve us well understand how these deposits form. Because of their importance to everything from flat panel computer displays and televisions to high performance magnets in wind turbines, electric cars, and speakers, we briefly examine rare earth ore deposits, which fall into many of the above categories.

Finally, we put our geochemical toolbox to use to understand how human activities can degrade environmental quality and how this can be addressed. Like ore deposits, this is an enormous topic and we have space to consider only a few examples. We begin with the problem of eutrophication and associated anoxia in fresh water lakes, using Lake Erie, one of the Great Lakes of North America, as an example. Eutrophication refers to situations where nutrient levels in water allow excessive growth of algae, usually cyanobacteria, which produce microcystin toxins. Lakes typically become temperature-stratified in summer such that oxygen in the deep water is not replenished. Bacteria consuming the remains of algae falling into the deep water can consume all available oxygen leading to anoxic conditions in the deep water and consequent fish kills. Persistent eutrophication in Lake Erie was successfully addressed by regulations in the 1970s that severely limited nutrients from sewage,

industrial effluents and particulate phosphorus in agricultural runoff and the lake was restored to health. In the late 1990s eutrophication is summer began to occasionally reoccur due to dissolved phosphorus from agricultural runoff. Solving the problem will require further modification of farming practices.

Toxic metals are another important environmental problem. One source is mining of sulfide deposits, such as the several types described above. Sulfides exposed to water and atmospheric oxygen quickly weather to produce sulfuric acid, resulting in a problem known as *acid mine drainage*. Not only is the acidity a problem with pH values as low as 2, but under these conditions many otherwise insoluble toxic metals become soluble. The solution is certainly not to simply shut down mines as when pumps are shut off, water penetrates in mine shafts, pits, and tailings ponds and the problem worsens. Indeed, the bigger problem is old, abandoned mines as a number of strategies are deployed in modern mining operations to prevent the problem. Lead and mercury are highly toxic metals and anthropogenic release of these elements to the atmosphere has polluted the entire surface of the planet. Lead, however, is an example of an environmental success story largely due to the efforts of one geochemist, Claire Patterson. Regulations that eliminated Pb from gasoline and emissions from smelters have dramatically reduced the amount of Pb in the environment. Regulations have also starkly reduced emissions of Hg, at least in developed countries, and local sources of extreme pollution, such as in Minamata, Japan, where mercury poisoning killed over 1700 people and disabled many more, have been eliminated in most cases. Nevertheless, levels in the atmosphere, soils, plants, the ocean, and many fish species remain high and will decrease only slowly in the future, even if all emissions are eliminated. An understanding of the unique geochemistry of Hg will enable us to understand why.

Finally, we examine the problem of *acid rain*. This results from burning of fossil fuels, particularly coal, which oxidizes sulfur and nitrogen ultimately to sulfuric and nitric acid, although use of nitrogen fertilizers also contributes. This can lower pH in rain to values as low as 4. Depending on the nature of the soil and bedrock this may or may not be a rocks with low acid neutralizing capacity, the low pH alone can have deleterious effects on trees, fish, and aquatic invertebrates, but that is not the main problem. Instead, the principal problems are loss of cations such as Ca^{2+} and aluminum toxicity. Aluminum is one of the most abundant elements in the Earth's crust, yet natural Al toxicity is rare. Once we understand the geochemistry of Al, we'll be able to understand why this is usually not an issue but can be when rain is acidic. Acid rain is another environmental success story, although a still unfolding one. Regulations have greatly reduced emissions in the developed world, but it will take decades before soils and stream chemistry returns to natural levels and for damaged ecosystems to heal.

REFERENCES AND SUGGESTIONS FOR FURTHER READING

Clarke, F. W. 1908. The Data of Geochemistry. *US Geological Survey Bulletin* 770. Washington, US Government Printing Office.

Vernadsky, V. I. 1926. *Biosfera*. Leningrad, Scientific Chemico-Technical Publishing.

Goldschmidt, V. M. 1937. The principals of distribution of chemical elements in mineral and rocks. *Journal of the Chemical Society of London* 1937: 655–73. doi: 10.1039/JR9370000655.

Gribbin, J. R. 1984. *In Search of Schrödinger's Cat: Quantum Physics and Reality*. New York, Bantam Books.

Lindley, D. 2001. *Boltzmann's Atom: The Great Debate that Launched a Revolution in Physics*, New York, The Free Press.

Morris, R. 2003. *The Last Sourcers: The Path from Alchemy to the Periodic Table*, Washington, DC, Joseph Henry Press.

Strathern, P. 2000. *Mendeleyev's Dream: The Quest for the Elements*, London, Berkley.

Chapter 2

Energy, entropy, and fundamental thermodynamic concepts

2.1 THE THERMODYNAMIC PERSPECTIVE

We defined geochemistry as the application of chemical knowledge and techniques to solve geologic problems. It is appropriate, then, to begin our study of geochemistry with a review of physical chemistry. Our initial focus will be on *thermodynamics*. Strictly defined, thermodynamics is the study of energy and its transformations. Chemical reactions and changes of states of matter inevitably involve energy changes. By using thermodynamics to follow the energy, we will find that we can predict the outcome of chemical reactions, and hence the state of matter in the Earth. In principle, at least, we can use thermodynamics to predict at what temperature a rock will melt and the composition of that melt, and we can predict the sequence of minerals that will crystallize to form an igneous rock from the melt. We can predict the new minerals that will form when that igneous rock undergoes metamorphism, and we can predict the minerals and the composition of the solution that forms when that metamorphic rock weathers and the nature of minerals that will ultimately precipitate from that solution. Thus, thermodynamics allows us to understand (in the sense that we defined understanding in Chapter 1) a great variety of geologic processes.

Thermodynamics embodies a *macroscopic* viewpoint, that is, it concerns itself with the properties of a system, such as temperature,

volume, and heat capacity, and it does not concern itself with how these properties are reflected in the internal arrangement of atoms. The *microscopic* viewpoint, which is concerned with transformations on the atomic and subatomic levels, is the realm of *statistical mechanics* and *quantum mechanics*. In our treatment, we will focus mainly on the macroscopic (thermodynamic) viewpoint, but we will occasionally consider the microscopic (statistical mechanical) viewpoint when our understanding can be enhanced by doing so. More detailed treatments of geochemical thermodynamics can be found in Anderson and Crerar (1993), Nordstrom and Munoz (1986), and Fletcher (1993).

In principle, *thermodynamics is only usefully applied to systems at equilibrium*. If an equilibrium system is perturbed, thermodynamics can predict the new equilibrium state, but cannot predict how, how fast, or indeed whether the equilibrium state will be achieved. (The field of *irreversible thermodynamics*, which we will not treat in this book, attempts to apply thermodynamics to nonequilibrium states. However, we will see in Chapter 5 that thermodynamics, through the *principle of detailed balancing* and *transition state theory*, can help us predict reaction rates.)

Kinetics is the study of rates and mechanisms of reaction. Whereas thermodynamics is concerned with the ultimate equilibrium state and not concerned with the pathway

Geochemistry, Second Edition. William M. White.
© 2020 John Wiley & Sons Ltd. Published 2020 by John Wiley & Sons Ltd.
Companion website: www.wiley.com/go/white/geochemistry

to equilibrium, kinetics concerns itself with the pathway to equilibrium. Very often, equilibrium in the Earth is not achieved, or achieved only very slowly, which naturally limits the usefulness of thermodynamics. Kinetics helps us to understand how equilibrium is achieved and why it is occasionally not achieved. Thus, these two fields are closely related, and together form the basis of much of geochemistry. We will treat kinetics in Chapter 5.

2.2   THERMODYNAMIC SYSTEMS AND EQUILIBRIUM

We now need to define a few terms. We begin with the term *system*, which we have already used. A thermodynamic system is simply that part of the universe we are considering. Everything else is referred to as the *surroundings*. A thermodynamic *system* is defined at the convenience of the observer in a manner so that thermodynamics may be applied. While we are free to choose the boundaries of a system, our choice must nevertheless be a careful one as the success or failure of thermodynamics in describing the system will depend on how we have defined its boundaries. Thermodynamics often allows us this sort of freedom of definition. This can certainly be frustrating, particularly for someone exposed to thermodynamics for the first time (and often even the second or third time). But this freedom allows us to apply thermodynamics successfully to a much broader range of problems than otherwise.

A system may be related to its environment in a number of ways. An *isolated* system can exchange neither energy (heat or work) nor matter with its surroundings. A truly isolated system does not exist in nature, so this is strictly a theoretical concept. An *adiabatic* system can exchange energy in the form of work, but not heat or matter, with its surroundings, that is to say it has thermally insulating boundaries. Though a truly adiabatic system is probably also a fiction, heat transport in many geologic systems is sufficiently slow that they may be considered adiabatic. *Closed* systems may exchange energy, in the form of both heat and work, with their surrounding but cannot exchange matter. An *open* system may exchange both matter and energy across it boundaries. The

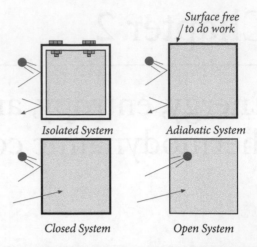

Figure 2.1   Systems in relation to their surroundings. The ball represents mass exchange, the arrow represents energy exchange.

various possible relationships of a system to its environment are illustrated in Figure 2.1.

Depending on how they behave over time, systems are said to be either in *transient* or *time-invariant* states. Transient states are those that change with time. Time-independent states may be either *static* or *dynamic*. A dynamic time-independent state, or *steady-state*, is one whose thermodynamic and chemical characteristics do not change with time despite internal changes or exchanges of mass and energy with its surroundings. As we will see, the ocean is a good example of a steady-state system. Despite a constant influx of water and salts from rivers and loss of salts and water to sediments and the atmosphere, its composition does not change with time (at least on geologically short time-scales). Thus, a steady-state system may also be an open system. We could define a static system as one in which nothing is happening. For example, an igneous rock or a flask of seawater (or some other solution) is static in the macroscopic perspective. From the statistical mechanical viewpoint, however, there is a constant reshuffling of atoms and electrons, but with no net changes. Thus, static states are generally also dynamic states when viewed on a sufficiently fine scale.

Let's now consider one of the most important concepts in physical chemistry, that of *equilibrium*. One of the characteristics of the *equilibrium state* is that it is static from a macroscopic perspective, that is, it does not change measurably with time. Thus, the equilibrium state is always time-invariant. However, while a reaction A→B may appear to have reached static equilibrium on a macroscopic scale, this reaction may still proceed on a microscopic scale but with the rate of reaction A→B being the same as that of B→A. Indeed, as we shall see in Chapter 5, a kinetic definition of equilibrium is that the forward and reverse rates of reaction are equal.

The equilibrium state is entirely independent of the manner or pathway in which equilibrium is achieved. Indeed, once equilibrium is achieved, no information about previous states of the system can be recovered from its thermodynamic properties. Thus a flask of CO_2 produced by combustion of graphite cannot be distinguished from CO_2 produced by combustion of diamond. In achieving a new equilibrium state, all records of past states are destroyed.

Time-invariance is a necessary but not sufficient condition for equilibrium. Many systems exist in metastable states. Diamond at the surface of the Earth is not in an equilibrium state, despite its time-invariance on geologic time-scales. Carbon exists in this metastable state because of kinetic barriers that inhibit transformation to graphite, the equilibrium state of pure carbon at the Earth's surface. Overcoming these kinetic barriers generally requires energy. If diamond is heated sufficiently, it will transform to graphite, or in the presence of sufficient oxygen, to CO_2.

The concept of equilibrium versus metastable or unstable (transient) states is illustrated in Figure 2.2 by a ball on a hill. The equilibrium state is when the ball is in the valley at the bottom of the hill, because its gravitational potential energy is minimized in this position. When the ball is on a slope, it is in an unstable, or transient, state and will tend to roll down the hill. However, it may also become trapped in small depressions on the side of the hill, which represent metastable states. The small hill bordering the depression represents a kinetic barrier. This kinetic barrier can only be overcome when the ball acquires enough energy to roll up and over it. Lacking that energy, it will exist in the metastable state indefinitely.

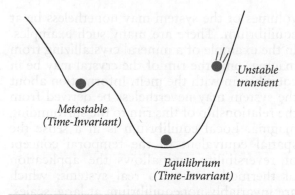

Figure 2.2   States of a system.

In Figure 2.2, the ball is at equilibrium when its (gravitational) potential energy is lowest (i.e., at the bottom of the hill). This is a good definition of equilibrium in this system, but as we will soon see, it is not adequate in all cases. A more general statement would be to say that the equilibrium state is the one toward which a system will change in the absence of constraints. So in this case, if we plane down the bump (remove a constraint), the ball rolls to the bottom of the hill. At the end of this chapter, we will be able to produce a thermodynamic definition of equilibrium based on the Gibbs free energy. We will find that, for a given pressure and temperature, the chemical equilibrium state occurs when the Gibbs free energy of the system is lowest.

Natural processes proceeding at a finite rate are irreversible under a given set of conditions; that is, they will only proceed in one direction. Here we encounter a problem in the application of thermodynamics: if a reaction is proceeding, then the system is out of equilibrium and thermodynamic analysis cannot be applied. This is one of the first of many paradoxes in thermodynamics. This limitation might at first seem fatal, but we get around it by imagining a comparable reversible reaction. Reversibility and local equilibrium are concepts that allow us to "cheat" and apply thermodynamics to nonequilibrium situations. A "reversible" process is an idealized one where the reaction proceeds in sufficiently small steps that it is in equilibrium at any given time (thus allowing the application of thermodynamics).

Local equilibrium embodies the concept that in a closed or open system, which may not be at equilibrium on the whole, small volumes of the system may nonetheless be at equilibrium. There are many such examples. In the example of a mineral crystallizing from magma, only the rim of the crystal may be in equilibrium with the melt. Information about the system may nevertheless be derived from the relationship of this rim to the surrounding magma. Local equilibrium is in a sense the spatial equivalent to the temporal concept of reversibility and allows the application of thermodynamics to real systems, which are invariably nonequilibrium at large scales. Both local equilibrium and reversibility are examples of simplifying assumptions that allow us to treat complex situations. In making such assumptions, some accuracy in the answer may be lost. Knowing when and how to simplify a problem is an important scientific skill.

2.2.1 Fundamental thermodynamic variables

In the next two chapters we will be using a number of variables, or properties, to describe thermodynamic systems. Some of these will be quite familiar to you, others less so. Volume, pressure, energy, heat, work, entropy, and temperature are the most fundamental variables in thermodynamics. As all other thermodynamic variables are derived from them, it is worth our while to consider a few of these properties.

Energy is the capacity to produce change. It is a fundamental property of any system, and it should be familiar from physics. By choosing a suitable reference frame, we can define an absolute energy scale. However, it is changes in energy that are generally of interest to us rather than absolute amounts. Work and heat are two of many forms of energy. Heat, or thermal energy, results from random motions of molecules or atoms in a substance and is closely related to kinetic energy. Work is done by moving a mass, M, through some distance, $x = X$, *against* a force F:

$$w = -\int_0^X F \, dx \qquad (2.1)$$

where w is work and force is defined as mass times acceleration:

$$F = M \frac{dv}{dt} \qquad (2.2)$$

(the minus sign is there because of the convention that work done on a system is positive, work done by a system is negative). This is, of course, Newton's first law. In chemical thermodynamics, pressure–volume work is usually of more interest. Pressure is defined as force per unit area:

$$P = \frac{F}{A} \qquad (2.3)$$

Since volume is area times distance, we can substitute eqn. 2.3 and $dV = A \, dx$ into eqn. 2.1 and obtain:

$$w = -\int_{x_0}^{x_1} \frac{F}{A} A \, dx = -\int_{V_0}^{V_1} P \, dV \qquad (2.4)$$

Thus, work is also done as a result of a volume change in the presence of pressure.

Potential energy is energy possessed by a body by virtue of its position in a force field, such as the gravitational field of the Earth, or an electric field. Chemical energy will be of most interest to us in this book. Chemical energy is a form of potential energy stored in chemical bonds of a substance. Chemical energy arises from the forces involved in the interaction between atoms and electrons. Internal energy, which we denote with the symbol U, is the sum of the potential energy arising from these forces as well as the kinetic energy of the atoms and molecules (i.e., thermal energy) in a substance. It is internal energy that will be of most interest to us.

We will discuss all these fundamental variables in more detail in the next few sections.

2.2.2 Properties of state

Properties or variables of a system that depend only on the present state of the system, and not on the manner in which that state was achieved, are called *variables of state* or *state functions*. *Extensive* properties depend on total size of the system. Mass, volume, and energy are all extensive properties. Extensive properties are additive, the value for the whole being the sum of values for the parts. *Intensive* properties are independent of the size of a system, for example temperature, pressure, and viscosity. They are not additive, for example, the temperature of a system is not the sum of the temperature of its parts.

In general, an extensive property can be converted to an intensive one by dividing it by some other extensive property. For example, density is the mass per volume and is an intensive property. It is generally more convenient to work with intensive rather than extensive properties. For a single component system not undergoing reaction, specification of three variables (two intensive, one extensive) is generally sufficient to determine the rest, and specification of any two intensive variables is generally sufficient to determine the remaining intensive variables.

A final definition is that of a *pure substance*. A pure substance is one that cannot be separated into fractions of different properties by the same processes as those considered. For example, in many processes, the compound H_2O can be considered a pure substance. However, if electrolysis were involved, this would not be the case.

2.3 EQUATIONS OF STATE

Equations of state describe the relationship that exists among the state variables of a system. We will begin by considering the ideal gas law and then very briefly consider two more complex equations of state for gases.

2.3.1 Ideal gas law

The simplest and most fundamental of the equations of state is the *ideal gas law*.* It states that pressure, volume, temperature, and the number of moles of a gas are related as:

$$\boxed{PV = NRT} \tag{2.5}$$

where P is pressure, V is volume, N is the number of moles, T is thermodynamic, or absolute temperature (which we will explain shortly), and R is the ideal gas constant[†] (an empirically determined constant equal to 8.314 J/mol-K, 1.987 cal/mol-K, or 82.06 cc-atm/deg-mol). This equation describes the relation between two extensive (mass-dependent) parameters, volume and the number of moles, and two intensive (mass-independent) parameters, temperature and pressure. We earlier stated

that if we defined two intensive and one extensive system parameter, we could determine the remaining parameters. We can see from eqn. 2.5 that this is indeed the case for an ideal gas. For example, if we know N, P, and T, we can use eqn. 2.5 to determine V.

The ideal gas law, and any equation of state, can be rewritten with intensive properties only. Dividing V by N we obtain the *molar volume*, $\overline{V}$. Substituting $\overline{V}$ for V and rearranging, the ideal gas equation becomes:

$$\overline{V} = \frac{RT}{P} \tag{2.6}$$

The ideal gas equation tells us how the volume of a given amount of gas will vary with pressure and temperature. To see how molar volume will vary with temperature alone, we can differentiate eqn. 2.6 with respect to temperature, holding pressure constant, and obtain:

$$\left(\frac{\partial V}{\partial T}\right)_P = \frac{\partial(NRT/P)}{\partial T} \tag{2.7}$$

which reduces to:

$$\left(\frac{\partial V}{\partial T}\right)_P = \frac{NR}{P} \tag{2.8}$$

It would be more useful to know the *fractional* volume change rather than the absolute volume change with temperature, because the result in that case does not depend on the size of the system. To convert to the fractional volume change, we simply divide the equation by V:

$$\frac{1}{V}\left(\frac{\partial V}{\partial T}\right)_P = \frac{NR}{PV} \tag{2.9}$$

Comparing eqn. 2.9 with eqn. 2.5, we see that the right-hand side of the equation is simply $1/T$, thus

$$\frac{1}{V}\left(\frac{\partial V}{\partial T}\right)_P = \frac{1}{T} \tag{2.10}$$

The left-hand side of this equation, the fractional change in volume with change in temperature, is known as the *coefficient of thermal expansion*, α:

$$\alpha \equiv \frac{1}{V}\left(\frac{\partial V}{\partial T}\right)_P \tag{2.11}$$

---

* Frenchman Joseph Gay-Lussac (1778–1850) established this law based on the earlier work of Englishman Robert Boyle and Frenchman Edme Mariotte.
[†] We will generally refer to it merely as the gas constant.

For an ideal gas, the coefficient of thermal expansion is simply the inverse of temperature.

The *compressibility* of a substance is defined in a similar manner as the fractional change in volume produced by a change in pressure at constant temperature:

$$\beta = -\frac{1}{V}\left(\frac{\partial V}{\partial P}\right)_T \qquad (2.12)$$

Geophysicists sometimes use the isothermal bulk modulus, K_T, in place of compressibility. The isothermal bulk modulus is simply the inverse of compressibility: $K_T = 1/\beta$. Through a similar derivation to the one we have just done for the coefficient of thermal expansion, it can be shown that the compressibility of an ideal gas is $\beta = 1/P$.

The ideal gas law can be derived from statistical physics (first principles), assuming the molecules occupy no volume and have no electrostatic interactions. Doing so, we find that R = N_Ak, where k is Boltzmann's constant (1.381×10–23 J/K), N_A is the Avogadro number (the number of atoms in one mole of a substance), and k is a fundamental constant that relates the average molecular energy, e, of an ideal gas to its temperature (in Kelvins) as e = 3kT/2.

Since the assumptions just stated are ultimately invalid, it is not surprising that the ideal gas law is only an approximation for real gases; it applies best in the limit of high temperature and low pressure. Deviations are largest near the condensation point of the gas.

The compressibility factor is a measure of deviation from ideality and is defined as

$$Z = PV/NRT \qquad (2.13)$$

By definition, $Z = 1$ for an ideal gas.

2.3.2 Equations of state for real gases

2.3.2.1 Van der Waals equation

Factors we need to consider in constructing an equation of state for a real gas are the finite volume of molecules and the attractive and repulsive forces between molecules arising from electric charges. The van der Waals equation is probably the simplest equation of state that takes account of these factors. The van der Waals equation is:

$$P = \frac{RT}{\overline{V} - b} - \frac{a}{\overline{V}^2} \qquad (2.14)$$

Here again we have converted volume from an extensive to an intensive property by dividing by N.

Let's examine the way in which the Van der Waals equation attempts to take account of finite molecular volume and forces between molecules. Considering first the forces between molecules, imagine two volume elements v_1 and v_2. The attractive forces will be proportional to the number of molecules or the concentrations, c_1 and c_2, in each. Therefore, attractive forces are proportional to $c_1 \times c_2 = c^2$. Since c is the number of molecules per unit volume, $c = n/V$, we see that attractive forces are proportional to $1/\overline{V}^2$. The a term is a constant that depends on the nature and strength of the forces between molecules, and will therefore be different for each type of gas.

In the first term on the right, $\overline{V}$ has been replaced by $\overline{V} - b$. b is the volume actually occupied by molecules, and the term $\overline{V} - b$ is the volume available for movement of the molecules. Since different gases have molecules of differing size, we can expect that the value of b will also depend on the nature of the gas. Table 2.1 lists the values of a and b for a few common gases.

2.3.2.2 Other equations of state for gases

The *Redlich-Kwong equation* (1949) expresses the attractive forces as a more complex function:

$$P = \frac{RT}{\overline{V} - b} - \frac{a}{T^{1/2}\overline{V}(\overline{V} + b)} \qquad (2.15)$$

The *Virial equation* is much easier to handle algebraically than the van der Waals equation and has some theoretical basis in statistical mechanics:

$$PV/RT = A + BP + CP^2 + DP^3 \ldots \qquad (2.16)$$

Table 2.1   Van der Waals constants for selected gases.

Gas	a liter-atm/mole^2	b liter/mole
Helium	0.034	0.0237
Argon	1.345	0.0171
Hydrogen	0.244	0.0266
Oxygen	1.360	0.0318
Nitrogen	1.390	0.0391
Carbon dioxide	3.592	0.0399
Water	5.464	0.0305
Benzene	18.00	0.1154

A, B, C, ... are empirically determined (temperature-dependent) constants.

2.3.3   Equation of state for other substances

The compressibility and coefficient of thermal expansion parameters allow us to construct an equation of state for any substance. Such an equation relates the fundamental properties of the substance: its temperature, pressure, and volume. The partial differential of volume with respect to temperature and pressure is such an equation:

$$dV = \left(\frac{\partial V}{\partial T}\right)_P dT + \left(\frac{\partial V}{\partial P}\right)_T dP \qquad (2.17)$$

Substituting the coefficient of thermal expansion and compressibility for $\partial V/\partial T$ and $\partial V/\partial P$ respectively we have:

$$dV = V(\alpha dT - \beta dP) \qquad (2.18)$$

Thus, to write an equation of state for a substance, our task becomes to determine its compressibility and coefficient of thermal expansion. Once we know them, we can integrate eqn. 2.18 to obtain the equation of state. These, however, will generally be complex functions of temperature and pressure, so the task is often not easy.

2.4   TEMPERATURE, ABSOLUTE ZERO, AND THE ZEROTH LAW OF THERMODYNAMICS

How do you define and measure temperature? We have discussed temperature with respect to the ideal gas law without defining it, though we all have an intuitive sense of what temperature is. We noted above that temperature of a gas is a measure of the average

(kinetic) energy of its molecules. Another approach might be to use the ideal gas law to construct a *thermometer* and define a temperature scale. A convenient thermometer might be one based on the linear relationship between temperature and the volume of an ideal gas. Such a thermometer is illustrated in Figure 2.3. The equation describing the relationship between the volume of the gas in the thermometer and our temperature, τ, is:

$$V = V_0 (1 + \gamma\tau) \qquad (2.19)$$

where V_0 is the volume at some reference point where $\tau = 0$ (Figure 2.3a) and γ is a scale factor. For example, we might choose $\tau = 0$ to be the freezing point of water and the scale factor such that $\gamma = 100$ (Figure 2.3b) occurs at the boiling point of water, as is the case in the centigrade scale. Rearranging, we have:

$$\tau = \frac{1}{\gamma}\left(\frac{V}{V_0} - 1\right) \qquad (2.20)$$

Then $\tau = 0$ at $V = V_0$. If V is less than the reference volume, then temperature will be negative on our scale. But notice that while any positive value of temperature is possible on this scale, there is a limit to the range of possible negative values. This is because V can never be negative. The minimum value of temperature on this scale will occur when V is 0. This occurs at:

$$\tau_0 = -\frac{1}{\gamma} \qquad (2.21)$$

Thus, implicit in the ideal gas law, which we used to make this thermometer, is the idea that there is an absolute minimum value, or an absolute zero, of temperature, which occurs when the volume of an ideal gas is 0. Notice

(a)

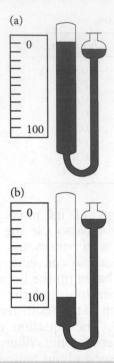

(b)

Figure 2.3 An ideal gas thermometer. The colored area is the volume occupied by the ideal gas.

that while the value $(-1/\gamma)$ of this absolute zero will depend on how we designed our thermometer (i.e., on V_0), the result, that a minimum value exists, does not. We should also point out that only an ideal gas can have a volume of 0. The molecules of real gases have a finite volume, and such a gas will have a finite volume at absolute zero.

The temperature scale used by convention in thermodynamics is the Kelvin* scale. The magnitude of units, called kelvins (not degrees kelvin) and designated K (not °K), on this scale are the same as the centigrade scale, so there are exactly 100 kelvins between the freezing and boiling points of water. There is some slight uncertainty (a very much smaller uncertainty than we need to concern ourselves with) concerning the value of absolute zero (i.e., the value of γ in eqns. 2.20 and 2.21). The scale has been fixed by choosing 273.16 kelvins to be the triple point of water (0.01°C). On this scale, the absolute zero of temperature occurs at 0 ± 0.01 kelvins. *The Kelvin scale should be used wherever temperature occurs in a thermodynamic equation.*

Temperature has another fundamental property, and this is embodied in the *zeroth law of thermodynamics*. It is sufficiently obvious from everyday experience that we might overlook it. It concerns thermal equilibrium and may be stated in several ways: *two bodies in thermal equilibrium have the same temperature* and *any two bodies in thermal equilibrium with a third are in equilibrium with each other.*

2.5 ENERGY AND THE FIRST LAW OF THERMODYNAMICS

2.5.1 Energy

The first law may be stated in various ways:

- *Heat and work are equivalent*[†].
- *Energy is conserved in any transformation.*
- *The change of energy of a system is independent of the path taken.*

All are restatements of the law of conservation of energy:

Energy can be neither created nor destroyed.

Mathematically:

$$\Delta U = Q + W \text{ or } \boxed{dU = dQ + dW} \quad (2.22)$$

---

* Named for Lord Kelvin. Born William Thomson in Scotland in 1824, he was appointed professor at Glasgow University at the age of 22. Among his many contributions to physics and thermodynamics was the concept of absolute temperature. He died in 1907.

† This may seem intuitively obvious to us, but it was not to James Joule (1818–1889), English brewer and physicist, who postulated it on the basis of experimental results. It was not obvious to his contemporaries either. His presentation of the idea of equivalence of heat and work to the British Association in 1843 was received with "entire incredulity" and "general silence". The Royal Society rejected his paper on the subject a year later. If you think about it a bit, it is not so obvious – in fact, there is no good reason why heat and work should be equivalent. This law is simply an empirical observation. The proof is a negative one: experience has found no contradiction of it. German physician Julius Mayer (1814–1878) formulated the idea of conservation of energy in 1842, but his writing attracted little attention. It was Joule's experiments with heat and work that conclusively established the principle of conservation of energy. By 1850, the idea of conservation of energy began to take hold among physicists, thanks to Joule's persistence and the support of a brilliant young physicist named William Thomson (later Lord Kelvin), who also had been initially skeptical.

Thermodynamics is concerned only with the internal energy of a system. We don't really care whether the system as a whole is in motion, i.e., whether it has kinetic energy (we do care, however, about the internal kinetic energy, or heat). For the most part, we also don't care whether it has potential energy, except to the extent that this influences the state of our system (e.g., pressure in the atmosphere is a function of the altitude, and hence would be of interest to us). In addition, we are almost always concerned *only* with energy *changes*, not with the absolute energy of a system. In thermodynamics, ΔU, not U, is the interesting quantity.

Of course, we now understand that it is not energy that is conserved, but rather mass energy. Albert Einstein proposed this important modification of Joule's result in 1905 (Einstein, 1905). Conversion of mass to energy fuels the Sun and the stars and, through radioactive decay, is an important source of energy in the Earth. Radioactive decay and nuclear fusion will be important topics in Chapters 8 and 10. For the geochemical processes we will be interested in the next few chapters, however, we can take the conservation of energy alone to be absolute.

Energy may be transferred between a system and its surroundings in several ways: heat, work, radiation, and advection (i.e., energy associated with mass gained or lost by the system). Whenever possible, we will want to choose our system such that it is closed and we don't have to worry about the latter. In most, but not all, instances of geochemical interest, radiation is not important. Thus in geochemical thermodynamics, *heat and work are the forms of energy flow of primary interest.*

2.5.2 Work

We have seen that work is the integral of force applied through a distance. Force times distance has units of energy (mass-velocity^2), thus work is a form of energy. The SI (*Système*

Internationale) unit of energy is the joule = $1\ kg\text{-}m^2/s^2$. Conversion factors for energy and other variables as well as values of important constants are listed in Appendix I.

There are several kinds of work of interest to thermodynamics, the most important of which is that involved in chemical reactions (later, when we consider oxidation and reduction reactions, we will be concerned with electrochemical work). One of the most important forms of work in classical thermodynamics is 'PV' work: expansion and contraction. Expressing eqn. 2.4 in differential form:

$$dW = -P_{ext}\,dV \qquad (2.23)$$

Pressure is force per unit area, and therefore has units of mass-distance^{-1}-time^{-2}, while volume has units of distance^3. The product of P and V therefore has units of energy: mass-(distance/time)^2.* The negative sign arises because, by convention, we define energy flowing into the system as positive. Work done by the system is thus negative, while work done on the system is positive. This conforms to a 1970 IUPAC (International Union of Pure and Applied Chemistry) recommendation.

While PV work is not as important in geochemistry as in other applications of thermodynamics, it is nevertheless of significant interest. There is, of course, a great range of pressures within the Earth.

Systems rising within the Earth, such as magma, a hydrothermal fluid, or upwelling water in the ocean or air in the atmosphere, will thus do work on their surroundings, and systems sinking, such as sediments being buried or lithosphere being subducted, will have work done on them.

We mentioned the concept of reversible and irreversible reactions and stated that a reversible reaction is one that occurs in sufficiently small steps that equilibrium is maintained. In an expansion or contraction reaction, equilibrium is maintained, and the reaction is reversible if the external pressure is equal to the internal pressure. The work done

---

* The pascal, the SI unit of pressure, is equal to $1\ kg/m\text{-}s^2$. Thus if pressure is measured in MPa (megapascals, 1 atm ≈ 1 bar $\approx 0.1\ MPa$) and volume in cc (= $10^{-6}\ m^{-3}$), the product of pressure times volume will be in Joules. This is rather convenient. It is named for French mathematician and physicist Blaise Pascal (1623–1662). Among his many contributions was the demonstration that atmospheric pressure was lower atop the Puy de Dome volcano than in the town of Clermont-Ferrand below it.

under these conditions is said to be reversible:

$$dW_{rev} = -PdV \qquad (2.24)$$

2.5.3 Path independence, exact differentials, state functions, and the first law

We said earlier that state functions are those that depend only on the present state of a system. Another way of expressing this is to say that state functions are path independent. Indeed, path independence may be used as a test of whether a variable is a state function or not. This is to say that if Y is a state function, then for any process that results in a change $Y1 \rightarrow Y2$, the net change in Y, ΔY, is independent of how one gets from $Y1$ to $Y2$. Furthermore, if Y is a state function, then the differential dY is said to be mathematically *exact*.

Let's explore what is meant by an exact differential. An exact differential is the familiar kind, the kind we would obtain by differentiating the function u with respect to x and y, and also the kind we can integrate. But not all differential equations are exact. Let's first consider the mathematical definition of an exact differential, then consider some thermodynamic examples of exact and inexact differentials.

Consider the first order differential expression:

$$Mdx + Ndy \qquad (2.25)$$

containing variables M and N, which may or may not be functions of x and y. Equation 2.25 is said to be an *exact differential* if there exists some function u of x and y relating them such that the expression:

$$du = Mdx + Ndy \qquad (2.26)$$

is the total differential of u:

$$du = \left(\frac{\partial u}{\partial x}\right)_y dx + \left(\frac{\partial u}{\partial y}\right)_x dy \qquad (2.27)$$

Let's consider what this implies. Comparing 2.26 and 2.27, we see that:

$$\frac{\partial u}{\partial x} = M \quad \text{and} \quad \frac{\partial u}{\partial y} = N \qquad (2.28)$$

A necessary, but not sufficient, condition for 2.25 to be an exact differential is that M and N must be functions of x and y.

A general property of partial differentials is the *reciprocity relation* or *cross-differentiation identity*, which states that the order of differentiation does not matter, so that:

$$\frac{\partial^2 u}{\partial x \partial y} = \frac{\partial^2 u}{\partial y \partial x} \qquad (2.29)$$

(The reciprocity relation is an important and useful property in thermodynamics, as we shall see at the end of this chapter.) If eqn. 2.26 is the total differential of u, it follows that:

$$\frac{\partial M}{\partial y} = \frac{\partial N}{\partial x} \qquad (2.30)$$

which is equivalent to:

$$\left(\frac{\partial M}{\partial y}\right)_x = \left(\frac{\partial N}{\partial x}\right)_y \qquad (2.31)$$

Equation 2.31 is a necessary and sufficient condition for 2.25 to be an exact differential; that is, if the cross-differentials are equal, then the differential expression is exact.

Exact differentials have the property that they can be integrated and an exact value obtained. This is true because they depend only on the initial and final values of the independent variables (e.g., x and y in eqn. 2.27).

Now let's consider some thermodynamic examples. Volume is a state function and we can express it as an exact differential in terms of other state functions, as in eqn. 2.17:

$$dV = \left(\frac{\partial V}{\partial T}\right)_P dT + \left(\frac{\partial V}{\partial P}\right)_T dP \qquad (2.17)$$

Substituting the coefficient of thermal expansion and compressibility for $\partial V/\partial T$ and $\partial V/\partial P$ respectively, equation 2.30 becomes equal to eqn. 2.18:

$$dV = \alpha VdT - \beta VdP \qquad (2.18)$$

According to eqn. 2.31, if V is a state function, then:

$$\frac{\partial(\alpha V)}{\partial P} = -\frac{\partial(\beta V)}{\partial T} \qquad (2.32)$$

You should satisfy yourself that eqn. 2.32 indeed holds for ideal gases and therefore that V is a state variable.

Work is not a state function, that is, the work done does not depend only on the initial and final states of a system. We would expect then that dW is not an exact differential, and indeed, this is easily shown for an ideal gas.

For PV work, d$W = -P$dV. Substituting eqn. 2.17 for dV and rearranging, we have:

$$dW = -P\left[\left(\frac{\partial V}{\partial T}\right)_P dT = \left(\frac{\partial V}{\partial P}\right)_T dP\right] \quad (2.33)$$

Evaluating $\partial V/\partial T$ and $\partial V/\partial P$ for the ideal gas equation and multiplying through by P, this becomes:

$$dW = -NRdT + \frac{NRT}{P}dP \quad (2.34)$$

but

$$\frac{\partial NR}{\partial P} \neq \frac{\partial(NRT/P)}{\partial T} \quad (2.35)$$

We cannot integrate eqn. 2.34 and obtain a value for the work done without additional knowledge of the variation of T and P because the amount of work done does not depend only on the initial and final values of T and P; it depends on the path taken. Heat is also not a state function, not an exact differential, and is also path dependent. Path dependent functions always have inexact differentials; path independent functions always have exact differentials.

On a less mathematical level, let's consider how the work and heat will vary in a transformation of a system, say from state 1 to state 2. Imagine that we burn gasoline in an open container. In this case, in the transformation from state 1 (gasoline) to state 2 (combustion products of gasoline), energy is given up by the system only as heat. Alternatively, we could burn the gasoline in an engine and recover some of the energy as work (expansion of the volume of the cylinder resulting in motion of the piston). The end states of these two transformations are the same, but the amount of heat released and work done varies, depending on the path we took. Thus, neither work nor heat can be state functions. Energy is a state function, is

path independent, and is an exact differential. Whether we burn the gasoline in an open container or an engine, the energy released will be the same. Herein lies the significance for thermodynamics of Joule's discovery: that *the sum of heat and work is independent of the path taken even though, independently, work and heat are not.*

2.6 THE SECOND LAW AND ENTROPY

2.6.1 Statement

Imagine a well-insulated box (an isolated system) somewhere in the universe (Figure 2.4). Imagine that within the box are two gases, separated by a removable partition. If we remove the partition, what happens? You know: the two gases mix completely. The process is entirely spontaneous. We have neither added energy to nor taken energy from the system, hence the first law says nothing about this process. Nor did removing the partition "cause" the reaction. This is apparent from the observation that if we reinsert the partition, the gases do not unmix. That you knew that the gases would mix (and knew as well that they would not unmix upon reinserting the partition) suggests there is something very fundamental and universal about this. We need a physical law that describes it. This is the second law.

The second law may be stated in a number of ways:

It is impossible to construct a machine that is able to convey heat by a cyclical process from one reservoir at a lower temperature to another at a higher temperature unless work

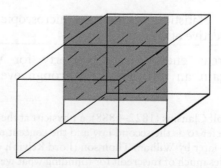

Figure 2.4 A gas-filled box with a removable partition. When the partition is removed, the gases mix. Entropy increases during this process.

*is done by some outside agency (i.e., air conditioning is never free).**

Heat cannot be entirely extracted from a body and turned into work (thus car engines always have cooling systems).

Every system left to itself will, on average, change toward a condition of maximum probability.

Introducing a new state function S called *entropy*, we may state the second law as:

The entropy of the universe always increases.

In colloquial terms we could say:

You can't shovel manure into the rear end of a horse and expect to get hay out its mouth.

The second law codifies some of our everyday experience. The first law would not prevent us from using a horse to manufacture hay from manure. It only says we cannot get more joules worth of hay out than we put in as manure. We would search in vain for any other physical law that prohibited this event. Yet our experience shows that it will not happen. Indeed, this event is so improbable that we find it comical. Similarly, we know that we can convert gasoline and oxygen to carbon dioxide and water in an internal combustion engine and use the resulting energy to drive a vehicle down the road. But adding CO_2 and water to the engine and pushing the car backwards down the street does not produce gasoline and oxygen, although such a result violates no other law of physics. *The second law states that there is a natural direction in which reactions will tend to proceed.* This direction is inevitably that of higher entropy of the system and its surroundings.

2.6.2 Statistical mechanics: a microscopic perspective of entropy

Whereas energy is a property for which we gain an intuitive feel through everyday experience, the concept of entropy is usually more difficult to grasp. Perhaps the best intuitive understanding of entropy can be obtained from the microscopic viewpoint of statistical mechanics. So for that reason, we will make the first of several brief excursions into the world of atoms, molecules, and quanta.

Let's return to our box of gas and consider what happens on a microscopic scale when we remove the partition. To make things tractable, we'll consider that each gas consists of only two molecules, so there are four all together, two red and two black. For this thought experiment, we will keep track of the individual molecules so we label them 1red, 2red, 1black, 2black. Before we removed the partition, the red molecules were on one side and the black ones on the other. Our molecules have some thermal energy, so they are free to move around. So by removing the partition, we are essentially saying that each molecule is equally likely to be found in either side of the box.

Before we removed the partition, there was only one possible arrangement of the system; this is shown in Figure 2.5a. Once we remove the partition, we have four molecules and two subvolumes, and a total of $2^4 = 16$ possible configurations (Figure 2.5b) of the system. *The basic postulate of statistical mechanics is: a system is equally likely to be found in any of the states accessible to it.* Thus, we postulate that each of these configurations is equally likely. Only one of these states corresponds to the original one (all red molecules on the left). Thus, the probability of the system being found in its original state is 1/16. That is not particularly improbable. However, suppose that we had altogether a mole of gas ($\approx 6 \times 10^{23}$ molecules). The probability of the system ever being found again in its original state is then $\approx 2 \times 10^{24}$, which is unlikely indeed.

Now consider a second example. Suppose that we have two copper blocks of identical

---

* Rudolf Clausius (1822–1888), a physicist at the Prussian military engineering academy in Berlin, formulated what we now refer to as the second law and the concept of entropy in a paper published in 1850. Similar ideas were published a year later by William Thomson (Lord Kelvin), who is responsible for the word *entropy*. Clausius was a theorist who deserves much of the credit for founding what we now call *thermodynamics* (he was responsible for, among many other things, the virial equation for gases). However, a case can be made that Sadi Carnot (1796–1832) should be given the credit. Carnot was a Parisian military officer (the son of a general in the French revolutionary army) interested in the efficiency of steam engines. The question of credit hinges on whether he was referring to what we now call entropy when he used the word *calorique*.

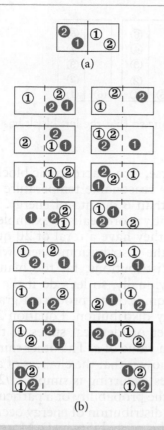

(a)

(b)

Figure 2.5 Possible distribution of molecules of a red and a black gas in a box before (a) and (b) after removal of a partition separating them.

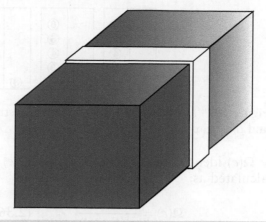

Figure 2.6 Two copper blocks at different temperatures separated by an insulator. When the insulator is removed and the blocks brought in contact, the blocks come to thermal equilibrium. Entropy increases in this process.

mass at different temperatures and separated by a thermally insulating barrier (Figure 2.6). Imagine that our system, which is the two copper blocks, is isolated in space so that the total energy of the system remains constant. What happens if we remove the insulating barrier? Experience tells us that the two copper blocks will eventually come into thermal equilibrium, and their temperatures will eventually be identical.

Now let's look at this process on a microscopic scale. We have already mentioned that temperature is related to internal energy. As we shall see, this relationship will differ depending on the nature and mass of the material of interest, but since our blocks are of identical size and composition, we can assume that temperature and energy are directly related in this case. Suppose that before we remove the insulation, the left block has one unit of energy and the right one has five units (we can think of these as quanta, but

this is not necessary). The question is, how will energy be distributed after we remove the insulation?

In the statistical mechanical viewpoint, we cannot determine how the energy will be distributed; we can only compute the possible ways it could be distributed. Each of these energy distributions is then equally likely according to the basic postulate. So let's examine how it can be distributed. Since we assume that the distribution is completely random, we proceed by randomly assigning the first unit to either the left or right block, then the second unit to either, and so on. With six units of energy, there are already more ways of distributing it ($2^6 = 64$) than we have space to enumerate here. For example, there are six ways energy can be distributed so that the left block has one unit and the right one has five units. This is illustrated in Figure 2.7. However, since we can't actually distinguish the energy units, all these ways are effectively identical. There are 15 ways, or combinations, to distribute the energy so that the left block has two units and the right has four units. Similarly, there are 15 combinations where the left block has four units and the right has two units. For this particular example, the rule is that if there are a total of E units of energy, e of which are assigned to the left block and ($E-e$) to the right, then there will

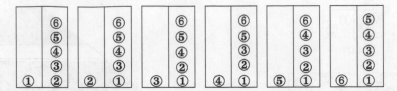

Figure 2.7 There are six possible ways to distribute six energy units so that the left block has one unit and the right block has five units.

be $\Omega(e)$ identical combinations where $\Omega(e)$ is calculated as:

$$\Omega(e) = \frac{E!}{e!(E - e!)} \qquad (2.36)^\dagger$$

Here we use $\Omega(e)$ to denote the function that describes the number of states accessible to the system for a given value of e. In this particular example, "states accessible to the system" refers to a given distribution of energy units between the two blocks. According to eqn. 2.36 there are 20 ways of distributing our six units of energy so that each block has three. There is, of course, only one way to distribute energy so that the left block has all of the energy and only one combination where the right block has all of it.

According to the basic postulate, any of the 64 possible distributions of energy are equally likely. The key observation, however, is that there are many ways to distribute energy for some values of e and only a few for other values. Thus the chances of the system being found in a state where each block has three units is 20/64 = 0.3125, whereas the chances of the system being in the state with the original distribution (one unit to the left, five to the right) are only 6/64 = 0.0938. So it is much more likely that we will find the system in a state where energy is equally divided than in the original state.

Of course, two macroscopic blocks of copper at any reasonable temperature will have far more than 6 quanta of energy. Let's take a just slightly more realistic example and suppose that they have a total of 20 quanta and compute the distribution. There will be 220 possible distributions, far too many to consider individually, so let's do it the easy way and use eqn. 2.36 to produce a graph of the probability distribution. Equation 2.36 gives the number of identical states of the system for a given value of e. The other thing that we need to know is that the chances of any one of these states occurring is simply $(1/2)^{20}$. So to compute the probability of a particular distinguishable distribution of energy occurring, we multiply this probability by Ω. More generally, the probability, P, will be:

$$P(e) = \frac{E!}{e!(E - e)!} p^e q^{E-e} \qquad (2.37)$$

where p is the probability of an energy unit being in the left block and q is the probability of it being in the right. This equation is known as the *binomial distribution*.[†] Since both p and q are equal to 0.5 in our case (if the blocks were of different mass or of different composition, p and q would not be equal), the product $p^e q^{E-e}$ is just p^E and eqn. 2.37 simplifies to:

$$P(e) = \frac{E!}{e!(E - e)!} p^E = \Omega(e) p^E \qquad (2.38)$$

[*] This is the equation when there are two possible outcomes. A more general form for a situation where there are m possible outcomes (e.g., copper blocks) would be:

$$\Omega = \frac{N!}{n_1! n_2! \ldots n_m!} \qquad (2.36a)$$

where there are n_1 outcomes of the first kind (i.e., objects assigned to the first block), n_2 outcomes of the second, etc. and $N = \sum n_i$ (i.e., N objects to be distributed).

[†] In Microsoft Excel™, you can use the BINOMDIST function to compute the outcome of this equation, which makes computing graphs such as Figure 2.8 much easier. In MATLAB™, you can compute the distribution using the probability distribution command: `pd=makedist ('Binomial',N,p)`, where N is the number of trials (e) and p is the probability of success (p). A plot similar to Figure 2.8a can be created using the *Probability Distribution Function App* with the `distool` command.

Since p^E is a constant (for a given value of E and configuration of the system), the probability of the left block having e units of energy is directly proportional to $\Omega(e)$. It turns out that this is a general relationship, so that for any system we may write:

$$\mathcal{P}(f) = C\Omega(f) \qquad (2.39)$$

where f is some property describing the system and C is some constant (in this case 0.5^{20}). Figure 2.8a shows the probability of the left block having e units of energy. Clearly, the most likely situation is that both will have approximately equal energy. The chance of one block having 1 unit and the other 19 units is very small (2×10^{-5} to be exact). In reality, of course, the number of quanta of energy available to the two copper blocks will be of the order of multiples of the Avogadro number. If one or the other block has 10 or 20 more units or even 10^{10} more quanta than the other, we wouldn't be able to detect it. Thus, energy will always appear to be distributed evenly between the two, once the system has had time to adjust.

Figure 2.8b shows Ω as a function of e, the number of energy units in the left block. Comparing the two, as well as eqn. 2.38, we see that the most probable distribution of energy between the blocks corresponds to the situation where the system has the maximal number of states accessible to it (i.e., to where $\Omega(e)$ is maximum).

According to our earlier definition of equilibrium, the state ultimately reached by this system when we removed the constraint (the insulation) is the equilibrium one. We can see here that, unlike the ball on the hill, we cannot determine whether this system is at equilibrium or not simply from its energy: the total energy of the system remained constant. In general, for a thermodynamic system, *whether the system is at equilibrium depends not on its total energy but on how that energy is internally distributed.*

Clearly, it would be useful to have a function that could predict the internal distribution of energy at equilibrium. The function that does this is the *entropy*. To understand this, let's return to our copper blocks. Initially, a thermal barrier separates the two copper blocks and we can think of each as an isolated system. We assume that each has an internal

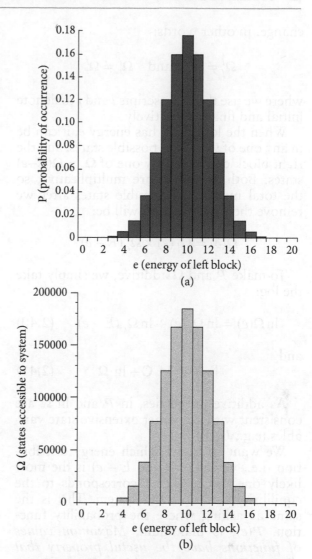

Figure 2.8 (a) Probability of one of two copper blocks of equal mass in thermal equilibrium having e units of energy when the total energy of the two blocks is 20 units. (b) Ω, number of states available to the system (combinations of energy distribution) as a function of e.

energy distribution that is at or close to the most probable one (i.e., each is internally at equilibrium). Each block has its own function Ω (which we denote as Ω_l and Ω_r for the left and right block, respectively) that gives the number of states accessible to it at a particular energy distribution. We assume that initial energy distribution is not the final one, so that when we remove the insulation, the energy distribution of the system will spontaneously change. In other words:

$$\Omega_l^i \neq \Omega_l^f \quad \text{and} \quad \Omega_r^i \neq \Omega_r^f$$

where we use the superscripts i and f to denote initial and final, respectively.

When the left block has energy e, it can be in any one of $\Omega_l = \Omega(e)$ possible states, and the right block can be in any one of $\Omega_r = \Omega(E-e)$ states. Both $\mathcal{P}$ and Ω are multiplicative, so the total number of possible states after we remove the insulation, Ω, will be:

$$\Omega(e) = \Omega_l(e) \times \Omega_r(E-e)$$

To make $\mathcal{P}$ and Ω additive, we simply take the log:

$$\ln \Omega(e) = \ln \Omega_l(e) + \ln \Omega_r(E-e) \quad (2.40)$$

and

$$\ln \mathcal{P} = \ln C + \ln \Omega \quad (2.41)$$

As additive properties, $\ln \mathcal{P}$ and $\ln \Omega$ are consistent with our other extensive state variables (e.g., U, V).

We want to know which energy distribution (i.e., values of e and $E - e$) is the most likely one, because that corresponds to the equilibrium state of the system. This is the same as asking where the probability function, $\mathcal{P}(e)$, is maximum. *Maximum values of functions have the useful property that they occur at points where the derivative of the function is 0.* That is, a maximum of function $f(x)$ will occur where $df(x)/d(x) = 0$.* Thus, the maximum value of $\mathcal{P}(e)$ in Figure 2.8 occurs where $d\mathcal{P}/de = 0$. The most probable energy distribution will therefore occur at:

$$\frac{\partial \mathcal{P}(e)}{\partial e} = 0 \quad \text{or equivalently} \quad \frac{\partial \ln \mathcal{P}(e)}{\partial e} = 0 \quad (2.42)$$

(we use the partial differential notation to indicate that, since the system is isolated, all other state variables are held constant).

Substituting eqn. 2.41 into 2.42, we have:

$$\frac{\partial \ln \mathcal{P}(e)}{\partial e} = \frac{\partial(\ln C + \ln \Omega(e))}{\partial e} = \frac{\partial \ln \Omega(e)}{\partial e} = 0 \quad (2.43)$$

(since C is a constant). Then substituting eqn. 2.40 into 2.43 we have:

$$\frac{\partial \ln \Omega(e)}{\partial e} = \frac{\partial \ln \Omega_l}{\partial e} + \frac{\partial \Omega_r \ln(E-e)}{\partial e} = 0 \quad (2.44)$$

so the maximum occurs at:

$$\frac{\partial \ln \Omega_l}{\partial e} = -\frac{\partial \ln \Omega_r(E-e)}{\partial e} \quad (2.45)$$

The maximum then occurs where the function $\partial \ln \Omega / \partial e$ for each of the two blocks are equal (the negative sign will cancel because we are taking the derivative $\partial f(-e)/\partial e$). More generally, we may write:

$$\frac{\partial \ln \Omega_l^f(E_l^f)}{\partial E_l^f} = \frac{\partial \ln \Omega_r^f(E_r^f)}{\partial E_r^f} \quad (2.46)$$

Notice two interesting things: *the equilibrium energy distribution is the one where $\ln \Omega$ is maximum (since it is proportional to P) and where the functions $\partial \ln \Omega / \partial E$ of the two blocks are equal.* It would appear that both are very useful functions. We define entropy, S, as:

$$\boxed{S = k \ln \Omega} \quad (2.47)^\dagger$$

and a function β such that:

$$\beta = \frac{\partial \ln \Omega}{\partial E} \quad (2.48)$$

where k is a constant (which turns out to be Boltzmann's constant or the gas constant; the choice depends on whether we work in units of atoms or moles, respectively). *The function S then has the property that it is maximum at equilibrium and β has the property that it is the same in every part of the system at equilibrium.*

---

* Either a maximum or minimum can occur where the derivative is 0, and a function may have several of both; so some foreknowledge of the properties of the function of interest is useful in using this property.

† This equation, which relates microscopic and macroscopic variables, is inscribed on the tombstone of Ludwig Boltzmann (1844–1906), the Austrian physicist responsible for it.

Entropy also has the interesting property that in any spontaneous reaction, the total entropy of the system plus its surroundings must increase. In our example, this is a simple consequence of the observation that the final probability, $\mathcal{P}(E)$, and therefore also Ω, will be maximum and hence never be less than the original one. Because of that, the final number of accessible states must exceed the initial number and:

$$\ln \Omega_l^f(E_l^f) + \ln \Omega_r^f(E_r^f) \geq \ln \Omega_l^i(E_l^i) + \ln \Omega_r^i(E_r^i) \tag{2.49}$$

rearranging:

$$[\ln \Omega_l^f(E_l^f) - \ln \Omega_l^i(E_l^i)] \geq [\ln \Omega_r^f(E_r^f) - \ln \Omega_r^i(E_r^i)]$$

The quantities in brackets are simply the entropy changes of the two blocks. Hence:

$$\Delta S_l \geq -\Delta S_r \tag{2.50}$$

In other words, any decrease in entropy in one of the blocks must be at least compensated for by an increase in entropy of the other block.

For an irreversible process, that is, a spontaneous one such as thermal equilibrium between two copper blocks, we cannot determine exactly the increase in entropy. Experience has shown, however, that the increase in entropy will always exceed the ratio of heat exchanged to temperature. Thus, the mathematical formulation of the second law is:

$$\boxed{dS \geq \frac{dQ}{T}} \tag{2.51}$$

Like the first law, eqn. 2.51 cannot be derived or formally proven; it is simply a postulate that has never been contradicted by experience. For a reversible reaction, that is, one that is never far from equilibrium and therefore one where dQ is small relative to T,

$$dS = \frac{dQ_{rev}}{T} \tag{2.52}$$

(see Example 2.1). In thermodynamics, we restrict our attention to systems that are close to equilibrium, so eqn. 2.52 serves as an operational definition of entropy.

2.6.2.1 Microscopic interpretation of temperature

Let's now return to our function β. The macroscopic function having the property of our new function β is temperature. The relation of temperature to β is

$$kT = 1/\beta \tag{2.53}$$

and

$$\frac{1}{kT} = \frac{\partial \ln \Omega}{\partial E} \tag{2.54}$$

Equation 2.53 provides a statistical mechanical definition of temperature. We can easily show that T *is the measure of the energy per degree of freedom*. To do this, though, we need one other relationship, which we introduce without proof. This is that Ω increases roughly with E as:

$$\Omega \propto E^f$$

where f is the number of degrees of freedom of the system (which, in turn, is proportional to the number of atoms or molecules in the system times the modes of motion, e.g., vibrational, rotational, translational, available to them). Hence:

$$\beta = \frac{\partial \ln \Omega}{\partial E} \propto \frac{f}{E} \tag{2.55}$$

Substituting $T = 1/\beta$, then

$$T \propto \frac{E}{kf}$$

In other words, the temperature of a system is proportional to its energy per degree of freedom.

2.6.2.2 Entropy and volume

Our discussion of entropy might leave the impression that entropy is associated only with heat and temperature. This is certainly not the case. Our first example, that of the gases in the box (Figure 2.5), is a good demonstration of how entropy changes can also accompany isothermal processes. When the partition is removed and the gases mix, there is an increase in the number of states accessible to the system. Before the partition is removed, there is only one state accessible to the system (here, *accessible states* means distribution of red and black molecules between the two sides of the box), so $\Omega = 1$. Suppose that after we remove the partition, we find the system state is one where there is one molecule of each kind on each side (the most probable case). There are four such possible configurations, so $\Omega = 4$. The entropy change has thus been

$$\Delta S = k(\ln 4 - \ln 1) = k\ln 4 = 2k\ln 2$$

From the macroscopic perspective, we could say that the red gas, initially confined to the left volume, expands into the volume of the entire box, and the black gas expands from the right half to the entire volume. Thus entropy changes accompany volume changes.

2.6.2.3 Summary

It is often said that entropy is a measure of the randomness of a system. From the discussion above, we can understand why. Entropy is a function of the number of states accessible to a system. Because there are more states available to a system when energy or molecules are "evenly" or "randomly" distributed than when we impose a specific constraint on a system (such as the thermal insulation between the blocks or the partition between the gases), there is indeed an association between randomness and entropy. When we remove the insulation between the copper blocks, we allow energy to be randomly distributed between them. In the example of the combustion of gasoline, before combustion all oxygen atoms are constrained to be associated with oxygen molecules. After combustion, oxygen is randomly distributed between water and CO_2 molecules.

More precisely, we may say that an increase in entropy of a system corresponds to a decrease in knowledge of it. In the example of our two gases in the box, before the partition is removed, we know all red molecules are located somewhere in the left half of the box and all black ones somewhere in the right half. After the partition is removed, we know only that the molecules are located somewhere within the combined volume. Thus our knowledge of the location of the molecules

decreases in proportion to the change in volume. Molecules in ice are located at specific points in the crystal lattice. When ice melts, or evaporates, molecules are no longer constrained to specific locations: there is an increase in entropy of H_2O and a corresponding decrease in our knowledge of molecular positions. When we allowed the two copper blocks to come to thermal equilibrium, entropy increased. There were more possible ways to distribute energy after the blocks equilibrated than before. As a result, we knew less about how energy is distributed after removing the insulation.

As a final point, we emphasize that the second law does not mean we cannot decrease the entropy of a "system." Otherwise, the organization of molecules we call life would not be possible. However, if the entropy of a system is to decrease, the entropy of its surroundings must increase. Thus, we can use air conditioning to cool a room, but the result is that the surroundings (the "outside") are warmed by more than the air in the room is cooled. Organisms can grow, but in doing so they inevitably, through consumption and respiration, increase the entropy of their environment. Thus we should not be surprised to find that the entropy of the manure is greater than that of hay plus oxygen.

2.6.3 Integrating factors and exact differentials

A theorem of mathematics states that any inexact differential that is a function of only two variables can be converted to an exact differential. dW is an inexact differential, and dV is an exact differential. Since $dW_{rev} = -PdV$, dW_{rev} can be converted to a state function by dividing by P since

$$\frac{dW_{rev}}{P} = -dV \qquad (2.56)$$

and V is a state function. Variables such as P that convert nonstate functions to state functions are termed *integrating factors*. Similarly, for a *reversible* reaction, heat can be converted to the state function entropy by dividing by T:

$$\frac{dQ_{rev}}{T} = dS \qquad (2.57)$$

Thus temperature is the integrating factor of heat. Entropy is a state function and therefore an exact differential. Therefore, eqn. 2.57 is telling us that although the heat gained or lost in the transformation from state 1 to state 2 will depend on the path taken, *for a reversible reaction the ratio of heat gained or lost to temperature will always be the same, regardless of path.*

If we return to our example of the combustion of gasoline above, the second law also formalizes our experience that we cannot build a 100% efficient engine: the transformation from state 1 to state 2 cannot be made in such a way that all energy is extracted as work; some heat must be given up as well. In this sense, the second law necessitates the automobile radiator.

Where P–V work is the only work of interest, we can combine the first and second laws as:

$$dU \leq TdS - PdV$$

The implication of this equation is that if equilibrium is approached at prescribed S and V, the energy of the system is minimized. For the specific situation of a reversible reaction

where $dS = dQ/T$, this becomes

$$\boxed{dU_{rev} = TdS_{rev} - PdV} \qquad (2.58)$$

This expresses energy in terms of its natural or *characteristic variables*, S and V. The characteristic variables of a function are those that give the simplest form of the exact differential. Since neither T nor P may have negative values, we can see from this equation that energy will always increase with increasing entropy (at constant volume) and that energy will decrease with increasing volume (at constant entropy). This equation also relates all the primary state variables of thermodynamics, U, S, T, P, and V. For this reason, it is sometimes called the *fundamental equation of thermodynamics*. We will introduce several other state variables derived from these five, but these will be simply a convenience.

By definition, an *adiabatic* system is one where $dQ = 0$. Since $dQ_{rev}/T = dS_{rev}$ (eqn. 2.52), it follows that for a reversible process, *an adiabatic change is one carried out at constant entropy, or in other words, an isoentropic change.* For adiabatic expansion or compression, therefore, $dU = -PdV$.

Example 2.1 Entropy in reversible and irreversible reactions

Air conditioners work by allowing coolant contained in a closed system of pipes to evaporate in the presence of the air to be cooled, then recondensing the coolant (by compressing it) on the warm or exhaust side of the system. Let us define our "system" as only the coolant in the pipes. The system is closed since it can exchange heat and do work but not exchange mass. Suppose our system is contained in an air conditioner maintaining a room at 20°C or 293 K and exhausting to outside air at 303 K. Let's assume the heat of evaporation of the coolant (the energy required to transform it from liquid to gas) is 1000 joules. During evaporation, the heat absorbed by the coolant, dQ, will be 1000 J. During condensation, –1000 J will be given up by the system. For each cycle, the *minimum* entropy change during these transformations is easy to calculate from eqn. 2.51:

$$\text{Evaporation} \qquad dS \geq \frac{dQ}{T} = 1000/293 = 3.413 \text{J/K}$$

$$\text{Condensation:} \qquad dS \geq \frac{dQ}{T} = -1000/303 = 3.300 \text{J/K}$$

The minimum net entropy change in this cycle is the sum of the two, or $3.413 - 3.300 = 0.113$ J/K. This is a "real" process and irreversible, so the entropy change will be greater than this.

If we performed the evaporation and condensation isothermally at the equilibrium condensation temperature (i.e., reversibly), then this result gives the exact entropy change in each case. In this imaginary reversible reaction, where equilibrium is always maintained, there would be no net entropy change over the cycle. But of course no cooling would be achieved either, so it would be pointless from a practical viewpoint. It is nevertheless useful to assume this sort of reversible reaction for the purposes of thermodynamic calculations, because exact solutions are obtained.

2.7   ENTHALPY

We have now introduced all the fundamental variables of thermodynamics, T, S, U, P and V. Everything else can be developed and derived from these functions. Thermodynamicists have found it convenient to define several other state functions, the first of which is called *enthalpy*. *Enthalpy is a composite function and is the sum of the internal energy plus the product PV*:

$$H = U + PV \qquad (2.59)$$

As is the case for most thermodynamic functions, it is enthalpy changes rather than absolute enthalpy that are most often of interest. For a system going from state 1 to state 2, the enthalpy change is:

$$H_2 - H_1 = U_2 - U_1 + P_2V_2 - P_1V_1 \qquad (2.60)$$

The first law states:

$$U_2 - U_1 = \Delta Q + \Delta W$$

so:

$$H_2 - H_1 = \Delta Q + \Delta W + P_2V_2 - P_1V_1$$

If pressure is constant, then:

$$\Delta H = \Delta Q_P + \Delta W + P\Delta V \qquad (2.61)$$

(we use the subscript P in ΔQ_P to remind us that pressure is constant). *If the change takes place at constant pressure and P–V work is the only work done by the system*, then the last two terms cancel and *enthalpy is simply equal to the heat gained or lost by the system*:

$$\Delta H = \Delta Q_P$$

or in differential form:

$$\boxed{dH = dQp} \qquad (2.62)$$

H is a state function because it is defined in terms of state functions U, P, and V. Because enthalpy is a state function, dQ must also be a state function under the conditions of constant pressure and the only work done being P–V work.

More generally, the enthalpy change of a system may be expressed as:

$$dH = dU + VdP + PdV$$

or at constant pressure as:

$$dH = dU + PdV \qquad (2.63)$$

In terms of its characteristic variables, it may also be expressed as:

$$dH \leq TdS + VdP \qquad (2.64)$$

From this it can be shown that H will be at a minimum at equilibrium when S and P are prescribed:

$$\boxed{dH_{rev} = TdS_{rev} + VdP} \qquad (2.65)$$

The primary value of enthalpy is measuring the energy consumed or released in changes of state of a system. For example, how much energy is given off by the reaction:

$$2H_2 + O_2 \rightleftharpoons 2H_2O$$

To determine the answer, we could place hydrogen and oxygen in a well-insulated piston-cylinder maintaining constant pressure. We would design it such that we could easily measure the temperature before and after reaction. Such an apparatus is known as a *calorimeter*. By measuring the temperature before and after the reaction and knowing the *heat capacity* of the reactants and our calorimeter, we could determine the enthalpy of this reaction. This enthalpy value is often also called the *heat of reaction* or *heat of formation* and is designated ΔH_r (or ΔH_f).

Similarly, we might wish to know how much heat is given off when NaCl is dissolved in water. Measuring temperature before and after reaction would allow us to calculate the heat of solution. The enthalpy change of a system that undergoes melting is known as the heat of fusion or heat of melting, ΔH_m (this quantity is sometimes denoted ΔH_f; we will use the subscript m to avoid confusion with heat of formation); that of a system undergoing boiling is known as the heat of vaporization, ΔH_v. As eqn. 2.65 suggests, measuring enthalpy change is also a convenient way of determining the entropy change.

At this point, it might seem that we have wandered rather far from geochemistry. However, we shall shortly see that functions such as entropy and enthalpy and measurements of such things as heats of solution and melting are essential to predicting equilibrium geochemical systems.

Example 2.2 Measuring enthalpies of reaction

Sodium reacts spontaneously and vigorously with oxygen to form Na_2O. The heat given off by this reaction is the enthalpy of formation ΔH_f of Na_2O. Suppose that you react 23 g of Na metal with oxygen in a calorimeter that has the effective heat capacity of 5 kg of water. The heat capacity of water is 75.3 J/mol K. If the calorimeter has a temperature of 20°C before the reaction and a temperature of 29.9°C after the reaction, what is ΔH_f of Na_2O? Assume that the Na_2O contributes negligibly to the heat capacity of the system.

Answer: The heat capacity of the calorimeter is

$$75.3 \, \text{J/mol K} \times 5000 \, \text{g} \div 18 \, \text{g/mol} = 20{,}917 \, \text{J/K}.$$

The heat required to raise its temperature by 9.9 K is then

$$9.9 \times 20{,}917 = 207.08 \, \text{kJ}$$

which is the enthalpy of this reaction. Our experiment created 0.5 moles of Na_2O, so ΔH is −414.16 kJ/mol.

2.8 HEAT CAPACITY

It is a matter of everyday experience that the addition of heat to a body will raise its temperature. We also know that if we bring two bodies in contact, they will eventually reach the same temperature. In that state, the bodies are said to be in thermal equilibrium. However, thermal energy will not necessarily be partitioned equally between the two bodies. It would require half again as much heat to increase the temperature of 1 g of quartz by 1°C as it would to increase the temperature of 1 g of iron metal by 1°C. (We saw that temperature is a measure of the energy per degree of freedom. It would appear then that quartz and iron have different degrees of freedom per gram, something we will explore below.) *Heat capacity* is the amount of heat (in joules or calories) required to raise the temperature of a given amount (usually a mole) of a substance by 1 K. Mathematically, we would say:

$$C = \frac{dQ}{dT} \qquad (2.66)$$

However, the heat capacity of a substance will depend on whether heat is added at constant volume or constant pressure, because some of the heat will be consumed as work if the volume changes. Thus, a substance will have two values of heat capacity: one for constant volume and one for constant pressure.

2.8.1   Constant volume heat capacity

Recall that the first law states:

$$dU = dQ + dW$$

If we restrict work to P–V work, this may be rewritten as:

$$dU = dQ - PdV$$

If the heating is carried out at constant volume (i.e., $dV = 0$), then $dU = dQ$ (all energy change takes the form of heat) and:

$$C_V = \left(\frac{dU}{dT}\right)_V \qquad (2.67)$$

In an ideal gas, each atom has three degrees of translational freedom. A mole of such gas will have N_A such atoms and $3N_A$ degrees of freedom. According to the kinetic theory of gases, the energy, U, of this gas is $^3/_2 N_A kT$. Thus $(dU/dT)_V = {}^3/_2 N_A k$, or $^3/_2 R$ where R is the gas constant. Molecular gases, however, are not ideal. Vibrational and rotational modes also come into play, and heat capacity of real gases, as well as solids and liquids, is a function of temperature.

For solids, motion is vibrational and heat capacities depend on vibrational frequencies, which in turn depend on temperature and bond strength (for stronger bonds there is less energy stored as potential energy, hence less energy is required to raise temperature), for reasons discussed below. For nearly incompressible substances such as solids, the difference between C_V and C_P is generally small.

2.8.2   Constant pressure heat capacity

While heat capacities at constant volume are readily measured for gases, they are difficult to measure for solids and liquids. In nature too, temperature changes tend not to take place at constant volume, so constant pressure heat capacities are of greater interest. Equation 2.61 states that $\Delta H = \Delta Q_P$. Substituting this expression in to eqn. 2.66 we have:

$$\boxed{C_p = \left(\frac{\partial H}{\partial T}\right)_P} \qquad (2.68)$$

Thus, enthalpy change at constant pressure may also be expressed as:

$$dH = C_P dT \qquad (2.69)$$

2.8.3   Energy associated with volume and the relationship between C_v and C_p

Constant pressure and constant temperature heat capacities are different because there is energy associated (work done) with expansion and contraction. Thus how much energy we must transfer to a substance to raise its temperature will depend on whether some of this energy will be consumed in this process of expansion. These energy changes are due to potential energy changes associated with changing the position of an atom or molecule in the electrostatic fields of its neighbors. The difference between C_V and C_p reflects this energy associated with volume. Let us now determine what this difference is.

We can combine relations 2.67 and 2.68 as:

$$C_P - C_V = \left(\frac{\partial H}{\partial T}\right)_P - \left(\frac{\partial U}{\partial T}\right)_V \qquad (2.70)$$

From this, we may derive the following relationship:

$$C_P - C_V = \left(\frac{\partial U}{\partial T}\right)_P - \left(\frac{\partial U}{\partial T}\right)_V + P\left(\frac{\partial V}{\partial T}\right)_P \qquad (2.71)$$

and further:

$$C_P - C_V = \left[\left(\frac{\partial U}{\partial V}\right)_T + P\right]\left(\frac{\partial V}{\partial T}\right)_P \qquad (2.72)$$

It can also be shown that, for a reversible process:

$$\left(\frac{\partial U}{\partial V}\right)_T = T\frac{\alpha}{\beta} - P \qquad (2.73)$$

$(\partial U/\partial V)_T$ is the energy associated with the volume occupied by a substance and is known as the *internal pressure* (P_{int}, which we introduced earlier in our discussion of the van der Waals law, e.g., eqn. 2.17). It is a measure of the energy associated with the forces holding molecules or atoms together. For real substances, energy changes associated with volume changes reflect potential energy increases associated with increased separation between charged molecules and/or atoms; there are no such forces in an ideal gas, so this term is 0 for an ideal gas. Substituting eqn. 2.73 into 2.72, we obtain:

$$C_P - C_V = T\overline{V}\frac{\alpha^2}{\beta} \qquad (2.74)$$

Thus, the difference between C_p and C_v will depend on temperature and pressure for real substances. The terms on the right will always be positive, so that C_p will always be greater than C_v. This accords with our expectation, since energy will be consumed in expansion when a substance is heated at constant pressure, whereas this will not be the case for heating at constant volume. For an ideal gas, $C_p - C_v = R$.

As it is impractical to measure C_v for solids and liquids, only experimentally determined values of C_p are available for them, and values of C_V must be obtained from eqn. 2.74 when required.

We found earlier that C_p is the variation of heat with temperature at constant pressure. How does this differ from the variation of energy with temperature at constant volume? To answer this question, we rearrange eqn. 2.71 and substitute C_V for $(\partial U/\partial T)_V$ and $V\alpha$ for $(\partial V/\partial T)_P$. After simplifying the result, we obtain (on a molar basis):

$$\left(\frac{\partial U}{\partial T}\right)_P = C_P = P\overline{V}\alpha \qquad (2.75)$$

For an ideal gas, the term $PV\alpha$ reduces to R, so that $(\partial U/\partial T)_P = C_p - R$. $C_p - R$ may be shown to be equal to C_V, so the energy change with temperature for an ideal gas is the same for both constant pressure and constant volume conditions. This is consistent with the notion that the difference between C_p and C_v reflects the energy associated with, and changing distances between, atoms and molecules in the presence of attractive forces between them. In an ideal gas, there are no such forces, hence $(\partial U/\partial T)_P = (\partial U/\partial T)_V$.

2.8.4 Heat capacity of solids: a problem in quantum physics

As we shall see, knowledge of the heat capacity of substances turns out to be critical to determining properties such as enthalpy and entropy, and, ultimately, to predicting chemical equilibrium. The heat capacity of a substance reflects the internal motion of its atoms. There are three kinds of motion available to atoms and molecules: translational, vibrational and rotational,* but often one or more of these modes will not be available and not contribute to the energy of a substance. For gases at low temperature, only rotational and translational motions are important (for a monatomic gas, only translational modes are available), while only vibrational motions are important for solids (translational modes are available to solids, which is why solids have finite vapor pressures, but they are extremely improbable, which is why vapor pressures of solids are very small and can usually be neglected). Twice as much energy is typically required to raise the temperature of a vibrational mode by 1 K as for a translational mode. This is because vibration involves both kinetic and potential energy of two or more atoms. Also, vibrational modes do not accept much energy at low temperatures.

This latter phenomenon is not predicted by classical physics; as a result, nineteenth-century physicists were puzzled by the temperature dependence of heat capacity. In 1869, James Maxwell referred to the problem as "the greatest difficulty yet encountered in molecular theory." The solution required a more radical revision to physics than Maxwell imagined: the heat capacity problem turned out to be one of the first indications of the inadequacy of classical physics.

An understanding of the dependence of heat capacity on temperature was only achieved in the twentieth century with the aid of quantum physics. A complete theoretical treatment of heat capacity of real substances is beyond the scope of this book. However, even the few statements we will make will require us to make another excursion into statistical mechanics, a closely related field,

---

* R. Clausius recognized the possibility that molecules might have these three kinds of motion in 1855.

to discover the Boltzmann distribution law. What we learn will be of considerable use in subsequent chapters.

2.8.4.1 The Boltzmann distribution law

Consider a mineral sample, A, in a heat bath, B (B having much more mass than A), and assume they are perfectly isolated from their surroundings. The total energy of the system is fixed, but the energy of A and B will oscillate about their most probable values. The question we ask is *what is the probability that system A is in a state such that it has energy E_A?*

We assume that the number of states accessible to A when it has energy E_A is some function of energy:

$$\Omega(a) = \Omega(E_A) \qquad (2.76)$$

Following the basic postulate, we also assume that all states are equally probable and that the probability of a system having a given energy is simply proportional to the number of states the system can assume when it has that energy:

$$\mathcal{P} = C\Omega \qquad (2.39)$$

where C is a constant. Thus, the probability of A being in state a with energy E_A is:

$$\mathcal{P}_a = C_A\Omega(E_A) \qquad (2.77)$$

Since the total energy of the two systems is fixed, system B will have some fixed energy E_B when A is in state a with energy E_A, and:

$$E_B = E - E_A$$

where E is the total energy of the system. As we mentioned earlier, Ω is multiplicative, so the number of states available to the total system, A + B, is the product of the number of states available to A times the states available to B:

$$\Omega_{Total} = \Omega_A\Omega_B$$

If we stipulate that A is in state a, then Ω_A is 1 and the total number of states available to the system in that situation is just Ω_B:

$$\Omega_{Total} = 1 \times \Omega_B = \Omega(E_B)$$

Thus, the probability of finding A in state a is equal to the probability of finding B in one of the states associated with energy E_B, so that:

$$\mathcal{P}_a = C_B\Omega(E_B) = C_B\Omega(E - E_A)$$
$$= C_B \exp[\ln \Omega(E - E_A)] \qquad (2.78)$$

We can expand $\ln \Omega (E - E_A)$ as a Taylor series about E:

$$\ln \Omega(E - E_A) = \ln \Omega(E) - E_A\left(\frac{d \ln \Omega(E)}{dE}\right) + \cdots \qquad (2.79)$$

and since B is much larger than A, $E \gg E_A$, higher-order terms may be neglected.

Substituting β for $\partial \ln\Omega(E)/dE$ (eqn. 2.48), we have:

$$\Omega(E - E_A) = \exp(\ln \Omega(E) - E_A\beta)$$
$$= \Omega(E)e^{-\beta E_A}$$

and

$$\mathcal{P}_A = C_B\Omega(E)e^{-\beta E_A} \qquad (2.80)$$

Since the total energy of the system, E, is fixed, $\Omega(E)$ must also be fixed, so:

$$\mathcal{P}_A = Ce^{-\beta E_A} \qquad (2.81)$$

Substituting $1/kT$ for β (eqn. 2.53), we have:

$$\mathcal{P}_A = Ce^{-E_A/kT}$$

We can deduce the value of the constant C by noting that $\sum \mathcal{P}_A = 1$, that is, the probabilities over all energy levels must sum to one (because the system *must always* be in one of these states). Therefore:

$$\sum \mathcal{P}_i = C \sum e^{-\beta E_i} = 1 \qquad (2.82)$$

so that

$$C = 1/{\textstyle\sum} e^{-\beta E_i} \qquad (2.83)$$

Generalizing our result, the probability of the system being in state i corresponding to energy ε_i is:

$$\boxed{\mathcal{P}_i = \frac{e^{-\varepsilon_i/kT}}{\sum\limits_n e^{-\varepsilon_n/kT}}} \qquad (2.84)$$

This equation is the *Boltzmann distribution law**, and one of the most important equations in statistical mechanics. Though we derived it for a specific situation and introduced an approximation (the Taylor series expansion), these were merely conveniences; the result is very general (see Feynman et al., 1989 for an alternative derivation). If we define our system as an atom or molecule, then this equation tells us the probability of an atom having a given energy value, ε_i. This is the statistical mechanical interpretation of this equation; it can also be interpreted in terms of quantum physics. The basic tenet of quantum physics is that energy is quantized: only discrete values are possible. The Boltzmann distribution law gives the probability of an atom having the energy associated with quantum level i.

The Boltzmann distribution law says that the population of energy levels decreases exponentially as the energy of that level increases (*energy among atoms is like money among men: the poor are many and the rich few*). A hypothetical example is shown in Figure 2.9.

2.8.4.2 *The partition function*

The denominator of eqn. 2.84, which is the probability normalizing factor or the *sum of the energy distribution over all accessible states*, is called the *partition function* and is denoted Q:

$$Q = \sum_i e^{-\varepsilon_i/kT} \qquad (2.85)$$

The partition function is a key variable in statistical mechanics and quantum physics. It is related to macroscopic variables with which we are already familiar, namely energy and entropy. Let's examine these relationships.

We can compute the total internal energy of a system, U, as the average energy of the atoms times the number of atoms, n. To do this we

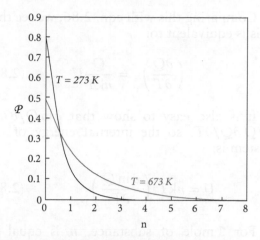

Figure 2.9 Occupation of vibrational energy levels calculated from the Boltzmann distribution. The probability of an energy level associated with the vibrational quantum number n is shown as a function of n for a hypothetical diatomic molecule at 273 K and 673 K.

need to know how energy is distributed among atoms. Macroscopic systems have a very large number of atoms ($\sim 10^{23}$, give or take a few in the exponent). In this case, the number of atoms having some energy ε_i is proportional to the probability of one atom having this energy. So to find the average, we take the sum over all possible energies of the product of energy times the possibility of an atom having that energy. Thus, the internal energy of the system is just:

$$U = n \sum_i \varepsilon_i P_i = \frac{n \sum \varepsilon_i e^{-\varepsilon_i/kT}}{Q} \qquad (2.86)$$

The derivative of Q with respect to temperature (at constant volume) can be obtained from eqn. 2.85:

$$\left(\frac{\partial Q}{\partial T} \right)_V = \frac{1}{kT} \sum \varepsilon_i e^{-\varepsilon_i/kT} \qquad (2.87)$$

---

* We now understand and interpret this law in terms of quantum physics, but Boltzmann formulated it 30 years before Planck and Einstein laid the foundations of quantum theory. Ludwig Boltzmann's work in the second half of the nineteenth century laid the foundations of statistical mechanics and paved the way for quantum theory in the next century. His work was heavily attacked by other physicists of the time, who felt physics should deal only with macroscopic observable quantities and not with atoms, which were then purely hypothetical constructs. These attacks contributed to increasingly frequent bouts of depression, which ultimately led to Boltzmann's suicide in 1906. Ironically and sadly, this was about the time that Perrin's experiments with Brownian motion, Millikan's oil drop experiment, and Einstein's work on the photoelectric effect confirmed the discrete nature of mass, charge, and energy, and thereby the enduring value of Boltzmann's work.

Comparing this with eqn. 2.86, we see that this is equivalent to:

$$\left(\frac{\partial Q}{\partial T}\right)_V = \frac{Q}{nkT}U \qquad (2.88)$$

It is also easy to show that $\partial \ln Q/\partial T = 1/Q \; \partial Q/\partial T$, so the internal energy of the system is:

$$U = nkT^2\left(\frac{\partial \ln Q}{\partial T}\right)_V \qquad (2.89)$$

For 1 mole of substance, n is equal to the Avogadro number, N_A. Since $R = N_A k$, eqn. 2.89, when expressed on a molar basis, becomes:

$$U = RT^2\left(\frac{\partial \ln Q}{\partial T}\right)_V \qquad (2.90)$$

We should not be surprised to find that entropy is also related to Q. This relationship, the derivation of which is left to you (Problem 13), is:

$$S = \frac{U}{T} + R \ln Q \qquad (2.91)$$

Since the partition function is a sum over all possible states, it might appear that computing it would be a formidable, if not impossible, task. As we shall see, however, *the partition function can very often be approximated to a high degree of accuracy by quite simple functions*. The partition function and Boltzmann distribution will prove useful to us in subsequent chapters in discussing several geologically important phenomena such as diffusion and the distribution of stable isotopes between phases, as well as in understanding heat capacities, discussed below.

2.8.4.3 *Energy distribution in solids*

According to quantum theory, all modes of motion are quantized. Consider, for example, vibrations of atoms in a hydrogen molecule. Even at absolute zero temperature, the atoms will vibrate at a ground state frequency. The energy associated with this vibration will be:

$$\varepsilon_0 = \frac{1}{2}h\nu_0 \qquad (2.92)$$

where h is Planck's constant and ν_0 is the vibrational frequency of the ground state. Higher quantum levels have higher frequencies (and hence higher energies) that are multiples of this ground state:

$$\varepsilon_n = \left(n + \frac{1}{2}\right)h\nu_0 \qquad (2.93)$$

where n is the quantum number (an integer ≥ 0).

Now consider a monatomic solid, such as diamond, composed of N identical atoms arranged in a crystal lattice. For each vibration of each atom, we may write an atomic partition function, q. Since vibrational motion is the only form of energy available to atoms in a lattice, the atomic partition function may be written as:

$$q = \sum_m e^{-\varepsilon_m/kT} = \sum_n e^{-\left(n+\frac{1}{2}\right)h\nu_0/kT} \qquad (2.94)$$

We can rewrite eqn. 2.94 as:

$$q = e^{-h\nu_0/2kT} \sum_n e^{-nh\nu_0/kT} \qquad (2.95)$$

The summation term can be expressed as a geometric series, $1 + x + x^2 + x^3 + ...$, where $x = e^{-h\nu_0/kT}$. Such a series is equal to $1/(1 - x)$ if $x < 1$. Thus, eqn. 2.95 may be rewritten in a simpler form as:

$$q = \frac{e^{-h\nu_0/2kT}}{1 - e^{-h\nu_0/kT}} \qquad (2.96)$$

At high temperature, $h\nu_0/kT \ll 1$, and we may approximate $e^{-h\nu_0/kT}$ in the denominator of eqn. 2.96 by $1 - h\nu_0/kT$, so that at high temperature:

$$q \cong \frac{kTe^{-h\nu_0/2kT}}{h\nu_0} \qquad (2.97)$$

Using this relationship, and those between constant volume heat capacity and energy and between energy and the partition function, it is possible to show that:

$$C_V = 3R \qquad (2.98)$$

This is called the *Dulong-Petit limit,* and it holds only where the temperature is high enough that the approximation $e^{-h\nu_0/kT} = 1 - h\nu_0/kT$ holds. For a solid consisting of N different kinds of atoms, the predicted heat capacity is 3NR. Observations bear out these predictions. For example, at 25°C the observed heat capacity for NaCl, for which N is 2, is 49.7 J/K, whereas the predicted value is 49.9 J/K. Substances whose heat capacity agrees with that predicted in this manner are said to be *fully activated.* The temperature at which this occurs, called the *characteristic* or *Einstein temperature,* varies considerably from substance to substance (for reasons explained below). For most metals, it is in the range of 100–600 K. For diamond, however, the Einstein temperature is in excess of 2000 K.

Now consider the case where the temperature is very low. In this case, $h\nu_0/kT \gg 1$ and the denominator of eqn. 2.96; therefore, tends to 1, so that eqn. 2.96 reduces to:

$$q \cong e^{-h\nu_0/2kT} \qquad (2.99)$$

The differential with respect to temperature of ln q is then simply:

$$\left(\frac{\partial \ln q}{\partial T}\right)_V = \frac{h\nu_0}{2kT} \qquad (2.100)$$

If we insert this into eqn. 2.90 and differentiate U with respect to temperature, we find that the predicted heat capacity at $T = 0$ is 0! In actuality, only a perfectly crystalline solid would have 0 heat capacity near absolute zero. Real solids have a small but finite heat capacity.

On a less mathematical level, the heat capacities of solids at low temperature are small because the spacings between the first few vibrational energy levels are large. As a result, energy transitions are large and therefore improbable. Thus, at low temperature, relatively little energy will go into vibrational motions.

We can also see from eqn. 2.93 that the gaps between energy levels depend on the fundamental frequency, ν_0. The larger the gap in vibrational frequency, the less likely will be the transition to higher energy states. The ground state frequency, in turn, depends on bond strength. Strong bonds have higher vibrational frequencies and, as a result, energy is less readily stored in atomic vibrations. In general, covalent bonds will be stronger than ionic ones, which, in turn, are stronger than metallic bonds. Thus, diamond, which has strong covalent bonds, has a low heat capacity until it is fully activated, and full activation occurs at very high temperatures. The bonds in quartz and alumina (Al_2O_3) are also largely covalent, and these substances also have low heat capacities until fully activated. Metals, on the other hand, tend to have weaker bonds and high heat capacities.

Heat capacities are more difficult to predict at intermediate temperatures and require some knowledge of the vibrational frequencies. One simple assumption, used by Einstein,[*] is that all vibrations have the same frequency. The Einstein model provides reasonable predictions of C_v at intermediate and high temperatures but does not work well at low temperatures. A somewhat more sophisticated assumption was used by Debye,[†] who assumed a range of frequencies up to a maximum value, ν_D, now called the *Debye frequency*, and then integrated the frequency spectrum. The procedure is too complex for us to treat here. At low temperature, the

[*] Albert Einstein (1879–1955), though best known for his relativity theories, was also the founder, along with Max Planck, of quantum physics. His work on the quantum basis of heat capacity of solids was published in 1907. Einstein was born in Ulm, Germany, and published some of his most significant papers while working as a patent clerk in Bern, Switzerland. He later joined the Prussian Academy of Sciences in Berlin. A dedicated and active pacifist, Einstein left Germany when Hitler came to power in 1933. He later joined the Center for Advanced Studies in Princeton, New Jersey.

[†] Peter Debye (1884–1966) was born in Maastricht, Netherlands (as Petrus Debije), but spent much of his early career in Germany, eventually becoming director of the Kaiser-Wilhelm-Institut in Berlin. While he was visiting Cornell University in 1940, Germany invaded Holland and Debye simply remained at Cornell, eventually becoming chairman of the Chemistry Department. Debye made numerous contributions to physics and physical chemistry; we shall encounter his work again in the next chapter.

Debye theory predicts:

$$C_v = \frac{12\pi^4}{5} NR \left(\frac{T}{\theta_D}\right)^3 \qquad (2.101)$$

where $\theta_D = h\nu_D / k$ and is called the *Debye temperature*.

Figure 2.10 shows an example of the variation in heat capacity. Consistent with predictions made in the discussion above, heat capacity becomes essentially constant at $T = h\nu/k$ and approaches 0 at $T = 0$. Together, the Debye and Einstein models give a reasonable approximation of heat capacity over a large range of temperature, particularly for simple solids.

Nevertheless, geochemists generally use empirically determined heat capacities. Constant pressure heat capacities are easier to determine, and therefore more generally available and used. For minerals, which are relatively incompressible, the difference between C_v and C_p is small and can often be neglected. Empirical heat capacity data is generally in the form of the coefficients of polynomial expressions of temperature. The *Maier-Kelley formulation* is:

$$C_P = a + bT - \frac{c}{T^2} \qquad (2.102)$$

where a, b, and c are the empirically determined coefficients. The *Haas–Fisher formulation* (Hass and Fisher, 1976) is:

$$C_P = a + bT - \frac{c}{T^2} + fT^2 + gT^{-1/2} \qquad (2.103)$$

with a, b, c, f, and g as empirically determined constants. The Hass–Fisher formulation is

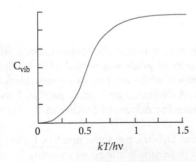

Figure 2.10 Vibrational contribution to heat capacity as a function of $kT/h\nu$.

more accurate and more widely used in geochemistry and heat capacity data are commonly tabulated this way (e.g., Helgenson, et al., 1978; Berman, 1988; Holland and Powell, 1998). We shall use the Maier–Kelly formulation because it is simpler, and we do not want to become more bogged down in mathematics than necessary.

Since these formulae and their associated constants are purely empirical (i.e., neither the equations nor constants have a theoretical basis), they should not be extrapolated beyond the calibrated range.

2.8.5 Relationship of entropy to other state variables

We can now use heat capacity to define the temperature dependency of entropy:

$$\left(\frac{\partial S}{\partial T}\right)_V = \frac{C_V}{T} \qquad (2.104)$$

$$\left(\frac{\partial S}{\partial T}\right)_P = \frac{C_P}{T} \qquad (2.105)$$

The dependencies on pressure and volume (at constant temperature) are:

$$\left(\frac{\partial S}{\partial P}\right)_T = -\alpha V \qquad (2.106)$$

$$\left(\frac{\partial S}{\partial V}\right)_T = \frac{\alpha}{\beta} \qquad (2.107)$$

2.8.6 Additive nature of silicate heat capacities

For many oxides and silicates, heat capacities are approximately additive at room temperature. Thus, for example, the heat capacity of enstatite, $MgSiO_3$, may be approximated by adding the heat capacities of its oxide components, quartz (SiO_2) and periclase (MgO). In other words, since:

$$SiO_2 + MgO \rightarrow MgSiO_3$$

then

$$C_{p-En} \approx C_{p-Qz} + C_{p-Pe}$$

Substituting values:

$$C_{p-En} \approx 44.43 + 37.78 = 82.21 \text{ J/mol-K}$$

The observed value for the heat capacity of enstatite at 300 K is 82.09 J/mol-K, which differs from our estimate by only 0.1%. For most silicates and oxides, this approach will yield estimates of heat capacities that are within 5% of the observed values. However, this is not true at low temperature. The same calculation for C_{p-En} carried out using heat capacities at 50 K differs from the observed value by 20%.

The explanation for the additive nature of oxide and silicate heat capacities has to do with the nature of bonding and atomic vibrations. The vibrations that are not fully activated at room temperature are largely dependent on the nature of the individual cation–oxygen bonds and not on the atomic arrangement in complex solids.

2.9 THE THIRD LAW AND ABSOLUTE ENTROPY

2.9.1 Statement of the third law

The entropies of substances tend toward zero as absolute zero temperature is approached, or as Lewis and Randall expressed it:

If the entropy of each element in some crystalline state may be taken as zero at the absolute zero of temperature, every substance has a finite positive entropy, but at absolute zero, the entropy may become zero, and does so become in the case of perfectly crystalline substances.

2.9.2 Absolute entropy

We recall that entropy is proportional to the number of possible arrangements of a system: $S = k\ln\Omega$. At absolute zero, a perfectly crystalline substance has only one possible arrangement, namely the ground state. Hence $S = k\ln 1 = 0$.

The implication of this seemingly trivial statement is that we can determine the absolute entropy of substances. We can write the complete differential for S in terms of T and P as:

$$dS = \left(\frac{\partial S}{\partial T}\right)_P dT + \left(\frac{\partial S}{\partial P}\right)_T dP \qquad (2.108)$$

Substituting eqns. 2.105 and 2.106 into this, we have:

$$dS = \frac{C_P}{T}dT - \alpha V dP \qquad (2.109)$$

The coefficient of thermal expansion is 0 at absolute zero; the choice of 1 atm for the heat capacity integration is a matter of convenience because C_P measurements are conventionally made at 1 atm.

Actually, the absolute entropies of real substances tend not to be zero at absolute zero, which is to say they are not "perfectly crystalline" in the third law sense. A residual entropy, S_0, which reflects such things as mixing of two or more kinds of atoms (elements or even isotopes of the same element) at crystallographically equivalent sites, must also be considered. This *configurational entropy* is important for some geologically important substances such as feldspars and amphiboles. Configurational entropy can be calculated as:

$$S_{conf} = -R \sum_j m_j \sum_i X_{ij} \ln X_{ij} \qquad (2.110)$$

where m_j is the total number of atoms in the jth crystallographic site (in atoms per formula unit) and $X_{i,j}$ is the mole fraction of the ith atom (element) in the jth site (see Example 2.3). We will return to this equation when we consider multicomponent systems.

Example 2.3 Configurational entropy

Olivine is an example of a solid solution, which we will discuss at length in Chapter 3. Fe and Mg may substitute for each other in the octahedral site. Assuming that the distribution of Fe and Mg within this site is purely random, what is the configurational entropy of olivine of the composition $(Mg_{0.8},Fe_{0.2})_2SiO_4$?

Answer: To solve this problem, we need to apply eqn. 2.110. We need only consider the octahedral site containing Fe and Mg, because O and Si are the only kinds of atoms occupying the tetrahedral and anion sites. The values for X for these two sites will therefore be 1, and $\ln(1) = 0$, so there is no contribution to configurational entropy.

For the octahedral site, $m = 2$, $X_1 = X_{Mg} = 0.8$, and $X_2 = X_{Fe} = 0.2$. Therefore, the configurational entropy will be:

$$S_{conf} = -8.314 \times 2(0.8\ln(0.8) + 0.2\ln(0.2)) = 8.32 \, \text{J mol}^{-1}\text{K}^{-1}$$

2.10   CALCULATING ENTHALPY AND ENTROPY CHANGES

2.10.1   Enthalpy changes due to changes in temperature and pressure

From eqn. 2.68, we can see that the temperature derivative of enthalpy is simply the isobaric heat capacity:

$$\left(\frac{\partial H}{\partial T}\right)_P = C_P$$

and hence:

$$dH = C_P dT \qquad (2.111)$$

Thus, the change in enthalpy over some temperature interval may be found as:

$$\Delta H = \int_{T_1}^{T_2} C_P dT \qquad (2.112)$$

C_p is often a complex function of temperature, so the integration is essential. Example 2.4 illustrates how this is done.

Example 2.4 Calculating isobaric enthalpy changes

How does the enthalpy of a 1 mol quartz crystal change if it is heated from 25°C to 300°C if the temperature dependence of heat capacity can be expressed as $C_p = a + bT - cT^{-2}$ J/K – mol, and a = 46.94, b = 0.0343, and c = 1129680? Assume pressure is constant.

Answer: The first step is to convert temperature to kelvins: all thermodynamic formulae assume temperature is in kelvins. So $T_1 = 298$ K and $T_2 = 573$ K. To solve this problem, we need to use eqn. 2.112. Substituting the expression for heat capacity into eqn. 2.112, we have:

$$\Delta H = \int_{273}^{373} (a + bT - cT^{-2}) dT = a\int_{273}^{373} dT + b\int_{273}^{373} TdT - c\int_{273}^{373} T^{-2}dT$$

Performing the integral, we have:

$$\Delta H = \left[aT + \frac{b}{2}T^2 + \frac{c}{T}\right]_{273}^{373} = \left[46.94 \times T + \frac{0.0343}{2}T^2 + \frac{1129680}{T}\right]_{273}^{373}$$

Now that we have done the math, all that is left is arithmetic. This is most easily done using a spreadsheet. Among other things, it is much easier to avoid arithmetical errors. In addition, we have a permanent record of what we have done. We might set up a spreadsheet to calculate this problem as follows:

	Values		Formulas & Results	
a_	46.94	H	(a_*Temp)+(b_*Temp^2)/2+c_/Temp	
b_	0.0343	H1	19301.98	J/mol
c_	1129680	H2	34498.98	J/mol
Temp1	273	ΔH	15.20	kJ/mol
Temp2	373			

This example is from Microsoft Excel™. On the left, we have written down the names for the various constants in one column, and their values in an adjacent one. Using the Create Names command, we assigned the names in the first column to the values in the second (to avoid confusion with row names, we have named T_1 and T_2 as Temp1 and Temp2 respectively; added an underscore to 'a', 'b', and 'c', so these constant appears as a_, etc. in our formula). In the column on the right, we have written the formula out in the second row, then evaluated it at T_1 and T_2 in the third and fourth rows respectively. The next row contains our answer, 15.2 kJ/mol, determined simply by subtracting 'H1' from 'H2' (and dividing by 1000). *Hint*: we need to keep track of units. Excel won't do this for us.

Isothermal enthalpy changes refer to those occurring at constant temperature, for example, changes in enthalpy due to isothermal pressure changes. Though pressure changes at constant temperature are relatively rare in nature, hypothetical isothermal paths are useful in calculating energy changes. Since enthalpy is a state property, the net change in the enthalpy of a system depends only on the starting and ending state: the enthalpy change is path independent. Imagine a system consisting of a quartz crystal that undergoes a change in state from 25°C and 1 atm to 500°C and 400 atm. How will the enthalpy of this system change? Though in actuality the pressure and temperature changes may have occurred simultaneously, because the enthalpy change is path independent, we can treat the problem as an isobaric temperature change followed by an isothermal temperature change, as illustrated in Figure 2.11. Knowing how to calculate isothermal enthalpy changes is useful for this reason.

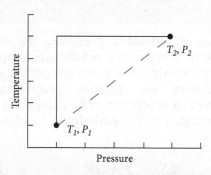

Figure 2.11 Transformations on a temperature–pressure diagram. Changes in state variables such as entropy and enthalpy are path independent. For such variables, the transformation paths shown by the solid line and dashed line are equivalent.

We want to know how enthalpy changes as a function of pressure at constant temperature. We begin from eqn. 2.63, which expresses the enthalpy change as a function of volume and pressure:

$$dH = dU + VdP + PdV \qquad (2.63)$$

By making appropriate substitutions for dU, we can derive the following of enthalpy on pressure:

$$dH = V(1 - \alpha T)dP \qquad (2.113)$$

If changes are large, α, β, and V must be considered functions of T and P and integration performed over the pressure change. The isothermal enthalpy change due to pressure change is thus given by:

$$\Delta H = \int_{P_1}^{P_2} V(1 - \alpha T)dP \qquad (2.114)$$

2.10.2 Changes in enthalpy due to reactions and change of state

We cannot measure the absolute enthalpy of substances, but we can determine the enthalpy *changes* resulting from transformations of a system, and they are of great interest in thermodynamics. For this purpose, a system of relative enthalpies of substances has been established. Since enthalpy is a function of both temperature and pressure, the first problem is to establish standard conditions of temperature and pressure to which these enthalpies apply. These conditions, by convention, are 298.15 K and 0.1 MPa (25°C and 1 bar). Under these conditions the elements are assigned enthalpies of 0. *Standard state enthalpy of formation*, or heat of formation, from the elements, $\Delta H°$, can then be determined for compounds by measuring the heat evolved in the reactions that form them from the elements (e.g., Example 2.2). For example, the heat of formation of water is determined from the energy released at constant pressure in the reaction: $H_2 + \frac{1}{2}O_2 \rightarrow H_2O$, which yields a $\Delta H°$ of -285.83 kJ/mol, where water is in the liquid state. *The minus sign indicates heat is liberated in the reaction*, that is, the reaction is *exothermic* (a reaction that consumes heat is said to be *endothermic*).

Having established such a system, the enthalpy associated with a chemical reaction is easily calculated using Hess's law, which is:

$$\Delta H_r = \sum_i v_i \Delta H^o_{f,i} \qquad (2.115)$$

where v_i is the stoichiometric coefficient for the ith species. In other words, the enthalpy of reaction is just the total enthalpy of the products less the total enthalpy of the reactants. The use of Hess's law is illustrated in Example 2.5 below.

The *heat of vaporization* of a substance is the energy required to convert that substance from liquid to gas, i.e., to boil it. If the reaction $H_2 + \frac{1}{2}O_2 \rightarrow H_2O$ is run to produce water vapor, the $\Delta H°$ turns out to be -241.81 kJ/mol. The difference between the enthalpy of formation of water and vapor, 44.02 kJ/mol, is the heat consumed in going from liquid water to water vapor. This is exactly the amount of energy that would be required to boil 1 mole of water. Analogously, the *heat of melting* (or fusion) is the enthalpy change in the melting of a substance. Because reaction rates are often very slow, and some compounds are not stable at 298 K and 1 MPa, it is not possible to measure the enthalpy for every compound. However, the enthalpies of formation for these compounds can generally be calculated indirectly.

Example 2.5 Enthalpies (or heats) of reaction and Hess's law

What is the energy consumed or evolved in the hydration of corundum (Al_2O_3) to form gibbsite ($Al(OH)_3$)? The reaction is:

$$\frac{1}{2}Al_2O_3 + \frac{3}{2}H_2O \rightarrow Al(OH)_3$$

Answer: We use *Hess's law*. To use Hess's law, we need the standard state enthalpies for water, corundum, and gibbsite. These are: Al_2O_3: -1675.70 kJ/mol, H_2O: -285.83 and $Al(OH)_3$: -1293.13. The enthalpy of reaction is $\Delta H_r = -1293.13 - (0.5 \times -1675.70) - (1.5 \times -285.83) = -26.53$ kJ.

This is the enthalpy of reaction at 1 bar and 298 K. Suppose you were interested in this reaction under metamorphic conditions such as 300°C and 50 MPa. How would you calculate the enthalpy of reaction then?

2.10.3 Entropies of reaction

Since

$$dH = dQ_P \qquad (2.62)$$

and

$$dS = \frac{dQ_{rev}}{T} \qquad (2.57)$$

then at constant pressure

$$dS_{rev} = \frac{dH}{T} \qquad (2.116)$$

Thus, at constant pressure, the entropy change in a reversible reaction is simply the ratio of enthalpy change to temperature.

Entropies are additive properties and entropies of reaction can be calculated in the same manner as for enthalpies, so Hess's law applies:

$$\Delta S_r = \sum_i v_i \Delta S^o_{f,i} \qquad (2.117)$$

The total entropy of a substance can be calculated as:

$$S_{298} = \int_0^{298} \frac{C_p dT}{T} + S_0 + \Delta S_\phi \qquad (2.118)$$

Table 2.2 Standard state thermodynamic data for some important minerals.

Phase/ Compound	Formula	ΔH_f^o kJ/mol	S^O J/K-mol	ΔG_f^o kJ/mol	$\overline{V}$ cc/mol*	a	C_p b	c
H₂O₉	H₂O (gas)	−241.81	188.74	−228.57	24789.00	30.54	0.01029	0
H₂O₁	H₂O (liquid)	−285.84	69.92	−237.18	18.10	29.75	0.03448	0
CO₂	CO₂	−393.51	213.64	−394.39	24465.10	44.22	0.00879	861904
Calcite	CaCO₃	−1207.30	92.68	−1130.10	36.93	104.52	0.02192	2594080
Graphite	C	0	5.740		5.298			
Diamond	C	1.86	2.37		3.417			
Aragonite	CaCO₃	−1207.21	90.21	−1129.16	34.15	84.22	0.04284	1397456
α-Qz	SiO₂	−910.65	41.34	−856.24	22.69	46.94	0.03431	1129680
β-Qz	SiO₂	−910.25	41.82	−856.24		60.29	0.00812	0
Cristobalite	SiO₂	−853.10	43.40	−853.10	25.74	58.49	0.01397	1594104
Coesite	SiO₂	−851.62	40.38	−851.62	20.64	46.02	0.00351	1129680
Periclase	MgO	−601.66	26.94	−569.38	11.25	42.59	0.00728	619232
Magnetite	Fe₃O₄	−1118.17	145.73	−1014.93	44.52	91.55	0.20167	0
Spinel	MgAl₂O₄	−2288.01	80.63	−2163.15	39.71	153.86	0.02684	4062246
Hematite	Fe₂O₃	−827.26	87.61	−745.40	30.27	98.28	0.07782	1485320
Corundum	Al₂O₃	−1661.65	50.96	−1568.26	25.58	11.80	0.03506	3506192
Kyanite	Al₂SiO₅	−2581.10	83.68	−2426.91	44.09	173.18	0.02853	5389871
Andalusite	Al₂SiO₅	−2576.78	92.88	−2429.18	51.53	172.84	0.02633	5184855
Sillimanite	Al₂SiO₅	−2573.57	96.78	−2427.10	49.90	167.46	0.03092	4884443
Almandine	Fe₃Al₂Si₃O₁₂	−5265.50	339.93	−4941.73	115.28	408.15	0.14075	7836623
Grossular	Ca₃Al₂Si₃O₁₂	−6624.93	254.68	−6263.31	125.30	435.21	0.07117	11429851
Albite	NaAlSi₃O₈	−3921.02	210.04	−3708.31	100.07	258.15	0.05816	6280184
K-feldspar	KAlSi₃O₈	−3971.04	213.93	−3971.4	108.87	320.57	0.01804	12528988
Anorthite	CaAl₂Si₂O₈	−4215.60	205.43	−3991.86	100.79	264.89	0.06190	7112800
Jadeite	NaAlSi₂O₆	−3011.94	133.47	−2842.80	60.44	201.67	0.04770	4966408
Diopside	CaMgSi₂O₆	−3202.34	143.09	−3029.22	66.09	221.21	0.03280	6585616
Enstatite	MgSiO₃	−1546.77	67.86	−1459.92	31.28	102.72	0.01983	2627552
Wollatonite	CaSiO₃	−1632.0	82.03	−1656.45	39.93	139.58	0.00236	1401200
Forsterite	Mg₂SiO₄	−2175.68	95.19	−2056.70	43.79	149.83	0.02736	3564768
Clinozoisite	Ca₂Al₃Si₃ O₁₂(OH)	−68798.42	295.56	−6482.02	136.2	787.52	0.10550	11357468
Tremolite	Ca₂MgSi₈O₂₂ (OH)₂	−12319.70	548.90	−11590.71	272.92	188.22	0.05729	4482200
Chlorite	MgAl(AlSi₃) O₁₀(OH)₈	−8857.38	465.26	−8207.77	207.11	696.64	0.17614	15677448
Pargasite	NaCa₂Mg₄Al₃ Si₈O₂₂(OH)₂	−12623.40	669.44	−11950.58	273.5	861.07	0.17431	21007864
Phlogopite	KMg₃AlSi₃O₁₀ (OH)₂	−6226.07	287.86	−5841.65	149.66	420.95	0.01204	8995600
Muscovite	KAl₃Si₃O₁₀ (OH)₂	−5972.28	287.86	−5591.08	140.71	408.19	0.110374	10644096
Gibbsite	Al(OH)₃	−1293.13	70.08	−1155.49	31.96	36.19	0.19079	0
Boehmite	AlO(OH)	−983.57	48.45	−908.97	19.54	60.40	0.01757	0
Brucite	Mg(OH)₂	−926.30	63.14	−835.32	24.63	101.03	0.01678	2556424

Data for the standard state of 298.15 K and 0.1 MPa. ΔH_f is the molar heat (enthalpy) of formation from the elements; S^o is the standard state entropy; V is the molar volume; a, b and c are constants for the heat capacity (C_p) computed as: $C_p = a + bT - cT^{-2}$ J/K-mol.

where S_0 is the entropy at 0 K (the configurational, or *third law* entropy) and $\Delta S\phi$ is the entropy change associated with any phase change. Compilations for S_{298} are available for many minerals. Table 2.2 lists some heat capacity constants for the power series formula as well as other thermodynamic data for a few geologically important minerals. Example 2.6 illustrates how entropy and enthalpy changes are calculated.

Example 2.6 Calculating enthalpy and entropy changes

If the heat capacity of steam can be represented by a three-term power series:

$$C_P = a + bT + cT^2$$

with constants $a = 36.37$ J/K – mol, $b = -7.84 \times 10^{-3}$ J/K^2 – mol, and $c = 9.08 \times 10^{-6}$ J/K^3 – mol, and the enthalpy of vaporization at 100°C is 40.6 kJ/mol, calculate the S and H changes when 1 mol of liquid water at 100°C and 1 atm is converted to steam and brought to 200°C and 3 atm. Assume that with respect to volume, steam behaves as an ideal gas (which, in reality, it is certainly not).

Answer: We need to calculate entropy and enthalpy associated with three changes: the conversion of water to steam, raising the steam from 100°C to 200°C, and increasing the pressure from 1 atm to 3 atm. Since both S and H are state variables, we can treat these three processes separately; our answer will be the sum of the result for each of these processes and will be independent of the order in which we do these calculations.

1. Conversion of water to steam. This process will result in ΔH of 40.6 kJ. For entropy, $\Delta S = \Delta H/T = 40.6/373 = 109$ J/K. We converted centigrade to Kelvin, or absolute, temperature.
2. Raising the steam from 100°C to 200°C (from 373 K to 473 K) isobarically. Since heat capacity is a function of temperature, we will have to integrate eqn. 2.112 over the temperature interval:

$$\int_{T_1}^{T_2} C_P dT = \int_{373}^{473} (a + bT + cT^2)dT = a\int_{373}^{473} dT + b\int_{373}^{473} TdT + c\int_{373}^{473} T^2dT$$

$$= \left[aT + \frac{b}{2}T^2 + \frac{c}{3}T^3 \right]_{373}^{473}$$

Evaluating this, we find that $\Delta H = (17.20 - 0.88 + 0.32) - (13.57 - 0.55 + 0.16) = 3.469$ kJ. The entropy change is given by:

$$\Delta S = \int_{T_1}^{T_2} \frac{C_P}{T}dT = \int_{373}^{473} \frac{a}{T}dT + \int_{373}^{473} bdT + \int_{373}^{473} cTdT = \left[a\ln T + bT + \frac{c}{2}T^2 \right]_{373}^{473}$$

Evaluating this, we find that $\Delta S = (224.01 - 3.71 + 1.02) - (215.37 - 2.93 + 0.63) = 8.24$ J/K.

3. Increasing pressure from 1 atm to 3 atm (0.1 MPa to 0.3 MPa) isothermally. We can use eqn. 2.114 to determine the enthalpy change associated with the pressure change. On the assumption of ideal gas behavior, we can substitute $1/T$ for α. Doing so, we find the equation goes to 0; thus, there is no enthalpy change associated with a pressure change for an ideal gas. This is in accord with assumptions about an ideal gas: namely, that there are no forces between molecules, hence no energy is stored as potential energy of attraction between molecules.

The isothermal pressure dependence of entropy is given by eqn. 2.106. We substitute $1/T$ for α and RT/P for V and integrate from P_1 to P_2:

$$\Delta S = \int_{P_1}^{P_2} -\frac{1}{T}\frac{RT}{P}dP = \int_{P_1}^{P_2} -\frac{R}{P}dP = -R[\ln P]_{0.1}^{0.3} = -8.315\left[\ln\frac{0.3}{0.1}\right] \text{ J/K} = 9.31\text{J/K}$$

The whole enthalpy and entropy changes are the sum of the changes in these three steps:

$$\Delta H = 40.6 + 3.5 + 0 = 44.1 \text{ kJ} \quad \Delta S = 108.8 + 8.2 - 9.1 = 107.9 \text{ J/K}$$

2.11 FREE ENERGY

We can now introduce two free energy functions, the Helmholtz free energy and Gibbs free energy. Gibbs free energy is one of the most useful functions in thermodynamics.

2.11.1 Helmholtz free energy

We can rearrange eqn. 2.58 to read $dU - TdS = -PdV$. The $-PdV$ term is the work term and the TdS term is the heat function. TdS is the energy unavailable for work. Therefore, $dU - TdS$ is *the amount of internal energy available for work*, or the *free energy*. We define it as A, the *Helmholtz free energy*:

$$A \equiv U - TS \qquad (2.119)$$

As usual, we are interested in the differential form (since we are more interested in changes than in absolutes):

$$dA = dU - d(TS) = dU - SdT - TdS \qquad (2.120)$$

or substituting eqn. 2.58 into 2.120:

$$dA = -SdT - PdV \qquad (2.121)$$

2.11.2 Gibbs free energy

2.11.2.1 Derivation

The Gibbs free energy is perhaps misnamed. By analogy to the Helmholtz free energy, it should be called the free enthalpy (but enthalpy is an energy), because it is derived as follows:

$$G \equiv H - TS \qquad (2.122)$$

and

$$dG = d(H - TS) = dH - d(TS) \qquad (2.123)$$

or

$$dG = TdS + VdP - d(TS)$$
$$= TdS + VdP - SdT - TdS$$

which reduces to:

$$\boxed{dG = VdP - SdT} \qquad (2.124)$$

Notice the similarity to the Helmholtz free energy; in that case we subtracted the TS term from the internal energy; in this case we subtracted the TS term from the enthalpy. *The Gibbs free energy is the energy available for nonPV work (such as chemical work)*. It has two other important properties: its independent variables are T and P, generally the ones in which we are most interested in geochemistry, and it contains the entropy term (as does the Helmholtz free energy), and hence can be used as an indication of the direction in which spontaneous reactions will occur.

2.11.2.2 Gibbs free energy change in reactions

For a finite change at constant temperature, the Gibbs free energy change is:

$$\boxed{\Delta G = \Delta H - T\Delta S} \qquad (2.125)$$

The free energy change of formation, ΔG_f, is related to the enthalpy and entropy change of reaction:

$$\Delta G_f^o = \Delta H_f^o - T\Delta S_f^o \qquad (2.126)$$

Like other properties of state, the Gibbs free energy is additive. Therefore:

$$\Delta G_r = \sum_i v_i \Delta G_{f,i} \qquad (2.127)$$

In other words, we can use Hess's law to calculate the free energy change of reaction. Values for ΔG_f at the standard state are available in compilations.

2.11.3 Criteria for equilibrium and spontaneity

The Gibbs free energy is perhaps the single most important thermodynamic variable in geochemistry because it provides this criterion for recognizing equilibrium. This criterion is:

> **Products and reactants are in equilibrium when their Gibbs free energies are equal.**

Another important quality of the Gibbs free energy is closely related:

> **At fixed temperature and pressure, a chemical reaction will proceed in the direction of lower Gibbs free energy (i.e., $\Delta G_r < 0$).**

The reverse is also true: a reaction will not proceed if it produces an increase in the Gibbs free energy.

On an intuitive level, we can understand the Gibbs free energy as follows. We know that transformations tend to go in the direction of the lowest energy state (e.g., a ball rolls down hill). We have also learned that transformations go in the direction of increased entropy (if you drop a glass it breaks into pieces; if you drop the pieces they don't re-assemble into a glass). We must consider both the tendency for energy to decrease and the tendency for entropy to increase in order to predict the direction of a chemical reaction. This is what the Gibbs free energy does. Example 2.7 illustrates how Gibbs free energy of reaction is used to predict equilibrium.

2.11.4 Temperature and pressure dependence of the Gibbs free energy

One reason why the Gibbs free energy is useful is that its characteristic variables are temperature and pressure, which are the "external" variables of greatest interest in geochemistry. Since it is a state variable, we can deduce its temperature and pressure dependencies from eqn. 2.124, which are:

$$\left(\frac{\partial \Delta G}{\partial P}\right)_T = \Delta V \quad (2.128)$$

$$\left(\frac{\partial \Delta G}{\partial T}\right)_P = -\Delta S \quad (2.129)$$

Example 2.7 Using Gibbs free energy to predict equilibrium

Using the thermodynamic data given in Table 2.2, calculate ΔG_r for the reaction:

$$CaAl_2Si_2O_8 + 2Mg_2SiO_4 \rightleftharpoons CaMgSi_2O_6 + MgAl_2O_4 + 2MgSiO_3$$

(Anorthite + 2 Fosterite $\rightleftharpoons$ Diopside + Spinel + 2 Enstatite)

at 298 K and 0.1 MPa. Which mineral assemblage is more stable under these conditions (i.e., which side of the reaction is favored)? Which assemblage will be favored by increasing pressure? Why? Which side will be favored by increasing temperature? Why?

Answer: We can calculate ΔG_r from ΔH_f and ΔS_f, values listed in Table 2.2:

$$\Delta G = \Delta H - \Delta T \Delta S$$

ΔH is calculated as: $\Delta H_{f,Di} + \Delta H_{f,Sp} + 2 \times \Delta H_{f,En} - (\Delta H_{f,An} + 2 \times \Delta H_{f,Fo})$. ΔS is calculated in a similar manner. Our result is -6.08 kJ/mol. Because ΔG_r is negative, the reaction will proceed to the right, so that the assemblage on the right is more stable under the conditions of 298 K and 1 atm.

To find out which side will be favored by increasing pressure and temperature, we use equations 2.128 and 2.129 to see how ΔG will change. For temperature, $\partial \Delta G / \partial T = -\Delta S$. ΔS_r is -36.37 J/K-mol, and $\partial \Delta G / \partial T = 36.37$. The result is positive, so that ΔG will increase with increasing T, favoring the left side. Had we carried out the calculation at 1000°C and 0.1 MPa, a temperature appropriate for crystallization from magma, we would have found that the anorthite–forsterite assemblage is stable. For pressure, $\partial \Delta G / \partial P = \Delta V$. ΔV for the reaction is -20.01 cc/mol (= J/MPa-mol), so will decrease with increasing pressure, favoring the right side. Reassuringly, our thermodynamic result is consistent with geologic observation. The assemblage on the left, which could be called *plagioclase peridotite,* transforms to the assemblage on the right, *spinel peridotite,* as pressure increases in the mantle.

Equations 2.128 and 2.129 allow us to predict how the Gibbs free energy of reaction will change with changing temperature and pressure. *Thus, we can predict how the direction of a reaction will change if we change temperature and pressure.* To obtain the ΔG_r at some temperature T' and pressure P' we integrate:

$$\Delta G_{T',P'} = \Delta G_{T_{ref}P_{ref}} + \int_{P_{ref}}^{P'} \Delta V_r dP - \int_{T_{ref}}^{T'} \Delta S_r dT \tag{2.130}$$

(See Example 2.8.) For liquids and particularly gases, the effects of pressure and temperature on ΔV are significant and cannot be ignored. The reference pressure is generally 0.1 MPa. For solids, however, we can often ignore the effects of temperature and pressure on ΔV so the first integral reduces to: $\Delta V(P' - P_{ref})$ (see Example 2.9). On the other hand, we cannot ignore the temperature dependence of entropy. Hence we need to express ΔS_r as a function of temperature. The temperature dependence of entropy is given by eqn. 2.105. Writing this in integral form, we have:

$$\Delta S(T) = \int_{T_{ref}}^{T} \frac{\Delta C_P}{T} dT$$

This is the change in entropy due to increasing the temperature from the reference state to T. The full change in entropy of reaction is then this plus the entropy change at the reference temperature:

$$\Delta S_r = \Delta S_{T_{ref}} + \int_{T_{ref}}^{T} \frac{C_P}{T} dT \tag{2.131}$$

Substituting this into 2.130, the second integral becomes:

$$-\int_{T_{ref}}^{T'} \Delta S_r dT = -\Delta S_{ref}(T' - T_{ref})$$

$$-\int_{T_{ref}}^{T'} \int_{T_{ref}}^{T'} \frac{\Delta C_P}{T} dT dT = \Delta G_{T'} \tag{2.132}$$

ΔG_T, as we have defined it here, is the change in free energy of reaction as a result of increasing temperature from the reference state to T'.

Example 2.8 Predicting the equilibrium pressure of a mineral assemblage

Using the thermodynamic reaction and data as in Example 2.7:

$$CaAl_2Si_2O_8 + 2Mg_2SiO_4 \rightleftarrows CaMgSi_2O_6 + MgAl_2O_4 + 2MgSiO_3$$

(Anorthite + Forsterite $\rightleftarrows$ Diopside + Spinel + 2 Enstatite)

determine the pressure at which these two assemblages will be in equilibrium at 1000°C. Assume that the volume change of the reaction is independent of pressure and temperature (i.e., α and $\beta = 0$).

Answer: These two assemblages will be in equilibrium if and only if the Gibbs free energy of reaction is 0. Mathematically, our problem is to solve eqn. 2.130 for P such that $\Delta G_{1273,P} = 0$.

Our first step is to find ΔG_r for this reaction at 1000°C (1273 K) using eqn. 2.130. Heat capacity data in Table 2.2 is in the form: $C_p = a + bT - cT^{-2}$. Substituting for ΔC_p, we have:

$$\Delta G_T = -\Delta S_{T_{ref}}(T' - T_{ref}) - \int_{T_{ref}}^{T'} \int_{T_{ref}}^{T'} \left(\frac{\Delta a}{T} + \Delta b - \frac{\Delta c}{T^3} \right) dT dT \tag{2.133}$$

Performing the double integral and collecting terms, and letting $\Delta T = T' - T_{ref}$, this becomes:

$$\Delta G_{T'} = -\Delta T \left[\Delta S_{T_{ref}} - \Delta a + \frac{\Delta b}{2}\Delta T - \frac{\Delta c \Delta T}{2T'T_{ref}^2} \right] - \Delta a T' \ln \frac{T'}{T_{ref}} \tag{2.134}$$

Equation 2.134 is a general solution to eqn. 2.130 when the Maier-Kelley heat capacity is used.

We found ΔS_{Tref} to be -36.37 J/K$-$mol in Example 2.7. Computing Δa as $(a_{di} + a_{Sp} + 2a_{En}) - (a_{An} + 2a_{Fo})$, we find $\Delta a = 15.96$ J/mol. Computing Δb and Δc similarly, they are -0.01732 J/(K-mol) and 1.66×10^6 J-K^2/mol, respectively. Substituting values into eqn. 2.136, we find $\Delta G_T = 36.74$ kJ/mol.

Since we may assume the phases are incompressible, the solution to the pressure integral is:

$$\Delta G_P = \int_{P_{ref}}^{P'} \Delta V_r dP = \Delta V_r (P' - P_{ref}) \tag{2.135}$$

Equation 2.130 may now be written as:

$$\Delta G_{T',P'} = 0 = \Delta G° + \Delta G_T + \Delta V_r(P'' - P_{ref})$$

Let $\Delta G_{1273,0.1} = \Delta G° + \Delta G_T$. $\Delta G°$ is -6.95 kJ/mol (calculated from values in Table 2.2), so $\Delta G_{1273,0.1} = 29.86$ kJ/mol. $\Delta G_{1273,0.1}$ is positive, meaning that the left side of the reaction is favored at 1000°C and atmospheric pressure, consistent with our prediction based on $\partial G/\partial T$.

Solving for pressure, we have

$$P' = \frac{-\Delta G_{T'P_{ref}}}{\Delta V_r} + P_{ref} \tag{2.136}$$

With $\Delta V = -20.01$ cc/mol, we obtain a value of 1.49 GPa (14.9 kbar). Thus, assemblages on the right and left will be in equilibrium at 1.49 GPa and 1000°C. Below that pressure, the left is stable, and above that pressure, the right side is the stable assemblage, according to our calculation.

The transformation from "plagioclase peridotite" to "spinel peridotite" actually occurs around 1.0 GPa in the mantle. The difference between our result and the real world primarily reflects differences in mineral composition: mantle forsterite, enstatite and diopside are solid solutions containing Fe and other elements. The difference does not reflect our assumption that the volume change is independent of pressure. When available data for pressure and temperature dependence of the volume change are included in the solution, the pressure obtained is only marginally different: 1.54 GPa.

Example 2.9 Volume and free energy changes for finite compressibility

The compressibility (β) of forsterite (Mg_2SiO_4) is 8.33×10^{-6} MPa^{-1}. Using this and the data given in Table 2.2, what is the change in molar volume and Gibbs free energy of forsterite at 100 MPa and 298 K?

Answer: Let's deal with volume first. We want to know how the molar volume (43.79 cc/mol) changes as the pressure increases from the reference value (0.1 MPa) to 1 GPa. The compressibility is defined as:

$$\beta = -\frac{1}{V}\left(\frac{\partial V}{\partial P}\right)_T \tag{2.12}$$

So the change in volume for an incremental increase in pressure is given by:

$$dV = -V\beta dP \tag{2.137}$$

To find the change in volume over a finite pressure interval, we rearrange and integrate:

$$\int_{V_o}^{V} \frac{dV}{V} = -\int_{P_o}^{P} \beta dP$$

Performing the integral, we have:

$$\ln \frac{V}{V^o} = -\beta(P - P^o) \tag{2.138}$$

This may be rewritten as:

$$V = V^o e^{-\beta(P-P^o)} \tag{2.139}$$

However, the value of $P-P^o$ is of the order of 10^{-2}, and in this case, the approximation $e^x \cong x + 1$ holds, so that eqn. 2.139 may be written as:

$$V \cong V^o(1 - \beta(P - P^o)) \tag{2.140}$$

Equation 2.140 is a general expression that expresses volume as a function of pressure when β is known, small, and is independent of temperature and pressure. Furthermore, in situations where $P \gg P^o$, this can be simplified to:

$$V \cong V^o(1 - \beta P) \tag{2.141}$$

Using equation 2.141, we calculate a molar volume of 43.54 cc/mol (identical to the value obtained using eqn. 2.139). The volume change, ΔV, is 0.04 cc/mol.

The change in free energy with volume is given by:

$$\left(\frac{\partial G}{\partial P}\right)_T = V$$

so that the free energy change as a consequence of a finite change is pressure can be obtained by integrating:

$$\Delta G = \int_{P^o}^{P} V \ dP \quad \Delta G = \int_{P^o}^{P} V \ dP$$

Into this we may substitute eqn. 2.141:

$$\Delta G = \int_{V^o}^{V} V(1 - \beta P)dP = V^o[P - \beta P^2]_{P^o}^{P} \tag{2.142}$$

Using eqn. 2.142 we calculate a value of ΔG of 4.37 kJ/mol.

2.12   THE MAXWELL RELATIONS*

The reciprocity relationship, which we discussed earlier, leads to a number of useful relationships. These relationships are known as the Maxwell relations. Consider the equation:

$$dU = TdS - PdV \tag{2.58}$$

If we write the partial differential of U in terms of S and V we have:

$$dU = \left(\frac{\partial U}{\partial S}\right)_V dS + \left(\frac{\partial U}{\partial V}\right)_S dV \tag{2.143}$$

From a comparison of these two equations, we see that:

$$\left(\frac{\partial U}{\partial S}\right)_V = T \tag{2.144a}$$

---

* The Maxwell relations are named for Scottish physicist James Clerk Maxwell (1831–1879), perhaps the most important figure in nineteenth-century physics. He is best known for his work on electromagnetic radiation, but he also made very important contributions to statistical mechanics and thermodynamics.

and

$$\left(\frac{\partial U}{\partial V}\right)_S = -P \qquad (2.144b)$$

And since the cross-differentials are equal, it follows that:

$$\left(\frac{\partial T}{\partial V}\right)_S = \left(\frac{\partial P}{\partial S}\right)_V \qquad (2.145)$$

The other Maxwell relations can be derived in an exactly analogous way from other state functions. They are:

- From dH (eqn 2.65):

$$\left(\frac{\partial T}{\partial P}\right)_S = \left(\frac{\partial V}{\partial S}\right)_P \qquad (2.146)$$

- from dA (eqn. 2.121)

$$\left(\frac{\partial P}{\partial T}\right)_V = \left(\frac{\partial S}{\partial V}\right)_T \qquad (2.147)$$

- from dG (eqn. 2.122)

$$\left(\frac{\partial V}{\partial T}\right)_P = -\left(\frac{\partial S}{\partial P}\right)_T \qquad (2.148)$$

2.13 SUMMARY

In this chapter, we introduced the fundamental variables and laws of thermodynamics.

- Temperature, pressure, volume, and energy are state variables who value depends only on the state of the system and not the path taken to that state. Two other fundamental variables, work and heat, are not state variables and their value is path dependent in transformations. Relationships between state variables are known as equations of state. Most often we are interested in changes in state variables rather than their absolute values and we often express these in terms of partial differential equations, for example, the dependence of volume on T and P is written as:

$$dV = \left(\frac{\partial V}{\partial T}\right)_P dT + \left(\frac{\partial V}{\partial P}\right)_T dP \qquad (2.17)$$

- The *first law* states the principle of conservation of energy: even though work and heat are path dependent, their sum is the energy change in a transformation and is path independent:

$$dU = dQ + dW \qquad (2.22)$$

- We introduced another important state variable, entropy, which is a measure of the randomness of a system and is defined as:

$$S = k \ln \Omega \qquad (2.47)$$

where Ω is the number of states accessible to the system and k is Boltzmann's constant.

- The *second law* states that in any real transformation the increase in entropy will always exceed the ratio of heat exchanged to temperature:

$$dS \geq \frac{dQ}{T} \qquad (2.51)$$

In the fictional case of a reversible reaction, entropy change equals the ratio of heat exchanged to temperature.

- The *third law* states that the entropy of a perfectly crystalline substance at the absolute 0 of temperature is 0. Any other substance will have a finite entropy at absolute 0, which is known as the configurational entropy:

$$S_{conf} = -R \sum_j m_j \sum_i X_{ij} \ln X_{ij} \qquad (2.110)$$

- We then introduced another useful variable, H, the enthalpy, which can be thought of as the heat content of a system and is related to other state variables as:

$$dH_{rev} = T dS_{rev} + V dP \qquad (2.65)$$

The value of enthalpy is in measuring the energy consumed or released in changes of state of a system, including phase changes such as melting.

- The heat capacity of a system, C, is the amount of heat required to raise its temperature. This will depend on whether volume or pressure is held constant. For the latter:

$$C_p = \left(\frac{\partial H}{\partial T}\right)_P \qquad (2.68)$$

- With these variables, we could then define a particularly useful state function called the *Gibbs free energy*, G:

$$\Delta G = \Delta H - T\Delta S \qquad (2.125)$$

Written in terms of its characteristic variables:

$$dG = VdP - SdT \qquad (2.124)$$

The Gibbs free energy is the amount of energy available to drive chemical transformations. It has two important properties:

Produces and reactants are in equilibrium when the Gibbs free energies are equal and At fixed temperature and pressure, a chemical reaction will proceed in the direction of lower Gibbs free energy.

REFERENCES AND SUGGESTIONS FOR FURTHER READING

Anderson, G.M. and Crerar, D.A. 1993. *Thermodynamics in Geochemistry*. New York, Oxford University Press.

Berman, R.G. 1988. Internally-consistent thermodynamic data for minerals in the system Na_2O-K_2O-CaO-MgO-FeO-Fe_2O_3-Al_2O_3-SiO_2-TiO_2-H_2O-CO_2. *Journal of Petrology* 29, 445–552.

Einstein, A. 1905. Ist die Trägheit eines Körpers von seinem Energieinhalt abhängig? *Annalen der Physik* 18, 639–643.

Feynman, R., Leighton, R.B. and Sands, M.L. 1989. *The Feynman Lectures on Physics Vol. I*. Pasadena: California Institute of Technology.

Fletcher, P. 1993. *Chemical Thermodynamics for Earth Scientists*. Essex: Longman Scientific and Technical.

Haas, J.L. and Fisher, J.R. 1976. Simultaneous evaluation and correlation of thermodynamic data. *American Journal of Science* 276: 525–45.

Helgeson, H.C., Delany, J.M., Nesbitt, H.W. and Bird, D.K. 1978. Summary and critique of the thermodynamic properties of rock-forming minerals. *American Journal of Science* 278A: 1–229.

Holland, T. J. B. and Powell, R. 1998. An internally consistent thermodynamic data set for phases of petrological interest. *Journal of Metamorphic Geology* 16(3), 309–343. doi: 10.1111/j.1525-1314.1998.00140.x.

Nordstrom, D.K. and Munoz, J.L. 1986. *Geochemical Thermodynamics*. Palo Alto, Blackwell Scientific.

PROBLEMS

1. For a pure olivine mantle, calculate the adiabatic temperature gradient $(\partial T/\partial P)_s$ at 0.1 MPa (1 atm) and 1000°C. Use the thermodynamic data in Table 2.2 for forsterite (Mg-olivine, Mg_2SiO_4), and $\alpha = 44 \times 10^{-6}$ K^{-1}, and $\beta = 8 \times 10^{-6}$ MPa^{-1}.

   Note that: 1 cc/mol = 1 J/MPa/mol.

2. Complete the proof that V is a state variable by showing that for an ideal gas:

$$\frac{\partial \alpha V}{\partial P} = -\frac{\partial \beta V}{\partial T}$$

3. A quartz crystal has a volume of 7.5 ml at 298 K and 0.1 MPa. What is the volume of the crystal at 840K and 12.3 MPa if

   (a) $\alpha = 1.4654 \times 10^{-5}$ K^{-1} and $\beta = 2.276 \times 10^{-11}$ Pa^{-1} and α and β are independent of T and P.

   (b) $\alpha = 1.4310 \times 10^{-5}$ K^{-1} + 1.1587×10^{-9} K^{-2}T
   $\beta = 1.8553 \times 10^{-11}$Pa^{-1} + 7.9453×10^{-8} Pa^{-1}

4. One mole of an ideal gas is allowed to expand against a piston at constant temperature of 0°C. The initial pressure is 1 MPa and the final pressure is 0.04 MPa. Assuming the reaction is reversible,

(a) What is the work done by the gas during the expansion?
(b) What is the change in the internal energy and enthalpy of the gas?
(c) How much heat is gained/lost during the expansion?

5. A typical eruption temperature of basaltic lava is about 1200°C. Assuming that basaltic magma travels from its source region in the mantle quickly enough so that negligible heat is lost to wall rocks, calculate the temperature of the magma at a depth of 40 km. The density of basaltic magma at 1200°C is 2610 kg/m³; the coefficient of thermal expansion is about 1 × 10⁻⁴/K. Assume a heat capacity of 850 J/kg-K and that pressure is related to depth as 1 km = 33 MPa (surface pressure is 0.1 MPa.).
(HINT: "Negligible heat loss" means the system may be treated as adiabatic.)

6. Show that the C_p of an ideal monatomic gas is 5/2 R.

7. Show that:

$$\left(\frac{\partial P}{\partial T}\right)_V = \frac{\alpha}{\beta}$$

8. Show that for a reversible process:

$$\left(\frac{\partial U}{\partial V}\right)_T = T\frac{\alpha}{\beta} - P \qquad (2.73)$$

(Hint: Begin with the statement of the first law (eqn. 2.58), make use of the Maxwell relations, and your proof in problem 7.)

9. Imagine that there are 30 units of energy to distribute among three copper blocks.

(a) If the energy is distributed completely randomly, what is the probability of the first block having all the energy?
(b) If n_1 is the number of units of energy of the first block, construct a graph (a histogram) showing the probability of a given value of n_1 occurring as a function of n_1.

(HINT: Use eqn. 2.37, but modify it for the case where there are three blocks.)

10. Consider a box partitioned into equal volumes, with the left half containing 1 mole of Ne and the right half containing 1 mole of He. When the partition is removed, the gases mix. Show, using a classical thermodynamic approach (i.e., macroscopic), that the entropy change of this process is $\Delta S = 2R \ln 2$. Assume that He and Ne are ideal gases and that temperature is constant.

11. Find expressions for C_p and C_v for a van der Waals gas.

12. Show that β (the compressibility, defined in eqn. 2.12) of an ideal gas is equal to $1/P$.

13. Show that $S = \frac{U}{T} + R\ln\Omega$
    Hint: Start with equations 2.47 and 2.36a using the approximation that $\ln N! = N \ln N{-}N$.

14. Show that $\Delta H = \int_{P_1}^{P_2} V(1 - \alpha T)dP$
    Hint: Begin with equation 2.63 and express dU as a function of temperature and volume change.

15. Helium at 298K and 1 atm has S° = 30.13 cal/K – mole. Assume He is an ideal gas.

(a) Calculate V, H, G, α, β, Cp, Cv, for He at 298K and 1 atm.
(b) What are the values for these functions at 600K and 100 atm?
(c) What is the entropy at 600 K and 100 atm?

16. Using the enthalpies of formation given in Table 2.02, find ΔH in Joules for the reaction:

$$Mg_2SiO_4 + SiO_2 \rightarrow 2MgSiO_3$$

17. Using the data in Table 2.2, calculate the enthalpy and entropy change of diopside as it is heated at constant pressure from 600 K to 1000 K.

18. Calculate the total enthalpy upon heating of 100g of quartz from 25° C to 900° C. Quartz undergoes a phase transition from α-quartz to β-quartz at 575° C. The enthalpy of this phase transition is $\Delta H_{tr} = 0.411$ kJ/mol. Use the Maier–Kelly heat capacity data in Table 2.2.

19. Calcite and aragonite are two forms of $CaCO_3$ that differ only their crystal lattice structure. The reaction between them is thus simply:

$$Calcite \rightleftharpoons Aragonite$$

Using the data in Table 2.2,

(a) Determine which of these forms is stable at the surface of the earth (25° C and 0.1 MPa).
(b) Which form is favored by increasing temperature?
(c) Which form is favored by increasing pressure?

20. Use the data in Table 2.2 to determine the pressure at which calcite and aragonite are in equilibrium at 300°C.

21. Suppose you found kyanite and andalusite coexisting in the same rock, that you had reason to believe this was an equilibrium assemblage, and that you could independently determine the temperature of equilibrium to be 400°C. Use the data in Table 2.2 to determine the pressure at which this rock equilibrated.

Chapter 3

Solutions and thermodynamics of multicomponent systems

3.1 INTRODUCTION

In the previous chapter, we introduced thermodynamic tools that allow us to predict the equilibrium mineral assemblage under a given set of conditions. For example, having specified temperature, we were able to determine the pressure at which the assemblage anorthite + forsterite is in equilibrium with the assemblage diopside + spinel + enstatite. In that reaction the minerals had unique and invariant compositions. In the Earth, things are not quite so simple: these minerals are present as solid solutions*, with substitutions of Fe^{2+} for Mg, Na for Ca, and Cr and Fe^{3+} for Al, among others. Indeed, most natural substances are solutions; that is, their compositions vary. Water, which is certainly the most interesting substance at the surface of the Earth and perhaps the most important, inevitably has a variety of substances dissolved in it. These dissolved substances are often of primary geochemical interest. More to the point, they affect the chemical behavior of water. For example, the freezing temperature of an aqueous NaCl solution is lower than that of pure water. You may have taken advantage of this phenomenon by spreading salt to de-ice sidewalks and roads.

In a similar way, the equilibrium temperature and pressure of the plagioclase + olivine ⇌ clinopyroxene + spinel + orthopyroxene reaction depends on the composition of these minerals. To deal with this compositional dependence, we need to develop some additional thermodynamic tools, which is the objective of this chapter. This may seem burdensome at first: if it were not for the variable composition of substances, we would already know most of the thermodynamics we need. However, as we will see in Chapter 4, we can use this compositional dependence to advantage in reconstructing conditions under which a mineral assemblage or a hydrothermal fluid formed.

A final difficulty is that the valance state of many elements can vary. Iron, for example, may change from its Fe^{2+} state to Fe^{3+} when an igneous rock weathers. The two forms of iron have very different chemical properties; for example, Fe^{2+} is considerably more soluble in water than is Fe^{3+}. Another example of this kind of reaction is photosynthesis, the process by which CO_2 is converted to organic carbon. These kinds of reactions are called *oxidation–reduction,* or *redox* reactions. The energy your brain uses to process the information you are now reading comes from oxidation of organic carbon – carbon originally reduced by photosynthesis in plants. To fully specify the state of a system, we must specify its "redox" state. We treat redox reactions in the final section of this chapter.

Though Chapter 4 will add a few more tools to our geochemical toolbox, and treat a number of advanced topics in thermodynamics, it is designed to be optional. With completion of this chapter, you will have a sufficient thermodynamic background to deal with a wide

---

* The naturally occurring minerals of varying composition are referred to as plagioclase rather than anorthite, olivine rather than forsterite, clinopyroxene rather than diopside, and orthopyroxene rather than enstatite.

---

range of phenomena in the Earth, and most of the topics in the remainder of this book.

3.2 PHASE EQUILIBRIA

3.2.1 Some definitions

3.2.1.1 Phase

Phases are real substances that are homogeneous, physically distinct, and (in principle) mechanically separable. For example, the phases in a rock are the minerals present. Amorphous substances are also phases, so glass and opal would be phases. The sugar that won't dissolve in your iced tea is a distinct phase from the tea, but the dissolved sugar is not. *Phase* is not synonymous with *compound*. Phases need not be chemically distinct: a glass of ice water has two distinct phases: water and ice. Many solid compounds can exist as more than one phase. Nor need they be compositionally unique: plagioclase, clinopyroxene, olivine, and so on, are all phases even though their composition can vary. A fossil in which the aragonite ($CaCO_3$) is partially retrograded into calcite (also $CaCO_3$) consists of two phases, which, although they might be chemically identical, have different crystal structures and hence different properties. Systems and reactions occurring within them that consist of a single phase are referred to as *homogeneous*; those systems consisting of multiple phases, and the reactions occurring within them, are referred to as *heterogeneous*.

3.2.1.2 Species

Species is somewhat more difficult to define than either *phase* or *component*. A species is a chemical entity, generally an element or compound (which may or may not be ionized). The term is most useful in the context of gases and liquids. A single liquid phase, such as an aqueous solution, may contain a number of species. For example, H_2O, H_2CO_3, HCO_3^-, CO_3^{2-}, H^+, and OH^- are all species commonly present in natural waters. The term *species* is generally reserved for an entity that actually exists, such as a molecule, ion, or solid, at least on a microscopic scale. This is not necessarily

the case with components, as we shall see. The term *species* is less useful for solids, although it is sometimes applied to the pure end-members of solid solutions and to pure minerals.

3.2.1.3 Component

In contrast to a species, a *component* need not be a real chemical entity; rather, it is simply an algebraic term in a chemical reaction. The *minimum number of components** of a system is rigidly defined as *the minimum number of independently variable entities necessary to describe the composition of each and every phase of a system*. Unlike species and phases, components may be defined in any convenient manner: what the components of your system are and how many there are depend on your interest and on the level of complexity you will be dealing with. Consider our aragonite–calcite fossil. If the only reaction occurring in our system (the fossil) is the transformation of aragonite to calcite, one component, $CaCO_3$, is adequate to describe the composition of both phases. If, however, we are also interested in the precipitation of calcium carbonate from water, we might have to consider $CaCO_3$ as consisting of two components: Ca^{2+} and CO_3^{2-}.

There is a rule to determine the minimum number of components in a system once you decide what your interest in the system is; the hard part is often determining your interest. The rule is:

$$c = n - r \qquad (3.1)$$

where *n is the number of species and r is the number of independent chemical reactions possible between these species*. Essentially, this equation simply states that if a chemical species can be expressed as the algebraic sum of other components, we need not include that species among our minimum set of components. Let's try the rule on the species we listed above for water. We have six species: H_2O, H_2CO_3, HCO_3^-, CO_3^{2-}, H^+, and OH^-.

We can write three reactions relating them:

$$HCO_3^- = H^+ + CO_3^{2-}$$
$$H_2CO_3 = H^+ + HCO_3^-$$
$$H_2O = H^+ + OH^-$$

---

* Caution: some books use the term *number of components* as synonymous with *minimum number of components*.

Equation 3.1 tells us we need $3 = 6 - 3$ components to describe this system: CO_3^{2-}, H^+, and OH^-. Put another way, we see that carbonic acid, bicarbonate, and water can all be expressed as algebraic sums of hydrogen, hydroxyl, and carbonate ions, so they need not be among our minimum set of components.

In igneous and metamorphic petrology, components are often the major oxides (though we may often choose to consider only a subset of these). On the other hand, if we were concerned with the isotopic equilibration of minerals with a hydrothermal fluid, ^{18}O might be considered as a different component than ^{16}O.

Perhaps the most straightforward way of determining the number of components is a graphical approach. If all phases can be represented on a one-dimensional diagram (that is, a straightline representing composition), we are dealing with a two-component system. For example, consider the hydration of Al_2O_3 (corundum) to form boehmite ($AlO(OH)$) or gibbsite $Al(OH)_3$. Such a system would contain four phases (corundum, boehmite, gibbsite, water), but is nevertheless a two-component system because all phases may be represented in one dimension of composition space, as shown in Figure 3.1. Because there are two polymorphs of gibbsite, one of boehmite, and two other possible phases of water, there are nine possible phases in this two-component system. Clearly, a system may have many more phases than components.

Similarly, if a system may be represented in two dimensions, it is a three-component system. Figure 3.2 is a ternary diagram illustrating the system Al_2O_3–H_2O–SiO_2. The graphical representation approach reaches its practical limit in a four-component system because of the difficulty of representing more than three dimensions on paper. A four-component system is a quaternary one, and can be represented with a three-dimensional quaternary diagram.

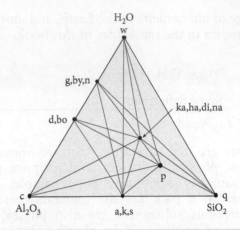

Figure 3.2 Phase diagram for the system Al_2O_3–H_2O–SiO_2. The lines are called *joins* because they join phases. In addition to the end-members, or components, phases represented are g: gibbsite, *by*: bayerite, *n*: norstrandite (all polymorphs of $Al(OH)_3$), d: diaspore, *bo*: boehmite (polymorphs of $AlO(OH)$), *a*: andalusite, *k*: kyanite, *s*: sillimanite (all polymorphs of Al_2SiO_5), *ka*: kaolinite, *ha*: halloysite, *di*: dickite, *na*: nacrite (all polymorphs of $Al_2Si_2O_5(OH)_4$), and *p*: pyrophyllite ($Al_2Si_4O_{10}(OH)_2$). There are also six polymorphs of quartz, *q* (coesite, stishovite, tridymite, cristobalite, α-quartz, and β-quartz).

It is important to understand that a component may or may not have chemical reality. For example in the exchange reaction:

$$NaAlSi_3O_8 + K^+ = KAlSi_3O_8 + Na^+$$

we could alternatively define the *exchange operator* KNa_{-1} (where Na_{-1} is -1 mol of Na ion) and write the equation as:

$$NaAlSi_3O_8 + KNa_{-1} = KAlSi_3O_8$$

In addition, we can also write the reaction:

$$K - Na = KNa_{-1}$$

Here we have four species and two reactions and thus a minimum of only two components. You can see that *a component is merely an algebraic term*.

There is generally some freedom in choosing components. For example, in the ternary (i.e., three-component) system SiO_2–Mg_2SiO_4–$MgCaSi_2O_6$, we could

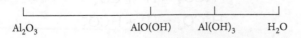

Figure 3.1 Graphical representation of the system Al_2O_3–H_2O.

choose our components to be quartz, diopside, and forsterite, or we could choose them to be SiO_2, MgO, and CaO. Either way, we are dealing with a ternary system (which contains $MgSiO_3$ as well as the three other phases).

3.2.1.4 Degrees of freedom

The number of degrees of freedom in a system is equal to the sum of the number of independent *intensive* variables (generally temperature and pressure) and independent concentrations (or activities or chemical potentials) of components in phases that must be fixed to define uniquely the state of the system. A system that has no degrees of freedom (i.e., is uniquely fixed) is said to be invariant, one that has one degree of freedom is univariant, and so on. Thus, in a univariant system, for example, we need specify the value of only one variable, for example, temperature or the concentration of one component in one phase, and the value of pressure and all other concentrations are then fixed and can be calculated (assuming the system is at equilibrium).

3.2.2 The Gibbs phase rule

The Gibbs* phase rule is a rule for determining the *degrees of freedom*, or *variance*, of a system *at equilibrium*. The rule is:

$$f = c - \phi + 2 \qquad (3.2)$$

where f is the degrees of freedom, c is the number of components, and ϕ is the number of phases. The mathematical analogy is that the degrees of freedom are equal to the number of variables minus the number of equations relating those variables. For example, in a system consisting of just H_2O, if two phases coexist, for example, water and steam, then the system is univariant. Three phases coexist at the triple point of water, so the system is said to be invariant, and T and P are uniquely fixed: there is only one temperature and one pressure at which the three phases of water

can coexist (273.15 K and 0.006 MPa). If only one phase is present, for example just liquid water, then we need to specify two variables to describe completely the system. It does not matter which two we pick. We could specify molar volume and temperature and from that we could deduce pressure. Alternatively, we could specify pressure and temperature. There is only one possible value for the molar volume if temperature and pressure are fixed. It is important to remember this applies to intensive parameters. To know volume, an extensive parameter, we would have to fix one additional extensive variable (such as mass or number of moles). And again, we emphasize that all this applies only to systems at equilibrium.

Now consider the hydration of corundum to form gibbsite. There are three phases, but there need be only two components. If these three phases (water, corundum, gibbsite) are at equilibrium, we have only one degree of freedom (i.e., if we know the temperature at which these three phases are in equilibrium, the pressure is also fixed).

Rearranging eqn. 3.2, we also can determine the *maximum* number of phases that can coexist at equilibrium in any system. The degrees of freedom cannot be less than zero, so for an invariant, one-component system, a maximum of three phases can coexist at equilibrium. In a univariant one-component system, only two phases can coexist. Thus, sillimanite and kyanite can coexist over a range of temperatures, as can kyanite and andalusite, but the three phases of Al_2SiO_5 coexist only at one unique temperature and pressure.

Let's consider the example of the three-component system Al_2O_3–H_2O–SiO_2 in Figure 3.2. Although many phases are possible in this system, for any given composition of the system only three phases can coexist at equilibrium over a range of temperature and pressure. Four phases (e.g, a, k, s, and p) can coexist only along a one-dimensional line or curve in P–T space. Such points are called univariant lines (or curves). Five phases can coexist at invariant points at which both temperature and pressure are uniquely fixed.

---

* J. Williard Gibbs (1839–1903) is viewed by many as the father of thermodynamics. He received the first doctorate in engineering granted in the US, from Yale in 1858. He was Professor of Mathematical Physics at Yale from 1871 until his death. He also helped to found statistical mechanics. The importance of his work was not widely recognized by his American colleagues, though it was in Europe, until well after his death.

Turning this around, if we found a metamorphic rock whose composition fell within the Al_2O_3–H_2O–SiO_2 system, and if the rock contained five phases, it would be possible to determine uniquely the temperature and pressure at which the rock equilibrated.

3.2.3 The Clapeyron equation

A common problem in geochemistry is to know how a phase boundary varies in P–T space, for example, how a melting temperature will vary with pressure. At a phase boundary, two phases must be in equilibrium, so ΔG must be 0 for the reaction Phase 1 ⇌ Phase 2. The phase boundary therefore describes the condition:

$$d(\Delta G_r) = \Delta V_r dP - \Delta S_r dT = 0$$

Thus, the slope of a phase boundary on a temperature-pressure diagram is:

$$\frac{dT}{dP} = \frac{\Delta V_r}{\Delta S_r} \qquad (3.3)$$

where ΔV_r and ΔS_r are the volume and entropy changes associated with the reaction. Equation 3.3 is known as the *Clausius–Clapeyron equation*, or simply the *Clapeyron*

equation. Because ΔV_r and ΔS_r are functions of temperature and pressure, this is, of course, only an instantaneous slope. For many reactions, however, particularly those involving only solids, the temperature and pressure dependencies of ΔV_r and ΔS_r will be small and the Clapeyron slope will be relatively constant over a large T and P range (see Example 3.1).

Because $\Delta S = \Delta H/T$, the Clapeyron equation may be equivalently written as:

$$\frac{dT}{dP} = \frac{T\Delta V_r}{\Delta H_r} \qquad (3.4)$$

Slopes of phase boundaries in P–T space are generally positive, implying that the phases with the largest volumes also generally have the largest entropies (for reasons that become clear from a statistical mechanical treatment). This is particularly true of solid–liquid phase boundaries, although there is one very important exception: water. How do we determine the pressure and temperature dependence of ΔV_r, and why is ΔV_r relatively T- and P-independent in solids?

We should emphasize that application of the Clapeyron equation is not limited to reactions between two phases in a one-component system but may be applied to any univariant reaction.

Example 3.1 The graphite–diamond transition

At 25°C the graphite–diamond transition occurs at 1600 MPa (megapascals, 1 MPa = 10 b). Using the standard state (298 K, 0.1 MPa) data below, predict the pressure at which the transformation occurs when temperature is 1000°C.

	Graphite	Diamond
α (K^{-1})	1.05×10^{-05}	7.50×10^{-06}
β (MPa^{-1})	3.08×10^{-05}	2.27×10^{-06}
$S°$ (J/K-mol)	5.74	2.38
V (cm^3/mol)	5.2982	3.417

Answer: We can use the Clapeyron equation to determine the slope of the phase boundary. Then, assuming that ΔS and ΔV are independent of temperature, we can extrapolate this slope to 1000°C to find the pressure of the phase transition at that temperature.

First, we calculate the volumes of graphite and diamond at 1600 MPa as (eqn. 2.140):

$$V = V°(1 - \beta\Delta P) \qquad (3.5)$$

where ΔP is the difference between the pressure of interest (1600 MPa in this case) and the reference pressure (0.1 MPa). Doing so, we find the molar volumes to be 5.037 for graphite and 3.405 for diamond, so ΔV_r is −1.6325 cc/mol. The next step will be to calculate ΔS at 1600 MPa. The pressure dependence of entropy is given by equation 2.106: $(\partial S/\partial P)_T = -\alpha V$. Thus, to determine the effect of pressure we integrate:

$$S_P = S° + \int_{P_{ref}}^{P_1} \left(\frac{\partial S}{\partial P}\right)_T dP = S° + \int_{P_{ref}}^{P_1} -\alpha V dP \tag{3.6}$$

(We use S_p to indicate the entropy at the pressure of interest and $S°$ the entropy at the reference pressure.) We need to express V as a function of pressure, so we substitute eqn. 3.5 into 3.6:

$$S_P = S° + \int_{P_{ref}}^{P_1} -\alpha V°(1 - \beta P)dP = S° - \alpha V° \left[\Delta P - \frac{\beta}{2}(P_1^2 - P_{ref}^2)\right] \tag{3.7}$$

The reference pressure, P_{ref}, is negligible compared with P_1 (0.1 MPa vs 1600 MPa), so that this simplifies to:

$$S_P = S° - \alpha V \left[\Delta P - \frac{\beta}{2}P_1^2\right]$$

For graphite, S_p is 5.66 J/K-mol, for diamond it is 2.34 J/K-mol, so ΔS_r at 1600 MPa is -3.32 J-K^{-1}-mol^{-1}.

The Clapeyron slope is therefore:

$$\frac{\Delta S}{\Delta V} = \frac{-3.322}{-1.63} = 2.035 \text{JK}^{-1}\text{cm}^{-3}$$

One distinct advantage of the SI units is that cm^3 = J/MPa, so the above units are equivalent to K/MPa. From this, the pressure of the phase change at 1000°C can be calculated as:

$$P_{1000} = P_{293} + \Delta T \times \frac{\Delta S}{\Delta V} = 1600 + 975 \times 2.035 = 3584 \text{MPa}$$

The Clapeyron slope we calculated (solid line) is compared with the experimentally determined phase boundary in Figure 3.3. Our calculated phase boundary is linear whereas the experimental one is not. The curved nature of the observed phase boundary indicates ΔV and ΔS are pressure- and temperature-dependent. This is indeed the case, particularly for graphite. A more accurate estimate of the volume change requires that β be expressed as a function of pressure.

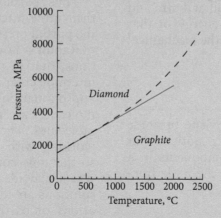

Figure 3.3   Comparison of the graphite–diamond phase boundary calculated from thermodynamic data and the Clapeyron slope (solid line) with the experimentally observed phase boundary (dashed line).

3.3 SOLUTIONS

Solutions are defined as homogeneous phases produced by dissolving one or more substances in another substance. In geochemistry we are often confronted by solutions: as gases, liquids, and solids. Free energy depends not only on T and P, but also on composition. In thermodynamics it is generally most convenient to express compositions in terms of mole fractions, X_i, the number of moles of i divided by the total moles in the substance (moles are weight divided by atomic or molecular weight). The sum of all the X_i fractions must, of course, total to 1.

Solutions are distinct from purely mechanical mixtures. For example, salad dressing (oil and vinegar) is not a solution. Similarly, we can grind anorthite ($CaAl_2Si_2O_8$) and albite ($NaAlSi_3O_8$) crystals into a fine powder and mix them, but the result is not a plagioclase solid solution. The Gibbs free energy of mechanical mixtures is simply the sum of the free energy of the components. If, however, we heated the anorthite–albite mixture to a sufficiently high temperature that the kinetic barriers were overcome, there would be a reordering of atoms and the creation of a true solution. Because this reordering is a spontaneous chemical reaction, there must be a decrease in the Gibbs free energy associated with it. This solution would be stable at 1 atm and 25°C. Thus, we can conclude that the solution has a lower Gibbs free energy than the mechanical mixture. On the other hand, vinegar will never dissolve in oil at 1 atm and 25°C because the Gibbs free energy of that solution is greater than that of the mechanical mixture.

3.3.1 Raoult's law

Working with solutions of ethylene bromide and propylene bromide, Raoult* noticed that the vapor pressures of the components in a solution were proportional to the mole fractions of those components:

$$P_i = X_i P_i^o \qquad (3.8)$$

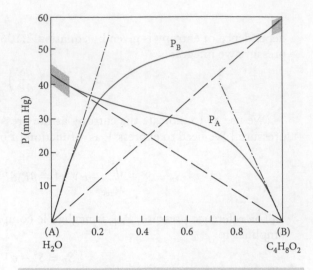

Figure 3.4 Vapor pressure of water and dioxane in a water–dioxane mixture showing deviations from ideal mixing. Shaded areas are areas where Raoult's law (dashed lines). Henry's law slopes are shown as dot-dashed lines. After Nordstrom and Munoz (1986).

where P_i is the vapor pressure of component i above the solution, X_i is the mole fraction of i in solution, and P_i^o is the vapor pressure of pure i under standard conditions. Assuming the partial pressures are additive and the sum of all the partial pressures is equal to the total gas pressure ($\Sigma P_i = P_{total}$):

$$P_i = X_i P_{total} \qquad (3.9)$$

Thus, partial pressures are proportional to their mole fractions. This is the definition of the **partial pressure** of the ith gas in a mixture.

Raoult's law holds only for ideal solutions, that is, substances where there are no intermolecular forces. It also holds to a good approximation where the forces between like molecules are the same as between different molecules. The two components Raoult was working with were very similar chemically, so that this condition held, and the solution was nearly ideal. As you might guess, not all solutions are ideal. Figure 3.4 shows the variations of partial pressures above a mixture of water and dioxane. Significant deviations

---

* Francois Marie Raoult (1830–1901), French chemist, chaired the Chemistry Department at the Université de Grenoble from 1867 until his death.

from Raoult's law are the rule except where X_i approaches 1.

3.3.2 Henry's law

Another useful approximation occurs when X_i approaches 0. In this case, the partial pressures are not equal to the mole fraction times the vapor pressure of the pure substance, but they do vary linearly with X_i. This behavior follows Henry's law,* which is:

$$P_i = hX_i \text{ for } X_i \ll 1 \qquad (3.10)$$

where h is known as the Henry's law constant.

3.4 CHEMICAL POTENTIAL

3.4.1 Partial molar quantities

Free energy and other thermodynamic properties are dependent on composition. We need a way of expressing this dependence. For any extensive property of the system, such as volume, entropy, energy, or free energy, we can define a *partial molar value*, which expresses how that property will depend on changes in amount of one component. For example, we define the partial molar volume of component i in phase ϕ as:

$$v_i^\phi = \left(\frac{\partial V}{\partial n_i}\right)_{T,P,n_{j,j\neq i}} \quad \text{such that} \quad V = \sum_i n_i v_i$$
$$(3.11)$$

(we will use small letters to denote partial molar quantities; the superscript refers to the phase and the subscript refers to the component). The plain language interpretation of eqn. 3.11 is that *the partial molar volume of component* i *in phase* ϕ *tells us how the volume of phase* ϕ *will vary with an infinitesimal addition of component* i, *if all other variables are held constant.* For example, the partial molar volume of Na in an aqueous solution such as seawater would tell us how the volume of that solution would change for an infinitesimal addition of Na. In this case i would refer to the Na component and ϕ would refer to the aqueous solution phase. In Table 2.2, we see that the molar volumes of the albite and anorthite end-members of the

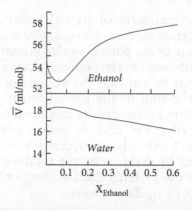

Figure 3.5 Variation of the partial molar volumes of water and ethanol as a function of the mole fraction of ethanol in a binary solution. This figure also illustrates the behavior of a very nonideal solution. After Nordstrom and Munoz (1986).

plagioclase solid solution are different. We could define v_{Ab}^{pl} as the partial molar volume of albite in plagioclase, which would tell us how the volume of plagioclase would vary for an infinitesimal addition of albite. (In this example, we have chosen our component as albite rather than Na. While we could have chosen Na, the choice of albite simplifies matters because the replacement of Na with Ca is accompanied by the replacement of Si by Al.)

The second expression in eqn. 3.11 says that the volume of a phase is the sum of the partial molar volumes of the components times the number of moles of each component present. Thus, the volume of plagioclase would be the sum of the partial molar volumes of the albite and anorthite components weighted by the number of moles of each.

Another example might be a solution of water and ethanol. The variation of the partial molar volumes of water and ethanol in a binary solution is illustrated in Figure 3.5. This system illustrates very clearly why the qualification "for an infinitesimal addition" is always added: the value of a partial molar quantity of a component may vary with the amount of that component present.

Equation 3.11 can be generalized to all partial molar quantities and also expresses an

---

* Named for English chemist William Henry (1775–1836), who formulated it.

important property of partial molar quantities: *an extensive variable of a system or phase is the sum of its partial molar quantities for each component in the system*. In our example above, this means that the volume of plagioclase is the sum of the partial molar volume of the albite and anorthite components.

Generally, we find it more convenient to convert extensive properties to intensive properties by dividing by the total number of moles in the system, Σn. Dividing both sides of eqn. 3.11 by Σn we have:

$$\overline{V} = \sum_i X_i v_i \qquad (3.12)$$

This equation says that the molar volume of a substance is the sum of the partial molar volumes of its components times their mole fractions. *For a pure phase, the partial molar volume equals the molar volume since* X = 1.

3.4.2 Definition of chemical potential and relationship to Gibbs free energy

We define μ as the *chemical potential*, which is simply the partial molar Gibbs free energy:

$$\boxed{\mu_i = \left(\frac{\partial G}{\partial n_i}\right)_{T,P,n_{j,j\neq i}} \qquad (3.13)}$$

The chemical potential thus tells us how the Gibbs free energy will vary with the number of moles, n_i, of component i holding temperature, pressure, and the number of moles of all other components constant. We said that the Gibbs free energy of a system is a measure of the capacity of the system to do chemical work. Thus, the chemical potential of component i is the amount by which this capacity to do chemical work is changed for an infinitesimal addition of component i at constant temperature and pressure. In a NiCd battery (common rechargeable batteries), for example, the chemical potential of Ni in the battery (our system) is a measure of the capacity of the battery to provide electrical energy per mole of additional Ni for an infinitesimal addition.

The total Gibbs free energy of a system will depend on composition as well as on temperature and pressure. The equations we introduced for Gibbs free energy in Chapter 2 fully describe the Gibbs free energy only for single component systems or systems containing only pure phases. The Gibbs free energy change of a phase of variable composition is fully expressed as:

$$dG = VdP + SdT + \sum_i \mu_i dn_i \qquad (3.14)$$

3.4.3 Properties of the chemical potential

We now want to consider two important properties of the chemical potential. To illustrate these properties, consider a simple two-phase system in which an infinitesimal amount of component i is transferred from phase β to phase α, under conditions where T, P, and the amount of other components is held constant in each phase. One example of such a reaction would be the transfer of Pb from a hydrothermal solution to a sulfide mineral phase. The chemical potential expresses the change in Gibbs free energy under these conditions:

$$dG = dG^\alpha + dG^\beta = \mu_i^\alpha dn_i^\alpha + \mu_i^\beta dn_i^\beta \qquad (3.15)$$

since we are holding everything else constant, atoms gained by α must be lost by β, so $-dn_i^\alpha = dn_i^\beta$ and:

$$dG = (\mu_i^\alpha - \mu_i^\beta)dn_i \qquad (3.16)$$

At equilibrium, $dG = 0$, and therefore

$$\mu_i^\alpha = \mu_i^\beta \qquad (3.17)$$

Equation 3.17 reflects a very general and very important relationship, namely:

> In a system at equilibrium, the chemical potential of every component in a phase is equal to the chemical potential of that component in every other phase in which that component is present.

Equilibrium is the state toward which systems will naturally transform. The Gibbs free energy is the chemical energy available to fuel these transformations. *We can regard differences in chemical potentials as the forces driving transfer of components between phases*. In this sense, the chemical potential is similar to other forms of potential energy, such as gravitational or electromagnetic. Physical systems spontaneously transform so as to minimize potential energy. Thus, for example, water on the surface of the Earth will move to a point where its gravitational potential energy is minimized – downhill. Just as gravitational potential energy drives this motion, the chemical potential drives chemical reactions, and just as water will come to rest when gravitational energy is minimized, chemical reactions will cease when chemical potential is minimized. So in our example above, the spontaneous transfer of Pb between a hydrothermal solution and a sulfide phase will occur until the chemical potentials of Pb in the solution and in the sulfide are equal. At this point, there is no further energy available to drive the transfer.

We defined the chemical potential in terms of the Gibbs free energy. However, in his original work, Gibbs based the chemical potential on the internal energy of the system. As it turns out, however, the quantities are the same:

$$\mu_i = \left(\frac{\partial G}{\partial n_i}\right)_{P,T,n_{j,j\neq i}} = \left(\frac{\partial U}{\partial n_i}\right)_{S,V,n_{j,j\neq i}} \quad (3.18)$$

It can be further shown (but we won't) that:

$$\mu_i = \left(\frac{\partial G}{\partial n_i}\right)_{P,T,n_{j,j\neq i}} = \left(\frac{\partial U}{\partial n_i}\right)_{S,V,n_{j,j\neq i}}$$

$$= \left(\frac{\partial H}{\partial n_i}\right)_{S,P,n_{j,j\neq i}} = \left(\frac{\partial A}{\partial n_i}\right)_{T,V,n_{j,j\neq i}}$$

3.4.4 The Gibbs–Duhem relation

Since μ is the partial molar Gibbs free passim energy, the Gibbs free energy of a system is the sum of the chemical potentials of each component:

$$G = \sum n_i \left(\frac{\partial G}{\partial n_i}\right)_{P,T,n_{j,j\neq i}} = \sum_i n_i \mu_i \quad (3.19)$$

The differential form of this equation (which we get simply by applying the chain rule) is:

$$dG = \sum_i n_i d\mu_i + \sum_i \mu_i dn_i \quad (3.20)$$

Equating this with eqn. 3.14, we obtain:

$$\sum_i n_i d\mu_i + \sum_i \mu_i dn_i = VdP - SdT + \sum_i \mu_i dn_i \quad (3.21)$$

Rearranging, we obtain the *Gibbs–Duhem relation*:

$$\boxed{VdP - SdT - \sum_i n_i d\mu_i = 0} \quad (3.22)$$

The Gibbs–Duhem equation describes the relationship between simultaneous changes in pressure, temperature, and composition in a single-phase system. In a closed system at equilibrium, net changes in chemical potential will occur only as a result of changes in temperature or pressure. *At constant temperature and pressure*, there can be no net change in chemical potential at equilibrium:

$$\boxed{\sum n_i d\mu_i = 0} \quad (3.23)$$

This equation further tells us that the chemical potentials do not vary independently but change in a related way. In a closed system, only one chemical potential can vary independently. For example, consider a two-component system. Then we have $n_1 d\mu_1 + n_2 d\mu_2 = 0$ and $d\mu_2 = -(n_1/n_2)d\mu_1$. If a given variation in composition produces a change in μ_1 then there is a concomitant change in μ_2.

For multiphase systems, we can write a version of the Gibbs–Duhem relation for each phase in the system. For such systems, the Gibbs–Duhem relation allows us to reduce the number of independently variable components in each phase by one. We will return to this point later in the chapter.

We can now state an additional property of chemical potential:

In spontaneous processes, components or species are distributed between phases so as to minimize the chemical potential of all components.

This allows us to make one more characterization of equilibrium: equilibrium is the point where the chemical potential of all components is minimized.

3.4.5 Derivation of the phase rule

Another significant aspect of the Gibbs–Duhem equation is that the phase rule can be derived from it. We begin by recalling that the variance of a system (the number of variables that must be fixed or independently determined to determine the rest) is equal to the number of variables minus the number of equations relating them. In a multicomponent single-phase system, consisting of c components, there are $c + 2$ unknowns required to describe the equilibrium state of the system: T, P, μ_1, μ_2, ...μ_c. But in a system of ϕ phases at equilibrium, we can write ϕ versions of eqn. 3.23, which reduces the independent variables by ϕ. Thus, the number of independent variables that must be specified to describe a system of c components and ϕ phases is:

$$f = c + 2 - \phi$$

which is the Gibbs phase rule.

Specification of f variables will completely describe the system, at least with the qualification that in thermodynamics we are normally uninterested in the size of the system, that is, in extensive properties such as mass and volume (though we are interested in their intensive equivalents), and outside forces or fields such as gravity, electric, or magnetic fields. Nevertheless, the size of the system is described as well, provided that only one of the f variables is extensive.

3.5 IDEAL SOLUTIONS

Having placed another tool, the chemical potential, in our thermodynamic toolbox, we are ready to continue our consideration of solutions. We will begin with *ideal solutions*, which, like ideal gases, are fictions that avoid some of the complications of real substances. For an ideal solution, we make an assumption similar to one of those made for an ideal gas, namely that there are no forces between molecules. In the case of ideal solutions, which may be gases, liquids, or solids, we can relax this assumption somewhat and require only that *the interactions between different kinds of molecules in an ideal solution are the same as those between the same kinds of molecules.*

3.5.1 Chemical potential in ideal solutions

How does chemical potential vary in an ideal solution? Consider the vapor pressure of a gas. The derivative of G with respect to pressure at constant temperature is volume:

$$\left(\frac{\partial G}{\partial P}\right)_T = V$$

Written in terms of partial molar quantities:

$$\left(\frac{\partial u}{\partial P}\right)_T = v$$

If the gas is ideal, then:

$$\left(\frac{\partial \mu}{\partial P}\right)_{T,ideal} = \frac{RT}{P} \qquad (3.24)$$

and if we integrate from $P°$ to P we obtain:

$$\mu^P - \mu^{P°} = RT \ln \frac{P}{P°} \qquad (3.25)$$

where $\mu^{P°}$ is the chemical potential of the pure gas at the reference (standard state) pressure $P°$. This is the standard-state chemical potential and is written as $\mu°$. If we let $P°$ be the vapor pressure of pure i and P be the vapor pressure of i in an ideal solution, then we may substitute X for $P/P°$ into Raoult's law (eqn. 3.8) to obtain the following:

$$\mu_{i,ideal} = \mu_i° + RT \ln X_i \qquad (3.26)$$

This equation describes the relationship between the chemical potential of component i and its mole fraction in an ideal solution.

3.5.2 Volume, enthalpy, entropy, and free energy changes in ideal solutions

We will be able to generalize a form of this equation to nonideal cases a bit later. Let's first consider some other properties of ideal mixtures. For real solutions, any extensive thermodynamic property such as volume can be considered to be the sum of the volume of the components plus a volume change due to mixing:

$$\overline{V} = \sum X_i \overline{V}_i + \Delta V_{mixing} \qquad (3.27)$$

The first term on the right reflects the volume resulting from mechanical mixing of the various components. The second term reflects volume changes associated with solution. For example, if we mixed 100 ml of ethanol and 100 ml of water (Figure 3.5), the volume of the resulting solution would be 193 ml. Here, the value of the first term on the right would be 200 ml, and the value of the second term would be −7 ml. We can write similar equations for enthalpy, and so on. But the volume change and enthalpy change due to mixing are both 0 in the ideal case. This is true because both volume and enthalpy changes of mixing arise from intermolecular forces, and, by definition, such intermolecular forces are absent in the ideal case. Thus:

$$\Delta V_{ideal\ mixing} = 0$$

therefore:

$$\overline{V}_{ideal} = \sum_i X_i v_i = \sum_i X_i \overline{V}_i$$

and

$$\Delta H_{ideal\ mixing} = 0$$

and therefore:

$$\overline{H}_{ideal} = \sum_i X_i h_i = \sum_i X_i \overline{H}_i$$

(where h is the partial molar enthalpy). This, however, is not true of entropy. You can imagine why: if we mix two substances on an atomic level, the number of possible arrangements of our system increases even if they are ideal substances. The entropy of ideal mixing is (compare eqn. 2.110):

$$\Delta S_{ideal\ mixing} = -R \sum_i X_i \ln X_i \quad (3.28)$$

$$\overline{S}_{ideal\ solution} = \sum X \overline{S} - R \sum X_i \ln X_i \quad (3.29)$$

Because $\Delta G_{mixing} = \Delta H_{mixing} - T\Delta S_{mixing}$ and $\Delta H_{mixing} = 0$, it follows that:

$$\overline{G}_{ideal} = RT \sum_i X_i \ln X_i \quad (3.30)$$

We stated above that the total expression for an extensive property of a solution is the sum of the partial molar properties of the pure phases (times the mole fractions), plus the mixing term. The partial molar Gibbs free energy is the chemical potential, so the full expression for the Gibbs free energy of an ideal solution is:

$$\boxed{\overline{G}_{ideal} = \sum_i X_i \mu_i^o + RT \sum_i X_i \ln X_i} \quad (3.31)$$

Rearranging terms, we can reexpress eqn. 3.31 as:

$$\overline{G}_{ideal} = \sum_i X_i(\mu_i^o + RT \ln X_i) \quad (3.32)$$

The term in parentheses is simply the chemical potential of component i, μ_i, as expressed in eqn. 3.26. Substituting eqn. 3.26 into 3.32, we have

$$\overline{G}_{ideal} = \sum_i X_i \mu_i \quad (3.33)$$

Note that for an ideal solution, μ_i is always less than or equal to $\mu_i^°$ because the term $RT \ln X_i$ is always negative (because the log of a fraction is always negative).

Let's consider ideal mixing in the simplest case, namely binary mixing. For a two-component (binary) system, $X_1 = (1 - X_2)$, so we can write eqn. 3.30 for the binary case as:

$$\Delta \overline{G}_{ideal\ mixing}$$
$$= RT[(1 - X_2) \ln(1 - X_2) + X_2 \ln X_2] \quad (3.34)$$

Since X_2 is less than 1, ΔG is negative and becomes increasingly negative with temperature, as illustrated in Figure 3.6. The curve is symmetrical with respect to X, that is, the minimum occurs at $X_2 = 0.5$.

Now let's see how we can recover information on μ_i from plots such as Figure 3.6, which we will call G-bar-X plots. Substituting $X_1 = (1 - X_2)$ into eqn. 3.33, it becomes:

$$\overline{G}_{ideal\ solution}$$
$$= \mu_1(1 - X_2) + \mu_2 X_2 = \mu_1 + (\mu_2 - \mu_1)X_2 \quad (3.35)$$

This is the equation of a straight line on such a plot with slope of $(\mu_2 - \mu_1)$ and intercept μ_1. This line is illustrated in Figure 3.7.

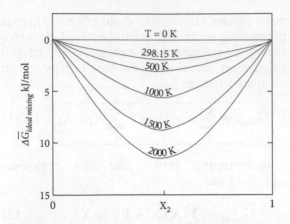

Figure 3.6   Free energy of mixing as a function of temperature in the ideal case. After Nordstrom and Munoz (1986).

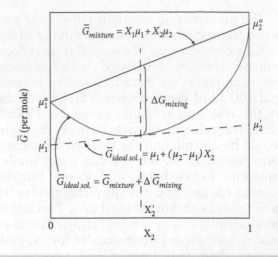

Figure 3.7   Molar free energy in an ideal mixture and a graphical illustration of eqn. 3.31. After Nordstrom and Munoz (1986).

The curved line is described by eqn. 3.31. The dashed line is given by eqn. 3.35. Both eqn. 3.31 and eqn. 3.35 give the same value of $\overline{G}$ for a given value of X_2, such as X'_2. Thus, the straight line and the curved one in Figure 3.7 must touch at X'_2. In fact, the straight line is the tangent to the curved one at X'_2. The intercept of the tangent at $X_2 = 0$ is μ_1 and the intercept at $X_2 = 1$ is μ_2. The point is, on a plot of molar free energy vs. mole fraction (a G–X diagram), *we can determine the chemical potential of component i in a two-component system by extrapolating a tangent of the free energy curve to $X_i = 1$.* We see that in Figure 3.7, as X_1 approaches 1 (X_2 approaches 0), the intercept of the tangent approaches μ°_1, i.e., μ_1 approaches μ°_1. Looking at eqn. 3.26, this is exactly what we expect. Figure 3.7 illustrates the case of an ideal solution, but the intercept method applies to nonideal solutions as well, as we shall see.

Finally, the solid line connecting the μ°'s is the Gibbs free energy of a mechanical mixture of components 1 and 2, which we may express as:

$$\Delta\overline{G}_{mixture} = \sum_i X_i\mu_i \qquad (3.36)$$

You should satisfy yourself that the ΔG_{mixing} is the difference between this line and the free energy curve:

$$\Delta\overline{G}_{ideal\,mixing} = \overline{G}_{ideal\,solution} - \overline{G}_{mixture} \qquad (3.37)$$

3.6   REAL SOLUTIONS

We now turn our attention to real solutions, which are somewhat more complex than ideal ones, as you might imagine. We will need to introduce a few new tools to help us deal with these complexities.

3.6.1   Chemical potential in real solutions

Let's consider the behavior of a real solution in view of the two solution models we have already introduced: Raoult's law and Henry's law. Figure 3.8 illustrates the variation of chemical potential as a function of composition in a hypothetical real solution. We can identify three regions where the behavior of the chemical potential is distinct:

1. The first is where the mole fraction of component X_i is close to 1 and Raoult's law holds. In this case, the amount of solute dissolved in i is trivially small, so molecular interactions involving solute molecules do not significantly affect the thermodynamic properties of the solution, and the behavior of μ_i is close to that in an ideal solution:

    $$\mu_{i,ideal} = \mu^\circ_i + RT\ln X_i \qquad (3.26)$$

2. At the opposite end is the case where X_i is very small. Here interactions between

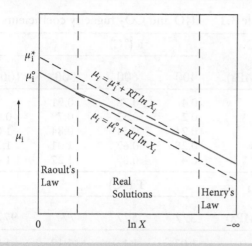

Figure 3.8 Schematic plot of the chemical potential of component i in solution as a function of $\ln X_i$. Here $\mu°$ is the chemical potential of pure i at the pressure and temperature of the diagram. After Nordstrom and Munoz (1986).

component i molecules are extremely rare, and the behavior of μ_i is essentially controlled by interactions between i and those of the solvent. While the behavior of μ_i is not ideal, it is nonetheless a linear function of $\ln X_i$. This is the region where Henry's law holds. Let's define the constant η (eta) such that for conditions where Henry's law holds:

$$\eta_i X_i = \frac{P_i}{P_i^o} \qquad (3.38)$$

Comparing this equation with eqn. 3.10, we see that η is merely another form of the Henry's law constant; whereas h has units of pressure, η is dimensionless. The compositional dependence of the chemical potential in the Henry's law region can be expressed as:

$$\mu_i = \mu_i^o + RT \ln \eta_i X_i \qquad (3.39)$$

This equation can be rewritten as:

$$\mu_i = \mu_i^o + RT \ln X_i + RT \ln \eta_i \qquad (3.40)$$

By definition, η is independent of composition at constant T and P and can be regarded as adding or subtracting a fixed amount to the standard state chemical potential (a fixed amount to the intercept in Figure 3.8). By *independent of composition*, we mean it is independent of X_i, the mole fraction of the component of interest. η will depend on the nature of the solution. For example, if Na is our component of interest, η_{Na} will not be the same for an electrolyte solution as for a silicate melt. We can define a new term, μ^*, as:

$$\mu_i^* \equiv \mu_i^o + RT \ln \eta_i \qquad (3.41)$$

Substituting eqn. 3.40 into 3.39 we obtain:

$$\mu_i = \mu_i^* + RT \ln X_i \qquad (3.42)$$

When plotted against $\ln X_i$, the chemical potential of i in the range of very dilute solutions is given by a straight line with slope RT and intercept μ^* (the intercept is at $X_i = 1$ and hence $\ln X_i = 0$ and $\mu_i = \mu^*$). Thus, μ^* can be obtained by extrapolating the Henry's law slope to $X = 1$. *We can think of μ^* as the chemical potential in the hypothetical standard state of Henry's law behavior at $X = 1$.*

3. The third region of the plot is that region of real solution behavior where neither Henry's law nor Raoult's law apply. In this region, μ is not a linear function of $\ln X$. We will introduce a new parameter, *activity*, to deal with this region.

3.6.2   Fugacities

The tools we have introduced to deal with ideal solutions and infinitely dilute ones are based on observations of the gaseous state: Raoult's law and Henry's law. We will continue to make reference to gases in dealing with real solutions that follow neither law. While this approach has a largely historical basis, it is nevertheless a consistent one. So following this pattern, we will first introduce the concept of fugacity, and derive from it a more general parameter, *activity*.

In the range of intermediate concentrations, the partial pressure of the vapor of component i above a solution is generally not linearly related to the mole fraction of component i in solution. Thus, the chemical potential of i cannot be determined from equations such as 3.26, which we derived on the assumption that the partial pressure was proportional to the mole fraction. To deal with this situation, chemists invented a fictitious partial pressure, *fugacity*. Fugacity may be thought of as the "escaping tendency" of a real gas from a solution. It is defined to have the same relationship to chemical potential as the partial pressure of an ideal gas:

$$\mu_i = \mu_i^o + RT \ln f_i/f_i^o \qquad (3.43)$$

where f^o is the standard-state fugacity, which is analogous to standard-state partial pressure. We are free to choose the standard state, but the standard state for f^o and μ^o must be the same. If we choose our standard state to be the pure substance, then f^o is identical to P^o, but we may wish to choose some other standard state where this will not be the case. Since the behavior of real gases approaches ideal at low pressures, the fugacity will approach the partial pressure under these circumstances. Thus the second part of the definition of fugacity is:

$$\lim_{P \to 0} \frac{f_i}{P_i} = 1$$

For an ideal gas, fugacity is identical to partial pressure. Since, as we stated above, fugacity bears the same relationship to chemical potential (and other state functions) of a nonideal substance as pressure of a nonideal gas, we substitute fugacity for pressure in thermodynamic equations.

The relationship between pressure and fugacity can be expressed as:

$$f = \phi P \qquad (3.44)$$

where ϕ is the *fugacity coefficient*. The fugacity coefficient expresses the difference in the pressure between a real gas and an ideal gas under comparable conditions. Kerrick and Jacobs (1981) fitted the Redlich–Kwong equation (eqn. 2.15) to observations on the volume, pressure and volume of H_2O and CO_2 to obtain values for the coefficients a and b in eqn. 2.15. From these, they obtained fugacity coefficients for these gases at a series of temperatures and pressures. These are given in Table 3.1; Example 3.2 illustrates their use.

Table 3.1 H_2O and CO_2 fugacity coefficients.

H_2O	T (°C)			
P (MPa)	400	600	800	1000
50	0.4	0.78	0.91	
200	0.2	0.52	0.79	0.94
400	0.21	0.54	0.84	1.03
600	0.28	0.67	1.01	1.22
800	0.4	0.89	1.27	1.49

CO_2	T (°C)			
P (MPa)	377	577	777	977
50	1.02	1.1	1.12	1.12
200	1.79	1.86	1.82	1.75
400	4.91	4.18	3.63	3.22
600	13.85	9.48	7.2	5.83
800	38.73	21.33	14.15	10.44

From Kerrick and Jacobs (1981).

3.6.3 Activities and activity coefficients

Fugacities are thermodynamic functions that are directly related to chemical potential and can be calculated from measured P–T–V properties of a gas, though we will not discuss how. However, they have meaning for solids and liquids as well as gases, since solids and liquids have finite vapor pressures. Whenever a substance exerts a measurable vapor pressure, a fugacity can be calculated. Fugacities are relevant to the equilibria between species and phase components, because if the vapor phases of the components of some solid or liquid solutions are in equilibrium with each other, and with their respective solid or liquid phases, then the species or phase components in the solid or liquid must be in equilibrium. One important feature of fugacities is that we can use them to define another thermodynamic parameter, the *activity*, a:

$$a_i \equiv \frac{f_i}{f_i^0} \qquad (3.45)$$

f^o is the standard state fugacity. Its value depends on the standard state you choose. You are free to choose a standard state convenient for whatever problem you are addressing.

If we substitute eqn. 3.45 into eqn. 3.43, we obtain the important relationship:

$$\mu_i = \mu_i^o + RT \ln a_i \qquad (3.46)$$

The "catch" on selecting a standard state for f°, and hence for determining a_i in eqn. 3.46, is that this state must be the same as the standard state for μ°. Thus, we need to bear in mind that standard states are implicit in the definition of activities, and that those standard states are tied to the standard-state chemical potential. Until the standard state is specified, activities have no meaning.

Comparing eqn. 3.46 with 3.26 leads to:

$$a_{i,ideal} = X_i \qquad (3.47)$$

Thus, in ideal solutions, the activity is equal to the mole fraction.

Chemical potentials can be thought of as driving forces that determine the distribution of components between phases of variable composition in a system. *Activities can be thought of as the effective concentration or the availability of components for reaction.* In real solutions, it would be convenient to relate all nonideal thermodynamic parameters to the composition of the solution,

because composition is generally readily and accurately measured. To relate activities to mole fractions, we define a new parameter, the *rational activity coefficient*, λ. The relationship is:

$$a_i = X_i \lambda_I \qquad (3.48)$$

The rational activity coefficient differs slightly in definition from the *practical activity coefficient*, γ, used in aqueous solutions. λ is defined in terms of mole fraction, whereas γ is variously defined in terms of moles of solute per moles of solvent, or more commonly, moles of solute per kg or liter of solution. Consider, for example, the activity of Na in an aqueous sodium chloride solution. For λ_{Na}, X is computed as:

$$X_{Na} = \frac{n_{Na}}{n_{Na} + n_{Cl} + n_{H_2O}}$$

whereas for γ_{Na}, X_{Na} is:

$$\frac{n_{Na}}{n_{H_2O}} \quad \text{or} \quad \frac{n_{Na}}{kg\ solution}$$

where n indicates moles of substance. γ is also used for other concentration units that we will introduce in section 3.7.

Example 3.2 Using fugacity to calculate Gibbs free energy

The minerals brucite ($Mg(OH)_2$) and periclase (MgO) are related by the reaction:

$$Mg(OH)_2 \rightleftharpoons MgO + H_2O$$

Which side of this reaction represents the stable phase assemblage at 600°C and 200 MPa?

Answer: We learned how to solve this sort of problem in Chapter 2: the side with the lowest Gibbs free energy will be the stable assemblage. Hence, we need only to calculate ΔG_r at 600°C and 200 MPa. To do so, we use eqn. 2.130:

$$\Delta G_{T'P'} = \Delta G^o - \int_{T_{ref}}^{T'} \Delta S_r dT + \int_{T_{ref}}^{T'} \Delta V_r dP \qquad (2.130)$$

Our earlier examples dealt with solids, which are incompressible to a good approximation so we could simply treat ΔV_r as independent of pressure. In that case, the solution to the first integral on the left was simply $\Delta V_r (P'-P_{ref})$. The reaction in this case, like most metamorphic reactions, involves H_2O, which is certainly not incompressible: its volume as steam or a supercritical fluid is very much a function of pressure. Let's isolate the difficulty by dividing ΔV_r into two parts: the volume change of reaction due to the solids, in this case the difference between molar volumes of periclase and brucite, and the volume change due to H_2O. We will denote the former as ΔV^S and assume that it is independent of pressure. The second integral in eqn. 2.132 then becomes:

$$\int_{P_{ref}}^{P'} \Delta V_r dP = \Delta V^s (P' - P_{ref}) + \int_{P_{ref}}^{P'} V_{H_2O} dP \tag{3.49}$$

How do we solve the pressure integral above? One approach is to assume that H_2O is an ideal gas. For an ideal gas:

$$V = \frac{RT}{P}$$

so that the pressure integral becomes:

$$\int_{P_{ref}}^{P'} \frac{RT}{P} dP = RT \ln \frac{P'}{P_{ref}}$$

Steam is a very nonideal gas, so this approach would not yield a very accurate answer. The concept of fugacity provides us with an alternative solution. For a nonideal substance, fugacity bears the same relationship to volume as the pressure of an ideal gas. Hence, we may substitute fugacity for pressure so that the pressure integral in eqn. 2.130 becomes:

$$\int_{f_{ref}}^{f'} \frac{RT}{f} df = RT \ln \frac{f'}{f_{ref}}$$

where we take the reference fugacity to be 0.1 MPa. Equation 3.49 thus becomes:

$$\int_{P_{ref}}^{P'} \Delta V_r dP = \Delta V^s (P' - P_{ref}) + \int_{f_{ref}}^{f'} V_{H_2O} df = \Delta V^s (P' - P_{ref}) + RT \ln \frac{f'}{f^{ref}} \tag{3.50}$$

We can then compute fugacity using eqn. 3.44 and the fugacity coefficients in Table 3.1.

Using the data in Table 2.2 and solving the temperature integral in 2.130 as usual (eqn. 2.132), we calculate the $\Delta G_{T,P}$ is 3.29 kJ. As it is positive, the left side of the reaction, i.e., brucite, is stable.

The ΔS of this reaction is positive, however, implying that at some temperature, periclase plus water will eventually replace brucite. To calculate the actual temperature of the phase boundary requires a trial and error approach: for a given pressure, we must first guess a temperature, then look up a value of ϕ in Table 2.1 (interpolating as necessary), and calculate ΔG_r. Depending on our answer, we make a revised guess of T and repeat the process until ΔG is 0. Using a spreadsheet, however, this goes fairly quickly. Using this method, we calculate that brucite breaks down at 660°C at 200 MPa, in excellent agreement with experimental observations.

3.6.4 Excess functions

The ideal solution model provides a useful reference for solution behavior. Comparing real solutions with ideal ones leads to the concept of *excess functions*, for example:

$$G_{excess} \equiv G_{real} - G_{ideal} \tag{3.51}$$

which can be resolved into contributions of excess enthalpy and entropy:

$$G_{excess} = H_{excess} - TS_{excess} \tag{3.52}$$

The excess enthalpy is a measure of the heat released during mixing the pure end-members to form the solution, and the excess entropy is a measure of all the energetic effects resulting from a nonrandom distribution of species in solution. We can express excess enthalpy change in the same way as excess free energy:

$$H_{excess} \equiv H_{real} - H_{ideal} \qquad (3.53)$$

But since $\Delta H_{ideal\ mixing} = 0$, $\Delta H_{excess} = \Delta H_{real}$; in other words, the enthalpy change upon mixing is the excess enthalpy change. Similar expressions may, of course, be written for volume and entropy (bearing in mind that unlike volume and enthalpy, ΔS_{ideal} is not zero).

Combining eqn. 3.46 with eqn. 3.48 leads to the following:

$$\mu_i = \mu_i^o + RT \ln X_i \lambda_i \qquad (3.54)$$

We can rewrite this as:

$$\mu_i = \mu_i^o + RT \ln X_i + RT \ln \lambda_i \qquad (3.55)$$

Equation 3.55 shows how activity coefficients relate to Henry's and Raoult's laws. Comparing eqn. 3.55 with eqn. 3.39, we see that in the region where Henry's law holds, that is, dilute solutions, the activity coefficient is equal to Henry's law constant. In the region where Raoult's law holds, the activity coefficient is 1 and eqn. 3.55 reduces to eqn. 3.26 since $RT \ln \lambda_i = 0$.

Since we know that

$$\left(\frac{\partial G}{\partial n_i} \right)_{T,P,n_{j,j \neq i}} = \mu_i = \mu_i^o + RT \ln X_i \lambda_i$$

comparing equations 3.51 and 3.55, we find that:

$$\left(\frac{\partial G_{excess}}{\partial n_i} \right)_{T,P,n_{j,j \neq i}} = RT \ln \lambda_i \qquad (3.56)$$

which is the same as:

$$\overline{G}_{excess,i} = RT \ln \lambda_i \qquad (3.56a)$$

so that the molar excess free energy associated with component i is simply RT times the log of the activity coefficient. The total molar excess free energy of the solution is then:

$$G_{excess} = RT \sum_i X_i \ln \lambda_i \qquad (3.57)$$

We will see the usefulness of the concept of excess free energy shortly when we consider activities in electrolyte solutions. It will also prove important in our treatment of nonideal solid solutions and exsolution phenomena in the next chapter.

Depression of the melting point

In northern climates, salting roads and sidewalks to melt snow and ice is a common practice in winter. We have now acquired the thermodynamic tools to show why salt melts ice, and that this effect does not depend on any special properties of salt or water. Depression of the melting point by addition of a second component to a pure substance is a general phenomenon. Suppose that we have an aqueous solution containing sodium chloride coexisting with pure ice. If the two phases are at equilibrium, then the chemical potential of water in ice must equal that of water in the solution:

$$\mu_{H_2O}^{ice} = \mu_{H_2O}^{aq} \qquad (3.58)$$

(we are using subscripts to denote the component, and superscripts to denote the phase; aq denotes the liquid aqueous solution). We define our standard state as that of the pure substance. According to eqn. 3.48, the chemical potential of water in the solution can be expressed as:

$$\mu_{H_2O}^{aq} = \mu_{H_2O}^o + RT \ln a_{H_2O}^{aq} \qquad (3.59)$$

where $\mu_{H_2O}^O$ denotes the chemical potential of pure liquid water. Substituting eqn. 3.59 into 3.58 and rearranging, we have:

$$\mu_{H_2O}^{ice} - \mu_{H_2O}^{o,aq} = RT \ln a_{H_2O}^{aq} \qquad (3.60)$$

Ice will incorporate very little salt; if we assume it is a pure phase, we may write eqn. 3.60 as:

$$\mu_{H_2O}^{o,ice} - \mu_{H_2O}^{o,aq} = RT \ln a_{H_2O}^{aq} \qquad (3.60a)$$

or

$$\mu_{H_2O}^{o,aq} - \mu_{H_2O}^{o,ice} = -RT \ln a_{H_2O}^{aq} \qquad (3.61)$$

(The order is important: eqn. 3.60a describes the freezing process, 3.61 the melting process. These processes will have equal and opposite entropies, enthalpies, and free energies.) The left-hand side of eqn. 3.61 is the Gibbs free energy of melting for pure water, which we denote as ΔG_m^o (ΔG_m^o is 0 at the melting temperature of pure water, which we denote $T^o{}_m$, but nonzero at any other temperature).

We may rewrite eqn. 3.61 as:

$$\Delta G_m^o = -RT \ln a_{H_2O}^{aq} \qquad (3.62)$$

If we assume that ΔH and ΔS are independent of temperature (which is not unreasonable over a limited temperature range) and we assume pressure is constant as well, the left-hand side of the equation may also be written as:

$$\Delta G_m^o = \Delta H_m^o - T\Delta S_m^o \qquad (3.63)$$

Substituting eqn. 3.63 into 3.62:

$$\Delta H_m^o - T\Delta S_m^o = -RT \ln a_{H_2O}^{aq} \qquad (3.64)$$

At the melting temperature of pure water, ΔG_m^o is zero, so that:

$$\Delta H_m^o = T_m^o \Delta S_m^o$$

Substituting this into eqn. 3.64 and rearranging:

$$\Delta S_m^o (T_m^o - T) = -RT \ln a_{H_2O}^{aq} \qquad (3.65)$$

Further rearrangement yields:

$$\frac{T_m^o}{T} - 1 = \frac{-R}{\Delta S_m^o} \ln a_{H_2O}^{aq}$$

For a reasonably dilute solution, the activity of water will approximately equal its mole fraction, so that:

$$\frac{T_m^o}{T} - 1 = \frac{-R}{\Delta S_m^o} \ln X_{H_2O}^{aq} \qquad (3.66)$$

The entropy of melting is always positive, and since X is always less than 1, the left-hand side of eqn. 3.66 must always be positive. Thus, the ratio T_m^o/T must always be greater than 1. So the temperature at which an aqueous solution will freeze will always be less than the melting point of pure water. Salting of roads is not a question of geochemical interest, but there are many examples of depression of the freezing point of geological interest. For example, the freezing point of the ocean is about $-2°C$, and this phenomenon is important in igneous petrology, as we shall see in the next chapter. A related phenomenon of geological interest is elevation of the boiling point of a liquid: for example, hydrothermal solutions boil at temperatures significantly above that of pure water. Can you demonstrate that elevation of the boiling point of an ideal solution depends only on the mole fraction of the solute?

3.7 ELECTROLYTE SOLUTIONS

Electrolyte solutions are solutions in which the solute dissociates to form ions, which facilitate electric conduction. Seawater is an obvious example of a natural electrolyte solution, but all natural waters are also electrolytes, though generally more dilute ones. These solutions, which Lavoisier* called the "rinsings of the Earth," are of enormous importance in many geologic processes.

3.7.1 The nature of water and water–electrolyte interaction

There is perhaps no compound more familiar to us than H_2O. Commonplace though it might be, H_2O is the most remarkable compound in nature. Its unusual properties include: the highest heat capacity of all solids and liquids except ammonia, the highest latent heat of vaporization of all substances, the highest surface tension of all liquids, its maximum density is at 4°C, with density decreasing below that temperature (negative coefficient of thermal expansion), the solid form is less dense than the liquid (negative Clapeyron slope), and finally, it is the best solvent known, dissolving more substances and in greater quantity than any other liquid. We will digress here briefly to consider the structure and properties of H_2O and the nature of water–electrolyte interactions from a microscopic perspective.

Many of the unusual properties of water arise from its nonlinear polar structure, which is illustrated in Figure 3.9a. The polar nature of water gives rise to van der Waals forces and the hydrogen bond discussed in Chapter 1. The hydrogen bond, which forms between hydrogen and oxygen atoms of adjacent molecules, imposes a dynamic partial structure on liquid water (Figure 3.9b). These bonds continually break and new ones reform, and there is always some fraction of unassociated molecules. On average, each water molecule is coordinated by four other water molecules. When water boils, all hydrogen bonds are broken. The energy involved in breaking these bonds accounts for the high heat of vaporization.

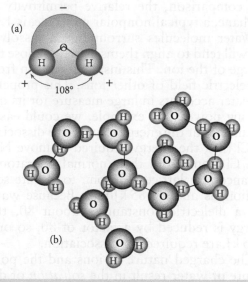

Figure 3.9 (a) Structure of the water molecule. Bond angle in the liquid phase is 108°, and 105° in the gas. The hydrogens retain a partial positive charge and the oxygen retains a partial positive charge. (b) Partial structure present in liquid water. Lines connecting adjacent molecules illustrate hydrogen bonds.

The dissolving power of water is due to its *dielectric* nature. A dielectric substance is one that reduces the forces acting between electric charges. When placed between two electrically charged plates (a capacitor), water molecules will align themselves in the direction of the electric field. As a result, the molecules oppose the charge on the plates and effectively reduce the transmission of the electric field. The *permittivity*, ε, of a substance is the measure of this effect. The *relative permittivity*, or *dielectric constant*, ε_r, of a substance is defined as the ratio of the capacitance observed when the substance is placed between the plates of a capacitor to the capacitance of the same capacitor when a vacuum is present between the plates:

$$\varepsilon_r = \frac{\varepsilon}{\varepsilon_0} \tag{3.67}$$

where ε_0 is the permittivity of a vacuum ($8.85 \times 10^{-12} \, C^2/J \, m$). The relative permittivity of water is 78.54 at 25°C and 1 atm.

---

* Antoine Lavoisier (1743–1794) laid the foundations of modern chemistry in his book, *Traité de Elémentaire de Chemie,* published in 1789. He died at the guillotine during the French Revolution.

For comparison, the relative permittivity of methane, a typical nonpolar molecule, is 1.7.

Water molecules surrounding a dissolved ion will tend to align themselves to oppose the charge of the ion. This insulates the ion from the electric field of other ions. This property of water accounts in large measure for its dissolving power. For example, we could easily calculate that the energy required to dissociate NaCl (i.e., the energy required to move Na^+ and Cl^- ions from their normal interatomic distance in a lattice, 236 pm, to infinite separation) is about 585 kJ/mol. Because water has a dielectric constant of about 80, this energy is reduced by a factor of 80, so only 7.45 kJ are required for dissociation.

The charged nature of ions and the polar nature of water result in the *solvation* of dissolved ions. Immediately adjacent to the ion, water molecules align themselves to oppose the charge on the ion, such that the oxygen of the water molecule will be closest to a cation (Figure 3.10). These water molecules are called the *first solvation shell* or layer and they are effectively bound to the ion, moving with it as it moves. Beyond the first solvation shell is a region of more loosely bound molecules that are only partially oriented, called the second solvation shell or layer. The boundary of this latter shell is diffuse: there is no sharp transition between

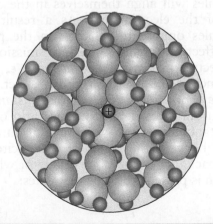

Figure 3.10 Solvation of a cation in aqueous solution. In the first solvation shell, water molecules are bound to the cation and oriented so that the partial negative charge on the oxygen faces the cation. In the second solvation shell, molecules are only loosely bound and partially oriented.

oriented and unaffected water molecules. The energy liberated in this process, called the *solvation energy*, is considerable. For NaCl, for example, it is −765kJ/mol (it is not possible to deduce the solvation energies of Na^+ and Cl^- independently). The total number of water molecules bound to the ion is called the *solvation number*. Solvation effectively increases the electrostatic radius of cations by about 90 pm and of anions by about 10 pm per unit of charge.

An additional effect of solvation is *electrostriction*. Water molecules in the first solvation sphere are packed more tightly than they would otherwise be. This is true, to a lesser extent, of molecules in the secondary shell. In addition, removal of molecules from the liquid water structure causes partial collapse of this structure. The net effect is that the volume occupied by water in an electrolyte solution is less than in pure water, which can lead to negative apparent molar volumes of solutes, as we shall see. The extent of electrostriction depends strongly on temperature and pressure.

A final interesting property of water is that some fraction of water molecules will *autodissociate*. In pure water at standard state conditions, one in every 10^{-7} molecules will dissociate to form H^+ and OH^- ions. Although in most thermodynamic treatments the protons produced in this process are assumed to be free ions, most will combine with water molecules to form H_3O^+ ions. OH^- is called the *hydroxyl* ion; the H_3O^+ is called *hydronium*.

3.7.2 Some definitions and conventions

The first two terms we need to define are solvent and solute. *Solvent* is the substance present in greatest abundance in a solution; in the electrolyte solutions that we will discuss here, water is always the solvent. *Solute* refers to the remaining substances present in solution. Thus, in seawater, water is the solvent and NaCl, $CaSO_4$, and so on, are the solutes. We may also refer to the individual ions as solutes.

3.7.2.1 Concentration units

Geochemists concerned with aqueous solutions commonly use a variety of concentration units other than mole fraction. The first is *molality* (abbreviated as lower-case *m*), which is *moles of solute per kg of solvent* (H_2O). Molality can be converted to moles solute per moles solvent units by dividing by 55.51 mol/kg. A second unit is *molarity* (abbreviated as uppercase *M*), which is *moles of solute per liter* of solution. To convert molality to mole fraction, we would divide by the molecular weight of solvent and use the rational activity coefficient. Natural solutions are often sufficiently dilute that the difference between molality and molarity is trivial (seawater, a relatively concentrated natural solution, contains only 3.5 weight percent dissolved solids). Another common unit is weight fraction (i.e., grams per gram solution), which may take several forms, such as weight percentage, parts per thousand, or parts per million (abbreviated %, ppt or ‰, ppm or mg/kg). To convert to mole fraction, one simply divides the weight of solute and H_2O by the respective molecular weights.

3.7.2.2 pH

One of the most common parameters in aqueous geochemistry is *pH*. *pH* is defined as the negative logarithm of the hydrogen ion activity:

$$pH \equiv - \log a_{H^+} \qquad (3.68)$$

3.7.2.3 Standard state and other conventions

The first problem we must face in determining activities in electrolyte solutions is specifying the standard state. With gases, the standard state is generally the pure substance (generally at 298 K and 1 atm), but this is generally not a reasonable choice for electrolytes. A NaCl solution will become saturated at about $0.1\, X_{NaCl}$, and crystalline NaCl has very different properties from NaCl in aqueous solution. *By convention, a hypothetical standard state of unit activity at 1 molal concentration is chosen*:

$$a° = m = 1 \qquad (3.69)$$

Activity is generally given units of molality in this case (it is dimensionless, as we defined it in eqn. 3.45), so that in this hypothetical standard state, activity equals molality. The

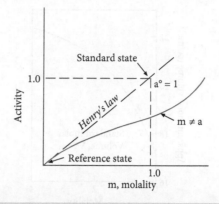

Figure 3.11 Relationship of activity and molality, reference state, and standard state for aqueous solutions. After Nordstrom and Munoz (1986).

standard state is hypothetical because, for most electrolytes, the activity will be less than 1 in a 1 m (molal) solution. Because the *standard state* generally is unattainable in reality, we must also define an attainable *reference state*, from which experimental measurements can be extrapolated. *By convention, the reference state is that of an infinitely dilute solution* – the Henry's law state. For multicomponent solutions, we also specify that the concentrations of all other components be held constant. Hence the reference state is:

$$\lim_{m \to 0} \frac{a_i}{m_i} = 1 \ (m_j \ constant) \qquad (3.70)$$

This convention is illustrated in Figure 3.11. In such solutions, the activity coefficient can be shown to depend on the charge of the ion, its concentration, and the concentration of other ions in the solution as well as temperature and other parameters of the solute. Comparing eqn. 3.70 with 3.46 and 3.48, we see that under these conditions, the activity coefficient is 1. By referring to infinite dilution, we are removing the effect of solute–solute interactions. The standard state properties of an electrolyte solution therefore only take account of solvent–solute interactions.

Clearly, it is impossible to measure the properties of the solute, such as chemical potential or molar volume, at infinite dilution. In practice, this problem is overcome by measuring properties at some finite dilution and extrapolating the result to infinite dilution. Indeed, even at finite concentrations, it is

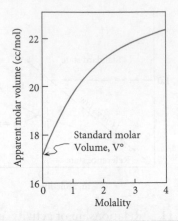

Figure 3.12   Apparent molar volume of NaCl in aqueous solution as a function of molality. The standard molar volume, V°, is the apparent molar volume at infinite dilution.

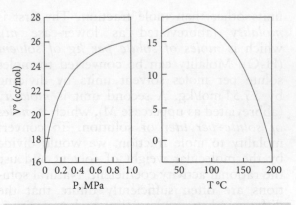

Figure 3.13   Standard molar volume of NaCl in aqueous solution as a function of temperature and pressure. Based on the data of Helgeson and Kirkham (1976).

not possible to measure directly many properties of electrolytes. Volume is a good example. One cannot measure the volume of the solute, but one can measure the volume change of the solution as a function of concentration of the solute. Then by assuming that the partial molar volume of water does not change, a partial molar volume of the solute can be calculated. This is called the *apparent molar volume*, $\overline{V_A}$. The apparent molar volume of NaCl as a function of molarity is shown in Figure 3.12. In essence, this convention assigns all deviations from nonideality to the solute and allows us to use the partial molar volume of pure water in the place of the true, but unknown, molar volume of water in the solution. Thus the volume of NaCl solution is given by:

$$V = n_w \overline{V}_w + n_{NaCl} \overline{V}_{NaCl}^{aq} \qquad (3.71)$$

This convention leads to some interesting effects. For example, the apparent molar volume of magnesium sulfate increases with pressure, and many other salts, including NaCl (Figure 3.13), exhibit the same behavior. Just as curiously, the apparent molar volume of sodium chloride in saturated aqueous solution becomes negative above ~200°C (Figure 3.13). Many other salts show the same effect. These examples emphasize the "apparent" nature of molar volume when defined in this way. Of course, the molar volume of NaCl does not actually become negative; rather, this

is the result of the interaction between Na+ and Cl− and H₂O (electrostriction) and the convention of assigning all nonideality to sodium chloride.

The concentration of a salt consisting of ν_A moles of cation A and ν_B moles of cation B is related to the concentration of its constituent ionic species as:

$$m_A = \nu_A m_{AB} \quad \text{and} \quad m_B = \nu_B m_{AB} \qquad (3.72)$$

By convention, the thermodynamic properties of ionic species A and B are related to those of the salt AB by:

$$\Psi_{AB} \equiv \nu_A \Psi_A + \nu_B \Psi_B \qquad (3.73)$$

where Ψ is some thermodynamic property. Thus the chemical potential of MgCl₂ is related to that of Mg²⁺ and Cl⁻ as:

$$\mu_{MgCl_2} = \mu_{Mg+} + 2 \times \mu_{Cl-}$$

The same holds for enthalpy of formation, entropy, molar volume, and so on.

A final important convention is that *the partial molar properties and energies of formation for the proton (H+) are taken to be zero under all conditions*.

3.7.3   Activities in electrolytes

The assumption we made for ideal solution behavior was that interactions between molecules (species might be a better term in the case of electrolyte solutions) of solute and molecules of solvent were not different from those interactions between solvent ions only. In light of the discussion of aqueous solutions earlier, we can see this is clearly not going to be the case for an electrolyte solution. We have seen significant deviations from ideality even where the components have no net charge (e.g., water–ethanol); we can expect greater deviations due to electrostatic interactions between charged species.

The nature of these interactions suggests that a purely macroscopic viewpoint, which takes no account of molecular and ionic interactions, may have severe limitations in predicting equilibria involving electrolyte solutions. Thus, chemists and geochemists concerned with the behavior of electrolytes have had to incorporate a microscopic viewpoint into electrolyte theory. On the other hand, they did not want to abandon entirely the useful description of equilibria based on thermodynamics. We have already introduced concepts, the activity and the activity coefficient, which allow us to treat non-ideal behavior within a thermodynamic framework. *The additional task imposed by electrolyte solutions, and indeed all real solutions, therefore, is not to rebuild the framework, but simply to determine activities from readily measurable properties of the solution.* The dependence of all partial molar properties of a solute on concentration can be determined once the activity coefficient and its temperature and pressure dependence are known.

3.7.3.1 The Debye–Hückel and Davies equations

Both solvent–solute and solute–solute interactions in electrolytes give rise to excess free energies and nonideal behavior. By developing a model to account for these two kinds of interactions, we can develop an equation that will predict the activity of ions in electrolyte solution.

In an electrolyte solution, each ion will exert an electrostatic force on every other ion. These forces will decrease with the increase in square of distance between ions. The forces between ions will be reduced by the presence of water molecules, due to its dielectric nature. As total solute concentration increases, the mean distance between ions will decrease.

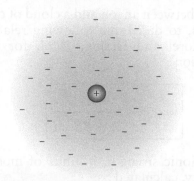

Figure 3.14 An ion surrounded by a cloud of oppositely charged ions, as assumed in Debye–Hückel theory.

Thus, we can expect that activity will depend on the total ionic concentration in the solution. The extent of electrostatic interaction will also obviously depend on the charge of the ions involved: the force between Ca^{2+} and Mg^{2+} ions will be greater at the same distance than between Na^+ and K^+ ions.

In the Debye–Hückel theory (Figure 3.14), a given ion is considered to be surrounded by an atmosphere or cloud of oppositely charged ions (this atmosphere is distinct from, and unrelated to, the solvation shell). If it were not for the thermal motion of the ions, the structure would be analogous to that of a crystal lattice, though considerably looser. Thermal motion, however, tends to destroy this structure. The density of charge in this ion atmosphere increases with the *square root* of the ionic concentrations but increases with the *square* of the charges on those ions. The dielectric effect of intervening water molecules will tend to reduce the interaction between ions. Debye–Hückel theory also assumes the following:

- All electrolytes are completely dissociated into ions.
- The ions are spherically symmetrical charges (hard spheres).
- The solvent is structureless; the sole property is its permittivity.
- The thermal energy of ions exceeds the electrostatic interaction energy.

With these assumptions, Debye and Hückel (1923) used the Poisson–Boltzmann equation, which describes the electrostatic interaction energy between an ion and a cloud of opposite charges, to derive the following relationship (see Morel and Hering, 1993, for the full derivation):

$$\log_{10}\gamma_i = \frac{-Az_i^2\sqrt{I}}{1 + B\mathring{a}_i\sqrt{I}} \qquad (3.74)$$

I is ionic strength, in units of molality or molarity, calculated as:

$$I = \frac{1}{2}\sum_j m_j z_j^2 \qquad (3.75)$$

where m is the concentration and z the ionic charge. The parameter $\mathring{a}$ is known as the *hydrated ionic radius, or effective radius* (significantly larger than the radius of the same ion in a crystal). A and B constants are known as solvent parameters and are functions of T and P. Equation 3.74 is known as the *Debye–Hückel extended law*; we will refer to it simply as the *Debye–Hückel equation*. Table 3.2a summarizes the Debye–Hückel solvent parameters over a range of temperatures and Table 3.2b gives values of $\mathring{a}$ for various ions (and see Example 3.3).

For very dilute solutions, the denominator of eqn. 3.74 approaches 1 (because I approaches 0), hence eqn. 3.74 becomes:

$$\log_{10}\gamma_i = -Az_i^2\sqrt{I} \qquad (3.76)$$

Table 3.2a Debye–Hückel solvent parameters.

T°C	A	B (10^8 cm)
0	0.4911	0.3244
25	0.5092	0.3283
50	0.5336	0.3325
75	0.5639	0.3371
100	0.5998	0.3422
125	0.6416	0.3476
150	0.6898	0.3533
175	0.7454	0.3592
200	0.8099	0.3655
225	0.8860	0.3721
250	0.9785	0.3792
275	1.0960	0.3871
300	1.2555	0.3965

From Helgeson and Kirkham (1974).

Table 3.2b Debye–Hückel effective radii.

Ion	$\mathring{a}$ (10^{-8} cm)
Rb^+, Cs^+, NH_4^+, Ag^+	2.5
K^+, Cl^-, Br^-, I^-, NO_3^-	3
OH^-, F^-, HS^-, BrO_3^{2-}, IO_4^-, MnO_4^-	3.5
Na^+, HCO_3^-, $H_2PO_4^-$, HSO_4^-, HPO_4^{2-}, PO_4^{3-}	4.0–4.5
Pb^{2+}, CO_3^{2-}, SO_4^{2-}	4.5
Sr^{2+}, Ba^{2+}, Cd^{2+}, Hg^{2+}, S^{2-}	5
Li^+, Ca^{2+}, Cu^{2+}, Zn^{2+}, Sn^{2+}, Mn^{2+}, Fe^{2+}, Ni^{2+}	6
Mg^{2+}, Be^{2+}	8
H^+, Al^{3+}, trivalent rare earths	9
Th^{4+}, Zr^{4+}, Ce^{4+}	11

From Garrels and Christ (1982).

This equation is known as the *Debye–Hückel limiting law* (so-called because it applies in the limit of very dilute concentrations).

Davies (1938, 1962) introduced an empirical modification of the Debye–Hückel equation. The Davies equation is:

$$\log_{10}\gamma_i = -Az_i^2\left[\frac{\sqrt{I}}{1 + \sqrt{I}} - bI\right] \qquad (3.77)$$

where A is the same as in the Debye–Hückel equation and b is an empirically determined parameter with a value of around 0.3. It is instructive to see how the activity coefficient of Ca^{2+} would vary according to Debye–Hückel and Davies equations if we vary the ionic strength of the solution. This variation is shown in Figure 3.15. The Davies equation predicts that activity coefficients begin to increase above ionic strengths of about 0.5 m. For reasons discussed below and in greater detail in Chapter 4, activity coefficients do actually increase at higher ionic strengths. On the whole, the Davies equation is slightly more accurate for many solutions at ionic strengths of 0.1–1 m. Because of this, as well as its simplicity, the Davies equation is widely used.

3.7.3.2 Limitations to the Debye–Hückel approach

None of the assumptions made by Debye and Hückel hold in the absolute. Furthermore, the Poisson–Boltzmann equation provides only an

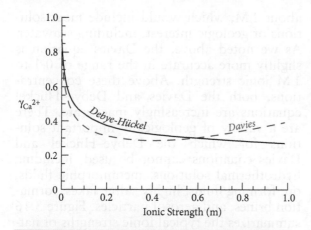

Figure 3.15 Variation of the Ca^{2+} activity coefficient with ionic strength according to the Debye–Hückel (black solid line) and Davies equations (red dashed line).

approximate description of ion interactions, and Debye and Hückel used an approximate solution of this equation. Thus, we should not expect the Debye–Hückel equations to provide an exact prediction of activity coefficients under all conditions.

Perhaps the greatest difficulty is the assumption of complete dissociation. When ions approach each other closely, the electrostatic interaction energy exceeds the thermal energy, which violates the assumption made in the approximate solution of the Poisson–Boltzmann equation. In this case, the ions are said to be associated. Furthermore, the charge on ions is not spherically symmetric and this asymmetry becomes increasingly important at short distances. Close approach is obviously more likely at high ionic strength, so not surprisingly the Debye–Hückel equation breaks down at high ionic strength.

We can distinguish two broad types of ion associations: ion pairs and complexes. These two classes actually form a continuum, but we will define a complex as an association of ions in solution that involves some degree of covalent bonding (i.e., electron sharing). Ion pairs, on the other hand, are held together purely by electrostatic forces. We will discuss formation of ion pairs and complexes in greater detail in subsequent chapters. Here, we will attempt to

convey only a very qualitative understanding of these effects.

An ion pair* can be considered to have formed when ions approach closer than some critical distance where the electrostatic energy, which tends to bind them, exceeds twice the thermal energy, which tends to move them apart. When this happens, the ions are electrostatically bound and their motions are linked. This critical distance depends on the charge of the ions involved and is therefore much greater for highly charged ions than for singly charged ones. As we will show in Chapter 4, ion pairs involving singly charged ions will never form, even at high ionic strengths. On the other hand, multiply charged ions will tend to form ion pairs even at very low ionic strengths.

Formation of ion pairs will cause further deviations from ideality. We can identify two effects. First, the effective concentration, or activity, of an ionic species that forms ionic associations will be reduced. Consider, for example, a pure solution of $CaSO_4$. If some fraction, α, of Ca^{2+} and SO_4^{2-} ions forms ion pairs, then the effective concentration of Ca^{2+} ions is:

$$[Ca^{2+}]_{eff} = [Ca_{2+}]_{tot}(1 - \alpha)$$

(here we follow the usual convention of using brackets to denote concentrations). The second effect is on ionic strength. By assuming complete dissociation, we similarly overestimate the effective concentration in this example by a factor of $(1 - \alpha)$.

A second phenomenon that causes deviations from ideality not predicted by Debye–Hückel is solvation. As we noted, an ion in aqueous solution is surrounded by a sphere of water molecules that are bound to it. Since those water molecules bound to the ion are effectively unavailable for reaction, the activity of water is reduced by the fraction of water molecules bound in solvation shells. This fraction is trivial in dilution solutions but is important at high ionic strength. The result of this effect is to increase the activity of ions.

Despite these problems, Debye–Hückel has proved to be remarkably successful in predict-

---

* The term *ion pair* is a bit of a misnomer because such associations can involve more than two ions. In concentrated solutions, ion pairs may consist of a cation plus several anions.

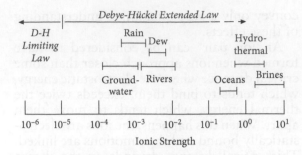

Figure 3.16 Ionic strength of natural electrolyte solutions and the applicability of the Debye–Hückel and Davies equations.

ing activity coefficients in dilute solution. The extended Debye–Hückel equation (eqn. 3.74) is most useful at concentrations less than 0.1 M, which includes many natural waters, and provides adequate approximation for activity coefficients up to ionic strengths of

about 1 M, which would include most solutions of geologic interest, including seawater. As we noted above, the Davies equation is slightly more accurate in the range of 0.1 to 1 M ionic strength. Above these concentrations, both the Davies and Debye–Hückel equations are increasingly inaccurate. There are a variety of geologically important solutions for which the Debye–Hückel and Davies equations cannot be used, including hydrothermal solutions, metamorphic fluids, ore-forming fluids, highly saline lakes, formation brines, and aerosol particles. Figure 3.16 summarizes the typical ionic strengths of natural solutions and the applicability of these equations. The Debye–Hückel limiting law is useful only for very dilute solutions, less than 10^{-5} mol/kg, which is more dilute than essentially all solutions of geologic interest. We will consider several methods of estimating activities in higher ionic strength solutions in Chapter 4.

Example 3.3 Calculating activities using the Debye–Hückel equation

Given the composition for the average river water in column A in the table below, calculate the activity of the Ca^{2+} ion at 25° C using the Debye–Hückel equation.

Answer: Our first step is to convert these concentrations to molality by dividing by the respective molecular weights. We obtain the molal concentrations in column B. We also need to compute z^2 (column C), and the product z^2m (column D). Using equation 3.75, we calculate the ionic strength to be 0.00202 m. (Note, one must use the ionic strength in molal or molar, and not millimolar, units in the Debye–Hückel Equation.

Ion	A g/kg	B mol/kg × 10^3	C z^2	D mz^2 × 10^3
Cl^-	0.0078	0.2201	1	0.2201
SO_4^{2-}	0.0112	0.1167	4	0.4667
HCO_3^-	0.0583	0.9557	1	0.9557
Mg^{2+}	0.0041	0.1687	4	0.6746
Ca^{2+}	0.015	0.3742	4	1.4970
K^+	0.0023	0.0588	1	0.0588
$Na+$	0.0041	0.1782	1	0.1782

We substitute this value for I, then find å = 6, A = 0.5092, and B = 0.3283 in Table 3.1, and obtain a value for the activity coefficient of 0.8237, and an activity of 0.308 ×10^{-3} m. If we did the calculation for other temperatures, we would see that for a dilute solution such as this, the activity coefficient is only a weak function of temperature, decreasing to 0.625 at 300° C.

3.8 IDEAL SOLUTIONS IN CRYSTALLINE SOLIDS AND THEIR ACTIVITIES

When we deal with solid solutions, we are again faced with the inadequacy of the purely macroscopic approach of classical thermodynamics. There is little disadvantage to this approach for gases, where the arrangement of molecules is chaotic. But the crystalline state differs from that of gases in that the arrangement of atoms in the crystal lattice is highly ordered, and the properties of the crystal depend strongly on the nature of the ordering. For this reason, we cannot afford to ignore the arrangement of atoms in solids, particularly with respect to solutions.

Solid solutions differ from those of gases and liquids in several respects. First, solution in the solid state inevitably involves substitution. While we can increase the concentration of HCl in water simply by adding HCl gas, we can only increase the concentration of Fe in biotite solid solution if we simultaneously remove Mg. Second, solid solutions involve substitution at crystallographically distinct sites. Thus, in biotite a solid solution between phlogopite ($KMg_3AlSi_3O_{10}(OH)_2$) and annite ($KFe_3AlSi_3O_{10}(OH)_2$) occurs as Fe^{2+} replaces Mg^{2+} in the octahedral site; the tetrahedral Si site and the anion (O) sites remain unaffected by this substitution. Third, substitution is often coupled. For example, the solid solution between anorthite ($CaAl_2Si_2O_8$) and albite ($NaAlSi_3O_8$) in plagioclase feldspar involves not only the substitution of Na^+ for Ca^{2+}, but also the substitution of Al^{3+} for Si^{4+}. The anorthite–albite solution problem is clearly simplified if we choose anorthite and albite as our components rather than Na^+, Ca^{2+}, Al^{3+} and Si^{4+}. Such components are known as *phase components*. Choosing pure phase end members as components is not always satisfactory either because substitution on more than one site is possible, leading to an unreasonably large number of components, or because the pure phase does not exist and hence its thermodynamic properties cannot be measured.

However we choose our components, we need a method of calculating activities that takes account of the ordered nature of the

crystalline state. Here we will discuss two ideal solution models of crystalline solids. We tackle the problem of nonideal solid solutions in Chapter 4.

3.8.1 Mixing-on-site model

Many crystalline solids can be successfully treated as ideal solutions. Where this is possible, the thermodynamic treatment and assessment of equilibrium are greatly simplified. A simple and often successful model that assumes ideality but takes account of the ordered nature of the crystalline state is the *mixing-on-site model*, which considers the substitution of species in sites individually. In this model, the activity of an individual species is calculated as:

$$a_{i,ideal} = (X_i)^\nu \qquad (3.78)$$

where X is the mole fraction of the ith atom and ν is the number of sites per formula unit on which mixing takes place. For example, $\nu = 2$ in the Fe–Mg exchange in olivine, $(Mg,Fe)_2SiO_4$. One trick to simplifying this equation is to pick the formula unit such that $\nu = 1$. For example, we would pick $(Mg,Fe)Si_{1/2}O_2$ as the formula unit for olivine. We must then consistently choose all other thermodynamic parameters to be half those of $(Mg,Fe)_2SiO_4$.

The entropy of mixing is given by:

$$\Delta S_{ideal\,mixing} = -R \sum_j \left(n_j \sum_i X_{i,j} \ln X_{i,j} \right) \qquad (3.79)$$

where the subscript j refers to sites and the subscript i refers to components, and n is the number of sites per formula unit. The entropy of mixing is the same as the configurational entropy, residual entropy, or "third law entropy" (i.e., entropy when $T = 0\,K$). For example, in clinopyroxene, there are two exchangeable sites, a sixfold-coordinated M1 site, (Mg, Fe^{2+}, Fe^{3+}, Al^{3+}), and an eightfold-coordinated M2 site (Ca^{2+}, Na^+). Here j ranges from 1 to 2 (e.g., 1 = M1, 2 = M2), but $n = 1$ in both cases (because both sites accept only one atom). i must range over all present ions in each site, so in this example, i ranges from 1 to 4 (1 = Mg, 2 = Fe^{2+}, etc.) when $j = 1$ and from 1 to 2 when $j = 2$. Since we have assumed an ideal solution, $\Delta H = 0$ and $\Delta G_{ideal} = -T\Delta S$. In other words, all we need is temperature and eqn. 3.79 to calculate the free energy of solution.

In the mixing-on-site model, the activity of a phase component in a solution, for example, pyrope in garnet, is the product of the activity of the individual species in each site in the phase:

$$a_\phi = \prod_i X^{v_i} \qquad (3.80)$$

where a_ϕ is the activity of phase component ϕ, i are the ion components of pure ϕ, and v_i is the stoichiometric proportion of i in pure ϕ. For example, to calculate the activity of aegirine ($NaFe^{3+}Si_2O_6$) in aegirine-augite ($[Na,Ca][Fe^{3+},Fe^{2+},Mg]Si_2O_6$), we would calculate the product: $X_{Na}X_{Fe^{3+}}$. Note that it would not be necessary to include the mole

fractions of Si and O, since these are 1 (see Example 3.4).

A slight complication arises when more than one ion occupies a structural site in the pure phase. For example, suppose we wish to calculate the activity of phlogopite ($KMg_3Si_3AlO_{10}(OH)_2$) in a biotite of composition $K_{0.8}Ca_{0.2}(Mg_{0.17}Fe_{0.83})_3Si_{2.8}Al_{1.2}O_{10}(OH)_2$. The tetrahedral site is occupied by Si and Al in the ratio of 3:1 in the pure phase end members. If we were to calculate the activity of phlogopite in pure phlogopite using eqn. 3.80, the activities in the tetrahedral site would contribute only $X_{Si}^2 X_{Al}^1 = (0.75)^3(0.25)^1 = 0.1055$ in the pure phase. So we would obtain an activity of 0.1055 instead of 1 for phlogopite in pure phlogopite. Since the activity of a phase component must be one when it is pure, we need to normalize the result. Thus, we apply a correction by multiplying by the raw activity we obtain from 3.92 by $1/(0.1055) = 9.481$, and thus obtain an activity of phlogopite of 1.

Example 3.4 Calculating activities using the mixing-on-site model

Sometimes it is desirable to calculate the activities of pure end-member components in solid solutions. Garnet has the general formula $X_3Y_2Si_3O_{12}$. Calculate the activity of pyrope, $Mg_3Al_2Si_3O_{12}$, in a garnet solid solution of composition:

$$(Mg_{.382}Fe_{2.316}^{2+}Mn_{.167}Ca_{.156})(Al_{1.974}Fe_{.044}^{2+})Si_3O_{12}$$

Answer: The chemical potential of pyrope in garnet contains mixing contributions from both Mg in the cubic site and Al in the octahedral site:

$$\mu_{py}^{gt} = \mu_{py}^o + 3RT \ln X_{Mg} + 2RT \ln X_{Al} = RT \ln(X_{Mg}^3 X_{Al}^2)$$

The activity of pyrope is thus given by:

$$a_{py}^{gt} = X_{py}^{gt} = X_{Mg}^3 X_{Al}^2$$

In the example composition above, the activity of Mg is:

$$a_{Mg} = X_{Mg}^3 = \left(\frac{[Mg]}{[Mg] + [Fe^{2+}] + [Mn] + [Ca]}\right) = 0.126^3 = 0.002$$

and that of Al is:

$$a_{Al} = X_{Al}^2 = \left(\frac{[Al]}{[Al] + [Fe^{3+}]} \right) = 0.976^2 = 0.956$$

The activity of pyrope in the garnet composition above is $0.002 \times 0.956 = 0.00191$. There is, of course, no mixing contribution from the tetrahedral site because it is occupied only by Si in both the solution and the pure pyrope phase.

3.8.2 Local charge balance model

Yet another model for the calculation of activities in ideal solid solutions is the *local charge balance* model. A common example is the substitution of Ca for Na in the plagioclase solid solution ($NaAlSi_3O_8$–$CaAl_2Si_2O_8$). To maintain charge balance, the substitution of Ca^{2+} for Na^+ in the octahedral site requires substitution of Al^{3+} for Si^{4+} in the tetrahedral site. In this model, the activity of the end-member of phase component is equal to the mole fraction of the component (see Example 3.5).

Example 3.5 Activities using the local charge balance model

Given the adjacent analysis of a plagioclase crystal, what are the activities of albite and anorthite in the solution?

Plagioclase Analysis

Oxide	Wt. percent
SiO_2	44.35
Al_2O_3	34.85
CaO	18.63
Na_2O	0.79
K_2O	0.05

Answer: According to the *local charge balance model*, the activity of albite will be equal to the mole fraction of Na in the octahedral site. To calculate this, we first must convert the weight percent oxides to formula units of cation. The first step is to calculate the moles of cation from the oxide weight percentages. First, we can convert weight percent oxide to weight percent cation using the formula:

$$wt.\%cation = wt.\%oxide \times \frac{atomic\;wt.cation \times formula\;units\;cation\;in\;oxide}{molecular\;wt.oxide}$$

Next, we calculate the moles of cation:

$$moles\;cation = \frac{wt.\%cation}{atomic\;wt.cation}$$

Combining these two equations, the *atomic wt. cation* terms cancel and we have:

$$moles\;cation = wt.\%cation \times \frac{formula\;units\;cation\;in\;oxide}{molecular\;wt.oxide}$$

Next, we want to calculate the number of moles of each cation per formula unit. A general formula for feldspar is: XY_4O_8, where X is Na, K, or Ca in the A site and Y is Al or Si in the tetrahedral site. So to calculate formula units in the A site, we divide the number of moles of Na, K, and Ca by the sum of moles of Na, K, and Ca. To calculate formula units in the tetrahedral site, we divide the number of moles of Al and Si by the sum of moles of Al and Si and multiply by 4, since there are four ions in this site. Since the number of oxygens is constant, we can refer to these quantities as the moles per eight oxygens. The following table shows the results of these calculations.

Cation formula units

	Mol. wt. oxide	Moles cation	Moles per 8 oxygens
Si	60.06	0.7385	2.077
Al	101.96	0.6836	1.923
Ca	56.08	0.3322	0.926
Na	61.98	0.0255	0.071
K	94.2	0.0011	0.003

The activity of albite is equal to the mole fraction of Na, 0.07; the activity of anorthite is 0.93.

3.9 EQUILIBRIUM CONSTANTS

Now that we have introduced the concepts of activity and activity coefficients, we are ready for one of the most useful parameters in physical chemistry: the equilibrium constant. Though we can predict the equilibrium state of a system, and therefore the final result of a chemical reaction, from the Gibbs free energy alone, the equilibrium constant is a convenient and succinct way express this. As we shall see, it is closely related to, and readily derived from, the Gibbs free energy.

3.9.1 Derivation and definition

Consider a chemical reaction such as:

$$aA + bB \rightleftharpoons cC + dD$$

carried out under isobaric and isothermal conditions. The Gibbs free energy change of this reaction can be expressed as:

$$\Delta G = c\mu_c + d\mu_d - a\mu_a - b\mu_b \quad (3.81)$$

At equilibrium, ΔG must be zero. A general expression then is:

$$\Delta G = \sum_i v_i \mu_i = 0 \quad (3.82)$$

where v_i is the stoichiometric coefficient of species i. Equilibrium in such situations need not mean that all the reactants (i.e., those phases on the left side of the equation) are consumed to leave only products. Indeed, this is generally not so. Substituting eqn. 3.46 into 3.82 we obtain:

$$\sum_i v_i \mu_i^o + RT \sum_i v_i \ln a_i = 0 \quad (3.83)$$

or:

$$\sum_i v_i \mu_i^o + RT \ln \prod_i a_i^{v_i} = 0 \quad (3.84)$$

The first term is simply the standard state Gibbs free energy change, $\Delta G°$, for the reaction. There can be only one fixed value of $\Delta G°$ for a fixed standard state pressure and temperature, and therefore of the activity products. The activity products are therefore called the *equilibrium constant* K, familiar from elementary chemistry:

$$\boxed{K = \prod_i a_i^{v_i}} \quad (3.85)$$

Substituting eqn. 3.85 into 3.84 and rearranging, we see that the equilibrium constant is related to the Gibbs free energy change of the reaction by the equation:

$$\boxed{\Delta G_r^o = -RT \ln K} \quad (3.86)$$

At this point, it is worth saying some more about *standard states*. We mentioned that one is free to choose a standard state, but there are pitfalls. In general, there are two kinds of standard states, fixed pressure–temperature standard states and variable *P–T* standard states. If you chose a fixed temperature standard state, then eqn. 3.86 is only valid at that standard-state temperature. If you chose a variable-temperature standard state, then eqn. 3.86 is valid for all temperatures, but $\Delta G°$ is then a function of temperature. The same goes for pressure. Whereas most thermodynamic quantities we have dealt with thus far are additive, equilibrium constants are multiplicative (see Example 3.6).

3.9.2 Law of mass action

Let's attempt to understand the implications of eqn. 3.85. Consider the dissociation of carbonic acid, an important geologic reaction:

$$H_2CO_3 \rightleftharpoons HCO_3^- + H^+$$

For this particular case, eqn. 3.85 is expressed as:

$$K = \frac{a_{HCO_3^-} a_{H^+}}{a_{H_2CO_3}}$$

The right side of the equation is a quotient, the product of the activities of the products divided by the product of the activities of the reactants and is called the *reaction quotient*. At equilibrium, the reaction quotient is equal to the equilibrium constant. The equilibrium constant therefore allows us to predict the relative amounts of products and reactants that will be present when a system reaches equilibrium.

Suppose now that we prepare a beaker of carbonic acid solution; it is not hard to prepare: we just allow pure water to equilibrate with the atmosphere. Let's simplify things by assuming that this is an ideal solution. This allows us to replace activities with concentrations (the concentration units will dictate how we define the equilibrium constant; see below). When the solution has reached equilibrium, just enough carbonic acid will have dissociated so that the reaction quotient will be equal to the equilibrium constant. Now let's add some H^+ ions, perhaps by adding a little HCl. The value of the reaction quotient increases above that of the equilibrium constant and the system is no longer in equilibrium. Systems will always respond to disturbances by moving toward equilibrium (how fast they respond is another matter, and one that we will address in Chapter 5). The system will respond by adjusting the concentrations of the three species until equilibrium is again achieved; in this case, hydrogen and bicarbonate ions will combine to form carbonic acid until the reaction quotient again equals the equilibrium constant. We can also see that had we reduced the number of hydrogen ions in the solution (perhaps by adding a base), the reaction would have been driven the other way (i.e., hydrogen ions would be produced by dissociation). Equation 3.85 is known as the *law of mass action*, which we can state more generally: *Changing the concentration of one species in a system at equilibrium will cause a reaction in a direction that minimizes that change.*

Example 3.6 Manipulating reactions and equilibrium constant expressions

Often we encounter a reaction for which we have no value of the equilibrium constant. In many cases, however, we can derive an equilibrium constant by considering the reaction of interest to be the algebraic sum of several reactions for which we do have equilibrium constant values. For example, the concentration of carbonate ion is often much lower than that of the bicarbonate ion. In such cases, it is more convenient to write the reaction for the dissolution of calcite as:

$$CaCO_3 + H_2O \rightleftharpoons Ca^{2+} + HCO_3^- + OH^- \tag{3.87}$$

Given the following equilibrium constants, what is the equilibrium constant expression for the above reaction?

$$K_2 = \frac{a_{H^+} a_{CO_3^{2-}}}{a_{HCO_3^-}} \qquad K_{cal} = \frac{a_{Ca^{2+}} a_{CO_3^{2-}}}{a_{CaCO_3}} \qquad K_{H_2O} = \frac{a_{H^+} a_{OH^-}}{a_{H_2O}}$$

Answer: Reaction 3.87 can be written as the algebraic sum of three reactions:

$$+ CaCO_3 \leftrightharpoons Ca^{2+} + CO_3^{2-}$$

$$+ H_2O \leftrightharpoons H^+ + OH^-$$

$$- HCO_3^- \leftrightharpoons H^+ + CO_3^{2-}$$

$$CaCO_3 - HCO_3^- + H_2O \leftrightharpoons Ca^{2+} + OH^-$$

The initial inclination might be to think that if we can sum the reactions, the equilibrium constant of the resulting reaction is the sum of the equilibrium constants of the components. However, this is not the case. Whereas we sum the reactions, we take the product of the equilibrium constants. Thus, our new equilibrium constant is:

$$K = \frac{K_{cal} K_{H_2O}}{K_2}$$

For several reasons (chief among them that equilibrium constants can be very large or very small numbers), it is often more convenient to work with the log of the equilibrium constant. A commonly used notation is pK. pK is the negative logarithm (base 10) of the corresponding equilibrium constant (note this notation is analogous to that used for pH). The pK's sum and our equilibrium constant expression is:

$$pK = pK_{cal} - pK_{H2O} - pK_2$$

3.9.2.1 Le Chatelier's principle

We can generalize this principle to the effects of temperature and pressure as well. Recall that:

$$\left(\frac{\partial \Delta G_r}{\partial P}\right)_T = \Delta V \qquad (2.128)$$

and

$$\left(\frac{\partial \Delta G_r}{\partial T}\right)_T = -\Delta S_r \qquad (2.129)$$

and that systems respond to changes imposed on them by minimizing G. Thus, a system undergoing reaction will respond to an increase in pressure by minimizing volume. Similarly, it will respond to an increase in temperature by maximizing entropy. The reaction ice → water illustrates this. If the pressure is increased on a system containing water and ice, the equilibrium will shift to favor the phase with the least volume, which is water (recall that water is unusual in that the liquid has a smaller molar volume than the solid). If the temperature of that system is increased, the phase with the greatest molar entropy is favored, which is also water.

Another way of looking at the effect of temperature is to recall that:

$$\Delta S \geq \frac{\Delta Q}{T}$$

Combining this with eqn. 2.129, we can see that if a reaction A + B → C + D generates heat, then increasing the temperature will retard formation of the products, that is, the reactants will be favored.

A general statement that encompasses both the law of mass action and the effects we have just discussed is then:

When perturbed, a system reacts to minimize the effect of the perturbation.

This is known as Le Chatelier's principle.

3.9.3 K_D values, apparent equilibrium constants, and the solubility product

It is often difficult to determine activities for phase components or species, and therefore it is more convenient to work with concentrations. We can define a new "constant," the distribution coefficient, K_D, as:

$$K_D = \prod_i X_i^{\nu_i} \quad (3.88)$$

K_D is related to the equilibrium constant K as:

$$K_D = \frac{K_{eq}}{K_\lambda} \quad (3.89)$$

where K_λ is simply the ratio of activity coefficients:

$$K_\lambda = \prod_i \lambda_i^{\nu_i} \quad (3.90)$$

Distribution coefficients are functions of temperature and pressure, as are the equilibrium constants, though the dependence of the two may differ. The difference is that K_D values are also functions of composition.

An alternative to the distribution coefficient is the *apparent equilibrium constant*, which we define as:

$$K^{app} = \prod_i m_i^{\nu_i} \quad (3.91)$$

$$K^{app} = \frac{K_{eq}}{K_\gamma} \quad (3.92)$$

with K_γ defined analogously to K_λ. The difference between the apparent equilibrium constant and the distribution coefficient is that we have defined the former in terms of molality and the latter in terms of mole fraction. Igneous geochemists tend to use the distribution coefficient, aqueous geochemists the apparent equilibrium constant.

Another special form of the equilibrium constant is the *solubility product*. Consider the dissolution of NaCl in water. The equilibrium constant is:

$$K = \frac{a_{Na^+_{aq}} a_{Cl^-_{aq}}}{a_{NaCl_s}}$$

where *aq* denotes the dissolved ion and *s* denotes solid. Because the activity of NaCl in pure sodium chloride solid is 1, this reduces to:

$$K = a_{Na^+_{aq}} a_{Cl^-_{aq}} = K_{sp} \quad (3.93)$$

where K_{sp} is called the *solubility product*. You should note that it is generally the case in dissolution reactions such as this that we take the denominator (i.e., the activity of the solid) to be 1 (see Example 3.7).

Example 3.7 Using the solubility product

The apparent (molar) solubility product of fluorite (CaF_2) at 25°C is 3.9×10^{-11}. What is the concentration of Ca^{2+} ion in groundwater containing 0.1 mM of F^- in equilibrium with fluorite?

Answer: Expressing eqn. 3.93 for this case we have:

$$K_{sp-Fl} = \frac{[Ca^{2+}][F^-]^2}{[CaF_2]} = [Ca^{2+}][F^-]^2$$

We take the activity of CaF_2 as 1. Rearranging and substituting in values, we have:

$$[Ca^{2+}] = \frac{K_{sp-Fl}}{[F^-]^2} = \frac{3.9 \times 10^{-11}}{[0.1 \times 10^{-3}]^2} = \frac{3.9 \times 10^{-11}}{1 \times 10^{-8}} = 3.9 \times 10^{-3} M = 3.9 mM$$

3.9.4 Henry's law and gas solubilities

Consider a liquid, water for example, in equilibrium with a gas, the atmosphere for example. Earlier in this chapter, we found that the partial pressure of component i in the gas could be related to the concentration of a component i in the liquid by Henry's law:

$$P_i = h_i X_i \qquad (3.10)$$

where h is Henry's law constant. We can rearrange this as:

$$h_i = \frac{P_i}{X_i} \qquad (3.94)$$

Notice that this equation is analogous in form to the equilibrium constant expression (3.88), except that we have used a partial pressure in place of one of the concentrations. A Henry's law constant is thus a form of equilibrium constant used for gas solubility: it relates the equilibrium concentration of a substance in a liquid solution to that component's partial pressure in a gas.

3.9.5 Temperature dependence of equilibrium constant

Since $\Delta G° = \Delta H° - T\Delta S°$ and $\Delta G°_r = -RT \ln K$, it follows that in the standard state, the equilibrium constant is related to enthalpy and entropy change of reaction as:

$$\ln K = -\frac{\Delta H^o_r}{RT} + \frac{\Delta S^o_r}{R} \qquad (3.95)$$

Equation 3.95 allows us to calculate an equilibrium constant from fundamental thermodynamic data (see Example 3.8). Conversely, we can estimate values for $\Delta S°$ and $\Delta H°$ from the equilibrium constant, which is readily calculated if we know the activities of reactants and products. Equation 3.95 has the form:

$$\ln K = \frac{a}{T} + b$$

where a and b are $\Delta H°/R$ and $\Delta S°/R$, respectively. If we can assume that ΔH and ΔS are constant over some temperature range (this is likely to be the case provided the temperature interval is small), then a plot of $\ln K$ vs. $1/T$

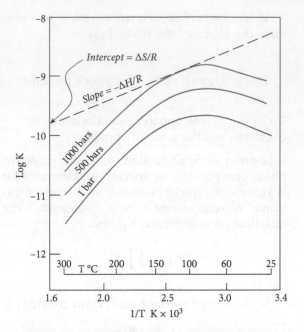

Figure 3.17 Log of the solubility constant of barite plotted against the inverse of temperature. The slope of a tangent to the curve is equal to $-\Delta H/R$. The intercept of the tangent (which occurs at $1/T = 0$ and is off the plot) is equal to $\Delta S/R$. After Blount (1977).

will have a slope of $\Delta H°/R$ and an intercept of $\Delta S°/R$. Thus, measurements of $\ln K$ made over a range of temperatures and plotted vs. $1/T$ provide estimates of $\Delta H°$ and $\Delta S°$. Even if ΔH and ΔS are not constant, they can be estimated from the instantaneous slope and intercept of a curve of $\ln K$ plotted against $1/T$. This is illustrated in Figure 3.17, which shows measurements of the solubility constant for barite ($BaSO_4$) plotted in this fashion (though in this case the $\log_{10}$ rather than natural logarithm is used). From changes of ΔH and ΔS with changing temperature and knowing the heat capacity of barite, we can also estimate heat capacities of the Ba^{2+} and SO_4^{2-} ions, which would obviously be difficult to measure directly. We can, of course, also calculate ΔG directly from eqn. 3.86. Thus, a series of measurements of the equilibrium constant for simple systems allows us to deduce the fundamental thermodynamic data needed to predict equilibrium in more complex systems.

Example 3.8 Calculating equilibrium constants and equilibrium concentrations

The hydration of olivine to form chrysotile (a serpentine mineral) may be represented in a pure Mg system as:

$$H_2O + 2H^+ + 2Mg_2SiO_4 \rightleftharpoons Mg_3Si_2O_5(OH)_4 + Mg^{2+}$$

If this reaction controlled the concentration of Mg^{2+} of the metamorphic fluid, what would the activity of Mg^{2+} be in that fluid if it had a pH of 4.0 at 300° C?

Answer: Helgeson (1967) gives the thermodynamic data shown in the table below for the reactants at 300° C. From these data, we use Hess's law to calculate ΔH_r and ΔS_r as −231.38 kJ and −253.01 J/K respectively. The equilibrium constant for the reaction may be calculated as:

$$K = \exp\left(-\frac{\Delta H_r^o}{RT} + \frac{\Delta S_r^o}{R}\right) = \exp\left(-\frac{-231.38 \times 10^3}{8.134 \times 573} + \frac{253.01}{8.314}\right) = 7.53 \times 10^7$$

Species	ΔH° kJ	S° J/K
$Mg_3Si_2O_5(OH)_4$	−4272.87	434.84
Mg^{2+}	−366.46	109.05
H^+	44.87	106.68
Mg_2SiO_4	−2132.75	186.02
H_2O	−232.19	211.50

The equilibrium constant for this reaction can be written as:

$$K = \frac{a_{Mg^{2+}}a_{Cry}}{a_{H^+}^2 a_{Fo}^2 a_{H_2O}}$$

which reduces to $K = \frac{a_{Mg^{2+}}}{a_{H^+}^2}$ if we take the activities of water, chrysotile, and forsterite as 1. Since pH $= -\log aH^+$, we may rearrange and obtain the activity of the magnesium ion as:

$$a_{Mg^{2+}} = K\, a_{H^+}^2 = 7.53 \times 10^7 \times 10^{-4 \times 2} = 7.53 \times 10^{-1}$$

Taking the derivative with respect to temperature of both sides of eqn. 3.95 (while holding pressure constant), we have:

$$\left(\frac{\partial \ln K}{\partial T}\right)_P = \frac{\Delta H_r^o}{RT^2} \quad (3.96)$$

This equation is known as the *van't Hoff equation*.

3.9.6 Pressure dependence of equilibrium constant

Since

$$\left(\frac{\partial \Delta G_r}{\partial P}\right)_T = \Delta V$$

and

$$\Delta G_r^o = -RT \ln K$$

then

$$\left(\frac{\partial \ln K}{\partial P}\right)_T = -\frac{\Delta V_r^o}{RT} \quad (3.97)$$

If ΔV_r does not depend on pressure, this equation can be integrated to obtain:

$$\ln K_{P_2} = \ln K_{P_1} - \frac{\Delta V_r^o}{RT}(P_1 - P_2) \quad (3.97a)$$

This assumption will be pretty good for solids because their compressibilities are very low, but slightly less satisfactory for reactions involving liquids (such as dissolution), because they are more compressible. This assumption will be essentially totally invalid for reactions involving gases, because their volumes are highly pressure-dependent.

3.10 PRACTICAL APPROACH TO ELECTROLYTE EQUILIBRIUM

With the equilibrium constant now in our geochemical toolbox, we have the tools necessary to roll up our sleeves and get to work on some real geochemical problems. Even setting aside nonideal behavior, electrolyte solutions (geologic ones in particular) often have many components and can be extremely complex. Predicting their equilibrium state can therefore be difficult. There are, however, a few rules for approaching problems of electrolyte solutions that, when properly employed, make the task much more tractable.

3.10.1   Choosing components and species

We emphasized at the beginning of the chapter the importance of choosing the components in a system. How well we choose components will make a difference to how easily we can solve a given problem. Morel and Hering (1993) suggested these rules for choosing components and species in aqueous systems:

1. All species should be expressible as stoichiometric functions of the components, the stoichiometry being defined by chemical reactions.
2. Each species has a unique stoichiometric expression as a function of the components.
3. H_2O should always be chosen as a component.
4. H^+ should always be chosen as a component.

H^+ activity, or pH, is very often the critical variable, also called the *master variable*, in problems in natural waters. In addition, recall that we define the free energy of formation of H^+ as 0. For these reasons, it is both convenient and important that H^+ be chosen as a component.

3.10.2   Mass balance

This constraint, also sometimes called *mole balance*, is a very simple one, and as such it is easily overlooked. When a salt is dissolved in water, the anion and cation are added in stoichiometric proportions. If the dissolution of the salt is the only source of these ions in the solution, then for a salt of composition $C_{\nu+}A_{\nu-}$ we may write:

$$\nu^-[C] = \nu^+[A] \qquad (3.98)$$

Thus, for example, for a solution formed by dissolution of $CaCl_2$ in water, the concentration of Cl^- ion will be twice that of the Ca^{2+} ion. Even if $CaCl_2$ is not the only source of these ions in solution, its congruent dissolution allows us to write the mass balance constraint in the form of a differential equation:

$$\frac{\partial Cl^-}{\partial Ca^{2+}} = 2$$

which just says that $CaCl_2$ dissolution adds two Cl^- ions to solution for every Ca^{2+} ion added.

By carefully choosing components and boundaries of our system, we can often write conservation equations for components. For example, suppose we have a liter of water containing dissolved CO_2 in equilibrium with calcite (for example, groundwater in limestone). In some circumstances, we may want to choose our system as the water plus the limestone, in which case we may consider Ca conserved and write:

$$\Sigma Ca = Ca^{2+}_{aq} + CaCO_{3s}$$

where $CaCO_{3s}$ is calcite (limestone) and Ca^{2+}_{aq} is aqueous calcium ion. We may want to avoid choosing carbonate as a component and choose carbon instead, since the carbonate ion is not conserved because of association and dissociation reactions such as:

$$CO_3^{2-} + H^+ \rightleftharpoons HCO_3^-$$

Choosing carbon as a component has the disadvantage that some carbon will be present as organic compounds, which we may not wish to consider. A wiser choice is to define CO_2 as a component. Total CO_2 would then include all carbonate species as well as CO_2 (very often, total CO_2 is expressed instead as total carbonate). The conservation equation for total CO_2 for our system would be:

$$\Sigma CO_2 = CaCO_{3s} + CO_2 + H_2CO_3 + HCO_3^-$$
$$+ CO_3^{2-}$$

Here we see the importance of the distinction we made between components and species earlier in the chapter. Example 3.9 illustrates the use of mass balance.

Example 3.9 Soil organic acid

Consider soil water with a pH of 7 containing a weak organic acid, which we will designate HA, at a concentration of 1×10^{-4} M. If the apparent dissociation constant of the acid is $10^{-4.5}$, what fraction of the acid is dissociated?

Answer: We have two unknowns: the concentration of the dissociated and undissociated acid, and we have two equations: the equilibrium constant expression for dissociation and the mass balance equation. We will have to solve the two simultaneously to obtain the answer. Our two equations are:

$$K_{dis} = \frac{[H^+][A^-]}{[HA]} = 10^{-4.5} \qquad \Sigma HA = [HA] + [A^-]$$

Solving the dissociation constant expression for $[A^-]$, we have:

$$[A^-] = \frac{[HA]K_{dis}}{[H^+]}$$

Then solving the conservation equation for [HA] and substituting, we have

$$[A^-] = \frac{(\Sigma HA - [A^-])K_{dis}}{[H^+]}$$

Setting H^+ to 10^{-7} and ΣHA to 10^{-4}, we calculate $[A^-]$ as 3.16×10^{-5} M, so 31.6% of the acid is dissociated.

3.10.3   Electrical neutrality

There is an additional condition that electrolyte solutions must meet: *electrical neutrality*. Thus, the sum of the positive charges in solutions must equal the sum of the negative ones, or:

$$\boxed{\sum_i m_i z_i = 0} \qquad (3.99)$$

While this presents some experimental obstacles, for example, we cannot add only Na^+ ion to an aqueous solution while holding other compositional parameters constant; it also allows placement of an additional mathematical constraint on the solution. It is often convenient to rearrange eqn. 3.99 so as to place anions and cations on different sides of the equation:

$$\sum_i m_i^+ z_i^+ = \sum_n m_n^- z_n^- \qquad (3.100)$$

As an example, consider water in equilibrium with atmospheric CO_2 and containing no other species. The charge balance equation in this case is:

$$[H^+] = [OH^-] + [HCO_3^-] + 2[CO_3^{2-}]$$

As Example 3.10 illustrates, the electrical neutrality constraint can prove extremely useful.

Example 3.10 Determining the pH of rainwater from its composition

Determine the pH of the two samples of rain in the adjacent table. Assume that sulfuric and nitric acid are fully dissociated and that the ions in the table, along with H^+ and OH^-, are the only ones present.

Analysis of rainwater		
	Rain 1(μM)	**Rain 2(μM)**
Na	9	89
Mg	4	16
K	5	9
Ca	8	37
Cl	17	101
NO_3^-	10	500
SO_4^{2-}	18	228

Answer: This problem is simpler than it might first appear. Given the stated conditions, there are no reactions between these species that we need to concern ourselves with. To solve the problem, we observe that this solution must be electrically neutral: any difference in the sum of cations and anions must be due to one or both of the two species not listed: OH^- and H^+.

We start by making an initial guess that the rain is acidic and that the concentration of H^+ will be much higher than that of OH^-, and that we can therefore neglect the latter (we will want to verify this assumption when we have obtained a solution). The rest is straightforward. We sum the product of charge times concentration (eqn. 3.99) for both cations and anions and find that anions exceed cations in both cases: the difference is equal to the concentration of H^+. Taking the log of the concentration (having first converted concentrations to M from μM by multiplying by 10^{-6}), we obtain a pH of 4.6 for the first sample and 3.14 for the second.

Now we need to check our simplifying assumption that we could neglect OH^-. The equilibrium between OH^- and H^+ is given by:

$$K = [H^+][OH^-] = 10^{-14}$$

From this we compute $[OH^-]$ as 10^{-10} in the first case and 10^{-11} in the second. Including these would not change the anion sum significantly, so our assumption was justified.

Charge balance for rainwater		
	Rain 1	**Rain 2**
Σ cations	38	204
Σ anions	63	1057
Δ	25	853
pH	4.60	3.07

3.10.4 Equilibrium constant expressions

For each chemical reaction in our system, we can write one version of eqn. 3.85. This allows us to relate the equilibrium activities of the species undergoing reaction in our system to one another.

Solution of aqueous equilibria problems often hinge on the degree to which we can simplify the problem by minimizing the number of equilibrium constant expressions we must solve. For example, H_2SO_4 will be completely dissociated in all but the most acidic natural waters, so we need not deal with reactions between H^+, SO_4^{2-}, HSO_4^-, and H_2SO_4, and need not consider the latter two in our list of species. Similarly, though many natural waters contain Na^+ and Cl^-, NaCl will precipitate only from concentrated brines, so we generally need not consider reactions between NaCl, Na^+, and Cl^-.

Carbonate is a somewhat different matter. Over the range of compositions of natural waters, H_2CO_3, HCO_3^-, and CO_3^{2-} may all be present. In most cases, however, one of these forms will dominate and the concentrations of the remaining ones will be an order of magnitude or more lower than that of the dominant one. In some cases, two of the above species may have comparable concentrations and we will have to consider equilibrium between them, but it is rarely necessary to consider equilibrium between all three. Thus, at most we will have to consider equilibrium between H_2CO_3 and HCO_3^-, or HCO_3^- and CO_3^{2-}, and we can safely ignore the existence of the remaining species. A successful solution of problems involving carbonate equilibria often requires correctly deciding which reactions to ignore. We will discuss carbonate equilibrium in greater detail in Chapter 6.

3.11 OXIDATION AND REDUCTION

An important geochemical variable that we have not yet considered is the *oxidation state* of a system. Many elements exist in nature in more than one valence state. Iron and carbon are the most important of these because of their abundance. Other elements, including transition metals such as Ti, Mn, Cr, Ce, Eu, and U, and nonmetals such as N, S, and As, are found in more than one valence state in nature. The valence state of an element can significantly affect its geochemical behavior. For example, U is quite soluble in water in its oxidized state, U^{6+}, but is much less soluble in its reduced state, U^{4+}. Many uranium deposits have formed when an oxidized, U-bearing solution was reduced. Iron is reasonably soluble in reduced form, Fe^{2+}, but much less

soluble in oxidized form, Fe^{3+}. The same is true of manganese. Thus, iron is leached from rocks by reduced hydrothermal fluids and precipitated when these fluids mix with oxidized seawater. Eu^{2+} in magmas substitutes readily for Ca in plagioclase, whereas Eu^{3+} does not. Nitrogen is the element most critical to life after C, H, and O and is abundant in the atmosphere and dissolved in natural waters as N_2. Plants and algae, however, can utilize N only in its reduced or oxidized states (such as ammonium or nitrate). The mobility of pollutants, particularly toxic metals, will depend strongly on whether the environment is reducing or oxidizing. Thus, the oxidation state of a system is an important geochemical variable.

The *valence number* of an element is defined as the electrical charge an atom would acquire if it formed ions in solution. For strongly electronegative and electropositive elements that form dominantly ionic bonds, valence number corresponds to the actual state of the element in ionic form. However, for elements that predominantly or exclusively form covalent bonds, valence state is a somewhat hypothetical concept. Carbon, for example, is never present in solution as a monatomic ion. Because of this, assignment of valence number can be a bit ambiguous. A few simple conventions guide assignment of valence number:

- The valence number of all elements in pure form is 0.
- The sum of valence numbers assigned to atoms in molecules or complex species must equal the actual charge on the species.
- The valence number of hydrogen is $+1$, except in metal hydrides, when it is -1.
- The valence number of oxygen is -2 except in peroxides, when it is -1.

The valence state in which an element will be present in a system is governed by the availability of electrons. Oxidation–reduction (*redox*) reactions involve the transfer of electrons and the resultant change in valence. *Oxidation is the loss of electrons; reduction* is the gain of electrons.* An example is the oxidation of magnetite (which consists of 1 Fe^{2+} and 2 Fe^{3+}) to hematite:

$$2Fe_3O_4 + \frac{1}{2}O_2 \rightleftharpoons 3Fe_2O_3 \qquad (3.101)$$

The Fe^{2+} in magnetite loses an electron in this reaction and is thereby oxidized; conversely, oxygen gains an electron and is thereby reduced.

We can divide the elements into *electron donors* and *electron acceptors*; this division is closely related to electronegativity, as you might expect. Electron acceptors are electronegative; electron donors are electropositive. Metals in 0 valence state are electron donors, nonmetals in 0 valence state are usually electron acceptors. Some elements, such as carbon and sulfur, can be either electron donors or receptors. Oxygen is the most common electron acceptor, hence the term *oxidation*. It is nevertheless important to remember that oxidation and reduction may take place in the absence of oxygen.

A reduced system is one in which the availability of electrons is high, due to an excess of electron donors over electron acceptors. In such a system, metals will be in a low valence state (e.g., Fe^{2+}). Conversely, when the availability of electrons is low, due to an abundance of electron acceptors, a system is said to be oxidized. Since it is the most common electron acceptor, the abundance of oxygen usually controls the oxidation state of a system, but this need not be the case.

To predict the equilibrium oxidation state of a system, we need a means of characterizing the availability of electrons, and the valence state of elements as a function of that availability. Low-temperature geochemists and high-temperature geochemists do this in different ways. The former use electrochemical potential while the latter use oxygen fugacity. We will consider both.

3.11.1 Redox in aqueous solutions

The simplest form of the chemical equation for the reduction of ferric iron would be:

$$Fe^{3+}_{aq} + e^- \rightleftharpoons Fe^{2+}_{aq} \qquad (3.102)$$

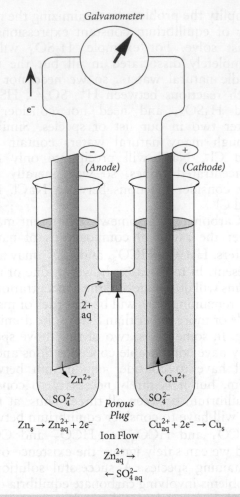

Figure 3.18 Electrode reactions in the Daniell cell.

where the subscript *aq* denotes the aqueous species. This form suggests that the energy involved might be most conveniently measured in an electrochemical cell.

The Daniell cell pictured in Figure 3.18 can be used to measure the energy involved in the exchange of electrons between elements, for example, zinc and copper:

$$Zn_s + Cu^{2+}_{aq} \rightleftharpoons Zn^{2+}_{aq} + Cu_s \qquad (3.103)$$

where the subscript *s* denotes the solid. Such a cell provides a measure of the *relative* preference of Zn and Cu for electrons. In practice, such measurements are made by applying a voltage to the system that is just sufficient to halt the flow of electrons from the zinc plate to the copper one. What is actually

---

* A useful mnemonic to remember this is **LEO** the lion says **GRR!** (**L**oss **E**quals **O**xidation, **G**ain **R**efers to **R**eduction.) Silly, perhaps, but effective. Try it!

measured, then, is a potential energy, denoted E, and referred to as the *electrode potential*, or simply the *potential* of the reaction.

If we could measure the potential of two separate half-cell reactions:

$$Zn_s \rightleftharpoons Zn_{aq}^{2+} + 2e^-$$

$$Cu_s \rightleftharpoons Cu_{aq}^{2+} + 2e^-$$

we could determine the energy gain/loss in the transfer of an electron from an individual element. Unfortunately, such measurements are not possible (nor would these reactions occur in the natural environment: electrons are not given up except to another element or species*). This requires the establishment of an arbitrary reference value. Once such a reference value is established, the potential involved in reactions such as 3.102 can be established.

3.11.1.1 Hydrogen scale potential, E_H

The established convention is to measure potentials in a standard hydrogen electrode cell (at standard temperature and pressure). The cell consists on one side of a platinum plate coated with fine Pt powder that is surrounded by H_2 gas maintained at a partial pressure of 1 atm and immersed in a solution of unit H^+ activity. The other side consists of the electrode and solution under investigation. A potential of 0 is assigned to the half-cell reaction:

$$\frac{1}{2}H_{2(g)} \rightleftharpoons H_{aq}^+ + e^- \tag{3.104}$$

where the subscript g denotes the gas phase. The potential measured for the entire reaction is then assigned to the half-cell reaction of interest. Thus, for example, the potential of the reaction:

$$Zn_{aq}^{2+} + H_{2(g)} \rightleftharpoons Zn_s + 2H^+$$

is -0.763 V. This value is assigned to the reaction:

$$Zn_{aq}^{2+} + 2e^- \rightleftharpoons Zn_s \tag{3.105}$$

Table 3.3 E_H° and $p\varepsilon^\circ$ for some half-cell reactions.

Half-cell reaction	E_H° (V)	$p\varepsilon^\circ$
$Li^+ + e^- \rightleftharpoons Li$	-3.05	-51.58
$Ca^{2+} + 2\,e^- \rightleftharpoons Ca$	-2.93	-49.55
$Th^{4+} + 4e^- \rightleftharpoons Th$	-1.83	-30.95
$U^{4+} + 4e^- \rightleftharpoons U$	-1.38	-23.34
$Mn^{2+} + 2e^- \rightleftharpoons Mn$	-1.18	-19.95
$Zn^{2+} + 2e^- \rightleftharpoons Zn$	-0.76	-12.85
$Cr^{3+} + 3e^- \rightleftharpoons Cr$	-0.74	-12.51
$CO_{2(g)} + 4H^+ + 4e^- \rightleftharpoons CH_2O^*$ $+ 2H_2O$	-0.71	-12.01
$Fe^{2+} + 2e^- \rightleftharpoons Fe$	-0.44	-7.44
$Eu^{3+} + e^- \rightleftharpoons Eu^{2+}$	-0.36	-6.08
$Ni^{2+} + 2e^- \rightleftharpoons Ni$	-0.26	-4.34
$Pb^{2+} + 2e^- \rightleftharpoons Pb$	-0.13	-2.2
$CrO_4^{2-} + 4H_2O + 3e^- \rightleftharpoons$ $Cr(OH)_3 + H_2O$	-0.13	-2.2
$2H^+ + 2e^- \rightleftharpoons H_{2(g)}$	0	0
$N_{2(g)} + 6H^+ + 6e^- \rightleftharpoons 2NH_3$	0.093	1.58
$Cu^{2+} + 2e^- \rightleftharpoons Cu$	0.34	5.75
$UO_2^{2+} + 2e^- \rightleftharpoons UO_2$	0.41	6.85
$S + 2e^- \rightleftharpoons S^{2-}$	0.44	7.44
$Cu^+ + e^- \rightleftharpoons Cu$	0.52	8.79
$Fe^{3+} + e^- \rightleftharpoons Fe^{2+}$	0.77	13.02
$NO^{3+} + 2H^+ + e^- \rightleftharpoons NO_{2(g)} +$ H_2O	0.80	13.53
$Ag^+ + e^- \rightleftharpoons Ag$	0.80	13.53
$Hg^{2+} + 2e^- \rightleftharpoons Hg$	0.85	14.37
$MnO_{2(s)} + 4H^+ + 2e^- \rightleftharpoons Mn^{2+} +$ $2H_2O$	1.22	20.63
$O_2 + 4H^+ + 4e^- \rightleftharpoons 2H_2O$	1.23	20.80
$MnO^{4-} + 8H^+ + 5e^- \rightleftharpoons Mn^{2+} +$ $4H_2O$	1.51	25.53
$Au^+ + e^- \rightleftharpoons Au$	1.69	28.58
$Ce^{4+} + e^- \rightleftharpoons Ce^{3+}$	1.72	29.05
$Pt^+ + e^- \rightleftharpoons Pt$	2.64	44.64

*CH_2O refers to carbohydrate, the basic product of photosynthesis.

and called the *hydrogen scale potential*, or E_H, of this reaction. Thus, the E_H for the reduction of Zn^{+2} to Zn^0 is -0.763 V. The hydrogen scale potentials of a few half-cell reactions are listed in Table 3.3. The sign convention for E_H is that the sign of the potential is positive when the reaction proceeds from left to right (i.e., from reactants to products). Thus, if a reaction has positive E_H, the metal ion will be reduced by hydrogen gas to the metal. If a reaction has negative E_H, the metal will be oxidized to the ion and H^+ reduced. The

---

* Ionization reactions, where free electrons are formed, do occur in nature at very high temperatures. They occur, for example, in stars or other very energetic environments in the universe.

standard state potentials (298 K, 0.1 MPa) of more complex reactions can be predicted by algebraic combinations of the reactions and potentials in Table 3.3 (see Example 3.11).

The half-cell reactions in Table 3.3 are arranged in order of increasing E_H^o. Thus, a species on the product (right) side of a given reaction will reduce (give up electrons to) the species on the reactant side in all reactions listed below it. Thus, in the Daniell cell reaction in Figure 3.18, Zn metal will reduce Cu^{2+} in solution. Zn may thus be said to be a stronger reducing agent than Cu.

Electrochemical energy is another form of free energy and can be related to the Gibbs free energy of reaction as:

$$\Delta G = -z\mathcal{F}E \qquad (3.106)$$

and

$$\Delta G^\circ = -z\mathcal{F}E^\circ \qquad (3.107)$$

where z is the number of electrons per mole exchanged (e.g., 2 in the reduction of zinc) and $\mathcal{F}$ is the Faraday constant ($\mathcal{F} = 96{,}485$ coulombs; 1 joule = 1 volt-coulomb). The free energy of formation of a pure element is 0 (by convention). Thus, the ΔG in a reaction that is opposite one such as 3.105, such as:

$$Zn_{(s)} \rightleftharpoons Zn^{2+} + 2e^-$$

is the free energy of formation of the ion from the pure element. From eqn. 3.106 we can calculate the ΔG for the reduction of zinc as 147.24 kJ/mol. The free energy of formation of Zn^{2+} would be -147.24 kJ/mol. Given the free energy of formation of an ion, we can also use eqn. 3.105 to calculate the hydrogen scale potential. Since

$$\Delta G = \Delta G^o + RT \ln \prod_i a_i^{v_i} \qquad (3.108)$$

we can substitute eqns. 3.106 and 3.107 into 3.108 and also write

$$E = E^\circ - \frac{RT}{z\mathcal{F}} \ln \prod_i a_i^{v_i} \qquad (3.109)$$

Equation 3.109 is known as the *Nernst equation.*[*] At 298 K and 0.1 MPa it reduces to:

$$E = E^o - \frac{0.0592}{z} \log \prod_i a_i^{v_i} \qquad (3.110)$$

We can deduce the meaning of this relationship from the relationship between ΔG and E in eqn. 3.106. At equilibrium ΔG is zero. Thus, in eqn. 3.108, *activities will adjust themselves such that E is 0.*

Example 3.11   Calculating the E_H of net reactions

We can calculate E_H values for reactions not listed in Table 3.3 by algebraic combinations of the reactions and potentials that are listed. There is, however, a catch. Let's see how this works.

Calculate the E_H for the reaction:

$$Fe^{3+} + 3e^- \rightleftharpoons Fe$$

Answer: This reaction is the algebraic sum of two reactions listed in Table 3.3:

$$Fe^{3+} + e^- \rightleftharpoons Fe^{2+}$$

$$Fe^{2+} + 2e^- \rightleftharpoons Fe$$

---

[*] Named for Walther Nernst (1864–1941). Nernst was born in Briesau, Prussia (now in Poland), and completed a PhD at the University of Würzburg in 1887. Nernst made many contributions to thermodynamics and kinetics, including an early version of the third law. He was awarded the Nobel Prize in 1920.

Since the reactions sum, we might assume that we can simply sum the E_H values to obtain the E_H of the net reaction. Doing so, we obtain an E_H of 0.33 V. However, the true E_H of this reaction is -0.037 V. What have we done wrong?

We have neglected to take into consideration the number of electrons exchanged. In the algebraic combination of E_H values, we need to multiply the E_H for each component reaction by the number of electrons exchanged. We then divide the sum of these values by number of electrons exchanged in the net reaction to obtain the E_H of the net reaction, i.e.,

$$E_{H(net)} = \frac{1}{z_{net}} \sum_i z_i E_{Hi} \tag{3.111}$$

where the sum is over the component reactions i. Looking at eqn. 3.106, we can see why this is the case. By Hess's law, the ΔG of the net reaction must be the simple sum of the component reaction ΔGs, but E_H values are obtained by multiplying ΔG by z. Equation 3.111 is derived by combining eqn. 3.106 and Hess's law. Using eqn. 3.111, we obtain the correct E_H of -0.037 V.

3.11.1.2   Alternative representation of redox state: $p\varepsilon$

Consider again the reaction:

$$Fe_{aq}^{3+} + e^- \rightleftharpoons Fe_{aq}^{2+} \tag{3.102}$$

If we were to express the equilibrium constant for this reaction, we would write:

$$K = \frac{a_{Fe^{2+}}}{a_{Fe^{3+}} a_{e^-}}$$

Thus, we might find it convenient to define an activity for the electron. For this reason, chemists have defined an analogous parameter to pH, called $p\varepsilon$, which is the negative log of the activity of electrons in solution:

$$\boxed{p\varepsilon \equiv -\log a_{e^-}} \tag{3.112}$$

The log of the equilibrium constant for eqn. 3.101 may then be written as:

$$\log K = \log \frac{a_{Fe^{2+}}}{a_{Fe^{3+}}} + p\varepsilon$$

Upon rearranging we have:

$$p\varepsilon = \log K - \log \frac{a_{Fe^{2+}}}{a_{Fe^{3+}}} \tag{3.113}$$

When the activities of reactants and products are in their standard states (i.e., $a = 1$), then:

$$p\varepsilon^o = \frac{1}{z} \log K \tag{3.114}$$

(where z again is the number of electrons exchanged: 1 in reaction 3.102). $p\varepsilon^\circ$ values are empirically determined and may be found in various tables. Table 3.3 lists values for some of the more important reactions. For any state other than the standard state, $p\varepsilon$ is related to the standard state $p\varepsilon^\circ$ by:

$$p\varepsilon = p\varepsilon^o - \log \frac{a_{Fe^{2+}}}{a_{Fe^{3+}}} \tag{3.115}$$

$p\varepsilon$ and E_H are related by the following equation:

$$p\varepsilon = \frac{\mathcal{F}E_H}{2.303RT} = \frac{5039E_H}{T} \tag{3.116}$$

(the factor 2.303 arises from the switch from natural log units to base 10 log units).

In defining electron activity and representing it in log units, there is a clear analogy between $p\varepsilon$ and pH. However, the analogy is purely mathematical, and not physical. Natural waters do not contain significant concentrations of free electrons. Also, although a system at equilibrium can have only one value for $p\varepsilon$, just as it will have only one value of pH, redox equilibrium is often not achieved in natural waters. The $p\varepsilon$ of a natural system is therefore often difficult to determine. *Thus, $p\varepsilon$ is a hypothetical unit*, defined for convenience of incorporating a representation of redox state that fits readily into established thermodynamic constructs such as the equilibrium constant. In this sense, eqn. 3.116

provides a more accurate definition of $p\varepsilon$ than does eqn. 3.112.

The greater the $p\varepsilon$, the greater the tendency of species to lose their transferable, or valence, electrons. In a qualitative way, we can think of the negative of $p\varepsilon$ as a measure of the availability of electrons. $p\varepsilon$ can be related in a general way to the relative abundance of electron acceptors. When an electron acceptor, such as oxygen, is abundant relative to the abundance of electron donors, the $p\varepsilon$ is high and electron donors will be in electron-poor valence states (e.g., Mn^{4+} instead of Mn^{2+}). $p\varepsilon$, and E_H, are particularly useful concepts when combined with pH to produce diagrams representing the stability fields of various species. We will briefly consider how these are constructed.

3.11.1.3 $p\varepsilon$–pH diagrams

$p\varepsilon$–pH and E_H–pH diagrams are commonly used tools of aqueous geochemistry, and it is important to become familiar with them. An example, the $p\varepsilon$–pH diagram for iron, is shown in Figure 3.19. $p\varepsilon$–pH diagrams look much like phase diagrams, and indeed there are many similarities. There are, however, some important differences. First, labeled regions do not represent conditions of stability for phases; rather they show which *species* will *predominate* under the $p\varepsilon$–pH conditions within the regions. Indeed, in Figure 3.19 we consider only a single phase: an aqueous solution. The bounded regions are called *predominance areas*. Second, species are stable beyond their region: *boundaries represent the conditions under which the activities of species predominating in two adjoining fields are equal*. However, since the plot is logarithmic, activities of species decrease rapidly beyond their predominance areas.

More generally, a $p\varepsilon$–pH diagram is a type of *activity* or *predominance diagram*, in which the region of predominance of a species is represented as a function of activities of two or more species or ratios of species. We will meet variants of such diagrams in later chapters.

Let's now see how Figure 3.19 can be constructed from basic chemical and thermodynamic data. We will consider only a very simple Fe-bearing aqueous solution. Thus, our solution contains only species of iron, the dissociation products of water and

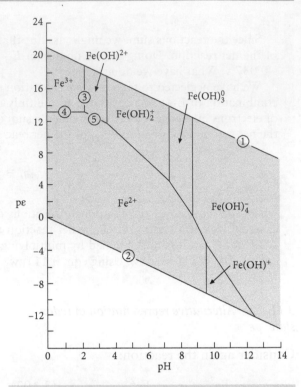

Figure 3.19 $p\varepsilon$–pH diagram showing predominance regions for ferric and ferrous iron and their hydrolysis products in aqueous solution at 25°C and 0.1 MPa.

species formed by reactions between them. Thermodynamics allow us to calculate the predominance region for each species. To draw boundaries on this plot, we will want to obtain equations in the form of $p\varepsilon = a + b \times$ pH. With an equation in this form, b is a slope and a is an intercept on a $p\varepsilon$–pH diagram. Hence we will want to write all redox reactions so that they contain e^- and all acid–base reactions so that they contain H^+.

In Figure 3.18, we are only interested in the region where water is stable. So to begin construction of our diagram, we want to draw boundaries outlining the region of stability of water. The upper limit is the reduction of oxygen to water:

$$\frac{1}{2}O_{2(g)} + 2e^- + 2H^+_{aq} \rightleftharpoons H_2O$$

The equilibrium constant for this reaction is:

$$K = \frac{a_{H_2O}}{P_{O_2}^{1/2} a_{e^-}^2 a_{H^+}^2} \qquad (3.117)$$

Expressed in log form:

$$\log K = \log a_{H_2O} - \frac{1}{2} \log P_{O_2} + 2p\varepsilon + 2pH$$

The value of log K is 41.56 (at 25°C and 0.1 MPa). In the standard state, the activity of water and partial pressure of oxygen are 1 so that 3.117 becomes:

$$p\varepsilon = 20.78 - pH \qquad (3.118)$$

Equation 3.118 plots on a $p\varepsilon$–pH diagram as a straight line with a slope of -1 intersecting the vertical axis at 20.78. This is labeled as line ① on Figure 3.19.

Similarly, the lower limit of the stability of water is the reduction of hydrogen:

$$H^+_{aq} + e^- \rightleftharpoons \frac{1}{2} H_{2(g)}$$

Because $\Delta G^°_r = 0$ and log K = 0 (by convention), we have $p\varepsilon = -pH$ for this reaction: a slope of 1 and intercept of 0. This is labeled as line ② on Figure 3.19. Water is stable between these two lines (region shown in gray on Figure 3.19).

Now let's consider the stabilities of a few simple aqueous iron species. One of the more important reactions is the hydrolysis of Fe^{3+}:

$$Fe^{3+}_{aq} + H_2O \rightleftharpoons Fe(OH)^{2+}_{aq} + H^+$$

The equilibrium constant for this reaction is 0.00631. The equilibrium constant expression is then:

$$\log K = \log \frac{a_{Fe(OH)^{2+}}}{a_{Fe^{3+}}} - pH = -2.2$$

Region boundaries on $p\varepsilon$–pH diagrams represent the conditions under which the activities of two species are equal. When the activities of $FeOH^{2+}$ and Fe^{3+} are equal, the equation reduces to:

$$-\log K = pH = 2.2$$

Thus, this equation defines the boundary between regions of predominance of Fe^{3+} and $Fe(OH)^{2+}$. The reaction is independent of $p\varepsilon$ (no oxidation or reduction is involved), and it plots as a straight vertical line pH

= 2.2 (line ③ on Figure 3.19). Boundaries between the successive hydrolysis products, such as $Fe(OH)^0_3$ and $Fe(OH)^-_4$, can be similarly drawn as vertical lines at the pH equal to their equilibrium constants, and occur at pH values of 3.5, 7.3, and 8.8. The boundary between Fe^{2+} and $Fe(OH)^-$ can be similarly calculated and occurs at a pH of 9.5.

Now consider equilibrium between Fe^{2+} and Fe^{3+} (eqn. 3.102). The $p\varepsilon^°$ for this reaction is 13.0 (Table 3.3), hence from eqn. 3.112 we have:

$$p\varepsilon = 13.0 - \log \frac{a_{Fe^{2+}}}{a_{Fe^{3+}}} \qquad (3.119)$$

When the activities are equal, this equation reduces to:

$$p\varepsilon = 13.0$$

and therefore plots as a horizontal line at $p\varepsilon$ = 13 that intersects the $FeOH^{2+}$–Fe^{3+} line at an invariant point at pH = 2.2 (line ④ on Figure 3.19).

The equilibrium between Fe^{2+} and $Fe(OH)^{2+}$ is defined by the reaction:

$$Fe(OH)^{2+}_{aq} + e^- + H^+ \rightleftharpoons Fe^{2+}_{aq} + H_2O$$

Two things are occurring in this reaction: reduction of ferric to ferrous iron, and reaction of H+ ions with the OH− radical to form water. Thus, we can treat it as the algebraic sum of the two reactions we just considered:

$$Fe^{3+}_{aq} + e^- \rightleftharpoons Fe^{2+}_{aq} \qquad p\varepsilon = 13.0$$

$$Fe(OH)^{2+}_{aq} + H^+ \rightleftharpoons Fe^{3+}_{aq} + H_2O \quad pH = 2.2$$

$$Fe(OH)^{2+}_{aq} + e^- + H^+ \rightleftharpoons \qquad p\varepsilon + pH$$

$$Fe^{2+}_{aq} + H_2O \qquad = 15.2$$

or:

$$p\varepsilon = 15.2 - pH$$

This boundary has a slope of -1 and an intercept of 15.2 (line ⑤ on Figure 3.19). Slopes and intercepts of other reactions may be derived in a similar manner.

Now let's consider some solid phases of iron as well, specifically hematite (Fe_2O_3) and magnetite (Fe_3O_4). First, let's consider the oxidation of magnetite to hematite. We could write this reaction as we did in eqn. 3.101, however, that reaction does not explicitly involve electrons, so that we would not be able to derive an expression containing $p\varepsilon$ or pH from it. Instead, we'll use water as the source of oxygen and write the reaction as:

$$2Fe_3O_4 + H_2O \leftrightharpoons 3Fe_2O_3 + 2H^+ + 2e^-$$
(3.120)

Assuming unit activity of all phases, the equilibrium constant expression for this reaction is:

$$\log K = -2pH - 2p\varepsilon \qquad (3.121)$$

From the free energy of formation of the phases ($\Delta G_f = -742.2$ kJ/mol for hematite, -1015.4 kJ/mol for magnetite, and -237.2 kJ/mol for water), we can calculate ΔG_r using Hess's law and the equilibrium constant using eqn. 3.86. Doing so, we find log K = -5.77. Rearranging eqn. 3.121 we have:

$$p\varepsilon = 2.88 - pH$$

The boundary between hematite and magnetite will plot as a line with a slope of -1 and an intercept of 2.88. Above this line (i.e., at higher $p\varepsilon$) hematite will be stable; below that magnetite will be stable (Figure 3.20). Thus, this line is equivalent to a phase boundary.

Next let's consider the dissolution of magnetite to form Fe^{2+} ions. The relevant reaction is:

$$Fe_3O_4 + 8H^+ + 2e^- \leftrightharpoons 3Fe^{2+} + 4H_2O$$

The equilibrium constant for this reaction is 7×10^{29}. Written in log form:

$$\log K = 3 \log a_{Fe^{2+}} - 8pH - 2p\varepsilon = 29.85$$

or:

$$p\varepsilon = 14.92 - 4pH - \frac{3}{2} \log a_{Fe^{2+}}$$

We have assumed that the activity of water is 1 and that magnetite is pure and therefore that its activity is 1. If we again assume unit

activity of Fe^{2+}, the predominance area of magnetite would plot as the line:

$$p\varepsilon = 14.92 - 4pH$$

that is, a slope of -4 and intercept of 0.58. However, such a high activity of Fe^{2+} would be highly unusual in a natural solution. A more relevant activity for Fe^{2+} would be perhaps 10^{-6}. Adopting this value for the activity of Fe^{2+}, we can draw a line corresponding to the equation:

$$p\varepsilon = 23.92 - 4pH$$

This line represents the conditions under which magnetite is in equilibrium with an activity of aqueous Fe^{2+} of 10^{-6}. For any other activity, the line will be shifted, as

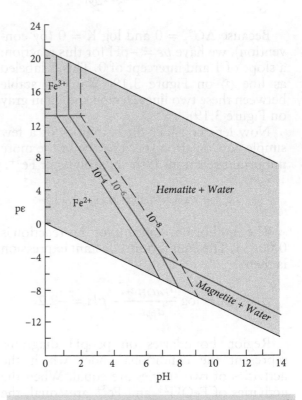

Figure 3.20 Stability regions for magnetite and hematite in equilibrium with an iron-bearing aqueous solution. Thick lines are for a Fe_{aq} activity of 10^{-6}, finer lines for activities of 10^{-4} and 10^{-8}. The latter is dashed.

illustrated in Figure 3.20. For higher concentrations, the magnetite region will expand, while for lower concentrations it will contract.

Now consider the equilibrium between hematite and Fe^{2+}. We can describe this with the reaction:

$$Fe_2O_3 + 6H^+ + 2e^- \leftrightharpoons 2Fe^{2+} + 3H_2O$$

The equilibrium constant (which may again be calculated from ΔG_r) for this reaction is 23.79.

Expressed in log form:

$$\log K = 2\log a_{Fe^{2+}} + 6pH + 2p\varepsilon = 23.79$$

Using an activity of 1 for Fe^{2+}, we can solve for $p\varepsilon$ as:

$$p\varepsilon = 11.9 - 3pH - \log a_{Fe^{2+}}$$

For an activity of Fe^{2+} of 10^{-6}, this is a line with a slope of 3 and an intercept of 17.9. This line represents the conditions under which hematite is in equilibrium with $a_{Fe^{2+}} = 10^{-6}$. Again, for any other activity, the line will be shifted as shown in Figure 3.20.

Finally, equilibrium between hematite and Fe^{3+} may be expressed as:

$$Fe_2O_3 + 6H^+ \leftrightharpoons 2Fe^{3+} + 3H_2O$$

The equilibrium constant expression is:

$$\log K = 2\log a_{Fe^{3+}} + 6pH = -3.93$$

For a Fe^{3+} activity of 10^{-6}, this reduces to:

$$pH = 1.34$$

Since the reaction does not involve transfer of electrons, this boundary depends only on pH.

The boundary between predominance of Fe^{3+} and Fe^{2+} is independent of the Fe concentration in solution and is the same as eqn. 3.119 and Figure 3.18, namely $p\varepsilon = 13$.

Examining this diagram, we see that for realistic dissolved Fe concentrations, magnetite can be in equilibrium only with a fairly reduced, neutral to alkaline solution. At pH of

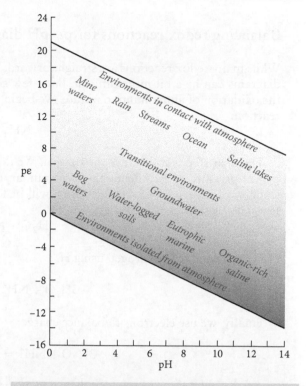

Figure 3.21 $p\varepsilon$ and pH of various waters on and near the surface of the Earth. After Garrels and Christ (1965).

about 7 or less, it dissolves and would not be stable in equilibrium with acidic waters unless the Fe concentration were very high. Hematite is stable over a larger range of conditions and becomes stable over a wider range of pH as $p\varepsilon$ increases. Significant concentrations of the Fe^{3+} ion ($>10^{-6}$ m) will be found only in very acidic, oxidizing environments.

Figure 3.21 illustrates the pH and $p\varepsilon$ values that characterize a variety of environments on and near the surface of the Earth. Comparing this figure with pH–$p\varepsilon$ diagrams allows us to predict the species we might expect to find in various environments. For example, Fe^{3+} would be a significant dissolved species only in the acidic, oxidized waters that sometimes occur in mine drainages (the acidity of these waters results from high concentrations of sulfuric acid that is produced by oxidation of sulfides). We would expect to find magnetite precipitating only from reduced seawater or in organic-rich, highly saline waters.

Balancing redox reactions for $p\varepsilon$–pH diagrams

While many redox reactions are straightforward, balancing more complex redox reactions for $p\varepsilon$–pH diagrams can be a bit more difficult, but a few simple rules make it easier. Let's take as an example the oxidation of ammonium to nitrate. We begin by writing the species of interest on each side of the reaction:

$$NH_4^+ = NO_3^-$$

The next step is to balance the oxygen. We don't want to use O_2 gas to do this. We used O_2 at a partial pressure of 1 to define the top boundary for the water stability region. Within the region of stability of water, the O_2 concentration will be lower and we don't necessarily know its value. This is usually best done using water:

$$3H_2O + NH_4^+ = NO_3^-$$

Next balance the hydrogen using H^+:

$$3H_2O + NH_4^+ = NO_3^- + 10H^+$$

Finally, we use electrons to balance charge:

$$3H_2O + NH_4^+ = NO_3^- + 10H^+ + 8e^-$$

As a check, we can consider the valance change of our principal species and be sure that our reaction makes sense. In ammonium, nitrogen is in the 3– state, while in nitrate it is in the 5+ state, a net change of 8. This is just the number of electrons exchanged in the reaction we have written.

3.11.2   Redox in magmatic systems

High-temperature geochemists use *oxygen fugacity* to characterize the oxidation state of systems. Consequently, we want to write redox reactions that contain O_2. Thus, equilibrium between magnetite and hematite would be written as:

$$4Fe_3O_4 + O_{2(g)} \leftrightharpoons 6Fe_2O_3 \quad (3.122)$$

(or alternatively, as we wrote in eqn. 3.101) rather than the way we expressed it in eqn. 3.120. We note, however, there is negligible molecular oxygen in magmatic systems, and other species are often responsible for transfer of electrons and O^{2-}. For example, the equilibrium between magnetite and hematite may be mediated by water:

$$2Fe_3O_4 + H_2O_{(g)} \leftrightharpoons 3Fe_2O_3 + H_2 \quad (3.123)$$

The above two reactions are thermodynamically equivalent in terms of magnetite oxidation. The first reaction is simpler, of course,

and hence preferred, but it may sometimes be necessary to consider the proportions of the actual gas species present.

If we can regard magnetite and hematite as pure phases, then their activities are equal to one and the equilibrium constant for reaction 3.122 is the inverse of the oxygen fugacity:

$$K_{MH} = \frac{1}{f_{O_2}} \quad (3.124)$$

We can rewrite eqn. 3.86 as:

$$K = e^{-\Delta G_f^o/RT} \quad (3.125)$$

and taking the standard state as 1000 K and 1 bar, we can write:

$$-\log K = \log f_{O_2}$$
$$= \left(\frac{6\Delta G_{f(Fe_2O_3,1000)}^o - 4\Delta G_{f(Fe_3O_4,1000)}^o}{2.303RT} \right)$$

Thus, oxygen fugacity can be calculated directly from the difference in the free energy of formation of magnetite and hematite at the appropriate T and P. Substituting appropriate values into this equation yields a value for log f_{O_2} of -10.86.

It is important to understand that the oxygen fugacity is fixed at this level (though the exact level at which it is fixed is still disputed because of uncertainties in the thermodynamic data) simply by the equilibrium coexistence of magnetite and hematite. The oxygen fugacity does not depend on the proportion of these minerals. For this reason, it is appropriately called a buffer. To understand how this works, imagine some amount of magnetite, hematite and oxygen present in a magma. If the oxygen fugacity is increased by the addition of oxygen to the system, equilibrium in the reaction in eqn. 3.121 is driven to the right until the log of the oxygen fugacity returns to a value of -10.86. Only when all magnetite is converted to hematite can the oxygen fugacity rise. A drop in oxygen fugacity would be buffered in exactly the opposite way until all hematite were gone. A number of other buffers can be constructed based on reactions such as:

$$3Fe_2SiO_4 + O_2 \rightleftharpoons 2Fe_3O_4 + 3SiO_2$$

(fayalite)   (magnetite) (quartz)

and

$$Fe + \frac{1}{2}O_2 \rightleftharpoons FeO$$

(iron)   (wüstite)

These can be used to construct the oxygen buffer curves in Figure 3.22.

3.12 SUMMARY

Natural systems often contain multiple phases, many of which are solutions of several components; in this chapter, we developed the thermodynamic tools to deal with them.

- We began by defining components, phases, and species. Together, the number of components and phases in a system determine the degrees of freedom of the system:

$$f = c - \phi + 2 \tag{3.2}$$

which are the number of independent variables we need to specify to completely

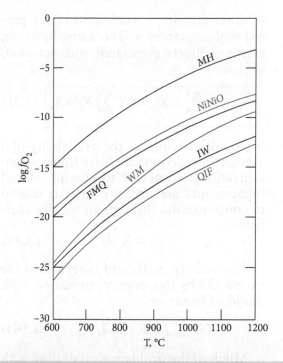

Figure 3.22 Oxygen buffer curves in the system Fe–Si–O at 1 bar. QIF, IW, WM, FMQ, and MH refer to the quartz–iron–fayalite, iron–wüstite, wüstite–magnetite, fayalite–magnetite–quartz and magnetite–hematite buffers, respectively.

describe the system. We derived the Clapeyron equation, which described the boundary between two phases, such as graphite and diamond, the $P-T$ space:

$$\frac{dT}{dP} = \frac{\Delta V_r}{\Delta S_r} \tag{3.3}$$

- We found the thermodynamic properties of solutions depend on their composition as well as T and P and to deal with this we introduced partial molar quantities, particularly the partial molar Gibbs free energy or chemical potential:

$$\mu_i = \left(\frac{\partial G}{\partial n_i}\right)_{T,P,n_{j,j \neq i}} \tag{3.13}$$

- The simplest solutions are ideal ones, where there are no energetic or volumetric solution are simply their sum of the partial molar quantities. There are, however, entropic effects associated with solution, so that

$$\overline{G}_{ideal} = \sum_i X_i \mu_i^o + RT \sum_i X_i \ln X_i \quad (3.31)$$

- In nonideal solutions, the availability of a species for reaction can differ from its concentration; to deal with this we introduced fugacity and activity; the latter is related to concentration through an activity coefficient:

$$a_i = X_i \lambda_I \quad (3.48)$$

The activity coefficient is related to the excess Gibbs free energy associated with nonideal behavior:

$$\overline{G}_{excess,i} = RT \ln \lambda_i \quad (3.56a)$$

Much of the problem with dealing with nonideal solutions is reduced to finding values for the activity coefficients.
- Electrolyte solutions, of which seawater is a good example, are common nonideal solutions. We reviewed the nature of these solutions and introduced approaches for calculating activity coefficients in them, such as the Debye–Hückel extended law:

$$\log_{10}\gamma_i = \frac{-Az_i^2\sqrt{I}}{1 + B\mathring{a}_i\sqrt{I}} \quad (3.74)$$

We then reviewed ways to calculate activities in ideal solid solutions.
- In section 3.9, we introduced the equilibrium constant:

$$K = \prod_i a_i^{\nu_i} \quad (3.85)$$

and found we could directly relate it to the Gibbs free energy of reaction.
- In section 3.11, we introduced the electrochemical potential to deal with changing valance states of elements, that is, oxidation–reduction reactions. This too we could relate to our thermodynamic framework:

$$\Delta G = -z\mathcal{F}E \quad (3.106)$$

A useful way to represent redox potential in low-temperature systems is the electron activity

$$p\varepsilon \equiv -\log a_{e^-} \quad (3.112)$$

which we could also directly relate to electrochemical potential. Since oxygen is the most common oxidant, in high-temperature systems, the redox state of the system is more commonly represented with oxygen fugacity, f_{O2}.

REFERENCES AND SUGGESTIONS FOR FURTHER READING

Anderson, G.M. and Crerar, D.A. 1993. *Thermodynamics in Geochemistry*. New York, Oxford University Press.

Blount, C.W. 1977. Barite solubilities and thermodynamic quantities up to 300°C and 1400 bars. *American Mineralogist* 62: 942–57.

Brookins, D.G. 1988. *Eh-pH Diagrams for Geochemistry*. New York, Springer Verlag.

Davies, C.W. 1938. The extent of dissociation of salts in water. VIII. An equation for the mean ionic activity coefficient in water, and a revision of the dissociation constant of some sulfates. *Journal of the Chemical Society* 2093–8.

Davies, C.W. 1962. *Ion Association*. London, Butterworths.

Debye, P. and Hückel, E. 1923. On the theory of electrolytes. *Phys. Z.* 24: 185–208.

Fletcher, P. 1993. *Chemical Thermodynamics for Earth Scientists*. Essex, Longman Scientific and Technical.

Garrels, R.M. and Christ, C.L. 1965. *Solutions, Minerals, and Equilibria*. San Francisco: Freeman Cooper.

Helgeson, H.C. 1967. Solution chemistry and metamorphism, in *Researches in Geochemistry* vol. 2 (ed. P.H. Abelson), pp. 362–402. New York, John Wiley and Sons, Ltd.

Helgeson, H.C. and Kirkham, D.H. 1974. Theoretical prediction of the thermodynamic behavior of aqueous electrolytes at high pressures and temperatures, Debye–Hückel parameters for activity coefficients and relative partial molar quantities. *American Journal of Science* 274: 1199–261.

Helgeson, H.C. and Kirkham, D.H. 1976. Theoretical prediction of the thermodynamic properties of aqueous electrolytes at high pressures and temperatures. III. Equation of state for aqueous species at infinite dilution. *American Journal of Science* 276: 97–240.

Kerrick, D.M. and Jacobs, G.K. 1981. A modified Redlich–Kwong equation for H_2O, CO_2 and H_2O–CO_2 mixtures at elevated pressures and temperatures. *American Journal of Science* 281: 735–67.

Morel, F.M.M. and Hering, J.G. 1993. *Principles and Applications of Aquatic Chemistry*, 2nd edn. New York, John Wiley and Sons, Ltd.

Nordstrom, D.K. and Munoz, J.L. 1986. *Geochemical Thermodynamics*. Palo Alto, Blackwell Scientific.

Saxena, S.K., Chaterjee, N., Fei, Y. and Shen, G. 1993. *Thermodynamic Data on Oxides and Silicates*. Berlin, Springer Verlag.

PROBLEMS

1. Consider the following minerals:

anhydrite:	$CaSO_4$	
bassanite:	$CaSO_4 \cdot \frac{1}{2}H_2O$	(plaster of Paris)
gypsum:	$CaSO_4 \cdot 2H_2O$	

   (a) If water vapor is the only phase of pure water in the system, how many phases are there in this system and how many components are there?

   (b) How many phases are present at invariant points in such a system? How many univariant reactions are possible? Write all univariant reactions, labeling each according the phase that does not participate in the reaction.

2. Consider a system consisting of olivine of variable composition (($Mg,Fe)_2SiO_4$) and orthopyroxene of variable composition (($Mg,Fe)SiO_3$). What is the *minimum* number of components needed to describe this system?

3. In section 3.2.1.3, we showed that a system containing H_2O, H_2CO_3, HCO_3^-, CO_3^{2-}, H^+, and OH^- could be described in terms of components CO_3^{2-}, H^+, and OH^-. Find a different set of components that describe the system equally well. Show that each of the species in the system is an algebraic sum of your chosen components.

4. Use the data in Table 2.2 to construct a temperature–pressure phase diagram that showing the stability fields of calcite and aragonite.

5. Consider the following hypothetical gaseous solution: gases 1 and 2 form an ideal binary solution; at 1000 K, the free energies of formation from the elements are −50 kJ/mol for species 1 and −60 kJ/mol for species 2.

   (a) Calculate ΔG_{mixing} for the solution at 0.1 increments of X_2. Plot your results.

   (b) Calculate $\overline{G}$ for an *ideal* solution at 0.1 increments of X_2. Plot your results.

   (c) Using the method of intercepts, find μ^1 and μ^2 in the solution at $X_2 = 0.2$

6. Using the thermodynamic data in Table 2.2, determine which side of this reaction is stable at 600°C and 400 MPa.:

$$2Al(OH)_3 \rightleftharpoons Al_2O_3 + 3H_2O$$

7. The following analysis of water is from the Rhine River as it leaves the Swiss Alps:

Ca^{2+}	40.7 ppm	HCO_3^-	113.5 ppm
Mg^{2+}	7.2 ppm	SO_4^{2-}	36.0 ppm
Na^+	1.4 ppm	NO_3^-	1.9 ppm
K^+	1.2 ppm	Cl^-	1.1 ppm

(a) Calculate the ionic strength of this water. (Recall that concentrations in ppm are equal to concentrations in $mmol\,kg^{-1}$ multiplied by formula weight.)

(b) Using the Debye–Hückel equation and the data in Table 3.2, calculate the practical activity coefficients for each of these species at 25°C.

8. Seawater has the following composition:

Na^+	0.481 M	Cl^-	0.560 M
Mg^{2+}	0.0544 M	SO_4^{2-}	0.0283 M
Ca^{2+}	0.0105 M	HCO_3^-	0.00238 M
K^+	0.0105 M		

(a) Calculate the ionic strength.

(b) Using the Davies equation and the data in Table 3.2, calculate the practical activity coefficients for each of these species at 25°C.

9. The following is an analysis of *Acqua di Nepi*, a spring water from the Italian province of Viterbo:

Ca^{2+}	82 ppm	HCO_3^-	451 ppm
K^+	50 ppm	SO_4^{2-}	38 ppm
Mg^{2+}	27 ppm	Cl^-	20 ppm
Na^+	28 ppm	NO_3^-	9 ppm
		F^-	1.3 ppm

(a) Calculate the ionic strength of this water.

(b) Using the Debye–Hückel equation and the data in Table 3.1, calculate the practical activity coefficients for each of these species at 25°C.

10. Water from Thonon, France, has the following composition:

Anions	mg/l	Cations	mg/l
HCO_3^-	332	Ca^{2+}	103.2
SO_4^{2-}	14	Mg^{2+}	16.1
NO_3^-	14	K^+	1.4
Cl^-	8.2	Na^+	5.1

(a) What is the ionic strength of this water?

(b) What are the activity coefficients for HCO_3^- and CO_3^{2-} in this water?

(c) Assuming an equilibrium constant for the dissociation of bicarbonate:

$$HCO_3^- \rightleftharpoons H^+ + CO_3^{2-}$$

of 4.68×10^{-11} and a pH of 7.3, what is the equilibrium concentration of CO_3^{2-} in this water?

11. The equilibrium constant for the dissolution of galena:

$$PbS_{solid} + 2H^+ \rightleftharpoons Pb^{2+}_{aq} + H_2S^{aq}$$

is 9.12×10^{-7} at 80°C. Using $\gamma_{Pb^{2+}} = 0.11$ and $\gamma_{H_2S} = 1.77$, calculate the equilibrium concentration of Pb^{2+} in aqueous solution at this temperature and at pHs of 6, 5 and 4. Assume the dissolution of galena is the only source of Pb and H_2S in the solution, and that there is no significant dissociation of H_2S. *Hint*: Mass balance requires that $[H_2S] = [Pb^{2+}]$.

12. The dissociation constant for hydrofluoric acid (HF) is $10^{-3.2}$ at 25°C. What would be the pH of a 0.1 M solution of HF? You may assume ideal behavior. (*Hint*: Ask yourself what addition constraints are imposed on the system. Your final answer will require solving a quadratic equation.)

13. The first dissociation constant for H_2S is $K_1 = 9.1 \times 10^{-3}$. Neglecting the second dissociation and assuming ideality (i.e., activity equals concentration), what is the pH of 1 liter of pure water if you dissolve 0.01 moles of H_2S in it? What fraction of H_2S has dissociated?
(*HINT*: Assume that the concentration of OH^- is negligible (in other words, no autodissociation of water) and use the quadratic equation for your final solution.)

14. Given the following analysis of biotite and assuming a *mixing-on-site model* for all sites, calculate the activities of the following components:

$$KMg_3Si_3AlO_{10}(F)_2 \text{ (fluorophlogopite)}$$

$$KFe_3^{2+}Si_3AlO_{10}(OH)_2 \text{ (annite)}$$

Site	Ion	Ions per site
Tetrahedral	Si	2.773
	Al	1.227
Octahedral	Al	0.414
	Ti	0.136
	Fe^{3+}	0.085
	Fe^{2+}	1.399
	Mg	0.850
Interlayer	Ca	0.013
	Na	0.063
	K	0.894
Anion	OH	1.687
	F	0.037

Hint: Check your result by making sure the activity of phlogopite in pure phlogopite is 1.

15. Given the following analysis of a pyroxene, use the mixing-on-site model of ideal activities to calculate the activity of *jadeite* ($NaAlSi_2O_6$) and *diopside* ($CaMgSi_2O_6$) in this mineral:

Site	Ion	Ions per site
Tetrahedral	Si	1.96
	Al	0.04
Octahedral M1	Al	0.12
	Mg	0.88
Octahedral M2	Fe^{2+}	0.06
	Ca	0.82
	Na	0.12

16. Write the equilibrium constant expression for the reaction:

$$CaCO_{3(s)} + 2H^+_{(aq)} + SO_4^{2-} + H_2O_{(liq)} \rightleftharpoons CaSO_4 \cdot 2H_2O + CO_{2(g)}$$

assuming the solids are pure crystalline phases and that the gas is ideal.

17. Assuming ideal solution behavior for the following:

    (a) Show that the boiling point of a substance is increased when another substance is dissolved in it, assuming the concentration of the solute in the vapor is small.
    (b) By how much will the boiling point of water be elevated when 10% salt is dissolved in it?

18. Find $\Delta \overline{G}$ for the reaction:

$$Pb^{2+} + Mn \rightleftharpoons Pb + Mn^{2+}$$

    Which side of the reaction is favored?
    (*Hint: Use the data in Table 3.3.*)

19. What is the $\Delta \overline{G}$ for the reaction:

$$Cu^{2+} + e^- \rightleftharpoons Cu^+$$

    What is the $p\varepsilon^\circ$ for this reaction?

20. Consider a stream with a pH of 6.7 and a total dissolved Fe concentration of 1 mg/l. Assume ideal behavior for this problem.

    (a) If the stream water is in equilibrium with the atmospheric O_2 (partial pressure of 0.2 MPa), what is the $p\varepsilon$ of the water?
    (b) Assuming they are the only species of Fe in the water, what are the concentrations of Fe^{3+} and Fe^{2+}? Use the $p\varepsilon$ you determined in part a.

21. Write reactions for the oxidation of nitrogen gas to aqueous nitrite and nitrate ions that contain the electron and hydrogen ion (i.e., reactions suitable for a $p\varepsilon$–pH diagram). Write the log equilibrium constant expression for these reactions. Using the data below, calculate the log equilibrium constant for these reactions under standard state conditions). Calculate $p\varepsilon^\circ$ and E^o_H for these reactions.

Species	ΔG^o_f kJ/mol
H_2O	−237.19
N_2 (gas)	0
NO_3^-	−111.3
NO_2^-	−32.2

Standard state is 25°C and 0.1 MPa. R = 8.314 J/mol-K.

Hint: remember, in the standard state, *all* reactants and products have activities of 1.

22. Construct a $p\varepsilon$–pH diagram for the following species of sulfur: HSO_4^-, SO_4^{2-}, H_2S, HS^-, and S^{2-} at 25°C and 0.1 MPa. The following free energies of formation should provide sufficient information to complete this task.

Species	ΔG^o_f	Species	ΔG^o_f
S^{2-} (aq)	+85.81	H_2O	−237.19
HS^- (aq)	+12.09	H^+	0
H_2S (aq)	−27.82	H_2 (g)	0
SO_4^{2-} (aq)	−744.54	O_2 (g)	0
HSO_4^- (aq)	−755.92		

Values are in kJ/mol, standard state is 25°C and 0.1 MPa. R = 8.314 J/mol-K.

23. Construct a $p\varepsilon$–pH diagram for dissolved species of uranium, UO_2^{2+} and $U(OH)_5^-$, and the two solid phases UO_2 and U_3O_8 at 25°C and 0.1 MPa. Assume the activity of dissolved uranium is fixed at 10^{-6}. The following free energies of formation should provide sufficient information to complete this task.

Species	ΔG_f^o
$U(OH)_5^-$ (aq)	−1630.80
UO_2^{2+} (aq)	−952.53
UO_2 (s)	−1031.86
U_3O_8 (s)	−3369.58
H_2O	−237.19

Values are in kJ/mol, standard state is 25°C and 0.1 MPa. R = 8.314 J/mol-K.

Chapter 4

Applications of thermodynamics to the Earth

4.1 INTRODUCTION

In the previous two chapters, we developed the fundamental thermodynamic relationships and saw how they are applied to geochemical problems. The tools now in our thermodynamic toolbox are sufficient to deal with many geochemical phenomena. They are not sufficient, however, to deal with all geochemical phenomena. In this chapter, we will add a final few thermodynamic tools. These allow us to deal with nonideal behavior and exsolution phenomena in solids and silicate liquids. With that, we can use thermodynamics to determine the pressure and temperature at which rock assemblages formed, certainly one of the most useful applications of thermodynamics to geology. Along the way, we will see how thermodynamics is related to one of the most useful tools in petrology: phase diagrams. We will then briefly consider how thermodynamics has been used to construct computer models of how magma compositions evolve during melting and crystallization. Finally, we return to the question of nonideal behavior in electrolyte solutions and examine the problems of ion association and solvation in more depth and how this affects ion activities. Deviations from ideal behavior tend to be greater in solutions of high ionic strength, which includes such geologically important solutions as hydrothermal and ore-forming fluids, saline lake waters, metamorphic fluids, and formation and oil field brines. We briefly examine methods of computing activity coefficients at ionic strengths relevant to such fluids.

4.2 ACTIVITIES IN NONIDEAL SOLID SOLUTIONS

4.2.1 Mathematical models of real solutions: Margules equations

Ideal solution models often fail to describe the behavior of real solutions; a good example is water and alcohol, as we saw in Chapter 3. Ideal solutions fail spectacularly when exsolution occurs, such as between oil and vinegar, or between orthoclase and albite, a phenomenon we will discuss in more detail shortly. In nonideal solutions, even when exsolution does not occur, more complex models are necessary.

Power, or Maclaurin, series are often a convenient means of expressing complex mathematical functions, particularly if the true form of the function is not known, as is often the case. This approach is the basis of Margules* equations, a common method of calculating excess state functions. For example, we could express the excess volume as a power series:

$$\overline{V}_{ex} = A + BX_2 + CX_2^2 + DX_2^3 + \dots \quad (4.1)$$

where X_2 is the mole fraction of component 2.

*Named for Max Margules (1856–1920), an Austrian meteorologist, who first used this approach in 1895.

Geochemistry, Second Edition. William M. White.
© 2020 John Wiley & Sons Ltd. Published 2020 by John Wiley & Sons Ltd.
Companion website: www.wiley.com/go/white/geochemistry

Following the work of Thompson (1967), Margules equations are used extensively in geochemistry and mineralogy as models for the behavior of nonideal solid solutions. It should be emphasized that this approach is completely empirical – true thermodynamic functions are not generally power series. The approach is successful, however, because nearly any function can be *approximated* as a power series. Thus, Margules equations are attempts to approximate thermodynamic properties from empirical observations when the true mathematical representation is not known. We will consider two variants of them: the symmetric and asymmetric solution models.

4.2.1.1 The symmetric solution model

In some solutions, a sufficient approximation of thermodynamic functions can often be obtained by using only a second-order power series (i.e., in eqn. 4.1, $D = E = \ldots = 0$). Now in a binary solution, the excess of any thermodynamic function should be entirely a function of mole fraction X_2 (or X_1, however we wish to express it). Put another way, where $X_2 = 0$, we expect $\overline{V}_{ex} = 0$. From this we can see that the first term in eqn. 4.1, A, must also be 0. Thus, eqn. 4.1 simplifies to:

$$\overline{V}_{ex} = BX_2 + CX_2^2 \qquad (4.2)$$

The simplest solution of this type would be one that is symmetric about the midpoint, $X_2 = 0.5$; this is called a *symmetric solution*. In essence, symmetry requires that:

$$BX_2 + CX_2^2 = BX_1 + CX_1^2 \qquad (4.3)$$

Substituting $(1 - X_2)$ for X_1 and expanding the right-hand side of eqn. 4.3, we have:

$$BX_2 + CX_2^2 = B - BX_2 + C - 2CX_2 + CX_2^2 \qquad (4.4)$$

Collecting terms and rearranging:

$$B(2X_2 - 1) = C(1 - 2X_2) \qquad (4.5)$$

which reduces to $B = -C$. Letting $W_V = B$ in eqn. 4.2, we have:

$$\overline{V}_{ex} = W_V X_2 - W_V X_2^2 = W_V X_2 (1 - X_2)$$
$$= X_1 X_2 W_V \qquad (4.6)$$

W is known as an *interaction parameter* because nonideal behavior arises from *interactions* between molecules or atoms and depends on temperature, pressure, and the nature of the solution, but not on X. Expressions similar to 4.2–4.6 may be written for enthalpy, entropy, and free energy; for example:

$$\overline{G}_{ex} = X_1 X_2 W_G \qquad (4.7)$$

The free energy interaction term, W_G, may be expressed as:

$$W_G = W_U + PW_V - TW_S \qquad (4.8)$$

Since the W_H term can be written as:

$$W_H = W_U + PW_V$$

then eqn. 4.8 may also be written:

$$W_G = W_H - TW_S \qquad (4.8a)$$

The temperature and pressure dependence of W_G are then

$$\left(\frac{\partial W_G}{\partial T}\right)_P = -W_S \qquad (4.9)$$

$$\left(\frac{\partial W_G}{\partial P}\right)_T = -W_V \qquad (4.10)$$

We will first make use of interaction parameters in the context of solid solutions, but as we shall see in subsequent sections, they are equally useful in silicate liquid and electrolyte solutions as well.

*Regular solutions** are a special case of symmetric solutions where:

$$W_s = 0 \qquad \text{and therefore} \qquad W_G = W_H$$

Regular solutions correspond to the case where $\Delta S_{ex} = 0$, i.e., where $\Delta S_{mixing} = \Delta S_{ideal}$,

---

* The term *regular solution* is often used to refer to symmetric solutions. In that case, what we termed a regular solution is called a strictly regular solution.

and therefore where $W_S = 0$. From eqn. 4.9, we see that W_G is independent of temperature for regular solutions. Examples of such solutions include electrolytes with a single, uncoupled, anionic or cationic substitution, such as $CaCl_2$–$CaBr_2$, or solid solutions where there is a single substitution in just one site (e.g., Mg_2SiO_4–Fe_2SiO_4).

Setting eqn. 4.7 equal to eqn. 3.57, we have for binary solutions:

$$\overline{G}_{ex} = X_1 X_2 W_G = RT[X_1 \ln \lambda_1 + X_2 \ln \lambda_2] \quad (4.11)$$

For a symmetric solution we have the additional constraint that at $X_2 = X_1$, $\lambda_1 = \lambda_2$. From this relationship it follows that:

$$RT \ln \lambda_i = X_j^2 W_G \quad (4.12)$$

This leads to the relationships:

$$\mu_1 = \mu_1^o + RT \ln X_1 + X_2^2 W_G \quad (4.13)$$

$$\mu_2 = \mu_2^o + RT \ln X_2 + X_1^2 W_G \quad (4.13a)$$

The symmetric solution model should reduce to Raoult's and Henry's laws in the pure substance and infinitely dilute solution, respectively. We see that as $X_1 \to 1$, eqns. 4.13 and 4.13a reduce, respectively, to:

$$\mu_1 = \mu_1^o + RT \ln X_1 \quad (4.14)$$

$$\mu_2 = \mu_2^o + RT \ln X_2 + W_G \quad (4.15)$$

Equation 4.14 is Raoult's law; letting:

$$\mu^* = \mu^\circ + W_G \quad \text{or} \quad W_G = RT \ln h$$

then eqn. 4.15 is Henry's law. Thus, the interaction parameter can be related to the parameters of Henry's law, and activity coefficient. In the Margules representation, a solution that is ideal throughout is simply the special case where $A = B = C = D = \ldots = 0$.

4.2.1.2 The asymmetric solution model

Many real solutions, for example mineral solutions with asymmetric solvi, are not symmetric. This corresponds to the case where D in eqn. 4.1 is nonzero, so we must carry the expansion to the third order. It can be

shown that in this case the excess free energy in binary solutions is given by:

$$\overline{G}_{ex} = (W_{G_1} X_2 + W_{G_2} X_1) X_1 X_2 \quad (4.16)$$

(You can satisfy yourself that this may be written as a power series to the third order of either X_1 or X_2.) The two coefficients are related to the Henry's law constants:

$$W_{G_1} = \mu_i^o - \mu_i^* = RT \ln h \quad (4.17)$$

Activity coefficients are given by:

$$\nu RT \ln \lambda_i = (2W_{G_j} - W_{G_i})X_j^2$$
$$+ 2(W_{G_i} + W_{G_j})X_j^3 \quad (4.18)$$

where $j = 2$ when $i = 1$ and vice versa, and ν is the stoichiometric coefficient. As for the symmetric solution model, the interaction parameters of the asymmetric model can be expressed as the sum of the W_U, W_V, and W_S interaction parameters to account for temperature and pressure dependencies (see Example 4.1).

The alkali feldspars ($NaAlSi_3O_8$–$KAlSi_3O_8$) are an example of a solid solution exhibiting asymmetric exsolution. Figure 4.1 shows a thin section of alkali feldspar that has undergone exsolution of an Na-rich component from orthoclase feldspar. Figure 4.2 shows the ΔG_{real}, ΔG_{ideal}, and ΔG_{excess} for the alkali feldspar solid solution computed for 600°C and 200 MPa using the asymmetric solution model of Thompson and Waldbaum (1969). ΔG_{excess} is computed from eqn. 4.16, ΔG_{ideal} is computed from eqn. 3.30. Figure 4.3 shows ΔG_{real} for a series of temperatures. Perhaps a clearer picture of how ΔG will vary as a function of both composition and temperature can be obtained by plotting all three variables simultaneously, as in Figure 4.4. The Thompson and Waldbaum approach uses end-member phases, $NaAlSi_3O_8$ and $KAlSi_3O_8$, as components, which simplify matters. So long as there is no substitution for Al, Si, or O, an equivalent result can be obtained by choosing elements as components.

For clarity, we have considered only binary solutions. Indeed, in many minerals, substitution occurs primarily or exclusively on one

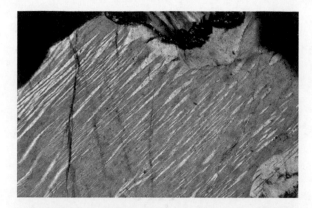

Figure 4.1 Exsolution lamellae of Na-rich feldspar (albite) from a K-rich (orthoclase) feldspar that developed during cooling. Photo by Alessandro Da Mommio. Used by permission.

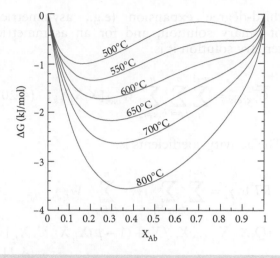

Figure 4.3 ΔG_{real} of alkali feldspar solution computed for a series of temperatures and 200 MPa.

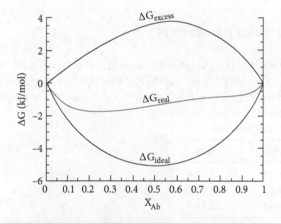

Figure 4.2 Alkali feldspar solid solution computed at 600°C and 200 MPa (2 kb) using the data of Thompson and Waldbaum (1969). $\Delta G_{real} = \Delta G_{ideal} + \Delta G_{excess}$.

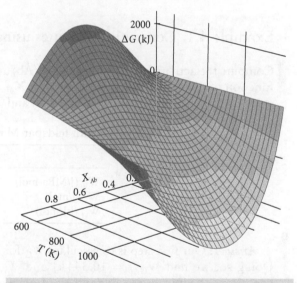

Figure 4.4 ΔG surface for the alkali feldspar solid solution as a function of the mole fraction of albite and temperature.

site, but in many other cases, multiple substitutions are possible. In particular amphiboles, for example hornblende with formula $(K,Na)_{0-1}(Ca,Na,Fe,Mg)_2(Mg,Fe,Al)_5(Al,Si)_8$ $O_{22}(OH)_2$, have an interlayer cation site (occupied by K, Na, or a vacancy), four crystallographically distinct octahedral cation sites occupied by Mg, Fe, Ca, Mn, Al, etc., and two distinct tetrahedral sites, which may contain Ti or Al as well as Si. In addition, O^{2+}, Cl^- and F^- may substitute for OH^-. For complex solutions such as these, the Margules expansion may need to be taken to a higher degree and can be computationally complex.

A general expression for excess free energy in nonideal solid solutions using Margules parameters is (Berman and Brown, 1984):

$$\overline{G}_{excess} = \sum_{j_1} \sum_{j_2, j_2 \neq j_1}^{n-1} \cdots \sum_{j_p, j_p \neq j_{p-1}}^{n}$$
$$W_{j_1, j_2 \ldots j_p} [X_{j_1} X_{j_2} \ldots X_{j_p}] \quad (4.19)$$

where j_i etc. denote components and p is the degree of the polynomial used in the Margules expansion. This reduces to eqn. 4.16 for a third-degree expansion (e.g., asymmetric) of binary solution, and for an asymmetric ternary solution it is:

$$\overline{G}_{excess} = \sum_{i_1}^{n-1} \sum_{j,j\neq i}^{n} \sum_{k,k\neq i,j}^{n} W_{i,j,k}[X_i X_j X_k] \quad (4.20)$$

The activity coefficients is:

$$RT\ln \gamma_i = \sum_{j_1}^{n-1} \sum_{j_2,j_2\neq j_1}^{n} \cdots \sum_{j_p,j_p\pm j_{p-1}}^{n} W_{j_1,j_2\ldots j_p}$$

$$[Q_i X_{j_1} X_{j_2} \ldots X_{j_p}/X_i + (1-p)X_{j_1} X_{j_2} \ldots X_{j_p}] \quad (4.21)$$

where Q_i is equal to the number of the j subscripts that are equal to i (0 or 1).

Phase component activities can then be calculated as the product of the activity in the ideal mixing-on-site model (eqn 3.80) times an activity coefficient:

$$a_\phi = \gamma_\phi \prod_j \prod_i \left(\frac{n_j}{\nu_{\phi ij}} X_i^j \right)^{\nu_{\phi ij}} \quad (4.22)$$

where γ_ϕ is calculated as:

$$\gamma_\phi = \prod_s \prod_i (\gamma_i)^{\nu_s} \quad (4.23)$$

where i refers to individual ions, s to crystallographic sites occupied by them and ν_s is the stoichiometric coefficient for that site.

Example 4.1 Computing activities using Margules parameters

Compute the activity of albite in an albite (Ab) and orthoclase (Or) solid solution (alkali feldspar) as a function of the mole fraction of albite from $X_{Ab} = 0$ to 1 at 600°C and 200 MPa. Use the asymmetric solution model and the data of Thompson and Waldbaum (1969) given below.

Alkali feldspar Margules parameters		
	Ab	**Or**
W_V (J/MPa-mol)	3.89	4.688
W_S (J/mol)	19.38	16.157
W_H (kJ/mol)	26.485	32.105

Answer: Our first step is to calculate W_G for each end -member where $W_G = W_H + W_v P - W_S T$. Doing so, we find $W_{GAb} = 10.344\,kJ$ and $W_{GOr} = 18.938\,kJ$. We can then calculate the activity coefficient as a function of X_{Ab} and X_{Or} from eqn. 4.18. The activity is then computed from $a = \lambda X_{Ab}$. The results are plotted in Figure 4.5.

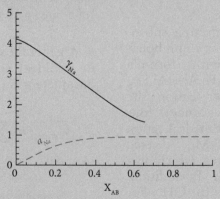

Figure 4.5 Activity and activity cofficient of albite in alkali feldspar solid solution computed at 600°C and 200 MPa using the asymmetric solution model of Thompson and Waldbaum (1969).

4.3   EXSOLUTION PHENOMENA

Now consider a binary system, such as $NaAlSi_3O_8$–$KAlSi_3O_8$ in the example above, of components 1 and 2, each of which can form a pure phase, but also together form a solution phase, which we will call c. The condition for spontaneous exsolution of components 1 and 2 to form two phases a and b is simply that $G_a + G_b < G_c$.

As we saw in Chapter 3, the free energy of a real solution can be expressed as the sum of an ideal solution and a nonideal or excess free energy term:

$$G_{real} = G_{ideal} + G_{ex}$$

The free energy of the ideal part is given by:

$$\overline{G}_{ideal} = \sum X_i \mu_i^o + RT \sum X_i \ln X_i \quad (3.31)$$

Further, the ideal part itself consists of two terms, the first term in eqn. 3.31 corresponding to the free energy of a mechanical mixture ($G_{mixture}$), and the second part being the free energy of ideal mixing ($\Delta G_{ideal\ mixing}$). Figure 4.6a illustrates the variation of G_{excess}, $G_{mixture}$, and G_{ideal} in this hypothetical system. $G_{mixture}$ is simply the free energy of a mechanical mixture of pure components 1 and 2 (e.g., orthoclase and albite). Figure 4.6b illustrates the variation of G_{real} in this system. So long as G_{real} is less than $G_{mixture}$, a solution is stable relative to pure phases 1 and 2. You can see that G_{ideal} is always less than $G_{mixture}$, as long as the G_{ex} term is not too great. In the hypothetical case illustrated in Figure 4.6, a solution is always stable relative to a mechanical mixture of the pure end-member phases. However, if we look carefully at Figure 4.6b, we see there is yet another possibility, namely that two phases a and b, each of which is a *limited* solid solution of components 1 and 2, are stable relative to a single solid solution. Thus, at equilibrium, two phases will exsolve from the single solution; this is just what occurs at lower temperatures in the alkali feldspar system. It would be useful if thermodynamics could predict when such exsolution will occur. Let's see if our thermodynamics tools are up to the task.

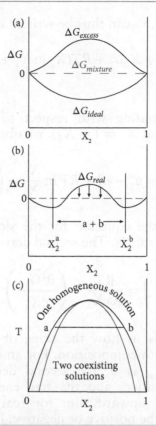

Figure 4.6   (a) Schematic isothermal, isobaric G–X plot for a real solution showing ΔG of mechanical mixing, ideal mixing and excess mixing. (b) Sum of ideal and excess mixing free energies shown in (a). Tangents to the minima give the chemical potentials in immiscible phases a and b. (c) T–X plot for same system as in (b). Solid line is the *solvus*, red dashed line is the *spinodal*. Exsolution may not occur between the spinodal and solvus because the free energy can locally increase during exsolution. After Nordstrom and Munoz (1986).

Looking at Figure 4.3, we see that at 800°C, ΔG_{real} defines a continuously concave upward path, while at lower temperatures, such as 600°C (Figure 4.2), inflections occur and there is a region where ΔG_{real} is concave downward. All this suggests we can use calculus to predict exsolution. For a binary solution of components 1 and 2, the $G_{mixture}$ and $\Delta G_{ideal\ mixing}$ terms are:

$$G_{mixture} = X_1 \mu_1^o + X_2 \mu_2^o$$

$$\Delta G_{ideal\ mixing} = RT(X_1 \ln X_1 + X_2 \ln X_2)$$

Equation 3.31 can thus be written as:

$$\overline{G} = X_1\mu_1^o + X_2\mu_2^o + RT(X_1 \ln X_1 + X_2 \ln X_2) + G_{ex} \qquad (4.24)$$

Differentiating with respect to X_2 (and recalling that $X_1 = 1 - X_2$), we obtain:

$$\left(\frac{\partial\overline{G}}{\partial X_2}\right) = \mu_2^o - \mu_1^o + RT \ln \frac{X_2}{X_i} + \left(\frac{\partial\overline{G}_{ex}}{\partial X_2}\right) \qquad (4.25)$$

This is the equation for the slope of the curve of G vs. X_2. The second derivative is:

$$\left(\frac{\partial^2\overline{G}}{\partial X_2^2}\right) = \frac{RT}{X_1 X_2} + \left(\frac{\partial^2\overline{G}_{ex}}{\partial X_2^2}\right) \qquad (4.26)$$

This tells us how the slope of the curve changes with composition. For an ideal solution, G_{excess} is 0, the second derivative is always positive, and the free energy curve is concave upward. But for real solutions G_{excess} can be positive or negative. If for some combination of T and X (and P), the second derivative of G_{excess} becomes negative and its absolute value is greater than the $RT/X_1 X_2$ term, inflection points appear in the G–X curve. Thus, exsolution is thermodynamically favored if for some composition:

$$\frac{RT}{X_1 X_2} + \left(\frac{\partial^2\overline{G}_{ex}}{\partial X_2^2}\right) \leq 0$$

The inflection points occur where the second derivative is 0; however, as may be seen in Figure 4.6b, the inflection points do not correspond with the thermodynamic limits of solubility, which in this diagram are between X_2^a and X_2^b.

We can draw a straight line that is at a tangent to the free energy curve at X_2^a and X_2^b. This line is the free energy of a mechanical mixture of the two limited solutions a and b. Phase a is mostly component 1 but contains X_2^a of component 2. Similarly, phase b is mostly component 2 but contains $1 - X_2^b$ of component 1. The mechanical mixture of a and b has less free energy than a single

solution phase everywhere between X_2^a and X_2^b. It is therefore thermodynamically more stable, so exsolution can occur in this region.

In Figure 4.3, we can see inflection points developing at about 650°C in the alkali feldspar solution. The inflection points become more marked and occur at increasingly different values of X_{Ab} as temperature decreases. The alkali feldspar system illustrates a common situation where there is complete solid solution at higher temperature but decreasing *miscibility* at lower temperature. This occurs because the free energy of ideal mixing becomes less negative with decreasing temperature (Figure 3.6).

Figure 4.6c shows a schematic drawing of a temperature–composition plot in which there is complete solution at higher temperature with a widening two-phase region at lower temperatures. The boundary between the two-phase and one-phase regions is shown as a solid line and is known as the *solvus*.

The analysis of exsolution above is relevant to immiscible liquids (e.g., oil and vinegar, silicate and sulfide melts) as well as solids. There is a difference, however. In solids, exsolution must occur through diffusion of atoms through crystal lattices, while in liquids both diffusion and advection serve to redistribute components in the exsolving phases. As exsolution begins, the exsolving phases begin with the composition of the single solution and must rid themselves of unwanted components. In a solid, this only occurs through diffusion, which is very slow. This leads to a kinetic barrier that often prevents exsolution even though two exsolved phases are more stable than a solution. This is illustrated in Figure 4.7. For example, consider a solution of composition C. It begins to exsolve protophases of A and B, which initially have compositions A′ and B′. Even though a mechanical mixture of A and B will have lower free energy than solution phase C, A′ and B′, the initial products of exsolution, have higher free energy than C. Furthermore, as exsolution proceeds and these phases move toward compositions A and B, this free energy excess becomes larger. Thus, exsolution causes a local increase in free energy and therefore cannot occur. This problem is not encountered at composition

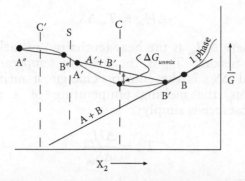

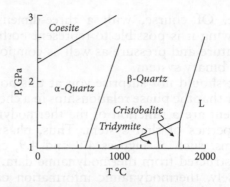

Figure 4.7 A small portion of a $\overline{G}$–X plot illustrating the origin of the spinodal. The process of exsolution of two phases from a single solid solution must overcome an energy barrier. As exsolution from a solution of composition C begins, the two exsolving phases have compositions that move away from C, e.g., A′ and B′. But the free energy of a mechanical mixture of A′ and B′ has greater free energy, by ΔG_{unmix}, than the original single solution phase. Exsolution will therefore be inhibited in this region. This problem does not occur if the original solution has composition C′.

Figure 4.8 P–T phase diagram for SiO_2. This system has one component but seven phases. L designates the liquid phase. The α–β quartz transition is thought to be partially second-order, that is, it involves only stretching and rotation of bonds rather than a complete reformation of bonds as occurs in first-order phase transitions.

C′, though, because a mixture of the exsolving protophases A″ and B″ has lower free energy than original solution at C′. Thus, the actual limit for exsolution is not tangent points such as B but at inflection points (where $\partial^2 G/\partial X^2 = 0$) such as S. The locus of such points is plotted in Figure 4.6c as the red line and is known as the *spinodal*.

4.4 THERMODYNAMICS AND PHASE DIAGRAMS

A *phase diagram* is a representation of the regions of stability of one or more phases as a function of two or more thermodynamic variables such as temperature, pressure, volume, or composition. In other words, if we plot two thermodynamic variables such as temperature and pressure or temperature and composition, we can define an area on this plot where a phase of interest is thermodynamically stable. Figure 4.8 is an example of a P–T phase diagram for a one-component system: SiO_2. The diagram shows the SiO_2 phase stable for a given combination of pressure and temperature.

Figure 4.9 is an example of a simple T–X diagram for the two-component system diopside–anorthite ($CaMgSi_2O_6$ or clinopyroxene and Ca-plagioclase, $CaAl_2Si_2O_8$; two of the more common igneous rock-forming minerals). In multicomponent systems we must always be concerned with at least three thermodynamic variables: P, T, and X. Thus, any T–X phase diagram will be valid for only one pressure, 0.1 MPa (1 bar $\approx$ 1 atm) in this

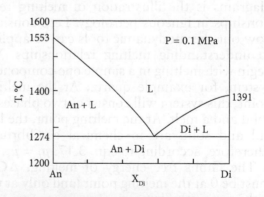

Figure 4.9 Phase diagram (T–X) for the two-component system diopside–anorthite at 1 atm. Four combinations of phases are possible as equilibrium assemblages: liquid (L), liquid plus diopside (L + Di), liquid plus anorthite (L + An), and diopside plus anorthite.

case. Of course, with a three-dimensional drawing it is possible to represent both temperature and pressure as well as composition in a binary system.

It should not surprise you at this point to hear that the phase relationships in a chemical system are a function of the thermodynamic properties of that system. Thus, phase diagrams, such as Figures 4.8 and 4.9, can be constructed from thermodynamic data. Conversely, thermodynamic information can be deduced from phase diagrams.

Let's now see how we can construct phase diagrams, specifically $T–X$ diagrams, from thermodynamic data. Our most important tool in doing so will be the $\overline{G}$ –X diagrams that we have already encountered. *The guiding rule in constructing phase diagrams from $\overline{G}$ –X diagrams is that the stable phases are those that combine to give the lowest $\overline{G}$.* Since a $\overline{G}$ –X diagram is valid for only one particular temperature, we will need a number of $\overline{G}$ –X diagrams at different temperatures to construct a single $T–X$ diagram (we could also construct $P–X$ diagrams from a number of $\overline{G} - X$ diagrams at different pressures). Before we begin, we will briefly consider the thermodynamics of melting in simple systems.

4.4.1 The thermodynamics of melting

One of the more common uses of phase diagrams is the illustration of melting relationships in igneous petrology. Let's consider how our thermodynamic tools can be applied to understanding melting relationships. We begin with melting in a simple one-component system, for example quartz. At the melting point, this system will consist of two phases: a solid and a melt. At the melting point, the liquid and solid are in chemical equilibrium. Therefore, according to eqn. 3.17: $\mu_l = \mu_s$.

The Gibbs free energy of melting, ΔG_m, must be 0 at the melting point (and only at the melting point). Since:

$$\Delta G_m = \Delta H_m - T_m \Delta S_m \qquad (4.27)$$

and $\Delta G_m = 0$ at T_m, then:

$$\Delta H_m = T_m \Delta S_m$$

where ΔH_m is the heat (enthalpy) of melting or fusion,* T_m is the melting temperature, and ΔS_m is the entropy change of melting. Thus, the melting temperature of a pure substance is simply:

$$T_m = \frac{\Delta H_m}{\Delta S_m} \qquad (4.28)$$

This is a very simple, but very important, relationship. This equation tells us that temperature of melting of a substance is the ratio of the enthalpy change to entropy change of melting. Also, if we can measure temperature and enthalpy change of the melting reaction, we can calculate the entropy change.

The pressure dependence of the melting point is given by the Clapeyron equation:

$$\frac{dT}{dP} = \frac{\Delta V_m}{\Delta S_m} \qquad (4.29)$$

Precisely similar relationships hold for vaporization (boiling). Indeed, the temperature and pressure boundaries between any two phases, such as quartz and tridymite, calcite and aragonite, and so on, depend on thermodynamic properties in an exactly analogous manner.

In eqn. 3.66 we found that addition of a second component to a pure substance depresses the melting point. Assuming ΔS_m and ΔH_m are independent of temperature, we can express this effect as:

$$\frac{T_{i,m}}{T} = 1 - \frac{R \ln a^\ell_{i,m}}{\Delta S_{i,m}} \qquad (4.30)$$

Since enthalpies of fusion, rather than entropies, are the quantities measured, eqn. 4.25 may be more conveniently expressed as:

$$\frac{T_{i,m}}{T} = 1 - \frac{T_{i,m} R \ln a^\ell_{i,m}}{\Delta H_{i,m}} \qquad (4.31)$$

Example 4.2 shows how the approximate phase diagram for the diopside–anorthite system (Figure 4.9) may be constructed using this equation.

---

*The heat of fusion is often designated by ΔH_f. I have chosen to use the subscript m to avoid confusion with heat of formation, for which we have already been using the subscript f.

Example 4.2   Calculating melting curves

Using the data given below and assuming (1) that the melt is an ideal solution and (2) diopside and anorthite solids are pure phases, calculate a T–X phase diagram for melting of an anorthite–diopside mixture.

	T_m °C	ΔH_m J/mol
Diopside	1391	138100
Anorthite	1553	136400

Data from Stebbins et al. (1983)

Answer: Solving eqn. 4.31 for T, and replacing activity with mole fraction (since we may assume ideality), we have:

$$T = \frac{\Delta H_{i,m}}{\Delta H_{i,m}/T_{i,m} - R \ln X_i^{\ell}}$$

(4.32)

We then calculate T for every value of X_{An} and X_{Di}. This produces two curves on a T–X plot, as shown in Figure 4.10. The curves intersect at the eutectic, or lowest point at which melt may exist in the system.

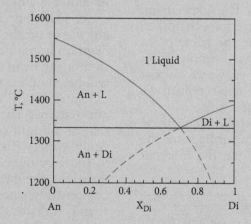

Figure 4.10   Computed phase diagram for the system anorthite–diopside ($CaAl_2Si_2O_8$–$CaMgSi_2O_6$). The eutectic occurs at $X_{Di} = 0.7$ and 1334°C. The dashed lines beyond the eutectic give the apparent melting points of the components in the mixture.

Comparing our result with the actual phase relationships determined experimentally (Figure 4.8), we see that while the computed phase diagram is similar to the actual one, our computed eutectic occurs at $X_{Di} = 0.70$ and 1335°C and the actual eutectic occurs at $X_{Di} \approx 0.56$ and 1274°C. The difference reflects the failure of the several assumptions we made. First, and most importantly, silicate liquids are not ideal solutions. Second, the entropies and enthalpies of fusion tend to decrease somewhat with decreasing temperature, violating the assumption we made in deriving eqn. 4.31. Third, the diopside crystallizing from anorthite–diopside mixtures is not pure, but contains some Al and an excess of Mg.

It must be emphasized that in deriving eqn. 3.66, and hence the eqns. 4.25 and 4.26, we made the assumption that the solid was a pure phase. This assumption is a reasonably good one for ice, and for the anorthite–diopside binary system, but it is not generally valid. Should the solid or solids involved exhibit significant solid solution, this assumption breaks down and these equations are invalid. In that case, melting phase diagrams can still be constructed from thermodynamic equations, but we need to model solid solution as well as the liquid one. Section 4.4.2.1 illustrates an example (anorthite–albite) where the two solutions can be modeled as ideal.

4.4.2   Thermodynamics of phase diagrams for binary systems

In a one-component system, a phase boundary, such as the melting point, is univariant since at that point two phases coexist and $f = c - p + 2 = 1 - 2 + 2 = 1$. Thus, specifying either temperature or pressure fixes the other. A three-phase point, such as the triple point of water, is invariant. Hence, simply from knowing that three phases of water coexist (i.e., knowing we are at the triple point), we know the temperature and pressure.

In binary systems, the following phase assemblages are possible according to the Gibbs phase rule (ignoring for the moment gas phases):

	Phases	Free compositional variables
Univariant	2 solids + liquid, 2 liquids + solid, 3 solids or liquids	0
Divariant	1 solid + 1 liquid, 2 solids, 2 liquids	0
Trivariant	1 solid or 1 liquid	1

When a $\overline{G}$–X diagram is drawn, it is drawn for a specific temperature and pressure, such that $\overline{G}$–X are isobaric and isothermal. Thus, we have already fixed two variables, and the compositions of all phases in univariant and divariant assemblages are fixed by virtue of

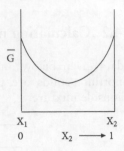

Figure 4.11 $\overline{G}$–X diagram for a two-component system exhibiting complete solution (either liquid or solid).

our having fixed T and P. Only in trivariant systems are we free to choose the composition of a phase on a $\overline{G}$–X diagram. Figure 4.11 is a schematic diagram of a two-component, one-phase (trivariant) assemblage, in which there is complete solution between component 1 and component 2. This phase might be either a liquid or a solid such as plagioclase. The composition of the phase may fall anywhere on the curve. Of course, since this diagram applies only to one temperature, we cannot say from this diagram alone that there will be complete solution at all temperatures.

Figure 4.12 illustrates four possible divariant systems. The first case (Figure 4.12a) is that of a liquid solution of composition L' in equilibrium with a solid of fixed composition S_2 (pure component 2). Because the system is divariant, there can be only one possible liquid composition since we have implicitly specified P and T. As usual, the equilibrium condition is described by $\mu_i^\ell = \mu_i^s$ (eqn. 3.17). For $i = 2$, this means the tangent to the free energy curve for the melt must intersect the $X_2 = 1$ line at μ_i^s as shown. In other words, the chemical potential of component 2 in the melt must be equal to the chemical potential of component 2 in the solid. Again, this diagram is valid for only one temperature; at any other temperature, the free energy curve for the liquid would be different, but the composition of this new liquid in equilibrium with solid S_2 would still be found by drawing a tangent from S_2 to the free energy curve of the liquid. At sufficiently high temperature, the tangent would always intersect below S_2. The temperature at which this first occurs is the melting temperature of S_2 (because it is the point at which the

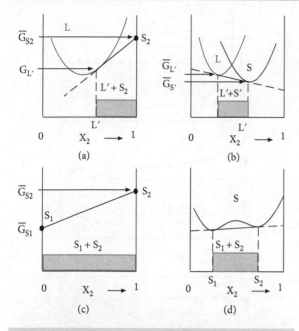

(a)

(b)

(c)

(d)

Figure 4.12 Plot of molar free energy vs. composition ($\overline{G}$–X_2) for two-phase divariant systems. (a) A liquid solution (L) is in equilibrium with a solid (S_2) of pure X_2. The shaded area shows the range of composition of systems for which L' and S_2 coexist as separate phases. (b) Here both solid and liquid have variable composition. Equilibrium compositions are determined by finding a tangent to both free energy curves. L' and S' will be the equilibrium phases for systems having compositions in the shaded area. (c) is the case of two immiscible solids, while (d) shows two limited solid solutions of composition S_1 and S_2. Here, the compositions of the solids are given by the point where a straight line is tangent to the curve in two places.

free energy of a liquid of pure 2 is less than the solid). The shaded region shows the compositions of systems that will have a combination of solid S_2 and liquid L' as their equilibrium phases at this temperature.

We can also think of the tangent line as defining the free energy of a mechanical mixture of S_2 and L'. In the range of compositions denoted by the shaded region, this mixture has a lower free energy than the liquid solution, hence at equilibrium we expect to find this mixture rather than the liquid solution.

Figure 4.12b illustrates a system with a liquid plus a solid solution, each of which has

its own G–X curve. Again, the equilibrium condition is $\mu_i^\ell = \mu_i^s$, so the compositions of the coexisting liquid and solid are given by a tangent to both curves. Since the system is divariant and we have fixed P and T, the compositions of the solutions are fixed. All system compositions in the shaded region can be accommodated by a mixture of liquid and solid. Compositions lying to the left of the region would have only a liquid; compositions to the right of the shaded region would be accommodated by a solid solution.

Figure 4.12c illustrates the case of two immiscible solids (pure components 1 and 2). The molar free energy of the system is simply that of a mechanical mixture of S_1 and S_2: a straight line drawn between the free energy points of the two phases.

Figure 4.12d illustrates the case of a limited solution. We have chosen to illustrate a solid solution, but the diagram would apply equally well to the case of two liquids of limited solubility.

Figure 4.13a shows the case of two solid solutions plus one liquid. The chemical potential of each component in each phase must be equal to the chemical potential of that component in every other phase, so chemical potentials are given by tangents to all three phases. This is a univariant system, so specifying either temperature, pressure, or the composition of a phase fixes other variables in the system. Because of this, if we move to a slightly higher or lower temperature at fixed pressure, one of the phases must be eliminated in a *phase elimination reaction*.

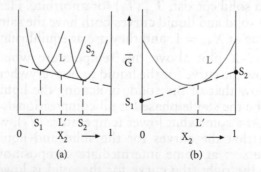

(a)

(b)

Figure 4.13 Two univariant systems: (a) a liquid plus two solid solutions, and (b) two pure solids and a liquid. Since these systems are univariant, they occur only at one fixed T if P is fixed.

If the phase is the liquid inbetween the two solids in composition, the reaction is known as a *eutectic*, which is the lowest temperature at which the liquid can exist. Moving to a higher temperature would result in elimination of one of the solids. If, alternatively, the liquid is not between two solids (for example, if the curves L and S_2 in Figure 4.13a were switched), the reaction would be known as a peritectic, and moving to lower temperature eliminates one of the solids. Thus, it is possible for a liquid to persist below a peritectic if the composition is right, but a liquid will never persist at equilibrium below a eutectic. Figure 4.13b is a eutectic in a system where the two solids are the phases of pure components 1 and 2. A line drawn between the free energies of the pure components is also tangential to the liquid curve.

4.4.2.1   *An example of a simple binary system with complete solution: albite–anorthite*

Phase diagrams in T–X space can be constructed by analyzing $\overline{G}$–X diagrams at a series of temperatures. Let's examine how this can be done in the case of a relatively simple system of two components, albite ($NaAlSi_3O_8$) and anorthite ($CaAl_2Si_2O_8$), whose solid (plagioclase) and liquid exhibit complete solid solution. Figure 4.14 shows $\overline{G}$–X diagrams for various temperatures as well as a T–X phase diagram for this system. Since both the solid and liquid exhibit complete solution, we need to consider $\overline{G}$–X curves for both.

We start at the highest point at which liquid and solid coexist, T_m (T_1) for anorthite. Here the solid and liquid curves both have the same value at $X_{An} = 1$, and they are at equilibrium. A $\overline{G}$–X plot above this temperature would show the curve for the liquid to be everywhere below that of the solid, indicating the liquid to be the stable phase for all compositions.

At a somewhat lower temperature (T_2), we see that the curves for the solid and liquid intersect at some intermediate composition. To the right, the curve for the solid is lower than that of the liquid, and tangents to the solid curve extrapolated to both $X_{Ab} = 1$ and $X_{An} = 1$ are always below the curve for the liquid, indicating the solid is the stable phase. As we move toward *Ab* (left) in composition,

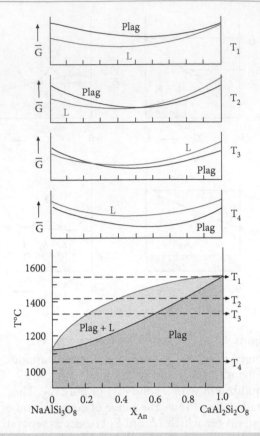

Figure 4.14 $\overline{G}$–X diagrams and a T–X phase diagram for the plagioclase–liquid system. After Richardson and McSween (1989).

tangents to the solid curve eventually touch the curve for the liquid. The point where the tangent touches each curve gives the compositions of the liquid and the solid stable at this temperature. In the compositional range between the points where the tangent touches the two curves, the tangent is below both curves; thus, a mechanical mixture of solid and liquid is stable over this compositional range at this temperature. For compositions to the left of the point where the tangent touches the liquid curve, the liquid curve is lower than both the solid curve and a tangent to both, so it is stable relative to both the solid and any mixture of solid and liquid.

Going to progressively lower temperatures (e.g., T_3), the points where a tangent intersects the two curves move toward Ab (to the left). Eventually, at a sufficiently low temperature (T_4), the curve for the solid is everywhere below that of the liquid and only solid solution is stable. By extracting information from

G–X curves at a number of temperatures, it is possible to reconstruct the phase diagram shown at the bottom of Figure 4.14.

Since both the solid and liquid show complete miscibility in this system, we will make the simplifying assumption that both solutions are ideal and do an approximate mathematical treatment. We recall that the condition for equilibrium was:

$$\mu_i^\alpha = \mu_i^\beta$$

We can express the chemical potential of each component in each phase as:

$$\mu_i^\alpha = \mu_i^{o\alpha} + RT \ln X_i^\alpha \qquad (4.33)$$

Combining these relationships, we have:

$$\mu_{Ab}^{o\ell} - \mu_{Ab}^{os} = RT \ln \left(\frac{X_{Ab}^s}{X_{Ab}^\ell} \right) \qquad (4.34)$$

and

$$\mu_{An}^{o\ell} - \mu_{An}^{os} = RT \ln \left(\frac{X_{An}^s}{X_{An}^\ell} \right) \qquad (4.35)$$

Here our standard states are the pure end-, members of the melt and solid. The left side of both of these equations corresponds to the standard free energy change of crystallization, thus:

$$\Delta \overline{G}_m^{Ab} = RT \ln \left(\frac{X_{Ab}^s}{X_{Ab}^\ell} \right) \qquad (4.36)$$

$$\Delta \overline{G}_m^{An} = RT \ln \left(\frac{X_{An}^s}{X_{An}^\ell} \right) \qquad (4.37)$$

Both sides of these equations reduce to 0 if and only if $X_i^\ell = X_i^s = 1$ and $T = T_m$. Rearranging:

$$X_{Ab}^s = X_{Ab}^\ell e^{\Delta \overline{G}_m^{Ab}/RT} \qquad (4.38)$$

$$X_{An}^s = X_{An}^\ell e^{\Delta \overline{G}_m^{An}/RT} \qquad (4.39)$$

Thus, the fraction of each component in the melt can be predicted from the composition of the solid and thermodynamic properties of the end-members. At equilibrium:

$$\mu_{An}^\ell = \mu_{An}^s$$

Since $X_{An}^s = 1 - X_{Ab}^s$ and $X_{An}^\ell = 1 - X_{Ab}^\ell$, we can combine eqns. 4.33 and 4.34 to obtain:

$$(1 - X_{Ab}^\ell)e^{\Delta \overline{G}_m^{An}/RT} = 1 - X_{Ab}^\ell e^{\Delta \overline{G}_m^{Ab}/RT} \qquad (4.40)$$

and rearranging yields:

$$X_{Ab}^\ell = \frac{1 - e^{\Delta \overline{G}_m^{An}/RT}}{e^{\Delta \overline{G}_m^{Ab}/RT} - e^{\Delta \overline{G}_m^{An}/RT}} \qquad (4.41)$$

The point is that the mole fraction of any component of any phase in this system can be predicted from the thermodynamic properties of the end-members. We must bear in mind that we have treated this as an ideal system, so we have ignored any G_{excess} term. Nevertheless, the ideal treatment is relatively successful for the plagioclase system. For nonideal systems, we merely replace mole fraction in the above equations with activity. Provided they are known, interaction parameters can be used to calculate activity coefficients (e.g., eqns. 4.18 or 4.12 as the case may be). Beyond that, nonideal systems can be treated in a manner exactly analogous to the treatment above.

4.4.3 Phase diagrams for multicomponent systems

For clarity and simplicity, we have focus on phase diagrams of binary systems. Nature, of course, is rarely so simple. For example, a minimum of seven major components are typically required to describe silicate rocks: SiO_2, Al_2O_3, FeO, MgO, CaO, Na_2O, and K_2O; most commonly, H_2O is also a major component, and a complete description would include TiO_2, Fe_2O_3, MnO, and in some cases CO_2. Many of the principal phases are, like alkali feldspar and plagioclase, solid solutions. Although the principles of computing phase relations are the same in these systems as the binary ones we have considered, calculation of equilibria is computationally intense.

Fortunately, computer codes for carrying out such calculations are available. One such

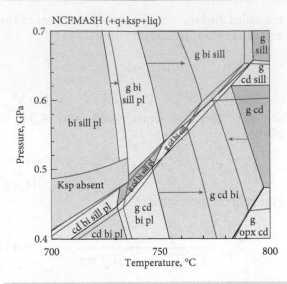

NCFMASH (+q+ksp+liq)

Figure 4.15 Phase diagram for a metapelite calculated using THERMOCALC. The composition is given in the text. The diagram is drawn for the system Na$_2$O–CaO–FeO–MgO–K$_2$O–Al$_2$O$_3$–SiO$_2$–H$_2$O using the THERMOCALC 2011 NCFMASH database (Holland and Powell, 2011). Phase boundaries for two slightly different compositions are shown; the one in red is richer in the anorthite component. Mineral abbreviations are as follows: bi (biotite), cd (cordierite: (Mg,Fe)$_2$Al$_4$Si$_5$O$_{18}$), g (garnet), opx (orthopyroxene), pl (plagioclase) sill (sillimanite). In addition to these phases, quartz, a silicate liquid and, except where noted, alkali feldspar (ksp) are present everywhere in the diagram.

program is THERMOCALC, developed by Roger Powell and Tim Holland (e.g., Powell et al., 1998), which is freely available at http://www.metamorph.geo.uni-mainz.de/thermocalc/index.html. THERMOCALC solves the set of simultaneous equations that describe the equilibria between phases. Components are generally phase components rather than oxides. Consider, for example, the phase boundary between biotite, sillimanite (Al$_2$SiO$_5$), plagioclase, garnet, alkali feldspar, quartz, and silicate liquid (the bold black line in the lower right of Figure 15). This is the reaction:

Garnet + Cordierite + Biotite + Alkali feldspar

+ SiO$_2$ + Liquid ⇌

Garnet + Cordierite + Orthopyroxene

+ Alkali feldspar + SiO$_2$ + Liquid

Except quartz, all of these phases are solutions, with biotite, cordierite, and orthopyroxene consisting of Mg and Fe end-members, alkali feldspar consisting of K-feldspar, albite, and anorthite end-members, and garnet consisting of Mg, Fe, and Ca end-members (pyrope, almandine, and grossular, respectively). In addition, the liquid composition can vary.

For each phase, THERMOCALC adopts a mixing on site, symmetric, or asymmetric solution model as appropriate. The boundaries between phase assemblages are the condition where $\Delta G_r = 0$. Solving for this condition involves simultaneously solving a series of nonlinear equations and requires a data set of thermodynamic parameters (V, S, C_p, etc. as well as interaction parameters for nonideal solutions). The THERMOCALC has been successively improved over several decades; most recent database versions are described by Green et al. (2016).

Figure 4.15 is an example of a P–T phase diagram for a metamorphic aluminous sediment (a metapelite) computed with THERMOCALC v3.33. The black lines show the version for a composition with the following weight percent oxides SiO$_2$ = 70.09, Al$_2$O$_3$ = 8.95, MgO = 3.64, FeO = 6.93, CaO = 0.28, Na$_2$O = 0.57, K$_2$O = 2.87, and H$_2$O = 6.66 (this is the composition used in the tutorial by R. W. White on the THERMOCALC website). Each such phase diagram is valid only for a specific composition. The red lines show the phase diagram for this composition slightly modified to have a greater proportion of anorthite and be more Fe-rich (SiO$_2$ = 68.83, Al$_2$O$_3$ = 9.06, MgO = 3.59, FeO = 8.80, CaO = 0.54, Na$_2$O = 0.56, K$_2$O = 2.70, and H$_2$O = 5.92). As we might anticipate, the major differences are that stability region of plagioclase increases and that of biotite decreases. In addition, a field where alkali feldspar (ksp) is absent is present in the second composition. THERMOCALC is also capable of computing mineral compositions (for a given rock composition) as functions of T and P (isopleths) and hence can be used for thermobarometry, a topic to which we now turn.

4.5   GEOTHERMOMETRY AND GEOBAROMETRY

An important task in geochemistry is estimating the temperature and pressure at which mineral assemblages equilibrate. The importance extends beyond petrology to tectonics and all of geology because it reveals the conditions under which geological processes occur. Here we take a brief look at the thermodynamics underlying geothermometry and geobarometry (since most reactions are both temperature and pressure dependent, it is perhaps more accurate to use the term *thermobarometer*). We have space to consider just a few relatively simple thermobarometers based on the composition of mineral pairs. Particularly in complex metamorphic systems, modern thermobarometry involves simultaneously solving for the equilibrium among many phases and requires relatively sophisticated computer algorithms such as PERPLEX (Connolly, 1989) or THERMOCALC (Powell and Holland, 2008). Those approaches, however, involve the same fundamental principles as the simple geobarometers we consider below. Here we focus on "chemical" geobarometers that depend on the distribution of chemical components between phases. In Chapter 9, we will see that temperatures can also be deduced from the distribution of isotopes of an element between phases.

Geothermometry and geobarometry involve two nearly contradictory assumptions. The first is that the mineral assemblage of interest is an equilibrium one, and the second is that the system did not re-equilibrate during the passage through lower P and T conditions that brought the rock to the surface where it could be collected. As we will see in the next chapter, reaction rates depend exponentially on temperature, hence these assumptions are not quite as contradictory as they might seem.

4.5.1   Theoretical considerations

In general, geobarometers and geothermometers make use of the pressure and temperature dependence of the equilibrium constant, K. In Section 3.9 we found that $\Delta G° = -RT \ln K$. Assuming that ΔC_p and ΔV of the reaction are independent of temperature and pressure,

we can write:

$$\Delta G° = \Delta H°_{T,P_{ref}} - T\Delta S°_{T,P_{ref}} + \Delta V°_{T,P_{ref}}(P - P_{ref})$$
$$= -RT \ln K \qquad (4.42)$$

where the standard state of all components is taken as the pure phase at the temperature and pressure of interest, and the enthalpy, entropy, and volume changes are for the temperature of interest and a reference pressure (generally 0.1 MPa).

Solving eqn. 4.42 for lnK and differentiating the resulting equation with respect to temperature and pressure leads to the following relations:

$$\left(\frac{\partial \ln K}{\partial T}\right)_P = \frac{\Delta H°_{T,P_{ref}} + \Delta V°_{T,P_{ref}}(P - P_{ref})}{RT^2}$$
$$\qquad (4.43)$$

$$\left(\frac{\partial \ln K}{\partial P}\right)_T = \frac{\Delta V°_{T,P_{ref}}}{RT} \qquad (4.44)$$

These equations provide us with the criteria for reactions that will make good geothermometers and geobarometers. For a good geothermometer, we want the equilibrium constant to depend heavily on T but be approximately independent of P. Looking at eqn. 4.43, we see this means the ΔH term should be as large as possible and the ΔV term as small as possible. A fair amount of effort was devoted to the development of a geothermometer based on the exchange of Fe and Mg between olivine and pyroxenes in the late 1960s. The effort was abandoned when it was shown that the ΔH for this reaction was very small. As a rule, a reaction should have a $\Delta H°$ of at least 1 kJ to be a useful geothermometer. For a good geobarometer, we want the ΔV term to be as large as possible. For example, even though the rhodonite ($[Mn,Fe,Ca]SiO_3$) and pyroxmangite ($[Mn,Fe]SiO_3$) pairs commonly occur in metamorphic rocks, the reaction rhodonite → pyroxmangite does not make a useful geobarometer because the ΔV of reaction is only 0.2 cc/mol. In general, a reaction should have a ΔV of greater than 2 cc/mol if it is to be used for geobarometry.

Since for most reactions both ΔH_r and ΔV_r are likely to be finite, the equilibrium constants will depend to some degree on both temperature and pressure. Consequently, a precise estimate of either also requires an estimate of the other. A good geothermometer will have a steep slope on a $P–T$ diagram. Conversely, a good geobarometer will have a shallow slope on such a diagram. While an independent estimate of either P or T is possible from one reaction alone, the most precise estimate can be obtained from the intersection of the two reaction lines.

The following discussion presents a few examples of useful chemical geothermometers and geobarometers. It is not an exhaustive treatment; examples have been chosen to demonstrate underlying thermodynamic principles rather than for their accuracy or utility. Reviews by Essene (1989), Anderson et al. (2008), Blundy and Cashman (2008), Powell and Holland (2008), and Putrika (2018) summarize a wide range of igneous and metamorphic thermobarometers.

There are two important caveats to thermobarometry. The first is that their successful application requires that chemical equilibrium was achieved and not subsequently disturbed. The second is metamorphic and intrusive igneous rocks may reequilibrate at temperatures below the maximum.

4.5.2   Practical thermobarometers

4.5.2.1   Univariant reactions and displaced equilibria

We can broadly distinguish three main types of thermobarometers. The first is the *univariant reaction*, in which the phases have fixed compositions. They are by far the simplest, and often make good geobarometers as the ΔV of such reactions is often large. Examples include the graphite–diamond transition, any of the SiO_2 transitions (Figure 4.8), and the transformations of Al_2SiO_5, shown in Figure 4.16. While such thermobarometers are simple, their utility for estimating temperature and pressure is limited. This is because exact temperatures and pressures can be obtained only if two or more phases coexist, for example, kyanite and andalusite in Figure 4.16. If kyanite and andalusite are both found in a rock, we can determine either temperature or pressure if we can independently determine the other. Where three phases, kyanite, sillimanite, and andalusite, coexist the system is invariant and P and

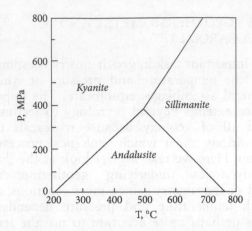

Figure 4.16   Phase diagram for Al_2SiO_5 (kyanite–sillimanite–andalusite) as determined by Holdaway (1971).

T are fixed. If only one phase occurs, for example sillimanite, we can only set a range of values for temperature and pressure. Unfortunately, the latter case, where only one phase is present, is the most likely situation. It is extremely rare that kyanite, sillimanite, and andalusite occur together.

The term *displaced equilibria* refers to variations in the temperature and pressure of a reaction that result from appreciable solution in one or more phases. Thermobarometers based on this phenomenon are more useful than univariant reactions because the assemblage can coexist over a wide range of P and T conditions. In the example shown in Figure 4.17, the boundaries between garnet-bearing, spinel-bearing, and plagioclase-bearing assemblages are curved, or "displaced" as a result of the solubility of Al in enstatite. In addition to the experimental calibration, determination of P and T from displaced equilibria requires (1) careful determination of phase composition and (2) an accurate solution model.

Geobarometers based on the solubility of Al in pyroxenes have been the subject of extensive experimental investigations. The general principle is illustrated in Figure 4.17, which shows the concentration of Al in orthopyroxene (opx) coexisting with olivine (forsterite) and an aluminous phase, anorthite, spinel, or garnet. The Al content of opx depends almost exclusively on pressure in the presence of anorthite, is essentially independent of pressure in the presence of spinel and depends

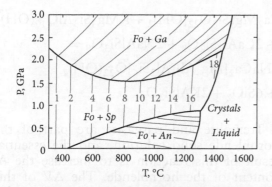

Figure 4.17 Isopleths of Al in orthopyroxene (thin red lines; weight percent) coexisting with forsterite plus an aluminous phase in the CMAS (Ca–Mg–Al–Si) system. After Gasparik (1984a).

on both temperature and pressure in the presence of garnet. Orthopyroxene–garnet equilibrium has proved to be a particularly useful geobarometer.

Garnet is an extremely dense phase. So we might guess that the ΔV of reactions that form it will be comparatively large, and therefore that it is potentially a good geobarometer. The concentration of Al in opx in equilibrium with garnet may be used as a geobarometer if temperature can be independently determined. Although there has been a good deal of subsequent work and refinement of this geobarometer, the underlying thermodynamic principles are perhaps best illustrated by considering the original work of Wood and Banno (1973).

Wood and Banno (1973) considered the following reaction:

$$Mg_2Si_2O_6 + MgAl_2SiO_6 \rightleftharpoons Mg_3Al_2Si_3O_{12}$$
$$(4.45)$$

opx solid solution $\rightleftharpoons$ pyrope garnet

In developing a geobarometer based on this reaction, they had to overcome a number of problems. First, the substitution of Al in orthopyroxene is a coupled substitution. For each atom of Al substituting in the M1* octahedral site, there must be another Al atom substituting for SiO_2 in the tetrahedral site. Second, there was a total lack of thermodynamic data on the $MgAl_2SiO_6$ phase

component. Data was lacking for a good reason: the phase does not exist and cannot be synthesized as a pure phase. Another problem was the apparent nonideal behavior of the system, which was indicated by orthopyroxenes in Fe- and Ca-bearing systems containing less alumina than in pure MgO systems at the same pressure.

The equilibrium constant for reaction 4.45 is:

$$K = \frac{a_{Mg_2Al_2Si_3O_{12}}}{a_{Mg_2Si_2O_6}a_{MgAl_2SiO_6}} \qquad (4.46)$$

where the activities in the denominator represent the activities of the enstatite and the hypothetical aluminous enstatite phase components in the enstatite solid solution. In the pure MgO system (i.e., no CaO, FeO, MnO, etc.), the numerator, the activity of pyrope, is 1 and we may write:

$$\Delta G° = RT \ln(a_{Mg_2Si_2O_6}a_{MgAl_2SiO_6})$$
$$= \Delta H° - T\Delta S° + (P - P_{ref})\Delta V° \quad (4.47)$$

(compare eqn. 4.42). For an ideal case, this may be rewritten as:

$$\Delta G° = RT \ln(X_{Mg_2Si_2O_6}X_{MgAl_2SiO_6})$$
$$= \Delta H° - T\Delta S° + (P - P_{ref})\Delta V° \quad (4.48)$$

Wood and Banno (1973) first estimated thermodynamic parameters (ΔH, ΔS, and ΔV for aluminous pyroxene) from experimental data. They dealt with the nonideality in two ways. First, they assumed ideal solution behavior at 1 bar and assumed all nonideality associated with substitution of Al in orthopyroxene at higher pressure could be accounted for in the volume term in eqn. 4.47, which they rewrote as:

$$\Delta \overline{V}° = \overline{V}°_{Mg_2Al_2Si_3O_{12}} - \overline{V}°_{MgAl_2SiO_6} - \overline{V}°_{Mg_2Si_2O_6}$$
$$(4.49)$$

As for nonideality related to substitution of Ca and Fe in the system, they noted that nonidealities of most silicate systems were of similar size and magnitude and hence the activity coefficients for garnet tend to cancel those for orthopyroxene. Furthermore, the ΔV and ΔH terms are both large and tend to reduce the errors due to nonideal behavior.

* The two octahedral sites in pyroxene normally occupied by metal ions such as Mg, Fe, and Ca are slightly different; the smaller of the two is labeled M1 and the larger of the two, occupied by Ca in diopside, is labeled M2.

Complexities arise because Fe, Ca, and Cr are also present in both garnet and orthopyroxene at significant concentrations. In addition, the available thermodynamic data (based on experiments) has greatly improved and, consequently, a number of studies have improved on and extended the work of Wood and Banno. Brey and Köhler (1990) developed equations that took advantage of the new data and took account of Cr and Ca concentrations; these were further refined by Brey et al. (2008). Some of the simplifying assumptions originally made by Wood and Banno, such as ideal mixing between Mg and Fe in garnet and pyroxene and between Mg and Al in orthopyroxene, were retained, but other aspects required more complex, asymmetric solution models to match experimental data. The equations of Brey et al. (2008) are somewhat complex. A spreadsheet for the calculations, PTEXL.xls, is available at http://www.uni-frankfurt.de/69525998/downloads?.

Since eqn. 4.47 contains temperature as well as pressure terms, it is obvious that the temperature must be known to calculate pressure of equilibration. In the same paper, Wood and Banno (1973) provided the theoretical basis for estimating temperature from the orthopyroxene–clinopyroxene miscibility gap, which we discuss below. Thus, in a system containing garnet, orthopyroxene, and clinopyroxene, both temperature and pressure of equilibration may be estimated from the composition of these phases.

This geobarometer–geothermometer is commonly used to estimate the temperature and pressure (depth) of equilibration of mantle-derived garnet lherzolite xenoliths. One of the first applications was by Boyd (1973), who calculated P and T for a number of xenoliths in South African kimberlites, and hence reconstructed the geotherm in the mantle under South Africa.

Another widely used geobarometer is based on the Al exchange between amphibole, mica and feldspars in reactions such as:

$$\text{tremolite} + \text{phlogopite} + 2\,\text{anorthite}$$
$$+\ 2\,\text{albite} \rightleftharpoons 2\,\text{pargasite} + 6\,\text{quartz}$$
$$+\ 2\text{K-feldspar}$$

$$Ca_2Mg_5Si_8O_{22}[OH]_2 + KMg_3Si_3AlO_{10}(OH)_2$$
$$+\ 2CaAl_2Si_2O_8 + 2NaAlSi_3O_8 \rightleftharpoons$$
$$2NaCa_2[Mg_4Al][Al_2Si_6]O_{22}(OH)_2$$
$$+\ 6SiO_2 + 2KAlSi_3O_8$$

Tremolite and pargasite are part of the hornblende solid solution, and the essential result of this reaction is to increase the Al content of the hornblende. The ΔV of this reaction is negative, so it is pressure sensitive while only moderately temperature sensitive. (We have written the reaction for a Fe-free system, but Fe substitutes for Mg in in both hornblende and phlogopite and the feldspars are, of course, solid solutions as well). Experimental work showed that the total amount of Al in the amphibole was indeed strongly pressure dependent. Mäder et al. (2004) and Ague (2007) provided a thermodynamic basis using a symmetric, second-degree Margules solution model. Pressure is calculated as (Anderson et al., 2008):

$$P(GPa) = 0.476Al - 0.301 - \left(\frac{T - 402}{8.5}\right)$$
$$[0.0530Al + 0.0005294(T - 402)] \quad (4.50)$$

where Al is the sum of tetrahedral and octahedral Al per 13 cations.

4.5.2.2 Solvus equilibria

Solvus equilibria provide a second kind of thermobarometer. Generally, these make better geothermometers than geobarometers. A good example is the ortho- and clinopyroxene system, illustrated in Figure 4.18. The two-pyroxene solvus has been the subject of particularly intensive experimental and theoretical work because ortho- and clinopyroxene coexist over a wide range of conditions in Mg,Fe-rich rocks of the crust and upper mantle.

One of the inherent thermodynamic difficulties with this type of geothermometer is that since it involves exsolution, ideal solution models will clearly be very poor approximations. Thus, considerable efforts have been made to develop solution models for the pyroxenes. Several factors further complicate

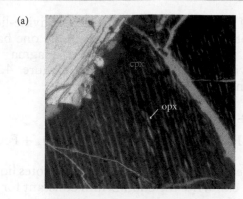

(a)

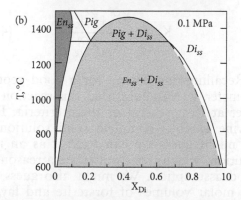

(b)

Figure 4.18 a. Lamellae of orthopyroxene (opx; bright lines) exsolved from a clinopyroxene crystal (cpx; dark) in a gabbric xenolith. Phase relationships in the $Mg_2Si_2O_6$ (enstatite)–$CaMgSi_2O_6$ (diopside) system (after Lindsley, 1983).

efforts to use the pyroxene solvus as a thermobarometer. The first is the existence of a third phase, pigeonite (a low-Ca clinopyroxene), at high temperatures and low pressures; the second is that the system is not strictly binary: natural pyroxenes in igneous rocks are solutions of Mg, Ca, and Fe components. The presence of iron is problematic because of the experimental difficulties encountered with Fe-containing systems. These difficulties include the tendency both for iron to dissolve in the walls of commonly used platinum containers and for Fe^{2+} either to oxidize to Fe^{3+} or to reduce to metallic iron, depending on the oxygen fugacity. In addition, other components, particularly Na, Ti, and Al, are often present in the pyroxenes.

Despite its complexities, the system has been modeled with some success using a symmetric solution model developed by Wood

(1987). Ignoring pigeonite and components other than Ca, Mg, and Fe, we can treat mixing in the M2 and M1 sites separately. Mixing in the M2 site can be treated as a ternary Mg, Fe, and Ca solution. In a symmetric ternary solution consisting of components A, B, and C, the activities of the components may be calculated from:

$$RT \ln \gamma_A = X_B^2 W_G^{AB} + X_C^2 W_G^{AC}$$
$$+ X_B X_C (W_G^{AB} + W_G^{AC} + W_G^{BC})$$
$$(4.51)$$

where W_G^{AB} is the A–B binary interaction parameter, and so on. Mixing of Fe and Mg between the M1 and M2 sites was treated as a simple exchange reaction:

$$Fe_{M2} + Mg_{M1} \rightleftharpoons Fe_{M1} + Mg_{M2}$$

with ΔH of 29.27 kJ/mol and ΔS of 12.61 J/mol. Using this approach, Wood calculated the temperature dependence of the solvus as shown in Figure 4.19. The model fits experimental observation reasonably well for the Mg-rich pyroxenes, but significant deviations occur for the Fe-rich pyroxenes.

The complexities of this system are sufficient that a full and accurate theoretical treatment has not been developed. The most widely used implementation of the

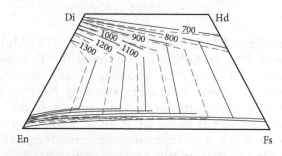

Figure 4.19 Comparison of calculated (solid lines) and experimentally observed (red dashed lines) phase relationships between clino- and orthopyroxene shown in the *pyroxene quadrilateral*, a part of the $CaSiO_3$–$MgSiO_3$–$FeSiO_3$ system. Di, diopside; En, enstatite; Hd, hedenbergite; Fs, ferrosilite. Lines show the limit of solid solution at the corresponding temperatures (°C).

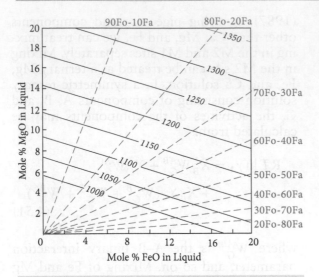

Figure 4.20 Olivine saturation surface constructed by Roeder and Emslie (1970). Fo and Fa denote the forsterite and fayalite components in olivine, respectively.

two-pyroxene geothermometer is an empirical one of Brey and Köhler (1990), where temperature is given by:

$$T = \frac{23664 + (249 + aX_{Fe}^{cpx})P}{13.38 + (\ln K_D)^2 bX_{Fe}^{opx}} \quad (4.52)$$

where

$$K_D = \frac{1 - Ca^{cpx}/(1 - Na^{cpx})}{1 - Ca^{opx}/(1 - Na^{opx})} \quad (4.53)$$

and $a = 1263$, $b = 11.59$, $X_{Fe} = $ Fe/(Fe+Mg), and P is in GPa. Since solving for P using eqn. 4.47 requires knowing T and solving eqn. 4.52 for T requires knowing P, solving for both requires an iterative approach: making an initial guess of P and T, solving the equations, and repeating the procedure with new values until two successive answers agree within error, which is roughly ± 15 K and ± 0.3 GPa.

4.5.2.3 Exchange reactions

Exchange reaction thermobarometers depend on the exchange of two species between phases. We will consider two examples of these.

The Roeder and Emslie (1970) olivine-liquid geothermometer is a rather simple one based on the equilibrium between magma and olivine crystallizing from it (Figure 4.20). Consider the exchange reaction:

$$MgO_{Ol} + FeO_{liq} \rightleftharpoons MgO_{liq}$$

$$+ FeO_{Ol} \quad MgO_{Ol} + FeO_{\ell} \rightleftharpoons MgO_{\ell} + FeO_{Ol}$$

where Ol denotes olivine and ℓ denotes liquid. We can write the equilibrium constant for this reaction as:

$$K_D = \frac{X_{FeO}^{Ol} X_{MgO}^{\ell}}{X_{Fe}^{\ell} X_{Mg}^{Ol}} \quad (4.54)$$

Recalling our criteria for a good geothermometer, we can guess that this reaction will meet at least several of these criteria. First, olivine exhibits complete solid solution, so we might guess we can treat it as an ideal solution, which turns out to be a reasonably good assumption. We might also guess that the molar volumes of forsterite and fayalite and of their melts will be similar, meaning the ΔV term, and hence pressure dependence, will be small, which is also true. As it turns out, however, the ΔH term, which is related to the difference in heats of fusion of forsterite and fayalite, is also relatively small, so the exchange reaction itself is a poor geothermometer. However, we can consider two separate reactions here:

$$MgO_{\ell} \rightarrow MgO_{Ol} \quad \text{and} \quad FeO_{\ell} \rightarrow FeO_{Ol}$$

and we can write two expressions for K_D. This was the approach of Roeder and Emslie (1970), who deduced the following relations from empirical (i.e., experimental) results:

$$\log K_D = \frac{3740}{T} - 1.87 \quad (4.55)$$

and

$$\log \frac{X_{FeO}^{Ol}}{X_{FeO}^{\ell}} = \frac{3911}{T} - 2.50 \quad (4.56)$$

These K_D's are much more temperature-dependent than for the combined exchange reaction. Subtracting eqn. 4.55 from 4.56 yields:

$$\log K_D = \frac{171}{T} - 0.63 \qquad (4.57)$$

This geothermometer has been reinvestigated by a number of workers since the work of Roedder and Emslie. Putrika et al. (2007) developed an expression that takes account of the effects of melt composition and pressure:

$$T(°C) =$$

$$\frac{\{15294.6 + 1318.8P + 2.4834P^2\}}{\left\{\begin{array}{l} 8.048 + 2.83252 \ln D_{Mg}^{ol/\ell} \\ +2.097 \ln(1.5 X_{NM}^{\ell}) + \\ 2.575 \ln(3 X_{SiO_2}^{\ell}) - 1.41 NF^{\ell} \\ +0.222 H_2O^{\ell} + 0.5P \end{array}\right\}} \qquad (4.58)$$

where P is in GPa, H_2O in weight percent, $D_{MgO}^{Ol/\ell} = X_{MgO}^{Ol}/X_{MgO}^{\ell}$, NF is a network former parameter:

$$NF = \frac{7}{2} \ln(1 - X_{AlO_{1.5}}^{\ell}) + 7 \ln(1 - X_{TiO_2}^{\ell})$$

and C_{NM} is the concentration of network modifiers:

$$X_{NM}^{\ell} = X_{MgO}^{\ell} + X_{MnO}^{\ell} + X_{FeO}^{\ell} + X_{CaO}^{\ell}$$
$$+ X_{Co}^{\ell} + X_{Ni}^{\ell}$$

In these equations, mole fractions are calculated on a single metal *cationic basis*; in practice, that means that mole fractions are calculated for Al_2O_3, Fe_2O_3, Na_2O, and K_2O as $AlO_{1.5}$, $FeO_{1.4}$, $NaO_{0.5}$, and $KO_{0.5}$; all other mole fractions are calculated as usual. (See section 4.6.1 for a discussion of network formers and modifiers in silicate melts.) In practice, this means that the mole fractions for those oxides are twice what they would otherwise be. At pressures corresponding to the near surface (i.e., $P \approx 0$) and form most basaltic compositions, temperatures calculated using 4.58 converge with those of Roeder and Emslie.

Example 4.3   Calculating magma temperatures using the olivine geothermometer

From the electron microscope analysis of a mid-ocean ridge basalt glass and its coexisting olivine microphenocrysts below, calculate the temperature at which the olivine and liquid equilibrated using the Roeder and Emslie olivine-liquid geothermometer. Compare your results with the temperature calculated using the equation of Putrika et al. (2007) (eqn. 4.58), assuming a pressure of 0.1 GPa and a water content of 0.15%.

SiO_2	50.3
Al_2O_3	14.3
ΣFeO	11.1
MgO	7.8
CaO	11.5
Na_2O	2.6
K_2O	0.23
MnO	0.20
TiO_2	1.71
Total	99.02
Mol % Fo in Ol	82

Answer: We will answer this assuming the glass composition represents that of the liquid and using eqns. 4.55 and 4.56. We first have to convert the analysis of the glass from weight percentage to mole fraction.

Let's set up a spreadsheet to do these calculations. First we deal with the Fe analysis. The analysis reports only iron as FeO. Generally, about 10% of the iron in a basalt will be present as ferric iron (Fe_2O_3), so we will have to assign 10% of the total iron to Fe_2O_3. To do this, we get the weight percent FeO simply by multiplying the total FeO by 0.9. To get weight percent Fe_2O_3, we multiply total FeO (11.1%) by 0.1, then multiply by the ratio of the molecular weight of Fe_2O_3 to FeO and divide by 2 (since there are 2 Fe atoms per oxide), which works out to multiplying by 1.11.

	wt%	w/10% ferric	Mol. wt.	Moles	Mol frac.	Cationic moles	Cation mol frac
SIO_2	50.3	50.3	60.09	0.8371	0.5237	0.8371	0.470
Al_2O_3	14.3	14.3	102	0.1402	0.0877	0.280	0.157
total FeO	11.1						
FeO		9.99	71.85	0.1390	0.0870	0.1390	0.078
Fe_2O_3		1.23	157.7	0.0078	0.0049	0.0156	0.009
MgO	7.8	7.8	40.6	0.1921	0.1202	0.1921	0.108
CaO	11.5	11.5	56.08	0.2051	0.1283	0.2051	0.115
Na_2O	2.6	2.6	61.98	0.0419	0.0262	0.0839	0.047
K_2O	0.23	0.23	94.2	0.0024	0.0015	0.0049	0.003
MnO	0.2	0.2	70.94	0.0028	0.0018	0.0028	0.002
TiO_2	1.71	1.71	79.9	0.0214	0.0134	0.0214	0.012
H_2O	0.15	0.15	18	0.0083	0.0052		
Total	99.89	100.01		1.5983	1.0000	1.7823	1.0000
C-NM					0.3024		
NF					-0.6836		
XMgO-Ol	0.82	TMgO	1383	K	1110	°C	
XFeO-Ol	0.18	TFeO	1389	K	1117	°C	
P	0.1			T Putrika	1121		

Now we calculate the mole fractions. We'll set up a column with molecular weights and divide each weight percent by the molecular weight to get the number of moles per 100 grams. To convert to mole fraction, we divide the number of moles by the sum of the number of moles.

Since the mole fraction of Mg in olivine is equal to the mole fraction of forsterite, we need only convert percent to fraction (i.e., divide by 100). The mole fraction of FeO in olivine is simply $1 - X_{MgO}$. Thus, $X_{MgO(ol)} = 0.82$ and $X_{FeO(ol)} = 0.18$. Now we are ready to calculate temperatures. We can calculate two temperatures: one from MgO and the other from FeO. The temperature based on the FeO exchange is:

$$T_{FeO} = \frac{3911}{\log\left(X_{FeO}^{ol}/X_{FeO}^{\ell}\right) + 250}$$

and that based on MgO is:

$$T_{MgO} = \frac{3740}{\log\left(X_{MgO}^{ol}/X_{FeO}^{\ell}\right) + 1.87}$$

The MgO and FeO difference is small, suggesting equilibrium in this case. Since assumptions about the fraction of ferric iron affect the FeO calculation more than the MgO one, the latter is more reliable.

A slightly different approach is needed to calculate mole fractions on a "cationic basis" for the Putrika equation. In effect, it means the number of moles for Al, Fe^3, Na, and K are twice as large. We then recalculate the total and recompute mole fractions. This equation produces a slightly higher temperature. If we set P to 0, it nearly converges with the Roeder and Emslie temperature.

A widely used geothermometer in silicic igneous and metamorphic rocks is the so-called TitaniQ titanium in quartz geothermometer (Wark and Watson, 2006, Ostapenko et al., 1987). In a system in which both quartz and rutile are present, we may write an equilibrium constant expression as:

$$K = \frac{a_{TiO_2}^{Qz}}{a_{TiO_2}^{Rut}} \qquad (4.59)$$

(Note that this does not require that rutile and quartz be in contact: if both are in equilibrium with a third phase such as a silicate or aqueous fluid, they must be in equilibrium with each other.) Since rutile can be considered pure TiO_2, the denominator in 4.59 is 1 and the equilibrium constant reduces to

$$K = a_{TiO_2}^{Qz} = \gamma_{TiO_2}^{Qz} X_{TiO_2}^{Qz} \qquad (4.60)$$

Since the concentration of TiO_2 in quartz is typically quite small, $\lesssim 1000\,ppm$, Henry's law behavior can be assumed, i.e., the activity coefficient is independent of the TiO_2 concentration, but dependent on T and P, as is the equilibrium constant (Equation 4.42). The T dependence is, of course, the basis of a geothermometer. In a series of experiments, Wark and Watson (2007) demonstrated a linear correlation between the log of the TiO_2 concentration in quartz and temperature in experiments where rutile was also present. A number of subsequent experimental calibrations of this system and cross calibrations with other thermobarometers have been performed that modified the original equation of Wark and Watson somewhat, the most recent being that of Thomas et al. (2015):

$$RT \ln X_{TiO_2}^{Qz} = -a + bT - cP + RT \ln a_{TiO_2}^{fluid} \qquad (4.61)$$

where $a = 60{,}952$, $b = 1.520$, and $c = 1741$ with P in GPa (the value of c differs from that of Thomas et al. because their equation uses P in kb).

The activity of TiO_2 in the fluid (magma or aqueous fluid) is taken to be 1 in systems in equilibrium with rutile. Where rutile is not present, the activity must be estimated. One approach is to assume that activity scales with rutile solubility; in other words:

$$a_{TiO_2}^{\ell} = \frac{X_{TiO_2}^{\ell}}{X_{TiO_2}^{sat.}} \qquad (4.62)$$

where X^{sat} is the concentration at which the liquid becomes rutile saturated. Hayden and Watson (2007) and Kularatne and Audétat (2014) describe methods for this. Nevertheless, the accuracy of the Ti in quartz geothermometer has been questioned (Wilson et al., 2012). The method of Ghiorso and

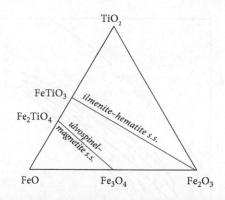

Figure 4.21   The TiO_2–FeO–Fe_2O_3 ternary system. Phases are: FeO, wüstite; Fe_2O_3, hematite; TiO_2, rutile; Fe_2TiO_4, ulvospinel; Fe_2O_4, magnetite; $FeTiO_3$, ilmenite. The system also includes the $FeTi_2O_5$–Fe_2TiO_5 solution, which is not shown.

Gualda (2013) is based on equilibrium in the iron-titanium oxide system, to which we now turn.

The iron-titanium oxide system evaluated by Buddington and Lindsley (1964) was one of the first means of obtaining quantitative estimates of crystallization temperatures of igneous rocks. It is important not only because it is useful over a wide range of temperatures and rock types, but also because it yields oxygen fugacity as well. Figure 4.21 shows the TiO_2–FeO–Fe_2O_3 (rutile–wüstite–hematite) ternary system. The geothermometer is based on the reaction:

$$yFe_2TiO_4 + (1 - y)Fe_3O_4 + \frac{1}{4}O_2 \rightleftharpoons yFe_2TiO_3$$
$$+ \left(\frac{3}{2} - y\right)Fe_2O_3 \qquad (4.63)$$

which describes equilibrium between the ulvospinel–magnetite (titanomagnetite) and ilmenite–hematite solid solution series. The equilibrium constant expression may be written as:

$$K = \frac{a_{FeTiO_3}^{y} a_{Fe_2O_3}^{3/2-y}}{a_{Fe_2TiO_4}^{y} a_{Fe_3O_4}^{1-y} f_{O_2}^{1/4}} \qquad (4.64)$$

The original Buddington and Lindsley geothermometer was based on empirical observations of compositional dependence on

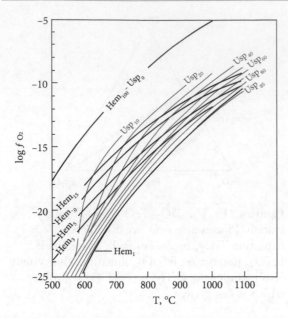

Figure 4.22 Relationship of composition of coexisting titanomagnetite and ilmenite to temperature and oxygen fugacity. Black lines show the percentage of hematite in the hematite–ilmenite phase (Hem) and red lines show the percentage of ulvospinel in the ulvospinel–magnetite phase (Usp). Intersection of two line occur at equilibrium T and fO_2 values.

oxygen fugacity and temperature, as shown in Figure 4.22. Having values for the compositions of the titanomagnetite and ilmenite phases, one simply read T and fO_2 from the graph. To understand the system from a thermodynamic perspective, it is better to consider the two fundamental reactions occurring separately in this system:

$$Fe_3O_4 + FeTiO_3 \rightleftharpoons Fe_2TiO_4 + Fe_2O_3 \quad (4.65)$$

magnetite + ilmenite $\rightleftharpoons$ ulvospinel + hematite

and $\quad 4Fe_3O_4 + O_2 \rightleftharpoons 6Fe_2O_3 \quad (4.66)$

The first reaction represents a temperature-dependent exchange between the titanomagnetite and ulvospinel solutions; the second reaction is the oxidation of magnetite to hematite.

Several investigators have studied the iron–titanium oxides attempting to improve upon the work of Buddington and Lindsley

(1964). The approach of Spencer and Lindsley (1981) was to consider two reactions 4.65 and 4.66. They modeled the ilmenite as a binary asymmetric Margules solution and titanomagnetite as a binary asymmetric Margules solution below 800°C and as an ideal binary solution above 800°C. They modeled configurational entropy-based ordering of Fe^{2+}, Fe^{3+}, and Ti^{4+} in the ilmenite lattice structure (they assumed Fe^{3+} mixed randomly with Ti^{4+} in A sites and Fe^{3+} and Fe^{2+} randomly in B sites). The ΔG of reactions above were written as:

$$-\frac{\Delta G_{ex}}{RT} = \ln \left[\frac{X_{Usp}^\alpha (1 - X_{Ilm})}{(1 - X_{Usp})^\alpha X_{ilm}^\alpha} \right] + \ln \left[\frac{\lambda_{Usp}^\alpha \lambda_{Hem}^\alpha}{\lambda_{Mt}^\alpha \lambda_{Ilm}^\alpha} \right] \quad (4.67)$$

and

$$-\frac{\Delta G_{ox}}{RT} = \ln \left[\frac{X_{-Hem}^{6-\alpha}}{X_{Mt}^{4-\alpha}} \right] + \ln \left[\frac{\lambda_{Hem}^{6-\alpha}}{\lambda_{Mt}^{4-\alpha}} \right] - \ln f_{O_2} \quad (4.68)$$

The α parameter is related to the number of sites involved in the exchange; Spencer and Lindsley assumed α was 2 for ilmenite and 1 for titanomagnetite. The excess free energy was expressed in the usual way for an asymmetric solution (eqn. 4.16):

$$\overline{G}_{ex} = (W_{G_1} X_2 + W_{G_2} X_1) X_1 X_2$$

for each solution series. When the effect of pressure is neglected, the free energy interaction parameter expression (eqn. 4.8) simplifies to:

$$W_G = W_H - TW_S \quad (4.69)$$

Values for W_H and W_S were obtained from least-squares fits to experimental data. The parameters obtained are listed in Table 4.1.

Substituting eqns. 4.69 and 4.16 into the free energy of solution expression ($\Delta G_{excess} = \Delta G_{ideal} - \Delta G_{real}$), the following equation can be obtained:

$$T(K) = \frac{\left\{ \begin{array}{c} AW_H^{Usp} - BW_H^{Mt} - CW_H^{Ilm} \\ + DW_H^{Hem} + \Delta H° \end{array} \right\}}{\left\{ \begin{array}{c} AW_S^{Usp} - BW_S^{Mt} - CW_S^{Ilm} \\ + DW_S^{Hem} + \Delta S° + R \ln K^{exch} \end{array} \right\}} \quad (4.70)$$

Table 4.1 Margules parameters for ilmenite and titanomagnetite solid solutions.

	Usp (<800°C)	Mag (<800°C)	Ilm	Hem
W_H (joules)	64835	20798	102374	36818
W_S (joules)	60.296	19.652	71.095	7.7714
W_G (>800°C) (joules)	0	0		
ΔS^o_{Usp} (joules/K)	4.192			
ΔH^o_{usp} (joules)	27799			

From Spencer and Lindsley (1981).

Oxygen fugacity is determined as:

$$\log f_{O_2} = \log MH$$

$$+ \left(\begin{array}{l} 12\ln(1 - X_{ilm}) - 4\ln(1 - X_{Usp}) + \\ \frac{1}{RT} \left[\begin{array}{l} 8X^2_{Usp}(X_{Usp} - 1)W^{Usp}_G \\ +4X^2_{Usp}(1 - 2X_{Usp})W^{Mt}_G + \\ 12X^2_{Ilm}(1 - X_{Ilm})W^{Ilm}_G \\ -6X^2_{Ilm}(1 - 2X_{Ilm})W^{Hem}_G \end{array} \right] \end{array} \right) / 2.303$$

$$(4.71)$$

where:

$$A = 2X^2_{Usp} - 4X_{Usp} + 1 \quad B = 3X^2_{Usp} - 2X_{Usp}$$

$$C = 3X^2_{Ilm} - 4X_{Ilm} + 1 \quad D = 3X^2_{Ilm} - 2X_{Ilm}$$

$$K^{exch} = \frac{X_{Usp}X^2_{Hem}}{X_{Mt}X^2_{Ilm}} \quad K^{exch} = \frac{X_{Usp}X^2_{Hem}}{X_{Mt}X^2_{Ilm}}, \quad \Delta H^o =$$

27.799 kJ/mol, $\Delta S^o = 4.1920$ J/K-mol and MH is the magnetite–hematite buffer (see Example 4.4): $\log MH = 13.966 - 24634/T$.

Example 4.4 Using the iron–titanium oxide geothermometer

An electron microprobe analysis of oxide phases in an andesite reveals that there is 68 mole percent of ulvospinel in an ulvospinel–magnetite phase, and 93.3% of ilmenite in an ilmenite–hematite phase. Calculate the temperature and fO_2 at which these phases equilibrated.

Answer: We can use eqns. 4.70 and 4.71 to answer this question. The data in Table 4.1 are relevant to the binary asymmetric solution model for the system below 800°C. Above 800°C, an ideal solution is assumed for the ulvospinel–magnetite phase, so the interaction parameters for this phase go to 0. But if we don't know the temperature, how do we know which equation to use? We begin by computing temperature using the parameters for less than 800°C. If the temperature computed in this way is greater than 800°C (1073 K), we set the W_H and W_S for ulvospinel and magnetite to 0 and recompute.

	X_{Usp}	X_{Ilm}		
	0.68	0.933		
ΔH		27799		
ΔS		4.192		
R		8.314		
Interaction Parameters				
WHU		64835	WSU	60.296
WHM		20798	WSM	19.652
WHI		102374	WSI	71.095
WHH		36818	WSH	7.7714
A		−0.3328		
B		0.0272		
C		−0.12053		
D		0.745467		
K		0.010958		

	X_{Usp}	X_{IIm}		
T=	$\underline{(A*WHU-B*WHM-C*WHI+D*WHH+\Delta H)}$			
	$(A*WSU-B*WSM-C*WSI+D*WSH+\Delta S-R*\ln(K))$			
T (<800)		1281 K		1008 °C
T (>800)		1250 K		932 °C
	WG = WH-T*WS			
	WGU	−7829.52	WGI	16695.29
	WGM	−2885.21	WGH	27452.45
			MH	−6.47
		LogfO_2 (<800)		−12.58
		Logf_2O_2 (>800)		−12.69

Once we have temperature, we can compute the W_G terms using the relationship $W_G = W_H - TW_S$, bearing in mind that $W_{Gusp} = W_{GMt} = 0$ if the temperature is greater than 800°C. With these values in hand, we can use eqn. 4.71 to calculate the fO_2. Our spreadsheet is shown here.

These data were taken from one of Spencer and Lindsley's (1981) experiments, performed at 938°C and log $fO_2 = -12.76$. Our calculations are in good agreement with the experimental observation.

A number of subsequent studies have further refined the iron–titanium oxide geothermometer based on new experimental data and more sophisticated thermodynamic models. The most recent of these efforts is that of Ghiorso and Evans (2008), whose formulation is too complex to describe here. Software for computing temperature and oxygen fugacity based on their formulation is available at www.ofm-research.org/publications.html.

We have reviewed just a few of the available thermobarometers in use. These were selected to illustrate the underlying principles. There are, however, many thermobarometers in use by geochemists and petrologists. Some of these are listed in Table 4.2.

4.6 THERMODYNAMIC MODELS OF MAGMAS

Magmas have played an extremely important role in the development of the Earth, as well as other bodies in the solar system. As we shall see, the Earth's core formed as iron melted and sank to the center, and the crust formed as melts from the mantle rose to the surface and cooled. Thus, an understanding of igneous processes is an essential part of earth science. Until a few decades ago, the primary approaches to igneous petrology were observational and experimental. Results of melting experiments in the laboratory were used to interpret observations of igneous rocks. This approach proved highly successful and is responsible for most of our understanding of igneous processes. However, such an approach has inherent limitations: virtually every magma is unique in its composition and crystallization history. Yet the experimental database is limited: it is not practical to subject every igneous rock to melting experiments in the laboratory. Realizing this, igneous petrologists and geochemists turned to thermodynamic models of silicate melts as a tool to interpret their evolution. With a proper "model" of the interaction of various components in silicate melts and adequate thermodynamic data, it should be possible to predict the equilibrium state of any magma under any given set of conditions. The obstacles in developing proper thermodynamic models of silicate liquids, however, have been formidable. Because they are stable only at high temperatures, obtaining basic thermodynamic data on silicate liquids is difficult. Furthermore, silicate liquids are very complex solutions, with eight or more elements present in high enough concentrations to affect the properties of the solution. Nevertheless, sufficient progress has been made on these problems that thermodynamics is now an important tool of igneous petrology.

Table 4.2   Commonly used thermobarometers.

Reaction	Type	Reference
Amphibole ⇌ plagioclase	Displaced equilibria	Holland and Blundy (1994)

$$NaCa_2Mg_5AlSi_7O_{22}(OH)_2 + 4SiO_2 \rightleftharpoons Ca_2Mg_5Si_8O_{22}(OH)_2 + NaAlSi_3O_8$$
$$NaCa_2Mg_5AlSi_7O_{22}(OH)_2 + NaAlSi_3O_8 \rightleftharpoons Na(NaCa)Mg_5Si_8O_{22}(OH)_2 + CaAl_2Si_2O_8$$

Calcite ⇌ dolomite	Solvus equilibria	Goldsmith and Newton (1969)

$$CaCO_3 \rightleftharpoons (Ca, Mg)CO_3$$

Garnet ⇌ biotite Fe – Mg	Exchange	Ferry and Spear (1978)

$$(Fe, Mg)_3Al_2Si_3O_{12} \rightleftharpoons K(Mg, Fe)AlSi_3O_{10}(OH)_2$$

Garnet + qtz ⇌ plag. + wollastonite	Displaced equilibria	Gasparik (1984b)

$$(Fe, Ca)_3Al_2Si_3O_{12} + SiO_2 \rightleftharpoons (Ca, Na)Al_2Si_2O_8 + 2CaSiO_3$$

Hercynite + qtz ⇌ garnet + sillimanite	Displaced equilibria	Bohlen et al. (1986)

$$FeAl_2O_4 + 5SiO_2 \rightleftharpoons Fe_3Al_2Si_3O_{12} + Al_2SiO_5$$

Ilmenite + Al₂SiO₅ ⇌ garnet + rutile + qtz	Displaced equilibria	Bohlen et al. (1983)

$$3FeTiO_3 + Al_2SiO_5 \rightleftharpoons 3TiO_2 + (Fe, Ca)_3Al_2Si_3O_{12} + SiO_2$$

Plagioclase ⇌ liquid	Exchange	Kudo and Weill (1970), Putirka (2005)

$$CaAl_2Si_2O_8 + NaO_{0.5}{}^{liq} + 2SiO_2{}^{liq} \rightleftharpoons NaAlSi_3O_8 + CaO^{liq} + 2AlO_{1.5}{}^{liq}$$

Plagioclase ⇌ garnet + kyanite + quartz	Displaced equilibria	Ghent (1976), Koziol and Newton (1988)

$$3(Ca, Na)Al_2Si_2O_8 \rightleftharpoons (Fe, Ca)_3Al_2Si_3O_{12} + 2Al_2SiO_5 + SiO_2$$

Ti in zircon–Zr in rutile	Exchange	Ferry and Watson (2007)

$$ZrSiO_4 + TiO_2 \rightleftharpoons ZrTiO_4$$
$$ZrSiO_4 + TiO_2 \rightleftharpoons ZrO_2 \text{ (in rutile)} + SiO_2$$

4.6.1   Structure of silicate melts

As was the case for silicate solids and electrolyte solutions, application of thermodynamics to silicate liquids requires some understanding of the interactions that occur on the atomic level. Thus, we will once again have to consider the microscopic viewpoint before developing a useful thermodynamic approach. In this section, we briefly consider the nature of silicate melts on the atomic level.

Most, though not all, of our knowledge of the structure has come from studies of glasses rather than melts. While the thermodynamic properties of silicate liquids and their respective glasses differ, other studies have confirmed the general structural similarities

of glasses and liquids. Spectral studies of glasses, which in some respects can be viewed as supercooled liquids, have revealed that silicate liquids have structures rather similar to those of silicate solids. In fact, the principal difference between silicate liquids and solids is the absence of long-range ordering in the former; short-range ordering is similar. As in silicate minerals, the primary structural element of silicate liquids is the silicon tetrahedron (Figure 1.11), consisting of a silicon atom surrounded by four oxygens. As in silicate minerals, tetrahedra may be linked by a shared oxygen, called a *bridging* oxygen; not surprisingly, unshared oxygens are termed *nonbridging* (Figure 4.23a). Unlinked

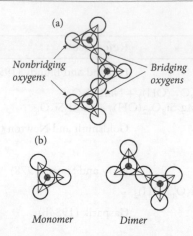

Figure 4.23 Silicate structures. (a) Short-range silicate structures in melts resemble those in solids. Individual tetrahedra may be linked by bridging oxygens and linked to two silicon atoms. (b) Units in silicate melts may include monomers, with no bridging oxygens, and dimers, where only one of four oxygens in each tetrahedra are "bridging."

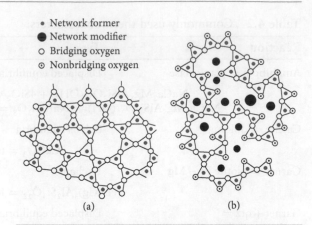

Figure 4.24 (a) Structure of pure silica glass and (b) a silica-rich glass with additional component ions.

silica tetrahedra, that is, those with no bridging oxygens, are termed monomers, SiO_4^{4-} (Figure 4.23b). Two tetrahedra linked by a single oxygen are termed *dimers* and have the formula $Si_2O_7^{6-}$. Tetrahedra may also be linked by two oxygens to form infinite chains; these have a chemical formula of SiO_3^{2-}. In silicates such as quartz and feldspar, the tetrahedra are all linked into a framework, and all oxygens are shared. All these structural elements can be present in silicate glasses.

The degree to which the silica tetrahedra are linked, or *polymerized*, in silicate liquids affects chemical and physical properties. The degree of polymerization in turn depends on other cations present. These may be divided into two groups, *network formers* and *network modifiers*. Relatively small, highly charged cations such Al^{3+} and Fe^{3+} (more rarely, Ti^{4+}, P^{3+}, and B^{3+} also) often substitute for silicon in tetrahedral sites and, along with Si, are termed *network formers*. The other common cations of natural silicate liquids, Ca^{2+}, Mg^{2+}, K^+, Na^+, and H^+, are network modifiers. These ions cannot substitute for silicon in tetrahedra and their positive charges can only be balanced by nonbridging oxygens. Addition of these ions disrupts the linkages between silica tetrahedra. Thus, as

silicate melts become richer in these network modifiers they become progressively depolymerized. This is illustrated in Figure 4.24, which compares the structure of pure silica glass (liquid) and a silica-rich glass (liquid). Melt structure in turn affects the physiochemical properties of the melt. For example, SiO_2-rich melts tend to have low densities and high viscosities. As ions such as MgO or CaO are added to the melt, viscosity decreases and density increases as the polymer structure is disrupted.

4.6.2 Magma solution models

Advances on several fronts have moved thermodynamic modeling of magmas from an academic curiosity to a useful petrological tool. First, spectroscopic (mainly Raman and infrared spectroscopy, both of which are sensitive to atomic and molecular vibrations) studies are revealing the structure of silicate melts, which provides the theoretical basis for thermodynamic models. Second, more sophisticated thermodynamic models more accurately reflect interactions in silicate melts. Third, the thermodynamic database has become more complete and more accurate. Finally, the wide accessibility and power of computers and appropriate programs have made the extensive matrix calculations involved in these models possible.

Several factors complicate the task of thermodynamic modeling of magmas. First, magmas are solutions of many components (typically eight or more). Second, the solids crystallizing from magmas are themselves solutions. Third, magmas crystallize over a substantial temperature range (as much as 400–500°C, and more in exceptional cases). Furthermore, crystallization may occur over a range of pressures as a magma ascends through the Earth, and crystallization may be accompanied by melting and assimilation of the surrounding country rock. Despite these complications, several models that are sufficiently accurate to be useful to petrologists have been published, most notably those of Ghiorso (Ghiorso et al., 1983; Ghiorso and Sack, 1995) and Nielsen and Dungan (1983). The goal of these models is to describe the phases and their proportions in equilibrium with a magma, and the resulting evolution of liquid composition. In particular, the models of Ghiorso and colleagues are applicable to both melting and crystallization. In the section below, we briefly consider the model of Ghiorso.

4.6.2.1 *The regular solution model of Ghiorso and others: "MELTS"*

Ghiorso and colleagues (e.g., Ghiorso et al., 1983; Ghiorso and Sack, 1995, Ghiorso et al., 2002; Ghiorso and Gualda, 2015)) noted that silicate liquids have substantial compositional regions in which immiscibility occurs and therefore argued that the simplest model that might be able to describe them is the regular solution model. As we saw earlier in the chapter, regular solution models attempt to describe excess functions with interaction, or Margules, parameters. The Margules equation for excess Gibbs free energy for many components is:

$$\overline{G}_{ex} = \frac{1}{2} \sum_i \sum_{j,j \neq i} X_i X_j W_G^{i,j} \qquad (4.72)^*$$

and the Gibbs free energy is:

$$\overline{G} = \sum_i X_i \mu_i^o + RT \sum_i X_i \ln X_i$$

$$+ \frac{1}{2} \sum_i \sum_{j,j \neq i} X_i X_j W_G^{i,j} \qquad (4.73)^\dagger$$

The chemical potentials of individual components are:

$$\mu_i = \mu_i^o + RT \sum_i X_i \ln X_i + \sum_{j,j \neq i} X_j W_G^{i,j}$$

$$- \frac{1}{2} \sum_{j,j \neq k} \sum_{k,k \neq j} X_i X_k W_G^{j,k} \qquad (4.74)$$

and the activity coefficients are:

$$RT \ln \lambda_i = \sum_{j,j \neq i} X_j W_G^{i,j} - \frac{1}{2} \sum_{j,j \neq k} \sum_{k,k \neq j} X_i X_k W_G^{j,k} \qquad (4.75)$$

Having chosen a general form for the solution model, the next step is to select the components. For practical reasons, Ghiorso et al. (1983) placed all components on an eight-oxygen basis. Ghiorso and Sack (1995) chose liquid components that were "mineral-like": SiO_2, TiO_2, Al_2O_3, Fe_2O_3, $MgCr_2O_4$, Fe_2SiO_4, Mg_2SiO_4, $CaSiO_3$, $KAlSiO_4$, and so on, and H_2O. For components of solid phases, they chose pure end-member *phase* components (e.g., $MgSiO_3$ in orthopyroxene). The problem with this approach is that the concentrations of these components varied greatly; for example, the mole fraction of SiO_2 is typically 0.4 in basaltic magmas, whereas that of Mg_2SiO_4 is typically less than 0.1 and that of $KAlSiO_4$ is typically less than 0.05. We can see from eqn. 4.60 that when X_i is small, the contribution of the interaction parameters for this component, $W_G^{i,j}$, to the free energy will also be small. Consequently, in the most recent version of this model, called *p*MELTS, Ghiorso et al. (2002) redefined the liquid components so that their mole fractions were more similar, e.g., $SiO_2 \rightarrow Si_4O_8$, $Na_2SiO_3 \rightarrow NaSi_{0.5}O_{1.5}$.

The next task is to find values for the interaction parameters. These can be calculated from solid–liquid equilibria experiments. The principle involved is an extension of that which we used in constructing phase diagrams: when a solid and liquid are in equilibrium, the chemical potential of each component in each phase must be equal. Since thermodynamic properties of the solids involved are available (determined using standard thermodynamics techniques), the thermodynamic properties of the coexisting

---

* The ½ term arises because the sum contains both $X_i X_j W_G^{ij}$ and $X_i X_j W_G^{ij}$ terms and $W_G^{ij} = W_G^{ji}$.

† For clarity, we have simplified Ghiorso's equation by neglecting H_2O, which they treated separately.

liquid may be calculated.

$$\phi_p \rightleftharpoons \sum_i v_{p,i} c_i \qquad (4.76)$$

where φ_p is the pth end-member component of phase φ, c_i refers to the formula for the ith component in the liquid, and $v_{p,i}$ refers to the stoichiometric coefficient of this component. Thus, for reaction of olivine with the liquid, we have two versions of eqn. 4.72:

$$(Mg_2SiO_4)_{Ol} \rightleftharpoons 2MgO_\ell + SiO_{2-\ell} \qquad (4.77a)$$

and

$$(Fe_2SiO_4)_{Ol} \rightleftharpoons 2FeO_\ell + SiO_{2-\ell} \qquad (4.77b)$$

We can express the Gibbs free energy change for each of these reactions as:

$$\Delta \overline{G}_r = \Delta \overline{G}_\phi^o + RT \sum v_{p,i} a_i^\ell + RT \ln a_{\varphi,p} \qquad (4.78)$$

where a_i^ℓ is the activity of the oxide component in the liquid and φ_p refers to phase component p in phase φ. $\Delta \overline{G}_r$ is, of course, 0 at equilibrium. For example, for reaction 4.77a above, we have:

$$\Delta \overline{G}_r = 0 = \Delta \overline{G}_{Fo}^o + RT[2 \ln a_{MgO}^\ell + \ln a_{SiO_2}^\ell] + RT \ln a_{Fo} \qquad (4.79)$$

where the subscript Fo refers to the forsterite (Mg_2SiO_4) component in olivine and the superscript ℓ refers to the liquid phase. Expanding the liquid activity term, we have:

$$0 = \Delta \overline{G}_{\varphi,p}^o + RT \sum_i v_{p,i} \ln X_i^\ell$$
$$+ RT \sum_i v_{p,i} \ln \lambda_i^\ell - RT \ln a_{\varphi,p} \qquad (4.80)$$

Substituting eqn. 4.75 for the activity coefficient term in eqn. 4.80 and rearranging to place the "knowns" on the left-hand side, we have:

$$-\Delta \overline{G}_{\varphi,p}^o + RT \ln a_{\varphi,p} - RT \sum_i v_{p,i} \ln X_i^\ell$$
$$= \sum_i v_{p,i} \sum_{j,j \neq i} X_j W_G^{i,j} - \frac{1}{2} \sum_{j,j \neq k} \sum_{k,k \neq j} X_j X_k W_G^{j,k} \qquad (4.81)$$

The quantities on the left-hand side of the equation are terms that can be calculated from the compositions of coexisting solids and liquids and solution models of the solids. The right-hand side contains the unknowns. One statement of eqn. 4.81 can be written for each component in each solid phase at a given temperature and pressure. With enough experiments, values for the interaction parameters can be extracted from the phase relations. Ghiorso et al. (1983) and Ghiorso and Sack (1995) used a statistical technique called least squares* to determine the interaction parameters from a large amount of published experimental data. Ghiorso and Sack (1995) also noted that the absence of a phase in an experiment provides thermodynamic information about that phase, in that its free energy must be higher than that of the phases that are present. Their approach made use of this information as well, though discussion of that aspect of their method would take us too far afield. In all, Ghiorso and Sack (1995) used data from 1593 published laboratory experiments. In constructing the pMELTS model, Ghiorso et al. (2002) used mineral–liquid equilibrium constraints derived from published results of 2439 different laboratory experiments.

One of the goals of pMELTS was to improve the thermodynamic predictions at higher pressures. Since many melting reactions involve significant volume changes, this required an improved equation of state for the liquid, that is, an improved description of the relationship between volume and pressure. Ghiroso et al. (2002) chose a third-order Birch-Murnaghan

---

* Least squares is a numerical technique that attempts to minimize the square of the difference between calculated and observed values of some parameter. The square is taken to give greater weight to large deviations. Thus, least squares techniques yield results where there are relatively few large deviations between the calculated and observed value of the parameter of interest. We discuss this technique further in Chapter 8.

equation:

$$P = \frac{3}{2}K\left[\left(\frac{V^\circ}{V}\right)^{\frac{1}{3}} - \left(\frac{V^\circ}{V}\right)^{\frac{5}{3}}\right]$$

$$\left\{1 = \frac{3}{4}(4 - K')\left[\left(\frac{V^\circ}{C}\right)^{\frac{3}{2}} - 1\right]\right\} \quad (4.82)$$

where K is the bulk modulus. New experimental data on density derived from new shockwave experiments and new experimental determinations of silicate liquid density (by suspending olivine crystals in the liquid and observing if they sink or float) were used to constrain the K' parameter. A new equation of state for water was also incorporated into pMELTS, and Ghiorso and Gualda (2015) developed algorithms for dealing with H_2O and CO_2 saturated silica-rich melts.

With values for the interaction parameters, the model can then be used to predict the assemblage of solids, their compositions, and the liquid composition that will be present in the system as a function of temperature and pressure. The equilibrium condition for a magma, as for any other system, is the condition where Gibbs free energy is at a minimum. Thus, the problem becomes finding compositions for the liquid and coexisting solids that minimize G at a particular temperature and pressure. In other words, we want to find values of G_ℓ and $G_{\varphi 1}$, $G_{\varphi 2}$, ... $G_{\varphi n}$ such that G_{sys} is minimal where:

$$G_{sys} = G_\ell + \sum_\varphi G_\varphi \quad (4.83)$$

Inherent in the problem is finding which solids will be in equilibrium with the liquid for a given bulk system composition at specified temperature and pressure. In Ghiorso's approach, an initial guess is made of the state of the system. This is done by taking the liquid composition as equal to the system composition and estimating what phases are likely to be in equilibrium with this liquid. Then G is expanded as a three-term Taylor Series* about that initial point, N', where N' is the composite vector containing the mole fractions describing the compositions of all phases in the system. The second term in the expansion is the matrix of first derivatives of G with respect to n_i, the moles of component i, which is simply the matrix of the chemical potentials. A minimum of G occurs where the first derivative is 0. Thus, the second term is set to 0 and the solution sought by successive iterations. After each iteration, N' is reset to the composition found in the most recent iteration. This approach clearly involves repetitive matrix calculations and would not be practical without a computer, but they can easily be performed on the current generation of computers.

The goal of a thermodynamic model such as MELTS is to predict both the composition of the melt and composition of coexisting solid phases if temperature, pressure and the composition of the system can be specified. Thus, such a program should be able to predict the composition of the melt generated in a region undergoing melting and how the composition of that melt evolves as it rises and cools nearer the surface. Figure 4.25 compares the predictions of the 1995 and 2002 versions of the model with experimentally determined compositions of the liquid produced by melting peridotite at 1 GPa. The agreement between the model and experimental observation is clearly improved in pMELTS, but it is also clear that the predictions still do not agree perfectly with observations.

Figure 4.26 compares the compositions of clinopyroxene crystals found in basaltic lavas of Cameroon Line volcanoes with the compositions predicted by pMELTS to crystallize from these magmas. Diamonds and circles are *megacrysts*, and are likely to have crystallized from the magmas. Stars are pyroxenes in ultamafic xenoliths, which are more likely pieces of mantle accidentally incorporated into the magma. The kink in the lower link

---

* A Taylor series expansion of a function $f(z)$ in the vicinity of some point z = a has the form:

$$f(z) = f(a) + \frac{(z-a)}{1!}f'(a) + \frac{(z-a)^2}{2!}f''(a) + \dots$$

where f' and f'' are the first and second derivatives of f with respect to z.

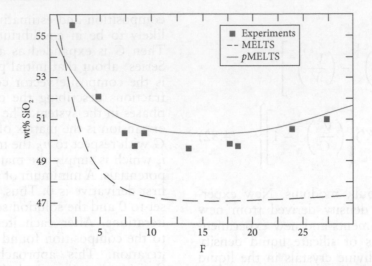

Figure 4.25 SiO$_2$ concentrations in a melt produced by melting of peridotite at 1 GPa as a function of F, the percent fraction of melt in the system. Figure compares the predictions of the earlier version of the MELTS model, the newer version, pMELTS, and experimentally determined composition. After Ghiorso et al. (2002).

reflects the onset of garnet crystallization. The agreement is not perfect but this diagram nevertheless shows the enormous value of this thermodynamic approach in igneous petrology. In this example, Rankenburg et al. (2004) were able to estimate the pressure and temperature of crystallization as 1400°C and 1.7–2.3 GPa. These pressures correspond to depths greater than the thickness of the crust in this area, hence the authors concluded the pyroxene megacrysts must have crystallized in the mantle. Future refinements of MELTS will undoubtedly close the gap between predictions and observations and enhance the value of this tool.

The latest versions of the model, rhyolite-MELTS and pMELTS, run on UNIX-based computers (including Mac OS X) and are available online at http://melts.ofm-research .org/index.html. This website also has an online MELTS calculator available. The website also has an Excel™ workbook that provides a graphical user interface to run MELTS (works only on Microsoft Windows™ and requires an internet connection). Another version of MELTS, alphaMELTS (Smith and Asimow, 2005), is available from https://magmasource.caltech.edu/alphamelts/ and can run on both Mac OS X and Windows (the later requires a virtual machine such as Linux Virtual Box or Perl). Magma Chamber Simulator (MCS) is an Excel virtual basic

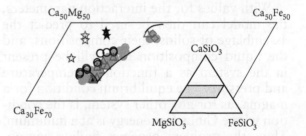

Figure 4.26 Compositions of pyroxenes found in lavas from the Cameroon Line. Diamonds, circles, and triangles are *megacrysts* and likely to have crystallized from these lavas. Stars are pyroxenes in *xenoliths* accidentally incorporated in the lavas. Lines show the compositions of the pyroxenes predicted by pMELTS to crystallize from these magmas as they cool and evolve. After Rankenburg et al. (2004).

program that uses MELTS to also simulate wallrock assimilation and magma recharge, including trace elements and isotope ratios (Bohrson et al., 2014). MCS is available at https://mcs.geol.ucsb.edu.

4.7 REPRISE: THERMODYNAMICS OF ELECTROLYTE SOLUTIONS

We discussed the nature of electrolyte solutions and introduced one approach to dealing with their nonideality; namely, the Debye–Hückel activity coefficients, in Section 3.7. We also noted a number of theoretical weaknesses in the Debye–Hückel approach and that this approach is restricted to fairly dilute solutions (ionic strengths less than 0.1 M). While most surface waters fall within this range and can be treated with this approach, many geochemically important solutions, including magmatic and metamorphic fluids, hydrothermal fluids, formation brines, and ore forming fluids are far more concentrated. In this section we will return to the problem of electrolyte solutions, examine the causes of nonideal behavior in high ionic strength solutions in more detail and introduce approaches for dealing with such fluids. Many of these fluids exist at temperatures and pressures far above the standard state of 298K and 0.1 MP; consequently, we will begin by reviewing the behavior at water under those conditions.

4.7.1   Equation of state for water

Water in the Earth's interior plays an important role in many important processes, such as metamorphism, ore formation, volcanic eruptions, etc. These processes occur over a large range of temperature and pressure. For water at the surface of the Earth, including shallow ground water, the effect of temperature and pressure on volume can often be neglected or calculated with the Kwong-Redlich or Virial equations. Volume changes at more extreme temperature and pressure are significant and cannot be neglected. They pose a particularly challenging problem because of the highly nonideal nature of water. Its equation of state is a consequently complex function of T and P. Since other thermodynamic properties are related to volume (e.g., $(\partial S/\partial P)_T = \alpha V$), they are similarly complex functions. This is illustrated in the phase diagrams for water in Figure 4.27. At 374.2°C and 22 MPa, pure water reaches a critical point. At the critical point:

$$\left(\frac{\partial V}{\partial P}\right)_T = 0 \quad \text{and} \quad \left(\frac{\partial V}{\partial T}\right)_P = 0$$

The second derivatives of volume with respect to T and P are also 0 (i.e., the critical point in an inflection point). Furthermore, the isothermal compressibility becomes infinite

at the critical point. Above this temperature and pressure, water becomes a supercritical fluid with a compressibility greater than that of liquid water but less than that of vapor. The T and P of the critical point is a strong function of composition, for example in a saturated NaCl solution it is ~600° C and ~390 MPa. CO_2 is similarly nonideal with similar behavior with a critical point at 31.1° C and 7.4 MPa.

Water exhibits other curious behaviors as well. It reaches maximum density at 4°C (277 K), and density decreases above and below this temperature, including when supercooled below 0°C. Experiments with supercooled water show that the coefficient of thermal expansion, α, and isothermal compressibility, β, extrapolate to $-\infty$ and $+\infty$, respectively at 228 K (however, it has not been possible to supercool liquid water to this temperature). This indicates behavior that can be described with an equation of the form:

$$\Psi = A\left(\frac{1}{T - T_s}\right)^\lambda \qquad (4.84)$$

where Ψ is some thermodynamic function, A and λ are constants, and T_s is this "singular temperature" of 228 K. Temperature and pressure dependencies of other thermodynamic properties can be fitted to similar equations. This suggests that the singular temperature of 228 K is an important transition point in the structure and behavior of water, presumably related to hydrogen bonding of individual water molecules.

4.7.2   Activities and mean ionic and single ion quantities

Consider an aqueous NaCl solution. In Chapter 3 we saw that the thermodynamic properties of a salt are related to those of its component ions by:

$$\Psi_{AB} \equiv \nu_A\Psi_A + \nu_B\Psi_B \qquad (3.73)$$

where Ψ is any thermodynamic property and ν is a stoichiometric coefficient. So, for example, the chemical potential of NaCl in solution is:

$$\mu_{NaCl} = \mu_{Na^+} + \mu_{Cl^-}$$

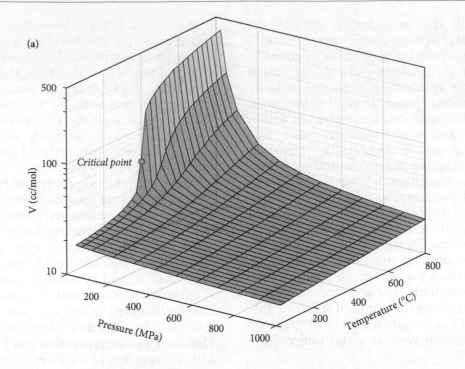

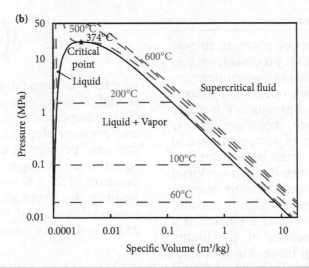

Figure 4.27 (a) Three-dimensional $P–T–V$ phase diagram for liquid and supercritical pure water based primarily on the data of Helgeson and Kirkham (1974a). (b) $P–V$ phase diagram for water illustrating the liquid, two-phase, and supercritical regions.

which we can express as:

$$\mu_{NaCl} = \mu^o_{Na^+} + \mu^o_{Cl^-} + RT(\ln a_{Na^+} + \ln a_{Cl^-}) \tag{4.85}$$

or

$$\mu_{NaCl} = \mu^o_{Na^+} + \mu^o_{Cl^-}$$
$$+ RT(\ln m_{Na^+} + \ln m_{Cl^-} + \ln \gamma_{Na^+} + \ln \gamma_{Cl^-})$$

4.7.2.1   Relationship between activity and molality of a salt

Let's consider the relationship between activity and molality of a salt in an electrolyte solution such as a NaCl solution. Figure 4.28a illustrates this relationship. What we immediately notice is that the slope in the Henry's law region is essentially zero, which is not at all what we expect for Henry's law behavior.

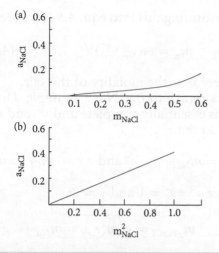

Figure 4.28 (a) Relationship between activity and molality of NaCl in aqueous solution. The activity is very low and the "Henry's law slope" is almost 0 at low concentrations. (b) Relationship between activity and the square of molality of NaCl in aqueous solution.

It can easily be shown that the relationship in Figure 4.28a is a simple consequence of the dissociation of NaCl into Na^+ and Cl^- ions. From eqn. 3.73 we have:

$$\mu_{NaCl} = \mu_{Na^+_{aq}} + \mu_{Cl^-_{aq}} \qquad (4.86)$$

Substituting this into eqn. 3.46, we obtain:

$$\mu_{NaCl} = \mu^o_{Na^+} + \mu^o_{Cl^-} + RT \ln a_{Na^+} + RT \ln a_{Cl^-}$$

In the reference state of infinitely dilute solution, $m_i = a_i$, so that:

$$\mu_{NaCl} = \mu^o_{Na^+} + \mu^o_{Cl^-} + RT \ln m_{Na^+} + RT \ln m_{Cl^-} \qquad (4.87)$$

Furthermore, charge balance requires that:

$$m_{Na^+} = m_{Cl^-} = m_{NaCl} \qquad (4.88)$$

Substituting eqn. 4.88 into 4.87 and rearranging:

$$\mu_{NaCl} = \mu^o_{Na^+} + \mu^o_{Cl^-} + 2RT \ln m_{NaCl} = \mu^o_{Na^+} + \mu^o_{Cl^-} + RT \ln m^2_{NaCl} \qquad (4.89)$$

Comparing this equation with eqn. 3.48, we see that:

$$a_{NaCl} \propto m^2_{NaCl}$$

When we plot activity versus the square of molality, we obtain a linear relationship (Figure 4.27b).

Generalizing this result for dissociation of a substance into a positive ion A and negative ion B, such as:

$$A_{v^+} B_{v^-} \rightleftharpoons v^+ A^+ + v^- B^-$$

then the relationship between activity of a salt and its molality is:

$$\boxed{a_{AB} \propto m^v_{AB}} \qquad (4.90)$$

For example, v is 3 for $CaCl_2$, 4 for $FeCl_3$, and so on. Recall that by convention the standard state is 1 molal solution where activity equals molality (and therefore $\gamma = 1$) and hence in the standard state:

$$a^o_{AB} = m_{AB} \qquad (4.91)$$

4.7.2.2 Mean ionic quantities

Although we can certainly determine the concentrations of Na and Cl in solution, how do we independently determine their activity coefficients? Since we cannot create a pure Na^+ solution or a pure Cl^- one, we cannot say what part of the nonideality of NaCl solution is due to Na^+ and what part is due to Cl^-. A practical solution then is to assign all nonideality equally to both ions. This leads to the concept of the *mean ion activity coefficient*:

$$\gamma_\pm = (\gamma_{Na^+}\gamma_{Cl^-})^{1/2} \qquad (4.92)$$

Thus, the mean activity coefficient of a salt is the geometric mean of the activity coefficients of its component ions. Equation 4.85 then becomes:

$$\mu_{NaCl} = \mu^o_{Na^+} + \mu^o_{Cl^-}$$
$$+ RT(\ln m_{Na^+} + \ln m_{Cl^-} + \ln \gamma^2_\pm)$$

Equation 4.92 is valid for 1:1 salts (i.e., 1 cation for each anion). A general expression for the mean activity coefficient of a salt of composition $A_v + B_v -$ is:

$$\boxed{\gamma_\pm = (\gamma^{v^+}_+ \gamma^{v^-}_-)^{1/v}} \qquad (4.93)$$

where v is the sum of the component positive and negative ions:

$$v = v^+ + v^- \qquad (4.94)$$

Mean activity coefficients have the advantage that they are readily measurable (through electrochemical means or solubility, for example).

We can extend the concept of mean ionic quantities to other thermodynamic variables as well. The *mean ionic potential*, $\mu_\pm$, is defined as:

$$\mu_\pm = \frac{v^+ \mu_+ + v^- \mu_-}{v} \qquad (4.95)$$

Thus, the mean ionic potential is simply the arithmetic mean of the potential of the individual ions weighted by their stoichiometric coefficients. We could also express the mean ionic potential as:

$$\mu_\pm = \mu_\pm^o + \frac{RT(\ln a_+^{v^+} + \ln a_-^{v^-})}{v} \qquad (4.96)$$

Rearranging once more, we obtain:

$$\mu_\pm = \mu_\pm^o + RT(\ln a_+^{v^+} a_-^{v^-})^{1/v} \qquad (4.97)$$

Comparing this relationship with eqn. 4.93, we define a *mean ionic activity* such that:

$$a_\pm = (a_+^{v^+} a_-^{v^-})^{1/v} \qquad (4.98)$$

We can also define mean ionic molalities such that $a_\pm = \gamma_\pm m_\pm$. Substituting $a_- = \gamma_- m_-$, and $a_+ = \gamma_+ m_+$, we find the *mean ionic molality* is then:

$$m_\pm = (m_+^{v^+} m_-^{v^-})^{1/v} \qquad (4.99)$$

Mass balance requires that:

$$m^+ = v^+ m \qquad \text{and} \qquad m^- = v^- m \qquad (4.100)$$

Substituting this into eqn. 4.93, we see that:

$$m_\pm = m(v_+^{v^+} v_-^{v^-})^{1/v} \qquad (4.101)$$

where m is the molality of the salt.

Let's return to our NaCl example. Dissociation is essentially complete and v^+ and v^- are unity, so that:

$$m_{Na^+} = m_{NaCl} \qquad \text{and} \qquad m_{Cl^-} = m_{NaCl}$$

Since $v^+ = v^- = 1$ and $v = 2$:

$$m_{\pm NaCl} = \sqrt{m_{NaCl}^2} = m_{NaCl}$$

Mean ionic molality is simply equal to molality for a completely dissociated salt consisting of monovalent ions such as NaCl. In general for strong electrolytes (salts that completely dissociate), the mean activity coefficient and mean activity of the salt are related to its activity coefficient and activity by:

$$\gamma = \gamma_\pm^v \qquad (4.102)$$

and

$$a = a_\pm^v \qquad (4.103)$$

Now let's return to the relationship between activity and molality and see what happens if we substitute the mean ion activity for activity. Since:

$$a_\pm^v = a_{AB}$$

We have: $a_\pm^v = m_{AB}^v$ or $a_\pm = m_{AB}$

This is the relationship that we observed in Figure 4.28, so we see that the mean ionic activity accounts for the effects of dissociation.

The mean ionic activity coefficient, or the *stoichiometric activity coefficient* as it is sometimes referred to, of NaCl would be the square root of the product of the component activity coefficients according to eqn. 4.93, as would the mean ionic activity. Example 4.5 illustrates the calculation for a 2:1 salt.

Example 4.5 Mean ionic parameters for a fully dissociated electrolyte

If the molality of a $CaCl_2$ solution is 0.3 M and the activity coefficients of Ca^{2+} and Cl^- calculated from the Debye–Hückel extended law are 0.21 and 0.56, respectively, calculate the activity and mean ionic molality of $CaCl_2$ in the solution. Assume that $CaCl_2$ fully dissociates.

Answer: For $CaCl_2$, $v^+ = 1$, $v^- = 2$, and $v = 3$. So we can use eqn. 4.99 to calculate mean ionic molality:

$$m_{\pm CaCl_2} = m_{CaCl_2}(1^1 \times 2^2)^{1/3}$$

Substituting 0.3 for m, we find that $m_{\pm} = 0.476$ M.
We then use eqn. 4.93 to calculate the mean ionic activity coefficient:

$$\gamma_{\pm CaCl_2} = (\gamma_{Ca^{2+}}^1 \gamma_{Cl^-}^2)^{1/3} = (0.21^1 \times 0.56^2)^{1/3} = 0.404$$

The mean ionic activity is then:

$$a_{\pm} = \gamma_{\pm} m_{\pm} = 0.513 \times 0.404 = 0.207$$

and the activity of $CaCl_2$ is:

$$a_{CaCl_2} = a_{\pm}^v = \gamma_{\pm} m_{\pm} \quad a_{CaCl_2} = a_{\pm CaCl_2}^v = (\gamma_{\pm} m_{\pm})^v$$

4.7.2.3 *Single ion properties*

The individual ion activities can be measured in a number of ways. Therefore, the above relationships allow calculation of the mean ionic activity coefficient from measurable quantities. One approach is the so-called MacInnes Convention, which references activities coefficients relative to those of KCl, which is a well-behaved salt consisting of ions that are similar in size and mobility in solution such that:

$$\gamma_{KCl} = \gamma_{\pm KCl} = \gamma_{K^+} = \gamma_{Cl^-} \quad (4.104)$$

If we measure the $CaCl_2$ and KCl activity coefficients under equivalent conditions, the activity coefficient of Cl^- can then be taken as equal to that of KCl. From this we may calculate the activity coefficient of Ca^{2+} in a $CaCl_2$ solution by expressing the $CaCl_2$ mean ion activity coefficient as:

$$\gamma_{\pm CaCl_2} = (\gamma_{Ca^{2+}} \gamma_{Cl^-}^2)^{1/3} = (\gamma_{Ca^{2+}} \gamma_{\pm KCl}^2)^{1/3}$$
$$(4.105)$$

and then solving for the Ca^{2+} activity coefficient:

$$\gamma_{Ca^{2+}} = \frac{\gamma_{\pm CaCl_2}^3}{\gamma_{\pm KCl}^2} \quad (4.106)$$

This approach works reasonably well for dilute solutions but becomes inaccurate at higher ionic strength.

An alternative method is based on a convention that we have already introduced, namely that ΔG_f of the hydrogen ion is taken as 0. Consider HCl. The standard state free energy of formation of HCl from H_2 and Cl_2 gas is –95.3 kJ/mol. The free energy of then dissolving this gas in water is –35.9 kJ/mol. These are additive, so the free energy of formation of HCl_{aq} is –131.2 kJ/mol. Taking the ΔG_f of H^+ as 0, then ΔG_f^o for the Cl^- ion is –131.2 kJ/mol.

Now consider NaCl. The free energy for the reaction:

$$Na_s + \tfrac{1}{2}Cl_{2(g)} \rightleftharpoons Na^+ + Cl^-$$

is –393.1 kJ/mol. Since the reactants are the elements in their standard states and we have determined the free energy of formation of the chloride ion to be –131.2 kJ/mol, the free energy of formation of the sodium ion is –393.2 + 131.1 = –261.9 kJ/mol. Other thermodynamic parameters may be determined in a similar manner. Free energy is related to the activity coefficient through eqn. 3.56, allowing us to calculate the activity coefficients as well.

Free energies of formation of ions calculated in this way are sometimes called *conventional free energies*. They differ from absolute free energy by the free energy of formation of the dissolved hydrogen ion from

H_2 gas. The standard state (298K, 0.1MPa) free energy of formation of gas phase H^+ from H_2 is 1514.4 KJ/mol and the free energy of hydration of the H^+ ion is ~ -400kJ/mol, so the differences between conventional free energies and absolute ones, $\sim$1114 kJ/mol, are significant. In most instances, the difference cancels out, but in some cases, the absolute free energies are needed. In general, the conventional and absolute thermodynamic properties of an ion, i, are related as

$$\Delta\Psi_{f,i}^{abs} = \Psi_i^{conv.} + \frac{1}{\nu_i}\Psi_i - z_i\left(\frac{1}{2}\Psi_{H_2} - \Psi_{H^+}\right)$$
(4.107)

where Ψ_i is the property of the pure element and ν_i is the stoichiometric coefficient in the pure element (e.g., 1 in Na metal, 2 in Cl_2 gas).

4.7.3 Activities in high ionic strength solutions

The Debye–Hückel equation becomes inaccurate at ionic strengths above about 0.1 m. This is illustrated in Figure 4.29, which shows the experimentally determined mean ion activity coefficient for NaCl as a function of ionic strength and temperature. At low temperatures, the activity begins to increase ionic strengths of about 1 m, whereas Debye-Hückel predicts continual decrease. The activities of many electrolytes eventually exceed 1 at high concentrations. The difference between the observed activity coefficients and those predicted by the Debye–Hückel equation is due to the effects of ion association and solvation. Debye and Hückel explicitly ignore any solute–solute interactions and assumed complete dissociation, i.e., no ion associations, and while their treatment included in a general way the dielectric properties of water, it neglected the effects of solvation. As we noted in Chapter 3, the effects of both ion association and solvation become increasingly important with increasing ionic strength. It should be no surprise then that the Debye–Hückel treatment breaks down at high ionic strength. Here we will consider these effects in greater detail and methods to predict activities at high ionic strength.

4.7.3.1 Correction for the concentration of water

At low and moderate ionic strength, we can assume that the mole fraction of water in

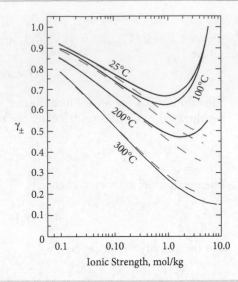

Figure 4.29 Observed mean ion activity coefficient, $\gamma_\pm$, of NaCl as a function of ionic strength and temperature (solid lines; data from Helgeson et al., 1981) compared with values predicted by the Debye–Hückel law (dashed red lines, computed as $(\gamma_{Na}+\gamma_{Cl^-})^{1/2}$). After Helgeson et al. (1981).

solution is 1. For example, in seawater with an ionic strength of 0.7, the mole fraction of water is about 0.99. Generally, activity coefficients and equilibrium constants are not known within 1%, so the error introduced by this assumption is still small compared with other errors. In higher ionic strengths, however, this assumption is increasingly invalid (for example, at a molality of 3, the mole fraction of water has decreased below 0.95), and this must be taken into account. A convenient way to do this is to incorporate it into the activity coefficient. The corrected activity coefficient is:

$$\gamma_{corr} = \frac{\gamma}{1 + 0.018\sum m_i}$$
(4.108)

where the sum is taken over all solutes.

4.7.3.2 Effects of solvation

Water molecules bound to ions in solvation shells have lost their independent translational motion and move with the ion as a single entity. These water molecules are effectively unavailable for reaction, hence solvation has the effect of reducing the activity of water, which increases the apparent concentration, or activity, of the solutes. In addition to solvation (i.e., the direct association of some water molecules with the ion), the charge of the ion causes collapse of the water structure beyond the solvation shell.

For a solution consisting of a single salt, e.g., NaCl, Robinson and Stokes (1959) proposed the contribution of solvation to the mean ion activity coefficient could be expressed as:

$$\log \gamma_{\pm}^{solv} = \frac{h}{\nu} \log a_w - \log(1 - 0.018hm)$$

$$(4.109)$$

where $\gamma_{\pm}^{solv}$ is the solvation contribution to the mean ion activity coefficient, h is the number of moles of water molecules bound to each mole of salt, a_w is the activity of water, m is the concentration of the salt in solution, and ν is as defined in eqn. 4.88 (i.e., total moles of ions produced upon dissolution of a mole of salt). Table 4.3 listed estimated values for the solvation number, that is, the number of water molecules in the solvation shell of each ion (h). The direct effect of the formation of the solvation shell on the activity of water can be adequately estimated as:

$$\Delta a_w \cong 0.04m$$

Figure 4.30 illustrates the effect of solvation on the activity coefficient. As may be seen, solvation substantially affects the activity coefficient at ionic strengths above about 0.5 m.

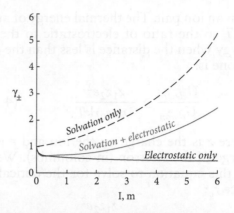

Figure 4.30   Comparison of the electrostatic contribution to the mean ion activity coefficient of NaCl (calculated by the Debye–Hückel extended law), the solvation contribution (calculated from eqn. 4.94 assuming $h = 4$) and the sum of the two.

4.7.3.3   Effects of ion association

The electrostatic energy between two ions separated by a distance r is:

$$U_{electro} = \frac{q_1 q_2}{4\pi\varepsilon_0\varepsilon r} \qquad (4.110)$$

where q is the electrostatic charge on the ion and ε is the permittivity of the medium and ε_0 is the permittivity in a vacuum. An ion pair can be considered to have formed when ions approach closer than some critical distance, r_c, where the electrostatic energy, which tends to bind them, exceeds twice the thermal energy, which tends to move them apart. When this happens, the ions are electrostatically bound and their motions are linked. They are said to

Table 4.3   Ion solvation numbers.

Species	h	Species	h
Li$^+$	2.3	OH$^-$	7.6
Na$^+$	3.3	F$^-$	6.7
K$^+$	2.3	Cl$^-$	2.7
Rb$^+$	2.3	Br$^-$	1.7
Mg^{2+}	8.9	CO$_3^{2-}$	14.4
Ca^{2+}	8.9	SO$_4^{2-}$	10.4
Cd^{2+}	6.3		
Ba^{2+}	9.2		

From Robinson and Stokes (1959)

form an ion pair. The thermal energy of an ion is kT so the ratio of electrostatic to thermal energy when the distance is less than the critical one is:

$$\frac{U_{electro}}{U_{therm}} \geq \frac{z_1 z_2 e^2}{4\pi\varepsilon\varepsilon_0 rkT} \qquad (4.111)$$

where z is the charge on the ion and e is the charge of the electron (in coulombs). We can use this equation to solve for the critical distance r_c:

$$r_c = \frac{z_1 z_2 e^2}{8\pi\varepsilon_0\varepsilon kT} \qquad (4.112)$$

For two singly charged ions at 25°C and 0.1 MPa, the critical distance is 0.357 nm (3.57 Å). In a 1 molal solution, the average separation between ions is about 1.2 nm, so even in such a relatively concentrated solution, ion pairs will not form between singly charged ions. Indeed, the critical distance is smaller than the combined Debye–Hückel radii of all pairs of singly charged ions. Thus, we do not expect ion associations to form from pairs of singly charged ions under most circumstances. In contrast, the critical distance for ion association between a singly and a doubly charged ion is 7 nm, considerably greater than the critical distance and the sum of their Debye–Hückel radii. It also exceeds the average separation of ions in a 0.01 m solution (about 5.5 nm), so that even in dilution solutions, we would expect significant ion pair formation for multiply charged ions.

As we saw earlier, all ions in solution are surrounded by a solvation shell of water molecules. This solvation shell may or may not be disrupted when ion pair formation occurs (Figure 4.31). If it is not disrupted, and the two solvation shells remain intact, an *outer sphere ion pair* (also called an outer sphere complex) is said to have formed. If water molecules are excluded from the space between the ions, an *inner sphere ion pair* (or complex) is said to have formed. We will discuss these complexes in more detail in Chapter 6.

For some purposes, ion pairs can be treated as distinct species having a charge equal to the algebraic sum of the charge of the ions involved. These can be included, for example, in calculation of ionic strength to obtain a somewhat more accurate estimate of activities. On the other hand, ion pairs, including neutral ones, can be highly dipolar and may behave as charge-separated ions.

Ion associations affect activities in two ways. First, associated ions are less likely to participate in reactions, thus reducing the activity of the ions involved. Second, ion association reduces the ionic strength of the solution, and hence reduces the extent of electrostatic interactions among ions. This has the effect of increasing activity. To understand the first effect, consider the case where a certain fraction of the free ions reassociates to form ion pairs:

$$v^+ A^{z+} + v^- A^{z-} \rightleftharpoons (A_{v^+} B_{v^-})^0_{aq}$$

where the superscript 0 indicates neutrality and the subscript aq a dissolved aqueous species. A salt that only partially dissociates in solution is called a weak electrolyte. Let α be the fraction of the ions that associate to form ion pairs or complexes. The association of this fraction of ions as ion pairs will be thermodynamically equivalent to that fraction of the substance not dissociating to begin with. The fraction of free ions is then $1 - \alpha$. Equation 4.100 becomes:

$$m^+ = (1-\alpha)v^+ m \quad \text{and} \quad m^- = (1-\alpha)v^- m \qquad (4.113)$$

where m is the molality of the solute. We can rewrite eqn. 4.98 as:

$$a_\pm = [(\gamma_+ m_+)^{v^+}(\gamma_- m_-)^{v^-}]^{1/v} \qquad (4.114)$$

Substituting 4.113 into 4.114 and rearranging, we obtain:

$$a_\pm = (\gamma_+^{v^+}\gamma_-^{v^-})^{1/v}$$
$$\{[(1-\alpha)v^+ m]^{v^+}[(1-\alpha)v^- m]^{v^-}\}^{1/v}$$

A little more rearranging and we have:

$$a_\pm = (\gamma_+^{v^+}\gamma_-^{v^-})^{1/v}$$
$$\{[(1-\alpha)m]^{(v^+ + v^-)}(v^+)^{v^+}(v^-)^{v^-}\}^{1/v}$$

Finally, since $v = v^+ + v^-$, we obtain:

$$a_\pm = (\gamma_+^{v^+}\gamma_-^{v^-})^{1/v}(1-\alpha)m\{(v^+)^{v^+}(v^-)^{v^-}\}^{1/v} \qquad (4.115)$$

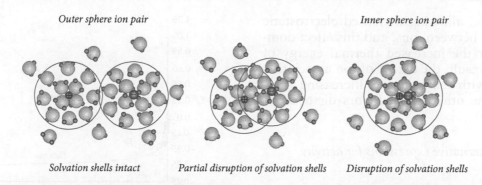

Outer sphere ion pair   *Inner sphere ion pair*

Solvation shells intact   *Partial disruption of solvation shells*   *Disruption of solvation shells*

Figure 4.31 In formation of ion pairs, the solvation shells may remain intact or be partially or totally disrupted. The former results in an outer sphere ion pair, the latter results in an inner sphere ion pair.

We can recognize the last term as $m_\pm$. Since $a_\pm = \gamma_\pm m_\pm$, we see that the mean ionic activity coefficient will be

$$\gamma_\pm = (1 - \alpha)(\gamma_+^{\nu^+} \gamma_-^{\nu^-})^{1/\nu} \qquad (4.116)$$

for an incompletely dissociated electrolyte. Thus, the mean ion activity coefficients are reduced by a factor of $1 - \alpha$. Provided we have appropriate stability constants (discussed in Chapter 6) for the ion pairs or complexes, α can be calculated and an appropriate correction applied.

Now consider a $CaSO_4$ solution of which some fraction of the Ca^{2+} and SO_4^{2-} ions, α, associate to form $CaSO_4^0$. Since $z = 2$ for both ions, the ionic strength of this solution would be

$$I = \frac{(1 - \alpha)}{2}(4m_{Ca^{2+}} + 4m_{SO_4^{2-}})$$

Thus, the ionic strength is reduced by a factor of $1 - \alpha$ as well.

Ion pairs and complexes need not be neutral species ($AlCl^{2+}$, for example). When they are not, they will contribute to ionic strength. A general expression for ionic strength taking account of ion associations must include charged ion pairs and complexes:

$$I = \frac{1}{2}\left[\sum_i (1 - \alpha)m_i z_i^2 + \sum_n c_n z_n^2\right] \qquad (4.117)$$

where α_i is the fraction of each ion involved in ion associations, c_n is the concentration of each ion pair or complex, and z_n is its charge. We could use this result directly in the

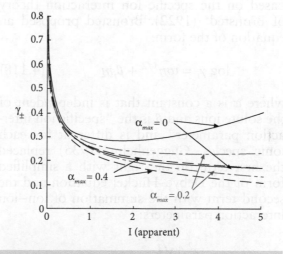

Figure 4.32 Effects of ion association on the activity coefficient. Mean ion activity coefficient of $CaCl_2$ for varying extents of ion association. Fraction of Ca^{2+} ions forming $CaCl^+$ was assumed to increase linearly with ionic strength up to a maximum value (α_{max}) at $I = 5$ m. Solid line shows electrostatic term (Debye–Hückel) after correction for ion association, dashed line shows the combined electrostatic and solvation term.

Debye–Hückel equation to make an improved estimate of ionic strength, and hence of the single ion activity coefficient.

Figure 4.32 illustrates the effect of ion pair formation for a hypothetical $CaCl_2$ solution in which some fraction of the ions combine to form ion pairs.

If the formation of ion pairs depends on the ratio of thermal to electrostatic energy, we might expect that ion pair formation will decrease with temperature. However, the relative permittivity of water decreases with temperature, allowing increased electrostatic interaction between ions, and this effect dominates over the increased thermal energy of ions. As a result, the extent of ion association increases with temperature. Increasing pressure, on the other hand, favors dissociation of ions.

4.7.3.4 Alternative expressions for activity coefficients

There have been a number of attempts to develop working equations that account for all the effects on activity coefficients at high ionic strength. Many of these are ultimately based on the specific ion interaction theory of Brønsted[*] (1922). Brønsted proposed an equation of the form:

$$\log \gamma_i = \alpha m^{1/2} + \beta_i m \qquad (4.118)$$

where α is a constant that is independent of the solute ions and β is the "specific ion interaction parameter" and is different for each ionic species. Guggenheim (1935) replaced the first term on the right with a simplified form of the Debye–Hückel equation and the second term with the summation of ion–ion interaction parameters:

$$\log \gamma_i = \frac{-z_i^2 A \sqrt{I}}{1 + \sqrt{I}} + 2 \sum_k \beta_{i,k} m_k \qquad (4.119)$$

where $\beta_{i,k}$ is a parameter describing the interactions between ions i and k and m_k is the molal concentration of k, very much

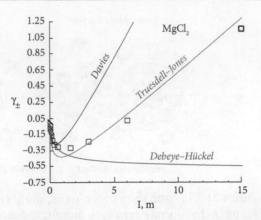

Figure 4.33 Measured mean ionic activity coefficients in $MgCl_2$ solution as a function of ionic strength, compared with values calculated from the Debye–Hückel, Davies and Truesdell–Jones equations.

analogous to the interaction parameters introduced for solid solutions earlier in this chapter. For natural waters with many species, the Guggenheim equation becomes complex. Also starting from Debye–Hückel, Truesdell and Jones (1974) proposed the following simpler equation:

$$\log \gamma_i = \frac{-z_i^2 A \sqrt{I}}{1 + B \mathring{a} \sqrt{I}} + b_i I \qquad (4.120)$$

The first term on the right is identical in form to Debye–Hückel; the second term is similar to the Brønsted specific ion interaction term. Truesdell and Jones determined parameters $\mathring{a}$ and b empirically. Table 4.4

Table 4.4 Truesdell–Jones parameters.

Ion	å	b
Na+	4.0	0.075
K+	3.5	0.015
Mg2+	5.5	0.20
Ca2+	5.0	0.165
Cl-	3.5	0.015
SO42-	5.0	−0.04
CO32-	5.4	0
HCO3-	5.4	0

From Truesdell and Jones (1974).

[*] Johannes Nicolaus Brønsted (1879–1947) was a Danish chemist who, along with English chemist Thomas Lowry, was responsible for the protonic acid–base theory, i.e., acids are proton donors and bases are proton acceptors.

lists these parameters for some common ions (see Example 4.6). Figure 4.33 compares mean activity coefficient calculated with the Debye–Hückel, Davies, and Truesdell–Jones equations with the actual measured values. Generally, the Truesdell–Jones equations fit these observations very well. This is not always the case, however. The fit for Na_2CO_3, for example, is little better than for Debye–Hückel.

Example 4.6 Activity coefficients in a brine

The following concentrations were measured in a pore water brine from Sudbury, Canada, at 22°C. Calculate the activity coefficients of these species using the Truesdell–Jones equation.

Species	Conc. g/kg
Na	18.9
K	0.43
Ca	63.8
Mg	0.078
SO_4	0.223
HCO_3	0.042
Cl	162.7

Answer: Our first task is to convert g/kg to molal concentrations. We do this by dividing by molecular weight. Next, we need to calculate ionic strength (eqn. 3.74), which we find to be 5.9 m. Calculation of activity coefficients is then straightforward using eqn. 4.120 and the parameters in Tables 3.2 and 4.4. Finally, we apply a correction for the decreased concentration of water (eqn. 4.108). Our final spreadsheet is shown below.

	‰	m	z	å_TJ	b_TJ	log (γ)	γ	γ (corr)
Na	18.9	0.822	1	5	0.165	0.7279	5.345	4.744
K	0.43	0.017	1	3.5	0.015	−0.2376	0.579	0.514
Ca	63.8	1.595	2	5	0.165	−0.0163	0.963	0.855
Mg	0.078	0.003	2	5.5	0.2	0.2642	1.837	1.631
SO_4	0.223	0.002	2	5	−0.04	−1.2289	0.059	0.052
HCO_3	0.058	0.001	1	5.4	0	−0.2332	0.585	0.519
Cl	162.7	4.59	1	3.5	0.015	−0.2376	0.579	0.514
	m	7.03	A	0.5092				
	I	5.915	B	0.3283				

4.7.3.5 Pitzer equations

A widely used approach to concentrated electrolyte solutions is based on the equations of Kenneth Pitzer* (1973, 1981). For an aqueous electrolyte, Pitzer used a Virial expansion (e.g., analogous to the Virial equation of state for gases) to express excess free energy (defined in eqn. 3.52) of electrolyte solutions:

$$
\frac{G_{ex}}{RT} = f(I) + \sum_i \sum_{j,j\pm1} m_j m_i W_{i,j}(I)
$$
$$
+ \sum_i \sum_j \sum_k m_i m_j m_k w_{i,j,k} \quad (4.121)
$$

The bar in this case denotes per kilogram of water; $f(I)$ is an ionic strength (I) dependent parameter, m_i, etc. is the molality of species i, $W_{i,j}(I)$ is the interaction parameter between species i and j and is a function of ionic strength as well as T and P, and $w_{i,j,k}$ is a third-order three-species interaction parameter assumed to be independent of ionic strength: exactly analogous to the interaction parameters we used for silicate solutions. From Equation 3.56:

$$
\left(\frac{\partial G_{excess}}{\partial n_i} \right)_{T,P,n_{j,j\neq i}} = RT \ln \lambda_i \quad (3.56)
$$

(Here we have replaced λ in eqn 3.56 with γ as we are working with molality rather than mole fraction.) The relationship between the activity coefficient and excess free energy is thus:

$$
\ln \gamma_i = \frac{1}{RT} \left(\frac{\partial G_{excess}}{\partial n_i} \right)_{T,P,n_{j,j\neq i}} \quad (4.122)
$$

and the single ion activity coefficients can be found by differentiating equation 4.121 with respect to molality which yields:

$$
\ln \gamma_i = z_i^2 + f' + \sum_j m_j \left(2B_{i,j} + C_{i,j} \sum_k z_k m_k \right) +
$$
$$
\sum_j m_j \left(2\Phi_{i,j} + \sum_k m_k \psi_{i,j,k} \right) + \sum_i \sum_k m_j m_k
$$
$$
(z_i^2 B'_{i,j} + z_j C_{j,k}) + \frac{1}{2} \sum_j \sum_k m_j m_k \psi_{i,j,k} \quad (4.123)
$$

where $B_{i,j}$. $C_{i,j,k}$, $B'_{i,k}$ $\psi_{i,j,k}$ and $\Phi_{i,j}$, are interaction parameters between the respective ions. $B_{i,j}$, $B'_{i,k}$, $C_{j,k}$ are interaction parameters for ions of different charge; they are 0 for ions of the same charge. $B_{i,j}$., $B'_{i,k}$ and $\Phi_{i,j}$, are functions of ionic strength and must be determined empirically. $C_{j,k}$ is a function of the charge of ions j and k. f' is derived from the modified Debye–Hückel equation:

$$
f' = -\frac{2.303A}{3} \left[\frac{\sqrt{I}}{1+b\sqrt{I}} + \frac{2}{b} \ln(1+b\sqrt{I}) \right] \quad (4.124)
$$

where A is the Debye–Hückel term and $b = 1.2$. The osmotic coefficient is given by:

$$
\phi - 1 = \frac{1}{\sum_i m_i} \left[\frac{-2A_\phi I^{3/2}}{1+bI^{1/2}} + 2 \sum_j \sum_k m_j m_j \right.
$$
$$
\left(B_{ij}^\phi + C_{ij}^\phi \sum_l \frac{z_l m_l}{z_j z_k} \right)
$$
$$
\left. + \sum_j \sum_k \Phi_{jk} + \sum_l m_l \psi_{ikl} \right] \quad (4.125)
$$

More details on these equations can be found in Fletcher (1993) and Anderson and Crerar (1993) as well as the original literature.

Clearly, calculating activity coefficients this way is computationally intensive for natural solutions, which typically have many ions. In addition, a database of the empirically determined parameters is required. Fortunately, computer codes are available to do these complex calculations. One such widely used program is EQ3/6 developed by Thomas Wolery at the Lawrence Livermore National Laboratory (Wolery and Jarek, 2003) and which is freely available (https://www-gs.llnl.gov/energy-cyber-and-infrastructure/geochemistry). Another is PHREEQC, developed by Pankhurst and Appelo of the US Geological Survey (Pankhurst and Appelo, 2013) and also freely available at https://wwwbrr.cr.usgs.gov/projects/GWC_coupled/phreeqc/. They are also incorporated in a commercial program, *Geochemist's Workbench*, which is rather expensive, but a free version is available to university students.

---

* Kenneth Pitzer (1914–1997) received his PhD from the University of California, Berkeley in 1937 and went on to become a professor and dean of the College of Chemistry there. Pitzer served as president of Rice University, where he was instrumental in integrating what was then a racially segregated institution and subsequently as president of Stanford University. In 1971 he returned to his position at UC Berkeley.

Example 4.7 Calculating activity coefficients using PHREEQC

Waters of the Great Salt Lake in the western United States are an example of a high ionic strength solution where the Debye–Hückel approach does not yield accurate activity coefficients. The briny water of the lake is quite variable in composition in both time and space, but the following is a typical composition. The pH is about the same as seawater, 8.2. Use PHREEQC to calculate activity coefficients.

	ppm
Cl^-	116667
Ca^{2+}	253
Mg^{2+}	6389
Na^+	68016
K^+	3611
SO_4^-	13810
HCO_3^-/CO_3^{2-}	260

Answer: First download and install the program from the USGS website and review the documentation. The program varies somewhat between operating systems: read the instructions on the website and in the download. This example is for MacOS operating system. The procedure is slightly different for Windows and LINUX.

We create the adjacent input file:

	log γ	γ
Cl^-	−0.226	0.59
Ca^{2+}	−0.008	0.98
Mg^{2+}	0.205	1.60
Na^+	0.069	1.17
K^+	−0.148	0.71
SO_4^{2-}	−1.994	0.01
HCO_3^{2-}	−0.54	0.29

PHREEQC comes with a number of databases; when prompted, specific the *Pitzer.dat* file. The results are shown below:

This output also calculates speciation and also shows that the brine is saturated in calcite, dolomite and several other phases. We will discuss speciation and precipitation in Chapter 6.

TITLE Great Salt Lake Activity Coefficients.
SOLUTION 1 Great Salt Lake Composition
units ppm
pH 8.2
temp 25.0
Cl 116667.0
Ca 253.0
Mg 6389.0
Na 68016.0
K 3611.0
S(6) 13810.0 as SO_4
C(4) 260.0 as HCO_3
END

4.7.4 Electrolyte solutions at elevated temperature and pressure

4.7.4.1 Born equation

Dissolution of ions in water also affects its volume and other thermodynamic properties through electrostriction. As we discussed in Chapter 3, the presence of ions in solution partially collapses the structure of water and further affects it through the solvation shell that forms around the ion. Max Born* proposed that the free energy change due to solvation could be expressed as:

$$\Delta \overline{G}_s^o = \frac{N_A(ze)^2}{2r}\left(\frac{1}{\varepsilon} - 1\right) \qquad (4.126)$$

where the subscript s denotes solvation, r is the effective ionic radius, z is the charge of the ion, e is the charge of the electron, ε is the relative permittivity of water, and N_A is Avogadro's number. This is known as the *Born equation*. Born assumed that the radii of anions are equal to their crystallographic ionic radii (for single atom ions, these are the radii shown in Figure 1.6) and the cations of cations are equal to crystallographic ionic radii plus $0.94 \times z$, where z is the cation charge. To a first approximation, these effective ionic radii can be taken as constant.

The first ratio on the right of eqn. 4.126 is a function of only charge and ionic radius and can be combined into a parameter, ω, called the Born coefficient:

$$\omega_i = \frac{N_A(z_ie)^2}{2r_i} \qquad (4.127)$$

A complication arises, however, from the convention that the free energy of formation of the hydrogen ion is 0 and that the thermodynamic properties of anions are taken as those of their corresponding acids (e.g., properties of Cl^- are taken as those of HCl_{aq}). To be consist with this convention, we subtract the absolute energy of solvation of the H^+ ion, so that Born parameters are calculated as:

$$\omega_i = \frac{N_A(z_ie)^2}{2r_i} - 0.5387z_i \qquad (4.128)$$

where z_i is the charge of the ion, r_i is the effective, or Born, radius of the ion, and ε is the permittivity of water.

The free energy change of solvation of ion i becomes:

$$\Delta \overline{G}_{s,i}^o = \omega_i \left(\frac{1}{\varepsilon} - 1\right) \qquad (4.129)$$

Free energies are additive so that the free energy of solvation of an electrolyte such as Na_2SO_4 is:

$$\Delta \overline{G}_{s,Na_2SO_4}^o = (2\omega_{Na} + \omega_{SO_4})\left(\frac{1}{\varepsilon} - 1\right) \qquad (4.130)$$

and

$$\omega_{Na_2SO_4} = 2\omega_{Na} + \omega_{SO_4} \qquad (4.131)$$

Other thermodynamic parameters can then be derived by differentiation of eqn 4.129, for example:

$$\Delta \overline{S}_s^o = -\left(\frac{\partial \overline{G}_s^o}{\partial T}\right)_P \qquad (4.132)$$

$$\Delta \overline{V}_s^o = \left(\frac{\partial \overline{G}_s^o}{\partial P}\right)_T \qquad (4.133)$$

$$\Delta C_{P,s}^o = \left(\frac{\partial \overline{H}_s^o}{\partial T}\right)_P = T\left(\frac{\partial \overline{S}_s^o}{\partial T}\right)_P \qquad (4.134)$$

If ionic radii are independent of T and P, then changes of volume, entropy, and heat capacities of solvation are functions only of the permittivity of water, i.e.:

$$\Delta \overline{S}_s^o = -\left(\frac{\partial \overline{G}_s^o}{\partial T}\right)_P = -\omega\left(\frac{\partial\left(\frac{1}{\varepsilon} - 1\right)}{\partial T}\right)_P \qquad (4.135)$$

---

* Max Born (1882–1970) received his PhD from the University of Göttingen in 1904 and held faculty positions there and in Berlin before taking up a position at the University of Edinburgh after the Nazi takeover and expulsion of Jewish professors in 1934. Over his career, he made many important contributions to quantum and relativity theory as well as to physical chemistry. He won the 1954 Nobel Prize in physics for his contributions to quantum mechanics.

$$\Delta \overline{V}_s^o = -\left(\frac{\partial \overline{G}_s^o}{\partial P}\right)_T = -\omega \left(\frac{\partial \left(\frac{1}{\varepsilon} - 1\right)}{\partial P}\right)_T \tag{4.136}$$

The permittivity of water varies substantially with pressure and temperature, decreasing with temperature from 78.5 at 25°C and 0.1 MPa to 55.3 to 100°C and increasing with pressure to 90.15 at 500 MPa. It is 15.83 at 600°C and 500 MPa. As we discussed in Chapter 3, the dielectric properties of water, quantified by the permittivity, plays a key role in solubility of ions, so we can expect these to change as well as a function of temperature and pressure.

4.7.4.2   The HKF model

In the 1970s Harold Helgeson* and colleagues developed a thermodynamic model of aqueous solutions at elevated temperature and pressure published in a series of papers in the *American Journal of Science* culminating in Helgeson Kirkham and Flowers (1981). This model, known as the HKF model, has been modified and improved over the decades as additional data has become available (e.g., Tanger and Helgeson (1989); Shock and Helgeson, 1990; Oelkers et al., 2009) and remains an often-used framework for thermodynamic modeling of aqueous solutions.

The HKF model is based on several fundamental properties of thermodynamic functions. First, the effects of temperature, pressure, and composition can be treated separately. We found in Chapter 3 that

$$dG = VdP + SdT + \sum_i \mu_i dn_i \tag{3.14}$$

Thus, thermodynamic properties of electrolyte solutions can be dealt with in three steps: variation temperature (integration of the first term on the right from the reference temperature to the temperature of interest) at standard state molality (1 m), variation in

pressure (integration of the second term on the right) again at standard state molality, and finally variation in composition at the T and P of interest (integration of the third term).

Second, the thermodynamic properties of the solution are the sum of the thermodynamic properties of its components. Thus, for example, we can solve separately for the chemical potentials or volumes of the constituents, then sum them to get molal free energy or volume of the solution. We must, however, consider the effects that the presence of each component has on the thermodynamic properties of every other component.

Third, fundamental thermodynamic variables, G, S, V, etc. are additive. Consider the dissolution of an electrolyte such as NaCl in water. It affects volume in three ways. The first is the intrinsic volume occupied by the ions themselves. The second is the local collapse of the structure of water (imposed by hydrogen bonding), resulting from the electrostatic field of the ion, and the third is the formation of the solvation shell surrounding the ion in which the water molecules are more tightly packed than otherwise. These latter effects are collectively termed *electrostriction*. In addition to volume effects, there are, of course, energetic and entropic effects as well. The total effect on a thermodynamic property such as volume is the sum of the three. The HKF model combines the intrinsic and structural contributions and then sums them with the solvation contribution, e.g.:

$$V_i^o = V_{n,j}^o + \Delta V_{s,j}^o \tag{4.137}$$

where the subscript n and s denote the non-solvation and solvation contributions to the molar volume of ion i, respectively.

The original HKF model used the Born equation to calculate the effect of solvation. The problem of solving finding the entropy and volume of solvation and their pressure and temperature dependencies thus comes down to solving

$$\left(\frac{\partial \left(\frac{1}{\varepsilon} - 1\right)}{\partial P}\right)_T \quad \text{and} \quad \left(\frac{\partial \left(\frac{1}{\varepsilon} - 1\right)}{\partial T}\right)_P$$

---

* Harold (Hal) Helgeson (1931–2007) received his PhD with Robert Garrels at Harvard University and then joined Garrels and Fred Mackenzie at Northwestern University where they together laid the foundation of the thermodynamics of water–rock interaction. Helgeson joined the faculty of the University of California at Berkeley in 1969 where he remained until his death.

or in other words, the temperature and pressure dependencies of the permittivity of water. Tanger and Helgeson (1989), however, found that Born ionic radii could not be taken as constant and must be considered functions of T and P, so, for example, equation 4.136 becomes, e.g.:

$$\Delta \overline{V}_{s,i}^{o} = -\omega_i \left(\frac{\partial \left(\frac{1}{\varepsilon} - 1 \right)}{\partial P} \right)_T - \left(\frac{1}{\varepsilon} - 1 \right) \left(\frac{\partial \omega_i}{\partial P} \right)_T$$

(4.138)

This imposes the additional task of determining the temperature and pressure dependencies of the Born coefficients. Since all other terms in the Born coefficient (eqn. 4.127) are constants, this boils down to determining how effective radii of solute ions vary with temperature and pressure. Helgeson et al. parameterized these equations by regressing experimental data on the behavior of electrolytes as functions of temperature and pressure. These results showed that the temperature and pressure dependence can be expressed as the product of an ion-specific constant and a second T and P dependent parameter, g, that is the same for all ions, and therefore considered a property of the solvent. This T and P dependence disappears at sufficiently elevated temperatures and pressures where the Born radii become constant.

Using a similar parameterization approach for nonsolvation effects, the modified HKF model accounted for nonsolvation effects with an equation of the form:

$$\overline{V}_{n,i}^{o} = a_i + \frac{b_i}{\Phi + P} + \frac{c_i}{T - T_s} + \frac{d_i}{(\Phi + P)(T - T_s)}$$

(4.139)

where the subscript n denotes the nonsolvation contribution to the molar volume of ion i, $a_i \ldots d_i$ are empirically determined constants specific to the ion, T_S is the singular temperature of 228 K discussed in section 4.7.3.1 and Φ is an empirically determined pressure of 260 MPa. Activity coefficients are the sum of intrinsic, structural, solvation, and ion–ion interaction terms:

$$\log \gamma_i = \frac{-A z_i^2 \sqrt{\overline{I}}}{1 + B\mathring{a}\sqrt{\overline{I}}} + \Gamma + \omega_i \sum_k b_k y_k \overline{I} + \sum_j \beta_{ij} m_j$$

(4.140)

The first term is the intrinsic/structural term where A and B are Debye–Hückel parameters (which are functions of T and P), $\mathring{a}$ is the effective electrostatic radius. Γ is simply a function that converts mole fraction to molality:

$$\Gamma = -2.303RT \log \left(1 + 0.180153 \sum_j m_j \right)$$

(4.141)

The third term is the ion solvation term and is calculated in terms of salts rather than ions. b is an empirical salt parameter determined by regression, y_k is the fractional contribution of the nth salt to the ionic strength:

$$y_k = \sum_j \frac{v_{jk}(z_j^2/2)m_j}{I}$$

(4.142)

$\overline{I}$ is the "true" ionic strength, that is corrected for the formation of complexes. The final term accounts for ion–ion interactions where β_{ij} is the interaction parameter between ions i and j. Figure 4.34 shows that the variation in molar volume and heat capacity of $CaCl_2$ as a function of temperature computed with the HKF model compares well with experimental data.

The HKF model is obviously complex and computationally intensive. Fortunately, the model was developed into a computer code and corresponding database, SUPCRT92 by Johnson et al. (1992). The program and database have in turn been progressively improved, most recently by Zimmer et al. (2013) and the current version is available at http://www.indiana.edu/~hydrogeo/supcrtbl.html.

4.7.4.3 Properties of ore-forming hydrothermal solutions

The majority of metal ores are precipitated from aqueous solutions at elevated temperatures and pressures. For this reason, as well as designing and operating geothermal power systems, an understanding of these solutions is useful. We will treat ore-forming processes in detail in Chapter 15. In this section, we will briefly summarize the solvent properties of water at elevated temperature and pressure.

Fluid inclusion studies show that ore forming solutions can be highly saline, ranging, for example, to >60% salinity in porphyry

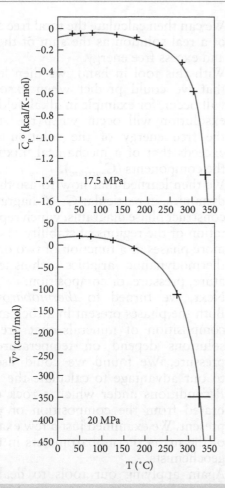

Figure 4.34 (a). Partial molar heat capacity of aqueous $CaCl_2$ at 17.5 MPa (a) and (b) partial molar volume at 20 MPa as a function of temperature computed with the revised HKF model compared to experimental data (crosses). Adapted from Tanger and Helgeson (1988).

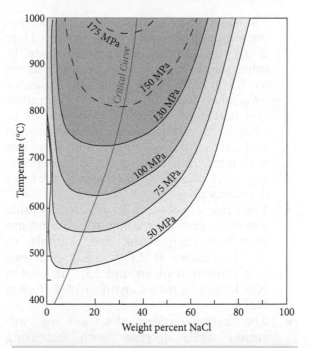

Figure 4.35 Temperature–composition phase diagram for aqueous solutions of NaCl at high temperatures and pressures. The presence of dissolved NaCl (and other ions) increases the two-phase field of water. Solutions plotting within the gray field at the specified pressures will undergo phase separation (boiling) into a high-salinity higher-density fluid and a low-salinity, lower-density fluid. Modified from Bodnar et al. (1985).

copper deposits (Bodnar et al., 2014). The dominant ions in these solutions are usually Na and Cl, so that to a rough, first approximation they can be treated as NaCl brines. A critical effect of salinity is to greatly expand the two-phase field of water (Figure 4.35). This is well documented in fluid-inclusions in ore deposits, as indicated, but among other things a lack of salinities between 20 and 30%, especially in inclusions trapped above 600°C (Bodnar et al., 2014). Consequently, some ore-bearing fluids can undergo boiling and separation into a high-salinity brine and a low-salinity, low density fluid or vapor.

Other important aspects of ore-forming fluids include:

- As pressure and temperature increase, the extent of hydrogen bonding diminishes leading to, among other things, decreased viscosity.
- As temperature increases, the dielectric constant decreases from 80.10 at 20°C to 7.22 at 373°C. This results in enhanced ion pairing/association and, consequently, metal complex formation. For example, in a 1M solution of NaCl at 380°C and density of 0.55 g/cm³, free ions of Na^+ and Cl^- constitute only 13% of species in solution (compared with ~100% at 20°C), with the remainder as species including not only NaCl but also Na_2Cl^+, Na_2Cl_2, etc. On the other hand, the dielectric constant increases with pressure.

- The value of pK_W decreases with increasing temperature, from 14 at 25°C to 11 at 250°C saturated water vapor pressure. In other words, the pH of pure water drops to 5.5 at 250°C. In detail, the value of pK_W also depends on the concentration and nature of solutes. As we will see in Chapter 6, pH is a controlling factor in mineral solubility. pK_W then increases with temperature above ~350°C, depending on pressure. Increasing pressure decreases pK_W.

- The value of pK_W is also affected by ionic strength, first decreasing to a minimum then increasing again. For example, in NaCl solutions at 25° C, pK_W decreases to a minimum of around 13.7 at ~0.4 m NaCl, then increases again to 14.4 at ~6m NaCl.

- The cation–oxygen distances in solvation shells decrease with increasing temperature; in other words, these shells shrink. Consequently, the number of water molecules surrounding a cation can decrease.

- While the solubility of most minerals, silicates in particular, increases with increasing temperature (*prograde* solubility), other minerals, most notably calcium salts, can exhibit retrograde solubility. Calcite solubility decreases with temperature over the entire relevant range while, gypsum, and anhydrite, and fluorite, show complex patterns of solubility as a function of temperature. Solubility also varies with pressure and pH and with the presence of other ions in solution; most notably the presence of alkali halides (e.g., NaCl) tends to increase solubility.

4.8 SUMMARY

In this chapter we introduced advanced tools to deal with nonideal solutions.

- We began by introducing *interaction parameters* to account for nonideal *interactions* between molecules or atoms. For example, for binary solutions with components 1 and 2 the excess Gibbs free energy is expressed as:

$$\overline{G}_{ex} = X_1 X_2 W_G$$
$$= RT[X_1 \ln \lambda_1 + X_2 \ln \lambda_2] \quad (4.11)$$

We can then calculate the total free energy of a real solution as the sum of the ideal and excess free energy.

- With this tool in hand, we then learned that we could predict when exsolution will occur, for example in alkali feldspars: exsolution will occur when the sum of the free energy of the solution ($\overline{G}_{real}$) exceeds that of a mechanical mixture of the components ($\overline{G}_{mixture}$).

- We then learned that how to use thermodynamics, in particular G–X diagrams, to construct phase diagrams, which representation of the regions of stability of one or more phases as a function of two or more thermodynamic variables such as temperature, pressure, or composition.

- Next, we turned to *thermobarometry*. Both the phases present in a rock and the composition of minerals that are solid solutions depend on temperature and pressure. We found we could use that to our advantage to calculate the T and P conditions under which a rock equilibrated from the composition of phases present. We examined just a few examples of the many thermobarometers in use by geochemists.

- Again applying our tools to deal with nonideal solutions, we turned to the problem of silicate magmas. While the principles are the same as for a simple two- component solution such as alkali feldspar, silicate magmas contain eight or more important components, so calculations require computer models such as MELTS. MELTS predicts the minerals crystallizing from a magma as a function of composition, T, and P by applying the fundamental principle that the stable state of a system is the one with the lowest Gibbs free energy.

- In the final part of the chapter, we returned to the problem another nonideal solution, aqueous electrolytes, to develop tools necessary to deal with them at high ionic strength and elevated T and P. We considered the relationship between the activity of a solute and its molality, finding that for a salt AB, which dissociates to produce ν ions that:

$$a_{AB} \propto m_{AB}^{\nu} \quad (4.90)$$

- Next, we introduced the concept of mean ionic quantities, for example:

$$\gamma_{\pm} = (\gamma_+^{v^+} \gamma_-^{v^-})^{1/v} \qquad (4.93)$$

and

$$m_{\pm} = (m_+^{v^+} m_-^{v^-})^{1/v} \qquad (4.101)$$

We reviewed the effects of ion association and introduced several approaches to calculate activities in high ionic strength solutions. Solvation is an important factor in electrolyte solubility and depends on the permittivity of water, which is a strong function of T and P. To deal with this we introduced the Born coefficient:

$$\Delta \overline{G}_s^o = \frac{N_A(ze)^2}{2r} \left(\frac{1}{\varepsilon} - 1 \right) \qquad (4.126)$$

The first term on the right is the Born coefficient:

$$\omega_i = \frac{N_A(z_i e)^2}{2r_i} \qquad (4.127)$$

This can be taken as constant in most circumstances, so the problem reduces to computing permittivity, ε. This led us to the widely used Helgeson, Kirkham, and Flowers (HKF) model for dealing with concentrated electrolytes at elevated T and P. Computations in the HKF model are complex but programs such as SUPERCRIT are available to deal with them.

REFERENCES AND SUGGESTIONS FOR FURTHER READING

Anderson, G.M. and Crerar, D.A. 1993. *Thermodynamics in Geochemistry*. New York, Oxford University Press.

Ague, J.J. 1997. Thermodynamic calculation of emplacement pressures for batholithic rocks, California: Implications for the aluminum-in-hornblende barometer. *Geology* 25: 563. doi: 10.1130/0091-7613(1997)025<0563:TCOEPF>2.3.CO;2.

Anderson, J.L., Barth, A.P., Wooden, J.L. and Mazdab, F. 2008. Thermometers and thermobarometers in granitic systems. *Reviews in Mineralogy and Geochemistry* 69: 121–42.

Anderson, J.L. and Smith, D.R. 1995. The effects of temperature and f_{O2} on the Al-in-hornblende barometer. *American Mineralogist* 80: 549–59.

Berman, R.G. and Thomas H, B., 1984. A thermodynamic model for multicomponent melts, with application to the system $CaO-A_{l2}O_3-SiO_2$. *Geochimica et Cosmochimica Acta* 48: 661–678. doi: 10.1016/0016-7037(84)90094-2.

Bodnar, R. J., Burnham, C. W. and Sterner, S. M. 1985. Synthetic fluid inclusions in natural quartz. III. Determination of phase equilibrium properties in the system $H_2O-NaCl$ to 1000°C and 1500 bars. *Geochimica et Cosmochimica Acta* 49: 1861–73. doi: 10.1016/0016-7037(85)90081-X.

Bodnar, R. J., Lecumberri-Sanchez, P., Moncada, D. and Steele-MacInnis, M. 2014. 13.5 - Fluid inclusions in hydrothermal ore deposits. In: Holland HD and Turekian KK (eds) *Treatise on Geochemistry* (Second Edition), Elsevier, Oxford, pp. 119–42. ISBN 119-142 978-0-08-098300-4.

Blundy, J. and Cashman, K. 2008. Petrologic reconstruction of magmatic system variables and processes. *Reviews in Mineralogy and Geochemistry* 69: 179–239.

Bohlen, S.R., Wall, V.J. and Boettcher, A.L. 1983. Experimental investigations and geological applications of equilibria in the system $FeO-TiO_2-Al_2O_3-SiO_2-H_2O$. *American Mineralogist* 68: 1049–58.

Bohlen, S.R., Dollase, W.A. and Wall, V.J. 1986. Calibration and application of spinel equilibria in the system $FeO-Al_2O_3-SiO_2$. *Journal of Petrology* 27: 1143–56.

Bohrson W. A., Spera, F. J., Ghiorso, M. S., Brown, G. A., Creamer, J. B. and Mayfield, A. 2014. Thermodynamic model for energy-constrained open-system evolution of crustal magma bodies undergoing simultaneous recharge, assimilation and crystallization: the magma chamber simulator. *Journal of Petrology* 55, 1685–1717. doi: 10.1093/petrology/egu036

Boyd, F.R. 1973. A pyroxene geotherm. *Geochimica Cosmochimica Acta* 37: 2533–46.

Brey, G.P. and Köhler, T. 1990. Geothermobarometry in four-phase lherzolites II. New thermobarometers, and practical assessment of existing thermobarometers. *Journal of Petrology* 31: 1353–78.

Brey, G.P., Bulatov, V.K. and Girnis, A.V. (2008) Geobarometry for peridotites: Experiments in simple and natural systems from 6 to 10 GPa. *Journal of Petrology* 49: 3–24.

Brønsted, J.N. 1922. Studies on solubility, IV. The principle of specific interaction of ions. *Journal of the American Chemical Society* 44: 877–98.

Buddington, A.F. and Lindsley, D.H. 1964. Iron-titanium oxide minerals and synthetic equivalents. *Journal of Petrology* 5: 310–57.

Connolly, J.A.D. 1989. Multivariable phase diagrams: An algorithm based on generalized thermodynamics. *American Journal of Science* 290, 666–718.

Essene, E. 1989. The current status of thermobarometry in metamorphic rocks, in *The Evolution of Metamorphic Belts* (eds J.S. Daly, R.A. Cliff and B.W.D. Yardley), pp. 1–44. London: Blackwell Scientific.

Ferry, J.M. and Spear, F.S. 1978. Experimental calibration of the partitioning of Fe and Mg between biotite and garnet. *Contributions to Mineralogy and Petrology* 25: 871–93.

Ferry, J. and Watson, E. 2007. New thermodynamic models and revised calibrations for the Ti-in-zircon and Zr-in-rutile thermometers. *Contributions to Mineralogy and Petrology* 154, 429–37.

Fletcher, P. 1993. *Chemical Thermodynamics for Earth Scientists*. Essex, Longman Scientific and Technical.

Garrels, R.M. and Christ, C.L. 1965. *Solutions, Minerals, and Equilibria*. New York, Harper and Row.

Gasparik, T. 1984a. Two-pyroxene thermobarometry with new experimental data in the system $CaO-MgO-Al_2O_3-SiO_2$. *Contributions to Mineralogy and Petrology* 87: 87–97.

Gasparik, T. 1984b. Experimental study of subsolidus phase relations and mixing properties of pyroxene and plagioclase in the system $CaO-Al_2O_3-SiO_2$. *Geochimica Cosmochimica Acta* 48: 2537–45.

Ghent, E.D. 1976. Plagioclase-garnet-Al_2O_5-quartz: a potential geobarometer-geothermometer. *American Mineralogist* 61: 710–14.

Ghiorso, M.S., Carmichael, I.S.E., Rivers, M.L. and Sack, R.O. 1983. The Gibbs free energy of mixing of natural silicate liquids; an expanded regular solution approximation for the calculation of magmatic intensive variables. *Contributions to Mineralogy and Petrology* 84: 107–45.

Ghiorso, M.S. and Evans, B.W. 2008. Thermodynamics of rhombohedral oxide solid solutions and a revision of the Fe-Ti two-oxide geothermometer and oxygen-barometer. *American Journal of Science* 308: 957–1039.

Ghiorso, M.S. and Gualda, G.A.R., 2013. A method for estimating the activity of titania in magmatic liquids from the compositions of coexisting rhombohedral and cubic iron–titanium oxides. *Contributions to Mineralogy and Petrology*. 165: 73–81. doi: 10.1007/s00410-012-0792-y.

Ghiorso, M.S. and Gualda, G.A. R. 2015. An $H_2O–CO_2$ mixed fluid saturation model compatible with rhyolite-MELTS. *Contributions to Mineralogy and Petrology* 169: 1–30. doi: 10.1007/s00410-015-1141-8.

Ghiorso, M.S., Hirshmann, M.M., Reiners, P.W. and Kress, V.C. 2002. The pMELTS: a revision of MELTS for improved calculation of phase relations and major element partitioning related to partial melting of the mantle to 3 GPa. *Geochemistry Geophysics Geosystems* 3: 2001GC000217.

Ghiorso, M.S. and Sack, R.O. 1995. Chemical mass transfer in magmatic processes IV. A revised and internally consistent thermodynamic model for the interpolation and extrapolation of liquid-solid equilibria in magmatic systems at elevated temperatures and pressures. *Contributions to Mineralogy and Petrology* 119: 197–212.

Goldsmith, J.R. and Newton, R.C. 1969. P-T–X relations in the system $CaCO_3-MgCO_3$ at high temperatures and pressures. *American Journal of Science* 267A: 160–90.

Green, E. C. R., White, R. W., Diener, J. F. A., Powell, R., Holland T.J.B., and Palin, R. M. 2016. Activity–composition relations for the calculation of partial melting equilibria in metabasic rocks. *Journal of Metamorphic Geology* 34: 845–869. doi: 10.1111/jmg.12211.

Guggenheim, E.A. 1935. The specific thermodynamic properties of aqueous solutions of strong electrolytes. *Philosophical Magazine* 19: 588.

Helgeson, H.C. 1969. Thermodynamics of hydrothermal systems at elevated temperatures and pressures. *American Journal of Science* 267: 729–804.

Helgeson, H. C., and Kirkham, D. H. 1974a. Theoretical prediction of the thermodynamic behavior of aqueous electrolytes at high pressures and temperatures; I, summary of the thermodynamic/electrostatic properties of the solvent. *American Journal of Science* 274: 1089–198. doi: 10.2475/ajs.274.10.1089.

Helgeson, H.C. and Kirkham, D.H. 1974b. Theoretical prediction of the thermodynamic behavior of aqueous electrolytes at high pressures and temperatures, Debye–Hückel parameters for activity coefficients and relative partial molar quantities. *American Journal of Science* 274: 1199–261.

Helgeson, H.C., Kirkham, D.H. and Flowers, G.C. 1981. Theoretical predictions of the thermodynamic behavior of electrolytes at high pressures and temperatures: IV. Calculation of activity coefficients, osmotic coefficients, and apparent molar and standard and relative partial molar properties to 600°C and 5 kbar. *American Journal of Science* 281: 1249–316.

Holdaway, M.J. 1971. Stability of andalusite and the aluminum silicate phase diagram. *American Journal of Science* 271: 91–131.

Holland, T.J.B. and Blundy, J. 1994. Nonideal interactions in calcic amphiboles and their bearing on amphibole–plagioclase thermometry. *Contributions to Mineralogy and Petrology* 116: 433–447.

Holland, T.J.B., Navrotsky, A. and Newton, R.C. 1979. Thermodynamic parameters of $CaMgSi_2O_6–Mg_2Si_2O_6$ pyroxenes based on regular solution and cooperative disordering models. *Contributions to Mineralogy and Petrology* 69: 337–44.

Holland, T. J. B. and Powell, R. 2011. An improved and extended internally consistent thermodynamic dataset for phases of petrological interest, involving a new equation of state for solids. *Journal of Metamorphic Geology* 29: 333–83. doi: 10.1111/j.1525-1314.2010.00923.x

Koziol, A.M. and Newton, R.C. 1988. Redetermination of the garnet breakdown reaction and improvement of the plagioclase-garnet-Al_2O_5-quartz geobarometer. *American Mineralogist* 73: 216–23.

Kudo, A.M. and Weill, D.F. 1970. An igneous plagioclase thermometer. *Contributions to Mineralogy and Petrology* 25: 52–65.

Kularatne, K. and Audétat, A. 2014. Rutile solubility in hydrous rhyolite melts at 750–900°C and 2kbar, with application to titanium-in-quartz (TitaniQ) thermobarometry. *Geochimica et Cosmochimica Acta* 125: 196–209. doi: 10.1016/j.gca.2013.10.020.

Mäder, U.K., Percival, J.A. and Berman, R.G., 1994. Thermobarometry of garnet–clinopyroxene–hornblende granulites from the Kapuskasing structural zone. *Canadian Journal of Earth Sciences* 31, 1134–45. doi: 10.1139/e94-101.

Morel, F.M.M. and Hering, J.G. 1993. *Principles and Applications of Aquatic Chemistry*. New York, John Wiley and Sons.

Nicholls, J. and Russell, J.K. 1990. *Modern Methods in Igneous Petrology, Reviews in Mineralology* vol. 24. Washington, Mineralogical Society of America.

Nielsen, R.L. and Dungan, M.A. 1983. Low pressure mineral-melt equilibria in natural anhydrous mafic systems. *Contributions to Mineralogy and Petrology* 84: 310–26.

Nordstrom, D.K. and Munoz, J.L. 1986. *Geochemical Thermodynamics*. Palo Alto: Blackwell Scientific.

Ostapenko, G., Gamarnik, M.Y., Gorogotskaya, L., Kuznetsov, G., Tarashchan, A. and Timoshkova, L. 1987. Isomorphism of titanium substitution for silicon in quartz: experimental data. *Mineralogicheskii Zhurnal* 9: 30–40.

Parkhurst, D. L., Appelo, C.A.J. (2013) Description of input and examples for PHREEQC version 3 – a computer program for speciation, batch-reaction, one-dimensional transport, and inverse geochemical calculations. *US Geological Survey Techniques and Methods*, book 6, chap A43, vol. US Geological Survey, Denver. 497p.

Pitzer, K. S. 1973. Thermodynamics of electrolytes. I. Theoretical basis and general equations. *Journal of Physical Chemistry* 77: 268–77.

Pitzer, K. S. 1981. Characteristics of very concentrated aqueous solutions. *Physics and Chemistry of the Earth* 13–14: 249–272. doi: 10.1016/0079-1946(81)90013-6.

Powell, R. and Holland, T.J.B. 2008. On thermobarometry. *Journal of Metamorphic Geology* 26: 155–79. doi: 10.1111/j.1525-1314.2007.00756.x, 2008.

Powell, R., Holland, T. and Worley, B. 1998. Calculating phase diagrams involving solid solutions via non-linear equations, with examples using THERMOCALC. *Journal of Metamorphic Geology* 16: 577–88. doi: 10.1111/j.1525-1314.1998.00157.x.

Putirka, K. 2005. Igneous thermometers and barometers based on plagioclase + liquid equilibria: tests of some existing models and new calibrations. *American Mineralogist* 90: 336–46.

Putirka, K. (2018) Geothermometry and geobarometry. In: White W.M. (Ed). *Encyclopedia of Geochemistry*, Springer, Dordrecht, pp. 602–20.

Putirka, K.D., Perfit, M., J., R.F., Jackson, M.G., 2007. Ambient and excess mantle temperatures, olivine thermometry, and active vs. passive upwelling. *Chemical Geology* 241: 177–206. doi: 10.1016/j.chemgeo.2007.01.014.

Rankenburg, K., Lassiter, J.C. and Brey, G. 2004. Origin of megacrysts in volcanic rocks of the Cameroon volcanic chain – constraints on magma genesis and crustal contamination. *Contributions to Mineralogy and Petrology* 147: 129–44.

Richardson, S.M. and McSween, H.Y. 1989. *Geochemistry: Pathways and Processes*. New York, Prentice Hall.

Robinson, R.A. and Stokes, R.H. 1959. *Electrolyte Solutions*. London, Butterworths.

Roeder, P.L. and Emslie, R.F. 1970. Olivine-liquid equilibrium. *Contributions to Mineralogy and Petrology* 29: 275–89.

Smith, P. M., Asimow, P. D. 2005. Adiabat_1ph: A new public front-end to the MELTS, pMELTS, and phMELTS models. *Geochemistry Geophysics Geosystems* 6: Q02004. doi: 10.1029/2004gc000816.

Spencer, K. and Lindsley, D.H. 1981. A solution model for coexisting iron-titanium oxides. *American Mineralogist* 66: 1189–201.

Stebbins, J.F., Carmichael, I.S.E. and Weill, D.F. 1983. The high temperature liquid and glass heat contents and the heats of fusion of diopside, albite, sanidine and nepheline. *American Mineralogist* 68: 717–30.

Tanger, J. C. and Helgeson, H. C. 1988. Calculation of the thermodynamic and transport properties of aqueous species at high pressures and temperatures; revised equations of state for the standard partial molal properties of ions and electrolytes. *American Journal of Science* 288: 19–98. doi: 10.2475/ajs.288.1.19.

Thompson, J.B. 1967. Thermodynamic properties of simple solutions, in *Researches in Geochemistry vol. 2* (ed. P.H. Abelson). New York, John Wiley and Sons.

Thompson, J.B. and Waldbaum, D.R. 1969. Mixing properties of sanidine crystalline solutions: III. Calculations based on two-phase data. *American Mineralogist* 54: 811–38.

Truesdell, A.H. and Jones, B.F. 1974. WATEQ, a computer program for calculating chemical equilibria in natural waters. *J. Res. US Geol. Surv.* 2: 233–48.

Thomas, J.B., Watson, E.B., Spear, F.S. and Wark, D.A., 2015. TitaniQ recrystallized: experimental confirmation of the original Ti-in-quartz calibrations. *Contributions to Mineralogy and Petrology* 169: 27. doi: 10.1007/s00410-015-1120-0.

Wark, D.A. and Watson, E.B., 2006. TitaniQ: a titanium-in-quartz geothermometer. *Contributions to Mineralogy and Petrology* 152, 743–754. doi: 10.1007/s00410-006-0132-1.

Wilson, C. J. N., Seward, T. M., Allan, A. S. R., *et al.* 2012. A comment on: 'TitaniQ under pressure: The effect of pressure and temperature on the solubility of ti in quartz', by Jay B. Thomas, E. Bruce Watson, Frank S. Spear, Philip T. Shemella, Saroj K. Nayak and Antonio Lanzirotti. *Contributions to Mineralogy and Petrology* 164: 359–368 doi: 10.1007/s00410-012-0757-1.

Wolery, T. W. and Jarek, R. L. (2003) Eq3/6, version 8.0 software user's manual. In, *vol 8. U.S. Department of Energy Office of Civilian Radioactive Waste Management Office of Repository Development*, Las Vegas, NV, p. 376.

Wood, B.J. 1987. Thermodynamics of multicomponent systems containing several solutions. In *Thermodynamic Modeling of Geological Materials: Minerals, Fluids and Melts* (eds. I.S.E. Carmichael and H.P. Eugster), pp. 71–96. Washington, Mineralogical Society of America.

Wood, B.J. and Banno, S. 1973. Garnet-orthopyroxene and orthopyroxene-clinopyroxene relationships in simple and complex systems. *Contributions to Mineralogy and Petrology* 42: 109–24.

PROBLEMS

1. Kyanite, andalusite, and sillimanite (all polymorphs of Al_2SiO_5) are all in equilibrium at 500°C and 376 MPa. Use this information and the table to construct an approximate temperature–pressure phase diagram for the system kyanite–sillimanite–andalusite. Assume ΔV and ΔS are independent of temperature and pressure. Label each field with the phase present.

ϕ	$\overline{V}$ (cm^2)	S (J/K-mol)
Kyanite	44.09	242.30
Andalusite	51.53	251.37
Sillimanite	49.90	253.05

2. Show that: $\overline{G}_{ex} = (W_{G_1}X_2 + W_{G_2}X_1)X_1X_2$ may be written as a four-term power expansion, i.e.:

$$\overline{G}_{excess} = A + BX_2 + CX_2^2 + DX_2^3$$

3. Construct G-bar–X diagrams for a regular solution with $W_G = 12$ kJ at 100°C temperature intervals of 200–700°C. Sketch the corresponding phase diagram.

4. Interaction parameters for the enstatite–diopside solid solution have been determined as follows: $W_{H-En} = 34.0$ kJ/mol, $W_{H-Di} = 24.74$ kJ/mol (assume W_V and W_S are 0).

   (a) Use the asymmetric solution model to calculate ΔG_{real} as a function of X_2 (let diopside be component 2) curves for this system at 100 K temperature intervals of 1000–1500 K. Label your curves.

   (b) What is the maximum mole fraction of diopside that can dissolve in enstatite in this temperature range?

   (c) Sketch the corresponding T–X phase diagram.

5. Sketch G-bar–X diagrams for 1600°C, 1500°C, 1300°C, and 1250°C for the system diopside–anorthite (Figure 4.9). Draw tangents connecting the equilibrium liquids and solids.

6. Suppose you conduct a 1 atm melting experiment on a plagioclase crystal. Predict the mole fractions of anorthite in the liquid and solid phases at a temperature of 1425°C. Assume both the liquid and solid behave as ideal solutions. Albite melts at 1118°C, anorthite at 1553°C. ΔH_m for albite is 54.84 kJ/mol; ΔH_m for anorthite is 123.1 kJ/mol.

7. Given the following two analyses of basaltic glass and coexisting olivine phenocrysts, determine the K_D for the MgO $\rightleftharpoons$ FeO exchange reaction, and calculate the temperatures at which the olivine crystallized using both MgO and FeO. Assume Fe_2O_3 to be 10 mole % of total iron (the analysis below includes only the total iron, calculated as FeO; you need to calculate from this the amount of FeO by subtracting an appropriate amount to be assigned as Fe_2O_3). Note that the mole % Fo in olivine is equivalent to the mole % Mg or MgO. (*HINT: You will need to calculate the mole fraction of MgO and FeO in the liquid.*)

Sample glass (liquid) composition	TR3D-1 (wt % oxide)	DS-D8A (wt % oxide)
SiO_2	50.32	49.83
Al_2O_3	14.05	14.09
ΣFe as FeO	11.49	11.42
MgO	7.27	7.74
CaO	11.49	10.96
Na_2O	2.3	2.38
K_2O	0.10	0.13
MnO	0.17	0.20
TiO_2	1.46	1.55
Olivine		
Mole % Fo (= mole % Mg)	79	81

8. Starting from equations 4.67, 4.68, and 4.16, use the fundamental relationships between free energy, entropy, enthalpy, and the equilibrium constant to derive the temperature dependence of the titanomagnetite–ilmenite exchange (eqn. 4.70).

9. Determine the temperature and oxygen fugacity of equilibration for the following set of coexisting iron-titanium oxides in lavas from the Azores:

	Titanomagnetite s.s. phase mole % magnetite	Ilmenite s.s. phase mole % hematite
G-4 groundmass	29.0	10.3
SJ-8 phenocrysts	41.9	13.0
SM-28 microphenocrysts	54.5	7.0
T-8 groundmass	33.7	8.1
F-29 microphenocrysts	36.2	6.0

Make a plot of f_{O2} vs. temperature using your results and compare with Figure 3.21. What buffer do the data fall near?

10. Average mid-ocean ridge basalt (MORB) has the composition in the table below. Use the "web applet" version of MELTS (http://melts.ofm-research.org/index.html) to answer the following questions.

   (a) At a pressure of 500 bars and f_{O2} of QFM-1, what is the liquidus temperature of this magma?
   (b) If this magma cools and undergoes fractional crystallization of solid phases to a temperature of 1100°C, what would be the composition of the remaining magma? What fraction of liquid would remain?
   (c) If instead the oxygen fugacity were QFM+1, what would be the composition of the remaining magma at 1100°C? How much liquid would remain?

Oxide	Weight percent
SiO_2	50.37
TiO_2	1.44
Al_2O_3	15.38
Fe_2O_3	1.10
FeO	8.94
MnO	1.10
MgO	7.92
CaO	11.51
Na_2O	2.70
K_2O	0.18
P_2O_5	0.15
H_2O	0.15

11. Derive the Thomas et al. (2015) TitaniQ equation (eqn. 4.61) of the temperature and pressure dependence of the equilibrium constant. What thermodynamic functions do the constants a, b, and c represent?

12. An analysis of an oil field brine from Mississippi with a temperature of 150°C is shown here. Calculate the *activities* of these species at that temperature using the Truesdell–Jones equation.

Species	Conc. (g/kg)
Na^+	63.00
K^+	6.15
Mg^{2+}	2.77
Ca^{2+}	44.6
Cl^-	200.4
SO_4^{2-}	0.13
HCO_3^-	0.03

13. Show that for a strong electrolyte, i.e., one in which dissociation is complete and:

$$m_- = v_-m \quad \text{and} \quad m_+ = v_+m$$

where m is the molality of the solute component A_v+B_v-, that:

$$m_\pm = (m_+^{v^+} m_-^{v^-})^{1/v}$$

where $v = v_+ + v_-$.

14. Mean ionic activity coefficients were measured for the following solutions at an ionic strength of 3: $\gamma_{KCl} = 0.569$, $\gamma_{NaCl} = 0.734$, $\gamma_{Na_2CO_3} = 0.229$. Assuming $\gamma_{Cl^-} = \gamma_{K^+} = \gamma_{\pm KCl}$, what is the activity coefficient of CO_3^{2-}?

15. Calculate the electrostatic, γ_{elect}, and solvation, γ_{solv}, contributions to the mean ionic activity coefficient of $MgCl_2$ at concentrations of 0.0033, 0.01, 0.033, 0.05, 0.1, 0.33, 0.5, and 1 using the Debye–Hückel (eqn. 3.74) and Robinson and Stokes (eqn. 4.109) equations, respectively. Plot your results, as well as $\gamma_{elect+solv} = \gamma_{elect} * \gamma_{solv}$ as a function of ionic strength (i.e., a plot similar to Figure 4.30).

16. Calculate the mean ionic activity coefficient for $NaCO_3$ in a 0.5 M solution using the Debye–Hückel (eqn. 3.74) and Truesdell–Jones (eqn. 4.120) equations and compare your results with the observed values shown here. Overall, which fits the data better?

I, m	$\gamma_{\pm}$ observed
0.001	
0.003	0.887
0.006	0.847
0.01	
0.015	0.78
0.03	0.716
0.06	0.644
0.1	
0.15	0.541
0.3	0.462
0.6	0.385
1	
1.5	0.292
3	0.229
6	0.182

17. Use PHREEQC to calculate activity coefficients of the following Dead Sea brine composition:

Dead Sea Brine Composition	
	mmol/kg
Cl	5294
Ca	391
Br	63
B	4.6
Li	3.4
Mg	1663
Na	1081
K	169
SO_4	4.2
HCO_3^-/CO_3^{2-}	0.86

Specify temperature as 25 °C and pH as 6.27 and use the Pitzer.dat database.

Chapter 5

Kinetics: the pace of things

5.1 INTRODUCTION

Thermodynamics concerns itself with the distribution of components among the various phases and species of a system at equilibrium. *Kinetics* concerns itself with the *path* the system takes in achieving equilibrium. Thermodynamics allows us to predict the equilibrium state of a system. Kinetics, on the other hand, tells us how and how fast equilibrium will be attained. Although thermodynamics is a macroscopic science, we found it often useful to consider the microscopic viewpoint in developing thermodynamics models. Because kinetics concerns itself with the path a system takes, what we will call *reaction mechanisms*, the microscopic perspective becomes essential, and we will very often make use of it.

Our everyday experience tells one very important thing about reaction kinetics: they are generally slow at low temperature and become faster at higher temperature. For example, sugar dissolves much more rapidly in hot tea than it does in iced tea. Good instructions for making iced tea might then incorporate this knowledge of kinetics and include the instruction to be sure to dissolve the sugar in the hot tea before pouring it over ice. Because of this temperature dependence of reaction rates, low-temperature geochemical systems are often not in equilibrium. A good example might be clastic sediments, which consist of a variety of phases. Some of these phases are in equilibrium with each

other and with porewater, but most are not. Another example of this disequilibrium is the oceans. The surface waters of the oceans are everywhere oversaturated with respect to calcite, yet calcite precipitates from seawater only through biological activity. At a depth of 2500 m, the ocean is undersaturated with calcite, yet calcite shells of microorganisms persist in sediments deposited at these depths (though they do dissolve at greater depths). Thus, great care must be used in applying thermodynamics to such systems. Even in the best of circumstances, thermodynamics will provide only a limited understanding of low-temperature geochemical systems. A more complete understanding requires the application of kinetic theory. Indeed, for such systems, kinetics is the deciding factor controlling their state and evolution. Even in metamorphic systems, with temperatures in the range of 300–700°C, kinetics factors are crucially important in determining their final states.

High-temperature geochemical systems, such as crystallizing magmas or high-grade metamorphic rocks, are more likely to be in equilibrium, and thermodynamics provides a reasonable understanding of these systems. However, even at high temperatures, kinetic factors remain important and can inhibit equilibrium. One obvious example of disequilibrium at high temperature is the formation of volcanic glasses. Thermodynamics predicts that magmas should crystallize as they cool. But where cooling is rapid enough, this

Geochemistry, Second Edition. William M. White.
© 2020 John Wiley & Sons Ltd. Published 2020 by John Wiley & Sons Ltd.
Companion website: www.wiley.com/go/white/geochemistry

does not occur. Glasses, which in many ways are simply extremely viscous liquids, form instead.

It is perhaps ironic that it is kinetic factors, and a failure to achieve equilibrium, that in the end allow us to use thermodynamics to make statements about the Earth's interior. As we pointed out in the preceding chapter, if equilibrium were always achieved, the only rocks we could collect at the surface of the Earth (which is, after all, the only place we can collect them) would consist of quartz, clays, serpentine, and so on; their petrology would tell us nothing about their igneous or metamorphic histories. Fortunately, kinetic factors allow the original minerals and textures of gneisses, peridotites, and lavas to be preserved for our study.

The foregoing might suggest that kinetics and thermodynamics are entirely unrelated subjects, and further, that what we have learned about thermodynamics is of little use in many instances. This is certainly not the case. As we shall see, transition state theory provides a very strong link between kinetics and thermodynamics. What we have learned about thermodynamics will prove very useful in our brief study of kinetics. Furthermore, chemical systems are always governed by a combination of thermodynamics and kinetics, so a full understanding of the Earth requires the use of both thermodynamic and kinetic tools. The goal of this chapter is to add the latter to our geochemical toolbox. This topic is vast and our coverage brief. Books by Lasaga (1997) and Zhong (2008) provide more thorough coverage.

5.2   REACTION KINETICS

5.2.1   Elementary and overall reactions

In thermodynamics, we found that the equilibrium state of a system is entirely independent of the path taken to reach that state. The goal of kinetics is a description of the way reactions proceed toward equilibrium. This description is inherently path-dependent. Consider, for example, the weathering of anorthite. We can write an *overall* reaction for this process as:

$$CaAl_2Si_2O_8 + 3H_2O + CO_{2(g)} \rightarrow CaCO_3$$
$$+ 2Al(OH)_3 + 2SiO_{2(qz)} \qquad (5.1)$$

In nature, however, this process will involve several intermediate steps. These intermediate steps can include:

$$H_2O + CO_{2(g)} \rightarrow H_2CO_{3(aq)} \qquad (5.2)$$

$$H_2CO_3 \rightarrow HCO_{3(aq)}^- + H^+ \qquad (5.3)$$

$$CaAl_2Si_2O_8 + H_2O + 2H^+$$
$$\rightarrow Si_2Al_2O_5(OH)_4 + Ca_{(aq)}^{2+} \qquad (5.4)$$

$$H_2O + Si_2Al_2O_5(OH)_4 \rightarrow 2SiO_{2(qz)}$$
$$+ 2Al(OH)_3 \qquad (5.5)$$

$$HCO_{3(aq)}^- \rightarrow CO_{3(aq)}^{2-} + H^+ \qquad (5.6)$$

$$CO_{3(aq)}^{2-} + CO_{(aq)}^{2+} \rightarrow CaCO_3 \qquad (5.7)$$

In thermodynamics, eqn. 5.1 is a perfectly adequate description of the reaction. In kinetics, a description of an *overall* reaction, such as 5.1, requires knowledge of the path taken, that is, a knowledge of the steps involved. Reactions 5.2 through 5.7 thus describe the overall reaction 5.1. Reactions 5.2, 5.3, and 5.6 are *elementary reactions* in that they involve only one step and the reaction as written describes what occurs on the microscopic level. The remaining reactions are not elementary in that they each consist of a number of more elementary steps.

5.2.2   Reaction mechanisms

Reaction 5.4 describes the breakdown of anorthite to form kaolinite plus a free calcium ion. This reaction involves profound structural changes in the solid phase that are not described by eqn. 5.4. A full kinetic description of 5.4 will require some knowledge of the steps involved in these structural changes. One possibility is that the first step in reaction 5.4 is the adsorption of hydrogen ions to the anorthite surface:

$$CaAl_2Si_2O_8 + 2H^+$$
$$\rightarrow CaAl_2Si_2O_8 = 2H^+ \qquad (5.4a)$$

where we are using the = sign to indicate surface adsorbed ions. The H$^+$ ions then replace calcium in the anorthite structure:

$$CaAl_2Si_2O_8 + 2H^+ \ CaAl_2Si_2O_8$$

$$= 2H^+ \rightarrow H_2Al_2Si_2O_8 + Ca^{2+} \quad (5.4b)$$

Reaction 5.5, the breakdown of kaolinite to quartz and gibbsite, could involve SiO_2 dissolving, subsequently precipitating as opaline silica, and later crystallizing to quartz:

$$Si_2Al_2O_5(OH)_4 + 5H_2O \rightarrow 2H_4SiO_{4(aq)}$$

$$+ 2Al(OH)_3 \quad (5.5a)$$

$$H_4SiO_{4(aq)} \rightarrow SiO_{2(opal)} + 2H_2O \quad (5.5b)$$

$$SiO_{2(opal)} \rightarrow SiO_{2(qz)} \quad (5.5c)$$

The description of an overall reaction in terms of elementary reactions is called the *reaction mechanism*. The rates of truly elementary reactions are path-independent because there is only one possible path. In this sense, elementary reactions are somewhat analogous to state functions in thermodynamics. Clearly then, an important step in any kinetic study is determination of the reaction mechanism, that is, to describe the process in terms of elementary reactions. As we shall see, there may be more than one possible path for an overall reaction, and several paths may be simultaneously involved. Kinetics can only provide an accurate description of a process if all these paths are known.

5.2.3 Reaction rates

Consider a reaction such as the precipitation of dolomite from a solution. We can describe this as:

$$Ca^{2+} + Mg^{2+} + 2CO_3^{2-} \rightleftharpoons CaMg(CO_3)_2$$

We *define* the *rate* of this reaction, $\mathfrak{R}$, as the rate at which dolomite is produced:

$$\mathfrak{R} \equiv \frac{d[CaMg(CO_3)_2]}{dt} \quad (5.8)$$

Clearly, if dolomite is to be formed, CO_3^{2-} must be consumed in this reaction twice as fast as Ca or Mg. For every mole of Ca or Mg consumed, exactly two moles of CO_3^{2-} will

also be consumed and one mole of dolomite produced. This being the case, we could equally well express the reaction rate as:

$$\mathfrak{R} = -\frac{1}{2}\frac{d[CO_3^{2-}]}{dt} \quad \text{or} \quad \mathfrak{R} = -\frac{d[Ca^{2+}]}{dt}$$

$$= -\frac{d[Mg^{2+}]}{dt}$$

We can now formulate the general rule. For any reaction such as:

$$aA + bB \rightarrow cC + dD \quad (5.9)$$

The *reaction rate*, $\mathfrak{R}$, is defined as the change in composition of the reaction mixture with time:

$$\mathfrak{R} \equiv -\frac{1}{a}\frac{d[A]}{dt} = -\frac{1}{b}\frac{d[B]}{dt} = \frac{1}{c}\frac{d[C]}{dt} = \frac{1}{d}\frac{[D]}{dt}$$

$$(5.10)$$

The brackets denote the concentrations of the species and the negative sign indicates that reactants are consumed as the reaction proceeds. Thus, *the rate of a reaction is simply the rate at which a reactant is consumed or product produced* divided by its stoichiometric coefficient.

5.2.3.1 The reaction rate for an elementary reaction: composition dependence

Reaction rates will, in general, depend on the concentration of the reactant. To understand this, consider the reaction:

$$N° + O_2 \rightleftharpoons NO + O \quad (5.11)$$

This reaction between free nitrogen atoms and oxygen molecules occurs in the stratosphere (where N° is produced by high-energy collisions involving N_2) and contributes to the production of nitrous oxide. Let's assume that reaction 5.8 is an adequate description of this reaction. In other words, we are assuming that 5.8 is an elementary reaction, and the reaction mechanism for the production of NO from nitrogen and oxygen gas is collision between a N° molecule and O_2 molecule. For the reaction to occur, the nitrogen and oxygen molecules must collide with enough kinetic energy that the mutual repulsion of the electron clouds is overcome and the electrons can

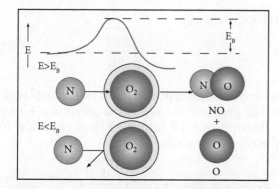

Figure 5.1 A nitrogen atom approaching an oxygen molecule must have enough kinetic energy to pass through the region where it is repelled by electrostatic repulsion of the electron cloud of the oxygen. Otherwise, it will not approach closely enough so that its electrons can combine with those of oxygen.

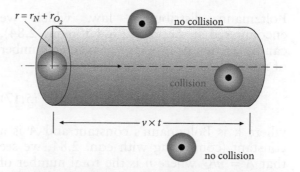

Figure 5.2 A nitrogen atom will sweep out a volume $V = v\pi(r_N + r_O)^2 t$ in time t. Whether a collision occurs will depend on whether the center (indicated by black dot) of an oxygen atom falls within this volume.

be redistributed into new covalent orbits. The repulsive force represents an energy barrier, E_B, which will prevent low-energy nitrogen and oxygen atoms from reacting. Figure 5.1 illustrates this point. The reaction rate will therefore depend on (1) the number of collisions per unit time, and (2) the fraction of N and O molecules having energy greater than the barrier energy.

Let's first consider the number of collisions per unit time. In order for a "collision" to occur, the electron clouds must overlap, that is, they must approach within $(r_N + r_O)$, where r_N and r_O are the radii of the nitrogen and oxygen molecules. To make things simple, imagine the oxygen to be fixed and the nitrogen in motion. In other words, our reference frame will be that of the oxygen molecules. We can imagine the nitrogen sweeping out a cross-section with radius $(r_N + r_O)$ as it travels. If the nitrogen is travelling at velocity v, in time t, it will sweep out a cylindrical volume (Figure 5.2):

$$V = v\pi(r_N + r_O)^2 t \qquad (5.12)$$

Whether a collision occurs will depend on whether the *center* of an oxygen molecule falls within this volume (Figure 5.2). The number of collisions that will occur in this time will be:

$$C = n_O v\pi(r_N + r_O)^2 t \qquad (5.13)$$

where n_O is the number of oxygen molecules per unit volume. The number of collisions per unit time is then simply:

$$\frac{C}{t} = n_O v\pi(r_N + r_O)^2 \qquad (5.14)$$

If there are n_N nitrogen atoms and the average velocity between nitrogen and oxygen molecules is $\bar{v}$, then the number of collisions per unit time is:

$$\dot{c} = n_N n_O \bar{v}\pi(r_N + r_O)^2 \qquad (5.15)$$

If we let

$$k = \bar{v}\pi(r_N + r_O)^2$$

then the rate at which collisions occur is:

$$\dot{c} = kn_N n_O \qquad (5.16)$$

Thus we see that the reaction rate in this case will depend on the concentration of nitrogen, oxygen and a constant that depends on the nature of the reactants. This is a general result.

5.2.3.2 The reaction rate for an elementary reaction: temperature dependence

We now need to estimate the fraction of nitrogen and oxygen atoms having at least the barrier energy, E_B. For simplicity, we will assume that oxygen and nitrogen molecules have an identical energy distribution. The

Boltzmann distribution law, which we encountered in Section 2.8.4.1 (eqn. 2.84), can be written to express the average number of molecules having energy level ε_i as:

$$n_i = Ae^{\varepsilon_i/kT} \qquad (5.17)$$

where k is Boltzmann's constant and A is a constant (comparing with eqn. 2.84, we see that $A = n/Q$ where n is the total number of molecules in the system and Q is the *partition function*). In plain English, this equation tells us that the number of molecules in some energy level i decreases exponentially as the energy of that level increases (Figure 2.9). We want to know the number of molecules with energy greater than E_B. In this case, we are dealing with translational energy. The quantum spacings between translational energy levels are so small that they essentially form a continuum, allowing us to integrate eqn. 5.17. Fortunately for us, the integration of 5.15 from $\varepsilon = E_B$ to infinity has a simple solution:

$$A \int_{E_B}^{\infty} e^{-\varepsilon_i/kT} d\varepsilon = AkTe^{-E_B/kT} \qquad (5.18)$$

The *fraction* of molecules with energy greater than E_B is just:

$$\frac{A \int_{E_B}^{\infty} e^{-\varepsilon_i/kT} d\varepsilon}{A \int_{0}^{\infty} e^{-\varepsilon_i/kT} d\varepsilon} = \frac{AkTe^{-E_B/kT}}{AkT} = e^{-E_B/kT}$$
$$(5.19)$$

The rate of reaction will be the rate of collision times the fraction of molecules having energy greater than E_B:

$$\mathfrak{R} = n_N n_O \bar{v} \pi (r_N + r_O)^2 e^{-E_B/kT} \qquad (5.20)$$

Now we just need to find a value for velocity. The average velocity can be calculated from the Maxwell–Boltzmann Distribution* of velocities of molecules in a gas. The average velocity is:

$$\bar{v} = \sqrt{\frac{8kT}{\pi\mu}} \qquad (5.19)$$

where μ is the reduced mass of the molecules in the gas; in this case $\mu = m_N m_O/(m_N + m_O)$. Substituting 5.18 into 5.17, our equation for the reaction rate is:

$$\mathfrak{R} = n_N n_O \pi (r_N + r_O)^2 \sqrt{\frac{8kT}{\pi\mu}} e^{-E_B/kT} \quad (5.22)$$

Redefining k as:

$$k = \pi(r_N + r_O)^2 \sqrt{\frac{8kT}{\pi\mu}} e^{-E_B/kT} \qquad (5.23)$$

our reaction rate equation is:

$$\mathfrak{R} = k n_N n_O \qquad (5.24)$$

Thus, the reaction rate in this case depends on the concentration of nitrogen and oxygen and a constant k, called the *rate constant,*[†] which depends on temperature, properties of the reactants, and the barrier energy.

In a more rigorous analysis we would have to take into consideration atoms and molecules not being spherically symmetric and that, as a result, some orientations of the molecules are more likely to result in a reaction than others. In addition, a head-on collision is more likely to result in reaction than a glancing blow, so the collision cross-section will be less than $\pi(r_N + r_O)^2$. These factors can, however, be accounted for by multiplying by a constant, called a *stearic factor*, so the form of our equation, and the temperature dependence, would not be affected. Values of stearic factors for various reactions range over many orders of magnitude and can be quite small. In rare

---

* So-called because Maxwell proposed it and Boltzmann proved it rigorously. James Clerk Maxwell (1831–1879) was a Scottish physicist and professor at Marischal College in Aberdeen and later Kings College of London. His contributions to physics rank with those of Newton and Einstein. He is best known for his electromagnetic theory of light. His contributions to physical chemistry include the kinetic theory of gases and the Maxwell relations introduced in Chapter 2.

† So many k's! To distinguish the rate constant, k, from Boltzmann's constant, k, we will always write the former in lower-case italics and the latter in roman typeface and the equilibrium constant, K, will be written in upper case roman typeface.

circumstances, they can be greater than 1 (implying an effective collision cross-section greater than the combined atomic radii).

Temperature occurs in two places in eqn. 5.23; however, the square-root dependence is slight compared with the exponential one. For example, consider a temperature change of 300 K to 325 K. For a reaction with an activation energy of 25 kJ, the exponential temperature dependence results in an increase in reaction rate of more than a factor of 2, whereas the square root dependence increases the reaction rate by only 4%. Hence the temperature dependence can be essentially expressed as:

$$k \propto e^{-E_B/kT}$$

The temperature dependence of the rate constant is most often written as:

$$\boxed{k = Ae^{-E_B/kT}} \qquad (5.25)$$

which is the important *Arrhenius relation*.* It expresses the rate constant in terms of the barrier, or *activation*, energy (also often written as E_A or E^*), and A, a proportionality constant sometimes called the *frequency factor* (because it depends on the frequency of collisions), and temperature. The gas constant, R, is the product of Boltzmann's constant and N_A, the Avagadro number. So we use R in place of k, when we deal in moles rather than atoms.

The temperature dependence of the rate constant is illustrated in Figure 5.3. We see that the reaction rate falls off by a factor of 10^2 as temperature is decreased from 500 to 200 K. This confirms our everyday experience that reaction rates are extremely temperature-dependent. Table 5.1 lists some examples of activation energies for geochemical reactions.

The pre-exponential factor, A, is often assumed to be independent of temperature.

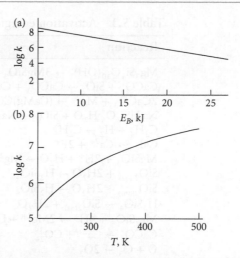

Figure 5.3 (a) Relative change in the reaction rate as a function of activation energy at 300 K. (b) Change in the reaction rate for the same as a function of temperature with an activation energy of 15 kJ.

Comparison of eqn. 5.25 with 5.23 shows, however, that it need not be. In the case of an elementary gas phase reaction, we would predict a dependence on the square root of temperature. Other kinds of reactions show other kinds of temperature dependencies of the frequency factor, however. A more accurate expression of temperature dependence of the reaction rate is:

$$k = AT^n e^{-E_B/kT} \qquad (5.26)$$

where the exponent n can be any number. Nevertheless, the temperature dependence of the frequency factor is usually small and it can often be safely neglected, as in our example above.

5.2.3.3 A general form of the rate equation

In general, the rate of a reaction such as:

$$aA + bB \rightarrow cC + dD$$

---

* Named for Svante August Arrhenius (1859–1927) because Arrhenius provided the theoretical justification for the equation which was originally proposed by Jacobus Van't Hoff's based on analogy to the Van't Hoff equation (eqn. 3.105). Arrhenius's PhD dissertation, completed in 1884 at the University of Uppsala in Sweden, was rated fourth class by the committee of examiners, implying great things were not expected of him. The old boys must have been a little surprised 19 years later when Arrhenius won the Nobel Prize for chemistry for his ionic theory of electrolyte solutions. Among Arrhenius's other contributions was the prediction that burning of fossil fuel would lead to rising atmospheric CO_2 concentrations and that global climate would warm as a consequence.

Table 5.1 Activation energies of some geochemical reactions.

Reaction	E_A kJ/mol
$Mg_3Si_4O_{10}(OH)_2 \rightarrow 3MgSiO_3 + SiO_2 + H_2O$	371.8
$CaCO_3 + SiO_2 \rightarrow CaCO_3 + CO_2$	225.0
$2CaCO_3 + Mg^{2+} \rightarrow (CaMg)CO_3 + Ca^{2+}$	117.1
$NaAlSi_2O_6 \cdot H_2O + SiO_2 \rightarrow NaAlSi_3O_8 + H_2O$	106.3
$C_2H_4 + H_2 \rightarrow C_2H_6$	102.8
$CaF_2 \rightarrow Ca^{2+} + 2F^+$	73.0
$MgSiO_3 + 2H^+ + H_2O \rightarrow Mg^{2+} + H_4SiO_4$	49.0
$SiO_{2\,(qz)} + 2H_2O \rightarrow H_4SiO_4$	40.6
$SiO_{2(am)} + 2H_2O \rightarrow H_4SiO_4$	35.8
$H_2SiO_4 \rightarrow SiO_{2(qz)} + 2H_2O$	28.4
$Mg_2SiO_4 + 4H^+ \rightarrow 2Mg^{2+} + H_2SiO_4$	21.7
$CaCO_3 \rightarrow Ca^{2+} + CO_{2+}^3$	20.1
$O + O_3 \rightarrow 2O_2$	13.4

can be expressed as:

$$\mathfrak{R} = k a_A^{n_A} a_B^{n_B} a_C^{n_C} a_D^{n_D} \qquad (5.27)^*$$

where k is the *rate constant* and a_A, etc. are activities (we will often use the simplifying assumption of ideality and replace these by concentrations). The exponents n_A, n_B, and so on, can be any number, including zero. The sum of the exponents n_A, n_B, ... is the *order of the reaction*. In general, the value of the exponents must be determined experimentally, though their values can be predicted if the reaction mechanism is known, as we saw in the above example.

Just as the mole fraction was the unit of choice for thermodynamics, *moles per volume, or moles per area in the case of reactions taking place on surfaces, is the unit of choice for kinetics*. Thus, wherever more than one phase is involved, one concentration should be expressed in moles per unit area or volume.

There are several simplifications of eqn. 5.25 for *elementary reactions*. First, the rate of reaction is independent of the concentration of the products, so the exponents of the products will be 0. Indeed, *one of the criteria for an elementary reaction is that the product does not influence the reaction rate*. Second, *the values of the exponents for the reactants are the stoichiometric coefficients*

of the reactant species. Thus, if the reaction can be written in terms of a series of elementary reactions, then the exponents for the rate equation can be deduced from those of the component elementary reactions. For elementary reactions, the order of reaction will be equal to the sum of the stoichiometric coefficients of the reactants. For complex reactions, however, the order of reaction must be deduced, either experimentally, or from the component elementary reactions.

A further simplification may be made where one of the reactants is in sufficient abundance that its concentration is not affected by the progress of the reaction of interest. For instance, the hydration of CO_2 through:

$$CO_2 + H_2O \rightarrow H_2CO_3$$

The rate of this reaction will be:

$$-\frac{d[CO_2]}{dt} = k[CO_2][H_2O]$$

which is a second-order reaction. However, in aqueous solution, H_2O will always be present in great excess over CO_2 and its abundance will not be significantly changed by this reaction. This allows us to treat the reaction as if it were first order and to

---

* Don't confuse this equation, which expresses the way in which reaction rates depend on concentrations, with eqn. 5.7, which is the definition of the reaction rate.

define a *pseudo-first-order* rate constant, k^*, as:

$$k^* = k\,[H_2O]$$

Since $[H_2O]$ is constant, it follows that k^* is as well. The reaction rate can then be written as:

$$-\frac{d[CO_2]}{dt} = k^*[CO_2]$$

In Examples 5.1 and 5.2 we have used just such a *pseudo-first- order* rate constant.

Example 5.1 Rate of hydration of $CO_{2(aq)}$

The rate for the hydration of CO_2 (i.e., $CO_2 + H_2O \leftrightharpoons H_2CO_3$) has been found to follow the first-order rate law:

$$-\frac{d[CO_{2(aq)}]}{dt} = k[CO_{2(aq)}] \tag{5.28}$$

At 25°C, k has been determined to be $0.014\ \text{sec}^{-1}$. Make a graph showing how the concentration of CO_2 will change with time as the reaction proceeds, assuming an equilibrium (i.e., final) CO_2 concentration of 0.

Answer: Since we are not given the absolute concentrations, we cannot determine the absolute change. We can, however, determine relative change. To do so, we just integrate 5.28:

$$-\int \frac{d[CO_{2(aq)}]}{[CO_2]}dt = k\int dt$$

With some rearranging, we obtain:

$$\frac{[CO_{2(aq)}]}{[CO_{2(aq)}]} = e^{-kt} \tag{5.29}$$

Figure 5.4 shows our result. It is apparent that this is a fast reaction. We can assume that equilibrium will prevail on most time scales of interest to us.

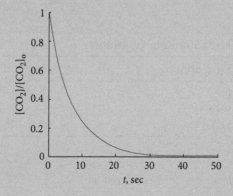

Figure 5.4   Progress in the reaction $CO_{2(aq)} + H_2O \rightarrow H_2CO_3$ with time, as measured by decrease in $[CO_2]$.

5.2.4 Rates of complex reactions

Deciding whether a reaction is elementary is not always straightforward. Consider the reaction:

$$2NO_2 \rightarrow 2NO + O_2 \qquad (5.30)$$

On a microscopic basis, we might describe this reaction as the collision of two NO_2 molecules to form two NO molecules and an O_2 molecule. Since no intermediate steps occur, this would appear to be an elementary reaction. The rate equation for this reaction has been experimentally determined to be:

$$= \frac{d[NO_2]}{dt} = 2k[NO_2]^2$$

This has the predicted form for an elementary reaction of second order; thus,

experiment confirms that reaction 5.28 is elementary.

Now consider the reaction:

$$2O_3 \rightarrow 3O_2$$

We might reason that this reaction requires only the collision of two ozone molecules with no intermediate products and that the reaction is therefore primary. However, the experimentally determined rate law is:

$$\frac{1}{3}\frac{d[O_2]}{dt} = -\frac{1}{2}\frac{d[O_3]}{dt} = k\frac{[O_3]^2}{[O_2]} \qquad (5.31)$$

Since the rate depends on the concentration of the product, the reaction is not elementary and must involve intermediate steps.

Example 5.2 Oxidation of Ferrous Iron

Given the equilibrium and pseudo-first order rate constants for the oxidation of three species of ferrous iron (Fe^{2+}, $Fe(OH)^+$, and $Fe(OH)_2$) to ferric iron in the adjacent table, calculate the overall rate of oxidation of ferrous iron at pH 2, 6, and 8 assuming a total Fe^{2+} concentration of 1×10^{-6} M.

Equilibrium Constants

Reaction	pK
$Fe^{2+} + H_2O \rightleftharpoons FeOH^+ + H^+$	4.5
$FeOH^+ + H_2O \rightleftharpoons Fe(OH)_2 + H^+$	7.4

Oxidation Rate Constants

Fe^{2+} Species	k (s^{-1})
Fe^{2+}	7.9×10^{-6}
$FeOH^+$	25
$Fe(OH)_2$	7.9×10^{6}

Answer: The overall oxidation rate can be written as:

$$\frac{d\Sigma Fe^{2+}}{dt} = k_1[Fe^{2+}] + k_2[FeOH^-] + k_3[Fe(OH)_2] \qquad (5.32)$$

Thus, to calculate the rate, we will have to calculate the concentrations of the various species. These are given by:

$$[FeOH^-] = K_1 \frac{[Fe^{2+}]}{[H^+]} \quad \text{and} \quad [Fe(OH)_2] = K_2 \frac{[Fe^{2+}]}{[H^+]^2}$$

Substituting these expressions into 5.32 we have:

$$\frac{d\Sigma Fe^{2+}}{dt} = [Fe^{2+}] \left(k_1 + \frac{k_2 K_1}{[H^+]} + \frac{k_3 K_2}{[H^+]^2} \right) \tag{5.33}$$

Since the total Fe^{2+} is the same at all three pH levels the concentration of the Fe^{2+} ion must vary. So we need to calculate the concentration of ionic Fe^{2+} at these pHs. The total Fe^{2+} is:

$$\Sigma Fe^{2+} = [Fe^{2+}] + [FeOH^-] + [Fe(OH)_2]$$

or:

$$\Sigma Fe^{2+} = [Fe^{2+}] \left(1 + \frac{K_1}{[H^+]} + \frac{K_2}{[H^+]^2} \right)$$

so that:

$$[Fe^{2+}] = \frac{\Sigma Fe^{2+}}{1 + \dfrac{K_1}{[H^+]} + \dfrac{K_2}{[H^+]^2}}$$

We can now calculate the rates. Substituting in appropriate values into eqn. 5.31, we find the rate is 0.0031 M/sec, 0.9371 M/sec, and 7.89 M/sec at a pH of 2, 4, and 8 respectively. Thus, the combination of the different rate constants and the pH dependency of the Fe speciation results in a very strong pH dependence of the oxidation rate.

5.2.4.1 Chain reactions and branching

Many overall reactions involve a series of sequential elementary reactions, or steps, each of which must be completed before a subsequent reaction can occur. Such reactions are termed *chain reactions*. It is also possible that the path of an overall reaction may include two or more alternative elementary reactions, or sequences of elementary reactions, that occur simultaneously. These alternative paths are called *branches*. The combustion of hydrogen is a good example because it is a chain reaction involving several branches.

Experiments have shown that the reaction rate for the combustion of hydrogen is not simply:

$$\frac{d[H_2O]}{dt} = k[H_2]^2[O_2] \tag{5.34}$$

and therefore $2H_2 + O_2 \rightarrow 2H_2O$ is not an elementary reaction. If it were an elementary reaction, eqn. 5.32 predicts that its rate should continuously decrease through the course of the reaction (provided temperature is held constant!) since the reactants will be consumed and their concentrations will decrease. In actuality, the rate of this reaction can increase rapidly, sometimes catastrophically (even at constant temperature), as it proceeds. Evidently, the reaction mechanism is more complex. Indeed, it appears to involve several steps. The final step of this reaction is:

$$OH + H_2 \rightarrow H_2O + H \tag{5.35}$$

This is an elementary reaction, depending only on the concentration of the two reactants. However, one of the reactants, OH, and one of the products, H, are not among the original constituents of the gas. Rather, they are created by intermediate steps. Species that do not appear in the overall reaction are termed *reactive intermediates*.

The first step in the combustion of hydrogen is breakup of the hydrogen molecule, forming highly reactive atomic H:

$$H_2 \rightarrow 2H \qquad (5.35a)$$

The next step is reaction of the atomic hydrogen with an oxygen molecule:

$$H + O_2 \rightarrow OH + O \qquad (5.35b)$$

Reactions 5.35a and 5.35b are an example of a *chain reaction*.

Since 5.35 is an elementary reaction, the reaction rate can be written as:

$$\frac{d[H_2O]}{dt} = k[OH][H_2]$$

This is also the rate of the overall reaction. Thus, the overall reaction will depend on the availability of OH. What makes the combustion of hydrogen particularly interesting is that there are several ways in which OH may be created. Reaction 5.35b is one way. The monatomic oxygen created in this reaction, however, provides two alternative mechanisms for the creation of the OH complex:

$$H_2 + O \rightarrow OH + H \qquad (5.35c)$$

and

$$H + O \rightarrow OH \qquad (5.35d)$$

Reactions 5.35b through 5.35d represent alternative reaction paths or *branches*. Notice that the final step also provides an alternative mechanism, or branch, for the production of monatomic hydrogen.

The branching that occurs provides the potential for a "runaway" or explosive reaction. This is apparent if we simply sum reactions 5.35 and 5.35b through 5.35d:

$$4 \times [OH + H_2 \rightarrow H_2O + H]$$

$$+ 2 \times [H + O_2 \rightarrow OH + O]$$

$$+ H_2 + O \rightarrow OH + H$$

$$+ H + O \rightarrow OH$$

$$5H_2 + 2O_2 \rightarrow 4H_2O + 2H \qquad (5.35e)$$

Each cycle of these reactions produces four water molecules plus two hydrogens. Since the rate of the overall reaction, i.e., the production of water, depends on [OH], which in turn depends on [H], the reaction will accelerate with time. (Actually, the combustion of hydrogen is a *very* complex reaction. When all the elementary reactions are written down, including the reverse reactions and reactions with the container wall, they fill an entire page. Interestingly, it displays this runaway behavior only under certain combinations of T, P, and container size and shape. The latter dependence results from reactions with, or catalyzed by, the container wall. Under certain conditions, it will become steady state; i.e., the creation and consumption of water balance to produce a constant concentration of water.)

5.2.4.2   Rate-determining step

It often happens that the reaction rate of a chain, or sequential, reaction, is controlled by a single step that is very much slower than the other steps. For example, how quickly you can buy a pencil at the campus bookstore on the first day of class will probably be controlled entirely by how quickly you can get through the checkout line. Such a step is called the *rate-determining step*. Once the rate of this step is determined, the rates of all other steps are essentially irrelevant.

Now consider a reaction that can occur through two branches. For example,

$$A \xrightarrow{1} B \quad \text{and} \quad A \xrightarrow{2} B$$

The reaction rate is then:

$$\frac{d[A]}{dt} = -(k_1 + k_2)[A] \qquad (5.36)$$

If one path is very much faster than the other, then the fastest of the two will always be taken. Thus for branched reactions, the fastest branch determines the reaction mechanism. Mathematically, we may say that if $k_1 \gg k_2$ then $(k_1 + k_2) \approx k_1$ and therefore:

$$\frac{d[A]}{dt} = -k_1[A]$$

In our analogy above, if an express checkout is available, you would certainly take it. In this case, the slowness of the regular checkout line becomes irrelevant for determining how quickly you can buy your pencil. To sum up, we may say that when reactions occur *in series*, then *the slowest reaction is the rate-determining step. When parallel*, or branched, reaction paths of very different speeds are available, then only *the fastest path is of interest*.

5.2.5   Steady state and equilibrium

Many geochemical systems are steady-state ones, that is, time-invariant systems, or approximately so. The equilibrium state is also a steady state, but not all steady-state systems are necessarily equilibrium ones. We may say, then, that steady state is a necessary, but not sufficient, condition for *equilibrium*. Let's consider how a system will approach the steady state and equilibrium.

Consider the elementary reaction:

$$A \rightarrow B$$

Suppose that this reaction does not entirely consume A but reaches a steady state where the concentration of A is $[A]_s$, the subscript s denoting the steady state. In this case, we can express the reaction rate as:

$$\frac{d[A]}{dt} = k([A]_s - [A]) \qquad (5.37)$$

where $[A]_s$ is the steady-state concentration of A. The reaction rate is 0 when $[A] = [A]_s$.

To see how the concentration will vary before steady state is achieved, we integrate eqn. 5.37:

$$\ln\left(\frac{[A]_s - [A]}{[A]_s - [A]^\circ}\right) = -kt$$

where $[A]^\circ$ is the initial concentration of A. This may be written as:

$$\left(\frac{[A]_s - [A]}{[A]_s - [A]^\circ}\right) = e^{-kt} \qquad (5.38)$$

The denominator is a constant (for a given set of initial conditions), so we can rewrite

eqn. 5.36 as:

$$[A]_s - [A] = Ce^{-kt}$$

The excess concentration of A, i.e., $[A] - [A]_s$ declines as e^{-t}, so that steady state is approached asymptotically. An effective steady state will be achieved when $t \gg 1/k$. As in Example 5.1, the reaction rate decreases exponentially with time, i.e.:

$$\frac{d[A]}{dt} = kCe^{-kt}$$

Now suppose that in addition to the reaction: $A \rightarrow B$, the reaction $B \rightarrow A$ also occurs and that both are first-order elementary reactions. The rates of reaction will be:

$$\frac{d[A]}{dt} = -k_+[A] + k_-[B] \qquad (5.39)$$

Here we are using k_+ for the rate constant of the forward reaction and k_- for the rate constant of the reverse reaction. Assuming the system is closed and that no other processes affect the concentrations of A and B, then:

$$[A] + [B] = \Sigma AB$$

where ΣAB is the total of A and B and is a constant. Equation 5.39 can therefore be written as:

$$\frac{d[A]}{dt} = -k_+[A] + k_-\left(\sum AB - [A]\right) \quad (5.40)$$

The concentration at some time τ, $[A]_\tau$, is obtained by integrating 5.40:

$$\int_{A^\circ}^{A_\tau} \frac{d[A]}{-(k_+ + k_-)[A] + k_-\Sigma AB} = \int_0^\tau dt$$

which yields:

$$\frac{-(k_+ + k_-)[A]_\tau + k_-\Sigma AB}{-(k_+ + k_-)[A]^\circ + k_-\Sigma AB} = e^{-(k_+ + k_-)\tau}$$

Since $[A]^\circ + [B]^\circ = \Sigma AB$, we can also express this as:

$$\frac{-(k_+ + k_-)[A]_\tau + k_-[B]_\tau}{-(k_+ + k_-)[A]^\circ + k_-[B]^\circ} = e^{-(k_+ + k_-)\tau} \quad (5.41)$$

Thus, in the general case, the concentrations of A and B will depend on their initial concentrations (see Example 5.3). However, for $\tau = \infty$, or as a practical matter when $\tau \gg$ $(k_+ + k_-)$, then a steady state will be achieved where eqn. 5.41 reduces to:

$$k_+[A]_\infty = k_-[B]_\infty \qquad (5.42)$$

Example 5.3 Racemization of amino acids

Amino acids are nitrogen-containing organic molecules that are essential to life. The chemical properties of amino acids depend not only on their composition, but also on their structure. Twenty different amino acids are used in building proteins. Amino acid comes in two forms, which can be distinguished by the direction in which they rotate polarized light. Interestingly enough, organisms synthesize only the form that rotates polarized light in a counterclockwise manner, labeled the L-form (Fig. 5.05a). After death of the organism, however, the amino acid can spontaneously convert to its mirror image, the D-form (Fig. 5.05b), corresponding to clockwise rotation of light. This process is termed *racemization*. Racemization is a first-order reaction, and rate constants for this process have been determined for a number of amino acids in various substances. This provides a means of dating sediment. Given that the rate constant k_+ for the L-isoleucine → D-alloisoleucine reaction is 1.2×10^{-7} y^{-1} and for the D-alloisoleucine → L-isoleucine is 9.6×10^{-8} y^{-1}, what is the age of a sediment whose D-alloisoleucine/L-isoleucine ratio is 0.1? Assume that the total isoleucine is conserved and an initial D-isoleucine concentration of 0.

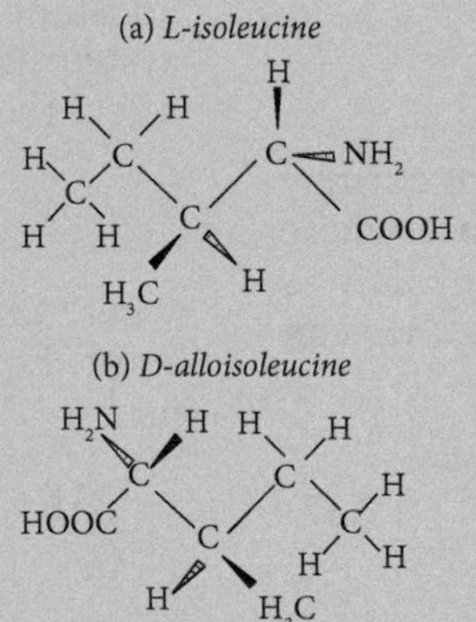

(a) *L-isoleucine*

(b) *D-alloisoleucine*

Figure 5.5 Structure of L-isoleucine and D-alloisoleucine. Solid wedge shapes indicate bonds coming out of the plane of the paper, hashed wedge shapes indicate bonds behind the paper.

Answer: This is a special case of equation 5.41 where $[B]°$ is 0 and $[A] + [B] = [A]°$. Letting γ be the ratio $[B]/[A]$ (D-alloisoleucine/L-isoleucine) and substituting into to 5.41, we obtain:

$$\frac{-k_+ + \gamma k_-}{-k_+(1+\gamma)} = e^{-(k_+ + k_-)t}$$

Substituting values and solving for t, we find the age is 8.27×10^3 yr. Of course, racemization rates, like all reaction rates, depend on temperature, so this result assumes constant temperature.

5.3 RELATIONSHIPS BETWEEN KINETICS AND THERMODYNAMICS

5.3.1 Principle of detailed balancing

Equation 5.42 describes the relation between the concentration of reactant and product of a reversible reaction after infinite time (i.e., in the steady state). This, then, is just the state the reaction will obtain in the absence of constraints and external disturbance. This is precisely the definition of equilibrium we decided on in Chapter 2. It follows that $[A]_\infty$ and $[B]_\infty$ are also the equilibrium concentrations. Thus we see, as we stated in Chapter 2, that equilibrium is not necessarily a static state on the microscopic scale. Rather, it is a steady state where the forward rate of reaction is equal to the reverse rate. Formally, we may say that for an elementary reaction such as:

$$A \rightleftharpoons B$$

at equilibrium the following relation must hold:

$$k_+[A]_{eq} = k_-[B]_{eq} \qquad (5.43)$$

where k_+ and k_- are the rate constants for the forward and reverse reactions respectively. This is known as the *principle of detailed balancing*, and it establishes an essential link between thermodynamics and kinetics. This link is apparent when we combine eqn. 5.43 with eqn. 3.85 to obtain:

$$\frac{k_+}{k_-} = \frac{[B]_{eq}}{[A]_{eq}} = K^{app} \qquad (5.44)$$

It is apparent from eqn. 5.43 that if the equilibrium constant and one of the rate constants for a reaction are known, the rate constant for the reverse reaction may be deduced. Furthermore, if the form of the rate law for either the forward or reverse reaction is known, the other can be deduced. This is a trivial point for elementary reactions since rate laws for such reactions are readily obtained in any case. The importance of this point is that it holds for overall reactions as well as elementary ones. For example, consider the serpentinization of olivine:

$$2Mg_2SiO_4 + H_2O + 2H^+$$
$$\rightleftharpoons Mg_3Si_2O_5(OH)_4 + Mg^{2+}$$

This is not an elementary reaction, as several intermediate steps are involved, as in the example of the weathering of anorthite discussed earlier. Nevertheless, if olivine, serpentine, and water can be assumed to be pure phases and have unit activity, the equilibrium constant for this reaction is:

$$K^{app} = \frac{[Mg^{2+}]}{[H^+]^2} \qquad (5.45)$$

The relation between the forward and reverse reaction rate constants must be:

$$k_-[Mg^{2+}] = k_+[H^+]^2$$

Suppose that experiments show that the rate law for the forward reaction is:

$$\frac{d[Mg^{2+}]}{dt} = k_+[Ol][H^+]^2$$

where $[Ol]$ is the specific area (area per solution volume) of olivine in the experiment. From eqn. 5.10, we can express the rate for the reverse reaction as:

$$\frac{d[Ol]}{dt} = -2\frac{d[Mg^{2+}]}{dt} = -k_+[Ol][H^+]^2$$

Using eqn. 5.44 to obtain a substitution for k_+, we find that the rate law for the reverse reaction, i.e., for the formation of olivine from serpentine, must be:

$$\frac{d[Ol]}{dt} = -2k_-\frac{[Mg^{2+}]}{[H^+]^2}[Ol][H^+]$$
$$= -2k_-\frac{[Mg^{2+}]}{[H^+]}[Ol] \qquad (5.46)$$

where k_- is the rate constant for the reverse reaction.

5.3.2 Enthalpy and activation energy

The principle of detailed balancing allows us to relate the activation energy in the Arrhenius relation (eqn. 5.25) to the heat (enthalpy) of reaction. Recall that the equilibrium constant is related to free energy change of reaction as:

$$K = e^{-\Delta G_r^o/RT} = e^{-\Delta H_r^o/RT + \Delta S/R} = e^{\Delta S/R}e^{-\Delta H_r^o/RT} \qquad (5.47)$$

(For simplicity and clarity, here, and in the subsequent discussion of transition state theory, we assume ideal behavior, and therefore that activities equal concentrations and that $K^{app} = K$.) If we write the Arrhenius relations for the forward and reverse reactions and combine them with eqn. 5.43 we obtain:

$$\frac{k_+}{k_-} = \frac{A_+ e^{-E_+/RT}}{A_- e^{-E_-/RT}}\frac{A_+}{A_-} e^{-(E_+ - E_-)/RT} = K \quad (5.48)$$

Comparing equations 5.47 and 5.48, we can see that:

$$E_+ - E_- = \Delta H_r^o \quad (5.49)$$

This relationship is illustrated for the example of an exothermic reaction in Figure 5.6. In the process of converting products to reactants, an amount of energy ΔH is released. To reach that state, however, an energy barrier of E_{B+} must be overcome. It is apparent then that the enthalpy change of reaction is just the difference between the barrier energies of the forward and reverse reactions. We also see that:

$$\frac{A_+}{A_-} = e^{\Delta S/R} \quad (5.50)$$

Indeed, it can be shown that the Arrhenius coefficient, or frequency factor, is related to entropy as:

$$A_+ = \frac{kT}{h}e^{\Delta S_+^o/R} \quad (5.51)$$

where ΔS_+^o is the entropy difference between the initial state and the activated state (discussed below) and h is Planck's constant. The ratio kT/h has units of time^{-1} and is called

the *fundamental frequency*. At 298 K, it has a value of 6.21×10^{12} sec^{-1}.

5.3.3 Aspects of transition state theory

In the above discussion, we have already made implicit use of *transition state theory*. Transition state theory postulates that an *elementary* reaction such as

$$A + BC \rightarrow AC + B \quad (5.52)$$

proceeds through the formation of an *activated complex* ABC*, also called a reactive intermediate. Thus, reaction 5.52 can be described by the mechanism:

$$A + BC \rightarrow ABC^* \quad (5.53)$$

and

$$ABC^* \rightarrow AC + B \quad (5.54)$$

The activated complex ABC^* is assumed to be in thermodynamic equilibrium with both reactants and products. Hence it is possible to define an equilibrium constant for reaction 5.53 (assuming ideal behavior) as:

$$K^* = \frac{[ABC^*]}{[A][BC]}$$

as well as a free energy change:

$$\Delta G^* = -RT \ln K^*$$

and enthalpy and entropy changes:

$$\Delta G^* = \Delta H^* - T\Delta S^*$$

Though to do so here would take us too far afield, it can be shown from a statistical mechanical approach that the rate constant for a reaction such as 5.53 is:

$$k = \kappa \frac{kT}{h}K^* \quad (5.55)$$

where kT/h is the fundamental frequency as we defined it above, and κ is a constant, called the transmission coefficient, whose value is often close to 1. Equation 5.55 is known as

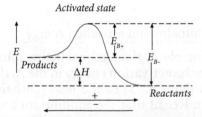

Figure 5.6 The relationship between enthalpy of reaction and the barrier energy for the forward and reverse reactions.

the *Eyring equation.*[*] It is then easily shown that the rate constant is:

$$k = \frac{kT}{h} e^{\Delta S^*/R} e^{-\Delta H^*/RT} = \frac{kT}{h} e^{-\Delta G^*/RT} \quad (5.56)$$

Thus, if the nature of the activated complex is understood, the rate constant can be calculated. For example, we saw that the partition function is related to entropy and energy (it is also easily shown that it is related to Gibbs free energy and enthalpy as well). The rate constant can be calculated from partition functions of the activated complex and reactants (see Example 5.4).

Now consider that reaction 5.49 is reversible so that:

$$A + BC \rightleftharpoons AC + B$$

and that the reverse reaction proceeds through the same activated complex ABC*.

The net rate of reaction is:

$$\mathfrak{R}_{net} = \mathfrak{R}_+ - \mathfrak{R}_- \quad (5.57)$$

If ΔG is the free energy difference between product and reactant, then the free energy difference between the product and the activated complex must be $\Delta G - \Delta G^*$. From this it is readily shown (Problem 3) that the ratio of the forward and reverse reaction rates is:

$$\frac{\mathfrak{R}_+}{\mathfrak{R}_-} = e^{-\Delta G/RT} \quad (5.58)$$

where ΔG is the actual free energy difference between products and reactants. The negative of ΔG in this context is often called the *affinity of reaction*, *reaction affinity*, or simply *affinity*, and is designated A_r (for clarity, however, we shall continue to designate this quantity as ΔG in most cases). Substituting 5.56 into 5.55 and rearranging, we have:

$$\mathfrak{R}_{net} = \mathfrak{R}_+(1 - e^{-\Delta G/RT}) \quad (5.59)$$

If the forward reaction is an elementary one, then $\mathfrak{R}_+$ will be:

$$\mathfrak{R}_+ = k_+[A][BC]$$

where k is as defined in 5.25.

It must be emphasized that *equations 5.58 and 5.59 apply to elementary reactions only*. However, a similar equation may be written for *overall* reactions:

$$\mathfrak{R}_{net} = \mathfrak{R}_+(1 - e^{-n\Delta G/RT}) \quad (5.60)$$

where n can be any real number. Using the Arrhenius expression for k (eqn. 5.25), eqn. 5.60 becomes:

$$\mathfrak{R}_{net} = A_+ e^{-E_{B+}/RT}(1 - e^{-n\Delta G/RT})[A]^{n_A}[B]^{n_B} \ldots \quad (5.61)$$

where $[A]$, $[B]$, ... are the concentrations (surface areas for solids) of the reactants and the n variables can be any real number but will generally be >1 for overall reactions (Lasaga, 1981b). Equation 5.58 links kinetics and thermodynamics and forms the basis of *irreversible thermodynamics*.

If the system is not far from equilibrium, then $\Delta G \ll RT$ and we may approximate e^x by $1 + x$, so that for an elementary reaction:

$$\mathfrak{R}_{net} \approx -\mathfrak{R}_+ \frac{\Delta G}{RT} \quad (5.62)$$

Thus close to equilibrium, the reaction rate will vary linearly with ΔG, slowing as equilibrium is approached. Substituting A_r for $-\Delta G$, eqn. 5.61 can also be written as:

$$\mathfrak{R}_{net} \approx \mathfrak{R}_+ \frac{A_r}{RT} \quad (5.63)$$

---

[*] Named for Henry Eyring (1901–1981), who formulated transition state theory in 1935. It was evidently an idea whose time had come, because M.G. Evans and M. Polanyi independently developed the same theory in a paper published the same year. Eyring, who was born in Juarez, Mexico, received his PhD from the University of California at Berkeley in 1929. He worked in the University of Wisconsin, the Kaiser Wilhelm Institute in Berlin (working with Polanyi), and Princeton University before becoming professor of chemistry at the University of Utah in 1946, where he remained for the rest of his life.

Example 5.4 Estimating ΔG^* for the aragonite–calcite transition

Aragonite is the high-pressure form of $CaCO_3$. Upon heating at 1 atm, it will spontaneously revert to calcite. Carlson (1980) heated aragonite crystals containing calcite nuclei to a series of temperatures for fixed times on the heating stage of a microscope, then measured the growth of the calcite nuclei and from that calculated growth rates shown in the table below. Using these data, determine the value of ΔG^* for this reaction.

Answer: This is a reversible reaction, so we have to consider that both the forward and reverse of the aragonite $\rightarrow$ calcite reaction will occur. According to transition state theory, the rate constant for the forward reaction is

$$k_+ = \frac{kT}{h}e^{-\Delta G^*/RT} \tag{5.64}$$

From eqn. 5.61, assuming this is an elementary reaction with $n = 1$, the rate of the net reaction is:

$$\Re_{net} = \frac{kT}{h}e^{-\Delta G^*/RT}(1 - e^{-\Delta G/RT})$$

This rate expression has units of time^{-1}, but Carlson's results are given in units of distance/time. How do we reconcile these? We might guess in this case that fundamental frequency, the pre-exponential term, ought to be multiplied by some sort of *fundamental distance*. A fundamental distance in this case would be lattice spacing, which for aragonite is about 5 Å (or 5×10^{-10} m). Thus, if λ is the lattice spacing, we have:

$$\Re_{net} = \frac{\lambda kT}{h}e^{-\Delta G^*/RT}(1 - e^{-\Delta G/RT})m/s$$

Solving for ΔG^*:

$$\Delta G^* = -RT\left[\ln \Re_{net} - \ln \frac{\lambda kT}{h} - \ln(1 - e^{-\Delta G/RT})\right]$$

To determine ΔG^*, we have to calculate ΔG, which we can do using the thermodynamic data in Table 2.2 and equation 2.133. Our spreadsheet is shown below. Calculating the average ΔG^* for the four measurements, we find $\Delta G^* = 161$ kJ. We can then use ΔG^* to predict the reaction rates. A comparison between the measured and predicted reaction rates is shown in Figure 5.7.

R	8.314	J/mol-K		h	6.63E-34	J-sec		
k	1.38E-23	J/K		λ	5.00E-10	m		
$\lambda k/h$	1.04E+01	m/K-sec						
					ln(1-exp			
T °C	R m/sec	ln R	T, K	ΔG, J	($\Delta G/RT$))	ln($\lambda kT/h$)	ΔG^* kJ	$-\ln R_{calc}$
455	7.45E-09	18.7152	728	−2828.8	−0.985	8.933	161.38	18.6485
435	3.63E-09	19.4352	708	−2709.3	−0.997	8.906	160.95	19.4397
415	1.61E-09	20.2452	688	−2592	−1.010	8.877	160.80	20.2756
395	6.24E-10	21.1952	668	−2476.6	−1.022	8.847	161.17	21.1605
375	2.72E-10	22.0252	648	−2363	−1.035	8.817	160.58	22.0986
						ΔG^*	160.98	ave

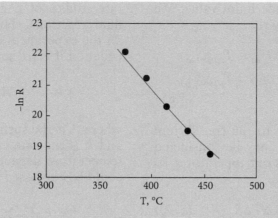

Figure 5.7   Comparison of observed and predicted rates of the aragonite ⇋ calcite reaction. Data (circles) from Carlson (1980).

At this point, you might think, "This is all fine and good, but how do I calculate ΔG?" There are several approaches to estimating the value of ΔG under nonequilibrium conditions. For the first method, let's return to the relationship between activities, $\Delta G°$, and K. In Chapter 3, we found we could express the relationship between chemical potential and activities *at equilibrium* as:

$$\sum_i v_i \mu_l^o + RT \ln \prod_i a_i^{v_i} = 0 \quad (3.84)$$

At equilibrium, the first term on the left is $\Delta G°$ and the second term is $RT \ln K$. *Under nonequilibrium conditions*, however, the product of activities will not be equal to K and eqn. 3.84 will not be equal to 0. Rather, it will have some finite value, which is ΔG. We define a quantity Q as:

$$Q \equiv \prod_i a_i^{v_i} \quad (5.65)$$

Q is called the *reaction quotient* (Chapter 3). Although eqn. 5.65 has the same form as our definition of the equilibrium constant (eqn. 3.85), there is an important difference. K defines the relationship between activities *at equilibrium*. In defining Q, we impose no such condition, so that Q is simply the product of activities. At equilibrium $Q = K$, but not otherwise. Under nonequilibrium conditions, we can express eqn. 3.86 as:

$$\Delta G° + RT \ln Q = \Delta G \quad (5.66)$$

Since $\Delta G°$ is equal to $-RT \ln K$, eqn. 5.66 can be written as:

$$RT \ln Q - RT \ln K = \Delta G$$

or:

$$\frac{Q}{K} = e^{\Delta G/RT} \quad (5.67)$$

In dissolution–precipitation reactions, the ratio Q/K is referred to as the saturation index or saturation ratio and sometimes designated as Ω: precipitation occurs when $Q/K > 1$, while dissolution occurs for $Q/K < 1$.

Substituting 5.67 into 5.59, we have for an elementary reaction:

$$\mathfrak{R}_{net} = \mathfrak{R}_+ \left(1 - \frac{Q}{K}\right) \quad (5.68)$$

Thus, we expect reaction rates to decrease as $Q \rightarrow K$. For a nonelementary reaction, 5.68 becomes:

$$\mathfrak{R}_{net} = \mathfrak{R}_+ \left(1 - \frac{Q}{K}\right)^n \quad (5.68a)$$

where n can be any number.

To arrive at the second method of estimating ΔG, we recall that ΔG may be written as $\Delta H - T\Delta S$. At equilibrium:

$$\Delta G_{eq} = \Delta H_{eq} - T_{eq}\Delta S_{eq} = 0$$

where the subscript *eq* denotes the quantity when products and reactants are at equilibrium. Under nonequilibrium conditions,

$\Delta H - T\Delta S$ will have some finite value. We can make use of this and write:

$$\Delta G = \Delta H - T\Delta S - (\Delta H_{eq} - T_{eq}\Delta S_{eq})$$
$$= \Delta H - \Delta H_{eq} - (T\Delta S - T_{eq}\Delta S_{eq}) \tag{5.69}$$

For temperatures close to the equilibrium temperature, ΔH and ΔS may be considered constant (i.e., independent of temperature), so that eqn. 5.69 simplifies to:

$$A_r = -\Delta G = \Delta T\Delta S \tag{5.70}$$

where $\Delta T = (T - T_{eq})$ and is sometimes called the temperature overstep (see Example 5.5). This may be substituted into eqn. 5.60, so that close to equilibrium we have

$$\mathfrak{R}_{net} = \frac{\mathfrak{R}_+ \Delta S(T - T_{eq})}{RT_{eq}} \tag{5.71}$$

Wood and Walther (1983) used this equation to analyze experimental reaction

rate studies of a variety of silicate–aqueous fluid reactions. They found that essentially all the experimental data could be fit to this equation if $\mathfrak{R}_+$ is given by:

$$\mathfrak{R}_+ = -kA$$

where A is the surface area of the solid phase and k is the rate constant. Furthermore, the temperature dependence of the rate constant could be expressed as:

$$\log k = -2900/T - 6.85 \tag{5.72}$$

This is illustrated in Figure 5.8. The data show a scatter of more than one order of magnitude about the line, so clearly the equation cannot be used for exact prediction of reaction rates. As Kerrick et al. (1991) pointed out, this approach has limits and cannot be applied to reactions involving carbonates. Nevertheless, Wood and Walther's work provides a useful way to estimate the order of magnitude of silicate dissolution rates.

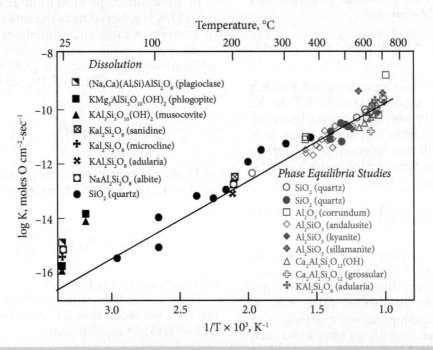

Figure 5.8 Log of the rate constant vs. inverse of temperature for a variety of silicate and aluminate dissolution reactions. Wood and Walther (1983) extracted reaction rate data from both studies of the rates of mineral dissolutions (labeled "Dissolution") and phase equilibria studies (labeled "Phase Equilibria Studies"). Notice that the rate constant has units of mole of oxygen per cc per second. After Wood and Walther (1983).

Example 5.5 Predicting rates of reversible metamorphic reactions

Consider the reaction:

$$Ca_2Mg_5Si_8O_{22}(OH) + 11CaMg(CO_3)_2 \rightleftharpoons 8Mg_2SiO_4 + 13CaCO_3 + 9CO_2 + H_2O$$

$$Tremolite + 11\ dolomite \rightleftharpoons 8\ forsterite + 13\ calcite + 9CO_2 + H_2O$$

Assume $T = 625°C$, $\Delta S_r = 1.140\ kJ/K$, $E_A = 579\ kJ$, $A = 1.54 \times 10^{27}\ sec^{-1}$ (Heinrich et al., 1989), an overstep of the equilibrium temperature of 5°C, and that the dolomite crystals are perfect cubes, i.e., $V_{Do} = (S_{Do})^{3/2}$ where V_{Do} and S_{Do} are the volume and surface area respectively of dolomite. Assume further that the initial assemblage contains only tremolite plus dolomite and that the reaction rate can be expressed as in eqn. 5.59, i.e.:

$$\frac{-dV_{Do}/V_{Do_0}}{dt} = k(1 - e^{\Delta G/RT})S_{Do}/S_{Do_0}$$

where V_{Do_0} and S_{Do_0} are the initial dolomite volume and surface areas respectively and k is the rate constant with the usual Arrhenius temperature dependence (the minus sign occurs because dolomite is a reactant). Calculate the extent of conversion of dolomite (i.e., volume relative to initial volume V_0) as a function of time.

Answer: To solve this problem, we need to integrate the equation above. First, we substitute eqn. 5.25 for k and $(V/V_{Do_0})^{2/3}$ for S_{Do}/S_{Do_0}. Upon integration, we obtain:

$$\frac{V_{Do}}{V_{Do_0}} = \left[1 - \frac{A}{3}e^{-E_A/RT}(1 - e^{\Delta G/RT})t\right]^3$$

Making use of $-\Delta T\Delta S \approx \Delta G$ (where ΔT is the temperature overstep; eqn. 5.70), we have:

$$\frac{V_{Do}}{V_{Do_0}} = \left[1 - \frac{A}{3}e^{-E_A/RT}(1 - e^{\Delta T\Delta S/RT})t\right]^3$$

The result is shown in Figure 5.9. On geological time scales, this reaction is clearly quite fast, going to completion within half a year (1.5×10^7 sec), even with a relatively small temperature overstep of 5°C. We also see in Figure 5.9 that the rate of reaction decreases as time progresses. This occurs because of the decreasing dolomite surface area.

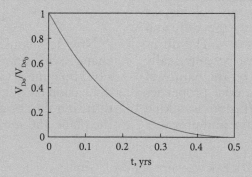

Figure 5.9 Relative volume of dolomite as a function of time predicted at 625°C.

The temperature conditions we chose for this example, a constant 5°C overstep, are not geologically realistic. A more realistic assumption would be that of steady temperature increase, such as would occur around an igneous intrusion or as a result of burial or underthrusting. That situation is addressed in Problem 8 at the end of this chapter.

5.4 DIFFUSION

In order for a reaction to occur, reactants must be brought together. In chemical laboratories, magnetic stirrers serve this purpose. In nature, fluid advection, driven ultimately by gravity, can serve to transport components on larger scales, but it is rarely effective on very small scales. On these scales, diffusion is usually the process responsible for transport of chemical components. Consequently, chemical transport generally involves both diffusion and advection: advection for large-scale transport, and diffusion for small-scale transport. In this section, we discuss the nature of diffusion and develop the tools necessary to treat it.

5.4.1 Diffusion flux and Fick's laws

Fick's law,[*] or *Fick's first law*, states that at steady state, the *flux, J*, of some species through a plane is proportional to the concentration gradient normal to that plane:

$$J = -D\frac{\partial c}{\partial x} \qquad (5.73)$$

The minus sign indicates diffusion is toward the region of lower concentration. The proportionality coefficient, D, is known as the *diffusion coefficient*. J has units of mass/area-time, e.g., moles/m^2-sec. If concentration is expressed per unit volume, as is often preferred in kinetics, the diffusion coefficient has units of m^2-sec^{-1}. The diffusion coefficient generally must be empirically determined and will depend on the nature of the diffusing species, the material properties of the system in which diffusion is taking place and, as usual, temperature and pressure.

Eqn. 5.73 is applicable to diffusion in only one dimension, and in our brief treatment here, we will consider only the one-dimensional case. A more general expression of Fick's first law, applicable in three-dimensional space is:

$$J = -D\nabla c \qquad (5.73a)$$

where ∇c is the concentration gradient vector:

$$\nabla c = \frac{\partial c}{\partial x}\mathbf{x} + \frac{\partial c}{\partial y}\mathbf{y} + \frac{\partial c}{\partial z}\mathbf{z}$$

and $\mathbf{x}, \mathbf{y}$, and $\mathbf{z}$ are unit vectors in the respective directions. In many cases, a three-dimensional case can be reduced to a one-dimensional one by choosing our x-direction to be the direction of the concentration gradient or by assuming that diffusion is radial. However, most minerals are anisotropic so that diffusion coefficient will vary with direction relative to crystallographic axes (this effect can be as large as several orders of magnitude). In that case, eqn. 5.70 becomes:

$$J = -\mathbf{D}\nabla c \qquad (5.73b)$$

where $\mathbf{D}$ is a *tensor* represented by a symmetric 3 × 3 matrix:

$$\mathbf{D} = \begin{bmatrix} D_{xx} & D_{xy} & D_{xz} \\ D_{yx} & D_{yy} & D_{yz} \\ D_{zx} & D_{zy} & D_{zz} \end{bmatrix} \qquad (5.73c)$$

Now let's consider how concentration will change with time as a consequence of diffusion. Imagine a volume enclosed in a cube of dimension dx (Figure 5.10). Further imagine a diffusion flux of a species of interest through the plane and into the volume at x and a flux

---

[*] Named for Adolf Fick (1829–1901). Fick was born in Kassel, Germany, and earned an MD from the University of Marburg in 1851. Fisk's interest in diffusion through cell membranes led him to formulate the laws that bear his name. It was actually the second law that was published first, in an 1855 paper titled *Über diffusion*. Fick deduced it by analogy to Fourier's equation for thermal diffusion. Fourier's equation is exactly analogous, with thermal diffusivity replacing the diffusion coefficient and the thermal gradient replacing the concentration gradient.

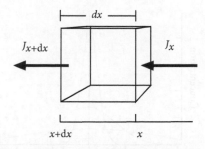

Figure 5.10 A volume of dimension dx with fluxes through the planes at x and dx.

out of the volume through the plane at $x + dx$. Suppose n_x atoms per second pass through the plane at x and n_{x+dx} atoms per second pass through the plane at $x + dx$. The fluxes at the two planes are thus $J_x = n_x/dx^2\text{-sec}$ and $J_{x+dx} = n_{x+dx}/dx^2\text{-sec}$. Conservation of mass dictates that the increase in the number of atoms, dn, within the volume is just what goes in minus what comes out. Over an increment of time dt this will be d$n = (J_x - J_{x+dx})dt$. The change in the concentration over this time is just this change in the number of atoms per unit volume:

$$dc = \frac{(J_x - J_{x+1})dt}{dx}$$

and the rate of change of concentration is:

$$\frac{dc}{dt} = \frac{(J_x - J_{x+1})}{dx}$$

If we specify that we are interested in the change in concentration at some fixed point, x, and some fixed time, t, in the limit of infinitesimal dt and dx, this equation can be written as:

$$\left(\frac{\partial c}{\partial t}\right)_x = -\left(\frac{\partial J}{\partial x}\right)_t \quad (5.74)$$

Equation 5.74 is called the *equation of continuity* since it follows from mass conservation. Now since the flux is given by Fick's first law, we can write:

$$\left(\frac{\partial c}{\partial t}\right)_x = -\left(\frac{\partial\,(-D\partial c/\partial x)}{\partial x}\right)_t$$

Simplifying, we arrive at *Fick's second law*:

$$\left(\frac{\partial c}{\partial t}\right)_x = D\left(\frac{\partial^2 c}{\partial x^2}\right)_t \quad (5.75)$$

In three dimensions, this becomes:

$$\left(\frac{\partial c}{\partial t}\right)_\xi = -\nabla J = -\frac{\partial J_x}{\partial x} - \frac{\partial J_y}{\partial y} - \frac{\partial J_z}{\partial z} \quad (5.76)$$

where the subscript ξ denotes a fixed location in space and J_x denotes the flux through a plane normal to x, etc. Where the diffusion coefficient varies with direction, Fick's second law in matrix notation is:

$$\left(\frac{\partial c}{\partial t}\right)_\xi = D \times \nabla^2 c \quad (5.77)$$

where D is the diffusion tensor. Equation 5.75 tells us that rate of change with time of the concentration at any point is proportional to the second differential of the diffusion profile. Fick's second law is illustrated in Figure 5.11.

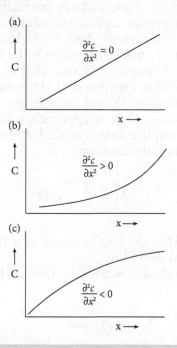

Figure 5.11 Three possible concentration gradients. In (a), $\partial^2 c/\partial x^2 = 0$ and therefore $\partial c/\partial t = 0$. Thus, for a gradient that is straight, the concentration at any point remains constant (even though there is diffusion along the gradient). This is therefore the steady-state case. In (b), $\partial^2 c/\partial x^2 > 0$ and hence the concentration at any point increases with time. In (c), $\partial^2 c/\partial x^2 < 0$ and therefore the concentration at any point decreases with time. Both (b) and (c) will tend, with time, toward the steady-state case (a).

5.4.1.1 Solutions to Fick's second law

There is no single solution for eqn. 5.75 (i.e., a function expressing $c(t,x)$); rather, there are a number of possible solutions, and the solution appropriate to a particular problem will depend on the boundary conditions. Let's consider a few of the simpler ones. In all cases, we assume that the system is uniform in composition in the y and z directions, so diffusion occurs only in the x direction.

As a first case, consider a thin film of some diffusing species sandwiched between layers of infinite length having concentration $c = 0$. This might represent a "doped" layer in a diffusion experiment in the laboratory. In nature, it might represent a thin sedimentary horizon enriched in some species (such as the iridium-enriched layer in many sediments at the Cretaceous–Tertiary boundary). Diffusion will cause the species to migrate away from $x = 0$ as time passes. Mathematically, this situation imposes certain boundary conditions on the solution of eqn. 5.75. We take the position of enriched horizon to be 0, and we seek a solution to 5.75 such that at $t = 0$, $c = 0$ everywhere except $x = 0$. At some time $t > 0$, our function should have the property that c approaches 0 as x approaches infinity. We further require that the total amount of the species remain constant:

$$M = \int_{-\infty}^{+\infty} c \, dx$$

where M is the total amount of substance in an initial thin film.

The solution is given by Crank (1975) as:

$$c(x,t) = \frac{M}{2(\pi Dt)^{1/2}} e^{-x^2/4Dt} \qquad (5.78)$$

Figure 5.12 shows how the concentration profile changes with time under these circumstances. It is interesting to note that these profiles are the same as those of a "normal" statistical distribution error curve with a standard deviation $\sigma = (2Dt)^{1/2}$.

Suppose a boundary condition is imposed such that diffusion can occur only in the positive direction. We can treat this case as if the diffusion in the negative direction is reflected at the plane $x = 0$. The solution is obtained

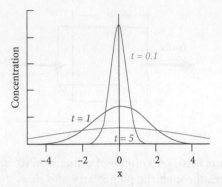

Figure 5.12 Concentration profiles at three different times resulting from outward diffusion from a thin film of the diffusing species. Note that the area under the curve remains constant through time.

by superimposing the solution for the negative case on the positive one:

$$c(x,t) = \frac{M}{(\pi Dt)^{1/2}} e^{-x^2/4Dt} \qquad (5.79)$$

Now consider a situation where the diffusing species has an initial uniform concentration C_o between $x = 0$ and $x = -\infty$, and 0 concentration between $x = 0$ and $x = \infty$. In the laboratory, this circumstance might arise if we place two experimental charges adjacent to one another, one having some finite concentration with the species of interest, the other none. In nature, a somewhat analogous situation might be a layer of fresh water overlying a formation brine in an aquifer, or river water overlying seawater in an estuary, or two adjacent crystals.

The solution to this case may be found by imagining the volume between $x = 0$ and $x = -\infty$ as being composed of an infinite number of thin films of thickness $\delta\xi$ (Figure 5.13). The concentration of the diffusing species at some point x_p at time t is then the sum of the contributions of each imaginary thin film (Crank, 1975). The mathematical solution is obtained by integrating the contribution of all such films:

$$c(x,t) = \frac{C_o}{2(\pi Dt)^{1/2}} \int_x^\infty e^{-\xi^2/4Dt} d\xi \qquad (5.80)$$

or defining $\eta = \xi/2\sqrt{Dt}$:

$$c(x,t) = \frac{C_o}{\pi^{1/2}} \int_{x/2\sqrt{Dt}}^\infty e^{-\eta^2} d\eta \qquad (5.81)$$

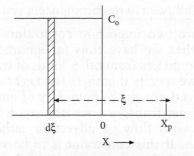

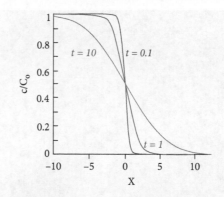

Figure 5.13 An extended initial distribution can be thought of as consisting of an infinite number of thin films of thickness $\delta\xi$. The concentration of a diffusing species at some point x_p is the contribution from each film from distances x to infinity (after Crank, 1975).

Figure 5.14 Distribution of a diffusing species initially confined to $-\infty < x < 0$ at three time intervals after diffusion begins.

The integral in eqn. 5.81 may be written as:

$$\int_{x/2\sqrt{Dt}}^{\infty} e^{-\eta^2}\,d\eta = \int_{0}^{\infty} e^{-\eta^2}\,d\eta - \int_{0}^{x/2\sqrt{Dt}} e^{-\eta^2}\,d\eta$$

$$(5.82)$$

This integral has the form of a standard mathematical function called the *error function*, which is defined as:

$$erf(x) = \frac{2}{\pi^{1/2}} \int_{0}^{x} e^{-\eta^2}\,d\eta \qquad (5.83)$$

Substituting eqn. 5.83 into 5.81, and since $erf(\infty) = 1$, eqn. 5.81 becomes:

$$c(x,t) = \frac{C_o}{2} \left\{ 1 - erf\left(\frac{x}{2\sqrt{Dt}}\right) \right\} \quad (5.84)$$

Values for the error function may be found in mathematical tables. The error function is also a standard function in MATLAB™ (erf) and Microsoft Excel™ (ERF)*. Alternatively, it may be approximated as:

$$erf(x) \cong \sqrt{1 - \exp(-4x^2/\pi}} \qquad (5.85)$$

Figure 5.14 shows how the concentration profile will appear at different times. Since

$erf(0) = 0$, the profiles have the interesting property that $c = C_o/2$ at $x = 0$ at all times.

A similar approach can be used for a diffusing species initially confined to a distinct region, for example: $-h < x < +h$. Examples might be sedimentary or metamorphic layers of finite thickness or a compositionally zoned crystal (Example 5.6). Again, the layer is treated as a series of thin films, but the integration in eqn. 5.75 is carried out from $-h$ to $+h$. The result is:

$$c(x,t) = \frac{C_o}{2} \left\{ erf\left(\frac{h-x}{2\sqrt{Dt}}\right) + erf\left(\frac{h+x}{2\sqrt{Dt}}\right) \right\}$$

$$(5.86)$$

Mineral grains in lavas sometimes exhibit zoning that can be observed in thin section. This can occur when the composition of the magma changes, and consequently, the equilibrium mineral composition changes during the crystal growth. Zoning results if temperature decreases before diffusion can homogenize the composition of the crystal. Distinct zones in crystals are an indication of a sudden change in magma composition, for example as a result of mixing with a new batch of magma arriving in the magma chamber. This is illustrated in Figure 5.15 and Example 5.6.

---

* The error function in Excel, ERF(), is an add-in function found among the analysis tools. ERF() does not properly treat the case where $x < 0$. The error function has the property that $erf(-x) = -erf(x)$. In working with Excel, test for a negative value of x and where x is <0 replace ERF(−X) with −ERF(X). IF functions in Excel have the format "IF(logical_test, value_if_true, value_if_false)." So, for example, use a statement such as "=IF(X<0,−ERF(−X),ERF(X))."

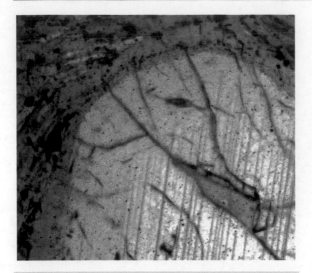

Figure 5.15 A zoned plagioclase crystal in a lava from the Azores viewed with crossed polarizers. Most of the crystal grew within a magma chamber while the thin, darker rim probably grew during eruption as the magma was cooling. Consequently, the rim is richer in albite component and is visible because of slightly different optical properties.

5.4.2 Diffusion in multicomponent systems

There are two important constraints on diffusion that we have thus far ignored. First, diffusion differs from other kinds of transport in that we specify that there is *no net transport of material across the boundary of interest*. If there is net transport, we are, by definition, dealing with flow or advection rather than diffusion. If this constraint is to be satisfied, movement of one species through a plane must be accompanied by movement of one or more other species in the opposite direction.

The second constraint is electrical neutrality. Diffusion of even small quantities of an ion will quickly lead to the development of a large electric potential. The force associated with the potential would prevent any further diffusion of that ion in that direction. Thus, diffusion of an ionic species must be coupled with diffusion of other charged species so as to maintain electrical neutrality. In addition to these constraints, we must recognize that diffusion in some cases will lead to nonideal mixing and the finite enthalpy and volume changes that accompany such situations. With this in

---

Example 5.6   Diffusion in a crystal

Igneous crystals are often zoned as a result of changes in the composition of the magma as they crystallize. Suppose an olivine crystal of 2 mm diameter with a concentration of 2000 ppm Ni comes into contact with a magma in which its equilibrium concentration should be 500 ppm Ni. How long would it take for diffusion to homogenize the crystal at a temperature of 1250°C, assuming instantaneous equilibration at the crystal–liquid boundary?

Answer: We can simplify things by treating the olivine crystal as a sphere and assume diffusion coefficients are independent of crystallographic orientation. Radial symmetry then allows us to consider the problem as a function of radius. We need only consider the variation of concentration along one radial direction with $0 < x < r$. Our boundary condition is that at $x = r$ (the edge of the crystal), concentration is held constant by reaction with the liquid. We'll call this concentration C_r. The initial distribution is $c = C_i$ for $0 < x < r$. According to Crank (1975), the solution is:

$$\frac{c - C_i}{C_r - C_i} = 1 + \frac{2r}{\pi x} \sum_{n=1}^{\infty} \frac{(-1)^n}{n} \sin\left(\frac{n\pi x}{r}\right) \exp\left(\frac{-Dn^2\pi^2 t}{r^2}\right) \qquad (5.87)$$

and the concentration at $x = 0$, C_0, is:

$$C_o = (C_r - C_i)\left\{ 1 + 2\sum_{n=1}^{\infty} (-1)^n \exp\left(\frac{-Dn^2\pi^2 t}{r^2}\right) \right\} + C_i \qquad (5.88)$$

From eqn. 5.71, we see that as the concentration gradient disappears, the rate of diffusion goes to 0. So the crystal approaches homogeneity only asymptotically, becoming homogenous only at $t = \infty$, but it will become *essentially* homogenous more quickly. *Essentially homogenous* implies we could not detect a gradient. If our analytical precision is only 5%, we could not detect a gradient of less than 5%. So let's rephrase the question to ask, how long will it take before the concentration gradient is less than 5%? We set $C_0/C_r = 1.05$, and substituting into eqn. 5.86 and rearranging, we obtain:

$$0.05 \leq \left(1 - \frac{C_i}{C_r}\right) 2 \sum_{n=1}^{\infty} (-1)^n \exp\left(\frac{-Dn^2\pi^2 t}{r^2}\right) \tag{5.89}$$

As it turns out, for relatively large values of t ($Dt/r^2 > 0.1$), the summation converges within 0.05% after the first term, so that eqn. 5.89 may be approximated by:

$$t \cong \frac{-r^2}{D\pi^2} \ln\left[\frac{0.05}{2(C_i/C_r - 1)}\right]$$

For the value of $D = 10^{-12}$ cm^2/sec given by Morioka and Nagasawa (1990), we find that about 154 years is required before the olivine homogenizes. If the olivine spent less than this time in contact with the magma, we would expect it to be zoned in Ni concentration. Figure 5.15 shows how the concentration profile of Ni would vary with time.

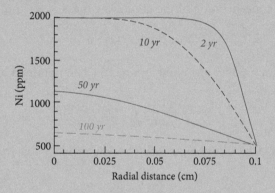

Figure 5.16   Distribution of Ni in a spherical olivine grain with an initial concentration of 2000 ppm and a rim concentration fixed at 500 ppm.

mind, we can recognize four classes of situations and four kinds of diffusion coefficients:

1. *Tracer, or self-diffusion*, in which the net mass and charge fluxes associated with the diffusing species are sufficiently small that they can be safely ignored. This situation occurs when, for example, an experimental charge is doped with a radioactive isotope in sufficiently small amounts such that the concentration of the element, and hence its chemical potential, does not vary significantly. This is the simplest situation, and the one that we have dealt with thus far.

2. *Chemical diffusion* refers to nonideal situations where chemical potential (μ) rather than concentration must be considered. In this circumstance, Fick's laws can be rewritten as:

$$J = -L\frac{\partial \mu}{\partial x} \tag{5.90}$$

and

$$\left(\frac{\partial c}{\partial t}\right)_x = L\left(\frac{\partial^2 \mu}{\partial x^2}\right) \tag{5.91}$$

L is called the *chemical* or *phenomenological coefficient*. These equations must be used in situations where there is a significant change in composition of the material through which diffusion is occurring, such as a chemically zoned liquid or solid, or across a phase boundary. For example, consider an olivine crystal in equilibrium with a surrounding basaltic liquid. There would be a significant change in the concentration of a species such as Mg at the phase boundary, and hence eqn. 5.75 would predict that, even at equilibrium, diffusion of Mg out of the olivine and into the liquid should occur, but this need not be the case. If the olivine crystal is in equilibrium with the surrounding melt, then the chemical potential of Mg is the same in both and thus eqn. 5.91 correctly predicts that no diffusion will occur at equilibrium.

Example 5.7 Equilibration between a mineral grain and pore water

Imagine a grain of calcite in an accumulating sediment surrounded by pore water. Assume that the distribution coefficient of Sr between calcite and water is 100, that the calcite has an initial Sr concentration of 2000 ppm, that a constant Sr concentration of 10 ppm is maintained in the water, and that the grain is spherical. If the diffusion coefficient for Sr in calcite is 10^{-15} cm^2/sec and the radius, r, of the calcite grain is 1 mm, how will the average concentration of Sr in the grain change with time?

Answer: This problem is similar to the previous example (Example 5.6). This time, however, we want to know the average concentration of the grain. The mass of Sr at time t in a spherical shell of thickness dr is:

$$M(t) = c(r, t)4\pi r^2 \mathrm{d}r$$

The average concentration of Sr in the grain at time t is then obtained by integrating and dividing by the volume:

$$\overline{C}(t) = \frac{1}{4/3\pi r^3} \int_0^r c(r, t)4\pi^2 \mathrm{d}r \tag{5.92}$$

where $c(r,t)$ is given by eqn. 5.87 in Example 5.6. The solution is (Albarède, 1995):

$$\overline{C}(t) = \frac{6C_i}{\pi^2} \sum_{n=1}^{\infty} \frac{1}{n} e^{-n^2 \pi^2 Dt/r^2} + C_o \tag{5.93}*$$

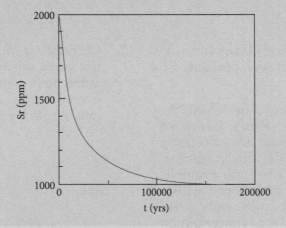

Figure 5.17 Change in the bulk concentration of Sr in a 2 mm diameter calcite grain assuming an initial concentration of 2000 ppm, a constant concentration in the pore water of 10 ppm and a calcite/water distribution coefficient of 100.

where[*] C_i is the initial concentration in the calcite and C_0 is the concentration at the edge of the crystal, which will be in equilibrium with the pore water. The solution is shown Figure 5.17. The grain reaches equilibrium with the pore water in 100,000–200,000 yrs.

*The summation in eqn. 5.93 is slow if Dt/a2 is small. An alternative solution to eqn. 5.93 is:

$$\overline{C}(t) = 1 - \frac{6\sqrt{Dt}}{r} \left[\sum_{n=1}^{\infty} \frac{e^{-nr/\sqrt{Dt}}}{\sqrt{\pi}} - \frac{nr}{\sqrt{Dt}} \left\{ 1 - erf\left(\frac{nr}{\sqrt{Dt}}\right) \right\} \right] + 3\frac{Dt}{r^2} \tag{5.94}$$

The two solutions give identical results. They differ only in the ease of computation. For large values of Dt/a^2 the summation in eqn. 5.93 is preferred. For small values of Dt/a^2 5.94 is preferred.

3. *Multicomponent diffusion* refers to situations where the concentration of the species of interest is sufficiently large that its diffusion must be coupled with diffusion of other species in the opposite direction to maintain electrical neutrality and/or constant volume. In such a circumstance, the diffusion of any one species is related to the diffusion of all other species.

Multicomponent diffusion computations are best carried out using matrices; hence we can express Fick's first law as:

$$\mathbf{J} = -\mathbf{D}\nabla\mathbf{C} \tag{5.95}$$

where $\mathbf{J}$ is the flux vector, $\mathbf{D}$ is the diffusion coefficient matrix, and $\nabla\mathbf{C}$ is the concentration gradient column vector. Written in full matrix notation:

$$\begin{pmatrix} J_1 \\ J_2 \\ \vdots \\ J_n \end{pmatrix} = - \begin{bmatrix} D_{11} & D_{12} & \cdots & D_{1n} \\ D_{21} & D_{22} & \cdots & D_{2n} \\ \cdots & \cdots & \cdots & \cdots \\ D_{n1} & D_{2n} & \cdots & D_{nn} \end{bmatrix} \begin{pmatrix} \frac{\partial c_1}{\partial x} \\ \frac{\partial c_2}{\partial x} \\ \vdots \\ \frac{\partial c_n}{\partial x} \end{pmatrix} \tag{5.95a}$$

The interdiffusion coefficient matrix should not be confused with the diffusion tensor in eqn. 5.73c; in the later the indices are directions; here they are components. Again, letting $\mathbf{C}$ be a column vector of concentrations, then Fick's

second law becomes:

$$\left(\frac{\partial \mathbf{C}}{\partial t}\right)_x = \mathbf{D}\left(\frac{\partial^2 \mathbf{C}}{\partial x^2}\right)_t \tag{5.96}$$

or:

$$\frac{\partial}{\partial t} \begin{pmatrix} c_1 \\ c_2 \\ \vdots \\ c_n \end{pmatrix} = \begin{bmatrix} D_{11} & D_{12} & \cdots & D_{1n} \\ D_{21} & D_{22} & \cdots & D_{2n} \\ \cdots & \cdots & \cdots & \cdots \\ D_{n1} & D_{2n} & \cdots & D_{nn} \end{bmatrix}$$

$$\frac{\partial^2}{\partial x^2} \begin{pmatrix} c_1 \\ c_2 \\ \vdots \\ c_n \end{pmatrix} \tag{5.96a}$$

Provided that we define our concentrations as fractions, either mole or mass fractions such that $\Sigma c_i = 1$, the stipulation of no net transport means that in a system of n components there are only $n - 1$ independent fluxes. Hence, the flux of species i can be computed as:

$$J_i = -\rho \sum_{k}^{n-1} D_{ik} \frac{\partial c_k}{\partial x} \tag{5.97}$$

where D_{ik} is the *interdiffusion coefficient* describing the interaction of species i and k, and n is the number of components in the system and ρ is density. If density is constant throughout the system, we can simplify things somewhat by ignoring the ρ term.

4. *Multicomponent chemical diffusion* refers to situations where both the chemical potential and the diffusion of other species must be considered. In this case,

the diffusion flux is calculated according to equation 5.91, but the diffusion coefficient matrix, **D**, must be calculated as:

$$D = -LG \qquad (5.98)$$

where **L** is the matrix of phenomenological coefficients and the elements of **G**, the thermodynamic matrix, are functions of the derivatives of activity with respect to concentration: $[G] = [\partial \mu_k / \partial c_i]$.

Just as for tracer diffusion, solutions to Fick's second law for multicomponent diffusion depend on boundary conditions. Solutions are generally best done using matrix algebra (see Example 5.8). For a given set of boundary conditions, the solutions are generally analogous to tracer diffusion solutions reviewed above but can produce nonintuitive results in some cases. For example, interdiffusion can result in diffusion up a concentration gradient, because the concentration of species i depends on the concentration gradients of all species, not just its own. This occurs when the net diffusive flux of all other components in eqn. 5.89 is larger than and has the opposite to the flux resulting from the concentration gradient of the species i. Liang (2010) provides an excellent review of the theory, and the experiments of Guo and Zhang (2016, 2018) provide a good example of the complexity and nonintuitive nature of multicomponent chemical diffusion, as does Al_2O_3 in Figure 5.18. Equation 5.98 is known as the *Fick–Onsager law*, or simply *Onsager's equation*.

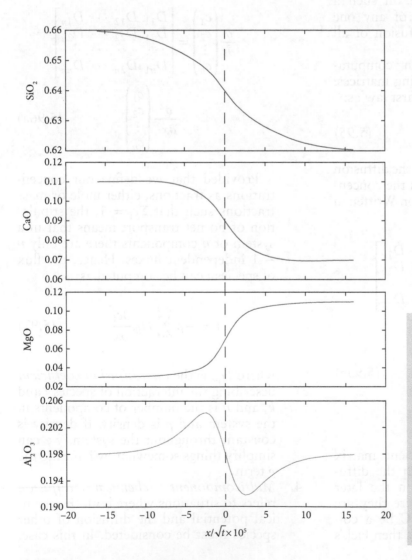

Figure 5.18 Electron microprobe traverses across the contact of basaltic liquids of different and K_2O concentrations but identical concentrations of SiO_2, CaO, MgO, and Al_2O_3 calculated from the diffusion experiments of Richter et al. (1998) in Example 5.8. Al_2O_3 appears to diffuse uphill against the gradient. The concentration scales differ and the entire range of Al_2O_3 is only about 1 wt. percent. The predicted profiles reproduce the ones observed by Richter et al. (1998) well.

Example 5.8   Calculating interdiffusion profiles

Richter et al. (1998) conducted diffusion experiments on pseudo-dacitic silicate liquid by placing two glasses of different SiO_2, MgO, and CaO concentrations but identical Al_2O_3 concentration, listed in the adjacent table, in contact, then holding them at 1500°C and 1 GPa for 2.5 hours. The liquids were quenched, and diffusion profiles measured with an electron microprobe. Interdiffusion coefficients with Al_2O_3 as the independent variable calculated by Liang (2010) from the measured profiles are listed as follows. From these data, calculate time-independent (units of x/t) diffusion profiles for the four components.

Experimental Compositions

	Comp. 1	Comp. 2
SiO_2	0.62	0.66
Al_2O_3	0.20	0.20
CaO	0.07	0.11
MgO	0.11	0.0.

Interdiffusion Coefficients

	SiO_2	CaO	MgO
SiO_2	6.44	0.89	1.12
CaO	−1.06	2.69	0.46
MgO	−2.69	−1.64	−0.52

From Liang (2010).

Al_2O_3-oxide interdiffusion coefficients (10^{-11} m^2/s)

Answer: The solution in this case is somewhat analogous to the "infinite thin films" solution for tracer diffusion described in eqns. 5.80 through 5.86. In this case, Liang (2010) gives the solution as:

$$c(x,t) = \frac{c_{+\infty} + c_{-\infty}}{2} + [B][E][B]^{-1} \times \frac{c_{-\infty} - c_{-\infty}}{2} \tag{5.99}$$

$c_{+\infty}$ and $c_{-\infty}$ are the concentrations at infinite distance (i.e., starting compositions). [B] is a matrix whose columns contain the eigenvectors of the interdiffusion coefficient matrix, [E] is the eigenvalue matrix with diagonal terms equal to:

$$E_{ii} = erf\left(\frac{x}{4\sqrt{\lambda_i t}}\right) \tag{5.100}$$

where x is distance from the boundary between the melts, t is time, λ_i are the eigenvalues of the interdiffusion coefficient matrix and $E_{ij} = 0$ for $i \neq j$.

As a matrix algebra problem, the computations are easiest done in MATLAB. The script below calculates eigenvalues and eigenvectors with the function *eig*, then uses eigenvalues to calculate the diagonal matrix E, and finally calculates the concentration matrix at 40 steps in each direction from the interface. Mass conservation requires the mass fractions of the four oxides must sum to 1, so that Al_2O_3 can be calculated by difference. The results are then plotted with the *plot* function. These are shown in Figure 5.18.

```
%Interdiffusion between two infinite rods
%define diffusion matrix
D=[6.44,0.89,1.12;-1.06,2.69,0.46;-2.69,-1.64,-0.52]
%specify end member comps (values at infinity); SiO2,CaO, MgO wt%
End1 = [0.62;0.07;0.11]
End2 = [0.66;0.11;0.03]
%Calculate eigenvectors (B) and eigenvalues as a diagonal matrix (Lam) of D
[B,Lam]=eig(D)
%compute the initial E matrix
%a loop is necessary to avoid undefined values in nondiagonal elements
E=zeros(3)
for k = 1:1:3
    E(k,k)=power(4*sqrt(Lam(k,k)),-1)
    end
%predefine concentration (C) and distance (z) arrays
C=zeros(3,81)
z=zeros(1,81)
for j= 1:1:81
    %z is x/sqrt(t) and will vary from 16 to -16
    z(j)=(164-j*4)/10
    C(1:3,j)=(End1+End2)/2+B*erf(z(j)*E)*inv(B)*(End1-End2)/2
    end
SiO2=C(1,1:81)
CaO=C(2,1:81)
MgO=C(3,1:81)
Al2O3=1-SiO2-CaO-MgO
plot(z,SiO2,'blue')
plot(z,CaO,'cyan')
plot(z,MgO,'red')
plot(z,Al2O3,'green')
```

5.4.3 Driving force and mechanism of diffusion

The motion of atoms dispersing along a concentration or chemical potential gradient increases the entropy of the system. In that sense, entropy and the second law can be thought of as the driving force for diffusion. The rate of entropy generation is related to the diffusion flux as:

$$\frac{dS}{dt} = \sum J_i \frac{\nabla \mu_i}{T} \qquad (5.101)$$

where, as usual, S is entropy and μ is the chemical potential.

The mechanism of diffusion on a microscopic scale is always the same: the random motion of atoms or molecules. To demonstrate this point, we can derive Fick's first law just from a consideration of random atomic motion. Consider two adjacent lattice planes in a crystal spaced a distance dx apart. Let the number of atoms of the element of interest at the first plane be n_1 and the number of atoms at the second be n_2. We assume that atoms can change position randomly by jumping to an adjacent plane and that this occurs with an average frequency v (i.e., 1 jump of distance dx every $1/v$ sec). We further assume that there are no external forces, so that a jump in any direction has equal probability. At the first plane there will be $vn_1/6$ atoms that jump to the second plane (we divide by 6 because there are six possible jump directions: up, down, back, front, right, left). At the second plane there will be $vn_2/6$ atoms that jump to first plane. The net flux from the first plane to the second is then:

$$J = \frac{vn_1/6 - vn_2/6}{dx^2} = \frac{v}{6}\frac{(n_1 - n_2)}{dx^2} \qquad (5.102)$$

The concentration, c, is the number of atoms of interest per unit volume (i.e., n/dx^3), so we may substitute cdx^3 for the number of atoms in eqn. 5.102:

$$J = \frac{v}{6}\frac{(c_1 - c_2)dx^3}{dx^2} = \frac{v}{6}(c_1 - c_2)dx$$

Letting $dc = -(c_1 - c_2)$ and multiplying by dx/dx, we can rewrite this equation as:

$$J = -\frac{vdx^2}{6}\frac{dc}{dx}$$

If we let $D = vdx^2/6$, then we have Fick's first law:

$$J = -D\frac{dc}{dx}$$

Hence D is related to the jump frequency, v, and square of the jump distance (dx).

We see that there is a net diffusion, not because of the presence of a force, but only because there are more atoms at one point than at an adjacent one. In the absence of a concentration gradient (i.e., n_1 and n_2 the same), there would be $nv/6$ atoms moving from the first to the second plane and $nv/6$ atoms moving from second to the first plane. But if we cannot distinguish atoms originally at the first plane from those originally at the second, these fluxes balance and go unnoticed. If we could distinguish the atoms, we could detect a flux even in the absence of a concentration gradient. Other factors, such as pressure or stress gradient, presence of an electromagnetic field, or concentration gradients of other species may make a jump in one direction more probable than another, as can differences in chemical potential between the two planes. These will affect the diffusion flux and terms must be added to the diffusion equations to account for them. Because both chemical potential and the diffusion gradient depend on temperature, a temperature gradient alone can induce diffusion, something known as the *Soret effect*. The point we are stressing here is that diffusion can occur in the absence of all such forces.

5.4.4 Diffusion in solids and the temperature dependence of the diffusion coefficient

We can imagine four ways in which diffusion might take place in solids (Figure 5.19):

1. *Exchange*: the interchange of position of two atoms in adjacent sites.

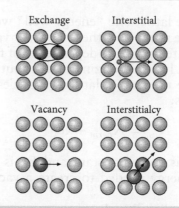

Figure 5.19 Four types of diffusion mechanisms in solids. After Henderson (1986).

2. *Interstitial*: in which an atom moves from one interstitial site to another.
3. *Interstitialcy*: in which an atom is displaced from a lattice site into an interstitial site.
4. *Vacancy*: in which an atom moves from a lattice site to a vacancy, creating a vacancy behind it.

Mechanisms 1 and 3 involve displacement of two atoms and therefore have high activation energies. Since interstitial sites are likely to be small, mechanism 2 will apply mainly to small atoms (H and He, for example). Thus, we are left with mechanism four as a principal mechanism of diffusion in solids.

Hence, diffusion in solids is a bit like a game of checkers: an atom can generally only travel by moving from lattice site to lattice site. Furthermore, it can only move to a vacant lattice site (and one of the correct types). In general, lattice vacancies are of two types: permanent and temporary. Permanent vacancies can arise from defects or through the presence of impurities, for example through substitution of a doubly charged ion for a singly charged one with a vacancy providing charge balance. Temporary sites arise from thermal agitation causing the volume of the solid to be slightly greater than the ideal volume by forcing atoms onto the surface. The number of the former is temperature-independent while the latter are temperature-dependent.

Let's attempt to calculate a diffusion coefficient *ab initio* for the simple one-dimensional case of tracer diffusion in a solid occurring through the vacancy mechanism. Since a certain minimum energy is required to get an ion out of the lattice site "energy well" we would expect the number of the temporary vacancies to have a temperature dependence of the form of eqn. 5.17, the Boltzmann distribution law. Thus, the number of lattice vacancies can be written as:

$$N_{vac} = N_{perm} + ke^{-E_H/RT}$$

where k is some constant and E_H is an activation energy needed to create a vacancy or "hole."

The probability, $\mathcal{P}$, of an atom making a successful jump to a vacant site is found by multiplying the number of attempts, $\aleph$, by the fractions of atoms having sufficient energy to get out of the well:

$$\mathcal{P} = \aleph e^{-E_B/RT} \qquad (5.103)$$

The number of attempts is simply the vibration rate, ν, times the number of holes:

$$\aleph = \nu[N_{vac}] = \nu[N_{perm} + ke^{-E_H/RT}] \quad (5.104)$$

Combining eqns. 5.103 and 5.104 we have:

$$\mathcal{P} = \nu N_{perm}e^{-E_B/RT} + \nu ke^{-(E_H+E_B)/RT}$$

The diffusion rate should be the number of jumps times the distance per jump, d:

$$\mathfrak{R} = \nu dN_{perm}e^{-E_B/RT} + \nu dke^{-(E_H+E_B)/RT}$$
$$(5.105)$$

or

$$\mathfrak{R} = mN_{perm}e^{-E_B/RT} + ne^{-(E_H+E_B)/RT} \quad (5.106)$$

where m and n are simply two constants replacing the corresponding terms in eqn. 5.105. Thus diffusion rates generally will have a temperature dependence similar to eqn. 5.25. At low temperature, the permanent vacancies will dominate and the diffusion rate equation will look like:

$$\mathfrak{R} \cong me^{-E_B/RT} \qquad (5.107)$$

At higher temperature where thermally generated vacancies come into play, the latter term in 5.104 dominates, and the diffusion rate equation will look like:

$$\mathfrak{R} \cong ne^{-(E_H+E_B)/RT} \qquad (5.108)$$

Diffusion that depends on thermally created vacancies is sometimes called *intrinsic* diffusion, while that depending on permanent vacancies is called *extrinsic* diffusion. The boundary between these regions will vary, depending on the nature of the material and the impurities present. For NaCl, the transition occurs around 500°C, while for silicates it generally occurs above 1000°C. Where the diffusion mechanism changes, a break in slope can be observed on a plot of ln D vs. 1/T. For example, Figure 5.20 shows how the diffusion coefficient might change based on eqns. 5.107 and 5.108.

Combining E_B and E_H into a single activation energy term, E_A, which is the energy necessary to create the vacancy and move another

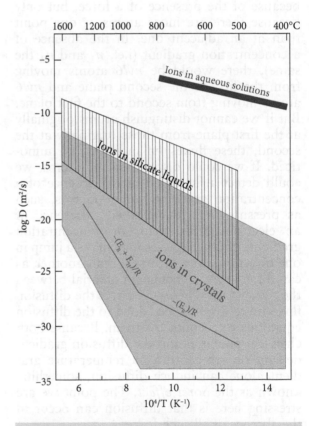

Figure 5.20 Schematic plot of log diffusion rate against inverse of temperature (Arrhenius plot) for different types of materials. Diffusion is faster in liquids than in solids due to the lack of long-range structure in the former. In solids, a change in diffusion mechanism from intrinsic to extrinsic can result in a break in slope. Adapted from Liang (2018) and Watson and Baxter (2007).

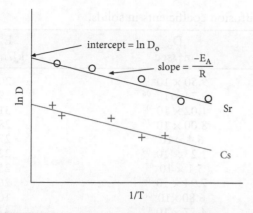

Figure 5.21 Schematic plot of log diffusion rate against inverse of temperature (Arrhenius plot) for two elements, Cs and Sr.

atom into it, a typical expression for temperature dependence of the diffusion coefficient in solids is:

$$D = D_o e^{-E_A/RT} \qquad (5.109)$$

where D_o is again called the frequency factor. As we have seen, it will depend on vibrational frequency and the distance of the interatomic jump.

Experimental observation supports our theoretical expectation of an exponential temperature dependence of diffusion, for example, in a series of measurements of the diffusion coefficient, D, at various temperatures (Figure 5.21). Taking the log of both sides of eqn. 5.107, we obtain:

$$\ln D = \ln D_o - \frac{E_A}{RT} \qquad (5.110)$$

Thus, on a plot of ln D versus reciprocal temperature, data for diffusion of a given element in a given substance should plot along a line with slope E_A/R and intercept ln D_o (Figure 5.21). Values for the activation energy are generally similar for most elements (typically 50–200 kJ), but the frequency factor varies widely. Table 5.2 list frequency factors and activation energies for several elements in various minerals.

The pressure dependence of the diffusion coefficient is:

$$D = D_o e^{-\{E_A + (P - P_{ref})\Delta V\}/RT} \qquad (5.111)$$

where ΔV is the *activation volume*.

5.4.5   Diffusion in liquids

In both liquids and solids, diffusion coefficients depend on both the nature of the diffusing species and the nature of the media. As you might expect, small atoms generally diffuse more rapidly than large ions. Since liquids lack the long-range order of crystalline solids, diffusion is more rapid in liquids than in solids (Figure 5.20).

The value of the diffusion coefficient in liquids can be estimated in a number of ways. Based on a model of molecular motion in a *nonionic* liquid, such as a hydrocarbon, composed of molecules of diameter d, and assuming a kinetic energy per atom of $3kT/2$ and a mean free path length of $2\alpha dT$, where α is the coefficient of thermal expansion, m is molecular mass, and k is Boltzmann's constant, the diffusion coefficient may be estimated as (Kirkaldy and Young, 1985):

$$D \cong \alpha d \sqrt{\frac{8k}{\pi m}} T^{3/2} \qquad (5.112)$$

This predicts a diffusion coefficient near the melting point of 10^{-4} cm^2/sec.

Diffusion coefficients in liquids are also commonly expressed in terms of viscosity. For uncharged species, the dependence of the diffusion coefficient on molecular radius and viscosity is expressed by the *Stokes–Einstein equation*:

$$D = \frac{kT}{6\pi\eta r} \qquad (5.113)$$

where r is molecular radius and η is viscosity. From this equation, we see that diffusion becomes more difficult as the liquid becomes more structured (polymerized) because the viscosity increases with increasing polymerization. Eyring replaced the $6\pi r$ term in equation 5.113 with an effective jump distance or effective volume term, λ:

$$D = \frac{kT}{\lambda\eta} \qquad (5.113a)$$

The Stokes–Einstein and Eyring equations work reasonably well for diffusion of neutral species, such as the heavier noble gases in aqueous solutions.

Because of the electrical neutrality effect, ion charge is important in diffusion of ions.

Table 5.2   Frequency factors and activation energies of diffusion coefficients in solids.

Species	Phase	D_o m^2/sec	E_A kJ/mol
Al	Qz (c axis)	9.50×10^{-11}	209
Ar	Orthoclase	3.74×10^{-06}	198
Ca	plag (An60)	1.02×10^{-05}	313
Dy	olivine	8.00×10^{-10}	289
Fe	Garnet	6.4×10^{-08}	275
Fe, Mg	Olivine	1.23×10^{-09}	220
Mn	Garnet	5.1×10^{-08}	253
Na	Plagioclase	7.94×10^{-04}	268
Na	Qz (c axis)	6.80×10^{-06}	103
O	Olivine	4.57×10^{-09}	338
O	Anorthite	1.50×10^{-10}	217
O	diopside	4.30×10^{-04}	457
O	Plag (An96)	8.40×10^{-13}	162
O	spinel	2.20×10^{-07}	404
Pb	K-feldspar	1.81×10^{-05}	309
Pb	zircon	7.76×10^{-02}	545
Rb	Orthoclase (Or94)	1.00×10^{-02}	339
Si	Qz (c axis)	6.40×10^{-06}	443
Sr	Diopside a-axis	6.4×10^{-04}	452
Sr	Diopside b-axis	1.2×10^{01}	565
Sr	Diopside c-axis	1.2×10^{-01}	511
Sr	Plagioclase	1.1×10^{-06}	295
Sr	Calcite	2.1×10^{-13}	132
Sr	Orthoclase (Or94)	1.0×10^{-11}	167
Sr	apatite	2.40×10^{-07}	271
Th	zircon	$8.63 \times 10^{+01}$	796
Ti	Qz	7.01×10^{-08}	273
U	zircon	1.63	726

Data from Zhang and Cherniak (2010)

In aqueous electrolytes, tracer diffusion coefficients depend on ion charge as:

$$D^\circ = \frac{RT\lambda^\circ}{|z|\mathcal{F}^2} \qquad (5.114)$$

where λ° is the limiting ionic conductance (conductance extrapolated to infinite dilution) of the ion (in cm^2/ohm-equivalent), $\mathcal{F}$ is Faraday's constant, and z is the charge of the ion. The limiting ionic conductance is itself a function of temperature, which leads to a strong dependence of D° on temperature. The nought ($\circ$) denotes the standard state of infinite dilution.

Table 5.3 lists values of D° for a few ions of geochemical interest. In dilute solution, diffusion coefficients depend approximately on the square root of ionic strength. In more concentrated solutions, diffusion coefficients show a complex dependence on ionic strength, the

treatment of which is beyond the scope of this book. Discussions of this problem may be found in Anderson (1981), Tyrell and Harris (1984), and Lasaga (1997).

None of these equations work well for silicate liquids. Instead, diffusion coefficients have a complex dependency on the nature of the diffusing species and the composition of the liquid. One of the most important chemical properties of silicate melts is the extent of polymerization, which is effectively controlled by the ratio of network forming ions, mainly oxide ions of Si and Al, to network modifiers, which includes most of the remaining components (described in section 4.6). The extent of polymerization controls viscosity. As an illustration of this complexity, diffusion noble gases, and alkali elements is faster in high-viscosity polymerized silica-rich

Table 5.3   Trace diffusion coefficients for ions in infinitely dilute aqueous solution at 25°C.

Cation	$D°$ $10^{-6}cm^{-2}sec^{-1}$	Anion	$D°$ $10^{-6}cm^{-2}sec^{-1}$
H^+	93.1	OH^-	52.7
Na^+	13.3	Cl^-	20.3
K^+	19.6	I^-	20.0
Mg^{2+}	7.05	SO_4^{2-}	10.7
Ca^{2+}	7.93	CO_3^{2-}	9.55
Sr^{2+}	7.94	HCO_3^{2-}	11.8
Ba^{2+}	8.48	NO_3^-	19.0
Fe^{2+}	7.19		
La^{3+}	6.17		

From Lasaga (1997).

liquids than in lower-viscosity weakly polymerized basaltic liquids, in contrast to the prediction of the Stokes–Einstein equation. This may reflect the denser packing of weakly polymerized melts, suggesting available space is the controlling factor. For ions of higher valence, diffusion is faster in weakly polymerized melts, suggesting the rigidity of the structure is more important.

A few generalizations of diffusion in silicate liquids are possible: network formers tend to diffuse more slowly than network modifiers and, in most cases, small ions and ions of low charge diffuse more rapidly than large ions or ions of high charge. This, of course, is expected, but it is not always the case. Table 5.4 provides some examples of the diffusion coefficients in silicate melts. Much more detailed on diffusion in silicate melts see the volume edited by Zhang and Cherniak (2010).

5.4.6 Diffusion in porous media

As we noted earlier in this chapter, when two parallel reaction pathways are available, the faster one is rate controlling and the slower one can become effectively irrelevant. Mineral grains in rocks are often surrounded by a fluid, particularly in soils, unconsolidated sediment, sediment undergoing diagenesis, rocks undergoing metamorphism, or during melting or crystallization of igneous rocks. Because diffusion is much more rapid in fluids than in solids (Figure 5.20), diffusion through the fluid occupying pore spaces dominates over diffusion through the soils, even though that pathway might be considerably longer

(the analogy is to driving a longer route to get to a destination on the other side of city rather than driving through city traffic). In porous media, Fick's first law is generally written as:

$$J = -\frac{\phi}{\tau}D\frac{dc}{dx}$$ (5.115)

where ϕ is the porosity (the fraction of volume filled by fluid) and τ is the tortuosity, a measure of the additional distance a diffusing species must take by driving the *fluid expressway*. For spherical grains, $\tau = 0.66$ as the diffusive pathway is on average 45° to the concentration gradient.

5.5 SURFACES, INTERFACES, AND INTERFACE PROCESSES

The properties of a phase at its surface are different from the bulk properties of the phase. This difference arises from the difference between the local environment of atoms on a surface or interface and those in the interior of a phase. An atom at the surface of a crystal is not surrounded by the same bonds and distribution of charges as it would be in the interior of the crystal lattice. Its potential energy must therefore be different. Here we define *surface* as the exterior boundary of a condensed phase (a solid or liquid) in a vacuum or gas. An *interface* is the boundary between two condensed phases, for example, between two crystals or between a mineral and water (the term *surface* is, however, often used for what we have just defined as an interface). Surfaces, surface energies, and interfaces play an important role in many

Table 5.4 Tracer diffusion coefficients in silicate melts.

Species	Phase	D_o m²/sec	E_A kJ/mol
Ar	Basaltic melt	8.05×10^{-02}	253
Ba	Basaltic melt	4.21×10^{-07}	130
Ba	Rhyolitic melt	3.80×10^{-06}	188
Ca	Basaltic melt	1.22×10^{-05}	159
Cl	Basaltic melt	7.54×10^{-04}	225
Cl	Rhyolitic melt	5.33×10^{-09}	92
Co	Basaltic melt	1.59×10^{-05}	168
H₂O	Basaltic melt	1.92×10^{-04}	159
H₂O	Dacitic melt	1.64×10^{-06}	131
Li	Basaltic melt	7.81×10^{-06}	116
Li	Rhyolitic melt	2.17×10^{-06}	91
Mg	Basaltic melt	7.79×10^{-01}	343
Na	Basaltic melt	8.96×10^{-05}	164
Na	Rhyolitic melt	1.01×10^{-06}	84
Nd	Basaltic melt	4.07×10^{-05}	200
Nd	Andesitic melt	1.08×10^{-05}	169
O	Basaltic melt	3.73×10^{-06}	170
Rb	Rhyolitic melt	2.01×10^{-07}	127
Sr	Basaltic melt	1.12×10^{-05}	169
Sr	Rhyolitic melt	3.74×10^{-05}	156
Yb	Basaltic melt	4.37×10^{-06}	169
Yb	Trachytic Melt	1.55×10^{-07}	187
Zr	Basaltic melt	1.40×10^{-05}	189

geochemical processes. All heterogeneous reactions (i.e., those involving more than one phase) must involve interfaces or surfaces. Dissolution, melting, exsolution, and precipitation are examples of processes that, on an atomic scale, occur entirely at or near the interface between two phases. Surfaces can also play important roles as catalysts in many geochemical reactions.

On a microscopic scale, the reactivity of mineral surfaces will vary locally for several reasons. The first is the microtopography of the surface (Figure 5.222). For example, a single growth unit (which might be a single atom, an ion, or molecule and called an *adatom*), located on an otherwise flat surface will be particularly unstable because it is bonded to other units on only one side. A *step* (which might be formed through growth, dissolution, or *screw dislocation*) provides a more favorable growth site because the new unit is bound to other units on two sides. An even better site for growth is at a *kink*, where bonds may be formed on three sides. Conversely, a unit at a kink (with three exposed sides) is less stable than one at a step (with

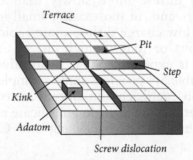

Figure 5.22 On a microscopic scale, the surface of a mineral exhibits a number of features. As a result, the local reactivity of the surface will be quite variable.

two exposed sides), which in turn is less stable than a unit on a flat surface, with only one exposed side. The point is, kinks and steps will be more reactive than other features, so surface reactions rates will depend in part on the density of these features.

Atoms are also freer to move alone a mineral surface rather than through its interior. This is process known as *grain boundary diffusion*, which can be many orders of magnitude more rapid than volume diffusion. This is a consequence of the abundance of unsatisfied bonds and defects on crystal surfaces.

Properties of mineral surfaces will also vary depending on the orientation of the surface relative to crystallographic axes. Most minerals grow or dissolve faster in one direction than in another. Most surface reactions involve the formation of new bonds between atoms of a mineral and atoms of the adjacent phase; the nature of the bonds that are possible will depend on the orientation of the surface relative to crystallographic axes. Reaction rates measured for one crystal face may not be representative of other faces.

Finally, most minerals have a variety of atoms and crystallographic sites, hence there will be a variety of bonds that are possible on any surface. We will discuss this aspect of surfaces in slightly more detail below.

5.5.1 The surface free energy

In Chapters 2 and 3, we introduced the concept of molar quantities and partial molar quantities. For example, the molar volume of a substance was:

$$\overline{V} = \frac{V}{n}$$

and we defined the partial molar volume as:

$$v_i^\phi = \left(\frac{\partial V}{\partial n}\right)_{T,P,n_j} \qquad (3.11)$$

We now define two new quantities, the *molar surface area*:

$$\overline{A} = \frac{A}{n} \qquad (5.118)$$

and the partial molar surface area:

$$\boxed{a_i^\phi = \left(\frac{\partial A}{\partial n}\right)_{T,P,n_j}} \qquad (5.117)$$

where A is the surface area of the phase and n is the number of moles of the component. A related quantity is the *specific surface area*, which is the area per unit mass. The molar

volume or the molar Gibbs free energy of pure quartz depends only on temperature and pressure, and the molar volume of each (pure) quartz crystal is the same as that of every other (pure) quartz crystal at that temperature and pressure. Unlike other molar quantities, the *molar or specific surface area and partial molar surface area depend on shape and size* and are therefore not intrinsic properties of the substance. For a sphere, for example, the partial molar surface area is related to molar volume as:

$$a = \frac{\partial A}{\partial V}\frac{\partial V}{\partial n} = \frac{2v}{r} \qquad (5.118)$$

For other shapes, the relationship between a and v will be different.

Finally, we define the *surface free energy* of phase ϕ as:

$$\boxed{\sigma^\phi \equiv \left(\frac{\partial G}{\partial A}\right)_{T,P,n}^\phi} \qquad (5.119)$$

The surface free energy represents those energetic effects that arise because of the difference in atomic environment on the surface of a phase. Surface free energy is closely related to surface tension. The total surface free energy of a phase is minimized by minimizing the phase's surface area. Thus, a water-drop in the absence of other forces will tend to form a sphere, the shape that minimizes surface area. When surface effects must be considered, we can revise the Gibbs free energy equation (eqn. 3.14) to be:

$$dG = VdP - SdT + \sum_i \mu_i dn_i + \sum_k \sigma_k dA_k \qquad (5.120)$$

where the last sum is taken over all the interfaces of a system. In this sense different crystallographic faces have different surface free energies. The last term in eqn. 5.120 increases in importance as size decreases. This is because the surface area for a given volume or mass of a phase will be greatest when particle size is small.

5.5.2 The Kelvin effect

When the size of phases involved is sufficiently small, surface free energy can have the effect of displacing equilibrium. For an equilibrium system at constant temperature and pressure, eqn. 5.120 becomes:

$$0 = \sum_i \nu_i \mu_i^o + RT \sum_i \nu_i \ln a_i + \sum_k \sigma_k dA_k$$

The first term on the right is ΔG°, which according to eqn. 3.85 is equal to $-RT \ln K$. This is the "normal" equilibrium constant, uninfluenced by surface free energy, so we'll call it K°. We will call the summation in the second term K^s, the equilibrium influenced by surface free energy. Making these substitutions and rearranging, we have:

$$\ln K^s = \ln K^o - \frac{\sum \sigma_k dA_k}{RT} \quad (5.121)$$

Thus, we predict that equilibrium can be shifted due to surface free energy, and the shift will depend on the surface or interfacial area. This is known as the *Kelvin effect*.

There are a number of examples of this effect. For example, fine, and therefore high surface area, particles are more soluble than coarser particles of the same composition. Water has a surface free energy of about $70 \, mJ/m^2$. Consequently, humidity in clouds and fogs can reach 110% when droplet size is small.

5.5.3 Nucleation and crystal growth

5.5.3.1 Nucleation

Liquids often become significantly oversaturated with respect to some species before crystallization begins. This applies to silicate liquids as well as aqueous solutions (surface seawater is several times oversaturated with respect to calcite). However, crystallization of such supersaturated solutions will often begin as soon as seed crystals are added. This suggests that nucleation is an important barrier to crystallization. This barrier arises because the formation of a crystal requires a local increase in free energy due to the surface free energy at the solid–liquid interface.

Let's explore a bit further how nucleation can be inhibited. For a crystal growing in a liquid, we can express the complete free energy change as:

$$dG_{tot} = \sigma dA + dG_{xt} \quad (5.122)$$

where dG_{xt} refers to the free energy change of the crystallization reaction that applies throughout the volume of the crystal (i.e., free energy in the usual sense, neglecting surface effects).

Let's consider a more specific example, that of a spherical crystal of phase ϕ growing from a liquid solution of component ϕ. The free energy change over some finite growth interval is:

$$\Delta G_{tot} = 4\pi r^2 \sigma + \frac{4}{3}\pi r^3 \frac{\Delta G_{xt}}{\overline{V}} \quad (5.123)$$

where r is the radius (we divide by $\overline{V}$ to convert joules per mole to joules per unit volume). The first term on the right is the surface free energy, and, although small, *is always positive*. At the point where the solution is exactly saturated, ΔG_{xt} will be 0. The net free energy, ΔG_{tot}, is thus positive, so the crystal will tend to dissolve. In order for spontaneous nucleation to occur, the second term on the right must be negative and its absolute value must exceed that of the first term on the right of 5.123 (i.e., the liquid must be supersaturated for nucleation to occur). Solving eqn. 5.123 for r, we find $\Delta G_{tot} \leq 0$ when $r \geq -3\sigma/\Delta G$.

How will ΔG vary with r up to this point? To answer this, we differentiate eqn. 5.123 with respect to r:

$$\frac{\partial \Delta G_{tot}}{\partial r} = 8\pi r\sigma + 4\pi r^2 \frac{\Delta G_{xt}}{\overline{V}} \quad (5.124)$$

Since the volume free energy term is proportional to r^2 and the surface free energy term to r, the latter necessarily dominates at very small values of r. For small values of r, ΔG_{tot} will increase with increasing r because σ is always positive. In other words, near the saturation point where ΔG is small, very small crystals will become increasingly unstable as they grow. The critical value of r, that is, the value at which ΔG will decrease upon further growth, occurs where $\partial G/\partial r = 0$. Solving eqn. 5.124, we find that

$$r_{crit} = -\frac{2\sigma \overline{V}}{\Delta G_{xt}} \quad (5.125)$$

For a solution that undergoes cooling and becomes increasingly saturated as a result

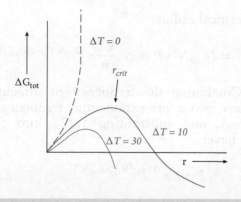

Figure 5.23 Free energy as a function of crystal radius for small crystals forming near the saturation point. ΔT is the amount of undercooling (difference between temperature and saturation temperature).

(e.g., a magma or a cooling hydrothermal solution), we can use eqn. 5.68 to approximate the ΔG term as $\Delta G \cong -\Delta T \Delta S$, where ΔT is the difference between actual temperature and the temperature at which saturation occurs, and ΔS is the entropy change of crystallization. For an aqueous solution near saturation, we can alternatively express ΔG_{xt} as $RT\ln(Q/K)$ from eqn. 5.67. Figure 5.23 shows the total free energy calculated in this way as a function of r for various amounts of undercooling.

The surface free energy term correlates with viscosity. Thus, nucleation should require less supersaturation for aqueous solutions than silicate melts. Among silicate melts, nucleation should occur more readily in basaltic ones, which have low viscosities, than in rhyolitic ones, which have high viscosities. This is what one observes. Also, we might expect rapid cooling to lead to greater supersaturation than slow cooling. This is because there is an element of chance involved in formation of a crystal nucleus (the chance of bringing enough of the necessary components together in the liquid so that r exceeds r_{crit}). Slow cooling provides time for this statistically unlikely event to occur and prevents high degrees of supersaturation from arising. With rapid cooling, crystallization is postponed until ΔG_r is large, when many nuclei will be produced. Let's briefly consider nucleation rates in more detail.

Surface free energy varies between different minerals and this explains some of the features we observe in metamorphic and igneous rocks. When surface free energy is large, nucleation is inhibited, and few grains will form but lack of competition for components allows them to grow to large size. The few but quite large garnets that can frequently be observed in metamorphosed pelitic sediments are a good example. In contrast, when surface free energy is low, many grains can nucleate but competition for components limits the size to which they can grow. The abundant but tiny grains of chrome spinel and other oxides that occur in basaltic lavas are a good example.

5.5.3.2 Nucleation rate

The first step in crystallization from a liquid is the formation of small clusters of atoms having the composition of the crystallizing phase. These so-called *heterophase fluctuations* arise purely because of statistical fluctuations in the distribution of atoms and molecules in the liquid. These fluctuations cause local variations in the free energy of the liquid, and therefore their distribution can be described by the Boltzmann distribution law:

$$N_i = N_\nu e^{-\Delta G_i/kT}$$

where N_i is the number of clusters per unit volume containing i atoms, N_ν is the number of atoms per unit volume of the cluster, and ΔG_i is the difference between the free energy of the cluster and that of the liquid as a whole. The number of clusters having the critical size (r_{crit}) is:

$$N_{crit} = N_\nu e^{-\Delta G_{crit}/kT}$$

where ΔG_{crit} is the total free energy (ΔG_{tot}) of clusters with critical radius obtained by solving eqn. 5.123 when $r = r_{crit}$. For spherical clusters, this is:

$$\Delta G_{crit} = \frac{16\pi}{3} \frac{\sigma^3 \overline{V}^2}{\Delta G_{xt}^2} \qquad (5.126)$$

Substituting eqn. 5.70 for ΔG_{xt}, we have:

$$\Delta G_{crit} = \frac{16\pi}{3} \frac{\sigma^3 \overline{V}^2}{(\Delta T \Delta S)^2} \qquad (5.127)$$

For a phase precipitating from solution at constant temperature, the equivalent equation is:

$$\Delta G_{crit} = \frac{16\pi}{3} \frac{\sigma^3 \overline{V}^2}{\left(RT \ln \frac{Q}{K}\right)^2} \quad (5.128)$$

If E_A is the activation energy associated with attachment of an additional atom to a cluster, then the probability of an atom having this energy is again given by the Boltzmann distribution law:

$$\mathcal{P} = e^{-E_A/kT}$$

Now according to transition state theory, the frequency of attempts, ν, to overcome this energy is simply the fundamental frequency $\nu = kT/h$. The attachment frequency is then the number of atoms adjacent to the cluster, N^*, times the number of attempts times the probability of success:

$$N^* \nu \mathcal{P} = N^* \frac{kT}{h} e^{-E_A/kT} \quad (5.129)$$

The nucleation rate, I, is then the attachment frequency times the number of clusters

of critical radius:

$$I = N_{crit} N^* \nu \mathcal{P} = N^* \frac{kT}{n} e^{-E_A/kT} e^{-\Delta G_{crit}/kT} \quad (5.130)$$

Combining the frequency of attachment terms into a pre-exponential frequency factor A, and substituting 5.128 into 5.130 we have:

$$I = Ae^{-16\pi\sigma^3 \overline{V}^2/3\Delta G^2 kT} \quad (5.131)$$

which is the usual expression for nucleation rate (e.g., McLean, 1965). If we substitute 5.70 into 5.131, we see that:

$$I = Ae^{-16\pi\sigma^3 \overline{V}^2/3(\Delta S \Delta T)^2 kT} \quad \text{or} \quad I \propto e^{-1/\Delta T^2} \quad (5.132)$$

This implies that nucleation rate will be a very strong function of "temperature overstepping" for relatively small values of ΔT, but will level off at higher values of ΔT. At low degrees of overstepping, nucleation rate will be nil, but will increase rapidly once a critical temperature is achieved, as is demonstrated in Example 5.9. A more detailed treatment of nucleation and growth of crystals in cooling magmas can be found in Toramaru (1991).

Example 5.9 Nucleation of Diopside

The enthalpy of fusion of diopside is 138 kJ/mol and its melting temperature is 1665 K. Assuming an activation energy of 1×10^{-18} J, how will the nucleation rate of diopside crystals in a diopside melt vary with temperature for surface free energies of 0.02, 0.06, and 0.12 J/m^2?

Answer: The one additional piece of information we need here is the molar volume, which we find to be 66 cc/mol from Table 2.02. We can calculate ΔS_m from the relation:

$$\Delta S_m = \frac{\Delta H_m}{T_m}$$

Assuming ΔS_m, σ, and E_A are independent of temperature, we can use equations 5.130 and 5.128 to calculate the nucleation rate. The calculation for the three surface free energies is shown in Figure 5.24a. Nucleation will be experimentally observable when the nucleation rate reaches $\approx 10^{-10}$ m^{-2}, which corresponds roughly to 1 nuclei/cm^2/hr. For a surface free energy of 0.02 J/m^2, the rate is reached only a few kelvins below the melting point. Further undercooling results in very high nucleation rates. For a surface energy of 0.06 J/m^2, an undercooling of 35 K is required, and an undercooling of 130 K is required at the highest value of surface energy. In the latter case, the rise in nucleation rate with undercooling is not nearly as steep.

In Figure 5.24b, we see that the nucleation rate passes through a maximum and as undercooling proceeds further, the rate decreases. This decrease reflects the $1/T$ dependence of both exponential terms in equation 5.130, i.e., the formation and growth of heterophase fluctuations will fall as temperature falls. Observed nucleation rates show this maximum, but the "bell" is generally more symmetric and considerably narrower. This reflects the increasing viscosity of the melt, and therefore the decreasing mobility of atoms, i.e., diffusion of atoms to the proto-nuclei slows.

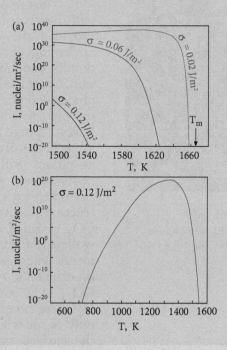

Figure 5.24   Calculated nucleation rate of diopside in diopside melt as a function of temperature for various values of the surface free energy, σ.

5.5.3.3   Heterogeneous nucleation

The nucleation of diopside crystals from diopside melt is an example of homogenous nucleation, that is nucleation in a system where initially only one phase is present. Heterogeneous nucleation refers to the nucleation of a phase on a preexisting one. Often the surface free energy between the nucleating phase and the preexisting surface is lower than between the nucleating phase and the phase from which it is growing. Hence, heterogeneous nucleation is often favored over homogenous nucleation. Perhaps the most familiar example is dew. Dew droplets appear on surfaces, such as those of grass, at significantly lower relative humidity than necessary for fog or mist to form. The reason is that the surfacefree energy between grass and water is lower than between water and air. Another example is the clusters of crystals

seen in igneous rocks. These result from one crystal nucleating on the other, again because the free energy of the crystal–crystal interface is lower than that of the crystal–magma interface. Figure 5.25 illustrates an example of crystals that have nucleated on each other in a magma.

Let's examine this in a more quantitative fashion. Consider a spherical cap of phase β nucleating from phase α on a flat surface, s (Figure 5.26), such as water (β) condensing from air (β) on a leaf (s). The balance of surface forces at the three-phase contact is:

$$\sigma_{\alpha-s} = \sigma_{\beta-s} + \sigma_{\alpha-\beta}\cos\theta \qquad (5.133)$$

and solving for θ:

$$\cos\theta = \frac{\sigma_{\alpha-s} - \sigma_{\beta-s}}{\sigma_{\alpha-\beta}}$$

Figure 5.25 Photomicrograph of an intergrowth of clinopyroxene and plagioclase grains in a basaltic lava that have nucleated on each other.

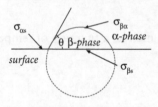

Figure 5.26 Illustration of the balance of forces as a spherical crystal or droplet of phase β crystallizes or condenses from phase α on a surface.

If the interfacial energy between the nucleating phase, β, and the surface ($\sigma_{\beta s}$) is smaller than that between phase α and the surface ($\sigma_{\beta s}$), then the angle of intersection, θ, will be small so as to maximize the interfacial surface area between β and s for a given volume of β. In the limit where $\sigma_{\beta s} \ll \sigma_{\alpha s}$ then θ will approach 0 and β will form a film coating the surface. As $\sigma_{\beta s}$ approaches $\sigma_{\alpha s}$ the nucleating phase will form more spherical droplets. If $\sigma_{\beta s} \geq \sigma_{\alpha s}$ then θ will be 90° or greater, and heterogeneous nucleation will not occur. To take account of the reduced interfacial energy

between β and s, eqn. 5.126 becomes:

$$\Delta G_{crit} = \frac{4\pi}{3} \frac{\sigma_{\alpha-\beta} \overline{V}}{\Delta G_{xt}^2} \qquad (5.134)$$

In metamorphic reactions, nucleation will necessarily always be heterogeneous. Provided the necessary components of the nucleating phase are available and delivered rapidly enough by fluid transport and diffusion, interfacial energy will dictate where new phases will nucleate, nucleation being favored on phases where the interfacial energy is lowest. Where transport of components limits growth, however, this may not be the case, as phases will nucleate where the components necessary for growth are available. For example, experimental investigation of the reaction calcite + quartz $\rightleftharpoons$ wollastonite + CO_2 revealed that, in the absence of water, wollastonite nucleated on quartz. In experiments where water was present, it nucleated on calcite. SiO_2 is not significantly soluble in CO_2, so it could not be transported in the H_2O free experiments, hence nucleation could only occur where SiO_2 was available (i.e., at the surface of quartz), despite a probable higher interfacial energy.

Unfortunately, agreement between observed and predicted nucleation rates is often poor (Kirkpatrick, 1981; Kerrick et al., 1991). Equation 5.134 and Figure 5.24 show that the nucleation rate is a very strong function of the surface free energy ($I \propto \exp(\sigma^3)$), and the poor agreement between theory and observation may reflect the lack of accurate data on surface free energy as well as the activation energy, E_A. However, it may also indicate that further work on the nucleation theory is required.

5.5.3.4 Diffusion-limited and heat-flow limited growth

Two other kinetic factors affect crystallization. These are the local availability of energy and local availability of components necessary for crystal growth. The latter can be important where the crystal is of different composition than the liquid (almost always the case in nature, except freezing of fresh water). Crystals can grow only as rapidly as the necessary chemical components are delivered to their surfaces. Where diffusion is not rapid enough to supply these components, diffusion will limit growth.

A second effect of slow diffusion is to change the *apparent* distribution coefficient, because the crystal "sees" the concentrations in the adjacent boundary layer rather than the average concentrations in the liquid. Thus, the crystal may become less depleted in elements excluded from the crystal, and less enriched in elements preferentially incorporated in it, than equilibrium thermodynamics would predict. For example, suppose a crystal of plagioclase is crystallizing from a silicate melt. Plagioclase preferentially incorporates Sr and excludes Rb. If diffusion of Sr and Rb to the crystal is slow compared with the crystal growth rate, the liquid in the boundary layer immediately adjacent to the crystal will become impoverished in Sr and enriched in Rb. The crystal will grow in equilibrium with this boundary layer liquid, not the average magma composition, thus will be poorer in Sr and richer in Rb than if it grew in equilibrium with the average magma. Figure 5.27 illustrates this point. If, however, growth rate of the crystal is very much slower than the transport of components to the crystal–liquid interface, this circumstance will not arise.

When crystals grow from a magma there will be a local increase in temperature at the crystal–liquid boundary, due to release of latent heat of fusion, ΔH_m, which will retard crystal growth. In most cases, however, advection and conduction of heat is probably sufficiently rapid that this is at best a minor effect. The effect is probably more important in prograde metamorphic reactions (e.g., dehydration reactions), which are usually endothermic and hence require a continuous supply of energy to maintain crystal growth. Where crystal growth and transport of components is sufficiently rapid, heat flow may limit rates of crystal growth. This is more likely to occur at high temperatures and in late stages of metamorphism when structures are already large (Fisher, 1978).

5.5.3.5 Grain coarsening, static annealing, and Ostwald ripening

When essentially monomineralic rocks such as limestone or quartz sandstone undergo metamorphism, the resulting metamorphic rocks can consist of the same minerals as the protolith but be texturally distinct in being coarser grained (illustrated in Figure 5.28). This, and the interlocking nature of the crystals, makes marble and quartzite much stronger rocks than their sedimentary equivalents. It's no accident that Michelangelo chose marble rather than limestone for his best work and Stone-Age humans sought out quartzite rather than sandstone for their stone tools. The coarsening is a consequence of interfacial free energy: for a given volume of crystals composing the rock, the total free energy of the system is lowered by decreasing the surface area. This is achieved by having fewer but larger crystals. The process involved is known as *static annealing*, which is a form of *Ostwald ripening*. Let's consider this in a bit more detail.

We begin by noting that sedimentary rocks form by accumulation of particles settling out of a fluid (water or air). The particles typically have a range of sizes and a variety of morphologies and there is initially substantial pore space between them which decreases upon burial. Once buried, this pore space will be filled with an aqueous solution, even if the sediment originated as desert sand. The pore water is important as it provides a pathway for diffusion. Over time, the pore fluid will come in the chemical equilibrium with the

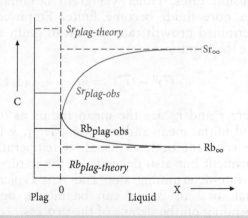

Figure 5.27 Variation of Sr and Rb concentrations from a plagioclase–liquid interface. Solid curves show the variation of concentration. The crystal–liquid interface is at 0. Dashed lines show the concentrations at infinite distance from the interface (Sr_∞, Rb_∞). Sr_{plag} and Rb_{plag} are the concentrations in the crystal.

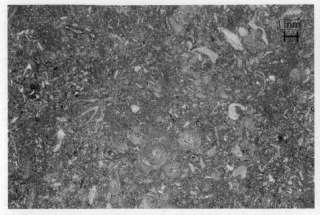

Figure 5.28 Left: Photomicrograph of a fossiliferous limestone from Texas (Ludders Bluff). Source: National Park Service Building stone database: Right: Photomicrograph of the Carrera Marble (Italy) viewed under crossed polarizers. Photo from Calia et al. (2006). Both consist primarily of calcite, but grain size is larger in the marble.

grains, which is to say it will be saturated in the components comprising the grains.

For simplicity, we assume the sediment consists of spheres of a single mineral (a quartz sandstone consisting of well-rounded grains, for example). We found that because of surface-free energy, equilibrium between a mineral and the solution from which it precipitates will depend on grain size. As a consequence, a solution in equilibrium with the assemblage as a whole will be undersaturated with grains smaller than some critical radius. The ratio of concentration at which a solution is saturated with respect to a grain of radius r, c_r, to the saturation concentration for an infinitely larger grain (i.e., one for which surface free energy is irrelevant), c_∞, is:

$$\frac{c_r}{c_\infty} = \exp\left[\frac{2\sigma\overline{V}}{RT}\left(\frac{1}{r} - \frac{1}{r_{crit}}\right)\right] \quad (5.135)$$

where r_{crit} is the critical radius (eqn. 5.125). Those grains with radius smaller than the critical radius will dissolve, the solution will consequently become supersaturated with respect to large ones, which will then grow. The rate at which a grain of radius r will grow with respect to the mean radius can be expressed as:

$$\frac{\partial r}{\partial t} = k(c_{\bar{r}} - c_r) \quad (5.136)$$

where k is the precipitation rate constant and $c_{\bar{r}}$ and c_r are the concentrations in the pore

fluid in equilibrium with the mean grain radius and a grain of radius r, respectively. The process eventually produces an asymmetric distribution of grain sizes about the mean, the form of which depends on the rate-determining step (Figure 5.29).

Nevertheless, the grains of a sandstone or limestone may persist in their original state for hundreds of millions of years. Metamorphism begins once temperatures rise to the point where precipitation, dissolution, and diffusion rates, either via grain boundaries or a pore fluid, become finite. Empirically determined growth rates for mean grain size have the form:

$$(\bar{r})^n - (\bar{r_0})^n = kt \quad (5.137)$$

where $\bar{r}$ and $\bar{r}_0$ are the mean radius at time t and initial mean radius, respectively, k is a rate constant that, as usual, is temperature dependent but also depends on the nature of the rate-determining step and n is typically equal to 2 or 3 but can be higher, again depending on the details of the process.

Figure 5.29a illustrates the equilibrium grain size about the mean for various rate-determining steps. Growth rate limited by grain boundary or volume diffusion results in a distribution with a mode slightly greater than the mean grain size but highly skewed to smaller grains. When reaction kinetics are rate-limiting, the mode is less than the mean

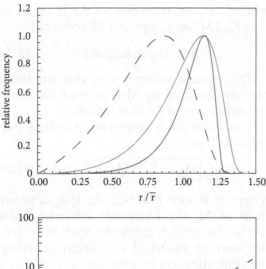

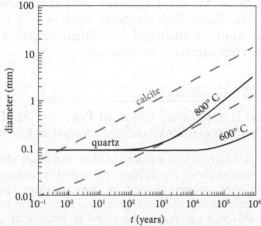

Figure 5.29 a. Distribution of grain sizes normalized to mean grain diameter based on equations in Joesten (1991). The distribution depends on the nature of the rate-controlling step: dashed green line: reaction kinetics; red line: grain boundary diffusion; solid blue line: volume diffusion. b. Increase in mean diameter with time for calcite (dashed red lines) at 600°C based on different experimentally determined rate constants summarized in Joesten (1991) and quartz (black lines) at 600° and 800°C and equation 5.135. The value of n is 3 for calcite and 2 for quartz. Modified from Joesten (1991).

and the distribution less skewed. Figure 5.29b illustrates how mean grain size will increase through time for calcite and quartz, depending on the rate-limiting step and temperature.

5.5.4 Adsorption

Many geochemically important reactions take place at the interface between solid and fluid

phases, and inevitably involve adsorption and desorption of species onto or from the surface of the solid. Two types of adsorption are possible: physical and chemical. Physical adsorption involves the attachment of an ion or molecule to a surface through intermolecular or van der Waals forces. Such forces are relatively weak, and heat of adsorption (ΔH_{ad}) relatively low (typically 4–12 kJ/mol). Chemical adsorption involves the formation of a new chemical bond between the adsorbed species and atoms on the surface of the solid. Heats of chemical adsorption are relatively large (>40 kJ/mol).

Adsorption of ions and molecules on a solid surface or interface affects the surface free energy. The relationship between surface free energy and adsorbed ions can be expressed as:

$$d\sigma = -\sum_i \frac{n_{i,s}}{A} d\mu_i = -\sum_i \Gamma_i d\mu_i \quad (5.138)$$

where $n_{i,s}$ is the number of mole of species i adsorbed at the surface, A is the surface area, and we define Γ_i as the *Gibbs adsorption density*. Because silicates and oxides generally have positive surface free energies, we can see that adsorption will decrease this energy and is therefore strongly favored (see Example 5.10).

5.5.4.1 *The relation between concentration and adsorption: Langmuir and Freundlich isotherms*

Consider the adsorption of aqueous species M at a surface site that we will denote as ≡S. The reaction may be written as:

$$M + \equiv S \rightleftharpoons M·S$$

We will denote the fraction of surface sites occupied by M as Θ_M, the rate constant for adsorption as k_+, and that for desorption as k_-. The fraction of free sites is then $(1 - \Theta_M)$, and we explicitly assume that M is the only species adsorbed from solution. Assuming the reaction is elementary, the rate of adsorption is then:

$$\frac{dM}{dt} = k_+[M](1 - \Theta_M) \quad (5.139)$$

The rate of desorption is:

$$\frac{d\Theta_M}{dt} = k_-\Theta_M \quad (5.140)$$

At equilibrium, the rate of adsorption and desorption will be equal, so

$$k_-\Theta_M = k_+[M](1 - \Theta_M) \quad (5.141)$$

Solving eqn. 5.141 for Θ_M, we obtain:

$$\Theta_M = \frac{k_+/k_-[M]}{1 + k_+/k_-[M]} \quad (5.142)$$

which expresses the fraction of site occupied by M as a function of the concentration of M. Since at equilibrium:

$$K_{ad} = \frac{k_+}{k_-} = \frac{[M]_{ad}}{[M]_{aq}} \quad (5.44)$$

where K_{ad} is the equilibrium constant for adsorption, eqn. 5.142 becomes:

$$\Theta_M = \frac{K_{ad}[M]}{1 + K_{ad}[M]} \quad (5.143)$$

Equation 5.141 is known as the *Langmuir isotherm*.* Since this is a chapter on kinetics, we have derived it using a kinetic approach, but it is a statement of thermodynamic equilibrium and can be readily derived from thermodynamics as well. From the definition of Θ_M, we may also write the Langmuir isotherm as:

$$\Gamma_M = \Gamma_M^{max}\frac{K_{ad}[M]}{1 + K_{ad}[M]} \quad (5.144)$$

where Γ_M^{max} is the maximum observed adsorption. Thus, the Langmuir isotherm predicts a maximum adsorption when all available sites are occupied by M. At large concentrations of M, then:

$$\Gamma_M = \Gamma_M^{max} \quad (5.145)$$

where the concentration of M is small such that $K_{ad}[M] \ll 1$, eqn. 5.128 reduces to:

$$\Theta_M \cong K_{ad}[M] \quad (5.146)$$

This equation simply says that the fraction of sites occupied by M is proportional to the concentration of M in solution.

The *Freundlich isotherm*, which is purely empirical, is:

$$\Theta_M = K_{ad}[M]^n \quad (5.147)$$

where n is any number. At low concentrations of M, the Langmuir isotherm reduces to the Freundlich isotherm with $n = 1$ (i.e., the amount adsorbed is a linear function of the concentration in solution).

5.5.5 Catalysis

The International Union of Pure and Applied Chemistry (IUPAC) defines *catalyst* as follows:

> *A catalyst is a substance that increases the rate without modifying the overall standard Gibbs energy change in the reaction; the process is called catalysis, and a reaction in which a catalyst in involved is known as a catalyzed reaction.*

Another definition of a catalyst is a *chemical species that appears in the rate law with a reaction order greater than its stoichiometric coefficient*. This latter definition makes it clear that a catalyst may be involved in the reaction as a reactant, a product, or neither. If it is a reactant or product, its presence affects the reaction rate to a greater extent than would be predicted from the stoichiometry of the reaction.

We can distinguish two kinds of catalysis. Homogenous catalysis refers to a situation in which the catalyst is present in the same phase in which the reaction is occurring (necessarily a solution). Examples of homogenous catalysts of geochemical reactions include acids and a collection of organic molecules called enzymes. Catalysis that occurs at the interface

---

*An admittedly odd name for this equation. It is named for Irving Langmuir (1881–1957). Langmuir obtained a PhD from the University of Göttingen and spent most of his career working for the General Electric Company. The lifetime of tungsten filaments in light bulbs will be shortened by reaction with oxygen absorbed onto the surface of the evacuated bulb. Langmuir's objective was to reduce the amount of adsorbed gas on the interior surface of the bulb and he developed this equation to describe his results. The term *isotherm* arises because such descriptions of adsorption are valid only for one temperature (i.e., K_{ads} is temperature-dependent, as we would expect); in this case, adsorption of gas on glass decreases with temperature. Langmuir won the Nobel Prize for Chemistry in 1932.

Example 5.10   The Langmuir isotherm

Consider a suspension of 1 mol/l of FeOH. Assuming an adsorption site density of 0.1 mol/mol and K for adsorption of Sr on FeOH of 10^5, how will the Sr adsorption density vary with the concentration of Sr in the solution? Assume that no other ions are present in the solution.

Answer: We can use equation 5.144 to solve this problem. Γ_M^{max} in this case is 0.1 mol/mol. Using this value and K_{ads} of 10^5 in this equation, we obtain the result shown in Figure 5.30. The inset shows that at concentrations less than about 4 µM, the adsorption density rises linearly with concentration. At higher concentrations, the adsorption density asymptotically approaches the maximum value of 0.1 mol.

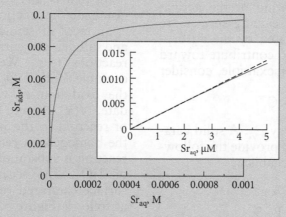

Figure 5.30   Variation of adsorption density of Sr on FeOH as a function of Sr concentration of the solution.

Adsorption and Determination of Mineral Surface Areas

By now you should have a sense that surface area is a key parameter in the kinetics of heterogeneous reactions and may be wondering how surface area is measured. Of course, if the geometry of grains is known, this is straightforward. For example, the surface area is a cubic pyrite grain of dimension l is simply $6l^2$. However, the geometry of mineral grains is rarely so simple: microscopic bumps, protrusions, pores, and other features are common on a surface so that the actual surface area is likely to be significantly greater than one would calculate based on its macroscopic geometry even if all the grains have the same size and geometry.

A commonly used approach to measuring surface area is the Brunauer–Emmett–Teller or BET method, which is based on the adsorption of gas on surfaces. This is done by placing a sample in a tube, heating it in a vacuum to drive off any adsorbed gas, then cooling it to near the boiling point of the gas being used in the analysis. The boiling point of N_2, the most commonly used gas, is 77 K. Krypton, another commonly used gas, has a boiling point of 120 K. The sample then is exposed to analysis gas in controlled increments. After each increment, the pressure is allowed to equilibrate and the quantity of gas adsorbed calculated from the pressure change. The quantity adsorbed at each pressure and temperature defines the adsorption isotherm, which in turn allows calculation of the specific surface area. Reproducibility of BET surface area measurements range from as good as 5% to as poor as 70% (Brantley and Mellot, 2000).

between two phases is referred to as heterogeneous catalysis. We will focus primarily on heterogeneous catalysis here. Heterogeneous catalysts are commonly simply surfaces of some substance. A familiar, but nongeochemical, example is the platinum in the catalytic converter of an automobile, which catalyzes the further oxidation of gasoline combustion products.

Catalysts work by providing an alternative reaction path with lower activation energy. In many cases, the lowering of the activation energy arises when reacting species are adsorbed. The heat liberated by the adsorption (ΔH_{ad}) is available to contribute toward the activation energy. For example, consider the reaction:

$$A + B \rightarrow C$$

having an activation energy E_A. A solid catalyst of this reaction would provide the following alternate reaction mechanism:

$$A + \equiv S \rightarrow A \cdot S$$
$$B + \equiv S \rightarrow B \cdot S$$
$$B \cdot \equiv S + A \cdot S \rightarrow C \cdot S$$
$$C \cdot S \rightarrow C + \equiv S$$

The net heat of adsorption for this process is:

$$\Delta H_{ad} = \Delta H_{ad}^A + \Delta H_{ad}^B - \Delta H_{ad}^C$$

Recalling that enthalpy is related to activation energy, we can write the activation energy for the catalyzed reaction as:

$$E_A^{cat} = E_A + \Delta H_{ad} \qquad (5.148)$$

If ΔH_{ad} is negative (i.e., heat liberated by adsorption), the activation energy is lowered and the reaction proceeds at a faster rate than it otherwise would.

As we noted earlier, a surface will have a variety of sites for adsorption/desorption and surface reactions on a microscopic scale. Each site will have particular activation energy for each of these reactions. The activation energies for these processes will, however, be related. Sites with large negative adsorption energies also will be sites with low activation energies for surface reactions. On the other hand, if a site has a large negative adsorption energy, the desorption energy will be large and positive and desorption inhibited. If either the activation energy or the desorption energy

is too large, catalysis of the overall reaction will be inhibited. What is required for fast overall reaction rates is a site where some compromise is achieved. In general, reaction and desorption energies will be related as:

$$\Delta G_r = -n\Delta G_d \qquad (5.149)$$

where n is some constant. The presence of several sites on a solid surface results in several possible reaction paths. The fastest reaction path, that is, the path that optimizes n, will dominate the reaction.

Surfaces of semiconductors (metal oxides and sulfides) can catalyze oxidation–reduction reactions (e.g., Wehrli et al., 1989). For example, both TiO_2 and Al_2O_3 can catalyze the oxidation of vanadyl, V(IV), to vanadate, V(V). Figure 5.31 compares the rate of reaction in the presence of TiO_2 solid to the homogenous reaction, demonstrating the reaction is substantially faster in the presence of TiO_2. The reaction mechanism for the surface catalyzed reaction may be described as follows (Figure 5.32):

$$2\equiv TiOH + VO^{2+} \rightleftharpoons \equiv(TiO)_2VO + 2H^+$$
$$\equiv(TiO)_2VO + TiOH \rightleftharpoons \equiv(Ti)_3VO_4^{2-} + H^+$$
$$\equiv(Ti)_3VO_4^{2-} + H_2O + 2OH^- \rightleftharpoons 3$$
$$\equiv TiOH + HVO_4^{2-}$$

where $\equiv Ti$ indicates that the Ti atom is part of a surface. The rate law for this reaction as determined by Wehrli and Stumm (1988) is:

$$-\frac{d\{V(IV)\}}{dt} = k\{VO(OTi \equiv)\}[O_2]$$

where the {} brackets denote surface concentrations. The reaction is thus second-order,

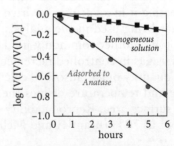

Figure 5.31 Oxygenation of vanadyl at pH 4 and $PO_2 = 1$ atm in experiments of Wehrli and Stumm (1988). After Wehrli et al. (1989).

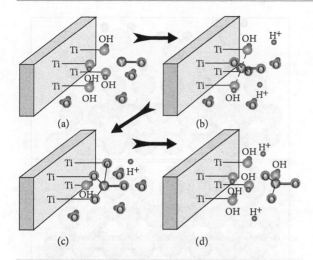

(a)   (b)

(c)   (d)

Figure 5.32 Mechanism of oxygenation of surface-bound vanadyl. In step (1) vanadyl is adsorbed at a TiO_2 surface (a → b). Note that the vanadium is bound to two surface TiO groups. In step (2), the vanadium binds to a third surface oxygen, releasing an H^+ ion (b → c). In step (3), the vanadate ion is replaced at the surface by three H^+ ions (c → d) (at intermediate pH, most vanadate will remain bound to the surface).

depending on the concentration of surface bound V(IV) and dissolved O_2. Wehrli and Stumm (1988) determined the rate constant for this reaction to be $0.051\,M^{-1}s^{-1}$ and the activation energy to be 56.5 kJ/mol at pH 7.

The surface-catalyzed reaction is essentially independent of pH, whereas the reaction in homogenous solution is strongly pH dependent. The rate law for the latter can be written as:

$$-\frac{d\{V(IV)\}}{dt} = k\{VO^+\}[O_2][H^+]$$

The apparent rate constant for this reaction is $1.87 \times 10^{-6}\,s^{-1}$ and the apparent activation energy is 140 kJ/mol. Part of the difference in the activation energies can be accounted for as the energy of the hydrolysis reaction:

$$VO^{2+} + H_2O \rightleftharpoons VO(OH)^+ + H^+$$

which is the first step in the homogenous reaction. This energy is 54.4 kJ/mol. Wehrli and Stumm (1988) speculated that the remainder

of the difference in activation energy is the energy required for the transition from the octahedral structure of the dissolved $VO(OH)^+$ ion to the tetrahedral structure of the dissolved vanadate ion.

5.6 KINETICS OF DISSOLUTION

Minerals formed under one set of conditions, for example an olivine crystal precipitating from a magma or a calcite crystal in the test formed by a planktonic foraminifer in the ocean surface waters, will not necessarily be stable under another set of conditions. Of particular interest in this respect is the Earth's solid surface, sometimes referred to as the *critical zone* – the region *"from the outer extent of vegetation down to the lower limits of groundwater"* (Brantley et al., 2007), which is the subject of Chapter 13. It is so named because all life ultimately depends on the chemistry of this zone. Here, minerals react with water and air, breaking down to produce the loose regolith, which combined with organic debris, forms the water-retaining soil that allows plants to take root and releasing nutrients essential for their growth.

Mineral dissolution reactions within the soil commonly involve hydrogen ions and are consequently pH dependent. The hydrogen ions, i.e., acidity, are largely supplied by carbonic acid derived from atmospheric CO_2 and respiration within the soil (reactions 5.2 and 5.3). The soluble products of the dissolution reactions are then carried to the ocean where they precipitate as carbonate (reactions 5.6 and 5.7). This series of reactions is sometimes referred to as the *Urey reactions,* as Harold Urey (1952) emphasized their importance in controlling atmosphere CO_2 concentrations (however, J. J. Ebelmen had pointed out the importance of these reactions in consuming atmospheric CO_2 in 1847 and T. C. Chamberlin (1899) pointed out their consequent importance for climate). That, in turn, has an important long-term control on the Earth's surface temperature and climate, a topic we will return to in Chapter 12. Organic acids also contribute hydrogen ions and furthermore some of these soil reactions are directly mediated by organisms ranging from single-celled bacteria to giant redwoods. Our brief coverage here will focus exclusively on kinetics of inorganic, abiotic reactions.

Readers are referred to books by Brantley et al. (2008) and Oelkers and Schott (2009) for more in-depth description of this topic.

5.6.1 Simple oxides

The rates of dissolution of nonionic solids are generally controlled by surface reactions at the solid–water interface. Absorption of ions to the surface of the solid plays a critical role in the dissolution process. In natural waters, adsorption of H^+ and OH^- ions at the surface dominate dissolution reactions, although adsorption of other species, particularly organic ones such as carboxylic acids, can be important as well. Adsorption of these aqueous species onto solid surfaces initiates the dissolution process.

Incompletely bonded oxygens on solid surfaces in contact with aqueous solution will be protonated under most circumstances, that is, an H^+ ion will attach to one of the surface O ligands to form a surface hydroxyl. Bonding of a single proton merely replaces the bond that the oxygen would have formed with a metal ion had it been located in the crystal interior. Addition of a second proton (i.e., protonation of the surface hydroxyl), however, has the consequence of weakening metal-oxide bonds. Figure 5.33 illustrates this process in which oxygens surrounding silicon atoms in quartz become progressively protonated, weakening the Si–O bonds until a $Si(OH)_4$ silicic acid unit is released to solution. Experiments on quartz dissolution rates in solutions of water plus various organic liquids found that the dissolution rate is proportional to the fourth power of water activity:

$$\Re_+ = k_{H_2O}a_{H_2O}^4 \tag{5.150}$$

indicating that the dissolution reaction proceeds as:

$$\equiv SiO + 4H_2O \rightleftharpoons \equiv Si\text{-}(OH)_4^{ads}$$

In the case of a trivalent ion such as Al_2O_3, protonation of three such bonds effectively frees the metal ion from the lattice structure. We can expect, therefore, that the dissolution rate will be proportional to finding three protonated ligands surrounding a single surface metal ion. The concentration of surface-bound protons, $[H^+]_s$, can be related

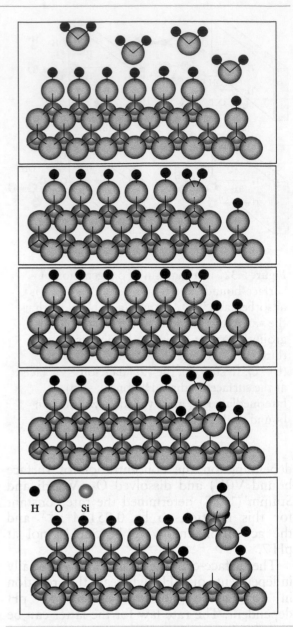

Figure 5.33 Cartoon of proton-promoted dissolution of an oxide such as quartz.

to the concentration of H^+ in solution through an absorption equilibrium constant K_{ad}, so that:

$$[H^+]_s/[\equiv S] = K_{ad}[H^+]_{aq}$$

where $[\equiv S]$ is the density of appropriate surface sites. The probability of finding a metal surrounded by three protonated ligands is then proportional to $([H^+]_s/[\equiv S])^3$. Thus, we expect the dissolution rate to be proportional to the third power of the surface protonation:

$$\Re \propto \{H_s^+/S\}^3 = \{K_{ad}[H_{ad}^+]\}^3 \tag{5.151}$$

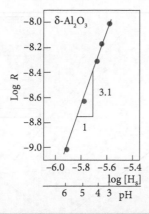

Figure 5.34 Log of the rate of Al_2O_3 dissolution plotted against the log of the concentration of surface protons. The slope of 3 indicates a rate law with third-order dependence on the surface concentration of protons. After Stumm and Wollast (1990).

Figure 5.34 shows that this is indeed the case for Al_2O_3.

At high pH deprotonation of surface OH groups will occur through the following reaction:

$$\equiv S-OH + OH^- \rightleftharpoons \equiv S-O^- + H_2O \quad (5.152)$$

where $\equiv S-O$ denotes a surface-bound oxygen. This deprotonation disrupts metal–oxygen bonds through polarization of electron orbitals. As a result, dissolution rates will also increase with increasing pH in alkaline solutions. Other ligands, particularly organic ones such as oxalates, can have a similar effect. In addition, water molecules can react with the surface in either protonation or deprotonation reactions.

In the context of transition state theory, formation of protonated or deprotonated surface species are considered precursors to the activated complex of the reaction. The forward rate of dissolution reactions should then be proportional to the concentrations of these precursor species and consequently is the sum of several independent reactions:

$$\mathfrak{R}_+ = k_{H^+}[\equiv MOH_2^+]^{n_H} + k_{OH^-}[\equiv MO^-]^{n_{OH}}$$
$$+ k_L[\equiv ML]^{n_L} + k_{H_2O}[\equiv MOH]^{n_{H_2O}}$$
$$(5.153)$$

where $\equiv M$ refers to a metal surface ion, L to a non-water-related ligand (such as PO_4), and

n is generally equal to the charge on the metal ion (Schott et al., 2009).

In the transition state theory developed by Oelkers and Schott (1995), protonation leads to the formation of a surface complex that serves as a precursor to the activated complex of the dissolving species. For a single metal oxide such as SiO_2, this precursor is formed by the reaction:

$$n_H\, H^+ + \equiv Si\text{-}OH \rightleftharpoons P^*$$

where $\equiv Si-OH$ is a surface-bound silicon, n_H is the number of protonation steps necessary to form the precursor, P^* (e.g., 4 in the case of Si). Assuming that the number of surface sites, S, is constant and $S = [\equiv Si--OH\} + P^*$, the concentration of the precursor species can be expressed by a version of the Langmuir isotherm:

$$[P^*] = S\frac{K^* \cdot a_{H^+}^n}{1 + K^* \cdot a_{H^+}^n} \quad (5.154)$$

where K^* is the absorption coefficient for the reaction above. Provided that $[P^*] \ll [\equiv Si-OH]$, this reduces to:

$$[P^*] = K^* \cdot a_{H^+}^n \quad (5.155)$$

and the rate of forward reaction is:

$$\mathfrak{R}_+ = k_+ K^* \cdot a_{H^+}^n \quad (5.156)$$

Although dissolution is a complex reaction involving successive steps of reactant transport, attachment of surface species, detachment of products from the surface and diffusion away from the surface, in most cases attachment and transport are quite fast. Detachment is the slowest step and hence determines the overall rate of reaction. In addition, at a given pH and the concentration of other ligands, one of the reactions in eqn. 5.153 will be much faster and therefore dominate. Consequently, quartz dissolution can approximate a simple elementary reaction. For example, in near-neutral pH, the overall reaction rate can be described by a

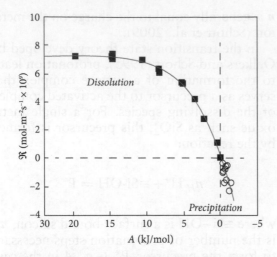

Figure 5.35 Quartz dissolution and precipitation rates at 200°C in near-neutral pH from Gautier (1999). The line shows the predicted rates as a function of reaction affinity, A, computed as $(1 - Q/K)$. From Schott et al. (2009).

version of equation 5.68:

$$\mathfrak{R}_{net} = \mathfrak{R}_+ \left(1 - \frac{Q}{K} \right)$$

$$= k_{H_2O}[\equiv SiOH] \left(1 - \frac{Q}{K} \right) \quad (5.157)$$

where [≡SiOH] is the concentration of SiOH surface species, and Q and K are the reaction quotient and equilibrium constant for quartz dissolution, respectively (Schott et al., 2009). Eqn. 5.157 (which in modified form is relevant to a wide range of minerals) predicts that the dissolution rate will slow as equilibrium between mineral and solution is approached (i.e., as the concentration of dissolved components increases). Figure 5.35 shows experimental data on quartz dissolution at near-neutral and 200°C is well explained by this equation.

5.6.2 Silicates

Figure 5.36 summarizes the pH dependence of dissolution rates for several silicates (as well as apatite). Surface protonation and deprotonation also play a dominant role in silicate dissolution. However, the dissolution of silicates is somewhat more complex than that of simple oxides because they typically contain

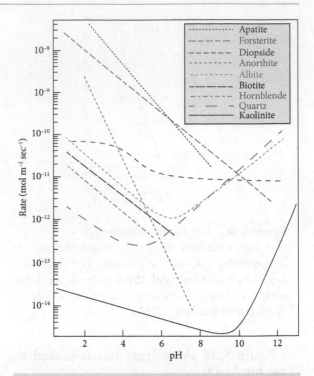

Figure 5.36 Experimentally determined dissolution rates of common minerals as a function of pH. From Brantley (2008).

several metals bound in different ways. These metal–oxygen bonds break at very different rates depending on their relative strength. In general, the strength of these bonds increases with the valence of the metal. Thus Na–O, Ca–O, Mg–O, etc. bonds break more readily than Al–O bonds and Si–O bonds break least readily. Consequently, in addition to sorption reactions reviewed above, the dissolution mechanism can also involve exchange reactions between protons and metals that are not essential to maintaining the mineral structure (Schott et al., 2009). This can result in incongruent dissolution, such that some metal ions may be released to solution more rapidly than others with formation of leached layers depleted in the most liable metals relative to silica.

A related, and particularly important, factor is lattice structure, in particular, the degree to which the individual silica tetrahedra share oxygens. There is a complete range among silicates in this respect, from orthosilicates, such as olivine, in which no oxygens are shared, to the tecto-, or framework-, silicates, such as quartz and the feldspars, in which all oxygens are shared. As we discussed in Chapter 4, shared oxygens are termed *bridging*, and non-shared ones *nonbridging* oxygens. Sharing of oxygens increases the degree of *polymerization* of the structure. The degree of polymerization is important in the context of dissolution because the nonbridging bonds are much more reactive than the bridging ones. Minerals with highly polymerized structures, such as feldspars, dissolve slowly and are subject to leaching, as components (particularly the network-modifiers) may be dissolved out, leaving the silicate framework still partially intact. Silicates with a low fraction of shared oxygens dissolve more rapidly and more uniformly. An example is olivine, whose structure is illustrated in Figure 5.37. Once the Mg ions surrounding it are removed, the individual silica tetrahedra are no longer bound to the mineral and are free to form H_4SiO_4 complexes in the solution (a more likely mode of dissolution is replacement of Mg^{2+} by $2H^+$; in essence, this produces a free H_4SiO_4 molecule). In contrast, removal of Na^+ by H^+ in albite (Figure 5.37) leaves the framework of tetrahedra largely intact. The rate of weathering can also be affected by the Al/Si ratio, as the silicate groups are less reactive than the aluminate ones except at high pH. Thus, the dissolution rate of plagioclase depends on the ratio of the anorthite ($CaAl_2Si_2O_8$) to albite ($NaAlSi_3O_8$) components, with calcic plagioclase weathering more rapidly.

Some idea of the role these factors play can be obtained from Table 5.5, which lists the mean lifetimes of a 1 mm crystal for a variety of minerals in contact with a solution of pH 5 based on experimentally determined dissolution rates. Estimates of weathering rates from field observations based on concentrations of weathering products in streams and reviews and rates of outcrop retreat are often orders of magnitude slower than laboratory determinations, but the order of stability of minerals based on field observations of weathering trends follows those calculated from laboratory experiments (Brantley and Olsen, 2014). We will discuss these field observations in Chapter 13.

In the context of the Oelkers and Schott transition state model, in multioxides such as silicates formation of the precursor complex is accompanied by removal of one or more metal ions to solution. A generalized reaction

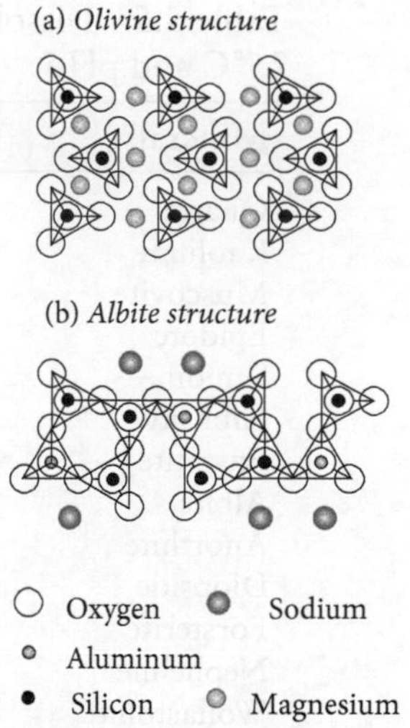

(a) *Olivine structure*

(b) *Albite structure*

○ Oxygen   ◉ Sodium
◎ Aluminum
● Silicon   ◉ Magnesium

Figure 5.37 Comparison of olivine (forsterite) and feldspar (albite) structures. In feldspar, all oxygens are shared by adjacent tetrahedra, in olivine none are; instead the excess charge of the SiO_4^{4-} units is compensated by $2Mg^{2+}$.

can be written as:

$$z \cdot n \cdot H^+ + \equiv M_1 - OH + M_2 - OH + \ldots \rightleftharpoons P^*$$
$$+ M^{z_1+} + M^{z_2+} + \ldots$$

The forward rate of reaction is:

$$\mathfrak{R}_+ = k_+ \prod_i \left[\frac{K_i \left(\dfrac{a_{H^+}^{z_i}}{a_{M_i^{z_i+}}} \right)^{n_i}}{1 + K_i \left(\dfrac{a_{H^+}^{z_i}}{a_{M_i^{z_i+}}} \right)^{n_i}} \right] \quad (5.158)$$

where K_i is the equilibrium constant for the proton exchange reaction with metal M_i, and the activities are those of the metals in solution. Where significant M_i remains in the surface leached layer, the product of the equilibrium constant and activity ratio will be small, and the denominator will approach 1. For example, in the case of a mineral composed of a single metal ion in addition to Si (e.g. enstatite, $MgSiO_3$) when Ka_H/a_M is $\ll 1$, eqn. 5.156 simplifies to:

$$\mathfrak{R}_+ \cong k_+ K^* \left(\frac{a_{H^+}^z}{a_{M^{z+}}} \right)^n \quad (5.159)$$

Table 5.5 Dissolution rates and mean lifetimes of crystals at 25°C and pH 5.

Mineral	Log rate (mol/m²/s)	Mean lifetime years
Quartz	−12.6	2,800,000
Kaolinite	−13.28	6,000,000
Muscovite	−13.3	2,300,000
Epidote	−12.61	923,000
Sanidine	−12.20	220,000
Gibbsite	−11.45	276,000
Enstatite	−11.9	440,100
Albite	−11.6	66,000
Anorthite	−11.6	58,000
Diopside	−10.7	11,000
Forsterite	−9.1	470
Nepheline	−8.55	211
Wollastonite	−8.00	79

Based on Lasaga et al. (1994) and Brantley and Olsen (2014).

In this case, the equation predicts that far from equilibrium (i.e., where ΔG is large) the rate will depend on both the concentration of the metal ion in solution and pH. Oelkers (2001) found that this was the case for a number of minerals, including enstatite (Figure 5.38), muscovite, and kaolinite, as well as basaltic glass. In contrast, where Ka_H/a_M is $\gg 1$, essentially all metal is removed from the mineral surface leading to maximum concentrations of the precursor complex concentration and the rate becomes independent of pH and the aqueous concentration of the metal. The case for olivine (Figure 5.35) and anorthosite.

Temperature and the presence of other dissolved species will also affect dissolution reaction rates. As for other reactions, we expect dissolution rates to have an Arrhenius-type temperature dependence of the rate constant. Values of such apparent activation energies for a few minerals are listed in Table 5.6.

Dissolution rates of silicates commonly show a sigmoidal dependence on reaction affinity ($-\Delta G_r$) such as the dissolution rate for albite shown in Figure 5.39a, in which rate is independent of ΔG_r far from equilibrium ($|\Delta G_r| \gg 0$) and depends on ΔG_r close to equilibrium with a transition zone inbetween. Hellmann and Tisserand (2006) used the transition state theory model developed by Burch et al. (1993) in which the dissolution rate is the sum of the rates of two parallel reactions:

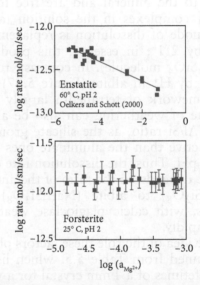

Figure 5.38 Enstatite (MgSiO₃) and forsterite (Mg₂SiO₄) dissolution rates as a function of aqueous Mg concentrations. The enstatite dissolution rate is a function of Mg^{2+} concentration whereas the forsterite dissolution rate is not. From Oelkers (2001).

$$\mathfrak{R} = k_1[1 - e^{-n(|\Delta G|/RT)^{m_1}}] + k_2[1 - e^{-|\Delta G|/RT}]^{m_2} \quad (5.160)$$

with $k_1 = 1.02 \times 10^{-8}$, $k_2 = 1.8 \times 10^{-10}$ mol/m²-s $n = 8 \times 10^{-5}$, $m_1 = 3.81$, and $m_2 = 1.17$. ΔG_r was calculated as $RT\ln(Q/K)$ (eqn. 5.67). The first term on the right expresses account for the plateau far from equilibrium and the sharp change in rates in the transition equilibrium region. The second

Table 5.6 Apparent activation energies for dissolution reactions in near-neutral pH.

Mineral	Ea kJ/mol
Alkali feldspar	51.7
Hydroxyapatite	250
Biotite	22
Calcite	23.5
Dolomite	52.2
Forsterite	79
Enstatite	80
Muscovite	22
Wollastonite	54.7

From Brantley and Osen (2014).

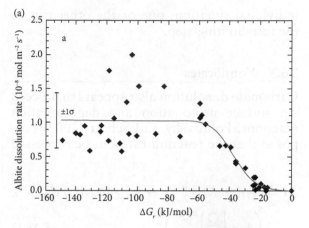

(a)

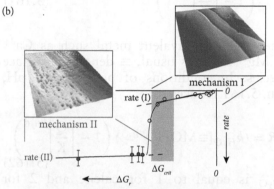

(b)

Figure 5.39 (a). Albite dissolution rates measured by Hellman and Tisserand (2006) in a continuously stirred flow-through reactor at 150°C and pH = 9.2 as a function of ΔG_r. The red line shows the predicted rates based on eqn. 5.160. Analytical uncertainty in the determined rates are comparatively large in experiments at high $|\Delta G_r|$ and can explain much of the scatter. From Hellman and Tisserand (2006). (b). Vertical scanning interferometry images of albite crystals undergoing dissolution at 80°C and pH 8.8 under far-from-equilibrium (left, mechanism II) and near-equilibrium (right, mechanism I) conditions. Under far-from-equilibrium conditions crystal surface is extensively pitted and dissolution is rapid and nearly independent of ΔG_r. Pits are absent under near-equilibrium conditions and dissolution rates vary with ΔG_e. From Arvidson and Luttge (2010).

term on the right expresses the dependence of reaction rate on ΔG_r close to equilibrium, with $m_2 \approx 1$; this term is consistent with transition state theory prediction for an elementary reaction (eqn. 5.59). They also determined an activation energy of the reaction of 60 kJ/mol.

The need for parallel rate equations can be explained by considering the details of the solid surfaces. Dissolution reactions occur more readily at discontinuities on the mineral surface: terraces, steps, kinks, defects, discontinuities, etc. (Figure 5.22). At such sites, metal ions will be coordinated by fewer oxygens than on smooth surfaces, consequently fewer metal–oxygen bonds need be broken to free the metal ion. Examination of mineral surfaces using techniques such as scanning electron microscopy (SEM), vertical scanning interferometry (VSI), and atomic force microscopy (AFM) reveals that minerals dissolving under far from equilibrium conditions have heavily pitted surfaces, while those dissolving under near equilibrium conditions have smooth surfaces (Figure 5.39). The differences are consistent with a theory in which a critical value of ΔG_r is required for etch pits to develop, which then spiral away from the pit across the mineral surface. Various studies have estimated the critical value of ΔG_r for pit development to be in the range of 40–70 kJ/mol. Above this energy, dissolution is rapid and the rate independent of reaction affinity, while below that energy dissolution occurs primarily at defects, steps, etc. at much lower rates.

In the steady state, mass balance requires that the rate of dissolution (i.e., the rate at which aqueous species are produced at the surface) and transport (the rate at which components are removed from the solution adjacent to the dissolving surface) must be equal. Thus, overall weathering rates are controlled by a combination of surface kinetics and transport kinetics. In each individual situation, one or the other can be the rate-limiting step.

5.6.3 Nonsilicates

Carbonate dissolution also appears to proceed via surface protonation and deprotonation reactions. Pokrovsky and Schott (2002) proposed that the reaction rate can be expressed as:

$$\Re = (k_H[\equiv CO_3H^0]^{n_H} + k_{H_2O}[\equiv MOH_2^+]^{n_{H_2O}})$$

$$\left(1 - \left[\frac{Q}{K}\right]^{n_{H_2O}}\right) \tag{5.161}$$

where M is a divalent metal such as Ca^{2+} or Mg^{2+} and, as usual, $\equiv$ denotes a surface species. In conditions of near neutral pH, eqn. 5.161 reduces to:

$$\Re = (k_{H_2O}[\equiv MOH_2^+]^{n_{H_2O}})\left(1 - \left[\frac{Q}{K}\right]^{n_{H_2O}}\right)$$
$$\tag{5.162}$$

n_{H_2O} is equal to 1 for calcite and 2 for dolomite but can range up to 4 for other carbonates. At 25°C log (k_{H_2O}) has a value of -1 mol-cm^{-2}sec^{-1} for calcite, but a value of 12.65 mol-cm^{-2}sec^{-1} for dolomite. Just as for silicates, reaction rates can be discontinuous functions of Q/K, reflecting different dominant surface reaction pathways under far-from- and near-equilibrium conditions.

Surface reactions are most often rate-limiting in dissolution and weathering of silicate minerals at low temperature (25°C). Dissolution of readily soluble minerals (e.g., halite) and even moderately soluble minerals (e.g., calcite, gypsum) are, by contrast, relatively fast and more likely to be limited by the rate at which the dissolving components can be transported away from the mineral–water interface by advection and diffusion. As temperature increases, transport is increasingly likely to become rate-limiting. This is because the activation energy of diffusion in aqueous solution, typically 5–10 kJ/mol, is generally less than the activation energy of surface reactions (typically >30 kJ/mol; Table 5.5). Thus, the diffusion rates increase more slowly with temperature than surface reaction rates. This point is illustrated for the case of calcite in Figure 5.40. At temperatures below

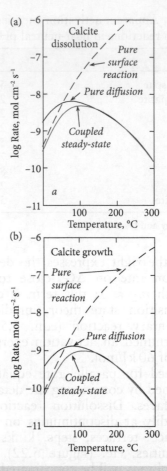

Figure 5.40 Log of steady-state dissolution (a) and growth (b) of calcite as a function of temperature, comparing diffusion-controlled and surface reaction-controlled kinetics. The model assumes a 1 μm hydrodynamic boundary layer and saturation in the case of growth. From Murphy et al. (1989).

75°C, growth and dissolution of calcite, a moderately soluble mineral, is effectively controlled by the surface reaction rate, while at temperatures greater than 125°C, diffusion is the rate-controlling step. Dissolution under hydrothermal and metamorphic conditions is most likely to be diffusion-controlled for most minerals (Guy and Schott, 1989).

5.7 DIAGENESIS

An introductory geology text might define *diagenesis* as the process through which sediment is converted to a sedimentary rock. We will use diagenesis to refer to a number of physical and chemical processes that occur subsequent to deposition of sediment, including compaction and expulsion of pore water, consumption of organic matter, and resulting changes in $p\varepsilon$. Some of the originally deposited phases dissolve in the pore water during diagenesis; other phases crystallize from the pore water. Some of these changes begin immediately after deposition, some only as a result of later deformation. Some occur as a result of moderately elevated temperature and pressure, though processes occurring at significantly higher temperature and pressure would be called metamorphism. Diagenesis and metamorphism form a continuum; though a geologist might volunteer a definite opinion on whether a particular specimen had been diagenetically or metamorphically altered, s/he would be hard pressed to come up with criteria to distinguish diagenesis from metamorphism that were not arbitrary. Here, we will briefly consider a few of these processes.

5.7.1 Compositional gradients in accumulating sediment

Let's turn our attention to the early stages of diagenesis in slowly accumulating sediment. Our first task is to decide on a reference frame. There are two choices: we could choose a reference frame fixed to a specific layer. In this case, the sediment–water interface will appear to move upward with time. Alternatively, we can choose a reference frame that is fixed relative to the sediment–water interface, thus depth always refers to distance downward from that interface. As sediment accumulates, a given layer of sediment will appear to move downward in this reference frame. In this reference frame, we can express the change in concentration at some depth, x, as the sum of changes in the composition due to diagenesis plus the change in the composition of sediment flowing downward past our fixed reference point:

$$\left(\frac{\partial C_i}{\partial t}\right)_x = \frac{dC_i}{dt} - \omega\left(\frac{\partial C_i}{\partial x}\right)_t \quad (5.163)$$

where C_i is concentration of some species i, and ω is the burial rate. The partial derivative on the left-hand side refers to changes *at some fixed depth,* and the total derivative refers to diagenetic changes occurring in a given layer,

or horizon, undergoing burial. $(\partial C_i/\partial x)_t$ is the concentration gradient of i at some fixed time t. This equation allows us to convert a reference frame that is fixed relative to the sediment–water interface to one that is fixed relative to some sedimentary layer.

Now let's consider two extremes where eqn. 5.163 is particularly simple. In the first, a steady state is reached and there is no change with time, hence:

$$\left(\frac{\partial C}{\partial t}\right)_x = 0 \quad (5.164)$$

In other words, the concentration of i at some fixed depth below the water–sediment interface is constant. Under these circumstances then,

$$\frac{dC_i}{dt} = \omega\left(\frac{\partial C_i}{\partial x}\right) \quad (5.165)$$

This case is illustrated in Figure 5.41.

In the second extreme, there is no diagenesis and the composition of a given layer is determined only by what is initially deposited, thus:

$$\frac{dC_i}{dt} = 0 \quad (5.166)$$

The concentration change with time at some fixed depth is then due to change in the composition of the sediment moving

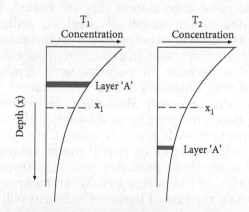

Figure 5.41 Steady-state diagenesis. Concentration at a fixed depth x_1 below the surface remains constant, but layer A, whose depth increases with time due to burial, experiences a decreasing concentration with time. After Berner (1980).

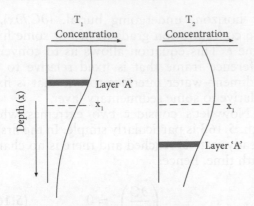

Figure 5.42 Concentration profiles in a sediment in which the composition of the material changes with time, but there is no diagenesis. The composition of any given layer is fixed, but the composition at some fixed depth relative to the water–sediment interface, such as x_1, changes with time. After Berner (1980).

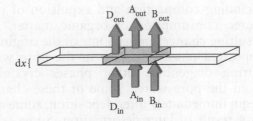

Figure 5.43 Fluxes through a box in a sedimentary layer of unit lateral dimensions and thickness dx. Arrows labeled A, B, and D indicate advective, biodiffusive, and molecular diffusive fluxes, respectively. Loss or gain by the box due to these processes depends on the difference in the flux into and out of the box: dF/dx.

downward past that point. Thus:

$$\left(\frac{\partial C_i}{\partial t}\right)_x = \omega \left(\frac{\partial C_i}{\partial x}\right)_t \quad (5.167)$$

This case is illustrated in Figure 5.42.

The sediment consists both of solid particles and the water buried with the particles, the pore water. Assuming no other fluid is present (e.g., gas, petroleum) then the volume fraction of water in the sediment is equal to the porosity ϕ. The volume fraction of solids is then simply $1 - \phi$. Most sediments will undergo compaction as they are buried. This is due to the weight of overlying sediment (gravitational compaction). Gravitational compaction results in expulsion of pore water and a decrease in porosity with depth. In addition, dissolution and cementation will also affect porosity. Since the molar volume of a phase precipitating or dissolving (the most important such phase is often $CaCO_3$) will be different from its partial molar volume in solution, these processes will also result in motion of the pore water. When compaction occurs, the rate of burial of sediment will not be equal to the sedimentation rate.

Now consider a box of sediment of thickness dx and unit length and width embedded within some sedimentary layer (Figure 5.43). We assume that the layer is of uniform composition in the lateral dimension, and therefore that there is no lateral diffusion, and that there

is also no lateral advection of fluid. Within the box there are C moles of species i. If we choose our concentration units to be moles per volume, then the concentration is simply C_i.

Let's consider the processes that can affect the concentration of species i within the box. First of all, reactions occurring within the box might affect i. For example, oxidation and reduction will affect species such as Fe^{3+}, SO_4^{2-}, and Mn^{2+}. If we are interested in the concentration of a dissolved species, then dissolution, crystallization, leaching, and so on, will all change this concentration.

In addition to reactions occurring within the box, diffusion, advection, and bioturbation will also affect the concentration of i if there is a difference between the fluxes into and out of the box. Bioturbation is the stirring effect produced by the activity of animals that live in the sediment (the infauna). From a geochemical perspective, bioturbation is much like diffusion in that it results from the random motion of particles (even if these particles are of very different size from atoms and ions) and acts to reduce compositional gradients. Mathematically, we can treat the effect of bioturbation in a way similar to diffusion, that is, we can define a bioturbation flux as:

$$J_B = D_B \left(\frac{\partial C_i}{\partial x}\right)_t \quad (5.168)$$

where D_B is the biodiffusion coefficient. Values of D_B for solid phases range from 10^{-6} cm^2/sec in nearshore clays to 10^{-11} in deep-sea pelagic sediments. The bioturbation coefficient will generally be different for solid species than for liquid ones. Since most animals live in the upper few centimeters of sediment, D_B will be a function of depth. In those circumstances, the time dependence of concentration is given by:

$$\left(\frac{\partial C_i}{\partial t}\right)_x = \left\{\frac{\partial(D_B(\partial C_i/\partial x))}{\partial x}\right\}_t \quad (5.169)$$

As we noted in section 5.4.6, diffusion through solids is much lower than through liquids, so we can neglect diffusion in the solid and deal only with diffusion through the pore water using eqn. 5.115. We can simplify that equation by incorporating the tortuosity term into the porosity, ϕ, so the diffusion of dissolved species will be:

$$J_M = -\phi D_M\left(\frac{\partial C}{\partial x}\right) \quad (5.170)$$

where we have adopted the subscript M to denote molecular diffusion. Fick's second law becomes:

$$\left(\frac{\partial C_i}{\partial t}\right)_x = \frac{\partial(\partial(\phi D_M(\partial C_i/\partial x))}{\partial x} \quad (5.171)$$

The advective flux is the product of the fluid (i.e., pore water) velocity, v, times the concentration:

$$J_A = v C_i \quad (5.172)$$

To describe the rate of change of concentration in the box, we want to know the rate of reactions within it and the change in flux across it, as it is the change in flux that dictates what is lost or gained by the box. Combining all the fluxes into a single term, F_i, the rate of change of species i in the box is:

$$\frac{dC_i}{dt} = -\frac{\partial F_i}{\partial x} + R_i \quad (5.173)$$

where the second term is the sum of the rates of all reactions affecting i. The flux term is negative because any decrease in flux over dx results in an increase in concentration within the box.

We can then use equation 5.163 to transform to a reference frame fixed relative to the sediment surface:

$$\left(\frac{\partial C_i}{\partial t}\right) = -\left(\frac{\partial F_i}{\partial x}\right)_t - \omega\frac{\partial C_i}{\partial x} + \sum R_i \quad (5.174)$$

The downward burial of sediment past point x can also be considered a flux. Combining this with the other flux terms, we have:

$$\boxed{\left(\frac{\partial C_i}{\partial t}\right) = -\left(\frac{\partial F_i}{\partial x}\right)_t + \sum R_i} \quad (5.175)$$

where F is the *net* flux of i in and out of the box and the last term is the rate of all internal changes, including chemical, biochemical, and radioactive, occurring within the box. Equation 5.175 is called the *diagenetic equation* (Berner, 1980). Let us now consider an example that demonstrates how this equation can be applied.

5.7.2 Reduction of sulfate in accumulating sediment

Organic matter buried with the sediment will be attacked by aerobic bacteria until all dissolved O_2 is consumed. When O_2 is exhausted, often within tens of centimeters of the surface, consumption will continue anaerobically, with sulfur in sulfate acting as the electron acceptor:

$$2\alpha CH_2O + SO_4^{2-} \rightleftharpoons H_2S + 2HCO_3^- \quad (5.176)$$

where CH_2O represents organic matter generally and α is the number of organic matter carbon atoms reduced per sulfur atom. Let's assume that sulfate is far more abundant than organic matter so that the rate of sulfate reduction depends only on the supply of organic matter and not on the abundance of sulfate. In this case:

$$\frac{d[CH_2O]}{dt} = -k[CH_2O] \quad (5.177)$$

We are greatly simplifying matters since there is a great variety of organic compounds in sediments, each of which will have a different rate constant. To further simplify matters, we will assume (1) that conditions become anaerobic at the sediment–water interface, (2) that all consumption of organic matter occurs anaerobically, (3) that steady-state is achieved (i.e., $(\partial C/\partial t)_x = 0$), and (4) that there is no compaction (and therefore no pore water advection) or bioturbation. Substituting 5.177 into 5.167, we have:

$$-k[CH_2O] = \omega\left(\frac{\partial[CH_2O]}{\partial x}\right)_t \qquad (5.178)$$

Integrating, we obtain the concentration of organic matter as a function of depth:

$$[CH_2O](x) = [CH_2O]^\circ e^{-kx/\omega} \qquad (5.179)$$

where $[CH_2O]^\circ$ is the organic matter concentration at the sediment–water interface $(x = 0)$.

We can now also solve for the variation in concentration sulfate in the pore water. According to eqn. 5.10, the rate of sulfate reduction is related to the organic matter consumption rate as:

$$\frac{d[SO_4^{2-}]}{dt} = \frac{1}{2\alpha}\frac{d[CH_2O]}{\partial t} = \frac{k}{2\alpha}[CH_2O] \qquad (5.180)$$

Whereas the organic matter can be considered fixed in sediment, the sulfate is a dissolved species, so we must also consider diffusion. Making appropriate substitutions into 5.173, we have:

$$\phi D\left(\frac{\partial^2[SO_4^{2-}]}{\partial x^2}\right) - \omega\left(\frac{\partial[SO_4^{2-}]}{\partial x}\right)_t -$$
$$\frac{k}{2\alpha}[CH_2O]^\circ e^{-kx\omega} = 0 \qquad (5.181)$$

This is a second-order differential equation and its solution will depend on the boundary conditions. Our boundary condition is that at $x = 0$, $C = C^\circ$. The solution under this condition is:

$$[SO_4^{2-}](x) = \frac{\omega^2[CH_2O]^\circ(e^{-kx\omega} - 1)}{2\alpha(\omega^2 + kD)\phi} + [SO_4^{2-}]^\circ \qquad (5.182)$$

where $[SO_4^{2-}]^\circ$ is the sulfate concentration at the surface.

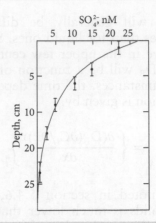

Figure 5.44   Dissolved sulfate concentrations in sediments from the Saanich Inlet. Data are shown as open circles with error bars. Curve is fitted using eqn. 5.182 and parameters given in the text. From Murray et al. (1978).

Murray et al. (1978) applied this model to data from sediment cores taken from the Saanich Inlet of British Columbia (Figure 5.44). Sedimentation rate, ω, was determined using ^{210}Pb (see Chapter 8) to be about 1 cm/yr; the factor α was independently estimated to be 0.5, with average porosity of 0.927. The value of D was taken to be 2.6×10^{-6} cm^2/sec. They fitted an exponential curve of the form $c = a\, e^{-bx}$ to the data and found $a = 26.6$ and $b = 0.184$. From this they determined the rate constant to be 6.1×10^{-9} sec^{-1}, and the initial concentration of metabolizable organic matter to be 380 mM/cm^3 total sediment. The latter was somewhat larger than the value determined from the profile of total organic carbon in the core. To explain the discrepancy, the authors suggested that methane is produced below the depth where sulfate is depleted. Methane then diffuses upward and is oxidized by sulfate-reducing bacteria.

5.8   SUMMARY

The chapter covered the fundamentals of kinetics and their application to geochemistry.

- We found that reaction rates generally depend on the concentrations of the species involved and can do so complexly, but elementary reactions depend only on the concentrations of reactants raised to the power of their stoichiometric coefficients. Reaction rates also depend exponentially on temperature, which can be expressed by the *Arrhenius relation*:

$$k = Ae^{-E_B/kT} \qquad (5.25)$$

- We found that equilibrium is a condition where the forward and reverse rates of reaction are equal. This allowed us to relate the rate constants (eqn. 5.25) of forward and reverse reactions to the equilibrium constant and the frequency factor *A* to the entropy change of reaction and the barrier energy (E_B) to the enthalpy change of reaction.
- Using transition state theory, we could show that the rate of reaction can be expressed as:

$$\mathfrak{R}_{net} = A_+ e^{-E_{B+}/RT}(1 - e^{-n\Delta G/RT})$$
$$[A]^{n_A}[B]^{n_B} \ldots \qquad (5.61)$$

In this case, ΔG is a measure of disequilibrium of the system and is often called the *reaction affinity,* A_r. Reaction affinity can be expressed in several ways, such as:

$$A_r = RT \ln (Q/K)$$

where Q is the reaction quotient and K the equilibrium constant or

$$A_r = \Delta T \Delta S$$

where ΔT is the temperature overstep.

- Reaction rates can be limited by the rate at which reactants are brought together, so our next topic was diffusion. Fick's first law states that the diffusion flux is proportional to the product of the concentration gradient and the diffusion coefficient:

$$J = -D\frac{\partial c}{\partial x} \qquad (5.73)$$

Fick's second law states that the diffusion gradient will evolve over time as:

$$\left(\frac{\partial c}{\partial t}\right)_x = D\left(\frac{\partial^2 c}{\partial x^2}\right)_t \qquad (5.75)$$

Since diffusion by definition involves no net mass transport, it can be necessary to simultaneously consider the diffusion of all species and in some cases the chemical potential replaces the concentration in Fick's laws.

- In solids, the diffusion coefficient has an Arrhenius-like temperature dependence:

$$D = D_o e^{-E_a/RT} \qquad (5.109)$$

- Heterogeneous reactions necessarily occur across interfaces, so we next considered the nature of surfaces and defined the partial molar surface area, which can between like phases and the partial molar free energy:

$$\sigma^\phi \equiv \left(\frac{\partial G}{\partial A}\right)^\phi_{T,P,n} \qquad (5.119)$$

Partial molar free energy plays an important role in the nucleation and growth of phases.

- We turned next to adsorption. The Langmuir isotherm describes the fraction of surface sites occupied by an absorbent, Θ_M, as:

$$\Theta_M = \frac{K_{ad}[M]}{1 + K_{ad}[M]} \qquad (5.143)$$

We considered the role of adsorption in catalysis and dissolution reactions.

- In the last section, we discussed diagenesis of sediments, which involves reactions between fluids and particles in a context of advection, chemical diffusion and biodiffusion (bioturbation). Berner's diagenetic equation accounts for these processes as:

$$\left(\frac{\partial C_i}{\partial t}\right) = -\left(\frac{\partial F_i}{\partial x}\right)_t + \sum R_i \qquad (5.175)$$

REFERENCES AND SUGGESTIONS FOR FURTHER READING

Aagaard, P. and Helgeson, H.C. 1982. Thermodynamic and kinetic constraints on reaction rates among minerals and aqueous solutions, I. Theoretical considerations. *American Journal of Science* 282: 237–85.

Albarède, F. 1995. *Introduction to Geochemical Modeling*. Cambridge, Cambridge University Press.

Anderson, D.E. 1981. Diffusion in electrolyte mixtures, in *Reviews in Mineralogy Volume 8: Kinetics in Geochemical Processes* (eds A. Lasaga and R. Kirkpatrick), pp. 211–60. Washington, Mineralogical Society of America.

Berner, R.A. 1980. *Early Diagenesis*. Princeton, Princeton University Press.

Berner, R.A. 1981. Kinetics of weathering and diagenesis, in *Reviews in Mineralogy, Volume 8: Kinetics in Geochemical Processes* (eds A. Lasaga and R. Kirkpatrick), pp. 111–34. Washington, Mineralogical Society of America.

Brantley, S. L. 2008. Kinetics of mineral dissolution. In: *Kinetics of Water–Rock Interaction* (eds Brantley, S. L., Kubicki, J. D. and White, A. F.), pp. 151–210. New York, Springer New York.

Brantley, S. L., Goldhaber, M. B. and Ragnarsdottir, K. V. 2007. Crossing disciplines and scales to understand the critical zone. *Elements* 3(5), 307–14. doi: 10.2113/gselements.3.5.307.

Brantley, S. L., Kubicki, J. D. and White, A. F. (2008) *Kinetics of Water–Rock Interaction*. New York, Springer Science+Business Media.

Brantley, S. L. and Mellott, N. P., 2000. Surface area and porosity of primary silicate minerals. *American Mineralogist* 85(11–12): 1767–83. doi: 10.2138/am-2000-11-1220.

Brantley, S. L. and Olsen, A. A., 2014. 7.3 - reaction kinetics of primary rock-forming minerals under ambient conditions. In: *Treatise on Geochemistry* (Second Edition) (eds Holland, H. D. and Turekian, K. K.), pp. 69–113, Elsevier, Oxford. doi: 10.1016/B978-0-08-095975-7.00503-9.

Burch, T. E., Nagy, K. L., Lasaga, A. C. 1993. Free energy dependence of albite dissolution kinetics at 80°C and pH 8.8. *Chemical Geology* 105: 137–62. doi: 10.1016/0009-2541(93)90123-Z

Calia, A., Giannotta, M. and Quarta, G., 2006. A contribution to the study of re-used architectural marbles in Troia Cathedral (Foggia province, Southern Italy): identification and determination of provenance. In: *ASMOSIA VIIIth International Conference*, pp. 739–58, Maisonneuve and Larose, Maison méditerranéenne des sciences de l'homme, Aix-en-Provence, France.

Carlson, W.D. 1980. Experimental Studies of Metamorphic Petrogenesis, PhD thesis, Los Angeles, UCLA.

Chamberlin, T. C. 1899. An attempt to frame a working hypothesis of the cause of glacial periods on an atmospheric basis. *Journal of Geology* 7: 545–84. doi: 10.1086/608449.

Crank, J. 1975. *The Mathematics of Diffusion*, 2nd ed. Oxford, Clarendon Press.

Fisher, G.W. 1978. Rate laws in metamorphism. *Geochimica et Cosmochimica Acta* 42: 1035–50.

Ganor, J., Mogollon, J.L. and Lasaga, A.C. 1995. The effect of pH on kaolinite dissolution rates and on activation energy. *Geochimica et Cosmochimica Acta* 59: 1037–52.

Gautier, J.-M., 1999. Etude expérimentale et modélisation de la cinétique de dissolution et de cristallisation des silicates en milieu hydrothermal: cas du quartz et du feldspath potassique. PhD Dissertation, 180 p. Université Paul Sabatier, Toulouse, France.

Guy, C. and J. Schott. 1989. Multisite surface reaction versus transport control during the hydrolysis of a complex oxide. *Chem. Geol.* 78: 181–204.

Guo, C., Zhang, Y. 2016. Multicomponent diffusion in silicate melts: SiO_2–TiO_2–Al_2O_3–MgO–CaO–Na_2O–k_2O system. *Geochimica et Cosmochimica Acta* 195: 126–41. doi: 10.1016/j.gca.2016.09.003.

Guo, C., Zhang, Y. 2018. Multicomponent diffusion in basaltic melts at 1350 °C. *Geochimica et Cosmochimica Acta* 228: 190–204. doi: 10.1016/j.gca.2018.02.043.

Heinrich, W., Metz, P. and Gottshalk, M. 1989. Experimental investigation of the kinetics of the reaction 1 tremolite + 11 dolomite ⇌ 8 fosterite + 13 calcite + $9CO_2$ + 1 H_2O. *Contributions to Mineralogy and Petrology* 102: 163–73.

Hellmann, R. and Tisserand, D. 2006. Dissolution kinetics as a function of the Gibbs free energy of reaction: An experimental study based on albite feldspar. *Geochimica et Cosmochimica Acta* 70: 364–83. doi: 10.1016/j.gca.2005.10.007.

Henderson, P. 1986. *Inorganic Geochemistry*. Oxford, Pergamon Press.

Hofmann, A.W., 1980. Diffusion in natural silicate melts, a critical review, in *Physics of Magmatic Processes* (ed. R.B. Hargraves). Princeton, Princeton University Press.

Joesten, R. L. 1991. Kinetics of coarsening and diffusion-controlled mineral growth. in *Contact Metamorphism* Reviews in Mineralogy and Geochemistry vol. 26 (ed. D.M. Kerrick), pp. 507–82. Washington, Mineralogical Society of America.

Kerrick, D.M., Lasaga, A.C. and Raeburn, S.P. 1991. Kinetics of heterogeneous reactions, in *Contact Metamorphism* vol. 26 (ed. D.M. Kerrick), pp. 583–722. Washington, Mineralogical Society of America.

Kirkaldy, J. S. and Young, D.J. 1985. *Diffusion in the Condensed State*. London, Institute of Metals.

Kirkpatrick, R.J. 1981. Kinetics of crystallization of igneous rocks, in *Reviews in Mineralogy, Volume 8: Kinetics in Geochemical Processes* (eds A. Lasaga and R. Kirkpatrick), pp. 321–98. Washington, Mineralogical Society of America.

Laidler, K.J. 1987. *Chemical Kinetics*. New York, Harper Collins.

Lasaga, A.C. 1981a. Rate laws of chemical reactions, in *Reviews in Mineralogy, Volume 8: Kinetics in Geochemical Processes* (eds A. Lasaga and R. Kirkpatrick), pp. 1–68. Washington, Mineralogical Society of America.

Lasaga, A.C. 1981b. Transition state theory, in *Kinetics of Geochemical Processes, Reviews in Mineralogy 8* (eds A.C. Lasaga and R.J. Kirkpatrick), pp. 135–70. Washington, Mineralogical Society of America.

Lasaga, A.C. 1997. *Kinetic Theory in the Earth Sciences*. Princeton NJ, Princeton University Press.

Lasaga, A.C., Soler, J.M., Burch, T.E. and Nagy, K.L. 1994. Chemical weathering rate laws and global geochemical cycles. *Geochimica et Cosmochimica Acta* 58: 2361–86.

Lennie, A.R. and Vaughan, D.J. 1992. Kinetics of the marcasite–pyrite transformation: an infrared study. *American Mineralogist* 77: 1166–71.

Liang, Y. 2010. Multicomponent diffusion in molten silicates: Theory, experiments, and geological applications. *Reviews in Mineralogy and Geochemistry* 72: 409–46. doi: 10.2138/rmg.2010.72.9.

Liang, Y. (2017) Diffusion. In: White WM (E.) *Encyclopedia of Geochemistry*. Springer International Publishing, Cham, pp. 1–13. doi: 10.1007/978-3-319-39193-9.

Liang, Y., Richter, F.M. and Chamberlin, L. 1997. Diffusion in silicate melts: III. Empirical models for multicomponent diffusion. *Geochimica Cosmochimica Acta* 61: 5295–312.

McLean, D. 1965. The science of metamorphism in metals, in *Controls of Metamorphism* (eds W.S. Pitcher and G.W. Flinn), pp. 103–18. Edinburgh, Oliver and Boyd.

Morioka, M. and Nagasawa, H. 1990. Ionic diffusion in olivine, in *Advances in Physical Geochemistry, Volume 8, Diffusion, Atomic Ordering, and Mass Transport* (ed. J. Ganguly), pp. 176–97. New York, Springer-Verlag.

Murray, J.W., Grundmanis, V. and Smethie, W.M. 1978. Interstital water chemistry in the sediments of Saanich Inlet. *Geochimica et Cosmochimica Acta* 42: 1011–26.

Oelkers, E. H. 2001. General kinetic description of multi-oxide silicate mineral and glass dissolution. *Geochimica et Cosmochimica Acta* 65, 3703–19. doi: 10.1016/S0016-7037(01)00710-4.

Oelkers, E. H. and Schott, J. 1995. Experimental study of anorthite dissolution and the relative mechanism of feldspar hydrolysis. *Geochimica et Cosmochimica Acta* 59, 5039–53. doi: 10.1016/0016-7037(95)00326-6.

Oelkers, E. H. and Schott, J. (2009) *Thermodynamics and Kinetics of Water–Rock Interaction. Reviews in Mineralogy and Geochemistry* v. 70. Mineralogical Society of America, Washington, DC. 549 p.

Pokrovsky, O. S. and Schott, J. 2002. Surface chemistry and dissolution kinetics of divalent metal carbonates. *Environmental Science and Technology* 36: 426–32. doi: 10.1021/es010925u.

Pearson, D.G., Davies, G.R. and Nixon, P.H. 1995. Orogenic ultramafic rocks of UHP (diamond facies) origin, in *Ultrahigh Pressure Metamorphism* (eds R.G. Coleman and X. Wang), pp. 456–510. Cambridge, Cambridge University Press.

Richter, F. M., Liang, Y. and Minarik, W. G. 1998. Multicomponent diffusion and convection in molten $MgO-Al_2O_3-SiO_2$. *Geochimica et Cosmochimica Acta* 62: 1985–991. doi: 10.1016/S0016-7037(98)00123-9.

Schott, J., Pokrovsky, O. S. and Oelkers, E. H. 2009. The link between mineral dissolution/precipitation kinetics and solution chemistry. *Reviews in Mineralogy and Geochemistry* 70(1): 207–58. doi: 10.2138/rmg.2009.70.6.

Sposito, G., 1989. *The Chemistry of Soils*. New York, Oxford University Press.

Stumm, W. (ed.) 1990. *Aquatic Chemical Kinetics: Reaction Rates of Processes in Natural Waters*. New York, Wiley-Interscience.

Stumm, W. and Wollast, R. 1990. Coordination chemistry of weathering: kinetics of the surface-controlled dissolution of oxide minerals. *Reviews in Geophysics* 28: 53.

Toramaru, A. 1991. Model of nucleation and growth of crystals in cooling magmas. *Contributions to Mineralogy and Petrology* 108: 106–17.

Tyrell, H.J.V. and Harris, K.R. 1984. *Diffusion in Liquids*. London, Butterworths.

Urey, H. C. 1952. *The Planets: Their Origin and Development*. Yale University. Press, New Haven, CT, 182 p.

Watson, E. B. and Baxter, E. F. 2007. Diffusion in solid-earth systems. *Earth and Planetary Science Letters* 253: 307–27 doi: 10.1016/j.epsl.2006.11.015.

Wehrli, R. and Stumm, W. 1988. Oxygenation of vanadyl(IV): Effect of coordinated surface hydroxyl groups and OH. *Langmuir* 4: 53–8.

Wehrli, B., Sulzberger, B. and Stumm, W. 1989. Redox processes catalyzed by hydrous oxide surfaces. *Chem. Geol.* 78: 167–79.

Wood, B.J. and Walther, J.V. 1983. Rates of hydrothermal reactions. *Science* 222: 413–15.

Zhang, Y. (2008) *Geochemical Kinetics*. Princeton University Press, Princeton NJ. 631 p.

Zhang, Y. and Cherniak, D. (eds) 2010. *Diffusion in Minerals and Melts*. Washington, Mineralogical Society of America.

Zhong, S. and Mucci, A. 1993. Calcite precipitation in seawater using a constant addition technique: a new overall reaction kinetic expression. *Geochimica et Cosmochimica Acta* 57: 1409–17.

PROBLEMS

1. (a) Assuming that the precipitation of calcite from aqueous solution occurs only through the reaction:

$$Ca^{2+} + CO_3^{2-} \rightarrow CaCO_{3(s)}$$

and that this reaction is *elementary*, write an equation for the rate of calcite precipitation.

(b) Assuming that the reaction above is reversible, i.e.:

$$Ca^{2+} + CO_3^{2-} \rightleftharpoons CaCO_{3(s)}$$

and still assuming that it is *elementary*, write an equation for the dependence of *net* rate of calcite precipitation on concentration and free energy change of reaction.

2. Zhong and Mucci (1993) found that at constant concentration of dissolved Ca^{2+} ($[Ca^{2+}] \approx$ 10.5 mmol/kg), the rate of calcite precipitation in seawater obeyed the following rate law:

$$\mathfrak{R} = K_f[CO_3^{2-}]^3 - k_-$$

where $K_f = k_+(a_{Ca2+})^n \gamma_{CO_3}^3 = 10^{3.5}$ and $k_- = 0.29$ ($\mathfrak{R}$ is in units of $\mu mol \ m^{-2} \ h^{-1}$).

(a) Is this rate law consistent with the mechanism of calcite precipitation in seawater being the elementary one described in problem 1 above, or with a more complex reaction mechanism? Justify your answer.

(b) Using this rate law, predict the rate of calcite precipitation for concentrations of CO_3^{2-} of 0.04, 0.066, and 0.3 mmol/kg.

3. Oxidation of methane in the atmosphere occurs through a number of mechanisms, including reaction with the hydroxyl radical:

$$OH + CH_4 \rightleftharpoons H_2O + CH_3$$

The rate of this reaction for a series of temperatures is shown in the table below. Based on these data, estimate the activation energy and frequency factor for this reaction. (*HINT*: Try using linear regression.)

Rates of methane hydroxyl reaction

T, °C	k
25	6.60×10^{-15}
10	4.76×10^{-15}
0	3.76×10^{-15}
−10	2.93×10^{-15}
−25	1.93×10^{-15}

4. Schott et al. (1981) found that dependence on pH of the rate of dissolution of enstatite could be expressed as:

$$\mathfrak{R} = k \, a_{H^+}^n$$

where k shows a typical Arrhenius temperature dependence.

(a) Reaction rates were measured at a series of pH values at constant temperature (22°C). These data are shown in the adjacent table. Using these data, estimate values of k and n for this temperature. (*HINT*: Try using linear regression.)

Rate of enstatite dissolution	
pH	Rate moles Si/g-sec
1	2.75×10^{-10}
2	7.08×10^{-11}
6	2.82×10^{-13}

(b) Reaction rates were also determined at various temperatures at constant pH (6). Using these data, estimate the activation energy and frequency factor for the rate constant.

T°C	Rate moles Si/g-sec
20	3.72×10^{-13}
50	2.34×10^{-12}
60	4.07×10^{-12}
75	8.13×10^{-12}

(c) Using your results from (a) and (b), estimate the rate of reaction (in moles Si released per sec per gram enstatite) at pH 4 and 30°C.

5. Marcasite and pyrite are polymorphs of FeS_2. Though pyrite has a lower ΔG_f than marcasite, the latter often forms metastably. Lennie and Vaughan (1992) found that the kinetics of the marcasite to pyrite transformation follows a simple first-order rate law:

$$-\frac{d\alpha}{dt} = k\alpha$$

where α is the volume fraction of marcasite and k has the usual Arrhenius temperature dependence with $A = 2.76 \times 10^{17}$ sec^{-1} and $E_A = 253$ kJ/mol. Assuming a system consisting initially of pure marcasite, calculate the time required for one half of the marcasite to convert to pyrite (i.e., $\alpha = 0.5$) at 300°C and 350°C.

6. If ΔG is the free energy of reaction for the reaction:

$$A + BC \rightarrow AC + B$$

and assuming (1) this is an elementary reaction, (2) ideal behavior, and (3) it is a reversible reaction, show that the ratio of the forward and reverse rates of the reaction is:

$$\frac{\mathcal{R}_+}{\mathcal{R}_-} = e^{-\Delta G/RT} \tag{5.55}$$

(HINT: Start with eqn. 5.43. Also consider that ΔG_r for this reaction can be written as:

$$\Delta G_r = \sum_1 v_i u_i^o + RT \sum_1 v_i \ln a_i = \Delta G_r^o + RT \sum_i v_i \ln a_i$$

The activity quotient term is equal to the equilibrium constant at equilibrium, but not otherwise.)

7. On a temperature–pressure diagram, draw a line such that the time required for complete conversion of a 1 mm aragonite crystal to calcite will be complete within 10^5 years. Assume spherically symmetric growth of calcite from a single nucleus in the center. Use the thermodynamic data in Table 2.2 and $\Delta G^* = 184\,kJ$ (see Example 5.4).

8. Using the data given in Example 5.5 for the reaction:

$$Ca_2Mg_5Si_8O_{22}(OH) + 11CaMg(CO_3)_2 \rightleftarrows 8Mg_2SiO_4 + 13\,CaCO_3 + 9CO_2 + H_2O$$

$$Tremolite + 11\,Dolomite \rightleftarrows 8\,Forsterite + 13\,Calcite + 9\,CO_2 + H_2O$$

make a plot of the relative volume of dolomite (V_{Do}/V_{Do0}) as a function of time assuming an initial temperature of 620°C (the equilibrium temperature) and a heating rate of 0.1°C per year. (*HINT*: Because the reaction is fast, the overall temperature change will be small, so you may assume that T (i.e. absolute temperature) is constant. However, because the temperature is close to the equilibrium temperature, the change in the temperature overstep, ΔT, will be significant. Approximate ΔG in eqn. 5.70 as $\Delta S \Delta T$ and express ΔT as a function of time, $\Delta T = R_H t$ where R_H is the heating rate, then integrate.)

9. The transformation of kaolinite to illite (muscovite) may be written as:

$$K^+ + 1.5Al_2Si_2O_5(OH)_4 \rightleftharpoons H^+ + KAl_3Si_3O_{10}(OH)_2 + 1.5H_2O$$

Chermack and Rimstidt (1990) determined that the forward rate of reaction was:

$$-\frac{d[K^+]}{dt} = k_+[K^+]$$

Forward and reverse rate constants for the reaction were determined to be:
$\ln k_+ = 12.90 - 1.87 \times 10^4/T$ and $\ln k_- = 6.03 - 1.21 \times 10^4/T$

(a) What are the activation energies for the forward and reverse reactions?
(b) What is the equilibrium constant for this reaction at 275°C ?
(c) Make a plot of log $\Re_{net}$ vs. log ($[H^+]/[K^+]$) at 250°C, assuming a K^+ concentration of 2.0×10^{-6} M, ideal solution, that muscovite and kaolinite are pure phases, and that the forward and reverse reactions are elementary.

10. Using the data in Table 5.2, determine along which crystallographic axis Sr diffuses fastest in diopside at 1000°C.

11. Use the data in the following table and concentration gradients of 0.022 mol/cm. -0.09 mol/cm, and -0.015 mol/cm for Mn^{2+}, Mg^{2+}, and Fe^{-2+}, respectively, to calculate the diffusion flux for these elements in garnet.

cm^2/sec	Mn	Mg	Fe
Mn	8.38×10^{-20}	-9.91×10^{-20}	-4.68×10^{-21}
Mg	-2.78×10^{-20}	7.76×10^{21}	-8.81×10^{-23}
Fe	-7.16×10^{-20}	-4.81×10^{-23}	1.19×10^{-20}

12. A remarkable feature of sediments recording the Cretaceous–Tertiary boundary is an enrichment in iridium (Ir) at the boundary, which is often marked by a boundary clay. Imagine a boundary clay 5 cm thick that initially has a uniform Ir concentration of 20 ppb. Assume that sediments above and below the boundary clay contain negligible Ir. If the detection limit for Ir is 2 ppb, how thick would the Ir-enriched layer be after 60 million years if the diffusion coefficient for Ir is 10^{-15} cm^2/sec?

13. Show that the partial molar surface area, a, of a cube is $4v/r$ where v is the partial molar volume and r is the length of a side.

14. Assuming a surface free energy of 10^{-4} J/cm^2, $\overline{V} = 101$ cc/mol, $\Delta H_m = 54.84$ kJ/mol, and $T_m = 1118°$C, what is the critical radius for a spherical albite crystal growing in a pure albite melt that has been undercooled by 10°, 20°, and 30°C? Make a plot of ΔG_{tot} as a function of crystal radius for each of these temperatures. (*HINT*: Your scale should span only 10 or 20 microns.)

15. Crystal growth and dissolution are reactions that involve both diffusion and surface reactions occurring in series (i.e., a component of a growing crystal must first be delivered to the surface, then incorporated into the growing crystal). Either of these processes can be the rate-limiting step at 25°C. Diffusion in aqueous solutions typically has an activation energy of 20 kJ/mol, whereas surface reactions in aqueous solution typically have activation energies of 60–80 kJ/mol. Assuming the rates of diffusion and surface reaction for growth of a certain mineral from aqueous solution are approximately equal at 25°C, will diffusion or surface reaction be rate-limiting at 200°C?

16. Some anaerobic bacteria can utilize Mn^{4+} to oxidize organic matter. The reaction may be represented as:

$$2MnO_2(s) + CH_2O + H_2O \rightleftharpoons 2Mn^{2+} + CO_2 + 4OH^-$$

In its oxidized form, Mn is highly insoluble and effectively immobile in sediment. However, in its reduced form, Mn is soluble and mobile. Imagine that at a depth of 50 cm in actively depositing marine sediments, conditions become sufficiently reducing so that the reaction above occurs. Furthermore, assume that reaction is such that a constant concentration of 0.02 mM of Mn^{2+} is maintained at this depth and below. Above this depth, Mn^{2+} is oxidized and precipitated through reactions such as:

$$Mn^{2+} + 2OH^- + \frac{1}{2} O_2 \rightleftharpoons MnO_2(s) + H_2O$$

Assuming: (1) that the rate of the above reaction may be written as:

$$-\frac{d[Mn^{2+}]}{dt} = k[Mn^{2+}]$$

(2) k for this reaction is 10^{-8} sec^{-1}, (3) the concentration of Mn^{2+} at the sediment–water interface is 0 and that diffusion from below is the sole source of Mn^{2+} between 0 and 50 cm, (4) D is 5×10^{-6} cm/sec, (5) a sedimentation rate of 1 cm/yr, (6) there is no advection, compaction, or bioturbation, and (7) a porosity of 0.85, make a plot of the concentration of dissolved Mn^{2+} vs. depth at steady state.

17. Diamond is remarkably stable at the surface of the Earth. Pearson et al. (1995) estimated that to convert 1 cc of diamond to graphite at 0.1 MPa and 1000°C would require 1 billion years, but only a million years would be required at 1200°C. From the difference in these rates, estimate the activation energy for the diamond–graphite transition.

18. Using eqn. 5.160 and the data given below it to make a plot showing how the two reaction rates and their sum varies as ΔG varies from 150 kJ/mol to 1 kJ/mol at 25°C.

Chapter 6

Aquatic chemistry

6.1 INTRODUCTION

Water continually transforms the surface of the Earth, through interaction with the solid surface and transport of dissolved and suspended matter. Beyond that, water is essential to life and central to human activity. Thus as a society, we are naturally very concerned with water quality, which in essence means water chemistry. Aquatic chemistry is therefore the principal concern of many geochemists.

In this chapter, we learn how the tools of thermodynamics and kinetics are applied to water and its dissolved constituents. We develop methods, based on the fundamental thermodynamic tools already introduced, for predicting the species present in water at equilibrium. We then examine the interaction of solutions with solids through precipitation, dissolution, and adsorption.

Most reactions in aqueous solutions can be placed in one of the following categories:

- Acid–base, e.g., dissociation of carbonic acid:

$$H_2CO_3 \rightleftharpoons H^+ + HCO_3^-$$

- Complexation, e.g., hydrolysis of mercury:

$$Hg^{2+} + H_2O \rightleftharpoons Hg(OH)^{1+} + H^+$$

- Dissolution/precipitation, e.g., dissolution of orthoclase:

$$KAlSiO_8 + H^+ + 7H_2O \rightleftharpoons Al(OH)_3 + K^+ + 3H_4SiO_4$$

- Adsorption/desorption, e.g., adsorption of Mn on a clay surface:

$$\equiv S + Mn^{2+} \rightleftharpoons \equiv S\text{—}Mn^{2+}$$

(where we are using $\equiv S$ to indicate the surface of the clay).

We'll put these methods to use when we examine water in soils, streams, rivers, and lakes in Chapter 13 and the ocean in Chapter 14.

6.2 ACID–BASE REACTIONS

The hydrogen and hydroxide ions are often participants in all the foregoing reactions. As a result, many of these reactions are pH-dependent. In order to characterize the state of an aqueous solution, that is, to determine how much $CaCO_3$ a solution will dissolve, the complexation state of metal ions, or the redox state of Mn, the first step is usually to determine pH. On a larger scale, weathering of rock and precipitation

Geochemistry, Second Edition. William M. White.
© 2020 John Wiley & Sons Ltd. Published 2020 by John Wiley & Sons Ltd.
Companion website: www.wiley.com/go/white/geochemistry

of sediments depends critically on pH. Thus, pH is sometimes called the *master variable* in aquatic systems. We note in passing that while pH represents the hydrogen ion, or proton concentration, the hydroxide ion concentration is easily calculated from pH since the proton and hydroxide concentrations are related by the dissociation constant for water by:

$$K_W = a_{H^+}a_{OH^-} \qquad (6.1)$$

The value of K_W, like all equilibrium constants, depends upon temperature, but is 10^{-14} at 25°C.

Arrhenius defined an *acid* as a substance that, upon solution in water, releases free protons. He defined a *base* as a substance that releases hydroxide ions in solution. These are useful definitions in most cases. However, chemists generally prefer the definition of Brønstead, who defined acid and base as proton donors and proton acceptors respectively. The strength of an acid or base is measured by its tendency to donate or accept protons. The dissociation constant for an acid or base is the quantitative measure of this tendency and thus is a good indication of its strength. For example, dissociation of HCl:

$$HCl \rightleftharpoons H^+ + Cl^-$$

has a dissociation constant:

$$K = \frac{a_{H^+}a_{Cl^-}}{a_{HCl}} = 10^3$$

HCl is a strong acid because only about 3% of the HCl molecules added will remain undissociated. The equilibrium constant for dissociation of hydrogen sulfide:

$$H_2S \rightleftharpoons H^+ + HS^-$$

is

$$K = \frac{a_{H^+}a_{HS^-}}{a_{H_2S}} = 10^{-7.1}$$

Thus, H_2S is a weak acid because very few of the H_2S molecules actually dissociate except at high pH.

Metal hydroxides can either donate or accept protons, depending on pH. For example, we can represent this in the case of aluminum as:

$$Al(OH) + H^+ \rightleftharpoons Al(OH)^{2+} + H_2O$$

$$Al(OH) + OH^- \rightleftharpoons Al(OH) + H_2O$$

Compounds that can either accept or donate protons are said to be *amphoteric*.

Metals dissolved in water are always surrounded by solvation shells. The positive charges of the hydrogens in the surrounding water molecules are to some extent repelled by the positive charge of the metal ion. For this reason, water molecules in the solvation shell are more likely to dissociate and give up a proton more readily than other water molecules. Thus, the concentration of such species will affect pH.

Most protons released by an acid will complex with water molecules to form hydronium ions, H_3O^+ or even $H_5O_2^+$. However, in almost all cases we need not concern ourselves with this and can treat the H^+ ion as if it were a free species. Thus we will use $[H^+]$ to indicate the concentration $H^+ + H_3O^+ + H_5O_2 + ...$

6.2.1 Proton accounting, charge balance, and conservation equations

6.2.1.1 Proton accounting

Knowing the pH of an aqueous system is the key to understanding it and predicting its behavior. This requires a system of accounting for the H^+ and OH^- in the system. There are several approaches to doing this. One such approach is the *proton balance equation* (e.g., Pankow, 1991). In this system, both H^+ and OH^- are considered components of the system, and the proton balance equation is written such that *the concentration of all species whose genesis through reaction with water caused the production of OH^- are written on one side, and the concentration of all species whose genesis through reaction with water caused the production of H^+ are written on the other side.* Because water dissociates to form one H^+ and one OH^-, $[H^+]$ always appears on the left side and OH^- always appears on the right side of the proton balance equation. The proton balance equation for pure water is thus:

$$[H^+] = [OH^-] \qquad (6.2)*$$

Thus in pure water the concentrations of H^+ and OH^- are equal.

Now, consider the example of a nitric acid solution. H^+ will be generated both by dissociation of water and dissociation of nitric acid:

$$HNO_3 \rightleftharpoons H^+ + NO_3^-$$

Since one NO_3^- ion is generated for every H^+, the proton balance equation becomes:

$$[H^+] = [OH^-] + [NO_3^-] \qquad (6.3)$$

Next consider a solution of calcium carbonate. We specify the calcium and carbonate ions as components. Hydrogen ions may be generated by hydrolysis of calcium:

$$Ca^{2+} + H_2O \rightleftharpoons H^+ + Ca(OH)^+$$

and hydroxide ions may be generated by:

$$CO_3^- + H_2O \rightleftharpoons HCO_3^- + H_2O$$

The proton balance equation for this reaction is:

$$[H^+] + [HCO_3^-] = [OH^-] + [Ca(OH)^+] \qquad (6.4)$$

Now consider a solution of a *diprotonic* acid such as H_2S. H_2S can undergo two dissociation reactions:

$$H_2S \rightleftharpoons H^+ + HS^- \qquad (6.5)$$

$$HS^- \rightleftharpoons H^+ + S^{2-} \qquad (6.6)$$

For every HS^- ion produced by dissociation of H_2S, one H^+ ion would have been produced. For every S^{2-} ion, however, two H^+ would have been produced, one from the first dissociation and one from the second. The proton balance equation is thus:

$$[H^+] = [OH^-] + [HS^-] + 2[S^{2-}] \qquad (6.7)$$

An alternative approach to the proton balance equation is the *TOTH proton mole balance equation* used by Morel and Hering (1993). In this system, *H^+ and H_2O are always chosen as components* of the system, but OH^- is not. The species OH^- is the algebraic sum of H_2O less H^+:

$$OH^- = H_2O - H^+ \qquad (6.8)$$

An implication of this selection of components is that when an acid, such as HCl, is present, we choose the conjugate anion as the component, so that the acid HCl is formed from components:

$$HCl = Cl^- + H^+$$

For bases, such as NaOH, we choose the conjugate cation as a component. The base, NaOH, is formed from components as follows:

$$NaOH = Na^+ + H_2O - H^+$$

Because aquatic chemistry almost always deals with dilute solutions, the concentration of H_2O may be considered fixed at a mole fraction of 1, or 55.4 M. Thus, in the Morel and Hering system, H_2O is made an *implicit* component, that is, its presence is assumed but not written, so that eqn. 6.8 becomes:

$$[OH^-] = -[H^+] \qquad (6.9)$$

The variable TOTH is the total amount of component H^+, rather than the total of species H^+. Every species containing the component H^+ contributes positively to *TOTH* while every species formed by subtracting component H^+ contributes negatively to *TOTH*. Because we create the species OH^- by subtracting component H^+ from component H_2O, the total of component H^+ for pure water will be:

$$TOTH = [H^+] - [OH^-]$$

---

* Be careful not to confuse algebraic expressions, written with an equal sign, such as the proton balance equation, with chemical reactions, written with the reaction symbol, $\rightleftharpoons$, or similar. In this case, it is obvious that this is not a balanced chemical reaction, but that will not always be the case.

Thus, TOTH in this case is the difference between the concentrations of H^+ and OH^-. Of course, in pure water, $[H^+] = [OH^-]$, so $TOTH = 0$.

Now let's consider our example of the $CaCO_3$ solution. In addition to H^+ and H_2O, we choose Ca^{2+} and CO_3^{2-} as components. In the proton mole balance equation, HCO_3^- counts positively (since $HCO_3^- = CO_3^{2-} + H^+$) and $CaOH^+$ (since $CaOH^+ = Ca^{2+} + H_2O - H^+$) negatively:

$$TOTH = [H^+] + [HCO_3^-] - [OH^-]$$
$$- [Ca(OH)^+] \qquad (6.10)$$

Comparing eqns. 6.10 and 6.4, we see that the TOTH is equal to the difference between the left- and right-hand sides of the proton balance equation, and that in this case $TOTH = 0$. This makes sense, because, having added neither $[H^+]$ nor $[OH^-]$ to the solution, the total of the component H the solution contains should be 0.

Now consider the dissolution of CO_2 in water to form carbonic acid:

$$CO_2 + H_2O \rightleftharpoons H_2CO_3 \qquad (6.11)$$

Under the right conditions of pH, this carbonic acid will dissociate to form bicarbonate ion:

$$H_2CO_3 \rightleftharpoons H^+ + HCO_3^- \qquad (6.12)$$

If we choose CO_2 as our component, bicarbonate ion would be made from components CO_2, H_2O, and H^+:

$$HCO_3^- = CO_2 + H_2O - H^+$$

Thus, in the TOTH proton mole balance equation, bicarbonate ion would count negatively, so TOTH is:

$$TOTH = [H^+] - [OH^-] - [HCO_3^-] \qquad (6.13)$$

Alternatively, had we defined CO_3^{2-} as a component, then species HCO_3^- is formed by the components:

$$HCO_3^- = H^+ + CO_3^{2-}$$

In this case, the proton mole balance equation is:

$$TOTH = [H^+] - [OH^-] + [HCO_3^-] \qquad (6.13a)$$

6.2.1.2 Conservation equations

A further constraint on the composition of a system is provided by *mass balance*. Acid–base reactions will not affect the total concentration of a substance. Thus, regardless of reactions 6.5 and 6.6, and any other complexation reactions, such as

$$Pb^{2+} + S^{2-} \rightleftharpoons PbS_{aq}$$

the total concentration of sulfide remains constant. We may then write the following:

$$\Sigma S = [H_2S] + [HS^-] + [S^{2-}] + [PbS_{aq}] + \cdots$$

We can write one mass balance, or *conservation*, equation for each component in solution. Of course, for components such as Na that form only one species, Na^+ in this case, the mass balance equation is trivial. Mass balance equations are useful for those components forming more than one species.

While the charge balance constraint is an absolute one and always holds, mass balance equations can be trickier because other processes, such a redox, precipitation, and adsorption, can affect the concentration of a species. We sometimes get around this problem by writing the mass balance equation for an element, since an elemental concentration is not changed by redox processes. We might also define our system such that it is closed to get around the other problems. Despite these restrictions, mass balance often provides a useful additional constraint on a system.

6.2.1.3 Charge balance

As we saw in Chapter 3, solutions are electrically neutral; that is, the number of positive and negative charges must balance:

$$\sum_i m_i z_i = 0 \qquad (6.14)$$

where m is the number of moles of ionic species i and z is the charge of species i.

Equation 6.14 is known as the *charge balance equation* and is identical to eqn. 3.99. This equation provides an important constraint on the composition of a system. Notice that in some cases, the proton balance and charge balance equations are identical (e.g., eqns. 6.2 and 6.7).

For each acid–base reaction, an equilibrium constant expression may be written. By manipulating these equilibrium constant expressions as well proton balance, charge balance, and mass balance equations, it is possible to predict the pH of any solution. In natural systems where there are many species present; however, solving these equations can be a complex task indeed. An important step in their solution is to decide which reactions have an insignificant effect on pH and neglect them.

6.2.2   The carbonate system

We now turn our attention to carbonate. Water at the surface of the Earth inevitably contains dissolved CO_2, either as a result of equilibration with the atmosphere or because of respiration by organisms. CO_2 reacts with water to form *carbonic acid*:

$$CO_2 + H_2O \rightleftharpoons H_2CO_3^- \qquad (6.15)$$

some of the carbonic acid dissociates to form bicarbonate and hydrogen ions:

$$H_2CO_3 \rightleftharpoons H^+ + HCO_3^- \qquad (6.16)$$

Some of the bicarbonate will dissociate to an additional hydrogen ion and a carbonate ion:

$$HCO_3 \rightleftharpoons H^+ + CO_3^{2-} \qquad (6.17)$$

We can write three equilibrium constant expressions for these reactions:

$$K_{sp} = \frac{a_{CO_2}}{f_{CO_2}} \qquad (6.18)$$

$$K_1 = \frac{a_{H^+} a_{HCO_3^-}}{a_{H_2CO_3}} \qquad (6.19)$$

$$K_2 = \frac{a_{H^+} a_{CO_3^{2-}}}{a_{HCO_3^-}} \qquad (6.20)$$

The equilibrium constants for these reactions are given in Table 6.1 as a function of temperature.

In the above series of reactions, we have simplified things somewhat and have assumed that dissolved CO_2 reacts completely with water to form H_2CO_3. This is actually not the case, and much of the dissolved CO_2 will actually be present as distinct molecular species, $CO_{2(aq)}$. Thus, reaction 6.15 actually consists of the two reactions:

$$CO_{2(g)} \rightleftharpoons CO_{2(aq)} \qquad (6.15a)$$

$$CO_{2(aq)} + H_2O \rightleftharpoons H_2CO_3 \qquad (6.15b)$$

The equilibrium for the second reaction favors $CO_{2(aq)}$. However, it is analytically

Table 6.1   Equilibrium constants for the carbonate system.

T (°C)	pK_{CO_3}[‡]	pK_1	pK_2	pK_{cal}	pK_{arag}	pK_{CaHCO_3}[*]	pK_{CaCO_3}[†]
0	1.11	6.58	10.63	8.38	8.22	−0.82	−3.13
5	1.19	6.52	10.55	8.39	8.24	−0.90	−3.13
10	1.27	6.46	10.49	8.41	8.26	−0.97	−3.13
15	1.34	6.42	10.43	8.43	8.28	−1.02	−3.15
20	1.41	6.38	10.38	8.45	8.31	−1.07	−3.18
25	1.47	6.35	10.33	8.48	8.34	−1.11	−3.22
30	1.52	6.33	10.29	8.51	8.37	−1.14	−3.27
45	1.67	6.29	10.20	8.62	8.49	−1.19	−3.45
60	1.78	6.29	10.14	8.76	8.64	−1.23	−3.65
80	1.90	6.34	10.13	8.99	8.88	−1.28	−3.92
90	1.94	6.38	10.14	9.12	9.02	−1.31	−4.05

[*]$K_{CaHCO_3^+} = a_{CaHCO_3^+}/a_{Ca^+} a_{HCO_3^-}$

[†]$K_{CaCO_3^0} = a_{CaCO_3^0}/a_{Ca^{2+}} a_{CO_3^{2-}}$

[‡]pressure in units of bars. From Drever (1988).

difficult to distinguish between the species $CO_{2(aq)}$ and H_2CO_3. For this reason, $CO_{2(aq)}$ is often combined with H_2CO_3 when representing the aqueous species. The combined total concentration of $CO_{2(aq)}$ and H_2CO_3 is sometimes written as $H_2CO_3^*$. We will write it simply as H_2CO_3. The equilibrium for the second reaction favors $CO_{2(aq)}$.

Example 6.1 Proton, mass, and charge balance equations for Na$_2$CO$_3$ solution

Write the proton, proton mass balance, charge balance, and carbonate conservation equations for a solution prepared by dissolving Na_2CO_3 in water. Assume that Na_2CO_3 dissociates completely and that the system is closed.

Answer: We begin with the proton balance equation. From the dissociation of water we have:

$$[H^+] = [OH^-]$$

In addition to this, hydroxide ions will also be generated by reaction between and water:

$$CO_3^{2-} + H_2O \rightleftharpoons HCO_3^- + OH^-$$

and

$$HCO_3^- + H_2O \rightleftharpoons H_2CO_3 + OH^-$$

Since for each HCO_3^- formed, one OH^- must have formed and for each H_2CO_3 present two OH^- must have formed, the proton balance equation is:

$$[H^+] + [HCO_3^-] + 2[H_2CO_3] = [OH^-] \tag{6.21}$$

Choosing CO_2 and sodium ions as components (in addition to H^+ and H_2O), the three carbonate species are made from components as follows:

$$H_2CO_3 = H_2O + CO_2$$

$$HCO_3^- = H_2O + CO_2 - H+$$

$$CO_3^{2-} = H_2O + CO_2 - 2H+$$

In this case, the proton mole balance equation is:

$$TOTH = [H+] - [HCO_3^-] - 2[CO_3^{2-}] - [OH^-] \tag{6.22}$$

The charge balance equation is:

$$[H^+] + [Na^+] = [OH^-] + [HCO_3^-] + 2[CO_3^{2-}] \tag{6.23}$$

The conservation equation for carbonate species is:

$$\Sigma CO_2 = [CO_3^{2-}] + [HCO_3^-] + [H_2CO_3] \tag{6.24}$$

Since the dissolution of Na_2CO_3 produces two moles of Na^+ for every mole of carbonate species, we may also write:
$$[Na^+] = 2\Sigma CO_2 = 2([CO_3^{2-}] + [HCO_3^-] + [H_2CO_3])$$

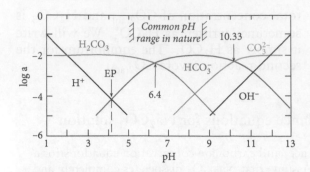

Figure 6.1   Activities of different species in the carbonate system as a function of pH, assuming $\Sigma CO_2 = 10^{-2}$. After Drever (1988).

The importance of the carbonate system is that by dissociating and providing hydrogen ions to solution, or associating and taking up free hydrogen ions, *it controls the pH of many natural waters*. Example 6.2 shows that pure water in equilibrium with atmospheric CO_2 will be slightly acidic. The production of free H^+ ions as a result of the solution of CO_2 and dissociation of carbonic acid plays an extremely important role in weathering.

Groundwaters may not be in equilibrium with the atmosphere but will nonetheless contain some dissolved CO_2 (see Example 6.3). Because of respiration of organisms in soil

Example 6.2   pH of water in equilibrium with the atmosphere

What is the pH of water in equilibrium with the atmospheric CO_2 at 25°C, assuming ideal behavior and no other dissolved solids or gases present? The partial pressure of CO_2 in the atmosphere is 3.7×10^{-4}.

Answer: In this case, the proton balance and charge balance equations are identical:

$$[H^+] = [OH^-] + [HCO_3^-] + 2[CO_3^{2-}] \tag{6.25}$$

We might guess that the pH of this solution will be less than 7 (i.e., $[H^+] > 10^{-7}$). Under those circumstances, the concentrations of the hydroxyl and carbonate ions will be much lower than that of the hydrogen and bicarbonate ions. Assuming we can neglect them, our equation then becomes simply:

$$[H^+] = [HCO_3^-] \tag{6.25a}$$

We can combine eqns. 6.18 and 6.19 to obtain an expression for bicarbonate ion in terms of the partial pressure of CO_2:

$$[HCO_3^-] = (K_1 K_2 P_{CO_2})/[H^+]$$

Substituting this into 6.25a and rearranging, we have:

$$[H^+]^2 \cong K_1 K_{CO_2} P_{CO_2} \tag{6.26}$$

Taking the negative log of this expression and again rearranging, we obtain:

$$pH \cong \frac{\log K_1 - \log K_{CO_2} - \log P_{CO_2}}{2}$$

Substituting values from Table 6.1, we calculate pH = 5.64. Looking at Figure 6.1, we can be assured that our assumption that carbonate and hydroxyl ion abundances are valid. Indeed, using the Solver in Excel™ differs from the approximate one by less than 0.0001 pH units.

(mainly plant roots and bacteria) through which they pass before penetrating deeper, groundwaters often contain much more CO_2 than water in equilibrium with the atmosphere. In addition, calcite and other carbonates are extremely common minerals in soils and in sedimentary, metamorphic, and altered igneous rocks. Groundwater will tend to approach equilibrium with calcite by either dissolving it or precipitating it:

$$CaCO_3 \rightleftharpoons Ca^{2+} + CO_3^{2-} \qquad (6.27)$$

Carbonate ions produced in this way will associate with hydrogen ions to form bicarbonate as in reaction 6.17 above, increasing the pH of the solution. Water containing high concentrations of calcium (and magnesium) carbonate is referred to as *hard water* and is generally somewhat alkaline.

Now suppose we have a known activity of all carbonate species in solution, say for example 10^{-2}:

$$a_{H_2CO_3} + a_{HCO_3^-} + a_{CO_3^{2-}} = \sum CO_2 = 10^{-2} \qquad (6.28)$$

From this, and the dissociation constants, we can calculate the amount of each species present as a function of pH and temperature. For example, we can use the equilibrium constant expressions to obtain substitutions for the carbonic acid and carbonate ion activities in eqn. 6.28 that are functions of bicarbonate ion activity and pH. We then solve eqn. 6.28 to obtain an expression for the activity of the bicarbonate ion as a function of total CO_2 and hydrogen ion activity:

$$a_{HCO_3^-} = \frac{\sum CO_2}{(a_{H^+}/K_1) + 1 + (K_2/a_{H^+})} \qquad (6.29)$$

Similar equations may be found for carbonic acid and carbonate ion. Carrying out these calculations at various pH values, we can construct the graph shown in Figure 6.1. In this figure, we see that carbonic acid is the

---

Example 6.3 pH of a solution with fixed total carbonate concentration

Groundwater moving through soil into a deep aquifer acquires a total dissolved CO_2 concentration of 10^{-2} M. Assuming the water does not exchange with surrounding rock, ideal behavior, and no other dissolved solids or gases, what is the pH of the water?

Answer: In this case, our charge and proton balance equations are the same as in Example 6.2, i.e., eqn. 6.25. Since the solution does not exchange with surrounding rock, it can be considered a closed system and we can write the following mass balance equation:

$$\sum CO_2 = [H_2CO_3] + [HCO_3^-] + [CO_3^{2-}] = 10^{-2} \qquad (6.31)$$

Simultaneously solving the change balance and mass balance equations, and using equilibrium constant expressions to eliminate carbonate and OH species, we obtain:

$$[H^+]^4 + K_1[H^+]^3 + \{K_2K_1 - K_W - K_1\sum CO_2\}[H^+]^2 - \{K_W + 2K_2\sum CO_2\}K_1[H^+] - K_2K_1K_W = 0$$

We might again guess that the concentration of the carbonate ion will be very low, and that we can therefore neglect all terms in which K_2 occurs. We might also guess that pH will be acidic so that $[H^+] \gg [OH^-]$, and therefore that we can neglect terms containing K_W. Our equation becomes:

$$K_1^{-1}[H^+]^2 + [H^+] = \sum CO_2$$

Solving this quadratic, we find that pH = 4.18.

dominant species at low pH, bicarbonate at intermediate pH, and carbonate at high pH.

6.2.2.1 Equivalence points

Particularly simple relationships occur when the activities of two species are equal. The pH where this occurs, known as an equivalence point, is determined by eqns. 6.19 and 6.20. For example, the point where carbonic acid concentration equals bicarbonate concentration can be determined by rearranging eqn. 6.19:

$$\frac{a_{HCO_3^-}}{a_{H_2CO_3}} = \frac{K_1}{a_{H^+}} = 1 \qquad (6.30b)$$

and therefore:

$$a_{H^+} = K_1 = 10^{-6.35} \qquad (6.30b)$$

The point labeled EP on Figure 6.1 is called the CO_2 equivalence point. At this point, the concentration of the carbonate ion is extremely low, and there is exactly enough H^+ to convert all HCO_3^- to H_2CO_3. From the perspective of the proton balance, then, the HCO_3^- concentration is equivalent to the same concentration of H_2CO_3. In a similar way, the point where the carbonic acid and carbonate ion concentrations are equal is called the bicarbonate equivalence point, and that where bicarbonate and hydroxyl concentrations are equal is called the carbonate equivalence point.

The exact concentrations of carbonate species depend on total carbonate concentration as well as the concentration of other ions in solution. Thus, the distribution shown in Figure 6.1 is unique to the conditions specified ($\Sigma CO_2 = 10^{-2}$, no other ions present). Nevertheless, the distribution will be qualitatively similar for other conditions.

6.2.3 Conservative and nonconservative ions

We can divide dissolved ions into conservative and nonconservative ones. The conservative ions are those whose concentrations are not affected by changes in pH, temperature, and pressure, assuming no precipitation or dissolution. In natural waters, the principal conservative ions are Na^+, K^+, Ca^{2+}, Mg^{2+}, Cl^-, SO_4^{2-} and NO_3^-. These ions are conservative because they are fully dissociated from their conjugate acids and bases over the normal range of pH of natural waters. Nonconservative ions are those that will undergo association and dissociation reactions in this pH range. These include the proton, hydroxyl, and carbonate species as well as $B(OH)_4^-$, $H_3SiO_4^-$, HS^-, NH_4OH, phosphate species, and many organic anions. Virtually all the nonconservative species are anions, the two principal exceptions being H^+ and NH_4OH (which dissociates to form NH_4^+ at low pH). Variations in the concentrations of nonconservative ions result from reactions between them, and these reactions can occur in the absence of precipitation or dissolution. For example, reaction of the carbonate and hydrogen ion to form bicarbonate will affect the concentrations of all three ions. Of course, if the system is at equilibrium, this reaction will not occur in the absence of an external disturbance, such as a change in temperature.

6.2.4 Total alkalinity and carbonate alkalinity

Alkalinity is a measure of acid-neutralizing capacity of a solution and is defined as the sum of the concentration (in equivalents) of bases that are titratable with strong acid. Mathematically, we define alkalinity as the negative of TOTH when the components are the principal species of the solution at the CO_2 equivalence point. The acidity can be defined as the negative of alkalinity, and hence equal to TOTH.

As a first example, let's consider a solution containing a fixed total dissolved concentration of $CaCO_3$. At the CO_2 equivalence point, H_2CO_3 is the principal carbonate species, so we could choose our components as H^+, CO_3^{2-}, and Ca^{2+}. (Since we always choose water as a component, we do not want to choose H_2CO_3 as a component, because it contains the component H_2O and hence is not fully independent). Instead, we choose CO_2 as the carbonate component in this case.) Species H_2CO_3, HCO_3^- and, CO_3^{2-} are made by combining these components.

The proton mole balance equation is then:

$$TOTH = [H^+] - [HCO_3^-] - 2[CO_3^{2-}] - [OH^-] \quad (6.32)$$

The alkalinity is then:

$$Alk = -TOTH = -[H^+] + [HCO_3^-]$$
$$+ 2[CO_3^{2-}] + [OH^-] \quad (6.33)$$

An analytical definition of alkalinity is that it is the quantity of acid that must be added to

---

Example 6.4   The Tableau method of Morel and Hering

Write an expression for the alkalinity of a solution containing $H_3SiO_4^-$, H_4SiO_4, $B(OH)_3$, $B(OH)_4$, H_2S, HS^-, $H_2PO_4^-$, HPO_4^-, H_2CO_3, HCO_3^-, and CO_3^{2-}, as well, of course, as OH^- and H^+.

Answer: The alkalinity will be the negative of TOTH when the components are the principal species of the solution at the CO_2 equivalence point, so the real problem is just choosing components and defining our species in terms of these. At the CO_2 equivalence point, the principal species will be H_4SiO_4 $B(OH)_3$, H_2S, $H_2PO_4^-$, and H_2CO_3. The problem now is simply defining the species in terms of components. Morel and Hering (1993) proposed a table, with the components listed across the top and the species listed vertically. Entries in the table are just the stoichiometric coefficients used to define each species in terms of its components. In this case, the Tableau will look like that shown above. The columns will show us what the coefficients will be in our TOTH equation. Our expression for alkalinity will thus be:

$$Alk = -TOTH = -\{[H^+] - [OH^-] - [HCO_3^-] - 2[CO_3^{2-}] - [H_2PO_4^-]$$
$$- [H_3SiO_4] - [B(OH)_4^-] - [HS^-] \quad (6.34)$$

The sum, $-[H^+] + [HCO_3^-] + 2[CO_3^{2-}] - [OH^-]$, is called the *carbonate alkalinity*. If no other non-conservative ions are present in solution, carbonate alkalinity and alkalinity are equal. To avoid confusion with carbonate alkalinity, alkalinity is sometimes called *total alkalinity*.

Tableau							
	H^+	H_2O	CO_2	$H_2PO_4^-$	H_2SiO_4	$B(OH)_3$	H_2S
H^+	1						
OH^-	−1	1					
H_2CO_3		1	1				
HCO_3^-	−1	1	1				
CO_3^{2-}	−2	1	1				
HPO_4^{2-}	−1			1			
$H_2PO_4^-$				1			
$H_3SiO_4^-$	−1				1		
H_4SiO_4					1		
$B(OH)_4^-$	−1	1				1	
$B(OH)_3$						1	
H_2S							1
HS^-	−1						1

the solution to bring the pH to the CO_2 equivalence point.

We can also express alkalinity in terms of conservative and non-conservative ions. The charge balance equation, eqn. 6.14, could be written as:

$$\Sigma_{\text{cations (in equivalents)}} - \Sigma_{\text{anions (in equivalents)}} = 0 \tag{6.35}$$

This can then be expanded to:

$$\Sigma_{\text{conserv.cations}} - \Sigma_{\text{conserv.anions}} + \Sigma_{\text{nonconserv.cations}}$$
$$- \Sigma_{\text{non-conserv.anions}} = 0$$

(all in units of equivalents)*. Rearranging, we have:

$$\Sigma_{\text{conserv.cations}} - \Sigma_{\text{conserv.anions}}$$
$$= -\Sigma_{\text{nonconserv.cations}} + \Sigma_{\text{nonconserv.anions}} \tag{6.36}$$

The right-hand side of eqn. 6.36 is equal to the *alkalinity*. Hence we may write:

$$Alk = \Sigma_{\text{conserv.cations}} - \Sigma_{\text{conserv.anions}}$$
$$= -\Sigma_{\text{nonconserv.cations}} + \Sigma_{\text{nonconserv.anions}} \tag{6.37}$$

This equation emphasizes an important point. The difference of the sum of conservative anions and cations is clearly a conservative property, i.e., they cannot be changed except by the addition or removal of components. Since alkalinity is equal to this difference, alkalinity is also a conservative quantity (i.e., independent of pH, pressure, and temperature). *Thus, total alkalinity is conservative, even though concentrations of individual species are not.*

6.2.4.1 Alkalinity determination and titration curves

If the concentrations of all major conservative ions in a solution are known, the alkalinity can be simply calculated from eqn. 6.37. It is often useful, however, to determine this independently. This is done, as the definition of alkalinity suggests, through titration. Titration is the process of progressively adding a strong acid or base to a solution until a specified pH, known as an end-point, is reached. In the case of the determination of alkalinity, this end-point is the CO_2 equivalence point.

Consider a solution containing a certain concentration of sodium bicarbonate (Na_2CO_3). Because the carbonate ion can act as a proton acceptor, $NaCO_3$ is a base. We can determine both the alkalinity and the total carbonate concentration of this solution by titrating with a strong acid, such as HCl. Let's examine the chemistry behind this procedure.

For clarity, we make several simplifying assumptions. First, we assume ideal behavior. Second, we assume the system is closed, so that all components are conserved, except for H^+ and Cl^-, which we progressively add. Third, we assume that the volume of our Na_2CO_3 solution is sufficiently large and our HCl sufficiently concentrated that there is no significant dilution of the original solution. Finally, we assume both Na_2CO_3 and HCl dissociate completely.

The charge balance equation during the titration is:

$$[Na^+] + [H^+] = [Cl^-] + [HCO_3^-]$$
$$+ 2[CO_3^{2-}] + [OH] \tag{6.38}$$

Since the Cl^- concentration is conservative, it will be equal to the total amount of HCl added. Into equation 6.38, we can substitute the following:

$$[HCO_3^-] = \frac{K_1[H_2CO_3]}{[H^+]} \tag{6.39a}$$

$$[CO_3^{2-}] = \frac{K_1K_2[H_2CO_3]}{[H^+]^2} \tag{6.39b}$$

---

*One *equivalent* of a species is defined as the number of moles multiplied by the charge of the species. Thus one equivalent of CO_3^{2-} is equal to 0.5 moles of CO_3^{2-}, but one equivalent of Cl^- is equal to 1 mole of Cl^-. For an acid or base, an equivalent is the number of moles of the substance divided by the number of hydrogen or hydroxide ions that can be potentially produced by dissociation of the substance. Thus, there are 2 equivalents per mole of H_2CO_3, but 1 equivalent per mole of Na(OH).

and

$$[OH^-] = \frac{K_W}{[H^+]} \qquad (6.39c)$$

Doing so and rearranging yields:

$$[Cl^-] = [Na^+] + [H^+] - \frac{K_1[H_2CO_3]}{[H^+]}$$
$$- \frac{K_1K_2[H_2CO_3]}{[H^+]^2} - \frac{K_W}{[H^+]} \qquad (6.40)$$

We may also write a conservation equation for carbonate species, which is the same as equation 6.24 in Example 6.1. Substituting equations 6.39a and 6.39b into 6.24 and rearranging, we have:

$$[H_2CO_3] = \frac{\Sigma CO_2}{1 + \frac{K_1}{[H^+]} + \frac{K_1K_2}{[H^+]^2}} \qquad (6.41)$$

Substituting this expression into 6.40, we obtain:

$$[Cl^-] = [Na^+] + [H^+]$$
$$- \frac{\Sigma CO_2}{[H^+] + K_1 + K_1K_2/[H^+]}$$
$$\left\{ K_1 - \frac{K_1K_2}{[H^+]} \right\} - \frac{K_W}{[H^+]} \qquad (6.42)$$

An analytical definition of alkalinity is its *acid-neutralizing capacity when the end-point of the titration is the CO_2 equivalence point* (Morel and Hering, 1993). We had previously defined alkalinity as the negative of *TOTH* when the principal components are those at the CO_2 equivalence point. Let's now show that these definitions are equivalent.

Our *TOTH* expression, written in terms of components at the CO_2 equivalence point, is identical to eqn 6.32:

$$TOTH = [H^+] - [HCO_3^-] - 2[CO_3^{2-}] - [OH^-] \qquad (6.32)$$

and the charge balance equation (before any HCl is added) is:

$$[Na^+] + [H^+] = [HCO_3^-] + 2[CO_3^{2-}] + [OH^-]$$

Combining the two we have:

$$TOTH = -[Na^+]$$

Since the alkalinity is the negative of *TOTH*, it follows that (before the addition of HCl):

$$Alk = [Na^+] \qquad (6.43)$$

We obtain exactly the same result from equation 6.37. It is easy to show that after titrating to the CO_2 equivalence point, the alkalinity is 0. The change in alkalinity is thus

Example 6.5 Calculating alkalinity of spring water

Calculate the alkalinity of spring water from Thonon, France, whose analysis is given at right (this is the same analysis as in Problem 3.8).

Anions	mM	Cations	mM
HCO$_3^-$	5.436	Ca^{2+}	2.475
SO$_4^{2-}$	0.146	Mg^{2+}	0.663
NO$_3^-$	0.226	K$^+$	0.036
Cl$^-$	0.231	Na$^+$	0.223

Answer: We can use equation 6.34 to calculate alkalinity. All the ions listed here are conservative with the exception of HCO$_3^-$. To calculate alkalinity, we first need to convert the molar concentrations to equivalents; we do so by multiplying the concentration of each species by its charge. We find the sum of conservative anion concentrations to be 0.749 meq (milliequivalents), and that of the conservative cation concentrations to be 6.733 meq. The alkalinity is the difference, 5.985 meq.

equal to the number of equivalents, or moles, of H^+ we have added to the solution. Since at the endpoint, $[H^+] = [HCO_3^-]$ and the concentrations of CO_3^{2-} and OH^- are negligible, our charge balance equation, 6.38, reduces to:

$$[Na^+] \simeq [Cl^-]$$

Comparing this with eqn. 6.42, we see that the alkalinity is equal to the amount of HCl added. In the example in Figure 6.2, the equivalence point occurs after the addition of 10 ml of 1 M HCl, or a total of moles of Cl^-. (Notice that since small additions of acid result in large changes in pH at the end-points, we do not have to determine pH particularly accurately during the titration for an accurate determination of alkalinity.) So the alkalinity is 0.01 equivalents. This is exactly the answer we obtain from eqn. 6.42 for 1 liter of 0.005 M Na_2CO_3, since there are 2 moles of Na^+ for each mole of Na_2CO_3.

By assuming that the concentration of H^+ contributes negligibly to charge balance, it is also easily shown (Problem 2) that at the bicarbonate equivalence point:

$$\Sigma CO_2 = [Cl^-] + [OH^-] \qquad (6.44)$$

Thus total carbonate is obtained by titrating to the bicarbonate equivalence point (knowing the pH of the end-point allows

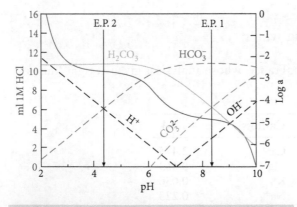

Figure 6.2 Titration curve (solid red line) for a one liter 0.005 M Na_2CO_3 solution titrated with 1 M HCl. The left axis shows the number of ml of HCl to be added to obtain a given pH. Also shown are the concentrations of carbonate species, H^+, and OH^- (dashed black lines, right axis gives scale). EP1 is the bicarbonate equivalence point, EP2 is the CO_2 equivalence point.

us to determine the ΣCO_2 exactly; however neglecting the $[OH^-]$ term in eqn. 6.44 results in only a 1% error in the example shown). In Figure 6.2, this occurs after the addition of 5ml 1 M HCl.

6.2.5 Buffer intensity

The carbonate system is a good example of a pH *buffer*. We define the *buffer intensity* of a solution as the inverse of change in pH per amount of strong base (or acid) added:

$$\beta \equiv \frac{dC_B}{dpH} = -\frac{dC_A}{dpH} \qquad (6.45)$$

where C_B and C_A are the concentrations, in equivalents, of strong base or acid respectively. The greater the buffer capacity of a solution, the less change in its pH as an acid or base is added. The buffer capacity of a solution can be calculated by differentiation of the equation relating base (or acid) concentration to pH, as is illustrated in Example 6.5.

A pH buffer acts to control pH within a narrow range as H^+ ions are added or removed from solution by other reactions. To understand how this works, imagine a solution containing carbonic acid, HCO_3^-, CO_3^{2-}, and H^+ ions in equilibrium concentrations. Now imagine that additional H^+ ions are added (for example, by addition of rainwater containing HNO_3). In order for the right-hand side of eqn. 6.19 to remain equal to K_1 despite an increase in the activity of H^+ (which it must at constant temperature and pressure), the bicarbonate activity must decrease and the carbonic acid activity increase. It is apparent then that reaction 6.16 must be driven to the left, taking free hydrogen ions from solution, hence driving the pH back toward its original value. Similarly, reaction 6.17, the dissociation of bicarbonate, will also be driven to the left, increasing the bicarbonate concentration and decreasing the hydrogen and carbonate ion concentrations.

The buffer capacity of the carbonate system depends strongly on pH and also on the concentration of the carbonate species and the concentration of other ions in solution (see Example 6.6). Natural solutions, however, can have substantial buffering capacity. Figure 6.3 illustrates three other examples of natural pH buffers. "Hard water" is an example of water with a substantial buffering capacity due to the presence of dissolved carbonates. In pure water containing no other ions and only carbonate in amounts in equilibrium with the atmosphere, the buffering capacity is negligible near neutral pH, as is shown in Figure 6.4. As we shall see, how adversely lakes and streams are impacted by "acid rain," depends on their buffering intensity.

6.3   COMPLEXATION

Ions in solution often associate with other ions, forming new species called complexes. Complex formation is important because it affects the solubility and reactivity of ions, as we will see in the following section. In some cases, complex formation is an intermediate step in the precipitation process. In other cases, ions form stable, soluble complexes

that greatly enhance the solubility of the one or both of the ions.

Complexation is usually described in terms of a central ion, generally a metal, and ions or molecules that surround it, or *coordinate* it, referred to as *ligands*. Perhaps the simplest and most common complexes are those formed between metals and water or its dissociation products. We learned in Section 3.7 that a solvation shell surrounds ions in aqueous solutions. The solvation shell (Figure 3.10) consists of water molecules, typically 6, though fewer in some cases, loosely bound to the ion through electrostatic forces. This solvation shell is referred to as an *aquo-complex*. Water molecules are the ligands in aquo-complexes. Aquo-complexes are ubiquitous: all charged species have a

Example 6.6   Calculating buffer intensity

How will pH change for given addition of a strong base such as NaOH for a solution of pure water in equilibrium with atmospheric CO_2? Calculate the buffer intensity of this solution as a function of pH. Assume that NaOH completely dissociates and behavior is ideal.

Answer: We want to begin with the charge balance equation in this case because it relates the two quantities of interest, $[Na^+]$ and $[H^+]$. The charge balance equation is the same as in Example 6.1:

$$[Na^+] + [H^+] = [OH^-] + [HCO_3^-] + 2[CO_3^{2-}] \tag{6.23}$$

Since Na^+ is a conservative ion, its concentration will depend only on the amount of NaOH added, so that $C_B = [Na^+]$. Substituting this into eqn. 6.14 and rearranging, we have:

$$C_B = [OH^-] + [HCO_3^-] + 2[CO_3^{2-}] - [H^+] \tag{6.46}$$

We can now use the equilibrium constant relations to substitute for the first three terms of the right-hand side of 6.46 and obtain:

$$C_B = \frac{K_W + K_1 K_{sp} P_{CO_2}}{[H^+]} + 2\frac{K_2 K_1 K_{sp} P_{CO_2}}{[H^+]^2} - [H^+]$$

Using the relation $pH = -\log[H^+]$ to replace $[H^+]$ in this equation with pH, we have:

$$C_B = \frac{K_W + K_1 K_{sp} P_{CO_2}}{10^{-pH}} + 2\frac{K_2 K_1 K_{sp} P_{CO_2}}{10^{-2pH}} - 10^{-pH}$$

Now differentiating with respect to pH, we obtain:

$$\frac{dC_B}{dpH} = \beta = \ln 10 \times \{(K_W + K_1 K_{sp} P_{CO_2}) \times 10^{pH} + 4K_2 K_1 K_{sp} P_{CO_2} 10^{2pH} + 10^{-pH}\} \tag{6.47}$$

Figure 6.4 shows a plot of this equation using the values in Table 6.1. Buffer intensity is negligible in neutral to slightly acidic conditions but increases rapidly with pH.

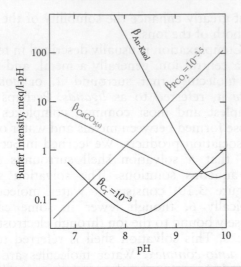

Figure 6.3 Buffer intensity as a function of pH for several ideal natural systems: β_{CT} fixed total dissolved CO_2, β_{CO_2} water in equilibrium with atmospheric CO_2 $\beta_{CaCO_3(s)}$ water in equilibrium with calcite, and $\beta_{An\text{-}Kaol.}$ water in equilibrium with anorthite and kaolinite. After Stumm and Morgan (1995).

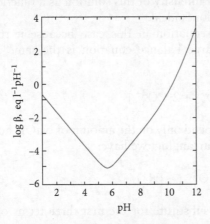

Figure 6.4 Buffer capacity of a carbonate solution in equilibrium with atmospheric CO_2.

solvation shell. Truly "free" ions do not exist: ions not otherwise complexed ("free ions") are in reality associated with surrounding water molecules and hence actually aquo-complexes. However, the existence of this type of complex is implicitly accounted for through the activity coefficient and not usually explicitly considered. Nevertheless, it is important to bear in mind that since all ions are complexed in some way to begin with, every complexation reaction in aqueous solution is essentially a ligand exchange reaction.

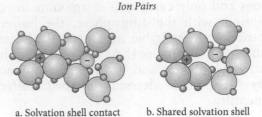

Figure 6.5 Illustration of ion pair and complex formation. Two types of ion pairs can be envisioned: (a) solvation shell contact and (b) solvation shell sharing. Ion pairs are sometimes referred to as outer sphere complexes. In formation of true complexes, ions are in contact (c) and there is some degree of covalent bonding between them. These are sometimes referred to as inner-sphere complexes.

Beyond aquo-complexes, we can distinguish two types of complexes:

- *Ion pairs*, where ions of opposite charge associate with one another through electrostatic attraction, yet each ion retains part or all of its solvation sphere. Figure 6.5 illustrates two possibilities: one where the two solvation spheres are merely in contact, the other where the water molecules are shared between the two solvation spheres. Ion pairs are also called *outer sphere complexes* (see also Figure 4.31).
- *Complexes (sensu stricto)*, where the two ions are in contact and a bond forms between them that is at least partly covalent in nature (Figure 6.5c). These are often called *inner sphere complexes*.

6.3.1 Stability constants

In its simplest form, the reaction for the formation of an ion pair or complex between a metal cation M and an anion or ligand L may be written as:

$$mM^+ + \ell L^- \rightleftharpoons M_mL_\ell$$

As with any other reaction, we may define an equilibrium constant as:

$$K = \frac{a_{M_mL_\ell}}{a_{M^+}^m a_{L^-}^\ell} \quad (6.48)$$

For example, the equilibrium constant for the reaction:

$$Zn(H_2O)_6^{2+} + OH^- \rightleftharpoons Zn(H_2O)_5OH^- + H_2O$$

is

$$K_1 = \frac{a_{Zn(H_2O)_5OH^-}}{a_{Zn(H_2O)_6^{2+}} a_{OH^-}}$$

(we omit the activity of water because we assume, as usual, that it is 1). As we noted, however, the aquo-complex is generally not explicitly expressed, so this same reaction would more often be written as:

$$Zn^{2+} + OH^- \rightleftharpoons ZnOH^+$$

and the equilibrium constant as:

$$\beta_1 = K_1 = \frac{a_{ZnOH^+}}{a_{Zn^{2+}} a_{OH^-}}$$

Equilibrium constants for complex formation reactions are often referred to as *stability constants*, since their value is an indication of the stability of the complex, and often denoted by β. Thus for the reaction above, β_1 and K_1 are synonymous. *By convention, stability constants are written so that the complex appears in the numerator (i.e., as a product of the reaction).*

The zinc ion might associate with a second hydroxyl:

$$ZnOH^- + OH^- \rightleftharpoons Zn(OH)_2$$

Here, however, the notation for the stability constant and the equilibrium constant differs. Whereas K_2 refers to the reaction above, β_2 refers to the reaction:

$$Zn^{2+} + 2OH^- \rightleftharpoons Zn(OH)_2$$

Hence

$$\beta_2 = \frac{a_{Zn(OH)_2}}{a_{Zn^{2+}} a_{OH^-}^2} = K_1 K_2$$

Finally, the notation $^*K$ and $^*\beta$ are sometimes uses for reaction when the complexation reaction is written so that the hydrogen ion occurs as a product, for example:

$$Zn^{2+} + HO \rightleftharpoons ZnOH^+ + H^+$$

and

$$^*K_1 = \frac{a_{Zn(OH)^+} a_{H^+}}{a_{Zn^{2+}}}$$

$^*K_1$ is then related to K_1 as:

$$^*K_1 = {}^*\beta_1 = K_1 K_W$$

where K_W is the dissociation constant of water ($=10^{-14}$ at 25°C).

We can define *apparent* equilibrium and stability constants, where the molar concentrations are used in place of activity. Indeed, as in other aspects of geochemistry, apparent equilibrium constants are more commonly encountered than true ones.

The equilibrium constant may, in turn, be related to the Gibbs free energy of the reaction, as in eqn. 3.85. Interestingly, the free energy changes involved in complexation reactions result largely from entropy changes. Indeed, the enthalpy changes of many complexation reactions are unfavorable, and the reaction proceeds only because of large positive entropy changes. These entropy changes result from the displacement of water molecules from the solvation shell.

The link between the equilibrium constant and the free energy change is particularly important and useful in complexation reactions, because it is in most instances difficult to determine the concentrations of individual complexes analytically. Thus, our knowledge of chemical speciation in natural waters derives largely from predictions based on equilibrium thermodynamics.

6.3.2 Water-related complexes

Let's further consider the types of complexes typically found in aqueous solution. Ferric iron, for example, can form a $Fe(H_2O)_6^{3+}$ complex. The positive charge of the central ion tends to repel hydrogen in the water molecules, so that water molecules in these aquo-complexes are more readily hydrolyzed than otherwise. Thus, these aquo-complexes can act as weak acids. For example:

$$Fe(H_2O)_6^{3+} \rightleftharpoons Fe(H_2O)_5(OH)^{2-} + H^+$$

$$\rightleftharpoons Fe(H_2O)_4(OH)_2^+ + 2H^+ \rightleftharpoons$$

$$Fe(H_2O)_3(OH)_3^0 + 3H^+$$

$$\rightleftharpoons Fe(H_2O)_2(OH)_4^- + 4H^+$$

As this suggests, equilibrium between these *hydroxo-complexes* depends strongly on pH.

The repulsion between the central metal ion and protons in water molecules of the solvation shell will increase with decreasing diameter of the central ion (decreasing distance between the protons and the metal) and with increasing charge of the central ion. For highly charged species, the repulsion of the central ion is sufficiently strong that all hydrogens are repelled, and it is surrounded only by oxygens.

Such complexes, for example, MnO_4^- and CrO_4^{2-} are known as *oxo-complexes*. Intermediate types in which the central ion is surrounded, or *coordinated*, by both oxygens and hydroxyls are also possible, for example $MnO_3(OH)$ and $CrO_3(OH)^-$, are known as hydroxo-oxo complexes. Figure 6.6

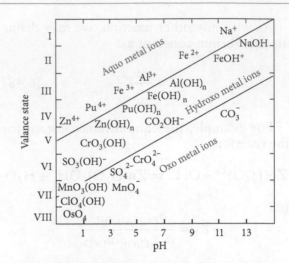

Figure 6.6 Predominant aquo-, hydroxo-, and oxo-complexes as a function of pH and valence state. After Stumm and Morgan (1995).

summarizes the predominance of aquo-, hydroxo-, hydroxo-oxo, and oxo-complexes as a function of pH and valence state. For most natural waters, metals in valence states I and II will be present as *free ions*, that is, aquo-complexes, valence III metals will be present as aquo- and hydroxo-complexes, and those with higher charge will be present as oxo-complexes.

Example 6.7 Complexation of Pb

Assuming an equilibrium constant for the reaction:

$$Pb^{2+} + H_2O \rightleftharpoons PbOH^+ + H^+$$

of $10^{-7.7}$, calculate the fraction of Pb that will be present as $PbOH^+$ from pH 6 to 9.

Answer: The equilibrium constant expression is:

$$K = \frac{[PbOH^+][H^+]}{[Pb^{2+}]} \qquad (6.49)$$

In addition to the equilibrium constant expression, we also need the conservation equation for Pb:

$$\Sigma Pb = Pb^{2+} + PbOH^+$$

Solving the conservation equation for Pb^{2+} and substitution in to the equilibrium constant expression, we obtain:

$$(\Sigma Pb - [PBOH^+])K = [PbOH^+][H^+]$$

With some rearranging, we eventually obtain the following expression:

$$\frac{[PbOH^+]}{\Sigma Pb} = \frac{K}{K + [H^+]}$$

The result is illustrated in Figure 6.7. Below pH 6, virtually all pH is present as Pb^+; above pH 9, virtually all Pb is present as $PbOH^+$.

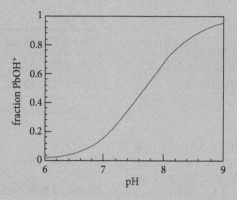

Figure 6.7   Fraction of Pb complexed as $PbOH^+$ as a function of pH.

Polynuclear hydroxo- and oxo-complexes, containing two or more metal ions, are also possible, for example:

$$Mn^{2+} - OH - Mn^{2+}$$

As one might expect, the extent to which such *polymeric* species form depends on the metal ion concentration: they become increasingly common as concentration increases. Most highly charged metal ions (3+ and higher oxidation states) are highly insoluble in aqueous solution. This is due in part to the readiness with which they form hydroxo-complexes, which can in turn be related to the dissociation of surrounding water molecules as a result of their high charge. When such ions are present at high concentration, formation of polymeric species such as those above quickly follows formation of the hydroxo-complex. At sufficient concentration, formation of these polymeric species leads to the formation of colloids and ultimately to precipitation. In this sense, these polymeric species can be viewed as intermediate products of precipitation reactions.

Interestingly enough, however, the tendency of metal ions to hydrolyze decreases with concentration. The reason for this is the effect of the dissociation reaction on pH. For example, increasing the concentration of dissolved copper decreases the pH, which in turn tends to drive the hydrolysis reaction to the left. To understand this, consider the following reaction:

$$Cu^{2+} + H_2O \rightleftharpoons CuOH^+ + H^+$$

for which the apparent equilibrium constant is $K^{app} = 10^{-8}$. We can express the fraction of copper present as $CuOH^+$, α_{CuOH+} as:

$$\alpha = \frac{[CuOH^+]}{Cu_T} = \frac{K}{[H^+] + K} \qquad (6.50)$$

where Cu_T is the total dissolved copper. At constant pH, the fraction of Cu complexed is constant. However, for a solution with a fixed amount of Cu ion dissolved, we can also write a proton balance equation:

$$[H^+] = [CuOH^+] + [OH^-] \qquad (6.51)$$

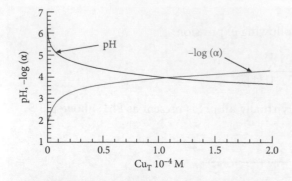

Figure 6.8 pH and $-\log \alpha$, as a function of total copper concentration in aqueous solution. α is the fraction of copper present as the hydroxo-complex.

Table 6.2 Classification of metal ions.

A-type metals
Li^+, Na^+, K^+, Rb^+, Cs^+, Be^{2+}, Mg^{2+}, Ca^{2+}, Sr^{2+}, Ba^{2+}, Al^{3+}, Sc^{3+}, Y^{3+}, REE, Ti^{4+}, Si^{4+}, Zr^{4+}, Hf^{4+}, Th^{4+}, Nb^{5+}, Ta^{5+}, U^{6+}

B-type metals
Cu^{2+}, Ag^+, Au^+, Tl^+, Ga^+, Zn^{2+}, Cd^{2+}, Hg^{2+}, Pb^{2+}, Sn^{2+}, Tl^{3+}, Au^{3+}, In^{3+}, Bi^{3+}

Transition metal ions
V^{2+}, Cr^{2+}, Mn^{2+}, Fe^{2+}, Co^{2+}, Ni^{2+}, Cu^{2+}, Ti^{3+}, V^{3+}, Cr^+, Mn^{3+}, Fe^{3+}, Co^{3+}

From Stumm and Morgan (1996).

and a mass balance equation. Combining these with the equilibrium constant expression, we can calculate both α and pH as a function of Cu_T (problem 11). When we do this, we see that as Cu_T increases, both pH and α decrease, as is demonstrated in Figure 6.8.

6.3.3 Other complexes

When nonmetals are present in solution, as they would inevitably be in natural waters, then other complexes are possible. In this respect, we can divide the elements into four classes (Table 6.2, Figure 6.9). The first is the nonmetals, which form anions or anion groups. The second group is the "A-type" or "hard" metals. These metals, listed

in Table 6.2, have spherically symmetric, inert-gas type outer electron configurations. Their electron shells are not readily deformed by electric fields and can be viewed as *hard spheres*. Metals in this group preferentially form complexes with fluorine and ligands having oxygen as the donor atoms (OH^-, CO_3^{2-}, PO_4^{3-}, SO_4^{2-}). Stability of the complexes formed by these metals increases with charge to radius ratio. Thus, the alkalis form only weak, unstable complexes, while elements such as Zr^{4+} form very strong, stable complexes (e.g., with fluorine). In complexes formed by A-type metals, anions and cations are bound primarily by electrostatic forces, i.e., ionic-type bonds. The A-type elements correspond approximately to the lithophile

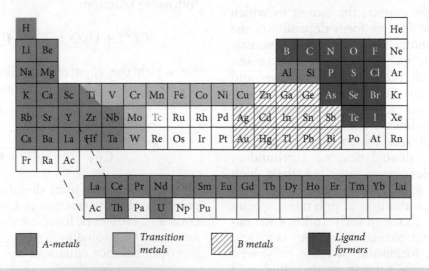

Figure 6.9 Classification of the elements with respect to complex formation in aqueous solution.

elements of Goldschmidt's classification presented in Chapter 7.

The third group is the B-type, or "soft," metal ions. Their electron sheaths are not spherically symmetric and are readily deformed by the electrical fields of other ions (hence the term soft). They preferentially form complexes with bases having S, I, Br, Cl, or N (such as ammonia; not nitrate) as the donor atom. Bonding between the metal and ligand(s) is primarily covalent and is comparatively strong. Thus, Pb forms strong complexes with Cl^- and S^{2-}. Many of the complexes formed by these elements are quite insoluble. The B-type elements consist primarily of the *chalcophile elements,* a term we will define in the next chapter.

The first series transition metals form the fourth group and correspond largely to the siderophile elements (see Chapter 7). Their electron sheaths are not spherically symmetric, but they are not so readily polarizable as the B-type metals. On the whole, however, their complex-forming behavior is similar to that of the B-type metals.

Among the transition metals, the sequence of complex stability is $Mn^{2+} < Fe^{2+} < Co^{2+} < Ni^{2+} < Cu^{2+} > Zn^{2+}$, a sequence known as the *Irving–Williams series.* This is illustrated in Figure 6.10. In that figure, all the sulfate complexes have approximately the same stability, a reflection of the predominantly

electrostatic bonding between sulfate and metal. Pronounced differences are observed for organic ligands. The figure demonstrates an interesting feature of organic ligands: although the absolute value of stability complexes varies from ligand to ligand, the relative affinity of ligands having the same donor atom for these metals is always similar.

Organic molecules can often have more than one functional group and hence can coordinate a metal at several positions, a process called chelation. Such ligands are called *multi-dentate* and organic compounds having these properties are referred to as chelators or chelating agents. We will explore this topic in greater detail in Chapter 12.

The kinetics of complex formation is quite fast in most cases, so that equilibrium can be assumed. There are exceptions, however. As we noted earlier in this section, all complexation reactions are ligand exchange reactions: water playing the role of ligand in "free ions." The rate at which complexes form is thus governed to a fair degree by the rate at which water molecules are replaced in the hydration sphere.

6.3.4   Complexation in fresh waters

Where only one metal is involved, the complexation calculations are straightforward, as exemplified in Example 6.7. Natural waters, however, contain many ions. The most abundant of these are Na^+, K^+, Mg^{2+}, Ca^{2+}, Cl^-, SO_4^{2-}, HCO_3^-, CO_3^{2-} and there are many possible complexes between them as well as with H^+ and OH^-. To calculate the speciation state of such solutions, an iterative approach is required. The calculation would be done as follows. First, we need the concentrations, activity coefficients, and stability constants (or apparent stability constants) for all species. Then we assume all ions are present as free ions and calculate the concentrations of the various possible complexes on this basis. In this pass, we need only consider the major ions (we can easily understand why with an example: formation of $PbCl^+$ when the concentration of Pb is 10^{-8} or less and the abundance of Cl^- is 10^{-4} or more will have an insignificant effect on the free ion Cl concentration). We then iterate the calculation, starting with the free ion concentrations corrected for abundances of complexes we

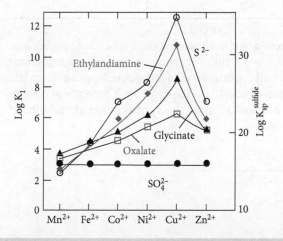

Figure 6.10 Stability constants for transition metal sulfate and organic complexes and their sulfide solubility constants, illustrating the Irving–Williams series. From Stumm and Morgan (1995).

calculated in the previous iteration. This process is repeated until two successive iterations produce the same result. Although it sounds difficult, such calculations typically converge within two to four iterations. Example 6.8 shows how this is done. Once free ion concentrations of the major ligands are known, the speciation of trace metals may be calculated.

As Example 6.8 demonstrates, the major metals in fresh waters are present mainly as free ions (aquo-complexes), as are the three most common anions, chloride, sulfate, and bicarbonate. The alkali and alkaline earth

trace elements are also largely uncomplexed. Co^{2+}, Ni^{2+}, Zn^{2+}, Ag^+, and Cd^{2+} are roughly 50% complexed. The remaining metals are present as primarily as complexes. B, V, Cr, Mo, and Re, as well as As and Se are present as anionic oxo-complexes. Other metals are usually present as hydroxide, carbonate, chloride, or organic complexes. Under reducing conditions, HS^- and S^{2-} complexes are important. In organic-rich waters such as swamps, organic complexes can be predominant. We will discuss organic complexes in more detail in Chapter 12.

Example 6.8 Speciation in fresh water

Using the water and stability constants given in the tables below, calculate the activities of the major species in this water.

Analysis of Stream Water (mM)

Na^+	0.32	Cl^-	0.22
K^+	0.06	SO_4^{2-}	0.12
Mg^{2+}	0.18	ΣCO_2	1.0
Ca^{2+}	0.36	pH	8.0

Log stability constants

	OH^-	HCO_3^-	CO_3^{2-}	SO_4^{2-}	Cl^-
H^+	14	6.35	10.33	1.99	–
Na^+	–	−0.25	1.27	1.06	–
K^+	–	–	–	0.96	–
Mg^{2+}	2.56	1.16	3.4	2.36	–
Ca^{2+}	1.15	1.26	3.2	2.31	–

Answer: The first two problems we need to address are the nature of the carbonate species and activity coefficients. At this pH, we can see from Figure 6.1 that bicarbonate will be the dominant carbonate species. Making the initial assumption that all carbonate is bicarbonate, we can calculate ionic strength and activity coefficients using the Debye–Hückel law (eqn. 3.77). These are shown in the following table. Having the activity coefficients, we can calculate the approximate abundance of the carbonate ion assuming all carbonate is bicarbonate:

$$[CO_3^{2-}] = \frac{\gamma_{HCO_3^-}\Sigma CO_2}{\beta a_{H^+}\gamma_{CO_3^{2-}}} \tag{6.51}$$

where β is the stability constant for the reaction:

$$CO_3^{2-} + H^+ \rightleftharpoons HCO_3^-$$

This equation is readily derived from conservation and the equilibrium constant expression. The bicarbonate ion is then calculated as:

$$[HCO_3^-] = \Sigma CO_2 = [CO_3^{2-}]$$

The result confirms our initial first-order assumption that all carbonate is present as bicarbonate.

Using the concentrations and stability constants given as well as the activity coefficients we calculated, we can then make a first pass at calculating the concentrations of the complexes. For example, $MgCO_3$ is calculated as

$$a_{MgCO_3} = \beta_{MgCO_3} a_{Mg^{2+}} a_{CO_3^{2-}}$$

The results of this first iteration are shown in a matrix. Chlorine does not complex with any of the major ions to any significant degree, so we can neglect it in our calculations.

We then correct the free ion activities by subtracting the activities of the complexes they form. Thus for example, the corrected free ion activity of Mg^{2+} is calculated as:

$$a_{Mg^{2+}}^{corr} = a_{Mg^{2+}}^{ini} - a_{MgOH^-} - a_{MgHCO_3^+} - a_{MgCO_3} - a_{MgSO_4}$$

We then repeat the calculation of the activities of the complexes using these corrected free ion activities. A second matrix shows the results of this second iteration. A third table shows the percent of each ion present as a free ion (aquo-complex). In fresh waters such as this one, most of the metals are present as free ions, the alkaline earths being 5% complexed by sulfate and carbonate.

Ion activities: Iteration 1

	free ion	OH^-	HCO_3^-	CO_3^{2-}	SO_4^{2-}
free ion		1×10^{-06}	9.17×10^{-04}	6.79×10^{-07}	1.62×10^{-04}
H^+	1×10^{-08}	—	2.12×10^{-05}	9.46×10^{-04}	1.75×10^{-10}
Na^+	3.03×10^{-04}	—	1.62×10^{-07}	2.51×10^{-08}	6.27×10^{-07}
K^+	5.69×10^{-05}	—	—	—	9.33×10^{-08}
Mg^{2+}	1.39×10^{-04}	5.09×10^{-08}	2.0310^{-06}	1.65×10^{-06}	6.10×10^{-06}
Ca^{2+}	2.77×10^{-04}	4.29×10^{-09}	5.08×10^{-06}	2.07×10^{-06}	1.08×10^{-05}

Ion activities: Iteration 2

	free ion	OH^-	HCO_3^-	CO_3^{2-}	SO_4^{2-}
free ion		1×10^{-06}	9.12×10^{-04}	1.12×10^{-06}	1.65×10^{-04}
H^+	1×10^{-08}	—	2.06×10^{-05}	9.12×10^{-04}	1.58×10^{-10}
Na^+	3.03×10^{-04}	—	1.57×10^{-07}	2.35×10^{-08}	5.64×10^{-07}
K^+	5.69×10^{-05}	—	—	—	8.40×10^{-08}
Mg^{2+}	1.40×10^{-04}	5.03×10^{-08}	1.84×10^{-06}	1.45×10^{-06}	5.14×10^{-06}
Ca^{2+}	2.80×10^{-04}	3.92×10^{-09}	4.64×10^{-06}	1.83×10^{-06}	9.16×10^{-06}

% free ion

Na^+	99.76%	Cl^-	100%
K^+	99.85%	SO_4^{2-}	91.7%
Mg^{2+}	94.29%	HCO_3^-	99.3%
Ca^{2+}	94.71%	CO_3^{2-}	25.3%

6.4 DISSOLUTION AND PRECIPITATION REACTIONS

6.4.1 Calcium carbonate in groundwaters and surface waters

Calcium carbonate is an extremely common component of sedimentary rocks and is present in weathered igneous and metamorphic rocks. It is also a common constituent of many soils. Water passing through such soils and rocks will precipitate or dissolve calcite until equilibrium is achieved. This process has a strong influence on carbonate concentrations, hardness, and pH as well as dissolved calcium ion concentrations. Let's examine calcite solubility in more detail.

The solubility product of calcite is:

$$K = a_{Ca^+} a_{CO_3^{2-}} \quad (6.52)$$

This can be combined with eqns. 6.18–20 to obtain the calcium concentration of water in equilibrium with calcite as a function of P_{CO_2}:

$$[Ca^+] = P_{CO_2} \frac{K_1 K_{sp-cal} K_{sp-CO_2}}{K_2 \gamma_{Ca^{2+}} \gamma^2_{HCO_3^-} [HCO_3^-]^2} \quad (6.53)$$

In a solution in equilibrium with calcite and a CO_2 gas phase and containing no other dissolved species, it is easy to modify equation 6.53 so that the calcium ion concentration is a function of P_{CO2} only. A glance at Figure 6.1 shows that we can neglect OH^-, H^+, and CO_3^{2-} if the final pH is less than about 9. The charge balance equation in this case reduces to:

$$2[Ca^{2+}] = [HCO_3^-] \quad (6.54)$$

Substituting this into 6.53, we obtain:

$$[Ca^+] = P_{CO_2} \frac{K_1 K_{sp-cal} K_{sp-CO_2}}{K_2 \gamma_{Ca^{2+}} \gamma^2_{HCO_3^-} [Ca^{2+}]^2} \quad (6.55)$$

or

$$[Ca^+] = \left\{ P_{CO_2} \frac{K_1 K_{sp-cal} K_{sp-CO_2}}{K_2 \gamma_{Ca^{2+}} \gamma^2_{HCO_3^-}} \right\}^{1/3} \quad (6.56)$$

There are two interesting aspects to this equation. First, the calcium ion concentration, and therefore calcite solubility, increases with increasing P_{CO_2}. This might seem counterintuitive at first, as one might think that increasing P_{CO_2} should produce an increase in carbonate ion concentration and therefore calcite solubility, and therefore drive the reaction toward precipitation. However, increasing P_{CO_2} decreases pH, which decreases CO_3^{2-} concentration, and therefore drives the reaction toward dissolution. Second, calcium ion concentration varies with the one-third power of P_{CO_2} (Figure 6.11). Because of this nonlinearity, mixing of two solutions, both of which are saturated in Ca^{2+} with respect to calcite, can result in the mixture being undersaturated with respect to Ca^{2+}. For example, consider the mixing of stream and groundwater. Stream water is in equilibrium with the atmosphere for which P_{CO_2} is $10^{-3.5}$. On the other hand, P_{CO_2} in soils is often as high as 10^{-2}. So mixing between calcite-saturated groundwater and calcite-saturated surface water would produce a solution that is undersaturated with calcite.

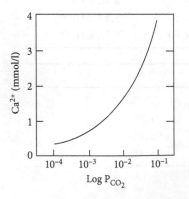

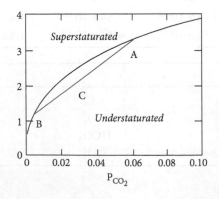

Figure 6.11 Concentration of calcium ion in equilibrium with calcite at 25°C and 1 atm as a function of P_{CO_2}. From Drever (1988).

Equation 6.56 describes calcite solubility for a system open to exchange with gaseous CO_2. For a P_{CO_2} of $10^{-3.5}$ (i.e., the atmosphere), this equation yields a calcium concentration of 1.39 mM. Water in pores and fractures in rocks does not exchange with a gas phase. Example 6.9 illustrates the calculation of calcite solubility in a closed system and shows that under those circumstances, less calcite will dissolve; in the case of $P_{CO_2} = 10^{-2}$, calcite saturation is reached at only 0.33 mM, or about a fourth as much. The difference is illustrated in Figure 6.12, which is a plot of log $[HCO_3^-]$ vs. pH. Systems in equilibrium with constant P_{CO_2} (open systems) evolve along straight lines on this plot and ultimately reach calcite saturation at higher pH and lower $[HCO_3^-]$ (and $[Ca^{2+}]$) than closed systems that initially equilibrate with the same P_{CO_2}.

6.4.2 Solubility of Mg

There are a number of compounds that can precipitate from Mg-bearing aqueous solutions, including brucite $(Mg(OH)_2)$, magnesite $(MgCO_3)$, and dolomite $(CaMg(CO_3)_2)$, as well as hydrated carbonates such as hydromagnesite $(MgCO_3(OH)_2 \cdot 3H_2O)$. The stability of these compounds may be described by the following reactions:

$$Mg(OH)_2 \rightleftharpoons Mg^{2+} + 2OH^- \qquad K = 10^{-11.6} \tag{6.57}$$

$$MgCO_3 \rightleftharpoons Mg^{2+} + CO_3^{2-} \qquad K_{mag} = 10^{-7.5} \tag{6.58}$$

$$CaMg(CO_3)_2 \rightleftharpoons Mg^{2+} + Ca^{2+} + 2CO_3^{2-} \qquad K_{dol} = 10^{-17} \tag{6.59}$$

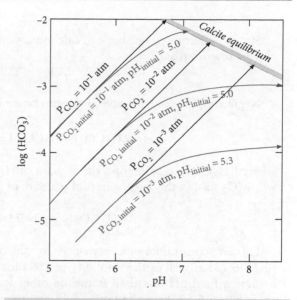

Figure 6.12 Comparison of the evolution of systems with constant P_{CO_2} (open systems) with those closed to gas exchange. After Stumm and Morgan (1995) and Deines et al. (1974).

(The solubility of dolomite is poorly known; values for this equilibrium constant vary between $10^{-16.5}$ and 10^{-20}.)

We can use these reactions and their equilibrium constants, together with the reactions for the carbonate system (eqn. 6.15–6.17) to construct predominance diagrams for Mg-bearing solutions in equilibrium with these phases. For example, for reaction 6.57, we may derive the following relationship (assuming the solid is pure):

$$\log a_{Mg2+} = -pK_{bru} + 2\,pK_W - 2\,pH$$
$$= 16.4 - 2\,pH \tag{6.60}$$

where we use the notation pK = −log (K).

Example 6.9 Calcite solubility in a closed system

Suppose groundwater initially equilibrates with a P_{CO_2} of 10^{-2} and thereafter is closed to gas exchange, so that there is a fixed ΣCO_2 initial. The water then equilibrates with calcite until saturation is reached. What will be the final concentration of calcium in the water? Assume ideal behavior and an initial calcium concentration of 0.

Answer: Since the system is closed, a conservation equation is a good place to start. We can write the following conservation equation for total carbonate:

$$\Sigma CO_2 = \Sigma CO_{2initial} + \Sigma CO_2 \text{ from calcite}$$

Since dissolution of one mole of calcite adds one mole of ΣCO_2 for each mole of Ca^{2+}, this equation may be rewritten as:

$$\Sigma CO_2 = \Sigma CO_{2\text{initial}} + [Ca^{2+}]$$

Neglecting the contribution of the carbonate ion to total carbonate, this equation becomes:

$$[H_2CO_3] + [HCO_3^-] = ([H_2CO_3]_{\text{initial}}) + [Ca^{2+}] \tag{6.61}$$

where $([H_2CO_3]_{\text{initial}})$ denotes the amount of H_2CO_3 calculated from equation 6.21 for equilibrium with CO_2 gas; in this case, a partial pressure of 10^{-2}. This can be rearranged to obtain:

$$([H_2CO_3]_{\text{initial}}) = [H_2CO_3] + [HCO_3^-] - [Ca^{2+}] \tag{6.62}$$

Further constraints are provided by the three carbonate equilibrium product expressions (6.21–6.23) as well as the solubility product for calcite (6.50), and the charge balance equation. We assume a final pH less than 9 and no other ions present, so the charge balance equation reduces to equation 6.54. From equation 6.18 and the value of K_{CO_2} in Table 6.1, $[H_2CO_3]_{\text{initial}} = 10^{-2} \times 10^{-1.47}$ M. Dividing equation 6.19 by 6.20 yields:

$$\frac{K_1}{K_2} = \frac{[HCO_3^-]^2}{[H_2CO_3][CO_3^{2-}]}$$

Then substituting equations 6.52, 6.54, and 6.62 gives:

$$\frac{K_1}{K_2} = \frac{4[Ca^{2+}]^3}{K_{sp-cal}\{[H_2CO_3]_{\text{initial}} - [Ca^{2+}]\}}$$

Into this equation we substitute $P_{CO_2}K_{CO_2} = [H_2CO_3]$ and rearrange to obtain:

$$[Ca^{2+}]^3 + \frac{K_1K_{sp-cal}}{4K_2}[Ca^{2+}] - \frac{K_1K_{sp-cal}K_{sp-CO_2}}{4K_2}P_{CO_2-\text{initial}} = 0 \tag{6.63}$$

This is a cubic equation that is readily solved for $[Ca^{2+}]$. For an initial P_{CO_2} of 10^{-2}, we calculate a calcium concentration of 0.334 mM.

For reaction 6.58, the equilibrium constant relationship may be written as:

$$\log \frac{a_{Mg_{aq}^{2+}}}{a_{MgCO_3}} = -\log a_{CO_3^{2-}} - pK_{mag} \tag{6.64}$$

However, the carbonate concentration will depend on both total carbonate (or the partial pressure of CO_2) and pH. To simplify things, let's specify that the solid is pure, the solution ideal, and $\Sigma CO_2 = 10^{-2.5}$ M. Then we can think of three limiting cases: where carbonic acid, bicarbonate ion, and carbonate ion predominate. In the latter case, $[CO_3^{2-}] \approx$ $\Sigma CO_2 = 10^{-2.5}$ M, so we have:

$$\log[Mg^{2+}] = 2.5 - pK = -5.0 \tag{6.64a}$$

When bicarbonate ion predominates, $[HCO_3^-] \approx \Sigma CO_2 = 10^{-2.5}$ M, and the carbonate ion concentration is:

$$\log[CO_3^{2-}] = pK_2 + \log[HCO_3^-] + pH$$
$$= 12.88 + pH$$

Substituting this into eqn. 6.64 we have:

$$\log[Mg^{2+}] = 5.33 - pH \tag{6.64b}$$

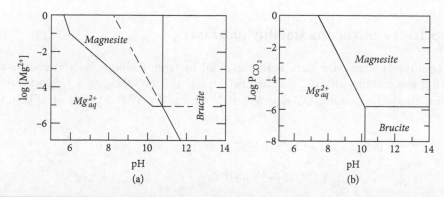

Figure 6.13 Predominance diagrams for Mg-bearing phases in equilibrium with aqueous solution. Total CO_2 is fixed at $10^{-2.5}$ M in (a). The concentration of Mg^{2+} is fixed at 10^{-4} M in (b). After Stumm and Morgan (1995).

Finally, when carbonic acid predominates, $[H_2CO_3] \approx \Sigma CO_2 = 10^{-2.5}$ M. The carbonate ion concentration as a function of $[H_2CO_3]$ is given by eqn. 6.39b. Taking the log and substituting into eqn. 6.64, we have:

$$\log [Mg^{2+}]$$

$$= pK_{mag} + pK_1 + pK_2 - \log [H_2CO_3] - 2pH$$

$$= 26.68 - 2pH \qquad (6.64c)$$

We can use these equations to construct *stability, or predominance diagrams* in a manner similar to that used to construct $p\varepsilon$–pH predominance diagrams (Chapter 3). Equations 6.63–6.64c express the Mg ion concentration as a function of pH and hence represent lines on a plot of $\log [Mg^{2+}]$ vs. pH. The lines divide the diagram (Figure 6.13) into three regions, where (1) only an Mg-bearing aqueous solution is stable; (2) magnesite is stable; and (3) brucite is stable. For example, on a plot of $\log [Mg^{2+}]$ vs. pH, the predominance boundary between Mg^{2+} and brucite plots as a line with a slope of −2 and an intercept of +16.4. Figure 6.13b shows a predominance diagram for this system, but where the Mg^{2+} concentration is fixed and P_{CO_2} and *pH* are the variables.

Virtually all natural solutions will contain dissolved calcium as well as magnesium. This being the case, we must also consider the stability of dolomite. We can construct similar predominance diagrams for these systems, but we must add an additional variable, namely the Ca^{2+} concentration. To describe the relative stability of dolomite and calcite, it is more convenient to express the solubility of dolomite as:

$$CaMg(CO_3)_2 + Ca^{2+} \rightleftharpoons Mg^{2+} + 2CaCO_3$$

because the reaction contains calcite as well as dolomite. Since this reaction can be constructed by subtracting two times the calcite dissolution (eqn. 6.27) from the dolomite dissolution (eqn. 6.59), the equilibrium constant for this reaction can be calculated from:

$$K = \frac{K_{dol}}{K_{cal}^2}$$

Figure 6.14 illustrates the stability of magnesite, dolomite, brucite, and calcite as a function of P_{CO_2} and the Ca^{2+}/Mg^{2+} concentration ratio. Whether any of these phases are stable relative to a Mg^{2+}-bearing solution depends on the Mg^{2+} concentration, which is not specified in the graph.

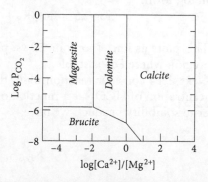

Figure 6.14 Stability of magnesite, dolomite, calcite, and brucite in equilibrium with a Mg- and Ca-bearing aqueous solution.

Example 6.10   Constructing stability diagrams

Using the equilibrium constant data below as well as from Table 6.1, construct a stability diagram showing the stability of FeS, siderite ($FeCO_{3(S)}$), and $Fe(OH)_2$ as a function of total sulfide concentration and pH assuming $\Sigma CO_2 = 5 \times 10^{-2}$ M, $Fe^{2+} = 10^{-6}$ M, and ideal behavior. Neglect any S^{2-}.

$$FeS_{(S)} + H^+ \rightleftharpoons Fe^{2+} + HS^- \qquad\qquad K_{FeS} = 10^{-4.2} \qquad\qquad (6.65)$$

$$FeCO_{3(S)} + H^+ \rightleftharpoons Fe^{2+} + HCO_3^- \qquad\qquad K_{FeCO_3} = 10^{-0.1} \qquad\qquad (6.66)$$

$$Fe(OH)_{2(S)} + 2H^+ \rightleftharpoons Fe^{2+} + 2H_2O \qquad\qquad K_{Fe(OH)_2} = 10^{12.9} \qquad\qquad (6.67)$$

$$H_2S_{(aq)} \rightleftharpoons H^+ + HS^- \qquad\qquad K_S = 10^{-7} \qquad\qquad (6.68)$$

Answer: Our first step is to set up an equation that describes the concentration of HS^- as a function of pH. From the conservation of sulfur, we have:

$$\Sigma S = [H_2S] + [HS^-]$$

From the equilibrium constant expression for the dissociation of H_2S, we may substitute:

$$[H_2S] = \frac{[H^+][HS^-]}{K_S}$$

and obtain:

$$\Sigma S = \frac{[H^+][HS^-]}{K_S} + [HS^-] \qquad\qquad (6.69)$$

Solving for $[HS^-]$, we have:

$$[HS^-] = \Sigma S \frac{K_S}{K_S + [H^+]}$$

We substitute this into the FeS (pyrrhotite) solubility product, and solving for ΣS we have:

$$\Sigma S = \frac{K_{FeS}[H^+](K_S + [H^+])}{[Fe^{2+}]K_S}$$

or in log form:

$$\log(\Sigma S) = \log(K_S + 10^{-pH}) - pH - \log([Fe^{2+}]) - pK_{FeS} + pK_S \qquad\qquad (6.70)$$

This plots as line ① on our ΣS vs. pH stability diagram (Figure 6.15). The area above the line is the region where FeS is stable.

Next, let's consider reaction 6.66, the solubility of siderite. We need an equation describing the concentration of HCO_3^- as a function of pH and ΣCO_2, which is eqn. 6.29. Substituting this into the siderite solubility product, we have:

$$K_{FeCO_3} = \frac{[Fe^{2+}]\Sigma CO_2}{\left\{ \dfrac{[H^+]}{K_1} + 1 + \dfrac{K_2}{[H^+]} \right\}[H^+]}$$

$$pH + \log[Fe^{2+}] + \log(\Sigma CO_2) - \log\left\{ \frac{10^{-pH}}{K_1} + 1 + \frac{K_2}{10^{-pH}} \right\} + pK_{FeCO_3} = 0 \qquad\qquad (6.71)$$

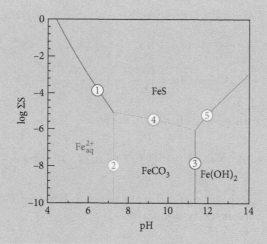

Figure 6.15   Stability diagram showing the stable solid Fe-bearing phases in equilibrium with a solution containing 10^{-6} M Fe^{2+} and 5×10^{-3} M ΣCO_2.

An approximate solution may be found by assuming $HCO_3^- = \Sigma CO_2$, which yields pH = 7.20. An exact solution requires an indirect method. Using the Solver in Microsoft Excel™, we find pH = 7.25, very close to our approximate solution (Solver uses a succession of "intelligent" guesses to find solutions to equations, such as 6.71, that have no direction solution). Thus siderite will precipitate when the pH is greater than 7.25 and, not surprisingly, this is independent of ΣS. The boundary between the Fe^{2+} and $FeCO_3$ field is then a vertical line (line ②) at pH = 7.25.

Now let's consider the solubility of ferrous iron hydroxide. The condition for precipitation of $Fe(OH)_2$ is described by the equation:

$$pH = \frac{pK_{Fe(OH)_2} - \log[Fe^{2+}]}{2} = 9.45 \tag{6.72}$$

Thus, $FeOH_2$ will not precipitate until pH reaches 9.45, i.e., above the point where $FeCO_3$ precipitates, so there is no boundary between a $Fe(OH)_2$ phase and a Fe^{2+}-bearing solution.

Next, we need equations that describe the reactions between solid phases. A reaction between $FeCO_3$ and $Fe(OH)_2$ can be obtained by subtracting reaction 6.67 from reaction 6.66. The corresponding equilibrium constant is obtained by dividing 6.66 by 6.67:

$$FeCO_3 + 2H_2O \rightleftharpoons (FeOH)_2 + H^+ + HCO_3^- \qquad K_{FeCO_3}/K_{Fe(OH)_2} = 10^{-13.0}$$

From this we derive:

$$pH = \log[HCO_3^-] + 13.0 \tag{6.73}$$

We can obtain an approximate solution, by simply assuming all carbonate is bicarbonate, in which case, we obtain pH = 11.69. Or we can substitute eqn. 6.29 for HCO_3^- and use the Solver in Excel™, which yields an exact solution of pH = 11.02. Not surprisingly, our approximate solution is less accurate than in the previous case because at this high pH, the carbonate ion makes up a significant fraction of the total carbonate (Figure 6.1). The boundary between the $FeCO_3$ and $Fe(OH)_2$ fields is thus a vertical line at pH = 11.02 (line ③).

The boundary between FeS and $FeCO_3$ is found by subtracting reaction 6.66 from 6.65 and dividing the corresponding equilibrium constants:

$$FeS + HCO_3 \rightleftharpoons FeCO_3 + HS^- \qquad K_{FeS}/K_{FeCO_3} = 10^{-4.1} \tag{6.74}$$

Substituting 6.68 and 6.29 into the corresponding equilibrium constant expression, we have:

$$-pK_{pry-sid} = \log[HS^-] - \log[HCO_3^-] = \log \Sigma S - pK_s - \log(K_s + [H^+]) -$$

$$\log \Sigma CO_2 + \log \left\{ \frac{[H^+]}{K_1} + 1 + \frac{K_2}{[H^+]} \right\}$$

Solving for ΣS and substituting 10^{-pH} for $[H^+]$, we have:

$$\log \Sigma S = 1 - pK_{pyr-sid} + pK_s + \log(K_s + 10^{-pH}) + \log \Sigma CO_2 - \log \left\{ \frac{10^{-pH}}{K_1} + 1 + \frac{K_2}{10^{-pH}} \right\}$$

This plots as line ④ on our diagram. Finally, the reaction between FeS and $Fe(OH)_2$ is obtained by subtracting reaction 6.67 from reaction 6.65:

$$FeS + 2H_2O \rightleftharpoons Fe(OH)_2 + H^+ + HS^- \qquad K_{FeS}/K_{Fe(OH)2} = 10^{-17.1}$$

From the equilibrium constant equation, we obtain an expression for ΣS as a function of pH:

$$\log \Sigma S = pH - pK_{FeS-Fe(OH)_2} + pK_S + \log(K_S + 10^{-pH})$$

This plots as line ⑤ on our diagram, which is now complete.

6.4.3 Solubility of SiO$_2$

Silicon is the most common element on the Earth's surface after oxygen. Its concentration in solution plays an important role in determining how weathering will proceed.

The dissolution of silica may be represented by the reaction:

$$SiO_{2(qtz)} + 2H_2O \rightleftharpoons H_4SiO_{4(aq)} \qquad (6.75)$$

The equilibrium constant expression is simply:

$$K_{qz} = a_{H_4SiO_{4(aq)}} = 10^{-4} \qquad (6.76)$$

at 25°C. This is to say, water is saturated with respect to quartz when the concentration of H_4SiO_4 is 10^{-4} moles per kilogram, or about 7.8 ppm by weight SiO_2.

However, there are some complicating factors. First, precipitation of quartz seems to be strongly kinetically inhibited. Equilibrium is more likely to be achieved with amorphous silica, the equilibrium constant for which is 2×10^{-3} (~115 ppm). Second, H_4SiO_4 is a weak acid and undergoes successive dissociation with increasing pH:

$$H_4SiO_4 \rightleftharpoons H_3SiO_4^- + H^+$$

$$K_1 = \frac{a_{H_4SiO_4^-} a_{H^+}}{a_{H_4SiO_4}}$$

$$H_3SiO_4^- \rightleftharpoons H_2SiO_4^{2-} + H^+$$

$$K_2 = \frac{a_{H_3SiO_4^{2-}} a_{H^+}}{a_{H_4SiO_4^-}}$$

the total dissolved silica concentration will be the sum of H_4SiO_4, $H_3SiO_3^-$, and $H_2SiO_3^{2-}$. Assuming activity coefficients of unity, the concentration of dissolved silica is then:

$$[SiO_T] = [H_4SiO_4] \left\{ 1 + \frac{K_1}{a_{H^+}} + \frac{K_1 K_2}{a_{H^+}^2} \right\}$$

$$(6.77)$$

From eqn. 6.77, we would expect silica solubility to be pH-dependent. This dependence is illustrated in Figure 6.16.

We could have defined the second dissociation reaction as:

$$H_4SiO_4 \rightleftharpoons H_2SiO_4^{2-} + 2H^+$$

in which case, the equilibrium constant would be:

$$K_2^* = \frac{a_{H_2SiO_4^{2-}} a_{H^+}^2}{a_{H_4SiO_4}} = 10^{-9.9} \times 10^{-11.7} = 10^{-21.6}$$

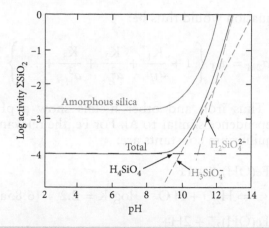

Figure 6.16 Log activity of dissolved silica in equilibrium with quartz and amorphous silica (dashed line) as a function of pH. After Drever (1988).

and eqn. 6.77 would have been:

$$[SiO_T] = [H_4SiO_4] \left\{ 1 + \frac{K_1}{a_{H^+}} + \frac{K_2^*}{a_{H^+}^2} \right\}$$
(6.77a)

The concentration of SiO_{2T} we calculate in this way would, of course, be identical. The point is, reactions and their corresponding equilibrium constants can be expressed in various ways, and we need to be alert to this.

6.4.4 Solubility of Al(OH)₃ and other hydroxides

The hydroxide is the least soluble salt of many metals. Therefore, it is the solubility of their hydroxides that controls the solubility of these

metals in natural waters. These are shown in Figure 6.17. Since these dissolution reactions involve OH⁻, they are pH-dependent, and the slope of the solubility curve depends on the valence of the metal (e.g., −3 for Fe^{3+}, −2 for Fe^{2+}, −1 for Ag^+). Let's consider in more detail the solubility of gibbsite, the hydroxide of aluminum.

Dissolution of gibbsite (Al(OH)₃) can be described by the reaction:

$$Al(OH)_3 + 3H^+ \rightleftharpoons Al^{3+} + 3H_2O$$

$$K_{gib} = \frac{a_{Al^{3+}}}{a_{H^+}^3}$$
(6.78)

However, a complication arises from hydrolyzation of the aluminum, which occurs in solutions that are not highly acidic (hydrolyzation is typical of many of the highly charged, ≥3, metal ions):

$$Al^{3+} + H_2O \rightleftharpoons Al(OH)^{2-} + H^+$$

$$K_1 = \frac{a_{Al(OH)^{2+}}a_{H^+}}{a_{Al^{3+}}} = 10^{-5}$$
(6.79)

$$Al^{3+} + 2H_2O \rightleftharpoons Al(OH)_2^- + 2H^+$$

$$K_2 = \frac{a_{Al(OH)_2^+}a_{H^+}^2}{a_{Al(OH)^{2+}}} = 10^{-9.3}$$
(6.80)

$$Al^{3+} + 3H_2O \rightleftharpoons Al(OH)_3^0 + 3H^+$$

$$K_3 = \frac{a_{Al(OH)_3^0}a_{H^+}^3}{a_{Al(OH)_2^-}} = 10^{-16}$$
(6.81)

$$Al^{3+} + 4H_2O \rightleftharpoons Al(OH)_4^+ + 4H^+$$

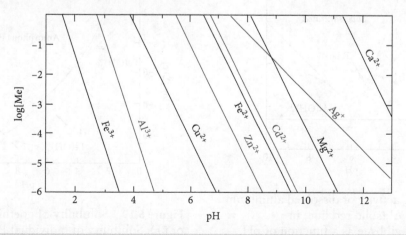

Figure 6.17 Solubility of metal hydroxides as a function of pH. After Stumm and Morgan (1995).

$$K_4 = \frac{a_{Al(OH)_4^-} a_{H^+}^4}{a_{Al(OH)_3^0}} = 10^{-22} \quad (6.82)$$

The total dissolved aluminum activity is given by:

$$a_{\Sigma Al} = a_{Al^{3+}} \left\{ 1 + \frac{K_1}{a_{H^+}} + \frac{K_2}{a_{H^+}^2} + \frac{K_3}{a_{H^+}^3} + \frac{K_4}{a_{H^+}^4} \right\}$$
$$(6.83)$$

Figure 6.18 shows the activities of the various aluminum species and total aluminum as a function of pH. The solubility of Al is low except at low and high pH, and as pH increases Al^{3+} becomes increasingly hydrolyzed. Also note that where positively charged species dominate (e.g., Al^{3+}), solubility increases with decreasing pH; where negatively charged species dominate, solubility increases with increasing pH. Minimum solubility occurs in the range of pH of natural waters (which is just as well because it has some toxicity).

Equation 6.83 can be generalized to other metals that undergo hydrolyzation reactions, *when the reactions are expressed in the same form as those above.* A general form of this

equation would thus be:

$$a_{\Sigma M} = a_M \left\{ 1 + \frac{K_1}{a_{H^+}} + \frac{K_2}{a_{H^+}^2} + \frac{K_3}{a_{H^+}^3} + \cdots \right\}$$

Thus iron and other metals show a pH dependence similar to Al. For Fe, the relevant equilibrium constants are:

$Fe(OH)^{2+} + H^+$

$\rightleftharpoons H_2O + HO \quad \log K = 2.2 \quad (6.85a)$

$Fe(OH)_2^+ + 2H+$

$\rightleftharpoons H_2O + 2H_2O \quad \log K = 5.7 \quad (6.85b)$

$Fe(OH)_3^- + 4H+$

$\rightleftharpoons H_2O + 4H_2O \quad \log K = 21.6 \quad (6.85c)$

$Fe(OH)_4 + 3H+$

$\rightleftharpoons H_2O + +3H_2O \quad \log K = 3.2 \quad (6.85d)$

Using these equilibrium constants, the solubility of amorphous goethite as a function of pH is readily calculated and is shown in Figure 6.19.

6.4.5 Dissolution of silicates and related minerals

The concentrations of Al and Si will usually not be controlled by equilibrium with quartz and gibbsite, but by equilibrium with other silicates. An example of this is shown in

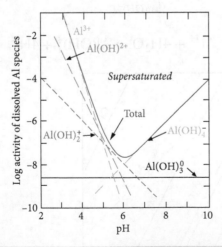

Figure 6.18   Log activity of dissolved aluminum species and total Al (solid red line) in equilibrium with gibbsite as a function of pH. After Drever (1988).

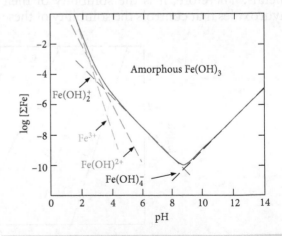

Figure 6.19   Solubility of goethite as a function of pH. Solubility of individual Fe-hydroxide species shown as dashed lines.

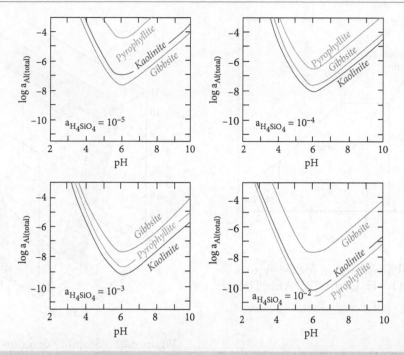

Figure 6.20 Total dissolved Al activity in equilibrium with gibbsite, pyrophyllite, and kaolinite as a function of pH at different dissolved silica activities. After Drever (1988).

Figure 6.20, which shows the concentration of dissolved Al in equilibrium with gibbsite, kaolinite, and pyrophyllite at four different activities of dissolved silica. Only at the lowest dissolved silica concentrations will gibbsite precipitate before one of the aluminosilicates.

For the most part, silicates do not dissolve in the conventional sense; rather, they react with water to release some ions to solution and form new minerals in place of the original ones. This phenomenon is known as *incongruent solution*. In considering such reactions, we can usually assume that all Al remains in the solid phase. If we consider only Al and Si, a simple reaction might be the breakdown of kaolinite to form gibbsite plus dissolved silica:

$$\tfrac{1}{2}Al_2Si_2O_5(OH)_{4(s)} + {}^5/_2H_2O$$
$$\rightleftharpoons Al(OH)_{3(s)} + H_4SiO_4 \quad (6.86)$$

Assuming the solid phases are pure, the equilibrium constant for this reaction is simply:

$$K_{kaol} = a_{H_4SiO_4} = 10^{-4.4} \quad (6.87)$$

which tells us that at H_2SiO_4 activities greater than $10^{-4.4}$, kaolinite is more stable than gibbsite. Similar reactions can, of course, be written between other phases, such as kaolinite and pyrophyllite. Introducing other ions into the system allows the possibility of other phases and other reactions, for example:

$$KAlSi_3O_8 + H^+ + 7H_2O$$
$$\rightleftharpoons Al(OH)_3 + K^+ + 3H_4SiO_4$$

$$K = \frac{a_{K^+}a^3_{H_4SiO_4}}{a_{H^+}} \quad (6.88)$$

$$2KAl_3Si_3O_{10}(OH)_2 + H^+ + {}^3/_2H_2O$$
$$\rightleftharpoons {}^3/_2Al_2Si_2O_5(OH)_4 + K^+$$

$$K_{mus-kao} = \frac{a_{K^+}}{a_{H^+}} \quad (6.89)$$

From a series of such reactions and their corresponding equilibrium constant expressions, we can construct stability diagrams such as the one in Figure 6.21. The procedure for constructing this and similar diagrams is essentially similar to that used in construction of $p\varepsilon$–pH diagrams and is illustrated in Example 6.10. In the case of Figure 6.21, we seek an equilibrium constant expression containing the a_{K^+}/a_{H^+} ratio and the activity

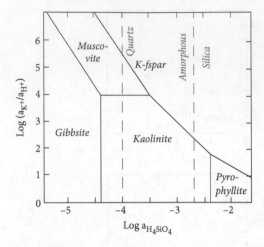

Figure 6.21 Stability diagram for the system $K_2O-Al_2O_3-SiO_2-H_2O$ at 25°C. After Drever (1988).

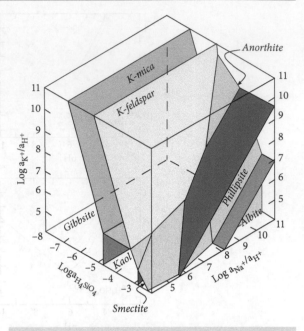

Figure 6.22 Stability diagram for the system $K_2O-Na_2O-CaO-Al_2O_3-SiO_2-H_2O$ at 25°C. After Garrels and Christ (1965).

of H_4SiO_4. From this expression we determine the slope and intercept, which allows us to plot the predominance boundary. For example, the boundary between the kaolinite and muscovite fields is given by eqn. 6.89. The equilibrium constant for the reaction is 10^4, so the boundary is a line with slope 0 at log $a_{K+}/a_{H+} = 4$. The boundary between gibbsite and kaolinite is reaction 6.86, and eqn. 6.87, written in log form, defines the line dividing the two regions. This boundary thus plots as a vertical line at log $a_{H_4SiO_4} = -4.4$. The kaolinite – K-feldspar boundary is the reaction:

$$\tfrac{1}{2}Al_2Si_2O_5(OH)_4 + K^+ + 2H_4SiO_4$$

$$\rightleftharpoons KAlSi_3O_8 + H^+ + \tfrac{9}{2}H_2O$$

$$K_{kao-kfs} = \frac{a_{H+}}{a_{K+}a_{H_2SiO_4}^2}$$

$$\log \frac{a_{K+}}{a_{H+}} = pK - 2\log a_{H_4SiO_4}$$

This boundary thus plots as a vertical line with a slope of −2 and an intercept equal to the negative of the log of the equilibrium constant. Boundaries for the remaining fields can be derived similarly.

The fields in Figure 6.21 show the phase that is the most stable of those we considered in constructing the diagram. (Strictly speaking, it does not tell us whether the phase can

be expected to precipitate, as this depends on the Al activity; however, because of the low solubility of Al, an aluminum-bearing phase can be expected to be stable in most instances.) By considering sodium and calcium as well as potassium, we can construct a three-dimensional stability diagram, such as that in Figure 6.22, in a similar manner.

Because many low-temperature reactions involving silicates are so sluggish, equilibrium constants are generally calculated from thermodynamic data rather than measured.

6.5 CLAYS AND THEIR PROPERTIES

Clays are ubiquitous on the surface of the Earth. Unfortunately, the term *clay* has two meanings in geology: in a mineralogical sense it refers to a group of sheet silicates, but in another sense, it refers to the finest fraction of a sediment or soil. In addition to clay minerals, fine-grained oxides and hydroxides are present in the *clay fraction*. Clays, in both senses, exert important controls on the composition of aqueous fluids, both because of chemical reactions that form them and because of their sorptive and ion exchange capacities. Generally, only clays in the mineralogical sense have true ion-exchange capacity, where ions in the clay can be exchanged for ions in the surrounding solution, but oxides and hydroxides can adsorb and desorb ions on their surfaces. We will first consider the mineralogy of the true clays, then consider their interaction with solution.

6.5.1 Clay mineralogy

Clay minerals (*sensu stricto*) are sheet silicates. We can think of each sheet in a clay mineral as consisting of layers of silica tetrahedra bound to a hydroxide layer in which the cation (most commonly Al, Mg, or Fe) is in octahedral coordination much as in pure hydroxide minerals such as gibbsite $(Al(OH)_3)$ or brucite $(Mg(OH)_2)$. Brucite and gibbsite are structurally similar to each other except that in gibbsite every third octahedral site is left empty to maintain charge balance. This is illustrated in Figure 6.23. Because only two out of three octahedral sites are occupied, gibbsite is said to have a *dioctahedral* structure, while brucite is said to be

trioctahedral. This terminology also applies in clay minerals in exactly the same sense.

6.5.1.1 Kaolinite group (1:1 clays)

The simplest clays consist of a tetrahedral silicate layer and an octahedral hydroxide layer, hence the term 1:1 clays. Kaolinite, $Al_4Si_4O_{10}(OH)_8$, is a good example. The structure of kaolinite is shown in Figure 6.24. Its unit cell consists of a layer of silica tetrahedra bound to an octahedral alumina layer whose structure is very similar to that of gibbsite, except that some hydroxyls are replaced by oxygens. Note that individual sheets are not bound together; they are held together only by van der Waals interactions, which are quite weak. The structure of serpentine, $Mg_6Si_4O_{10}(OH)_8$, is similar to that of kaolinite, with Mg replacing Al, and every octahedral site is occupied. In both minerals, a mismatch in the spacing of octahedra and tetrahedra results in a curvature of the lattice. Also, successive layers of kaolinite are generally stacked in a random manner.

6.5.1.2 Pyrophyllite group (2:1 clays)

This is a large group of clay minerals consisting of a *hydroxide layer* sandwiched between two layers of silica tetrahedra, hence the

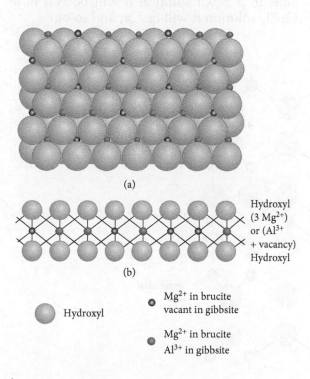

(a)

(b)

Hydroxyl
$(3\ Mg^{2+})$
or $(Al^{3+}$
+ vacancy)
Hydroxyl

- Hydroxyl
- Mg^{2+} in brucite vacant in gibbsite
- Mg^{2+} in brucite Al^{3+} in gibbsite

Figure 6.23 Structure of gibbsite and brucite. (a) Plan (vertical) view. (b) Expanded cross-sectional view.

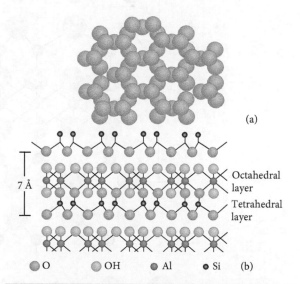

(a)

7 Å

Octahedral layer

Tetrahedral layer

● O ◐ OH ● Al ● Si   (b)

Figure 6.24 Structure of kaolinite. (a) Plan view of the tetrahedral layer. (b) Cross-sectional view.

term 2:1 clays. More of the hydroxyls in the hydroxide layer are replaced by oxygen than in 1:1 clays. The simplest two such clays are pyrophyllite, $Al_2Si_4O_{10}(OH)_2$, and talc, $Mg_3Si_4O_{10}(OH)_2$. In pyrophyllite, every third octahedral site is vacant, so that it, like kaolinite, is said to be dioctahedral, while serpentine and talc are trioctahedral. The structure of pyrophyllite is shown in Figure 6.25. Other, more complex clays, including those of the smectite group, the biotite group, the vermiculite group, saponite, and muscovite, have structures similar to those of pyrophyllite and talc. These are derived in the following ways:

1. Substitution of Al^{+3} for Si^{+4} in the tetrahedral sites, resulting in a charge deficiency that is balanced by the presence of a cation between the layers (interlayer positions). When the number of interlayer cations is small, they are generally exchangeable; when the number is large, K is typically the cation and it is not very exchangeable.

2. Substitution of Mg^{2+}, Fe^{2+}, Fe^{3+} or a similar cation for Al^{3+} or substitution of Fe^{2+}, Fe^{3+}, or Al^{3+} for Mg^{2+} in the octahedral layer. Where charge deficiency results, it is balanced by the presence of exchangeable cations in the interlayer sites, or by vacancies in the octahedral sites.

Smectites are distinguished by their expansion to a unit cell thickness of 14 Å upon treatment with ethylene glycol. This expansion results from entry of the ethylene glycol molecule into the interlayer position of the clays. Smectites generally also have water present in the interlayer space, the amount of water being determined by the cation present. Generally, there is little water present when the interlayer cation is divalent Mg or Ca, but can be very considerable when the cation is Na. The amount of water also depends on the humidity; as a result, smectites, and sodium-bearing ones in particular, will swell on contact with water, affecting permeability. The most common smectite is montmorillonite, $X_{1/3}(Mg_{1/3}Al_{5/3})Si_4O_{10}$ $(OH)\cdot nH_2O$. The interlayer cation of smectites is exchangeable: in a NaCl solution it will be Na, in a $CaCl_2$ solution it will be Ca, and so on.

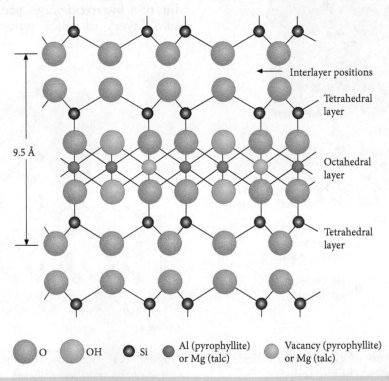

9.5 Å

Interlayer positions

Tetrahedral layer

Octahedral layer

Tetrahedral layer

⬤ O   ⬤ OH   ⬤ Si   ⬤ Al (pyrophyllite) or Mg (talc)   ⬤ Vacancy (pyrophyllite) or Mg (talc)

Figure 6.25 Structure of pyrophyllite.

Vermiculites have a higher net charge on the 2:1 layer (i.e., there is greater cation deficiency). As a result, the electrostatic forces holding the layers together are greater, so that the interlayer space is less expandable and the interlayer cations less exchangeable.

Micas (biotite and muscovite) are related to pyrophyllite and talc by substitution of an Al^{3+} for a Si^{4+} in a tetrahedral site. The structure of muscovite is illustrated in Figure 6.26. The result is that the silicate layers are relatively strongly bound to the interlayer K, which is not normally exchangeable. *Illite* is a name applied to clay-sized micas, though it is sometimes restricted to the dioctahedral mica (muscovite). Generally, illite has less K and Al and more Si than igneous or metamorphic muscovite; in this sense it can be viewed as a solid solution of muscovite and pyrophyllite.

Because of the structural similarity of various 2:1 clays, they can form crystals that consist of layers of more than one type, for example illite–smectite. In addition, layers of gibbsite or brucite may occur in smectite. Different layers may be distributed randomly or may be ordered. These are called, for example, mixed-layer chlorite–smectite or hydroxy-interlayer smectite.

6.5.1.3 *Chlorite group (2:2 clays)*

This group is characterized by having a unit cell consisting of two tetrahedral layers and two (hydroxide) octahedral layers. The ideal formula of chlorite is $(Mg,Fe,Al)_6(SiAl)_4O_{10}(OH)_8$, where the elements in parentheses can be in any proportions. The structure of chlorite is shown in Figure 6.27. Chlorites with unit cells consisting of a single tetrahedral and single octahedral layer also occur and are called septechlorite (because they have a 7 Å spacing; true chlorites have 14 Å spacing), but are less stable and uncommon.

6.5.2 Ion-exchange properties of clays

One of the most important properties of clays is their capacity for ion exchange. In soil science, the term *ion exchange* refers specifically to replacement of an ion adsorbed to the surface by one in solution. However, we shall use the term in a more general sense here, and also include exchange reactions between ions in solution and ions bound within the solid. The ability of a substance to exchange ions is called the *ion exchange capacity* and is generally measured in equivalents or milli-equivalents (meq). Ion exchange capacities of clays are listed in Table 6.3.

The exchange reaction of two monovalent ions between clay and solution may be written:

$$X_{clay} + Y^+ \rightleftharpoons Y_{clay} + X^+$$

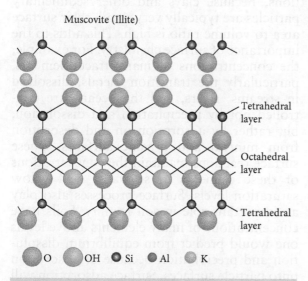

Muscovite (Illite)

Tetrahedral layer

Octahedral layer

Tetrahedral layer

○ O ◯ OH ● Si ● Al ● K

Figure 6.26 Structure of muscovite $(KAl_3Si_3O_{10}(OH)_2)$. The structure of the clay illite is similar, but illite typically has less K and Al and more Si than muscovite.

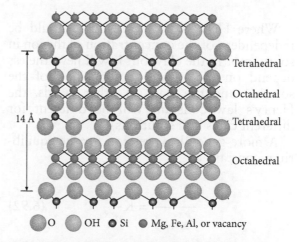

Tetrahedral

Octahedral

14 Å

Tetrahedral

Octahedral

○ O ◯ OH ● Si ● Mg, Fe, Al, or vacancy

Figure 6.27 Structure of chlorite.

Table 6.3 Ion exchange capacity of clays.

Clay	Exchange capacity (meq/100g)
Smectite	80–150
Vermiculite	120–200
Illite	10–40
Kaolinite	1–10
Chlorite	<10

From Drever (1988).

The corresponding equilibrium constant expression is written as:

$$K = \frac{a_{Y-clay}a_{X^+}}{a_{X-clay}a_{Y^+}} \quad \text{or} \quad \frac{a_{Y-clay}}{a_{X-clay}} = K\frac{a_{Y^+}}{a_{X^+}}$$

(6.90)

If we express this with molar concentrations in solution and mole fractions (X) in the solid rather than activities:

$$\frac{X_{Y-clay}}{X_{X-clay}} = K'\frac{[Y^+]}{[X^+]}$$

K' is called the *selectivity constant*. It expresses the selectivity of the clay the Y ion over the X ion. Because we have expressed it in mole fraction rather than activity, we can expect its value to depend on the composition of both the clay and the solution. We may also define a *distribution coefficient K_d* as:

$$K_d = \frac{[X_{clay}]}{[X^+]}$$

(6.91)

Where Henry's law holds, K_d should be independent of the concentration of the ion in solution or in the clay, but it will nevertheless depend on the overall composition of the solution and the clay (in other words, the Henry's law constant will be different for different clays and solutions).

A more general expression for the equilibrium constant is:

$$\frac{a_{Y-clay}^{v_Y}}{a_{X-clay}^{v_X}} = K\frac{a_Y^{v_Y}}{a_X^{v_X}}$$

(6.92)

where v is the stoichiometric coefficient.

The power term is important. Consider the case of exchange between Na^+ and Ca^{2+}. The reaction is:

$$2Na_{clay}^+ + Ca_{aq}^{2+} \rightleftharpoons Ca_{clay}^{2+} + 2Na_{aq}^+ \quad (6.93)$$

If we assume that: (1) the mole fractions of Na^+ and Ca^{2+} in the clay must sum to one (i.e., they are the only ions in the exchanging site); (2) molar concentrations of 1 for Na and Ca in solution; and (3) $K' = 1$, solving eqn. 6.94 yields $X_{Na} = 0.62$ and $X_{Ca} = 0.38$. If we kept the ratio of Ca and Na in solution constant, but dilute their concentrations 1000-fold, we obtain $X_{Na} = 0.03$ and $X_{Ca} = 0.97$. Thus, by diluting the solution, the divalent cation has almost entirely replaced the monovalent ion. The composition of the clay will thus depend on the ionic strength of the solution. The dominant exchangeable cation is Ca^{2+} in fresh water, but Na^+ in seawater.

6.6 MINERAL SURFACES AND THEIR INTERACTION WITH SOLUTIONS

Reactions between solutions and solids necessarily involve the interface between these phases. The details of interface processes thus govern equilibria between solids and solutions. Because clays and other sedimentary particles are typically very small, their surface area to volume ratio is high. This adds to the importance of surface chemistry. For example, the concentrations of many trace elements, particularly the transition metals, dissolved in streams, rivers, and the oceans are controlled not by precipitation and dissolution, but rather by adsorption on and desorption from mineral and organic surfaces. These surface reactions maintain the concentrations of these elements in seawater well below saturation levels. Surface processes also play an important role in soil fertility. Soils have concentrations of many elements above levels one would predict from equilibrium dissolution and precipitation because of adsorption onto particle surfaces. Surface adsorption will also strongly affect the dispersion of pollutants in soils, ground and surface waters. We discussed some aspects of surface chemistry in Chapter 5 within the context of kinetic fundamentals. We return to it in this chapter in a broader context.

6.6.1 Adsorption

Adsorption is a remarkably important process in natural waters: adsorption onto particle surfaces retains nutrients in soils and often controls the abundance of trace metals in natural waters, topics we'll return to in Chapters 13 and 14. We can define adsorption as attachment of an ion in solution to a preexisting solid surface, for example a clay particle. Adsorption involves one or more of the following:

- *Surface complex formation:* The formation of coordinative bonds between metals and ligands at the surface. Considered in isolation, this process is very similar to the formation of complexes between dissolved components.
- *Electrostatic interactions:* As we shall see, solid surfaces are typically electrically charged. This electrostatic force, which is effective over greater distances than purely chemical forces, affects surface complex formation and loosely binds other ions to the surface. For solutions, we were able to make the simplifying assumption of electrical neutrality. We cannot make this assumption about surfaces.
- *Hydrophobic adsorption:* Many organic substances, most notably lipids, are highly insoluble in water due to their nonpolar nature. These substances become adsorbed to surfaces, not because they are attracted to the surface, but rather because they are repelled by water.

The interaction of the three effects makes quantitative prediction of adsorption behavior more difficult than prediction of complexation in solution. The functional groups of mineral and organic surfaces have properties similar to those of their dissolved counterparts. In this sense, surface complexation reactions are similar to complexation reactions in solution. However, reactions between these surface groups and dissolved species are complicated by the proximity of surface groups to each other, making them subject to long-range electrostatic forces from neighboring groups. For example, the surface charge

will change systematically as the adsorbed surface concentration of a positive species such as H^+ increases. This change in surface charge will decrease the attraction between H^+ ions and the surface. As a result, the equilibrium for the surface protonation reaction will change as the surface concentration of H^+ increases.

We found in Chapter 5 that adsorption is usually described in terms of adsorption isotherms. We introduced two such isotherms, the *Langmuir* isotherm:

$$\Theta_M = \frac{K_{ad}[M]}{1 + K_{ad}[M]} \qquad (5.143)$$

(where Θ_M is the fraction of surface sites occupied by species M, $[M]$ is the dissolved concentration of M, and K_{ad} is the adsorption equilibrium constant), and the *Freundlich* isotherm:

$$\Theta_M = K_{ad}[M]^n \qquad (5.147)$$

where n is an empirical constant. We derived the Langmuir isotherm from kinetic fundamentals, but we could have also derived it from thermodynamics. Inherent in its derivation are the assumptions that (1) the free energy of adsorption is independent of the number of sites available, and therefore that (2) the law of mass action applies, and that (3) only a monolayer of adsorbate can form. The Langmuir isotherm thus shows a decrease in the fraction of M adsorbed when the concentration of M in solution is high, reflecting saturation of the surface. In contrast, the Freundlich isotherm, which is merely empirical, shows no saturation. We also found that at low relative saturation of the surface, the Freundlich isotherm with $n = 1$ approximates the Langmuir isotherm.

Adsorption phenomena can be treated with the *surface complexation model*, which is a generalization of the Langmuir isotherm (Stumm and Morgan, 1995; Morel and Hering, 1993). The model incorporates both *chemical bonding of solute species to surface atoms* and *electrostatic interactions between the surface and solute ions*. The model assumes that these two effects can be treated separately. Thus the free energy of adsorption is the sum of a complexation, or intrinsic, term and an electrostatic, or coulombic term:

$$G_{ad} = \Delta G_{intr} + \Delta G_{coul} \qquad (6.95)$$

From this it follows that the adsorption equilibrium constant can be written as:

$$K_{ad} = K_{intr} K_{coul} \qquad (6.96)$$

Letting $\equiv S$ denote the surface site and M denote a solute species, we may write the adsorption reaction as:

$$\equiv S + M \rightleftharpoons \equiv SM$$

Let's begin by considering comparatively simple surfaces: those of metal oxides. Although silicates are likely to be more abundant than simple oxides, the properties of silicate surfaces approximate those of mixtures of their constituent oxides' surfaces. Hence, what we learn from consideration of oxides can be applied to silicates as well. We will initially focus just on the intrinsic terms in eqns. 6.95 and 6.96. We will return to the coulombic term at the end of this section.

Oxygen and metal atoms at an oxide surface are incompletely coordinated; i.e., they are not surrounded by oppositely charged ions as they would be in the interior of a crystal (Figure 6.28a). Consequently, mineral surfaces immersed in water attract and bind water molecules (Figure 6.28b). These water molecules can then dissociate, leaving a hydroxyl group bound to the surface metal ion. We may write this reaction as:

$$\equiv M^+ + H_2O \rightleftharpoons \equiv MOH + H^+$$

where $\equiv M$ denotes a surface metal ion.

In a similar fashion, incompletely coordinated oxygens at the surface can also bind water molecules, which can then dissociate, again creating a surface hydroxyl group:

$$\equiv O^- + H_2O \rightleftharpoons \equiv OH + OH^-$$

Thus the surface of an oxide immersed in water very quickly becomes covered with hydroxyl groups (Figure 6.28c), which we can write as $\equiv SOH$ and which are considered to constitute part of the surface rather than the

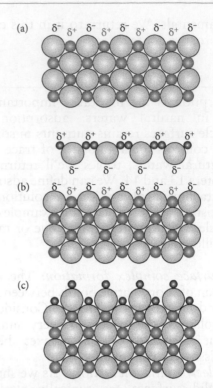

Figure 6.28 (a) Metal ions (small red spheres) and oxygens (large gray spheres) on a mineral surface are incompletely coordinated, leading to a partial charge on the surface (indicated by δ^+ and δ^-). (b) When the mineral surface is immersed in water, water molecules coordinate metal ions on the surface. (c) Water molecules will dissociate leaving hydroxyl groups coordinating metal ions. Protons (small dark spheres) will associate with surface oxygens, forming additional hydroxyl groups.

solution. These hydroxyl groups can then act as either proton acceptors or proton donors through further association or dissociation reactions, for example:

$$\equiv SOH + H^+ \rightleftharpoons \equiv SOH^+$$

or

$$\equiv SOH \rightleftharpoons \equiv SO^- + H^+$$

We should not be surprised to find that these kinds of reactions are strongly pH-dependent.

Adsorption of metals to the surface may occur through replacement of a surface proton, as is illustrated in Figure 6.29a, while

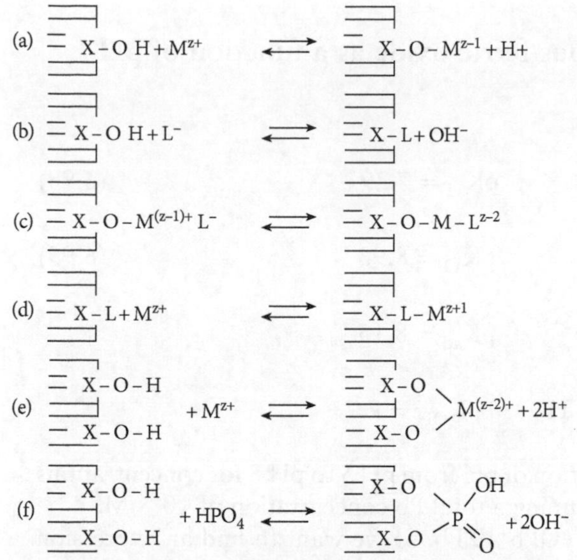

(a) $\equiv X-O\,H + M^{z+}$ ⇌ $\equiv X-O-M^{z-1} + H^+$

(b) $\equiv X-O\,H + L^-$ ⇌ $\equiv X-L + OH^-$

(c) $\equiv X-O-M^{(z-1)+}\,L^-$ ⇌ $\equiv X-O-M-L^{z-2}$

(d) $\equiv X-L + M^{z+}$ ⇌ $\equiv X-L-M^{z+1}$

(e) $\begin{matrix}\equiv X-O-H \\ \equiv X-O-H\end{matrix} + M^{z+}$ ⇌ $\begin{matrix}\equiv X-O \\ \equiv X-O\end{matrix}\Big\rangle M^{(z-2)+} + 2H^+$

(f) $\begin{matrix}\equiv X-O-H \\ \equiv X-O-H\end{matrix} + HPO_4$ ⇌ $\begin{matrix}\equiv X-O \\ \equiv X-O\end{matrix}\Big\rangle P\Big\langle\begin{matrix}OH \\ O\end{matrix} + 2OH^-$

Figure 6.29 Complex formation of solid surfaces may occur when (a) a metal replaces a surface proton, or (b) a ligand replaces a surface OH group. The adsorbed metal (c) may bind an additional ligand, and the ligand (d) may bind an additional metal. Multidentate adsorption involves more than one surface site (e, f).

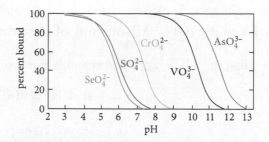

Figure 6.30 Binding of ligands (anions) on the surface of hydrous ferric oxide ($\Sigma Fe = 10^{-3}$ M) from dilute solution (5×10^{-7} M; $I = 0.1$) as a function of pH. After Stumm and Morgan (1995).

ligands may be absorbed by replacement of a surface OH group (Figure 6.29b). The adsorbed metal may bind an additional ligand (Figure 6.29c), and the adsorbed ligand may bind an additional metal (Figure 6.29d).

An additional possibility is multidentate adsorption, where a metal or ligand is bound to more than one surface site (Figures 6.29e and f). This raises an interesting dilemma for the Langmuir isotherm. Where x sites are involved, we could write the reaction:

$$x \equiv S + M \rightleftharpoons \equiv S_x M$$

and the corresponding equilibrium constant expression as:

$$K_{ad} = \frac{[\equiv S_x M]}{[\equiv S]^x [M]} \qquad (6.97)$$

where x is the number of sites involved and M is the species being adsorbed. This assumes, however, that the probability of finding x sites together is proportional to the xth power of

concentration, which is not the case. A better approach is to assume that the reaction occurs with a multidentate surface species, $\equiv S_x$ and that its concentration is $[\equiv S]/x$. The equilibrium constant is then:

$$K_{ad} = \frac{[\equiv S_x M]}{[M][\equiv S]/x}$$

Alternatively, the $1/x$ can be contained within the equilibrium constant.

Since surface-bound H^+ and OH^- are almost inevitably involved in adsorption, we would expect that adsorption of metals and ligands will be strongly pH-dependent. This is indeed the case, as may be seen in Figures 6.30 and 6.32: adsorption of cations increases with increasing pH, while adsorption of anions decreases with increasing pH. Figure 6.32 shows that adsorption of metals on goethite goes from insignificant to nearly complete over a very narrow range of pH. This strong dependence on pH certainly reflects protonation of the surface as we have discussed above, but it also reflects the extent of hydrolysis of the ion in solution. We also see that metals vary greatly in how readily they are adsorbed. At a pH of 7, for example, and a solution containing a 1 μM concentration of the metal of interest, the fraction of surface sites occupied by Ca, Ag, and Mg is trivial and only 10% of surface sites would be occupied by Cd. At this same pH, however, 97% of sites would be occupied by Pb and essentially all sites would be occupied by Hg and Pd.

Example 6.11 Adsorption of Pb^{2+} on hydrous ferric oxide as a function of pH

Using the following apparent equilibrium constants:

$$\equiv FeOH_2^+ \rightleftharpoons \equiv FeOH + H^+ \qquad pK_{a1} = 7.29 \qquad (6.98)$$

$$\equiv FeOH \rightleftharpoons \equiv FeO^- + H^+ \qquad pK_{a2} = 8.93 \qquad (6.99)$$

$$\equiv FeOPb^+ \rightleftharpoons \equiv FeO^- + Pb^{2+} \qquad pK_{ad} = 8.15$$

$$Pb^{2+} + H_2O \rightleftharpoons PbOH^+ + H^+ \qquad pK_{OH} = 7.7$$

calculate the fraction of surface adsorbed Pb as a function of pH from pH 5 to pH 8 for concentrations of surface sites of 10^{-3} M, 10^{-4} M, and 10^{-5} M assuming a total Pb concentration of 10^{-8} M.

Answer: The quantity we wish to calculate is $[\equiv FeOPb^+]/\Sigma Pb$, so we want to find an expression for $[\equiv FeOPb^+]$ as a function of pH. We chose our components to be H^+, Pb^{2+}, and $\equiv FeOH_2^+$ and begin by writing the two relevant conservation equations:

$$\Sigma Pb = [Pb^{2+}] + [PbOH^+] + [\equiv FeOPb^+] \qquad (6.100)$$

$$\Sigma \equiv Fe = [\equiv FeOH_2^+] + [\equiv FeOH] + [\equiv FeO^-] + [\equiv FeOPb^+] \qquad (6.101)$$

From the equilibrium constant expressions, we have the following:

$$[\equiv FeOH] = \frac{[FeOH_2^+]}{[H^+]} K_{a1} \qquad (6.102)$$

$$[\equiv FeO^-] = \frac{[\equiv FeOH]}{[H^+]} K_{a2} = \frac{[FeOH_2^+]}{[H^+]^2} K_{a1} K_{a2} \qquad (6.103)$$

$$[\equiv FeOPb^+] = \frac{[\equiv FeO^-][Pb^{2+}]}{K_{ad}} = \frac{[FeOH_2^+]}{[H^+]^2}[Pb^{2+}]\frac{K_{a1} K_{a2}}{K_{ad}} \qquad (6.104)$$

Substituting equations 6.102–6.104 into 6.101 we have:

$$\Sigma \equiv Fe = [FeOH_2^+] \left\{ 1 + \frac{K_{a1}}{[H^+]} + \frac{K_{a1} K_{a2}}{[H^+]^2} + \frac{[Pb^{2+}] K_{a1} K_{a2}}{K_{ad}} \right\} \qquad (6.105)$$

Since the $[Pb^{2+}]$ is small, the last term on the right can be neglected so we have:

$$\Sigma \equiv Fe = [FeOH_2^+] \left\{ 1 + \frac{K_{a1}}{[H^+]} + \frac{K_{a1} K_{a2}}{[H^+]^2} \right\}$$

In a similar way, we obtain:

$$\sum Pb = [Pb^{2+}] \left\{ 1 + \frac{[FeOH_2^+] K_{a1} K_{a2}}{[H^+]^2 K_{ad}} + \frac{K_{OH}}{[H^+]} \right\}$$

Solving this pair of equations for $[\equiv FeOH_2^+]$ and $[Pb^{2+}]$, and substituting these into 6.104, we obtain:

$$[\equiv FeOPb^+] = \frac{[FeOH_2^+]}{[H^+]^2 + [H^+]K_{a1} + K_{a1}K_{a2}}\left\{\frac{\Sigma Pb}{\dfrac{\Sigma\equiv FeK_{a1}K_{a2}/K_{ad}}{[H^+]^2 + [H^+]K_{a1} + K_{a1}K_{a2}} + \dfrac{K_{OH}}{[H^+]}}\right\}\frac{K_{a1}K_{a2}}{K_{ad}}$$

Dividing by ΣPb and simplifying, we have:

$$\frac{[\equiv FeOPb^+]}{\Sigma Pb} = \left\{\frac{\Sigma\equiv FeK_{a2}}{K_{ad}([H^+]^2/K_{a1} + [H^+] + K_{a2}) + ([H^+]/K_{a1} + 1 + K_{a2}/[H^+])K_{OH}K_{ad} + \Sigma\equiv FeK_{a2}}\right\}$$

The result is shown in Figure 6.31. For the highest concentration of surface sites, Pb goes from virtually completely in solution to virtually completely adsorbed within 2 pH units.

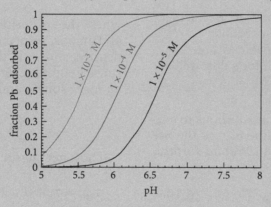

Figure 6.31 Calculated adsorption of Pb^{2+} on hydrous ferric oxide for three different concentrations of surface sites: 10^{-3} M, 10^{-4} M, and 10^{-5} M.

As is the case with soluble complexes, surface complexes may be divided into inner sphere and outer sphere complexes (Figure 6.33). Inner sphere complexes involve some degree of covalent bonding between the adsorbed species and atoms on the surface. In an outer sphere complex, one or more water molecules separate the adsorbed ion and the surface; in this case, adsorption involves only electrostatic forces. The third possibility is that an ion may be held within the diffuse layer (see following section) by long-range electrostatic forces.

6.6.2 Development of surface charge and the electric double layer

Mineral surfaces develop electrical charge for three reasons:

1. *Complexation* reactions between the surface and dissolved species, such as those we discussed in the previous section. Most important among these are protonation and deprotonation. Because these reactions depend on pH, this aspect of surface charge is pH-dependent. This pH dependence is illustrated in Figure 6.34.

2. *Lattice imperfections* at the solid surface as well as substitutions within the crystal lattice (e.g., Al^{3+} for Si^{4+}). Because the ions in interlayer sites of clays are readily exchangeable, this mechanism is particularly important in the development of surface charge in clays.

3. *Hydrophobic adsorption*, primarily of organic compounds, and "surfactants" in particular. We will discuss this effect in Chapter 12.

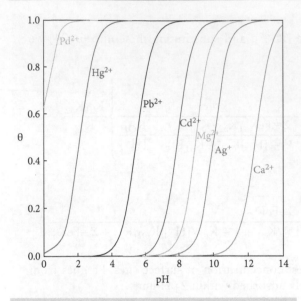

Figure 6.32 Calculated adsorption (Θ, fraction of sites occupied) of metals on goethite (FeOOH) using adsorption coefficients of Mathur and Dzombak (2006) for a dissolved metal concentration of 10^{-6} M.

Thus, there are several contributions to surface charge density. We define σ_{net} as the *net density of electric charge on the solid surface*, and express it as:

$$\sigma_{net} = \sigma_0 + \sigma_H + \sigma_{SC} \qquad (6.106)$$

where σ_0 is the *intrinsic* surface charge due to lattice imperfections and substitutions, σ_H is the net proton charge (i.e., the charge due to binding H^+ and OH^-), σ_{SC} is the charge due to other surface complexes. σ is usually measured in coulombs per square meter (C/m^2). σ_H is given by:

$$\sigma_H = \mathcal{F}\,(\Gamma_H - \Gamma_{OH}) \qquad (6.107)$$

where $\mathcal{F}$ is the Faraday constant and Γ_H and Γ_{OH} are the adsorption densities (mol/m^2) of H^+ and OH^-, respectively. In a similar way, the charge due to other surface complexes is given by

$$\sigma_{SC} = \mathcal{F}\,(Z_M\Gamma_M + Z_A\Gamma_A) \qquad (6.108)$$

where the subscripts M and A refer to metals and anions, respectively, Γ is again adsorption density, and Z is the charge of the ion. The surface complex term may also be broken into an inner sphere and outer sphere component:

$$\sigma_{SC} = \sigma_{IS} + \sigma_{OS} \qquad (6.109)$$

Thus, net charge on the mineral surface is:

$$\sigma_{net} = \sigma_0 + (\Gamma_H - \Gamma_{OH} + Z_M\Gamma_M + Z_A\Gamma_A) \qquad (6.110)$$

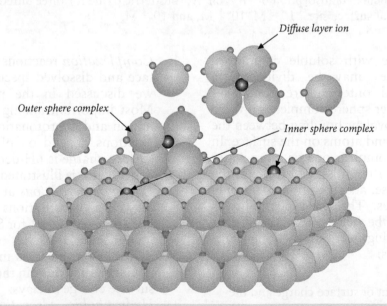

Figure 6.33 Inner sphere surface complexes involve some degree of covalent bonding between the surface and the ion; outer sphere complexes form when one or more water molecules intervenes between the surface and the ion. Ions may also be held in the diffuse layer by electrostatic forces.

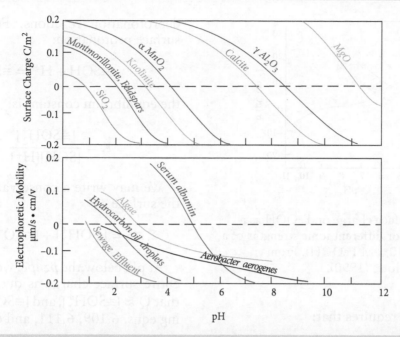

Figure 6.34 (a) Surface charge of some common sedimentary materials as a function of pH. (b) Electrophoretic mobility, which is related to surface charge, of representative organic substances as a function of pH. The pH dependence of surface charge reflects the predominance of attached protons at low pH and the predominance of attached hydroxyls at higher pH. From Stumm (1992).

Figure 6.34 shows that at some value of pH the surface charge, σ_{net}, will be zero. The pH at which this occurs is known as the iso-electric point, or *zero point of charge* (ZPC). *The ZPC is the pH at which the charge on the surface of the solid caused by binding of all ions is 0*, which occurs when the charge due to adsorption of cations is balanced by charge due to adsorption of anions. A related concept is the *point of zero net proton charge* (pznpc), which is the point of zero charge when the charge due to the binding of H^+ and OH^- is 0; that is, pH where $\sigma_H = 0$. Table 6.4 lists values of the point of zero net proton condition for some important solids. Surface charge depends on the nature of the surface, the nature of the solution, and the ionic strength of the latter. An important feature of the point of zero charge, however, is that it is independent of ionic strength, as illustrated in Figure 6.35.

6.6.2.1 Determination of surface charge

The surface charge due to binding of protons and hydroxyls is readily determined by

Table 6.4 Point of zero net proton charge of common sedimentary particles.

Material	pH
SiO_2 (quartz)	2.0
SiO_2 (gel)	1.0–2.5
α-Al_2O_3	9.1
$Al(OH)_2$ (gibbsite)	8.2
TiO_2 (anatase)	7.2
Fe_3O_4 (magnetite)	6.5
α-Fe_2O_3 (hematite)	8.5
$FeO(OH)$ (goethite)	7.8
$Fe_2O_3 \cdot nH_2O$	8.5
δ-MnO	2.8
β-MnO	7.2
Kaolinite	4.6
Montmorillonite	2.5

From Stumm (1992).

titrating a solution containing a suspension of the material of interest with strong acid or base. The idea is that any deficit in H^+ or OH^- in the solution is due to binding with the surface. For example, consider a simple hydroxide surface with surface species $\equiv SOH_2^+$ and $\equiv SO^-$ (as well as $\equiv SOH^0$).

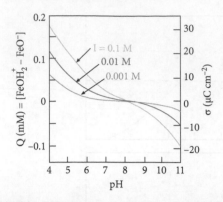

Figure 6.35 Surface charge on FeOOH as a function of pH for different ionic strengths of a 1:1 electrolyte (10^{-3} M FeOOH). From Dzombak and Morel (1990).

Charge balance requires that:

$$C_A - C_B + [OH^-] - [H^+]$$
$$= [\equiv SOH_2^+] - [\equiv SO^-]$$

where C_A and C_B are the concentrations of conjugate of the acid or base added (e.g., Na^+ is the conjugate of the base NaOH) and $[\equiv SOH_2^+]$ and $[\equiv SO^-]$ are the concentrations (in moles per liter) of the surface species. The surface charge, Q (in units of moles of charge per liter), is simply:

$$Q = [\equiv SOH_2^+] - [\equiv SO^-]$$

So that the surface charge is determined from:

$$Q = C_A - C_B - [H^+] + \frac{10^{-14}}{[H^+]} \quad (6.111)$$

The surface charge *density*, σ, is calculated from Q as:

$$\sigma = \frac{QF}{A[\equiv S]} \quad (6.112)$$

where A is the specific surface area (m²/mol) and $[\equiv S]$ is the concentration of solid (in moles/l).*

We can write equilibrium constant expressions for the surface protonation and deprotonation reactions. For example, for surface protonation:

$$\equiv SOH + H^+ \rightleftharpoons \equiv SOH_2^+$$

the equilibrium constant is:

$$K = \frac{[\equiv SOH_2^+]}{[SOH][H^+]} \quad (6.113)$$

We may write a conservation equation for the surface as:

$$\Sigma \equiv S = [SOH_2^+] + [\equiv SO^-] + [\equiv SOH^0] \quad (6.114)$$

At pH below the *pznpc*, we can consider the entire surface charge as due to $[\equiv SOH_2^+]$, so that $Q \approx [\equiv SOH_2^+]$, and $[\equiv SO^-] \approx 0$. Combining eqns. 6.109, 6.111, and 6.112, we have:

$$K = \frac{Q}{(\Sigma \equiv S - Q)[H^+]} \quad (6.115)$$

In eqn. 6.111, we see that if the amount of acid (or base) added is known, the surface charge can be determined by measuring pH (from which $[OH^-]$ may also be calculated). This is illustrated in Figure 6.36.

Thus, the value of the protonation reaction equilibrium constant may be calculated from the surface charge and pH. The equilibrium constant for the deprotonation reaction may be obtained in a similar way. These equilibrium constants are also known as *surface acidity constants*, and sometimes denoted (as in Example 6.11) as K_{a1} and K_{a2} for the protonation and deprotonation reaction respectively.

6.6.2.2 *Surface potential and the double layer*

The charge on a surface exerts a force on ions in the adjacent solution and gives rise to an electric potential, Ψ (measured in volts), which will in turn depend on the nature and distribution of ions in solution, as well as intervening water molecules. The surface charge results in an excess concentration of oppositely charged ions (and a deficit of like charged ions) in the immediately adjacent solution.

* If the concentration of solid is expressed in kg/l, as it commonly is, then the specific surface area should be in units of m²/kg.

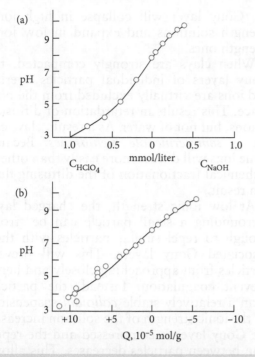

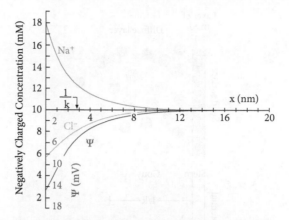

Figure 6.37 Variation in electrical potential and ions with distance from a negatively charged surface based on the Gouy–Chapman model. Electrical potential varies exponentially with distance, as do ion concentrations. $1/\kappa$ is the distance where the potential has decreased by $1/e$. From Morel and Hering (1993).

Figure 6.36 (a) Titration of a suspension of α–FeOOH (goethite) (6 g/liter) by $HClO_4$ and NaOH in the presence of 0.1 M $NaClO_4$. (b) Charge calculated by charge balance (eqn. 6.109) from the titration. From Stumm (1992).

The surface charge, σ, and potential, Ψ_0, can be related by *Gouy–Chapman theory**, which is conceptually and formally similar to Debye–Hückel theory (Chapter 3). The relationship between surface charge and the electric potential is:

$$\sigma = (8RT\varepsilon_r\varepsilon_0 I)^{1/2} \sinh\left(\frac{z\Psi F}{2RT}\right) \quad (6.116)$$

where z is the valence of a symmetrical background electrolyte (e.g., 1 for NaCl), Ψ_0 is the potential at the surface, F is the Faraday constant, T is temperature, R is the gas constant, I is ionic strength of the solution in contact with the surface, ε_r is the dielectric constant of water, and ε_0 is the permittivity of a vacuum (see Chapter 3). At 25°C, eqn. 6.116 may be

written as:

$$\sigma = \alpha I^{1/2} \sinh(\beta z \Psi_0) \quad (6.117)$$

where α and β are constants with values of 0.1174 and 19.5, respectively.

Where the potential is small, the potential as a function of distance from the surface is:

$$\Psi(x) = \Psi_0 e^{-\kappa x} \quad (6.118)$$

where κ has units of inverse length and is called the Debye parameter or Debye length, and is given by:

$$\kappa = \sqrt{\frac{2F^2 I}{\varepsilon_r \varepsilon_0 RT}} \quad (6.119)$$

From eqn. 6.118, we see that the inverse of κ is the distance at which the electrostatic potential will decrease by $1/e$. The variation in potential, the Debye length, and the excess concentration of counter-ions with distance from the surface is illustrated in Figure 6.37.

* Gouy–Chapman theory assumes an infinite flat charge plane in one dimension. The electrostatic interaction between the surface and a cloud of charged particles is described by the Poisson–Boltzmann equation, as in Debye–Hückel theory. Unlike Debye–Hückel, the Poisson–Boltzmann equation has an exact solution in this case. The theory was developed by Gouy and Chapman around 1910, a decade before Debye and Hückel developed their theory. See Morel and Hering (1993) for the details of the derivation.

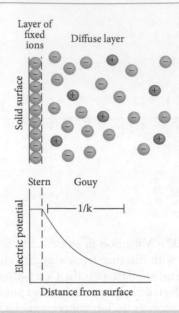

Figure 6.38 The double layer surrounding clay particles.

An addition simplification occurs where the potential is small, namely that eqn. 6.116 reduces to:

$$\sigma = \varepsilon\varepsilon_0\kappa I^{2/3}\Psi_0$$

As is illustrated in Figure 6.37, an excess concentration of oppositely charged ions develops adjacent to the surface. Thus, an *electric double layer* develops adjacent to the mineral surface. The inner layer, or Stern layer*, consists of charges fixed to the surface; the outer diffuse layer, or Gouy layer, consists of dissolved ions that retain some freedom of thermal movement. This is illustrated in Figure 6.38. The Stern layer is sometimes further subdivided into an inner layer of specifically adsorbed ions (inner sphere complexes) and an outer layer of ions that retain their solvation shell (outer sphere complexes), called the inner and outer Helmholtz planes, respectively. Hydrogens adsorbed to the surface are generally considered to be part of the solid rather than the Stern layer. The thickness of the Gouy (outer) layer is considered to be the Debye length, $1/\kappa$. As is apparent in eqn. 6.119, this thickness will vary inversely with the square root of ionic strength. Thus,

the Gouy layer will collapse in high ionic strength solutions and expand in low ionic strength ones.

When clays are strongly compacted, the Gouy layers of individual particles overlap and ions are virtually excluded from the pore space. This results in retardation of diffusion of ions, but not of water. As a result, clays can act as *semipermeable membranes*. Because some ions will diffuse more easily than others, a chemical fractionation of the diffusing fluid can result.

At low ionic strength, the charged layer surrounding a small particle can be strong enough to repel similar particles with their associated Gouy layers. This will prevent particles from approaching closely and hence prevent coagulation. Instead, the particles form a relatively stable *colloidal* suspension. As the ionic strength of the solution increases, the Gouy layer is compressed and the repulsion between particles decreases. This allows particles to approach closely enough that they are bound together by attractive van der Waals forces between them. When this happens, they form larger aggregates and settle out of the solution. For this reason, clay particles suspended in river water will flocculate and settle out when river water mixes with seawater in an estuary – a topic we will take up in Chapter 14.

6.6.2.3 *Effect of the surface potential on adsorption*

The electrostatic forces also affect complexation reactions at the surface, as we noted at the beginning of this section. An ion must overcome the electrostatic forces associated with the electric double layer before it can participate in surface reactions. We can account for this effect by including it in the Gibbs free energy of reaction, as in eqn. 6.95:

$$\Delta G_{ads} = \Delta G_{intr} + \Delta G_{coul} \qquad (6.95)$$

where ΔG_{ads} is the total free energy of the adsorption reaction, ΔG_{intr} is the *intrinsic* free energy of the reaction (i.e., the value the reaction would have in the absence of electrostatic forces; in general, this will be

---

* This fixed layer is also sometimes called the Helmholtz layer, after Herman von Helmholtz (1821–1894), who first proposed it (and for whom Helmholtz free energy is also named).

similar to the free energy of the same reaction taking place in solution), and ΔG_{coul} is the free energy due to the electrostatic forces and is given by:

$$\Delta G_{coul} = \mathcal{F}\Delta Z\Psi_0 \qquad (6.120)$$

where ΔZ is the change in molar charge of the surface species due to the adsorption reaction. For example, in the reaction:

$$\equiv SOH + Pb^{2+} \rightleftharpoons \equiv SOPb^+ + H^+$$

the value of ΔZ is $+1$ and $\Delta G_{coul} = \mathcal{F}\Psi$.

Example 6.12 Effect of surface potential on surface speciation of ferric oxide

Using the surface acidity constants given in Example 6.11, calculate the surface speciation of hydrous ferric oxide as a function of pH in a solution with a background electrolyte concentration of $I = 0.1$ M. Assume the concentration of solid is 10^{-3} mol/l, the specific surface area is 5.4×10^4 m^2/mol and that there are 0.2 mol of active sites per mole of solid.

Answer: The concentration of surface sites, $\Sigma\equiv Fe$, is 0.2 mol sites/mol solid $\times$ 10^{-3} mol solid/l $= 2 \times 10^{-4}$ mol sites/l. Our conservation equation is:

$$\Sigma\equiv Fe = [\equiv FeOH_2^+] + [\equiv FeOH] + [\equiv FeO^-] = 2 \times 10^{-4}$$

We define P as:

$$P = e^{-\mathcal{F}\Delta Z\Psi_0/RT} \qquad (6.121)$$

so that our equilibrium constant expressions (6.98 and 6.99) become:

$$[\equiv FeOH] = \frac{[\equiv FeOH_2^+]}{[H^+]}K_{a1}P^{-1} \qquad (6.122)$$

and

$$[\equiv FeO^-] = \frac{[\equiv FeOH_2^+]}{[H^+]}K_{a2}P^{-1} = \frac{[\equiv FeOH_2^+]}{[H^+]}K_{a1}K_{a2}P^{-2} \qquad (6.123)$$

Substituting into our conservation equation, and solving for $[\equiv FeOH_2^+]$ we have:

$$[\equiv FeOH_2^+] = \Sigma\equiv Fe\left\{1 + \frac{K_{a1}P^{-1}}{[H^+]} + \frac{K_{a1}K_{a2}P^{-2}}{[H^+]^2}\right\} \qquad (6.124)$$

The concentration of surface charge, Q, is simply:

$$Q = [\equiv FeOH_2^+] - [\equiv FeO^-]$$

and the surface charge density is:

$$\sigma = \frac{\mathcal{F}}{A[\equiv S]}([\equiv FeOH_2^+] - [\equiv FeO^-])$$

(P enters the equations as the inverse because we have defined the equilibrium constants in Example 6.11 for the *desorption* reactions.) Substituting into the surface charge density equation, we have:

$$\sigma = \frac{\mathcal{F}}{A[\equiv S]}[\equiv FeOH_2^+]\left(1 - \frac{K_{a1}K_{a2}P^{-2}}{[H^+]^2}\right) \qquad (6.125)$$

Substituting eqn. 6.114 into 6.125, we have:

$$\sigma = \frac{\mathcal{F}}{A[\equiv S]}\Sigma \equiv Fe\left\{1 + \frac{K_{a1}P^{-1}}{[H^+]} + \frac{K_{a1}K_{a2}P^{-2}}{[H^+]^2}\right\}^{-1}\left(1 - \frac{K_{a1}K_{a2}P^{-2}}{[H^+]^2}\right) \qquad (6.126)$$

Finally, substituting equation 6.117 for σ, and 6.123 for P, we have:

$$\sinh(\beta z \Psi_0) = \frac{\mathcal{F}}{A[\equiv S]\alpha I^{1/2}}\Sigma \equiv Fe\frac{\left(1 - \dfrac{K_{a1}K_{a2}e^{2\mathcal{F}\Delta Z\Psi_0/RT}}{[H^+]^2}\right)}{\left\{1 + \dfrac{K_{a1}e^{\mathcal{F}\Delta Z\Psi_0/RT}}{[H^+]} + \dfrac{K_{a1}K_{a2}e^{2\mathcal{F}\Delta Z\Psi_0/RT}}{[H^+]^2}\right\}} \qquad (6.127)$$

A pretty intimidating equation, and one with no direct solution. It can, however, be solved by indirect methods (i.e., iteratively) on a computer. A quick and easy way is to use the Solver feature in Microsoft Excel™. Figure 6.39 shows the results and compares them to the surface speciation when surface potential is to not considered. The effect of including the surface potential term is to reduce the surface concentration of $\equiv FeO^-$ and $\equiv FeOH_2^+$ and broaden the pH region where $\equiv FeO$ dominates.

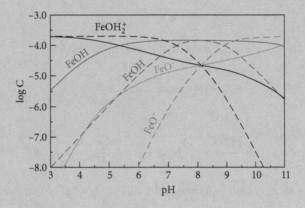

Figure 6.39 Surface speciation of hydrous ferric oxide for $I = 0.1$ M calculated in Example 6.12. Solid lines show speciation when surface potential term is included, dashed lines show the calculated speciation with no surface potential.

Thus, if we can calculate ΔG_{coul}, this term can be added to the intrinsic ΔG for the adsorption reaction (ΔG_{intr}) to obtain the effective value of ΔG (ΔG_{ads}). From ΔG_{ads} it is a simple and straightforward matter to calculate K_{ads}. From eqn. 3.86 we have:

$$K = e^{-\Delta G_{ads}/RT}$$

Substituting eqn. 6.96, we have:

$$K = e^{-\Delta G_{intr}/RT}e^{-\Delta G_{coul}/RT} \qquad (6.128)$$

Since $K_{intr} = e^{-\Delta G_{intr}/RT}$ and $\Delta G_{coul} = \mathcal{F}\Delta Z\Psi_0$, we have:

$$K = K_{intr}e^{-\mathcal{F}\Delta Z\Psi_0/RT} \qquad (6.129)$$

Thus, we need only find the value of Ψ_0, which we can calculate from σ using eqn. 6.114. Example 6.12 illustrates the procedure.

The effect of surface potential on a given adsorbate will be to shift the adsorption curves to higher pH for cations and to lower pH for anions. Figure 6.40 illustrates the

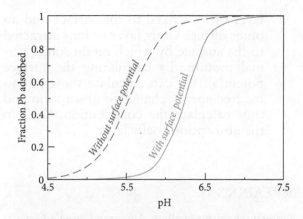

Figure 6.40 Comparison of calculated adsorption of Pb on hydrous ferric oxide with and without including the effect of surface potential.

example of adsorption of Pb on hydrous ferric oxide. When surface potential is considered, adsorption of a given fraction of Pb occurs at roughly 1 pH unit higher than in the case where surface potential is not considered. In addition, the adsorption curves become steeper.

6.7   SUMMARY

In this chapter, we reviewed methods to predict the species present in aqueous solutions at the surface of the Earth and their interaction with solids through dissolution, precipitation, and adsorption.

- Almost everything about natural solutions depends on their pH, consequently we began by developing methods to deal with acid–base reactions and how to combine *proton accounting* and mass and charge balance to predict the speciation of the major nonconservative ions in solution. Among these the carbonate species are arguably most important and often control the pH state of natural waters. *Alkalinity*, which is a measure of *acid-neutralizing capacity* of a solution, is an important parameter in water chemistry and it can be readily determined analytically. It can also be calculated in several ways; among these the parameter *TOTH* is the negative of alkalinity.

- The *buffer intensity* of a solution is a measure of how much acid or base must be added to the solution to change pH and is defined as:

$$\beta \equiv \frac{dC_B}{dpH} = -\frac{dC_A}{dpH} \qquad (6.45)$$

"Hard water" is an example of water with a substantial buffering capacity due to the presence of dissolved carbonates.

- In the following section, we examined how ions can associate to form *complexes* and *ion pairs*. *Stability constants*, which are a type of equilibrium constant but with rules for their expression, allow us to predict when ions will form complexes. The most common complexes are water-related, and whether an ion forms an aquo-, hydroxo- or oxo-complex depends on pH and the charge on the ion. Metals with valences of 3 and 4 typically form hydroxo-complexes which are sparingly soluble.

- Complex formation is often a precursor to precipitation, which we discussed in our next section. Solubility is almost always pH- dependent the solubility of a metal is often the sum of the solubility of the complexes it forms. We learned to construct *stability diagrams* to predict the solubility of phases containing several metal ions.

- Next, we returned to the topic of *adsorption* and the *surface complexation model*, which incorporates both chemical bonding and electrostatic interactions between surfaces and ions in solution. The net free energy change is the sum of these effects to the absorption coefficient is the product:

$$K_{ad} = K_{intr} K_{coul} \qquad (6.96)$$

Electrostatic interactions occur because mineral surfaces generally have a surface charge as a result of surface complexation and lattice imperfections, with the total charge, σ_{sc} being the sum of these effects. The surface charge gives rise to a surface potential, Ψ_0:

$$\sigma = \alpha I^{1/2} \sinh(\beta z \Psi_0) \qquad (6.117)$$

which decreases with distance from the surface as:

$$\Psi(x) = \Psi_0 e^{-x\sqrt{\frac{2F^2 I}{\varepsilon_r \varepsilon_0 RT}}} \quad (6.118)$$

This gives rise to the *electric double layer*, which consists of an inner Stern layer of ions fixed to the surface and an outer diffuse Gouy layer of ions attracted to the surface by which retain some thermal motion. By calculating the surface potential, we can calculate the coulombic free energy change of absorption and thus calculate the contribution, K_{coul} to the absorption coefficient.

REFERENCES AND SUGGESTIONS FOR FURTHER READING

Chou, L. and Wollast, R. 1985. Steady-state kinetics and dissolution mechanisms of albite. *American Journal of Science* 285: 965–93.

Deines, P., Langmuir, D. and Harmon, R.S. 1974. Stable carbon isotope ratios and the existence of a gas phase in the evolution of carbonate ground waters. *Geochimica et Cosmochimica Acta* 38: 1147–64.

Drever, J.I. 1988. *The Geochemistry of Natural Waters*. Englewood Cliffs, Prentice Hall.

Dzombak, D.A. and Morel, F.M.M. 1990. *Surface Complexation Modelling: Hydrous Ferric Oxide*. New York, Wiley Interscience.

Garrels, R.M. and Christ, C.L. 1965. *Solutions, Minerals and Equilibria*. New York, Harper and Row.

James, R.O. and Healy, T.W. 1972. Adsorption of hydrolyzable metal ions at the oxide–water interface. *Journal of Colloid Interface Science* 40: 42–52.

Mathur, S.S. and Dzombak, D.A. 2006. Surface complexation modeling: goethite, in *Surface Complexation Modelling* (ed. J. Lützenkirchen), pp. 443–68. Amsterdam, Elsevier.

Morel, F.M.M. and Hering, J.G. 1993. *Principles and Applications of Aquatic Chemistry*. New York, John Wiley and Sons.

Oelkers, E.H. and Schott, J. 2009. *Thermodynamics and Kinetics of Water–Rock Interaction*. Reviews in Mineralogy and Geochemistry, 70. Washington, Mineralogical Society of America.

Pankow, J.F. 1991. *Aquatic Chemistry Concepts*. Chelsea, MI, Lewis Publishers.

Richardson, S.M. and McSween, H.Y. 1988. *Geochemistry: Pathways and Processes*. New York, Prentice Hall.

Schlesinger, W.H. 1991. *Biogeochemistry*. San Diego, Academic Press.

Sposito, G. 1989. *The Chemistry of Soils*. New York, Oxford University Press.

Stumm, W. 1992. *Chemistry of the Solid–Water Interface*. New York, Wiley Interscience.

Stumm, W. and Morgan, J.J. 1995. *Aquatic Chemistry*. New York, Wiley and Sons.

White, A.F. and Brantley, S.L. (eds) 1995. *Chemical Weathering Rates in Silicate Minerals*. Washington, Mineralogical Society of America.

PROBLEMS

1. Make a plot similar to Figure 6.1, but for water in equilibrium with atmospheric CO_2 ($P_{CO_2} = 10^{-3.4}$). Assume ideality and that there are no other species present in solution but those shown on the graph. What is the pH of the CO_2 equivalence point in this case? What is the pH of the bicarbonate equivalence point?

2. Using the composition given in Problem 3.8, calculate the alkalinity of seawater at 25°C.

3. For a sodium carbonate solution titrated with HCl to the bicarbonate equivalence point, show that:

$$\Sigma CO_2 \cong [Cl^-] + [OH^-]$$

4. Calculate the pH of a solution containing $\Sigma CO_2 = 10^{-2}$ at 25°C at the bicarbonate and carbonate equivalence points. Assume ideality and use the equilibrium constants in Table 6.1.

5. Consider a 0.005 M solution of Na_2CO_3 at 25°C. Assuming ideality and that the system is closed:

(a) What is the pH of this solution?

(b) What is the pH of this solution when titrated to the bicarbonate equivalence point?

(c) What is the pH of this solution when titrated to the CO_2 equivalence point?

6. Consider a 0.01 M solution of $NaHCO_3$ (sodium bicarbonate) at 20°C. Assuming ideality:

(a) What is the pH of this solution?

(b) Plot the titration curve for this solution (i.e., moles of HCl added vs. pH).

(c) What is the pH of the CO_2 equivalence point of this solution?

7. Explain why pH changes rapidly near the bicarbonate and CO_2 equivalence points during titration.

8. Mars probably once had a more substantial atmosphere and water on its surface. Suppose that it had a surface atmospheric pressure of 1 bar (0.1 MPa) and that the partial pressure of CO_2 was the same as it is today, 6×10^{-3}. Further suppose the surface temperature was 5°C. Assume ideal behavior and use the equilibrium constants in Table 6.1 for this problem. Under these conditions, at what concentration of Ca^{2+} ion would an ancient Martian stream become saturated with $CaCO_3$? What would the pH of that stream be? Assume that calcium, carbonate species, and the dissociate products of water are the only ions present.

9. Calculate the buffer capacity of a solution produced by equilibrating pure water with calcite for pH between 6 and 9 at 25°C.

10. Calculate the calcium ion concentration for a solution in equilibrium with calcite and fixed ΣCO_2 of 10^{-2} M at 25°C.

11. Show that the fraction of copper complexed as $CuOH^+$, as defined in eqn. 6.50, will decrease with increasing concentration of total copper in solution. Assume ideal behavior and that H^+, OH^-, Cu^{2+} and $CuOH^+$ are the only ions present in solution, and that the stability constant of $CuOH^+$ is 10^{-8}.

12. Using the following equilibrium constants and reactions, make a plot of Zn^{2+}, $ZnOH^+$, $Zn(OH)_2$, $Zn(OH)_3^+$, $Zn(OH)_4^{2-}$, and total zinc concentration as a function of pH from pH 1 to pH 14. Assume ideal behavior and that H^+, OH^-, and various species of Zn are the only ions in solution.

(*Hint:* Use a log scale for the Zn concentrations.)

$$ZnO + 2H^+ \rightleftharpoons Zn^{2+} + H_2O \qquad \log K_{Zn} = 11.2$$

$$Zn^{2+} + H_2O \rightleftharpoons ZnOH^+ + H^+ \qquad \log K_1 = -9$$

$$Zn^{2+} + 2H_2O \rightleftharpoons Zn(OH)_2 + 2H^+ \qquad \log K_2 = -16.9$$

$$Zn^{2+} + 3H_2O \rightleftharpoons Zn(OH)_3^- + 3H^+ \qquad \log K_3 = -28.1$$

$$Zn^{2+} + 4H_2O \rightleftharpoons Zn(OH)_4^{2-} + 4H^+ \qquad \log K_4 = -40.2$$

13. Using the reactions and equilibrium constants given by eqns. 6.57–59, derive the equations used to construct the stability diagram shown in Figure 6.13b. Assume a fixed Mg^{2+} concentration of 10^{-7} M, that the solution is ideal, and that the solids are pure phases.

14. For the adsorption of Zn^{2+} on hydrous ferric oxide:

$$\equiv FeOH^0 + Zn^{2+} \rightleftharpoons \alpha FeOZn^+ + H^+$$

the apparent equilibrium constant is $10^{0.99}$. For this problem, use the surface acidity constants (i.e., equilibrium constants for adsorption and desorption of H^+) given in eqns. 6.98 and 6.99 (Example 6.11).

(a) Make a plot of Θ_{Zn} (fraction of sites occupied by Zn) vs. the aqueous concentration of Zn^{2+} (use log $[Zn^{2+}]$) at pH 7 and a total concentration of surface sites of 10^{-3} M. Assume that Zn^{2+} forms no complexes in solution.

(b) Ignoring electrostatic effects and any aqueous complexation of Zn^{2+}, make a plot of the fraction of Zn^{2+} adsorbed as a function of pH (from pH 2 to pH 6), assuming a total Zn^{2+} concentration of 10^{-8} M.

(c) Do the same calculation as in (b) but take into consideration the aqueous complexation reactions and equilibrium constants in Problem 11.

15. For the adsorption of Pb^{2+} on aluminum oxide:

$$\equiv SOPb^+ \rightleftharpoons \alpha SO^- + Pb^{2+}$$

the apparent equilibrium constant, K_{ad}, is $10^{-6.1}$. In addition, consider the reactions:

$$\equiv SOH^+ \rightleftharpoons \equiv SOH + H^+ \qquad K = 10^{-6}$$

$$\equiv SOH \rightleftharpoons \equiv SO^- + H^+ \qquad K = 10^{-7.7}$$

Make a plot of the fraction of Pb adsorbed as a function of pH from pH $= 4$ to pH $= 7$, assuming a total concentration of alumina of 10^{-2} M, a surface site density of 10^{-2} moles/mole Al_2O_3, and total Pb concentration of 10^{-9} M, ignoring electrostatic effects and any complexation in solution. (Hint: Θ_{Pb}, the fraction of sites occupied by Pb, will be negligible.)

16. Consider a 10^{-3} M suspension of aluminum oxide (Al_2O_3) in a 1:1 electrolyte having a specific surface area of 500 m^2/g. At pH 7, the surface charge, Q, is found to be 7.93×10^{-5} moles/l. What is the surface charge density, σ? If the temperature is 25°C and ionic strength, I, is 10^3 M, what is the surface potential, Ψ_0? What is the potential at a distance of 1 Debye length from the surface? Make a plot of how Ψ_0 and the Debye length change as I varies from 10^{-3} to 1 (equivalent to going from river water to seawater).

Chapter 7

Trace elements in igneous processes

7.1 INTRODUCTION

In this chapter we will consider the behavior of trace elements, particularly in magmas, and introduce methods to model this behavior. Though trace elements, by definition, constitute only a small fraction of a system of interest, they provide geochemical and geological information out of proportion to their abundance. There are several reasons for this. First, variations in the concentrations of many trace elements are much larger than variations in the concentrations of major components, often by many orders of magnitude. Second, in any system there are far more trace elements than major elements. In most geochemical systems, there are ten or fewer major components that together account for 99% or more of the system. This leaves 80 trace elements. Each element has chemical properties that are to some degree unique, hence there is unique geochemical information contained in the variation of concentration for each element. Third, the range in behavior of trace elements is large, and collectively they are sensitive to processes to which major elements are insensitive. One example is the depth at which partial melting occurs in the mantle. When the mantle melts, it produces melts whose composition is only weakly dependent upon pressure – it always produces basalt. Certain trace elements, however, are highly sensitive to the depth of melting (because the phase assemblages are functions of pressure). Furthermore, on a large scale, the composition of the Earth's mantle appears to be relatively uniform, or at least those parts of it that give rise to basaltic magmas. Indeed, it has proved very difficult to demonstrate any heterogeneity in the mantle based on the major element chemistry of the magmas it has produced. In contrast, it has been amply demonstrated that trace element concentrations of the mantle are quite variable. Trace elements, particularly when combined with isotope ratios, which we shall discuss in the next chapter, thus provide a chemical fingerprint of different mantle reservoirs. Finally, the behavior of trace elements is almost always simpler than that of major elements because trace elements obey Henry's law (Chapter 3).

7.1.1 Why care about trace elements?

Trace element geochemistry has been of enormous use in understanding the evolution of the Earth. As we shall see in subsequent chapters, a fair amount of what we know about the evolution of the core, the mantle, and the crust has come from the study of trace element abundances. For example, the abundance of certain *siderophile* (a term we shall define shortly) trace elements in the mantle and mantle-derived rocks provides reason to believe that segregation of the Earth's iron–nickel core must have been largely complete before the Earth had entirely

Geochemistry, Second Edition. William M. White.
© 2020 John Wiley & Sons Ltd. Published 2020 by John Wiley & Sons Ltd.
Companion website: www.wiley.com/go/white/geochemistry

accreted from the cloud of gas and dust surrounding the early Sun (the solar nebula). We also know, for example, that much of the upper mantle has undergone partial melting at some point in the past. These partial melts of the upper mantle have, through time, created the continental crust. From the abundances of trace gases in the mantle and their isotopic composition, we conclude that the solid portion of the Earth must have undergone extensive outgassing within the first few hundred million years of Earth history. As we shall see in subsequent chapters, magmas from a given tectonic setting tend to share patterns of trace element abundances. This allows the tectonic setting of anciently erupted magmas to be deduced.

Though our focus here will be on igneous processes, trace elements are equally useful in other geologic problems as well. For example, trace elements can provide useful clues as to the origin of sulfide ore deposits. The concentrations of trace elements such as cadmium in the fossil shells of microorganisms provide information about the biological productivity and circulation patterns of ancient oceans, and the concentration of Sr in corals provides a measure of temperature of ancient seas. Indeed, throughout the earth sciences, trace element geochemistry has become a powerful tool.

Our purpose in this chapter is to add this tool to our geochemical toolbox. We will begin by considering the chemical properties of the various groups of trace elements, with particular emphasis on how they behave in nature. We then introduce quantitative means of describing trace element distribution. Our primary tool here will be the distribution, or partition, coefficient, which we first introduced in Chapter 3 (as K_D). We will examine how the distribution coefficient depends on temperature, pressure, composition, and the fundamental chemical properties of the element of interest. Finally, we develop equations to predict the behavior of trace elements during melting and crystallization. The knowledge of trace element behavior that we gain in this chapter will be useful in the following one where we discuss radiogenic isotope geochemistry, because all the radioactive elements and their daughter products, with a single exception, are trace elements. We will apply the tools we acquire here to

understanding the evolution of the core, the mantle, and the crust in subsequent chapters.

7.1.2 What is a trace element?

The term *trace element* is a bit hard to define. For igneous and metamorphic systems (and sedimentary rocks for that matter), an operational definition might be as follows: trace elements are those elements that are not stoichiometric constituents of phases in the system of interest. Clearly this definition is a bit fuzzy: a trace element in one system is not one in another. For example, potassium never forms its own phase in mid-ocean ridge basalts (MORB), its concentration rarely exceeding 1500 ppm, but K is certainly not a trace element in granites. For most silicate rocks, O, Si, Al, Na, Mg, Ca, and Fe are major elements. H, C, S, K, P, Ti, Cr, and Mn are sometimes "major elements" in the sense that they can be stoichiometric constituents of phases. These are often referred to as minor elements. All the remaining elements are always trace elements, apart from a few rare, but important, circumstances such as pegmatites and ore deposits.

The above definition breaks down entirely for fluids and natural waters since there is only one phase, namely the fluid, and it is not stoichiometric. In seawater, anything other than Mg^{2+}, Ca^{2+}, K^+, Na^+ Cl^-, SO_4^{2-}, and HCO_3^- (and H_2O, of course) can be considered a trace constituent, though Sr^{2+}, $B(OH)_4^-$, and Br^- are sometimes considered major constituents also (constituents or species is a better term here than elements). These, including the last three, constitute over 99.99% of the total dissolved solids in seawater. Trace elements in seawater and in rocks do have one thing in common: neither affect the chemical or physical properties of the system as a whole to a significant *extent*. This might serve as a definition. However, trace (or at least minor) elements can determine the color of a mineral (e.g., the green color of chrome diopside), so even this definition has problems. And CO_2, with a concentration in the atmosphere of only 410 ppm, profoundly affects the transparency of the atmosphere to infrared radiation, and, as a result, Earth's climate. At even lower concentrations, ozone in the upper atmosphere controls the atmospheric transparency to ultraviolet radiation. So this definition is not satisfactory, either.

Yet another possible definition of a trace element is an element whose activity obeys Henry's law in the system of interest. This implies sufficiently dilute concentrations that, for trace element A and major component B, A–A interactions are not significant compared to A–B interactions, simply because A–A interactions will be rare.

There is perhaps no satisfactory quantitative definition of a trace element that will work in every situation. For our present purposes, any of these definitions might do, but bear in mind that a trace element in one system need not be a trace element in another.

7.2 BEHAVIOR OF THE ELEMENTS

7.2.1 Goldschmidt's classification

No matter how we define the term *trace element,* most elements will fall into this category, as is illustrated in Figure 7.1. That being the case, this is a good place to consider the geochemical characteristics of the elements. Victor Goldschmidt recognized four broad categories: atmophile, lithophile, chalcophile,

and siderophile (Figure 7.2, Table 7.1). *Atmophile* elements are highly volatile, forming gases or liquids at the surface of the Earth, and are concentrated in the atmosphere and hydrosphere. Lithophile, siderophile, and chalcophile refer to the tendency of the element to partition into a silicate, metal, or sulfide *liquid* respectively. *Lithophile* elements are those showing an affinity for silicate phases and are concentrated in the silicate portion (crust and mantle) of the Earth. *Siderophile* elements have an affinity for a metallic liquid phase. They are depleted in the silicate portion of the Earth and presumably concentrated in the core. *Chalcophile* elements have an affinity for a sulfide liquid phase. They are also depleted in the silicate earth and may be concentrated in the core. Many sulfide ore deposits originated from aqueous fluids rather than sulfide liquid. A chalcophile element need not necessarily be concentrated in such deposits. As it works out, however, they generally are. Most elements that are siderophile are usually also somewhat chalcophile and vice versa.

There is some basis for Goldschmidt's classification in the chemistry of the elements. Figure 7.2 shows that the lithophile elements

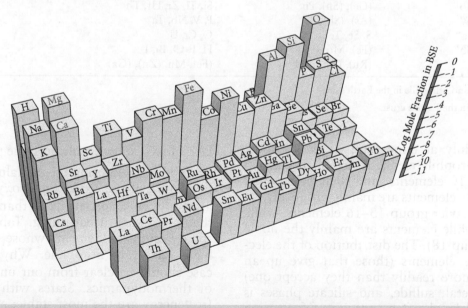

Figure 7.1 Three-dimensional histogram illustrating the abundance of the elements (as the log of mole fraction) in the silicate portion of the Earth (the bulk silicate Earth, BSE). Just six elements, oxygen, magnesium, silicon, iron, aluminum, and calcium make up 99.1% of the silicate Earth. If we include the core and consider the composition of the entire Earth, then only nickel, and perhaps sulfur, need be added to this list. The remaining elements, though sometimes locally concentrated (e.g., in the crust, in the hydrosphere, in ores) can be considered *trace elements.*

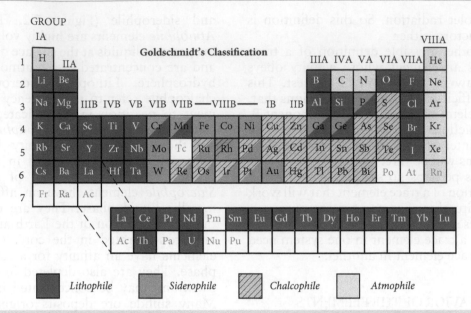

Figure 7.2   Goldschmidt's classification of the elements.

Table 7.1   Goldschmidt's classification of the elements.

Siderophile	Chalcophile	Lithophile	Atmophile
Fe*, Co*, Ni*	(Cu), Ag	Li, Na, K, Rb, Cs	(H), N, (O)
Ru, Rh, Pd	Zn, Cd, Hg	Be, Mg, Ca, Sr, Ba	He, Ne, Ar, Kr, Xe
Os, Ir, Pt	Ga, In, Tl	B, Al, Sc, Y, REE	
Au, Re†, Mo†	(Ge), (Sn), Pb	Si, Ti, Zr, Hf, Th	
Ge*, Sn*, W‡	(As), (Sb), Bi	P, V, Nb, Ta	
C‡, Cu*, Ga*	S, Se, Te	O, Cr, U	
Ge*, As†, Sb†	(Fe), Mo, (Os)	H, F, Cl, Br, I	
	(Ru), (Rh), (Pd)	(Fe), Mn, (Zn), (Ga)	

*Chalcophile and lithophile in the Earth's crust.

†Chalcophile in the Earth's crust.

occur mainly at either end of the periodic table, siderophile elements are mainly group 8, 9, and 10 elements (and their neighbors), chalcophile elements are mainly group 11, 12, and the heavier group 13–16 elements, while the atmophile elements are mainly the noble gases (group 18). The distribution of the electropositive elements (those that give up an electron more readily than they accept one) among metal, sulfide, and silicate phases is controlled by the free energies of formation of the corresponding sulfides and silicates. By comparing the free energies of formation with those of ferrous sulfide and ferrous silicate, it is possible to deduce which elements are siderophile, which are chalcophile and which are lithophile. For historical reasons, namely lack of ΔG_f° data on silicates, the point is generally illustrated using the enthalpy of formation, ΔH_f, of the oxide. Since *oxyphile* could arguably be a better term than lithophile, this is not such a bad thing. Table 7.2 gives some examples. Elements whose oxides have high $-\Delta G_f$ are lithophile. Why this is the case should be clear from our understanding of thermodynamics. States with the lowest free energy are the most stable: a high $-\Delta G_f$ indicates the oxide is much more stable than the metal. Elements whose oxides have $-\Delta G_f$ similar to that of FeO combine with oxygen only about as readily as Fe and are generally siderophile. Those elements whose oxides have low $|\Delta G^{\circ}|$ are generally chalcophile.

Table 7.2   Free energy of formation of some oxides.

Oxide	$-\Delta G^o_f$(kJ/ mole/oxygen)	Oxide	$-\Delta G^o_f$(kJ/ mole/oxygen)
CaO	604.0	In_2O_3	304.2
ThO_2	584.6	SnO_2	260.0
MgO	569.4	FeO	245.9
Al_2O_3	527.3	WO_3	247.3
ZrO_2	521.6	CdO	221.9
CeO_2	512.6	NiO	211.6
TiO_2	444.7	MoO_3	215.4
SiO_2	428.1	Sb_2O_3	207.9
Na_2O	398.3	PbO	189.3
Ta_2O_3	374.0	As_2O_3	180.1
MnO	362.8	Bi_2O_3	168.8
Cr_2O_3	353.1	CuO	127.6
ZnO	318.4	Ag_2O_3	10.9

Lithophile elements also have either very low electronegativities or very high ones and tend to form ionic bonds (although the basic silicate bond, the Si–O bond, is only about 50% ionic, metal–oxygen bonds in silicates are dominantly ionic). Siderophile and chalcophile elements have intermediate electronegativities and tend to form covalent or metallic bonds.

7.2.2   The geochemical periodic table

Goldschmidt's classification is relevant mainly to distribution of elements in meteorites and to how elements distribute themselves between the Earth's major geochemical reservoirs: the core, the mantle and crust, and the hydrosphere and atmosphere. Since there is an overabundance of O in the outer part of the Earth, metallic liquids do not form, and siderophile elements have little opportunity to behave as such. Similarly, sufficient S is rarely available to form more than trace amounts of sulfides. As a result, siderophile elements such as Ni and chalcophile elements such as Pb occur mainly in silicate phases in the crust and mantle.

We can, however, group the elements based on how they behave in the silicate portion of the Earth, the mantle and crust. Figure 7.3 illustrates this grouping. We first note that we have added sodium to those six elements whose molar abundance exceeds 1 percent, to form the group called major elements, and which we will not discuss in this chapter. Although the elements K, Ti, Mn, and P are often reported in rock analyses as major elements, we will include them in our discussion of trace elements. Let's now briefly examine the characteristics of the remaining groups.

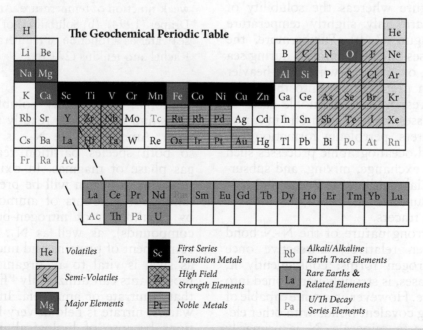

Figure 7.3   The geochemical periodic table, in which elements are grouped according to their geochemical behavior.

7.2.2.1 The volatile elements

The defining feature of the noble gases is their filled outer electron shell, making them chemically inert as well as volatile. Hence, they are never chemically bound in rocks and minerals. Furthermore, except for He, they have rather large radii and cannot easily be accommodated in either cationic or anionic lattice sites of many minerals. Thus, they are typically present at very low concentrations. Their concentrations are usually reported in STP cm^3/g at (i.e., cm^3/g at standard temperature and pressure: 273 K and 0.1 MPa; 1 cm^3/g = 4.46 × 10^{-5} moles/g). Concentrations in silicate rocks and minerals are typically 10^{-4} to 10^{-12} STP cm^3/g (10^{-1} to 10^{-9} ppm). Their solubility in silicate melts is a strong function of pressure, as well as both atomic radius (the heavy noble gases are more soluble) and melt composition as is illustrated in Figure 7.4a. Although they cannot form true chemical bonds with other atoms, they can be strongly adsorbed to crystal surfaces through van der Waals forces.

Solubilities of the noble gases in water range over more than an order of magnitude with the light noble gases being the least soluble and the heavy ones the most. The solubilities of the heavier noble gases (Ar, Kr, and Xe) exhibit a strong, nonlinear decrease with temperature whereas the solubility of the He, Ne are only slightly temperature dependent (Figure 7.4b). Furthermore, the light noble gases partition into ice during sea ice formation or melting while the heavier gases partition preferentially into the water. This varying response of the noble gases to physical processes makes then useful in studying groundwaters, petroleum formation and migration, and oceanographic processes such as air–sea gas exchange, mixing, and subsurface basal glacial melting. See Burnard (2013) for a useful summary of the application of noble gases as tracers.

The very strong nature of the N–N bond makes nitrogen relatively unreactive once molecular nitrogen forms; consequently it, like the rare gases, is strongly partitioned into the atmosphere. However, it is quite capable of forming strong covalent bonds with other elements. In silicate minerals, N is primarily present as the ammonia ion rather than N$_2$. As such, it readily substitutes for K$^+$. As

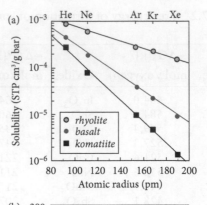

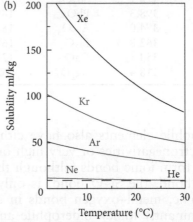

Figure 7.4 (a) Solubility of the rare gases in melts of varying composition at 1200° to 1400°C. Solubility is a strong function of atomic radius and melt composition, but only a weak function of temperature. After Carroll and Draper (1994). (b) Solubilities of noble gases in seawater as a function of temperature. After Rachel and Jenkins (2013).

ammonia, it is highly soluble in aqueous fluids and is therefore readily transported by them. Ammonia, like N$_2$, is quite volatile, so both species partition readily into the gas phase of magmas. In oxidized aqueous solutions, nitrogen will be present as nitrate (and trace amounts of ammonia, produced by breakdown of nitrogen-bearing organic compounds), as well as N$_2$. Nitrogen is a component of proteins and nucleic acids, and as such is vital to all organisms. However, most plants can utilize only "fixed" nitrogen, that is nitrate or ammonia. In many natural waters, nitrate is held at very low concentrations because of biological utilization, and this often limits primary productivity (i.e., photosynthesis).

7.2.2.2   The semivolatiles

The shared characteristic of this group is that they partition readily into a fluid or gas phase (e.g., Cl, Br) or form compounds that are volatile (e.g., SO_2, CO_2). Not all are volatile in a strict sense (volatile in a strict sense means having a high vapor pressure or low boiling point; indeed, carbon is highly refractory in the elemental form).

The partitioning of sulfur between liquid and gas phases is a strong function of f_{O_2}. At high oxygen fugacities, sulfur is present primarily as SO_2, but at low f_{O_2} it is present primarily as sulfide. The solubility of sulfide in silicate liquids is, however, low. At sufficiently high sulfur concentrations in magmas, sulfide and silicate liquids will exsolve. Sulfide liquids are rich in Fe and Ni and other chalcophile metals and are the source of many economically important ore deposits. Large volumes of sulfide liquid are rare, but microscopic droplets of sulfide liquids commonly occur in mid-ocean ridge magmas.

Similarly, the solubility of CO_2 in silicate magmas is limited and is a strong function of pressure. At low CO_2 concentrations, CO_2 exsolves from magmas to form a CO_2–H_2O gas phase. However, at higher CO_2/H_2O ratios and total CO_2 concentrations, carbonatite magmas can form in which $CaCO_3$ is the dominant component. On the whole, carbonatites are rare, but over the course of geologic history they have erupted on every continent. In certain localities, such as the modern East African Rift, they can be fairly common.

The remaining elements in this group are always present in trace concentrations and never reach saturation in magmas, and hence never exsolve as independent gas or fluid phases. Rather, they partition into the gas phase formed by exsolution of CO_2 and H_2O.

7.2.2.3   Alkali and alkaline earth elements

The alkali and alkaline earth elements have electronegativities less than 1.5 and a single valence state (+1 for the alkalis, +2 for the alkaline earths). The difference in electronegativity between these elements and most anions is 2 or greater, so the bonds these elements form are strongly ionic in character (Be is an exception, as it forms bonds with a more covalent character). Ionic bonds are readily disrupted by water due to its polar nature (see Chapter 3). The low ionic potential (ratio of charge to ionic radius) makes these elements relatively soluble in aqueous solution. Because of their solubility, they are quite mobile during metamorphism and weathering.

Because bonding is predominantly ionic, the atoms of these elements behave approximately as hard spheres containing a fixed point charge at their centers (these are among the group A or hard ions discussed in Chapter 6). Thus, the factors that most govern their behavior in igneous rocks are *ionic radius and charge*. K, Rb, Cs, Sr, and Ba, are often collectively termed the *large-ion lithophile (LIL) elements*. As the name implies, these elements all have large ionic radii, ranging from 118 picometers (pm) for Sr to 167 pm for Cs. As we found in Chapter 1, lattice sites in crystals are determined by cation ionic radii. The major minerals in basaltic and ultramafic rocks have two kinds of cationic lattice sites: small tetrahedral sites occupied by Si and Al (and less often by Fe^{3+} and Ti^{4+}), and larger octahedral ones usually occupied by Ca, Mg, or Fe, and more rarely by Na. The ionic radii of the heavy alkali and alkaline earth elements are larger than the radii of even the larger octahedral sites. As a result, substitution of these elements in these sites results in local distortion of the lattice, which is energetically unfavorable. These elements thus tend to be concentrated in the melt phase when melting or crystallization occurs. Such elements are called *incompatible elements*. Incompatible elements are defined as those elements that partition readily into a melt phase when the mantle undergoes melting. Compatible elements, conversely, remain in the residual minerals when melting occurs. Over the history of the Earth, partial melting of the mantle and eruption or intrusion of the resulting magmas on or in the continental crust has enriched the crust in incompatible elements.

In contrast to the heavy alkaline earths, Be has an ionic radius smaller than most octahedral sites. Substitution of a small ion in a large site is also energetically unfavorable as the bond energy is reduced. Thus, Be is also an incompatible element, though only moderately so. While Li has an ionic radius similar to that of Mg and Fe^{2+}, its substitution for one of these elements creates a charge imbalance that requires a coupled substitution. This is also energetically unfavorable, hence Li is also an incompatible element, though again only moderately so.

7.2.2.4   *The rare earth elements and Y*

The rare earths are the two rows of elements commonly shown at the bottom of the periodic table. The first row is the *lanthanide* rare earths, the second is the *actinide* rare earths (the International Union of Pure and Applied Chemistry recommends the terms *lanthanoid* and *actinoid* rather than *lanthanide,* since in chemistry the word-ending "ide" generally refers to anionic species, but the -ide usage remains the most common). However, the term "rare earth elements" (commonly abbreviated REE) is most often used in geochemistry to refer to only to the lanthanide rare earths. We will follow that practice in this book, though we will discuss both the actinide and lanthanides in this section. Only two of the actinides, U and Th, have nuclei stable enough to survive over the history of the Earth. Yttrium shares the same chemical properties, including charge and ionic radius, as the heavier rare earths, and as a result behaves much like them.

As the alkalis and alkaline earths, the REE and Y are strongly electropositive; the lanthanides have electronegativities of 1.2 or less, the actinides U and Th have slightly higher electronegativities. As a result, they form predominantly ionic bonds, and the hard, charged sphere again provides a good model of their behavior. The lanthanide rare earths are in the +3 valence state over a wide range of oxygen fugacities. At the oxygen fugacity of the Earth's surface, however, Ce can be partly or wholly in the +4 state, and Eu can be partly in the +2 state at the low oxygen fugacities of the Earth's interior. Th is always in a +4 valence state, but U may be in a +4 or +6 valence state, depending on oxygen fugacity (or $p\varepsilon$, if we choose to quantify the redox state that way). Unlike the alkali and alkaline earth elements, they are relatively insoluble in aqueous solutions, a consequence of their higher charge and high ionic potential and resulting need to be coordinated by anions. The one exception is U in its fully oxidized U^{6+} form, which forms a soluble oxyanion complex, UO_2^{-2}.

The REE are transition metals. In the transition metals, the *s* orbital of the outermost shell is filled before filling of lower electron shells is complete (Chapter 1). In atoms of the period 6 transition elements, the *6s* orbital is filled before the *5d* and *4f* orbitals. In the lanthanide rare earths, it is the *4f* orbitals that are being filled, so the configuration of the valence electrons is similar in all the REE, hence all exhibit similar chemical behavior. Ionic radius, which decreases progressively from La^{3+} (115 pm) to Lu^{3+} (93 pm), illustrated in Figures 7.5 and 1.6, is thus the characteristic that governs their relative behavior.

Because of their high charge and large radii, the rare earths are incompatible elements, as are U and Th. However, the heavy rare earths have sufficiently small radii that they can be accommodated to some degree in many common minerals. The heaviest rare earths readily substitute for Al^{3+} in garnet, and hence can be concentrated by it. Eu, when in its 2+ state, substitutes for Ca^{2+} in plagioclase feldspar more readily than the other rare earths. Thus, plagioclase is often anomalously rich in Eu compared with the other rare earths, and other phases in equilibrium with plagioclase become relatively depleted in Eu as a consequence.

The systematic variation in lanthanide rare earth behavior is best illustrated by plotting the log of the *relative abundances* as a function of atomic number (this sort of plot is sometimes called a Masuda, Masuda–Coryell, or Coryell plot, but most often is simply termed a rare earth plot or diagram). Relative

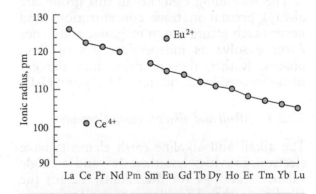

Figure 7.5   Ionic radii of the lanthanide rare earth elements (3+ state except where noted). Promethium (Pm) has no isotope with a half-life longer than 5 years.

abundances are calculated by dividing the concentration of each rare earth by its concentration in a set of normalizing values, such as the concentrations of rare earths in chondritic meteorites. Why do we use relative abundances? As we shall see in Chapter 10, the abundances of even-numbered elements in the solar system (and most likely the cosmos) are greater than those of neighboring odd-numbered elements. Furthermore, because of the way the elements have been created, abundances generally decrease with increasing atomic number. Thus, a simple plot of abundances produces a saw-tooth pattern of decreasing abundances. This can be seen in Figure 7.6, which shows rare earth abundances in the CI chondrite Orgueil (CI chondrites are a class of meteorites that are taken to be the best representative of the average concentrations of condensable elements in the solar system; see Chapter 10). "Normalizing" the rare earth abundances to those of chondritic meteorites eliminates effects related to nuclear stability and nucleosynthesis, and produces a smooth pattern, such as those seen in Figure 7.7.

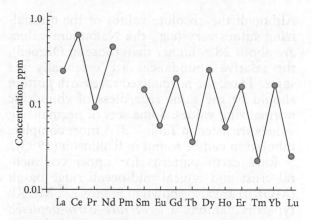

Figure 7.6 Concentrations of the rare earths in the carbonaceous chondritic meteorite Orgueil.

Though all igneous geochemists normalize rare earth abundances to some set of chondritic values, there is no uniformity in the set chosen. Popular normalization schemes include the CI chondrite Orgueil, an average of 20 ordinary chondrites reported by Nakamura (1974), and the chondritic meteorite Leedy (Masuda and Nakamura, 1973).

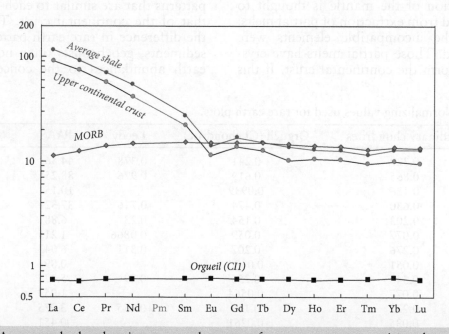

Figure 7.7 A rare earth plot showing rare earth patterns for average upper continental crust (Rudnick and Fountain, 1995), average shale (McClennan, 2018), typical mid-ocean ridge basalt (White and Klein, 2014), and the meteorite Orgueil. As a carbonaceous chondrite, Orgueil is richer in volatile elements and has lower rare earth concentrations than the average of ordinary chondrites used for normalization (see Chapter 10).

Although the absolute values of the normalizing values vary (e.g., the Nakamura values are about 28% higher than those of Orgueil), the relative abundances are essentially the same. Thus, the normalized rare earth *pattern* should be the same regardless of chondritic normalizing values. Some sets of normalizing values are listed in Table 7.3. A more complete tabulation can be found in Rollinson (1993).

Rare earth patterns for upper continental crust and typical mid-ocean ridge basalt (MORB) are also shown in Figure 7.7. MORB typically exhibits a *light rare earth-depleted* pattern; upper continental crust is *light rare earth-enriched* with a negative "Eu anomaly." This anomaly is quantified as follows:

$$Eu/Eu^* = Eu_N/\sqrt{Sm_N \times Gd_N} \qquad (7.1)$$

where the subscript N denotes the "normalized" concentration and Eu^* is the interpolated concentration; i.e., the concentration if there were no anomaly.

The light rare earth depletion of MORB reflects the incompatible element-depleted nature of the upper mantle from which these magmas are derived. This incompatible element depletion of the mantle is thought to have resulted from extraction of partial melts, in which the incompatible elements were concentrated. Those partial melts have crystallized to form the continental crust. If this

is so, the complementary nature of the rare earth patterns of MORB and continental crust is not coincidental. We expect that the relative abundances of REE in the whole Earth are strongly similar to those of chondrites (i.e., the rare earth pattern of the Earth should be flat or nearly so). Mass balance therefore requires the sum of all the various rare earth reservoirs in the Earth to have an approximately flat rare earth pattern. If we assume the continental crust and the mantle are the only two reservoirs with significant concentrations of rare earth elements, and if the continental crust is light rare earth-enriched, then the mantle should be light rare earth-depleted.

A negative Eu anomaly is typical of many continental rocks, as well as most sediments and seawater. The Eu anomaly arises because many crustal rocks of granitic and granodioritic composition were produced by intracrustal partial melting. The residues of those melts were rich in plagioclase, hence retaining more Eu in the lower crust and creating a complementary Eu-depleted upper crust. Sediments and seawater inherit this Eu anomaly from their source rocks in the upper continental crust.

Many sedimentary rocks have rare earth patterns that are similar to each other, and to that of the continental crust. To accentuate the difference in rare earth patterns between sediments, geochemists often normalize rare earth abundances to the concentrations in

Table 7.3 Normalizing values used for rare earth plots.

	Ordinary chondrites	Orgueil (CI chondrite)	Leedy	PAAS	Average shale
La	0.329	0.241	0.378	44.56	38.2
Ce	0.865	0.619	0.976	88.25	79.6
Pr	0.126	0.0939		10.15	8.83
Nd	0.630	0.474	0.716	37.32	33.9
Sm	0.203	0.154	0.23	6.88	5.55
Eu	0.077	0.059	0.0866	1.215	1.08
Gd	0.276	0.207	0.311	6.043	4.66
Tb	0.051	0.0380		0.891	0.774
Dy	0.343	0.256	0.39	5.325	4.68
Ho	0.077	0.0564		1.053	0.99
Er	0.225	0.166	0.255	3.075	2.85
Tm	0.035	0.0261		0.451	0.405
Yb	0.220	0.169	0.249	3.012	2.82
Lu	0.034	0.0250	0.0387	0.439	0.433

"Ordinary chondrites" is from Nakamura (1974) (modified to interpolate missing values), Orgueil are the values tabulated by Lodders (2010), Leedy, an ordinary chondrite, is from Masuda and Nakamura (1973), PAAS is Post-Archean Australian Shale composite from Poumand et al. (2012) and "average shale" is from McClennan (2018).

average shale. Again, there are several sets of normalizing values (two sets are given in Table 7.3, others may be found in Rollinson, 1993), but the relative abundances are all similar. Figure 7.8 shows examples of shale-normalized rare earth patterns.

Because the rare earths are highly insoluble and immobile, rare earth patterns often remain unchanged during weathering and metamorphism. Hence rare earth patterns can provide information on the pre-metamorphic history of a rock. Indeed, even during the production of sediment from crystalline rock, the rare earth patterns often remain little changed, and rare earth patterns have been used to identify the provenance (i.e., the source) of sedimentary rocks. Although their solubility in aqueous solutions is low, the REE rarely precipitate from solution. Instead, their abundance is limited in solutions, particularly seawater, by adsorption onto particle surfaces (Chapter 6a); they are said to be *particle reactive*. Rare earth patterns have also become useful tools in chemical oceanography, now that modern analytical techniques allow their accurate determination despite concentrations in the parts per trillion range, as we shall see in Chapter 14. As Figure 7.8 shows, seawater is strongly depleted in Ce, a consequence of it being partly or entirely in the IV valence state and being much more readily removed from seawater by reaction with particle surfaces

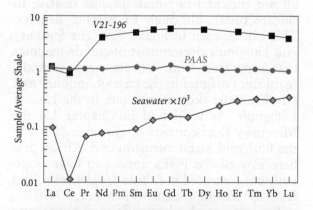

Figure 7.8 Shale-normalized REE patterns of a Pacific pelagic sediment (V21-196; Ben Othman et al., 1989), the Post-Archean Australian Shale composite, and typical seawater (Table 14.1). Both the pelagic sediment and seawater display a negative Ce anomaly, a consequence of Ce being in the 4+ oxidation state.

than the other REE. This Ce "anomaly" can be quantified in an analogous way to the Eu anomaly:

$$Ce/Ce^* = Ce_N/\sqrt{La_N \times Pr_N} \qquad (7.2)$$

7.2.2.5 The HFS elements

The *high field strength (HFS) elements* are so called because of their high ionic charge: Zr and Hf have +4 valence states and Ta and Nb have +5 valence states. Th and U are sometimes included in this group. As we noted, Th has a +4 valence state and U either a +6 or +4 valence state. Because of their high charge, all are relatively small cations, with ionic radii of 64 pm for Nb^{5+} and Ta^{5+}, and 72 and 76 pm for Zr^{4+} and Hf^{4+} respectively (U^{4+} and Th^{4+} are larger, however). Although they are of appropriate size for many cation sites in common minerals, their charge is too great and requires one or more coupled substitutions to maintain charge balance. As we noted earlier, such substitutions are energetically unfavorable. Thus, Hf and Zr are moderately incompatible elements while Nb and Ta are highly incompatible elements. These elements are less electropositive than the alkalis, and alkaline and rare earths. That, as well as their high charge and the involvement of *d* orbitals (which are highly directional) in bonding in the case of Ta and Nb, means that there is a greater degree of covalency in the bonds they form. Thus, the simple charged sphere is a less satisfactory model of their behavior.

As a consequence of their high ionic potential, or ionic charge to ionic radius ratio, the HFS elements are particularly insoluble, and like the REE they are particle-reactive. As a result, these elements tend to be very immobile during weathering and metamorphism. They are therefore valuable in the study weathering and soil development and in the study of ancient igneous rock suites as they can sometimes provide insights into the environment in which those rocks formed. Ta and Nb are present in anomalously low concentrations in magmas associated with subduction zones (indeed, this is considered a diagnostic feature of subduction-related volcanism). Although this depletion is not well understood, it is probably at least in part a consequence of the low solubility of these elements and the consequent failure of aqueous fluids generated by dehydration of the subducting oceanic crust to transport these elements into the magma genesis zone.

7.2.2.6 The first series transition metals

The chemistry of the transition elements is considerably more complex than that of the elements we have discussed thus far. There are several reasons for this. First, many of the transition elements have two or more valence states in nature. Second, the transition metals have higher electronegativity than the alkali and alkaline earths, so that covalent bonding plays a more important role in their behavior. Bonding with oxygen in oxides and silicates is still predominantly ionic, but bonding with other nonmetals, such as sulfur, can be largely covalent. A final complicating factor is the geometry of the d-orbitals, which are highly directional and thus bestow upon the transition metals specific preferences for the geometry of coordinating anions, or ligands. We will discuss this aspect of their behavior in more detail in a subsequent section.

The solubility of the transition metals, though generally lower than that of the alkalis and alkaline earths, is quite variable and depends on valence state and the availability of anions with which they can form soluble coordination complexes. In general, however, their solubility in most natural waters under oxidizing conditions and near-neutral pH is limited and they are considered particle-reactive. Their behavior in magmas is also variable. They range from moderately incompatible (e.g., Ti, Cu, Zn) to very compatible (e.g., Cr, Ni, Co), but their exact behavior is generally a stronger function of composition (of both solid and melt phases) than that of the highly incompatible elements. With the exception of Mn, the first transition series metals are also siderophile and/or chalcophile.

7.2.2.7 The noble metals

The platinum group elements (Rh, Ru, Pd, Os, Ir, Pt), often called PGEs or platinoid metals, are transition metals of periods 5 and 6. Because of the lanthanide contraction, the radii of the heavy PGEs, Os, Ir, and Pt, are similar to those of the light ones, Rh, Ru, and Pd. Together with gold and rhenium, these elements are sometimes collectively called the *highly siderophile elements* or *noble metals*. They are so-called for several reasons: first, they are the most strongly siderophile of elements: experiments show that they concentrate by factors of 10^4 or more into liquid metal in equilibrium with a silicate one. Second, they are unreactive and quite stable in metallic form. Finally, they are rare: their concentrations in the silicate Earth are <1% of their concentrations in chondritic meteorites (Figure 7.9). This depletion reflects their highly siderophilic character: most of the Earth's inventory of these elements is in the core. Because of their low concentrations, their behavior is still incompletely understood.

In a manner analogous to the rare earths, PGE data are sometimes presented as plots of the chondrite-normalized abundance, as in Figure 7.9. In this case, however, elements are not ordered by atomic number, but rather by decreasing melting point (which ranges from downward from 3186°C for Re to 1555°C for Pd and 1064°C for Au). This, as it turns out, places them in order of decreasing compatibility, with the exception of Re. Within the mantle, the noble metals decrease in compatibility in the order Os>Ir>Ru>Rh>Pt>Pd>Re>Au with Os and Ir being strongly compatible at the one extreme and Re and Au, which are moderately to strongly incompatible, at the other.

The noble metals are also chalcophile (i.e., all are enriched in sulfide liquids relative to silicate ones), although to varying degrees. This can be seen in Figure 7.9. The Sudbury and Langmuir ores consist of sulfide magmas that separated from silicate ones: a mafic (gabbroic) magma in the case of Sudbury and an ultramafic (komatiitic) one in the case of Langmuir. As we'll find in Chapter 15, the Merensky Reef consists of chromite layers in the Bushveld mafic intrusion of S. Africa, and here as well the PGEs appear to have originally concentrated in sulfide droplets exsolved from the silicate magma (Hiemstra, 1979).

In mafic and ultramafic rocks such as gabbros and peridotites, Ir, Os, Ru, and Rh (the Ir-group) are hosted in Fe-Ni monosulfide solid solutions that form rounded inclusions in olivine while the Pt and Pd (the Pd-group) are associated with Cu-Ni sulfides. During partial melting of the mantle, the Cu-Ni sulfides enter the melt, whereas the Fe-Ni monosulfides

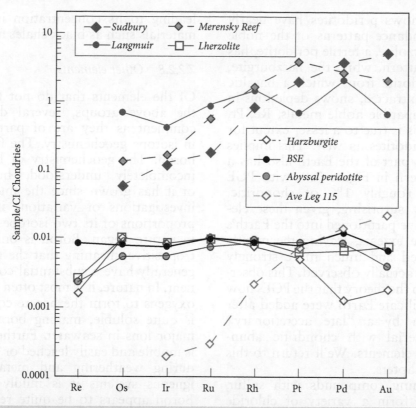

Figure 7.9 Chondrite-normalized abundances of the noble metals in ore deposits, peridotites, and basalts. "Sudbury" is the magmatic sulfide from the Little Strobie #1 Mine of the Sudbury intrusion, Canada (Naldrett et al., 1979); "Langmuir" is the sulfide of the Langmuir komatiite, Canada (Naldrett et al., 1979); "Merensky Reef" is a composite of the Merensky Reef chromite layer of the Bushveld Complex, South Africa (Hiemstra, 1979). "Lherzolite" is an average of lherzolitic peridotites from Lherz, France (Fischer-Goode et al., 2011), "Harzburgite" is a harzburgitic peridotite from Lherz (Fischer-Goode et al., 2011), "BSE" is the estimated bulk silicate earth or primitive mantle composition from Puchtel (2018), "Abyssal Peridotite" is the average of peridotites recovered from mid-ocean ridges and associated fracture zones (Snow and Schmidt, 1998), and "Ave Leg 115" is the average of basalts drilled from the Indian Ocean on ODP Leg 115 (Fryer and Greenough, 1992). CI chondrite normalizing values are from Fisher-Goode et al. (2010) and are 41, 491, 462, 688, 130, 943, 563, and 149 *ppb* for Re, Os, Ir, Ru, Rh, Pt, Pd, and Au, respectively.

remain in the residue until the degree of melting reaches ~25%, at which point the latter increasingly enter the melt as the degree of melting continues to increase. This partly accounts for the variable compatibility of these elements. Basalts produced by low degrees of partial melting are particularly strongly depleted in the Ir-group relative to the Pd-group (for example, the average ODP Leg 115 basalts in Figure 7.9). Picrites and komatiites produced by higher degrees of partial melting have higher PGE contents and are only slightly to moderately depleted in the Ir-group relative to the Pd-group (Puchtel, 2018). This partitioning also explains some

of the differences in sulfide deposits seen in Figure 7.9. Sudbury is relatively enriched in Pt, Pd, and Au and depleted in Os and Ir because of the way these elements partitioned into the original basaltic magma during melting. In contrast, the original Langmuir magma was komattitic, produced by a large extent of melting and is nearly as enriched in the Ir-group elements as the Pd-group ones. In residual peridotites, such as harzburgites, that have experience melt extraction and consequent sulfur depletion, the Ir-group may be hosted in native metal grains, such as osmiridium (an approximately 50:50 Os–Ir alloy).

Figure 7.9 shows peridotites have nearly chondritic abundance patterns of the noble metals. The lherzolite, a fertile peridotite, has a nearly flat pattern, while the harzburgite, a residual peridotite from which a basaltic melt has been extracted, shows depletions in the more incompatible noble metals, Re, Pt, Pd, and Au. This is true to a lesser extent for the abyssal peridotites as well. This implies that the silicate part of the Earth also has a nearly flat pattern in Figure 7.9, with PGE concentrations roughly 1% of chondritic. This is actually surprising, given these elements would have partitioned into the Earth's core to variable extents, leaving the mantle variably depleted and much more strongly depleted than is actually observed. This observation has led to the theory that the PGEs now present in the silicate Earth were added after core formation by a "late accretionary" veneer of material with chondritic abundances of these elements. We'll return to this topic in later chapters.

Besides forming compounds with sulfur, these elements form a variety of chloride and other halide complexes, which play an important role in the formation of some PGE deposits and certainly in many gold deposits (Chapter 15). These elements are transition elements and, like the first transition series, can exist in multiple valence states, ranging from 0 to +8, and have bonding behavior influenced by the d-orbitals. Thus, their chemistry is complex. When in high valence states, some, for example Os and Ru, form oxides that are highly volatile. Under most conditions, however, they are highly insoluble in aqueous solutions.

Rhenium is of interest because it decays radioactively to Os, as we'll see in the following chapter. Though it is not one of the platinum group elements, Re is adjacent to them in the periodic table and shares many of their properties, including being highly refractory under reducing conditions, being both highly siderophile and chalcophile, having a variety of potential valence states, having a large E_H, so that the metal is relatively resistant to oxidation, and forming a volatile (though only moderately so) oxide species. In oxidized aqueous solutions at the Earth's surface it is present as the perrhenate ion, ReO_4^-, and is quite soluble. However, it is readily adsorbed from solution under reducing conditions,

leading to its concentration in organic-rich materials such as black shales and coals.

7.2.2.8 Other elements

Of the elements that do not fit into any of the above groups, several deserve special comment as they are of particular interest in isotope geochemistry. The first of these is boron. The geochemistry of boron remains incompletely understood, but knowledge of it has grown since the mid-1980s when investigations of variations in the relative proportions of its two isotopes, ^{10}B and ^{11}B, in nature began. Boron is only mildly electropositive, meaning that the bonds it forms generally have a substantial covalent component. In nature, it is most often bound to three oxygens to form the borate complex. Borate is quite soluble, making borate one of the major ions in seawater. Furthermore, borate is mobile and easily leached or added to rocks during weathering and metamorphism. In igneous systems it is mildly incompatible. Boron appears to be quite readily removed from subducting oceanic crust and sediment by fluids and enriched in subduction-related magmas.

Lead is of great interest, not only because of its economic importance but also because it is the product of radioactive decay of ^{232}Th, ^{235}U, and ^{238}U and is quite toxic. The latter is of concern because its widespread use, particularly in paint and gasoline, has led to widespread pollution. Pb is chalcophile, though not so much as the PGEs and Re, and perhaps slightly siderophile. It is also quite volatile. Pb is in the +2 state throughout virtually the entire range of natural redox conditions. Pb has an electronegativity of 1.9, indicating a greater degree of covalency in its bonding than for the alkalis, alkaline earths, and rare earths. The solubility of Pb is reasonably low under most conditions, but it can form strong complexes with elements such as Cl and it can be readily transported in metamorphic and hydrothermal solutions. Its ionic radius is 119 pm in octahedral coordination, virtually identical to that of Sr. In igneous systems it is moderately incompatible, as might be expected from its ionic radius and charge. It is typically highly insoluble in natural waters at near-neutral pH and is highly particle-reactive. Lead, like many heavy metals, is toxic and its widespread use has led to environmental and health problems that we'll explore these in Chapter 15.

The concentration of phosphorus is sometimes high enough that it is treated as a major element. With a valence of +5 and being moderately electropositive, it is generally present in nature as the oxyanion. PO_4^{3-}, in which it is doubly bound to one of the oxygens and singly bound to the others. In rocks, it forms the common mineral apatite $(Ca_3(PO)_4(OH,F,Cl))$ as well as rarer minerals such as monazite. It behaves as a moderately incompatible element in mafic and ultramafic igneous systems. Phosphorus is, of course, an essential element of life: it is part of the structural framework of RNA and DNA, it transports energy within cells as adenosine triphosphate (ATP), in phospholipids provides the main structural components of all cellular membranes, and is a key structural component of bones in mammals. In both marine and terrestrial ecosystems, the availability of phosphorus can limit primary productivity (i.e., photosynthesis) and it is consequently a key ingredient in fertilizers critical to high-productivity modern agriculture. Marine life requires 1 phosphorus atom for every 106 carbon atoms; in terrestrial life the ratio varies between 100 and 1000.

The elements of the U and Th decay series have no stable nuclei. They exist on the Earth only because they are continually created through the decay of U and Th. They are of interest in geochemistry only because of their radioactivity. As we shall see in the next chapter, they can be useful geochronological tools. As radioactive substances, they represent potential environmental hazards and are of interest for this reason as well. Radon is perhaps the element that causes the greatest concern. As a noble gas, it is quite mobile. Relatively high levels of this gas can accumulate in structures built on uranium-rich soils and rocks. In this group we could also include the two "missing" elements, Tc and Pm. Like the U-decay series elements, they have no long-lived nuclei. They are, however, present in the Earth in exceedingly minute amounts, created by natural fission of U as well as by capture of fission-produced neutrons. Merely detecting them is a considerable challenge.

Ga and Ge can substitute, albeit with difficulty because of their larger radii, for their more abundant neighbors directly above them in the periodic table, Al and Si. Both are thus moderately incompatible elements. Their concentrations in the silicate Earth are somewhat low because of their siderophile nature. Germanium's greatest contribution to geochemistry may, however, be in experimental studies. Germanates created by substituting Ge for Si are used to simulate the properties of silicates at high pressures. This approach works because the oxygen ion is far more compressible than are cations. Thus, the ratio of the ionic radius of oxygen to silicon at high pressure is similar to that of oxygen to germanium at low pressure. Such studies of "germanium analogs" considerably advanced the understanding of the Earth's deep interior decades before the technology existed to reproducing the pressure of the deep Earth in the laboratory. Studies of germanium analogs continue in parallel with ultrahigh pressure experiments.

The geochemistry of the remaining elements, particularly in igneous processes, is poorly understood. Because of their low abundances, there have been few analyses published of these elements. Progress is being made on this front, however. For example, Newsom et al. (1986) demonstrated that the behavior of the chalcophile element Mo closely follows that of the rare earth Pr. W appears to be highly incompatible and its behavior mimics that of Ba. In oxidizing solutions, Mo forms a very soluble oxyanion complex, MoO_4^{2-}, so that its concentration in seawater is relatively high. Mo, despite its scarcity. place a key biological role in that it is part of the enzyme that bacteria use to break apart the N_2 molecule, the first step in converting nitrogen to biological useful forms.

Tin (Sn) and antimony (Sb) usually behave as moderately incompatible elements, with behaviors in igneous systems that are similar to that of the rare earth Sm (Jochum et al., 1985, 1993). However, these elements appear to form soluble species so, unlike Sm, they can be relatively mobile. It appears, for example, that they are readily stripped from subducting oceanic crust and carried into the magma genesis zones of island arc volcanoes.

7.3 DISTRIBUTION OF TRACE ELEMENTS BETWEEN COEXISTING PHASES

7.3.1 The partition coefficient

Geochemists find it convenient to define a *partition or distribution coefficient* as:

$$D_i^{\alpha-\beta} = \frac{C_i^\alpha}{C_i^\beta} \qquad (7.3)$$

where C is concentration, i refers to an element (or species) and α and β are two phases. By convention, if one phase is a liquid, the concentration ratio is written solid over liquid:

$$D_i^{s/\ell} = \frac{C_i^s}{C_i^\ell} \qquad (7.4)$$

where s refers to some solid phase and ℓ refers to the liquid phase. The distribution coefficient is a convenient concept for relating the concentration of some element in two different phases. It is also readily measured, either experimentally or "observationally." In the former, two phases are equilibrated at the temperature and pressure of interest and the concentration of i is subsequently measured in both. In the latter, the concentration of element i is simply measured in two natural phases thought to be in equilibrium. In both cases, the catch is, of course, equilibrium. As we found in Chapter 5, kinetic effects can lead to apparent partition coefficients that differ from equilibrium ones.

Having introduced the partition coefficient, we can now quantitatively define two terms we have already introduced: compatible and incompatible. *Incompatible* elements are those with $D^{s/l} \ll 1$. *Compatible* elements are those with $D^{s/l} \geq 1$. The partition coefficient for a given element will vary considerably between phases and can be less than 1 for one phase and greater than 1 for another. Hence the terms compatible and incompatible have meaning only when the phases are specified. These terms refer to partitioning *between silicate melts and phases common to mafic or ultramafic (i.e., basaltic or peridotitic) rocks*. It is this phase assemblage that dictates whether lithophile trace elements are concentrated in the Earth's crust, hence the significance of these terms.

7.3.2 Thermodynamic basis

Though trace elements are often treated differently from major elements, we should remember that the principle governing their distribution is the same as that governing the distribution of major elements. We are already familiar with this principle: *at equilibrium, elements will distribute themselves between coexisting phases so that the chemical potential of that element is the same in every phase in the system*. The chemical potential of element i in phases α and β is given by eqn. 3.54:

$$\mu_i^\alpha = \mu_i^{o\alpha} + RT \ln X_i^\alpha \lambda_i^\alpha \qquad (7.5)$$

and

$$\mu_i^\beta = \mu_i^{o\beta} + RT \ln X_i^\beta \lambda_i^\beta \qquad (7.6)$$

At equilibrium, $\mu^\alpha = \mu^\beta$ so:

$$\mu_i^{o\beta} - \mu_i^{o\alpha} = RT \ln(X_i^\alpha \lambda_i^\alpha / X_i^\beta \lambda_i^\beta) \qquad (7.7)$$

While eqn. 7.7 is expressed in mole fraction, the equation is identical when expressed in ppm because any difference in the molecular weights of the phases can be incorporated into the activity coefficients. A little algebra shows that the distribution coefficient is related to chemical potential as:

$$D_i^{\alpha-\beta} = \frac{C_i^\alpha}{C_i^\beta} = \frac{\lambda_i^\beta}{\lambda_i^\alpha} \exp\left(\frac{\mu_i^{o\beta} - \mu_i^{o\alpha}}{RT}\right) \qquad (7.8)$$

The standard chemical potentials on the right-hand side are the chemical potentials of i in pure i versions of phases α and β. Suppose, for example, we were interested in the partitioning of nickel between olivine and a silicate melt. In that case, $\mu_i^{o\alpha}$ would be the chemical potential of Ni in pure Ni olivine (i.e., Ni_2SiO_4) and $\mu_i^{o\beta}$ would be the chemical potential of Ni in Ni-silicate melt. This difference in chemical potential is the standard-state Gibbs free energy change resulting from transfer of i between these two phases, so:

$$D_i^{\alpha-\beta} = \frac{\lambda_i^\beta}{\lambda_i^\alpha} \exp\left(\frac{-\Delta G_i^{o\alpha-\beta}}{RT}\right) \qquad (7.9)$$

It is reasonable to expect that species present in trace quantities will obey Henry's law. In that case, we can rewrite eqn. 7.9 as:

$$D_i^{\alpha-\beta} = \frac{h_i^\beta}{h_i^\alpha} \exp\left(\frac{\mu_i^{o\beta} - \mu_i^{o\alpha}}{RT}\right) \quad (7.10)$$

where h is the Henry's law constant (Chapter 3). Drake and Weill (1975) found that Sr and Ba obeyed Henry's law in the system plagioclase–liquid up to concentrations of 5% or so of the element in the liquid (i.e., well above most natural concentrations). Many subsequent studies have confirmed that trace elements generally obey Henry's law over their entire range of natural concentration.

Comparing our derivation of the partition coefficient with that of the equilibrium constant shows that the two are closely related. Indeed, the partition coefficient is a form of apparent equilibrium constant and effectively identical with the K_D as defined in eqns. 3.89 and 3.90. The terms partition coefficient, distribution coefficient, and K_D are synonymous and often used interchangeably.

In a system with three phases, α, β, and γ, if α and β are in equilibrium and α and γ are in equilibrium, then β and γ must also be in equilibrium. It follows that:

$$D^{\beta/\gamma} = D^{\alpha/\gamma}/D^{\alpha/\beta} \quad (7.11)$$

This relationship has practical uses. For example, if we can determine the partition coefficient for an element between pyroxene and melt and between garnet and pyroxene, we can then calculate the garnet–melt partition coefficient for this element from eqn. 7.11.

7.4 FACTORS GOVERNING THE VALUE OF PARTITION COEFFICIENTS

7.4.1 Temperature and pressure dependence of the partition coefficient

In ideal solutions, the temperature dependence of the partition coefficient is the same as that of the equilibrium constant:

$$D_i = \exp\left(\frac{-\Delta G_i^o}{RT}\right) \quad (7.12)$$

ΔG in eqn. 7.12 can be expanded into entropy and enthalpy terms, as in eqn. 4.43:

$$\left(\frac{\partial \ln D_i}{\partial T}\right)_P = \frac{\Delta H^o + (P - P^o)\Delta V}{RT^2} \quad (7.13)$$

In ideal solution, and assuming ΔV is independent of temperature and pressure, the pressure dependence is also the same as that of the equilibrium constant, that is:

$$\left(\frac{\partial \ln D_i}{\partial P}\right)_T = \frac{-\Delta V}{RT} \quad (7.14)$$

From eqn. 7.14, we would predict a strong pressure dependence when the ionic radius of an element differs greatly from that of the available crystal lattice site. Thus, for example, we would predict the partition coefficient for K between pyroxene and melt would be strongly pressure-dependent since the ionic radius of K is 150 pm and is much larger than the size of the M2 site in clinopyroxene, which is normally occupied by Ca, with a radius of about 100 pm. Conversely, where the size difference is small (e.g., Mn (83 pm) substituting for Fe (78 pm)), we would expect the pressure dependence to be smaller.

For nonideal solutions (which will be the usual case), the temperature and pressure dependence will be more complex than given in eqns. 7.13 and 7.14 because the ratio of Henry's law constants in eqn. 7.10 will also be pressure- and temperature-dependent. The form of that dependence is usually difficult to deduce.

7.4.2 Ionic size and charge

Much of the interest in trace elements in igneous processes centers on the elements located in the lower left portion of the periodic table (alkalis K, Rb, Cs; alkaline earths Sr and Ba; the rare earths, Y, Zr, Nb, Hf, and Ta). The reason for this focus of attention is in part due to the relative ease with which these elements can be analyzed. These elements are all lithophile and therefore present at *relatively* high abundance in the Earth's crust and mantle. There is another reason, however: their chemical behavior is comparatively simple. All have electronegativities of less than 1.5 (Nb is the sole exception, with electronegativity of 1.6), and most have only a single valence state over the range of oxygen fugacity in the Earth. The difference in electronegativity between these elements and oxygen, effectively the only electronegative element in common igneous rocks, is 2 or greater, so the bonds these elements will form are predominantly ionic. To a reasonable approximation, then, the atoms of these elements behave as hard spheres containing a fixed point-charge at their centers. Thus, the two factors that most govern their chemical behavior are ionic radius and ionic charge. As we shall see, these elements have partition coefficients of less than 1 for most common minerals. The term *incompatible elements* often refers specifically to these elements.

The other trace elements that receive the most attention from igneous geochemists are the first transition series elements. Though their electronic structures and bonding behavior are considerably more complex (as we shall see in a subsequent section), charge and size are also important in the behavior of these elements. Many of these elements, particularly Ni, Co, and Cr, have partition coefficients greater than 1 in many Mg–Fe silicate minerals. Hence, the term *compatible elements* often refers to these elements.

The effect of ionic charge and size is illustrated in Figure 7.10, which shows a plot of ionic radius versus charge contoured for the clinopyroxene/melt partition coefficient for elements substituting into the M1 and M2 sites normally occupied by Mg, Fe, and Ca. Those elements whose charge and ionic radius most closely matches that of the major elements present in the cation sites have partition coefficients close to 1, while those whose charge or radius differs significantly have lower partition coefficients. Thus, Ba, even though its charge is the same (2+), has an ionic radius of 135 pm and fits only with difficulty into the lattice site normally occupied by Ca (100 pm), Mg (72 pm), or Fe (78 pm). This strains the lattice, and hence additional energy is required for the substitution to occur. On the other hand, Zr, which has an ionic radius identical to that of Mg, is not accepted in this site because its charge, 4+, is too great. Substitution of Zr^{4+} for Mg^{2+} would require either leaving one cation site vacant or one or more coupled substitutions (e.g., Al^{3+} for Si^{4+}) to maintain charge balance. Both of these are energetically unfavorable. In addition, ions having a radius much smaller than that of the element normally occupying the site also

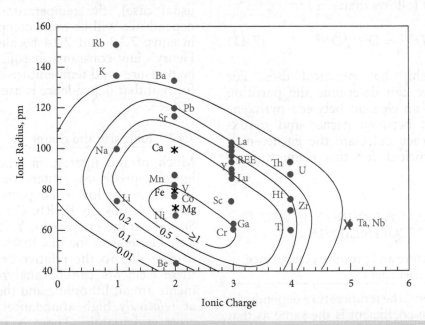

Figure 7.10 Ionic radius (picometers) vs. ionic charge contoured for clinopyroxene/liquid partition coefficients. Cations normally present in clinopyroxene M1 and M2 sites are Ca^{2+}, Mg^{2+}, and Fe^{2+}, shown by * symbols. Elements whose charge and ionic radius most closely match that of the major elements have the highest partition coefficients.

have low partition coefficients because their substitution also induces strain in the lattice.

7.4.2.1 Goldschmidt's rules of substitution

The importance of ionic radius and charge have long been known. Indeed, Goldschmidt developed the following rules regarding substitutions of elements into crystal structures:

1. If two ions have the same radius and the same charge, they will enter a given lattice site with equal facility.
2. If two ions have similar radii and the same charge, the smaller ion will enter a given site more readily.
3. If two ions have similar radii, the ion with the higher charge will enter a given site more readily.

Ringwood* noted the need also to consider the electronegativity in substitution. His rule is:

4. Whenever a substitution is possible between two elements with significantly different electronegativities ($\Delta > 0.1$), the one with the lower electronegativity will be preferentially incorporated.

However, the bottom-line is, the stronger the bond, the more likely the substitution.

7.4.2.2 Quantitative treatment of ionic size and charge

Goldschmidt's rules are simple qualitative statements based on empirical observation. Can we turn these into quantitative tools? The answer is yes. We can take advantage of the thermodynamic framework developed over the past century and incorporate into it the energetic effects of substituting a misfit ion into a specific lattice site to develop quantitative predictions of distribution coefficients. As we have seen, for example with the surface complexation model, thermodynamics generally, and the Gibbs free energy in particular, provide a wonderfully flexible framework into which we can incorporate new knowledge and understanding, often developed on the microscopic scale, to make quantitative predictions about the macroscopic behavior of chemical systems.

In brief, the idea is that when a trace element of different ionic radius is substituted for the ion that normally occupies a lattice site, the lattice must adjust. This adjustment is accomplished through the expenditure of strain energy, which can be calculated by measurable parameters of the mineral. Since this strain energy contributes to the Gibbs free energy of reaction, and since the partition coefficient is related to ΔG, the effect of the lattice strain energy on the partition coefficient can be calculated.

Clinopyroxenes are important because they are nearly ubiquitous in mafic and ultramafic rocks. Clinopyroxene has three cation sites, the tetrahedral site normally occupied by silicon, with the tetrahedra linked in chains by shared oxygens, the octahedral M1 site which lies at the apices of tetrahedra and the six- or eight-fold coordinated M2 site that lies between the tetrahedra bases (Figure 1.11). In the pure magnesian end-member, diopside, the M1 site is occupied by Mg and the M2 site is occupied by Ca. The M2 site is larger, enables clinopyroxene to accommodate a range of trace element cations more readily than other common mafic minerals. Smaller cations such as Co^{2+} can be accommodated in the M1 site while small trivalent cations such as Al^{3+} and Fe^{3+} can substitute either in the M1 or tetrahedral sites, which can maintain charge balance for a nondivalent substitution in the M2 site. Consequently, clinopyroxene exerts a strong control on trace element fractionation during melting and crystallization in these rocks and related magmas.

Consider the formation of diopside containing a divalent metal ion M^{2+}, such as Sr^{2+}

---

* Alfred (Ted) E. Ringwood (1930–1993). After receiving his BSc and PhD degrees at the University of Melbourne, Ringwood did a postdoctoral fellowship at Harvard University. His subsequent career was spent at the Australian National University, from where he exerted a strong influence on the development of earth science in Australia. Ringwood pioneered the post-war development of many aspects of geochemistry, including phase relationships of the mantle at very high pressures, the origin of basaltic magmas, the origin of the Earth and the Moon, as well as safe methods of long-term nuclear waste storage. Among his other contributions, he demonstrated that seismic discontinuities in the mantle were due to phase transitions, the existence of which he predicted using germanate analogs.

or Ba^{2+}, in the M2 site normally occupied by Ca^{2+}. We could write this reaction as:

$$MO^\ell + MgO^\ell + 2SiO_2^\ell \rightleftharpoons MMgSi_2O_6 \tag{7.15}$$

where ℓ denotes the liquid phase. However, the mineral $MMgSi_2O_6$ may not exist in nature and may not be synthesizable in the laboratory, so it might not be possible to determine its thermodynamic properties. Instead, we can imagine two reactions. The first would be the crystallization of diopside from a melt:

$$CaO^\ell + MgO^\ell + 2SiO_2^\ell \rightleftharpoons CaMgSi_2O_6 \tag{7.15a}$$

The second would be an exchange reaction such as the replacement of Ca^{2+} in the M2 site of diopside by metal ion M^{2+}:

$$M^{2+} + CaMgSi_2O_6 \rightleftharpoons Ca^{2+} + MMgSi_2O_6 \tag{7.15b}$$

We can see that reaction 7.15 is simply the difference between reaction 7.15b and 7.15a. Hence, the Gibbs free energy change of this reaction can be expressed as:

$$\Delta G_r = \Delta G_{exchange}^{M-Ca} - \Delta G_{melting}^{Di} \tag{7.16}$$

The first term is the free energy change involved in transferring an M^{2+} ion from the melt into the crystal lattice and simultaneously transferring a Ca^{2+} ion from the lattice site to the liquid. The second term is the free energy change associated with the melting of diopside, and in multicomponent systems this governs the distribution of Ca between diopside and the liquid. The distribution coefficient for element M then depends on these two components of free energy:

$$D_M^{Di/\ell} = \exp\left(\frac{\Delta G_{melting}^{Di} - \Delta G_{exchange}^{M-Ca}}{RT}\right) \tag{7.17}$$

According to the Bryce (1975) lattice strain energy theory adopted by Blundy and Wood (1994, 2003), $\Delta G_{exchange}$ is dominated by the energy associated with the lattice strain that results from the M^{2+} ion being a different size from the Ca^{2+} ion normally occupying the lattice site. Because the melt (at least at low pressure) has a far less rigid structure and is more compressible than the solid, any strain in the melt is essentially negligible compared with the strain in the solid. Hence:

$$\Delta G_{exchange}^{M-Ca} \cong \Delta G_{strain} \tag{7.18}$$

The strain energy, ΔG_{strain}, may be calculated from:

$$\Delta G_{strain} = 4\pi E N_A\left[\frac{r_0}{2}(r_M - r_0)^2 + \frac{1}{3}(r_M - r_0)^3\right] \tag{7.19}$$

where r_0 is the optimal radius of the lattice site, r_M is the ionic radius of M, N_A is Avogadro's number, and E is *Young's modulus*. The Young's modulus of a substance is the ratio of stress applied to the resulting strain and may be calculated as:

$$E = \frac{L_0 F}{\Delta L A} \tag{7.20}$$

where L_0 is the original length of the object undergoing strain, ΔL is the length change under stress, and F is the force applied over area A. E has units of pressure. Young's modulus is related to the bulk modulus, K, (which, as we found in Chapter 2, is the inverse of compressibility) through Poisson's ratio. Poisson's ratio is the ratio of the transverse contraction per unit dimension to elongation per unit length of a bar of uniform cross-section when subjected to tensile stress, and in most silicates is ~ 0.25. Consequently, $E \cong 1.5\ K$.

The $\Delta G_{melting}$ term in eqn. 7.17 governs the distribution of Ca between liquid and crystal:

$$D_{Ca}^{Di/\ell} = \exp\left(\frac{\Delta G_{melting}^{Di}}{RT}\right) \tag{7.21}$$

Substituting this and eqn. 7.19 into eqn. 7.17 we obtain:

$$D_M^{Di/\ell} = D_{Ca}^{Di/\ell} \exp\left(\frac{-\Delta G_{strain}^{Di}}{RT}\right) = D_{Ca}^{Di/\ell} \exp$$

$$\times\left(\frac{-4\pi E N_A\left[\frac{r_0}{2}(r_M - r_0)^2 + \frac{1}{3}(r_M - r_0)^3\right]}{RT}\right)$$

More generally, this equation may be written as:

$$D_i^{s/\ell} = D^o \exp$$

$$\left(\frac{-4\pi E N_A \left[\frac{r_0}{2}(r_M - r_0)^2 + \frac{1}{3}(r_M - r_0)^3 \right]}{RT} \right) \tag{7.22}$$

where D^o is the partition coefficient of an ion of radius r_o, which has the same charge as i and enters the lattice site without strain (Blundy and Wood, 1994).

Although the T term in eqn. 7.22 results in an increase in partition coefficients with temperature, partition coefficients generally decrease with increasing temperature. Decreasing T raises the free energy of melting ($\Delta G_{melting}$ in eqn. 7.21) and results in a decrease in D^o. The addition of water lowers the activity and activity coefficients of the trace components in it resulting in a decrease in partition coefficients. These effects are illustrated in Figure 7.11.

Now consider the case where the substituting trace element has a different charge from the ion normally occupying the site, for example:

$$La^{3+\ell} + CaMgSi_2O_6 \rightleftharpoons Ca^{2+\ell} + LaMgSi_2O_6 \tag{7.23}$$

Placing a trivalent ion into the site normally occupied by a divalent ion requires doing electrochemical work, which can be expressed as:

$$W = \frac{N_A(z_i - z_o)^2 e^2}{2\varepsilon r} \tag{7.24}$$

where z_i is the charge on the substituted ion, z_o is charge of the ion normally occupying the site, ε is the dielectric constant of the substance, and r is the effective radius of the charge defect. There's a tradeoff, however, between the electrostatic work done by placing the ion in the crystal versus placing in the melt. The electrostatic free energy of the substitution is thus the difference in the work done in these two cases:

$$\Delta G_{elec} = W_{xtl} - W_{melt} \tag{7.25}$$

The effect is to change the value of D_o:

$$D_o^{\Delta z} = e^{-\Delta G_{elec}/RT} \tag{7.26}$$

The charge imbalance must be compensated in one of two ways: either through a coupled substitution, such as:

$$La^{3+\ell} + Na^{+\ell} + 2CaMgSi_2O_6$$
$$\rightleftharpoons 2Ca^{2+\ell} + 2La_{0.5}Na_{0.5}MgSi_2O_6 \tag{7.27a}$$

or

$$La^{3+\ell} + Al^{3+\ell} + CaMgSi_2O_6$$
$$\rightleftharpoons Ca^{2+\ell} + Si^{4+} + LaMgAlSiO_6 \tag{7.27b}$$

or through a creation of a vacancy, for example:

$$2La^{3+\ell} + 3CaMgSi_2O_6$$
$$\rightleftharpoons 3Ca^{2+\ell} + 2LaMgSi_2O_6 + MgSi_2O_6 \tag{7.28}$$

If the charge balance mechanism is known, then the free energy of exchange may be calculated as the sum of free energies associated with melting, strain, and the charge balance mechanism. From this, the distribution coefficient can be calculated. For example, if charge is balanced by the coupled substitution in reaction 7.27b, then the distribution can be calculated as:

$$D_{La}^{Di/\ell}$$
$$= \exp \left(\frac{\Delta G_{melting}^{Di} - \Delta G_{strain}^{La-Ca} - \Delta G_{exchange}^{Al-Si}}{RT} \right) \tag{7.29}$$

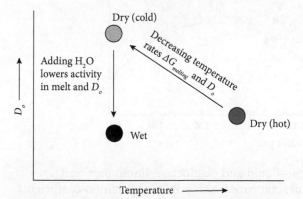

Figure 7.11 Effects of temperature and water in the melt on trace element partition coefficients. From Wood and Blundy (2014).

Equation 7.24 may be written as:

$$D_{La}^{Di/\ell} = \exp\left(\frac{\Delta G_{melting}^{Di} - \Delta G_{exchange}}{RT}\right)$$

$$\exp\left(\frac{-\Delta G_{strain}^{La-Ca}}{RT}\right) \quad (7.29a)$$

The first term on the right may be replaced by D^o, so that eqn. 7.22 may also be applied to heterovalent exchanges.

Of the possible charge-compensating substitutions, that of Al in the tetrahedral site (denoted ^{IV}Al) appears to be dominant. Wood and Blundy (2001) found that ΔG_{elec} (eqn. 7.25) was about 19 kJ/mol per unit charge for aluminous clinopyroxene (^{IV}Al = 0.195) and about 28 kJ/mol for low aluminum pyroxene.

Equation 7.22 suggests that the dependence of the partition coefficient on ionic radius should be highly nonlinear, and experiments have proved this to be the case. Figure 7.12 compares clinopyroxene–liquid and plagioclase–liquid partition coefficients for the rare earths determined experimentally at 1225°C and 1.5 GPa with partition coefficients predicted using eqn. 7.22. Plots

of partition coefficient versus ionic radius are known as Onuma diagrams after Onuma et al. (1968).

Partition coefficients depend on pressure as expressed in eqn. 7.14, so if we know the volume change of the exchange reaction, we can assess pressure effects by integrating eqn. 7.14 to the pressure of interest. However, pressure also affects the equilibrium composition of clinopyroxene crystallizing from a melt and the composition in turn affects the site radius, r_o. Clinopyroxene incorporates an increasing amount of jadeite component ($NaAlSi_2O_6$); i.e., Na in the M2 site and Al in the M1 site, as pressure increases, which results in a decrease in the radius of the M2 site. This can be expressed as:

$$r_o = 97.4 + 0.67X_{Ca}^{M2} - 0.51X_{Al}^{M1} \text{ pm} \quad (7.30)$$

In addition, Young's modulus is also pressure- and temperature-dependent, and this can be expressed as:

$$E_{M2}^{3+} = 318.6 + 6.9P - 0.036T \text{ GPa} \quad (7.31)$$

with pressure in GPa and temperature in Kelvins (Wood and Blundy, 2014). This shifts the partition coefficient parabola on the

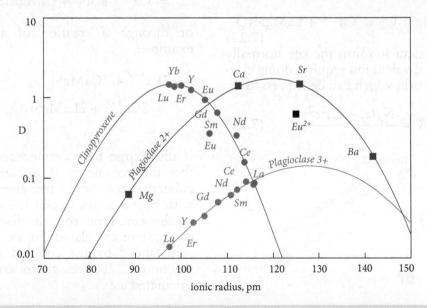

Figure 7.12 Experimentally determined clinopyroxene–liquid and plagioclase–liquid partition coefficients for rare earths as a function of ionic radius, compared with predicted partition coefficients using eqn. 7.20. For clinopyroxene, the model assumes D^o = 1.47, E = 262 GPa, and r_0 = 98.1 pm. For trivalent cations in plagioclase, these parameters are D^o = 0.14, E = 82 GPa, and r_0 = 128 pm and for divalent cations in plagioclase they are D^o = 1.71, E = 124 GPa, and r_0 = 120 pm. Based on Blundy et al. (1998) and Sun et al. (2017).

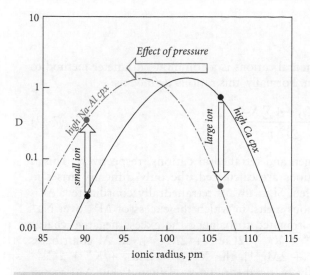

Figure 7.13 Because increasing pressure increases the amount of jadeite (NaAlSi$_2$O$_6$) in clinopyroxene, the radius of the M2 site decreases and partition coefficients for large ions decrease while those for small ions increase. After Wood et al. (1999).

Onuma diagram. As illustrated in Figure 7.13, partition coefficients of large ions will decrease with increasing pressure, while those of small ions will increase. For mantle pyroxenes, Sun and Liang (2012) found the effect of pressure was relatively small up to 2 or 3 GPa.

7.4.3 Compositional dependency

Equation 7.9 predicts that partition coefficients will depend on the composition of the phases involved, since the Henry's law constant will depend on the concentration of the major species. Experience has substantiated this prediction and shown that partition coefficients are composition-dependent. For example, elements that are incompatible in mafic (SiO$_2$-poor) systems often have partition coefficients equal to or greater than 1 in silicic (SiO$_2$-rich) systems. This behavior is illustrated by the solubility of zircon, ZrSiO$_4$. In basaltic liquids, zircon will not precipitate until the Zr concentration reaches 10,000 ppm (1% Zr) or more. In granitic liquids, however, zircon crystallizes at Zr levels of around 100 ppm. In another example, Watson (1976) conducted two-liquid partitioning experiments in which the concentrations of various elements were measured in coexisting immiscible mafic and silicic silicate liquids. P,

the rare earth elements (REEs), Ba, Sr, Mg, Zr, Mn, Ti, Cr, Ca, and Ta were found to be enriched in the mafic melt by factors of 1.5 to 10 (in the above order). Cs was enriched in the silicic melt by a factor of about 3.

Some of this compositional dependence can be related to structural changes in the melt phase. As we saw in Chapter 4, as the SiO$_2$ concentration of the melt increases, the polymerization of the melt increases with SiO$_2$ content as the Si tetrahedra become increasingly linked through shared, or *bridging*, oxygens. Si-poor melts will be largely depolymerized as more nonbridging oxygens are needed to charge-balance the network-modifying cations. In addition to Si, Al, Ti, P, and some or all of the Fe^{3+} are tetrahedrally coordinated in the melt phase and can contribute to polymerization. The ratio of *nonbridging oxygens* to these tetrahedral cations, written NBO/T, provides a measure of the degree of melt depolymerization (see Example 7.1). NBO/T ranges from >1 for silica-poor magmas such as picritic basalts to <0.1 for rhyolites.

Kohn and Schofield (1994) carried out a series of experiments demonstrating the effect of melt polymerization on forsterite/melt partitioning of Zn and Mn. Figure 7.14 shows that the Zn partition coefficient decreases from about 4.5 in highly polymerized melts (low NBO/T) to about 0.8 in depolymerized melts (high NBO/T). Zn is present in octahedral sites (i.e., surrounded by six oxygens) in silicate melts, though this remains uncertain. Changes in the partition coefficient most likely reflect the availability of suitably coordinated sites in the melt (see Example 7.1).

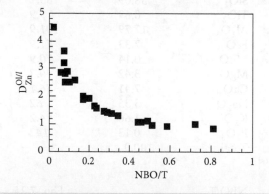

Figure 7.14 Variation of the zinc olivine/liquid partition coefficient as a function of the number of nonbridging oxygens per tetrahedral cation (NBO/T) in experiments by Kohn and Schofield (1994).

Example 7.1   Calculating NBO/T

NBO/T, the ratio of nonbridging oxygens to tetrahedral cations is a common parameter method to quantify the extent of silicate melt parameterization. Formally, this ratio is defined as:

$$NBO/T = \frac{2X_O - 4\sum X_T}{\sum X_T} \qquad (7.32)$$

where X_O and X_T are the mole fractions of oxygen and tetrahedral cations, respectively. This is a straightforward calculation once the mole fractions are calculated, the only difficulty arises in deciding which cations are tetrahedrally coordinated. Si is always tetrahedrally coordinated; Al is tetrahedrally coordinated except in rare peraluminous melts, in which the excess of Al^{3+} over Na^+ $+ K^+ + 2Ca^{2+} + 2Mg^{2+}$ will be octahedrally coordinated to balance the excess negative charge of tetrahedral oxygens. Thus, when $Al^{3+} < \{Na^+ + K^+ + 2Ca^{2+} + 2Mg^{2+}\}$, all Al contributes to $\sum X_T$, and when $Al^{3+} > \{Na^+ + K^+ + 2Ca^{2+} + 2Mg^{2+}\}$, the $X_{T-Al} = \{Na^+ + K^+ + 2Ca^{2+} + 2Mg^{2+}\}$. Ferric iron may be either tetrahedrally or octahedrally coordinated, but Mössbauer spectroscopy indicates that for $Fe^{3+}/\Sigma Fe < 0.3$, iron is octahedrally coordinated and becomes increasingly tetrahedrally coordinated for $Fe^{3+}/\Sigma Fe > 0.3$ and becomes predominantly tetrahedrally coordinated for $Fe^{3+}/\Sigma Fe > 0.5$ (Mysen, 1990). For most igneous rocks $Fe^{3+}/\Sigma Fe < 0.3$. Phosphorus is also tetrahedrally coordinated, but it is generally a minor constituent and has little effect on NBO/T.

An alternative equation for NBO/T is:

$$NBO/T = \frac{\sum nX_O - (X_{Al_T^{3+}} + X_{Fe_T^{3+}}) + X_{P_T^{5+}}}{\sum X_T} \qquad (7.33)$$

where X_{Oc} is the mole fraction of octahedrally coordinated cations and n is the charge on the cation, e.g., 2 for Ca^{2+} (all cations not in tetrahedral coordination are assumed to be in octahedral coordination). Subtraction of terms for Al, Fe^{3+}, and P accounts for the difference in charge on tetrahedra centered by tri- or quin-valent cations rather than the quadrivalent cations.

Problem: Calculate NBO/T for the Aleutian andesite, KS12-1, whose composition is given below.

	wt. %	mol. wt.	moles	molescations	moles O
SiO_2	58.09	60.1	0.97	0.967	1.933
TiO_2	0.69	79.9	0.01	0.009	0.017
Al_2O_3	17.79	101.9	0.17	0.349	0.524
FeO	7.33	71.9	0.10	0.102	0.102
MnO	0.14	70.9	0.00	0.002	0.002
MgO	3.42	40.3	0.08	0.085	0.085
CaO	7.41	56.1	0.13	0.132	0.132
Na_2O	3.35	62	0.05	0.108	0.054
K_2O	1.66	94.2	0.02	0.035	0.018
P_2O_5	0.13	142	0.00	0.002	0.005
Total	100.0			1.791	2.872
			ΣX_T	1.326	
			ΣX_{Oc}	0.785	
NBO/T		Eqn. 7.25		0.330	
		Eqn. 7.26		0.330	

Answer: Our first step is to convert the wt. percent values to moles by dividing by the molecular weights listed in the second column. The moles of each oxide component are listed in the third column. To calculate the number of moles of each cation, we multiply by the stoichiometric coefficient of the cation in the formula unit (e.g., 1 for Si, 2 for Al). If we use eqn. 7.32, we must also calculate the number of moles of O, multiplying by the O stoichiometric coefficient (e.g., 3 for Al_2O_3). We'll assume that any ferric iron is in octahedral coordination and all Al is in tetrahedral coordination, thus the ΣX_T is the mole sum of Si + Al + Ti + P, and X_O is the mole sum of oxygen. NBO/T is then calculated using equation 1. Notice that it is not necessary to divide the sums by the total moles to calculate mole fraction as the sum terms will cancel.

Alternatively, we can use eqn. 7.33, which is somewhat simpler as it does not require calculation of the number of moles of oxygen. Both approaches give the same result.

Another effect is a decrease in octahedral or other sites for transition and highly charged cations as melts become more polymerized. High field strength elements, such as Zr^{4+} and Ta^{5+}, need to be coordinated by a large number of anions. For example, Zr is coordinated by eight oxygens in zircon. Cations with large radii, such as Cs, do not easily fit in tetrahedral sites. Lacking suitable sites in silicic melts, they partition preferentially into solid phases.

Example 7.2 Parameterizing partition coefficients

One approach to accounting for compositional variability when the underlying thermodynamics is not fully understood is to parameterize the partition coefficient, i.e., to express it as a function of other measurable or predictable variables. The Zn partition coefficients determined by Kohn and Schofield are clearly strong functions of the NBO/T parameter, suggesting we might usefully try to express the Zn partition coefficient as a function of NBO/T. From eqn. 7.7, we can expect that the partition coefficient will be a function of temperature as well. Furthermore, we can use this equation to predict the form of the equation. Taking the log of both sides of eqn. 7.9, we have:

$$\ln D = \ln\left(\frac{\lambda_i^\beta}{\lambda_i^\alpha}\right) - \frac{\Delta G^\circ}{RT} \tag{7.9}$$

The first term on the right expresses compositional dependence and the second term expresses the temperature dependence. Thus, we might expect an equation of the form:

$$\ln D = a\ln\left(\frac{NBO}{T}\right) + \frac{b}{T}$$

We can then use a technique called *polynomial regression*, a multivariable extension of linear regression, to determine the constants *a* and *b*. Using this approach, we obtain the following equation:

$$\ln D = -0.405\ln\left(\frac{NBO}{T} + 6.077\right) + \frac{3594}{T} \tag{7.34}$$

Figure 7.15 compares the observed Zn partition coefficients with those predicted by equation 7.34.

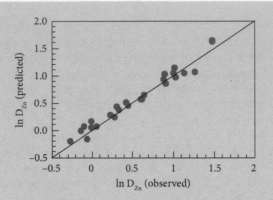

Figure 7.15 Comparison of Zn olivine/melt partition coefficients of Kohn and Schofield with values predicted by eqn. 7.34.

This approach does have limitations and dangers, however. In the experiments of Kohn and Schofield (1994), for example, temperature and NBO/T were highly correlated, thus some of the temperature dependence may be hidden in the first term of the equation. Second, it is unlikely that all compositional effects on the partition coefficient can be adequately expressed by a single parameter. Furthermore, our equation does not take variations in the composition of olivine into account. In addition, Fe–Mg solid solution in natural olivines of the solid may also affect partitioning behavior. Nevertheless, eqn. 7.32 is highly successful and accounts for over 96% of the variation in the partition coefficient observed by Kohn and Schofield (1994).

In contrast to olivine, whose composition is independent of melt composition other than the Fe-Mg substitution, clinopyroxene composition varies with melt composition, making it difficult to distinguish between the effects of clinopyroxene composition and melt composition on partition coefficients. Based on new experiments as well as previous ones by Huang et al., (2006) and Gaetani (2004), Michely et al. (2017) found that dependence of REE clinopyroxene-melt could be expressed as:

$$D_{REE}^{cpx/melt} = a \left[\frac{NBO}{T} \right]^{-1} \qquad (7.35)$$

The factor a varies with ionic radius and fits the parabolic function of eqn. 7.22 with E and r_0 values identical, within error, to those found by Wood and Bundy (1994) (258 kJ vs 262 kJ and 98.5 vs 98.1 pm, respectively), but the value of D_o differs (0.384 vs the value of 1.47 of Blundy and Wood). Thus, the two equations can be united as:

$$D_{REE}^{cpx/melt} = D^o \exp$$

$$\times \left(\frac{-4\pi E N_A \left[\frac{r_0}{2}(r_M - r_0)^2 + \frac{1}{3}(r_M - r_0)^3 \right]}{RT} \right)$$

$$\times \left[\frac{NBO}{T} \right]^{-1} \qquad (7.36)$$

After correcting for melt structure in this way, Michely et al. (2017) found that partition coefficients increased linearly with the Al content of the tetrahedral site (^{IV}Al) in clinopyroxene. The increase results in a general increase in the D_0 value in eqn. 7.36, consistent with the work of Wood and Blundy (2001), with a slight shift of a few pm in the site radius (r_o) (Figure 7.16).

They also found that REE partition coefficients increased with the Na_2O content of the melt (Figure 7.16) and speculated that this resulted because Na^+ could breakup REE^{3+}–Al complexes in the melt, which inhibit incorporation of REE in pyroxene.

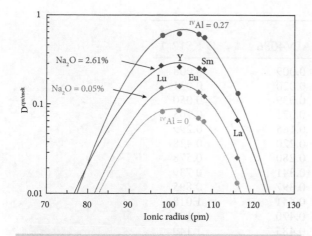

Figure 7.16 Effect of tetrahedral aluminum mole fraction and Na_2O melt concentration on clinopyroxene REE partition coefficients. Based on Michely et al. (2017).

Na_2O melt content had no discernable effect on either r_o or E and can be entirely described as an increase in D_o in eqn. 7.33. Example 7.3 illustrates this.

7.4.4 Mineral–liquid partition coefficients for mafic and ultramafic systems

As we have seen, partition coefficients depend on temperature, pressure, and the composition of the phases involved. There are nevertheless circumstances when a general set of partition coefficients is useful. For example, temperature or melt composition may not be precisely known, or great accuracy may not be needed. Table 7.4 is a set of mineral–melt partition coefficients for mafic and ultramafic magmas over temperature ranges of 1100°C (for amphibole, which is not stable above that temperature) to 1450°C and pressures generally <3GPa. They have been fitted to the Bryce lattice strain model where possible. Even within these constraints, absolute partition coefficients will vary substantially, but the partitioning patterns will generally be similar. Additionally, experimental and analytical uncertainties are considerable for some very small partition coefficients (e.g. in olivine and spinel), but once the value of a partition coefficient is less than about 0.01, its exact value will make very little difference in modeling.

Figure 7.18 illustrates the rare earth partition coefficients from this dataset. In general, the minerals clinopyroxene, garnet and plagioclase and, when present, amphibole (amphibole is not usually present in basalts because it is not stable at low pressure or above 1100°C) will control the patterns of incompatible element partitioning during melting and crystallization of basaltic

Example 7.3 Calculating partition coefficients

Problem: Compare the REE clinopyroxene-melt partition coefficients between the Aleutian andesite KS12-1, whose composition is given in Example 7.1 and a mid-ocean ridge basalt, ALV4086-1925, whose composition is listed in the table below. Assume $D_o = 0.384$, $E = 258\,kJ$, and $r_o = 98.5$ pm and the ionic radii from Rollinson and Adetunji (2018) listed below Assume a temperature of 1000 K for the basalt and 800 K for the andesite.

	ALV4086-1925
SiO_2	51.12
TiO_2	1.66
Al_2O_3	14.17
FeO	10.93
MnO	0.22
MgO	7.83
CaO	11.03
Na_2O	2.61
K_2O	0.06
P_2O_5	0.35

	r_i (pm)	D-ALV4086	D KS12-1
La	116	0.009	0.008
Ce	114.3	0.020	0.021
Pr	112.6	0.039	0.050
Nd	110.9	0.071	0.105
Sm	107.9	0.165	0.299
Eu	106.6	0.220	0.428
Gd	105.3	0.280	0.578
Tb	104	0.341	0.739
Dy	102.7	0.397	0.895
Ho	101.5	0.441	1.018
Er	100.4	0.470	1.103
Tm	99.4	0.485	1.149
Yb	98.5	0.490	1.163
Lu	97.7	0.486	1.152

Answer: In Example 7.1 we found that NBO/T was 0.330 for the andesite; using the same procedure, we calculate NBO/T = 0.784 for the basalt. Using eqn. 7.36, we calculate partition coefficients listed in the adjacent table. Partition coefficients are comparable for the lightest rare earths but a factor of 2 or more greater for the heavy earths. The difference reflects the combined effect of temperature and melt structure. We have, however, not considered the effects of melt Na_2O and clinopyroxene composition on partition coefficients, which could further increase D values in the andesite. The partition coefficients are plotted in Figure 7.17. We have also not considered the effect of temperature on D_o; the lower temperature would further increase partition coefficients in the andesite.

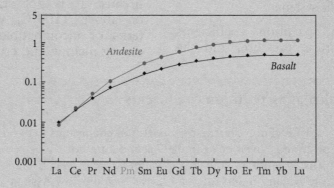

Figure 7.17   Comparison of calculated REE patterns for a representative andesite and basalt using the lattice strain model corrected for melt structure using the NBO/T parameter.

magmas because they have the highest partition coefficients. Olivine, though by far the most abundant mineral in the upper mantle, will produce little fractionation* of incompatible elements because its partition coefficients are so low. Spinel, which is usually not present in more than a few volume percent, will also have little effect on most relative trace element abundances. On the other hand, olivine largely controls the fractionation of the compatible transition metals.

---

* *Fractionation*, in this context, refers to a change in the relative abundances, or ratios, of elements. For example, if the ratio of La to Sm changes during a process, these elements are said to have been fractionated by that process.

Table 7.4 Mineral-melt partition coefficients.

	Olivine	Opx	Cpx	Plag	Spinel	Garnet	Amph
Li	0.29	0.22	0.25	0.34	0.13	0.04	0.1
K	0.00027	0.003	0.004	0.2	< 0.0001	0.05	0.37
Sc	0.133	0.25	2.2	0.34	0.3	2.6	4.2
Rb	0.0002	0.001	0.0004	0.043	< 0.0001	0.007	0.08
Sr	0.0006	0.001	0.17	1.59	0.05	0.01	0.44
Y	0.0047	0.04	0.54	0.016	0.0003	1.34	1.17
Zr	0.0048	0.0035	0.14	0.015	0.002	0.6	0.48
Nb	1.0×10^{-4}	0.001	0.024	0.044	0.08	0.03	0.38
Cs	< 0.0002	0.0009	0.0001	0.0025	< 0.0001	0.0005	0.1
Ba	< 0.0002	0.0001	0.0001	0.164	0.0001	0.0007	0.42
La	4.32×10^{-7}	0.0001	0.06	0.077	0.0003	0.0001	0.16
Ce	3.30×10^{-6}	0.0005	0.09	0.072	0.0003	0.001	0.28
Pr	1.75×10^{-5}	0.0017	0.14	0.065	0.0003	0.003	0.42
Nd	3.81×10^{-5}	0.0028	0.2	0.057	0.0003	0.010	0.64
Sm	0.00018	0.0073	0.32	0.041	0.0003	0.082	1.01
Eu	0.00037	0.0111	0.38	1.67	0.0003	0.173	1.11
Gd	0.00072	0.0152	0.44	0.029	0.0003	0.338	1.3
Tb	0.0015	0.0240	0.48	0.023	0.0003	0.609	1.36
Dy	0.0027	0.0321	0.53	0.018	0.0003	1.014	1.29
Ho	0.0045	0.0412	0.55	0.015	0.0003	1.517	1.20
Er	0.0073	0.0509	0.57	0.012	0.0003	2.076	1.11
Tm	0.0110	0.0597	0.57	0.0096	0.0003	2.637	1.01
Yb	0.0171	0.0694	0.57	0.0079	0.0003	3.154	0.91
Lu	0.0217	0.0743	0.56	0.0066	0.0003	3.595	0.82
Hf	0.0064	0.01	0.31	0.013	0.002	1	0.8
Ta	6.0×10^{-4}	0.001	0.071	0.023	0.001	0.033	0.3
Pb	0.00062	0.009	0.067	0.134	0.0005	0.005	0.05
Th	1.6×10^{-6}	0.0008	0.023	0.022	0.0004	0.013	0.01
U	1.0×10^{-6}	0.002	0.026	0.07	0.0004	0.029	0.01

Values from Sun and Liang (2012), Sun et al. (2017), Bedard (2005), Wijbrans et al. (2015), Tiepolo et al. (2017), and Wood and Bundy (2014) and references therein.

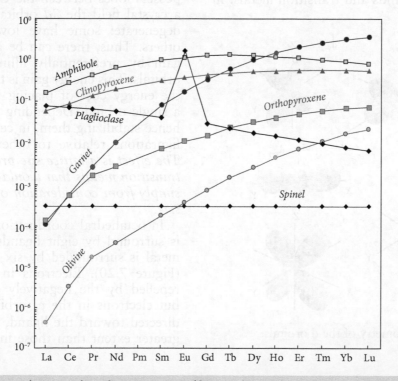

Figure 7.18 Rare earth mineral–melt partition coefficients for mafic magmas. Data from Table 7.4.

7.5   CRYSTAL-FIELD EFFECTS

We pointed out earlier that the ions of the alkalis, alkaline earths and rare earths can be satisfactorily modeled as hard spheres containing point charges at their centers. This model of ionic behavior is notably less satisfactory for many of the transition metals, because of the complex geometry of the bonding electron orbitals, illustrated in Figure 7.19. A more accurate prediction of bonding and substitution requires consideration of the electrostatic field of surrounding ions on the electron structure of transition elements.

7.5.1   Crystal-field theory

Crystal-field theory, which was developed in 1929 by physicist Hans Bethe, describes the effects of electrostatic fields on the energy levels of a transition-metal ion in a crystal structure. These electrostatic forces originate from the surrounding negatively charged anions or dipolar groups or ligands. While originally derived from cations in a crystal lattice, the same formalism can be applied to cations in silicate melts and aqueous solutions because metal ions in these liquids are also coordinated by ligands in structures analogous to those in crystals. Crystal-field theory is the simplest of several theories that attempt to describe the interaction and bonding between ligands and transition metals. In

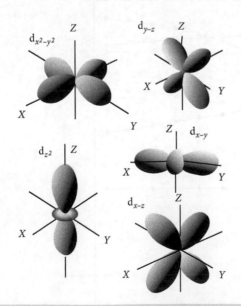

Figure 7.19   Geometry of the d orbitals.

crystal-field theory, the ligands are regarded simply as negative point charges about the transition metal ion. The electromagnetic field produced by these ligands, the "crystal field," destroys the spherical symmetry possessed by an isolated transition metal. The changes induced depend on the type, position, and symmetry of the coordinating ligands, as well as the nature of the transition metal.

In the usual case, electron orbitals are filled successively from inner to outer as one proceeds "up" the periodic table to heavier elements. In transition metals, however, filling of the outermost s orbital is begun before the d orbitals are completely filled. Ions are formed when the outermost s and, in some cases, some of the outermost d electrons (outermost will be $4s$ and $3d$ for the first transition series, $5s$ and $4d$ for the second) are removed from the metal.

In an isolated first series transition metal, the five $3d$ orbitals (each containing up to two electrons: a total of 10 electrons are possible in the d orbitals) are energetically equivalent and have equal probability of being filled: they are said to be *degenerate*. They possess, however, different spatial configurations (Figure 7.19). One group, the e_g orbitals, consisting of the d_{z2} and the d_{x2-y2} orbitals, has lobes directed along the Cartesian coordinates, while the t_{2g} group, consisting of the d_{xy}, d_{yz}, and d_{xz}, possess lobes between the Cartesian axes. In a crystal field the $3d$ orbitals are no longer degenerate: some have lower energy than others. Thus, there can be a relative energy gain by preferentially filling low-energy d orbitals. This energy gain is traded off against the energy cost of placing two electrons in a single orbital. Depending on this tradeoff, hence stabilizing them, in certain lattice configurations relative to other configurations. *The effect is a lattice site preference of some transition metals that would not be predicted simply from consideration of ion charge and size.*

In octahedral coordination (i.e., the metal is surround by eight ligands), the transition metal is surrounded by six identical ligands (Figure 7.20). Electrons in all orbitals are repelled by the negatively charged ligands, but electrons in the e_g orbitals, the orbitals directed toward the ligand, are repelled to a greater extent than those in the t_{2g} orbitals.

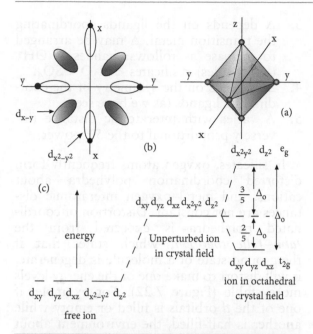

Figure 7.20 (a) Orientation of ligands and Cartesian coordinates for a metal ion in octahedral coordination. (b) Orientation of ligands (points) and d_{xy} (shaded) and $d_{x^2-y^2}$ (unshaded) orbitals in the x–y plane of a metal in octahedral coordination. (c) Energy-level diagram for d orbitals of a free transition metal ion, an ion in a spherically symmetric crystal field and an octahedral crystal field. In an octahedral crystal field, the energy of the d orbitals projected between the coordinates and the ligands (the t_{2g} orbitals) are lowered relative to the energy of the orbitals projected toward the ligands (e_g orbitals). After Burns (1970).

The energy separation between the t_{2g} and the e_g orbitals is termed octahedral crystal-field splitting and is denoted by Δ_o. The t_{2g} orbitals are lowered by $^2/_5\Delta_o$ while the e_g orbitals are raised by $^3/_5\Delta_o$ relative to the mean energy of an unperturbed ion. Therefore, each electron in a t_{2g} orbital stabilizes a transition metal ion by $^2/_5\Delta_o$ and each electron in an e_g orbital diminishes stability by $^3/_5\Delta_o$. The resultant net stabilization energy (i.e., $\Sigma\Delta_o$) is called the *octahedral crystal field stabilization energy* or octahedral CFSE. Crystal field stabilization energies can be derived from the electromagnetic spectra of minerals.

How electrons are distributed in an octahedrally coordinated transition metal is governed by two opposing tendencies. Coulomb forces between electrons cause them to be distributed over as many orbitals as possible, but crystal-field splitting favors the occupation of lowest energy orbitals. These two factors, in turn, depend on the number of d electrons possessed by the metal and the strength of the crystal field.

In ions having 1, 2, or 3 d electrons, all electrons will be in only t_{2g} orbitals, regardless of the strength of the crystal field, since there is only one electron per orbit. In ions having 8, 9, or 10 electrons in d orbitals, each orbital must contain at least 1 electron and three orbitals must contain 2 electrons, so all the 3 t_{2g} orbitals will be filled even for weak ligands. But in ions having 4, 5, 6, and 7 d electrons, there is a choice. If the crystal-field splitting is large, as in the case of strong field ligands, all electrons are in t_{2g} orbitals. This is the *low-spin* case, because the spins of electrons are anti-aligned (recall the Pauli exclusion principle that electrons can only occupy the same orbit if their spins are opposite), so net spin is low. When the crystal-field splitting is small, the energy cost of placing two electrons in the same orbit is greater than the energy gain from the octahedral CFSE, and electrons are distributed over both t_{2g} and e_g orbitals. This is known as the *high-spin* case because the electrons will preferentially occupy different orbitals with their spins parallel. On the Earth's surface in equilibrium with the atmosphere all first series transition metals apart from Co^{3+} and Ni^{3+} exist in the high-spin state either because they in their oxidized form (e.g., Fe^{3+}, Mn^{4+}) and do not have more than 3 d electrons or because they have 8 or more d electrons. Cr^{3+}, Ni^{2+}, and Co^{3+} have particularly high CFSE in octahedral coordination. Under more reducing conditions such as in magmas, these metals can also be in the low-spin state.

The distinction between high- and low-spin configurations is important in understanding magnetic properties of transition metal compounds because magnetism relates to spin alignment of electrons. Also, the crystal-field splitting energies are in the visible light band and hence relate to the coloration of transition-metal-bearing minerals. For example, consider titan-augite (a variety of clinopyroxene containing appreciable amounts of Ti). In Ti^{3+}, the single d electron is normally in the

t_{2g} orbital. Absorption of light of appropriate frequency ($v = \Delta_o/h$) excites the electron into an e_g orbital. This energy corresponds to violet light, which is emitted when the electron returns to the t_{2g} orbital.

In tetrahedral coordination where the metal is surrounded by six ligands (Figure 7.21), the e_g orbitals become more stable than the t_{2g} orbitals, but the tetrahedral crystal-field splitting parameter, Δ_t, is smaller than Δ_o. Other things being equal, $\Delta_t = 4/9\Delta_o$.

The crystal field splitting parameter, Δ, depends on a–number of things, but some generalizations can be made:

1. Values of Δ are higher for +3 ions than +2 ions.
2. The values are 30% higher for each succeeding transition element (left to right in the periodic table).

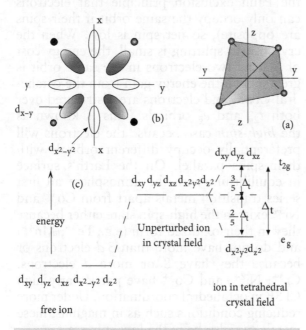

Figure 7.21   (a) Orientation of ligands and Cartesian coordinates for a metal ion in tetrahedral coordination. (b) Orientation of ligands (points) and d_{xy} (shaded) and $d_{x^2-y^2}$ (unshaded) orbitals in the $x - y$ plane of a metal in tetrahedral coordination. Black points are in front of plane of orbitals, gray points are behind the plane. (c) Energy-level diagram for d orbitals of a free transition metal ion, and an ion in a spherically symmetric crystal field and a tetrahedral crystal field. After Burns (1970).

3. Δ depends on the ligands coordinating the transition metal. Δ may be arranged to increase as follows: halides < OH$^-$ (hydroxides) < silicates < H$_2$O < SO$_4$.
4. Δ depends on the symmetry of the coordinating ligands (as we have seen).
5. Δ varies with interatomic distance (inversely proportional to the 5th power).

In silicates, oxygen atoms frequently form distorted coordination polyhedra about cations, and metal–oxygen interatomic distances are not constant. Distortion of coordinated polyhedra is expected from the *Jahn–Teller theorem*, which states that if the ground state of a molecule is degenerate, it will distort to make one of the energy levels more stable (Figure 7.22). For example, if one of the d orbitals is filled or empty while another is half-filled, the environment about the transition metal ion is predicted to distort spontaneously to a different geometry in which a more stable electronic configuration is attained by making the occupied orbital lower in energy. Further splitting of d orbital energy levels occurs when transition metal ions are in distorted coordination.

This can be illustrated for the case of Mn^{3+} in octahedral coordination with oxygen. The Mn^{3+} ion has the high-spin configuration (Table 7.5) in which each t_{2g} orbital is occupied by one electron and the fourth electron may occupy either of the e_g orbitals. If the four oxygen atoms in the x–y plane move toward the central Mn^{3+} ion and the two oxygens along the z-axis move away from it (Figure 7.22a), then the one e_g electron will favor the d$_{z^2}$ orbital in which the repulsion by the O ions is smaller than in the d$_{x^2-y^2}$ orbital. Thus, the e_g orbital group separates into two energy levels with the d$_{z^2}$ becoming more stable. The t_{2g} orbital group is also split, with the d$_{xz}$ and d$_{yz}$ becoming more stable than the d$_{xy}$ orbital. If the two O ions along the z-axis move closer to the Mn^{3+} ion (Figure 7.20c), the d$_{x^2-y^2}$ becomes more stable than the d$_{z^2}$. In either of the distorted environments, the Mn^{3+} becomes more stable than in an undistorted octahedral site. Transition metals subject to Jahn–Teller distortions in octahedral coordination are those with d^4, d^9, and low-spin d^7 configurations, in which one or three electrons occupy e_g orbitals. Looking

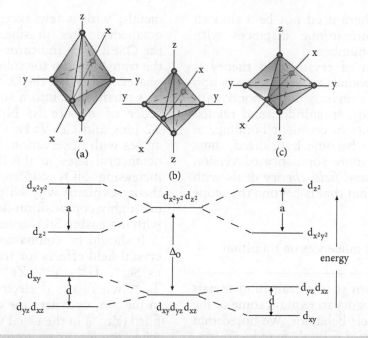

Figure 7.22 Arrangement of ligands and energy levels for (a) an octahedral site distorted along the z-axis; (b) an undistorted site; and (c) an octahedral site distorted along the y-axis. After Burns (1970).

Table 7.5 Electronic configurations and crystal-field stabilization energies of first transition series metal ions in octahedral configuration. From Burns (1970).

Number of 3d electrons	Ion	High spin state				Low spin state			
		Electronic configuration		Unpaired electrons	CFSE	Electronic configuration		Unpaired electrons	CFSE
		t_{2g}	e_g			t_{2g}	e_g		
0	Sc^{3+}, Ti^{4+}			0	0			0	0
1	Ti^{3+}	↑		1	$2/5\Delta_o$	↑		1	$2/5\Delta_o$
2	Ti^{2+}, V^{3+}	↑↑		2	$4/5\Delta_o$	↑↑		2	$4/5\Delta_o$
3	V^{2+}, Cr^{3+}, Mn^{4+}	↑↑↑		3	$6/5\Delta_o$	↑↑↑		3	$6/5\Delta_o$
4	Cr^{2+}, Mn^{3+}	↑↑↑	↑	4	$3/5\Delta_o$	↑↓↑↑		2	$8/5\Delta_o$
5	Mn^{2+}, Fe^{3+}	↑↑↑	↑↑	5	0	↑↓↑↓↑		1	$10/5\Delta_o$
6	Fe^{2+}, Co^{3+}, Ni^{3+}	↑↓↑↑	↑↑	4	$2/5\Delta_o$	↑↓↑↓↑↓		0	$12/5\Delta_o$
7	Co^{2+}, Ni^{3+}	↑↓↑↓↑	↑↑	3	$4/5\Delta_o$	↑↓↑↓↑↓	↑	1	$9/5\Delta_o$
8	Ni^{2+}	↑↓↑↓↑↓	↑↑	2	$6/5\Delta_o$	↑↓↑↓↑↓	↑↑	2	$6/5\Delta_o$
9	Cu^{2+}	↑↓↑↓↑↓	↑↓↑	1	$3/5\Delta_o$	↑↓↑↓↑↓	↑↓↑	1	$3/5\Delta_o$
10	Zn^{2+}	↑↓↑↓↑↓	↑↓↑↓	0	0	↑↓↑↓↑↓	↑↓↑↓	0	0

at Table 7.5, we can see that these are Cr^{2+}, Mn^{3+}, Cu^{2+}, Co^{2+}, and Ni^{3+} ions.

As noted, electronic transition energies are related to color. Because of the distortion, additional electronic transitions become possible. The differing probabilities of the various electronic transitions in polarized radiation is one of the causes of *pleochroism** in minerals.

Crystal-field effects lead to irregularities in the interatomic distances, or ionic radii of transition metals. As you might expect, they depend on the nature of the site, and

---

* Pleochroism refers to the property possessed by some crystals of exhibiting different colors when viewed along different axes in polarized light.

for a given site, there need not be a smooth contraction of interatomic distances with increasing atomic number.

The application of crystal-field theory is restricted to compounds where the transition metal forms a dominantly ionic bond with surrounding ligands. In sulfides, and related minerals, the effects of covalent bonding, in which the orbitals become hybridized, must be considered. A more sophisticated version of this theory, *ligand field theory* deals with covalent bonding but that is beyond the scope of our treatment.

7.5.2 Crystal-field influences on transition metal partitioning

We can now return to the transition metals and crystal-field theory to explain some of the peculiarities of their behavior. We noted that the energy of some *d* orbitals is reduced (stabilized) by the effects of the electrostatic field of coordinating ligands in both octahedral and tetrahedral sites, and that the octahedral CFSE is always greater than the tetrahedral CFSE. We now introduce one more quantity: the octahedral site preference energy (OSPE), which is defined as the octahedral CFSE minus the tetrahedral CFSE. Table 7.6 lists these energies for various first transition series metals. Silicate melts contain both octahedral and tetrahedral sites, but transition

metals, with a few exceptions, occupy only octahedral sites in silicate minerals. Thus, the OSPE is an indicator of the preference of the transition ion for solid phases over liquid phases: the higher the OSPE, the more readily it is partitioned into a solid phase. Predicted order of uptake is: $Ni>Co>Fe^{2+}>Mn$ for $+2$ ions and $Cr>V>Fe^{3+}$ for $+3$ ions, which agrees with observation. Since the number of octahedral sites in the liquid decrease with increasing SiO_2 concentration, crystal-field theory explains why Ni partition coefficients are highly composition-dependent, increasing with increasing SiO_2 concentration.

It should be emphasized that there are no crystal-field effects for transition metals such as Sc^{2+}, Ti^{4+}, Y^{3+}, Zr^{4+}, Nb^{5+}, Hf^{4+}, and Ta^{5+}, where the *d* electrons are not present in the ion, or where the d shell is completely filled (Zn^{2+}) in the usual valence state, at least when the electrons are in their ground state. However, color, which arises from excitation of electrons into higher orbitals and subsequent decay to the ground state, may still relate to crystal-field effects even when the *d* orbitals are not filled in the ground state. The second and third transition series metals for which crystal-field effects are expected are all highly siderophile or chalcophile and highly depleted in the Earth's crust and mantle. Little information is available on their behavior in silicate systems.

Table 7.6 Crystal-field splittings and stabilization energies in transition-metal ions.

Number of 3d electrons	Ion	Electron configuration		Δ (cm^{-1}) $M(H_2O)_{aq}^{6+}$	CFSE hydrate (kJ/mol)	Octahedral CFSE (kJ/mol)	Tetrahedral CFSE (kJ/mole)	Octahedral site preference energy (kJ)
1	Ti^{3+}	$(t2_g)^1$		20,300	$^2/_5\Delta = 97$	87.4	58.6	28.9
2	V^{3+}	$(t2_g)^2$		17,700	$^4/_5\Delta = 169$	160.2	106.7	53.6
3	Cr^{3+}	$(t2_g)^3$		17,400	$^6/_5\Delta = 249$	224.7	66.9	157.7
4	Cr^{2+}	$(t2g)^3$	$(e_g)^1$	13,900	$^3/_5\Delta = 100$	100.7	29.3	71.1
4	Mn^{3+}	$(t2g)^3$	$(e_g)^1$	21,000	$^3/5\Delta = 151$	135.6	40.2	95.4
5	Mn^{2+}	$(t2g)^3$	$(e_g)^2$	7,800	0	0	0	0
5	Fe^{3+}	$(t2g)^3$	$(e_g)^2$	13,700	0	0	0	0
6	Fe^{2+}	$(t2_g)^4$	$(e_g)^2$	10,400	$^2/_5\Delta = 50$	49.8	33.1	16.7
6	Co^{3+}	$(t2_g)^6$		18,600	$^{12}/_5\Delta = 536^*$	188.3	108.8	79.5
7	Co^{2+}	$(t2_g)^5$	$(e_g)^2$	9,300	$^4/_5\Delta = 89$	92.9	61.9	31.0
7	Ni^{3+}	$(t2_g)^6$	$(eg)^1$	–	$^9/_5\Delta =$			
8	Ni^{2+}	$(t2_g)^6$	$(e_g)^2$	8,500	$^6/_5\Delta = 29.6$	122.2	36.0	86.2
9	Cu^{2+}	$(t2_g)^6$	$(e_g)^3$	12,600	$^3/_5\Delta = 21.6$	90.4	26.8	63.6

*Low-spin complexes. The calculated CFSE must be reduced by the energy required to couple two electrons in a t_{2g} orbital. Data from Orgel (1966) and McClure (1957).

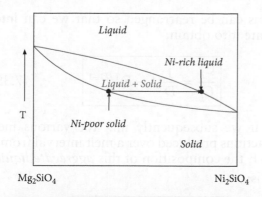

Figure 7.23 Schematic phase diagram for the system forsterite–Ni olivine showing Ni-poor olivine in equilibrium with Ni-rich liquid.

An understanding of crystal-field theory solves an interesting dilemma. A phase (T–X) diagram for the binary system Mg_2SiO_4–Ni_2SiO_4 is shown schematically in Figure 7.23. It is apparent from a quick glance that for any coexisting liquid and solid in the system, the solid will be poorer in Ni than the silicate liquid, i.e., $(Ni/Mg)_{ol}$ < $(Ni/Mg)_{liq}$. However, olivine crystallizing from basaltic liquids is always richer in Ni than the liquid. The reason for this is that in the pure olivine system, only octahedral sites are available in the melt and the solid, and thus Ni has no particular preference for the solid due to crystal-field effects. But basaltic melts have both tetrahedral and octahedral sites, while olivine has only octahedral sites (available to Ni). The greater availability of octahedral sites in the solid provides an added incentive for Ni to partition into olivine relative to basaltic liquid.

7.6   TRACE ELEMENT DISTRIBUTION DURING PARTIAL MELTING

In igneous geochemistry, trace elements are useful in understanding magmatic processes and in evaluating the composition of magma sources such as the mantle and lower crust. To make use of trace elements in such studies, we need to understand how magmatic processes such as partial melting and fractional crystallization will affect trace element abundances.

The task of the igneous geochemist is often to make inferences about the sources of magma, the mantle and lower crust, from the composition of the magmas themselves. This can be done through a mathematical model of melting. In the following sections, we will begin by considering two simple alternative models of melting: batch, or equilibrium, melting, and fractional melting. In fractional melting, the melt is extracted as soon as it is created, and only an infinitesimal increment of melt will be in equilibrium with the solid residue at any given time. In batch melting, a finite amount of melt, for example 5% or 10%, is produced and equilibrates completely with the solid residue.

We will go on to consider more complex models, such as continuous melting where melt is part of the melt is only partly extracted as it is produced. Once a melt is created and begins to rise, it may further interact with the surrounding "wallrock." We will also consider one possible model of this interaction: zone refining. Choosing between alternative models of partial melting requires a knowledge of how melting and melt extraction actually occurs. Unfortunately, melting and melt extraction in the Earth remain poorly understood because we are unable to observed them directly. Although melting experiments are useful in determining phase relationships, melting temperatures, and distribution coefficients, they do not provide much direct information on how melt is extracted. By and large, our knowledge of the melt extraction process comes from indirect inferences. Rarely, we can identify partial melting residues that have been tectonically emplaced at the surface of the Earth, and studies of these have provided some insights into the melting process. We will consider some of these insights in a subsequent section.

7.6.1   Equilibrium or batch melting

Equilibrium crystallization or melting implies complete equilibration between solid and melt. This means that the entire batch equilibrates with the residue before it is removed. From mass balance we may write:

$$C_i^o = C_i^s(1 - F) + C_i^\ell F \qquad (7.37)$$

where i is the element of interest, C^o is the original concentration in the solid phase (and the concentration in the whole system), C^ℓ is the concentration in the liquid, C^s is the concentration remaining in the solid and F is the melt fraction (i.e., mass of melt/mass of system). Since $D = C^s/C^\ell$, and rearranging:

$$C_i^o = C_i^\ell D_i^{s/\ell}(1 - F) + C_i^\ell F$$

or:

$$\boxed{\frac{C_i^\ell}{C_i^o} = \frac{1}{D^{s/\ell}(1 - F) + F}} \qquad (7.38)$$

This equation is an extremely useful one and describes the relative enrichment or depletion of a trace element in the liquid as a function of degree of melting. Two approximations are often useful and give us a feel for this equation:

1. Consider the case where $D \ll F$. In this case $C^\ell/C^o \approx 1/F$, that is, the enrichment is inversely proportional to the degree of melting. This is the case for highly incompatible elements at all but the smallest degrees of melting.
2. Consider the case where F approaches 0. In this case $C^\ell/C^o \approx 1/D$, the enrichment is inversely proportional to the partition coefficient.

Thus, the maximum enrichment possible in a partial melt is $1/D$. For highly compatible elements, that is, those with large D such as Ni, the depletion in the melt is $1/D$ when F is small and is relatively insensitive to F.

7.6.2 Fractional melting

Now consider the case where only an infinitesimally small amount of melt equilibrates with the solid residue, in other words, imagine we remove the liquid as fast as we make it. If i^s is the mass of element i in the solid phase being melted, S the mass of the solid phase, L the mass of the liquid phase, i^l the mass of i in the liquid, S° the original mass of the solid (and mass of the system), and $i°$ the original mass of i in the solid (and system), then:

$$C_i^s = \frac{i^s}{S} = \frac{i° - i^l}{S° - L} \quad \text{and} \quad C_i^\ell = \frac{1}{D_i}\frac{i° - i^\ell}{S° - L} = \frac{di^\ell}{dL}$$

This can be rearranged so that we can integrate it to obtain:

$$\boxed{\frac{C_i^\ell}{C_i^o} = \frac{1}{D}(1 - F)^{1/D-1}} \qquad (7.39)$$

If we subsequently mix the various melt fractions produced over a melt interval from 0 to F, the composition of this *aggregate liquid*, $\overline{C}$ is:

$$\boxed{\frac{\overline{C_i^\ell}}{C_i^o} = \frac{1}{F}(1 - (1 - F)^{1/D_i})} \qquad (7.40)$$

Figure 7.24 illustrates the variation of the liquid enrichment (C^ℓ/C^o) with degree of melting for both batch and fractional melting. The aggregate liquid of fractional melting, which may be the most realistic of the three equations we have considered so far, follows a trend close to that of batch melting.

7.6.3 Zone refining

If melt percolates slowly through the source region, trace element fractionation may be best approximated by equations governing zone refining. The term *zone refining* comes from the industrial purification process in which a melt zone is caused to migrate along a column of material. Several passes of this process efficiently extract low melting-temperature components. The relevant equation is:

$$\frac{C_i^\ell}{C_i^o} = \frac{1}{D_i} - \left(\frac{1}{D_i} - 1\right)e^{-D_iR} \qquad (7.41)$$

where R is the ratio of host, or wallrock, to melt. Note that for large R, $C^\ell/C^o \sim 1/D$.

7.6.4 Multiphase solids

The above equations are relevant when the solid undergoing melting is homogenous. If it consists of several phases, we need to replace D with a *bulk* distribution coefficient, which is simply the weighted mean of the individual solid/liquid partition coefficients:

$$\overline{D_i} = \sum_\phi m_\phi D_i^{\phi/\ell} \qquad (7.42)$$

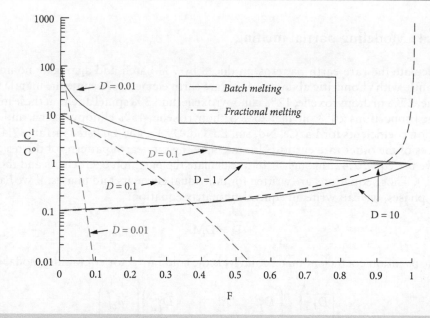

Figure 7.24 Variation in C^ℓ/C^o with degree of melting, F, for various partition coefficients for batch and fractional melting.

where m is simply the *modal mass fraction* of phase ϕ, that is, the fraction of phase ϕ as it exists in the rock.

In general, minerals do not enter the melt in the proportion in which they exist in a rock (i.e., their modal proportions). Thus, a realistic melting model requires that we modify our equations to account for this. We need to define a new parameter P, which is simply the average individual partition coefficients weighted according to the proportions in which the minerals enter the melt:

$$\overline{P}_i = \sum_\phi p_\phi D_i^{\phi/\ell} \qquad (7.43)$$

where p_ϕ is the proportion in which phase ϕ enters the melt. It is often assumed that the p_ϕ values are constants. In reality, they will be functions of F. The equations for batch and fractional melting in this *nonmodal melting* case become:

Nonmodal batch melting:

$$\frac{C_i^\ell}{C_i^o} = \frac{1}{F(1 - \overline{P}_i) + \overline{D}_i} \qquad (7.44)$$

Nonmodal fractional melting

$$\frac{C_i^\ell}{C_i^o} = \frac{1}{\overline{D}_i}\left(1 - \frac{\overline{P}_i F}{\overline{D}_i}\right)^{1/\overline{P}_i - 1} \qquad (7.45)$$

The aggregate liquid for nonmodal fractional melting is given by:

Aggregate:

$$\frac{\overline{C}_i^\ell}{C_i^o} = \frac{1}{F}\left[1 - \left(1 - \frac{\overline{P}_i F}{\overline{D}_i}\right)^{1/\overline{P}_i}\right] \qquad (7.46)$$

These equations are from Shaw (1970); their use is illustrated in Example 7.4.

7.6.5 Continuous melting

In most circumstances, the way in which rock melts in the Earth is probably intermediate between our batch and fractional melting models: only part of the melt is extracted continuously, while some fraction remains to fill the pore spaces between the mineral grains. This process has been called *continuous melting*. Let's look at how we can modify

Example 7.4 Modeling partial melting

Problem: Calculate the rare earth patterns produced by 7% batch and aggregate nonmodal partial melting of mantle with chondritic abundances of rare earth elements. Assume the mantle is composed of 58% olivine, 27% orthopyroxene, 12% clinopyroxene, and 3% spinel and that these minerals enter the melt in the proportions 20% olivine, 25% orthopyroxene, 45% clinopyroxene, and 10% spinel. Use the partition coefficients for La, Ce, Nd, Sm, Eu, Gd, Dy, Er, and Lu listed in Table 7.4. Interpolate the abundances of the other rare earths. Also calculate the rare earth patterns of the residual mantle.

Answer: We use eqns. 7.42 through 7.46 to calculate the partition coefficients and the enrichment factors (C^ℓ/C^o). Those equations are written for individual elements and phases. If we have a suite of elements and phases, we can write an equivalent matrix equation:

$$\overline{\mathbf{D}} = \mathbf{D}\mathbf{M} \tag{7.47}$$

where **D** is the partition coefficient matrix and **M** is the column vector containing modal abundances, e.g:

$$\begin{pmatrix} \overline{D}_{La} \\ \overline{D}_{Ce} \\ \vdots \\ \overline{D} \end{pmatrix} = \begin{pmatrix} D_{La}^{ol} & D_{Ce}^{ol} & \cdots & D_{Lu}^{ol} \\ D_{La}^{opx} & D_{Ce}^{opx} & \cdots & D_{Lu}^{opx} \\ \vdots & \vdots & \ddots & \vdots \\ D_{La}^{sp} & D_{Ce}^{sp} & \cdots & D_{Lu}^{sp} \end{pmatrix} \begin{pmatrix} M_{ol} \\ M_{opx} \\ \vdots \\ M_{sp} \end{pmatrix} \tag{7.47a}$$

Eqn. 7.43 can be recast in similar matrix notation. This suggests this is a problem most easily carried out using MATLAB, although it can also be done in Excel.

Once we have calculated for these nine rare earths, we can calculate the missing ones by a nonlinear interpolation, similar to that used in eqns. 7.1 and 7.2 should be used; for example, we calculate the praseodymium enrichment factor as

$$C_{Pr} = \sqrt{C_{Ce} \times C_{Nd}} \tag{7.48}$$

The interpolation of Tm and Yb is a bit trickier; for Tm for example:

$$C_{Tm} = (C_{Er}^2 \times C_{Lu})^{1/3} \tag{7.49}$$

The composition of the residual mantle can be calculated from mass balance. Rearranging eqn. 7.37, we have:

$$\frac{C_i^r}{C_i^o} = \frac{1 - (C_i^\ell/C_i^o)F}{(1-F)} \tag{7.50}$$

where *r* denotes the residual solid. Because we are working with relative abundances, $C^o = 1$ in this case. Since we are assuming a chondritic composition for the mantle the calculated enrichment factors, C^ℓ/C^o, are precisely the values we wish to plot: we do not need to know the chondritic values.

In MATLAB, we define our partition coefficient matrix and M and P vectors. We then compute the bulk partition coefficients and define *F*. Then we use eqns. 7.42 and 7.44 to compute enrichment factors. Our MATLAB solution is shown below.

The results are plotted in Figure 7.25. We see there is little difference between the batch and aggregate fractional melts. For this reason, may geochemists prefer the batch melt model for its simplicity, even though the aggregate fractional melt is likely more realistic. We also see that the mantle becomes highly depleted in light rare earths, particularly in the aggregate fractional melt model, as a consequence of melting.

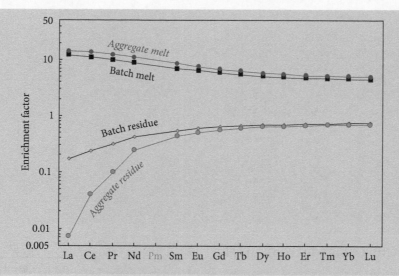

Figure 7.25 Rare earth patterns of 7% batch and aggregate fractional partial melts and of the respective residual solids calculated using partition coefficients in Table 7.4.

```
D  =                                              M=          P=
0.0000     0.0001     0.0600     0.0003     0.5800      0.2000
0.0000     0.0006     0.0900     0.0003     0.2700      0.2500
0.0000     0.0028     0.2000     0.0003     0.2500      0.4500
0.0002     0.0073     0.3200     0.0003     0.0300      0.1000
0.0004     0.0111     0.3800     0.0003
0.0007     0.0162     0.4400     0.0003
0.0027     0.0321     0.5300     0.0003
0.0073     0.0509     0.5700     0.0003
0.0217     0.0743     0.5600     0.0003
>>Dbar=D*M
>>Pbar=D*P
>>Dbar=                    Pbar=
    0.0150                    0.0271
    0.0227                    0.0407
    0.0508                    0.0907
    0.0821                    0.1459
    0.0982                    0.1739
    0.1148                    0.2022
    0.1427                    0.2471
    0.1605                    0.2701
    0.1727                    0.2750
>>F=0.07
>>Cbatch=(F*(1-Pbar)+Dbar).^-1
>>Cagg=(1/F)*(1-(1- ((Pbar*F)./Dbar)^(Pbar.^-1)))
Cbatch=                    Cagg=
    12.0265                    14.1868
    11.1343                    13.7595
     8.7393                    11.0101
     7.0488                     8.5396
     6.4083                     7.6106
     5.8601                     6.8336
     5.1174                     5.8170
     4.7272                     5.3046
     4.4760                     4.9913
```

```
>> Rbatch=(1-Cbatch*F)/(1-F)
>> Ragg=(1-Cagg*F)/(1-F)
Rbatch=                Ragg =
    0.1701                 0.0074
    0.2372                 0.0396
    0.4175                 0.2466
    0.5447                 0.4325
    0.5929                 0.5024
    0.6342                 0.5609
    0.6901                 0.6374
    0.7195                 0.6760
    0.7384                 0.6996
```

our fractional melting equation (eqn. 7.37) for this situation.

Consider a rock undergoing melting. We assume that it has a melt-filled porosity of ϕ, where ϕ is defined by mass. We can replace the partition coefficient in eqn. 7.37 with an effective partition coefficient, D', which takes account of a fraction of liquid, ϕ, in the rock with a partition coefficient of 1 (Albarède, 1995). Equation 7.39 thus becomes:

$$\frac{C_i^\ell}{C_i^o} = \frac{1}{D_i'}(1 - F)^{1/D_i' - 1} \qquad (7.51)$$

F in this case is the fraction of melt extracted, which differs from the total amount of melt produced by an amount equal to ϕ. D' is related to the usual partition coefficient, D, (eqn. 7.3) as:

$$D_i' = (1 - \phi)D_i + \phi \qquad (7.52)$$

The exponential term in eqn. 7.51, $1/D' - 1$, is related to D by:

$$\frac{1}{D_i'} - 1 = \frac{1 - \phi}{(1 - \phi)D_i + \phi}(1 - D) \qquad (7.53)$$

Substituting these back into eqn. 7.51, our expression for continuous melting written in terms of the usual partition coefficient is:

$$\frac{C_i^\ell}{C_i^o} = \frac{1}{(1 - \phi)D_i + \phi}(1 - F)^{\frac{(1-\phi)(1-D_i)}{(1-\phi)D_i+\phi}} \qquad (7.54)$$

Porosity is normally defined in terms of volume, but the above equations use a porosity defined in terms of mass. The relationship between the mass porosity and the volume porosity is:

$$\phi = \frac{\rho_\ell \varphi}{\rho_s(1 - \varphi) + \rho_\ell \varphi} \qquad (7.55)$$

where ϕ is the mass porosity, φ volume porosity, ρ_s is the density of the solid, and ρ_ℓ is the density of the liquid.

We can also derive an equation for an aggregate continuous melt simply by replacing D with D' in eqn. 7.40. Figure 7.26 compares continuous and fractional melting for $D = 0.0001$ and $\phi = 0.001$. Leaving residual melt in the pores has the effect of buffering the depletion of the solid, so that the concentration of an incompatible element does not decrease as fast in the case of continuous melting as for fractional melting. As Figure 7.26 shows, for high values of F,

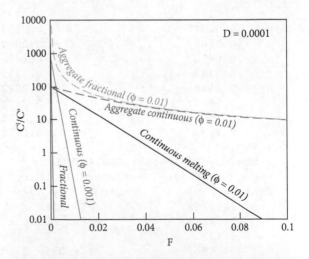

Figure 7.26 Comparison of continuous and fractional melting for D = 0.0001 and ϕ = 0.01. The aggregate melt is similar in both cases when F is greater than about 2%. A separate curve for continuous melting is shown for ϕ = 0.001.

the aggregate melts produced by fractional and continuous melting have almost identical compositions. The compositions of the residual solids, however, will be far different, with the residue of fractional melting being far more depleted in incompatible elements than the residue of batch melting.

7.6.6 Constraints on melting models

To summarize the previous discussion, we may say that the concentration of a trace element in a melt is a function of: (1) the solid phases (i.e., minerals) present in the system before and during melting; (2) the extent of melting (i.e., F); (3) the manner of melting (e.g., fractional vs. batch); (4) the concentration of the element in the original solid (i.e., $C°$); and (5) the partition coefficients. The partition coefficients, as we have seen, are functions of temperature, pressure, and composition of the phases involved. Two tasks of trace element geochemistry are to deduce something about the melting process and about the composition of the source of magmas. If we are to use trace elements for these purposes, it is essential we independently constrain some of these variables.

Most magmas are generated by partial melting of the upper mantle. Although temperature increases with depth in the mantle, the solidus temperature (i.e., the temperature where melting begins) increases more rapidly, so that the deep mantle is generally well below its solidus.* Though they have played a very important role in the evolution of the Earth, magmas produced by melting of the deep mantle are very much rarer. So our discussion here will be limited to the melting process in the upper mantle. The phases present in the upper mantle, and their compositions, are discussed in more detail in Chapter 11, so we will omit that topic from the discussion here.

7.6.6.1 Relationship between melt fraction and temperature and pressure

We can shorten our list of variables if we can somehow relate the degree of melting to temperature and ultimately to pressure. We can do this through a simplified thermodynamic analysis.

Most melting in the mantle, with the notable and important exception of subduction zones, appears to result from decompression: packets of mantle moving upward. Pressure in the Earth is related to depth, (h), by the simple relationship:

$$\frac{dP}{dh} = \rho g \qquad (7.56)$$

where ρ is density and g is the acceleration of gravity. For a typical upper mantle density, pressure increases by about 1 GPa for every 35 km depth.

Because of the scales generally involved (kilometers to hundreds of kilometers) and the low thermal conductivity of rock, it is reasonable to assume that a rising packet of mantle is adiabatic. As we learned in Chapter 2, this means it can do work or have work done on it, but it does not exchange heat with its surroundings (i.e., $dQ = 0$). We also learned in Chapter 2 that an adiabatic system is an isoentropic one (i.e., $dS = 0$). The constraint that the system is isoentropic allows us to relate the amount of melting that will occur to the temperature and pressure of the rising mantle.

The variation of entropy with temperature and pressure can be expressed as:

$$dS = \left(\frac{\partial S}{\partial T}\right)_P dT + \left(\frac{\partial S}{\partial P}\right)_T dP \qquad (7.57)$$

Substituting eqns. 2.105 and 2.106 into eqn. 7.57, we have:

$$dS = \frac{C_P}{T} dT - \alpha V dP \qquad (7.58)$$

Since the system is isoentropic, $dS = 0$, and we can solve eqn. 7.58 to obtain:

$$\left(\frac{\partial T}{\partial P}\right)_S = \frac{T\alpha V}{C_P} \qquad (7.59)$$

This equation describes the *adiabat*, the *P–T* path that adiabatically rising mantle follows. (By the way, we can see that the adiabat

---

* Recent seismic studies suggest the possible presence of melt pockets in the lowermost mantle, near the core–mantle boundary.

will be curved, since its slope depends on T). The solidus temperature will also change with pressure, and its slope is given by the *Clapeyron equation* (eqn. 3.3):

$$\left(\frac{dT}{dP}\right)_{sol} = \frac{\Delta V_m}{\Delta S_m} \qquad (3.3)$$

The slope of the solidus is steeper than that of the adiabat, so that rising hot mantle will eventually intersect the solidus (Figure 7.27). For simplicity, let's assume the solid consists of a single phase. When the solidus is reached, the system will consist of two phases, solid and liquid, and we can write one version of eqn. 7.57 for the solid and one version for any melt that has formed. Now let's specify that the two phases (melt and solid in this case) coexist at equilibrium along a univariant reaction curve, whose slope in P–T space is $(dT/dP)_{2\phi}$. We can solve eqn. 7.57 to determine how entropy of each phase changes with pressure, for example, for the solid:

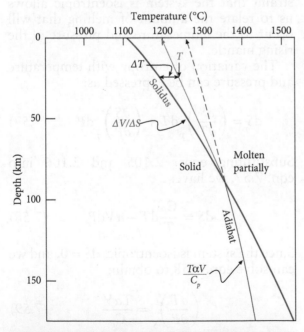

Temperature (°C)

Figure 7.27 Representation of melting of an ascending packet of mantle in temperature and pressure space. Below the solidus, the mantle rises along the adiabat. Once the packet intersects the solidus, the P–T path of the mantle packet is deflected, as shown by the solid line marked T.

$$\frac{dS^s}{dP} = \frac{C_P^s}{T}\left(\frac{dT}{dP}\right)_{2\phi} - \alpha^s V^s \qquad (7.60)$$

The total specific entropy (i.e., entropy per unit mass) of the system, S_o, can be expressed as the sum of the entropy of the solid and the melt.

$$FS^\ell + (1 - F)S^s = S_0 \qquad (7.61)$$

where S^ℓ and S^s are specific entropies of the melt and solid respectively and F is the fraction of melt. If we solve eqn. 7.61 for F, we have:

$$F = \frac{S_0 - S^s}{S^\ell - S^s} \qquad (7.62)$$

The term $S^\ell - S^s$ is just entropy of melting, ΔS_m, so eqn. 7.62 can be written as:

$$F = \frac{S_0 - S^s}{\Delta S_m} \qquad (7.63)$$

As long as the melt is not extracted, the system remains isoentropic and S_0 is a constant; however, neither S^ℓ nor S^s are necessarily constant. Let's assume for the moment that ΔS_m is also constant (equivalent to assuming that S^ℓ and S^s change in an identical way). If we now differentiate eqn. 7.63 with respect to pressure, we have:

$$\left(\frac{\partial F}{\partial P}\right)_S = \frac{1}{\Delta S_m}\left(\frac{C_P^s}{T}\left(\frac{dT}{dP}\right)_{2\phi} - \alpha^s V^s\right) \qquad (7.64)$$

Equation 7.64 shows that even assuming that all the thermodynamic parameters therein are constant, the amount of melt produced by rising mantle will be a function of its temperature.

Once melting begins, the rising mantle follows a P–T path that is steeper than adiabatic (Figure 7.27), since some energy is consumed in melting. Let's call the temperature that the system would have attained, had melting not occurred, the potential temperature, T_p. The difference between that temperature and the actual temperature T is related to the entropy change during melting, ΔS_m. We can determine the entropy change due to the difference between T and T_p by integrating eqn. 2.105:

$$\Delta S = \int_T^{T_p} \frac{C_P}{T} dT \qquad (7.65)$$

Since we are interested in a simple, approximate analysis, let's assume that C_p is constant. In that case, eqn. 7.65 becomes:

$$\Delta S = C_p \ln \frac{T_p}{T} \qquad (7.66)$$

To find a simple linear solution, let's approximate eqn. 7.66 with a Taylor series expansion about T_{act}, which yields:

$$\Delta S \cong \frac{C_p}{T}(T_p - T) \qquad (7.67)$$

As long as melt has not been lost, the system remains isoentropic, so the entropy difference in eqn. 7.67 must simply be the entropy consumed in melting:

$$\Delta S = \Delta S_m F \qquad (7.68)$$

Equating the two, we have:

$$\Delta S_m F \cong \frac{C_p}{T}(T_p - T) \qquad (7.69)$$

Rearranging, and letting $\Delta T = (T_{pot} - T)$, we have:

$$T_{pot} - T \cong \frac{T}{C_p}\Delta S_m F \qquad (7.70)$$

This difference, $T_{pot} - T$, is the temperature deflection due to melting in the P–T path and is shown in Figure 7.27 as ΔT. If we differentiate eqn. 7.70 with respect to P (and still holding S constant), we have:

$$\left(\frac{\partial(T_p - T)}{\partial P}\right)_S \cong \frac{\Delta S_m}{C_p}\left(\frac{\partial(TF)}{\partial P}\right)_S$$

$$= \frac{\Delta S_m}{C_p}\left[F\left(\frac{\partial T}{\partial P}\right)_S + T\left(\frac{\partial F}{\partial P}\right)_S\right]$$

and finally:

$$\left(\frac{\partial \Delta T}{\partial P}\right)_S \cong \frac{\Delta S_m}{C_p}F\left(\frac{\partial T}{\partial P}\right)_S + \frac{\Delta S_m}{C_p}T\left(\frac{\partial F}{\partial P}\right)_S \qquad (7.71)$$

Solving for $(\partial T/\partial P)_S$, we have:

$$\left(\frac{\partial T}{\partial P}\right)_S \cong \left(\frac{\partial T_p}{\partial P}\right)_S - \frac{T}{C_p}\Delta S_m\left(\frac{\partial F}{\partial P}\right)_S \qquad (7.72)$$

The term $(\partial T_{pot}/\partial P)_s$ is just the adiabatic gradient, given by eqn. 7.58, and substituting that into eqn. 7.72 we have:

$$\left(\frac{\partial T}{\partial P}\right)_S \cong \frac{T\alpha V}{C_P} - \frac{T}{C_p}\Delta S_m\left(\frac{\partial F}{\partial P}\right)_S \qquad (7.73)$$

Equation 7.73 describes the P–T path that a system undergoing isoentropic melting will follow as it rises.

The degree of melting will be a function of excess temperature, that is, the difference between the solidus temperature and the actual temperature, which we shall call ΔT. In Figure 7.24, ΔT can be found by subtracting the solidus temperature from the temperature path of the mantle packet:

$$\left(\frac{\partial \Delta T}{\partial P}\right)_S \cong \left[\frac{T\alpha V}{C_P} - \frac{T}{C_p}\Delta S_m\left(\frac{\partial F}{\partial P}\right)_S\right] - \left(\frac{\partial T}{\partial P}\right)_{2\phi} \qquad (7.74)$$

There have been many attempts to determine the relationship between melting and temperature for mantle materials. Such melting curves are notoriously difficult to determine. Figure 7.28 shows an experimentally determined melting curve for a peridotite composition at 3.5 GPa. The curve has several breaks in slope that correspond to elimination of phases. Despite the kinks, one can extract from this kind of experiment a relationship between degree of melting and excess

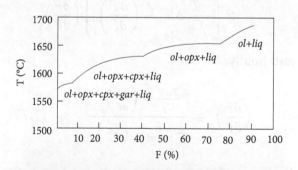

Figure 7.28 Relationship between extent of melting, F, and temperature in peridotite at 3.5 GPa determined experimentally in graphite capsules by Harrison (1979). Kinks in the curve correspond to consumption of phases, in the order garnet (gar), clinopyroxene (cpx), and orthopyroxene (opx).

temperature, that is, a value for $(\partial F/\partial T)_P$. For example, Langmuir et al. (1993) adopted a value of about 0.00285 for $(\partial F/\partial T)_P$ below 22% melting, and 0.0015 for $(\partial F/\partial T)_P$ above 22% melting. We want to incorporate this information into our analysis. We do this as follows. First, we express the variation in temperature as a function of melt fraction and pressure:

$$dT = \left(\frac{\partial T}{\partial P}\right)_F dP + \left(\frac{\partial T}{\partial F}\right)_P dF \quad (7.75)$$

If we differentiate eqn. 7.75 with respect to pressure, specifying that entropy be held constant, we can derive the following relationship:

$$\left(\frac{\partial T}{\partial P}\right)_S = \left(\frac{\partial T}{\partial P}\right)_F + \left(\frac{\partial T}{\partial F}\right)_P \left(\frac{\partial F}{\partial P}\right)_S$$

This can be substituted into eqn. 7.72 to obtain:

$$\left(\frac{\partial T}{\partial P}\right)_F + \left(\frac{\partial T}{\partial F}\right)_P \left(\frac{\partial F}{\partial P}\right)_S$$
$$\cong \frac{T\alpha V}{C_P} - \frac{T}{C_p}\Delta S_m \left(\frac{\partial F}{\partial P}\right)_S \quad (7.76)$$

Rearranging:

$$\left(\frac{\partial T}{\partial P}\right)_F - \frac{T\alpha V}{C_P}$$
$$\cong \left[-\frac{T}{C_p}\Delta S_m - \left(\frac{\partial T}{\partial F}\right)_P\right]\left(\frac{\partial F}{\partial P}\right)_S$$

and finally:

$$\left(\frac{\partial F}{\partial P}\right)_S \cong \frac{\frac{T\alpha V}{C_P} - \left(\frac{\partial T}{\partial P}\right)_F}{\frac{T}{C_p}\Delta S_m + \left(\frac{\partial T}{\partial F}\right)_P} \quad (7.77)$$

Equation 7.77 gives the melt fraction as a function of pressure above the pressure where the mantle intersects the solidus.

Let's now attempt to evaluate eqn. 7.77 by substituting some real values into it. The term $(\partial T/\partial P)_F$ is the slope in P–T space of lines of constant melt fraction. We can make two simplifying assumptions: (1) the lines of constant

melt fraction are parallel to the solidus; and (2) the solidus can be adequately described by a Clapeyron slope, eqn. 7.3.3 (because the composition of both melt and solid can vary in the real mantle, the solidus will not be a simple univariant curve described by the Clapeyron equation), so eqn. 7.77 becomes:

$$\left(\frac{\partial F}{\partial P}\right)_S \approx \frac{\frac{T\alpha V}{C_P} - \frac{\Delta V_m}{\Delta S_m}}{\frac{T}{C_p}\Delta S_m + \left(\frac{\partial T}{\partial F}\right)_P} \quad (7.78)$$

The coefficient of thermal expansion, α, is about $3 \times 10^{-5}\,K^{-1}$, V is about $0.3175\,cm^3/g$ ($= 0.3175\,JMPa^{-1}g^{-1}$), and C_p is about $1.15\,JK^{-1}g^{-1}$. Thus, the adiabatic gradient at $1673\,K$ ($1400°C$) is about $12\,K/GPa$. The term $(\partial T/\partial F)_P$ is, of course, just the inverse of $(\partial F/\partial T)_P$ and has a value of $1/0.00285 = 350.88\,K$. ΔV_m is about $0.0434\,cm^3/g$ ($0.0434\,JMPa^{-1}g^{-1}$) and ΔS_m is about $0.362\,JK^{-1}\,g^{-1}$, which corresponds to a slope of the solidus of about $120\,K/GPa$. From this we calculate a value for $(\partial F/\partial P)_S$ of about $-0.12\,GPa^{-1}$, or about $-1.2\%/kbar$ (it is negative because the extent of melt *increases* as pressure *decreases*). Of course, we have greatly simplified matters here by neglecting the pressure and temperature dependencies of all thermodynamic functions. Thus, this relationship is only approximate, and considering the uncertainty in our assumptions and the thermodynamic parameters, this value could lie anywhere between $-0.08/GPa$ and $-0.2/GPa$. So, for example, if a rising packet of mantle intersects the solidus at 100 km depth ($\approx 3\,GPa$), upon reaching a depth of 30 km ($\sim 1\,GPa$) that packet would have undergone 24% melting.

The solidus temperature of silicates can be substantially lowered by the addition of water and, at high pressures, of CO_2. In the presence of either H_2O or CO_2, the melting curve will be different from that shown in Figure 7.28, and the relationship we deduced between melt, temperature, and pressure will also be different.

A final point to make is that once melt is extracted, the system is no longer isoentropic because the extracted melt carries away some of the entropy of the system. Thus, our analysis would be strictly limited to batch melting, where the melt remains in equilibrium with the solid. A complete treatment of the thermodynamics of melting, including fractional melting, can be found in Azimov et al. (1997). Morgan (2001) discusses the situation where the material undergoing melting is lithologically heterogeneous.

7.6.6.2 Mantle permeability and melt distribution and withdrawal

Whether the melting process approximates the batch (equilibrium) model or the fractional model depends on the permeability of the source region. If the source region is highly permeable, melt will flow out as it is created, approximating the fractional melting model; if it is impermeable, it will build up in place before ascending, approximating the equilibrium model. Permeability depends on the degree to which the melt is interconnected, and this in turn depends on the crystal–liquid interfacial energy.

We explored the effects of interfacial energy on nucleation in Chapter 5 (section 5.5.3.3). We found that the difference in interfacial energy determined the geometry of nucleation. Here we wish to consider the case of how a liquid will distribute itself between grains of a solid undergoing melting. We assume that the solid consists of a single phase (e.g., olivine) and that the interfacial energy between these grains is σ_{ss}. Now consider the intersection between three such grains (Figure 7.29a). When melt is present, there will also be an interfacial energy between the grains and the melt, σ_{sm}. If θ is the angle formed by a melt pocket at a grain triple junction, the balance of forces may be described as:

$$\sigma_{ss} = 2\sigma_{s\ell} \cos \frac{\theta}{2} \qquad (7.79)$$

Rearranging, we have:

$$\cos \frac{\theta}{2} = \frac{\sigma_{ss}}{2\sigma_{s\ell}} \qquad (7.80)$$

and

$$\cos \frac{\theta}{2} = \frac{\sigma_{ss}}{2\sigma_{s\ell}} \qquad (7.80a)$$

The bottom line is that the lower the solid–liquid interfacial energy, the more extensively melt will interconnect (the more extensively grains will "wet") and the more readily

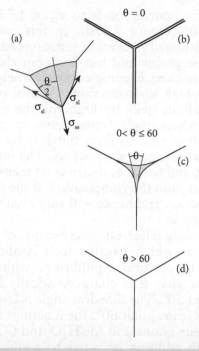

Figure 7.29 Relationship between dihedral angle, θ, and melt distribution at grain triple junctions. (a) The balance of solid–solid and solid–liquid interfacial energies, σ_{ss} and σ_{sl}, at the junction. (b) $\theta = 0$, and the melt (shaded) is distributed along grain–grain boundaries as well as triple junctions. (c) The melt forms an interconnected network along grain triple junctions. (d) θ is greater than 60° and melt is present only at four-grain junctions. After Kingery (1960) and Kohlstedt (1993).

melt will flow. Considering eqn. 7.80 in greater detail reveals that, depending on the value of θ, the melt can distribute itself in a number of ways. These are illustrated in Figure 7.29. The first case occurs when the solid–solid interfacial energy is twice that of the solid–melt interfacial energy; if so, then $\cos \theta/2 = 1$ and $\theta \approx 0$. In this case, solid–solid interfaces are energetically unfavorable and melt will form a thin film that coats all grain boundaries (Figure 7.29b). The second case is where the solid–solid interfacial energy is more than 1.73 times but less than twice that of the solid–melt interfacial energy ($2\sigma_{sl} > \sigma_{ss} > 1.73\sigma_{sm}$), which corresponds to $0 < \theta < 60°$ (Figure 7.29c). In this case, the melt will form interconnected channels along grain triple junctions, as is illustrated in Figure 7.29, but is absent from grain–grain surfaces. The third case corresponds to $\sigma_{ss} < 1.73\sigma_{sl}$ and $\theta > 60°$ (Figure 7.29d). In this case, melt forms isolated pockets at junctions between 4 or more grains and but is absent elsewhere. These pockets become connected only at relatively high melt fractions (several percent). Permeability will be high for the first two cases where melt forms films or channels that allow melt to flow, but low for the last case of isolated melt pockets. The interfacial energy, and hence θ, depends on temperature, pressure, and the composition of the melt and solid phase, and hence will vary even within a single rock.

Scanning electron microscopy of experiments in which basaltic melt is allowed to come to textural equilibrium with olivine indicate that θ is characteristically between 25° and 50°. The dihedral angle is larger, typically greater than 60°, for junctions between pyroxene grains and for H_2O and CO_2 fluids (though addition of water to a silicate rock reduces θ). Since the upper mantle consists of over 60% olivine, however, it is likely that melt forms an interconnected network such as that illustrated in Figure 7.30, resulting in high permeability. Experiments in which melt is induced to migrate, either as a result of a gradient in melt fraction in the experimental charge or as a result of stress, confirm that permeability of mantle material undergoing melting will be high. From our perspective, this means melt is likely to be extracted fairly quickly after it is created, and that very small melt fractions, perhaps as low as 0.1%, can segregate from the mantle. Thus, the fractional melting model may more closely approximate melting in the mantle.

Laporte (1994) carried out similar experiments with quartz and hydrous silicate melts and found that the dihedral is in the range of 12–18°, indicating a high ratio of σ_{ss}/σ_{sl}. This in turn indicates that the permeability within regions of the crust undergoing melting will be relatively high. However, the rate at which melt segregates from its source depends on melt viscosity as well as permeability. Because the viscosity of even hydrous granitic magmas is four orders of magnitude greater than that of basalt, segregation of granitic melt requires a higher melt fraction than does segregation of basaltic melt. Nevertheless, Laporte argued that melt fractions as low as 10% will segregate on time scales of 10^5

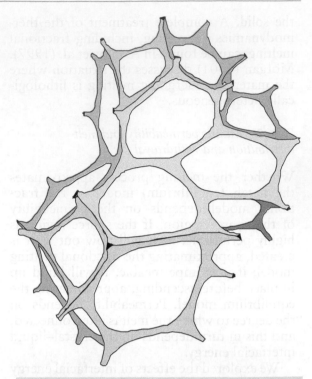

Figure 7.30 Three-dimensional network formed by melt along triple junctions of olivine grains. From Riley and Kohlstedt (1990).

yrs, whereas it had been previously believed that melt fractions as high as 30% would not segregate on reasonable geologic time scales.

7.6.6.3 Realistic models of mantle melting

As we pointed out above, in most circumstances melting in the mantle occurs because of decompression. (A possible exception is in subduction zones; here the generation of melt is still poorly understood but may ultimately be due to hydration of the mantle wedge. Addition of water lowers the melting temperature, so this is a form of flux melting.) Decompression melting is necessarily a dynamic process: a parcel of mantle will begin to melt at some depth and will continue to melt as it rises. The fraction of melt produced will increase with height above the initial melting depth. If, as we have argued above, melt segregates readily, melt will rise faster than the solid. As a result, once the parcel of mantle has risen above the depth where melting begins, melt from below will continually stream through it. The melt entering the parcel from below will initially not be in equilibrium with the solid within the parcel, having been produced as a smaller melt fraction at greater depth (and hence greater pressure and temperature). Thus, melt passing through the parcel will react with the solid in an attempt to reach equilibrium with it. This is similar to the process we described above as zone refining.

The situation then is analogous to diagenesis in sediments, which we discussed in Chapter 5. There are some differences, however. In diagenesis, the fraction of solid relative to fluid does not change, except through expulsion of fluid. In the melting process, solid is converted to fluid by nature of the process. In the melting process, length scales are such that diffusion does not significantly contribute to the flux and bioturbation does not exist, so advection is the only significant flux. Furthermore, our reduction of the problem to one dimension by assuming lateral uniformity will not be valid for the melting process. This is because the extent of melt will also decrease with distance from some central point (a point under a volcano or under a spreading mid-ocean ridge), and because melt will be focused in from these peripheral regions toward the center. With these caveats, however, the diagenetic equation (eqn. 5.175) may be directly applicable to the melting process.

Unfortunately, a truly thorough quantitative treatment of the melting process has not yet been undertaken. In one of the more thorough discussions, Langmuir et al. (1993) concluded that despite the complexity of the melting process, the batch melting equation gives a reasonably good approximation of incompatible element concentrations in the melt as a function of the *average* degree of melting. Beneath mid-ocean ridges, the average degree of melting will be less than the maximum degree of melting, because different parcels of mantle follow different paths. Only mantle directly beneath the ridge is able to rise the maximum amount, and hence melt the maximum amount. In the simple case illustrated in Figure 7.31a, the average extent of melting is one half the maximum extent. Other ratios are possible for other models of mantle flow.

There are two situations where batch melting may not be a good approximation of incompatible element concentrations. The first is where there is a large volume of mantle

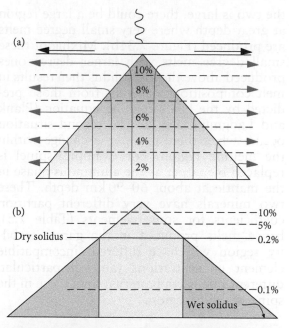

Figure 7.31 (a) Melting regime under a mid-ocean ridge. Red lines show the flow of mantle induced by passive spreading of overlying plates. Since melting results from decompression, no further melting occurs once motion becomes horizontal. Only those parts of the mantle directly under the ridge reach the maximum extent of melting. The melting regime along-ridge can be assumed to be uniform parallel to the ridge, hence the process is two-dimensional. The cartoon is, however, readily adapted to mantle-plume related volcanism by assuming radial symmetry. (b) Low-degree melts generated between the "wet" and "dry" solidi could enrich higher degree melts from the normal mantle column (light shading) in highly incompatible elements. However, the volume of this region must be large (typically 10 times that of the normal mantle column) for this to be effective, requiring efficient transport and focusing of melt over scales of hundreds of kilometers. After Plank and Langmuir (1992).

from which only very low degree melts are extracted. This situation might arise as a result of suppression of the solidus by H_2O or CO_2 fluid, a well-established phenomenon. If melting is such that a small fraction of melt, say 0.1% or so, is generated between the *wet solidus* (i.e., H_2O or CO_2 present) and the *dry solidus* and the temperature range between the two is large, there could be a large region at great depth where very small degree melts are produced (Figure 7.31b). Mixing of these small degree melts with larger degree ones produced above the dry solidus then results in melt compositions different from those predicted by the batch melting equation (Plank and Langmuir, 1992). The second situation occurs when there is a phase change within the melting region. For example, spinel is replaced by garnet as the aluminous phase in the mantle at about 60–90 km depth. These two minerals have very different partition coefficients for some elements (Table 7.5), hence melts produced in the garnet stability region will have different incompatible element concentrations (and in particular, different rare earth patterns) than those in the spinel stability region.

7.7 TRACE ELEMENT DISTRIBUTION DURING CRYSTALLIZATION

7.7.1 Equilibrium crystallization

Equilibrium crystallization occurs when the total liquid and total solid remain in equilibrium throughout the crystallization. If we define X as the fraction of material crystallized, then

$$\frac{C_i^l}{C_i^0} = \frac{1}{DX + (1 - X)} \qquad (7.81)$$

where C^ℓ is the concentration in the remaining liquid and $C°$ is the concentration in the original liquid (we derive this equation in a manner exactly analogous to eqn. 7.38). The limit of trace element enrichment or depletion occurs when $X = 1$, when $C_l/C_o = 1/D$. Equilibrium crystallization requires the liquid keeps in contact with all crystals. Crystal interiors would have to maintain equilibrium through solid state diffusion, a slow process. Thus, equilibrium crystallization is probably relevant only to a limited range of situations, such as the slow crystallization of an intrusion.

7.7.2 Fractional crystallization

Fractional crystallization, which assumes only instantaneous equilibrium between solid and liquid, is a more generally applicable model of crystallization. In this case, trace element concentrations in the melt are governed by:

$$\frac{C_i^\ell}{C_i^o} = (1 - X)^{D-1} \qquad (7.82)$$

There is no limit to the enrichment or depletion of the liquid in this case. If D is very large, C^ℓ/C^o approaches 0 as X approaches 1; it approaches ∞ as X approaches 1 if D is very small. What happens when $D = 0$?

For multiphase crystallization, we need to replace D in eqns. 7.81 and 7.82 with the bulk distribution coefficient as we defined it in eqn. 7.42, where m_ϕ in that equation would become the fraction of phase ϕ in the crystallizing mass.

Though fractional crystallization can, in principle, produce extreme trace element enrichment, this rarely occurs. A melt that has crystallized 90% or more (which would produce a tenfold enrichment of a perfectly incompatible element in the melt) would have major element chemistry very different from its parent. From our knowledge of the compositional dependence of partition coefficients, we could predict that incompatible elements would have partition coefficients close to 1 for such an acid melt.* This limits the enrichment of incompatible elements. However, highly *compatible* elements (elements with solid/liquid partition coefficients greater than 1, such as Ni) do have concentrations that approach 0 in fractionated melts (generally they disappear below detection limits). Variation of relative trace element concentration as a function of the fraction of liquid remaining is shown in Figure 7.32.

In summary, for moderate amounts of fractionation, crystallization has only a moderate effect on trace element concentrations, given that these concentrations vary by orders of magnitude.

---

* Silica-rich, or silicic, melts are sometimes referred to as "acidic" and Mg- and Fe-rich ones as "basic." The reason is historical: it was once thought that silica was present in melts as H_4SiO_4. This is not the case, but the terminology persists.

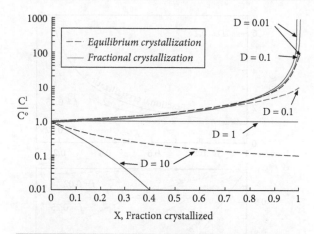

Figure 7.32 Variation of relative trace element concentration in a liquid undergoing crystallization.

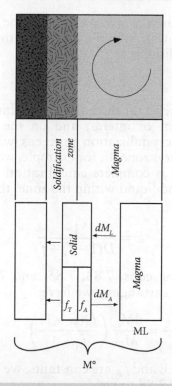

Figure 7.33 Magma chamber undergoing *in situ* crystallization. The solidification zone is the transition region between consolidated cumulates (that nevertheless retain some trapped liquid) and the magma chamber. As crystallization proceeds, some liquid will be expelled from the solidification zone back into the magma. After Langmuir (1989).

7.7.3 *In situ* crystallization

A magma chamber is likely to be substantially hotter than the rock surrounding it, which will result in a substantial thermal gradient at the margin of the chamber. Thus, the margins, particularly the roof and walls, are likely to be cooler than the interior, and it is here where crystallization will primarily occur. Crystals sloughed from the walls and roof would accumulate on the floor. When crystallization is restricted to marginal zones of the chamber, magma composition will evolve in a somewhat different manner than for simple fractional crystallization. Langmuir (1989) called this process *in situ crystallization*.

Imagine a magma chamber that is bounded by a zone of well-consolidated crystals at the margin. There may be some liquid within this zone, but we assume that the permeability is sufficiently low that it will never return to the magma chamber. Between this *cumulate zone* and the free magma is a transition zone of higher permeability, which we call the *solidification zone* (Figure 7.33). Magma is added to the solidification zone as crystallization advances into the chamber. As crystallization and compaction proceed within the solidification zone, liquid is expelled back into the central magma. The flux of magma to the solidification zone we will designate dM_I and the return flux as dM_A. We let f be the fraction of interstitial liquid remaining after crystallization within the solidification zone, and $(1 - f)$ be the fraction of liquid crystallized within this zone. Some fraction f_T of

the liquid remains to form the trapped liquid within the cumulate zone, and some fraction f_A returns to the magma, so that $f = f_T + f_A$. Hence:

$$dM_A = f_A dM_I \qquad (7.83)$$

If the magma plus cumulates form a closed system, then the change in mass of liquid within the chamber is the difference between the flux into and out of the solidification zone:

$$dM_L = dM_A - dM_I = dM_I(f_A - 1) \qquad (7.84)$$

If C_L is the concentration of some element in the magma and C_f is the concentration in the liquid returning from the solidification zone, then the change in mass of the element is:

$$d(M_L C_L) = C_L dM_L + M_L dC_L$$
$$= C_f dM_A - C_L dM_I \qquad (7.85)$$

We define the parameter E as the ratio of the concentration in the magma to that in the returning liquid:

$$E = C_f/C_L \qquad (7.86)$$

E will depend on the partition coefficient for the element of interest and on the manner in which crystallization proceeds within the solidification zone. If, for example, we assume that there is complete equilibration between crystals and liquid within the zone, then from eqn. 7.81:

$$E = \frac{1}{D(1-f)+f} \qquad (7.87)$$

Substituting eqns. 7.83, 7.85, and 7.86 into 7.84 and rearranging, we have:

$$\frac{dC_L}{C_L} = \frac{dM_L}{M_L}\left(\frac{f_A(E-1)}{f_A-1}\right) \qquad (7.88)$$

Assuming E and f_A are constants, we can integrate eqn. 7.88 to yield:

$$\frac{C_L}{C^0} = \left(\frac{M_L}{M^0}\right)^{f_A(E-1)/(f_A-1)} \qquad (7.89)$$

Figure 7.34 compares the change in concentration due to *in situ* crystallization and fractional crystallization for two values of D and several values of f. In general, *in situ* crystallization results in less enrichment of incompatible elements and less depletion of compatible elements for a given degree of crystallization of a magma body than does fractional crystallization. The degree of enrichment depends on f, the fraction of liquid remaining when liquid is expelled from the solidification zone. For very small values of f, that is, complete crystallization within the solidification zone, the composition of the magma remains nearly constant. For $f = 1$, that is, for no crystallization within the solidification zone, eqn. 7.89 reduces to eqn. 7.82, and the *in situ* curves in Figure 7.34 coincide with the fractional crystallization curves.

7.7.4 Crystallization in open system magma chambers

Thus far, we have treated crystallization only in closed systems, that is, where a certain volume of magma is intruded and subsequently cools and crystallizes without withdrawal or

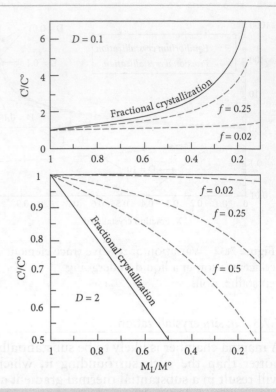

Figure 7.34 Comparison of the effects of *in situ* and fractional crystallization on concentration for two values of the distribution coefficient, D, and several values of f. f_A is assumed to equal f (i.e., no trapped liquid remains in the cumulate zone). After Langmuir (1989).

further addition of magma. This is certainly not a very realistic model of magmatism and volcanism. In the well-studied Hawaiian volcanoes for example, injections of new magma into crustal magma chambers are fairly frequent. Indeed, it appears that addition of magma to the Kilauea's magma chamber is nearly continuous. Furthermore, many igneous rocks show petrographic evidence of mixing of differentiated magmas with more primitive ones. In this section, we consider the concentrations of trace elements in open magma chambers; that is, the case where new "primary" magma mixes with a magma that has already undergone some fractional crystallization. Magma chambers where crystallization, eruption and addition of new magma occur are sometimes called RTF magma chambers, the RTF referring to *refilled, tapped,* and *fractionated.*

The extreme case of an RTF magma chamber is a *steady-state* system, where the magma resupply rate equals the rate of crystallization

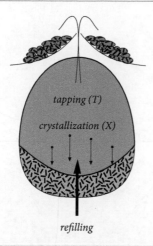

tapping (T)

crystallization (X)

refilling

Figure 7.35 Schematic illustration of a steady-state and periodically refilled, fractionally crystallized, and tapped magma chamber beneath a mid-ocean ridge.

and eruption, thus maintaining a constant volume of liquid (Figure 7.35). In such a magma chamber, the concentrations of all elements eventually reach steady state after many cycles of refilling, eruption, and fractional crystallization. Steady state occurs when the rate of supply of the elements (due addition of new magma) becomes equal to the rate of loss (due to crystallization and withdrawal and eruption of magma).

To understand how steady-state is achieved, consider a cyclic process where a volume C is lost by crystallization and a volume T is lost by eruption and a volume $(T + C)$ is added to the magma chamber during each cycle. For incompatible elements, the concentration in the liquid initially increases because a greater mass of these elements is added by refilling than is lost by crystallization and eruption. As the concentration in the liquid increases, so does the concentration in the solid since $C^s = D^{s/l}C^l$ ($D^{s/l}$ would be a bulk distribution coefficient if more than one phase is crystallizing). Eventually a point is reached where the concentration in the solid is so great, that loss of the element by crystallization and eruption equals the gain resulting from refilling.

We can quickly derive an expression for the steady-state concentration of an element in the *equilibrium crystallization* case. In steady-state, the losses of an element must equal gains of that element, so:

$$C^\circ(X + T) = TC^{ssl} + XC^s \qquad (7.90)$$

where C° is the concentration in the primary magma being added to the chamber, C^{ssl} is the concentration in the steady-state liquid, and C^s is the concentration in the solid. Since $C^s = D^{s/l}C^{ssl}$:

$$C^\circ(X + T) = TC^{ssl} + XDC^{ssl} \qquad (7.91)$$

We can rearrange this to obtain the enrichment in the steady-state liquid relative to the primary magma:

$$\frac{C^{ssl}}{C^0} = \frac{X + T}{X + TD^{s/l}} \qquad (7.92)$$

For the fractional crystallization case, the enrichment factor is:

$$\frac{C^{ssl}}{C^0} = \frac{(X + T)(1 - X)^{D-1}}{1 - (1 - X - T)(1 - X)^{D-1}} \qquad (7.93)$$

These equations are from O'Hara (1977).

Compatible element concentrations reach steady-state after fewer cycles than do incompatible elements. This is illustrated in Figure 7.36, which shows how the concentrations of La, an incompatible element, and Ni, a compatible element, vary as a function of the total amount crystallized in a steady-state system. The Ni concentration, which is 200 ppm in the primary magma, reaches a steady-state concentration of about 20 ppm after the equivalent of five magma chamber masses has crystallized. After the equivalent of 30 magma chamber masses crystallization, La has not quite reached its steady-state concentration of around 25 ppm.

Unfortunately, it is never possible to measure the fraction of magma that has crystallized. In place of our parameter, X, petrologists often use the concentration of some *index species*, i.e., some species whose behavior is relatively well understood and whose concentration should vary smoothly as a function of the fraction crystallized. In basalts, MgO is commonly used as the index species, while SiO_2 is a more common index in more acidic magmas such as andesites and dacites. Figure 7.37 shows the La and Ni concentrations against the MgO concentration in the same steady-state system. Since MgO is a compatible element (though not a trace element, we could treat it using the same

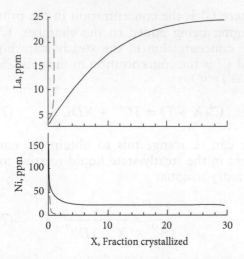

Figure 7.36 Concentration of Ni and La in closed system fractional crystallization (dashed line) and open system crystallization (solid line) as a function of the fraction crystallized. In the open system case the mass injected into the chamber is equal to the mass crystallized (i.e., $Y = 0$).

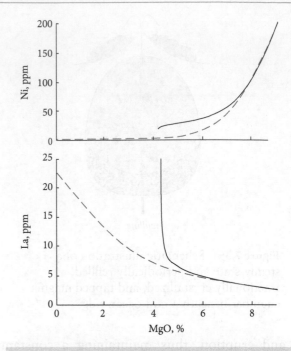

Figure 7.37 La and Ni concentrations plotted against MgO concentration in a basalt undergoing closed system fractional crystallization (dashed red line) and in a steady-state magma chamber where the mass of new magma equals the mass crystallized in each cycle (solid line).

equations we have derived for trace elements), it quickly reaches a steady-state concentration at around 4.25%, while the La concentration continues to increase. Lavas erupted from this magma chamber could thus have essentially constant MgO and Ni concentrations, but variable La concentrations. This apparent *decoupling* of compatible and incompatible element concentrations is a feature of open magmatic systems.

7.7.5 Comparing partial melting and crystallization

For moderate amounts of crystallization, fractional crystallization does not have dramatic effects on *incompatible* element concentrations. Concentrations of highly *compatible* elements are, however, dramatically affected by fractional crystallization. The RTF model does have significantly greater effects on incompatible element concentrations than simpler models, however.

Partial melting has much more dramatic effects on *incompatible* element concentrations. It is likely that much of the incompatible element variations observed in magmas and magmatic rocks are related to variations in degree of melting. Depth of melting also has an effect, in that the phases with which melts

equilibrate vary with depth. For example, the presence of garnet dramatically affects rare earth element (REE) abundances as the heavy rare earths are accepted into the garnet structure. Highly *compatible* elements are depleted in a partial melt. But the degree of depletion is rather insensitive to the degree of melting.

Consequently, compatible elements are good qualitative indicators of the extent of fractional crystallization and incompatible elements are good indicators of the degree of melting.

Ratios of incompatible elements are generally less sensitive to fractional crystallization and partial melting than are absolute abundances, particularly if they are of similar incompatibility. For relatively large extents of melting, the ratio of two incompatible elements in a magma will be similar to that ratio in the magma source. For this reason, trace element geochemists are often more interested in the ratios of elements than in absolute abundances.

One approach commonly used is to plot the ratio of two incompatible elements against the abundance of the least compatible of the two. This kind of plot is sometimes referred to as a *process identification plot* because fractional crystallization and partial melting result in very different slopes on such a diagram. Figure 7.38 is a schematic version of a process identification plot. Crystallization, both fractional and equilibrium, produce rather flat slopes on such a diagram, as does crystallization in an open system magma chamber. Partial melting produces a much steeper slope and the slope produced by aggregates of fractional melts is similar to that of equilibrium partial melting. *In situ* crystallization can produce a range of slopes depending on the value of f. Langmuir (1989) found that a value for f of 0.25 best matched the

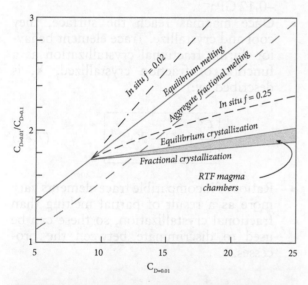

Figure 7.38 Plot of the ratio of two incompatible elements (one with $D = 0.01$, the other with $D = 0.1$) versus the concentration of the more incompatible element. The plot shows calculated effects of equilibrium partial melting and aggregate partial melting, assuming concentrations of 1 in the source for both elements. Other lines show the effect of crystallization on the composition of a liquid produced by 10% equilibrium melting. Fractional crystallization, equilibrium crystallization, and open system crystallization (RTF magma chambers) produce less variation of the ratio than does partial melting. *In situ* crystallization can mimic the effect of partial melting if the value of f, the fraction of liquid returned to the magma, is sufficiently small.

variation observed in the Kiglapait Intrusion in Labrador, a classic large layered intrusion. Assuming this value of f is typical, then the change in a trace element ratio due to *in situ* crystallization should be only moderately greater than for fractional crystallization.

7.8 SUMMARY OF TRACE ELEMENT VARIATIONS DURING MELTING AND CRYSTALLIZATION

In this chapter we reviewed the behavior of trace elements in igneous systems. A trace element in one circumstance might not be one in another and we defined a trace element as one which obeyed Henry's law in a given system.

- We reviewed the behavior of elements in the Earth generally and introduced Goldschmidt's classification: *lithophile*, *siderophile*, *chalcophile*, and *atmophile*. We also introduced the important terms *incompatible* and *compatible*, describing elements that partition into the magma and solid residue during mantle melting, respectively. The former are enriched in the Earth's crust.

- We defined the partition coefficient between phases α and β as:

$$D_i^{\alpha-\beta} = \frac{C_i^\alpha}{C_i^\beta} \quad (7.3)$$

We found we could relate these to chemical potential and Gibbs free energy and consequently predict temperature and pressure dependence. The primary factors governing partition coefficients are the size and charge of the ion relative to the size and charge of an ion normally occupying a crystallographic site. A general expression for this is:

$$D_i^{s/\ell} = D^o \exp$$

$$\left(\frac{-4\pi E N_A \left[\frac{r_0}{2}(r_M - r_0)^2 + \frac{1}{3}(r_M - r_0)^3 \right]}{RT} \right)$$

$$(7.22)$$

where r_M and r_0 are the size of the ion and the size of the crystallographic site and E is Young modulus, effectively the pressure produced by expanding or contracting the site to accommodate the ion.

- For transition metals, the orientation of d-orbitals relative to coordinating ligands is also an important control on partition coefficients, and effect described by *crystal-field theory*.
- We can use the partition coefficient and simple mass balance to describe how a trace element behaves during batch partial melting:

$$\frac{C_i^\ell}{C_i^o} = \frac{1}{D^{s/\ell}(1 - F) + F} \tag{7.38}$$

where F is the fraction of melt produced. Fractional melting may be a more realistic description of the process; the governing equation is:

$$\frac{C_i^\ell}{C_i^o} = \frac{1}{D}(1 - F)^{1/D-1} \tag{7.39}$$

However, if fractional melts aggregate the resulting variation with extent of melting approximates batch melting. Two simplifications are important for partial melting (both batch and fractional). *When $D \ll F$, the enrichment is 1/F.* Thus, for small D, the enrichment is highly dependent on the degree of melting. If D is large (i.e., $D > 1$ and $D \gg F$), the depletion of the element in the melt is rather insensitive to F. In either case, *when F approaches 0, the maximum enrichment or depletion is 1/D.*

- The extent to which melting approximates batch or fractional models depends on the extent of grain boundary wetting, which depends on interfacial energy between mineral grains.
- Most melting in the Earth results from decompression: as mantle rises convectively along an adiabat, it can intersect the solidus and begin to melt as it rises. We used thermodynamics to predict that the extent of melting should increase with decreasing pressure according to:

$$\left(\frac{\partial F}{\partial P}\right)_S \approx \frac{\frac{T\alpha V}{C_P} - \frac{\Delta V_m}{\Delta S_m}}{\frac{T}{C_p}\Delta S_m + \left(\frac{\partial T}{\partial F}\right)_P} \tag{7.78}$$

which works out to be roughly $-0.12\,\text{GPa}^{-1}$.

- Once magmas reach the surface, they cool and crystallize. Trace element behavior during fractional crystallization as a function of fraction crystallized, X, is described by:

$$\frac{C_i^\ell}{C_i^o} = (1 - X)^{D-1} \tag{7.82}$$

- Ratios of incompatible trace elements vary more as a result of partial melting than fractional crystallization, so these can be used to discriminate between the processes.

REFERENCES AND SUGGESTIONS FOR FURTHER READING

Albarède, F. 1995. *Introduction to Geochemical Modeling.* Cambridge, Cambridge University Press.

Azimov, P.D., Hirschmann, M.M. and Stolper, E. 1997. An analysis of variations in isoentropic melt productivity. *Philosophical Transactions of the Royal Society London A* 355: 255–81.

Bédard, J.H. 2005. Partitioning coefficients between olivine and silicate melts. *Lithos* 83: 394–419. doi 10.1016/j.lithos.2005.03.011.

Ben Othman, D., White, W.M. and Patchett, J. 1989. The geochemistry of marine sediments, island arc magma genesis, and crust-mantle recycling. *Earth and Planetary Science Letters* 94: 1–21.

Blundy, J.D., Robinson, J.C. and Wood, B.J. 1998. Heavy REE are compatible in clinopyroxene on the spinel lherzolite solidus. *Earth and Planetary Science Letters* 160: 493–504.

Blundy, J. and Wood, B. 1994. Prediction of crystal-melt partition coefficients from elastic moduli. *Nature* 372: 452–4.

Blundy, J. and Wood, B. 2003. Partitioning of trace elements between crystals and melts. *Earth and Planetary Science Letters* 210: 383–97.

Brice, J.C. 1975. Some thermodynamic aspects of the growth of strained crystals. *Journal of Crystal Growth* 28: 249–53.

Burnard, P. 2003 *The Noble Gases as Geochemical Tracers.* Springer, Berlin, Heidelberg, 391p.

Burns, R.G. 1970. *Mineralogical Applications of Crystal Field Theory.* Cambridge, Cambridge University Press.

Burns, R.G. 1973. The partitioning of trace transition elements in crystal structures: a provocative review with applications to mantle geochemistry. *Geochimica et Cosmochimica Acta* 37: 2395–403.

Carroll, M.R. and Draper, D.S. 1994. Noble gases as trace elements in magmatic processes. *Chemical Geology* 117: 37–56.

Drake, M.J. and Weill, D.F. 1975. Partition of Sr, Ba, Ca, Y, Eu^{2+}, Eu^{3+}, and other REE between plagioclase feldspar and magmatic liquid: an experimental study. *Geochimica et Cosmochimica Acta* 39, 689–712.

Fischer-Gödde, M., Becker, H. and Wombacher, F. 2010. Rhodium, gold and other highly siderophile element abundances in chondritic meteorites. *Geochimica et Cosmochimica Acta* 74(1), 356–79. doi: 10.1016/j.gca .2009.09.024.

Fischer-Gödde, M., Becker, H. and Wombacher, F. 2011. Rhodium, gold and other highly siderophile elements in orogenic peridotites and peridotite xenoliths. *Chemical Geology* 280(3), 365–83. doi: 10.1016/j.chemgeo .2010.11.024.

Fryer, B.J. and Greenough, J.D. 1992. Evidence for mantle heterogeneity from platinum-group element abundances in Indian Ocean basalts. *Canadian Journal of Earth Sciences* 29: 2329–40.

Gaetani, G. A. 2004. The influence of melt structure on trace element partitioning near the peridotite solidus. *Contributions to Mineralogy and Petrology* 147: 511–27. doi: 10.1007/s00410-004-0575-1.

Harrison, W.J. 1979. Rare earth elements partitioning between garnet lherzolite minerals and melts during partial melting. *Carnegie Institute of Washington Yearbook* 78: 562–8.

Hart, S.R. 1993. Equilibration during mantle melting: a fractal tree model. *Proceedings of the National Academy of Sciences USA* 90: 11914–18.

Hiemstra, S.A. 1979. The role of collectors in the formation of platinum deposits in the Bushveld Complex. *Canadian Mineralogist* 17: 469–82.

Huang, F., Lundstrom, C. and McDonough, W. 2006. Effect of melt structure on trace-element partitioning between clinopyroxene and silicic, alkaline, aluminous melts. *American Mineralogist* 91: 1385–400. doi: 10.2138/am .2006.1909.

Jochum, K.-P., Hofmann, A.W. and Seifert, H.M. 1985. Sn and Sb in oceanic basalts and the depletion of the siderophile elements in the primitive mantle (abs). *EOS* 66: 1113.

Jochum, K.-P., Hofmann, A.W. and Seifert, H.M. 1993. Tin in mantle-derived rocks: Constraints on Earth evolution. *Geochimica et Cosmochimica Acta* 57: 3585–95.

Kingery, W.D. 1960. *Introduction to Ceramics*. New York, John Wiley and Sons, Ltd.

Kohlstedt, D.L. 1993. Structure, rheology and permeability of partially molten rocks at low melt fractions, in *Mantle Flow and Melt Generation at Mid-ocean Ridges, Geophysical Monograph 71* (eds J.P. Morgan, D.K. Blackman and J.M. Sinton), pp. 103–21. Washington, American Geophysical Union.

Kohn, S.C. and Schofield, P.F. 1994. The implication of melt composition in controlling trace-element behavior: an experimental study of Mn and Zn partitioning between forsterite and silicate melts. *Chemical Geology* 117: 73–87.

Langmuir, C.H. 1989. Geochemical consequences of *in situ* crystallization. *Nature* 340: 199–205.

Langmuir, C.H., Klein, E.M. and Plank, T. 1993. Petrological systematics of mid-ocean ridge basalts: constraints on melt generation beneath oceanic ridges, in *Mantle Flow and Melt Generation at Mid-ocean Ridges, Geophysical Monograph 71* (eds J.P. Morgan, D.K. Blackman and J.M. Sinton), pp. 183–280. Washington, American Geophysical Union.

Laporte, D. 1994. Wetting behavior of partial melts during crustal anatexis: the distribution of hydrous silicic melts in polycrystalline aggregates of quartz. *Contributions to Mineralogy and Petrology* 116: 486–99.

Masuda, A. and Nakamura, N. 1973. Fine structures of mutually normalized rare-earth patterns of chondrites. *Geochimica et Cosmochimica Acta* 37: 239–48.

McClure, D.S. 1957. The distribution of transition metal cations in spinel. *J. Phys. Chem. Solids* 3: 311–17.

McLennan, S.M. (2018) Lanthanide rare earths. In: White WM (ed) *Encyclopedia of Geochemistry*, Springer International Publishing, Cham, pp. 792–99.

Michely, L. T., Leitzke, F. P., Speelmanns, I. M., Fonseca, R. O. C. 2017. Competing effects of crystal chemistry and silicate melt composition on trace element behavior in magmatic systems: insights from crystal/silicate melt partitioning of the REE, HFSE, Sn, In, Ga, Ba, Pt and Rh. *Contributions to Mineralogy and Petrology* 172: 39. doi: 10.1007/s00410-017-1353-1.

Morgan, J.P. 2001. Thermodynamics of pressure release melting of a veined plum pudding mantle. *Geochem. Geophys. Geosyst.*, 2: paper no. 2002GC000049.

Nakamura, N. 1974. Determination of REE, Ba, Fe, Mg, Ba, and K in carbonaceous and ordinary chondrites. *Geochimica et Cosmochimica Acta* 38: 7577–775.

Naldrett, A.J., Hoffman, E.L., Green, A.H., Chou, C.L. and Naldrett, S.R. 1979. The composition of Ni-sulfide ores with particular reference to their content of PGE and Au. *Canadian Mineralogist* 17: 403–15.

Newsom, H.E., White, W.M., Jochum, K.P. and Hofmann, A.W. 1986. Siderophile and chalcophile element abundances in oceanic basalts, Pb isotope evolution and growth of the Earth's core. *Earth and Planetary Science Letters* 80: 299–313.

O'Hara, M.J. 1977. Geochemical evolution during fractional crystallization of a periodically refilled magma chamber. *Nature* 266: 503–7.

O'Hara, M.J. 1985. Importance of the 'shape' of the melting regime during partial melting of the mantle. *Nature* 314: 58–62.

Onuma, N., Higuchi, H., Wakita, H. and Nagasawa, H. 1968. Trace element partition between two pyroxenes and the host lava. *Earth and Planetary Science Letters* 5: 47–51.

Orgel, L.E. 1966. *An Introduction to Transition Metal Chemistry: Ligand Field Theory*. London, Methuen.

Pourmand, A., Dauphas, N., Ireland, T. J. 2012. A novel extraction chromatography and MC-ICP-MS technique for rapid analysis of REE, Sc and Y: Revising CI-chondrite and Post-Archean Australian Shale (PAAS) abundances. *Chemical Geology* 291: 38–54. doi: 10.1016/j.chemgeo.2011.08.011.

Plank, T. and Langmuir, C.H. 1992. Effects of the melting regime on the composition of the oceanic crust. *Journal of Geophysical Research* 97: 19749–70.

Puchtel, I. S. (2018) Platinum group elements. In: White W.M. (Ed.) *Encyclopedia of Geochemistry*, Springer International Publishing, Cham, pp. 1236–9. ISBN: 978-3-319-39312-4.

Riley, G.N. and Kohlstedt, D.L. 1990. Kinetics of melt migration in upper mantle-type rocks. *Earth and Planetary Science Letters* 105: 500–21.

Rollinson, H. 1993. *Using Geochemical Data: Evaluation, Presentation, Interpretation*. Essex, Longman Scientific and Technical.

Rollinson, H., Adetunji, J. (2017) Ionic radii. In: White W.M. (Ed.) *Encyclopedia of Geochemistry*, vol. Springer International Publishing, Cham, pp. 1–6. 978-3-319-39193-9.

Rudnick, R.R. and Fountain, D.M. 1995. Nature and composition of the continental crust: a lower crustal perspective. *Reviews in Geophysics* 33: 267–309.

Shaw, D.M. 1970. Trace element fractionation during anatexis. *Geochimica et Cosmochimica Acta* 34: 237–43.

Snow, J. E. and Schmidt, G., 1998. Constraints on Earth accretion deduced from noble metals in the oceanic mantle. *Nature* 391, 166–169. doi: 10.1038/34396.

Stanley, R. H. R., Jenkins, W. J. (2013) Noble gases in seawater as tracers for physical and biogeochemical ocean processes. In: Burnard, P. (Ed.) *The Noble Gases as Geochemical Tracers*, Springer, Berlin, Heidelberg, pp. 55–79.

Sun, C., Liang, Y. 2012. Distribution of REE between clinopyroxene and basaltic melt along a mantle adiabat: effects of major element composition, water, and temperature. *Contributions to Mineralogy and Petrology* 163: 807–23. doi: 10.1007/s00410-011-0700-x.

Sun, C., Graff, M. and Liang, Y. 2017. Trace element partitioning between plagioclase and silicate melt: The importance of temperature and plagioclase composition, with implications for terrestrial and lunar magmatism. *Geochimica et Cosmochimica Acta* 206, 273–95. doi: 0.1016/j.gca.2017.03.003.

Tiepolo, M., Oberti, R., Zanetti, A., Vannucci, R. and Foley, S. F. 2007. Trace-Element Partitioning Between Amphibole and Silicate Melt. *Reviews in Mineralogy and Geochemistry* 67: 417–52. doi: 10.2138/rmg.2007.67.11.

Watson, E.B. 1976. Two-liquid partition coefficients: experimental data and geochemical implications. *Contributions to Mineralogy and Petrology* 56: 119–34.

Wijbrans, C. H., Klemme, S., Berndt, J. and Vollmer, C. 2015. Experimental determination of trace element partition coefficients between spinel and silicate melt: the influence of chemical composition and oxygen fugacity. *Contributions to Mineralogy and Petrology* 169: 45. doi: 10.1007/s00410-015-1128-5.

Wood, B. J. and Blundy, J. D. (2014) 3.11 – Trace Element Partitioning: The Influences of Ionic Radius, Cation Charge, Pressure, and Temperature. In: Turekian K.K. and Holland, H.D. (eds) *Treatise on Geochemistry (Second Edition)*, vol. Elsevier, Oxford, pp. 421–48, doi: 978-0-08-098300-4.

Wood, B.J., Blundy, J.D. and Robinson, J.C. 1999. The role of clinopyroxene in generating U-series disequilibrium during mantle melting. *Geochimica et Cosmochimica Acta* 63: 1613–20.

PROBLEMS

1. For one element not in the groups described in Section 7.2.2 (Mo, W, Ag, Cd, P, In, Sn, Tl, or Bi), write several paragraphs on its geochemistry. Include answers to the following: what valence state (or states) will it have in nature? What is its ionic radius in its most common valence state (preferably in octahedral coordination)? What is its electronegativity? What kinds of bonds will it most likely form? How will it behave, in particular what is its solubility, in aqueous solution? What element will it most easily substitute for in silicate rocks? Is it volatile or does it form volatile compounds? Is it siderophile or chalcophile? Will it behave as a compatible or incompatible element? What are its uses? What are the primary sources of the element for use by humans?

2. Make rare earth plots for the following two samples. For the granite, plot it normalized to one of the chondritic values in Table 7.3. For the Mn nodule, make one plot normalizing it to chondrites and one plot normalizing it to average shale. Describe the features of the REE patterns.

	La	Ce	Pr	Nd	Sm	Eu	Gd	Tb	Dy	Ho	Er	Tm	Yb	Lu
Granite	41	75		25	3.6	0.6	2.7		1.3		0.6		0.27	0.04
Mn nodule	110	858		116	24	4.9	24		24.1		14.4		13	1.92

3. Using the Blundy and Wood model, calculate the partition coefficients for the alkali metals in plagioclase at 1250°C. Assume that the site radius in 124 nm and that an ion of this radius would have a partition coefficient of 1. Assume that the value of Young's modulus in plagioclase is 64 GPa and that ionic radii for the alkalis are as follows: Li: 76 pm, Na 102 pm, K 138 pm, Rb 152 pm, and Cs 167 pm.

4. Relatively Ca-rich garnets (i.e., grossular-rich) appear to accept high field strength elements in both the X (normally occupied by Ca^{2+}, Mg^{2+}, Fe^{2+}, etc.) and Y (normally occupied by Al^{3+}, Fe^{3+}, etc.) crystallographic sites. The garnet/liquid partition coefficient should be the sum of the individual partition coefficients for each site. Assume that D_0, r_o, and E for the X site are 7, 91 pm, and 1350 GPa, respectively, and the corresponding values for the Y site are 1.9, 67 pm, and 920 GPa, respectively. Ions in the X site will be in eightfold coordination, while ions in the Y site will be in sixfold coordination. In eightfold coordination, the ionic radii of Zr^{4+}, Hf^{4+}, Th^{4+}, and U^{4+} are 84 pm, 83 pm, 105 pm, and 100 pm respectively; in sixfold coordination they will be 72 pm, 71 pm, 94 pm, and 89 pm, respectively. Calculate the garnet/liquid partition coefficients for these four elements at 1250°C using the Blundy and Wood model.

5. Calculate NBO/T for the two sample compositions listed in the adjacent table. Assuming the values of a in eqn. 7.35 are the clinopyroxene (cpx) partitions coefficients listed in Table 7.5, calculate the corrected cpx/melt partition coefficients for La, Sm, and Yb for these two samples based on eqn. 7.35.

	L1104	STL227
SiO_2	52.68	63.08
TiO_2	0.93	0.55
Al_2O_3	16.8	18.05
FeO	9.54	4.25
MgO	5.63	1.83
CaO	10.38	6.47
Na_2O	2.6	2.83
K_2O	0.51	1.43
MnO	0.17	0.09
P_2O_5	0.17	0.13
Total	99.41	98.71

6. Construct a table similar to Table 7.6 showing electronic configuration and CFSE (in terms of Δ_t) for both high-spin and low-spin states in *tetrahedral* coordination for Ti^{2+}, V^{2+}, Fe^{2+}, Co^{2+}, Ni^{2+}, and Cu^{2+}.

7. The following table gives the spectroscopically measured octahedral crystal-field splitting parameter (Δ_o) of oxides for several transition metal ions.

(a) For each ion below, calculate the octahedral CFSE (crystal-field stabilization energy), in joules per mole, for the high spin state. The data are given in terms of wavenumber, which is the inverse of wavelength, λ. Recall from your physics that $\lambda = c/v$, and that $e = hv$. Useful constants are $h = 6.626 \times 10^{-34}$ joule-sec, $c = 2.998 \times 10^{10}$ cm/sec, $N_A = 6.023 \times 10^{23}$ atoms/mole.

(b) Assuming $\Delta_t = {}^4/_9\Delta_o$, use the table you constructed in problem 7 to calculate the tetrahedral CFSE for the high-spin state.

(c) Calculate the OSPE (octahedral site preference energy) for each (OSPE = Octa. CFSE − tetra. CFSE).

Ion	Δ_o (cm^{-1})
Ti^{2+}	16100
V^{2+}	13550
Fe^{2+}	11200
Co^{2+}	8080
Ni^{2+}	7240
Cu^{2+}	12600

8. Calculate the relative concentrations (i.e., C^l/C^o) in a partial melt at increments of $F = 0.1$, under the following assumptions:
   (a) Homogenous solid phase, equilibrium (batch) melting for $D = 0.01$ and $D = 10$.
   (b) Calculate the relative concentration in the aggregate liquid for fractional melting for $D = 0.01$ and $D = 10$.
   (c) Plot C^l/C^o vs. F for a and b on the same graph, labeling each curve (use different colors or line types as well).

9. Calculate the enrichment of rare earth elements in an equilibrium partial melt of a mantle consisting of 10% cpx, 5% gar, 25% opx, and 60% ol, assuming *modal* (phases enter the melt in the same proportion as they exist in the solid) melting for $F = 0.02$ and $F = 0.10$. Assume the concentrations in the mantle are chondritic. Use the partition coefficients given in Table 7.5. *Where data are missing in this table, interpolate the values of partition coefficients.* Use only the following eight rare earths: La, Ce, Nd, Sm, Eu, Gd, Dy, and Lu. Plot the results on a semi-log plot of chondrite-normalized abundance versus atomic number (i.e. typical REE plot). Draw a smooth curve through the REE, interpolating the other REE. (*Hint:* Work only with chondrite-normalized abundances, don't worry about absolute concentrations, so the C^o values will all be 1.)

10. Calculate the relative enrichments of La and Sm and the La/Sm ratio in an aggregate melt produced by continuous melting. Assume bulk distribution coefficients for these two elements of 0.01 and 0.05, respectively. Do the calculation for porosities (ϕ) of 0.001 and 0.01.
    (a) Plot your results as a function of extent of melting, F, letting F vary from 0.001 to 0.1.
    (b) Plot your results on a process identification diagram, i.e., plot La/Sm vs. La, assuming initial La and Sm concentrations of 1.
    (c) Do the same calculation for and batch melting. Compare the two processes, aggregate continuously and batch on a plot of plot La/Sm vs. La.

11. Calculate the change in the La/Sm ratio of a melt undergoing *in situ* crystallization assuming bulk partition coefficients for La and Sm of 0.05 and 0.2, respectively. Assume the melt initially has a La/Sm ratio of 1. Do the calculation for values of f of 0.05 and 0.25 and assume that $f_A = f$.
    (a) Plot your results as a function of M^l/M^o.
    (b) Plot your results on a process identification diagram, i.e., plot La/Sm vs. La, assuming initial La and Sm concentrations of 1.

Chapter 8

Radiogenic isotope geochemistry

8.1 INTRODUCTION

Radiogenic isotope geochemistry had an enormous influence on geologic thinking in the twentieth century. The story began, however, in the late nineteenth century. At that time Lord Kelvin (born William Thomson, and who profoundly influenced the development of physics and thermodynamics in the nineteenth century) estimated the age of the solar system to be about 100 million years, based on the assumption that the Sun's energy was derived from gravitational collapse. In 1897 he revised this estimate downward to the range of 20–40 million years. A year earlier, another Englishman, John Jolly, estimated the age of the Earth to be about 100 million years based on the assumption that salts in the ocean had built up through geologic time at a rate proportional to their delivery by rivers. Geologists were particularly skeptical of Kelvin's revised estimate, feeling the Earth must be older than this, but had no quantitative means of supporting their arguments. They did not realize it, but the key to the ultimate solution of the dilemma, radioactivity, had been discovered about the same time (1896) by Frenchman Henri Becquerel. Only 11 years elapsed before Bertram Boltwood, an American chemist, published the first *radiometric age*. He determined the lead concentrations in three samples of pitchblende, a uranium ore, and concluded they ranged in age from 410 to 535 million years.

In the meantime, Jolly also had been busy exploring the uses of radioactivity in geology and published what we might call the first book on isotope geochemistry in 1908. When the dust settled, the evidence favoring an older Earth was deemed conclusive.

Satisfied though they might have been with this victory, geologists remained skeptical of radiometric age determinations. One exception was Arthur Holmes, who in 1913 estimated that the oldest rocks were at least 1600 million years old (Holmes was exceptional also in his support for Alfred Wegener's hypothesis of continental drift). Many geologists were as unhappy with Holmes's age for the Earth as they had been with Kelvin's.

Until the end of World War II, the measurement of isotope ratios was the exclusive province of physicists. One name, that of Alfred Nier, stands out over this period. Nier determined the isotopic compositions of many elements and made the first measurements of geologic time based on isotope ratios rather than elemental abundances. Modern mass spectrometers, while vastly more sophisticated than those of a half century ago, have evolved from Nier's 1940 design. After World War II, mass spectrometers began to appear in geologic laboratories. Many of these laboratories were established by former students and associates of Nier or Harold Urey of the University of Chicago. As this occurred, isotope geochemistry expanded well beyond geochronology, ultimately to have an impact in almost every branch of earth science and beyond.

Beyond providing precise ages of geologic events, radioactive decay is important because it provides natural tracers of geologic processes and because it provides information on the rates and pathways of geologic evolution. To understand the first point, consider a biologist who wishes to know how a nutrient, phosphorus for example, is utilized by an organism, a horse for example. The biologist can feed the horse grain doped with a small amount of radioactive phosphorus. Then by taking tissue and fluid samples and determining the amount of radioactive phosphorus present, he can trace phosphorus through various metabolic pathways. Similarly, an engineer might test a new automobile design by placing a model in a wind tunnel and using smoke as a tracer to follow the path of air around it. In principle at least, we could do a similar thing with the Earth. We might add dye to downwelling ocean water to trace deep ocean currents or add a radioactive tracer to subducting lithosphere to trace mantle convection currents. In practice, however, even the contemplation of such experiments is a bit absurd. We would need far too much dye or radioactive tracer: the scales of distance and mass are simply too large for this kind of experiment. Even if we could overcome that obstacle, we would be long dead before any useful results came from our experiment: the rates of geologic processes are simply too slow.

Nature, however, has provided natural tracers in the form of the radiogenic isotopes, which are products of natural radioactivity, and these tracers have been moving through the Earth since its beginning. For example, subducting oceanic crust has a different ratio of ^{87}Sr to ^{86}Sr than does the mantle, so we can use the ^{87}Sr/^{86}Sr ratio to trace the flow of subducting lithosphere through the mantle. Similarly, Atlantic water has a lower ^{143}Nd/^{144}Nd ratio than does Pacific water, so we can trace the flow of North Atlantic Deep Water into the Pacific using the ^{143}Nd/^{144}Nd ratio.

To understand the second point, consider the continental crust, which has a much higher ratio of Rb to Sr than does the mantle. Through time, this has led to a higher ratio of ^{87}Sr, the product of radioactive decay of ^{87}Rb, to ^{86}Sr in the crust than the mantle. However, the ^{87}Sr/^{86}Sr ratio in crustal rocks is lower than it should be had these rocks had their present Rb/Sr ratio for 4500 million years. From this observation we can conclude that the crust has not existed, or at least has not had its present composition, for the full 4500 million year history of the Earth. The situation is just the opposite for the mantle: had the mantle had its present Rb/Sr ratio for 4500 million years, it should have a lower ^{87}Sr/^{86}Sr than it does. Apparently, the mantle has had a higher Rb/Sr ratio in the past. From these simple observations we can draw the inference that the crust has evolved from the mantle through time. With more quantitative observations we can use isotope ratios to estimate the *rate* of crustal evolution.

Two fundamental assumptions are involved in virtually all geologic uses of radiogenic isotope ratios. The first is that the rate of radioactive decay is independent of all external influences, such as temperature and pressure. The second is that the isotopes of the same element are chemical identical and therefore that chemical processes cannot change, or fractionate, the ratio of two isotopes of the same elements. Neither of these assumptions holds in the absolute.* Nevertheless, all available evidence indicates that violations of these assumptions are entirely negligible.

This and the following chapter provide a brief introduction to the application of isotope studies in geochemistry. A companion book, *Isotope Geochemistry* (White, 2015) covers these topics in far more detail. *Geochronology and Thermochronology* by Reiners et al. (2017) covers geochronology in far more detail than we have space for here.

---

*There is a slight dependence of the rate of electron capture on pressure, and at extreme temperatures where nuclei become thermally excited there could be a dependence of decay rate on temperature (such temperatures, however, will only occur in the interiors of stars). Furthermore, there are subtle differences in the chemical behavior of the different isotopes of an element, which can be exploited for geologic use, which is the subject of the next chapter. For most radiogenic elements, *isotopic fractionations* are small and corrections for them are easily and routinely made.

8.2 PHYSICS OF THE NUCLEUS AND THE STRUCTURE OF NUCLEI

Nuclear physics is relevant to geochemistry for two reasons. First, the study of the distribution of isotopes forms an increasingly important part of geochemistry as well as earth science generally. Second, geochemistry concerns itself not only with the distribution of elements, but also with their origin, and the elements originated through nuclear processes, a topic we will consider in Chapter 10.

Nuclei are made up of various numbers of neutrons and protons. We'll use N to represent the number of neutrons, the *neutron number*, and Z to represent the number of protons, or *proton number*. Z is also the *atomic number* of the element, because the chemical properties of elements depend almost exclusively on the number of protons (since in the neutral atom the number of electrons equals the number of protons). The sum of Z and N is the mass number A.

8.2.1 Nuclear structure and energetics

Not all possible combinations of protons and neutrons result in stable nuclei. Typically for stable nuclei, $N \approx Z$. Thus, a significant portion of the nucleus consists of protons, which tend to repel each other. From the observation that nuclei exist at all, it is apparent that another force must exist that is stronger than coulomb repulsion at short distances. It must be negligible at larger distances, otherwise all matter would collapse into a single nucleus. This force, called the nuclear force, is a manifestation of one of the fundamental forces of nature (or a manifestation of the single force in nature if you prefer unifying theories), called the *strong* force. If this force is assigned a strength of 1, then the strengths of other forces are: electromagnetic 10^{-2}; weak force 10^{-5}; gravity 10^{-39} (we will discuss the weak nuclear force shortly). Just as electromagnetic forces are mediated by a particle, the photon, the nuclear force is mediated by the pion. The photon carries one quantum of electromagnetic force field; the pion carries one quantum of nuclear force field. The strong force also binds quarks together to form hadrons, a class of particles that includes neutrons and protons. The intensity of the strong force decreases rapidly with distance, so that at distances of more than about 10^{-14} m it is weaker than the electromagnetic force.

One of the rules of thermodynamics is that the configuration with the lowest Gibbs free energy is the most stable. This is really just one example of the general physical principle that the lowest energy configuration is the most stable; this rule applies to electron orbital configurations, as we saw in crystal field theory, and to nuclei. We would thus expect that if ^{4}He is stable relative to two free neutrons and two free protons, ^{4}He must be a lower energy state compared with the free particles. If this is the case, then from Einstein's mass–energy equivalence:

$$E = mc^2 \qquad (8.1)$$

we can predict that the ^{4}He nucleus will have less mass than two free neutrons and two free protons. It does in fact have less mass. From the principle that the lowest energy configurations are the most stable and the mass–energy equivalence, we should be able to predict the relative stability of various nuclei from their masses alone.

We begin by calculating the nominal weight of an atom from the sum of the mass of the constituent particles:

Proton: 1.007276 u or Da* = 1.6726218 × 10^{-27} kg = 938.272 MeV/c²

Neutron: 1.008664 u = 1.67492735 × 10^{-27} kg = 939.765 MeV/c²

Electron: 0.00054858 u = 9.1093829 × 10^{-31} kg = 0.5109889 MeV/c²

---

* The atomic unit of mass was formerly called an atomic mass unit or amu, but is now called the *unified atomic mass unit* and abbreviated u. It is defined as a twelfth of the mass of a ^{12}C atom. 1 u = 1.6605389 × 10^{-27} kg. In organic chemistry and biochemistry the unit of atomic mass is generally called the Dalton, Da. The international organization in charge of units and measures, the *Comité international des poids et mesures* (CIPM), has that u and Da are equivalent and either may be used. In this chapter we will use u but use Da in Chapter 12. Physicists sometimes find it useful to speak of the equivalent rest energy of elementary particles, calculated from eqn. 8.1. Strictly speaking, these have units of energy divided by the speed of light, e.g., MeV/c². Physicists, however, often just refer to these masses as MeV.

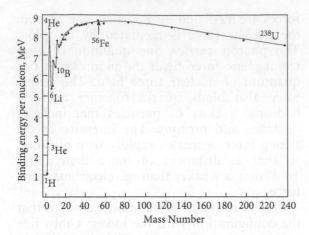

Figure 8.1 Binding energy per nucleon versus mass number.

Then we define the *mass decrement* of an atom as:

$$\delta = W - M \qquad (8.2)$$

where W is the sum of the mass of the constituent particles and M is the actual mass of the atom. For example, W for ^{4}He is: W $= 2m_p + 2m_n + 2m_e = 4.03297\,u$. The mass of ^{4}He is 4.002603 u, so $\delta = 0.030306\,u$. Converting this to energy using eqn. 8.1 yields 28.29 MeV. This energy is known as the *binding energy*. Dividing by A, the mass number, or number of nucleons, gives the *binding energy per nucleon*, E_b:

$$E_b = \frac{W - M}{A}c^2 \qquad (8.3)$$

E_b is a measure of nuclear stability: those nuclei with the largest binding energy per nucleon are the most stable. Figure 8.1 shows E_b as a function of mass. Note that the nucleons of intermediate mass tend to be the most stable.

Example 8.1 Calculating binding energies

Calculate the binding energies of ^{50}V, ^{50}Cr, and ^{50}Ti. Which of these three nuclei is the least stable? Which is the most stable?

Answer: The nucleus of ^{50}V consists of 23 protons and 27 neutrons, that of ^{50}Cr consists of 24 protons and 26 neutrons, that of ^{50}Ti consists of 22 protons and 28 neutrons. Atoms of these elements also have, respectively, 23, 24, and 22 electrons. First we calculate W for each:

$$W(^{50}V) = 23 \times 1.007276 + 27 \times 1.008665 + 23 \times 0.0005458 = 50.413856\,u$$

$$W(^{50}Cr) = 24 \times 1.007276 + 26 \times 1.008665 + 24 \times 0.0005458 = 50.413013\,u$$

$$W(^{50}Ti) = 22 \times 1.007276 + 28 \times 1.008665 + 22 \times 0.0005458 = 50.41470\,u$$

The actual masses (M) of these nuclides are: ^{50}V: 49.947163 u; ^{50}Cr: 49.94805 u; ^{50}Ti: 49.944792 u. Using equation 8.2 we calculate the mass decrement, and then divide by 50 to calculate the mass decrement per nucleon. We convert the result to kg using the conversion factor $1\,u = 1.6605 \times 10^{-27}$ kg. We then multiply by the square of the speed of light (2.99792×10^8 m/sec) to obtain the binding energy in kg-m/sec or joules. We use $1\,J = 6.2415 \times 10^{12}$ MeV to convert our answer to MeV. The results are shown in the following table.

Nuclide	δ u	δ kg	δ/A kg/nucleon	E_b J/nucleon	E_b MeV
^{50}V	0.46669	7.7496×10^{-28}	1.5499×10^{-29}	1.3930×10^{-12}	8.6944
^{50}Cr	0.46493	7.7209×10^{-28}	1.5442×10^{-29}	1.3878×10^{-12}	8.6622
^{50}Ti	0.43991	7.8030×10^{-28}	1.5606×10^{-29}	1.4023×10^{-12}	8.7543

Our results indicate that of the three, ^{50}V is the least stable and ^{50}Ti the most stable, though the difference is not that great. ^{50}V is an example of a branched decay: it can decay by β^- to ^{50}Cr or ^{50}Ti by electron capture.

Some indication of the relative strength of the nuclear binding force can be obtained by comparing the mass decrement associated with it to that associated with binding an electron to a proton in a hydrogen atom. The mass decrement above is of the order of 1%, 1 part in 10^2. The mass decrement associated with binding an electron to a nucleus is of the order of 1 part in 10^8, so bonds between nucleons are about 10^6 times stronger than bonds between electrons and nuclei.

Why are some combinations of N and Z more stable than others? The answer has to do with the forces between nucleons and how nucleons are organized within the nucleus. The structure and organization of the nucleus are questions still being actively researched in physics, and full treatment is certainly beyond the scope of this book, but we can gain some valuable insight to nuclear stability by considering two of the simplest models of nuclear structure. The simplest model of the nucleus is the *liquid-drop model*, proposed by Niels Bohr in 1936. This model assumes all nucleons in a nucleus have equivalent states. As its name suggests, the model treats the binding between nucleons as similar to the binding between molecules in a liquid drop. According to the liquid-drop model, three effects influence the total binding of nucleons: a volume energy, a surface energy, and a coulomb energy. The variation of these three forces with mass number and their total effect is shown in Figure 8.2.

Looking again at Figure 8.1, we see that, except for very light nuclei, the binding energy per nucleon is roughly constant, or that total binding energy is roughly proportional to the number of nucleons. Similarly, for a drop of liquid, the energy required to evaporate it, or unbind the molecules, would be proportional to the volume of the liquid. So the volume effect contributes a constant amount of energy per nucleon.

The surface effect arises from saturation of the strong nuclear force: a nucleon in the interior of the nucleus is surrounded by other nucleons and exerts no force on more distance nucleons. But at the surface, the force is unsaturated, leading to a force similar to surface tension in liquids. This force tends to minimize the surface area of the nucleus. The surface force is strongest for light nuclei and becomes rapidly less important for heavier nuclei.

The third effect on nuclear stability considered by the liquid-drop model is the repulsive force between protons. This force is a longer range one than the strong force and does not show saturation. It is proportional to the total number of proton pairs ($Z(Z - 1)/2$) and inversely proportional to radius. Figure 8.3 shows the stable combinations of N and Z on a plot of N against Z. Clearly, for light isotopes, N must roughly equal Z for a nucleus to be stable. For heavier isotopes, the field of stability moves in the direction of N > Z. This effect may also be explained by the repulsive coulomb force of the protons. The additional neutrons act to dilute the charge density (increase the radius) and thereby increase stability.

The liquid-drop model can account for the general pattern of binding energy in Figure 8.1 and the general distribution of stable nuclei in Figure 8.3, but not the details. The model predicts a smooth variation in binding energy with mass number, but it is apparent from Figure 8.1 that this is not the case: certain maxima occur and some configurations are more stable than others. From this, we might guess that the nucleus has some internal structure.

Another interesting observation is the distribution of stable nuclei. Nuclei with even numbers of protons and neutrons are more stable than those with odd numbers of protons or neutrons. As Table 8.1 shows, stable even–even configurations are most common; stable odd–odd configurations are particularly rare. In addition, as can be seen

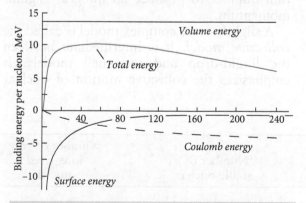

Figure 8.2 Binding energy per nucleon versus mass number as calculated using Bohr's liquid-drop experiment.

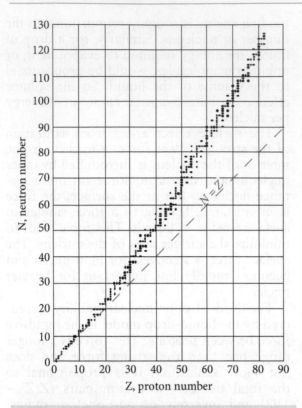

Figure 8.3 Neutron number versus proton number for stable nuclides.

These states can be described by quantum numbers. One of these quantum numbers is spin. Two protons can have the same spatial quantum numbers if their spins are anti-aligned (the situation is analogous to electrons sharing orbits). This is also true of neutrons. Apparently, nuclei are more stable when spins cancel (i.e., even number of protons or neutrons). The first proton and neutron shells are filled when occupied by two nucleons each. As in the atomic model, filling these shells produces a particularly stable configuration. The next shells are filled when six additional protons and neutrons are added for a total of eight (each). This configuration is ^{16}O. And so on, shells being filled with 2, 8, 20, 28, 50, 82, and 126 nucleons. These numbers of nucleons, which correspond to particularly stable nuclei, were called *magic numbers* and were an important clue leading to the shell model.

Another important aspect of the shell model is its prediction of nuclear angular momentum. Even–even nuclei have no angular momentum because spins of the neutrons cancel by anti-alignment, as do the proton spins. And the angular orbital momentum is zero because the nucleons are in closed shells. In even–odd and odd–even nuclides, one odd nucleon combines its half-integral spin with the integral orbital angular momentum quantum number of the nucleus, yielding half-integral angular momentum. In odd–odd nuclei, the odd proton and odd neutron each contribute a spin of ½, yielding an integral angular momentum, which can combine with an integral orbital angular momentum quantum number to produce an integral angular momentum.

A slightly more complex model is called the *collective model*. It is intermediate between the liquid-drop and the shell models. It emphasizes the collective motion of nuclear

in Figure 8.3, stable nuclei seem to be particularly common at *magic numbers*, that is, when either N or Z equals 2, 8, 20, 28, 50, 82, and 126. These observations, the even number and magic number effects, led to the *shell model of the nucleus*. It is similar to the shell model of electron structure and is based on the same physical principles, namely the Pauli exclusion principle and quantum mechanics. The Pauli exclusion principle says that no state occupied by one nucleon can be occupied by another nucleon; a nucleon added to a nucleus must occupy a new state, or niche.

Table 8.1 Numbers of stable odd and even nuclei.

Z	N	A(Z + N)	Number of stable nuclei	Number of very long-lived nuclei
Odd	Odd	Even	4	5
Odd	Even	Odd	50	3
Even	Odd	Odd	55	3
Even	Even	Even	165	11

matter, particularly the vibrations and rotations, both quantized in energy, in which large groups of nucleons can participate. Even–even nuclides with Z or N close to magic numbers are particularly stable with nearly perfect spherical symmetry. Spherical nuclides cannot rotate because of a dictum of quantum mechanics that a rotation about an axis of symmetry is undetectable, and hence cannot exist, and in a sphere every axis is a symmetry axis. The excitation of such nuclei (i.e., when their energy rises to some quantum level above the ground state) may be ascribed to the vibration of the nucleus as a whole. On the other hand, even–even nuclides far from magic numbers depart substantially from spherical symmetry and the excitation energies of their excited states may be ascribed to rotation of the nucleus.

8.2.2   The decay of excited and unstable nuclei

Just as an atom can exist in any one of a number of excited states, so too can a nucleus have a set of discrete, quantized, excited nuclear states. The behavior of nuclei in transforming to more stable states is somewhat similar to atomic transformation from excited to more stable sites, but there are some important differences. First, energy level spacing is much greater; second, the time an unstable nucleus spends in an excited state can range from 10^{-14} sec to 10^{11} years, whereas atomic lifetimes are usually about 10^{-8} sec; third, excited atoms emit photons, but excited nuclei may emit other particles as well as photons. The photon emitted through the decay of unstable nuclei is called a gamma ray. Nuclear reactions must obey general physical laws, conservation of momentum, mass–energy, spin, and so on, and conservation of nuclear particles. In addition to the decay of an excited nucleus to a more stable state, it is also possible for an unstable nucleus to decay to an entirely different nucleus, through the emission or absorption of a particle of nonzero rest mass.

Nuclear decay takes place at a rate that follows the law of radioactive decay. Interestingly, the decay rate is dependent only on the nature and energy state of the nuclide. It is independent of the past history of the nucleus, and essentially independent of external influences such as temperature, pressure,

and so on. Also, it is impossible to predict when a given nucleus will decay. We can, however, predict the probability of its decay in a given time interval. The probability of decay of a nucleus in some time interval, dt, is λ, where λ is called the decay constant. The probability of a decay among some number, N, of nuclides within dt is λN. Therefore, the rate of decay of N nuclides is:

$$\frac{dN}{dt} = -\lambda N \qquad (8.4)$$

The minus sign simply indicates that N decreases with time. Equation 8.4 is a first-order rate law that we will call the *basic equation of radioactive decay*. It is very much analogous to rate equations for chemical reactions (Chapter 5), and in this sense λ is exactly analogous to k, the rate constant, for chemical reactions, except that λ is independent of all other factors.

8.2.2.1   Gamma decay

Gamma emission occurs when an excited nucleus decays to a more stable state. A gamma ray is simply a high-energy photon (i.e., electromagnetic radiation). Its frequency, ν, is related to the energy difference by:

$$h\nu = E_u - E_l \qquad (8.5)$$

where E_u and E_l are the energies of the upper (excited) and lower (ground) states and h is Planck's constant. The nuclear reaction is written as:

$$^N Z^* \rightarrow {}^N Z + \gamma \qquad (8.6)$$

where Z is the element symbol, N is the mass number, and γ denotes the gamma ray.

8.2.2.2   Alpha decay

An α-particle is simply a helium nucleus. Since the helium nucleus is particularly stable, it is not surprising that such a group of particles might exist within the parent nucleus before α-decay. Emission of an alpha particle decreases the mass of the nucleus by the mass of the alpha particle plus a mass equivalent to the energy lost during the decay, which includes the kinetic energy of the alpha particle (constant for any given decay) and the remaining nucleus (because of the conservation of momentum, the remaining nucleus recoils from the decay reaction), and any gamma ray emitted.

The escape of the α particle is a bit of a problem, because it must overcome a very substantial energy barrier, a combination of the strong force and the coulomb repulsion, to get out. For example, α particles with energies below 8 MeV are scattered from the ^{238}U nucleus. However, the α particle emerges from the decaying ^{238}U with an energy of only about 4 MeV. This is an example of a quantum mechanical effect called *tunneling* and can be understood as follows. Quantum mechanics holds that we can never know exactly where the α particle is (or any other particle, or you or I, for that matter), we only know the probability of its being in a particular place. This probability is determined by the square of the particle's wave function, ψ. Although the wave is strongly attenuated through the potential energy barrier, it nevertheless has a small but finite amplitude outside the nucleus, and hence there is a small but finite probability of the α particle being located outside the nucleus. Anything that can occur ultimately will, so sooner or later the alpha particle escapes the nucleus.

The daughter may originally be in an excited state, from which it decays by γ decay.

Figure 8.4 shows an energy-level diagram for such a decay.

Alpha-decay occurs in nuclei with masses above the maximum in the binding energy curve of Figure 8.1, located at ^{56}Fe. Quite possibly, all such nuclei are unstable relative to alpha-decay, but the half-lives of most of them are immeasurably long.

8.2.2.3 Beta decay

Beta decay is a process in which the charge of a nucleus changes, but not the number of nucleons. If we plotted Figure 8.3 with a third dimension, namely energy of the nucleus, we would see that stable nuclei are located in an energy valley. Alpha-decay moves a nucleus down the valley axis; beta decay moves a nucleus down the walls toward the valley axis. Beta-decay results in the emission of an electron or positron (a positively charged electron), depending on which side of the valley the parent lies. Consider the three nuclei in Figure 8.5. These are known as *isobars*, since they have the same number of nucleons (12; *isotopes* have the same number of protons, *isotones* have the same number of neutrons). From what we know of nuclear structure, we can predict that the ^{12}C nucleus is the most stable of these three, because the spins of the neutrons and protons cancel each other. This is the case: ^{12}B decays to ^{12}C by

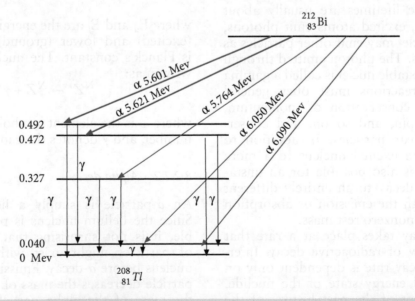

Figure 8.4 Nuclear energy-level diagram showing decay of bismuth-212 by alpha emission to the ground and excited states of thallium-208.

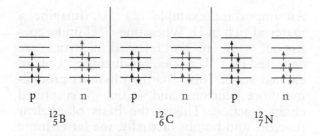

Figure 8.5 Proton and neutron occupation levels of boron-12, carbon-12, and nitrogen-12.

the creation and emission of a β^- particle and the conversion of a neutron to a proton. ^{12}N decays by emission of a β^+ and conversion of a proton to a neutron.

The discovery of beta decay presented physicists with a problem. Angular momentum must be conserved in the decay of nuclei. The ^{12}C nucleus has integral spin, as do ^{12}B and ^{12}N. But the beta particle (an electron or positron) has ½ quantum spin units, hence β decay apparently resulted in the loss of ½ spin units. The solution, proposed by Enrico Fermi,* was another, essentially massless particle called the *neutrino*, with ½ spin to conserve angular momentum. (It is now known that neutrinos do have mass, albeit a very small one that has not yet been directly measured. Other theoretical constraints require the neutrino mass to be somewhere between 0.5 eV and 0.05 eV; comparing this with the electron mass of about 0.5 MeV, we see it is roughly 10^{-6} to 10^{-7} smaller.) It is also needed to balance energy.

The kinetic energies of alpha particles are discrete. Not so for betas: they show a spectrum with a characteristic maximum energy for a given decay. Since energy must be conserved, and the total energy given off in any decay must be the same, it is apparent that the neutrino must also carry away part of the energy. The exact distribution of energy between the beta and the neutrino is random: it cannot be predicted in an isolated case, though there tends to be a fixed statistical distribution of energies, with the average observed beta energies being about a third the

maximum value (the maximum value is the case where the beta carries all the energy).

Beta decay involves the weak nuclear force. The weak force transforms a neutral particle into a charged one and vice versa. Both the weak and the electromagnetic force are now thought to be simply a manifestation of one force that accounts for all interactions involving charge (in the same sense that electrical and magnetic forces are manifestations of electromagnetism). This force is called electroweak. In β^+ decay, for example, a proton is converted to a neutron, giving up its +1 charge to a neutrino, which is converted to a positron. This process occurs through the intermediacy of the W+ particle in the same way that electromagnetic processes are mediated by photons. The photon and W particles are members of a class of particles called *bosons* that mediate forces between the basic constituents of matter. However, W particles differ from photons in having a substantial mass.

8.2.2.4 Electron capture

Another type of reaction is electron capture. This is sort of the reverse of beta decay and has the same effect, more or less, as β^+ decay. Interestingly, this is a process in which an electron is added to a nucleus to produce a nucleus with less mass than the parent! The missing mass is carried off as energy by an escaping neutrino, and in some cases by a γ. In some cases, a nucleus can decay by either electron capture, β^-, or β^+ emission. An example is the decay of ^{40}K, which decays to ^{40}Ar by β^+ or electron capture and to ^{40}Ca by β^-. In Example 8.1, we found that ^{50}V was less stable than its two isobars: ^{50}Cr and ^{50}Ti. In fact, a ^{50}V atom will eventually decay to either a ^{50}Cr atom by β^- decay or to ^{50}Ti by electron capture. The half-life for this decay is 1.4×10^{17} years, so that the decay of any single atom of ^{50}V is extremely improbable.

β decay and electron capture often leave the daughter nucleus in an excited state. In this case, it will decay to its ground state (usually very quickly) by the emission of a

* Enrico Fermi (1901–1954) had the unusual distinction of being both an outstanding theorist and an outstanding experimentalist. He made many contributions to quantum and nuclear physics and won the Nobel Prize in 1938. Interestingly, the journal *Nature* rejected the paper in which Fermi made this proposal!

γ-ray. Thus, γ rays often accompany β decay. A change in charge of the nucleus necessitates a rearrangement of the electrons in their orbits. This rearrangement results in X-rays being emitted from electrons in the inner orbits.

8.2.2.5 Spontaneous fission

Fission is a process by which a nucleus splits into two or more fairly heavy daughter nuclei. In nature, this is a very rare process, occurring only in the heaviest nuclei, ^{238}U, ^{235}U, and ^{232}Th (it is most likely in ^{238}U). This particular phenomenon is perhaps better explained by the liquid-drop model than the shell model. In the liquid-drop model, the surface tension tends to minimize the surface area while the repulsive coulomb energy tends to increase it. We can visualize these heavy nuclei as oscillating between various shapes. The nucleus may very rarely become so distorted by the repulsive force of 90 or so protons that the surface tension cannot restore the shape. Surface tension is instead minimized by splitting the nucleus entirely. Since there is a tendency for the N/Z ratio to increase with A for stable nuclei, the parent is neutron-rich. When fission occurs, some free neutrons are produced and nuclear fragments (the daughters, which may range from A = 30, zinc, to A = 64, terbium) are too rich in neutrons to be stable. The immediate daughters will decay by β^- decay until enough neutrons have been converted to protons that it has reached the valley of energy stability. It is this tendency to produce unstable nuclear byproducts, rather than fission itself, which makes fission in bombs and nuclear reactors such radiation hazards.

Some unstable heavy nuclei and excited heavy nuclei are particularly subject to fission.

An important example is ^{236}U. Imagine a material rich in U. When one ^{238}U undergoes fission, some of the released neutrons are captured by ^{235}U nuclei, producing ^{236}U in an excited state. This ^{236}U then fissions producing more neutrons, and so on – a sustained chain reaction. This is the basis of nuclear reactors and bombs (actually, the latter more often use some other nuclei, like Pu). The concentration of U, and ^{235}U in particular, is not high enough for this sort of thing to happen naturally – fission chain reactions require U enriched in ^{235}U. However, the concentration of ^{235}U was higher in the ancient Earth and at least one sustained natural chain reaction is known to have occurred about 2 billion years ago in the Oklo uranium deposit in Gabon, Africa. This deposit was found to have an anomalously low $^{235}U/^{238}U$ ratio, indicating some of the ^{235}U had been "burned" in a nuclear chain reaction. Anomalously high concentrations of fission-produced nuclides confirmed that this had indeed occurred.

Individual fission reactions are less rare. When fission occurs, there is a fair amount of kinetic energy produced, the fragments literally flying apart. These fragments damage the crystal structure through which they pass, producing "tracks" whose visibility can be enhanced by etching. This is the basis of *fission track dating*.

Natural fission also can produce variations in the isotopic abundance of the daughter elements. In general, however, the amount of the daughter produced is so small relative to that already present in the Earth that these isotopic variations are immeasurably small. An important exception is xenon, whose isotopic composition can vary slightly due to contributions from fission of U and the extinct radionuclide ^{244}Pu.

Measuring isotope ratios: mass spectrometry

Although determining the isotopic composition of elements can be done in a number of ways, for example spectroscopically in stars or by counting the decays of short-lived radionuclides, isotope geochemistry relies almost exclusively on mass spectrometers to determine isotope ratios because only this technique can produce the requisite precision. There are a variety of types of mass spectrometers, but all work by ionization and separating ions based on their mass to charge ratio m/z in a vacuum.

The most common mass spectrometers are quadrupole mass spectrometers used to determine the components of organic compounds rather than determine isotope ratios. Quadrupoles work by sending a stream of ions between four parallel aligned metal rods whose electric charge is rapidly oscillated. The frequency of oscillation is adjusted so that only ions of one m/z successfully pass through the instrument to reach the detector without colliding with one of the rods. Quadrupole mass spectrometers are used in geochemistry for elemental analysis. In these inductively coupled plasma–mass spectrometers (ICP-MS), a solution containing the sample is introduced into a stream of Ar gas, which is then excited to a plasma with radio frequency. With some exceptions, such quadrupole instruments are not used to measure isotope ratios.

Most isotope ratio mass spectrometers are of the magnetic-sector, or Nier-type* design, a schematic of which is shown in Figure 8.6. It consists of three essential parts: an ion source, a mass analyzer, and a detector. Ions are generated and accelerated by the ion source traveling down a tube into a magnetic field, where they experience a force, **F**, according to:

$$\mathbf{F} = q\mathbf{v} \times \mathbf{B} \tag{8.6}$$

where **B** is the magnetic field strength, **v** is the ion's velocity, which varies with mass for a given energy, and q is its charge (bold is used to denote vector quantities). The force is applied perpendicular to the direction of motion (hence, it is more properly termed a torque), and to the magnetic field vector, which changes the trajectory of the ion. The extent of this change depends on the mass to charge ratio and the strength of the magnetic field is adjusted such that only if ions of specific m/z are focused into a detector, or, as is common in most modern mass spectrometers and illustrated in Figure 8.6, a set of detectors separated such that each detects ions of a separate m/z.

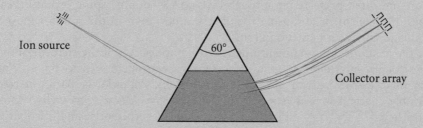

Ion source
60°
Collector array

Figure 8.6 Schematic of a magnetic sector mass spectrometer. Ions produced and accelerated in the ion source and then move through a evacuated tube passing through the field on an electromagnetic. The magnetic field is adjusted to deflect ions of one or more m/z ratios into one or more collectors that either measure the ion current or detectors that count the ions.

From this basic design, mass spectrometers vary based on how ions are produced, how they are detected, and whether and how the ion beam is filtered. The most important variation is in how ions are produced. Traditionally, this has been either thermal ionization or gas-source ion bombardment. In thermal ionization, most commonly used for radiogenic isotope ratio analysis, a solution of the purified element of interested is dried on a filament, introduced into the mass spectrometer, and the filament heated to a temperature (typically ~1000–2000°C) where the sample evaporates and ionizes. In gas-source instruments, used for noble gases and stable isotope analyses (Chapter 9), a gas containing the element of interest, for example, Ar, CO_2, or H_2O, is ionized by bombardment with electrons. Secondary ion mass spectrometers (SIMS), also called ion microprobes, which are used for *in situ* isotopic analysis, ionize atoms on the polished surface of a sample by focusing an ion beam on them. With this technique it is possible to analyze areas of the order of a few tens μm^2, allowing for analysis of individual parts of mineral grains. An important development over the past two decades has been the adaptation of ICP-MS instruments for isotopic analysis by replacing the quadrupole with a magnetic sector mass spectrometer and using multiple detectors or ion collectors. These are known as MC-ICP-MS instruments. A laser can be used to ablate material from the polished surface of a sample and introduced into the Ar gas stream of ICP-MS instruments, providing an another means of *in situ* analysis (LA-MC-ICP-MS), although on spots of slightly larger areas than ion probes. Because the temperature of the plasma in ICP-MS instruments reaches ~6000 K, ionization is virtually complete for all elements, which is an important advantage as many elements cannot readily be converted to gas and/or are difficult to ionize thermally (Hf being a good example).

A disadvantage is that the plasma produces a spectrum of energies, which leads to a spectrum of velocities for a given *m/z* in equation 8.6. This requires the ion beam to be energy-filtered, which is done by deflecting the beam electrostatically before it passes through the magnetic field. Instruments equipped with energy filters are referred to as *double focusing*.

Accelerator mass spectrometers are used for analysis of very rare isotopes such as ^{14}C and ^{10}Be. The principles are largely the same, but ions are accelerated to ~2.5 MeV compared with accelerations of the order of 10 kV in conventional mass spectrometers and additional steps are taken for energy filtration and to strip out interfering ions.

Some extent of mass fractionation occurs in all mass spectrometers resulting in either light or heavy isotopes being disproportionately detected. A correction for this effect must be made and is done either by measuring an additional isotope ratio of the element and correcting the isotope ratio of interest based on the observed fractionation in that additional isotope ratio. For example, ^{143}Nd/^{144}Nd ratios are routinely corrected for mass fractionation by comparing the measured ^{146}Nd/^{144}Nd with the known ratio and assuming the fractionation is a simple function of mass difference. When this cannot be done, for example for stable isotopes ratios, a correction can be made by analyzing standards of known isotopic composition immediately before and after the sample and making a correction based on the observed fractionation of the standard (this is known as sample-standard bracketing, or SSB). An alternative, and often preferred, approach is to add a "double spike" solution enriched in two minor isotopes of the element and using the observed deviation in the ratio of those isotopes from the known value to correct for fractionation.

Except for the *in situ* techniques mentioned above, purification of the element of interest is necessary to remove isobaric interferences because most mass spectrometers do not have the mass resolution to distinguish them (SIMS instruments are an exception). Potential isobaric interferences include singly and doubly charged species (e.g., ^{87}Sr$^+$ –^{87}Rb$^+$, ^{88}Sr^{2+}–^{44}Ca$^+$), as well as molecular species such as oxides (e.g., ^{160}Dy^{18}O$^+$ and ^{176}Lu$^+$), although if sufficiently small, isobaric interferences can be accurately corrected (e.g., by measuring ^{85}Rb and correcting based on the known ^{87}Rb/^{85}Rb ratio). Purification generally involves processing the sample chemically, for example by ion exchange, or cryogenically in the case of gases. By pairing high-performance chromatography with gas source mass spectrometers, it is possible to separately analyze the isotopic composition of different organic molecules in a sample, something known as *compound specific isotopic analysis*.

*This design was developed by Alfred Nier (1911–1994) of the University of Minnesota in the 1930s. Nier used his instrument to determine the isotopic abundances of many of the elements and was partly responsible for the fields of stable and radiogenic isotope geochemistry. He produced the first radiometric age based on isotopic measurements. In World War II, he participated in the Manhattan project, building mass spectrometers among other contributions. In the 1980s he was still designing mass spectrometers, this time miniature ones that could fly on spacecraft on interplanetary voyages. These instruments provided measurements of the isotopic composition of atmospheric gases of Venus and Mars.

8.3 BASICS OF RADIOGENIC ISOTOPE GEOCHEMISTRY AND GEOCHRONOLOGY

The basic equation of radioactive decay is:

$$\frac{dN}{dt} = -\lambda N \qquad (8.4)$$

λ is the decay constant, which we defined as the probability that a given atom would decay in some time dt. It has units of time^{-1}. Let's rearrange eqn. 8.4 and integrate:

$$\int_{N_0}^{N} \frac{dN}{N} = \int_{0}^{t} -\lambda dt \qquad (8.7)$$

where N_0 is the number of atoms of the radioactive, or parent, isotope present at time $t = 0$. Integrating, we obtain:

$$\ln \frac{dN}{N} = -\lambda t \qquad (8.8)$$

This can be expressed as:

$$\frac{N}{N_0} = e^{-\lambda t} \text{ or } N = N_0 e^{-\lambda t} \qquad (8.9)$$

Suppose we want to know the amount of time for the number of parent atoms to decrease to half the original number, that is, t when $N/N_0 = \frac{1}{2}$. Setting N/N_0 to $\frac{1}{2}$, we can rearrange eqn. 8.8 to get:

$$\ln \frac{1}{2} = -\lambda t_{1/2} \text{ or } \ln 2 = \lambda t_{1/2}$$

and finally:

$$t_{1/2} = \frac{\ln 2}{\lambda} \qquad (8.10)$$

This is the definition of the *half-life*, $t_{1/2}$.

Now the decay of a parent produces a daughter, or *radiogenic*, nuclide. The number of daughters produced is simply the difference between the initial number of parents and the number remaining after time t:

$$D = N_0 - N \qquad (8.11)$$

Rearranging eqn. 8.9 to isolate N_0 and substituting that into 8.11, we obtain:

$$D = Ne^{\lambda t} - N = N(e^{\lambda t} - 1) \qquad (8.12)$$

This tells us that the number of daughters produced is a function of the number of parents present and time. Since in general there will be some atoms of the daughter nuclide around to begin with, when $t = 0$, a more general expression is:

$$D = D_0 + N(e^{\lambda t} - 1) \qquad (8.13)$$

where D_0 is the number of daughters originally present.

An exponential function can be expressed as a Taylor series expansion:

$$e^{\lambda t} = 1 + \lambda t + \frac{(\lambda t)^2}{2!} + \frac{(\lambda t)^3}{2!} + \cdots \qquad (8.14)$$

Provided $\lambda t \ll 1$, the higher-order terms become very small and can be ignored; hence for times that are short compared with the decay constant (i.e., for $t \ll 1/\lambda$), eqn. 8.13 can be written as:

$$D \cong D_0 + N \lambda t \qquad (8.15)$$

Let's now write eqn. 8.13 using a concrete example, such as the decay of ^{87}Rb to ^{87}Sr:

$$^{87}Sr = {}^{87}Sr_0 + {}^{87}Rb(e^{\lambda t} - 1) \qquad (8.16)$$

As it turns out, it is generally much easier, and usually more meaningful, to measure the ratio of two isotopes than the absolute abundance of one. We therefore measure the ratio of ^{87}Sr to a nonradiogenic isotope, which by convention is ^{86}Sr. Thus, the useful form of eqn. 8.16 is:

$$\frac{^{87}Sr}{^{86}Sr} = \left(\frac{^{87}Sr}{^{86}Sr}\right)_0 + \frac{^{87}Rb}{^{86}Sr}(e^{\lambda t} - 1) \qquad (8.17)$$

Similar expressions can be written for other decay systems.

Equation 8.17 is a concise statement of Sr isotope geochemistry: the $^{87}Sr/^{86}Sr$ ratio in a system depends on: (1) the $^{87}Sr/^{86}Sr$ at time $t = 0$, (2) the $^{87}Rb/^{86}Sr$ ratio of the system (in most cases, we can assume the $^{87}Rb/^{86}Sr$ ratio is directly proportional to the Rb/Sr ratio), and (3) the time elapsed since $t = 0$. Actually, the best summary statement of isotope geochemistry was given by Paul Gast (Gast, 1960):

In a given chemical system the isotopic abundance of ^{87}Sr is determined by four parameters: the isotopic abundance at a given initial time, the Rb/Sr ratio of the system, the decay constant of ^{87}Rb, and the time elapsed since the initial time. The isotopic composition of a particular sample of strontium, whose history may or may not be known, may be the result of time spent in a number of such systems or environments. In any case the isotopic composition is the time-integrated result of the Rb/Sr ratios in all the past environments. Local differences in the Rb/Sr will, in time, result in local differences in the abundance of ^{87}Sr. Mixing of material during processes will tend to homogenize these local variations. Once homogenization occurs, the isotopic composition is not further affected by these processes.

This statement can, of course, be equally well applied to the other decay systems.

Table 8.2 lists the radioactive decay schemes of principal geologic interest. The usefulness and significance of each of the decay schemes are different and depend on the geochemical behavior of the parent and daughter, the half-life and the abundance of the parent. We will shortly consider the geologic significance of each one.

Geochronology is one of the most important applications of isotope geochemistry and the two are often closely intertwined. Let's rewrite eqn. 8.17 in more general terms:

$$R = R_0 + R_{P/D}(e^{\lambda t} - 1) \qquad (8.18)$$

where R_0 is the initial ratio and $R_{P/D}$ is the parent/daughter ratio. Measurement of geologic time is based on this equation or various derivatives of it. First let's consider the general case. Given a measurement of an isotope ratio, R, and a parent/daughter ratio, $R_{P/D}$, two unknowns remain in eqn. 8.18: t and the initial ratio. We can calculate neither from a single pair of measurements. But if we can measure R and $R_{P/D}$ on a second system for which we believe t and R_0 are the same, we have two equations and two unknowns, and subtracting the two equations yields:

$$\Delta R = \Delta R_{P/D}(e^{\lambda t} - 1) \qquad (8.19)$$

which we can solve for t. Rearranging:

$$\frac{\Delta R}{\Delta R_{P/D}} = (e^{\lambda t} - 1) \qquad (8.20)$$

t may then be solved for as:

$$t = \frac{\ln\left(\frac{\Delta R}{\Delta R_{P/D}}\right) + 1}{\lambda} \qquad (8.21)$$

We can obtain the ratio $\Delta R/\Delta R_{P/D}$ from any two data points, regardless of whether they are related or not. To overcome this problem in practice, many pairs of measurements of R and $R_{P/D}$ are made. How do we calculate the age when many pairs of measurements are made? Note that eqn. 8.18 has the form $y = a + bx$, where y is R, a is the intercept, R_0, b is the slope, $e^{\lambda t} - 1$, and x is $R_{P/D}$. An equation of this form is, of course, a straight line on a plot of R versus $R_{P/D}$, such as Figure 8.7. The slope of the line passing through the data is thus related to the age of the system and is called an *isochron*. When multiple measurements of the daughter isotope ratio and parent/daughter ratio are available, the slope, $\Delta R/\Delta R_{P/D}$, can be calculated by the statistical technique of *linear regression*, which we mentioned several times previously. The age is then obtained by substituting the value of the slope into eqn. 8.21 (see Example 8.2). Regression also yields an intercept, which is

Table 8.2 Long-lived radioactive decay systems of geochemical interest.

Parent	Decay mode	λ	Half-life	Daughter	Ratio
^{40}K	β^+, e.c., β^-	5.5492×10^{-10} y^{-1*}	1.25×10^9 y	^{40}Ar, ^{40}Ca	^{40}Ar/^{36}Ar
^{87}Rb	β^-	1.42×10^{-11} y^{-1}	4.88×10^{10} y	^{87}Sr	^{87}Sr/^{86}Sr
^{138}La	β^-, e.c.	1.55×10^{-12} y$^{-1\dagger}$	4.49×10^{11} y	^{138}Ce, ^{138}Ba	^{138}Ce/^{142}Ce, ^{138}Ce/^{136}Ce
^{147}Sm	α	6.54×10^{-12} y^{-1}	1.06×10^{11} y	^{143}Nd	^{143}Nd/^{144}Nd
^{176}Lu	β^-	1.867×10^{-11} y$^{-1\ddagger}$	3.71×10^{10} y	^{176}Hf	^{176}Hf/^{177}Hf
^{187}Re	β^-	1.67×10^{-11} y^{-1}	4.16×10^{10} y	^{187}Os	^{187}Os/^{186}Os
^{232}Th	α	4.948×10^{-11} y^{-1}	1.40×10^{10} y	^{208}Pb, ^{4}He	^{208}Pb/^{204}Pb, ^{3}He/^{4}He
^{235}U	α	9.849×10^{-10} y$^{-1\S}$	7.04×10^8 y	^{207}Pb, ^{4}He	^{207}Pb/^{204}Pb, ^{3}He/^{4}He
^{238}U	α	1.55125×10^{-10} y^{-1}	4.47×10^9 y	^{206}Pb, ^{4}He	^{206}Pb/^{204}Pb, ^{3}He/^{4}He

Note: the branching ratio, i.e. ratios of decays to ^{40}Ar to total decays of ^{40}K is 0.117. The production of ^{4}He from ^{147}Sm decay is insignificant compared with that produced by decay of U and Th.

*Value suggested by Renne et al. (2010). The conventional value is 5.543×10^{-10} y^{-1}

$\dagger$The branching ratio, i.e. ratios of decays to ^{138}Ce to total decays of ^{138}La is 0.348; this gives an effective half-life for decay to ^{138}Ce of 2.925×10^{11} yr and an effective decay constant of 2.37×10^{-12} y^{-1}.

$\ddagger$Value recommended by Söderlund et al. (2004).

$\S$Value suggested by Mattinson (2010). The conventional value is 0.98485×10^{-10} y^{-1}.

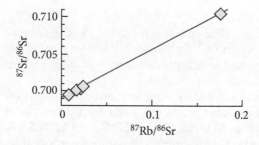

Figure 8.7 A Rb-Sr isochron. Five analyses from a clast in the Bholghati meteorite fall on an isochron, whose slope is related to the age of the system. Data from Nyquist et al. (1990).

simply the initial ratio R_0, since, as may be seen from eqn. 8.18, $R = R_0$ when $R_{P/D} = 0$. Example 8.2 illustrates this technique using the Rb-Sr system.

This approach is known as the *isochron method* of dating and geochronology with most decay systems is done using this method. There are several cases where variations on this technique are used. In limited special cases, we can calculate t from a single pair of measurements. One such case is where there is no significant amount of the daughter element present at $t = 0$, and we can drop the initial ratio term from eqn. 8.18. An example is K-Ar dating of volcanic rocks. Ar, the daughter, is lost upon eruption, and the only Ar present is radiogenic Ar. Another example is minerals that concentrate the parent and exclude the daughter. Zircon, for example, accepts U but not Pb; micas accept Rb but not Sr. Even in these cases, however, precise age determination generally requires we make some correction for the very small amounts of the daughter initially present. But since $R \gg R_0$, simple assumptions about R_0 can suffice. We'll explore K-Ar and zircon dating techniques in greater detail in subsequent sections.

There are two important assumptions built into the use of eqn. 8.18.

1. The system of interest was in isotopic equilibrium at time $t = 0$. Isotopic equilibrium in this case means the system had a homogenous, uniform value of R_0.

2. The system as a whole and each analyzed part of it was closed between $t = 0$ and time t (usually the present time). Violation of these conditions is the principal source of error in geochronology.

In conventional linear regression, the x value is assumed to be known absolutely (as would be, for example, the age or height of an individual). In geochronology, there is some analytical error associated with the x value (the parent/daughter ratio). It is important to take these analytical errors into account. In practical geochronology, an approach called two-error regression (e.g., York, 1969), which takes account of measurement errors in both R and $R_{P/D}$, is generally used. The details of this method are, however, beyond the scope of this book, but a Microsoft Excel™ plug-in for these calculations called *Isoplot* is available from the Berkeley Geochronology Center (http://www.bgc.org/).

Assumption 1 generally requires a thermal event to isotopically homogenize the system. In igneous rocks, melting accomplishes this, in metamorphic rocks, this is accomplished through diffusion. As we found in Chapter 5, diffusion rates increase exponentially with temperature. Once a rock undergoing metamorphism reaches the critical temperature for diffusion to achieve isotopic equilibrium, it will remain in this state until temperatures drop to the point where diffusion rates are no longer rapid enough to maintain isotopic equilibrium. The temperature is known as the *closure temperature,* which also depends on cooling rate and varies between different isotopic systems and between minerals (a detailed consideration of closure temperatures would take us too far afield, but can be found in White, 2015). Closure temperatures can be as high as ~1000°C for Sm-Nd in ultramafic rocks and as low as ~150°C for K-Ar in sedimentary rocks.

Example 8.2 Calculating isochrons and ages

The following $^{87}Rb/^{86}Sr$-$^{87}Sr/^{86}Sr$ data were measured by Nyquist et al. (1990) on a clast within the achondritic meteorite *Bholghati*. These data are plotted in Figure 8.6. What is the age and error on the age of this clast? What is the initial $^{87}Sr/^{86}Sr$ ratio of the clast and the error on this ratio?

Answer: We can calculate the age using eqn. 8.21 above. The value of $\Delta R / \Delta R_{P/D}$ is the slope of the isochron, i.e., the slope of a line through the data on a plot of $^{87}Sr/^{86}Sr$ vs. $^{87}Rb/^{86}Sr$. We will use simple least squares linear regression to obtain the slope and intercept; the latter will be the initial $^{87}Sr/^{86}Sr$ ratio.

Least squares regression is a calculation intensive procedure. Many computer programs are available for calculating regression, and regression functions are built into some hand calculators. Some spreadsheets also have built-in regression functions, such as SLOPE, INTERCEPT, and LINEST in Microsoft Excel™. In MATLAB, use the **regress** function. However, to illustrate the procedure, we will use a spreadsheet to calculate regression using basic spreadsheet functions.

Our spreadsheet is shown below. To make it easier to read, we have defined names for those cells containing calculated parameters; the names are shown to the left of the cell. We first calculate the parameters that we will need to use several times in our equations. These are the mean of x and y, the number of values, n, the sum of squares of x and y $(\Sigma x_i^2, \Sigma y_i^2)$ and the sum of the cross products, $\Sigma x_i y_i$. From these parameters, we also calculate the sum of squares of deviations for x $(\Sigma x_i^2 - \overline{x}^2 n)$. From these parameters, it is fairly straightforward to calculate the slope, b, the intercept, a, and the errors on the slope and intercept. Our result is that the age of the clast is 4.57 ± 0.02 Ga (Ga is the abbreviation for gigayears, 10^9 years). At that time, the $^{87}Sr/^{86}Sr$ ratio of the clast was 0.69892 ± 0.00003.

$^{87}Rb/^{86}Sr$	$^{87}Sr/^{86}Sr$
0.01015	0.69966
0.17510	0.71052
0.02166	0.70035
0.02082	0.70029
0.01503	0.69988

	$^{87}Rb/^{86}Sr$		$^{87}Sr/^{86}Sr$	
	0.01015		0.69966	
	0.1751		0.71052	
	0.02166		0.70035	
	0.02082		0.70029	
	0.01503		0.69988	
meanx	0.048552	meany	0.70214	
n	5			
SSQx	0.031892	SSQy	2.46509	
CP	0.171782			
SSQDev	0.020105	(SSQX-x-bar^2*n)		
b	0.066204	(CP-meanx*meany*n)/SSQDev		
a (initial)	0.69892	meany-b*meanx		
sigma b	0.000320	SQRT(((SSQy-meany^2*n) -(CP-meanx*meany*n)^2/SSQDev) *1/(SSQDev*(n-2)))		
sigma a	0.000026	SQRT(((SSQy-meany^2*n) -(CP-meanx*meany*n)^2/SSQDev) *((1/n)+meanx^2/SSQDev)*1/(n-2))		
t	4.57237	±	0.0228	Ga

8.4 DECAY SYSTEMS AND THEIR APPLICATIONS

Decay systems commonly used in geochemistry are listed in Table 8.2. We'll consider each of these in further detail in the following sections. Solar system initial ratios and chondritic parent/daughter ratios are listed in Table 3.

8.4.1 Rb-Sr

This decay system was one of the first to be widely used in geochronology and remains one of the most useful geochemical tracers. An important advantage of the system is the relatively large variations of the Rb/Sr ratio in rocks. Because of the difference in geochemical properties of the two elements, Rb/Sr can vary by several orders of magnitude. Since the accuracy of an age determination depends heavily on the spread of measured ratios, this makes the Rb-Sr system a useful geochronological tool, particularly in granitic and metasedimentary rocks. As we noted in Chapter 7, Rb is a highly soluble, highly incompatible element. Sr is also relatively soluble and is fairly incompatible in mafic and, particularly, ultramafic systems. However, it is relatively compatible in silica-rich igneous systems, partitioning preferentially into plagioclase. The result is that the mantle has a relatively uniform and low $^{87}Sr/^{86}Sr$ ratio, and the continental crust has a much more variable, and, on average, higher ratio. The mobility of these elements, however, is a distinct disadvantage in geochronology as the closed system requirement is often violated.

The Sr isotopic evolution of the Earth and its major silicate reservoirs (the continental crust and mantle) is illustrated in Figure 8.8, which is a plot of $^{87}Sr/^{86}Sr$ versus time (in giga-annum, or billions of years, abbreviated Ga). Such a plot is called an *isotope evolution diagram*. A characteristic of such diagrams is that a closed reservoir will evolve along a line whose slope is proportional to the parent/daughter ratio, in this case $^{87}Rb/^{86}Sr$. That this is the case is easy to show from eqn. 8.17. Where $t \ll 1/\lambda$, then eqn. 8.17 becomes:

$$\frac{^{87}Sr}{^{86}Sr} = \left(\frac{^{87}Sr}{^{86}Sr}\right)_0 + \frac{^{87}Rb}{^{86}Sr}\lambda t \qquad (8.22)$$

This equation has the form:

$$^{87}Sr/^{86}Sr = a + bt$$

This is the equation of a straight line on a plot of $^{87}Sr/^{86}Sr$ versus t with slope λ $^{87}Rb/^{86}Sr$ and intercept $(^{87}Sr/^{86}Sr)_0$ (to the degree that the Taylor series approximation is not exact, the line will actually be curved, but the approximation is fairly good in this case). In Figure 8.8 we have plotted geologic age rather than the time since $t = 0$; in other words, we have transformed t to $(4.55 - t)$, where 4.55 Ga is the assumed age of the Earth. Thus, the intercept, a, the initial value, occurs at $t = 4.55$ rather than 0.

The initial $^{87}Sr/^{86}Sr$ ratio of the Earth can be estimated from the initial ratio in meteorites under the assumption that the entire solar system had a uniform $^{87}Sr/^{86}Sr$ at the time it formed. Once it forms, the Earth is a closed system, so it evolves along a straight line with slope proportional to the bulk Earth $^{87}Rb/^{86}Sr$. This ratio has been estimated in various ways to be about 0.075 ± 0.015 (as we noted, the $^{87}Rb/^{86}Sr$ is proportional to the Rb/Sr ratio; the bulk Earth Rb/Sr is about 0.025 ± 0.005).

Now suppose that a portion of the mantle partially melts at 3.8 Ga to form a segment of continental crust. The new crust has a higher Rb/Sr ratio than the mantle because Rb is more incompatible than Sr. Thus, this crust evolves along a steeper line than the bulk Earth. The residual mantle is left with a lower Rb/Sr ratio than the bulk Earth and evolves along a shallow slope. If the melting in the mantle and creation of crust were a continuous process, the mantle would evolve along a continuously decreasing slope, i.e., a concave-downward path.

The relative mobility of these elements, particularly Rb, can be a disadvantage for geochronology because the closed system assumption is violated. Even very young rocks can be contaminated during weathering or hydrothermal activity. Mobility of Rb or Sr can result in imprecise ages or incorrect initial ratios. On the other hand, the large range in Rb/Sr ratios in siliceous igneous and metamorphic rocks means that Rb-Sr ages are somewhat insensitive to variations in initial $^{87}Sr/^{86}Sr$ ratios.

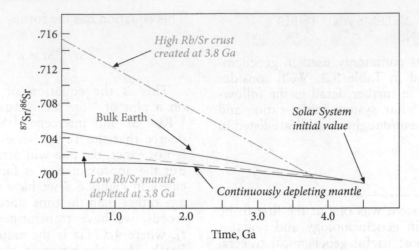

Figure 8.8 Sr isotopic evolution of the bulk Earth, evolution of high Rb/Sr crust created at 3.8 Ga, evolution of the resulting residual mantle, and the evolution of a mantle being continuously depleted.

Because Rb is volatile, we can only guess at the Rb/Sr ratio of the Earth. The present-day chondritic $^{87}Sr/^{86}Sr$ ratio is about 0.725 whereas the $^{87}Sr/^{86}Sr$ ratio of the Earth is thought to be around 0.7045 ± 0.001.

One interesting and useful feature of this system arises from the long residence time of Sr in seawater and its easy substitution in calcium carbonate. Because of its long residence time ($\sim 10^6$ years), the $^{87}Sr/^{86}Sr$ ratio of seawater is homogenous in the modern open ocean and presumably was so in the past. It does vary with time, however, as is shown in Figure 8.9. The $^{87}Sr/^{86}Sr$ ratio of seawater is controlled to a first approximation by the relative input of Sr from the weathering of the continents and sea-floor hydrothermal activity. The ratio of these will vary with mean spreading rate, erosion rates, and plate geometry. The variation of $^{87}Sr/^{86}Sr$ in seawater through the Phanerozoic has been determined from the analysis of carbonate and phosphate marine fossils. The age of marine sediments can thus be determined simply by determining the $^{87}Sr/^{86}Sr$ ratio of carbonate precipitated from seawater and comparing it with curve in Figure 8.9. This is particularly true of the Cenozoic, where evolution of $^{87}Sr/^{86}Sr$ in seawater has been particularly precisely determined and there has been a rapid and steady rise in $^{87}Sr/^{86}Sr$. This dating technique, referred to as *Sr isotope chronostratigraphy*, is widely used in paleo-oceanography and petroleum exploration.

8.4.2   Sm-Nd

^{143}Nd is produced by α-decay of ^{147}Sm. The Sm-Nd system is in many ways the opposite of the Rb-Sr system. First, the Sm/Nd ratio of the mantle is higher than that of the crust and hence the $^{143}Nd/^{144}Nd$ ratio is higher in the mantle than in the crust. Second, the Sm/Nd ratios of the crust and siliceous igneous rocks are relatively uniform, making this system often unsuitable for dating such rocks, but variable in mafic and ultramafic rocks (the opposite is true of Rb/Sr). Third, neither Sm nor Nd is particularly mobile, so ages and initial ratios are relatively insensitive to weathering and metamorphism, meaning the close system requirement is less likely to be violated. Fourth, Nd has a short seawater residence time and $^{143}Nd/^{144}Nd$ is not uniform in seawater.

Like all rare earth elements, Sm and Nd are "refractory lithophile elements" and their *relative* abundances in chondritic meteorites are uniform. It is generally assumed that the Earth also had the same relative abundances of these elements and therefore that the Sm/Nd ratio of the Earth is the same as in chondritic meteorites. More recently, differences in the ratio $^{142}Nd/^{144}Nd$ between the Earth and chondrites suggests the possibility that Sm/Nd of the Earth might be as much as a few percent different than chondrites. Nevertheless, $^{143}Nd/^{144}Nd$ ratios are often reported relative to the chondritic ratio using the ε notation. ε_{Nd} is simply the relative

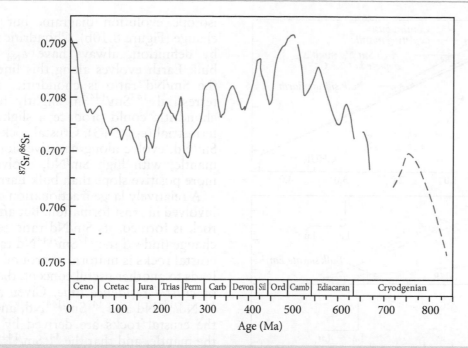

Figure 8.9 ^{87}Sr/^{86}Sr in seawater over the last 800 Ma, determined from the analysis of phosphate and carbonate fossils. Grayed areas represent uncertainties. From McArthur et al. (2012).

deviation of the ^{143}Nd/^{144}Nd ratio from the chondritic ratio in parts in 10^4:

$$\varepsilon_{Nd} = \frac{(^{141}\text{Nd}/^{144}\text{Nd}) - (^{141}\text{Nd}/^{144}\text{Nd})_{chon}}{(^{141}\text{Nd}/^{144}\text{Nd})_{chon}}$$

$$\times 10^4 \qquad (8.23)$$

The epsilon notation was introduced specifically for the Sm-Nd system by DePaolo and Wasserburg (1976) in light of the small variations in ^{143}Nd/^{144}Nd. The notation has subsequently been expanded and is now used more generally in isotope geochemistry to denote relative deviation of an isotope ratio in parts per 10,000 from some reference value.

The present-day chondritic value of ^{143}Nd/^{144}Nd is 0.512630 (when corrected for mass fractionation to a ^{146}Nd/^{144}Nd ratio of 0.7219). One value of the ε_{Nd} notation is that most rocks will have ε_{Nd} values in the range of −20 to +10, and it is certainly easier to work with a one- or two-digit number than a five- or six-digit one. Furthermore, we learn something about the history of a rock just by knowing whether its ε_{Nd} is positive or negative. A negative value of ε_{Nd} implies that on average over the history of the Earth the Sm/Nd ratio of that rock or its precursors

has been lower than chondritic. This, in turn, implies the rare earth pattern of the rock or its precursors was light rare earth-enriched. We can draw the opposite inference from a positive ε_{Nd} value. It is often useful to compare the ^{143}Nd/^{144}Nd of some ancient rock with the chondritic value at the time of its formation, that is, its initial ^{143}Nd/^{144}Nd or ε_t. We can calculate ε_t simply by substituting the chondritic value appropriate to that time into eqn. 8.23.

Figure 8.10a illustrates the gross features of the Nd isotope evolution of the Earth. The initial ^{143}Nd/^{144}Nd of the Earth is assumed to be the same as that of meteorites. If the ^{147}Sm/^{144}Nd ratio of the Earth is a little higher than that of chondrites (0.1960), the bulk Earth evolves along a curve a little steeper than the chondritic one (CHUR). Because Nd is more incompatible than Sm and will therefore be more enriched in a partial melt than Sm, the crust, formed by partial melting of the mantle, has a low Sm/Nd ratio and evolves along a line of lower slope. Partial melting depletes the mantle more in Nd than in Sm, so the mantle is left with a high Sm/Nd ratio and evolves along a steeper slope.

By replacing ^{143}Nd/^{144}Nd with ε_{Nd} we do not change the fundamental features of the

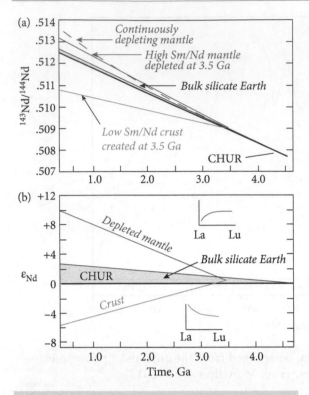

Figure 8.10 (a) Nd isotope evolution in mantle and crust. Bold line shows the evolution of a chondritic uniform reservoir (CHUR); also shown is the evolution of crust created at ~3.5 Ga, the corresponding residual mantle, and the evolution of a continuously depleted mantle. (b) Evolution of CHUR, crust, and mantle when $^{143}Nd/^{144}Nd$ is transformed to ε_{Nd}. Gray areas show the range of possible compositions of the bulk silicate Earth.

isotope evolution diagram, but the slopes change (Figure 8.10b). Chondritic meteorites, by definition, always have $\varepsilon_{Nd} = 0$. The bulk Earth evolves along this line as well if the Sm/Nd ratio is chondritic, although a terrestrial $^{147}Sm/^{144}Sm$ slightly higher than chondritic could produce a slightly positive terrestrial ($\varepsilon_{Nd} \lesssim +3$). Crustal rocks, with low Sm/Nd, evolve along negative slopes and the mantle, with high Sm/Nd, evolves along a more positive slope than bulk Earth.

A relatively large fractionation of Sm/Nd is involved in crust formation. But after a crustal rock is formed, its Sm/Nd ratio tends not to change (indeed the $^{147}Sm/^{144}Nd$ ratio of most crustal rocks is uniform at around 0.13). This leads to another useful concept, the *model age* or *crustal residence time*. Given a measured $^{143}Nd/^{144}Nd$ and $^{147}Sm/^{144}Nd$, and assuming the crustal rocks are derived by melting of the mantle, and that the $^{147}Sm/^{144}Nd$ ratio of crustal rocks has not changed since the time it formed by melting, we can estimate the "age" or crustal residence time, that is, the time the rock has spent in the crust. To do this, however, we need to make an assumption about how the mantle $^{143}Nd/^{144}Nd$ ratio has evolved through time. DePaolo and Wasserburg (1976) assumed that it has evolved like chondrites. In this case, we project the $^{143}Nd/^{144}Nd$ ratio back along a line whose slope corresponds to the measured present $^{147}Sm/^{144}Nd$ ratio until it intersects the chondritic growth line (Figure 8.11). The model age

Example 8.3 Calculating Nd model ages

DePaolo (1981) reported that the Pike's Peak granitic batholith has a $^{147}Sm/^{144}Nd$ ratio of 0.1061 and $^{143}Nd/^{144}Nd$ of 0.51197. What are its CHUR model and depleted mantle model ages?

Answer: The closed system isotopic evolution of any sample can be expressed as:

$$(^{143}Nd/^{144}Nd)_{sam} = (^{143}Nd/^{144}Nd)_0 + (^{147}Sm/^{144}Nd)_{sam}(e^{\lambda t} - 1) \qquad (8.24)$$

The chondritic evolution line is:

$$(^{143}Nd/^{144}Nd)_{chon} = (^{143}Nd/^{144}Nd)_0 + (^{147}Sm/^{144}Nd)_{chon}(e^{\lambda t} - 1) \qquad (8.25)$$

The CHUR model age, τ_{CHUR}, of a system is the time elapsed, $t = \tau$, since it had a chondritic $^{143}Nd/^{144}Nd$ ratio, assuming the system has remained closed. We can find τ by subtracting eqn. 8.25 from 8.24:

$$(^{143}Nd/^{144}Nd)_{sam} - (^{143}Nd/^{144}Nd)_{chon} = [(^{147}Sm/^{144}Nd)_{sam} - (^{147}Sm/^{144}Nd)_{chon}](e^{\lambda\tau} - 1) \quad (8.26)$$

Another way of thinking about this problem is to imagine a $^{143}Nd/^{144}Nd$ vs. time plot: on that plot, we want the intersection of the sample's evolution curve with the chondritic one. In terms of the above equations, this intersection occurs at $(^{143}Nd/^{144}Nd)_0$.

Solving eqn. 8.26 for τ:

$$\tau_{CHUR} = \frac{1}{\lambda} \ln\left(\frac{(^{143}Nd/^{144}Nd)_{sam} - (^{143}Nd/^{144}Nd)_{chon}}{(^{147}Sm/^{144}Nd)_{sam} - (^{147}Sm/^{144}Nd)_{chon}} + 1\right) \quad (8.27)$$

The chondritic $^{143}Nd/^{144}Nd$ and $^{147}Sm/^{144}Nd$ ratios are 0.512630 and 0.1960, respectively. Substituting these values, the values for the Pike's Peak batholith given above, and the value for λ_{147} from Table 8.2, we have:

$$\tau_{CHUR} = \frac{1}{6.54 \times 10^{-12}} \ln\left(\frac{0.51197 - 0.512630}{0.1067 - 0.1960} + 1\right) = 1.12 \ Ga$$

To calculate the depleted mantle model age, τ_{DM}, we use the same approach, but this time we want the intersection of the sample evolution line and the depleted mantle evolution line. So eqn. 8.27 becomes:

$$\tau_{CHUR} = \frac{1}{\lambda} \ln\left(\frac{(^{143}Nd/^{144}Nd)_{sam} - (^{143}Nd/^{144}Nd)_{DM}}{(^{147}Sm/^{144}Nd)_{sam} - (^{147}Sm/^{144}Nd)_{DM}} + 1\right) \quad (8.28)$$

The depleted mantle (as sampled by MORB) has an average ε_{Nd} of about 10, or $^{143}Nd/^{144}Nd = 0.51315$. The simplest possible evolution path, and the one we shall use, would be a closed system evolution since the formation of the Earth, 4.55 Ga ago (i.e., a straight line on a $^{143}Nd/^{144}Nd$ vs. time plot). This evolution implies $^{147}Sm/^{144}Nd$ of 0.2132. Substituting these values into eqn. 8.28:

$$\tau_{DM} = \frac{1}{6.54 \times 10^{-12}} \ln\left(\frac{0.51197 - 0.5132}{0.1067 - 0.2137} + 1\right) = 1.68 \ Ga$$

The age of the Pike's Peak batholith has been dated through conventional geochronology as 1.02 Ga (the "crystallization age"), only slightly younger than the τ_{CHUR}. If we assume that the mantle is chondritic in its $^{143}Nd/^{144}Nd$ and $^{147}Sm/^{144}Nd$ ratios, then we would conclude that Pike's Peak batholith could have formed directly from mantle-derived material, perhaps by fractional crystallization of basalt. In that case, it represents a new addition of material from the mantle to continental crust. However, as DePaolo (1981) pointed out, the mantle has not maintained chondritic $^{143}Nd/^{144}Nd$ and $^{147}Sm/^{144}Nd$ ratios, its evolution being more closely approximated by the depleted mantle evolution model described above. Since the τ_{DM} is much older than the crystallization age, he concluded that Pike's Peak batholith originated by melting of preexisting crustal material: material that had already resided in the crust for some 0.67 Ga before the batholith formed.

is this age at which these lines intersect and in this case is called a chondritic or *CHUR model age* (τ_{CHUR}). We have subsequently come to realize that a better assumption is that the mantle has evolved along a steeper slope to a higher $^{143}Nd/^{144}Nd$ value. For example, we can take the present average mid-ocean ridge basalt (MORB) value of about $\varepsilon_{Nd} = +10$. In this case the term *depleted mantle model age* is used (τ_{DM}). In either case, the model age is calculated by extrapolating the $^{143}Nd/^{144}Nd$ ratio back to the intersection with the mantle growth curve, as illustrated in Figure 8.11. Example 8.3 illustrates just how such model ages are calculated.

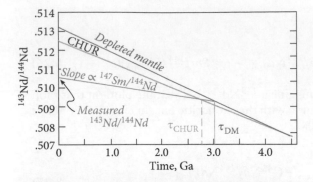

Figure 8.11 Sm-Nd model ages. $^{143}Nd/^{144}Nd$ is extrapolated backward (slope depending on Sm/Nd) until it intersects a mantle or chondritic growth curve. In this example, the *CHUR model age* is 3.05 Ga while the *depleted mantle model age* is 3.3 Ga.

Combined use of Sr and Nd isotope ratios can be a particularly powerful geochemical tool. Figure 8.12 shows the Sr and Nd isotopic characteristics of the major terrestrial reservoirs. Sr and Nd isotope ratios tend to be anticorrelated in the mantle and in mantle-derived magmas. This part of the diagram is sometimes called the *mantle array*. The inverse correlation reflects the relative compatibilities of the parent–daughter pairs. Rb is less compatible than Sr, while Sm is more compatible than Nd. Thus, magmatic processes affect the Rb/Sr and Sm/Nd ratios in opposite ways. In time, this leads to an inverse relationship between $^{87}Sr/^{86}Sr$ and ε_{Nd}. The anticorrelation between Sr and Nd isotope ratios in the mantle reflects the dominance of magmatic processes in the chemical evolution of the mantle. Most of the mantle has higher ε_{Nd} and low $^{87}Sr/^{86}Sr$. This in turn implies that Sm/Nd ratios have been high and Rb/Sr ratios have been low in the mantle. Mid-ocean ridge basalts (MORB) tend to have lowest $^{87}Sr/^{86}Sr$ and highest ε_{Nd}. This implies that the part of the mantle from which MORB is derived must be highly depleted in incompatible elements, presumably through past episodes of partial melting.

The continental crust has, on average, high $^{87}Sr/^{86}Sr$ and low ε_{Nd}. The anticorrelation between $^{87}Sr/^{86}Sr$ and ε_{Nd} is much weaker in crustal rocks. In part, this reflects the important role that nonmagmatic processes have played in evolution of the crust. It also reflects a weaker inverse relationship

in Rb/Sr and Sm/Nd in siliceous igneous rocks. The lower continental crust appears to have somewhat lower $^{87}Sr/^{86}Sr$ than does the upper continental crust, but it should be emphasized that there is no clear division in isotopic composition between the two, and that lower crustal rocks may plot anywhere within the crustal field. Continental basalts range to much higher $^{87}Sr/^{86}Sr$ and lower ε_{Nd} than do oceanic basalts. For the most part, this is due to assimilation of continental crust by these magmas. In some cases, however, it may reflect derivation of these magmas from incompatible element-enriched regions of the subcontinental lithospheric mantle.

Because $^{143}Nd/^{144}Nd$ ratios are not uniform in seawater, $^{143}Nd/^{144}Nd$ ratios in ancient sediments cannot be used for dating in the way $^{87}Sr/^{86}Sr$ ratios can. However, $^{143}Nd/^{144}Nd$ ratios have been used to identify different water masses and to attempt to constrain ocean circulation.

8.4.3 Lu-Hf

The Lu-Hf system is in many respects similar to the Sm-Nd system: (1) in both cases the elements are relatively immobile; (2) in both cases they are refractory lithophile elements; and (3) in both cases the daughter is preferentially enriched in the crust, so both $^{143}Nd/^{144}Nd$ and $^{176}Hf/^{177}Hf$ ratios are lower in the crust than in the mantle.

By analogy to the Sm-Nd system, we define ε_{Hf} as:

$$\varepsilon_{Hf} = \left[\frac{^{176}Hf/^{177}Hf_{sample} - {}^{176}Hf/^{177}Hf_{chon}}{^{176}Hf/^{177}Hf_{chon}} \right]$$
$$\times 10^4 \qquad (8.29)$$

The present-day chondritic value is 0.282785 ± 0.000011. As for the Sm-Nd system, it is generally assumed that the Lu/Hf ratio of the bulk silicate Earth is the same as that of chondrites. In a way analogous to the Sm-Nd system we can also define chondritic (τ_{CHUR}) and depleted mantle (τ_{DM}) model ages or crustal residence times.

The correlation between Nd and Hf isotope ratios in mantle-derived rocks (Figure 8.13), the *mantle array*, is somewhat better than between Sr and Nd isotope ratios, as one

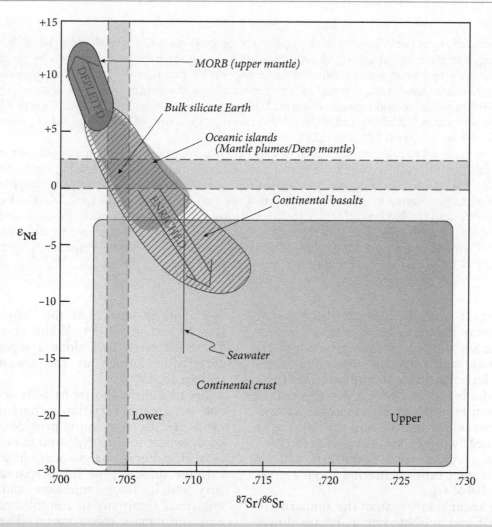

Figure 8.12 Sr and Nd isotope ratios in major geochemical reservoirs. The isotopic composition of the bulk Earth is in the range of $\varepsilon_{Nd} \approx 0-3$ and the dashed vertical gray area corresponds to the $^{87}Sr/^{86}Sr$ of the bulk Earth ($\approx 0.7045 \pm 0.001$). Arrows labeled "enriched" and "depleted" show where incompatible element enriched and depleted reservoirs would plot.

Example 8.4 Modeling the Sr and Nd isotope evolution of the crust and mantle

A simple model for the evolution of the crust and mantle might be that the crust was created early in Earth's history by partial melting of the mantle (the mass of the crust is 0.55% that of the mantle). Let's briefly explore the isotopic implications of this model. Assume that the crust was created as a 0.55% equilibrium partial melt of the mantle 2.5 Ga ago. Using partition coefficients of 0.00175, 0.0145, 0.081, and 0.046 for Rb, Sr, Sm, and Nd, respectively, calculate the present $^{87}Sr/^{86}Sr$ and $^{143}Nd/^{144}Nd$ of the crust and mantle. Assume that the silicate Earth (i.e., crust plus mantle) $^{87}Rb/^{86}Sr$ and $^{147}Sm/^{144}Nd$ ratios are 0.083 and 0.1960, respectively, and that the silicate Earth $^{87}Sr/^{86}Sr$ and $^{143}Nd/^{144}Nd$ ratios are at present 0.70450 and 0.51260, respectively. Also assume that the silicate Earth evolved as a single, uniform, closed system up to 2.5 Ga.

Answer: We first calculate the Rb/Sr and Sm/Nd fractionations during partial melting. Using eqn. 7.38, we find that the Rb, Sr, Sm, and Nd concentrations in the melt (the crust) are 137, 49, 11.8, and

19.5 times those in the original source respectively. Since $D = C^s/C^\ell$, we calculate the (relative) concentrations in the residual solid (the mantle) by multiplying the melt concentrations by the partition coefficients. From these, we calculate the Rb/Sr and Sm/Nd ratios in the melt to be, respectively, 2.8 and 0.602 times those in the original source; these ratios in the residual solid are, respectively, 0.336 and 1.048 times those in the original source. Multiplying these values by the silicate Earth Rb/Sr and Sm/Nd, we obtain ^{87}Rb/^{86}Sr and ^{147}Sm/^{144}Nd ratios in the crust of 0.213 and 0.118, respectively; those in the mantle are 0.026 and 0.205, respectively.

The next step is to calculate the initial ratios at 2.5 Ga. We can calculate the initial ratios at 2.5 Ga by noting that the bulk silicate Earth is still a closed system and hence by solving eqn. 8.18 for R_0 at $t = 2.5$ Ga and substituting the values given above for bulk silicate earth parent/daughter ratios and the decay constants in Table 8.2. Doing this, we find the initial ratios (at 2.5 Ga) to be 0.7018 for ^{87}Sr/^{86}Sr and 0.50937 for ^{143}Nd/^{144}Nd. Finally, we calculate present ratios for crust and mantle using eqn. 8.18 using these initial ratios and the $R_{P/D}$ ratios we calculated above. In this way, we find present ^{87}Sr/^{86}Sr ratios of 0.70945 and 0.70268 for the crust and mantle, respectively, and present ^{143}Nd/^{144}Nd ratios of 0.51132 and 0.51275 for crust and mantle, respectively.

might expect from the chemical behavior of the elements involved: Lu is after all a rare earth like Sm and Nd, and the behavior of Hf is somewhat similar to that of the rare earths in being incompatible and immobile. As Lu is only modestly incompatible, the fractionation of Lu from Hf tends to be somewhat greater than between Sm and Nd and the half-life of ^{176}Lu is nearly three times shorter, so the variation in ε_{Hf} is greater than that of ε_{Nd}, with Hf and Nd isotope ratios scattering about a trend of $\varepsilon_{Hf} = 1.4 \times \varepsilon_{Nd}$.

In the sedimentary system the similarity of the Lu-Hf and Sm-Nd systems breaks down. Hf is geochemically very similar to Zr and consequently concentrated in zircon in crustal rocks. Zircon characteristically has very low ^{176}Lu/^{177}Hf ratios so that the Hf it contains is unradiogenic (low ε_{Hf}). Zircon is also a dense, very mechanically and chemically stable mineral. It does not break down during weathering, consequently releasing little or none of its Hf to solution, and is deposited predominantly in fluvial or coastal settings; thus, the dissolved load of rivers is more radiogenic than continents as a whole. In contrast, most of the REE budget (including Lu) is contained in more readily weathered minerals, which release some of the REE to solution as they break down to form clays that can be carried into and deposited in deeper water as pelagic sediment. Consequently, fine-grained sediments, seawater, and sediments formed by precipitation from seawater such as Mn crusts and nodules contain Hf that is more radiogenic than igneous rocks with the same

ε_{Nd}. This is known as the "zircon effect" (Patchett et al., 1984; White et al., 1986). These materials plot along a separate weak correlation known as the *seawater array* (Albarede, 1998).

Because of the shorter half-life of the parent and the greater variation in parent/daughter ratios, Hf isotope ratios provide a superior geochemical tool to Nd isotope ratios. However, Hf isotopic analysis was hindered by the extreme difficulty of the preparative chemistry and its poor ionization, and therefore analytical sensitivity in conventional thermal ionization mass spectrometry. The development of multi-collector inductively coupled plasma mass spectrometers (MC-ICP-MS), which overcomes the ionization issue, has resulted in an enormous flood of Hf isotope data over the last 20 years.

The Lu-Hf decay system also provides a useful geochronological tool, although in limited circumstances. One of the most successful geochronologic applications has been garnet geochronology. Lu partitions strongly into garnet (Chapter 7), whereas Hf is excluded, and the closure temperature for garnet is particularly high. The Hf isotope ratios in zircon have also proved exceptionally useful in unraveling crustal histories. As we'll see in a subsequent section, U–Pb dating of zircons is a particularly powerful geochronological tool, but this tells us little of the origin and provenance of the zircon other than its age. Using *in situ* analytical techniques, the geochronological power of U–Pb zircon dating can be combined with Hf

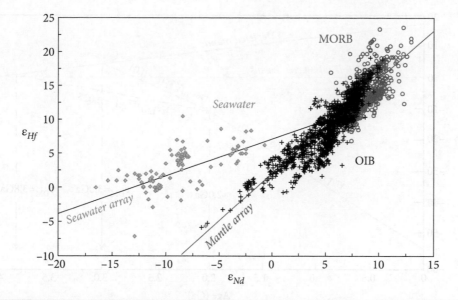

Figure 8.13 ε_{Hf} vs. ε_{Nd} in oceanic basalts and seawater.

isotope geochemistry to address the questions of origin and provenance. *In situ* techniques include laser ablation MC-ICP-MS, in which part of a zircon a few microns in diameter is ablated and carried into the mass spectrometer, and ion probes (also called secondary ionization mass spectrometers), in which a similarly small spot is ionized and the ions carried into a mass spectrometer. The very low $^{176}Lu/^{177}Hf$ of zircon as well as its chemical inertness allows for precise determination of the initial ε_{Hf}. Making inferences about the $^{176}Lu/^{177}Hf$ of the rock's source (e.g., upper continental crust, oceanic crust?), we can project this backward to obtain τ_{CHUR} and τ_{DM} model ages (Figure 8.14). This is particularly useful for detrital zircons, i.e., zircons found in sediments or metasediments whose original host rock is unknown. Indeed, the oldest known terrestrial materials are just such detrital zircons in metasediments of the Jack Hills in Western Australia. The initial ε_{Hf} in them suggests they were derived from crust created 4.4 to 4.5 billion years ago.

8.4.4 Re-Os

^{187}Re decays to ^{187}Os by β^- decay with a half-life of 42 billion years. Unlike the other decay systems of geological interest, rhenium and osmium are both siderophile elements. Thus, they are depleted in the silicate Earth by about two orders of magnitude compared

with chondritic meteorites. The missing Re and Os are presumably in the core. The resulting very low concentration levels (parts per billion to parts per trillion) make analysis extremely difficult. Although the first Re-Os isotope studies were published in the early 1960s, isotopic analysis of Os proved so difficult that there were no follow-up studies until 1980, when an analytical technique using the ion probe was developed (Luck et al., 1980). However, interest in this system only blossomed with the development of a sensitive and accurate technique employing thermal ionization mass spectrometry to analyze OsO_3^- (Creaser et al., 1991).

Several aspects of this decay system make it particularly useful in addressing a variety of geological processes. The first is the siderophile/chalcophile nature of these elements, making this a useful system to address questions of core formation and ore genesis. Second, Re can be concentrated in organic-rich sediments, making this system useful in studying such rocks and petroleum derived from them. Third, whereas all the other radioactive and radiogenic elements are incompatible, and hence enriched in melts, Os is a highly compatible element (bulk $D \sim 10$) and is enriched in the residual solid. This makes Os isotope ratios particularly useful in studies of the mantle. In addition, while Os is highly compatible, Re is moderately

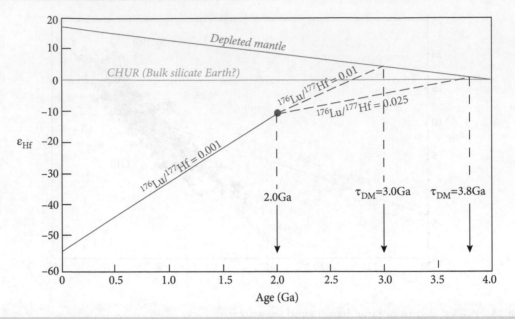

Figure 8.14   Low $^{176}Lu/^{177}Hf$ in zircons allows for calculation of their ε_{Hf} at their U–Pb age, in the case 2.0 Ga. Making inferences about the $^{176}Lu/^{177}Hf$ of the source of the magma from which the zircons originally crystallized, we can calculate depleted mantle model ages, τ_{DM}, which indicates the time this continental crust first formed. After Vervoort (2013).

incompatible and is slightly enriched in mantle melts. For example, mantle peridotites have average Re/Os close to the chondritic value of 0.08 whereas the average Re/Os in basalts is ~10. Thus, partial melting appears to produce an increase in the Re/Os ratio by a factor of ~10^3. As a consequence, the range of Os isotope ratios in the Earth is enormous compared with other radiogenic elements. (Thus, analytical precision need not be as high as for elements such as Sr, Nd, and Hf.) The mantle has a $^{187}Os/^{188}Os$ ratio close to the chondritic value of 0.1270 (Allègre and Luck, 1980), whereas the crust has a $^{187}Os/^{188}Os$ ratio of $\approx$1. By contrast, the difference in $^{143}Nd/^{144}Nd$ ratios between crust and mantle is only about 0.5%. Figure 8.15 illustrates the evolution of Os isotope ratios in the crust and mantle.

In the early work on this system, Hirt et al. (1963) reported the isotope ratio as $^{187}Os/^{186}Os$ (normalized for fractionation to $^{192}Os/^{188}Os$ of 3.08271). However, ^{186}Os is itself radiogenic, being the product of α-decay of ^{190}Pt. ^{190}Pt is sufficiently rare, and its half-life sufficiently long (462 billion years), that in most cases the amount of radiogenic ^{186}Os is insignificant. Significant radiogenic ^{186}Os was, however, observed by Walker

et al. (1991) in copper ores from Sudbury, Ontario, which prompted them to adopt the convention of using ^{188}Os as the normalizing isotope. Today Os isotope ratios are reported as $^{187}Os/^{188}Os$. Older $^{187}Os/^{186}Os$ ratios may be converted to $^{187}Os/^{188}Os$ ratios by multiplying by 0.12035.

Perhaps the first significant result from Re-Os studies was the realization the Re/Os ratio of the mantle was within the range of chondritic values (Allègre and Luck, 1980) (unlike Sm/Nd and Lu/Hf, Re/Os varies considerably within and between classes of chondritic meteorites). This is a bit surprising if most of the Re and Os have been extracted to the core. If the core and mantle are in equilibrium, then mantle concentrations will be determined by metal–silicate partition coefficients, which are large but quite different for both elements. This should lead to the ratio of the two in the mantle to be quite different from chondritic. The leading explanation for this is that the present inventory of highly siderophile elements such as Re and Os were derived from a "late accretionary veneer" of chondritic material subsequent to core formation (Walker, 2016). We will discuss this further in Chapter 11.

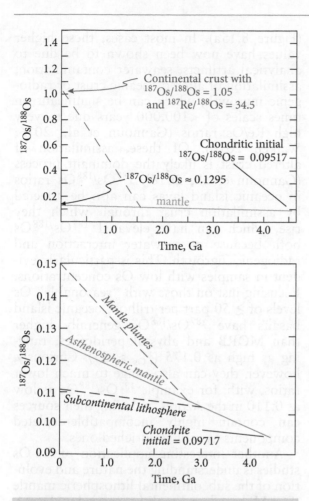

Figure 8.15 (a) Schematic evolution of Os isotope ratios in the mantle and crust. (b) $^{187}Os/^{188}Os$ evolution in the mantle as deduced from analyses of osmiridium and laurite grains in ancient mantle peridotites, initial ratios in peridotites, and modern oceanic peridotites and basalts. The pattern is consistent with other isotopic systems: the asthenospheric mantle is more incompatible element-depleted than mantle plumes, which produced oceanic island basalts. The mantle root of continents (lithospheric mantle) appears to have been severely depleted of its basaltic component and Re by melt extraction. The continental crust is strongly enriched in Re over Os and has an average $^{187}Os/^{188}Os$ of 1.05 (Peucker-Ehrenbrink and Jahn, 2001).

Since the silicate Earth appears to have a near-chondritic $^{187}Os/^{188}Os$ ratio, it is useful to define a parameter analogous to ε_{Nd} and ε_{Hf} that measures the deviation from chondritic.

Walker et al. (1989) defined γ_{Os} as:

$$\gamma_{Os} = \frac{^{187}Os/^{188}Os - {}^{187}Os/^{188}Os}{^{187}Os/^{188}Os_{chon}} \times 100$$

(8.30)

where the present-day chondritic value is 0.1270. Thus, γ_{Os} is the *percent* deviation from the chondritic value. Just as with ε_{Nd}, we can calculate values of γ_{Os} for times other than the present by using initial $^{187}Os/^{188}Os$ and a chondritic value appropriate to that time. However, the analogy to Nd and Hf has it limits because unlike $^{143}Nd/^{144}Nd$ and $^{176}Hf/^{177}Hf$, chondrites do not have a uniform $^{187}Os/^{188}Os$ ratio. $^{187}Os/^{188}Os$ averages 0.1262 ± 0.0006 in carbonaceous chondrites, 0.1283 ± 0.001 in ordinary chondrites and 0.1281 ± 0.0004 in enstatite chondrites, reflecting their variable Re/Os ratios (Walker, 2016). The estimated $^{187}Os/^{188}Os$ of the bulk silicate Earth (or primitive upper mantle: PUM) appears to be somewhat higher at 0.1296 ± 0.0008, corresponding to a $^{187}Re/^{188}Os$ of 0.435, but within the range of ordinary chondrites.

Because of the extremely low concentrations of Os in most rocks, Os isotopic analysis remains analytically challenging. As Os is a highly compatible element, its concentrations are higher in peridotites than in crustal rocks and most studies have focused on them and other PGE-rich rocks such as sulfide deposits and organic-rich sediments. An additional challenge in analyzing basaltic rocks is contamination though assimilation as the magmas rise through crust, and weathering reactions with seawater or groundwater.

Nevertheless, a picture is emerging from studies of mantle-derived materials that is consistent with the one painted by Sr, Nd, and Hf isotope systematics, showing that MORB and related abyssal peridotites have lower $^{187}Os/^{188}Os$ ratios than basalts from oceanic islands (summarized by Day, 2013). These results are consistent with the Nd and Sr isotope evidence we have already discussed that the source of MORB is more incompatible-element depleted than mantle plumes, which are thought to produce oceanic island volcanoes such as Hawaii. Abyssal peridotites recovered from mid-ocean ridges and the transform faults offsetting

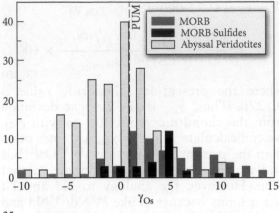

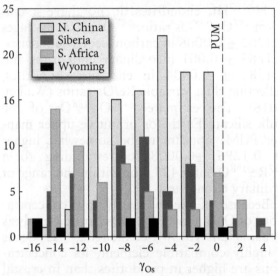

Figure 8.16 (a) Histogram comparing γ_{Os} in abyssal peridotites, MORB, and sulfide grains separated from MORB. Based on Walker (2016). (b) Histogram showing γ_{Os} in peridotites from the subcontinental lithospheric mantle beneath cratonic regions of North China (Liu et al., 2011), Siberia (Pearson et al., 1995; Ionov et al., 2014, 2015), Southern Africa (Pearson et al., 1995, 2004) and Wyoming, USA (Carlson and Irving, 1994).

them have $^{187}Os/^{188}Os$ ratios averaging about 0.125 (Figure 8.16a), lower than chondrites or PUM, consistent with long-term depletion in incompatible elements (recall that Re is incompatible with Os is compatible).

$^{187}Os/^{188}Os$ ratios reported in MORB range from values similar to those in abyssal peridotites, ~0.125, to ratios as high as 0.155

(Figure 8.15a). In most cases, these higher values have now been shown to be due to analytical artifacts, seawater contamination, assimilation of altered oceanic crust or radiogenic ingrowth, which can be significant on times scales of <100,000 years due to very high Re/Os ratios (Gannoun et al., 2016; Walker, 2016). Of these, assimilation of altered crust is likely the dominant process (Gannoun et al., 2016). $^{187}Os/^{188}Os$ ratios in oceanic island lavas can also be affected by assimilation crust through which they rise, which can have elevated $^{187}Os/^{188}Os$ both because of seawater interaction and radiogenic ingrowth. This is particularly evident in samples with low Os concentrations. Focusing just on those with "reasonable" Os levels of $\gtrsim$ 50 part per trillion, oceanic island basalts have $^{187}Os/^{188}Os$ generally higher than MORB and abyssal peridotites, ranging as high as 0.175 ($\gamma_{Os} \approx 35$). Curiously, however, they can also range to much lower ratios, with, for example $^{187}Os/^{188}Os$ as low as 0.110 in the Azores, indicting their sources can contain highly incompatible-depleted components as well as enriched ones.

Another interesting application of Re-Os studies is understanding the nature and evolution of the subcontinental lithospheric mantle (SCLM) based on *xenoliths*[*] carried to the surface by magmas. Kimberlites, better known for the diamond *xenocrysts* they sometimes carry to the surface, are the most prodigious carriers of such xenoliths. These xenoliths reveal that the SCLM, particularly beneath old "cratonic" regions of the continents, is generally poor in clinopyroxene and garnet and hence depleted in its basaltic component (the term *infertile* is often used to refer to such compositions), as a result of previous episodes of melting. They are also strongly depleted in Re but not in Os, consistent with partitioning of these elements during melting, and consequently have systematically low $^{187}Os/^{188}Os$ (Figure 8.16b). Indeed, much of the Re in the xenoliths appears to have been introduced just prior to or during eruption of the host magma. Assuming all such Re originates in that way lead to a rhenium depletion model age, T_{RD}, that is analogous to Sm-Nd model ages discussed earlier. It assumes that prior to

[*] Xenolith is a word derived from the Greek words for "foreign" and "rock." A xenolith is simply a foreign rock in an igneous one, ripped from the walls of the conduit or magma chamber. Xenocryst is an analogous single mineral grain.

rhenium depletion, the sample's $^{187}Re/^{188}Os$ was chondritic. The T_{RD} is calculated as:

$$T_{RD} = 1/\lambda \times \ln\left\{ \frac{(^{187}Os/^{186}Os)_{chon} - (^{187}Os/^{186}Os)_{sample(EA)}}{(^{187}Re/^{186}Os)_{chon}} + 1 \right\}$$

(8.31)

where $(^{187}Os/^{188}Os)_{sampleEA}$ is the $(^{187}Os/^{188}Os)$ at the eruption age, i.e., corrected for any ^{187}Re decay since eruption An alternative is the model age, T_{MA}, which is analogous to the Sm-Nd model age:

$$T_{MA} = 1/\lambda \times \ln\left\{ \frac{(^{187}Os/^{186}Os)_{chon} - (^{187}Os/^{186}Os)_{sample}}{(^{187}Re/^{186}Os)_{chon} - (^{187}Re/^{186}Os)_{sample}} + 1 \right\}$$

(8.32)

These studies have demonstrated that depletion age of the SCLM under cratonic regions matches the age of the overly crust and as old as 3–3.5 billion years.

Surprisingly, these xenoliths, although petrologically infertile, often show evidence of incompatible element enrichment, including high $^{87}Sr/^{86}Sr$ and low ε_{Nd}. This latter feature is often attributed to reaction of the mantle lithosphere with very small degree melts percolating upward through it (a process termed *mantle metasomatism*). This process, however, apparently leaves the Re-Os system unaffected, so that $^{187}Re/^{188}Os$ and $^{187}Os/^{188}Os$ remain low. Thus, Re-Os studies are proving to be important in understanding the evolution of the subcontinental lithosphere, and its role in volcanism. We shall return to these topics in Chapter 11.

There has also been considerable interest in the Os isotope composition of seawater. The $^{187}Os/^{188}Os$ ratio of modern seawater is ~1.026 to 1.046 (Gannoun et al., 2016). Like that of $^{87}Sr/^{86}Sr$, $^{187}Os/^{188}Os$ depends on the balance of continental crustal fluxes (e.g., rivers, with $^{187}Os/^{188}Os \sim 1$) and oceanic crustal fluxes (e.g., hydrothermal activity, with $^{187}Os/^{188}Os \sim 0.125$). In addition, however, cosmic fluxes ($^{187}Os/^{188}Os \sim 0.13$), which include both cosmic dust (which continually settles through the atmosphere into

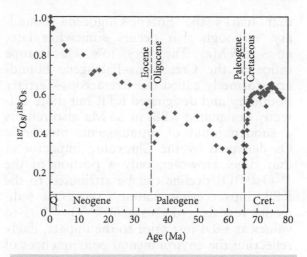

Figure 8.17 Os isotope composition of seawater over the last 80 Ma. Based on Peucker-Ehrenbrink and Ravizza, 2012).

the oceans) and large meteorite impacts, can be significant for Os.

The variation in seawater $^{187}Os/^{188}Os$ through time can be assessed through analysis of leachable Os in pelagic clays and of ferro-manganese crusts. Like Sr, the Os isotopic composition of seawater has changed over time (Figure 8.17). There are obvious similarities between the Os isotopic and Sr isotopic evolution of seawater (Figure 8.9), most notably the increase of both through the Tertiary period (the last 66 million years). This may in part reflect a decreasing hydrothermal flux resulting from decreasing sea floor spreading rates. There are also differences that reflect the differing geochemical behavior of Sr and Os. The geochemical behavior of both at the surface of the Earth is related to carbon, but while Sr is concentrated in carbonates, Os is concentrated in organic-rich sediments.

Also apparent in the marine Os isotope record are major asteroid impacts, as impactors with chondritic or iron meteorite compositions would deliver significant amounts of unradiogenic Os to the Earth's surface. Minimum values in the late Eocene are associated with elevated Os and Ir concentrations and appear to be related to impacts that produced the 100 km diameter Popigai crater in Siberia at 35.7 Ma and the 85 km Chesapeake Bay impact crater at 35.5 Ma. There is speculation that these impacts are related to the extinction that marks the Eocene-Oligocene boundary, although that occurs somewhat later at ~33.9 Ma. The very low Os isotope ratios at the Cretaceous–Paleogene boundary (formerly called the Cretaceous–Tertiary Boundary and designated KTP, but more correctly designated KPB) at 66 Ma also reflects a sudden input of unradiogenic meteoritic Os delivered by the Chicxulub impactor at that time. However, only a portion of the $^{187}Os/^{188}Os$ decline can be attributed to the KPB impact event. Latest Cretaceous sediments reveals a decrease in $^{187}Os/^{188}Os$ to values at ~0.4 just prior to the impact, likely reflecting the environmental consequences of the main eruptive phase of the Deccan Traps (Peucker-Ehrenbrink and Ravizza, 2012). This is consistent with emerging evidence that the Deccan Traps flood basalt event was a significant contribution to the KPB mass extinction.

8.4.5 La-Ce

^{138}La is a branched decay system: it decays by β^- to ^{138}Ce (34.5%) and to ^{138}Ba by β^+ and electron capture. Since ^{138}La is a rare isotope (an odd–odd), and has a long half-life, this is another system that is analytically challenging. ^{138}Ba is among the most abundant heavy isotopes (because it has 82 neutrons, a magic number), and the amount of radiogenic ^{138}Ba is relatively small, so attention has focused on ^{138}Ce. The rarity of ^{138}La, its slow decay ($t_{1/2}$ = 449 Ga), and most of this decay (65.2%) going to ^{138}Ba means very high analytical precision is required. Furthermore, ^{138}Ce is also a rare isotope of Ce (0.251%) and that, together with the presence of a very abundant isobar (^{38}Ba), makes Ce isotope analysis challenging. Consequently, few analyses have been reported in the literature. Dickin (1987) reported some analyses of relatively low precision, but only recently has a significant amount of high precision data been reported. Most studies have reported the $^{138}Ce/^{142}Ce$ ratio; because the abundance of ^{138}Ce is ~50 times lower than that of ^{142}Ce, Willig and Stracke (2019) have measured the $^{138}Ce/^{136}Ce$ ratio instead. $^{138}Ce/^{136}Ce$ can be converted to $^{138}Ce/^{142}Ce$ by multiplying with $^{136}Ce/^{142}Ce$ = 0.01688 if the same fractionation normalization is used.

Like Nd, Ce is a rare earth element and it is reasonable to assume that the Earth has a chondritic $^{138}Ce/^{142}Ce$ ratio or nearly so. We therefore define an ε_{Ce} in an analogous manner to ε_{Nd}:

$$\varepsilon_{Ce} = \left[\frac{(^{138}Ce/^{142}Ce)_{sample} - (^{138}Ce/^{142}Ce)_{chon}}{(^{138}Ce/^{142}Ce)_{chon}} \right] \times 10000$$

(8.33a)

or:

$$\varepsilon_{Ce} = \left[\frac{(^{138}Ce/^{136}Ce)_{sample} - (^{138}Ce/^{136}Ce)_{chon}}{(^{138}Ce/^{136}Ce)_{chon}} \right] \times 10000$$

(8.33b)

where $(^{138}Ce/^{142}Ce)_{chon}$ = 0.02256643 and $(^{138}Ce/^{136}Ce)_{chon}$ = 1.336897 (Willig and Stracke, 2019).

Figure 8.18 summaries the available high-precision data. Overall, there is a strong negative correlation between ε_{Ce} and ε_{Nd}, as we would expect because the parent, La, is more incompatible than the daughter, Ce, but the daughter of the Nd–Sm system is more incompatible than the parent. Much of the recent work has focused on island arc lavas to assess the subducted sediment contribution to them and to answer the question as to whether Ce anomalies (Chapter 7) are inherited from those sediments. Island arc magmas do appear to be offset to higher ε_{Ce} in Figure 8.18 compared with MORB and OIB, and Bellot et al. (2018) concluded that this is indeed the case. Doucelance et al. (2014) concluded from Ce and Nd isotopic analyses of carbonatites that the carbonate in these magmas originated from surface carbonate that had been recycled into the mantle through subduction.

Tazoe et al. (2011) analyzed Ce and Nd isotope ratios in Pacific seawater. The data form a weak correlation of lower slope than in mantle-derived rocks, as shown in Figure 8.18. Tazoe et al. found that water in the western North Pacific, including the East and South China Seas, had ε_{Ce} of +0.7 to 1.4, reflecting input of radiogenic Ce from continents, while ε_{Ce} ranged from −0.4 to +0.3 in Pacific Equatorial Water near volcanic

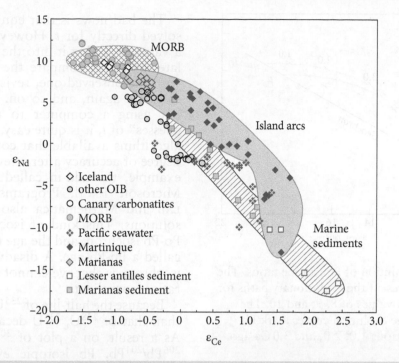

Figure 8.18 High-precision Ce and Nd isotope data on MORB, OIB, island arc volcanics (Bellot et al., 2015, 2018; Doucelance et al., 2014; Boyet et al., 2019; and Willig and Stracke, 2019), and marine sediments (Bellot et al., 2015, 2018), and Pacific seawater (Tazoe et al., 2011).

islands such as New Guinea, indicating input of Ce recently derived from the mantle.

8.4.6 U-Th-Pb

The U-Th-Pb system is somewhat of a special case since there are three decay schemes producing isotopes of Pb. In particular, two U isotopes decay to two Pb isotopes, and since the two parents and two daughters are chemically identical, we get two decay systems for the price of one, and together they provide a particularly powerful tool.

Let's explore the mathematics of this. Following convention, we will designate the $^{238}U/^{204}Pb$ ratio as μ, and the $^{232}Th/^{238}U$ ratio as κ. Now, we can write two versions of eqn. 8.18:

$$^{207}Pb/^{204}Pb = {}^{207}Pb/^{204}Pb_0 + \mu \frac{^{235}U}{^{238}U}(e^{\lambda_{235}t} - 1)$$

(8.34)

and

$$^{206}Pb/^{204}Pb = {}^{206}Pb/^{204}Pb_0 + \mu(e^{\lambda_{238}t} - 1)$$

(8.35)

Equations 8.34 and 8.35 can be rearranged by subtracting the initial ratio from both sides. For example, using Δ to designate the difference between the initial and the present ratio, eqn. 8.35 becomes:

$$\Delta^{206}Pb/^{204}Pb = \mu(e^{\lambda_{238}t} - 1)$$

(8.36)

Dividing the equivalent equation for $^{235}U/^{207}Pb$ by eqn. 8.36 yields:

$$\frac{\Delta^{207}Pb/^{204}Pb}{\Delta^{206}Pb/^{204}Pb} = \frac{^{235}U}{^{238}U}\frac{(e^{\lambda_{235}t} - 1)}{(e^{\lambda_{238}t} - 1)}$$

(8.37)

Conventionally, the $^{235}U/^{238}U$ was assumed to have a constant, uniform value of 1/137.88 in the Earth. Recent studies, however, have demonstrated that this ratio varies. Although this variation appears to result from mainly kinetic chemical fractionation (discussed in the following chapter) during oxidation and reduction, variation of a few tenths of a percent has also been demonstrated in high-temperature minerals such as zircon and monazite (Heiss et al., 2012). Heiss et al. found the average value to be 1/137.82.

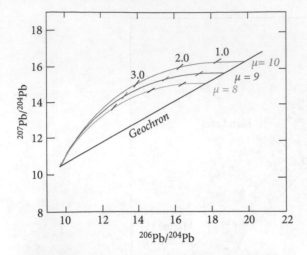

Figure 8.19 Evolution of Pb isotope ratios. The curved lines represent the evolutionary paths for systems having μ values of 8, 9, and 10. The hashmarks on the evolution curves mark Pb isotope compositions 1.0, 2.0, and 3.0 Ga ago.

Except where the highest precision is necessary, use of this value is probably adequate. Where the highest precision is necessary, the $^{235}U/^{238}U$ ratio must be measured. In subsequent discussions and derivations, we will assume this ratio has a constant value of 1/137.82.

The left-hand side of this equation is the slope of a line on a plot of $^{207}Pb/^{204}Pb$ versus $^{206}Pb/^{204}Pb$ such as Figure 8.19. The slope depends only on time and three constants (λ_{238}, λ_{235}, and $^{238}U/^{235}U$). Because its slope depends on time, a line on a plot of $^{207}Pb/^{204}Pb$ versus $^{206}Pb/^{204}Pb$ is an isochron, analogous to the isochrons in a plot such as Figure 8.6. We derived eqn. 8.37 by subtracting the initial ratio from the present one, but we could derive the identical equation by subtracting equations for two separate samples that share common initial ratios and time. Equation 8.37 differs from the conventional isochron equation (eqn. 8.18) in that terms for the initial values and the parent/daughter ratios do not appear. The point is, if we measure the Pb isotope ratios in a series of samples that all started with the same initial isotopic composition of Pb at some time t_0, and which all remained closed to gain or loss of Pb or U since then, we can determine the age of this system without knowing the parent/daughter ratio.

The bad news is that eqn. 8.37 cannot be solved directly for t. However, we can guess a value of t, plug it into the equation, calculate the slope, compare the calculated slope with the observed one, revise our guess of t, calculate again, and so on. Pretty laborious but using a computer to make "educated guesses" of t, it is quite easy. If fact, there are algorithms available that converge to a high degree of accuracy after a few iterations – for example, the add-in called the "Solver" in Microsoft Excel. Programs such as MATLAB and Mathematica also readily provide solutions. This kind of isochron is called a Pb-Pb isochron, and the age derived from it is called a Pb-Pb age. A disadvantage of Pb-Pb isochrons is that we cannot determine initial ratios from them.

Because the half-life of ^{235}U is much shorter than that of ^{238}U, ^{235}U decays more rapidly. As a result, on a plot of $^{207}Pb/^{204}Pb$ versus $^{206}Pb/^{204}Pb$, Pb isotopic evolution follows curved paths; the exact path depends on μ. Three such evolution curves are shown in Figure 8.19. All systems that begin with a common initial isotopic composition at time t_0 lie along a straight line at some later time t. This line is the Pb–Pb isochron.

As an example, let's consider the various bodies of the solar system. We assume that (and there has been no evidence to the contrary) when the solar system formed 4.55 billion years ago, it had a single, uniform Pb isotope composition, listed in Table 8.3, which we will refer to as primordial Pb. Planets formed shortly (in a geologic sense) after the solar system formed, and we can reasonably assume that they have remained closed since their formation (i.e., they have neither gained nor lost Pb or U). There is no reason to assume that all planetary bodies started with the same ratio of U to Pb, that is, the same value of μ. Indeed, there is good evidence that they did not. So Pb in each planetary body would evolve along a separate path that would depend on the value of μ in that body. However, at any later time t, the $^{207}Pb/^{204}Pb$ and $^{206}Pb/^{204}Pb$ ratios of all planetary bodies plot on a unique line. This line, called the *Geochron*, has a slope whose age corresponds to the age of the solar system, and it passes through the composition of primordial Pb.

Equation 8.37, in which the parent/daughter ratio term does not appear, turns out

Table 8.3 Initial isotopic compositions and parent/daughter ratios.

System	R_0	$R_{P/D}$*	Source
^{87}Rb-^{87}Sr	0.69898	~0.076	Basaltic achondrites
^{147}Sm-^{143}Nd	0.50670	0.1960[†]	Chondrites
^{176}Lu-^{176}Hf	0.27979[†]	0.0336[†]	Chondrites
^{136}La-^{138}Ce	0.0225322	0.00306	Chondrites
^{187}Re-^{187}Os	0.09517	0.402[‡]	Chondrites
^{235}U-^{207}Pb	10.294	~0.06	Canyon Diablo troilite
^{238}U-^{206}Pb	9.314	~8.5	Canyon Diablo troilite
^{232}Th-^{208}Pb	29.476	~36	Canyon Diablo troilite

*The parent daughter ratio is given as the present-day value.

[†]Based on the chondritic value of Bouvier et al. (2008).

[‡]The silicate Earth value is ~0.4343.

to be very useful for several reasons. First, for geochronological applications, U is a rather mobile element, so the closed system assumption is often violated. Often this mobility occurs late (perhaps occurring as the rock is unburied and reaches the surficial weathering zone). So a normal U–Pb isochron often gives erroneous results. But if the loss (or gain) has been sufficiently recent, Pb isotope ratios will not be significantly affected. Because of this, Pb-Pb ages calculated with eqn. 8.37 are robust with respect to recent mobility of U, and good ages can be obtained this way even when they cannot be obtained from a conventional U–Pb isochron (i.e., from the slope of ^{206}Pb/^{204}Pb vs. ^{238}U/^{204}Pb). Zircons and other U-rich, Pb-poor minerals are particularly useful in geochronology and specialized approaches have been developed to deal with them, as explained in the box below.

U–Pb zircon dating

Zircon, as we noted earlier, is extremely resistant to mechanical and chemical weathering; indeed, the oldest surviving mineral grains on the planet are zircons. Zr and U have the same charge, +4, and similar size (Zr = 72 pm; U = 89 pm), whereas Pb has a +2 charge and 119 pm radius) and consequently zircons accept U and reject Pb. Although never a major phase in any rock, it is a common accessory phase in silicic igneous rocks as well as metamorphic rocks. This makes it the almost perfect time capsule. "Almost" because some zircons contain so much U that they can suffer radiation damage; such radiation-damaged crystals are referred to as *metamict*. Furthermore, Pb is so incompatible in zircon that it is prone to diffuse out, particularly in metamict zircons. Zircons are consequently subject to partial resetting as well as to overgrowth of new crystal rims in metamorphism (Figure 8.20). Fortunately, it is often possible to recognize these effects and "see through" them to obtain original crystallization ages.

Zircons typically contain so little initial Pb that it is relatively easily corrected for based on the amount of ^{204}Pb present and simple assumptions about the ^{206}Pb/^{204}Pb and ^{207}Pb/^{204}Pb at the time the zircon formed (this can be done iteratively, once we have an approximate age). Having corrected for initial Pb, we denote the radiogenic Pb as ^{206}Pb* and ^{207}Pb*. From this and the amount of U parent we can calculate two separate ages:

$$t_{238} = \ln\left[\frac{^{206}Pb*}{^{238}U}+1\right]/\lambda_{238} \qquad (8.38)$$

and

$$t_{235} = \ln\left[\frac{^{207}Pb*}{^{235}U}+1\right]/\lambda_{235} \qquad (8.39)$$

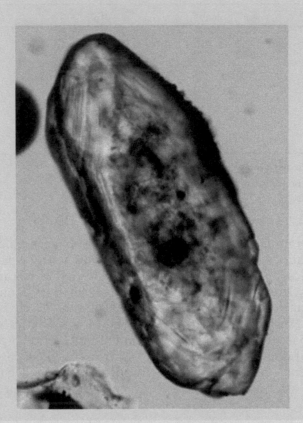

Figure 8.20   Zircon crystal seen under the microscope. This crystal is approximately 250 microns in length. Complex zoning and dark metamict areas are apparent. Photograph by Thomas Becker, Technische Universität München. Reproduced under Wikipedia Commons license.

If the ages agree, they are said to be concordant. To assess this, we use a *concordia diagram* first introduced by Wetherill (1956). A concordia diagram (Figure 8.21) is simply a plot of the locus of $^{206}Pb^*/^{238}U$ and $^{207}Pb^*/^{235}U$ values for which the two ages agree. It is curved because ^{235}U decays much faster than ^{238}U.

   If analytical uncertainty is taken into considerations, individual zircons plot as ellipses rather than points (because the errors in the two ratios are correlated). If the ages agree within error and our zircon plots on the concordia curve (point ① on Figure 8.1) we can be confident our ages record a real event and we're done. Techniques have also been developed to physically abrade or chemically leach rims and metamict parts of zircons, which can enhance concordancy, e.g., Mattinson (2012).

   Old zircons, however, which have often suffered through metamorphic events, can be quite discordant. Suppose that a zircon formed at 3.25 Ga suffered metamorphism and lost all its Pb at 2.0 Ga. This would result in complete resetting and would give us an age of 2.0 Ga. Now suppose the zircon lost only some of its Pb. Because the two isotopes of Pb would have been lost in exactly the proportions in which they were present in the zircon, its $^{206}Pb^*/^{238}U$ and $^{207}Pb^*/^{235}U$ would have shifted downward toward 0 on the plot *at that time*. Today, it would plot somewhere on a chord intercepting the concordia curve at 3.25 and 2.0 Ga (point ② on Figure 8.21). If we have analyzed only a single sample (which nonetheless might consist of multiple zircons), we can draw no quantitative conclusions as to its age. But if we analyze several zircons that crystallized at the same time and had simultaneously lost variable amounts of Pb during metamorphism (points ③ and ④ on Figure 8.21), we can draw a chord between the two intercepts and determine both the original crystallization age and the age of metamorphism. If instead of Pb loss during metamorphism, overgrowth rims had developed, the rims would have plotted on the concordia curve at 2.0 Ga and the zircons as a whole would have also plotted on a chord between 3.25 and 2.0, just as in the case of Pb loss.

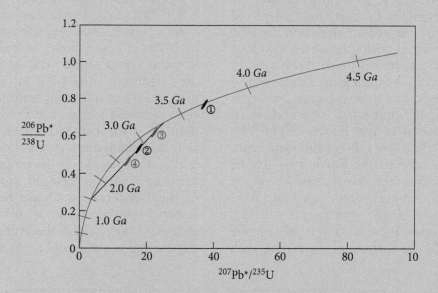

Figure 8.21 A concordia diagram illustrating U–Pb dating of zircons and other U-rich, Pb-poor minerals. The concordia curve is the locus of points where the $^{206}Pb*/^{238}U$ and $^{207}Pb*/^{235}U$ agree (asterisk denotes radiogenic Pb). Zircons plotting on the concordia curve (e.g., point ①) are said to be concordant. Pb loss or overgrowths during metamorphism will cause zircons to be discordant (point ②) and lie on a chord between crystallization age and metamorphic age. By analyzing multiple zircons (points ③ and ④), the chord can be defined, with the lower and upper intercepts giving the metamorphic and crystallization ages, respectively.

Using *in situ* techniques such as ion probes and laser ablation MC-ICP-MS, it is possible to analyze parts of zircons, avoiding metamict areas and metamorphic overgrowths.

This same approach can also be used for other minerals that strongly incorporate U and exclude Pb, such as titanite and apatite.

Pb isotope ratios can help constrain, but not date exactly, the age of mantle reservoirs. It is reasonable to assume that the isotope ratios in a volcanic rock will be the same as the isotope ratios of its source because the different isotopes of an element, for example ^{206}Pb and ^{204}Pb, are chemically identical and cannot be fractionated by magmatic processes (which is to say that their partition coefficients are identical). But isotopes of different elements, for example ^{238}U and ^{206}Pb, are not chemically identical and can be fractionated. Thus, the U/Pb or $^{238}U/^{204}Pb$ ratio of a volcanic rock is not necessarily the same as that of its source. As a result, we cannot calculate an "age" of the source of a series of volcanic

rocks using the normal isochron equation, 8.34 or 8.35. Since the parent/daughter ratio does not occur in eqn. 8.34, we can calculate an age using eqn. 8.37. Such "mantle isochron" ages for oceanic basalt sources are typically 1 to 2 Ga. Again, these should not be interpreted as literal ages as recent mixing between two distinct reservoirs can produce identical correlations, but nonetheless provide qualitative constraints on the age of mantle heterogeneity. As we shall see in subsequent sections, other isotopes indicate distinct mantle reservoirs have existed for the entire age of the Earth.

Cosmochemical constraints on the U/Pb ratio of the silicate Earth (and the bulk Earth) are particularly weak because, while

U is refractory and lithophile, Pb is both somewhat volatile, chalcophile, and somewhat siderophile. Consequently, the Earth is depleted in Pb relative to chondrites, and the silicate Earth might be even more depleted if Pb has concentrated in the core. Isotope ratios of terrestrial rocks, however, allow us to place some constraints on the U/Pb ratio of the silicate Earth. Assuming the Earth formed at the same time and from the same materials as meteorites, and that it has remained closed to gain or loss of U and Pb, its $^{207}Pb/^{204}Pb$ and $^{206}Pb/^{204}Pb$ must fall on the Geochron. (This applies only to the Earth as a whole; reservoirs within the Earth, such as the crust, mantle, and individual rock units, have clearly not been closed systems for 4.55 Ga, so their isotope ratios need not fall on the Geochron.) Major terrestrial reservoirs, such as the upper mantle (represented by MORB), upper and lower continental crust, plot near the Geochron between growth curves for $\mu = 8$ and $\mu = 8.8$ (Figure 8.21). This suggests the $^{238}U/^{204}Pb$ of the Earth is around 8.5.

If a system has experienced a decrease in U/Pb at some point in the past, its Pb isotopic composition will lie to the left of the Geochron; if its U/Pb ratio increased, its present Pb isotopic composition will lie to the right of the Geochron. U is more incompatible than Pb. This being the case, incompatible element depleted reservoirs should plot to the left of the Geochron, incompatible element enriched reservoirs should plot to the right of the Geochron in Figure 8.22. From the isotopic composition of other radiogenic elements, we would therefore predict that continental crust should lie to the right of the

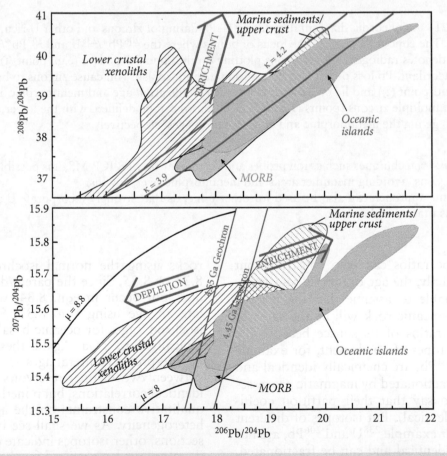

Figure 8.22 Pb isotope ratios in major terrestrial reservoirs. Typical lower continental crust and upper continental crust are represented by lower crustal xenoliths and modern marine sediments respectively (these somewhat underestimate the total variance in these reservoirs). MORB and oceanic islands represent the isotopic compositions of upper mantle and deep mantle, respectively.

Geochron and the mantle to the left. Upper continental crustal rocks do plot mainly to the right of the Geochron, as expected, but surprisingly, Pb isotope ratios of mantle-derived rocks also plot mostly to the right of the Geochron. This indicates the U/Pb ratio in the mantle has increased, not decreased as expected. This phenomenon is known as the *Pb paradox* and it implies that a simple model of crust–mantle evolution that involves only transfer of incompatible elements from crust to mantle through magmatism is inadequate.

The only large geochemical reservoir known to plot to the left of the Geochron is the lower continental crust. It is unlikely that the four reservoirs represented on Figure 8.22, the upper crust, the lower crust, the upper mantle (as represented by MORB), and the deep mantle (as represented by oceanic islands) average to a composition that lies on the Geochron, which is another part of the Pb paradox. Part of the explanation is that the Earth may be as much as 100 Ma younger than the solar system; indeed, there is now good reason to believe that it is (Chapter 10). Thus, it might be more appropriate to construct a Geochron based on an age of 4.45 Ga, which is also illustrated in Figure 8.22. The 4.45 Ga Geochron is shifted to higher $^{206}Pb/^{204}Pb$ and passes through the middle of the MORB field and to the right of the lower continental crust field, relieving much of the Pb isotope mass balance problem.

We can combine the growth equations for $^{208}Pb/^{204}Pb$ and $^{206}Pb/^{204}Pb$ in a way similar to eqn. 8.37. We end up with:

$$\frac{\Delta^{208}Pb/^{204}Pb}{\Delta^{206}Pb/^{204}Pb} = \kappa \frac{(e^{\lambda_{232}t} - 1)}{(e^{\lambda_{238}t} - 1)} \qquad (8.40)$$

where, again, κ is the $^{232}Th/^{238}U$ ratio. Provided κ is constant in a series of cogenetic rocks (by cogenetic we mean that at $t = 0$ they all had identical Pb isotope ratios), the slope of an array on a plot of $^{208}Pb/^{204}Pb$ and $^{206}Pb/^{204}Pb$ will depend only on t and κ, and not on the parent/daughter ratios. We can calculate κ from the slope of the data on a plot of $^{208}Pb/^{204}Pb$–$^{206}Pb/^{204}Pb$. U and Th are both highly incompatible elements. They have similar, but not identical, partition coefficients in most minerals. It seems unlikely that they will behave identically, and hence unlikely that κ will be strictly constant. It can

be shown that if κ varies and is positively correlated with variations in µ, then eqns. 8.35 and 8.35, when solved for κ, will actually overestimate it.

8.4.7   U and Th decay series isotopes

U and Th do not decay directly to Pb; rather, the transition from U and Th to Pb passes through many intermediate radioactive daughters (Figure 8.22). Many of these daughters have very short half-lives, ranging from milliseconds to minutes, and are of little use to geochemists. However, a number of these intermediate daughters have half-lives ranging from days to hundreds of thousands of years and do provide useful information about geologic processes. Table 8.4 lists half-lives and decay constants of some of the most useful of these nuclides.

The half-lives of these daughter isotopes are short enough so that any atoms present when the Earth formed have long since decayed (to Pb). They exist in the Earth (and in all other bodies of the solar system) only because they are continually produced by the decay of U and Th. The abundance of such an isotope depends on the balance between its own radioactive decay and its production by the decay of its parent:

$$\frac{dN}{dt} = \lambda_p N_P - \lambda_D N_D \qquad (8.41)$$

where subscripts P and D refer to parent and daughter, respectively, and *the derivative expresses the change in abundance (number of atoms or moles) of the daughter with time.* This equation says simply that the rate of change of the abundance of the daughter isotope is equal to the rate of production less the rate of decay. This can be integrated to give:

$$N_D = \frac{\lambda_P}{\lambda_D - \lambda_P} N_P^0 (e^{-\lambda_P t} - e^{-\lambda_D t}) + N_D^0 e^{-\lambda_D t}$$
$$(8.42)$$

Scientists dealing with the intermediate daughters of U and Th (and it is the daughters of ^{238}U that are of the most interest) generally work with activities, measured in the number of decays per unit time, rather than atomic abundances. One reason for this is that the abundances of these isotopes were historically determined by detecting their decay. Indeed,

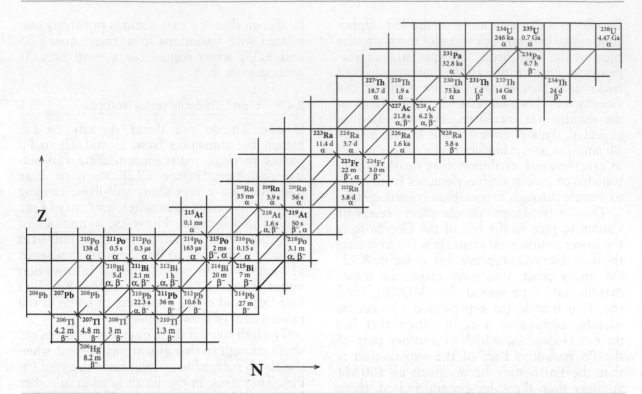

Figure 8.23 Part of the chart of the nuclides showing the series of decays that occur as 238U, 235U, and 232Th are transformed to 206Pb, 207Pb, and 208Pb, respectively. Nuclides of the 238U series are shown in red, the 232Th series in gray, and the 235U series in black.

Table 8.4 Half-lives and decay constants of long-lived U and Th daughters.

Nuclide	Half-life, yrs	Decay constant, yr^{-1}	Parent
^{234}U	246,000	2.794×10^{-6}	^{238}U
^{231}Pa	32,480	2.134×10^{-5}	^{235}U
^{230}Th	75,200	9.217×10^{-6}	^{238}U
^{226}Ra	1,622	4.272×10^{-4}	^{238}U
^{228}Ra	6.7	1.06×10^{-1}	^{232}Th
^{210}Pb	22.26	3.11×10^{-2}	^{238}U

the shorter-lived ones are so rare they cannot be detected any other way. The other reason will become apparent shortly. Activities are related to atomic (or molar) abundances by the basic equation of radioactive decay:

$$\frac{dN_i}{dt} = -\lambda_i N_i \qquad (8.43)$$

where dN_i/dt is the rate of decay or *activity* of i. Thus, so long as we know the decay constant, we can calculate activity from concentration and vice versa. The SI unit of activity is the *becquerel* (Bq), which is defined as 1 decay per second. An older unit of activity is the curie (Ci), which is equal to 3.7×10^{10} Bq.* In geochemistry, however, activity is often simply expressed in units of *disintegrations per minute* (dpm) per gram or per mole (1 dpm = 60 Bq/g) and is denoted by writing parentheses around the isotope (or the isotope ratio). Thus (^{234}U) is the activity of ^{234}U. Indeed, U-decay series isotopes were originally measured by alpha-counting, so what was measured was activity, not abundance. This required fairly large quantities of material. Improvements in mass spectrometry made it possible to measure ^{234}U, ^{230}Th, and other key nuclides such as ^{231}Pa and ^{226}Ra with smaller quantities of material and much better precision than α-counting. This has led to a considerable expansion of applications of U-decay series isotopes.

The state of *radioactive equilibrium* is the condition where the *activities* of the daughter and the parent are equal:

$$\boxed{\frac{dN_D}{dt} = \frac{dN_P}{dt}} \qquad (8.44)$$

where the derivatives refer to the activities of parent and daughter. It shares the same fundamental characteristic as the chemical equilibrium state: it is the state that will eventually be achieved if the system is not perturbed (remains closed). We can demonstrate this is so in two ways. The first is a simple mathematical demonstration. The equilibrium state is the steady state where the abundance of the daughter does not change, and so the left-hand side of eqn. 8.41 is zero:

$$0 = \lambda_p N_p - \lambda_D N_D \qquad (8.45)$$

We substitute the dN/dt terms for the λN terms in eqn. 8.45, rearrange, and we obtain eqn. 8.44 as predicted.

The second demonstration is a thought experiment. Imagine a hopper, such as a grain hopper with an open top and a door in the bottom. The door is spring-loaded such that the more weight that is placed on the door, the wider it opens. Suppose we start dropping marbles into the hopper at a constant rate. The weight of marbles accumulating in the hopper will force the door open slightly and marbles will start falling out at a slow rate. Because the marbles are falling out more slowly than they are falling in, the number and weight of marbles in the hopper will continue to increase. As a result, the door will continue to open. At some point, the door will be open so wide that marbles are falling out as fast as they are falling in. This is the steady state. Marbles can no longer accumulate in the hopper and hence the door is not forced to open any wider. The marbles falling into the door are like the decay of the parent isotope. The marbles in the hopper represent the population of daughter isotopes. Their decay is represented by their passing through the bottom door. Just as the number of marbles passing through the door depends on the number of marbles in the hopper, the activity (number of decays per unit time) of an isotope depends on the number of atoms present.

If the rate of marbles dropping into the hopper decreases for some reason, marbles will fall out of the hopper faster than they fall in. The number of marbles in the hopper will decrease; as a result, the weight on the door

---

* The Becquerel is named in honor of French physicist and engineer Henri Becquerel (1952–1908) who serendipitously discovered radioactivity in 1896 when he left uranium ore on black-paper covered photograph plates and found the plates exposed. The curie is named in honor of Marie Curie (1867–1934) whose pioneering work in radioactivity won her the 1903 Nobel Prize in physics jointly with her husband Pierre Curie and Henri Becquerel. She was the first woman to win the prize and in 1900 became the first woman faculty member at the École Normale Supérieure of Paris. In 1911 she won the Nobel Prize in chemistry for having discovered the elements radium and polonium (the latter named for her native Poland). Marie Curie remains the first and only person to have won Nobel prizes in two fields of science. She died in 1934 from radiation poisoning, a year before her daughter Irène Joliot-Curie won the Nobel Prize for chemistry jointly with her husband.

decreases and it starts to close. It continues to close (as the number of marbles decreases) until the rate at which marbles fall out equals the rate at which marbles fall in. At that point, there is no longer a change in the number of marbles in the hopper and the position of the door stabilizes. Again, equilibrium has been achieved, this time with fewer marbles in the hopper, but nevertheless at the point where the rate of marbles going in equals the rate of marbles going out. The analogy to radioactive decay is exact.

Thus, when a system is disturbed it will ultimately return to equilibrium. The rate at which it returns to equilibrium is determined by the decay constants of the parent and daughter. If we know how far out of equilibrium the system was when it was disturbed, we can determine the amount of time that has passed since it was disturbed by measuring the present rate of decay of the parent and daughter. Equilibrium is approached asymptotically, and these dating schemes are generally limited to time scales of less than 5–10 times the half-life of the daughter. At longer times, the difference between actual activities and equilibrium becomes too small to measure reliably.

There are several useful dating methods based on the degree of disequilibria between U decay series nuclides. Our first example is ^{234}U-^{238}U dating of sediments. As may be seen from Figure 8.22, ^{234}U, which has a half-life of 246,000 years, is the "great-granddaughter" of ^{238}U. For most purposes, the half-lives of the two intermediate daughters ^{234}Th and ^{234}Pa are so short that they can be ignored (because they quickly come into equilibrium with ^{238}U). As it turns out, ^{234}U and ^{238}U in seawater are not in equilibrium – the ^{234}U/^{238}U activity ratio is not 1. It is fairly constant, however, at about 1.15. The reason for this disequilibrium is that ^{234}U is preferentially leached from rocks because ^{234}U is present in minerals in damaged sites. It occupies the site of a ^{238}U atom that has undergone α-decay. The α particle and the recoil of the nucleus damage this site. Since it occupies a damaged site, it is more easily leached or dissolved from the crystal during weathering than ^{238}U. The oceans collect this leachate; hence they are enriched in ^{234}U. When U enters a sediment, it is isolated from seawater (not necessarily immediately) and

^{234}U decays faster than it is created by decay of ^{238}U, so it slowly returns to the equilibrium condition where $(^{234}$U/^{238}U) = 1.

Let's now consider the problem from a mathematical perspective and derive an equation describing this return to equilibrium. We can divide the ^{234}U activity in a sample into that which is *supported* by ^{238}U (i.e., that amount in radioactive equilibrium with ^{238}U), and that amount that is excess (i.e., *unsupported* by ^{238}U):

$$(^{234}U) = (^{234}U)_s - (^{234}U)_u \qquad (8.46)$$

Subscripts s and u denote supported and unsupported. The activity of the excess ^{234}U decreases with time according to eqn. 8.4, which we can rewrite as:

$$(^{234}U)_u = (^{234}U)_u^o e^{-\lambda_{234}t} \qquad (8.47)$$

where the superscript naught denotes the initial unsupported activity (at $t = 0$). We can also write:

$$(^{234}U)_u^o = (^{234}U)^o - (^{234}U)_s \qquad (8.48)$$

which just says that the initial unsupported activity of ^{234}U is equal to the total initial activity of ^{234}U minus the (initial) supported activity of ^{234}U. *Since to a very good approximation the activity of the parent, ^{238}U, does not change over time on the order of the half-life of ^{234}U or even 10 half-lives of ^{234}U, the present ^{238}U activity is equal to the activity at $t = 0$* (we make the usual assumption that the system is closed). By definition, the supported activity of ^{234}U is equal to the activity of ^{238}U, both now and at $t = 0$, hence eqn. 8.46 can be expressed as:

$$(^{234}U) = (^{238}U) + (^{234}U)_u \qquad (8.49)$$

and eqn. 8.48 becomes

$$(^{234}U)_u^o = (^{238}U) + (^{234}U)_u \qquad (8.50)$$

Substituting eqn. 8.50 into 8.47 yields:

$$(^{234}U)_u = [(^{234}U)^o - (^{238}U)]e^{-\lambda_{234}t} \qquad (8.51)$$

Substituting eqn. 8.51 into 8.49, we have:

$$(^{234}U) = (^{238}U) + [(^{234}U)^o - (^{238}U)]e^{-\lambda_{234}t} \qquad (8.52)$$

Just as for other isotope systems, it is generally most convenient to deal with ratios rather than absolute activities (among other things, this allows us to ignore detector efficiency, provided the detector is equally efficient at all energies of interest); hence we divide by the activity of ^{238}U:

$$\left(\frac{^{234}U}{^{238}U}\right) = 1 + \left[\frac{(^{234}U)^o - (^{238}U)}{(^{238}U)}\right] \cdot e^{-\lambda_{234}t} \qquad (8.53)$$

or since $^{238}U = {}^{238}U^0$:

$$\left(\frac{^{234}U}{^{238}U}\right) = 1 + \left[\left(\frac{^{234}U}{^{238}U}\right)^o - 1\right]e^{-\lambda_{234}t} \qquad (8.54)$$

Corals, for example, concentrate U. If we measure the $(^{234}U/^{238}U)$ ratio of an ancient coral and could make a first-order assumption that the seawater in which that coral grew had a $(^{234}U/^{238}U)$ the same as modern seawater (1.15), then the age of the coral can be obtained by solving eqn. 8.54 for t.

Because the disequilibrium between ^{230}Th and ^{238}U can be much larger than between ^{234}U and ^{238}U, ^{230}Th-^{238}U disequilibria is a more commonly used dating scheme than is ^{234}U-^{238}U. ^{230}Th is the daughter of ^{234}U (the decay chain is $^{238}U \rightarrow {}^{234}Th + \alpha$, $^{234}Th \rightarrow {}^{234}Pa + \beta^-$, $^{234}Pa \rightarrow {}^{234}U + \beta^-$, $^{234}U \rightarrow {}^{230}Th + \alpha$). To simplify the math involved, let's assume ^{234}U and ^{238}U are in radioactive equilibrium. In high-temperature systems, this is a very good assumption because α-damaged sites, which cause the disequilibrium noted above, are quickly annealed. With this assumption, we can treat the production of ^{230}Th as if it

were the direct decay product of ^{238}U. We write an equation analogous to 8.46 and from it derive an equation analogous to 8.52:

$$(^{230}Th) = (^{238}U) + [(^{230}Th)^o - (^{238}U)]e^{-\lambda_{230}t} \qquad (8.55)$$

We divide by the activity of ^{232}Th:

$$\left(\frac{^{230}Th}{^{232}Th}\right) = \left(\frac{^{238}U}{^{232}Th}\right)$$

$$+ \left[\left(\frac{^{230}Th}{^{232}Th}\right)^o - \left(\frac{^{238}U}{^{232}Th}\right)\right]e^{-\lambda_{230}t} \qquad (8.56)$$

and rearranging:

$$\left(\frac{^{230}Th}{^{232}Th}\right) = \left(\frac{^{230}Th}{^{232}Th}\right)^o e^{-\lambda_{230}t}$$

$$+ \left(\frac{^{238}U}{^{232}Th}\right)(1 - e^{-\lambda_{230}t}) \qquad (8.57)^*$$

$^{230}Th/^{238}U$ is commonly used to date sediments and to determine sedimentation rates. However, unlike the case of $(^{234}U/^{238}U)^0$, $(^{230}Th/^{232}Th)^0$ is not known *a priori*, so there are two unknowns in eqn. 8.57. As was the case for isochrons, however, we can solve for these two unknowns if we have a series of two or more measurements on sediments with the same initial $(^{230}Th/^{232}Th)$. Example 8.5 demonstrates how this is done for the case of a Mn nodule. In other cases (e.g., corals), we can assume that $(^{230}Th/^{232}Th)^0$ is the same as in modern seawater.

Example 8.5 Determining the growth rate of a Mn nodule

The tops of manganese nodules grow by precipitation of Mn-Fe oxides and hydroxides from seawater. They are known to grow very slowly, but how slowly? If we assume the rate of growth is constant, then depth in the nodule should be proportional to time. If z is the depth in the nodule, and s is the growth (sedimentation) rate, then:

$$t = z/s \qquad (8.58)$$

and eqn. 8.57 becomes:

$$\left(\frac{^{230}Th}{^{232}Th}\right) = \left(\frac{^{230}Th}{^{232}Th}\right)^o e^{-\lambda_{230}z/s} + \left(\frac{^{238}U}{^{232}Th}\right)(1 - e^{-\lambda_{230}z/s}) \qquad (8.59)$$

---

* This equation may also be derived directly from eqn. 8.42 since $\lambda_{230} - \lambda_{238} \cong \lambda_{230}$, $^{238}U \cong {}^{238}U^o$, and $e^{-\lambda_{238}t} \cong 1$ for any value of t over the range in which this method is useful (~500,000 years).

At the surface of the nodule, $z = 0$, so the exponential terms both go to 1 and the measured activity ratio is the initial activity ratio. Having a value for $(^{230}Th/^{232}Th)^0$, eqn. 8.59 can then be solved for s, the growth rate, if measurements are made at some other depth.

In practice, however, it is difficult to obtain a sample exactly at the surface: a finite amount of material is required for analysis, and this often translates into a layer of several mm thickness. Equation 8.59 is solved in that instance by less direct means. For example, consider the data shown in Figure 8.24 on a Pacific manganese nodule reported by Huh and Ku (1984). In this plot of $(^{230}Th/^{232}Th)$ vs. depth, the initial ratio is the intercept of the best-fit line through the data. A growth rate was obtained by making an initial guess of the initial $(^{230}Th/^{232}Th)$ ratio, then iteratively refining the solution to eqn. 8.54 by minimizing the difference between computed and observed activity ratios. A growth rate of 49.5 mm/Ma and $(^{230}Th/^{232}Th)$ of 77.7 was found to best fit the observations.

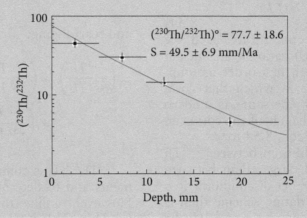

Figure 8.24 $(^{230}Th/^{232}Th)$ as a function of depth in a manganese nodule from MANOP Site H. After Huh and Ku (1984).

One of the most successful applications of U-Th dating has been dating corals to determine sea level rise at the end of the last glaciation. Reef-building zoanthid corals incorporate symbiotic algae in their tissues and thus always live within a few meters of sea level. As sea level rose at the end of the last glaciation, coral reefs grew to keep up. By drilling into these reefs and determining age as a function of depth, the timing of sea level rise can be determined. Coral structures are composed of calcium carbonate, and the original approach was to date layers using ^{14}C dating. As we will see in a subsequent section, however, variable initial ^{14}C in the atmosphere requires calibration of ^{14}C against some other chronometer to determine exact ages. As it happens, coral carbonate also incorporates U from seawater and excludes Th, making

Th-U dating a superb chronometer for dating corals.

A first-order approach would be to assume that the $(^{234}U/^{238}U)$ ratio has been constant through time at 1.15. Accurate ages, however, require that we take account of variable ^{234}U-^{238}U disequilibrium. To do so, we use the following equation:

$$\left(\frac{^{230}Th}{^{234}U}\right) = 1 - e^{-\lambda_{230}t} + \left[\left(\frac{^{230}U}{^{238}U}\right) - 1\right]$$
$$\frac{\lambda_{230}}{\lambda_{230} - \lambda_{234}}(1 - e^{-(\lambda_{230} - \lambda_{234})t})$$

$$(8.60)$$

Because time occurs in two separate exponents, we cannot solve this equation directly, but we can solve it with the same iterative methods we would use for eqn. 8.37. In addition, it is generally necessary to take account of initial small amounts of ^{230}Th, which are generally associated with detrital material incorporated in the reef structure. This can be done by also measuring ^{232}Th (see Reiners et al., 2017 for details).

Another advantage of the U-series dating is that, as for U–Pb, there are two radioactive isotopes of U. ^{235}U decays to ^{231}Pa through the intermediate nuclide ^{231}Th, which has only a one-day half-life. ^{231}Pa has a 5+ valence state and ionic radius of 92 pm and like Th is strongly excluded from corals. Assuming, as we did for ^{238}U, that the abundance of ^{235}U does not change on the time scales of interest, we derive the following relationship for ^{231}Pa/^{235}U:

$$\left(\frac{^{231}Pa}{^{235}U}\right) = (1 - e^{-\lambda_{231}t}) + \left(\frac{^{231}Pa}{^{235}U}\right)^0 e^{-\lambda_{231}t}$$

(8.61)

The ^{235}U-^{231}Pa decay has disadvantages that both ^{235}U and ^{231}Pa are far less abundant that ^{238}U and ^{230}Th and the decay constant of ^{231}Pa is less well-known than that of ^{230}Th. Nevertheless, using both systems we can create a concordia diagram analogous to that for U–Pb and use it to provide a "double-check" for the accuracy of our ages (see Edwards et al., 2003). Using this approach, a variety of workers beginning with Edwards et al. (1987) have produced the sea level curve shown in Figure 8.25. In addition to determining this curve, by simultaneously measuring ^{14}C, the ^{14}C chronometer has been calibrated through the last 50,000 years.

Another important application of U-Th dating has been dating cave art. Cave painting provides clear evidence of the emergence of human symbolic thinking. Many of the paintings were done in limestone caves where carbonate, "flowstone" precipitates from water running down cave walls. Any flowstone over the paintings must necessarily be younger than the paintings, so dating this provides a minimum age of the painting. By carefully removing the calcite covering the paintings and analyzing it using mass spectrometry, some of these paintings have now been dated. In a recent example, Hofmann et al. (2018) dated paintings and hand stencils in Spanish caves to ~70 ka (Figure 8.26). This predates the arrival of *Homo sapiens*

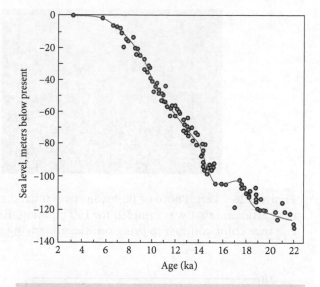

Figure 8.25 Sea level rise at the end of the last glaciation based on Th-U dating of coral reefs from New Guinea, Tahiti, and Barbados. After Edwards et al. (2003).

in Europe, suggesting the artists were Neanderthals capable of abstract and symbolic thinking.

Th–U disequilibria may also be used for dating lavas, and we now turn our attention briefly to this application. Equation 8.57 has the form of a straight line on a (^{230}Th/^{232}Th)–(^{238}U/^{232}Th) plot; the first term is the intercept and $(1 - e^{-\lambda_{230}t})$ is the slope. Since the slope is proportional to time, a line on a (^{230}Th/^{232}Th)–(^{238}U/^{232}Th) plot is an isochron, though unlike a conventional isochron the intercept also changes with time (Figure 8.27).

To understand how this works, imagine a crystallizing magma with homogenous (^{230}Th/^{232}Th) and (^{238}U/^{232}Th) ratios. Th and U partition into different minerals differently, so the minerals will have variable (^{238}U/^{232}Th) ratios but constant (^{230}Th/^{232}Th) ratios (assuming crystallization occurs quickly compared with the half-life of Th) since these two isotopes are chemically identical. Thus, the minerals will plot on a horizontal line in Figure 8.27 at $t = 0$. After the system closes, ^{238}U and ^{230}Th will begin to come to radioactive equilibrium (either ^{230}Th will decay faster than it is produced or vice versa, depending on whether (^{230}Th/^{238}U) is greater than or less than 1, the equilibrium value).

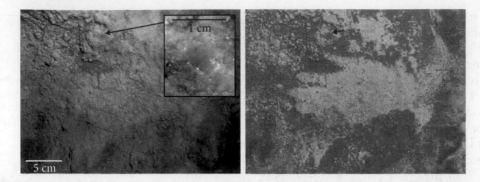

Figure 8.26   Left. Photo of flowstone over a hand stencil in Maltravieso Cave in Spain. The inset shows the carbonate that was sampled for U-Th dating. Right shows the same picture digitally processed to enhance color contrast to bring out the underlying image. From Hofmann et al. (2018).

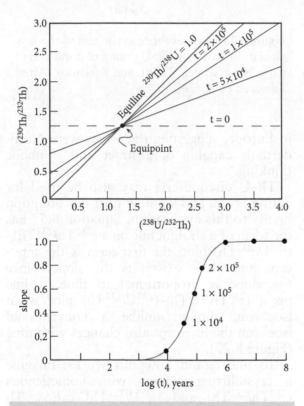

Figure 8.27   (a) ^{230}Th–^{238}U isochron diagram. The (^{238}U/^{232}Th) of the source is given by the intersection of the isochron with the equiline. (b) This shows how the slope changes as a function of time. After Faure (1986).

Thus, the original horizontal line will rotate, as in a conventional isochron diagram, but unlike the conventional isochron diagram, the intercept also changes. The rotation occurs about the point where (^{230}Th/^{232}Th) = (^{238}U/^{232}Th), which is known as the *equipoint*.

As t approaches infinity, the exponential term approaches 1 and:

$$\lim_{t \to \infty} \left(\frac{^{230}Th}{^{232}Th} \right) = \left(\frac{^{238}U}{^{232}Th} \right) \qquad (8.62)$$

Thus, the equilibrium situation, at $t = \infty$, is (^{230}Th/^{232}Th) = (^{238}U/^{232}Th). In this case, all the minerals will fall on a line, having a slope of 1. This line is known as the *equiline*. Figure 8.28 shows an example of a Th-U isochron for a Seguam Island (Aleutians) dacite from Jicha et al. (2005). The age, although a bit older, is within error of the ^{40}Ar/^{39}Ar age of 142 ka.

^{226}Ra, the daughter of ^{230}Th, is another relatively long-lived nuclide ($t_{1/2} = 1600$ yr) in the ^{238}U decay chain. Unfortunately, Ra has no stable isotope to which one can ratio ^{226}Ra. In some cases, ^{226}Ra has been ratioed to Ba and used in an analogous way to the Th-U isochron approach mentioned above. However, Ra and Ba partition between magma and minerals differently, so this has its limits. While deriving precise ^{226}Ra/^{230}Th ages is difficult, as the initial ratio cannot generally be defined, with assumptions as to this ratio, "model ages" can be computed that can give at least qualitative information on the time scales of melting and magma transport.

8.4.8   Isotopes of He and other rare gases

8.4.8.1   *Helium*

Alpha particles are, of course, simply ^{4}He nuclei, and therefore ^{4}He is produced by

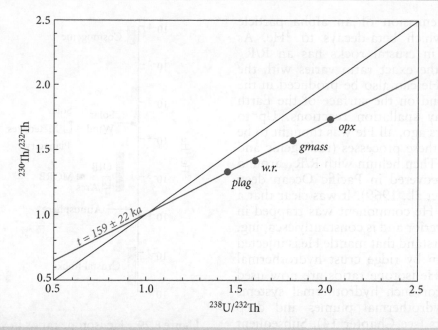

Figure 8.28 (a) ^{230}Th–^{238}U mineral isochron on a dacite lava from Seguam Island in the Aleutians. This and other ^{230}Th–^{238}U ages on Seguam lavas agree within error with ^{40}Ar/^{39}Ar ages, implying a short residence time of crystals in the magma chamber. plag: plagioclase, w.r.: whole rock, gmass groundmass, opx: orthopyroxene Data from Jicha et al. (2005).

alpha decay of U and Th. Thus, the ratio of ^{4}He/^{3}He varies in the Earth. Unlike virtually all other elements, He is not conserved on the Earth: much of the He present when the Earth formed has been subsequently lost. Being a rare gas, He does not combine chemically with anything, and it also diffuses rapidly (ever had a helium balloon? How long did it last?) Helium brought to the Earth's surface by magmatism eventually finds its way into the atmosphere. Once there, it can escape from the top of the atmosphere because of its low mass.* (H also escapes from the atmosphere, but most H is bound up as water or hydrous minerals, so relatively little is lost.) Since ^{4}He is continually produced and, for all practical purposes, ^{3}He is not, it should not be surprising that the ^{4}He/^{3}He ratio in the Earth is a very large number. Perhaps because geochemists like to deal with very small numbers rather than very large ones, or perhaps because it is actually ^{3}He that is

most interesting, the He isotope ratio is often expressed as ^{3}He/^{4}He, in contradiction to the normal convention of placing the radiogenic isotope in the numerator. We will adhere to this particular convention of not adhering to the convention and use the ^{3}He/^{4}He ratio.

The ^{3}He/^{4}He ratio of the atmosphere is 1.384×10^{-6}. Since this ratio is uniform and atmospheric He is available to all laboratories, it is a convenient standard and moreover provides the basis for a convenient normalization. He isotope ratios are often reported and discussed relative to the atmospheric value. The units are called simply R/R_A, where R is the ^{3}He/^{4}He in the sample of interest and R_A indicates the atmospheric ratio.

Actually, it is not quite true that ^{3}He is not produced in the Earth. It is produced in very small quantities through the nuclear reaction: ^{6}Li(n,α)→^{3}H(β)→^{3}He, which is to say ^{6}Li is excited by the capture of a neutron that has been produced by U fission, and decays

---

*At equilibrium the average kinetic energy of all varieties of molecules in a gas will be equal. Since kinetic energy is related to velocity by $E = \frac{1}{2}mv^2$, and since the mass of He is lower than for other species, the velocities of He atoms will be higher and more likely to exceed the escape velocity. In actuality, however, escape of He is more complex than simple thermal acceleration and involves other processes such as acceleration of He ions by the Earth's magnetic field.

through the emission of an alpha particle to tritium, which beta-decays to ^{3}He. As a result, He in crustal rocks has an R/R_A ≈ 0.1–0.01 (the exact ratio varies with the Li/U ratio). ^{3}He can also be produced in the atmosphere and on the surface of the Earth by cosmic ray spallation reactions. Up to about 50 years ago, all He was thought to be a product of these processes (radiogenic and cosmogenic). Then helium with R/R_A around 1.22 was discovered in Pacific Ocean deep water (Clark et al., 1969). It was clear that a "primordial" He component was trapped in the Earth's interior and is constantly escaping. We now understand that mantle He is injected into the ocean by ridge crust hydrothermal activity, and He isotope ratios are now used to prospect for such hydrothermal systems and trace hydrothermal plumes and deep ocean currents (see Chapter 14). Subsequent measurements beginning with Lupton and Craig (1975) have shown that the ^{3}He/^{4}He of MORB erupted at ocean ridges is quite uniform around R/R_A ≈ 8.8 ± 2.5. This is only primordial in a relative sense: the early solar system had R/R_A ≈ 200. Thus, even in MORB, He is 95% radiogenic.

Figure 8.29 illustrates the He isotopic composition of various terrestrial reservoirs. Oceanic islands and other hotspots often have even higher ratios, up to about R/R_A ≈ 40 (Sano, 2018) and ratios up to 50 have been measured in early Tertiary basalts on Baffin Island, which are thought to have been produced by the Iceland mantle plume. Other islands, Tristan da Cunha for example, have lower ratios: R/R_A ≈ 5 (e.g., Kurz et al., 1982). This suggests that most oceanic island basalts are derived from a less degassed, and in that sense more primordial, reservoir than MORB. This is consistent with the mantle plume hypothesis.

Island arc volcanics (IAV) also seem to be fairly uniform, with R/R_A ≈ 6 (Lupton, 1983). Ratios lower than MORB suggest the presence of a slab-derived component, but also indicate most of the He in IAV comes from the mantle wedge (a conclusion similar to the one reached from other radiogenic isotopes).

Two other interesting developments were the discovery of cosmogenic He in Hawaiian

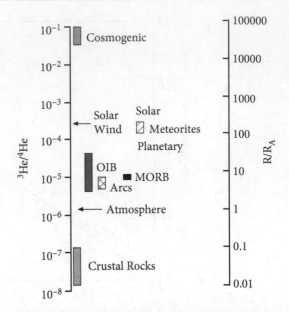

Figure 8.29 He isotope ratios in various terrestrial and solar system materials. "Solar" and "Planetary" refer to the solar and planetary components in meteorites. "Crustal rocks" shows the values expected for *in situ* production from α-decay and neutron-induced nuclear reactions.

basalts (Kurz, 1987), and very high R/R_A in some very slowly accumulating marine sediments. The former is a result of the ^{6}Li(n,α)→^{3}H(β)→^{3}He reaction instigated by cosmic ray produced neutrons and spallation.* The cosmogenic component decreases rapidly with depth of the rock (it is largely restricted to the top meter) and increases with elevation above sea-level (the effect was noticed at the summit of Haleakala at 2000 m). Thus, one should be suspicious of high ^{3}He/^{4}He ratios in old subaerial rocks. This has, however, been subsequently turned into a tool to determine exposure times and erosion rates (e.g., Kurz, 1990).

Very high R/R_A found in some very slowly accumulating marine sediments is an effect of accumulation of cosmic dust in sediment. ^{40}Ar/^{36}Ar ratios lower than atmospheric have also been observed in deep-sea sediment, which also suggest a cosmic origin. He ratios in magma can change through the combined effects of diffusion out of the magma and

---

* Spallation is the process by which a nucleus breaks into smaller nuclei as a result of a collision with a very high-energy particle such as a cosmic ray.

radiogenic growth of He, on the time scale of the residence time of magma in a magma chamber. This means we must be cautious of low ^{3}He/^{4}He ratios as well.

Finally, the production of ^{4}He by U and Th decay provides a useful geochronological tool. ^{4}He accumulates at predictable rates in U- and Th-rich minerals such as apatite and zircon. However, the high diffusion rates of He means that diffusive loss becomes significant even at relatively low temperatures ($\sim$50°C for apatite, $\sim$120°C for zircon). Diffusion rates can be determined as a function of temperature experimentally (see Chapter 5), so this can be turned to advantage for determining the rates at which rocks have cooled, for example as they are tectonically uplifted. In combination with other techniques such as the K-Ar decay and fission track dating, U-He dating is part of the important field of *thermochronology*. Reiners et al. (2017) cover this topic in detail.

8.4.8.2 Neon

Ne has three isotopes, ^{20}Ne, ^{21}Ne, and ^{22}Ne. Though none of them are radiogenic, the isotopic composition of neon varies in the Earth as well as extraterrestrial materials (Figure 8.30). The Earth's atmosphere has ^{20}Ne/^{22}Ne and ^{21}Ne/^{22}Ne ratios of 9.8 and 0.0290, respectively. This is displaced to lower ^{20}Ne/^{22}Ne and ^{21}Ne/^{22}Ne ratios from the solar wind values of 14.0 and 0.0336, respectively (Moreira, 2013), along a slope of 2, which implies the process responsible is mass-dependent fractionation (which we cover in the next chapter). The "B" component (^{20}Ne/^{22}Ne = 12.73, ^{21}Ne/^{22}Ne = 0.0321) isolated in meteorites (see Chapter 10) also plots on this line. In addition to mass dependent fractionation, Ne isotopes also vary due to production by nuclear reactions such as ^{18}O(α,n)$\rightarrow$ ^{21}Ne, ^{24}Mg(n,α)$\rightarrow$ ^{21}Ne, and ^{25}Mg(n,α)$\rightarrow$ ^{22}Ne, with the α particles and neutrons coming from α-decay and fission, respectively. Production rates are quite small, for example about 4.2×10^{-22} cc/g of ^{21}Ne in the mantle, but because Ne is rare in the Earth this production can be significant. The production rate of ^{21}Ne is about an order of magnitude higher than that of ^{20}Ne and ^{22}Ne, and ^{21}Ne is the least abundant of the Ne isotopes (0.27%), so the effect of these reactions is to increase the ^{21}Ne/^{22}Ne ratio

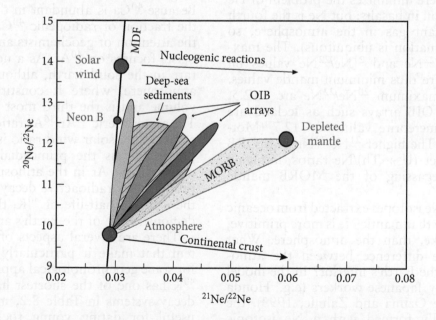

Figure 8.30 Ne isotope ratios in terrestrial materials. The line marked "MDF" is the mass dependent fractionation line. The line has a slope of 2 because the change in ^{20}Ne/^{22}Ne resulting from fractionation is twice that of ^{21}Ne/^{22}Ne because the mass difference between ^{20}Ne and ^{22}Ne is twice that between ^{21}Ne and ^{22}Ne. OIB arrays are data fields for basalts and mantle xenoliths.

with little change the $^{20}Ne/^{22}Ne$ ratio. This "nucleogenic" component builds up more rapidly in the Earth's crust (due to the high levels of U and Th and low intrinsic Ne concentrations) and can be used to determine groundwater residence times. Neon isotopes, particularly ^{21}Ne, can also be produced by cosmic ray spallation both in meteorites and on the Earth's surface. Because ^{21}Ne is such a rare nuclide in rocks, the amount of cosmogenic ^{21}Ne is significant so that, like ^3He, cosmogenic ^{21}Ne can be used to determine *exposure ages*: the time a rock has been exposed on the Earth's surface or a meteorite, ejected from its parent body by a collision, has been exposed in interplanetary space (Niedermann, 2002).

Neon trapped in basalts from mid-ocean ridges and basalts and mantle xenoliths from oceanic islands define arrays that diverge from atmospheric composition to compositions with higher $^{20}Ne/^{22}Ne$ and $^{21}Ne/^{22}Ne$ (e.g., Sarda et al., 1988; Poreda and Farley, 1992). These arrays reflect contamination of mantle Ne by atmospheric Ne: they are mixing arrays between mantle components with high $^{20}Ne/^{22}Ne$ and variable $^{21}Ne/^{22}Ne$ ratios (the very low concentrations of He in the atmosphere minimizes the problem of He contamination in basalts; but Ne is the fourth most abundant gas in the atmosphere, so this contamination is ubiquitous). The maximum $^{20}Ne/^{22}Ne$ and $^{21}Ne/^{22}Ne$ values on each array are thus minimum mantle values. In MORB maximum $^{20}Ne/^{22}Ne$ are ~12.5, while some OIB arrays such as Iceland are as high as meteorite value of ~12.7 (Moreira, 2013). The higher $^{21}Ne/^{22}Ne$ in MORB reflects higher (U + Th)/Ne ratios, implying extensive degassing of the MORB mantle source.

Overall, Ne isotopes extracted from oceanic basalts show that mantle Ne is more primitive, i.e., solar-like, than the atmosphere. What explains the difference between the atmosphere and the Earth's interior? In the model advocated by Japanese workers (e.g., Honda et al., 1991; Ozima and Zahnle, 1993), the Earth initially formed with a Ne isotopic composition similar to that of solar wind. During the Earth's very early history, much of Earth's atmophile element inventory escaped from the interior, perhaps during the existence of a magma ocean, to form a hot, massive atmosphere. This, together with a potentially much more unstable young, outburst-prone young Sun could have led to the escape of the lightest gases, hydrogen, helium, and a fraction of neon from this primitive atmosphere during this time. The efficiency of escape will depend on the mass of the molecule, so that the proportional loss of ^{20}Ne is greatest and that of ^{22}Ne the least. Other gases such as N_2, CO_2, and Ar would have been too heavy to reach escape velocity and hence would not have experienced significant isotope fractionation. Marty (2012) proposed an alternative that atmospheric Ne was added after the Earth accreted by a late veneer of chondritic material having the "planetary" component ("Neon A") or "Q" observed in meteorites (Moreira, 2013), which is depleted in light Ne isotopes (e.g., $^{20}Ne/^{22}Ne$ in "Q" is ~10.67).

8.4.8.3   K-Ar-Ca

^{40}K is an odd–odd nuclide and consequently subject to both β^- decay to ^{40}Ca and electron capture decay to ^{40}Ar, with ~88% of decays going to ^{40}Ca. ^{40}Ca, however, with 20 neutrons and 20 protons ^{40}Ca is a doubly magic nuclide and hence very abundant. Because ^{40}Ca is abundant in the solid Earth, the fraction of radiogenic ^{40}Ca is small and the attention of geochemists and geochronologists focuses on ^{40}Ar. As a noble gas, Ar is rare in the solid Earth, although not in the atmosphere, where it constitutes ~1% by volume and is the third most abundant gas. The atmospheric $^{40}Ar/^{36}Ar$ ratio is 296.16. By comparison, solar wind ratio is 0.003, which presumably is the primordial value. Thus, nearly all the Ar in the atmosphere has been produced by radioactive decay of ^{40}K. Given the ~1.4 Ga half-life of ^{40}K, this is a potent demonstration of the Earth's antiquity.

There are several aspects of the K-Ar system that make it particularly advantageous for some geochronological applications. First, ^{40}K has one of the shortest half-lives of the decay systems in Table 8.2, making it more useful for dating young rocks than other systems. In favorable cases, K-Ar ages on rocks as young as 30,000 yrs can be obtained with useful precision. Second, Ar is a rare gas and it is virtually totally lost upon eruption in most lavas. With little or no initial component, an age can be obtained from a single K-Ar determination. Nonradiogenic Ar in such samples is often (but not always) of atmospheric origin, and since the $^{40}Ar/^{36}Ar$ ratio in the atmosphere is constant, this can be readily accounted for. Finally, due to the ease with which Ar diffuses through minerals, the closure temperature for the K-Ar system is low, making it a useful system for study of relatively low-temperature phenomena (such as petroleum genesis). This, of course, can be a disadvantage, as it means the K-Ar system is reset rather easily. Most modern geochronological K-Ar studies, however, use the $^{40}Ar/^{39}Ar$ technique, which is described in the accompanying box.

$^{40}Ar/^{39}Ar$ dating

If you look at the chart of the nuclides, you'll find that ^{39}Ar has a 269-year half-life and does not occur naturally. You might wonder then how this can be used to date anything. The *40-39 method* is actually a variant of ^{40}K-^{40}Ar dating employing a unique analytical technique for the potassium first described by Merrihue and Turner (1966). The key is the production of ^{39}Ar by a nuclear reaction on ^{39}K:

$$^{39}K + n \longrightarrow {}^{39}Ar + p$$

The reaction is produced by irradiating a sample with fast neutrons in a nuclear reactor. The amount of ^{39}Ar produced is then a function of the amount of ^{39}K present, the reaction probability or cross-section, the neutron flux, and the irradiation time. Since the $^{40}K/^{39}K$ ratio in the Earth can be assumed to be constant (at any given time), the amount of ^{40}K can be calculated from ^{39}Ar produced. In practice, the analysis is performed by simultaneously irradiating and analyzing a standard of known age because the reaction cross-section is a function of neutron energy and there typically is a spectrum of energies present. The flux, capture cross-section, and decay constant terms will be the same for the standard as for the unknown sample, so that the flux and capture cross-section terms can be combined into a single constant, c. The relationship between age, t, and the $^{40}Ar/^{39}Ar$ ratio is then:

$$\frac{^{40}Ar^*}{^{39}Ar} = \frac{\lambda_e}{\lambda} \frac{^{40}K(e^{\lambda t} - 1)}{c^{39}K} \tag{8.63}$$

where, as usual, the asterisk denoted the radiogenic component. The constant c is conventionally combined with the other constants in this equation, the branching ratio and the $^{40}K/^{39}K$ ratio, into a constant, J, so that the age is given by:

$$t = \frac{1}{\lambda} \ln \left[1 + J \frac{^{40}Ar^*}{^{39}Ar} \right] \tag{8.64}$$

The advantage of this technique is that both K and Ar can be determined simultaneously on the same sample aliquot, whereas in conventional K-Ar dating, the two are determined by different methods on different aliquots. Perhaps more importantly, a series of measurements can be made on the same sample aliquot by heating the sample in steps and analyzing the $^{40}Ar/^{39}Ar$ ratio released in each heating step. The process terminates in fusion and complete release of all Ar.

Computing the age still requires subtraction of the nonradiogenic ^{40}Ar to determine $^{40}Ar^*$. A first-order approach is to assume it is atmospheric, so that the amount can be determined from the $^{40}Ar/^{36}Ar$ ratio. However, the nonradiogenic Ar can be inherited from the magma, xenocrysts, or the metamorphic protolith. Because step-heating provides a number of individual measurements on the sample, the initial ratio can be determined using a variation of the isochron approach described in Example 8.2. ^{36}Ar is a minor isotope and subject to greater analytical uncertainty, so that rather than plotting $^{40}Ar/^{36}Ar$ vs. $^{39}Ar/^{36}Ar$, $^{36}Ar/^{40}Ar$ is plotted against $^{39}Ar/^{40}Ar$. In this case, the intercept with $^{39}Ar/^{40}Ar = 0$ gives the initial $^{36}Ar/^{40}Ar$ and the intercept with $^{36}Ar/^{40}Ar = 0$ gives the age.

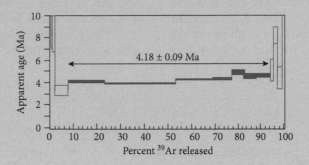

Figure 8.31 A $^{40}Ar/^{39}Ar$ step-heating release spectrum of a basalt from Grenada. Each step is plotted as a bar whose thickness is proportional to uncertainty. The middle temperature steps form a "plateau" (filled bars) that comprises 86% of the total Ar released. The first few and the last few steps (unfilled bars) are not included in the age computation. Analysis by P. Copeland. From White et al. (2017).

Another common practice is to plot the apparent age (eqn. 8.64) vs. cumulative Ar released. Doing so can reveal issues such as Ar loss or inherited Ar. In the example shown in Figure 8.31, Ar in the first low temperature release steps has variable apparent ages, likely reflecting a combination of adsorption and diffusion from crystal rims. The last few high temperature steps have higher apparent ages, perhaps reflecting inherited Ar. But more than 86% of the Ar released has the same apparent age, forming a *plateau*. The lowest and highest temperature steps are thus rejected and the "plateau age" is taken to be the actual eruption age.

From a geochemical viewpoint, the K-Ar system is of most interest with respect to the degassing of the Earth and evolution of the Earth's atmosphere. The most important questions in this regard are: what proportion of the Earth has been degassed, how extreme has this outgassing been, and when did this degassing occur? To briefly illustrate how K-Ar systematics can address these problems, consider the second question. We can think of two extreme possibilities: outgassing occurred yesterday; outgassing occurred simultaneously with the formation of the Earth. If outgassing occurred yesterday, then the $^{40}Ar/^{36}Ar$ ratio of the Earth's interior will be the same as that of the atmosphere. If outgassing occurred early in Earth's history, then the $^{40}Ar/^{36}Ar$ ratio of the atmosphere should be close to the initial ratio for the solar system and the $^{40}Ar/^{36}Ar$ of the Earth's interior should be much higher because of the decay of ^{40}K over the history of the Earth. In actuality, neither of these models matches observation. The $^{40}Ar/^{36}Ar$ ratio measured in MORB, in which noble gases are trapped by the pressure prevailing on the ocean floor, is very much higher than in the atmosphere. Ratios in MORB can be as high as 40,000, compared with the atmospheric ratio of 296.16. So the Earth did not outgas yesterday. On the other hand, the atmospheric ratio is much higher than the solar system initial $^{40}Ar/^{36}Ar$ ratio. Consequently, we can conclude that substantial degassing occurred after the Earth formed.

Now let's consider the first question: how much of the Earth has degassed? As we mentioned, essentially all the Ar in the atmosphere is radiogenic. Since both the concentration of Ar and its isotopic composition are uniform in the atmosphere, it is fairly easy to estimate the amount of ^{40}Ar in the atmosphere, which works out to be 1.65×10^{18} moles. The K concentration in the silicate part of the Earth is about 200 ± 50 ppm, which corresponds to $2.05 \pm 0.5 \times 10^{22}$ moles of K. Over the 4.5 Ga of Earth's history, this amount of K should have produced about $2.8 \pm 0.7 \times 10^{18}$ moles of Ar. Thus, about $60 \pm 15\%$ of the Ar that we expect has been produced by radioactive decay in the present atmosphere; the remainder must still be in the solid Earth. As Allègre et al. (1996) showed, most of this must be in the mantle rather than the crust.

Ar isotopes also reveal that the extent of degassing of the Earth's mantle is not uniform. Just as for neon, there is inevitable atmospheric contamination of mantle samples, so it is the maximum $^{40}Ar/^{36}Ar$ ratio in a suite of samples from a given locality that is of interest. While ratios in MORB can be as high as 40,000, ratios in OIB do not exceed about 8,000. $^{40}Ar/^{36}Ar$ ratios also correlate with $^{3}He/^{4}He$ and $^{20}Ne/^{22}Ne$. These correlations suggest $^{40}Ar/^{36}Ar$ ratios of $\leq 10,000$ for OIB sources.

8.5 "EXTINCT" AND COSMOGENIC NUCLIDES

8.5.1 "Extinct" radionuclides and their daughters

There is evidence that many short-lived radionuclides were present in the early solar system. They must have been created shortly before the solar system formed, implying a nucleosynthetic process, such as a red giant star or a supernova event shortly before the formation of the Earth (nucleosynthesis, the processes by which the elements are created, is discussed in Chapter 10). Some of these isotopes (e.g., ^{26}Al) may have been sufficiently abundant that they could have been significant heat sources early in solar system history. Evidence of their existence comes from nonuniform distribution of their daughter products. For example, if ^{26}Mg, the daughter of ^{26}Al, were found to be more abundant in an Al-rich phase such as plagioclase or spinel than in olivine, one might conclude that the excess ^{26}Mg was produced by the decay of ^{26}Al. With a couple of exceptions, evidence of these short-lived radionuclides is restricted to meteorites: they had decayed completely before the Earth formed. These extinct, or "fossil" radionuclides nevertheless provide

very important constraints on the early history of the solar system, and we will consider them in detail in Chapter 10. Here we will briefly discuss four short-lived radionuclides that provide important constraints on the composition and early history of the Earth: ^{129}I, which decays to ^{129}Xe with a half-life of 16 Ma, ^{244}Pu, which decays through fission to several heavy Xe isotopes with a half-life of 82 Ma, ^{146}Sm, which decays to ^{142}Nd with a half-life of 103 Ma, and ^{182}Hf, which decays to ^{182}W with a half-life of 6 Ma.

It has long been known that $^{129}Xe/^{130}Xe$ and $^{136}Xe/^{130}Xe$ ratios are higher in some well gases from volcanic areas in the western US than in the atmosphere (e.g., Butler et al., 1963). The most reasonable explanation of this is that Xe in the mantle contains more ^{129}Xe, produced by beta decay of ^{129}I, and ^{136}Xe, produced by fission of ^{244}Pu and ^{238}U, than does the atmosphere. Let's focus first on ^{129}Xe. The atmosphere, of course, contains no iodine whereas the mantle does. In this sense, we should not be surprised to find the decay product of ^{129}I in greater amounts in the mantle than the atmosphere. A rule of thumb is that a radionuclide will decay below detection levels after 5–10 half-lives. Thus, essentially all ^{129}I must have been decayed within $16 \times 10 = 160$ Ma of the beginning of the solar system. One explanation is that the Earth's atmosphere formed early, before all the ^{129}I had decayed to ^{129}Xe, implying that the Earth had formed and differentiated, and atmosphere had formed by at least that time. This leads to the idea of formation of the atmosphere by "catastrophic degassing" early in Earth's history. Of course, subsequent degassing must also have occurred as well to explain the atmospheric ^{40}Ar. An alternative hypothesis is that, as suggested for Ne, the Earth's atmosphere derived from a late veneer or chondritic or cometary material.

In addition to this, studies beginning with Staudacher and Allègre (1982) have revealed variations in Xe isotopic compositions in oceanic basalts implying Xe isotopic heterogeneity in the mantle. MORB are enriched in both ^{129}I and heavy Xe isotopes such as ^{131}Xe and ^{136}Xe. Once again, atmospheric contamination means that the maximum values measured values are the minimum mantle values; maximum measured $^{129}Xe/^{130}Xe$ in

MORB are > 6.9 while those in OIB are <6.7. Both ^{238}U and ^{244}Pu produce ^{131}Xe through ^{136}Xe isotopes by fission, but in different proportions (e.g., ^{244}Pu produces proportionally more ^{131}Xe). Working with subglacially erupted samples that had retained much of their noble gases, Mukhopadhyay (2012) demonstrated that the Iceland plume source, in addition to lower ^{129}Xe/^{130}Xe, also has a higher proportion of Pu- to U-derived fission Xe, requiring the plume source to be less degassed than the MORB source. Mukhopadhyay concluded that the Earth had accreted material from at least two separate sources and that processes over the last 4.45 billion years of Earth history has not erased this signature of heterogeneous accretion and early differentiation.

Small variations in ^{142}Nd/^{144}Nd were observed in meteorites, and since these correlated with the Sm/Nd ratio of the material analyzed, it was reasonable to assume these variations were produced by α-decay of ^{146}Sm. Because of the relatively long half-life of ^{146}Sm (103 Ma), some geochemists thought that ^{142}Nd/^{144}Nd variations might also be found in the Earth's oldest rocks. However, only tiny amounts of ^{146}Sm were ever present in the solar system (the solar system initial ^{142}Sm/^{144}Sm ratio was only 0.009, and ^{144}Sm is the least abundant stable isotope of Sm!), so very precise measurements were required. In the 1990s, Harvard researchers reported ^{142}Nd/^{144}Nd ratios in early Archean rocks from Isua, Greenland, that were higher than the ratio in laboratory standards (Harper and Jacobsen, 1992), but this work could not initially be reproduced in other laboratories. Eventually, however, other workers confirmed anomalously high ^{142}Nd/^{144}Nd in Isua rocks (e.g., Caro et al., 2003). Subsequent studies have found other instances of high ^{142}Nd/^{144}Nd in rocks as young as 2.7 Ga. O'Neil et al. (2008) reported low ^{142}Nd/^{144}Nd in meta-igneous rocks of the early Archean Nuvvuagittuq greenstone belt of Labrador. The variations are illustrated in Figure 8.32, where the ^{142}Nd/^{144}Nd ratio is expressed in epsilon notation, ε_{142Nd}[*]. This is analogous to our ε_{Nd} notation, but it expresses deviations in parts per 10,000

from *a terrestrial standard* rather than the chondritic value. The clear implication is that the Earth had begun to differentiate into light REE-enriched reservoirs, such as continental crust and light REE-depleted reservoirs such as the mantle with a few hundred million years of solar system formation. The absence of detectable variation of ^{142}Nd/^{144}Nd in post-Archean rocks suggests the original heterogeneities in the mantle have been subsequently homogenized.

Another surprise came in 2005 when it was demonstrated ordinary chondrites have a ^{142}Nd/^{144}Nd ratio that was 20 parts per million higher than in post-Archean terrestrial rocks (Boyet and Carlson, 2005). This suggested that either the Earth, or the observable part of it including the crust and all mantle magma sources, had Sm/Nd ratio a few percent higher than chondritic or that ^{142}Nd and ^{146}Sm were somehow nonuniformly distributed in the early solar system. Boyet and Carlson (2005) suggested an early LREE-enriched, low Sm/Nd reservoir, such as a basaltic crust, formed early but because of density was subsequently sequestered in the deep mantle leaving a LREE-depleted high Sm/Nd reservoir which then gave rise to the continental crust and modern mantle reservoirs. Caro and Bourdon (2010) suggested this early basaltic crust was lost by collisional erosion as the Earth accreted through collisions of small bodies, leaving the Earth slightly LREE depleted relative to chondrites.

It subsequently became clear that the ^{142}Nd/^{144}Nd varied between classes of chondritic meteorites, despite uniform Sm/Nd ratios. Furthermore, other, nonradiogenic isotopes of Nd, such as ^{148}Nd and ^{150}Nd, also vary among chondrites and between chondrites and the Earth (Burkhardt et al., 2016) and these variations could be related to the cosmic environment in which these nuclides were produced (red giant stars vs. supernovae or kilonovae; we'll discuss this in Chapter 10). It thus appears that most of the difference in ^{142}Nd/^{144}Nd can be attributed to nonuniform distribution of these isotopes in the solar nebula. It remains unclear whether all of the difference can be attributed to this, but any difference in Sm/Nd between the

---

[*] We will use this symbol to denote variation of the ^{142}Nd/^{144}Nd and continue to use ε_{Nd} to denote variations of the ^{143}Nd/^{144}Nd ratio.

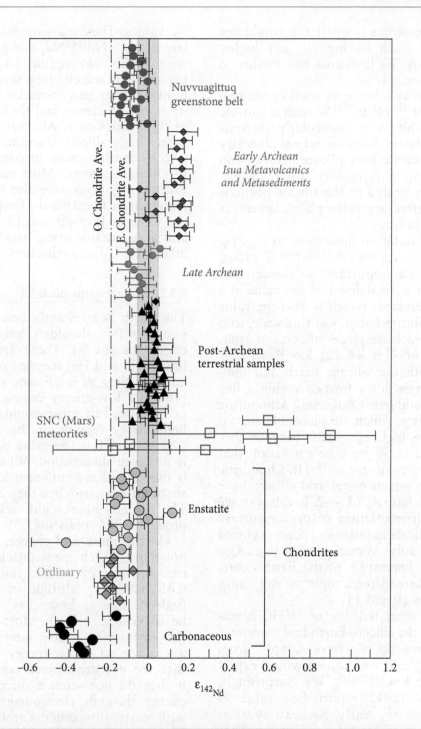

Figure 8.32 Variation of $^{142}Nd/^{144}Nd$ in terrestrial materials and meteorites. All terrestrial rocks younger than about 2.7 billion years have $^{142}Nd/^{144}Nd$ uniform within 5 ppm and that are about 20 ppm (0.2 parts in 10,000) higher than in ordinary chondrites. Solid line show the value for laboratory standards and modern terrestrial rocks, dashed line show the average of enstatite chondrites, and dashed-dot line shows the average of ordinary chondrites. Higher $^{142}Nd/^{144}Nd$ ratios are found in early Archean rocks from Isua, Greenland, and ratios lower than the modern terrestrial value have been found in the Nuvvuagittuq greenstone belt, Labrador. ε_{142Nd} is the deviation in parts per 10,000 of $^{142}Nd/^{144}Nd$ from a terrestrial laboratory standard.

Earth and chondrites is small and would not result in the Earth having ε_{Nd} any higher than +2 to +6. We'll discuss this further in Chapters 10 and 11.

The final decay scheme we want to mention is the decay of ^{182}Hf to ^{182}W with a half-life of 8.9 Ma. This is a particularly interesting decay scheme because when planetary bodies differentiate into silicate mantles and iron cores, Hf, a refractory lithophile element, is concentrated in the silicate portions while W, a refractory siderophile element, is concentrated in cores.

An epsilon notation analogous to ε_{Nd}, ε_W is often used to represent $^{182}W/^{184}W$ ratios, but there is an important difference. The zero value for ε_w is defined as the value in a laboratory standard, which is also the value of the bulk silicate Earth. On this scale, iron meteorites, which are pieces of cores of asteroids, have ε_w of −3 to −4, i.e., low $^{182}W/^{184}W$ compared with the silicate Earth and indicating that these cores formed within a few million years and less. Conversely, achondritic stony meteorites, which are pieces of asteroid mantles, have high ε_w ~+20 because they were depleted in W by core formation that preceded complete decay of ^{182}Hf. Chondritic meteorites, in which metal and silicate have not separated, have ε_w of ~−2. In this way, we know that differentiation of protoplanetary bodies into silicate mantles and cores occurred very early in solar system history (e.g., Qin et al., 2008). Formation of the Earth's core, however, occurred later, a topic we will return to in Chapters 10 and 11.

Given the short half-life of ^{182}Hf, it was assumed that the silicate Earth had a uniform $^{182}W/^{184}W$ (we do not have samples from the Earth's core; if we did, we would expect them to have low $^{182}W/^{184}W$). Surprisingly, Willbold et al. (2012) reported ε_W values of +0.13±0.06 in the early Archean (4.0 Ga) Acasta gneisses of Northwest Canada. They interpreted these as a result of an incomplete mixing of a late accretionary veneer of chondritic material. Subsequently, Touboul et al. (2012) reported ε_W of +0.15 in 2.8 billion-year-old komatiites from Siberia. Since then, numerous other examples of similar tungsten isotope anomalies have been reported in rocks ranging in age from 3.8 to 2.4 Ga, including some examples of negative

ε_W values. These ε_W variations do not correlate with $^{142}Nd/^{144}Nd$, nor do they correlate with highly siderophile abundances, making the late accretionary veneer explanation unlikely. They may instead reflect heterogeneity in the efficiency and timing of separation of the Earth's core. Alternatively, they could reflect early silicate fractionation because W is considerably more incompatible than Hf in silicate systems. Most recently, tungsten isotope variations have also been reported in Phanerozoic flood basalts from Baffin Bay and the Ontong Java Plateau (Rizo et al., 2016) and some oceanic island basalts (Mundl et al., 2017). We'll discuss this further in Chapter 11.

8.5.2 Cosmogenic nuclides

The Earth is constantly bombarded by cosmic rays (we shouldn't feel picked on; the entire cosmos is). These are atomic nuclei (mostly H and He) stripped of their electrons and traveling at relativistic velocities. Some, primarily low-energy cosmic rays, originate in the Sun, but most originate elsewhere in high-energy regions of the cosmos such as supernovae. For the most part, their origin is not well understood. What is understood is that they have sufficient kinetic energy to shatter a nucleus when they collide with one. The nucleus breaks into several parts in a process called *spallation*.

Having mass and charge, cosmic rays do not have much penetrating power. Thus, the intensity of cosmic radiation increases with increasing altitude in the atmosphere (indeed, this is how they were shown to be of cosmic origin). Most are stopped in the atmosphere, their interactions creating a cascade of lower-energy particles; those that are not are slowed considerably. Even if they do not score a direct hit, they lose energy through electromagnetic interaction with matter (ionizing the atoms they pass by). Thus, cosmic rays have their greatest effect in the atmosphere and somewhat less of an effect in the uppermost meter or so of rock.

The nuclear effects of cosmic radiation are on the whole pretty trivial. Nitrogen and oxygen, being the principal constituents of the atmosphere, are the most common targets, yet there is no change in the isotopic abundances of these elements. Cosmogenic production is only significant for those nuclides whose

Table 8.5 Some radioactive cosmogenic nuclides of geologic interest.

Nuclide	Half-life	λ (yr^{-1})
^{7}Be	53 days	4.777
^{10}Be	1.39×10^{6} yr	4.99×10^{-5}
^{14}C	5730 yr	1.209×10^{-4}
^{26}Al	7.02×10^{5} yr	9.87×10^{-7}
^{36}Cl	300×10^{5} yr	2.31×10^{-6}
^{41}Ca	102×10^{5} yr	6.80×10^{-6}

half-lives are so short they would not exist otherwise or for stable but exceedingly rare nuclides, such as ^{3}He and ^{21}Ne in rocks on the Earth's surface, as we have already mentioned. The radionuclides of greatest interest are listed in Table 8.5, along with their half-lives and their decay constants. These nuclides are created either directly through spallation (e.g., ^{10}Be), or by nuclear reactions with particles produced by spallation (e.g., ^{14}C is produced by secondary neutrons: ^{14}N (n,p)$\rightarrow$^{14}C).

When cosmogenic dating was introduced by Libby et al. (1949), it was assumed that the cosmic ray flux was constant. This has proved to be not the case. Variations in the Earth's geomagnetic field and the solar wind, both of which tend to deflect galactic cosmic rays away from the Earth, result in variable production rates through time. The Earth's geomagnetic field also causes latitudinal variations in the cosmic ray flux. And, of course, the flux decreases with depth in the atmosphere so that surfaces at high elevations are exposed to a greater flux than those at sea level. For example, production rates at sea level are some 15–20 times lower than at 4000 m altitude at the same latitude, and production rates are about 2.5 times lower at the equator than at latitudes >60°. These variations in space have little effect on ^{14}C because carbon has a long enough residence time in the atmosphere as CO_2 that the distribution is more or less uniform (slow mixing across the equator does result in slight variations). It is also not significant for ^{10}Be in marine deposits because of mixing in the oceans.

Below we briefly review a few of the applications of cosmogenic nuclides.

8.5.2.1 ^{14}C geochronology

The best known of the cosmogenic nuclide dating methods is of course ^{14}C, a technique first introduced by Willard F. Libby in 1949 (and for which he won the Nobel Prize in Chemistry in 1960, one of only two such prizes awarded for geochemistry). It is useful in dating archeological, climatological, volcanological, seismological, and paleontological "events." It is also useful in oceanography in determining the age of deep water (see Chapter 14) and circulation time scales. Present technology utilizing accelerator mass spectrometry rather than traditional β-counting can provide useful information on samples as old as ~40,000 years. The underlying principle of this method is quite simple. One assumes a constant production of ^{14}C in the atmosphere. The atmosphere is well mixed and has a uniform ^{14}C/^{12}C ratio. ^{14}C isolated from the atmosphere in organic matter or in carbonate will decrease with time according to:

$$^{14}C = \,^{14}C^{o}e^{-\lambda t} \qquad (8.65)$$

If the production rate is constant, the "naught" value is the present-day atmospheric value.

^{14}C was originally measured by beta counting, and often still is, and reported as the *specific activity*, either in disintegrations per minute (dpm) or becquerels per gram carbon. The reference "modern" atmospheric value, $^{14}C^{o}$ in eqn. 8.56, is 13.56 dpm/gC. Today, accelerator mass spectrometry is the method of choice as it can be performed with greater accuracy on considerably smaller samples but is considerably more expensive. Accelerator mass spectrometry measures the ^{14}C/^{12}C ratio, for which the reference modern atmospheric value is $1.176 \pm 0.010 \times 10^{-12}$. Although the ^{14}C half-life is now known to be 5730 ± 40 yrs, "conventional" radiocarbon ages are calculated using a half-life of 5568 yrs ($\lambda = 1.245 \times 10^{-4}$ yr^{-1}) for consistency with older data. In its simplest form, an age can be calculated directly from these values and eqn. 8.65. Accurate ages, however, must take into consideration a number of other factors, making the calculation more complex.

The first of these is mass dependent fractionation, which we'll discuss in detail in the following chapter. Because the $^{14}C/^{12}C$ ratio can vary due to the mass difference in these isotopes, a correction must be applied. This is done by measuring the $^{13}C/^{12}C$ ratio and correcting to a standard value; to a first approximation, the variation in $^{14}C/^{12}C$ ratio is twice that of $^{13}C/^{12}C$ ratio. This effect can amount to several percent. The second is correcting for the variable $^{14}C/^{12}C$ ratio in the atmosphere through time. This has several causes. First is the variation in the cosmic ray flux and hence production rate of ^{14}C mentioned above. The second factor is what are called reservoir effects, changes in the amount of noncosmogenic atmospheric CO_2, which dilutes the ^{14}C. This can happen when "old" dissolved CO_2 in the oceans shifts into the atmosphere (the oceans contain $\sim$50 times as much CO_2 as the atmosphere). An example is the end of the last ice age when atmospheric CO_2 concentration increased from 180 ppm to 280 ppm as a result of this shift. In addition, in the last $\sim$150 years, burning of fossil fuels has further increased atmospheric CO_2 to >400 ppm, and this carbon is ^{14}C-free, resulting in further dilution of atmospheric ^{14}C. Finally, atmospheric nuclear weapons tests in the 1950s produced a large pulse of ^{14}C, much of which ended up in the atmosphere. This bomb-produced ^{14}C is now slowly being removed from the atmosphere by the oceans and biosphere.

To get around these problems, ^{14}C ages have been calibrated by comparison with absolute chronometers. This was originally done with tree rings and been extended back about 12,000 years. Subsequently, the calibration has been extended back to 50,000 yrs BP by comparing ^{14}C ages with $^{230}Th/^{238}U$ ages of corals as described in section 8.4.7. As seawater has slightly older carbon than the atmosphere at any given time, this calibration is relevant to marine samples. A separate calibration has been done using cave carbonate (speleothems) and there are slightly different curves for the northern and southern hemisphere (Reimer et al., 2013). By using the appropriate calibration curve and the correct decay constant, conventional ^{14}C ages can be converted to calendar years. An online-calculator is available for this purpose at http://calib.org/calib/ (Stuiver et al., 2018).

8.5.2.2 ^{36}Cl in hydrology

^{36}Cl has been applied to hydrological problems for some time. The general principle is that ^{36}Cl is produced at a constant rate in the atmosphere, carried to the surface of the Earth in rain and incorporated into hydrological systems. Cl is an excellent candidate element for such studies because it is generally "conservative" in groundwater, i.e., neither lost nor gained from solution once it leaves the atmosphere. Imagine a simple system in which rainwater is incorporated into an aquifer at some unknown time in the past. In this case, if we can specify the number of ^{36}Cl atoms per liter in rain, $^{36}Cl^0$ and assume this value is time-invariant, then we can determine the age of water in the aquifer by measuring the number of ^{36}Cl atoms per liter simply from:

$$^{36}Cl = {}^{36}Cl^0\, e^{-\lambda t} \qquad (8.66)$$

Dealing with just the number, or concentration, of ^{36}Cl atoms can have disadvantages, and can be misleading. Evaporation, for example, would increase the number of ^{36}Cl atoms. Thus, the $^{36}Cl/Cl$ ratio (Cl has two stable isotopes: ^{35}Cl and ^{37}Cl) is often used.

As for ^{14}C, there are additional factors that must be considered. First, the production rate varies with time. Second, nuclear tests also produced ^{36}Cl (this bomb-produced ^{36}Cl was quickly removed from the atmosphere by rain and the resulting pulse of ^{36}Cl can also be used as a tracer). Third, unlike carbon, the residence time of chlorine in the atmosphere is short, so variations with latitude and elevation are important. Fourth, the $^{36}Cl/Cl$ ratio in rain varies with distance from the ocean (because the ocean is the source of atmospheric Cl). Fifth, some ^{36}Cl can be

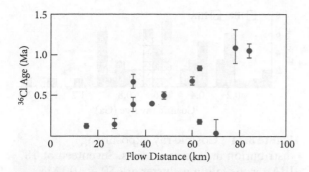

Figure 8.33 ^{36}Cl/Cl age of groundwater in the Milk River aquifer in Alberta, Canada as a function of flow distance. From Phillips (2000).

produced when neutrons produced by ^{238}U fission are captured by ^{35}Cl. Finally, while Cl is approximately conservative in solution, it is not absolutely so, and ^{36}Cl-free chlorine can be added by leaching from rocks or mixing with formation brines.

Determining the age of water in underground aquifers is an important problem because of the increasing demands placed in many parts of the world on limited water resources. A prudent policy for water resource management is to withdraw from a reservoir at a rate no greater than the recharge rate. Determination of recharge rate is thus prerequisite to wise management. Figure 8.33 shows an example of groundwater age (time since leaving the atmosphere) as a function of distance from the recharge zone for the Milk River aquifer in Alberta, Canada.

8.5.2.3 Be isotopes

^{7}Be and ^{10}Be are produced by spallation reactions on O and N but are quickly washed out of the atmosphere by rain and find their way into streams, soils, groundwater and eventually the ocean and marine sediments. Because of its short half-life, ^{7}Be has only a few limited applications such as monitoring short-term sediment transport on and in drainage networks and as a tracer of air movement in the atmosphere and water movement in the oceans. In addition to determining exposure ages, discussed below, ^{10}Be has several applications, including studying soil development, erosion, and sediment transport and dating ice cores (Bierman and Portenga, 2018). In the marine realm, it has been used to estimate

terrestrial terrigenous sediment fluxes to the oceans as well as dating sediments, including manganese nodules.

One unique application of ^{10}Be has been to trace sediment subduction into the mantle (Tera et al., 1986). Since ^{10}Be does not exist in the Earth's interior, its presence there could result only from subduction of sediment (which concentrates cosmogenic ^{10}Be). ^{10}Be has been identified in some island arc volcanics, but not in other volcanic rocks (at least not above the background level of 10^6 atoms per gram, i.e., 1 atom in 10^{18}). This is strong evidence that subducted sediment plays a role in island arc magma genesis, something suspected on the basis of other geochemical evidence. We will discuss this in more detail in Chapter 11.

8.5.2.4 Surface exposure ages

Rocks and soil at the Earth's surface are also exposed to cosmic rays that induce spallation reactions creating new nuclides both stable and unstable. In the simplest case of a rare stable nuclide such as ^{3}He or ^{21}Ne, the number of atoms produced over exposure time t is simply:

$$N = Pt \qquad (8.67)$$

where P is the production rate (dN/dt). If we are a bit more sophisticated and take account of the variable production rate, P becomes an integral over time t. We must also take account of any of these nuclides inherently present in the rock. The production also decreases exponential with depth such that the production rate at some depth, z, is:

$$P = P_0\, e^{-lz} \qquad (8.68)$$

where l is the penetration scale length that depends on composition and density but is typically $\sim$60 cm^{-1} for most rocks, so that at a depth of 60 cm, the production rate is $\sim$37% of the surface rate.

For a radionuclide, the number of atoms present is a balance of production and decay rates:

$$\frac{dN}{dt} = P - \lambda N \qquad (8.69)$$

Assuming the rock initially contained none of these atoms, the number of atoms present after an exposure time of t is:

$$N = \frac{P}{\lambda}(1 - e^{-\lambda t}) \qquad (8.70)$$

^{10}Be, ^{36}Cl, and ^{26}Al have all been used in exposure dating. Exposure ages have been used to date lava flows, impact craters, landslides, and fault surfaces. One of the more important applications has been in glacial chronology, such as dating moraines and glacial erratics left by receding ice sheets (Balco, 2011). Using nuclides in combination, such as ^{10}Be/^{26}Al, is particularly powerful.

Erosion rates can also be determined using cosmogenic nuclides. Assuming that production and decay have reached steady-state, the number of atoms of a radionuclide now at the surface will be:

$$N = \frac{P_{surf}}{\lambda + \varepsilon l} \qquad (8.71)$$

where ε is the erosion rate.

8.5.3 Cosmic ray exposure ages of meteorites

The surfaces of meteorites in space are subject to a fairly high cosmic ray flux because there is no atmosphere or geomagnetic field to protect them. Here, rare stable isotopes are used rather than radioactive ones because of the long times involved. For example, potassium is not present naturally in iron meteorites, but is produced by cosmic ray interactions. Knowing the production rate of ^{41}K and its abundance, it is possible to calculate how long a meteorite has been exposed to cosmic rays. Two important results of such studies are worth mentioning: (1) exposure ages are much younger than formation ages; and (2) meteorites that are compositionally and petrologically similar tend to have similar exposure ages (Figure 8.34). This means meteorites now colliding with the Earth have not existed as small bodies since the solar system formed. Instead, they are probably more or less continually (on a time scale of 10^9 yr) produced by the breakup of larger bodies through collisions. Also, similar meteorites probably come from the same parent body.

8.6 SUMMARY

Radioactive decay of unstable nuclei leads to variation in the isotopic composition of

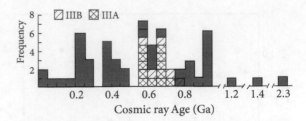

Figure 8.34 Cosmic ray exposure age distribution in iron meteorites. Seventeen of 18 IIIAB irons fall in a cluster at 650 ± 100 Ma (after Voshage, 1967).

a number of elements. This provides both a method of determining geological ages and natural tracers to help us understand fundamental geological processes such as crustal evolution, mantle structure, ore formation, and ocean circulation that we will discuss in subsequent chapters.

- Unstable nuclei decay to stable ones by emitting beta or alpha particles or capturing an electron. This decay occurs according to a first-order rate equation:

$$\boxed{\frac{dN}{dt} = -\lambda N} \qquad (8.4)$$

where λ is the decay constant. Unlike the rate constant in chemical reactions, λ is independent of all external influences, making radioactive decay a perfect natural clock. From this equation we can derive one that describes the evolution of a radiogenic isotope ratio such as ^{87}Sr/^{86}Sr through time:

$$R = R_0 + R_{P/D}(e^{\lambda t} - 1) \qquad (8.18)$$

where R_0 is the initial ratio at time t and $R_{P/D}$ is the parent/daughter ratio (e.g., ^{87}Rb/^{86}Sr in the Rb-Sr system). This equation also forms the basis of *isochron dating*: it has the form $y = a + bx$ such that if a number of cogenetic samples are analyzed for R and $R_{P/D}$, they define an *isochron* of slope $e^{\lambda t} - 1$ and an intercept of R_0 on a plot of R versus $R_{P/D}$. Deriving an age in this way requires that all samples had an identical value of R_0 at time t and the system has remained closed since t.

- We reviewed the properties and applications of the principal radiogenic systems: Rb-Sr, Nd-Sm, Lu-Hf, La-Ce, Re-Os, and U-Th-Pb. Each has unique advantages and disadvantages in geochronology and geochemistry. The decay at different ratios of two isotopes of U to two isotopes of Pb provides a particularly valuable geochronological tool. Because zircons incorporate U and exclude Pb and are robust with respect to mechanical and chemical weathering, U–Pb zircon dating is often considered the *gold standard* of geochronology.

- Because the rare earth elements and Hf are refractory lithophile elements, their relative abundances in the Earth should be identical to those of chondrites; this leads to the epsilon representation of Nd, Hf, and Ce isotope ratios, which are deviations in parts in 10,000 from the chondritic ratio:

$$\varepsilon = \frac{R_{sample} - R_{chon}}{R_{chon}} \times 10^4 \qquad (8.29)$$

- Uranium and thorium decay to isotopes of Pb by alpha and beta decays through a series of progressively lighter nuclei. These intermediate daughters provide an additional geochronological tool because once disturbed, the decay rates of a parent

and daughter in a given chemical system returns to the equilibrium condition:

$$\boxed{\frac{dN_D}{dt} = \frac{dN_P}{dt}} \qquad (8.44)$$

at a predictable rate. ^{238}U-^{230}Th dating has proved particularly valuable in determining post-glacial sea-level rise through dating of coral reefs.

- We then reviewed the variation of isotope ratios of the noble gases He, Ne, Ar, and Xe produced by nuclear processes. These have proven valuable not only in geochronology, for example ^{40}Ar–^{39}Ar dating, but also in understanding the degassing history of the Earth.

- Radioactive nuclides can be produced by spallation induced by cosmic rays. These cosmogenic isotopes, most notably ^{14}C but also others such as ^{10}Be, ^{21}Ne, and ^{36}Cl, provide important geochronological tools on geologically short time scales.

- Finally, we turned to extinct radionuclides: those present in the early solar system but with half-lives too short to survive in the modern Earth. Some, however, such as ^{129}I and ^{146}Sm, were still extant in the early Earth and produced measurable variations in their daughter products such as ^{129}Xe and ^{142}Nd that provide important constraints on the Earth's early evolution.

REFERENCES AND SUGGESTIONS FOR FURTHER READING

Albarède, F., Simonetti, A., Vervoort, J. D., Blichert-Toft, J. and Abouchami, W. 1998. A Hf-Nd isotopic correlation in ferromanganese nodules. *Geophysical Research Letters* 25: 3895–8. doi:10.1029/1998gl900008.

Allègre, C.J. and Luck, J.M. 1980. Osmium isotopes as petrogenic and geologic tracers. *Earth and Planetary Science Letters* 48: 148–54.

Allègre, C.J., Hofmann, A.W. and O'Nions, K. 1996. The argon constraints on mantle structure. *Geophysical Research Letters* 23: 3555–7.

Balco, G. 2011. Contributions and unrealized potential contributions of cosmogenic-nuclide exposure dating to glacier chronology, 1990–2010. *Quaternary Science Reviews*, 30: 3–27. doi: 10.1016/j.quascirev.2010.11.003.

Ballentine, C. J., Burgess, R. and Marty, B., 2002. Tracing fluid origin, transport and interaction in the crust. *Reviews in Mineralogy and Geochemistry* 47(1), 539–614. doi: 10.2138/rmg.2002.47.13.

Bellot, N., Boyet, M., Doucelance, R., Pin, C., Chauvel, C. and Auclair, D. 2015. Ce isotope systematics of island arc lavas from the Lesser Antilles. *Geochimica et Cosmochimica Acta* 168: 261–79. doi: 10.1016/j.gca.2015.07.002.

Bellot, N., Boyet, M., Doucelance, R., et al. 2018. Origin of negative cerium anomalies in subduction-related volcanic samples: Constraints from Ce and Nd isotopes. *Chemical Geology* 500: 46–63. doi: 10.1016/j.chemgeo.2018.09.006.

Bierman, P. R. and Portenga, E. W. 2018. Beryllium isotopes. In: White, W.M. (Ed.) *Encyclopedia of Geochemistry*. Springer International Publishing, Cham, pp 95–99. 978-3-319-39312-4.

Black, S., MacDonald, R. and Kelly, M.R. 1997. Crustal origin for peralkaline rhyolites from Kenya: evidence from U-series disequilibria and Th-isotopes. *Journal of Petrology* 38: 277–97. doi: 10.1093/petroj/38.2.277.

Bouvier, A., Vervoort, J.D. and Patchett, P.J. 2008. The Lu-Hf and Sm-Nd isotopic composition of CHUR; constraints from unequilibrated chondrites and implications for the bulk composition of terrestrial planets. *Earth and Planetary Science Letters* 273: 48–57.

Boyet, M. and Carlson, R.W. 2005. ^{142}Nd evidence for early (>4.53 Ga) global differentiation of the silicate Earth. *Science* 309: 576–81.

Burkhardt, C., Borg, L.E., Brennecka, G.A., Shollenberger, Q.R., Dauphas, N. and Kleine, T. 2016. A nucleosynthetic origin for the Earth's anomalous ^{142}Nd composition. *Nature* 537: 394–8. doi:10.1038/nature18956.

Butler, W.A., Jeffery, P.M., Reynolds, J.H. and Wasserburg, G.J. 1963. Isotopic variations in terrestrial xenon. *Journal of Geophysical Research* 68: 3283–91.

Caro, G. and Bourdon, B. 2010. Nonchondritic Sm/Nd ratio in the terrestrial planets: Consequences for the geochemical evolution of the mantle–crust system. *Geochimica et Cosmochimica Acta* 74: 3333–49.

Caro, G., Bourdon, B., Birck, J.-L. and Moorbath, S. 2003. ^{146}Sm-^{142}Nd evidence from Isua metamorphosed sediments for early differentiation of the Earth's mantle. *Nature* 423: 428–32.

Clark, W.B., Beg, M.A. and Craig, H. 1969. Excess ^{3}He in the sea: evidence for terrestrial primordial helium. *Earth and Planetary Science Letters* 6: 213–30.

Creaser, R.A., Papanastassiou, D.A. and Wasserburg, G.J. 1991. Negative thermal ion mass spectrometry of osmium, rhenium, and iridium. *Geochimica et Cosmochimica Acta* 55: 397–401.

Day, J. M. D. 2013. Hotspot volcanism and highly siderophile elements. *Chemical Geology*, 341(0), 50–74. doi: 10.1016/j.chemgeo.2012.12.010.

DePaolo, D.J. 1981. Neodymium isotopes in the Colorado Front Range and crust mantle evolution in the Proterozoic. *Nature* 291: 193–6.

DePaolo, D.J. and Wasserburg, G.J. 1976. Nd isotopic variations and petrogenetic models. *Geophysical Research Letters* 3: 249–52.

Doucelance, R., Bellot, N., Boyet, M., Hammouda, T. and Bosq, C. 2014. What coupled cerium and neodymium isotopes tell us about the deep source of oceanic carbonatites. *Earth and Planetary Science Letters* 407: 175–86. doi: 10.1016/j.epsl.2014.09.042

Dickin, A.P. 1987. Cerium isotope geochemistry of ocean island basalts. *Nature*, 326: 283–4.

Edwards, R. L., Chen, J. H., Ku, T.-L. and Wasserburg, G. J., 1987. Precise timing of the last interglacial period from mass spectrometric determination of thorium-230 in corals. *Science* 236(4808): 1547–53. doi: 10.1126/science.236.4808.1547.

Edwards, R. L., Gallup, C. D. and Cheng, H., 2003. Uranium-series dating of marine and lacustrine carbonates. *Reviews in Mineralogy and Geochemistry* 52(1), 363–405. doi: 10.2113/0520363.

Esser, B.K. and Turekian, K.K. 1993. The osmium isotopic composition of the continental crust. *Geochimica et Cosmochimica Acta* 57: 3093–104.

Faure, G. 1986. *Principles of Isotope Geology*, 2nd ed. New York, John Wiley & Sons, Ltd.

Gannoun, A., Burton, K. W., Day, J. M. D., Harvey, J., Schiano, P. and Parkinson, I. 2016. Highly siderophile element and Os isotope systematics of volcanic rocks at divergent and convergent plate boundaries and in intraplate settings. *Reviews in Mineralogy and Geochemistry* 81: 651–724. doi:10.2138/rmg.2016.81.11.

Gast, P.W. 1960. Limitations on the composition of the upper mantle. *Journal of Geophysical Research* 65: 1287–97.

Harper, C.L. and Jacobsen, S.B. 1992. Evidence from coupled ^{147}Sm-^{143}Nd and ^{146}Sm-^{142}Nd systematics for very early (4.5-Gyr) differentiation of the Earth's mantle. *Nature* 360: 728–32.

Hawkesworth, C.J., Blake, S., Evans, P., et al. 2000. Time scales of crystal fractionation in magma chambers – integrating physical, isotopic and geochemical perspectives. *Journal of Petrology* 41: 991–1006. doi: 10.1093/petrology/41.7.991.

Heiss, J., Condon, D.J., McLean, N. and Noble, S.R. 2012. ^{238}U/^{235}U systematics in terrestrial uranium-bearing minerals. *Science* 335: 1610–14.

Hirt, B., Herr, W. and Hoffmester, W. 1963. Age determinations by the rhenium-osmium method, in *Radioactive Dating*, pp. 35–44. Vienna, International Atomic Energy Agency.

Hoffmann, D. L., Standish, C. D., García-Diez, M., Pettitt, P. B., Milton, J. A., Zilhão, J., Alcolea-González, J. J., Cantalejo-Duarte, P., Collado, H., de Balbín, R., Lorblanchet, M., Ramos-Muñoz, J., Weniger, G.-C. and Pike, A. W. G., 2018. U-Th dating of carbonate crusts reveals Neandertal origin of Iberian cave art. *Science* 359(6378): 912–15. doi: 10.1126/science.aap7778.

Honda, M., Patterson, D., Doulgeris, A. and Clague, D. 1991. Possible solar noble-gas component in Hawaiian basalts. *Nature* 349: 149–51.

Huh, C.-A. and Ku, T.-L. 1984. Radiochemical observations on manganese nodules from three sedimentary environments in the north Pacific. *Geochimica et Cosmochimica Acta* 48: 951–64.

Ionov, D. A., Carlson, R. W., Doucet, L. S., Golovin, A. V. and Oleinikov, O. B. 2015a. The age and history of the lithospheric mantle of the Siberian craton: Re–Os and PGE study of peridotite xenoliths from the Obnazhennaya kimberlite. *Earth and Planetary Science Letters* 428: 108–19. doi: 10.1016/j.epsl.2015.07.007.

Ionov, D. A., Doucet, L. S., Carlson, R. W., Golovin, A. V. and Korsakov, A. V. 2015b. Post-Archean formation of the lithospheric mantle in the central Siberian craton: Re–Os and PGE study of peridotite xenoliths from the Udachnaya kimberlite. *Geochimica et Cosmochimica Acta* 165: 466—83. doi: 10.1016/j.gca.2015.06.035

Jicha, B. R., Singer, B. S., Beard, B. L. and Johnson, C. M., 2005. Contrasting timescales of crystallization and magma storage beneath the Aleutian Island arc. *Earth and Planetary Science Letters*, 236(1): 195–210. doi: 10.1016/j.epsl.2005.05.002.

Kurz, M.D. 1987. In situ production of terrestrial cosmogenic helium and some applications to geochronology. *Geochimica et Cosmochimica Acta* 50: 2855–62.

Kurz, M.D., Jenkins, W.J. and Hart, S.R. 1982. Helium isotopic systematics of oceanic islands and mantle heterogeneity. *Nature* 297: 43–7.

Kurz, M. D., Colodner, D., Trull, T. W., Moore, R. B. and O'Brien, K. 1990. Cosmic ray exposure dating with in situ produced cosmogenic He: results from young Hawaiian lava flows. *Earth and Planetary Science Letters* 97, 177–89. doi: 10.1016/0012-821X(90)90107-9.

Libby, W. F., Anderson, E. C. and Arnold, J. R. 1949. Age determination by radiocarbon content: world-wide assay of natural radiocarbon. *Science* 109: 227–8. doi: 10.1126/science.109.2827.227.

Liu, J., Rudnick, R. L., Walker, R. J., et al. 2011. Mapping lithospheric boundaries using Os isotopes of mantle xenoliths: An example from the North China Craton. *Geochimica et Cosmochimica Acta* 75: 3881–902. doi: 10.1016/j.gca.2011.04.018

Luck, J.-M., Birck, J.-L. and Allegre, C.J. 1980. ^{187}Re-^{187}Os systematics in meteorites: Early chronology of the solar system and age of the Galaxy. *Nature* 283: 256–9.

Lupton, J.E. 1983. Terrestrial inert gases: isotopic tracer studies and clues to primordial components in the mantle. *Annual Review of Earth and Planetary Science*, 11: 371–414.

Lupton, J.G. and Craig, H. 1975. Excess ^{3}He in oceanic basalts: evidence for terrestrial primordial helium. *Earth and Planetary Science Letters* 26: 133–9.

Makishima, A. and Masuda, A. 1994. Ce isotope ratios of N-type MORB. *Chemical Geology* 118: 1–8. doi 10.1016/0009-2541(94)90166-X.

Marty, B. 2012. The origins and concentrations of water, carbon, nitrogen and noble gases on Earth. *Earth and Planetary Science Letters* 313, 56–66. doi: 10.1016/j.epsl.2011.10.040.

Mattinson, J.M. 2010. Analysis of the relative decay constants of ^{235}U and ^{238}U by multi-step CA-TIMS measurements of closed-system natural zircon samples. *Chemical Geology* 275: 186–98.

McArthur, J. M., Howarth, R. J. and Shields, G. A. (2012) Chapter 7 - Strontium isotope stratigraphy. In: Gradstein FM, Ogg JG, Schmitz MD and Ogg GM (eds) *The Geologic Time Scale*, Elsevier, Boston, pp 127–44. 978-0-444-59425-9.

Merrihue, C. and Turner, G. 1966. Potassium-argon dating by activation with fast neutrons. *Journal of Geophysical Research* 71: 2852–57 doi:10.1029/JZ071i011p02852.

Moreira, M. 2013. Noble gas constraints on the origin and evolution of Earth's volatiles. *Geochemical Perspectives* 2(2): 229–30. doi: 10.7185/geochempersp.2.2.

Mukhopadhyay, S. 2012. Early differentiation and volatile accretion recorded in deep-mantle neon and xenon. *Nature* 486: 101–4. doi: 10.1038/nature11141.

Mundl, A., Touboul, M., Jackson, M. G., et al. 2017. Tungsten-182 heterogeneity in modern ocean island basalts. *Science* 356: 66–9. doi:10.1126/science.aal4179.

Nyquist, L.E., Bogard, D.D., Wiesmann, H., et al. 1990. Age of a eucrite clast from the Bholghati howardite. *Geochimica et Cosmochimica Acta* 54: 2195–206.

O'Neil, J., Carlson, R.L., Francis, D. and Stevenson, R.K. 2008. Neodymium-142 evidence for Hadean mafic crust. *Science* 321: 1828–31.

O'Neill, H.S.C. and Palme, H. 2008. Collisional erosion and the nonchondritic composition of the terrestrial planets. *Philosophical Transactions of the Royal Society A* 366: 4205–38. doi: 10.1098/rsta.2008.0111, 2008.

Ozima, M. and Zahnle, K. 1993. Mantle degassing and atmospheric evolution: noble gas view. *Geochemical Journal* 27: 185–200.

Patchett, P.J. 1983. Importance of the Lu-Hf isotopic system in studies of planetary chronology and chemical evolution. *Geochimica et Cosmochimica Acta* 47: 81–91.

Patchett, P.J., White, W.M., Feldmann, H., Kielinczuk, S. and Hofmann, A.W. 1984. Hafnium/rare earth element fractionation in the sedimentary system and crustal recycling into the Earth's mantle. *Earth and Planetary Science Letters* 69: 365–78.

Pearson, D. G., Carlson, R. W., Shirey, S. B., Boyd, F. R. and Nixon, P. H. 1995a. Stabilisation of Archaean lithospheric mantle: A ReOs isotope study of peridotite xenoliths from the Kaapvaal craton. *Earth and Planetary Science Letters* 134: 341–57. doi: 10.1016/0012-821X(95)00125-V.

Pearson, D. G., Irvine, G. J., Ionov, D. A., Boyd, F. R. and Dreibus, G. E. 2004. Re-Os isotope systematics and platinum group element fractionation during mantle melt extraction: a study of massif and xenolith peridotite suites. *Chemical Geology* 208: 29–59.

Pearson, D. G., Shirey, S. B., Carlson, R. W., Boyd, F. R., Pokhilenko, N. P., Shimizu, N. 1995b. Re-Os, Sm-Nd, and Rb-Sr isotope evidence for thick Archaean lithospheric mantle beneath the Siberian craton modified by multistage metasomatism. *Geochimica et Cosmochimica Acta* 59: 959–77.

Peucker-Ehrenbrink, B. and Jahn, B.-M. 2001. Rhenium-osmium isotope systematics and platinum group element concentrations: Loess and the upper continental crust. *Geochemistry, Geophysics, Geosystems* 2: 1061 doi:10.1029/2001gc000172.

Peucker-Ehrenbrink, B. and Ravizza, G. (2012) Chapter 8 – Osmium isotope stratigraphy. In: Gradstein FM, Ogg JG, Schmitz MD and Ogg GM (eds) *The Geologic Time Scale*. Elsevier, Boston, pp 145–66. 978-0-444-59425-9.

Poreda, R.J. and Farley, K.A. 1992. Rare gases in Samoan xenoliths. *Earth and Planetary Science Letters* 113: 129–44.

Reiners, P. W., Carlson, R. W., Renne, P. R., Cooper, K. M., Granger, D. E., McLean, N. M. and Schoene, B., 2017. *Geochronology and Thermochronology*. Wiley.

Renne, P.R., Mundil, R., Balco, G., Min, K. and Ludwig, K.R. 2010. Joint determination of ^{40}K decay constants and ^{40}Ar*/^{40}K for the Fish Canyon sanidine standard, and improved accuracy for ^{40}Ar/^{39}Ar geochronology. *Geochimica et Cosmochimica Acta* 74: 5349–67. doi: 10.1016/j.gca.2010.06.017.

Reimer, P., Bard, E., Bayliss, A., Beck, J., Blackwell, P., Ramsey, C., . . . Van der Plicht, J. (2013). IntCal13 and marine13 radiocarbon age calibration curves 0–50,000 Years cal BP. *Radiocarbon* 55(4): 1869–87. doi:10.2458/azu_js_rc.55.16947.

Rizo, H., Walker, R. J., Carlson, R. W., et al. 2016. Preservation of Earth-forming events in the tungsten isotopic composition of modern flood basalts. *Science* 352: 809–12. doi:10.1126/science.aad8563.

Sano, Y., 2018. Helium isotopes. In: *Encyclopedia of Geochemistry* (ed White, W. M.), pp. 659–63. Springer International Publishing, Cham.

Sarda, P., Staudacher, T. and Allègre, C.J. 1988. Neon isotopes in submarine basalts. *Earth and Planetary Science Letters* 91: 73–88.

Stuiver, M., Reimer, P.J., and Reimer, R.W. 2018, CALIB 7.1 [WWW program] at http://calib.org, accessed 2018-11-4.

Staudacher, T. and Allègre, C.J. 1982. Terrestrial xenology. *Earth and Planetary Science Letters* 60: 389–406. doi:10.1016/0012-821X(82)90075-9.

Tazoe, H., Obata, H. and Gamo, T. 2011. Coupled isotopic systematics of surface cerium and neodymium in the Pacific Ocean. *Geochemistry, Geophysics, Geosystems*. 12. doi: 10.1029/2010GC003342.

Tera, F., Brown, L., Morris, J., et al. 1986. Sediment incorporation in island-arc magmas: inferences from ^{10}Be. *Geochimica et Cosmochimica Acta* 50: 535–50.

Touboul, M., Puchtel, I.S. and Walker, R.J. 2012. ^{182}W evidence for long-term preservation of early mantle differentiation products. *Science* 335: 1065–9.

Vervoort, J. (2013) Lu-Hf Dating: The Lu-Hf Isotope System. In: Rink WJ, Thompson J (eds) *Encyclopedia of Scientific Dating Methods*, Springer Netherlands, Dordrecht, pp. 1–20. 978-94-007-6326-5.

Vervoort, J.D. and Blichert-Toft, J. 1999. Evolution of the depleted mantle; Hf isotope evidence from juvenile rocks through time. *Geochimica et Cosmochimica Acta* 63: 533–56.

Voshage, H. 1967. Bestrahlungsalter und Herkunft der Eisenmeteorite. *Zeitschrift. Naturforschung*, 22a: 477–506.

Walker, R. J. 2016. Siderophile elements in tracing planetary formation and evolution. *Geochemical Perspectives* 5: 1–145. doi:10.7185/geochempersp.5.1.

Walker, R.J., Carlson, R.W., Shirey, S.B. and Boyd, F.R. 1989. Os, Sr, Nd, and Pb isotope systematics of southern African peridotite xenoliths: implications for the chemical evolution of the subcontinental mantle. *Geochimica et Cosmochimica Acta* 53: 1583–95.

Walker, R.J., Morgan, J.W., Naldrett, A.J., Li, C. and Fassett, J.D. 1991. Re–Os isotope systematics of Ni-Cu sulfide ores, Sudbury igneous complex: evidence for a major crustal component. *Earth and Planetary Science Letters* 105: 416–29.

Wetherill, G. W. 1956. Discordant U–Pb ages. *Transactions of the American Geophysical Union* 37: 320.

Willbold, M., Elliott, T. and Moorbath, S. 2011. The tungsten isotopic composition of the Earth's mantle before the terminal bombardment. *Nature* 477: 195–8. doi:10.1038/nature10399

White, W. M., 2015. *Isotope Geochemistry*. John Wiley & Sons, Chichester.

White, W. M., Copeland, P., Gravatt, D. R. and Devine, J. D. 2017. Geochemistry and geochronology of Grenada and Union islands, Lesser Antilles: The case for mixing between two magma series generated from distinct sources. *Geosphere* 5: 1359–91. doi:10.1130/GES01414.1.

White, W.M., Patchett, P.J. and BenOthman, D. 1986. Hf isotope ratios of marine sediments and Mn nodules: evidence for a mantle source of Hf in seawater. *Earth and Planetary Science Letters* 79: 46–54.

York, D. 1969. Least squares fitting of a straight line with correlated errors. *Earth and Planetary Science Letters* 5: 320–4.

PROBLEMS

1. What are the binding energies per nucleon of ^{147}Sm (mass $= 146.914907$ u) and ^{143}Nd (mass $= 142.909823$ u)?

2. Calculate the maximum β^- energy in the decay of ^{87}Rb to ^{87}Sr. The mass of ^{87}Rb is 86.9091836 u; the mass of ^{87}Sr is 86.9088902 u.

3. What is the decay constant (λ) of ^{152}Gd if its half-life is 1.1×10^{14} yr?

4. The following data were measured on whole rock gneiss samples from the Bighorn Mountains of Wyoming. Use linear regression to calculate the age and initial ^{87}Sr/^{86}Sr for this gneiss.

Sample	^{87}Rb/^{86}Sr	^{87}Sr/^{86}Sr
4173	0.1475	0.7073
3400	0.2231	0.7106
7112	0.8096	0.7344
3432	1.1084	0.7456
3422	1.4995	0.7607
83	1.8825	0.7793

5. The following were measured on a komatiite flow in Canada. Use simple linear regression to calculate the slope. Plot the data on isochron diagrams.

	^{147}Sm/^{144}Nd	^{143}Nd/^{144}Nd
M654	0.2427	0.513586
M656	0.2402	0.513548
M663	0.2567	0.513853
M657	0.2381	0.513511
AX14	0.2250	0.513280
AX25	0.2189	0.513174
M666	0.2563	0.513842
M668	0.2380	0.513522

(a) Calculate the Sm-Nd age and the error on the age.
(b) Calculate the initial ε_{Nd} (i.e., the ε_{Nd} at the age calculated above) and the error on the initial ε_{Nd}.

6. A sample of granite has ^{143}Nd/^{144}Nd and ^{147}Sm/^{144}Nd of 0.51215 and 0.1342, respectively. The present chondritic ^{143}Nd/^{144}Nd and ^{147}Sm/^{144}Nd are 0.512630 and 0.1960, respectively. The decay constant of ^{147}Sm is 6.54×10^{-12} yr^{-1}. Calculate τ_{CHUR} (i.e., the crustal residence time relative to a chondritic mantle) for this granite.

7. Imagine that an initially uniform silicate Earth underwent melting at some time in the past to form continental crust (melt) and mantle (melting residue). Calculate the present-day Sr and Nd isotopic composition of 1%, 2%, and 5% partial melts and respective melting residues, assuming the bulk partition coefficients given in Example 8.4. Assume that the present-day ^{87}Rb/^{86}Sr, ^{87}Sr/^{86}Sr, ^{147}Sm/^{144}Nd, ^{143}Nd/^{144}Nd ratios of the bulk silicate Earth are 0.085, 0.7045, 0.1960, and 0.51263, respectively. Perform the calculation assuming the melting occurred at 4.0 Ga, 3.0 Ga, and 2.0 Ga. Plot your results on a Sr-Nd isotope diagram (i.e., ^{143}Nd/^{144}Nd vs. ^{87}Sr/^{86}Sr).

8. Given the data on a series of whole rocks below:

   (a) Use linear regression to calculate the age of the rocks and the error on the age.
   (b) Use linear regression to calculate their initial $^{143}Nd/^{144}Nd$, and the error on the initial ratio.
   (c) From (b), calculate the initial ε_{Nd}, that is, ε_{Nd} at the time calculated in (a). Take the present-day chondritic $^{143}Nd/^{144}Nd$ to be 0.512630 and the (present-day) chondritic $^{147}Sm/^{144}Nd$ to be 0.1960 (you need to calculate the chondritic value at the time the rock formed to calculate initial ε_{Nd}).
   (d) Calculate the depleted mantle model age τ_{DM}. Assume that the present $^{143}Nd/^{144}Nd$ of the depleted mantle is 0.51310, and that the depleted mantle has evolved from the chondritic initial with a constant $^{147}Sm/^{144}Nd$ since 4.56 Ga. How does it compare with the age you calculated in (a)?

Sample	$^{147}Sm/^{144}Nd$	$^{143}Nd/^{144}Nd$
Clinopyroxene	0.1886	0.512360
Garnet	0.6419	0.513401
Whole rock	0.1146	0.512205

Linear regression functions are available on some scientific calculators and in Microsoft Excel and MATLAB.

9. The following were measured on a komatiite flow in Canada. Use simple linear regression to calculate slopes. Plot the data on isochron diagrams.

	$^{206}Pb/^{204}Pb$	$^{207}Pb/^{204}Pb$	$^{208}Pb/^{204}Pb$
M665	15.718	14.920	35.504
M654	15.970	14.976	35.920
M656	22.563	16.213	41.225
M663	16.329	15.132	35.569
M657	29.995	17.565	48.690
AX14	32.477	17.730	49.996
AX25	15.869	14.963	35.465
M667	14.219	14.717	33.786
M666	16.770	15.110	35.848
M668	16.351	15.047	36.060
M658	20.122	15.700	39.390

   (a) Calculate the Pb-Pb age and the error on the age.
   (b) Calculate the Th/U ratio of the samples.

10. Calculate the $^{207}Pb/^{204}Pb–^{206}Pb/^{204}Pb$ age for the rocks below. (*Hint*: First calculate the slope using linear regression, then use Excel's Solver to calculate the age.)

Sample	$^{206}Pb/^{204}Pb$	$^{207}Pb/^{204}Pb$
NPA5	15.968	14.823
NPA12	17.110	15.043
NPA15	17.334	15.085
NPA15HF	17.455	15.107

11. **(a)** Calculate the isotopic evolution ($^{207}Pb/^{204}Pb$ and $^{206}Pb/^{204}Pb$ only) of Pb in a reservoir having a $^{238}U/^{204}Pb$ of 8. Do the calculation at 0.5 Ga intervals from 4.5 Ga to present. Assume the reservoir started with Canyon Diablo initial Pb isotopic composition. Plot your results on a $^{207}Pb/^{204}Pb$ vs. $^{206}Pb/^{204}Pb$ graph.

   (b) Calculate the isotopic evolution ($^{207}Pb/^{204}Pb$ and $^{206}Pb/^{204}Pb$ only) of Pb in a reservoir having a $^{238}U/^{204}Pb$ of 7 from 4.5 to 2.5 Ga and $^{238}U/^{204}Pb$ of 9 from 2.5 Ga to present. Do the calculation at 0.5 Ga intervals from 4.5 Ga to present. Assume the reservoir started with Canyon Diablo initial Pb isotopic composition. Plot your results on a $^{207}Pb/^{204}Pb$ vs. $^{206}Pb/^{204}Pb$ graph.

   HINT: The equation:

$$^{206}Pb/^{204}Pb = (^{206}Pb/^{204}Pb)_0 + \mu(e^{\lambda_{238}t} - 1)$$

   is valid for calculating the growth of $^{206}Pb/^{204}Pb$ only between the present and the initial time; i.e., the time when $^{206}Pb/^{204}Pb = (^{206}Pb/^{204}Pb)_0$ (because the $^{238}U/^{204}Pb$ ratio used is the present ratio). The growth of $^{206}Pb/^{204}Pb$ between two other times, t_1 and t_2, where t_1 is older than t_2, may be calculated by taking the growth of $^{206}Pb/^{204}Pb$ between t_2 and the present, and between t_1 and the present, and subtracting the latter from the former. If μ_1 is the value of μ between t_1 and t_2, and μ_2 is the value of μ between t_2 and the present, the relevant equation is then:

$$^{206}Pb/^{204}Pb = (^{206}Pb/^{204}Pb)_{t_1} + \mu_1(e^{\lambda_{238}t_1} - 1) + \mu_2(e^{\lambda_{238}t_2} - 1)$$

   if $\mu_1 = \mu_2$, then the equation simplifies to:

$$^{206}Pb/^{204}Pb = (^{206}Pb/^{204}Pb)_{t_1} + \mu(e^{\lambda_{238}t_1} - e^{\lambda_{238}t_2})$$

12. Show that when $\lambda_D \gg \lambda_P$ and $t \ll 1/\lambda_P$, eqn. 8.42 reduces to eqn. 8.57 (with D referring to daughter ^{230}Th, and P referring to parent ^{238}U).

13. A basalt from Réunion has a $(^{230}Th/^{232}Th)$ ratio of 0.93 and a $(^{238}U/^{232}Th)$ ratio of 0.75. Assuming the age of the basalt is 0:

   (a) What is the $(^{238}U/^{232}Th)$ ratio of the source of the basalt?
   (b) What is the $^{232}Th/^{238}U$ *atomic* ratio of the source of the basalt?
   (c) Assuming bulk distribution coefficients of 0.01 for U and 0.005 for Th and equilibrium melting, what is the percent melting involved in generating this basalt?

   (Remember, the parentheses denote *activity* ratios).

14. Given the following data on the Cheire de Mazaye flow in the Massif Central, France, calculate the age of the flow and the initial $^{232}Th/^{238}U$ (atomic). Use simple linear regression in obtaining your solution.

	$(^{238}U/^{232}Th)$	$(^{230}Th/^{232}Th)$
Whole rock	0.744	0.780 ± 0.012
Magnetite M1 (80–23μ)	0.970	0.864 ± 0.017
Magnetite M2 (23–7μ)	1.142	0.904 ± 0.017
Clinopyroxene	0.750	0.791 ± 0.019
Plagioclase	0.685	0.783 ± 0.018

Chapter 9

Stable isotope geochemistry

9.1 INTRODUCTION

Stable isotope geochemistry is concerned with variations in the isotopic compositions of elements arising from physicochemical processes rather than nuclear processes. Isotope fractionation, meaning change in an isotope ratio that arises as a result of some chemical or physical process, might at first seem to be an oxymoron. After all, in the last chapter we stated that the value of radiogenic isotopes was that the various isotopes of an element had identical chemical properties and therefore that isotope ratios such as $^{87}Sr/^{86}Sr$ are not changed measurably by chemical processes. In this chapter we will find that this is not quite true, and that the very small differences in the chemical behavior of different isotopes of an element can provide a very large amount of useful information about chemical (both geochemical and biochemical) processes.

The origins of stable isotope geochemistry are closely tied to the development of modern physics in the first half of the twentieth century. The discovery of the neutron in 1932 by H. Urey and the demonstration of variable isotopic compositions of light elements by A. Nier in the 1930s and 1940s were the precursors to this development. The real history of stable isotope geochemistry begins in 1947 with the publication of Bigeleisen and Mayer's paper, *Calculation of Equilibrium Constants for Isotopic Exchange Reactions,* and Harold Urey's paper, *The Thermodynamic Properties of Isotopic Substances.* These works not only showed why, on theoretical grounds, isotope fractionations could be expected, but also suggested that these fractionations could provide useful geologic information. Urey then set up a laboratory to determine the isotopic compositions of natural substances and the temperature dependence of isotopic fractionations, in the process establishing the field of stable isotope geochemistry.

What has been learned in the 70 years since those papers were published would undoubtedly astonish even Urey. Stable isotope geochemistry, like radiogenic isotope geochemistry, has become an essential part of not only geochemistry, but also the earth sciences as a whole and has found its way into many other fields as well, such as biochemistry, ecology, archeology, and forensics. In this chapter, we will attempt to gain an understanding of the principles underlying stable isotope geochemistry and then briefly survey its many-fold applications in the earth sciences. In doing so, we add the final tool to our geochemical toolbox. Space will allow for only a relatively brief review of the fundamentals of this field and some of its applications. More detail can be found in the companion book, White (2015), a *Reviews in Mineralogy* monograph (Valley and Cole, 2001), and texts by Sharp (2017) and Hoefs (2017).

Geochemistry, Second Edition. William M. White.
© 2020 John Wiley & Sons Ltd. Published 2020 by John Wiley & Sons Ltd.
Companion website: www.wiley.com/go/white/geochemistry

9.1.1   Scope of stable isotope geochemistry

Traditionally, the principal elements of interest in stable isotope geochemistry were H, C, N, O, and S. These elements are shown in dark background in Figure 9.1. Over the last two decades, Li and B have also become "staples" of isotope geochemistry. These elements have several common characteristics:

1. They have low atomic mass.
2. The relative mass difference between their isotopes is large.
3. They form bonds with a high degree of covalent character.
4. The elements exist in more than one oxidation state (C, N, and S), form a wide variety of compounds (O), or are important constituents of naturally occurring solids and fluids.
5. The abundance of the rare isotope is sufficiently high (generally at least tenths of a percent) to facilitate analysis.

In contrast to these light elements, most elements of interest in radiogenic isotope geochemistry are heavy (Sr, Nd, Ce, Hf, Pb), most form dominantly ionic bonds, generally exist in only one oxidation state, and there are only small relative mass differences between the isotopes of interest. Thus, isotopic fractionation of these elements is quite small. More importantly natural fractionation

is routinely corrected for in the process of correcting much larger fractionations that typically occur during analysis. Thus is it that one group of isotope geochemists make their living by measuring isotope fractionations while the other group makes their living by ignoring them!

It was once thought that elements not meeting these criteria would not show measurable natural variation in isotopic composition. However, as new techniques offering greater sensitivity and higher precision have become available (particularly use of the MC-ICP-MS), geochemists have begun to explore isotopic variations of many more elements (shown in red on Figure 9.1), indeed isotopic variations in the vast majority of elements with two or more isotopes have now been investigated. The isotopic variations observed in these elements are generally much smaller than the five traditional elements, but analytical precision now allows geologically and environmentally useful information to be obtained from them. We have space to briefly consider only a few of them here; the application of these nontraditional stable isotopes is reviewed in a recent book (Teng et al., 2017).

Stable isotope geochemistry has been applied to a great variety of problems, and we will see several examples in this chapter. One of the most common is geothermometry. Another is process identification. For

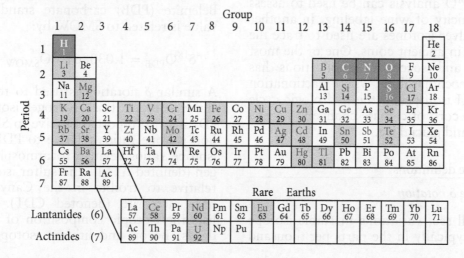

Figure 9.1   Periodic Table of the Elements illustrating the traditional stable isotopes (shown in bold with dark background) and the "nontraditional" ones (shown in red) that geochemists have begun to study in the last 20 years.

Table 9.1   Isotope ratios of light stable isotopes.

Element	Notation	Ratio	Standard	Absolute ratio
Hydrogen	δD	D/H (^{2}H/^{1}H)	SMOW	1.557×10^{-4}
Lithium	δ^7Li	^{7}Li/^{6}Li	NIST 8545 (L-SVEC)	12.285
Boron	δ^{11}B	^{11}B/^{10}B	NIST 951	4.044
Carbon	δ^{13}C	^{13}C/^{12}C	PDB	1.122×10^{-2}
Nitrogen	δ^{15}N	^{15}N/^{14}N	Atmosphere	3.613×10^{-3}
Oxygen	δ^{18}O	^{18}O/^{16}O	SMOW, PDB	2.0052×10^{-3}
	δ^{17}O	^{17}O/^{16}O	SMOW	3.76×10^{-4}
Sulfur	δ^{34}S	^{34}S/^{32}S	CDT	4.416×10^{-2}
	δ^{33}S	^{33}S/^{32}S	CDT	7.877×10^{-3}
	δ^{36}S	^{36}S/^{32}S	CDT	1.535×10^{-4}

instance, plants that produce "C$_4$" hydro-carbon chains (i.e., hydrocarbon chains four carbons long) as their primary photosynthetic product will fractionate carbon differently from plants that produce C$_3$ chains. This fractionation is retained up the food chain. This allows us, for example, to draw some inferences about the diet of fossil mammals from the stable isotope ratios in their bones. Sometimes stable isotope ratios are used as "tracers", much as radiogenic isotopes are. So, for example, we can use oxygen isotope ratios in igneous rocks to determine whether they have assimilated crustal material, as crust generally has different O isotope ratios than does the mantle. The oxygen isotope composition of wine reflects that of rainfall in the area, the terroir, where the grapes grew, so δ^{18}O analysis can be used to assess the authenticity of wine labeling. In another example, silver isotopes are used to trace the metal used in ancient coins. One of the most important and successful applications has been in paleoclimatology, where fractionation of O and H isotopes in the hydrologic cycle has left a record of climate change in marine sediments and glacial ice.

9.1.2   Some definitions

9.1.2.1   The δ notation

As we shall see, variations in stable isotope ratios are typically in the parts per thousand

to parts per hundred range and are most conveniently and commonly reported as *per mil* deviations, δ, from some standard. For example, O isotope ratios are often reported as per-mil deviations from SMOW* (standard mean ocean water) as:

$$\delta^{18}O = \left[\frac{(^{18}O/^{16}O)_{sample} - (^{18}O/^{16}O)_{SMOW}}{(^{18}O/^{16}O)_{SMOW}} \right] \times 1000 \tag{9.1}$$

Unfortunately, a dual standard developed for reporting O isotopes. While O isotopes in most substances are reported relative to SMOW, the oxygen isotope composition of carbonates is reported relative to the Pee Dee Belemite (PDB) carbonate standard.† This value is related to SMOW by:

$$\delta^{18}O_{PDB} = 1.03086\ \delta^{18}O_{SMOW} + 30.86 \tag{9.2}$$

A similar δ notation is used to report other stable isotope ratios. Hydrogen isotope ratios, δD, are also reported relative to SMOW, carbon isotope ratios relative to PDB, nitrogen isotope ratios relative to atmospheric nitrogen (denoted ATM), and sulfur isotope ratios relative to troilite in the Canyon Diablo iron meteorite (denoted CDT). Table 9.1 lists the isotopic composition of these standards. Boron and lithium isotope ratios as

---

* In practice, this and other standards are now artificially prepared and distributed by the International Atomic Energy Commission in Vienna. These working standards are sometimes referred to as "V-SMOW", "V-CDT", and so on.

† There is, however, a good historical reason for this: analytical difficulties in the measurement of carbonate oxygen prior to 1965 required it be measured against a carbonate standard.

well as most of the nontraditional stable isotopes are reported relative to laboratory standards.

9.1.2.2 The fractionation factor

The *fractionation factor, α*, is the ratio of isotope ratios in two phases:

$$\alpha_{A-B} \equiv \frac{R_A}{R_B} \qquad (9.3)$$

The fractionation of isotopes between two phases is also often reported as:

$$\Delta_{A-B} = \delta_A - \delta_B \qquad (9.4)$$

The relationship between Δ and α is:

$$\Delta \approx (\alpha - 1) \times 10^3 \quad \text{or} \quad \Delta \approx 10^3 \ln \alpha \quad (9.5)^*$$

The proof is left as an exercise (Problem 9.1).

As we will see, *at equilibrium, α* may be related to the equilibrium constant of thermodynamics by:

$$\alpha_{A-B} = (K/K_\infty)^{1/n} \qquad (9.6)$$

where n is the number of atoms exchanged, K_∞ is the equilibrium constant at infinite temperature, and K is the equilibrium constant written in the usual way (except that molar concentrations are used rather than activities because the ratios of the activity coefficients are equal to 1, i.e., there are no isotopic effects on the activity coefficient).

9.2   THEORETICAL CONSIDERATIONS

Isotope fractionation can originate from both *kinetic* effects and *equilibrium* effects. The former may be intuitively expected, but the latter may at first seem somewhat surprising. After all, we have been taught that the chemical properties of an element are dictated by its electronic structure, and that the nucleus plays no real role in chemical interactions. In the following sections, we will see that quantum mechanics predicts that the mass of an atom affects its vibrational motion, and therefore the strength of its chemical bonds. It also affects rotational and translational motions. From an understanding of these effects of atomic mass, it is possible to predict the small differences in the chemical properties of isotopes quite accurately.

9.2.1   Equilibrium isotope fractionations

Most isotope fractionations arise from equilibrium effects. Equilibrium fractionations arise from translational, rotational, and vibrational motions of molecules in gases and liquids and atoms in crystals because the energies associated with these motions are mass-dependent. We found in Chapter 2 that systems tend to adjust themselves so as to minimize energy. Thus, isotopes will be distributed so as to minimize the vibrational, rotational, and translational energy of a system. Of the three types of energies, vibrational energy makes by far the most important contribution to isotopic fractionation. Vibrational motion is the only mode of motion available to atoms in a solid. These effects are, as one might expect, small. For example, the equilibrium constant for the reaction:

$$\tfrac{1}{2}C^{16}O_2 + H_2{}^{18}O \rightleftharpoons \tfrac{1}{2}C^{18}O_2 + H_2{}^{16}O \qquad (9.7)$$

is only about 1.04 at 25°C and the ΔG of the reaction, given by $-RT \ln K$, is only -100 J/mol (you'll recall most ΔGs for reactions are typically in the kJ/mol range).

9.2.1.1 The quantum mechanical origin of isotopic fractionations

It is fairly easy to understand, at a qualitative level at least, how some isotope fractionations can arise from vibrational motion. Consider the two hydrogen atoms of a hydrogen molecule: they do not remain at a fixed distance from one another; rather they continually oscillate toward and away from each other, even at absolute zero. The

---

* Unfortunately, there are two other uses of the upper-case delta notation in isotope geochemistry, namely for mass independent fractionations and isotopic clumping, topics covered in subsequent sections. These can generally be distinguished by a following subscript or superscript.

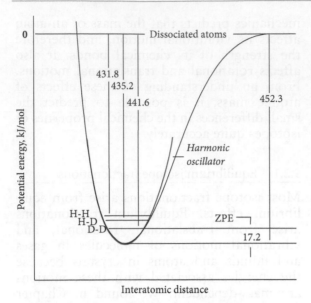

Figure 9.2 Energy-level diagram for the hydrogen molecule. Fundamental vibration frequencies are 4405 cm^{-1} for H_2, 3817 cm^{-1} for HD, and 3119 cm^{-1} for D_2. The zero point energy of H_2 is greater than that of HD, which is greater than that of D_2. Arrows show the energy, in kJ/mol, required to dissociate the three species. After O'Neil (1986).

frequency of this oscillation is quantized, that is, only discrete frequency values are possible. Figure 9.2 is a schematic diagram of energy as a function of interatomic distance in the hydrogen molecule. As the atoms vibrate back and forth, their potential energy varies as shown by the curved line. The *zero point energy* (ZPE) is the energy level at which the molecule will vibrate in its ground state, which is the state in which the molecule will be at low temperature. The zero point energy is always some finite amount above the minimum potential energy of an analogous harmonic oscillator.

The potential energy curves for various isotopic combinations of an element are identical, but as the figure shows, the zero point vibrational energies differ, as do vibration energies at higher quantum levels, being lower for the heavier isotopes. Vibration

energies and frequencies are directly related to bond strength. Because of this, the energy required for the molecule to dissociate will differ for different isotopic combinations. For example, 440.6 kJ/mol is necessary to dissociate a D_2 ($^2H_2$) molecule, but only 431.8 kJ/mol is required to dissociate the $^1H_2$ molecule. Thus, the bond formed by the two deuterium atoms is 9.8 kJ/mol stronger than the H–H bond. The differences in bond strength can also lead to kinetic fractionations, since molecules that dissociate more easily will react more rapidly. We will discuss kinetic fractionations later.

9.2.1.2 Predicting isotopic fractionations from statistical mechanics

Now let's attempt to understand the origin of isotopic fractionations on a more quantitative level. We have already been introduced to several forms of the *Boltzmann distribution law* (e.g., eqn. 2.84), which describes the distribution of energy states. It states that the probability of a molecule having internal energy E_i is:

$$P_i = \frac{g_i e^{-E_i/kT}}{\sum_n g_i e^{-E_n/kT}} \qquad (9.8)$$

where g is a statistical weight factor,* k is Boltzmann's constant, and the sum in the denominator is taken over all possible states. The average energy (per molecule) is:

$$\overline{E} = \sum_n E_n P_n = \frac{\sum_n g_i E_n e^{-E_n/kT}}{\sum_n g_i e^{-E_n/kT}} \qquad (9.9)$$

As we saw in Chapter 2, the *partition function, Q*, is the denominator of this equation:

$$Q = \sum_n g_n e^{-E_n/kT} \qquad (9.10)$$

The partition function is related to thermodynamic quantities U and S (eqns. 2.90

---

* This factor comes into play where more than one state corresponds to an energy level E_i (the states are said to be "degenerate"). In that case g_i is equal to the number of states having energy level E_i.

and 2.91). Since there is no volume change associated with isotope exchange reactions, we can use the relationship:

$$\Delta G = \Delta H - T\Delta S \qquad (9.11)$$

to derive the relationship Gibbs free energy and the partition function:

$$\Delta G_r = -RT \ln \prod_i Q_i^{\xi i} \qquad (9.12)$$

and comparing with eqn. 3.86, that the equilibrium constant is related to the partition function as:

$$K = \prod_i Q_i^{vi} \qquad (9.13)$$

for isotope exchange reactions. In the exchange reaction above (eqn. 9.7) this is simply:

$$K = \frac{Q_{C^{18}O_2}^{0.5} Q_{H_2{}^{16}O}}{Q_{C^{16}O_2}^{0.5} Q_{H_2{}^{18}O}} \qquad (9.14)$$

The usefulness of the partition function is that it can be calculated from quantum mechanics, and from it we can calculate equilibrium fractionations of isotopes.

There are three modes of motion available to gaseous molecules: vibrational, rotational, and translational (Figure 9.3). Only one vibrational mode is available to diatomic molecules. Possible vibrational and rotational modes of motion of polyatomic molecules are more complex. The partition function can be written as the product of the translational, rotational, vibrational, and electronic partition functions:

$$Q_{total} = Q_{vib}Q_{trans}Q_{rot}Q_{elec} \qquad (9.15)$$

The electronic configurations and energies of atoms are unaffected by the isotopic differences, so the last term can be ignored in the present context.

The vibrational motion is the most important contributor to isotopic fractionations, and it is the only mode of motion available to atoms in solids. So we begin by calculating the vibrational partition function. We can

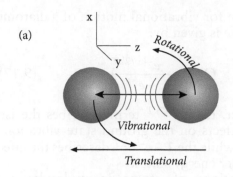

(a)

Rotational

Vibrational

Translational

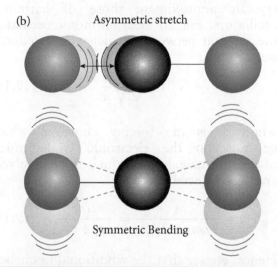

(b)   Asymmetric stretch

Symmetric Bending

Figure 9.3 (a) The three modes of motion for a diatomic molecule. Rotations can occur about both the y- and x-axes; only the rotation about the y-axis is illustrated. Since radial symmetry exists about the z-axis, rotations about that axis are not possible according to quantum mechanics. Three modes of translational motion are possible: in the x, y, and z directions. (b) Additional vibrational modes available to a triatomic molecule such as CO_2.

approximate the vibration of atoms in a simple diatomic molecule such as CO or O_2 by that of a harmonic oscillator. The energy of a "quantum" oscillator is:

$$E_{vib} = (n + 1/2)h\nu_0 \qquad (9.16)$$

where ν_0 is the ground-state vibrational frequency, h is Planck's constant, and n is the vibrational quantum number. The partition function for vibrational motion of a diatomic molecule is given by:

$$Q_{vib} = \frac{e^{-h\nu/2kT}}{1 - e^{-h\nu/kT}} \qquad (9.17)^*$$

In effect, the $e^{-h\nu/2kT}$ term describes the isotopic effects on the ground state vibrational energy while the $1 - e^{-h\nu/kT}$ describes the effect on higher energy states.

Vibrations of atoms in molecles and crystals approximate those of harmonic oscillators. For an ideal harmonic oscillator, the relation between frequency and reduced mass is:

$$\nu = \frac{1}{2}\sqrt{\frac{k}{\mu}} \qquad (9.18)$$

where k is the forcing constant, which depends on the electronic configuration of the molecule, but not the isotope involved, and μ is reduced mass:

$$\mu = \frac{1}{1/m_1 + 1/m_2} = \frac{m_1 m_2}{m_1 + m_2} \qquad (9.19)$$

Hence, we see that the vibrational frequency, and therefore also vibrational energy and partition function, depends on the mass of the atoms involved in the bond.

Rotational motion of molecules is also quantized. We can approximate a diatomic molecule as a dumbbell rotating about the center of mass. The rotational energy for a quantum rigid rotator is:

$$E_{\text{rot}} = \frac{j(j+1)h^2}{8\pi^2 I} \qquad (9.20)$$

where j is the rotational quantum number and I is the moment of inertia, $I = \mu r^2$, where r is the interatomic distance. The statistical weight factor, g, in this case is equal to $(2j + 1)$ since there two axes about which rotations are possible. For example, if $j = 1$, then there are $j(j + 1) = 2$ quanta available to the molecule, and $(2j + 1)$ ways of distributing these two quanta: all to the x-axis, all to the y-axis or one to each. Thus:

$$Q_{rot} = \sum_j (2j + 1)e^{-j(j+1)h^2/8\pi^2 IkT} \qquad (9.21)$$

Because the rotational energy spacings are small, eqn. 9.21 can be integrated to yield:

$$Q_{rot} = \frac{8\pi^2 IkT}{\sigma h^2} \qquad (9.22)^\dagger$$

where σ is a symmetry factor whose value is 1 for heteronuclear molecules such as ^{16}O-^{18}O and 2 for homonuclear molecules such as ^{16}O-^{16}O. This arises because in homonuclear molecules, the quanta must be equally distributed between the rotational axes, i.e., the js must be all even or all odd. This restriction does not apply to heterogeneous molecules, hence the symmetry factor.

Finally, translational energy associated with each of the three possible translational motions (x, y, and z) is given by the solution to the Schrödinger equation for a particle in a box:

$$E_{trans} = \frac{n^2 h^2}{8Ma^2} \qquad (9.23)$$

---

* Polyatomic molecules have many modes of vibrational motion available to them. The partition function is calculated by summing the energy over all available modes of motion; since energy occurs in the exponent, the result is a product:

$$Q_{vib} = \prod_i \frac{e^{-h\nu_i/2kT}}{1 - e^{-h\nu_i/kT}} \qquad (9.17a)$$

† Equation 9.22 also holds for linear polyatomic molecules such as CO_2. The symmetry factor is 1 if it has plane symmetry, and 2 if it does not. For nonlinear polyatomic molecules, eqn. 9.21 is replaced by:

$$Q_{rot} = \frac{8\pi^2(8\pi^3 ABC)^{1/2}(kT)^{3/2}}{\sigma h^3} \qquad (9.22a)$$

where A, B, and C are the three principal moments of inertia.

where n is the translational quantum number, M is molecular mass, and a is the dimension of the box. This expression can be inserted into eqn. 9.10. Above about 2 K, spacings between translational energy levels are small, so eqn. 9.10 can also be integrated. Recalling that there are three degrees of freedom, the result is:

$$Q_{trans} = \frac{(2\pi MkT)^{3/2}}{h^3} V \qquad (9.24)$$

where V is the volume of the box (a^3). Thus, the full partition function is:

$$Q_{total} = Q_{vib} Q_{rot} Q_{trans}$$

$$= \frac{e^{-hv/2kT}}{1 - e^{-hv/kT}} \frac{8\pi^2 IkT}{\sigma h^2} \frac{(2\pi MkT)^{3/2}}{h^3} V \qquad (9.25)$$

It is the ratio of partition functions that occurs in the equilibrium constant expression, so that many of the terms in eqn. 9.25 eventually cancel. Thus, the partition function ratio for two different isotopic species of the same diatomic molecule, A and B, reduces to:

$$\frac{Q_A}{Q_B} = \frac{\dfrac{e^{-hv_A/2kT}}{1 - e^{-hv_A/kT}} \dfrac{I_A}{\sigma_A} M_A^{3/2}}{\dfrac{e^{-hv_B/2kT}}{1 - e^{-hv_B/kT}} \dfrac{I_B}{\sigma_B} M_B^{3/2}}$$

$$= \frac{e^{-hv_A/2kT}(1 - e^{-hv_B/kT}) I_A \sigma_B M_A^{3/2}}{e^{-hv_B/2kT}(1 - e^{-hv_A/kT}) I_B \sigma_A M_B^{3/2}} \qquad (9.26)$$

This equation can be simplified through use of the Teller–Redlich spectroscopic theorem[*] to:

$$\frac{Q_A}{Q_B} = \frac{e^{-hv_A/2kT}(1 - e^{-hv_B/kT}) v_A \sigma_B m_A^{3r_A/2}}{e^{-hv_B/2kT}(1 - e^{-hv_A/kT}) v_B \sigma_A m_B^{3r_B/2}} \qquad (9.27)$$

and for a polyatomic molecule:

$$\frac{Q_A}{Q_B} = \left(\frac{m_A}{m_B}\right)^{3r/2} \frac{\sigma_B}{\sigma_A}$$

$$\prod_i \frac{e^{-hv_A^i/2kT}(1 - e^{-hv_B^i/kT}) v_B^i}{e^{-hv_B^i/2kT}(1 - e^{-hv_A^i/kT}) v_A^i} \qquad (9.28)$$

where the product is over all frequencies, m_i is the mass of the atom exchanged, and r is the number of atoms of the element being exchanged present in the molecule, i.e., m is 16 for ^{16}O and 18 for ^{18}O and r is 2 for CO_2 and 1 for H_2O in reaction 9.7.

Notice that all the temperature terms cancel except for those in the vibrational contribution. Thus, vibrational motion alone is responsible for the temperature dependency of isotopic fractionations.

As we see from eqn. 9.6, we need to calculate K_∞ to calculate the fractionation factor α from the equilibrium constant. For a reaction such as:

$$aA_1 + bB_2 \rightleftharpoons aA_2 + bB_1$$

where A_1 and A_2 are two molecules of the same substance differing only in their isotopic composition, e.g., $H_2^{16}O$ and $H_2^{18}O$, and a and b are the stoichiometric coefficients, the equilibrium constant is:

$$K_\infty = \frac{(\sigma_{A_2}/\sigma_{A_1})^a}{(\sigma_{B_1}/\sigma_{B_2})^b} \qquad (9.29)$$

Thus, for a reaction where only a single isotope is exchanged, K_∞ is simply the ratio of the symmetry factors.

---

[*] The Teller–Redlich theorem relates the products of the frequencies for each symmetry type of the two isotopes to the ratios of their masses and moments of inertia:

$$\left(\frac{I_1}{I_2}\right)\left(\frac{M_1}{M_2}\right)^{3/2} = \left(\frac{m_1}{m_2}\right)^{3/2} \frac{hv_1/kT}{hv_2/kT}$$

where m is the isotope mass and M is the molecular mass. We need not concern ourselves with its details.

Example 9.1　Predicting isotopic fractionations

Consider the exchange of ^{18}O and ^{16}O between carbon monoxide and oxygen:

$$C^{16}O + {}^{16}O^{18}O \rightleftharpoons C^{18}O + {}^{16}O_2 \tag{9.30}$$

The frequency for the C–^{16}O vibration is 6.505×10^{13} sec^{-1}, and the frequency of the $^{16}O_2$ vibration is 4.738×10^{13} sec^{-1}. How will the fractionation factor, $\alpha = (^{18}O/^{16}O)_{CO}/^{18}O/^{16}O)_{O2}$, vary as a function of temperature?

Answer: The equilibrium constant for this reaction is:

$$K = \frac{Q_{C^{18}O}Q_{^{16}O_2}}{Q_{C^{16}O}Q_{^{18}O^{16}O}} \tag{9.31}$$

The rotational and translational contributions are independent of temperature, so we calculate them first. The rotational contribution is:

$$K = \left(\frac{Q_{C^{18}O}Q_{^{16}O_2}}{Q_{C^{16}O}Q_{^{18}O^{16}O}}\right)_{rot} = \frac{\mu_{C^{18}O}\sigma_{C^{16}O}}{\mu_{C^{16}O}\sigma_{C^{18}O}}\frac{\mu_{^{16}O_2}\sigma_{^{18}O^{16}O}}{\mu_{^{18}O^{16}O}\sigma_{^{16}O_2}} = \frac{1}{2}\frac{\mu_{C^{18}O}\mu_{^{16}O_2}}{\mu_{C^{16}O}\mu_{^{18}O^{16}O}} \tag{9.32}$$

(the symmetry factor, σ, is 2 for $^{16}O_2$ and 1 for the other molecules). Substituting values $\mu_{C^{16}O} = 6.857$, $\mu_{C^{18}O} = 7.20$, $\mu_{^{18}O^{16}O} = 8.471$, $\mu_{C^{16}O_2} = 8$, we find: $K_{rot} = 0.9917/2$.

The translational contribution is:

$$K_{trans} = \frac{m_{C^{18}O}^{3/2}\, m_{^{16}O_2}^{3/2}}{m_{C^{16}O}^{3/2}\, m_{^{18}O^{16}O}^{3/2}} \tag{9.33}$$

Substituting $m_{C^{16}O} = 28$, $m_{C^{18}O} = 30$, $m_{^{18}O^{16}O} = 34$, $m_{^{16}O_2} = 32$ into 9.33, we find $K_{trans} = 1.0126$.

The vibrational contribution to the equilibrium constant is:

$$K_{vib} = \frac{e^{-h(\nu_{C^{18}O}-\nu_{C^{16}O}+\nu_{^{16}O_2}-\nu_{^{18}O^{16}O})/2kT}(1-e^{-h\nu_{C^{16}O}/kT})(1-e^{-h\nu_{^{18}O^{16}O}/kT})}{(1-e^{-h\nu_{C^{18}O}/kT})(1-e^{-h\nu_{^{16}O_2}/kT})} \tag{9.34}$$

To obtain the vibrational contribution, we can assume the atoms vibrate as harmonic oscillators and, using experimentally determined vibrational frequencies for the ^{16}O molecules, solve equation 9.18 for the forcing constant, k and calculate the vibrational frequencies for the ^{18}O-bearing molecules. These turn out to be 6.348×10^{13} sec^{-1} for $C^{18}O$ and 4.605×10^{13} sec^{-1} for $^{18}O^{16}O$, so that 9.34 becomes:

$$K_{vib} = \frac{e^{5.580/T}(1-e^{-3119/T})(1-e^{-2208/T})}{(1-e^{-3044/T})(1-e^{-2272/T})}$$

If we carry the calculation out at $T = 300$ K, we find:

$$K = K_{rot}K_{trans}K_{vib} = \frac{1}{2}0.9917 \times 1.0126 \times 1.1087 = \frac{1.0229}{2}$$

To calculate the fractionation factor α from the equilibrium constant, we need to calculate K_∞

$$K_\infty = \frac{(1/1)^1}{(2/1)^1} = \frac{1}{2}$$

so that

$$\alpha = K/K_\infty = 2K$$

At 300 K, α = 1.0233. The variation of α with temperature is shown in Figure 9.4.

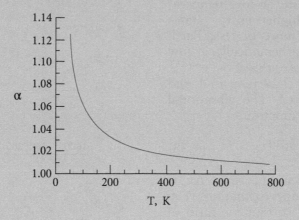

Figure 9.4   Fractionation factor, $\alpha = (^{18}O/^{16}O)_{CO}/(^{18}O/^{16}O)_{O2}$, calculated from partition functions as a function of temperature.

9.2.1.3   Reduced partition function ratios: β factors

Returning to our reaction between CO_2 and H_2O, we can rearrange the equilibrium constant expression (9.7) as:

$$K = \frac{\left(Q_{^{18}O}/Q_{^{16}O}\right)_{CO_2}^{1/2}}{\left(Q_{^{16}O}/Q_{^{18}O}\right)_{H_2O}} \quad (9.35)$$

Thus, the equilibrium constant is the ratio of partition function ratios of the two isotopic versions of the two substances. From this, we can see that the mass terms in equation 9.28 will also cancel in the computation and consequently we can omit them. The partition function with the mass terms omitted is sometimes referred to as the *reduced partition function*.

When theoretical calculations such as these are involved, isotope fractionation is also often expressed in terms of the predicted equilibrium ratio of reduced partition functions, β, of the isotope ratio of the substance of interest to the isotope ratio of dissociated atoms, for example:

$$\beta_{H_2O}^{^{18}O/^{16}O} \equiv \frac{(^{18}O/^{16}O)_{H_2O}}{(^{18}O/^{16}O)_O} \quad (9.36)$$

The denominator is the isotope ratio in a gas of monatomic oxygen and the numerator is the isotope ratio in water in equilibrium with it. The β factor for oxygen isotopes in CO_2 would also have the isotope ratio of dissociated atoms in the denominator and thus in our example of CO_2–water exchange, the denominators cancel so that:

$$\alpha = \frac{\beta_{CO_2}}{\beta_{H_2O}} \quad (9.37)$$

β-factors are closely related to the reduced partition function and are equal to it where the molecule of interest contains only one atom of the element of interest (this would be the case for water, for example, which contains only 1 oxygen atom). Where this is not the case, CO_2 for example, the β-factor differs from the reduced partition function by what Richet et al. (1987) termed an *excess factor*, which arises from our interest in atomic isotopic ratios rather than of isotopic molecular abundances.

9.2.1.4   Temperature dependence of the fractionation factor

As we noted above, the temperature dependence of the fractionation factor depends only on the vibrational contribution. As we noted earlier, the $e^{-hv/2kT}$ term describes the isotopic effects on the ground state vibrational energy while the $1 - e^{-hv/kT}$ describes the effect on higher energy states. Atomic vibrations tend to be in the ground state at low temperatures, only occupying higher energy states at temperatures above, more or less, Earth surface temperatures. Mathematically, at low temperatures $T \ll hv/k$, so the $1 - e^{-hv/kT}$ terms in eqns. 9.17 and 9.27 tend to 1 and can therefore be neglected. The vibrational partition function approximates:

$$Q_{vib} \cong e^{-hv/kT} \qquad (9.38)$$

In a further simplification, since Δv is small, we can use the approximation $e^x \approx 1 + x$ (valid for $x \ll 1$), so that the ratio of vibrational energy partition functions *at low temperature* becomes:

$$Q^A_{vib}/Q^B_{vib} \cong 1 - h\Delta v/2kT$$

Since the translational and rotational contributions are temperature-independent, we expect a relationship of the form:

$$\alpha \cong A + \frac{B}{T} \qquad (9.39)$$

In other words, α should vary inversely with temperature at low temperatures.

As temperature increases and higher energy vibrational states are occupied, the $1 - \exp(-hv/kT)$ term departs significantly from 1. Furthermore, at higher vibrational frequencies, the harmonic oscillator approximation breaks down (as suggested in Figure 9.1), as do several of the other simplifying assumptions we have made, so that *at high temperature* the relation between the fractionation factor and temperature approximates:

$$\ln \alpha \propto \frac{1}{T^2} \qquad (9.40)$$

Since α is generally small, $\ln \alpha \approx 1 + \alpha$, so that $\alpha \propto 1 + 1/T^2$. At infinite temperature, the fractionation is unity, since $\ln \alpha = 0$. This is illustrated in Figure 9.5 for distribution of ^{18}O and ^{16}O between CO_2 and H_2O. The $\alpha \propto 1/T$ relationship breaks down around 200°C; above

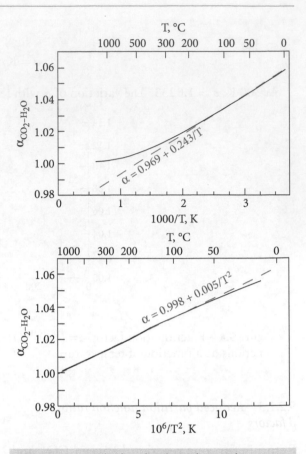

Figure 9.5 Calculated value of α_{18O} for CO_2–H_2O, shown vs. $1/T$ and $1/T^2$. Dashed lines show that up to ~200°C, $\alpha \approx 0.969 + 0.0243/T$. At higher temperatures, $\alpha \approx 0.9983 + 0.0049/T^2$. Calculated from the data of Richet et al. (1977).

that temperature, the relationship follows $\alpha \propto 1/T^2$.

It must be emphasized that the simple calculations performed in Example 9.1 are applicable only to a gas whose vibrations can be approximated by a simple harmonic oscillator. Real gases often show fractionations that are complex functions of temperature, with minima, maxima, inflections, and crossovers. Furthermore, solids have many vibrational modes.

9.2.1.5 Composition and pressure dependence

The nature of the chemical bonds in the phases involved is most important in determining isotopic fractionations. *A general rule of thumb is that the heavy isotope goes into the phase in which it is most strongly bound.*

Bonds to ions with a high ionic potential and low atomic mass are associated with high vibrational frequencies and have a tendency to incorporate the heavy isotope preferentially. For example, quartz, SiO_2, is typically the most ^{18}O rich mineral and magnetite the least. Oxygen is dominantly covalently bonded in quartz, but dominantly ionically bonded in magnetite. The O is bound more strongly in quartz than in magnetite, so the former is typically enriched in ^{18}O.

In both minerals and aqueous solution, bond strength is inversely related to coordination number and to bond length (shorter bonds tend to be stronger). For example, metal–oxygen bonds are stronger when the metal atom is surrounded by four oxygens (tetrahedral coordination) than six oxygens (octahedral coordination). As we found in Chapter 6, metal ions are bound to water molecules of the solvation shell forming aquo-complexes. A metal ion is more strongly held in solution when, for example, it is coordinated by six water molecules than with seven water molecules. Consequently, there will be a tendency for metal ions to partition into the phase in which it is coordinated by the smallest number of ligands.

Substitution of cations in a dominantly ionic site (typically the octahedral sites) in silicates has only a secondary effect on the O bonding, so that isotopic fractionations of O isotopes between similar silicates are generally small. Substitutions of cations in sites that have a strong covalent character (generally tetrahedral sites) result in greater O isotope fractionations. Thus, for example, we would expect the fractionation between the end-members of the alkali feldspar series and water to be similar, since only the substitution of K^+ for Na^+ is involved. We would expect the fractionation factors between end-members of the plagioclase series and water to be greater, since this series involves the substitution of Al for Si as well as Ca for Na, and the bonding of O to Si and Al in tetrahedral sites has a large covalent component. Substitution of Fe for Mg, common in many minerals, appears to have almost no effect on fractionation factors. Substitution of Ca for Mg (or Fe) in pyroxenes and garnet has a small effect of ~0.5‰.

Carbonates tend to be very ^{18}O-rich because O is bonded to a small, highly charged atom, C^{4+}. The fractionation, $\Delta^{18}O_{cal-water}$, between calcite and water is about 30 per mil at 25°C. The cation in the carbonate has a secondary role, with the fractionation factor decreasing with both increasing mass and ionic radius. Since the latter tend to be correlated, it is not entirely clear which is the dominant effect. The $\Delta^{18}O_{carb-H2O}$ decreases by about 1.5‰ when Ba replaces Ca at 240°C.

Crystal structure plays a secondary role. The $\Delta^{18}O$ between aragonite and calcite is of the order of 0.5 per mil. However, there is a large fractionation (10 per mil) of C between graphite and diamond.

Pressure effects on fractionation factors turn out to be small, no more than 0.1 per mil over 1 GPa in most cases. We can understand the reason for this by recalling that $\partial \Delta G / \partial P = \Delta V$. The volume of an atom is entirely determined by its electronic structure, which does not depend on the mass of the nucleus. Thus, the volume change of an isotope exchange reaction will be small, and hence there will be little pressure dependence. There will be a minor effect because vibrational frequency and bond length change as crystals are compressed. The compressibility of silicates is of the order of 1 part in 10^4, so we can expect effects of the order of 10^{-4} or less, which are generally insignificant.

Hydrogen isotope fractionation involving water is a notable exception. For example, the fractionation factor between brucite $(Mg(OH)_2)$ and water increases by >10‰ over a pressure range of 0.8 GPa. Similar pressure effects have been found for fractionation between water and other hydrous minerals with the effect of pressure increasing the partitioning of deuterium into the mineral.

9.2.2   Kinetic isotope fractionations

Kinetic isotope fractionations are normally associated with fast, incomplete, or unidirectional processes like evaporation, diffusion, dissociation reactions, and biologically mediated reactions. As an example, recall that temperature is related to the average kinetic energy. In an ideal gas, the average kinetic energy of all molecules is the same. The kinetic energy is given by:

$$E = \frac{1}{2}mv^2 \qquad (9.41)$$

Consider two molecules of carbon dioxide, $^{12}C^{16}O_2$ and $^{13}C^{16}O_2$, in such a gas. If their energies are equal, the ratio of their velocities is $(45/44)^{1/2}$, or 1.011. Thus, $^{12}C^{16}O_2$ can diffuse 1.1% further in a given amount of time than $^{13}C^{16}O_2$. This result, however, is largely limited to ideal gases, with low pressures where collisions between molecules are infrequent and intermolecular forces negligible. For the case where molecular collisions are important, the ratio of their diffusion coefficients is the inverse ratio of the square roots of the reduced masses of CO_2 and air (mean molecular weight 28.8):

$$\frac{D_{^{12}CO_2}}{D_{^{13}CO_2}} = \frac{\sqrt{\mu_{^{13}CO_2-air}}}{\mu_{^{12}CO_2-air}} = \frac{4.1906}{4.1721} = 1.0044 \tag{9.42}$$

Hence, we predict that gaseous diffusion will lead to a 4.4‰ rather than 11‰ fractionation.

Molecules containing the heavy isotope are more stable and have higher dissociation energies than those containing the light isotope. This can be readily seen in Figure 9.2. The energy required to raise the D_2 molecule to the energy where the atoms dissociate is 441.6 kJ/mol, whereas the energy required to dissociate the H_2 molecule is 431.8 kJ/mol. Therefore, it is easier to break the H-H than the D-D bond. Where reactions attain equilibrium, isotopic fractionations will be governed by the considerations of equilibrium discussed above. *Where reactions do not achieve equilibrium, the lighter isotope will usually be preferentially concentrated in the reaction products*, because of this effect of the bonds involving light isotopes in the reactants being more easily broken. Large kinetic effects are associated with biologically mediated reactions (e.g., photosynthesis, bacterial reduction), because such reactions generally do not achieve equilibrium and do not go to completion (e.g., plants do not convert all CO_2 to organic carbon). Thus, ^{12}C is enriched in the products of photosynthesis in plants (hydrocarbons) relative to atmospheric CO_2, and ^{32}S is enriched in H_2S produced by bacterial reduction of sulfate.

We can express this in a more quantitative sense. In Chapter 5, we found the reaction rate constant could be expressed as:

$$k = Ae^{-E_B/kT} \tag{5.23}$$

where k is the rate constant, A the frequency factor, and E_b is the barrier energy. For example, in a dissociation reaction, the barrier energy is the difference between the dissociation energy, ε, and the zero point energy when the molecule is in the ground state, or some higher vibrational frequency when it is not (Figure 9.2). The frequency factor is independent of isotopic composition, thus the ratio of reaction rates between the HD molecule and the H_2 molecule is:

$$\frac{k_D}{k_H} = \frac{e^{-(\varepsilon-1/2h\nu_D)/kT}}{e^{-(\varepsilon-1/2h\nu_H)/kT}} \tag{9.43}$$

or

$$\frac{k_D}{k_H} = e^{(\nu_H-\nu_D)h/kT} \tag{9.44}$$

Substituting for the various constants, and using the wavenumbers given in the caption to Figure 9.2 (remembering that $\omega = c\nu$, where c is the speed of light) the ratio is calculated as 0.24; in other words, we expect the H_2 molecule to react four times faster than the HD molecule, a very large difference. For heavier elements, the rate differences are smaller. For example, the same ratio calculated for $^{16}O_2$ and $^{18}O^{16}O$ shows that the $^{16}O_2$ will react about 15% faster than the $^{18}O^{16}O$ molecule.

The greater translational velocities of lighter molecules also allow them to break through a liquid surface more readily and hence evaporate more quickly than a heavy molecule of the same composition. Thus, water vapor above the ocean is typically around $\delta^{18}O = -13$ per mil, whereas at equilibrium the vapor should only be about 9 per mil lighter than the liquid.

Let's explore this a bit further. An interesting example of a kinetic effect is the fractionation of O isotopes between water and water vapor. This is an example of Rayleigh distillation (or condensation) and is similar to fractional crystallization. Let A be the amount of the species containing the major isotope, for example $H_2^{16}O$, and B be the amount of the species containing the minor isotope ($H_2^{18}O$). The rate at which these species evaporate is proportional to the amount present:

$$dA = k_A A \tag{9.45a}$$

and

$$dB = k_B B \qquad (9.45b)$$

Since the isotopic composition affects the reaction, or evaporation, rate, $k_A \neq k_B$. Earlier we saw that for equilibrium fractionations, the fractionation factor is related to the equilibrium constant. For kinetic fractionations, the fractionation factor is simply the ratio of the rate constants, so that:

$$\frac{k_A}{k_B} = \alpha \qquad (9.46)$$

and

$$\frac{dB}{dA} = \alpha \frac{B}{A} \qquad (9.47)$$

Rearranging and integrating, we have:

$$\ln \frac{B}{B^0} = \alpha \ln \frac{A}{A^0}$$

or

$$\frac{B}{B_0} = \left(\frac{A}{A_0} \right)^{\alpha} \qquad (9.48)$$

where A^0 and B^0 are the amount of A and B originally present. Dividing both sides by A/A^0:

$$\frac{B/A}{B_0/A_0} = \left(\frac{A}{A_0} \right)^{\alpha-1} \qquad (9.49)$$

Since the amount of B makes up only a trace of the total amount of H_2O present, A is essentially equal to the total water present, and A/A^0 is essentially identical to f, the fraction of the original water remaining. Hence:

$$\frac{B/A}{B_0/A_0} = f^{\alpha-1} \qquad (9.50)$$

Subtracting 1 from both sides, we have:

$$\frac{B/A - B_0/A_0}{B_0/A_0} = f^{\alpha-1} - 1 \qquad (9.51)$$

The left side of eqn. 9.51 is the relative deviation from the initial ratio. The per-mil difference from the original isotope ratio, δ_0, is simply:

$$\delta - \delta_0 = 1000(f^{\alpha} - 1) \qquad (9.52)$$

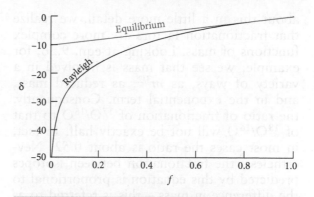

Figure 9.6   Fractionation of isotope ratios during Rayleigh and equilibrium condensation. δ is the per-mil difference between the isotopic composition of original vapor and the isotopic composition when fraction f of the vapor remains.

Of course, the same principle applies when water condenses from vapor. Assuming a value of α of 1.01, δ will vary with f, the fraction of vapor remaining, as shown in Figure 9.6.

Even if the vapor and liquid remain in equilibrium through the condensation process, the isotopic composition of the remaining vapor will change continuously. The relevant equation is:

$$\Delta = \left(1 - \frac{1}{(1-f)/\alpha + f} \right) \times 1000 \qquad (9.53)$$

where Δ is the change in the isotope ratio when fraction f of the vapor remains. The effect of equilibrium condensation is also shown in Figure 9.6.

9.2.3   Mass-dependent and mass-independent fractionations

Oxygen and sulfur both have more than two isotopes, as do, of course, many other elements. How do we expect the fractionations between various isotope ratios to be related? For example, if a 4‰ fractionation of $\delta^{18}O$ is observed in a particular sample, what value of $\delta^{17}O$ do we predict? A first guess might be that we would expect the difference in the fractionation to be proportional to the mass difference. In other words, if $\delta^{18}O$ increases by 4‰ in some process, we would expect $\delta^{17}O$ to increase by about 2‰. If we think about this in a little more detail, we realize that fractionation factors are more complex functions of mass. Looking at eqn. 9.26, for example, we see that mass is involved in a variety of ways, as $m^{3/2}$, as reduced mass, and in the exponential term. Consequently, the ratio of fractionation of $^{17}O/^{16}O$ to that of $^{18}O/^{16}O$ will not be exactly half. In fact, in most cases the ratio is about 0.52. Nevertheless, the fractionation between isotopes predicted by this equation is proportional to the difference in mass – this is referred to as *mass-dependent fractionation*.

There are, however, some exceptions where the observed ratio of fractionation of $^{17}O/^{16}O$ to that of $^{18}O/^{16}O$ is close to 1. Since the extent of fractionation in these cases seems independent of the mass difference, this is called *mass-independent fractionation*. Mass-independent fractionation is quantified using a Δ (delta) notation, for example for oxygen:

$$\Delta^{17}O = \delta^{17}O - 1000(1 + \delta^{18}O/1000)^{\theta} - 1 \tag{9.54}$$

where θ is the predicted ratio of fractionation of $^{17}O/^{16}O$ to that of $^{18}O/^{16}O$, i.e., ~0.52. For small values of $\delta^{17}O$, which is the usual case,

this reduces to:

$$\Delta^{17}O = \delta^{17}O - \theta \times \delta^{18}O \tag{9.55}$$

θ is the slope on a three-isotope diagram such as Figure 9.7 and Δ is the deviation from that slope. A similar notation is used for $\Delta^{33}S$ with $\theta \approx 0.515$ and $\Delta^{36}S$ with $\theta \approx 1.91$.

Mass-independent fractionation is rare. It was first observed in oxygen isotope ratios in meteorites (we will consider these variations in Chapter 10) and has subsequently been observed in oxygen isotope ratios of atmospheric gases, most dramatically in stratospheric ozone (Figure 9.7), and in sulfur isotope ratios of Archean sediments and modern sulfur-bearing aerosols in ice. Mass-independent fractionations can have several causes, none of which are completely understood as yet.

The cause of mass-independent O isotope fractionation in atmospheric ozone involves the dependence of reaction rates on molecular symmetry (Gao and Marcus, 2001). This theory can be roughly explained as follows. Formation of ozone in the stratosphere typically involves the energetic collision of monatomic and molecular oxygen:

$$O + O_2 \rightarrow O_3^*$$

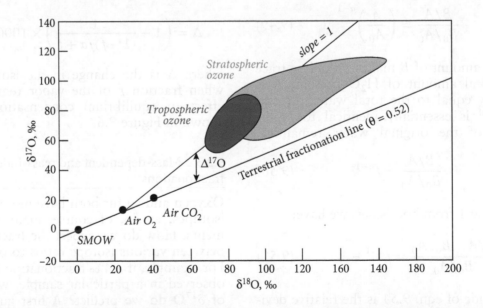

Figure 9.7   Oxygen isotopic compositions in the stratosphere and troposphere show the effects of mass-independent fractionation. A few other atmospheric trace gases show similar effects. Essentially all other material from the Earth and Moon plot on the *terrestrial fractionation line* whose slope, θ, is ≈0.52. After Johnson et al. (2001).

The ozone molecule thus formed is in an excited transition state, denoted by the * and, consequently, subject to dissociation if it cannot lose this excess energy. Excess vibrational energy can be lost either by collisions with other molecules, or by partitioning to rotational energy, with collision being the most common. In the stratosphere, collisions are less frequent, hence repartitioning of vibrational energy represents an important alternative pathway to stability. Because there are more possible energy transitions available to asymmetric species such as $^{16}O^{16}O^{18}O$ and $^{16}O^{16}O^{17}O$ than symmetric ones such as $^{16}O^{16}O^{16}O$, the former can more readily repartition its excess energy and form a stable molecule. At higher pressures, such as prevail in the troposphere, the symmetric molecule can lose energy through far more frequent collisions, lessening the importance of the vibrational to rotational energy repartition. Gao and Marcus (2001) were able to closely match observed experimental fractionations, but their approach was in part empirical because a full quantum mechanical treatment is not yet possible.

The cause of mass-independent isotope fractionations of oxygen in the solar nebula appears to be due to photolysis and to "self-shielding" of UV radiation that produces it. In the solar nebula, the wavelengths necessary to dissociate $C^{16}O$ were completely absorbed by gas in the inner solar system, while, because $C^{17}O$ and $C^{18}O$ are far less abundant, wavelengths necessary to dissociate those species were not. Consequently, dissociation made disproportionate amounts of ^{17}O and ^{18}O available to react and form other compounds, silicates, for example, that were incorporated into planets. Self-shielding appears also to be involved in mass-independent fractionation of sulfur isotopes, particularly in the early Earth. We'll consider mass independent fractionation in the Archean in more detail in section 9.8.

Two other factors can be responsible for mass-independent fractionations. The first is the so-called the *nuclear field shift* or *nuclear volume effect* and arises from the overlap between electronic and nuclear wave functions that results in *s* electrons being less tightly bound to a large nucleus (Bigeleisen, 2006). Furthermore, odd-numbered nuclei have slightly smaller volumes than even-numbered ones. This in turn lead to slightly different bond strengths and chemical behavior of odd- and even-number nuclei and consequently different fractionation factors and different temperature dependencies of them. This effect is restricted to large nuclei, i.e., those with high atomic number, and is most obviously manifested in Hg, Tl, and U isotope ratios.

The second factor is the *magnetic isotope effect*. As we noted in Chapter 8, protons and neutrons have "spin" or a "nuclear magnetic moment," and these cancel in even-numbered nuclei, leaving only odd-numbered nuclei with a magnetic moment. The moments of electrons can couple with this nuclear magnetic moment and again lead to slightly different behavior between odd- and even-numbered isotopes. This effect is purely kinetic and has so far been observed only in Hg isotope fractionations in photochemical reactions (Estrade et al., 2009).

9.2.4 Isotopic clumping

In Example 9.1, we calculated the distribution of ^{18}O between CO and O_2. However, the CO and O_2 will consist of a 12 isotopically distinct molecules or *isotopologues*, such as $^{12}C^{16}O$, $^{12}C^{17}O$, $^{13}C^{18}O$, $^{16}O^{17}O$, etc. The distribution of isotopes between these species will not be random but, rather, some of these isotopologues will be thermodynamically favored because the reduction in free energy achieved by *clumping* the heavy isotopes in molecules exceeds the free energy increase of clumping light isotopes. To understand this, consider just the isotopologues of CO and reactions relating them:

$$^{12}C^{16}O + {}^{13}C^{17}O \rightleftharpoons {}^{13}C^{16}O + {}^{12}C^{17}O$$

$$^{12}C^{16}O + {}^{13}C^{18}O \rightleftharpoons {}^{13}C^{16}O + {}^{12}C^{18}O$$

The equilibrium constant for first reaction can be calculated as:

$$K = \frac{q_{^{13}C^{16}O}q_{^{12}C^{17}O}}{q_{^{12}C^{16}O}q_{^{13}C^{17}O}} \tag{9.56}$$

and a similar equation can be written for the second reaction. The individual partition functions can be calculated as described in the previous section. Doing so, we find that the two heaviest species, $^{13}C^{17}O$ and $^{13}C^{18}O$, will be more abundant than if isotopes were merely randomly distributed among the six isotopologues, i.e., the heavy isotopes tend to *clump*. Wang et al. (2004) introduced a delta notation to describe this effect:

$$\Delta_i = \left(\frac{R_{i-e}}{R_{i-r}} - 1 \right) \times 1000 \qquad (9.57)$$

where R_{i-e} is the isotope ratio in the observed or calculated equilibrium abundance of isotopologue i to the isotopologue containing no rare isotopes and R_{i-r} is that same ratio if isotopes were distributed among isotopologues randomly. Thus, for example, in the system above:

$$\Delta_{^{13}C^{18}O} = \left\{ \frac{\left([^{13}C^{18}O]/[^{12}C^{16}O] \right)_e}{\left([^{13}C^{18}O]/[^{12}C^{16}O] \right)_r} - 1 \right\} \times 1000$$

$$(9.58)$$

Since R_{i-r} is the random distribution, it can be calculated directly as the probability of choosing isotopes randomly to form species. In the case of $^{13}C^{18}O$, it is:

$$R_{^{13}C^{18}O-r} = \left(\frac{[^{13}C^{18}O]}{[^{12}C^{16}O]} \right)_r = \frac{[^{13}C][^{18}O]}{[^{12}C][^{16}O]}$$

$$(9.59)$$

(where, as usual, brackets denote concentrations). It gets a little more complex for molecules with more than two isotopes. In most cases, we are interested in combinations of isotopes rather than permutations, which is to say we don't care about order. This will not be the case for highly asymmetric molecules such as nitrous oxide, N_2O. The structure of this molecule is N–N–O and $^{14}N^{15}N^{16}O$ will have different properties front $^{15}N^{14}N^{16}O$, so in that case, order does matter. The CO_2 molecule is, however, symmetric and we cannot distinguish $^{16}O^{12}C^{18}O$ from $^{18}O^{12}C^{16}O$. Its random abundance would be calculated as:

$$R_{^{13}C^{16}O^{18}O-r} = \left(\frac{[^{13}C^{16}O^{18}O]}{[^{12}C^{16}O_2]} \right)_r$$

$$= \frac{2[^{13}C][^{16}O][^{18}O]}{[^{12}C][^{16}O]^2}$$

$$= \frac{2[^{13}C][^{18}O]}{[^{12}C][^{16}O]} \qquad (9.60)$$

As Wang et al. (2004) showed, the value of Δ as defined in eqn. 9.57 is related to the equilibrium constant (eqn. 9.56) for the exchange reaction as:

$$\Delta \approx -1000 \ln \frac{K}{K_r} \qquad (9.61)$$

Since K_r refers to the case of random distribution of isotopes, it is equal to 1 and since K will have a value close to 1, we may use the approximation $\ln x \approx 1 - x$, so that eqn. 9.61 reduces to

$$\Delta \approx (K - 1) \times 1000 \qquad (9.62)$$

Just as isotope fractionations depend on temperature due to the vibrational partition coefficient temperature dependency, so too does the extent of isotope clumping, expressed in eqn. 9.57. Thus, this effect can be used as a geothermometer, with the advantage that is independent of the isotopic composition of other phases.

Carbonates are among the most common sedimentary minerals. Most form by precipitation from water (often biologically mediated), and the oxygen isotopic fractionation resulting from this precipitation reaction is temperature dependent and $\delta^{18}O$ values of carbonates are useful geothermometer and paleoclimatic tools. The isotopic ratios of carbonates are determined by gas-source mass spectrometry of CO_2 released by digestion with acid. Carbon has two isotopes and oxygen 3, so they can combine to form 12 distinct isotopologues. Ghosh et al. (2006) showed that the abundance of $^{13}C^{18}O^{16}O$ isotopologue, with mass 47, of analyzed CO_2 gas was proportional to the abundance of $^{13}C-^{18}O$ bonds in the carbonate. Thus, the extent of clumping, in particular the Δ_{47} value of CO_2 released from biogenic carbonates, depends on the temperature at which the carbonate precipitated (Figure 9.8). Cave speleothems exhibit a similar correlation, but it is offset to slightly to lower Δ_{47} (~0.05‰) for kinetic reasons. Thus, clumped isotope analysis is a useful geothermometer just as $\delta^{18}O$. However, the $\delta^{18}O$ of carbonates also depends on the isotopic composition of the water from which they precipitated; calculating temperature thus requires knowing that isotopic composition. In contrast, the relative abundance

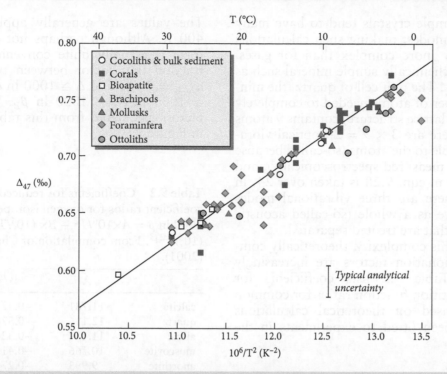

Figure 9.8 Variation of Δ_{47} of biogenic carbonates with the temperature in which they precipitated from water demonstrating the potential of clumped isotope analysis of carbonates as a geothermometer. Line is the original calibration of Ghorsh et al. (2006). Based on Eiler (2011).

of carbonate isotopologues depends only on temperature and the isotopic composition of the carbonate. Since the latter is readily measured, temperature can be calculated from the abundance of the isotopologues, provided those can be measured. We will briefly consider application of clumped isotope analysis in a subsequent section. More details can be found in Eiler (2011) and Eiler (2013).

9.3 ISOTOPE GEOTHERMOMETRY

One important use of stable isotope geochemistry is geothermometry. Like "conventional" chemical geothermometers, stable isotope geothermometers are based on the temperature dependence of the equilibrium constant. As we have seen, this dependence may be generally expressed as:

$$\ln \alpha \propto \frac{1}{T^2} \qquad (9.63)$$

From this expression we see that α tends 1 at infinite temperature, corresponding to no isotope separation, and to infinity at absolute zero, corresponding to complete separation

(but reaction rates tend to 0 at absolute 0, so this can never be achieved!). We can obtain a qualitative understanding of why this is so by recalling that the entropy of a system increases with temperature. At infinite temperature, there is complete disorder, hence isotopes would be mixed randomly between phases (ignoring for the moment the slight problem that at infinite temperature there would be neither phases nor isotopes). At absolute zero, there is perfect order, hence no mixing of isotopes between phases. A and B are, however, sufficiently invariant over a limited range of temperatures that they can be viewed as constants. We have also noted that at low temperatures, the form of eqn. 9.63 changes to $\alpha \propto 1/T$.

Fractionation factors for mineral pairs, the basis of isotope geothermometry, have generally been determined experimentally. Just as for gases, however, they can also be calculated from partition functions. The good news is that neither rotational nor translational modes of motion are available to atoms in a crystal lattice, so these terms are dropped from eqn. 9.28. The bad news is that even simple crystals tend to have many vibrational modes, making such calculations considerably more complex than for gases. Consider a chemically simple mineral such as quartz, SiO_2. The unit cell of quartz (the minimum number of atoms needed to completely describe the lattice structure) contains 9 atoms (Si_3O_6). There are $3 \times 9 = 24$ optical vibrations available to the atoms (so-called because they can be measured spectroscopically), and the product in eqn. 9.28 is taken over 24. In addition, there are three vibrational modes of the lattice as a whole (so-called acoustic vibrations) that are treated separated.

Despite this complexity, theoretically computed fractionation factors are increasingly available. Table 9.2 lists coefficients for reduced partition function ratios for common minerals based on theoretical calculations that have been fitted to experiments in the form of

$$1000 \ln \beta = A \times \left(\frac{10^6}{T^2}\right) + B \times \left(\frac{10^6}{T^2}\right)^2$$
$$+ C \times \left(\frac{10^6}{T^2}\right)^3 \qquad (9.64)$$

The values are generally applicable above 400 K. Although perhaps not obvious, this form is actually quite convenient since the fractionation factor between two phases is $\alpha_{A-B} = \beta_A/\beta_B$ and $\Delta \cong 1000 \ln \alpha$. Thus, $\Delta_{A-B} \cong 1000 \ln \beta_A - 1000 \ln \beta_B$. Fractionation factors calculated from this table are shown in Figure 9.9.

Table 9.2 Coefficients for reduced partition coefficient ratios for oxygen isotope fractionation $1000 \ln \beta = A \times 10^6/T^2 + B \times (10^6/T^2)^2 + C \times (10^6/T^2)^3$. From compilation of Chacko et al. (2001).

	A	B	C
calcite	11.781	−0.42	0.0158
quartz	12.116	−0.370	0.0123
albite	11.134	−0.326	0.0104
muscovite	10.766	−0.412	0.0209
anorthite	9.993	−0.271	0.0082
phlogopite	9.969	−0.382	0.0194
diopside	9.237	−0.199	0.0053
forsterite	8.236	−0.142	0.0032
rutile	7.258	−0.125	0.0033
magnetite	5.674	−0.038	0.0003

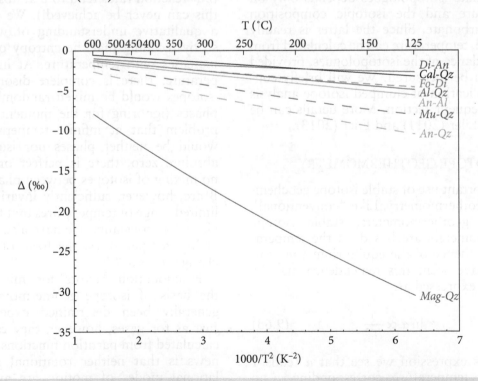

Figure 9.9 Oxygen isotope fractionation for several mineral pairs as a function of temperature based in partition coefficient ratios listed in Table 9.2.

Because of the dependence of the equilibrium constant on the inverse square of temperature, stable isotope geothermometry is employed primarily below magmatic temperatures. At temperatures in excess of 1000°C or so, the fractionations are generally small, making accurate temperatures difficult to determine from them, although increased analytical precision makes high temperature geothermometry more tractable

than in the past. Even at temperatures of the upper mantle, fractionations, although small, remain significant. As temperature increases, the latter two terms in eqn. 9.64 become increasingly small and fractionation factors become linear functions of $1/T^2$; Table 9.3 lists these high temperature mineral–mineral fractionation factors. In addition, systems at high temperature are more likely to achieve equilibrium than at lower temperatures.

Example 9.2   Oxygen isotope geothermometry

A granite-gneiss contains coexisting quartz, muscovite and magnetite with the following $\delta^{18}O$: quartz, 11.1; magnetite, 1.9, muscovite 9.2. Find the temperature of equilibration based on the Δ_{mag-qz} calculated from β-factors listed in Table 9.2.

Answer: The fractionation factor can be found as $\Delta_{mag-qz} \cong 1000 \ln \beta_{mag} - 1000 \ln \beta_{qz}$. First, we calculate the observed $\Delta^{18}O$ for the quartz-magnetite pair, which is $\Delta_{mag-qz} = 1.9{-}11.1 = -9.2$. Next, we express the fractionation factors for the mineral pair as a function of temperature using the β-factors listed in Table 9.2. For quartz-magnetite:

$$\Delta_{mag-qz} = (A_{mag} - A_{qz}) \times \frac{10^6}{T^2} + (B_{mag} - B_{qz}) \times \left(\frac{10^6}{T^2}\right)^2 + (C_{mag} - C_{qz}) \times \left(\frac{10^6}{T^2}\right)^3 \quad (9.65)$$

substituting values from Table 9.2 for quartz, this becomes:

$$\Delta_{mag-qz} = 6.442 \times \frac{10^6}{T^2} - 0.332 \times \left(\frac{10^6}{T^2}\right)^2 + 0.009 \times \left(\frac{10^6}{T^2}\right)^3 \quad (9.65a)$$

Representing $10^6/T^2$ as x and a as $A_{mag}-A_{qz}$, etc. and rearranging eqn. 9.65 as:

$$ax + bx^2 + cx^3 - \Delta = 0$$

we recognize eqn. 9.65 as a cubic equation, which can be solved directly, albeit with quite an ugly formula. Alternatively, there are several online calculators to solve cubic equations or it can be solved numerically in Excel or MATLAB (cubic equations can have several roots, but this equation should yield only one reasonable one). A third alternative is to note that in this case the c term is quite small, and we might guess we can drop it with little loss of accuracy so that 9.65a becomes a readily solved quadratic equation.

Our solution to the cubic equation is $10^6/T^2 = 1.54$ and $T = 805$ K or 532°C. Where we have data on multiple mineral pairs, we would also want to compute temperatures for those. Only when multiple pairs give reasonable agreement can we be confident that equilibrium has been achieved and the calculated temperatures are reasonable. Determining whether magnetite-muscovite and muscovite-quartz yield the same temperatures is left to the reader as Problem 9.4.

Table 9.4 lists similar coefficients for the sulfur fractionation between H_2S and sulfur-bearing compounds. Recall that if phases α and γ, and α and β are in equilibrium with each other, then γ is also in equilibrium

with β. Thus, this table may be used to obtain the fractionation between any two of the phases listed. For example, the fractionation between covellite (CuS) and sphalerite (ZnS) can be calculated as: $\Delta_{CuS-ZnS} = \Delta_{H2S-CuS} -$

Table 9.3 Mineral pair oxygen isotope fractionation factors for temperatures >600°C. From compilation of Chacko et al. (2001).

	calcite	albite	muscovite	anorthite	phlogopite	diopside	forsterite	rutile	magnetite
quartz	0.38	0.94	1.37	1.99	2.16	2.75	3.67	4.69	6.29
calcite		0.56	0.99	1.61	1.78	2.37	3.29	4.31	5.91
albite			0.43	1.05	1.22	1.81	2.37	3.75	5.35
muscovite				0.62	0.79	1.38	2.30	3.32	4.92
anorthite					0.17	0.76	1.68	2.7	4.3
phlogopite						0.59	1.51	2.53	4.13
diopside							0.92	1.91	3.54
forsterite								1.02	2.62
rutile									1.6

Table 9.4 Coefficients for sulfur isotope fractionation: $\Delta_{\phi-H_2S} = A \times 10^6/T^2 + B \times 10^3/T + C$ (T in kelvin).

ϕ	A	B	C	T (°C) range
HS⁻	−0.06±0.15		−0.6	50–350
S²⁻	−0.21	−1.23	−1.23	
S	−0.16		±0.5	200–400
CaSO₄	5.26	6.0 ± 0.5		200–350
SO₂	4.7	0	0.5±−0.5	350–1050
FeS₂	0.4 ± 0.08			200–700
FeS	0.25			
ZnS	0.10			50–705
CuS	0.04			280–490
CuFeS₂	0.05			
MoS₂	0.45 ± 0.1			
Ag₂S	−0.8 ± 0.1			
PbS	−0.64			50–700

From Ohmoto and Rye (1979) and Li and Liu (2006).

$\Delta_{H2S-ZnS}$. These fractionation factors are illustrated in Figure 9.10.

All geothermometers are based on the apparently contradictory assumptions that complete equilibrium was achieved between phases during, or perhaps after, formation of the phases, but that the phases did not re-equilibrate when they subsequently cooled. The reason these assumptions can be made and geothermometry works at all is the exponential dependence of reaction rates on temperature that we discussed in Chapter 5. Isotope geothermometers have the same implicit assumptions about the achievement of equilibrium as other geothermometers.

The importance of the equilibrium basis of geothermometry must be emphasized. Because most stable isotope geothermometers (though not all) are applied to relatively low-temperature situations, violation of the assumption that complete equilibrium was achieved is not uncommon. We have seen that isotopic fractionations may arise from kinetic as well as equilibrium effects. If reactions do not run to completion, the isotopic differences may reflect kinetic effects as much as equilibrium effects. There are other problems that can result in incorrect temperature as well; for example, the system may partially re-equilibrate at some lower temperature during cooling. A further problem with isotope geothermometry is that free energies of the exchange reactions are relatively low, meaning there is little chemical energy available to drive the reaction. Indeed, isotopic equilibrium probably often depends

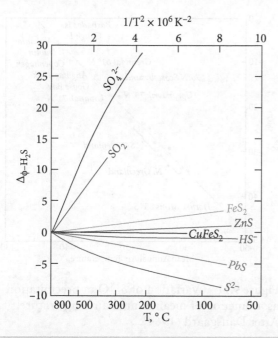

Figure 9.10 Sulfur isotope fractionation between $H_2S_{(aq)}$ and various sulfur compounds.

on other reactions occurring that mobilize the element involved in the exchange. Solid-state exchange reactions will be particularly slow at temperatures well below the melting point. Equilibrium between solid phases will thus generally depend on reaction of these phases with a fluid. This latter point is true of "conventional" geothermometers as well, and metamorphism, one of the important areas of application of isotope geothermometry, generally occurs in the presence of a fluid.

Isotope geothermometers do have several advantages over conventional chemical ones. First, as we have noted, there is no volume change associated with isotopic exchange reactions and hence no pressure-dependence of the equilibrium constant. However, Rumble has suggested an indirect pressure-dependence, wherein the fractionation factor depends on fluid composition, which in turn depends on pressure. Second, whereas conventional chemical geothermometers are generally based on solid solution, isotope geothermometers can make use of pure phases such as SiO_2. Generally, any dependence on the composition of phases involved is of relatively second-order importance (there are, however, exceptions). For example, isotopic exchange between calcite

and water is independent of the concentration of CO_2 in the water. Compositional effects can be expected only where they affect bonds formed by the element involved in the exchange. For example, we noted that substitution of Al for Si in plagioclase affects O isotope fractionation factors because of the nature of the bond with oxygen.

9.4 ISOTOPIC FRACTIONATION IN THE HYDROLOGIC SYSTEM

As we noted above, isotopically light water has a higher vapor pressure, and hence lower boiling point, than isotopically heavy water. Let's consider this in a bit more detail. Raoult's law (eqn. 3.8) states that the partial pressure of a species above a solution is equal to its molar concentration in the solution times the partial pressure exerted by the pure solution. So for the two isotopic species of water (we will restrict ourselves to O isotopes for the moment), Raoult's law is:

$$p_{H_2^{16}O} = p^o_{H_2^{16}O}[H_2^{16}O] \qquad (9.66a)$$

and

$$p_{H_2^{18}O} = p^o_{H_2^{18}O}[H_2^{18}O] \qquad (9.66b)$$

where p is the partial pressure and, as usual, the square brackets indicate the aqueous concentration. Since the partial pressure of a species is proportional to the number of atoms of that species in a gas, we can define the fractionation factor, α, between liquid water and vapor as:

$$\alpha_{v/\ell} = \frac{p_{H_2^{18}O}/p_{H_2^{16}O}}{[H_2^{18}O]/[H_2^{16}O]} \qquad (9.67)$$

Substituting eqns. 9.66a and 9.66b into 9.67, we arrive at the relationship:

$$\alpha_{v/\ell} = \frac{p^o_{H_2^{18}O}}{p^o_{H_2^{16}O}} \qquad (9.68)$$

Thus, interestingly enough, the fractionation factor for oxygen between water vapor and liquid turns out to be just the ratio of the standard state partial pressures.

The next question is how the partial pressures vary with temperature. Thermodynamics provides the answer. The temperature dependence of the partial pressure of a species may be expressed as:

$$\frac{d \ln p}{dT} = \frac{\Delta H}{RT^2} \qquad (9.69)$$

where T is temperature, ΔH is the enthalpy or latent heat of evaporation, and R is the gas constant. Over a sufficiently small temperature range, we can assume that ΔH is independent of temperature. Rearranging and integrating, we obtain:

$$\ln p = \frac{\Delta H}{RT} + \text{const} \qquad (9.70)$$

We can write two such equations, one for $[H_2^{16}O]$ and one for $[H_2^{18}O]$. Dividing one by the other we obtain:

$$\ln \frac{p^o_{H_2^{18}O}}{p^o_{H_2^{16}O}} = A - \frac{B}{RT} \qquad (9.71)$$

where A and B are constants. This can be rewritten as:

$$\alpha = ae^{B/RT} \qquad (9.72)$$

where $a = e^A$. Over a larger range of temperature, ΔH is not constant. The log of the fractionation factor in that case depends on the inverse square of temperature, so that temperature dependence of the fractionation factor can be represented as:

$$\ln \alpha = a - \frac{B}{T^2} \qquad (9.73)$$

Given the fractionation between water and vapor, we might predict that there will be considerable variation in the isotopic composition of water in the hydrologic cycle, and indeed there is. Figure 9.11 shows the global variation in $\delta^{18}O$ in *precipitation*.

Precipitation of rain and snow from clouds is a good example of Rayleigh condensation. Isotopic fractionations will therefore follow eqn. 9.52. Thus, in addition to the temperature dependence of α, the isotopic composition of precipitation will also depend on f, the fraction of water vapor remaining in

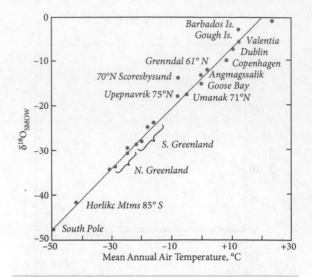

Figure 9.11 Variation of $\delta^{18}O$ in precipitation as a function of mean annual air temperature. After Dansgaard (1964).

the air. The farther air moves from the site of evaporation, the more water is likely to have condensed and fallen as rain, and therefore, the smaller the value of f in eqn. 9.52. Thus, fractionation will increase with distance from the region of evaporation (principally tropical oceans). Topography also plays a role as mountains force air up, causing it to cool and water vapor to condense, hence decreasing f. Precipitation from air that has passed over a mountain range will be isotopically lighter than precipitation on the ocean side of a mountain range. These factors are illustrated in Figure 9.12.

Hydrogen as well as oxygen isotopes will be fractionated in the hydrologic cycle. Indeed, $\delta^{18}O$ and δD are reasonably well correlated in precipitation, as shown in Figure 9.13. The fractionation of hydrogen isotopes, however, is greater because the mass difference is greater. This correlation is known as the *meteoric water line*. When meteoric water undergoes significant evaporation, such as rivers and lakes in arid regions, the residual water will form a trend of lower slope, that is, enrichment in ^{18}O relative to ^2H that intersects the meteoric water line at the composition of unevaporated water (Gat, 1996), as illustrated by the "evaporative trend" shown in Figure 9.13.

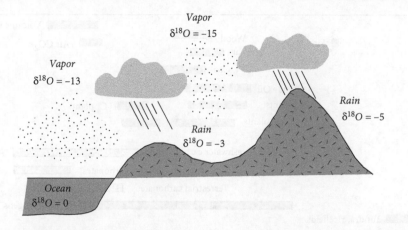

Figure 9.12 Process of Rayleigh fractionation and the decreasing $\delta^{18}O$ in rain as it moves inland.

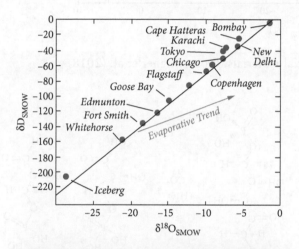

Figure 9.13 Northern hemisphere variation in δD and $\delta^{18}O$ in precipitation and meteoric waters. The relationship between δD and $\delta^{18}O$ is approximately $\delta D = 8 \times \delta^{18}O + 10$. After Dansgaard (1964).

9.5 ISOTOPIC FRACTIONATION IN BIOLOGICAL SYSTEMS

Biological processes often involve large isotopic fractionations. Indeed, for carbon, nitrogen, and sulfur, biological processes are the most important causes of isotopic fractionations. The largest fractionations of carbon occur during the initial production of organic matter by the so-called primary producers, or *autotrophs*. These include all plants and many kinds of bacteria. The most important means of production of organic matter is photosynthesis, but organic matter may also be produced by chemosynthesis, for example at mid-ocean ridge hydrothermal vents. Large fractionations of both carbon and nitrogen isotopes occur during primary production. Additional fractionations also occur in subsequent reactions and up through the food chain as *heterotrophs* consume primary producers, but these are generally smaller.

9.5.1 Carbon isotope fractionation during photosynthesis

Biological processes are the principal cause of variations in carbon isotope ratios, as is apparent in Figure 9.14: isotopic variation in biogenic organic matter vastly exceeds that of inorganic carbon. The most important of these processes is photosynthesis. As we noted earlier, photosynthetic fractionation of carbon isotopes is primarily kinetic. The early work of Park and Epstein (1960) suggested that fractionation occurred in several steps. Subsequent work has elucidated the fractionations involved in these steps, which we will consider in slightly more detail.

For terrestrial plants (those utilizing atmospheric CO_2), the first step is diffusion of CO_2 into the boundary layer surrounding the leaf, through the stomata, and internally in the leaf. On theoretical grounds, a fractionation of $-4.4‰$ is expected, as we found in eqn. 9.42. Marine algae and aquatic plants can utilize either dissolved CO_2 or HCO_3^- for photosynthesis:

$$CO_{2(g)} \rightarrow CO_{2(aq)} + H_2O$$
$$\rightarrow H_2CO_3 \rightarrow H^+ + HCO_3^-$$

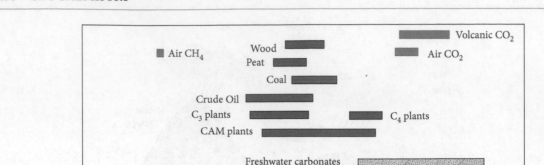

Figure 9.14 Carbon isotopic composition in terrestrial materials. From Wagner et al. (2018).

An equilibrium fractionation of +0.9‰ is associated with dissolution ($^{13}CO_2$ will dissolve more readily), and an equilibrium +7 to +8‰ fractionation occurs during hydration and dissociation of H_2CO_3.

At this point, there is a divergence in the chemical pathways. Most plants use an enzyme called *ribulose bisphosphate carboxylase oxygenase* (RUBISCO) to catalyze a reaction in which *ribulose bisphosphate carboxylase* reacts with one molecule of CO_2 to produce two molecules of 3-phosphoglyceric acid, a compound containing three carbon atoms, in a process called *carboxylation* (Figure 9.15). The carbon is subsequently reduced, carbohydrate formed, and the ribulose bisphosphate regenerated. Such plants are called C_3 plants, and this process is called the Benson-Calvin cycle, or simply the Calvin cycle. C_3 plants constitute about 90% of all plants today and include algae and autotrophic bacteria and comprise the majority of cultivated plants, including wheat, rice, and nuts. There is a kinetic fractionation associated with carboxylation of ribulose bisphosphate that has been determined by several methods to be −29.4‰ in higher terrestrial plants. Bacterial carboxylation has different reaction mechanisms and a smaller fractionation of about −20‰. However, for terrestrial plants a fractionation of about

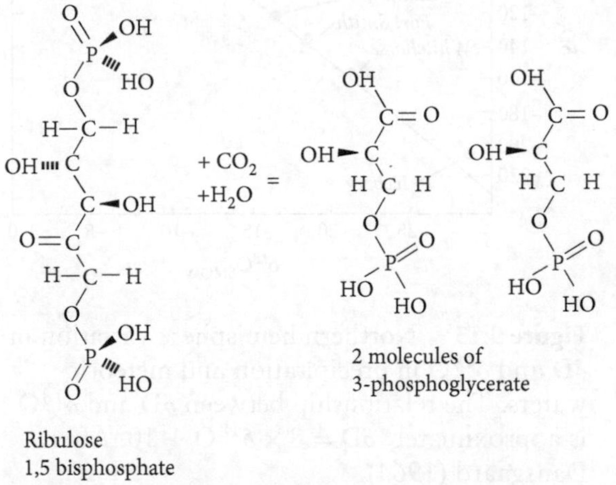

Figure 9.15 Ribulose bisphosphate (RuBP) carboxylation, the reaction by which C_3 plants fix carbon during photosynthesis.

−34‰ is expected from the sum of the individual fractionations. The actual observed total fractionation is in the range of −20 to −30‰ with a mean of about −27‰. This range and this difference between it and the expected fractionation reflect depletion of CO_2 inside plants. As photosynthesis preferentially consumes $^{12}CO_2$ in the plant interior, the remaining CO_2 becomes enriched in ^{13}C, so there is less net fractionation. This can be expressed by the following equation (e.g.,

Farquhar et al., 1982):

$$\Delta_{net} = \Delta_{diff} + (c_{in}/c_{ex})(\Delta_{carbox} - \Delta_{diff}) \quad (9.74)$$

where Δ_{diff} and Δ_{carbox} are the fractionation factors for diffusion and carboxylation, respectively, and c_{in} and c_{ex} are the interior and exterior CO_2 concentrations, respectively.

The other photosynthetic pathway is the Hatch–Slack cycle, used by the C_4 plants that include hot-region grasses and related crops such as maize and sugarcane. These plants use *phosphenol pyruvate carboxylase* (PEP) to fix the carbon initially and form oxaloacetate, a compound that contains four carbons (Figure 9.16). A much smaller fractionation, about −2.0 to −2.5‰, occurs during this step. In phosphoenol pyruvate carboxylation, the CO_2 is fixed in outer mesophyll cells as oxaloacetate and carried as part of a C_4 acid, either malate or aspartate, to inner bundle sheath cells where it is decarboxylated and refixed by RuBP (Figure 9.17). The environment in the bundle sheath cells is almost a closed system, so that virtually all the carbon carried there is refixed by RuBP, so there is less fractionation during this step. Thus, C_4 plants have an average $\delta^{13}C$ of −13‰, much less than C_3 plants. As in the case of RuBP

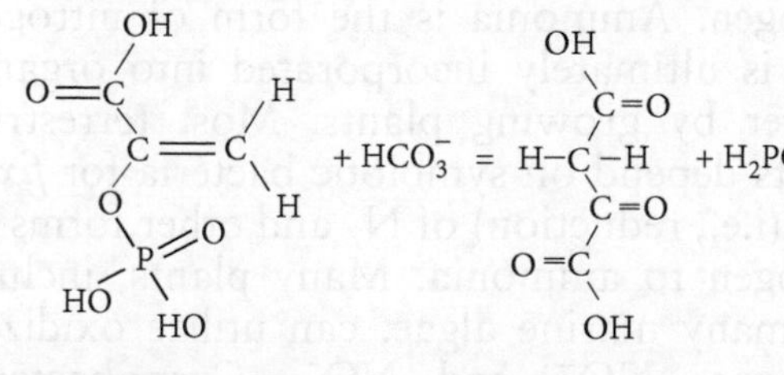

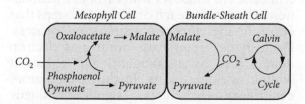

Figure 9.16 Phosphoenolpyruvate carboxylation, the reaction by which C_4 plants fix CO_2 during photosynthesis.

Figure 9.17 Chemical pathways in C_4 photosynthesis.

photosynthesis, the fractionation appears to depend on the ambient concentration of CO_2.

A third group of plants, the CAM plants, has a unique metabolism called the *Crassulacean acid metabolism*. These plants generally use the C_4 pathway but can use the C_3 pathway under certain conditions. These plants are generally succulents adapted to arid environments and include pineapple and many cacti; they have $\delta^{13}C$ intermediate between C_3 and C_4 plants.

Terrestrial plants, which utilize CO_2 from the atmosphere, generally produce greater fractionations than marine and aquatic plants, which utilize dissolved CO_2 and HCO_3^-, together referred to as *dissolved inorganic carbon* or DIC. As we noted above, there is about a +8‰ equilibrium fractionation between dissolved CO_2 and HCO_3^-. Since HCO_3^- is about two orders of magnitude more abundant in seawater than dissolved CO_2, marine algae utilize this species, and hence tend to show a lower net fractionation between dissolved carbonate and organic carbon during photosynthesis. Diffusion is slower in water than in air, so diffusion is often the rate-limiting step. Most aquatic plants have some membrane-bound mechanism to pump DIC, which can be turned on when DIC is low. When DIC concentrations are high, fractionation in aquatic and marine plants is generally similar to that in terrestrial plants. When it is low and the plants are actively pumping DIC, the fractionation is less because most of the carbon pumped into cells is fixed. Thus, carbon isotope fractionations between dissolved inorganic carbon and organic carbon can be as low as 5‰ in algae. $\delta^{13}C$ in marine algae organic matter averages about −21‰.

Not surprisingly, the carbon isotope fractionation in C fixation is also temperature-dependent. Thus, higher fractionations are observed in cold-water phytoplankton than in warm-water species. However, this observation also reflects a kinetic effect: there is generally less dissolved CO_2 available in warm waters because of the decreasing solubility with temperature. As a result, a larger fraction of the CO_2 is utilized and there is consequently less fractionation. Surface waters of the ocean are generally enriched in ^{13}C (relative to ^{12}C) because of uptake of ^{12}C during photosynthesis (Figure 9.18). The

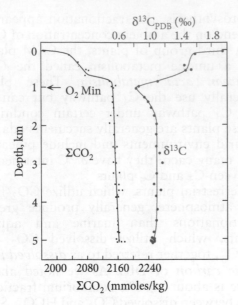

Figure 9.18 Depth profile of total dissolved inorganic carbon and $\delta^{13}C$ in the North Atlantic. Data from Kroopnick et al. (1972).

degree of enrichment depends on the productivity: biologically productive areas show greater depletion. Deep water, on the other hand, is enriched in ^{12}C. Organic matter falls through the water column and is decomposed and *remineralized* (i.e., converted to inorganic carbon) by the action of bacteria, enriching deep water in ^{12}C and total DIC. Thus, biological activity acts to "pump" carbon, and particularly ^{12}C from surface to deep waters.

Nearly all organic matter originates through photosynthesis. Subsequent reactions convert the photosynthetically produced carbohydrates to a variety of other organic compounds utilized by organisms. Within organisms, additional fractionations occur between these compounds. Most lipids are ^{13}C–depleted compared with starch ($\Delta\delta^{13}C_{Lipid-Starch} \approx -6.0‰$), sugars ($\Delta\delta^{13}C_{Lipid-Sugar} \approx -5.5‰$), and proteins ($\Delta\delta^{13}C_{Lipid-Protein} \approx -5.0‰$) with some fatty acids being as much as 10‰ lighter than carbohydrates. These fractionations are thought to be kinetic in origin and may partly arise from organic C–H bonds being enriched in ^{12}C and organic C–O bonds being enriched in ^{13}C. ^{12}C is preferentially consumed in respiration (again, because bonds are weaker

and it reacts more rapidly), which enriches residual organic matter in ^{13}C. The carbon isotopic composition of organisms becomes slightly more positive moving up the food chain.

The principal exception to the creation of organic matter through photosynthesis is *chemosynthesis*. In chemosynthesis, chemical reactions rather than light provide the energy to "fix" CO_2. Regardless of the energy source, however, fixation of CO_2 involves the Benson–Calvin cycle and RUBISCO. Not surprisingly, then, chemosynthesis typically results in carbon isotope fractionations similar to those of photosynthesis. Thus, large carbon isotope fractionations are the signature of both photosynthesis and chemosynthesis.

9.5.2 Nitrogen isotope fractionation in biological processes

Nitrogen is another important element in biological processes, being an essential component of all amino acids, proteins, and other key compounds such as RNA and DNA. There are five important forms of inorganic nitrogen: N_2, NO_3^-, NO_2^-, NH_3, and NH_4^+. Equilibrium isotope fractionations occur between these five forms, and kinetic fractionations occur during biological assimilation of nitrogen. Ammonia is the form of nitrogen that is ultimately incorporated into organic matter by growing plants. Most terrestrial plants depend on symbiotic bacteria for *fixation* (i.e., reduction) of N_2 and other forms of nitrogen to ammonia. Many plants, including many marine algae, can utilize oxidized nitrogen, NO_3^- and NO_2^-. Cyanobacteria (blue-green algae) can utilize N_2 directly and legumes harbor nitrogen-fixing bacteria in their root nodules, effectively allowing them to utilize N_2. Except in these latter cases, nitrogen must first be reduced by the action of reductase enzymes. As with carbon, fractionation may occur in each of the several steps that occur in the nitrogen assimilation process. Denitrifying bacteria use nitrates as electron donors (i.e., as an oxidant) and reduce it without assimilating it. In this dissimilatory denitrification, there is a significant kinetic fractionation, with the light isotope, ^{14}N, being preferentially reduced, leaving residual nitrate enriched in ^{15}N by 6–7‰.

While isotope fractionations during assimilation of ammonium are still poorly understood, it appears there is a strong dependence on the concentration of the ammonium ion. Such dependence has been observed, as for example in Figure 9.19. The complex dependence in Figure 9.19 is interpreted as follows. The increase in fractionation from highest to moderate concentrations of ammonium reflects the switching on of active ammonium transport by cells. At the lowest concentrations, essentially all available nitrogen is transported into the cell and assimilated, so there is little fractionation observed.

The isotopic compositions of marine particulate nitrogen and non-nitrogen-fixing plankton are typically –3‰ to +12‰ δ^{15}N. Non-nitrogen-fixing terrestrial plants unaffected by artificial fertilizers generally have a narrower range of +6‰ to +13 per mil. Marine blue-green algae range from –4 to +2, with most in the range of –4 to –2‰. Most nitrogen-fixing terrestrial plants fall in the range of –2 to +4‰, and hence are typically lighter than non-nitrogen-fixing plants.

9.5.3 Oxygen and hydrogen isotope fractionation by plants

Oxygen is incorporated into biological material from CO_2, H_2O, and O_2. However, both CO_2 and O_2 are in oxygen isotopic equilibrium with water during photosynthesis, and water is the dominant source of O. Therefore, the isotopic composition of plant water determines the oxygen isotopic composition of plant material. The oxygen isotopic composition of plant material seems to be controlled by exchange reactions between water and carbonyl oxygens (oxygens doubly bound to carbon):

$$C{=}^{16}O + H_2{}^{18}O \rightarrow C{=}^{18}O + H_2{}^{16}O$$

Fractionations of +16 to +27‰ (i.e., the organically bound oxygen is heavier) have been measured for these reactions. Consistent with this, cellulose from most plants has δ^{18}O of +27 ± 3‰. Other factors, however, play a role in the oxygen isotopic composition of plant material. First, the isotopic composition of water varies from δ^{18}O ≈ –55‰ in Arctic regions to δ^{18}O ≈ 0‰ in the oceans. Second, less than complete equilibrium may be achieved if photosynthesis is occurring at a rapid pace, resulting in less fractionation. Finally, some fractionation of water may occur during transpiration, with residual water in the plant becoming heavier.

Hydrogen isotope fractionation during photosynthesis occurs such that the light isotope is enriched in organic material. In marine algae, isotope fractionations of –100 to –150‰ have been observed, which is a little more than that observed in terrestrial plants (–86 to –120‰). Among terrestrial plants, there appears to be a difference between C_3 and C_4 plants. The former show fractionations of –117 to –121‰, while fractionations of –86 to –109‰ have been observed in C_4 plants. However, little is known in detail about the exact mechanisms of fractionation.

As for oxygen, variations in the isotopic composition of available water and fractionation during transpiration are important in controlling the hydrogen isotopic composition of plants. This is illustrated in Figure 9.20.

9.5.4 Biological fractionation of sulfur isotopes

Though essential to life, sulfur is a minor component in living tissue (C:S atomic ratio is about 200). Plants take up sulfur as sulfate and subsequently reduce it to sulfide and incorporate it into cysteine, an amino acid. There is apparently no fractionation of sulfur

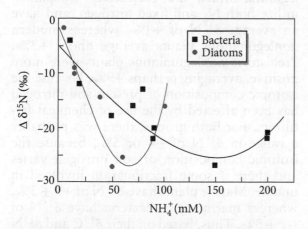

Figure 9.19 Dependence of nitrogen isotope fractionation by bacteria and diatoms on dissolved ammonium concentration. After Fogel and Cifuentes (1993).

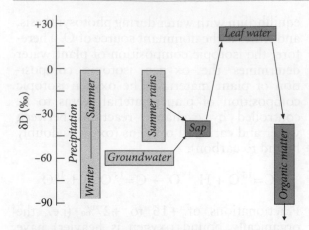

Figure 9.20 Isotopic fractionations of hydrogen during primary production in terrestrial plants. After Fogel and Cifuentes (1993).

isotopes in transport across cell membranes and incorporation into cysteine, but there is a fractionation of +0.5 to −4.5‰ in the reduction process, referred to as *assimilatory sulfate reduction*. This is substantially less than the expected fractionation of about −20‰, suggesting that most sulfur taken up by primary producers is reduced and incorporated into tissue.

Sulfur, however, plays two other important roles in biological processes. First, sulfur, in the form of sulfate, can act as an electron acceptor or oxidant, and is utilized as such by sulfur-reducing bacteria. This process, in which H_2S is liberated, is called *dissimilatory sulfate reduction* and plays an important role in biogeochemical cycles, both as a sink for sulfur and a source for atmospheric oxygen. A large fractionation of +5 to −46‰ is associated with this process. This process produces by far the most significant fractionation of sulfur isotopes, and thus governs the isotopic composition of sulfur in the exogene. Sedimentary sulfate typically has $\delta^{34}S$ of about +17‰, which is similar to the isotopic composition of sulfate in the oceans (+20‰), while sedimentary sulfide has a $\delta^{34}S$ of around −18‰. The living biomass has a $\delta^{34}S$ of ≈ 0.

The final important role of sulfur is as a reductant. Sulfide is an electron acceptor used by anaerobic photosynthetic green sulfur bacteria as well as other microbes in the reduction of CO_2 to organic carbon. Among these are the chemosynthetic archaea of submarine hydrothermal vents. They utilize H_2S emanating from the vents as an energy source and form the base of the food chain in these unique ecosystems. A fractionation of +2 to −18‰ is associated with this process.

9.5.5 Isotopes and diet: you are what you eat

As we have seen, the two main photosynthetic pathways, C_3 and C_4, lead to organic carbon with different carbon isotopic compositions. Terrestrial C_3 plants have $\delta^{13}C$ values that average about −27‰, C_4 plants an average $\delta^{13}C$ of about −13‰. Marine plants (which are all C_3) utilize dissolved bicarbonate rather than atmospheric CO_2. Seawater bicarbonate is about 8.5‰ heavier than atmospheric CO_2, and marine plants average about 7.5‰ heavier than terrestrial C_3 plants. In addition, because the source of the carbon they fix is isotopically more variable, the isotopic composition of marine plants is also more variable. Finally, marine cyanobacteria (blue-green algae) tend to fractionate carbon isotopes less during photosynthesis than do true marine plants, so they tend to average 2–3‰ higher in $\delta^{13}C$.

Plants may also be divided into two types based on their source of nitrogen: those that can utilize N_2 directly, and those that utilize only "fixed" nitrogen in ammonia and nitrate. The former include the legumes (e.g., beans, peas, etc.) and marine cyanobacteria. The legumes, which are exclusively C_3 plants, utilize both N_2 and fixed nitrogen, and have an average $\delta^{15}N$ of +1‰, whereas modern nonleguminous plants average about +3‰. Prehistoric nonleguminous plants were more positive, averaging perhaps +9‰, because the isotopic composition of present soil nitrogen has been affected by the use of chemical fertilizers. For both groups, there was probably a range in $\delta^{15}N$ of ±4 or 5‰, because the isotopic composition of soil nitrogen varies and there is some fractionation involved in uptake. Marine plants have $\delta^{15}N$ of +7 ± 5‰, whereas marine cyanobacteria have $\delta^{15}N$ of −1 ± 3‰. Thus, based on their $\delta^{13}C$ and $\delta^{15}N$ values, autotrophs can be divided into several groups, which are summarized in Figure 9.21.

DeNiro and Epstein (1978) studied the relationship between the carbon isotopic composition of animals and their diet. They

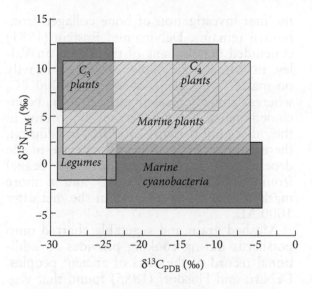

Figure 9.21 Relationship between δ^{13}C and δ^{15}N among the principal classes of autotrophs. Adapted from DeNiro (1987).

found that there is only slight further fractionation of carbon by animals, and that the carbon isotopic composition of animal tissue closely reflects that of the animal's diet. Typically, carbon in animal tissue is about 1‰ heavier than their diet. The small fractionation between animal tissue and diet is a result of the slightly weaker bond formed by ^{12}C compared with ^{13}C. The weaker bonds are more readily broken during respiration, and, not surprisingly, the CO_2 respired by most

animals investigated was slightly lighter than their diet. Thus, only a small fractionation in carbon isotopes occurs as organic carbon passes up the food web. Terrestrial food chains usually do not have more than three trophic levels, implying a maximum further fractionation of +3‰; marine food chains can have up to seven trophic levels, implying a maximum carbon isotope difference between primary producers and top predators of 7‰. These differences are smaller than the range observed in primary producers.

In another study, DeNiro and Epstein (1981) found that δ^{15}N of animal tissue reflects the δ^{15}N of the animal's diet but is typically 3–4‰ higher than that of the diet. Thus, in contrast to carbon, significant fractionation of nitrogen isotopes will occur as nitrogen passes up the food chain. These relationships are summarized in Figure 9.22. *The significance of these results is that it is possible to infer the diet of an animal from its carbon and nitrogen isotopic composition.*

Schoeninger and DeNiro (1984) found that the carbon and nitrogen isotopic composition of bone collagen in animals was similar to that of body tissue as a whole. Apatite in bone appears to undergo isotopic exchange with meteoric water once it is buried, but bone collagen and tooth enamel appear to be robust and retain their original isotopic compositions. This means that the nitrogen and carbon isotopic composition of fossil bone

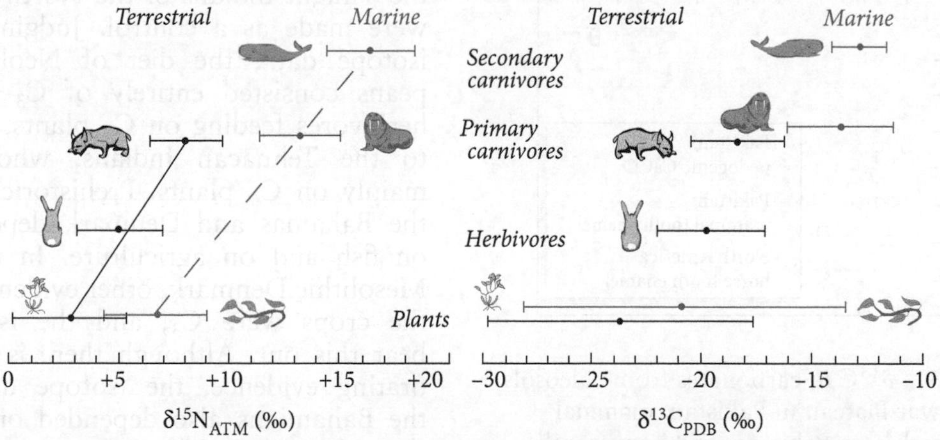

Figure 9.22 Values of δ^{13}C and δ^{15}N in various marine and terrestrial organisms. After Schoeninger and DeNiro (1984).

collagen and teeth can be used to reconstruct the diet of fossil animals.

Plant photosynthesis can also influence the isotopic composition of soil carbonate: when the plant dies, its organic carbon is incorporated into the soil and oxidized to CO_2 by bacteria and other soil organisms. In arid regions, some of this CO_2 precipitates as soil calcium carbonate. In an area of Pakistan now dominated by C_4 grasses, Cerling et al. (1993) found that a sharp shift in $\delta^{13}C$ in soil carbonate occurred about 7 million years ago (Figure 9.23). The same shift is seen in mammalian tooth enamel from fossils collected in the Siwaliks of Pakistan and fossil horse teeth from North America. Cerling et al. concluded that the shift marks the transition from C_3-dominant to C_4-dominant grasslands. This synchronicity suggests a global cause. C_4 photosynthesis is more efficient at lower concentrations of CO_2 than the C_3 pathway. Cerling et al., 1993 argued that this Miocene global expansion of C_4 grasslands occurred in response to a decrease in atmospheric CO_2.

9.5.5.1 Isotopes in archaeology

The differences in nitrogen and carbon isotopic composition of various foodstuffs and the preservation of these isotope ratios in bone collagen provides a means of determining what ancient peoples ate. In

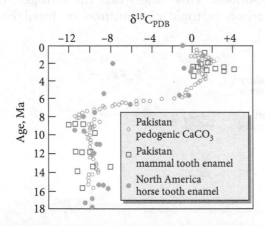

Figure 9.23 $\delta^{13}C$ in carbonates from paleosols of the Potwar Plateau in Pakistan, mammal tooth enamel from Pakistan, and horse tooth enamel from North America. The change in $\delta^{13}C$ reflects the global expansion of C_4 grasslands. From Quade et al. (1989).

the first investigation of bone collagen from human remains, DeNiro and Epstein (1981) concluded that Indians of the Tehuacan Valley in Mexico probably depended heavily on maize (a C_4 plant) as early as 4000 BC, whereas archeological investigations had concluded maize did not become important in their diet until perhaps 1500 BC. In addition, there seemed to be a steady increase in the dependence on legumes (probably beans) from 6000 BC to 1000 AD and a more marked increase in legumes in the diet after 1000 AD.

Mashed grain and vegetable charred onto potsherds during cooking provides an additional record of the diets of ancient peoples. DeNiro and Hasdorf (1985) found that vegetable matter subjected to conditions similar to burial in soil underwent large shifts in $\delta^{15}N$ and $\delta^{13}C$, but that vegetable matter that was burned or charred did not. The carbonization (charring, burning) process itself produced only small (2 or 3‰) fractionations. Since these fractionations are smaller than the range of isotopic compositions in various plant groups, they are of little significance. Since potsherds are among the most common artefacts recovered in archeological sites, this provides a second valuable means of reconstructing the diets of ancient peoples.

Figure 9.24 summarizes the results obtained in a number of studies of bone collagen and potsherds (DeNiro, 1987). Studies of several modern populations, including Eskimos and the Tlinglit Indians of the Northwestern US, were made as a control. Judging from the isotope data, the diet of Neolithic Europeans consisted entirely of C_3 plants and herbivores feeding on C_3 plants, in contrast to the Tehuacan Indians, who depended mainly on C_4 plants. Prehistoric peoples of the Bahamas and Denmark depended both on fish and on agriculture. In the case of Mesolithic Denmark, other evidence indicates the crops were C_3, and the isotope data bear this out. Although there is no corroborating evidence, the isotope data suggest the Bahamians also depended on C_3 rather than C_4 plants. The Bahamians had lower $\delta^{15}N$ because the marine component of their diet came mainly from coral reefs. Nitrogen fixation is particularly intense on coral reefs,

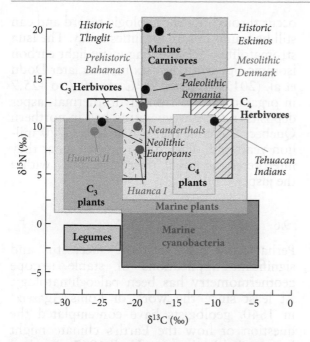

Figure 9.24 $\delta^{13}C$ and $\delta^{15}N$ of various foodstuffs and of diets reconstructed from bone collagen and vegetable matter charred onto pots by DeNiro and colleagues. The Huanca people were from the Upper Mantaro Valley of Peru. Data from potsherds of the Huanca I period (AD 1000–1200) suggest both C_3 and C_4 plants were cooked in pots, but only C_3 plants during the Huanca II period (AD 1200–1470). Adapted from DeNiro (1987).

which leads to ^{15}N depletion of the water, and consequently of reef organisms.

Since these pioneering studies, the use of stable isotopes in archeology has exploded and there are many examples of how stable isotopes have been able to tell the life histories of peoples no longer alive to tell them. One approach uses analysis of dental enamel and bones. Dental enamel forms in childhood and is not replaced during a subject's lifetime; its isotopic composition reflects the child's diet and environment. Bone on the other hand, is constantly replaced with some bones doing so more rapidly than others. Dense bone such as that of the femur remodels slowly and its isotopic composition reflects an average of at least the last 10 years of life, which other bones such as ribs regenerate roughly every 2–5 years.

An interesting example is that of King Richard III of England, who reigned from 1483 until he was killed in the Battle of Bosworth in 1485. His body was exhumed from beneath a car park in Leicester in 2012. Because Richard was not expected to become king, relatively little is known about his early life. Based on Sr, Pb, O, C, and N isotopic analysis of his bones and dentine, Lamb et al. (2014) concluded Richard moved from eastern England to the west, possibly Wales. His diet became richer in luxury foods such as fish and game birds near the end of his life and a positive shift in $\delta^{18}O$ later in life suggests an increase in consumption of wine imported from France.

A significant advance in such studies has been the ability to analyze $\delta^{13}C$ is specific organic compounds, known as *compound specific isotopic analysis*. This is done by using chromatography to separate the compounds before mass spectrometric analysis. Copley et al. (2018) found that C_{18} fatty acids (those based on 18-long carbon chains; see Chapter 12) in plants were about 8‰ lighter than the carbohydrates. C_{18} fatty acids in the ruminant milk are derived directly from diet and match the $\delta^{13}C$ of that diet (–27‰ for ruminants grazing C_3 plants), while adipose (muscle tissue) C_{18} fatty acids are synthesized by the animals and are ~2‰ heavier than milkfat.

It had long been known that Neolithic peoples of Europe raised domesticated ruminants (cattle, sheep, and goats) but it was unclear whether they were raised for meat or for milk. Although shorter fatty acids degrade, longer fatty acids, such as C_{18} ones, can be preserved on pot sherds. Extracting fatty acids from early Neolithic pot sherds from Britain, Copley et al. (2003) found all contained either ruminant adipose fats or dairy fats, but not both and none contained porcine fats (derived from swine). Sherds containing predominantly dairy fats were the most common (78%). They concluded that dairy farming was well established in Britain by 6000 years ago. Subsequent studies have demonstrated the early presence of dairy farming in Ireland and cheese-making in Poland as early as 7000 years ago.

9.5.6   Isotopic "fossils" and the earliest life

As noted earlier, large carbon isotope fractionations are common to all autotrophs.

Consequently, $\delta^{13}C$ values of $-20‰$ or less are generally interpreted as evidence of biologic origin of those compounds. Schidlowski (1988) first reported $\delta^{13}C$ as low as $-26‰$ in carbonate rocks from West Greenland that are ostensibly older than 3.5 Ga. Mojzsis et al. (1996) reported $\delta^{13}C$ between -20 to $-50‰$ for graphite inclusions in grains of apatite in 3.85 Ga banded-iron formations (BIFs) from the same area. In 1999, Rosing reported $\delta^{13}C$ of $-19‰$ from graphite in turbiditic and pelagic metasedimentary rocks from the Isua greenstone belt in the same area. These rocks are thought to be older than 3.7 Ga. In each case, these negative $\delta^{13}C$ values were interpreted as evidence of a biogenic origin of the carbon, and therefore that life existed on Earth at this time.

This interpretation has been controversial. There are several reasons for the controversy, but all ultimately relate to the extremely complex geological history of the area. The geology of the region includes not only the early Archean rocks, but also rocks of middle and late Archean age. Most rocks are multiply and highly deformed and metamorphosed, and the exact nature, relationships, and structure of the precursor rocks are difficult to decipher. Indeed, Rosing (1999) argued that at least some of the carbonates sampled by Schidlowski (1988) are veins deposited by metamorphic fluid flow rather than metasediments. Others have argued that the graphite in these rocks formed by thermal decomposition of siderite ($FeCO_3$) and subsequent reduction of some of the carbon. Yet that argument has been unconvincing because the required large isotopic fractionations have not been observed outside the laboratory.

Two new studies have now significantly strengthened the case for early life. Nutman et al. (2016) reported stromatolite-like structures from Isua rocks that are roughly 3.7 billion years old. Stromatolites are layered bacteria mats composed of carbonates that occur throughout the geologic record and can still be found in rare localities today. The Isua stromatolites lack the diagnostic light carbon isotopic signature, but six months later Dodd et al. (2017) reported $\delta^{13}C$ of -19.7 to -25.7 in organic carbon from hydrothermal jasper deposits in the Nuvvuagittuq belt in northern Quebec, Canada, that are at least 3770 million years old. They also reported what they interpreted as fossil microstructures within the jasper.

9.6 PALEOCLIMATOLOGY

Perhaps one of the most successful and significant applications of stable isotope geothermometry has been paleoclimatology. At least since the work of Louis Agassiz[*] in 1840, geologists have contemplated the question of how the Earth's climate might have varied in the past. Until 1947, they had no means of quantifying paleotemperature changes. In that year, Harold Urey initiated the field of stable isotope geochemistry. In his classic 1947 paper, Urey calculated the temperature dependence of oxygen isotope fractionation between calcium carbonate and water and proposed that the isotopic composition of carbonates could be used as a paleothermometer. Urey's students and postdoctoral fellows empirically determined temperature dependence of the fractionation between calcite and water as:

$$T = 16.5 - 4.3 \times (\delta^{18}O_{calcite} - \delta^{18}O_{water})$$
$$+ 0.14 \times (\delta^{18}O_{calcite} - \delta^{18}O_{water})^2$$
$$(9.75)$$

For example, a change in Δ_{cal-H_2O} from 30 to 31 per mil implies a temperature change from 8° to 12°C. Urey suggested that the Earth's climate history could be recovered from oxygen isotope analyses of ancient marine carbonates. Although the problem has turned out to be

---

[*] Louis Agassiz (1807–1873) was born in Haut-Vully, Switzerland. He earned PhD and MD degrees from the Universities of Erlangen and Munich, respectively, in Germany before becoming professor of natural history at the University of Neuchâtel, specializing in ichthyology and paleontology. Based on the studies of glaciers and geomorphology in his native Switzerland, in 1840 Agassiz proposed that the Earth had experienced far colder temperatures, an *Ice Age*, in the recent past. In 1847, he became professor of zoology and natural history at Harvard University and also served as a visiting professor at Cornell University from its opening in 1865 until his death. By that time, Agassiz was perhaps the most famous scientist in the world.

much more complex than Urey anticipated, this has proved to be an extremely fruitful area of research. Deep-sea carbonate oozes contained an excellent climatic record, and, as we shall see, several other paleoclimatic records are available as well.

9.6.1 The marine Quaternary $\delta^{18}O$ record and Milankovitch cycles

The principles involved in paleoclimatology are fairly simple. As Urey formulated it, the isotopic composition of calcite secreted by organisms should provide a record of paleo-ocean temperatures because the fractionation of oxygen isotopes between carbonate and water is temperature-dependent. In actual practice, the problem is somewhat more complex because the isotopic composition of the shell, or test, of an organism will depend not only upon temperature, but also upon the isotopic composition of the water in which the organism grew, *vital effects* (i.e., different species may fractionate oxygen isotopes somewhat differently), salinity, and post-burial isotopic exchange with sediment pore water. As it turns out, the vital effects, salinity, and post-burial exchange are usually not very important for late Tertiary/Quaternary carbonates, but the isotopic composition of water is.

The first isotopic work on deep-sea sediment cores with the goal of reconstructing the temperature history of Pleistocene glaciations was by Cesare Emiliani (1955), who was then a student of Urey at the University of Chicago. Emiliani analyzed $\delta^{18}O$ in foraminifera from piston cores from the world's oceans. Remarkably, many of Emiliani's findings are still valid today, though they have been revised to various degrees. Rather than just the five glacial periods that geomorphologists had recognized, Emiliani found 15 glacial–interglacial cycles over the last 600,000 years. He found that these were global events, with notable cooling even in low latitudes, and concluded that the fundamental driving force for Quaternary climate cycles was variations in the Earth's orbit and consequent variations in the solar energy flux: *insolation*.

Emiliani had realized that the isotopic composition of the ocean would vary between glacial and interglacial times as isotopically light water was stored in glaciers, thus enriching the oceans in ^{18}O. He estimated that this factor accounted for about 20% of the observed variations. The remainder he attributed to the effect of temperature on isotope fractionation. Subsequently, Shackleton and Opdyke (1973) concluded that storage of isotopically light water in glacial ice was the main effect causing oxygen isotopic variations in biogenic marine carbonates, and that the temperature effect was only secondary.

The question of just how much of the variation in deep-sea carbonate sediments is due to ice build-up and how much is due to the effect of temperature on fractionation is an important one. The resolution depends, in part, on the isotopic composition of glacial ice and how much this might vary between glacial and interglacial times. It is fairly clear that the average $\delta^{18}O$ of glacial ice is probably less than $-15‰$, as Emiliani had assumed. Typical values for Greenland ice are -30 to $-35‰$ (relative to SMOW) and as low as $-50‰$ for Antarctic ice. If the exact isotopic composition of ice and the ice volume were known, it would be a straightforward exercise to calculate the effect of continental ice build-up on ocean isotopic composition. For example, the present volume of continental ice is $27.5 \times 10^6 \, km^3$, while the volume of the oceans is $1350 \times 10^6 \, km^3$. Assuming glacial ice has a mean $\delta^{18}O$ of $-30‰$ relative to SMOW, we can calculate the $\delta^{18}O$ of the total hydrosphere as $-0.6‰$ (neglecting freshwater reservoirs, which are small). At the height of the Wisconsin Glaciation (the most recent one), the volume of glacial ice is thought to have increased by $42 \times 10^6 \, km^2$, corresponding to a lowering of sea level by $125 \, m$. If the $\delta^{18}O$ of ice was the same then as now $(-30‰)$, we can readily calculate that the $\delta^{18}O$ of the ocean would have increased by $1.59‰$. This is illustrated in Figure 9.25.

We can use eqn. 9.75 to see how much the effect of ice volume on seawater $\delta^{18}O$ affects estimated temperature changes. According to this equation, at 20°C, the fractionation between water and calcite should be 33‰. Assuming a present water temperature of 20°C, Emiliani would have calculated a temperature change of 6°C for an observed increase in the $\delta^{18}O$ of carbonates between glacial times and a present $\delta^{18}O$ of 2‰, after correction for 0.5‰ change in the isotopic

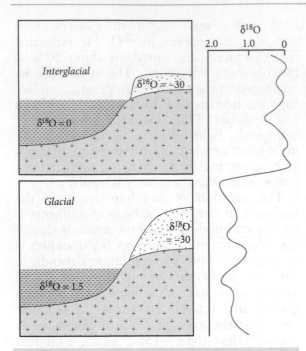

Figure 9.25 Cartoon illustrating how $\delta^{18}O$ of the ocean changes between glacial and interglacial periods.

composition of water. In other words, he would have concluded that surface ocean water in the same spot would have had a temperature of 14°C. If the change in the isotopic composition of water were actually 1.5‰, the calculated temperature difference would be only about 2°C. Thus, the question of the volume of glacial ice and its isotopic composition must be resolved before $\delta^{18}O$ in deep-sea carbonates can be used to calculate paleotemperatures.

Comparison of sea-level curves derived from ^{14}C and U-Th dating of terraces and coral reefs (Chapter 8) indicates that each 0.011‰ variation in $\delta^{18}O$ represents a 1 m change in sea-level. Based on this and other observations, it is now generally assumed that the $\delta^{18}O$ of the ocean changed by ~1.75‰ between glacial and interglacial periods in the Late Pleistocene.

By now, thousands of deep-sea sediment cores have been analyzed for oxygen isotope ratios. Figure 9.26 shows the global benthic foraminifera $\delta^{18}O$ record constructed by averaging analyses from 57 cores over the last 800,000 years (Lisiecki and Raymo, 2005). A careful examination of the global curve shows a periodicity of approximately 100,000 years. The same periodicity was apparent in Emiliani's initial work and led him to conclude that the glacial–interglacial cycles were due to variations in the Earth's orbital parameters. These are often referred to as the Milankovitch cycles, after Milutin Milanković, a Serbian astronomer who, beginning in 1912 and culminating in 1941, made a detailed calculation of just how these orbital changes would redistribute solar energy in time over the surface of the Earth (Milankovitch, 1941).

Let's explore Milankovitch's idea in a bit more detail. The *eccentricity* (i.e., the degree to which the orbit differs from circular) of the Earth's orbit about the Sun and the degree of tilt, or *obliquity*, of the Earth's rotational axis vary slightly, about 5%. In addition, the direction in which the Earth's rotational axis points, relative to the stars,

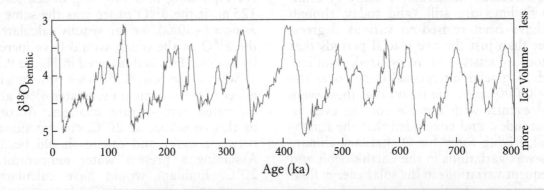

Figure 9.26 Late Pleistocene average $\delta^{18}O_{VPDB}$ of benthic foraminifera in 57 globally distributed deep-sea piston and ODB drilling cores. This variation primarily reflects ice volume because bottom water temperatures should be nearly invariant. From Lisiecki and Raymo (2005).

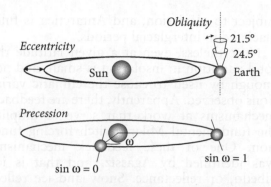

Obliquity

Eccentricity

21.5°
24.5°
Earth
Sun

Precession

ω

sin ω = 0

sin ω = 1

Figure 9.27 Illustration of the *Milankovitch parameters*. The eccentricity is the degree to which the Earth's orbit departs from circular. Obliquity is the tilt of the Earth's rotational axis with respect to the plane of the ecliptic and varies between 21.5° and 24.5°. Precession is the variation in the direction of tilt at the Earth's closest approach to the Sun (perihelion). The parameter ω is the angle between the Earth's position on June 21 (summer solstice), and perihelion.

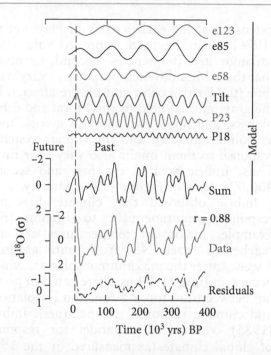

Figure 9.28 Gain and phase model of Imbrie, relating variations in eccentricity, tilt, and precession to the oxygen isotope curve. Top shows the variation in these parameters over the past 400,000 and next 25,000 yrs. Bottom shows the sum of these functions with appropriated gains and phases applied and compares them with the observed data. After Imbrie (1985).

varies, a phenomenon called *precession*. These variations are illustrated in Figure 9.27. Though variation in these "Milankovitch parameters" has negligible effect on the *total* radiation the Earth receives, they do affect the *pattern* of incoming radiation. For example, tilt of the rotational axis determines seasonality and the latitudinal gradient of insolation. The gradient is extremely important because it drives atmospheric and oceanic circulation. If the tilt is small, seasonality will be reduced (cooler summers, warmer winters). Precession relative to the eccentricity of the Earth's orbit also affects seasonality. For example, the Earth is presently closest to the Sun in January. As a result, northern hemisphere winters (and southern hemisphere summers) are somewhat warmer than they would be otherwise. For a given latitude and season, precession will result in a ±5% difference in insolation. While the Earth's orbit is only slightly elliptical and variations in eccentricity are small, these variations are magnified because insolation varies with the inverse square of the Earth–Sun distance.

Variation in tilt approximates to a simple sinusoidal function with a period of 41,000 yrs. Variations in eccentricity can be approximately described with a period of 100,000 yrs. In actuality, however, eccentricity variation is more complex, and is more accurately described with periods of 123,000 yrs, 85,000 yrs, and 58,000 yrs. Similarly, variation in precession has characteristic periods of 23,000 and 18,000 yrs.

Although Emiliani suggested that $\delta^{18}O$ variations were related to variations in these orbital parameters, the first quantitative approach to the problem was that of Hayes et al. (1976). They applied Fourier analysis to the $\delta^{18}O$ curve (Fourier analysis is a mathematical tool that transforms a complex variation such as that in Figure 9.28 to the sum of a series of simple sine functions). Hayes et al. then used spectral analysis to show that much of the spectral power of the $\delta^{18}O$ curve occurred at frequencies similar to those of the Milankovitch parameters. This was further refined by Imbrie (1985). Imbrie's treatment involved several refinements and extension of the earlier work of Hayes et al. (1976). First, he used improved values for Milankovitch frequencies. Second, he noted that these Milankovitch parameters vary with time (the Earth's orbit and tilt are affected by the gravitational field of the Moon and other planets, and other astronomical events, such as bolide impacts), and the climate system's response to them might also vary over time. Thus, Imbrie treated the first and second 400,000 years of Figure 9.28 separately.

Imbrie observed that climate does not respond instantaneously to forcing. For example, maximum temperatures are not reached in Ithaca, New York, until late July, 4 weeks after the maximum insolation, which occurs on June 21. Thus, there can be a *phase lag* between the forcing function (insolation) and climatic response (temperature). Imbrie (1985) constructed a model for response of global climate (as measured by the $\delta^{18}O$ curve) in which each of the six Milankovitch forcing functions was associated with a different gain and phase. The values of gain and phase for each parameter were found statistically by minimizing the residuals of the power spectrum of the $\delta^{18}O$ curve. The resulting *gain and phase model* is shown in comparison with the data for the past 400,000 yrs and the next 25,000 yrs in Figure 9.28. The model has a correlation coefficient, r, of 0.88 with the data. Thus, about r^2, or 77%, of the variation in $\delta^{18}O$, and therefore presumably in ice volume, can be explained by Imbrie's Milankovitch model. The correlation for the period 400,000–782,000 yrs is somewhat poorer, around 0.80, but nevertheless impressive. This picture has changed little with subsequent analysis (e.g., Lisiecki and Raymo, 2005).

Since variations in the Earth's orbital parameters do not affect the average total annual insolation the Earth receives, but only its pattern in space and time, one might ask how this could cause glaciation? Milankovitch concluded it was the insolation received during summer at high northern latitudes (65° N) in June. If summers are not warm enough to melt the accumulated snow of the previous winter, this snow accumulates to form glaciers and ice sheets. Northern latitudes are the area where large continental ice sheets develop. The southern hemisphere, except for Antarctica, is largely ocean, and therefore not subject to glaciation, and Antarctica is fully glaciated in interglacial periods.

Nevertheless, even at a given latitude the total variation in insolation is small, and not enough by itself to cause the climatic variations observed. Apparently, there are feedback mechanisms at work that serve to amplify the fundamental Milankovitch forcing function. One of these feedback mechanisms was identified by Agassiz, and that is ice albedo, or reflectance. Snow and ice reflect much (~90%) of the incoming sunlight back into space. Thus, as glaciers advance, they will cause further cooling. Any additional accumulation of ice in Antarctica, however, does not result in increased albedo, because the continent is fully ice covered even in interglacial times – hence, the dominant role of northern hemisphere insolation in driving climate cycles. Other feedback mechanisms include carbon dioxide and ocean circulation. The role of atmospheric CO_2 in controlling global climate is a particularly important issue because burning of fossil fuels has resulted in a significant increase in atmospheric CO_2 concentration over the last 150 years; we'll discuss the role of CO_2 in Chapters 12 and 14.

9.6.2   The record in glacial ice

Climatologists recognized early on that continental ice preserves a stratigraphic record of climate change. Some of the first ice cores recovered for the purpose of examining the climatic record and analyzed for stable isotopes were taken from Greenland in the 1960s (e.g., Camp Century Ice Core). Subsequent cores have been taken from Greenland, Antarctica, and various alpine glaciers. Very long ice cores that covered 150,000 yrs were recovered by the Russians from the Vostok station in Antarctica in the 1980s and were deepened over the next 20 years, eventually reaching back 400,000 years. In 2003, drilling began on the EPICA (European Project for Ice Coring in Antarctica) project Dome C Ice Core in Antarctica, which eventually drilled through 3260 m, corresponding to 800 ka of ice before halting just short of bedrock. Jouzel et al. (1996) focused on δD variations in these cores because it is more sensitive to temperature than is $\delta^{18}O$. In their initial work, temperatures were calculated from δD based on an empirical relationship between measured 0 and δD in Antarctic snow and after subtracting the effect of changing ice volume on δD of the oceans. Subsequently, they used the *deuterium excess* in combination with global circulation models to obtain the relationship between δD and temperature (Jouzel et al., 2007). The deuterium excess parameter is the deviation of δD from the meteoritic δD–$\delta^{18}O$ correlation in Figure 9.13. This correlation has a slope of ~8, and the deuterium excess is defined as $d = \delta D - 8 \times \delta^{18}O$. This provides a measure of the kinetic fractionation occurring during evaporation snow formation. These recent results showed temperature variations of 15°C between glacial and interglacial times, and these variations track the marine isotope record well, as can be seen in Figure 9.29, except in the lowermost 60 m or so of ice, which showed evidence of being deformed.

Figure 9.30 compares temperatures calculated from the EPICA Antarctic core (lower line) with $\delta^{18}O$ in ice from the Greenland Ice Core Project (GRIP) (upper line). The Greenland ice record agrees well with the Antarctic one, although the former is limited to the last 120 ka, largely because ice accumulates much faster in Greenland. The high ice accumulation rates in Greenland limit the range of time covered by the record but do provide high resolution. They reveal sharp changes in temperature and ice volume on

time-scales of a millennium and less. These rapid fluctuations in climate are known as Dansgaard–Oeschager events and can be correlated to $\delta^{18}O$ variations in high-resolution sediment cores from the North Atlantic. The warming events correlate with temperature maxima in Antarctica, suggesting coupling of climate between the two hemispheres. The cause of these Dansgaard–Oeschager events is still unclear, but changes in the North Atlantic Ocean circulation, perhaps triggered by an influx of fresh water from melting glacier and icebergs, are suspected.

9.6.3 Soils and paleosols

As we found in Chapter 6, the concentration of CO_2 in soils is very much higher than in the atmosphere, reaching 1% by volume. As a result, soil water can become supersaturated with respect to carbonates. In soils where evaporation exceeds precipitation, soil carbonates form. The carbonates form in equilibrium with soil water, and there is a strong correlation between $\delta^{18}O$ in soil carbonate and local meteoric water, though soil carbonates tend to be about 5‰ more enriched than expected from the calcite–water fractionation. There are two reasons why soil carbonates are heavier. First, soil water is enriched in ^{18}O relative to meteoric water due to preferential evaporation of isotopically

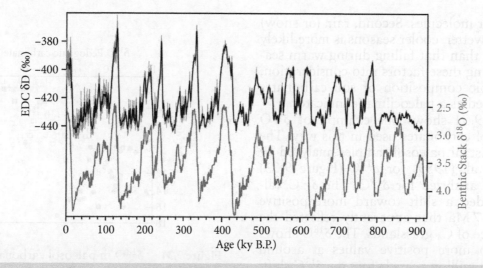

Figure 9.29 δD of ice in the EPICA Dome C ice core (upper line) compared with $\delta^{18}O$ of benthic foraminifera in deep-sea cores (lower line). Because the temperature variation in deep ocean water is small, the latter record mainly variations in ice volume. There is good overall agreement between the two cores. From Jouzel et al. (2007).

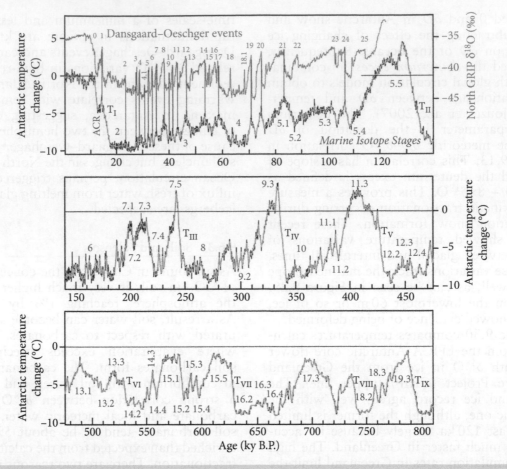

Figure 9.30   Antarctic temperature variation calculated from the EPICA ice core (lower line) compared with the $\delta^{18}O$ record from the GRIP ice core from Greenland (upper line). The former suggests a 15°C temperature range from the coldest glacial to the warmest interglacial period. From Jouzel et al. (2007).

light water molecules. Second, rain (or snow) falling in wetter, cooler seasons is more likely to run off than that falling during warm seasons. Taking these factors into consideration, the isotopic composition of soil carbonates may be used as a paleoclimatic indicator.

Figure 9.31 shows one example of $\delta^{18}O$ in paleosol carbonates used in this way. The same Pakistani paleosol samples analyzed by Quade et al. (1989) for $\delta^{13}C$ (Figure 9.23) were also analyzed for $\delta^{18}O$. The $\delta^{13}C$ values recorded a shift toward more positive values at 7 Ma that apparently reflected the appearance of C_4 grasslands. The $\delta^{18}O$ shows a shift to more positive values at around 8 Ma, or a million years before the $\delta^{13}C$ shift. Quade et al. interpreted this as due to an intensification of the monsoon system at that time, an interpretation consistent with marine paleontological evidence.

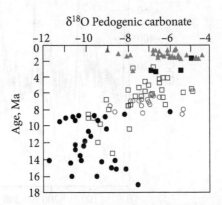

Figure 9.31 $\delta^{18}O$ in paleosol carbonate nodules from the Potwar Plateau in northern Pakistan. Different symbols correspond to different, overlapping sections that were sampled. After Quade et al. (1989).

Clays, such as kaolinites, are another important constituent of soil. Lawrence and Taylor (1972) showed that during soil formation, kaolinite and montmorillonite form in approximate equilibrium with meteoric water so that their $\delta^{18}O$ values are systematically shifted to higher values relative the local meteoric water, while δD values are shifted to lower values. Thus, kaolinites and montmorillonites define a line parallel to the meteoric water line (Figure 9.32), the so-called *kaolinite line*. From this observation, Sheppard et al. (1969) and Lawrence and Taylor (1972) reasoned that one should be able to deduce the isotopic composition of rain at the time ancient kaolinites formed from their δD values. Since the isotopic composition of precipitation is climate-dependent, ancient kaolinites provide another continental paleoclimatic record.

Many additional studies have followed up on this original work. Sheppard and Gilg (1996) determined the temperature dependence of the water–kaolinite fractionation of δD as:

$$1000 \ln \alpha_{kaol-H_2O} = -\frac{2.2 \times 10^6}{T^2} - 7.7 \tag{9.76a}$$

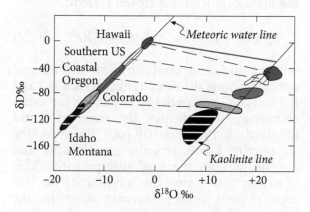

Figure 9.32 Relationship between δD and $\delta^{18}O$ in modern meteoric water and kaolinites. Dashed lines connect kaolinites with local meteoritic water. Bold red line is the calculated fraction between water and kaolinite at 20°C based on Sheppard and Gilg (1996). This corresponds to a fractionation of ^{18}O by about 25.4‰ and ^2H by about 33.3‰. Modified from Lawrence and Taylor (1972).

and that for $\delta^{18}O$ as:

$$1000 \ln \alpha_{kaol-H_2O} = \frac{2.76 \times 10^6}{T^2} - 6.75 \tag{9.76b}$$

which is only slightly different from the earlier work of Lawrence and Taylor (1972).

The temperature at which kaolinite recrystallizes can be determined from:

$$T = \left(\frac{3.06 \times 10^6}{\delta^{18}O_{kaol} - 0.125 \times \delta D_{kaol} + 7.04} \right)^{1/2} \tag{9.77}$$

Thus, kaolinites can be used as paleothermometer. When temperatures calculated in this way from modern kaolinites, they tend to be about 3°C than local mean annual temperature. This is thought to be due to crystallization primarily in the warm summer months. Consistent with other paleothermometers, kaolinites indicate that the Neogene has had temperatures several degrees cooler than in the Cretaceous and Eocene (Sheldon and Tabor, 2009).

9.7 HYDROTHERMAL SYSTEMS AND ORE DEPOSITS

When large igneous bodies are intruded into high levels of the crust, they heat the surrounding rock and the water in the cracks and pores in this rock, setting up convection systems. The water in these hydrothermal systems reacts with the hot rock and undergoes isotopic exchange; the net result is that both the water and the rock change their isotopic compositions.

9.7.1 Water in hydrothermal systems

One of the first of many important contributions of stable isotope geochemistry to understanding hydrothermal systems was the demonstration by Harmon Craig (another student of Harold Urey) that water in these systems is meteoric, not magmatic (Craig, 1963). The argument is based on the data shown in Figure 9.33. For each geothermal system, the δD of the "chloride" type geothermal waters is the same as the local precipitation and groundwater, but the $\delta^{18}O$ is shifted to higher values. The shift in

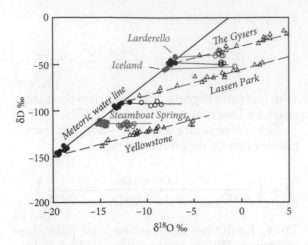

Figure 9.33 δD and $\delta^{18}O$ in meteoric hydrothermal systems. Closed circles show the composition of meteoric water in Yellowstone, Steamboat Springs, Mt. Lassen, Iceland, Larderello, and The Geysers, and open circles show the isotopic composition of chloride-type geothermal waters at those locations. Open triangles show the acidic, sulfide-rich geothermal waters at those locations. Solid lines connect the meteoric and chloride waters, dashed lines connect the meteoric and acidic waters. The "Meteoric Water Line" is the correlation between δD and $\delta^{18}O$ observed in precipitation in Figure 9.13. After Craig (1963).

$\delta^{18}O$ results from high-temperature reaction ($\lesssim 300°C$) of the local meteoric water with hot rock. However, because the rocks contain virtually no hydrogen, there is little change in the hydrogen isotopic composition of the water. If the water involved in these systems were magmatic, it would not have the same isotopic composition as local meteoric water (it is possible that these systems contain a few percent magmatic water).

Acidic, sulfur-rich waters from hydrothermal systems can have δD that is different from local meteoric water. This shift occurs when hydrogen isotopes are fractionated during boiling of geothermal waters. The steam produced is enriched in sulfide. The steam mixes with cooler meteoric water, condenses, and the sulfide is oxidized to sulfate, resulting in their acidic character. The mixing lines observed reflect mixing of the steam with meteoric water as well as fractionation during boiling.

Estimating temperatures at which ancient hydrothermal systems operated is a fairly straightforward application of isotope geothermometry, which we have already discussed. If we can measure the oxygen (or carbon or sulfur) isotopic composition of any two phases that were in equilibrium, and if we know the fractionation factor as a function of temperature for those phases, we can estimate the temperature of equilibration. We will focus now on water–rock ratios, which may also be estimated using stable isotope ratios.

9.7.2 Water–rock ratios

For a closed hydrothermal system, we can write two fundamental equations. The first simply describes isotopic equilibrium between water and rock:

$$\Delta = \delta_w^f - \delta_r^f \qquad (9.78)$$

where we use the subscript w to indicate water, and r to indicate rock. The superscript f indicates the final value. So eqn. 9.78 just says that the difference between the final isotopic composition of water and rock is equal to the fractionation factor (we implicitly assume equilibrium). The second equation is the mass balance for a closed system:

$$c_w W \delta_w^i + c_r R \delta_r^i = c_w W \delta_w^f + c_r R \delta_r^f \qquad (9.79)$$

where c indicates concentration (we assume concentrations do not change, which is valid for oxygen, but perhaps not valid for other elements), W indicates the mass of water involved, R the mass of rock involved, the superscript i indicates the initial value and f again denotes the final isotope ratio. This equation just states the amount of an isotope present before reaction must be the same as after reaction. We can combine these equations to produce the following expression for the water–rock ratio:

$$\frac{W}{R} = \frac{\delta_r^f - \delta_r^i}{\delta_w^i - \delta_r^f - \Delta} \left(\frac{c_r}{c_w} \right) \qquad (9.80)$$

The initial $\delta^{18}O$ can generally be inferred from unaltered samples and from the $\delta D–\delta^{18}O$ meteoric water line, and the final isotopic composition of the rock can be measured. The fractionation factor can be estimated if we know the temperature and the phases in the rock. For oxygen, the ratio of concentration in the rock to water will be close to 0.56 in all cases.

Equation 9.78 is not very geologically realistic because it applies to a closed system. A completely open system, where water makes one pass through hot rock, would be more realistic. In this case, we might suppose that a small parcel of water, dW, passes through the system and induces an incremental change in the isotopic composition of the rock, $d\delta_r$. In this case, we can write:

$$Rc_r d\delta_r = \delta_w^i - [\Delta + \delta_r]c_w dW \qquad (9.81)$$

This equation states that the mass of an isotope exchanged by the rock is equal to the mass of that isotope exchanged by the water (we have substituted $\Delta + \delta_r$ for δ_w^f). Rearranging and integrating, we have:

$$\frac{W}{R} = \ln\left(\frac{\delta_w^i - \delta_r^i + \Delta}{\delta_w^i - \delta_r^f + \Delta}\right)\frac{c_w}{c_r} \qquad (9.82)$$

Thus, it is possible to deduce the water–rock ratio for an open system as well as a closed one.

Oxygen isotope studies can be a valuable tool in mineral exploration. Mineralization is often (though not exclusively) associated with the region of greatest water flux, such as areas of upward-moving hot water above intrusions. Such areas are likely to have the lowest values of $\delta^{18}O$. To understand this, let's solve eqn. 9. for δ_r^f, the final value of $\delta^{18}O$ in the rock:

$$\delta_r^f = \delta_w^i + \Delta - (\delta_r^i - \delta_w^i + \Delta)e^{-W/R \times c_w/c_r} \qquad (9.83)$$

Assuming a uniform initial isotopic composition of the rocks and the water, all the terms on the right-hand side are constants except W/R and Δ, which is a function of temperature. Thus, the final values of $\delta^{18}O$ are functions of the temperature of equilibration, and an exponential function of the W/R ratio. Figure 9.34 shows δ_r^f plotted as a function of W/R and Δ, where δ_r^f is assumed to be +6 and δ_w^i is assumed to be −13. In a

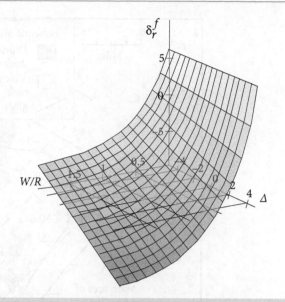

Figure 9.34 $\delta^{18}O$ as a function of W/R and Δ computed from eqn. 9.83.

few cases, such as porphyry copper deposits, ores apparently have precipitated from fluids derived from the magma itself. In those cases, this kind of approach does not apply.

Figure 9.35 shows an example of the $\delta^{18}O$ imprint of an ancient hydrothermal system: the Bohemia Mining District in Lane County, Oregon, where tertiary volcanic rocks of the western Cascades have been intruded by a series of dioritic plutons. Approximately $1,000,000 worth of gold was removed from the region between 1870 and 1940. $\delta^{18}O$ contours form a bull's-eye pattern, and the region of low $\delta^{18}O$ corresponds roughly with the area of propylitic (i.e., greenstone) alteration. Notice that this region is broader than the contact metamorphic aureole. The primary area of mineralization occurs within the $\delta^{18}O < 0$ contour. In this area, relatively large volumes of gold-bearing hydrothermal solution cooled, perhaps due to mixing with groundwater, and precipitated gold. This is an excellent example of the value of oxygen isotope studies to mineral exploration. Similar bull's-eye patterns are found around many other hydrothermal ore deposits.

9.7.3 Sulfur isotopes and ore deposits

Of the various stable isotope systems we will consider here, sulfur isotopes are the

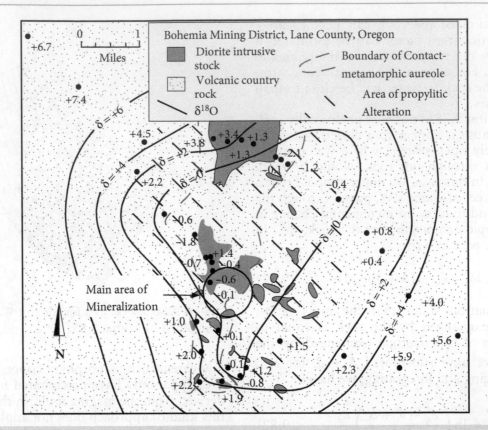

Figure 9.35 $\delta^{18}O$ variations in the Bohemia Mining District, Oregon. Note the bull's-eye pattern of the $\delta^{18}O$ contours. After Taylor (1974).

most complex. This complexity arises in part because there are five common valence states in which sulfur can occur in the Earth, +6 (e.g., $BaSO_4$), +4 (e.g., SO_2), 0 (e.g., S), −1 (e.g., FeS_2) and −2 (H_2S). Significant equilibrium fractionations occur between these valence states. Each of these valence states forms a variety of compounds, and fractionations can occur between these as well (e.g., Figure 9.10). Finally, sulfur is important in biological processes, and fractionations in biologically mediated oxidations and reductions are often different from fractionations in the abiological equivalents.

Sulfide ore deposits are important sources of a wide variety of metals particularly the base metals, Cu, Pb, and Zn, but also Co, Ni, Ag, Au, and platinum group metals. These have formed in a great variety of environments and under a great variety of conditions (Chapter 15). Sulfur isotopic compositions provides information on the source of these ores. Before discussing these ores, let's briefly

consider how sulfur isotopes vary in the Earth.

MORB, in which sulfur is primarily present in reduced form, have mean $\delta^{34}S$ of −0.88 ± 0.23‰. High-quality data on oceanic island basalts (OIB) is scarce, but available data show a somewhat wider range with most values being isotopically heavier than MORB. As discussed in subsequent sections, this likely reflects recycling of crustal S into the mantle. Ecologic diamonds have even more variable $\delta^{34}S$, again likely reflecting subduction of surficial sulfur into the mantle. Island arc volcanics, particularly andesitic ones have distinctly heavier sulfur than MORB or OIB. In contrast to oceanic basalts, much of this sulfur is present in oxidized form (such as SO_2). The most positive values likely reflect a combination of (1) fractionation during oxidation of sulfide to sulfate, (2) their sources containing some subducted sedimentary sulfur, and (3) combinations of assimilation and fractional crystallization.

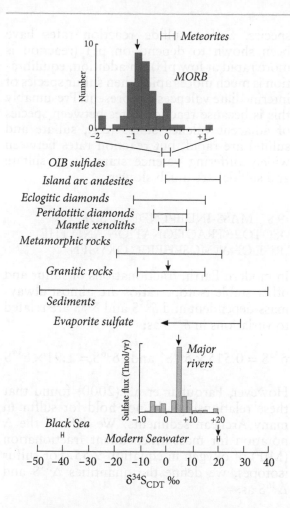

Figure 9.36 $\delta^{34}S_{CDT}$ in various terrestrial materials. Arrows show the mean values: MORB: −0.88‰, seawater: +21.2‰, rivers: +4.4‰. River water histogram from Burke et al. (2018).

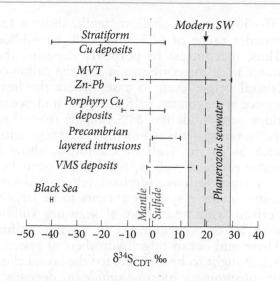

Figure 9.37 $\delta^{34}S_{CDT}$ of sulfide in ore deposits. Volcanogenic massive sulfide (VMS), porphyry copper, and Precambrian layered intrusions have $\delta^{34}S$ close to the mantle value, indicating the sulfur is mainly magmatically derived. Mississippi Valley type (MVT) deposits and stratiform deposits show a large range of $\delta^{34}S$ consistent with sulfate derived from seawater or anoxic seas and reduced either by thermal sulfate reduction or bacterial sulfate reduction. Dashed vertical line is mantle $\delta^{34}S$.

Seawater of the open ocean, in which sulfur is present as SO_4^{2-}, has nearly uniform $\delta^{34}S$ of +21.2 ±0.88* (Tostevin et al., 2014). However, in enclosed basins, such as the Black Sea and Curacao Trench where the water is anoxic at depth, marine sulfate is bacterially reduced to sulfide with a quite large fractionation and this sulfide can be very isotopically light ($\delta^{34}S \approx -40‰$). Sulfur in sedimentary and metamorphic rocks of the continental crust also have a large range of $\delta^{34}S$ (Figure 9.36).

Figure 9.37 shows $\delta^{34}S_{CDT}$ of sulfide in various types of ore deposits. Sulfides in sulfur-poor magmatic systems, such as the Merensky Reef of the Bushveld Complex, South Africa, generally have $\delta^{34}S$ close to the mantle value, while sulfide-rich magmas, such as at Duluth, Minnesota, have a wide range of positive values reflecting crustal assimilation. Most porphyry copper deposits, which account for most of the world's copper production and which typically are associated with subduction zone volcanism, have $\delta^{34}S$ between +4 and −6, indicating that most of the sulfur (as well as the water) is magmatically derived with lesser contributions from meteoric water and continental crust. Epithermal deposits, which are hydrothermal deposits mineralized in Cu, Pb, Zn, Ag, and Au and formed at low temperature

* This is the modern value for seawater. The sulfur isotopic composition of seawater is uniform at any one time but has varied over geologic time.

(50–300°C) and shallow depth, show a far broader range of $\delta^{34}S$ from +8‰ to +30‰. Thus, in contrast to porphyry deposits, the water is of meteoric origin and the sulfur of crustal origin even in cases where the heat source is magmatic. $\delta^{34}S$ in H_2S of mid-ocean ridge vent fluids has $\delta^{34}S$ ranging from 0 to +6‰, but systems sedimented ridge segments, such as in the Gulf of California, show a wider range of values, reflecting a contribution from sediment-derived sulfur. Reduced sulfur in these systems appears to be largely derived from reduction of seawater sulfate with lesser contributions of basaltic sulfur. These mid-ocean ridge hydrothermal systems are thought to be a model for the broad class of *volcanogenic massive sulfide* ore deposits.

In sediment-hosted ores, such as the Permian Kupferschiefer stratiform Cu deposit of Poland and Germany developed at the interface between organic-rich shales and oxidized red sandstones, bacterially reduced sulfides in sediments deposited in euxinic basins (for which the Black Sea is a modern analog) appears to be the sulfur source. These can have $\delta^{34}S$ as light as –40‰. In the case of Mississippi Valley type deposits, which are carbonate-hosted Pb-Zn sulfide ores formed at <200°C, $\delta^{34}S$ is typically positive, ranging from +30 to –5, indicative of evaporite sulfate or sulfate in formation brines as a primary source, which was subsequently reduced to sulfide resulting in precipitation of the ores. In other instances, the sulfur sources appear to be sulfides associated with organic matter and petroleum. Omoto (1986), Marini et al. (2011), and Shanks (2014) provide more details of sulfur isotopes in ore deposits. We explore ore deposits and their origins in greater detail in Chapter 15.

In addition to tracing ore sources, sulfur isotopic compositions can be used to determine the temperatures at which the ores formed, although with caveats. At magmatic temperatures, reactions generally occur rapidly and sulfur phases within these systems are generally found to be close to isotopic equilibrium. Below 200°C, however, sulfur isotopic equilibration is slow even on geologic time-scales, hence equilibration is rare and kinetic effects often dominate. Isotopic equilibration between two sulfide species or between two sulfate species is achieved more readily than between sulfate and sulfide

species. Sulfate–sulfide reaction rates have been shown to depend on pH (reaction is more rapid at low pH). In addition, equilibration is much more rapid when sulfur species of intermediate valences are present. Presumably this is because reaction rates between species of adjacent valence states (e.g., sulfate and sulfite) are rapid, but reaction rates between widely differing valence states (e.g., sulfate and sulfide) are much slower.

9.8 MASS-INDEPENDENT SULFUR ISOTOPE FRACTIONATION AND THE RISE OF ATMOSPHERIC OXYGEN

In modern Earth, fractionations of sulfur and other stable isotope ratios are almost always mass-dependent and $\delta^{33}S$ and $\delta^{36}S$ are related to variations in $\delta^{34}S$ as:

$$\delta^{33}S = 0.515 \times \delta^{34}S \quad \text{and} \quad \delta^{36}S = 1.91 \times \delta^{34}S$$

However, Farquhar et al. (2000) found that these relationships do not hold for sulfur in many Archean sediments. We defined the Δ notation for mass-independent fractionation (MIF) of oxygen in equation 9.55. For sulfur isotopes, we define the quantities $\Delta^{33}S$ and $\Delta^{36}S$ as:

$$\Delta^{33}S = \delta^{33}S - 0.515 \times \delta^{34}S \qquad (9.84)$$

and

$$\Delta^{36}S = \delta^{36}S - 1.90 \times \delta^{34}S \qquad (9.85)$$

Farquhar et al. found that many sulfides in Archean (>2.5 Ga) sediments and metasediments have positive $\Delta^{33}S$ and negative $\Delta^{36}S$, while hydrothermal sulfide ores and sedimentary sulfates (mainly barite) have negative $\Delta^{33}S$ and positive $\Delta^{36}S$ (Figure 9.38). This discovery has been amply confirmed and while Farquhar et al. found $\Delta^{33}S$ as great as 3‰, subsequent work has revealed $\Delta^{33}S$ as high as +14.‰ and $\Delta^{36}S$ as low as –13‰ in rocks formed at 3.3 Ga. Strongly positive $\Delta^{33}S$ are most common in late Archean rocks with ages between 2.7 and 2.5 Ga, and then disappear entirely in the Paleoproterozoic around 2.3 Ga.

The disappearance of MIF sulfur in the sedimentary record coincides with the time,

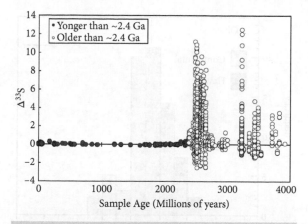

Figure 9.38 Δ^{33}S in sulfur through time. Mass independently fractionated sulfur isotopes are common in Archean rocks and early Proterozoic rocks but disappear around 2.3 Ga, coincident with other geochemical evidence for the atmosphere becoming oxidizing at this time. From Farquhar (2018).

2.3 Ga, for which there is other geochemical evidence that the Earth's atmosphere first became oxidizing: an abundance of banded iron formations formed by oxidation of ferrous to ferric iron, the first appearance of Fe-rich paleosols, the disappearance of detrital sulfides and uraninites, the appearance of *red beds,* sandstones with lithic grains coated with ferric iron, etc. The disappearance of MIF sulfur isotope around this time can be understood in this context and indeed provides further evidence of what has become known as the *Great Oxidation Event* or *GOE* (Farquhar and Wing, 2003; Kasting and Catling, 2003).

Sulfur dioxide lofted into the atmosphere by volcanic eruptions and other processes can be photolyzed by UV photons with wavelengths <219 nm:

$$SO_2 \xrightarrow{\nu} SO + O \qquad (9.86)$$

In modern Earth, this energic UV radiation is almost entirely absorbed in the stratosphere by ozone and only rare, large volcanic eruptions are capable of injecting SO_2 into the stratosphere, so such sulfur photolysis is rare. In the absence of atmospheric oxygen in the early Earth, UV radiation could penetrate through the entire atmosphere. In addition, in modern Earth, any SO and O produced

in this manner simply recombine to form SO_2. However, in the absence of significant atmospheric oxygen, the SO and O produced in 9.86 undergo further reactions such that the net reaction is:

$$3SO_2 \xrightarrow{\nu} 2SO_3 + S^0 \qquad (9.87)$$

The SO_3 then reacts with water to form H_2SO_4, dissolves in rain, and ultimately finds its way into the oceans. Some of this would precipitate as barite ($BaSO_4$), and some reduced to sulfide. The S would form particulate S_8 and be swept out of the atmosphere by rain and ultimately incorporated into sediments, where it would form sedimentary sulfides.

Precisely why this process results in mass dependent fractionation is still not fully understood and while several factors may be involved (Ono, 2017), there is a near consensus that the primary cause is self-shielding similar to that responsible for MIF oxygen in the solar nebula. The different isotopologues of SO_2 are photolyzed by slightly different UV wavelengths. In the Archean atmosphere, wavelengths necessary to photolyze the most abundant isotopogues, those of ^{32}S, were completely absorbed in the upper atmosphere while those necessary to photolyze isotopologues of ^{33}S, ^{34}S, and ^{36}S were not. MIF sulfur has subsequently been discovered in diamonds and some modern oceanic island basalts. We'll discuss this in more detail in Chapter 11.

Subsequently, sulfate possessing a MIF isotopic signature was found in Antarctic ice cores, most notably in layers formed in years of large volcanic eruptions, such as Mt. Pinatubo in 1992, that lofted large amounts of SO_2 into the stratosphere (Savarino et al., 2003). The fractionations are smaller than in the Archean with Δ^{33}S < 0.7‰ and Δ^{36}S > −4‰. The Δ^{33}S–Δ^{36}S pattern is different from in the Archean sulfur, suggesting a different mechanism. The proposed mechanism also involves UV radiation in a way analogous to that discussed for ozone in Section 9.2.3. Less energetic UV radiation with wavelengths of 240−340 nm can excite SO_2 into unstable higher energy states, which, if the vibrational energy is not repartitioned, can result in dissociation. Some isotopologues can repartition this vibrational energy into rotation energy more readily and thus avoid dissociation.

9.9 STABLE ISOTOPES IN THE MANTLE AND MAGMATIC SYSTEMS

9.9.1 Stable isotopic composition of the mantle

The mantle is the largest reservoir for oxygen, as well as for S and perhaps H, C, and N. Thus, we need to know the isotopic composition of the mantle to know the isotopic composition of the Earth. Variations in stable isotope ratios in mantle and mantle-derived materials also provide important insights on mantle and igneous processes.

9.9.1.1 Oxygen

Assessing the oxygen isotopic composition of the mantle, and particularly the degree to which its oxygen isotope composition might vary, has proved to be more difficult than expected. One approach has been to use basalts as samples of mantle, as is done for radiogenic isotopes. Isotope fractionation occurring during partial melting is small, so the oxygen isotopic composition of basalt should be similar to that of its mantle source. However, assimilation of crustal rocks by magmas and oxygen isotope exchange during weathering complicate the situation. An alternative is to use direct mantle samples such as xenoliths occasionally found in basalts although these are considerably rarer than are basalts.

Figure 9.39 shows the oxygen isotope compositions of olivines and clinopyroxenes in 76 peridotite xenoliths analyzed by Mattey et al. (1994) using the laser fluorination technique. The total range of values observed is only about twice that expected from analytical error alone, suggesting the mantle is fairly homogenous in its isotopic composition. The difference between coexisting olivines and clinopyroxenes averages about 0.5 per mil, which is consistent with the expected fractionation between these minerals at mantle temperatures. Mattey et al. (1994) estimated the bulk composition of these samples to be about +5.5 per mil. Subsequent work has confirmed these observations.

Figure 9.40 shows the distribution of $\delta^{18}O$ in basalts from four different groups.

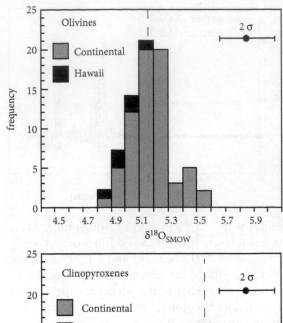

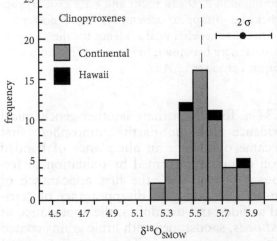

Figure 9.39 Oxygen isotope ratios in olivines and clinopyroxenes from mantle peridotite xenoliths. Data from Mattey et al. (1994).

To avoid weathering problems, Harmon and Hoefs (1995) included only submarine basaltic glasses and basalts having less than 0.75% water or have erupted historically in their compilation. There are several points worth noting in these data.

• MORB has a mean $\delta^{18}O_{SMOW}$ of +5.7‰, with a relatively smaller variation about this mean, suggesting that the depleted upper mantle appears is relatively homogenous. The small difference between MORB and the estimated mantle composition (~0.2‰ is consistent with a small fractionation occurring during partial melting. Cooper et al. (2004) reported a range of $\delta^{18}O$ of 0.47‰ in MORB

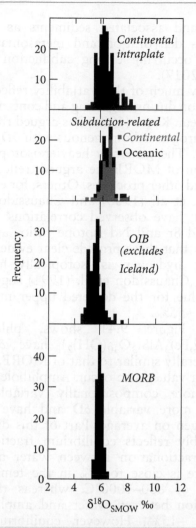

Figure 9.40 $\delta^{18}O$ in young, fresh basalts. Dashed line is at the mean of MORB (+5.7). After Harmon and Hoefs (1995).

from the N. Atlantic and a correlation between $\delta^{18}O$ and La/Sm, $^{87}Sr/^{86}Sr$, and $^{143}Nd/^{144}Nd$ indicative of real O isotope heterogeneity in the depleted upper mantle. Subsequent studies of MORB in other regions indicate similar variability, but no systematic variation between ocean basins (Cooper et al., 2009).

- Harmon and Hoefs originally excluded Iceland from their compilation of oceanic island basalts (OIB) because these basalts are characterized by anomalous $\delta^{18}O$ and there was a suspicion this reflected assimilation of crust that had interacted with the particularly low $\delta^{18}O$ meteoric water in this region. Subsequent studies (e.g.,

Thirlwall et al., 2006) indicate that while such assimilation has indeed occurred in some instances, the Icelandic mantle source does have notably low $\delta^{18}O$, as low or lower than +4.5‰. Olivines in OIB $\delta^{18}O$ range from this to at least +5.7. The lowest values tend to be associated with high $^3He/^4He$ (Starkey et al., 2016) and in some cases there are clear correlations with radiogenic isotope ratios. This variability most likely reflects subduction of material that has interacted with water at the Earth's surface into the mantle.

- Oxygen isotopes in olivines from subduction-related basalts (i.e., island arc basalts and their continental equivalents) vary between 4.88‰ and 6.78‰ (e.g., Bindeman et al., 2006), corresponding to calculated melt values of 6.36‰ to 8.17‰, and on average have more positive $\delta^{18}O$ than MORB. This also reflects subduction of material that has interacted with water at the Earth's surface.

9.9.1.2 Hydrogen

Estimating the isotopic composition of mantle hydrogen, carbon, nitrogen, and sulfur is more problematic than for oxygen isotopes. These are all volatile elements and are present at low concentrations in mantle materials. They partition partially or entirely into the gas phase of magmas upon their eruption. This gas phase is lost, except in deep submarine eruptions. Furthermore, C, N, and S have several oxidation states, and isotopic fractionations occur between the various compounds these elements form (e.g., CO_2, CO, CH_4, H_2, H_2O, H_2S, SO_2). This presents two problems. First, significant fractionations can occur during degassing, even at magmatic temperatures. Second, because of loss of the gas phase, the concentrations of these elements are low. Among other problems, this means their isotope ratios are subject to disturbance by contamination. Thus, the isotopic compositions of these elements in igneous rocks do not necessarily reflect those of the magma or its mantle source.

Hydrogen, which is primarily present as water but also as H_2 and H_2S, is often lost from magmas during degassing. However, basalts erupted beneath a kilometer or more of ocean retain most of their dissolved water.

Thus, mid-ocean ridge basalts and basalts erupted on seamounts are important sources of information of the abundance and isotopic composition of hydrogen in the mantle.

The first attempt to assess the hydrogen isotopic composition of the mantle materials was that of Sheppard and Epstein (1970), who analyzed hydrous minerals in xenoliths and concluded that δD varied in the mantle. Since then, many additional studies have been carried out. Based on analysis of fresh glasses, MORB has a mean δD_{SMOW} of about $-70.1‰$ and a standard deviation of $\pm 13‰$ (Figure 9.41). Atlantic MORB are more dispersed (standard deviation of 15 vs. 11 for the Pacific) and slightly lighter (-70.1 vs. -72.6 for the Pacific); the differences are not statistically significant. Although some of the variation may be due to magmatic fractionation, regional differences and regional correlations with radiogenic isotope ratios show that most of this variation reflects mantle heterogeneity. This appears to be the result of recycling of water within the oceanic crust and associated sediments as well as various dehydration and metasomatic processes occurring during subduction (Dixon et al., 2017).

How much of this variability reflects fractionation during degassing and contamination is unclear. Kyser (1986) has argued that mantle hydrogen is homogenous with δD_{SMOW} of $-80‰$. The generally heavier isotopic composition of MORB, he argued, reflects H_2O loss and other processes. Others, for example, Poreda et al. (1986) and Chaussidon et al. (1991), have observed correlations between δD and Sr and Nd isotope ratios and have argued that these provide clear evidence that mantle hydrogen is isotopically heterogeneous. Chaussidon et al. (1991) suggested a δD value for the depleted upper mantle of about $-55‰$.

As Figure 9.41 shows, phlogopites ($K(Mg,Fe)_3AlSi_3O_{10}(OH)_2$) have δD that is generally similar to that of MORB, though heavier values also occur. Amphiboles, which are more compositionally variable, have much more variable δD and have heavier hydrogen on average. Part of this difference probably reflects equilibrium fractionation. The fractionation between water and phlogopite is close to 0‰ in the temperature range of 800–1000°C, whereas the fractionation between water and amphibole is about $-15‰$. However, equilibrium fractionation alone cannot explain either the variability of amphiboles or the difference between the mean δD of phlogopites and amphiboles. Complex processes that might include Rayleigh distillation may be involved in the formation of mantle amphiboles. This would be consistent with the more variable water content of amphiboles compared with phlogopites. There are also clear regional variations in δD in xenoliths that argue for large-scale heterogeneity.

9.9.1.3 Carbon

Most carbon in basalts is in the form of CO_2, which has more limited solubility in basaltic liquids than water. As a result, basalts begin to exsolve CO_2 well before they erupt. Thus, virtually every basalt, including those erupted at mid-ocean ridges, has lost some carbon, and subaerial basalts have lost virtually all carbon (as well as most other volatiles). Therefore

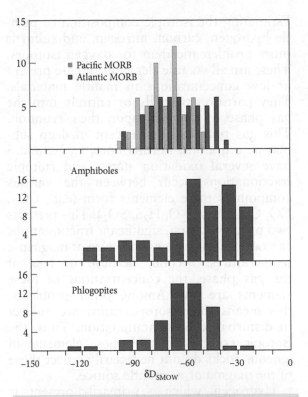

Figure 9.41 δD in MORB and in mantle phlogopites and amphiboles. The MORB and phlogopite data suggest the mantle has δD_{SMOW} of about -60 to -90.

only basalts erupted beneath several km of water provide useful samples of mantle carbon, so the basaltic dataset is essentially restricted to MORB and samples recovered from seamounts and the submarine part of Kilauea's East Rift Zone. The question of the isotopic composition of mantle carbon is further complicated by fractionation during degassing and contamination. There is a 3.3‰ fractionation between CO_2 dissolved in basaltic melts and the gas phase, with ^{13}C enriched in the gas phase (Graham et al., 2018). These submarine samples are invariably contaminated with organic material (predominantly derived from bacteria), which is isotopically light. This requires that analyses be done by step-heating, which removes the organic matter in the lower temperature steps (and these low temperature steps almost invariably have $\delta^{13}C < -20$). An alternative technique is to crush the samples, releasing the CO_2 in vesicles.

MORB have a mean $\delta^{13}C$ of -5.8 ± 1.6‰ (Figure 9.42), however, almost all these lavas have experienced some CO_2 degassing, which would decrease the $\delta^{13}C$ of the remaining CO_2; thus this value is more likely a lower limit of the mean $\delta^{13}C$ of the MORB mantle. The most CO_2-rich MORB samples have $\delta^{13}C$ of about -4‰. Since they are the least degassed, they presumably best represent the isotopic composition of the depleted mantle (Javoy and Pineau, 1991). Ocean island basalts erupted under sufficient water depth to preserve some CO_2 in the vesicles have slightly heavier carbon, with a mean of -6.58 ± 1.8‰. To what extent this difference is real is not yet clear, given the effects of fractionation. Aubaud (2006) concluded the mantle source of Pitcairn Island basalts had slightly lighter carbon ($\delta^{13}C \sim -6$‰) than MORB. Subglacial basalts from Iceland show a particularly large range of $\delta^{13}C$, including quite negative values, a result of both fractionation and crustal assimilation; nevertheless Barry et al. (2014) estimated the Iceland mantle had slightly heavier $\delta^{13}C$ (~ -2‰) than MORB. Subduction zone volcanics and back-arc basin basalts, which erupt behind subduction zones and are often geochemically similar to island arc basalts, show an even greater range of $\delta^{13}C$ with a standard deviation of 3, although the mean value, -7.65‰, is modestly lighter

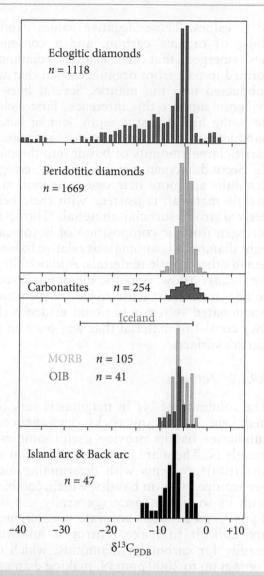

Figure 9.42 Carbon isotope ratios in mantle and mantle-derived materials. Diamond histograms are from Cartigny (2014).

than MORB. Again, whether this difference is real is unclear.

Carbonatites also have a mean $\delta^{13}C$ close to -4‰. Diamonds show a large range of carbon isotopic compositions (Figure 9.42). $\delta^{13}C$ of peridotitic diamonds, those with inclusions of typical peridotite minerals, have a mode of -5‰± 1 (Cartigny, 2014), hence similar to MORB. Thus, the $\delta^{13}C$ of the mantle is likely in the range of -4 to -5‰.

Eclogitic diamonds, those with inclusions of eclogitic minerals such as omphacitic pyroxene and garnet, are much more isotopically variable and are skewed to quite negative

$\delta^{13}C$ values. These negative values overlap those of organic carbon, and a consensus has emerged that the eclogitic diamonds formed in part from organic carbon that was subducted into the mantle. Several lines of evidence support this inference. First, eclogite is the high-pressure equivalent of basalt. Subduction of oceanic crust continuously carries large amounts of basalt into the mantle. Second, oxygen isotopes in some eclogite xenoliths are more heterogeneous than most mantle material, consistent with their being derived from surficial material. Third, the nitrogen isotopic composition of isotopically light diamonds is anomalous relative to nitrogen in other mantle materials. Additionally, as we discuss below, eclogitic diamonds contain sulfide inclusions with mass independently fractionated sulfur: additional evidence that they consist of material that was once on the Earth's surface.

9.9.1.4 Nitrogen

The solubility of N_2 in magmas is very limited, hence of volcanic rocks, once again only submarine basalts provide useful samples of mantle N. There are both contamination and analytical problems with determining nitrogen isotope ratios in basalts, which, combined with its low abundance (generally less than 1 ppm), mean that accurate measurements are difficult to make. Nitrogen substitutes readily for carbon in diamonds, which can contain up to 2000 ppm N, making diamonds important samples of mantle N. Nitrogen may be present as the ammonium ion in many kinds of rocks. NH_4^+ substitutes readily for K in many minerals; consequently, N concentrations in sediments and metasediments may reach several hundred ppm. In all, a fair amount of data is now available for a variety of mantle and crustal materials, and these data are summarized in Figure 9.43. Measurements of $\delta^{15}N_{ATM}$ in MORB range from about −10 to +8‰, with a mean value of about −3 to −5‰. Peridotitic diamonds show a similar range, but are a little lighter on average. Eclogitic diamonds show considerably greater scatter in $\delta^{15}N_{ATM}$, with the most common values in the range of −12 to −6‰. Interestingly, ocean island basalts, which presumably sample mantle plumes, have distinctly more positive $\delta^{15}N_{ATM}$, with

a mean in the range of +3 to +4‰. The N isotopic composition of mantle plumes thus appears to match well with that of organic matter in post-Archean sediments, metamorphic rocks, and subduction-related volcanic rocks. This observation led Marty and Dauphas (2003) to propose that nitrogen in mantle plumes is largely recycled from the surface of the Earth. They argue that because nitrogen can be bound in minerals as ammonium, it is more readily subducted and recycled into the mantle than other gases.

Figure 9.43 also illustrates an interesting shift in $\delta^{15}N$ in organic matter in sediments in the late Archean. Marty and Dauphas (2003) suggest two possible causes. First, the absence of atmospheric oxygen in the Archean meant that there was no nitrate, and hence no dissimilatory denitrification and consequently no fractionation associated with that process. Second, chemosynthetic life may have dominated the Archean. Modern chemosynthetic bacteria from hydrothermal vent ecosystems have a lower $\delta^{15}N$ than most plants, meaning the N isotopic composition of organic matter in Archean sediments would have been lighter than in post-Archean sediments.

9.9.1.5 Sulfur

Although sulfur is more soluble in magma than H_2O, CO_2, or N_2, H_2S and other sulfur species are also lost from magma at low pressure, consequently the reliable data on sulfur isotopes is somewhat limited. The available data is shown in Figure 9.36. As we noted earlier, MORB have mean $\delta^{34}S$ of 0.88 ± 0.23‰. There is negligible fractionation of sulfur isotopes during partial melting and Labibi et al. (2016) concluded this value approximately represents the mantle value. However, $\delta^{34}S$ in MORB varies regionally and correlates with radiogenic isotope ratios (e.g., Labidi et al., 2014), clearly demonstrating that the sulfur isotopic composition of the mantle is variable.

There is also little reliable data on oceanic island basalts as yet. Labidi et al. (2015) found that $\delta^{34}S$ of sulfides in Samoan submarine lavas varied between +0.11‰ and +2.79‰ and correlated with $^{87}Sr/^{86}Sr$. Roughly 10–20% of sulfur in these lavas is present as sulfate, and the sulfate is isotopically heavier, with $\delta^{34}S$ ranging from +4.19‰ to +9.71‰,

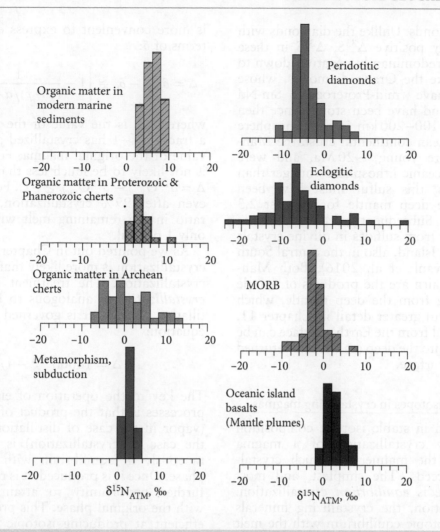

Figure 9.43 Isotopic composition of nitrogen in rocks and minerals of the crust and mantle. Modified from Marty and Dauphas (2003).

consistent with the fractionation during oxidation. They concluded that the variation in $\delta^{34}S$ primarily reflects recycled sediment in the Samoan mantle plume, as earlier inferred from radiogenic isotope ratios (White and Hofmann, 1982).

Dramatic evidence of recycling of material from the Earth's surface into the mantle came with the discovery of mass independently fractionated sulfur in sulfide inclusions in diamonds from the Orapa kimberlite mine in Botswana (Farquhar et al., 2002). As we discussed in Section 9.9, MIF sulfur is almost entirely confined to the Archean and earliest Proterozoic when UV radiation could penetrate deeply into the oxygen-free atmosphere. $\Delta^{33}S$ values in the diamonds

are smaller than in Archean sediments, range from +0.6‰ to –0.2‰, but nevertheless provide evidence that sulfur that was once in the atmosphere had been carried into the mantle to depths of 100 km or more. These diamonds are of eclogitic paragenesis, and the MIF sulfur is consistent with the carbon and nitrogen isotopic evidence that they formed from material once at the Earth's surface.

Dramatic evidence of sulfur recycling into the deep mantle came with the discovery of MIF sulfur in sulfide inclusions in olivines from lavas of Mangaia Island in the Cook-Austral chain located in the central South Pacific (Cabral et al., 2013). The sulfides are present as inclusions in olivine crystals and the $\Delta^{33}S$ values are smaller than those in diamonds. Unlike the diamonds with predominantly positive $\Delta^{33}S$, $\Delta^{33}S$ in these sulfides are predominantly negative, down to –0.4‰. Unlike the Orapa diamonds, whose inclusions have mid-Proterozoic Sm-Nd model ages and have been stored since then at depths of 100–200 km in the lithosphere beneath Archean continental crust, the Mangaia lavas are young, ~20 Ma, and were erupted on oceanic lithosphere younger than 70 Ma. Thus, this sulfur must have been stored in the deep mantle for at least 2.3 billion years. Subsequently, MIF sulfur was also reported from sulfides in olivine crystals from Pitcairn Island, also in the central South Pacific (Delavault et al. 2016). Both Mangaia and Pitcairn are the products of mantle plumes rising from the deep mantle, which we'll discuss in greater detail in Chapter 11. Thus, material from the Earth's surface can be transported into the deep mantle and returned again to the surface.

9.9.2 Stable isotopes in crystallizing magmas

The variation in stable isotope composition produced by crystallization of a magma depends on the manner in which crystallization proceeds. The simplest, and most unlikely, case is *equilibrium* crystallization. In this situation, the crystallizing minerals remain in isotopic equilibrium with the melt until crystallization is complete. At any stage during crystallization, the isotopic composition of a mineral and the melt will be related by the fractionation factor, α. Upon complete crystallization, the rock will have precisely the same isotopic composition as the melt initially had. At any time during the crystallization, the isotope ratio in the remaining melt will be related to the original isotope ratio as:

$$\frac{R_\ell}{R_0} = \frac{1}{f + \alpha(1 - f)} \quad (9.88)$$

where α is the fraction factor, R_ℓ is the isotope ratio in the liquid, R_s is the isotope ratio in the solid, R_0 is the isotope ratio of the original magma, and f is the fraction of liquid remaining. This equation is readily derived from mass balance, the definition of α, and the assumption that the element concentration in the magma is equal to that in the crystals; an assumption valid to within about 10%. It

is more convenient to express eqn. 9.88, in terms of δ:

$$\Delta = \delta_{melt} - \delta_0 \cong \left[1 - \frac{1}{(1 - f)/\alpha + f}\right] \times 100 \quad (9.89)$$

where δ_{melt} is the value of the magma after a fraction $f-1$ has crystallized and δ_o is the value of the original magma. For silicates, α is not likely to be much less than 0.998 (i.e., $\Delta = \delta^{18}O_{xtals} - \delta^{18}O_{melt} \geq -2$). For $\alpha = 0.999$, even after 99% crystallization, the isotope ratio in the remaining melt will change by only 1 per mil.

As we pointed out in Chapter 7, fractional crystallization is more likely than equilibrium crystallization. The treatment of *fractional crystallization* is analogous to Rayleigh distillation. Indeed, it is governed by the same equation:

$$\Delta = 1000(f^{\alpha-1} - 1) \quad (9.52)$$

The key to the operation of either of these processes is that the product of the reaction (vapor in the case of distillation, crystals in the case of crystallization) is only instantaneously in equilibrium with the original phase. Once it is produced, it is removed from further opportunity to attain equilibrium with the original phase. This process is more efficient at producing isotopic variations in igneous rocks, but its effect remains limited because α is generally not greatly different from 1. Figure 9.44 shows the calculated change in the oxygen isotopic composition of melt undergoing fractional crystallization for various values of Δ ($\approx 1000(\alpha-1)$). In reality, Δ will change during crystallization because of (1) changes in temperature, (2) changes in the minerals crystallizing, and (3) changes in the liquid composition. The changes will generally mean that the effective Δ will increase as crystallization proceeds. We would expect the greatest isotopic fractionation in melts crystallizing nonsilicates such as magnetite, and melts crystallizing at low temperature, such as rhyolites, and the least fractionation for melts crystallizing at highest temperatures, such as basalts.

Figure 9.45 shows observed $\delta^{18}O$ as a function of temperature in two suites: one from a propagating rift on the Galapagos Spreading Center, the other from the island

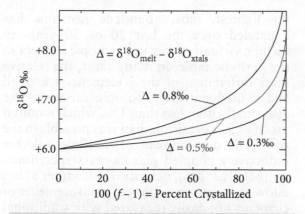

Figure 9.44 Plot of $\delta^{18}O$ versus fraction of magma solidified during fractional crystallization, assuming the original $\delta^{18}O$ of the magma was +6. After Taylor and Sheppard (1986).

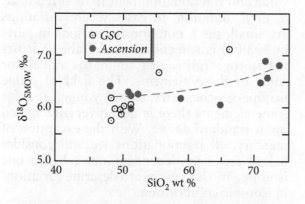

Figure 9.45 $\delta^{18}O$ as a function of SiO_2 in a tholeiitic suite from the Galapagos Spreading Center (GSC) (Muehlenbachs and Byerly, 1982) and an alkaline suite from Ascension Island (Sheppard and Harris, 1985). Dashed line shows model calculation for the Ascension suite.

of Ascension. There is a net change in $\delta^{18}O$ between the most and least differentiated rocks in the Galapagos of about 1.3‰; the change in the Ascension suite is only about 0.5‰. These, and other suites, indicate the effective Δ is generally small, of the order of 0.1–0.3‰. Consistent with this is the similarity of $\delta^{18}O$ in peridotites and MORB, which suggests a typical fractionation during melting of ~0.2‰.

We can generalize the temperature dependence of oxygen isotope fractionations by saying that at low temperature (i.e., Earth

surface temperatures up to the temperature of hydrothermal systems, 300–400°C), oxygen isotope ratios are changed by chemical processes. The amount of change can be used as an indication of the nature of the process involved, and, under equilibrium conditions, of the temperature at which the process occurred. At high temperatures (temperatures of the interior of the Earth or magmatic temperatures), oxygen isotope ratios are minimally affected by chemical processes and can be used as tracers much as radiogenic isotope ratios are.

These generalizations lead to an axiom: *Igneous rocks whose oxygen isotopic compositions show significant variations from the primordial value ($\delta^{18}O \sim +5.6$) must either have been affected by low-temperature processes, or must contain a component that was at one time at the surface of the Earth* (Taylor and Sheppard, 1986).

9.9.3 Combined fractional crystallization and assimilation

Because oxygen isotope ratios of most mantle-derived magmas are reasonably uniform (5.6‰ ± 1‰) and generally different from rocks that have equilibrated with water at the surface of the Earth, oxygen isotopes are a useful tool in identifying and studying the assimilation of country rock by intruding magma. We might think of this process as simple mixing between two components: magma and country rock. In reality, it is always at least a three-component problem, involving country rock, magma, and minerals crystallizing from the magma. Magmas are essentially never superheated; hence the heat required to melt and assimilate surrounding rock can only come from the latent heat of crystallization of the magma. Approximately 1 kJ/g would be required to heat rock from 150–1150°C and another 300 J/g would be required to melt it. If the latent heat of crystallization is 400 J/g, crystallization of 3.25 g of magma would be required to melt and assimilate 1 g of country rock. Since some heat will be lost by simple conduction to the surface, we can conclude that the amount of crystallization will inevitably be greater than the amount of assimilation (the limiting case where mass crystallized equals mass assimilated could occur only at great depth

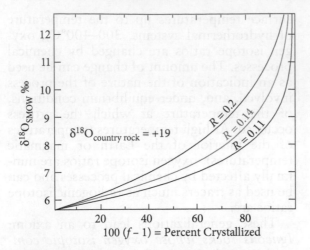

Figure 9.46 Variation in $\delta^{18}O$ of a magma undergoing AFC versus amount crystallized. Initial $\delta^{18}O$ of the magma is +5.7. After Taylor (1980).

in the crust where the rock is at its melting point to begin with). The change in isotopic composition of a melt undergoing the combined process of assimilation and fractional crystallization (AFC) is given by:

$$\delta_m - \delta_0 = \left([\delta_a - \delta_0] - \frac{\Delta}{R} \right) \{1 - f^{-R/(R-1)}\}$$

(9.90)

where R is the ratio of mass assimilated to mass crystallized, Δ is the difference in isotope ratio between the crystals and the magma ($\delta_{crystal} - \delta_{magma}$), f is the fraction of liquid remaining, δ_m is the $\delta^{18}O$ of the magma, δ_0 is the initial $\delta^{18}O$ of the magma, and δ_a is the $\delta^{18}O$ of the material being assimilated (Taylor and Sheppard, 1986). The assumption is made that the concentration of oxygen is the same in the crystals, magma and assimilant, which is a reasonable assumption. This equation breaks down at $R = 1$, but, as discussed above, R is generally significantly less than 1. Figure 9.46 shows the variation of $\delta^{18}O$ of a magma with an initial $\delta^{18}O = 5.7$ as crystallization and assimilation proceed.

9.10 NONTRADITIONAL STABLE ISOTOPES

As we mentioned in the introduction to this chapter, stable isotope geochemistry, once almost entirely restricted to a half-dozen of

the lightest, most abundant elements, has expanded over the last 20 or 30 years to include virtually every polyisotopic element in the periodic table. In many cases, the relative mass differences of these elements are small and consequently the isotopic variations are also small, often less than 1‰, which required much higher precision than was possible in the past. The development of multiple-collector inductively coupled plasma-mass spectrometry (MC-ICP-MS), described in Chapter 8, has allowed for precise isotopic measurements of elements not easily measured with traditional gas source or thermal ionization instruments (Halliday et al., 1995; Maréchal et al., 1999).

The notation and conventions of nontraditional stable isotopes are the same as the traditional ones, with most ratios reported in delta-notation relative to some standard value and fractionation factors reported in Δ, α, or β, notation. In cases where variations are small, the ε notation (deviations in parts in 10,000) is sometimes used. Table 9.5 lists the isotope ratios and standards values for some of these elements. The field of stable isotope geochemistry is still young, and for some elements there is no universally agreed upon standard as yet. With the exception of mercury, all fractionations we will consider below are mass-dependent ones, hence one isotope ratio is sufficient to describe variations in isotopic composition.

9.10.1 Boron isotopes

Boron has two stable isotopes: ^{10}B (19.9%) and ^{11}B (80.1%). The $^{11}B/^{10}B$ ratio is conventionally reported as $\delta^{11}B$ in per mil variations from the NBS951 standard. In nature, boron has a valence of +3 and is almost always bound to oxygen or hydroxyl groups in either trigonal (e.g., BO_3) or tetrahedral (e.g., $B(OH)_4^-$) coordination. Since the bond strengths and vibrational frequencies of trigonal and tetrahedral forms differ, we can expect that isotopic fractionation will occur between these two forms and this is confirmed by experiments which show a roughly 20‰ fractionation between $B(OH)_3$ and $B(OH)_4^-$, with ^{11}B preferentially found in the $B(OH)_3$ form. Boron is relatively abundant in seawater, with a concentration of 4.5 ppm. In silicate rocks, its concentration of boron ranges from a few tenths of a ppm or less in fresh basalts and

Table 9.5 Reference values of nonconventional stable isotope ratios.

Element	Notation	Ratio	Standard	Absolute ratio
Lithium	δ^7Li	$^7Li/^6Li$	NIST L-SVEC	12.1735
Boron	$\delta^{11}B$	$^{11}B/^{10}B$	NIST 951	4.0436
Magnesium	$\delta^{26}Mg$	$^{26}Mg/^{24}Mg$	DSM3	0.13979
Silicon	$\delta^{30}Si$	$^{30}Si/^{28}Si$	NBS28 (NIST-RM8546)	0.033532
	$\delta^{29}Si$	$^{29}Si/^{28}Si$		0.050804
Chlorine	$\delta^{37}Cl$	$^{37}Cl/^{35}Cl$	seawater (SMOC)	0.31963
			NIST-SRM 975	0.31977
Calcium	$\delta^{44/42}Ca$	$^{44}Ca/^{42}Ca$	NIST SRM 915a	0.310163
	$\delta^{44/40}Ca$	$^{44}Ca/^{40}Ca$		0.021518
	$\delta^{43/42}Ca$	$^{43}Ca/^{42}Ca$		0.208655
Titanium	$\delta^{49}Ti$	$^{49}Ti/^{47}Ti$	OL-Ti	0.74977
Iron	$\delta^{56}Fe$	$^{56}Fe/^{54}Fe$	IRMM-14	15.698
	$\delta^{57}Fe$	$^{57}Fe/^{54}Fe$		0.363255
Copper	$\delta^{65}Cu$	$^{65}Cu/^{63}Cu$	NIST 976	0.44562
Zinc	$\delta^{68}Zn$	$^{68}Zn/^{64}Zn$	JMC3-0749L	0.37441
	$\delta^{66}Zn$	$^{66}Zn/^{64}Zn$		0.56502
Mercury	$\delta^{199}Hg$	$^{199}Hg/^{198}Hg$	NIST SRM3133	1.6872
	$\delta^{200}Hg$	$^{200}Hg/^{198}Hg$		2.3047
	$\delta^{201}Hg$	$^{201}Hg/^{198}Hg$		1.3121
	$\delta^{202}Hg$	$^{202}Hg/^{198}Hg$		2.9614
	$\delta^{204}Hg$	$^{204}Hg/^{198}Hg$		0.68012

fresh peridotites, to several tens of ppm in clays.

While there were some earlier studies, modern study of boron isotopes began with the work of Spivack and Edmond (1987), preceding development of ICP-MS instruments. The range of $^{11}B/^{10}B$ variation in terrestrial materials is ~100‰ (Figure 9.47), reflecting the large mass difference between the two isotopes (~10%). Variation is also large in chondritic meteorites, reflecting both fractionation and nucleosynthetic variations. Seawater lies near the heavy extreme of this spectrum, with $\delta^{11}B$ of +39.6 ±0.04‰ and is uniform within analytical error. Evaporites of continental saline lakes define the light extreme. Boron is readily incorporated into the alteration products, so that even slightly altered basalts show a dramatic increase in B concentration and an increase in $\delta^{11}B$, with altered oceanic crust having $\delta^{11}B$ in the range of –2 to +26‰. Smith et al. (1995) estimated that average altered oceanic crust contains 5 ppm B and $\delta^{11}B$ of +3.4‰.

$\delta^{11}B$ in fresh MORB range from –9.4‰ to –2.2 ‰, but much of this range appears to reflect assimilation of altered oceanic crust or brines stored within it. Marschall et al. (2017) found that MORB with Cl/K ratios <0.08 had nearly homogenous $\delta^{11}B$ of –7.1 ± 0.9‰ and inferred this likely represented the mantle and bulk silicate Earth composition. Oceanic island basalts (OIB) have a larger range of $\delta^{11}B$ with an average of –4.4‰. The heaviest values in this range have been shown to result from assimilation of seawater-altered crust, most notably in the Azores (Genske et al., 2014) and Iceland. Excluding these samples, the OIB mean is slightly lighter than MORB, –8.2‰, but with an apparent bimodal distribution. One possible explanation is that OIB sources contain recycled sediment containing heavy boron and recycled oceanic crust containing light boron held in phengite (Palmer, 2017), a metamorphic mineral sometimes found in eclogites.

The boron isotopic composition of island arc volcanics (IAV) is heavier and more variable than OIB or MORB, with a mode of ~+5‰. $\delta^{11}B$ in island arc volcanics correlates negatively with Na_2O, Nb and Sr concentrations, and La/Sm and Nb/B ratios, and positively with slab dip, B and Sc concentrations and B/Ce and $^{87}Sr/^{86}Sr$ ratios. As

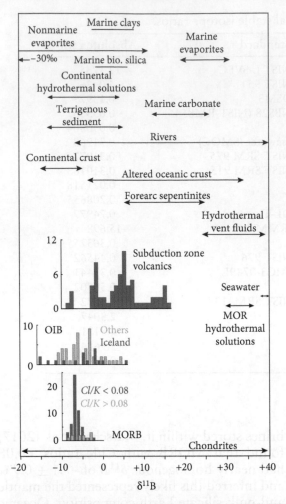

Figure 9.47 Boron isotopic composition of terrestrial materials and chondrites.

the oceanic crust and seawater is to decrease the $\delta^{11}B$ of seawater (Spivack and Edmond, 1987). The isotopic composition of these fluids is slightly lower than that of seawater (typically +24‰ to +37‰), suggesting that the B in these fluids is a simple mixture of seawater- and basalt-derived B, and that little or no isotopic fractionation was involved.

One of the more interesting applications of boron isotopes has been determining the paleo-pH of the oceans. Boron is readily incorporated into carbonates, with modern marine carbonates having B concentrations in the range of 15–60 ppm. The incorporation of B in carbonate is preceded by surface adsorption of $B(OH)_4^-$ (Vengosh et al., 1991; Heming and Hanson, 1992), and consequently it is primarily $B(OH)_4^-$ that is incorporated rather than $B(OH)_3$. This provides a means of determining pH in ancient seawater.

We noted above that boron is present in seawater both as $B(OH)_3$, and $B(OH)_4^-$. Since the reaction between the two may be written as:

$$B(OH)_3 + H_2O \rightleftharpoons B(OH)_4^- + H^+ \qquad (9.91)$$

the equilibrium constant for this reaction is:

$$pK^{app} = \log \frac{B(OH)_4^-}{B(OH)_3} - pH \qquad (9.92)$$

The relative abundance of these two species is thus pH-dependent. Furthermore, we can easily show that the isotopic composition of these two species must vary if the isotopic composition of seawater is constant. From mass balance we have:

$$\delta^{11}B_{sw} = \delta^{11}B_4 f + \delta^{11}B_3(1 - f) \qquad (9.93)$$

where f is the fraction of $B(OH)_3$ (and $1 - f$ is therefore the fraction of $B(OH)_4^-$), $\delta^{11}B_3$ is the isotopic composition of $B(OH)_3$, and $\delta^{11}B_4$ is the isotopic composition of $B(OH)_4^-$. If the isotopic compositions of the two species are related by a constant fractionation factor, Δ_{3-4}, we can write eqn. 9.93 as:

$$\delta^{11}B_{sw} = \delta^{11}B_4 f + \delta^{11}B_3 - \delta^{11}B_3 f$$
$$= \delta^{11}B_3 + \Delta_{3-4} f \qquad (9.94)$$

summarized by De Hoog and Savov (2018) and Palmer (2017), $\delta^{11}B$ is so high in IAV that its only plausible source is serpentinized mantle, either in the subducting slab, the hydrated mantle above the slab or in the forearc. As Figure 9.47 shows, such forearc peridotites have quite heavy $\delta^{11}B$ and are B-rich as a result of reaction with seawater.

Perhaps the most remarkable aspect of B isotope geochemistry is the very large fractionation of B isotopes between the oceans and the silicate Earth. This difference partly reflects the fractionation that occurs during adsorption of boron on clays (e.g., Schwarcz et al., 1969). However, this fractionation is only about 30‰ or less, whereas the difference between the continental crust and seawater is close to 50‰. Furthermore, the net effect of hydrothermal exchange between

Solving for $\delta^{11}B_4$, we have:

$$\delta^{11}B_4 = \delta^{11}B_{sw} - \Delta_{3-4}f \qquad (9.95)$$

Thus, assuming a constant fractionation factor and isotopic composition of seawater, the $\delta^{11}B$ of the two B species will depend only on f, which, as we can see in eqn. 9.92, will depend on pH. The proof is left to the reader (Problem 9.12).

Klochko et al. (2006) experimentally determined α_{3-4} to be 1.0272 ± 0.0006 (Δ_{3-4} = 27.2). There are some additional factors that must be considered: (1) there are species-specific fractionations, perhaps because they alter the pH of their micro-environment, or perhaps because $B(OH)_3$ is also incorporated to varying degrees; (2) the fractionation factor may be temperature-dependent (this appears to be small); and (3) the B isotopic composition of seawater may vary with time. Considerable research has been done to constrain these additional factors.

The pH of seawater, in turn, is largely controlled by the carbonate equilibrium, and depends therefore on the partial pressure of CO_2 in the atmosphere (see Example 6.2). Thus, if the pH of ancient seawater can be determined, it should be possible to estimate p_{CO_2} of the ancient atmosphere. Given the concern about the relation of future climate to future p_{CO_2}, it is obviously interesting to know how these factors related in the past.

Pearson and Palmer (2000) measured $\delta^{11}B$ in foraminiferal carbonate extracted from Ocean Drilling Program (ODP) cores, and from this calculated pH. The results indicated dramatically lower seawater pH and dramatically higher p_{CO_2} in the Paleogene. The apparent variation in p_{CO_2} is qualitatively consistent with what is known about Cenozoic climate change – namely, that a long-term cooling trend began in the early to middle Eocene. Since that initial study, a number of other, more detailed studies have been carried out for the Cenozoic as a whole and the last glacial cycle (Figure 9.48). These estimates are in reasonable agreement with ice core CO_2 data and the C_{37} alkenone $\delta^{13}C$ CO_2 proxy we will discuss in Chapter 12.

On a more limited time-scale, Hönisch and Hemming (2005) and Foster (2008)

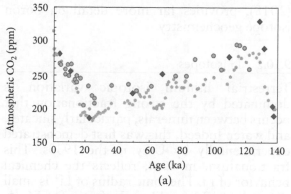

(a)

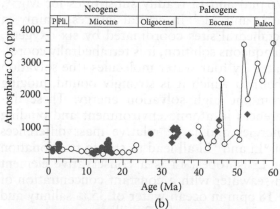

(b)

Figure 9.48 (a) Comparison of calculated atmospheric CO_2 based on ocean pH derived $\delta^{11}B$ in shells of planktonic foraminifera with CO_2 measured in the EPICA Antarctic ice core (Bereiter et al., 2015; small gray circles). Red diamonds are from Hönish and Hemming (2005); Blue circles are from Foster et al. (2015). (b) Concentration of atmospheric CO_2 derived from seawater pH calculated from $\delta^{11}B$. Circles are the original results of Pearson and Palmer (2000); other symbols are from subsequent studies.

investigated $\delta^{11}B$ during the Pleistocene. They controlled for temperature by analyzing the Mg/Ca ratio of the carbonate shells, which is known to be strongly temperature-dependent. For the last two glacial cycles, their calculated pH values ranged from 8.11 to 8.32, corresponding to a p_{CO_2} range of ~180 to ~325 ppm. As Figure 9.48 shows, these results track the CO_2 measured in Antarctic ice, although the $\delta^{11}B$-based values tend to be offset by a few tens of ppm to higher CO_2.

The recent book *Boron Isotopes, The Fifth Element* edited by Marschall and Foster

(2018), provides far more detail on boron isotope geochemistry.

9.10.2 Li isotopes

Terrestrial lithium isotopic variation is dominated by the strong fractionation that occurs between minerals, particularly silicates, and water. Indeed, this was first demonstrated experimentally by Urey in the 1930s. This fractionation, in turn, reflects the chemical behavior of Li. The ionic radius of Li^+ is small (76 pm) and Li readily substitutes for Mg^{2+}, Fe^{2+}, and Al^{3+} in crystal lattices, mainly in octahedral sites coordinated by six oxygens. In aqueous solution, it is tetrahedrally coordinated by four water molecules (the solvation shell) to which it is strongly bound, judging from the high solvation energy. These differences in atomic environment and binding energies and large relative mass differences of ^6Li and ^7Li all lead to strong fractionation of Li isotopes. Li is a conservative element in seawater with a constant concentration of 0.18 ppm in ocean water of 35‰ salinity and a residence time of 1.3 million years.

Li isotope ratios are today reported as δ^7Li, per-mil deviations from the NIST 8545 L-SVEC standard $^7Li/^6Li$. However, prior to 1996, it was reported as δ^6Li, that is, deviations from the $^6Li/^7Li$ ratio of that standard. For variations of less than about 10‰, $\delta^7Li \approx -\delta^6Li$. While it remains the reference standard for the δ^7Li value, L-SVEC has been exhausted and standard IRMM-16 has become the practical analytical standard.

Modern study of Li isotope ratios began with the work of Chan and Edmond (1988). They found that the isotopic composition of seawater was uniform within analytical error; subsequent higher precision analysis revealed a seawater $\delta^7Li = +31.3‰$. This compares to the mean values for rivers (~+23‰), groundwaters (~+15‰), hydrothermal fluids (~+7‰), and upper continental crust (~+0.6).

As is the case for boron, seawater represents one extreme of the spectrum of isotopic compositions in the Earth (Figure 9.49). During mineral–water reactions, the heavier isotope, ^7Li, is preferentially partitioned into the solution: this reflects the lower coordination number and stronger bonds formed in solution than in most minerals. There appears to be little or no fractionation during mineral dissolution, but strong fractionations occur during secondary mineral formation that accompanies weathering with the lighter isotope, ^6Li, is preferentially absorbed onto oxides and incorporated in clays, leaving the water ^7Li-rich. Thus, weathering on the continents results in river water being isotopically heavy compared with average continental crust (Teng et al., 2004). Seawater is some 10 per mil heavier than average river water, so additional fractionation must occur in the marine environment. This includes authigenic clay formation and low-temperature alteration of oceanic crust, which are the primary sinks for Li in the oceans. Chan et al. (1992) estimated the fractionation factor for clay-solution exchange at −19‰. Reported δ^7Li values in marine noncarbonate sediments range from −1 to +15‰. Marine carbonate sediments, which tend to be Li-poor, typically have higher δ^7Li than noncarbonate sediments.

Bulk chondrites range from +3‰ to +4‰ and unmetasomatized mantle peridotites range from +2.5‰ to +5‰, thus the Earth appears to have a chondritic Li isotopic composition. There appears to be little isotopic fractionation, <0.5‰, during fractional crystallization, and perhaps also partial melting (Tomascak et al., 2016). Fresh MORB have δ^7Li of +3.6 ± 0.8‰, a range not much larger than that expected from analytical error alone (Figure 9.49) and indistinguishable from peridotites. Oceanic island basalts (OIB) have on average higher δ^7Li, +4.4 ± 1.3‰. There are clear correlations with radiogenic isotope ratios in individual island chains, demonstrating that the mantle is heterogeneous with respect to δ^7Li (White, 2014). The highest δ^7Li occurs on islands characterized by particularly radiogenic Pb (the so-called HIMU OIB group), such as on some islands of the Cook-Austral chain. Here, δ^7Li may be as high as +8‰. This may reflect a recycled crustal component in their sources (e.g., Vlastelic et al., 2009).

The average δ^7Li of island arc volcanics, ~+3.7‰, is indistinguishable from that of MORB, but they are considerably more variable with a standard deviation of 2.4‰. Lithium isotopic signatures in arc magmas can vary from the mantle value for a number of reasons: a contribution from subducted seawater-altered oceanic crust, contribution

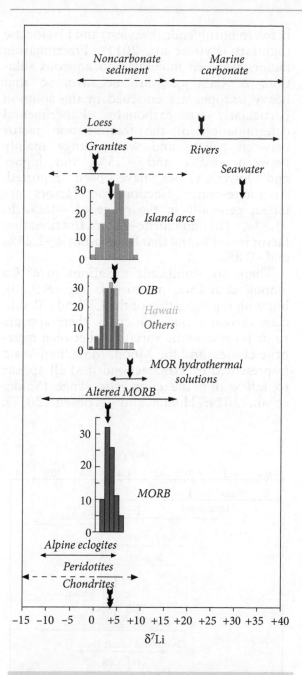

Figure 9.49 Li isotopic composition of terrestrial materials and chondrites. Arrows indicate mean values.

the large potential for diffusional isotopic fractionation. As we noted earlier in the chapter, lighter isotopes of an element will diffuse faster than heavier isotopes. The relatively large mass difference between ^7Li and ^6Li (17%) results in substantially faster diffusion of ^6Li. Richter et al. (1999) found that the ratio of diffusion coefficients for isotopes of an ion diffusing in molten silicates obeyed the relation:

$$\frac{D_1}{D_2} = \left(\frac{m_2}{m_1}\right)^{\beta} \tag{9.96}$$

where β is an experimentally determined empirical parameter. Richter et al. (2003) found that the value of β was 0.215 for Li, implying 3% faster diffusion of ^6Li. Teng et al. (2006) observed that the isotopic composition of Li diffusing out of a pegmatite into surrounding amphibolite decreased from +7.6‰ to −19.9‰ over a distance of 10 m.

δ^7Li in oceanic crust altered by seawater at low temperatures takes up Li from solution and has high δ^7Li compared with fresh basalt. In hydrothermal reactions, however, Li is extracted from basalt into solution and hydrothermal fluids can have Li concentrations up to 50 times greater than seawater. ^7Li is extracted more efficiently than ^6Li during this process, so hydrothermally altered basalt can have δ^7Li as low as −2‰. Serpentinites (hydrothermally altered peridotite) can have even lower δ^7Li. Because they extract Li from oceanic crust so completely, hydrothermal solutions have Li isotopic compositions intermediate between MORB and seawater despite this fractionation.

Misra and Froelich (2012) found that δ^7Li of seawater has risen by 9‰ over the Cenozoic, a change that mirrors, albeit inexactly, those of $^{87}Sr/^{86}Sr$ (Figure 8.9) and $^{187}Os/^{188}Os$ (Figure 8.17). Misra and Froelich interpreted this as due to an increase in weathering rates, but a concomitant decrease in weathering intensity; i.e., greater dissolution but less secondary mineral formation. This is what would be expected with rapid tectonic uplift: increasing the importance of mechanical over chemical weathering. The Cenozoic has indeed been a period of intense tectonic activity, most notably the Himalayan orogeny. Some confirmation of

from subducted sediment, and fractionation during dehydration as oceanic crust and sediment subducts, which should produce and isotopically heavy fluid and light residue.

Diffusion represents another mechanism by which Li isotopes may fractionate. Because of its small ionic size and charge, Li diffuses more readily than most other elements. This rapid diffusion also contributes to this interpretation was reported by Pogge von Strandmann et al. (2015), who found that δ^7Li in New Zealand rivers was inversely correlated with local uplift rates. Misra and Froelich also found a ~5‰ drop in δ^7Li across the Cretaceous–Paleogene Boundary. Neither of the suspected causes of the extinction that defines the boundary, the Chicxulub bolide and the Deccan Traps eruption, could have caused this drop directly. Instead, Misra and Froelich suggested it resulted from "massive continental denudation and acid rain weathering of continental soils that were partially incinerated and deforested by the impact aftermath and washed into the sea."

More details on Li isotope geochemistry can be found in the book *Advances in Lithium Isotope Geochemistry* (Tomascak et al., 2016) and the review by Pennison-Dorland et al. (2017).

9.10.3 Calcium isotopes

Because Ca is a biochemically important element and linked to the carbon cycle, particularly through carbonate precipitation, there is great interest in using calcium isotopes to understand biogeochemical cycles. Calcium has six naturally occurring isotopes, ^{40}Ca (96.941), ^{42}Ca (0.647%), ^{43}Ca (0.135%), ^{44}Ca (2.086%), ^{46}Ca (0.004%), and ^{48}Ca (0.187%), a consequence of it having a magic number of protons (Section 8.2.1). Because it also has a magic number of neutrons ^{40}Ca is particularly abundant. The $^{44}Ca/^{40}Ca$ ratio is the one generally measured and reported as $\delta^{44}Ca$ or as $\delta^{44/40}Ca$ per-mil deviations of from the NIST-SRM-915b standard. However, some data have been reported relative to seawater or estimates of bulk silicate Earth. Because $^{40}Ar^+$ interferes with the $^{40}Ca^+$ signal in MC-ICP-MS analysis, some geochemists have chosen to measure and report the $^{44}Ca/^{42}Ca$ as $\delta^{44/42}Ca$. For small values of δ (and almost all Ca isotope variations are small), the two reporting conventions are related as:

$$\delta^{44}Ca = \delta^{44/40}Ca = 2.01 \times \delta^{44/42}Ca \quad (9.97)$$

Here we will use $\delta^{44}Ca_{NIST-SRM-915a}$ exclusively in our discussion of Ca isotopes.

Minerals separated from a granodiorite have a range of $\delta^{44/40}Ca$ of ~0.6‰

between hornblende (heaviest) and plagioclase (lightest) (Ryu et al., 2011). Fractionation factors between minerals and aqueous solution, $\Delta^{44}Ca_{min-Ca^{2-}}$, are negative, so that heavy isotopes are enriched in the solution particularly for carbonates. Experimental determination of the fractionation factor between calcite and water range mainly between –0.5‰ and –1.5‰, but higher and lower values have been reported. Aragonite–water fractionation factors are larger, generally in the range of –1.2‰ to –1.8‰. The anhydrite–water fractionation factor is ~–1‰ and that for gypsum is –2.25‰ and –0.8‰.

There are significant variations in $\delta^{44}Ca$ among chondritic meteorites (Figure 9.50), but with the exception of the CV and CR subclass carbonaceous chondrites, there appears to be no systematic variations between meteorite classes, and the Moon, Mars, and Vesta (represented by HED achondrites) all appear to fall within the chondritic range (Valdes et al., 2014; Huang and Jacobsen, 2017).

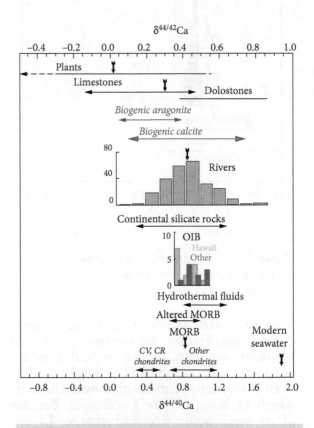

Figure 9.50 Calcium isotope ratios in terrestrial materials and meteorites.

Unmetasomatized peridotites have a limited range of $\delta^{44}Ca$ from +0.9‰ to +1.06‰, leading Huang et al. (2010) to estimate an upper mantle $\delta^{44}Ca$ of 0.97‰. Based on only a few analyses, MORB have an average $\delta^{44/40}Ca$ of +0.81±0.04‰. Oceanic island basalts are more variable, with $\delta^{44}Ca$ ranging from 0.7‰ to 1.07‰ with an average and standard deviation of 0.86‰ and 0.11‰, respectively. This variation seems unlikely to be due to fractional crystallization as Zhang et al. (2018) found no detectable variation with fractional crystallization in the Kilauea Iki lava lake and calculated $\Delta^{40}Ca_{cpx-melt} = 0.04±0.03‰$. Hawaiian basalts appear to be lighter on average (0.83±0.11‰) than other OIB (0.91±0.1‰). Huang et al. (2011) found that $\delta^{40/44}Ca$ ratios in lavas of Koolau series from Oahu, Hawaii, were particularly low: ~0.75; furthermore, they found that the calcium isotope ratios correlated inversely with $^{87}Sr/^{86}Sr$ and with Sr/Nb ratios. Huang et al. interpreted this as evidence of the presence anciently subducted marine carbonate in the Hawaiian mantle plume.

Seawater has a $\delta^{44}Ca$ of +1.9±0.18‰, with most of the variation likely to be entirely analytical; given its long residence time in seawater, calcium should be isotopically uniform. As was the case with B, and Li, seawater represents the heavy extreme of terrestrial Ca isotopic compositions. Rivers are the primary source of Ca in the oceans, and they have a mean $\delta^{44}Ca$ of 0.86‰. As carbonate weathering is thought to contribute 75–90% of the riverine Ca flux, this value is somewhat surprising because it is closer to average silicate rock (~+0.94‰) than average carbonate rock (0.63‰).

Several studies have shown that there is apparently little fractionation during mineral dissolution and weathering. In laboratory dissolution experiments Ryu et al. (2011) found that $\delta^{42/44}Ca$ of solutions produced in the experiments varied due to incongruent dissolution, but that little mass fractionation occurred. However, precipitation of secondary carbonates as travertines and soil carbonates, could increase the $\delta^{44}Ca$ of dissolved calcium in rivers, which could explain the relatively high mean $\delta^{40}Ca$ of rivers. Indeed, rivers draining carbonate terrains have higher $\delta^{40}Ca$ than those draining silicate terrains, suggesting some precipitation of isotopically light carbonate may be occurring (Tippler et al., 2016).

In contrast to the minimal fractionation in weathering, there is significant fractionation in uptake by plants. Plant themselves are isotopically variable, with roots and woody tissue being isotopically light compared with the nutrient sources and to leaves, although with considerable variability. On average, the mean $\Delta^{44}Ca_{plant-source}$ is about –0.7‰. In animals, Ca isotope systematics are complex. The data suggest that in animals there is a large fractionation in mineralized tissue formation. The isotopic composition of soft tissues (muscle, blood) appears to track that of diet, although with small fractionation between different types of tissues. In mammals, bones are the largest Ca reservoir and are enriched in the lighter isotopes of Ca by ~1‰ compared with blood, and urine is enriched in the heavy isotopes compared with blood by ~2.4‰ in humans. The biosphere, which is dominated by plants, represents the isotopically lightest Ca in the global Ca cycle (Figure 9.50).

High-temperature hydrothermal fluids on mid-ocean ridges are another source of Ca to the oceans and they have $\delta^{44}Ca$ ranging only from +0.95‰ to +0.80‰, which is up to ~0.15‰ heavier than fresh MORB. However, anhydrite precipitates as these fluids exit the seafloor, which has $\Delta^{44}Ca_{An-fluid}$ of –0.55‰, increasing $\delta^{44}Ca$ of the fluid by about 0.14‰. From this, Amini et al. (2008) concluded that Ca is leached from the oceanic crust without isotopic fractionation. Anhydrite precipitated in vents later dissolves at low temperature. Thus, hydrothermal activity produces a net flux of isotopically light Ca (~+0.80‰) to the oceans and does not directly affect the Ca isotopic composition of the oceanic crust. However, significant amounts of calcite and aragonite precipitate in the oceanic crust at low temperature with an apparent $\Delta^{44}Ca_{min-Ca^{2-}}$ of –1.54‰, which tends to increase in $\delta^{44/40}Ca$ of seawater.

Since fluxes into the oceans are isotopically light, they must be balanced by an isotopically light flux from the oceans. This flux is carbonate precipitation, which is the principal way in which Ca is removed from the oceans. Autotrophs such as the alga *Emiliania huxleyi* construct their tests, or shells, of calcite as do planktonic foraminifera; together with other open ocean calcite- secreting organisms, coccolithophorids and foraminifera account for about 55% of the calcium flux from the ocean. Although the exact fractionation factors between organisms and temperature, CO_3^{2-} concentration, precipitation rate, and Mg^{2+} concentration, there is a collective well-defined mean fractionation, $\Delta^{44}Ca_{min-Ca^{2-}}$, of open ocean removal of −1.30‰ (Blätter et al., 2012). The remaining 45% is removed by shallow water organisms. While some of these, such as coralline algae secrete calcite with a relatively small fractionation from seawater (∼ −0.9‰), most, such as reef-building corals, secrete aragonite with a much larger average fractionation on (∼ −1.6‰). The isotopic composition of the net flux out of seawater thus depends on the relative amounts removed by open ocean versus shallow water calcifiers. Planktonic calcifiers became important only in the Jurassic, so this balance could have been different in the past.

Farkas et al. (2007) attempted to deduce the Ca isotopic history of seawater over Phanerozoic time from the isotopic composition of marine carbonates and phosphates. The results suggest that $\delta^{44}Ca$ has increased by ∼0.7‰ over the last 500 Ma, although the path has been bumpy. Additional data has supported these results while supplying somewhat clearer picture (Blätter et al., 2012). Seawater $\delta^{44/40}Ca$ was about 0.65‰ lower than present in the early Paleozoic and rose dramatically in the Carboniferous to near-modern values before declining through the Permian to values ∼0.4‰ lower than modern. $\delta^{44}Ca$ then began to rise again in the Jurassic reaching near modern values in the Cretaceous. Farkaš et al. concluded these variations reflected a change in nature the dominant Ca output fluxes from dominantly aragonitic to dominantly calcic. Higgins et al. (2018) suggested that instead, diagenesis and dolomitization of shallow water carbonate might be the controlling factor.

Because of the close coupling of the Ca and carbon cycles, Ca isotopes have the potential to elucidate changes in the biogeochemical carbon cycle recorded by $\delta^{13}C$ in marine carbonates such as the decline in $\delta^{13}C$ of roughly 3‰ over several hundred thousand years at the Permian–Triassic boundary, defined by the largest mass extinction, and in which $\delta^{44/40}Ca$ declines by 0.3‰ (Payne et al., 2010). This $\delta^{13}C$ excursion is indicative of a major disruption of the global biogeochemical carbon cycle including the collapse of ocean primary productivity and extinction of many calcifiers, but Komar and Zeebe (2016) concluded that other factors must have been involved an increase in the carbonate weathering flux, and a change in $\Delta^{44/}Ca_{min-Ca^{2-}}$ due to changing $[HCO_3^-]$, all of which could have been secondary effects of the Siberian Traps eruption. A more extreme $\delta^{13}C$ excursion, −12‰ occurs in ∼560 Ma old Ediacaran strata, when multicellular animals (metazoans) were first appearing in the fossil record. There has been debate as to whether this excursion was merely diagenetic, but the $\delta^{13}C$ excursion is also observed in most pristine samples preserve distinctly aragonitic values of $\delta^{44}Ca$, indicating $\delta^{13}C$ does indeed record major disruption of the global biogeochemical carbon cycle.

9.10.4 Silicon isotopes

Silicon, which is the third or fourth most abundant element on Earth (depending on how much may be in the core), has three isotopes: ^{28}Si (92.23%), ^{29}Si (4.67%), and ^{30}Si (3.10%). By convention, $^{30}Si/^{28}Si$ ratios are reported as $\delta^{30}Si$ relative to the standard NBS28 (NIST-RM8546) (Table 9.5), although $^{29}Si/^{28}Si$ is also sometimes reported as $\delta^{29}Si$. The study of silicon isotope geochemistry began in the 1950s, but the field has rapidly expanded only since the advent of the high-precision multi-collector-ICP-MS in the last two decades. Silica is surprisingly important in biogeochemical cycles and is utilized by terrestrial plants as well as freshwater and marine algae and protists. Although abundant in the solid Earth, its concentration in solution is limited due in part to its low solubility but also to extensive bioutilization. Silicon isotopes thus have potential for understanding biogeochemical cycling.

Figure 9.51 provides an overview of the Si isotopic composition of terrestrial and extraterrestrial materials. In addition to biogeochemistry, silicon isotope geochemistry is also relevant to assessing the parental material of the Earth and the composition of the

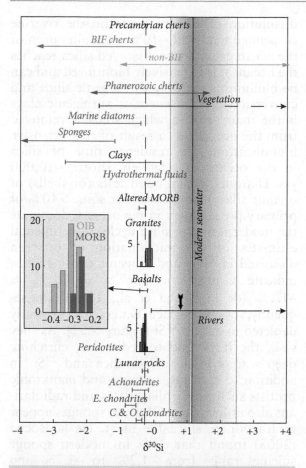

Figure 9.51 Silicon isotopic composition of terrestrial and solar system materials. Dashed line is the estimated $\delta^{30}Si$ of the bulk silicate Earth. "C & O chondrites" are carbonaceous and ordinary chondrites; "E chondrites" are enstatite chondrites; gray area shows the range of $\delta^{30}Si$ in seawater; solid line is estimated average seawater composition. Arrow shows estimated average composition of rivers.

Mg/Si is indicative of ^{28}Si depletion related to volatile loss (Dauphas et al., 2015). Second, it results from Si isotopic fractionation between silicate and iron, during core formation. Experiments (e.g., Hin et al., 2014) and theoretical calculations (Schauble et al., 2007) show that some Si partitions into the metal phase at sufficiently low oxygen fugacity and high pressure. Geophysical constraints require that the core contain several percent of one or more light element. Considering both volatile fractionation and metal-silicate partitioning, Dauphas et al. (2015) concluded the Earth's core contains 3.6 ± 3% Si.

The average $\delta^{30}Si$ of mantle-derived ultramafic xenoliths is –0.30±0.04‰. MORB appear to be isotopically uniform within analytical error with a mean of –0.28±0.03‰. OIB are slightly lighter on average and more variable: –0.32±0.05‰. While these differences are small, OIB are clearly skewed to lower values (Figure 9.51) and this is a result of particularly light compositions of basalts from just a few island chains (Pringle et al., 2016). This suggests, as do other stable isotopes, that OIB and the mantle plumes from which they are produced, contain a component of surficial material that has been recycled through subduction from the Earth's surface.

There is a slight fractionation associated with melting and fractional crystallization, with lighter silicon isotopes partitioning preferentially into SiO_2-poor mafic minerals such as olivine and consequently granites are heavier than basalts: –0.23±0.13‰. Savage et al. (2012) found that $\delta^{30}Si$ in igneous rocks depended on SiO_2 composition approximately as:

$$\delta^{30}Si(‰) = 0.0056 \times [SiO_2] - 0.567 \quad (9.98)$$

Compared with other processes, however, Si isotope fractionation in igneous processes is quite small. At lower temperatures and where silicic acid (H_4SiO_4) is involved, fractionations are greater. Igneous minerals at the surface of the Earth undergo weathering reactions in which Si from crystalline rock partitions between residual clays such as kaolinite, quartz, and silicic acid in aqueous solution. Méhuet et al. (2007) used a theoretical approach to calculate a $\Delta^{30}Si$ kaolinite–quartz fractionation factor of

Earth's core. The bulk silicate Earth appears to have a $\delta^{30}Si$ of –0.29±0.07‰ (Savage et al., 2010; Fitoussi et al., 2009), which is higher than that of carbonaceous chondritic meteorites (–0.36‰ to –0.56‰) and ordinary chondrites (–0.41‰ to –0.49‰) and much higher than enstatite chondrites (–0.52‰ to –0.82‰). The Moon appears to have the same Si isotopic composition as the silicate Earth. The difference in Si isotopic composition between the silicate Earth and chondrites results from two factors. The first nebular fractionation: volatile-poor meteorites have high Mg/Si and high $\delta^{30}Si$. The Earth's high

−1.6‰ at 25°C. Combined with other studies, the fractionation between kaolinite and dissolved silica is inferred be ∼ −2.5‰. In addition, some of dissolved silica released can be absorbed on the surface of oxides and hydroxides in soils such as ferrihydrite and goethite; experiments indicate ^{30}Si fractionation factors of $\Delta^{30}Si_{ferri\text{-}solu.} \approx -1.59‰$ and $\Delta^{30}Si_{geoth.\text{-}solu.} \approx -1.06‰$, making dissolved silica even isotopically heavier. A strong negative correlation between $\delta^{30}Si$ and the chemical index of alteration (CIA; the ratio of Al to Al plus Ca and alkali oxides; see Chapter 13) confirms that weathering of silicate rocks produces isotopically light weathering products and an isotopically heavy solution.

Roots of terrestrial plants take-up silicic acid from the soil solution with a fractionation of $\Delta^{30}Si_{roots\text{-}soil} \approx -1.2‰$ and incorporate it in various tissues, where it increases rigidity, increases photosynthetic efficiency, limits loss of water by evapotranspiration, and increases the resistance to pathogens and grazing by herbivores, including insects. The 'phytoliths' formed in this process consist of hydrated opaline silica and vary widely in silicon isotopic composition due to Rayleigh fractionation during precipitation of the phytoliths with an apparent fractionation factor of $\Delta^{30}Si_{phyto\text{-}solu.} \approx +2‰$. Within a watershed, silica may cycle between dissolved form, incorporation in or adsorption on secondary minerals, the living biota and dead organic matter before finally being exported, with isotopic fractionations associated with all these transitions.

Uptake by diatoms in streams, rivers and lakes results in additional fractionations. In view of the complexity of these fractionations, it is not surprising to find that rivers have highly variable $\delta^{30}Si$ ranging from −0.17‰ to +4.6‰. Isotopic compositions vary not only between different rivers, but also seasonally in individual rivers depending on discharge and biological productivity. Rivers nonetheless tend to be isotopically heavier than the rock in the watershed they drain, but given this variability, the mean value is difficult to define with estimates ranging from +0.8‰ (De la Rocha et al., 1997) to +1.26‰ (Sutton et al., 2018a).

Rivers and submarine groundwater supply about ∼65% of the dissolved silica flux to the oceans, with ∼25% coming from dissolution of aeolian dust and the riverine suspended load and ∼10% from alteration of the ocean crust. Once dissolved silica reaches the oceans it is extensively bioutilized and can be biolimiting. Settling of biogenic silica and conversion of it to chert or authigenic clays is the main way in which silica is removed from the oceans. As a result of this extensive bioutilization, the residence time of silica in the oceans is relatively short: ∼10,000 yrs. Diatoms, which build tests (or shells) of opaline silica and account for some >40% of primary productivity in the oceans, dominate the modern marine silica cycle. Experimental estimates of the fractionation of between dissolved H_4SiO_4 and biogenic opaline silica indicate a near constant $\Delta^{30}Si_{opal\text{-}diss.\ Si} \approx -1.1‰$ (de la Rocha et al., 1997). Consequently, ocean surface waters are typically depleted in Si and ^{28}Si (Figure 9.52). As they sink, the tests tend to redissolve, enriching deep waters in dissolved silica and ^{29}Si. In addition to diatoms, sponges and planktonic protists such as silicoflagellates and radiolarians also utilize silica. Of these, sponges appear to fractionate silica the most. De la Rocha (2003) found that $\delta^{30}Si$ in modern sponge spicules range from −1.2‰ to −3.7‰ and fractionation factors $\Delta^{30}Si_{sponge\text{-}sw}$, although variable, averaged −3.8±0.8‰.

Because $\delta^{30}Si$ in oceans is tightly coupled with biological productivity and hence the carbon cycle, there is much interest in using silicon isotopes in understanding the modern and ancient silica cycle and using $\delta^{30}Si$ in siliceous biogenic sediments to reconstruct productivity and carbon export variations in the oceans. For example, assuming simply Rayleigh fractionation, high levels of photosynthesis and silica utilization in the surface waters should lead to increasing $\delta^{30}Si$ in the diatoms tests as Si utilization increases, such that at complete utilization, $\delta^{30}Si_{diatom} = \delta^{30}Si_{sw}$. This will then be recorded in siliceous oozes on the ocean floor. Sponge spicules record bottom water $\delta^{30}Si$. The fractionation factor between seawater and sponge spicules depends on the silica concentration (probably related to the dependence of growth rate) so that those growing silica-poor water would have the lowest $\delta^{30}Si$.

Deep-sea cores record changes in $\delta^{30}Si$ of biogenic silica of radiolarians, diatoms,

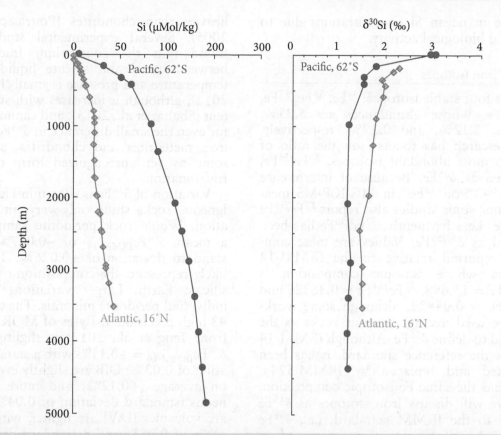

Figure 9.52 Variation of $\delta^{30}Si$ and dissolved silica concentration with depth in the Sub-Antarctic Pacific (data from de Souza et al., 2012) and the Sargasso Sea, Central North Atlantic (data from Sutton et al., 2018b). Silica, like other nutrients, is more depleted in Atlantic deep water and has lower $\delta^{30}Si$ than other oceans as a consequence of the vertical circulation of the oceans: Atlantic deep water is younger (see Chapter 14).

and sponges in the ~25,000 years since the last glacial maximum that are interpreted to reflect changes in the relative abundances of biolimiting nutrients, notably Fe, N, and Si (Hendry and Brzezinski, 2014). These changes are in turn thought to reflect changes in delivery of nutrients to surface waters both from the atmosphere (wind-derived dust during glacial periods enhanced Fe concentrations) and ocean circulation (upwelling returning N and Si to the surface). Alternatively, Frings et al. (2016) argue that the cooler, drier glacial climate, with larger continental shelves, large ice sheets and altered vegetation resulted in substantially lower mean $\delta^{30}Si$ of dissolved Si delivered to the ocean from the continents. Improved understanding of the silica cycle is needed before $\delta^{30}Si$ can be fully utilized as a paleo-oceanographic proxy.

On longer time scales, Fontorbe et al. (2016) used $\delta^{30}Si$ in sponge spicules and radiolarians to document that the low dissolved silica concentrations in the Atlantic have persisted for at least 60 million years. Tazel et al. (2017) found that $\delta^{30}Si$ in cherts decreased across the Ediacaran–Cambrian (Proterozoic–Phanerozoic) boundary, which they attribute to the expansion of the siliceous sponges and speculate that associated nutrient balances may have led to the rise in atmospheric oxygen that occurred then. Prior to the evolution of sponges, and radiolarians, which first appeared in the Early Paleozoic, most cherts appear to be abiotic and have a large range of $\delta^{30}Si$, reaching a peak in the Mid-Proterozoic of +2.2 to 3.9‰ followed by a decrease to eventual modern levels. Ding et al. (2017) speculated that this reflected a decrease in ocean Si concentrations due to increased biological activity.

9.10.5 Iron isotopes

Iron has four stable isotopes: ^{54}Fe, ^{56}Fe, ^{57}Fe, and ^{58}Fe, whose abundances are 5.85%, 91.75%, 2.12%, and 0.29%, respectively. Most research has focused on the ratio of the two most abundant isotopes, $^{56}Fe/^{54}Fe$, expressed as $\delta^{56}Fe$. Because of interference of $^{40}Ar^{16}O^+$ on $^{56}Fe^+$ in MC-ICP-MS measurements, some studies also report $^{57}Fe/^{54}Fe$ as $\delta^{57}Fe$. Less frequently, $^{57}Fe/^{56}Fe$ has been reported as $\delta^{57/56}Fe$. Values are most commonly reported relative to the IRMM-14 standard whose isotopic composition is $^{56}Fe/^{54}Fe = 15.698$, $^{57}Fe/^{54}Fe = 0.36325$ and $^{58}Fe/^{54}Fe = 0.04823$, although some workers have used average igneous rocks as the standard to define $\delta^{56}Fe$. Although IRMM-14 remains the reference standard, it has been exhausted and replaced by IRMM-524a, which has the same Fe isotopic composition. Here we will discuss iron isotopes as $\delta^{56}Fe$ relative to the IRMM standard, i.e., $\delta^{56}Fe_{IRMM-14}$. As all fractionations measured to date are mass dependent, $\delta^{57}Fe$ values can be converted as $\delta^{56}Fe = \delta^{57}Fe \times 0.668$.

Iron exists in the Earth in three oxidation states: Fe^{3+}, Fe^{2+}, and Fe^0, the latter being extremely rare in nature outside of the Earth's core. Not surprisingly, the largest Fe isotope fractionations are between redox states. Quite often, these redox are biologically mediated; that and the importance of Fe as a nutrient lead to considerable interest in iron isotopes in biogeochemical cycling. Fractionations involving biologically mediated reduction can be as great as 3‰. In contrast, nonredox fractionations between igneous minerals are generally <0.1‰. Dauphas (2017) provides a comprehensive summary of Fe isotope fractionation factors.

Carbonaceous, ordinary and enstatite chondrites have uniform $\delta^{56}Fe$ values of -0.005 ± 0.006‰ (Craddock and Dauphas, 2011). The three other planetary bodies from which we have samples, the Moon, Vesta, and Mars, have Fe isotopic compositions that are nearly indistinguishable from those of chondrites. Iron meteorites have a mean and standard deviation $\delta^{56}Fe$ of $+0.050\pm0.101$‰ and are thus just slightly heavier than chondrites (Poitrasson et al., 2005). Several experimental studies have found that the equilibrium fractionation between metal and silicate liquid at high temperature and pressure is small (Hin et al., 2012), although it increases with sulfur content (Shahar et al., 2015), and cannot account for even the small difference in $\delta^{56}Fe$ between iron meteorites and chondrites, suggesting some as yet unrecognized form of kinetic fractionation.

Variation of $\delta^{56}Fe$ is shown in Figure 9.53. Igneous rocks show only very limited variation. Whole rock peridotite samples have a mean $\delta^{56}Fe_{IRMM-14}$ of -0.027‰ with a standard deviation of ±0.026‰. This value likely represents the composition of the bulk silicate Earth. Larger variations occur in individual peridotite minerals. The average of 43 high precision analyses of MORB (mainly from Teng et al., 2013) is slightly heavier, $\delta^{56}Fe_{IRMM-14} = +0.11$‰ with a standard deviation of 0.02‰; OIB are slightly even heavier on average (+0.12‰) and more heterogeneous (standard deviation of 0.04‰). Island arc volcanics (IAV) are lighter, with a mean $\delta^{56}Fe$ of 0.05‰ and a standard deviation of 0.003‰ (Foden et al., 2018). The difference between basalts and peridotites is consistent with a small ($\Delta^{56}Fe \approx -0.2$‰) fractionation between olivine and silicate liquid during partial melting and fractional crystallization and with the observation that the most depleted peridotites typically have the lightest Fe (Williams and Bizimis, 2014). The lighter compositions of IAV reflects the more oxidized environment of subduction zones; the greater variability of OIB reflects recycling of surficial material into the deep mantle. Granitoid igneous rocks are slightly heavier and more variable, $\delta^{56}Fe = +0.16\pm0.07$‰. These small fractionations are consistent with quite small measured and calculated fractionation factors among igneous minerals.

The largest iron isotopic variation is observed in sediments and low-temperature fluids and is principally due to the relatively large equilibrium fractionation ($\sim+3$‰) associated with oxidation of aqueous Fe^{2+} to Fe^{3+}. However, Fe^{3+} produced by oxidative weathering of igneous and high-grade metamorphic rocks is immobile, consequently, isotopically heavy Fe^{3+} remains bound in the solid phase in minerals such as magnetite,

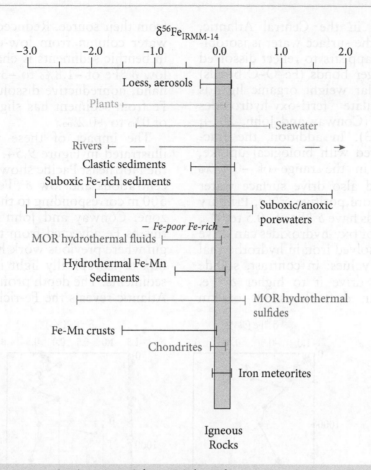

Figure 9.53 Iron isotopes ratios in terrestrial materials and meteorites.

iron-bearing clays, and iron oxy-hydroxides. Thus, weathering in an oxidative environment produces little net change in Fe isotopic composition. Instead, the largest fractionations evident in Figure 9.53 are often biologically mediated redox reactions (Johnson et al., 2008). Two biological processes are important in reducing ferric iron in anoxic environments: dissimilatory iron reduction (DIR) and bacterial sulfate reduction (BSR). Both of these reactions occur only in the absence of free oxygen. In DIR, iron is the electron receptor in the oxidation of organic carbon; in BSR, sulfur is the electron receptor in organic carbon oxidation. Iron is then reduced by reaction with sulfide and precipitated as iron sulfide (e.g., pyrite). Of these pathways, DIR produces the largest decrease in $\delta^{56}Fe$.

Some of the greatest interest in iron isotopes is in tracking Fe in the marine realm because Fe can be a biolimiting nutrient in as much as a quarter of the world's surface oceans. As such, it plays a role in the carbon cycle and, ultimately, climate. While rivers are the main source of most dissolved salts in the ocean, this is not the case for iron, in part due to the very low solubility of ferric iron and in part due to its removal by particle adsorption in estuaries (see Chapter 14). Instead, the main sources of iron are atmospheric dust, hydrothermal fluids, and sediments. Within the oceans, iron has a complex chemistry and is cycled between a variety of forms: the living biota, dissolved (mainly complexed by organic ligands), colloidal, organic and inorganic particles, and adsorbed on these surfaces.

Geochemists are attempting to use Fe isotopes to distinguish between sources of Fe to the oceans and its cycling within the oceans. Mineral dust has an isotopic composition close to +0.1‰ but in regions of the ocean where the dust flux is the greatest, such as from the Sahara in the Central Atlantic, dissolved iron in the surface water is isotopically heavy. This appears to reflect dissolved Fe forming stronger bonds (Fe–O–C bonds) with low molecular weight organic ligands than in particulate ferri-oxy-hydroxides (Fe–O–Fe bonds) (Conway and John, 2014; Ilina et al., 2013). In addition, the fractionation associated with biological uptake, $\Delta^{56}Fe_{phyto-diss}$, is in the range of –0.13‰ to –0.25‰ would also drive surface water dissolved Fe to more positive $\delta^{56}Fe$. Primary hydrothermal fluids have $\delta^{56}Fe$ of –0.5 to 0‰, but precipitation of oxy-hydroxides can drive the remaining dissolved iron in hydrothermal plumes to lower values; in contrast, sulfide precipitation can drive it to higher $\delta^{56}Fe$. These plumes can drift thousands of km

from their source. Reduced Fe released to the water column from low-oxygen porewaters in benthic sediments is characterized by very low $\delta^{56}Fe$ of –1.8‰ to –3.5‰. On the other hand, nonreductive dissolution of releases of Fe from sediment has slightly positive $\delta^{56}Fe$ of 0‰ to +0.2‰.

The impact of these various sources is illustrated in Figure 9.54. A depth profile in the Northeast Pacific shows enhanced Fe concentrations and low $\delta^{56}Fe$ at depths around 500 m corresponding to the oxygen minimum zone. Conway and John (2015) interpreted this as Fe advected from the California margin where previous work had demonstrated a flux of isotopically light iron from reducing sediments. The depth profile from the Central Atlantic reveals the Fe-rich, isotopically light

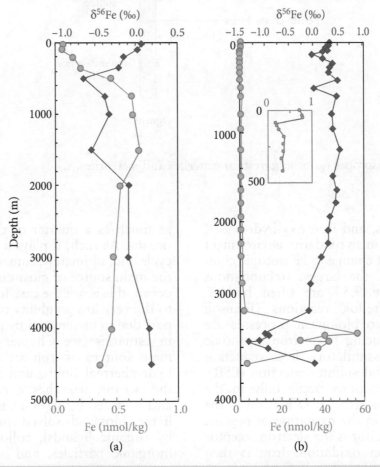

(a)   (b)

Figure 9.54 $\delta^{56}Fe$ (red) and Fe concentration profiles in the ocean. (a) Northeast Pacific Ocean (30°N, 140°W) shows a mid-depth increase in Fe and a decrease in $\delta^{56}Fe$ reflecting advection of Fe from reducing sediments on the California margin. (b) Central Atlantic (26°N, 45°W) shows a large increase in Fe and decrease in $\delta^{56}Fe$ resulting from hydrothermal input from the Mid-Atlantic Ridge. Inset shows the enrichment in iron in the surface water reflecting Saharan dust. Data from Conway and John (2014, 2015).

nature of the hydrothermal plume from the TAG hydrothermal field on the Mid-Atlantic Ridge. The inset shows enrichment of the surface water in isotopically light Fe derived from Saharan dust.

Because of the large fractionation of Fe isotopes associated with redox reactions, they can help constrain the redox history of the Earth's surface. δ^{56}Fe in various sedimentary deposits can help unravel this history. Interestingly, δ^{56}Fe in sedimentary rocks, including banded iron formations, were particularly variable around the time of the Great Oxidation Event in the early Proterozoic. We will discuss Fe isotope ratios in this context in Chapter 15.

Dauphas et al. (2017) provide a more detailed review of iron isotope geochemistry.

9.10.6 Mercury isotopes

Mercury has a large number of stable isotopes (7): ^{196}Hg (0.15%), ^{198}Hg (9.97%), ^{199}Hg (16.87%), ^{200}Hg (23.1%), ^{201}Hg (13.18%), ^{202}Hg (29.86%), and ^{204}Hg (6.87%); it is also highly toxic, and considerable amounts of anthropogenic mercury have entered the environment. We will discuss Hg pollution in more detail in Chapter 15. Mercury is redox sensitive metal and is present in the solid Earth in the Hg^{2+} or Hg^{+1} valence state but can be present in aqueous solution as Hg^0. Mercury is also volatile and a variety of processes release Hg^0 and organically bound Hg^{2+} to the atmosphere. This combination of redox sensitivity and volatility results in comparatively large isotopic fractionations. Furthermore, unlike most other elements we have considered, Hg experiences mass independent (MIF) as well as mass dependent (MDF) fractionations. Because of the former, Hg isotopic variations cannot be fully characterized by a single isotope ratio.

By convention, ^{198}Hg is taken as the normalizing isotope and variations are expressed as:

$$\delta^{xxx}Hg$$

$$= \left[\frac{(^{xxx}Hg/^{198}Hg)_{sample} - (^{xxx}Hg/^{198}Hg)_{std}}{(^{xxx}Hg/^{198}Hg)_{std}} \right]$$

$$\times 1000 \qquad (9.99)$$

where xxx = 199, 200, 201, 202, or 204. The conventional standard is NIST-SRM3133 (Table 9.5). MDF are generally reported as δ^{202}Hg. The ^{199}Hg/^{198}Hg ratio is generally used to characterize MIF, which are reported using the same Δ notation as for oxygen and sulfur, which is the difference between the observed fractionation and the predicted mass dependent one:

$$\Delta^{199}Hg = \delta^{199}Hg - 0.2520 \times \delta^{202}Hg$$
$$(9.100)$$

Similar equations for other isotopes can be written with constants: ^{200}Hg: 0.5024, ^{201}Hg: 0.7520, and ^{204}Hg: 1.493.

Fractionations in the solid Earth and extraterrestrial materials are dominantly mass dependent. δ^{202}Hg in chondrites and achondrites range from −7.13‰ to −0.73‰ but with Δ^{199}Hg ranging only from −0.26‰ to +0.31‰ (Meier et al., 2016). Terrestrial igneous rocks show a smaller range: δ^{202}Hg: −1.70‰ to 1.61‰ and Δ^{199}Hg: −0.13‰ to +0.13‰. Sedimentary rocks, excepting coal, have δ^{202}Hg ranges from −2.68‰ to 0.23‰ and Δ^{199}Hg ranges from −0.46‰ to +0.34‰; coal shows larger ranges: δ^{202}Hg: −4.00‰ to 0.91‰, Δ^{199}Hg: −0.63‰ to +0.34‰ (Blum and Johnson, 2017).

Mass independent fractionations of Hg isotopes result from the nuclear magnetic and nuclear volume effects discussed in Section 9.2.3. Both these effects ultimately arise from the effects of nucleon pairing in the nucleus, and hence they primarily produce fractionations between odd- and even-numbered nuclei in heavy elements. While mass-dependent fractionations of Hg isotopes are ubiquitous, MIF occur in only a limited number of reactions, primarily photochemical ones.

The largest mass independent fractionations result from different reaction rates of odd and even numbered isotopes as a consequence of the magnetic isotope effect. Bergquist and Blum (2007) found that kinetic fractionations during photochemical reduction of Hg^{2+} to Hg^0 in aqueous solution result in enrichments of odd-numbered isotopes (Figure 9.55). Photochemical reduction of methyl mercury ($HgCH_3^+$ and $Hg(CH_3^+)_2$) in aqueous solution result in kinetic fractionations with $\Delta^{199}Hg/\delta^{2021}Hg$ of 2.4 and $\Delta^{199}Hg/\Delta^{201}Hg$ of 1.36. The nuclear volume

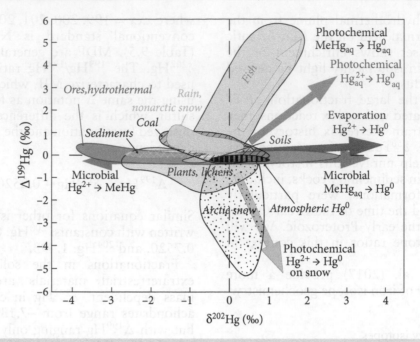

Figure 9.55 Mercury isotope fractionations and isotope compositions in terrestrial materials. δ^{202}Hg tracks mass dependent fractionations while Δ^{199}Hg tracks mass independent ones. Based on data in Blum and Johnson (2017) and Blum et al. (2014).

(or nuclear field shift) effect on bond strength results in kinetic and equilibrium fractionations during evaporation with enrichment of odd-numbered isotopes but these tend to be smaller and have Δ^{199}Hg/δ^{202}Hg of ~0.1 and Δ^{199}Hg/Δ^{201}Hg $\approx$ 2. Arctic snow cover shows depletions in odd-number Hg isotopes with Δ^{199}Hg/δ^{2021}Hg of –3.3 and Δ^{199}Hg/Δ^{201}Hg $\approx$ 1 that result from oxidation of atmospheric Hg^0 followed by deposition on the snow surface and subsequent photoreduction and evaporation of Hg^0 back to the atmosphere, leaving the snow depleted in odd-numbered isotopes (Sherman et al., 2010). Finally, mass independent fractionations observed in rain and nonArctic snow affect even-numbered isotopes with Δ^{200}Hg/^{204}Hg of –0.5 to –0.6 as well as odd-numbered isotopes. These are thought to be generated by photochemical redox reactions in the upper atmosphere but the cause of these not yet understood (e.g., Chen et al., 2016; Sun et al., 2016). As mercury cycles through atmospheric, marine, and terrestrial reservoirs it acquires MIF signatures through these processes (Berqquist, 2018).

Mercury enters the atmosphere as primarily as Hg^0 (~90%) and as such it has an atmospheric residence time of ~0.5 years. Mercury is also emitted to the atmosphere as methyl mercury produced by sulfur- and iron-reducing bacteria in reducing environments. Natural sources of atmospheric mercury include volcanoes, geothermal areas, forest fires, erosion, evasion from saline and fresh waterbodies, and soil degasification. However, anthropogenic sources now account for roughly 80% of emissions to the atmosphere. These include fossil fuel burning (primarily coal), artisanal gold mining and refining, and other industrial sources.

Mercury is removed from the atmosphere primarily by rain and photochemical reactions on cloud droplets leaves atmospheric Hg^0 with on average slightly positive δ^{202}Hg and slightly negative Δ^{199}Hg; atmospheric Hg^{2+} has the opposite signature. Plants take up atmospheric Hg^0 directly through these stomata and hence also have negative Δ^{199}Hg. Organic-rich soils have slightly negative odd-MIF signatures as plant material is transferred to the soil through litterfall. Sediments derive Hg from both runoff of these soils and wet Hg^{2+} deposition (rain and snow) and have Δ^{199}Hg $\approx$ 0 but with a wide range of δ^{202}Hg.

Much of the interest in mercury stems from its toxicity and widespread pollution and its bioaccumulation in aquatic food webs, leading to dangerous exposure at the top of the food chain, including in humans. Natural sources and pathways of Hg in the environment can be traced using Hg isotope ratios. We'll return to this topic in Chapter 15.

9.11 SUMMARY

In this chapter we reviewed why isotope fractionations can result from chemical processes and how these can be applied to further our understanding of the Earth.

- Stable isotope ratios are reported in the delta notation, e.g., $\delta^{18}O$, which is the per-mil difference of the ratio from that of a standard. Fractionations between phases are represented either in the Δ_{A-B} notation, which is the difference in δ values between phases A and B or in the α_{A-B} notation, which is the ratio of isotope ratios in these phases.

- Stable isotope ratios vary because the mass of an atom affects the strength of chemical bonds, which in turn influences the distribution of isotopes of an element between phases. As a rule, the heavier isotope partitions preferentially into the strongest bonds. Such equilibrium isotope fractionations can be predicted from statistical mechanics and partition functions. These predictions are represented as the ratio of reduced partition functions, β, which is the predicted ratio of the isotope ratio in a substance of interest to that isotope ratio in dissociated atoms.

- Equilibrium isotope fractionations are inversely proportional to temperature at temperatures corresponding to the Earth's surface and inversely proportional to the square of temperature at higher temperatures.

- Because bond strength influences reaction rates and because lighter atoms and molecules diffuse faster, differences in atomic mass also results in kinetic fractionations.

- Fractionations between two isotopes generally depend on the mass difference between the isotopes such that, for example, the fractionation between ^{17}O

and ^{16}O is approximately half (~ 0.52) that between ^{18}O and ^{18}O. A variety of essentially kinetic factors, as well as nuclear influences on electronic structure in heavy elements, can result in less common *mass independent fractionations* (MIF). These are expressed in a capital delta notation, for example, $\Delta^{33}S$, which is the difference between the observed $^{33}S/^{32}S$ and that predicted from $^{34}S/^{32}S$. For example, photolysis and UV self-shielding produced mass independent fractionation of sulfur isotopes in the oxygen-free Archean atmosphere. The absence of such fractionations after 2.3 Ga provides further evidence of the oxygenation of the atmosphere around that time.

- Variations of O and H isotopes in the hydrological cycles result from both kinetic and equilibrium fractionations such both O and H in precipitation become isotopically lighter with decreasing temperature and distance from source, defining the *meteoric water line*. The build-up of isotopically light ice on continents as well as the temperature-dependent fractionation of $\delta^{18}O$ between water and carbonate has allowed reconstruction of the Pleistocene ice ages, revealing that the pacemaker of these climate swings are the Milankovitch variations in the Earth's orbit and rotation. This small signal was amplified by changes in ocean circulation and the distribution of CO_2 between the atmosphere and the ocean.

- Carbon isotope fractionations of -15 to -25 per mil ratios occur during photosynthesis, with the largest fractionations produced by C_3 plants and the smallest produced by marine autotrophs. Only small additional fractionations occur in other biological reactions, so the carbon isotope signature is maintained through the food chain. Carbon isotopes have thus proved quite useful in archeology and paleo-ecology.

- Over the past several decades, stable isotope geochemistry has expanded to include most multi-isotope elements. The fundamental principles involved, however, remain the same as for traditional stable isotopes.

- Isotope fractionations at igneous and mantle temperatures approach 0. As a consequence, igneous and mantle rocks whose stable isotopic compositions show significant variation from the primordial values must either have been affected by low-temperature processes or must contain a component that was at one time at the surface of the Earth. Variations in carbon, sulfur, oxygen, and a variety of

nontraditional stable isotopes allows us to identify surficial material in the mantle and mantle-derived materials. The presence of organic carbon in diamonds and of MIF sulfur in mantle plume-derived lavas demonstrates that geochemical cycles can operate on truly planetary scales: from the top of the atmosphere to the base of the mantle.

REFERENCES AND SUGGESTIONS FOR FURTHER READING

Amini, M., Eisenhauer, A., Böhm, F., Fietzke, J., Bach, W., Garbe-Schönberg, D., Rosner, M., Bock, B., Lackschewitz, K. S. and Hauff, F., 2008. Calcium isotope ($\delta^{44}/^{40}$Ca) fractionation along hydrothermal pathways, Logatchev field (Mid-Atlantic Ridge, 14°45'N). *Geochimica et Cosmochimica Acta* 72(16), 4107–4122. doi: 10.1016/j.gca.2008.05.055.

Bereiter, B., Eggleston, S., Schmitt, J., et al. 2015. Revision of the EPICA Dome C CO_2 record from 800 to 600 kyr before present. *Geophysical Research Letters* 42: 542–549. doi: 10.1002/2014GL061957

Bergquist, B. A. 2018. Mercury isotopes. In: White, WM (ed) *Encyclopedia of Geochemistry*, Springer International Publishing, Cham, pp 900–906. doi: 978-3-319-39312-4.

Bergquist, B. A., Blum, J. D. 2007. Mass-dependent and -independent fractionation of Hg isotopes by photoreduction in aquatic systems. *Science* 318: 417–420. doi: 10.1126/science.1148050

Bigeleisen, J. 1996. Nuclear size and shape effects in chemical reactions. Isotope chemistry of the heavy elements. *Journal of the American Chemical Society* 118: 3676–3680. doi: 10.1021/ja954076k

Bigeleisen, J. and Mayer, M.G. 1947. Calculation of equilibrium constants for isotopic exchange reactions. *Journal of. Chemical. Physics.* 15: 261–7.

Bindeman, I. N., Eiler, J. M., Yogodzinski, G. M., et al. 2005. Oxygen isotope evidence for slab melting in modern and ancient subduction zones. *Earth and Planetary Science Letters* 235: 480–496. doi: 10.1016/j.epsl.2005.04.014

Blättler, C. L., Henderson, G. M., Jenkyns, H. C. 2012. Explaining the Phanerozoic Ca isotope history of seawater. *Geology* 40: 843–846. doi: 10.1130/g33191.1

Blum, J. D., Johnson, M. W. 2017. Recent developments in mercury stable isotope analysis. *Reviews in Mineralogy and Geochemistry* 82: 733–757. doi: 10.2138/rmg.2017.82.17

Blum, J. D., Sherman, L. S., Johnson, M. W. 2014. Mercury isotopes in Earth and environmental sciences. *Annual Reviews of Earth and Planetary Sciences* 42: 249–269. doi: 10.1146/annurev-earth-050212-124107

Burke, A., Present, T. M., Paris, G., Rae, E. C. M., Sandilands, B. H., Gaillardet, J., Peucker-Ehrenbrink, B., Fischer, W. W., McClelland, J. W., Spencer, R. G. M., Voss, B. M. and Adkins, J. F., 2018. Sulfur isotopes in rivers: Insights into global weathering budgets, pyrite oxidation, and the modern sulfur cycle. *Earth and Planetary Science Letters* 496, 168–77. doi: 10.1016/j.epsl.2018.05.022.

Cabral, R. A., Jackson, M. G., Rose-Koga, E. F., Koga, K. T., Whitehouse, M. J., Antonelli, M. A., Farquhar, J., Day, J. M. D. and Hauri, E. H., 2013. Anomalous sulphur isotopes in plume lavas reveal deep mantle storage of Archaean crust. *Nature* 496(7446), 490–3. doi: 10.1038/nature12020.

Cartigny, P., Palot, M., Thomassot, E., Harris, J. W. 2014. Diamond formation: A stable isotope perspective. *Annual Reviews of Earth and Planetary Science* 42: 699–732. doi: 10.1146/annurev-earth-042711-105259.

Cerling, T.E., Wang, Y. and Quade, J. 1993. Expansion of C_4 ecosystems as an indicator of global ecological change in the late Miocene. *Nature* 361: 344–5.

Chan, L.-H. and Edmond, J.M. 1988. Variation of lithium isotope composition in the marine environment: a preliminary report. *Geochimica et Cosmochimica Acta* 52: 1711–17.

Chan, L.H., Edmond, J.M., Thompson, G. and Gillis, K. 1992. Lithium isotopic composition of submarine basalts: implications for the lithium cycle of the oceans. *Earth and Planetary Science Letters* 108: 151–60.

Chaussidon, M., Sheppard, S.M.F. and Michard, A. 1991. Hydrogen, sulfur and neodymium isotope variations in the mantle beneath the EPR at 12°50'N, in *Stable Isotope Geochemistry: A Tribute to Samuel Epstein* (eds H.P. Taylor, J.R. O'Neil and I.R. Kaplan), pp. 325–38. San Antonio, Geochemical Society.

Chen, J., Hintelmann, H., Feng, X., Dimock, B. 2012. Unusual fractionation of both odd and even mercury isotopes in precipitation from Peterborough, ON, Canada. *Geochimica et Cosmochimica Acta* 90: 33–46. doi: 10.1016/j.gca.2012.05.005.

Chiba, H., Chacko, T., Clayton, R.N. and Goldsmith, J.R. 1989. Oxygen isotope fractionations involving diopside, forsterite, magnetite, and calcite: application to geothermometry. *Geochimica et Cosmochimica Acta* 53: 2985–95.

Conway, T. M. and John, S. G., 2014. Quantification of dissolved iron sources to the North Atlantic Ocean. *Nature* 511, 212. doi: 10.1038/nature13482.

Conway, T. M. and John, S. G., 2015. The cycling of iron, zinc and cadmium in the North East Pacific Ocean – Insights from stable isotopes. *Geochimica et Cosmochimica Acta* 164, 262–283. doi: 10.1016/j.gca.2015.05.023.

Cooper, K. M., Eiler, J. M., Asimow, P. D., Langmuir, C. H. 2004. Oxygen isotope evidence for the origin of enriched mantle beneath the mid-Atlantic ridge. *Earth and Planetary Science Letters* 220: 297–316. doi: 10.1016/s0012-821x(04)00058-5.

Cooper, K. M., Eiler, J. M., Sims, K. W. W., Langmuir, C. H. 2009. Distribution of recycled crust within the upper mantle: Insights from the oxygen isotope composition of MORB from the Australian-Antarctic Discordance. *Geochemistry, Geophysics, Geosystems* 10. doi: 10.1029/2009GC002728.

Copley, M. S., Berstan, R., Dudd, S. N., et al. 2003. Direct chemical evidence for widespread dairying in prehistoric Britain. 100: 1524–1529. doi:10.1073/pnas.0335955100 *Proceedings of the National Academy of Sciences* 100: 1524–1529.

Craddock, P. R. and Dauphas, N., 2011. Iron isotopic compositions of geological reference materials and chondrites. *Geostandards and Geoanalytical Research* 35(1), 101–123. doi: 10.1111/j.1751-908X.2010.00085.x.

Craig, H. 1963. The isotopic composition of water and carbon in geothermal areas, in *Nuclear Geology on Geothermal Areas* (ed. E. Tongiorgi), pp. 17–53. Pisa, CNR Lab. Geol. Nucl.

Dansgaard, W. 1964. Stable isotopes in precipitation. *Tellus* 16: 436–63.

Dauphas, N., John, S. G. and Rouxel, O., 2017. Iron isotope systematics. In: *Non-traditional Stable Isotopes* (eds Teng, F.-Z., Dauphas, N. and Watkins, J. M.) Reviews in Mineralogy and Geochemistry, pp. 415–51, Mineralogical Society of America, Washington DC.

Dauphas, N., Poitrasson, F., Burkhardt, C., Kobayashi, H., Kurosawa, K. 2015. Planetary and meteoritic Mg/Si and δ^{30}Si variations inherited from solar nebula chemistry. *Earth and Planetary Science Letters* 427: 236–248. doi: 0.1016/j.epsl.2015.07.008.

De Hoog, J. C. M., Savov, I. P. (2018) Boron isotopes as a tracer of subduction zone processes. In: Marschall H, Foster G (eds) *Boron Isotopes: The Fifth Element*, Springer International Publishing, Cham, pp 217–247. doi: 978-3-319-64666-4.

De la Rocha, C. L. 2003. Silicon isotope fractionation by marine sponges and the reconstruction of the silicon isotope composition of ancient deep water. *Geology* 31: 423–426. doi: 10.1130/0091-7613(2003)031.

De la Rocha, C. L., Brzezinski, M. A., DeNiro, M. J. 2000. A first look at the distribution of the stable isotopes of silicon in natural waters. *Geochimica et Cosmochimica Acta* 64: 2467–2477. doi: 10.1016/s0016-7037(00)00373-2.

de Souza, G. F., Reynolds, B. C., Rickli, J., et al. 2012. Southern Ocean control of silicon stable isotope distribution in the deep Atlantic Ocean. *Global Biogeochemical Cycles* 26:. doi: 10.1029/2011GB004141.

Delavault, H., Chauvel, C., Thomassot, E., Devey, C. W. and Dazas, B., 2016. Sulfur and lead isotopic evidence of relic Archean sediments in the Pitcairn mantle plume. *Proceedings of the National Academy of Sciences* 113(46), 12952–12956. doi: 10.1073/pnas.1523805113.

DeNiro, M.J. 1987. Stable isotopy and archaeology. *American Scientist* 75: 182–91.

DeNiro, M.J. and Epstein, S. 1978. Influence of diet on the distribution of carbon isotopes in animals. *Geochimica et Cosmochimica Acta* 42: 495–506.

DeNiro, M.J. and Epstein, S. 1981. Influence of diet on the distribution of nitrogen isotopes in animals. *Geochimica et Cosmochimica Acta* 45: 341–51.

DeNiro, M.J. and Hasdorf, C.A. 1985. Alteration of $^{15}N/^{14}N$ and $^{13}C/^{12}C$ ratios of plant matter during the initial stages of diagenesis: studies utilizing archeological specimens from Peru. *Geochimica et Cosmochimica Acta* 49: 97–115.

Ding, T. P., Gao, J. F., Tian, S. H., et al. 2017. The δ^{30}Si peak value discovered in middle Proterozoic chert and its implication for environmental variations in the ancient ocean. *Scientific Reports* 7: 44000. doi: 10.1038/srep44000.

Ding, T. P., Zhou, J. X., Wan, D. F., Chen, Z. Y., Wang, C. Y., Zhang, F. 2008. Silicon isotope fractionation in bamboo and its significance to the biogeochemical cycle of silicon. *Geochimica et Cosmochimica Acta* 72: 1381–1395. doi: 10.1016/j.gca.2008.01.008.

Dodd, M. S., Papineau, D., Grenne, T., et al. 2017. Evidence for early life in Earth's oldest hydrothermal vent precipitates. *Nature* 543: 60–64. doi: 10.1038/nature21377

Eiler, J. M., 2011. Paleoclimate reconstruction using carbonate clumped isotope thermometry. *Quaternary Science Reviews* 30, 3575–3588. doi: 10.1016/j.quascirev.2011.09.001.

Eiler, J.M. 2007. "Clumped-isotope" geochemistry – the study of naturally-occurring, multiply-substituted isotopologues. *Earth and Planetary Science Letters* 262: 309–27.

Eiler, J. M., 2013. The isotopic anatomies of molecules and minerals. *Annual Review of Earth and Planetary Sciences* 41, 411–441. doi:. doi: 10.1146/annurev-earth-042711-105348.

Eldridge, C.S., Compston, W., Williams, I.S., Harris, J.W. and Bristow, J.W. 1991. Isotope evidence for the involvement of recycled sediment in diamond formation. *Nature* 353: 649–53.

Emiliani, C. 1955. Pleistocene temperatures. *Journal of Geology* 63: 538–78.

Estrade, N., Carignan, J., Sonke, J. E., Donard, O. F. X. 2009. Mercury isotope fractionation during liquid–vapor evaporation experiments. *Geochimica et Cosmochimica Acta* 73: 2693–2711. doi: 10.1016/j.gca.2009.01.024.

Farquhar, J. (2018) Sulfur Isotopes. In: White WM (ed) *Encyclopedia of Geochemistry*, Springer International Publishing, Cham, pp 1402–1409. doi: 978-3-319-39312-4.

Farquhar, J. and Wing, B.A. 2003. Multiple sulfur isotopes and the evolution of atmospheric oxygen. *Earth and Planetary Science Letters* 213: 1–13.

Farquhar, J., Bao, H. and Thiemens, M. 2000. Atmospheric influence of Earth's earliest sulfur cycle. *Science* 289: 756–8.

Farquhar, J., Wing, B.A., McKeegan, K.D., *et al.* 2002. Mass-independent fractionation sulfur of inclusions in diamond and sulfur recycling on early Earth. *Science* 298: 2369–72.

Faure, G. 1986. *Principles of Isotope Geology*, 2nd ed. New York, John Wiley and Sons, Ltd.

Ferronsky, V.I., and Polyakov, V.A. 1982. *Environmental Isotopes in the Hydrosphere*. Chichester, John Wiley and Sons Ltd.

Fitoussi, C., Bourdon, B., Kleine, T., Oberli, F., Reynolds, B. C. 2009. Si isotope systematics of meteorites and terrestrial peridotites: implications for Mg/Si fractionation in the solar nebula and for Si in the Earth's core. *Earth and Planetary Science Letters* 287: 77–85. doi: 10.1016/j.epsl.2009.07.038.

Foden, J., Sossi, P. A. and Nebel, O., 2018. Controls on the iron isotopic composition of global arc magmas. *Earth and Planetary Science Letters* 494, 190–201. doi: 10.1016/j.epsl.2018.04.039.

Fogel, M.L., and Cifuentes, M.L. 1993. Isotope fractionation during primary production, in *Organic Geochemistry: Principles and Applications* (eds M.H. Engel and S.A. Macko), pp. 73–98. New York, Plenum Press.

Fontorbe, G., Frings, P. J., De La Rocha, C. L., Hendry, K. R., Conley, D. J. 2016. A silicon depleted North Atlantic since the Palaeogene: Evidence from sponge and radiolarian silicon isotopes. *Earth and Planetary Science Letters* 453: 67–77. doi: 10.1016/j.epsl.2016.08.006.

Foster, G.L. 2008. Seawater pH, pCO_2 and $[CO_3^{2-}]$ variations in the Caribbean Sea over the last 130 kyr: A boron isotope and B/Ca study of planktic foraminifera. *Earth and Planetary Science Letters* 271: 254–66.

Gao, Y.Q. and Marcus, R.A. 2001. Strange and unconventional isotope effects in ozone formation. *Science* 293: 259–63.

Gat, J. R., 1996. Oxygen and hydrogen isotopes in the hydrologic cycle. *Annual Review of Earth and Planetary Sciences* 24: 225–262. doi: 10.1146/annurev.earth.24.1.225.

Genske, F. S., Turner, S. P., Beier, C., et al. 2014. Lithium and boron isotope systematics in lavas from the Azores islands reveal crustal assimilation. *Chemical Geology* 373: 27–36. doi: 10.1016/j.chemgeo.2014.02.024.

Ghosh, P., Adkins, J., Affek, H., Balta, B., Guo, W., Schauble, E. A., Schrag, D. and Eiler, J. M., 2006. ^{13}C–^{18}O bonds in carbonate minerals: A new kind of paleothermometer. *Geochimica et Cosmochimica Acta* 70(6), 1439–1456. doi: 10.1016/j.gca.2005.11.014.

Graham, D. W., Michael, P. J., Rubin, K. H. 2018. An investigation of mid-ocean ridge degassing using He, CO_2, and $\delta^{13}C$ variations during the 2005–06 eruption at 9°50′N on the East Pacific Rise. *Earth and Planetary Science Letters* 504: 84–93. doi: 10.1016/j.epsl.2018.09.040.

Halliday, A. N., Lee, D.-C., Christensen, J. N., et al. 1995. Recent developments in inductively coupled plasma magnetic sector multiple collector mass spectrometry. *International Journal of Mass Spectrometry and Ion Processes* 146–147: 21–33. doi: 10.1016/0168-1176(95)04200-5.

Harmon, R.S. and Hoefs, J. 1995. Oxygen isotope heterogeneity of the mantle deduced from global ^{18}O systematics of basalts from different geotectonic settings. *Contributions to Mineralogy and Petrology* 120: 95–114. doi: 10.1007/bf00311010.

Hayes, J.D., Imbrie, J. and Shackleton, N.J. 1976. Variations in the Earth's orbit: pacemaker of the ice ages. *Science* 194: 1121–32.

Heming, N.G. and Hanson, G.N. 1992. Boron isotopic composition and concentration in modern marine carbonates. *Geochimica et Cosmochimica Acta* 56: 537–43.

Hendry, K. R., Robinson, L. F., McManus, J. F., Hays, J. D. 2014. Silicon isotopes indicate enhanced carbon export efficiency in the North Atlantic during deglaciation. *Nature Communications* 5: 3107. doi: 10.1038/ncomms4107.

Higgins, J. A., Blättler, C. L., Lundstrom, E. A., et al. 2018. Mineralogy, early marine diagenesis, and the chemistry of shallow-water carbonate sediments. *Geochimica et Cosmochimica Acta* 220: 512–534. doi: 10.1016/j.gca.2017.09.046.

Hin, R. C., Fitoussi, C., Schmidt, M. W., Bourdon, B. 2014. Experimental determination of the Si isotope fractionation factor between liquid metal and liquid silicate. *Earth and Planetary Science Letters* 387: 55–66. doi: 10.1016/j.epsl.2013.11.016.

Hin, R. C., Schmidt, M. W. and Bourdon, B., 2012. Experimental evidence for the absence of iron isotope fractionation between metal and silicate liquids at 1 GPa and 1250–1300 °C and its cosmochemical consequences. *Geochimica et Cosmochimica Acta* 93(0), 164–81. doi: 10.1016/j.gca.2012.06.011.

Hoefs, J. 19872015. *Stable Isotope Geochemistry*, 3rd 7rd edn. Berlin, Springer-Verlag.

Hofmann, A.W. and White, W.M. 1982. Mantle plumes from ancient oceanic crust. *Earth and Planetary Science Letters* 57: 421–36.

Hönisch, B. and Hemming, N.G. 2005. Surface ocean pH response to variations in pCO_2 through two full glacial cycles. *Earth and Planetary Science Letters* 236: 305–14. doi: 10.1016/j.epsl.2005.04.027.

Hönisch, B., Hemming, N.G., Archer, D., Siddall, M. and McManus, J.F. 2009. Atmospheric carbon dioxide concentration across the Mid-Pleistocene transition. *Science* 324: 1551–4. doi: 10.1126/science.1171477.

Huang, S., Farkaš, J. and Jacobsen, S. B., 2010. Calcium isotopic fractionation between clinopyroxene and orthopyroxene from mantle peridotites. *Earth and Planetary Science Letters* 292, 337–344. doi: 10.1016/j.epsl.2010.01.042.

Huang, S., Farkaš, J. and Jacobsen, S. B., 2011. Stable calcium isotopic compositions of Hawaiian shield lavas: Evidence for recycling of ancient marine carbonates into the mantle. *Geochimica et Cosmochimica Acta* 75, 4987–4997. doi: 10.1016/j.gca.2011.06.010.

Ilina, S. M., Poitrasson, F., Lapitskiy, S. A., Alekhin, Y. V., Viers, J. and Pokrovsky, O. S., 2013. Extreme iron isotope fractionation between colloids and particles of boreal and temperate organic-rich waters. *Geochimica et Cosmochimica Acta* 101, 96–111. doi: 10.1016/j.gca.2012.10.023.

Imbrie, J. 1985. A theoretical framework for the Pleistocene ice ages. *Journal of the Geological Society London* 142: 417–32.

Imbrie, J., Hayes, J.D., Martinson, D.G., et al. 1985. The orbital theory of Pleistocene climate: support from a revised chronology of the marine $\delta^{18}O$ record, in *Milankovitch and Climate, Part 1* (eds A.L. Berger, J. Imbrie, J. Hayes, G. Kukla and B. Saltzman), pp. 269–305. Dordrecht, D. Reidel.

Javoy, M. and Pineau, F. 1991. The volatile record of a "popping" rock from the Mid-Atlantic Ridge at 14°N: chemical and isotopic composition of the gas trapped in the vesicles. *Earth and Planetary Science Letters* 107: 598–611.

Johnson, C. M., Beard, B. L. and Roden, E. E., 2008. The iron isotope fingerprints of redox and biogeochemical cycling in modern and ancient earth. *Annual Review of Earth and Planetary Sciences* 36(1), 457–493. doi: 10.1146/annurev.earth.36.031207.124139.

Johnson, D.G., Jucks, K.W., Traub, W.A. and Chance, K.V. 2001. Isotopic composition of stratospheric ozone. *Journal of Geophysical Research* 105: 9025–31.

Jouzel, J., Masson-Delmotte, V., Cattani, O., et al. 2007. Orbital and millennial Antarctic climate variability over the last 800,000 years. *Science* 317: 793–6.

Jouzel, J., Waelbroeck, C., Malaizé, B., et al. 1996. Climatic interpretation of the recently extended Vostok ice records. *Clim. Dyn.* 12: 513–21.

Kasting, J.F. and Catling, D. 2003. Evolution of a habitable planet. *Annual Reviews of Astronomy Astrophysics* 41: 429–63.

Klochko, K., Kaufman, A.J., Yao, W., Byrne, R.H. and Tossell, J.A. 2006. Experimental measurement of boron isotope fractionation in seawater. *Earth and Planetary Science Letters* 248: 276–85. doi: 10.1016/j.epsl.2006.05.034, 2006.

Komar, N., Zeebe, R. E. 2016. Calcium and calcium isotope changes during carbon cycle perturbations at the end-Permian. *Paleoceanography* 31: 115–30. doi: 10.1002/2015PA002834

Kroopnick, P., Weiss, R.F. and Craig, H. 1972. Total CO_2, ^{13}C and dissolved oxygen-^{18}O at GEOSECS II in the North Atlantic. *Earth and Planetary Science Letters* 16: 103–10.

Kyser, K.T. 1986. Stable isotope variations in the mantle, in *Stable Isotopes in High Temperature Geologic Processes* (eds J.W. Valley, H.P. Taylor and J.R. O'Neil), pp. 141–64. Washington, Mineralogical Society of America.

Lamb, A. L., Evans, J. E., Buckley, R., Appleby, J. 2014. Multi-isotope analysis demonstrates significant lifestyle changes in King Richard III. *Journal of Archaeological Science* 50: 559–565. doi: 10.1016/j.jas.2014.06.021

Lawrence, J.R. and Taylor, H.P. 1972. Hydrogen and oxygen isotope systematics in weathering profiles. *Geochimica et Cosmochimica Acta* 36: 1377–93.

Li, Y., Liu, J. 2006. Calculation of sulfur isotope fractionation in sulfides. *Geochimica et Cosmochimica Acta* 70: 1789–1795. doi: 10.1016/j.gca.2005.12.015.

Lisiecki, L. E., Raymo, M. E. 2005. A Pliocene-Pleistocene stack of 57 globally distributed benthic $\delta^{18}O$ records. *Paleoceanography.* 20: 1. doi: 10.1029/2004PA001071.

Maréchal, C. N., Télouk, P., Albarède, F. 1999. Precise analysis of copper and zinc isotopic compositions by plasma-source mass spectrometry. *Chemical Geology* 156: 251–273. doi: 10.1016/s0009-2541(98)00191-0.

Marini, L., Moretti, R., Accornero, M. 2011. Sulfur isotopes in magmatic-hydrothermal systems, melts, and magmas. *Reviews in Mineralogy and Geochemistry* 73: 423–492. doi: 10.2138/rmg.2011.73.14.

Marschall, H. R., Wanless, V. D., Shimizu, N., Pogge von Strandmann, P. A. E., Elliott, T., Monteleone, B. D. 2017. The boron and lithium isotopic composition of mid-ocean ridge basalts and the mantle. *Geochimica et Cosmochimica Acta* 207: 102–138. doi 10.1016/j.gca.2017.03.028.

Marschall, H., Foster, G. (eds.) 2018. *Boron Isotopes, The Fifth Element*. Springer International Publishing Cham, p 289.

Marty, B. and Dauphas, N. 2003. The nitrogen record of crust-mantle interaction and mantle convection from Archean to present. *Earth and Planetary Science Letters* 206: 397–410.

Mattey, D., Lowry, D. and Macpherson, C. 1994. Oxygen isotope composition of mantle peridotite. *Earth and Planetary Science Letters* 128: 231–41.

Mattey, D.P. 1987. Carbon isotopes in the mantle. *Terra Cognita* 7: 31–8.

Mauersberger, K. 1987. Ozone isotope measurements in the stratosphere. *Geophysical Research Letters* 14: 80–83.

Méheut, M., Lazzeri, M., Balan, E., Mauri, F. 2007. Equilibrium isotopic fractionation in the kaolinite, quartz, water system: Prediction from first-principles density-functional theory. *Geochimica et Cosmochimica Acta* 71: 3170–3181. doi: 10.1016/j.gca.2007.04.012.

Milankovitch, M. (1941) *Kanon der Erdbestrahlung und seine Anwendung auf das Eiszeitenproblem*. Académie royale serbe éditions speciales 132. 633pp. Académie royale serbe. Belgrade.

Misra, S. and Froelich, P. N., 2012. Lithium isotope history of Cenozoic seawater: changes in silicate weathering and reverse weathering. *Science* 335(6070), 818–23. doi: 10.1126/science.1214697.

Mojzsis, S.J., Arrhenius, G., McKeegan, K.D., et al. 1996. Evidence for life on Earth before 3800 million years ago. *Nature* 384: 55–9.

Muehlenbachs, K. and Byerly, G. 1982. ^{18}O enrichment of silicic magmas caused by crystal fractionation at the Galapagos Spreading Center. *Contributions to Mineralogy and Petrology* 79: 76–9.

Nutman, A. P., Bennett, V. C., Friend, C. R. L., Van Kranendonk, M. J., Chivas, A. R. 2016. Rapid emergence of life shown by discovery of 3,700-million-year-old microbial structures. *Nature*, advance online publication: 537: 535–538. doi: 10.1038/nature19355

O'Leary, M. H. 1981. Carbon isotope fractionation in plants, *Phytochemistry* 20, 553–67.

O'Neil, J. R. 1986. Theoretical and experimental aspects of isotopic fractionation, in *Stable Isotopes in High Temperature Geological Processes, Reviews in Mineralogy* 16 (eds J.W. Valley, H.P. Taylor and J.R. O'Neil), pp. 1–40. Washington, Mineralogical Society of America.

Ohmoto, H. 1986. Stable isotope geochemistry of ore deposits, in *Stable Isotopes in High Temperature Geological Processes, Reviews in Mineralogy* 16 (eds J.W. Valley, H.P. Taylor and J.R. O'Neil), pp. 491–560. Washington, Mineralogical Society of America.

Ohmoto, H. and Rye, R.O. 1979. Isotopes of sulfur and carbon, in *Geochemistry of Hydrothermal Ore Deposits* (ed. H. Barnes). New York, John Wiley and Sons, Ltd.

Ono, S. 2017. Photochemistry of sulfur dioxide and the origin of mass-independent isotope fractionation in Earth's atmosphere. *Annual Review of Earth and Planetary Sciences* 45: 301–29. doi: 10.1146/annurev-earth-060115-012324.

Palmer, M. R. 2017. Boron cycling in subduction zones. *Elements* 13: 237–42. doi: 10.2138/gselements.13.4.237.

Park, R. and Epstein, S. 1960. Carbon isotope fractionation during photosynthesis. *Geochimica et Cosmochimica Acta* 21: 110–26.

Payne, J. L., Turchyn, A. V., Paytan, A., et al. 2010. Calcium isotope constraints on the end-Permian mass extinction. *Proceedings of the National Academy of Sciences* 107: 8543–48. doi: 10.1073/pnas.0914065107.

Pearson, P.N. and Palmer, M.R. 2000. Atmospheric carbon dioxide concentrations over the past 60 million years. *Nature* 406: 695–9.

Penniston-Dorland, S., Liu, X.-M. and Rudnick, R. L., 2017. Lithium isotope geochemistry. *Reviews in Mineralogy and Geochemistry* 82(1), 165–217. doi: 10.2138/rmg.2017.82.6.

Pogge von Strandmann, P. A. E. and Henderson, G. M., 2015. The Li isotope response to mountain uplift. *Geology* 43(1), 67–70. doi: 10.1130/G36162.1.

Poitrasson, F., Levasseur, S. and Teutsch, N., 2005. Significance of iron isotope mineral fractionation in pallasites and iron meteorites for the core–mantle differentiation of terrestrial planets. *Earth and Planetary Science Letters* 234(1–2), 151–64. doi: 10.1016/j.epsl.2005.02.010.

Poreda, R., Schilling, J.G. and Craig, H. 1986. Helium and hydrogen isotopes in ocean-ridge basalts north and south of Iceland. *Earth and Planetary Science Letters* 78: 1–17.

Pringle, E. A., Moynier, F., Savage, P. S., Jackson, M. G., Moreira, M., Day, J. M. D. 2016. Silicon isotopes reveal recycled altered oceanic crust in the mantle sources of ocean island basalts. *Geochimica et Cosmochimica Acta* 189: 282–95. doi: 10.1016/j.gca.2016.06.008.

Quade, J., Cerling, T.E. and Bowman, J.R. 1989. Development of Asian monsoon revealed by marked ecological shift during the latest Miocene in northern Pakistan. *Nature* 342: 163–6.

Richet, P., Bottinga, Y. and Javoy, M. 1977. A review of hydrogen, carbon, nitrogen, oxygen, sulphur, and chlorine stable isotope fractionation among gaseous molecules. *Annual Review Earth and Planetary Science Letters* 5: 65–110.

Richter, F.M., Davis, A.M., DePaolo, D.J. and Watson, E.B. 2003. Isotope fractionation by chemical diffusion between molten basalt and rhyolite. *Geochimica et Cosmochimica Acta* 67: 3905–23. doi: 10.1016/s0016-7037(03)00174-1.

Richter, F.M., Liang, Y. and Davis, A.M. 1999. Isotope fractionation by diffusion in molten oxides. *Geochimica et Cosmochimica Acta* 63: 2853–61. doi: 10.1016/s0016-7037(99)00164-7.

Rosing, M. 1999. ^{13}C-depleted carbon microparticles in >3700 Ma sea-floor sedimentary rocks from West Greenland. *Science* 283: 674–6.

Ryu, J.-S., Jacobson, A. D., Holmden, C., Lundstrom, C. and Zhang, Z., 2011. The major ion, $\delta^{44/40}$Ca, $\delta^{44/42}$Ca, and $\delta^{26/24}$Mg geochemistry of granite weathering at pH=1 and T=25°C: power-law processes and the relative reactivity of minerals. *Geochimica et Cosmochimica Acta* 75(20), 6004–6026. doi: 10.1016/j.gca.2011.07.025.

Savage, P. S., Georg, R. B., Armytage, R. M. G., Williams, H. M., Halliday, A. N. 2010. Silicon isotope homogeneity in the mantle. *Earth and Planetary Science Letters* 295: 139–46. doi: 10.1016/j.epsl.2010.03.035.

Savage, P. S., Georg, R. B., Williams, H. M., Turner, S., Halliday, A. N., Chappell, B. W. 2012. The silicon isotope composition of granites. *Geochimica et Cosmochimica Acta* 92: 184–202. doi: 10.1016/j.gca.2012.06.017.

Savarino, J., Bekki, S., J., C.-D., H., T. M. 2003. UV induced mass-independent sulfur isotope fractionation in stratospheric volcanic sulfate. *Geophysical Research Letters* 30: 11–14. doi: 10.1029/2003GL018134.

Schauble, E. A., Georg, R. B., Halliday, A. N. 2007. Silicate-metal fractionation of silicon isotopes at high pressure and temperature. *Eos Trans AGU Fall Meet Suppl* 88 (52): V42A–07.

Schidlowski, M. 1988. A 3800 million-year isotopic record of life from carbon in sedimentary rocks. *Nature* 333: 313–18.

Schoeninger, M.J. and DeNiro, M.J. 1984. Nitrogen and carbon isotopic composition of bone collagen from marine and terrestrial animals. *Geochimica et Cosmochimica Acta* 48: 625–39.

Schwarcz, H.P., Agyei, E.K. and McMullen, C.C. 1969. Boron isotopic fractionation during clay adsorption from sea-water. *Earth and Planetary Science Letters* 6: 1–5.

Shackleton, N.J. and Opdyke, N.D. 1973. Oxygen isotope and paleomagnetic stratigraphy of an equatorial Pacific core V28-238: oxygen isotope temperatures and ice volumes on a 10^5 and 10^6 year time scale. *Quaternary Research* 3: 39–55.

Shahar, A., Hillgren, V. J., Horan, M. F., Mesa-Garcia, J., Kaufman, L. A. and Mock, T. D., 2015. Sulfur-controlled iron isotope fractionation experiments of core formation in planetary bodies. *Geochimica et Cosmochimica Acta* 150, 253–64. doi: 10.1016/j.gca.2014.08.011.

Shanks, W. C., 2014. 13.3 - Stable isotope geochemistry of mineral deposits. In: *Treatise on Geochemistry* (Second Edition) (eds Holland, H. D. and Turekian, K. K.), pp. 59–85, Elsevier, Oxford.

Sharp, Z. D., 2007. *Principles of Stable Isotope Geochemistry*. Pearson Prentice Hall, Upper Saddle River, NJ.

Sheppard, S.M.F. and Epstein, S. 1970. D/H and ^{18}O/^{16}O ratios of minerals of possible mantle or lower crustal origin. *Earth and Planetary Science Letters* 9: 232–9.

Sheppard, S. M. F., Gilg, H. A. 1996. Stable isotope geochemistry of clay minerals. *Clay Minerals* 31: 1–24.

Sheppard, S.M.F. and Harris, C. 1985. Hydrogen and oxygen isotope geochemistry of Ascension Island lavas and granites: variation with fractional crystallization and interaction with seawater. *Contributions to Mineralogy and Petrology* 91: 74–81.

Sheppard, S.M.F., Nielsen, R.L. and Taylor, H.P. 1969. Oxygen and hydrogen isotope ratios of clay minerals from porphyry copper deposits. *Economic Geology* 64: 755–77.

Sherman, L. S., Blum, J. D., Johnson, K. P., Keeler, G. J., Barres, J. A., Douglas, T. A. 2010. Mass-independent fractionation of mercury isotopes in Arctic snow driven by sunlight. *Nature Geoscience* 3: 173. doi: 10.1038/ngeo758.

Smith, H.J., Spivack, A.J., Staudigel, H. and Hart, S.R. 1995. The boron isotopic composition of altered oceanic crust. *Chemical Geology* 126: 119–35.

Spivack, A. and Edmond J.M. 1987. Boron isotope exchange between seawater and the oceanic crust. *Geochimica et Cosmochimica Acta* 51: 1033–44.

Starkey, N. A., Jackson, C. R. M., Greenwood, R. C., et al. 2016. Triple oxygen isotopic composition of the high-^{3}He/^{4}He mantle. *Geochimica et Cosmochimica Acta* 176: 227–38. doi: 10.1016/j.gca.2015.12.027.

Sun, G., Sommar, J., Feng, X., et al. 2016. Mass-dependent and -independent fractionation of mercury isotope during gas-phase oxidation of elemental mercury vapor by atomic Cl and Br. *Environmental Science and Technology* 50: 9232–41. doi: 10.1021/acs.est.6b01668.

Sutton, J. N., André, L., Cardinal, D., et al. 2018a. A review of the stable isotope bio-geochemistry of the global silicon cycle and its associated trace elements. *Frontiers in Earth Science* 5:. doi: 10.3389/feart.2017.00112.

Sutton, J. N., de Souza, G. F., García-Ibáñez, M. I., De La Rocha, C. L. 2018b. The silicon stable isotope distribution along the GEOVIDE section (GEOTRACES GA-01) of the North Atlantic Ocean. *Biogeosciences* 15: 5663–5676. doi: 10.5194/bg-15-5663-2018.

Taylor, H.P. 1974. The application of oxygen and hydrogen studies to problems of hydrothermal alteration and ore deposition. *Economic Geology* 69: 843–83.

Taylor, H.P. 1980. The effects of assimilation of country rocks by magmas on $^{18}O/^{16}O$ and $^{87}Sr/^{86}Sr$ systematics in igneous rocks. *Earth and Planetary Science Letters* 47: 243–54.

Taylor, H.P. and Sheppard, S.M.F. 1986. Igneous rocks: I. Processes of isotopic fractionation and isotope systematics, in *Stable Isotopes in High Temperature Geological Processes* (eds J.W. Valley, H.P. Taylor and J.R. O'Neil), pp. 227–71. Washington, Mineralogical Society of America.

Teng, F.-Z., Dauphas, N., Huang, S. and Marty, B., 2013. Iron isotopic systematics of oceanic basalts. *Geochimica et Cosmochimica Acta* 107, 12–26. doi: 10.1016/j.gca.2012.12.027.

Teng, F.-Z., McDonough, W.F., Rudnick, R.L. and Walker, R.J. 2006. Diffusion-driven extreme lithium isotopic fractionation in country rocks of the Tin Mountain pegmatite. *Earth and Planetary Science Letters* 243: 701–10. doi: 10.1016/j.epsl.2006.01.036.

Teng, F.-Z., McDonough, W.F., Rudnick, R.L., et al. 2004. Lithium isotopic composition and concentration of the upper continental crust. *Geochimica et Cosmochimica Acta* 68: 4167–78.

Thirlwall, M. F., Gee, M. A. M., Lowry, D., Mattey, D. P., Murton, B. J., Taylor, R. N. 2006. Low $\delta^{18}O$ in the Icelandic mantle and its origins: Evidence from Reykjanes Ridge and Icelandic lavas. *Geochimica et Cosmochimica Acta* 70: 993–1019. doi: 10.1016/j.gca.2005.09.008.

Tomascak, P. B., Magna, T. and Dohmen, R., 2016. *Advances in Lithium Isotope Geochemistry*. Springer, Cham.

Urey, H. 1947. The thermodynamic properties of isotopic substances. *Journal of the Chemical Society (London)* 1947: 562–81.

Valley, J. W. and Cole, D. R., 2001. *Stable Isotope Geochemistry. Reviews in Mineralogy* v. 43, pp. 662, Mineralogical Society of America, Washington, D.C.

Valley, J.W., Taylor, H.P. and O'Neil, J.R. (eds) 1986. *Stable Isotopes in High Temperature Applications, Reviews in Mineralogy Volume 16*. Washington, Mineralogical Society of America.

Vengosh, A., Kolodny, Y., Starinsky, A., Chivas, A. and McCulloch, M. 1991. Coprecipitation and isotopic fractionation of boron in modern biogenic carbonates. *Geochimica et Cosmochimica Acta* 55: 2901–10.

Vlastelic, I., Koga, K., Chauvel, C., Jacques, G. and Telouk, P. 2009. Survival of lithium isotopic heterogeneities in the mantle supported by HIMU-lavas from Rurutu Island, Austral Chain. *Earth and Planetary Science Letters* 286: 456–66. doi: 10.1016/j.epsl.2009.07.013.

Wagner, T., Magill, C. R., Herrle, J. O. (2017) Carbon Isotopes. In: White WM (ed) *Encyclopedia of Geochemistry*. Springer International Publishing, Cham, pp 1–11 978-3-319-39193-9.

Wang, Y., Cerling, C.E. and McFadden, B.J. 1994. Fossil horse teeth and carbon isotopes: new evidence for Cenozoic dietary, habitat, and ecosystem changes in North America. *Paleogeogr., Paleoclim., Paleoecol.* 107: 269–79.

Wang, Z., Schauble, E. A. and Eiler, J. M., 2004. Equilibrium thermodynamics of multiply substituted isotopologues of molecular gases. *Geochimica et Cosmochimica Acta* 68, 4779–97. doi: 10.1016/j.gca.2004.05.039.

White, W. M. and Hofmann, A. W., 1982. Sr and Nd isotope geochemistry of oceanic basalts and mantle evolution. *Nature* 296, 821–825. doi: 10.1038/296821a0.

White, W. M., 2015. *Isotope Geochemistry*. John Wiley and Sons.

Williams, H. M. and Bizimis, M., 2014. Iron isotope tracing of mantle heterogeneity within the source regions of oceanic basalts. *Earth and Planetary Science Letters* 404(0), 396–407. doi: 10.1016/j.epsl.2014.07.033.

Zhang, H., Wang, Y., He, Y., Teng, F.-Z., Jacobsen, S. B., Helz, R. T., Marsh, B. D. and Huang, S., 2018. No Measurable Calcium Isotopic Fractionation During Crystallization of Kilauea Iki Lava Lake. *Geochemistry, Geophysics, Geosystems* 19(9), 3128–3139. doi: 10.1029/2018GC007506.

PROBLEMS

1. Show that $\Delta \approx (\alpha - 1) \times 10^3$ and $\Delta \approx 10^3 \ln \alpha$.

2. Using the procedure and data in Example 9.1, calculate the fractionation factor $\Delta_{CO\text{-}O2}$ for the $^{17}O/^{16}O$ ratio at 300 K. What is the expected ratio of fractionation of $^{17}O/^{16}O$ to that of $^{18}O/^{16}O$ at this temperature?

3. What would you predict would be the ratio of the diffusion coefficients of H_2O and D_2O in air?

4. In Example 9.2, we calculated the temperature of equilibration of a granite-gneiss as 805 K based on $\delta^{18}O$ of 11.1 for quartz and magnetite, 1.9. This sample also contains muscovite with a $\delta^{18}O$ of 9.2. Using the data in Table 9.2, calculate the separate equilibration temperatures for muscovite-quartz and magnetite muscovite. Do your temperatures agree with that calculated from quartz-magnetite?

5. Sphalerite (ZnS) and galena (PbS) from a certain Mississippi Valley deposit were found to have $\delta^{34}S_{CDT}$ of +13.2‰ and +9.8‰ respectively.
   (a) Using the information in Table 9.4, find the temperature at which these minerals equilibrated.
   (b) Assuming they precipitated from an H_2S-bearing solution, what was the sulfur isotopic composition of the HS⁻?

6. Glaciers presently constitute about 2.1% of the water at the surface of the Earth and have a $\delta^{18}O_{SMOW}$ of ≈ −30. The oceans contain essentially all remaining water. If the mass of glaciers were to increase by 50%, how would the isotopic composition of the ocean change (assuming the isotopic composition of ice remains constant)?

7. Consider the condensation of water vapor in the atmosphere. Assume that the fraction of vapor remaining can be expressed as a function of temperature (in kelvins):

$$f = \frac{T - 223}{50}$$

Also assume that the fractionation factor can be written as:

$$\ln \alpha = 0.0018 + \frac{12.8}{RT}$$

Assume that the water vapor has an initial $\delta^{18}O_{SMOW}$ of −9‰. Make a plot showing how the $\delta^{18}O$ of the remaining vapor would change as a function of f, fraction remaining.

8. Calculate the $\delta^{18}O$ of raindrops forming in a cloud after 80% of the original vapor has already condensed, assuming (1) the water initially evaporated from the ocean with $\delta^{18}O = 0$, (2) the liquid–vapor fractionation factor $\alpha = 1.0092$.

9. A saline lake that has no outlet receives 95% of its water from river inflow and the remaining 5% from rainfall. The river water has $\delta^{18}O$ of −10‰ and the rain has $\delta^{18}O$ of −5‰. The volume of the lake is steady-state (i.e., inputs equal outputs) and has a $\delta^{18}O$ of −3‰. What is the fractionation factor, α, for evaporation?

10. Consider a sediment composed of 70 mol % detrital quartz ($\delta^{18}O = +10‰$) and 30 mol % calcite ($\delta^{18}O = +30‰$). If the calcite/quartz fractionation factor, α, is 1.005 at 300°C, determine the O isotopic composition of each mineral after metamorphic equilibrium at 300°C. Assume that the rock is a closed system during metamorphism, i.e., $\delta^{18}O$ of the whole rock does not change during the process.

11. Assume that α is a function of the fraction of liquid crystallized such that $\alpha = 1 + 0.008(1-f)$. Make a plot showing how the $\delta^{18}O$ of the remaining liquid would change as a function of $1-f$, fraction crystallized, assuming a value for Δ of 0.5‰.

12. A basaltic magma with a $\delta^{18}O$ of +6.0 assimilates country rock with $\delta^{18}O$ of +20 as it undergoes fractional crystallization. Assuming a value of $\alpha = 0.998$, make a plot of $\delta^{18}O$ vs. f for R = 0.2 and R = 0.1, going from $f = 1$ to $f = 0.05$.

13. Derive a relationship between pH and $\delta^{11}B$ in calcite from eqns. 9.92 through 9.95.

Chapter 10

The big picture: cosmochemistry

10.1 INTRODUCTION

In the previous nine chapters we acquired a full set of geochemical tools. In this and subsequent chapters, we will apply these tools to understanding the Earth. Certainly, any full understanding of the Earth includes an understanding of its origin and its relationship to its neighboring celestial bodies. That is our focus in this chapter.

The question of the origin of the Earth is closely tied to that of the composition of the Earth, and certainly the latter is central to geochemistry. Indeed, one of the primary objectives of early geochemists was to determine the abundance of the elements in the Earth. It is natural to wonder what accounts for these abundances and to ask whether the elemental abundances in the Earth are the same as the abundances elsewhere in the solar system and in the universe. We might also ask why the Earth consists mainly of oxygen, iron, magnesium, and silicon? Why not titanium, fluorine, and gold? Upon posing these questions, the realm of geochemistry melds smoothly into the realms of cosmochemistry and cosmology. Cosmochemistry has among its objectives an understanding of the distribution and abundance of elements in the solar system and the cosmos, although our knowledge of the composition of objects beyond the solar system is much less complete. Perhaps more importantly, cosmochemistry seeks an understanding of how the solar system formed and evolved.

The composition of the Earth is unique, at least within our solar system but nevertheless similar to that of the other terrestrial planets: Mercury, Venus, Mars, and the Moon. The Earth also shares a common geochemical heritage with the remainder of the solar system, and all bodies in the solar system have similar isotopic compositions of most elements. What we know of the composition of the remainder of the universe suggests that it has a composition that is grossly similar to our solar system: it is dominated by hydrogen and helium, with lesser amounts of carbon, oxygen, magnesium, silicon, and iron, but there are local differences, particularly in the abundances of elements heavier than hydrogen and helium.

The unique composition of the Earth is the product of three sets of processes. These include those processes responsible for the creation of elements, that is, nucleosynthetic processes, the creation of the solar system, and finally the formation of the Earth within the solar system. How the elements are distributed within the Earth reflects its subsequent evolution.

We will begin by considering nucleosynthesis. Meteorites are the principal record of formation of the solar system and of the planetary bodies within it, so we devote considerable space to understanding these objects. Perhaps ever since we acquired the capacity to contemplate the abstract, mankind has wondered about how and when the Earth formed. Combining the tools of geochemistry with astronomical observations can provide the answer to this question. We close the chapter by attempting to construct a history of solar system and planetary formation.

10.2 IN THE BEGINNING ... NUCLEOSYNTHESIS

10.2.1 Astronomical background

Nucleosynthesis is the process of creation of the elements. While we could simply take for granted the existence of the elements, such an approach is somehow intellectually unsatisfactory. The origin of the elements is a cosmological question but also both a geochemical and astrophysical one. Our understanding of nucleosynthesis comes from a combination of observations of the abundances of the elements and their isotopes in meteorites and from observations of stars and related objects. Thus, to understand how the elements formed, we need to review a few astronomical observations and concepts. The universe began about 13.8 Ga ago with the Big Bang; since then the universe has been expanding, cooling, and evolving. This hypothesis follows from two observations: the relationship between red-shift and distance, and the cosmic background radiation, but particularly the former.

Stars shine because of exothermic nuclear reactions occurring in their cores. The energy released by these processes creates a tendency for thermal expansion that, in general, balances the tendency for gravitational collapse. Surface temperatures are very much cooler than temperatures in stellar cores. For example, the Sun has a surface temperature of 5700 K and a core temperature thought to be 14,000,000 K.

Stars are classified based on their color (and spectral absorption lines), which is, in turn, related to their surface temperature. From hot to cold, the classification is: O, B, F, G, K, M, with subclasses designated by numbers, such as F5. (The mnemonic is "*O Be a Fine Girl, Kiss Me!*".) The Sun is class G. Stars are also divided into Populations. Population I stars are second- or later generation stars and have greater heavy element contents than Population II stars. Population I stars are generally located in the main disk of the galaxy, whereas the old first-generation stars

of Population II occur mainly in globular clusters that circle the main disk.

On a plot of luminosity versus wavelength of their principal emissions (i.e., color), called a Hertzsprung-Russell diagram (Figure 10.1), most stars (about 90%) fall along an array defining an inverse correlation, called the *main sequence,* between these two properties. Since wavelength is inversely related to the fourth power of temperature, this correlation means simply that hot stars give off more energy (are more luminous) than cooler stars. Mass and radius are also simply related to temperature for these so-called main sequence stars: hot stars are big, cool stars are small. Thus, O and B stars are large, luminous and hot; K and M stars are small, faint, and cool. The relationship between mass, luminosity, and temperature is nonlinear, however. For example, an O star that is 30 times as massive as the Sun will have a surface temperature 7 times as hot and a luminosity 100,000 times brighter. Stars on the main sequence produce energy by *hydrogen burning,* fusion of hydrogen to produce helium. Since the rate at which this reaction occurs depends on

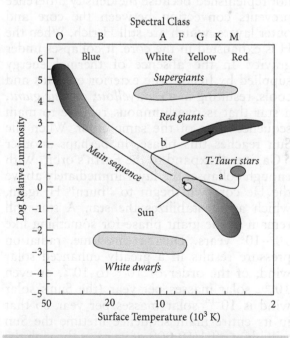

Figure 10.1 The Hertzsprung–Russell diagram of the relationship between luminosity and surface temperature. Arrows show the evolutionary path for a star the size of the Sun in pre- (a) and post- (b) main sequence phases.

temperature and density, hot, massive stars release more energy than smaller ones. As a result, they exhaust the hydrogen in their cores much more rapidly. Thus, there is an inverse relationship between the lifetime of a star, or at least the time it spends on the main sequence, and its mass. The most massive stars, up to 100 solar masses, have life expectancies of only about 10^6 years, whereas small stars, as small as 0.01 solar masses, remain on the main sequence for more than 10^{10} years. Most stars are small. While the Sun is often referred to as an average star, it is more massive than nearly 80% of all stars and half of all stars have masses less than 25% of the Sun's mass. Truly massive stars are exceedingly rare. B-type stars, which have masses more than twice that of the Sun, make up less than 1% of all stars and O stars, those more than 15 times as massive as the Sun, are far rarer.

The two most important exceptions to the main sequence stars, the red, yellow, and blue giants and the white dwarfs, represent stars that have burned all the H fuel in the cores and have moved on in the evolutionary sequence. In all but the smallest stars, H in the core is not replenished because the density difference prevents convection between the core and outer layers, which are still H-rich. When the H is exhausted in the core, it collapses under gravity. In the absence of thermal energy supplied by fusion, the exterior expands and cools, resulting in a *red, yellow, or blue giant*, a star that is overluminous relative to main sequence stars of the same color. When the Sun reaches this phase, in perhaps another 5 Ga, it will expand to the Earth's orbit. With enough collapse, the layer immediately above the He core will begin to "burn" H again, which again stabilizes the star. A star will remain in the giant phase for something like 10^6–10^8 years. During this time, radiation pressure results in a greatly enhanced solar wind, of the order of 10^{-6} to 10^{-7}, or even 10^{-4}, solar masses per year (the Sun's solar wind is 10^{-14} solar masses per year, so that in its entire main sequence lifetime the Sun will blow off 1/10,000 of its mass through the solar wind).

For stars of at least 0.5 solar masses, the inner part of the core eventually collapses to the point where temperature and pressure are sufficient to ignite He burning, in which three He atoms fuse to form carbon. Such stars are called *asymptotic giant branch*, or AGB stars. The fate of stars after this phase depends on their mass. Nuclear reactions in stars less than about eight solar masses cease and they begin to contract. In a dying gasp, they may expel a significant fraction of their mass as a "planetary nebula". Following that, they simply contract, their exteriors heating up as they do so, to become *white dwarfs*. This is the fate of the Sun. White dwarfs are underluminous relative to stars of similar color on the main sequence. They can be thought of as little more than glowing ashes, radiating energy produced by previous nuclear reactions and gravitational potential energy.

Unless they blow off sufficient mass during their post-main sequence evolution, large stars die explosively, in supernovae. (Novae are entirely different events that occur in binary systems when mass from a main sequence star is pulled by gravity onto a white dwarf companion.) Supernovae are incredibly energetic events. The energy released by a supernova can exceed that released by an entire galaxy (which, it will be recalled, consists of over 10^{10} stars) for days or weeks!

10.2.2   The polygenetic hypothesis of Burbidge, Burbidge, Fowler, and Hoyle

Our understanding of nucleosynthesis comes from three sets of observations: (1) the abundance of isotopes and elements in the cosmos; (2) experiments on nuclear reactions that determine what reactions are possible (or probable) under given conditions; and (3) inferences about the conditions that prevail in possible sites of nucleosynthesis. The abundances of the elements and their isotopes in primitive meteorites are by far our most important source of information. Additional information can be obtained from spectral observations of stars. The abundances of the elements and isotopes in the solar system are shown in Figure 10.2. Any successful theory of nuclear synthesis must explain these abundance patterns. Therefore, the chemical and isotopic composition of meteorites is a matter of keen interest, not only to geochemists, but to astronomers and physicists as well.

The cosmology outlined above provides two possibilities for formation of the elements: (1) they were formed in the Big Bang

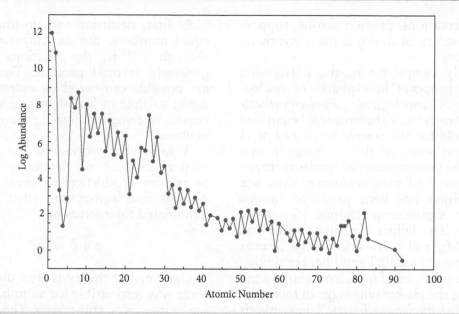

Figure 10.2 Solar system abundance of the elements relative to silicon as a function of atomic number.

itself, or (2) they were subsequently produced. One key piece of evidence comes from looking back into the history of the cosmos. Astronomy is a bit like geology in that just as we learn about the evolution of the Earth by examining old rocks, we can learn about the evolution of the cosmos by looking at old stars. The old stars of Population II are considerably poorer in heavy elements than are young stars. In particular, Population II stars have a Fe/H ratio typically a factor of 100 lower than the Sun. This suggests that much of the heavy element inventory of the galaxy has been produced since these stars formed more than 10 Ga ago. There are also significant variations in the Fe/H ratio between galaxies. In particular, dwarf spheroidal galaxies appear to be deficient in Fe, and sometimes in C, N, and O, relative to our own galaxy. Other galaxies show distinct radial compositional variations. For example, the O/H ratio in the interstellar gas of the disk of the spiral galaxy M81 falls by an order of magnitude with distance from the center. Finally, one sees a systematic decrease in the Fe/H ratio of white dwarfs (the remnants of small to medium sized stars) with increasing age. On the other hand, old stars seem to have about the same He/H ratio as young stars. Indeed, ^{4}He seems to have an abundance of $25 \pm 2\%$ by mass everywhere in the universe.

Thus, the observational evidence suggests that (1) H and He are everywhere uniform, implying their creation and fixing of the He/H ratio in the Big Bang, and (2) elements heavier than Li were created by subsequent processes. The production of these heavier elements seems to have occurred over time, but the efficiency of the process varies between galaxies and even within galaxies.

Early attempts (~1930–1950) to understand nucleosynthesis focused on single mechanisms. Failure to find a single mechanism that could explain the observed abundance of nuclides, even under varying conditions, led to the present view that relies on a number of mechanisms operating in different environments and at different times for creation of the elements in their observed abundances. This view, which has been called the polygenetic hypothesis, was first proposed by Burbidge, Burbidge, Fowler, and Hoyle (1957). The abundance of the elements and their isotopic compositions in meteorites were perhaps the most critical observations in development of the theory. An objection to the polygenetic hypothesis was the apparent uniformity of the isotopic composition of the elements, but variations in the isotopic composition of many elements have now been found in meteorites. Furthermore, there are significant variations in isotopic composition of some elements, such as carbon, among stars.

These observations provide strong support for the existence of multiple nucleosynthetic environments.

To briefly summarize it, the polygenetic hypothesis proposes four phases of nucleosynthesis. *Cosmological nucleosynthesis* occurred shortly after the universe began and is responsible for the cosmic inventory of H and He, and some of the Li. Even though helium is the main product of nucleosynthesis in the interiors of main sequence stars, not enough helium has been produced in this manner to significantly change its cosmic abundance. The lighter elements, up to and including Mg and a fraction of the heavier elements, but excluding Li and Be, are synthesized in the interiors of medium to large sized stars during the He-burning stage of their evolution (*stellar nucleosynthesis*). The synthesis of the remaining elements occurs as large stars exhaust the nuclear fuel in their interiors and explode in nature's grandest spectacle, the supernova, and mergers between their neutron star remnants (*explosive nucleosynthesis*). Finally, Li and Be are continually produced in interstellar space by interaction of cosmic rays with matter (*galactic nucleosynthesis*). Burbidge and others outlined this scenario in a rather general way. In the following decades, astronomical observations, theoretical calculations and computer simulations based on them have been filling in the details, but our understanding of nucleosynthesis remains incomplete. As research continues, we can expect our understanding to evolve over the coming decades.

10.2.3 Cosmological nucleosynthesis

Immediately after the Big Bang, the universe was too hot for any matter to exist. But within a microsecond or so, it had cooled to 10^{11} K so that matter began to condense. At first electrons, positrons, and neutrinos dominated, but as the universe cooled and expanded, quarks coalesced to form hadrons, including protons and neutrons. These existed in an equilibrium dictated by the following reactions:

$$^1H + e^- \rightleftharpoons n + \nu \quad \text{and} \quad n + e^+ \rightleftharpoons {}^1H + \bar{\nu}^*$$

At first, neutrons and protons existed in equal numbers. But as temperatures cooled through 10^{10} K, the reactions above progressively favored protons. These reactions are possible only at these extreme temperatures, so that in less than two seconds they ceased, freezing in a 6 to 1 ratio of protons to neutrons.

A somewhat worrisome question is why matter exists at all. Symmetry would seem to require production of equal numbers of protons and anti-protons that would have annihilated themselves by:

$$p + \bar{p} \rightarrow 2\gamma^*$$

One current theory is that the hyperweak force was responsible for an imbalance favoring matter over anti-matter. The same theory predicts a half-life of the proton of $\sim 10^{32}$ yrs, a prediction not yet verified.

It took another 10 seconds or so for the universe to cool to 10^9 K, which is cool enough for ^2H to form:

$$^1H + {}^1n \rightarrow {}^2H + \gamma$$

At about the same time, the following reactions could also occur:

$$^2H + {}^1n \rightarrow {}^3H + \gamma; \quad {}^2H + {}^1H \rightarrow {}^3He + \gamma$$
$$^2H + {}^1H \rightarrow {}^3H + \beta^+ + \gamma; \quad {}^3He + n \rightarrow {}^4He + \gamma$$

and

$$^2H + {}^2H \rightarrow {}^3He + n;$$
$$^3He + {}^4He \rightarrow {}^7Be + \gamma;$$
$$^7Be + e^- \rightarrow {}^7Li + \gamma$$

Formation of elements heavier than Li, however, was inhibited by the instability of nuclei of masses 5 and 8. Figure 10.3 illustrates the compositional evolution of the early cosmos. Two factors bring this process to a halt: (1) falling temperatures and (2) the decay of free neutrons: outside the nucleus, neutrons have a half-life of only 10 minutes, so nucleosynthetic reactions that consume neutrons eventually cease. Within 20 minutes or so, the universe cooled below $\sim 10^8$ K and these

---

* γ (gamma) is used here to indicate energy in the form of a gamma ray. ν denotes the neutrino; an overbar denotes an anti-particle.

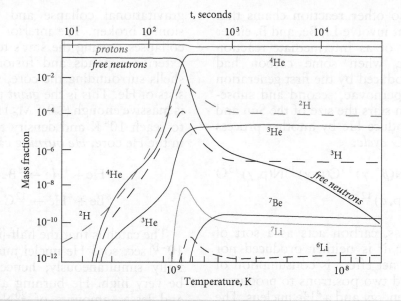

Figure 10.3 Compositional evolution during cosmological nucleosynthesis. ^{7}Be and ^{3}H are unstable and decay to ^{7}Li and ^{3}He with half-lives of 53 days (4.6×10^6 s) and 12.3 years (4×10^8 s), respectively. After Ned Wright's Cosmology Tutorial (http://www.astro.ucla.edu/~wright/cosmolog.htm).

nuclear reactions were no longer possible. Thus, the Big Bang created H, He, and a little Li (^{7}Li/H $= 10^{-9}$). Some 400,000 years later, the universe had cooled to about 3000 K, cool enough for electrons to be bound to nuclei, forming atoms. With this *recombination*, the universe became transparent, that is, radiation could freely propagate through it. The faint radiation known as the cosmic background radiation provides a snapshot of the universe at that time.

The first seconds and minutes of the cosmos are surprisingly well understood, in part because the conditions that prevailed can be reproduced, at least on a micro-scale, in high-energy accelerators/colliders. Experiments in these accelerators provide tests and calibrations of cosmological theory. The ratios of hydrogen, helium, and lithium created by these reactions depend on density, the ratio of particles to photons, and temperature and in part because of the decay of free neutrons, they are also a function of the rate at which these fall as the cosmos expands. An indication of the success of cosmological theory is that it predicts the H and He abundances, and their isotopic composition, in the cosmos. On the other hand, the very earliest history of the cosmos, the first femtosecond or so, is less well understood because the conditions that

prevailed are too energetic to be reproduced in accelerators.

10.2.4   Nucleosynthesis in stellar interiors

10.2.4.1   Hydrogen, helium, and carbon burning stars

For quite some time after the Big Bang, the universe was a more or less homogenous, hot gas. "Less" turns out to be critical. Inevitably (according to fluid dynamics), inhomogeneities developed in the gas, and indeed, very slight inhomogeneities are observed in the cosmic background radiation. These inhomogeneities enlarged in a sort of runaway process of gravitational attraction and collapse. Thus were formed protogalaxies, perhaps 0.5 Ga after the Big Bang. Instabilities within the protogalaxies collapsed into stars. Once this collapse proceeded to the point where density reached 6 g/cm and temperature reached 10 to 20 million K in the core of the star, nucleosynthesis began in the interior of stars by *hydrogen burning*, or the *pp process*, which involves reactions such as:

$$^1H + {}^1H \rightarrow {}^2H + \beta^+ + \nu; \quad {}^2H + {}^1H \rightarrow {}^3He + \gamma$$

$$\text{and} \quad {}^3He + {}^3He \rightarrow {}^4He + 2{}^1H + \gamma$$

There are also other reaction chains that produce ^{4}He that involve Li, Be, and B, either as primary fuel or as intermediate reaction products. Later, when some carbon had already been produced by the first generation of stars and supernovae, second and subsequent generation stars the size of the Sun and larger could produce He by another process as well, the *CNO cycle*:

$$^{12}\text{C}(p, \gamma)\,^{13}\text{N}(\beta^+, \gamma)\,^{13}\text{C}(p, \gamma)\,^{14}\text{N}(p, \gamma)\,^{15}\text{O}$$

$$(\beta^+, \nu)\,^{15}\text{N}(p, \alpha)\,^{12}\text{C}^*$$

In this process, carbon acts as a sort of nuclear catalyst: it is neither produced nor consumed. The net effect is consumption of four protons and two positrons to produce a neutrino, some energy and a ^{4}He nucleus. The $^{14}\text{N}(p,\gamma)^{15}\text{O}$ is the slowest in this reaction chain, so there tends to be a net production of ^{14}N as a result. Also, although both ^{12}C and ^{13}C are consumed in this reaction, ^{12}C is consumed more rapidly, so this reaction chain should increase the $^{13}\text{C}/^{12}\text{C}$ ratio. The CNO cycle is strongly temperature dependent, so that while only a small fraction of He is produced this way in the Sun, it becomes the dominant mechanism for stars larger than about $1.3\,\text{M}_\odot$[†].

The heat produced by these reactions counterbalances gravitational collapse in main sequence stars (Figure 10.1). However, as more of the stellar core is converted to helium, the core slowly contracts. As a result, the fusion reactions speed up and the star becomes very slightly brighter over time: the Sun now radiates about 30% more energy than it did 4.5 billion years ago. Eventually, the hydrogen in the stellar core is consumed. How quickly this happens depends, as we noted earlier, on the mass of the star.

Once the H is exhausted in the stellar core, fusion ceases, and the balance between gravitational collapse and thermal expansion is broken. The interior of the star thus collapses, raising the star's temperature. The exterior expands and fusion begins in the shells surrounding the core, which now consists of He. This is the *giant* phase. If the star is massive enough ($\gtrsim 0.5\,\text{M}_\odot$) for temperatures to reach 10^8 K and density to reach 10^4 g/cc in the He core, *He burning* can occur:

$$^4\text{He} + {}^4\text{He} \rightarrow {}^8\text{Be} + \gamma \quad \text{and}$$

$$^8\text{Be} + {}^4\text{He} \rightarrow {}^{12}\text{C} + \gamma$$

The catch is that the half-life of ^{8}Be is only 10^{-16} sec, so 3 He nuclei must collide essentially simultaneously; hence densities must be very high. He burning also produces O, and lesser amounts of ^{20}Ne and ^{24}Mg, in the red giant phase by successive exothermic α-capture reactions such as:

$$^{12}\text{C} + {}^4\text{He} \longrightarrow {}^{16}\text{O} + \gamma$$

$$^{16}\text{O} + {}^4\text{He} \longrightarrow {}^{20}\text{Ne} + \gamma;$$

$$^{20}\text{Ne} + {}^4\text{He} \longrightarrow {}^{24}\text{Mg} + \gamma$$

However, Li, Be, and B are skipped: they are not synthesized in these phases of stellar evolution. These nuclei are unstable at the temperatures of stellar cores. Rather than being produced, they are consumed in stars.

There is a division of evolutionary paths once helium in the stellar core is consumed. Densities and temperatures necessary to initiate further nuclear reactions cannot be achieved by stars with masses less than about 8 to $11\,\text{M}_\odot$ so their evolution ends after the red giant-AGB phase and the star becomes a white dwarf. Evolution during the preceding He burning can be unsteady, particularly in intermediate mass stars, with convection dredging up material from the stellar core.

---

[*] Here we are using a notation commonly used in nuclear physics. The notation:

$$^{12}\text{C}(p, \gamma)\,^{13}\text{N}$$

is equivalent to:

$$^{12}\text{C} + p \rightarrow {}^{13}\text{N} + \gamma$$

[†] The symbol $\odot$ is the astronomical symbol for the Sun. $\text{M}_\odot$ therefore indicates the mass of the Sun.

Ejection of outer layers leads to the formation of a carbon–oxygen white dwarf.

Massive stars undergo further collapse and the initiation of *carbon and oxygen burning* when temperatures reach 600 million K and densities 5×10^5 g/cc. Evolution in these stars now proceeds at an exponentially increasing pace (Figure 10.4) with reactions of the type:

$$^{12}C + {}^{12}C \rightarrow {}^{20}Ne + {}^{4}He + \gamma$$

$$^{16}O + {}^{16}O \rightarrow {}^{28}Si + {}^{4}He + \gamma$$

and

$$^{12}C + {}^{16}O \rightarrow {}^{24}Mg + {}^{4}He + \gamma$$

Also, ^{14}N created in the CNO cycle of second-generation stars can be converted to ^{22}Ne. A number of other less abundant nuclei, including Na, Al, P, S, and K, are also synthesized at this time, and in the subsequent process, *Ne burning*.

During the final stages of evolution of massive stars, a significant fraction of the energy released is carried off by neutrinos created by electron–positron annihilations in the core of the star. If the star is sufficiently oxygen-poor that its outer shells are reasonably transparent, the outer shell of

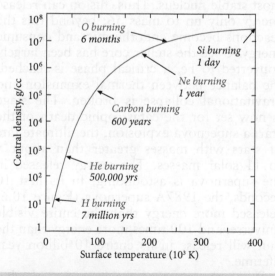

Figure 10.4 Evolutionary path of the core of star of 25 solar masses (after Bethe and Brown, 1985). The period spent in each phase depends on the mass of the star: massive stars evolve more rapidly.

these "supergiants" may collapse during last few 10^4 years of evolution to form a *blue giant*.

10.2.4.2 The e-process

Eventually, a new core consisting mainly of ^{28}Si is produced. At temperatures near 10^9 K and densities above 10^7 g/cc a process known as *silicon burning*, or the *e-process* (for equilibrium), begins, and lasts for a week or less, again depending on the mass of the star. The process of silicon burning is really a variety of reactions that can be summarized as the photonuclear rearrangement of a gas originally consisting of ^{28}Si nuclei into one that, via intervening beta decays, consists mainly of ^{56}Ni, which then decays with a half-life of six days to ^{56}Fe, the most stable of all nuclei. At 10^9 K, thermal photons have energies near 1 MeV; absorbing such photons overcomes the energy barriers between nuclei, allowing the system to evolve towards a minimum energy state by making the most stable nuclei. This is analogous to the rapid approach to chemical equilibrium that occurs once temperatures become high enough to overcome kinetic barriers.

The e-process includes reactions such as:

$$^{28}Si + \gamma \rightleftarrows {}^{24}Ne + {}^{4}He$$

$$^{28}Si + {}^{4}He \rightleftarrows {}^{32}S + \gamma$$

$$^{32}S + {}^{4}He \rightleftarrows {}^{36}Ar + \gamma$$

While these reactions can proceed in either direction, there is a tendency for the build-up of heavier nuclei with masses 32, 36, 40, 44, 48, 52, and 56. Partly as a result of the e-process, these nuclei are unusually abundant in nature. In addition, because a variety of nuclei are produced during C and Si burning phases, other reactions are possible, synthesizing a number of minor nuclei.

The star is now a cosmic onion of sorts, consisting of a series of shells of successively heavier nuclei and a core of Fe (Figure 10.5). Though temperature increases toward the interior of the star, the structure is stabilized with respective to convection and mixing because each shell is denser than the one overlying it.

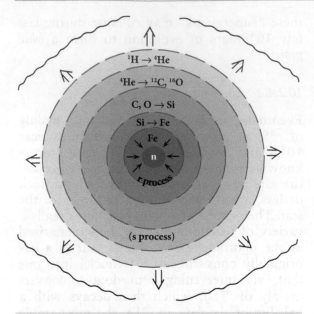

Figure 10.5 Schematic diagram of stellar structure at the onset of the supernova stage. Nuclear burning processes are illustrated for each stage.

Fe-group elements are also synthesized by the e-process in Type I supernovae.* Type I supernovae occur when white dwarfs of intermediate mass (3–10 solar masses) in binary systems accrete material from their companion. When their cores reach the Chandrasekhar limit of 1.4 solar masses, explosive C-burning (called carbon detonation) can occur, which in turn triggers further fusion reactions producing nuclides up to ^{56}Ni and the star explodes. Type I supernovae are responsible for a substantial fraction of the so-called *iron-peak* elements (chromium through nickel) in the universe.

10.2.4.3 The s-process

In second and later generation stars containing heavy elements, yet another nucleosynthetic process can operate. This is the slow neutron capture or *s-process*. It is so called because the rate of capture of neutrons is slow compared with the *r-process*, which we will discuss below. It operates mainly in the giant phase (as evidenced by the existence of ^{99}Tc and

enhanced abundances of several s-process nuclides in such stars) where neutrons are produced by reactions such as:

$$^{13}C + {}^4He \rightarrow {}^{16}O + n$$

$$^{22}Ne + {}^4He \rightarrow {}^{25}Mg + n$$

$$^{17}O + {}^4He \rightarrow {}^{20}Ne + n$$

(but even H burning produces neutrons, so that the s-process operates to some degree even in main sequence stars). These neutrons are captured by nuclei to produce successively heavier elements. The principal difference between the s-process and r-process, discussed below, is the rate of capture relative to the decay of unstable isotopes. In the s-process, a nucleus may only capture a neutron every year or so. If the newly produced nucleus is not stable, it will decay before another neutron is captured. As a result, nuclear instabilities cannot be bridged; the s-process path closely follows the valley of stability on the chart of the nuclides.

10.2.5 Explosive nucleosynthesis

The e-process stops at mass 56. In Chapter 8 we noted that ^{56}Fe had the highest binding energy per nucleon, that is, it is the most stable nucleus. Thus, fusion can release energy only up to mass 56; beyond this the reactions become endothermic and consume energy. Once the stellar core has been largely converted to Fe, a critical phase is reached: the balance between thermal expansion and gravitational collapse is broken. The stage is now set for the catastrophic death of the star: a supernova explosion, the ultimate fate of stars with masses greater than about 8 to 11 solar masses. The energy released in the supernova is astounding. In its first 10 seconds, the 1987A supernova (Figure 10.6) released more energy than the entire visible universe, and 100 times more energy than the Sun will release in its entire 10 billion year lifetime.

When the mass of the iron core reaches 1.4 solar masses (the Chandrasekhar mass),

---

* Astronomers recognize two kinds of supernovae: Type I and Type II. A Type I supernova occurs when a white dwarf in a binary system accretes mass from its sister star. Its mass reaches the point where carbon burning initiates and the star is explosively disrupted. The explosions of massive stars that we are considering are the Type II supernovae.

Figure 10.6 Rings of glowing gas surrounding the site of the supernova explosion named Supernova 1987A photographed by the wide field planetary camera on the Hubble Space Telescope in 1994. The nature of the rings is uncertain, but they may be debris of the supernova illuminated by high-energy beams of radiation or particles originating from the supernova remnant in the center. NASA photo.

further gravitational collapse cannot be resisted even by coulomb repulsion. The supernova begins with the collapse of this stellar Fe core, which would have a radius of several thousand km (similar to the Earth's radius) before collapse, to a radius of 100 km or so. The collapse occurs on a time scale of milliseconds. When matter in the center of the core is compressed beyond the density of nuclear matter (3×10^{14} g/cc), it rebounds, sending a massive shock wave back out. As the shock wave travels outward through the core, the temperature increase resulting from the compression produces a breakdown of nuclei by photodisintegration, for example:

$$^{56}Fe + \gamma \rightarrow 13\,^4He + 4\,^1n;$$

$$^4He + \gamma \rightarrow 2\,^1H + 2\,^1n$$

The extreme densities also favor electron captures and conversion of protons to neutrons. Thus, much of what took millions of years to produce is undone in an instant. However, photodisintegration produces a large number of free neutrons (and protons), which leads to another important nucleosynthetic process, the *r-process*, discussed below.

When the shock wave reaches the surface of the core, the outer part of the star is blown apart in an explosion of unimaginable violence. But amidst the destruction, new nucleosynthetic processes occur. As the shock wave passes through outer layers, it may "reignite" them, producing explosive Si, Ne, O, and C burning. These processes produce isotopes of S, Cl, Ar, Ca, Ti, and Cr, and some Fe.

While the outer layers of the star are blown out into space, the inner core collapses into a neutron star that might have a mass of 1 to $2\,M_\odot$ and a radius of 10 km or so. Like a ballerina pulling in her arms, it conserves angular momentum by spinning faster as it collapses, reaching as much as 700 revolutions per second. Neutron stars emit radiation in beacon-like fashion: a *pulsar*. If the original star was sufficiently massive, perhaps greater than $30\,M_\odot$, the collapse may not stop at all, collapsing to a diameter of zero and infinite density. Such an object is called a *singularity*. Its gravitational attraction is so great even light cannot escape, creating a *black hole*.

Another important effect during core collapse is the creation of huge numbers of neutrinos by positron–electron annihilations, these particles having "condensed" as pairs from gamma rays. The energy carried away by neutrinos leaving the supernova exceeds the kinetic energy of the explosion by a factor of several hundred and exceeds the visible radiation by a factor of some 30,000. The neutrinos leave the core at nearly the speed of light. Both neutrinos and photons leaving the core are delayed by interaction with the dense matter of core, but the neutrinos are delayed less than the electromagnetic radiation. Thus, neutrinos from the 1987A supernova arrived at Earth (some 160,000 years after the event) a few hours before the supernova became visible.

10.2.5.1 The r-process

The *r-process* (rapid neutron capture) is the principal mechanism for building up the nuclei heavier than the iron peak (in later generation stars, some heavy nuclides are also produced by the s-process). It occurs in an environment where the abundance of neutrons is so high that nuclei capture them at an extremely rapid rate – so rapid that even an unstable nucleus will capture multiple neutrons before it decays. The result is a build-up of neutron-rich unstable nuclei. Eventually the nuclei capture enough neutrons that they are not stable even for a small fraction of a second. At that point, they undergo β^- decay to new nuclides, which are more stable

and capable of capturing more neutrons. It reaches a limit when nuclei beyond A ≈ 90 are reached and these heavy nuclei fission into several lighter fragments. Figure 10.7 illustrates this process.

Type II supernovae described above were long thought to be the environment in which the r-process occurs. However, computer simulations of supernovae had generally not been successful in reproducing the observed abundances of r-process nuclides (e.g., Freiburghaus et al., 1999a; Thielemann et al., 2011). Consequently, astronomers searched for another high energy, neutron-rich environment where the r-process might occur. Lattimer and Schram (1974) had suggested that tidal disruption of a neutron star by a neighboring black hole might be an appropriate environment. Subsequent research suggested neutron star mergers, events known as *kilonovae*, might be suitable environments (e.g., Li and Bohdan, 1998; Freiburghaus et al., 1999b). An event that literally shook the cosmos appears to have confirmed this suspicion.

On August 17, 2017, gravitational wave detectors in the US and Italy (the US LIGO detector had been operational for less than 2 years and the Italian Virgo detector for only weeks) detected gravity waves in an area of roughly 28 deg² of the sky, an event labeled

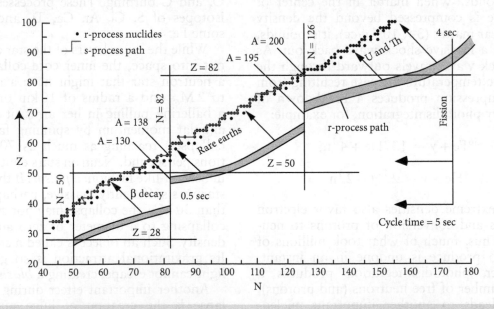

Figure 10.7 Diagram of the r-process path on a Z versus N diagram. The r-process path is indicated; the solid line through stable isotopes shows the s-process path.

GW170817. In this same area of sky, a gamma ray burst was detected 1.7 seconds later in the galaxy NGC4993 located roughly 100 million light years away. These features were consistent with the merger of two neutron stars in the mass range of 1.17–1.60 $M_\odot$ (LIGO Scientific Collaboration and Virgo Collaboration, 2017). In the days following this event, telescopes around the work were trained on this object, analyzing the entire electromagnetic spectrum. Although optically bright, 10^8 times more luminous than the Sun, it was less bright and dimmed more rapidly than a supernova but remained luminous in the infrared. Analysis of the spectrum confirmed an ejected mass of 0.03–0.05 $M_\odot$ expanding at a rate close to 0.2 times the speed of light. The infrared spectrum revealed this outflow was rich in heavy elements such as the rare earths, platinum and gold (Pian et al., 2017). That, together with the gamma-ray burst, confirmed that r-process nucleosynthesis had occurred. Some of this is relatively "cold" neutron capture as tidal forces fling material from the approaching stars and synthesizes mainly heavy nuclides ($A > 140$). As the stars merge, additional matter is squeezed out of polar regions by shock heating, synthesizing lighter nuclides ($A < 140$). The time scales for these processes are several seconds (Kasen et al., 2017).

Although neutron star merges might seem unlikely, most stars are binaries, so a pair of large stars will eventually produce a pair of neutron stars. As they revolve around their common center of gravity, tidal forces and emission of gravitational waves steal angular momentum and they slowly spin inward over hundreds of millions of years until they finally merge. Kilonovae are now thought to be the source of mysterious short gamma-ray bursts long recognized by astronomers (long gamma-ray bursts are produced primarily by supernovae). Based on the amount of heavy elements ejected during GW170817 and the estimate rate of neutron star mergers in our galaxy (about 1 every 2–60 million years), Kasen et al. (2017) concluded, "such mergers are a dominant mode of r-process production in the universe."

The r-process produces broad peaks in elemental abundance around the neutron magic numbers (50, 82, 126, corresponding to the elements Sr and Zr, Ba, and Pb). This occurs because nuclei with magic numbers of neutrons are particularly stable, have very low cross-sections for capture of neutrons, and because nuclei just short of magic numbers have particularly high capture cross-sections. Thus, the magic number nuclei are both created more rapidly and destroyed more slowly than other nuclei. When they decay, the sharp abundance peak at the magic number becomes smeared out.

10.2.5.2 The p-processes

There are some heavy proton-rich nuclides that cannot be produced by neutron capture. Typically, they are very much less abundant than other isotopes of the same element, for example, ^{84}Sr, ^{138}La, ^{144}Sm, ^{180}Ta, ^{190}Pt. These were initially thought to be products of proton capture. The probability of proton capture is low, and it is easy to understand why: the proton must have sufficient energy to overcome the coulomb repulsion and approach to within 10^{-14} cm of the nucleus where the strong force dominates over the electromagnetic one. In contrast, there is no such barrier to neutron capture. Thus, other processes are now thought to be responsible for these nuclides. In high energy environments, photodissociation reactions such as (γ, n), (γ, α), (γ, p) can occur, which may be more likely sources of these nuclides than proton capture. Both Type I and Type II supernovae as well as neutron star mergers may be appropriate astrophysical sites for such reactions. In addition, in the massive neutrino outflows from Type II supernovae, neutrino-induced reactions such as:

$$^{85}\text{Rb}\,(\nu_e, \beta^-, \text{n})\,^{84}\text{Sr}$$

and

$$^{181}\text{Hf}\,(\nu_e, \beta^-, \text{n})\,^{180}\text{Ta}$$

(where ν_e is the electron neutrino) may also contribute to p-process nuclides (Rauscher et al., 2013).

Figure 10.8 illustrates how these three processes, the s-, r-, and p-processes, create different nuclei. Notice the shielding effect. If an isotope with z protons and n neutrons has a stable isobar with $n + x$ neutrons and $p - x$ protons, this isotope is shielded from production by the r-process because

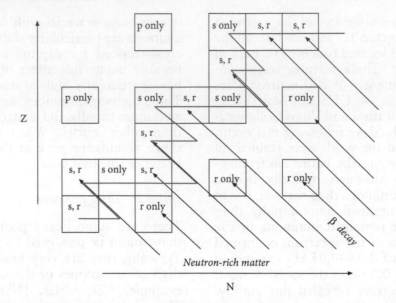

Figure 10.8   Z versus N diagram showing production of isotopes by the r-, s- and p-processes. Squares are stable nuclei; diagonal lines are the beta-decay path of neutron-rich isotopes produced by the r-process; a solid red line through stable isotopes shows the s-process path.

β-decay will cease when that stable isobar is reached. The most abundant isotopes of an element tend to be those created by all processes; the least abundant are those created by only one, particularly only by the p-process. The exact abundance of an isotope depends on a number of factors, including its neutron-capture cross-section,* and the neutron-capture cross-sections and stabilities of neighboring nuclei.

10.2.6   Nucleosynthesis in interstellar space

Except for production of ^{7}Li in the Big Bang, Li, Be, and B are not produced in any of the above situations. One clue to the creation of these elements is their abundance in galactic cosmic rays: they are overabundant by a factor of 10^6, as is illustrated in Figure 10.9. They are believed to be formed by interactions of cosmic rays with interstellar gas and dust, primarily reactions of ^{1}H and ^{4}He with carbon, nitrogen and oxygen nuclei. These

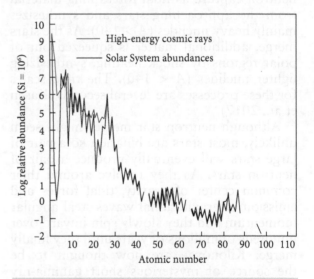

Figure 10.9   Comparison of relative abundances in cosmic rays and the solar system.

reactions occur at high energies (higher than the Big Bang and stellar interiors), but at low temperatures where the Li, B and Be can survive.

---

* In a given flux of neutrons, some nuclides will be more likely to capture and bind a neutron than others, just as some atoms will be more likely to capture and bind an electron than others. The neutron-capture cross-section of a nuclide is a measure of the affinity of that nuclide for neutrons, i.e., a measure of the probability of that nuclide capturing a neutron in a given neutron flux.

10.2.7 Summary

Figure 10.10 is a part of the Z versus N plot showing the abundance of the isotopes of elements 68 through 73. It is a useful region of the chart of the nuclides for illustrating how the various nucleosynthetic processes have combined to produce the observed abundances. First, we notice that even-numbered elements tend to have more stable nuclei than odd-numbered ones – a result of the greater stability of nuclides with even Z. We also notice that nuclides having a neutron-rich isobar (recall that isobars have the same value of A, but a different combination of N and Z) are underabundant, for example ^{176}Lu and ^{170}Yb. This under-abundance results from these nuclides being "shielded" from production by β^- decay of r-process neutron-rich nuclides. In this example, ^{170}Er and ^{176}Yb would be the ultimate product of neutron-rich unstable nuclides of mass number 170 and 176 produced during the r-process. Also notice that ^{168}Yb, ^{174}Hf, and ^{180}Ta are very rare. These nuclides are shielded from the r-process and are also off the s-process path. They are produced only by the p-processes. Finally, those nuclides that can be produced by both the s-process and the r-process tend to be the most abundant; for example, ^{176}Yb is about half as abundant as ^{174}Yb because the former is produced by the r-process only while the latter can be produced by both the s-process and the r-process. ^{176}Yb cannot be produced by the s-process because, during the s-process, the flux of neutrons is sufficiently low that any ^{175}Yb produced decays to ^{175}Lu before it can capture a neutron and become a stable ^{176}Yb.

The heavy element yield of stellar and explosive nucleosynthesis will vary tremendously with the mass of the star. A star of 60 solar masses will convert some 40% of its mass to heavy elements. The bulk of this is apparently ejected into the interstellar medium. Stars too small to become supernovae will convert relatively small fractions of their mass to heavy elements, and a smaller fraction of this will be ejected. On the whole, stars in the mass range of 20–30 solar masses probably produce the bulk of the heavy elements in the galaxy. While such stars, which are already quite large compared with the mean stellar mass, convert a smaller fraction of their mass to heavy elements than truly massive stars, they are much more abundant.

Novae may also make a significant contribution to the cosmic inventory of a few relatively light elements such as ^{19}F and ^{7}Li, as well as the rarer isotopes of C, N, and O. Novae occur when mass is accreted to a white dwarf from a companion red giant. If the material is mainly hydrogen and accretion is relatively slow, H burning may be ignited on the surface of the white dwarf, resulting in an explosion that ejects a relatively small fraction of the mass of the star.

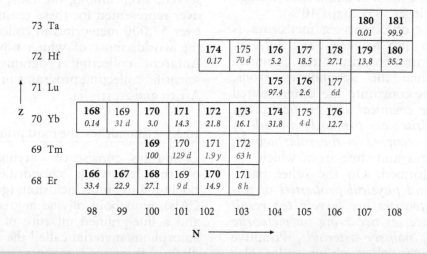

Figure 10.10 View of part of the chart of the nuclides. Mass numbers of stable nuclides are shown in bold, their isotopic abundance is shown in italics as percent. Mass numbers of short-lived nuclides are shown in plain text with their half-lives also given.

10.3 METEORITES: ESSENTIAL CLUES TO THE BEGINNING

In subsequent sections we want to consider the formation of the Earth and its earliest history. The Earth is a dynamic body; its rock formations are continually being recycled into new ones. As a result, old rocks are rare. The oldest rocks are 4.0 Ga; some zircon grains as old as 4.4 Ga have been found in coarse-grained, metamorphosed sediments. The geologic record ends there: the Earth's earliest history is not recorded in terrestrial rocks. So to unravel Earth's early history, we have to turn to other bodies in the solar system. So far, we have samples only of the Moon and meteorites, and some analyses of the surface of Venus and Mars.* The Moon provides some clues to the early history of planets, but meteorites provide the best clues as to the formation of planets and the solar system. We now turn our attention to them.

Meteorites are traditionally classified according to their composition, mineralogy, and texture. The first-order division is between *stones* and *irons*. You can pretty well guess what this means: stones are composed mainly of silicates, while irons are mainly metal. An intermediate class is the *stony-irons*, a mixture of silicate and metal. Stones are subdivided into *chondrites* and *achondrites* depending on whether they contain *chondrules*, which are small spherical particles that were once molten and can constitute up to 80% of the mass of chondrites (though the average is closer to perhaps 40%).

Another way of classifying meteorites is to divide them into *primitive* and *differentiated*. Chondrites constitute the primitive meteorites, while the achondrites, irons, and stony-irons constitute the differentiated meteorites. *The chemical and physical properties of chondrites are primarily a result of processes that occurred in the solar nebula*, the cloud of gas and dust from which the solar system formed. On the other hand, *the chemical and physical properties of differentiated meteorites are largely the result of igneous processes occurring on meteorite parent bodies, namely asteroids.* Primitive meteorites contain clues about early solar system formation, whereas differentiated meteorites contain clues about early planetary differentiation.

Meteorites are also divided into *falls* and *finds*. Falls are meteorites recovered after observation of a fireball whose trajectory can be associated an impact site. Finds are meteorites found but not observed falling. Some finds have been on the surface of the Earth for considerable time and consequently can be weathered. Thus, the compositional information they provide is less reliable than that of falls. An exception of sorts to this is the Antarctic and Saharan meteorites. Meteorites have been found in surprising numbers in the last 40 years or so in areas of low snowfall in Antarctica where ice is eroded by evaporation and wind. Meteorites are concentrated in such areas by glaciers. Because of storage in the Antarctic deep freeze, they are little weathered. This is largely true as well of meteorites found in desert areas, and Northwest Africa has been a particularly productive region for well-preserved meteorite finds. Meteorites from these areas have greatly expanded the range of classes of meteorites and therefore our knowledge of the early solar system.

Figure 10.11 illustrates the relative abundance of the various meteorite types among falls. Stones, and ordinary chondrites in particular, predominate among falls. Irons are over-represented among finds because they are more likely to be recognized as meteorites, and because they are more likely to be preserved. Even among the falls, irons may be over-represented for these reasons. There are over 57,000 meteorites in collections around the world, most of which now come from Antarctic collecting programs and similar scientific collecting programs in the deserts of Africa and Australia.

10.3.1 Chondrites: the most primitive objects

Chondrites consist of varying proportions of the following: chondrules, refractory calcium-aluminum inclusions (generally called CAIs), amoeboid olivine aggregates (AOAs), and a fine-grained mixture of minerals and amorphous material called the matrix – we'll discuss these components in more detail

* As we shall see, a few meteorites come from Mars, providing additional information on the composition of that planet.

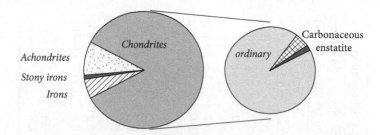

Figure 10.11   Relative abundance of major types of meteorite falls. The smaller pie chart on the right shows the relative proportions of different chondrite classes.

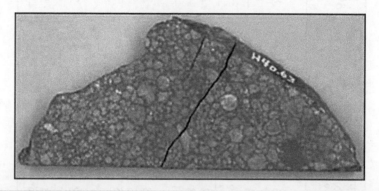

Figure 10.12   NWA869, a brecciated L-4 ordinary chondrite. Both chondrules and metal grains are apparent in this photo. The NWA designation indicates it was found in the deserts of Northwest Africa. Fragments totaling at least 2 tons of NWA869 have been recovered. Photo by H. Raab, Wikipedia creative commons.

below. An example is shown in Figure 10.12. The mineralogical, textural, and compositional (including isotopic composition) features of these components indicate that they formed while dispersed in the solar nebula and were subsequently aggregated to form the meteorite parent bodies. These components were subsequently processed in the parent bodies through aqueous alteration or thermal metamorphism. In addition, many are highly brecciated as a result of collisions and impacts on the surface of the parent bodies. Nevertheless, all chondrite classes have concentrations of *condensable* elements, that are closer to solar values than differentiated meteorites, terrestrial materials, lunar materials, and so on, all of which differ strongly from the solar composition. Since the Sun comprises more than 99% of the mass of the solar system, its composition is effectively identical to the composition of the solar system, and to the solar nebula from which the solar system formed. *The importance of chondrites is thus clear: they represent samples of the cloud of gas and dust from which all bodies in the solar system formed.* To the degree that they have not been obscured by subsequent processing in asteroidal bodies, details of their compositions, their mineralogy, and their textures provide insights into the conditions and processes in the solar nebula that led to the solar system we observe and inhabit today.

10.3.1.1   Chondrite classes and their compositions

Table 10.1 summarizes the general characteristics of the various chondrite groups. There are three main classes: *Carbonaceous (C)*, *Ordinary*, and *Enstatite (E)* chondrites. These classes are further divided into groups based on composition and texture. The *ordinary chondrites*, which as Figure 10.11 shows are by far the most common, are composed primarily of olivine, orthopyroxene and lesser amounts of Ni-Fe alloy. They are subdivided into classes H (high iron or bronzite), L (low

Table 10.1   Characteristics of chondrite groups.

Principal ferromagnesian silicates		R.I.s (vol.%)	Fe/(Fe+Mg) of silicate (mole %)	Metal (vol. %)	Mean Mg/Si (molar)	Mean Al/Si (mol %)	Mean Ca/Si (mol %)	δ18O ‰	δ17O ‰	Chondrules Size (mm)	Chondrules Volume (%)
Carbonaceous											
CI	serpentine	<0.01	45	<0.01	1.05	8.6	6.2	16.4	8.8	—	—
CM	serpentine	1.2	43	0.1	1.05	9.7	6.8	12.2	4	0.3	20
CO	olivine	1.0	9–23	1–5	1.05	9.3	6.8	−1.1	−5.1	0.15	40
CV	olivine	3.0	6–14	0–5	1.07	11.6	8.4	0	−4	1.0	45
CK	olivine	0.2	29–33	<0.01	1.08	10.2	7.6	−0.8	−4.6	0.8	15
CR	chlorite	0.12	37–40	5–8	1.05	8.2	5.6	6.3	2.3	0.7	50–60
CH	olivine	0.1	2.5	20	1.06	8.3	6	1.5	−0.7	0.02–0.09	70
CB	olivine	<0.1	3.5	60–70	1.08	11.1	7.2	1.7	−1.4	0.5–5	30–40
Ordinary											
H	olivine	0.01–0.2	17	15–19	0.96	6.8	4.9	4.1	2.9	0.3	60–80
L	olivine	<0.1	22	4–9	0.93	6.6	4.7	4.6	3.5	0.5	60–80
LL	olivine	<0.1	27	0.3–3	0.94	6.5	4.7	4.9	3.9	0.6	60–80
Enstatite											
EH	enstatite	<0.1	0.05	8	0.73	5	3.6	5.6	3	0.2	60–80
EL	enstatite	<0.1	0.05	15	0.87	5.8	3.8	5.3	2.7	0.6	60–80
Other											
R	olivine	<0.1	38	<0.3	0.77	6.4	4.1	4.5	5.2	0.4	>40
K	enstatite	<0.1	2–4	6–10	0.95	6.9	5	2.7	−1.3	0.6	20–30

Based on Wasson (1974) and Scott and Krot (2014).

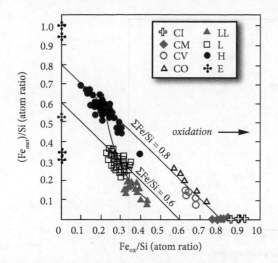

Figure 10.13 Ratio of reduced and oxidized iron to Si in chondrites. Carbonaceous and H group chondrites have approximately equal ΣFe/Si ratios; L and LL groups are iron-depleted. After Wasson (1974).

iron or hypersthene), and LL. The name LL reflects low total iron and low metallic iron. H chondrites contain 25–31% total iron, of which 15–19% is reduced, metallic iron. L chondrites contain 20–25% iron, of which 4–10% is metallic. LL chondrites contain about the same total iron as L chondrites, but only 1–3% is metallic. The enstatite chondrites are highly reduced, with virtually all the iron present as metal. Reduction of iron increases the $Si/(Fe^{2+}+Mg)$ ratio in silicates and results in enstatite, rather than olivine, being the dominant mineral in these objects, hence the name of the class. The E chondrites can be further subdivided into EH (high iron) and EL (low iron) classes. Besides enstatite, metal and sulfides, enstatite chondrites contain a number of other exotic minerals, such as phosphides, carbides and a oxynitride of Si, that indicate they formed under highly reducing conditions.

The amount of Fe and the degree of oxidation are two of the features that differentiate the various chondrite classes. This is illustrated in Figure 10.13. The diagonal lines are lines of constant total iron content. H and C chondrites have similar total iron contents,

but their oxidation state differs (carbonaceous chondrites are more oxidized). L and LL chondrites have lower total iron contents (and variable oxidation state). E chondrites are highly reduced, and may have high (EH) or low (EL) total iron. The variation in Fe content extends to other siderophile elements as well, reflecting general fractionation between siderophile and lithophile elements in the solar nebula.

As their name implies, *carbonaceous chondrites* differ from other chondrite classes in being rich in carbon compounds, including a variety of organic compounds, most notably amino acids (indeed, 70 different amino acids have been identified in *Murchison* – there are only 20 biological amino acids), water, and other volatiles. They are also enriched in hydrogen (present mainly in hydrated silicates) and nitrogen and somewhat poorer in Si compared with other chondrites. The composition of carbonaceous chondrites matches that of the Sun even more closely than chondrites as a whole. They are subdivided into eight groups, with the name of each subgroup derived from a type example: CI (*Ivuna*), CM (*Mighei*), CV (*Vigarano*) and CO (*Ornans*), CK (*Karoonda*), CR (*Renazzo*), CB (*Benccubbin*), CH (High Iron). Of these, meteorites from the CM, CV, and CO groups are the most common. CI chondrites are rare but are nevertheless of great significance. Perhaps ironically, CI chondrites lack chondrules and CAIs. Nevertheless, they are rich in carbonaceous matter and are compositionally similar to other chondrites and hence classed with the carbonaceous chondrites. Figure 10.14 compares the abundances in CI chondrites with those in the solar photosphere. As may be seen, the CI chondrite compositions match that of the Sun remarkably well for all but H (not shown), C, N, the rare gases, and Li.* CI chondrites seem to be collections of bulk nebular dust that escaped the high-temperature processing and attendant chemical fractionations that affected material in other chondrites.

Because it matches the composition of the Sun so well, the composition of CI chondrites is taken to represent the composition of the

---

* These elements will never fully condense from the solar nebula; instead large fractions will remain in the gas phase. These elements are thus termed *noncondensable*. Li is depleted in the Sun compared with chondrites because, as we saw, it is consumed in nuclear reactions in the Sun.

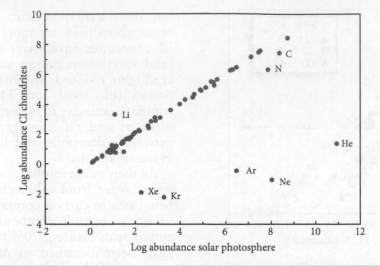

Figure 10.14   Abundances of the elements in CI carbonaceous chondrites versus their abundances in the Sun's photosphere. Abundances for most elements agree within analytical error except noncondensing elements and Li. Data from Lodders (2010).

solar system as a whole for the condensable elements. Table 10.2 lists the concentrations of the elements in the CI chondrites and in the solar system as a whole. The latter are derived from a combination of CI chondrites and spectroscopic measurements of the Sun's photosphere. You might ask, why use the CI chondritic composition for the solar system when data for the composition of the Sun are available? The answer is that while the composition of the Sun can be determined spectrographically, this technique is not very accurate for trace elements (indeed, there are no reliable photosphere measurements for some trace elements), and, as we learned in Chapter 7, most elements are trace elements. The terms *chondritic* and *solar composition* are thus nearly synonymous. Unfortunately, meteorites of this group are rare, likely a consequence of their fragile nature, and few falls, most notably *Orgueil* (which fell in the French town of that name in 1864) exists in enough quantity for complete and detailed chemical analysis.

In addition to these groups, there are two minor classes of chondrites, sometimes grouped together as "other chondrites." These include the R chondrites, of which there are only 19 known specimens. The R chondrites are so named for the type example *Rumuruti*, which fell in Kenya in

1934. The R chondrites are the opposite of the E chondrites in the sense that they are highly oxidized with practically no free metal. They typically contain fewer chondrules than ordinary chondrites and many are highly brecciated, suggesting they came from near the surface of their asteroid parent body. The K group consists of just three specimens. They are rich in the iron sulfide, troilite, show numerous primitive, armored chondrules, and have a unique chemical and oxygen isotope composition.

Other compositional differences between chondrite classes, in addition to metal contents and oxidation state, are the concentrations of *refractory* elements and the concentrations of *volatile* elements. Let's pause here to define these terms. Si, Mg, and Fe are the most abundant condensable elements in chondrites. In a hot gas of solar composition, these three elements would condense at similar temperatures (50% of the Si, Mg, and Fe would condense between 1340 and 1311 K at 0.1 Pa). Fifty percent of Al will condense at 1650 K, whereas 50% of Na will not condense until 970 K. We refer to elements that condense at temperatures higher than Si, Mg, and Fe, such as Al, as *refractory* and elements that condense at lower temperatures, such as Na, as *volatile*. This classification is sometimes further refined into

Table 10.2 Abundances of the elements.

Element		Solar system abundance*	Mean CI abundance†	σ (%)	Element		Solar system abundance*	Mean CI abundance†	σ (%)
1	H	1×10^{12}	1.97%	10	44	Ru	60.3	0.69	5
2	He	8.41×10^{10}	0.00917		45	Rh	12.6	0.127	5
3	Li	1.91×10^{3}	1.45	10	46	Pd	46.8	0.560	4
4	Be	2.09×10^{1}	0.0219	7	47	Ag	16.6	0.201	9
5	B	6.46×10^{2}	0.775	13	48	Cd	53.7	0.674	7
6	C	2.45×10^{8}	3.48%	10	49	In	6.0	0.078	5
7	N	7.24×10^{7}	0.295%	15	50	Sn	123	1.63	15
8	O	5.37×10^{8}	45.9%	10	51	Sb	10.7	0.145	14
9	F	2.75×10^{4}	58.2	16	52	Te	158.7	2.28	7
10	Ne	1.12×10^{8}	0.00018		53	I	37.2	0.53	20
11	Na	1.95×10^{6}	4692	9	54	Xe	186.2	1.74×10^{-4}	
12	Mg	3.47×10^{7}	9.54%	4	55	Cs	12.6	0.188	6
13	Al	2.88×10^{6}	0.840%	6	56	Ba	151.4	2.42	5
14	Si	3.39×10^{7}	10.70%	3	57	La	15.5	0.241	3
15	P	2.82×10^{5}	985	8	58	Ce	39.8	0.619	3
16	S	1.45×10^{7}	5.35%	5	59	Pr	5.9	0.0939	3
17	Cl	1.78×10^{5}	698	15	60	Nd	29.5	0.474	3
18	Ar	3.16×10^{6}	0.00133		62	Sm	9.1	0.154	3
19	K	1.29×10^{5}	546	9	63	Eu	3.4	0.0588	3
20	Ca	2.04×10^{6}	0.911%	6	64	Gd	12.3	0.207	3
21	Sc	1.17×10^{3}	5.81	6	65	Tb	2.2	0.0380	3
22	Ti	8.51×10^{4}	447	7	66	Dy	13.8	0.256	3
23	V	9.77×10^{3}	54.6	6	67	Ho	3.1	0.0564	3
24	Cr	4.478×10^{5}	2623	5	68	Er	8.9	0.166	3
25	Mn	3.16×10^{5}	1916	6	69	Tm	1.4	0.0261	3
26	Fe	2.88×10^{7}	18.66%	4	70	Yb	8.7	0.169	3
27	Co	7.94×10^{4}	513	4	71	Lu	1.3	0.0250	3
28	Ni	1.66×10^{6}	1.09%	7	72	Hf	5.4	0.107	3
29	Cu	1.86×10^{4}	133	14	73	Ta	0.72	0.015	10
30	Zn	4.47×10^{4}	309	4	74	W	4.68	0.096	104
31	Ga	1.26×10^{3}	9.62	6	75	Re	1.91	0.040	5
32	Ge	3.89×10^{3}	32.6	9	76	Os	23.4	0.495	5
33	As	2.09×10^{2}	1.74	9	77	Ir	22.9	0.469	5
34	Se	2.29×10^{3}	20.3	7	78	Pt	43.7	0.925	5
35	Br	3.62×10^{2}	3.26	15	79	Au	6.61	0.148	12
36	Kr	1.91×10^{3}	5.22×10^{-5}		80	Hg	15.5	0.35	50
37	Rb	2.40×10^{2}	2.32	8	81	Tl	6.17	0.140	11
38	Sr	7.94×10^{2}	7.79	7	82	Pb	114.8	2.62	8
39	Y	1.58×10^{2}	1.46	5	83	Bi	4.68	0.110	9
40	Zr	3.72×10^{2}	3.63	5	90	Th	1.20	0.030	7
41	Nb	26.3	0.283	10	92	U	0.30	0.0081	7
42	Mo	87.1	0.961	10					

*Atoms relative to $H = 10^{12}$, derived from a combination of CI meteorites and solar photosphere measurements. Data from Lodders (2010).

†Concentrations in ppm by mass unless otherwise indicated. Data from Palme (2017).

σ is the estimated uncertainty in the concentrations in CI chondrites.

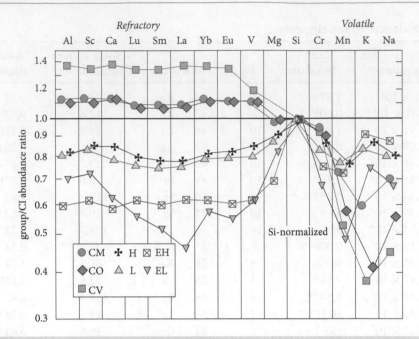

Figure 10.15   Silicon- and CI-normalized abundances of key elements in the main classes of chondrites. Elements are arranged from left to right in order of decreasing condensation temperature. Si concentration plots at 1 in each case. CI chondrites would plot as a horizontal line at 1. After Wasson and Kallemeyn (1988).

moderately volatile, highly refractory, and so on. Figure 10.15 illustrates the compositional variation between chondrite classes for several lithophile elements, which are arranged going from most refractory (Al) to most volatile (Na). Concentrations are shown normalized both to Si and to CI chondrites. Thus, Si plots at 1 for all classes, and CI chondrites would plot as a horizontal line at 1. We can see that all other chondrite groups are depleted in volatile elements compared with CI chondrites. We can also see that the other carbonaceous chondrites are enriched in refractory elements compared with CI chondrites, and ordinary and enstatite chondrites are depleted in refractory elements compared with CI chondrites. Carbonaceous chondrites are enriched in the most volatile elements (H, C, N, etc. – not shown on this plot) compared with other chondrites. Thus, ordinary and enstatite chondrites are depleted in both the most refractory and the most volatile elements compared with CI chondrites. An interesting and important feature of this diagram is that despite variations in the absolute levels of refractory elements, the ratios of refractory elements to each other remain nearly constant across all chondrite classes except EL. This is

an important observation generally taken as evidence that nebular processes were generally not able to fractionate the most refractory elements.

The compositional variations among chondrite classes illustrated in Figures 10.14 and 10.15 reflect variations in conditions in the solar nebula. Nevertheless, all meteorites have undergone subsequent processing on their parent bodies. Van Schmus and Wood (1967) devised a simple way to indicate the degree of parent body processing by dividing chondrites into petrologic types 1 through 6, based on increasing degree of metamorphism and decreasing volatile content. Types 1 and 2 have experienced low-temperature (T ≤ 50° C) aqueous alteration, while types 4–5 have experienced metamorphism at temperatures up to 800° C and up to 1000° C for type 6. Type 3 objects have undergone the least parent body processing and are in this sense the most primitive, but even they may have experienced metamorphic temperatures up to 300° C in some cases (Scott and Krot, 2014). Petrologic types are sometimes further subdivided with an additional digit, for example, 3.7. Table 10.3 summarizes the Van Schmus and Wood classification scheme.

Table 10.3 Van Schmus and Wood petrographic classification of chondrites (after Van Schmus and Wood, 1967).

Type	1	2	3	4	5	6
	←Aqueous alteration			Increasing temperature→ — Thermal metamorphism →		
I. Olivine and pyroxene homogeneity		Greater than 5% mean deviation		Less than 5% mean deviation		Uniform
II. Structural state of low-Ca pyroxene		Predominantly monoclinic		Abundant monoclinic crystals		Orthorhombic
III. Development of secondary feldspar		Absent		Predominantly as micro-crystalline aggregates		Clear, interstitial grains
IV. Igneous glass		Clear and isotropic primary glass; variable abundance		Turbid if present		Absent
V. Metallic minerals (max. Ni content)		Taenite absent or minor (<20%)		Kamacite and taenite present (>20%)		
VI. Average Ni of sulfide minerals		>0.5%		<0.5%		
VII. Chondrules	No chondrules	Very sharply defined chondrules		Well-defined chondrules	Chondrules rarely seen	Poorly defined chondrules
VIII. Texture of matrix	All fine-grained	Much opaque matrix	Opaque matrix	Transparent microcrystalline matrix		Recrystallized matrix
IX. Bulk carbon content	3–5%	0.8–2.6%	0.2–1%		<0.2%	
X. Bulk water content	18–22%	2–16%	0.3–3%		1.5%	

The petrologic types are used together with the above groups to classify meteorites as to origin and metamorphic grade, e.g., CV3. The petrologic type is correlated to a certain degree with chondrite class. Type 1 is restricted to CI and a few rare CM and CR chondrites, and petrologic grades above 1 are not found among CI chondrites. Type 2 is restricted to CM and CR chondrites. Petrologic types 5 and 6 occur only in CK, ordinary, E, and R chondrites. Thermal metamorphism results in equilibration of the various minerals present in meteorites; consequently, petrologic types 4–6 are sometimes termed *equilibrated*, while those of low petrologic type are sometimes referred to as *unequilibrated*.

Chondrites can also be classified according to the degree of shock they have experienced. Class *S1* indicates no shock, class *S6* indicates very strong shock, with some shock melting present.

To gain a better understanding of what chondrites are and how they help us to understand the processes in the solar nebula, we briefly review the principal components of chondrites in the following sections.

10.3.1.2 Chondrules

Chondrules are usually a few tenths of a mm to a few mm in diameter (Figure 10.12). Mean size varies between chondrite classes (Table 10.1), but is typically around 0.5 mm. In the least metamorphosed meteorites, they consist of mixture of crystals and glass. Most are porphyritic, with relatively large olivine or pyroxene crystals in a fine-grained or glassy matrix. Nonporphyritic chondrules can consist of cryptocrystalline material or radial pyroxene or barred olivine, all of which suggest rapid crystallization. Olivine and Ca-poor pyroxene (enstatite, hypersthene) are by far the dominant minerals, with troilite (FeS), kamacite (FeNi alloy), Ca-rich pyroxene (pigeonite, diopside), Mg-Al spinel, chromite, and feldspar being less abundant. Some rare Al-rich chondrules have compositions similar to CAIs, which are discussed below, and consist of some of the same refractory minerals (plus glass). Some chondrules contain relict mineral grains and a few contain relict CAIs. Chondrules have remnant magnetism that was acquired as they cooled through their Curie point in the presence of a magnetic field,

indicating the presence of such a field in the solar nebula. From the number of compound chondrules (two chondrules fused together) and those having indentations suggestive of collisions with other chondrules, the chondrule density was as high as a few per m^3 at times and places in the solar nebula. While "dents" are observed in chondrules, microcraters produced by high-velocity impact are absent. Many chondrules are compositionally zoned, and most chondrules contain nuclei of relict crystals. Many are rimmed with fine-grained dark secondary coatings of volatile-rich material broadly similar in composition to the chondrite matrix.

Chondrules make up nearly half the mass, on average, of primitive meteorites. Therefore, understanding their origin is critical to understanding processes in the solar nebula because the dust in the nebula, which is the raw material for terrestrial planets, was apparently processed into chondrules. The presence of glass and their spheroidal shape indicates that chondrules represent melt droplets, as has been realized for at least 100 years. How these melts formed has been more difficult to understand. The main problem is that at the low pressures that must have prevailed in the solar nebula, liquids are not stable: solids should evaporate rather than melt. Chondrules seem to have been heated quite rapidly, at rates of 10^4 K/hr or more to peak temperatures of 1650 to 1850 K, and then cooled rapidly as well, at rates of 100–1000 K/hr. In most cases, peak temperatures were maintained for only minutes and they apparently cooled completely in hours to days. These inferences are based on compositional zonation of minerals and experimental reproduction of textures, strengthened by other experiments that show chondrules would have evaporated if they existed in the liquid state any longer than this. Though cooling was rapid, it was considerably slower than the rate that would have resulted from radiative cooling in open space. All these observations indicate they formed very quickly and may never have reached equilibrium. Some chondrules show evidence of having experienced multiple melting events; a few contain relict CAI grains, indicating that CAIs made up part of "dust" that was ultimately recycled into chondrules.

Over the past 100 years or so, many mechanisms for chondrule formation have been proposed. These include formation through volcanism on planets or asteroids, impact melting resulting from collisions of planets or planetesimals, condensation from hot nebular gas, and transitory heating of preexisting nebular or interstellar dust. At present, there is a consensus that they formed by transitory heating of "cool" (<1000 K) nebular dust. Possibilities include collisions of small (<1 m) bodies, frictional heating of dust traveling through gas during infall, lightning, energy released by magnetic flares or reconnection of magnetic field lines, and radiative heating resulting from high-velocity outflows during the T-Tauri phase (see below) of the protosun. At present, the leading hypothesis is that most chondrules were produced in shock waves in the solar nebula. Such shock waves could have been produced by accretion shocks, bow shocks from planetesimals, infalling clumps, interactions with passing stars, or spiral density waves. The latter result from uneven distribution of mass in the nebula and resulting gravitational torques. They can be thought of as somewhat analogous to spiral arms of galaxies. Shocks produce heating because gas is accelerated in the shocks more rapidly than dust, so that the dust is heated by gas drag. Numerical modeling by Desch and Connolly (2002) has shown that shock waves can produce the rapid heating and cooling that chondrules apparently experienced. Such shock waves must have been common in the inner solar system because 40% or so of the dust that ultimately formed the asteroids and the terrestrial planets was processed into chondrules. Some chondrules might have formed by other mechanisms: for example, unusual features of chondrules in CB chondrites indicate they formed in a high-energy collision between planetary embryos (Krot et al., 2005).

10.3.1.3 Calcium–aluminum inclusions

Ca-Al inclusions or CAIs are submillimeter to centimeter-sized clasts consisting primarily of calcium- and aluminum-rich minerals. They were first described only in 1968 and were at first thought to be restricted to just the CO, CV, and CM chondrites. However, they have now been recognized in essentially all chondrite classes, except CI, although they are rare, and typically very small, in all except the carbonaceous chondrites. The principal minerals are spinel ($MgAl_2O_4$), melilite ($Ca_2Al_2SiO_7$–$Ca_2Mg_2Si_2O_7$), perovskite ($CaTiO_3$), hibonite ($CaAl_{12}O_{19}$), anorthite ($CaAl_2Si_2O_8$), and calcic pyroxene ($CaMg_2Si_2O_6$). Forsteritic olivine is also common in one subtype (forsterite, of course, is not a Ca–Al mineral *per se*, although some forsterites in CAIs are relative rich in the Ca olivine end member, monticellite). Ni–Fe alloys (taenite, kamacite, awaruite) and a wide variety of other minerals may also be present as minor or trace phases.

CAIs have attracted great interest for several reasons. First, they consist of those minerals thermodynamically predicted to condense first as hot (>1200 to 1300 K at 1 to 100 Pa) gas of solar composition cools or the last solids to evaporate as it heats up. Consistent with this, CAIs are remarkably poor in more volatile elements (except where they have been altered by secondary processes on parent bodies) and they are rich in refractory trace elements such as Ba, Th, Zr, Hf, Nb, Ta, Y, and the rare earths. They sometimes contain microscopic metallic nuggets, called *fremdlinge,* that consist of metals, such as Re, Os, Re, Pt, Ir, W, and Mo, which condense at temperatures even higher than the Ca–Al minerals. Second, as we shall see in a subsequent section, they are the oldest dated objects in the solar system, pre-dating other chondritic components by up to several million years.

A number of different types of CAIs have been recognized. The most common type is the so-called *spinel–pyroxene inclusions,* which are typically much smaller than 1 mm in diameter (except in CV3 meteorites). *Type A* CAIs are less than 1 mm in diameter (except in CV3 meteorites) and consist primarily of melilite ($CaMg(Al_2SiO_7)$) intergrown with hibonite ($Ca(Al,Ti,Mg)_{12}O_{19}$), spinel ($MgAlSiO_4$), perovskite ($CaTiO_3$), and noble metal nuggets. Both the Type A and the spinel–pyroxene inclusions can be more than 1 cm in diameter in CV3 meteorites. *Type B* inclusions are typically larger, up to 1 cm in diameter, more varied in composition, and are restricted to CV3 meteorites. *Type C* CAIs, which are rare, consist mostly of spinel, calcic pyroxene, and anorthite. A fifth type, the so-called *hibonite-rich CAIs,* consist of hibonite, sometimes accompanied by spinel and perovskite. A sixth type, the *hibonite–silicate spherules*, consist of hibonite intergrown with aluminous pyroxene and perovskite embedded in glass of aluminous pyroxene composition. All types of CAIs are typically surrounded by an accretionary rim several tens of μm thick, typically consisting of the same minerals as are present in the interior, that appears to have resulted from high-temperature gas–solid or gas–melt interaction.

Although CAIs as a whole have compositions that approximate that of the highest temperature condensate of a gas of solar composition, their compositions do not match a condensation trend exactly. Furthermore, the textures of most types of CAIs indicate that they have experienced complex histories, including episodes of melting, evaporation, reaction with nebular gas and finally aqueous alteration and/or metamorphism on parent bodies (Grossman, 2010). However, some CAIs, notably the fluffy Type A inclusions, do have compositions, including fractionated rare earth patterns, and textures suggesting they are indeed high-temperature condensates of nebular gas. It is possible that many other CAIs began as such condensates and they experienced subsequent episodes of transient high temperatures. Other CAIs may have formed as evaporative residues. Regardless of the details, CAIs provide evidence that some nebular dust experienced transient heating events with temperatures reaching 1700 K. This is much hotter than the nebula should have ever been at the position of the asteroid belt. For this and other reasons, there is an emerging consensus that CAIs formed close to the Sun and were subsequently cycled back out into the deeper nebula, perhaps by *X-winds,* which we will discuss later in the chapter.

10.3.1.4 Amoeboid olivine aggregates

Amoeboid olivine aggregates (AOAs) are, as their name implies, aggregates of anhedral forsteritic olivine with lesser amounts of Fe–Ni metal, spinel, aluminous diopside, and rare anorthite and melilite. They are fine-grained (5–20 μm) and the aggregates have dimensions similar to those of chondrules in the same meteorite. Some contain melted CAIs. In some cases, the olivine is partially replaced by enstatite. Some have igneous textures, suggesting they have been partially melted. AOAs most likely represent aggregates of grains that condensed from nebular gas at high temperature. They may well have formed in the same environment, albeit at lower temperature, as CAIs.

10.3.1.5 The chondrite matrix

The matrix of chondrites is dark, FeO- and volatile-rich material that is very fine-grained (typical grain size is about 1 μm). It can be quite heterogeneous, even on a 10 μm scale. It also varies between meteorite classes, with an order of magnitude variation in Mg/Si, Al/Si and Na/Si. The primary constituents appear to be Fe- and Ca-poor pyroxene and olivine and amorphous material, but magnetite, Fe-metal, and a wide variety of silicates, sulfides, carbonates, and other minerals are also present. In the most volatile-rich meteorites the olivine and pyroxene have been altered to serpentine and chlorite; in the carbonaceous chondrites, carbonaceous material is present in substantial quantities. On the whole, the composition of the matrix is complementary to that of the chondrules: whereas the latter are depleted in Fe and volatiles, the former are enriched in them. Very significantly, the matrix includes grains of SiC, graphite, diamond, and other phases of anomalous isotopic composition. These "presolar grains" are of great significance and we will discuss these isotopic variations in greater detail in a subsequent section.

10.3.2 Differentiated meteorites

The differentiated meteorites are products of melting on asteroid parent bodies. They are igneous rocks with igneous textures, although in some cases, brecciation may be the dominant texture.

10.3.2.1 Achondrites

While all the chondrites seem reasonably closely related, the achondrites are a more varied group. The *Acapulcoites, Lodranites, Winonaites,* and *Urelites* form a group called the *primitive achondrites* because they resemble chondrites in composition and mineralogy. Beyond that, they are quite diverse. Meteorites of the first three groups are extremely rare. They represent chondritic material that has experienced extreme metamorphism and low-degree partial melting. In a few cases, relict chondrules have been identified, providing further evidence of their primitive nature. *Urelites*, which are both more common and more diverse than the other primitive achondrites, consist of olivine, pyroxene, and a few metal grains plus a percent or so carbon, present as graphite and diamond, the latter a product of shock metamorphism produced by impacts. Their origin is problematic; it is possible they have several origins. Some are partial melting residues like other primitive achondrites; others appear to be highly fractionated igneous rocks. *Brachinites* are also sometimes included in the primitive achondrites.

Upon heating, chondritic material will first form a metal-sulfide melt, which is much denser than the mainly silicate residual solid. This metal will drain out of the matrix and ultimately form a core at the center of the body. Upon further heating, a silicate partial melt will form, into which the more incompatible elements will partition. Thus, primitive achondrites are variously depleted in metal and incompatible elements relative to chondrites.

Figure 10.16 illustrates the compositional relationship between the various achondrite groups. Primitive achondrites tend to have Mn/Mg ratios similar to those of chondrites, while the remaining achondrites have superchondritic Mn/Mg ratios and Fe/Mg ratios higher than primitive achondrites. This, as well as their texture and mineralogy, indicates that those remaining achondrite groups originated through igneous processes. During partial melting of chondritic material, these three elements partition into the melt in the order Mn > Fe > Mg. Subsequent fractional crystallization increases the Fe/Mg ratio but has little further effect on the Mn/Fe ratio.

Howardites eucrites, and *diogenites*, collectively termed HED meteorites, are among the most common achondrites (there are nearly 2000 specimens), comprising about 4% of all falls, and as described below are very likely derived from the large asteroid 4 Vesta. Diogenites are Ca-poor and consist principally of hypersthene (orthopyroxene), which is accompanied by minor amounts of olivine and plagioclase. They are

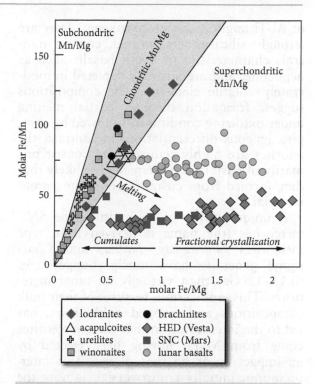

Figure 10.16 Variation in Fe/Mn and Fe/Mg ratios in achondrites and lunar basalts. Primitive achondrites are variably depleted in Fe, but have Mn/Mg ratios similar to chondrites. The HED and SNC chondrites, like lunar basalts, have superchondritic Mn/Mg ratios and higher Fe/Mg ratios than primitive achondrites as a consequence of their igneous origin. After Goodrich and Delaney (2000).

coarse-grained and their texture suggests they formed as cumulates in a magma chamber. The Ca-rich achondrites include the *eucrites* and *howardites*. The eucrites are Ca-rich and resemble lunar and, to a lesser extent, terrestrial basalts, and are also called *basaltic achondrites*, and like terrestrial basalts they consist primarily of plagioclase and pyroxene. The howardites are extremely brecciated and are a heterogeneous mixture of eucrite and diogenite material. They also contain clasts of carbonaceous chondritic material, other xenolithic inclusions, and impact melt clasts. Their brecciated character suggests they were part of the regolith, or surface, of their asteroid parent body.

Two final groups are the *angrites* and the *aubrites*. The angrites (of which there are only 12 specimens, the name of the class being derived from *Angra dos Reis*) consist mostly of Al-Ti augite (Ca-rich pyroxene). They are strongly silica-undersaturated, contain minerals characteristic of alkali basalts such as nepheline, and are strongly depleted in moderately volatile elements. The compositions suggest formation through partial melting under oxidizing conditions, followed by complex igneous differentiation. In contrast, the aubrites are highly reduced and consist primarily of enstatite. It seems highly likely that they formed from enstatite chondrite parent material.

A unique group of achondrites, the *SNC* meteorites (the name is derived from type meteorites *Sherogotty, Nakla,* and *Chassigny*) generally have much younger ages (0.15–1.5 Ga) than virtually all other meteorites. This, and certain features of their bulk compositions and trapped noble gases, has led to the interpretation that these meteorites come from Mars, having been ejected by an impact event on that planet. This interpretation, initially controversial, is now the consensus view. There are 201 specimens of this class. Finally, there are 326 meteorites of lunar origin. A handful of achondrites are unique and cannot be assigned to any class.

10.3.2.2 Irons

Iron meteorites were originally classified based largely on phase and textural relationships. Compositionally, they all consist primarily of Fe–Ni alloys with lesser amounts of (mainly Fe–Ni) sulfides. Octahedral taenite, one of the Fe–Ni alloys, is the stable Fe-Ni metal phase at $T > 900°C$ (Figure 10.17). At lower temperature, kamacite, a Ni-poor Fe–Ni alloy, exsolves on the crystal faces of the octahedron. If the Ni content falls below 6%, all the metal converts to kamacite at lower temperature. Thus, the phases and textures of iron meteorites are related to their composition and cooling history. Iron meteorites consisting only of kamacite are named *hexahedrites*. If Ni exceeds 6%, some taenite persists and the overall pattern is octahedral (= *octahedrites*), producing what is known as a *Widmanstätten* pattern. At low Ni contents, kamacite dominates and forms large crystals (*coarse octahedrites*). At higher Ni, kamacite and crystal size diminish (*fine and medium octahedrites*). *Ataxites* are Ni-rich (>14%) iron meteorites consisting of a fine-grained

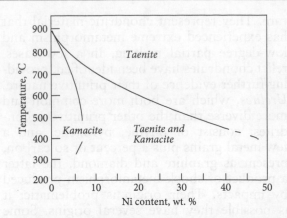

Figure 10.17 Phase diagram for iron–nickel alloy. After Wasson (1974).

intergrowth of kamacite and taenite. The 20% or so of irons with silicate inclusions form a separate class.

Wasson (1985) reclassified the irons based on composition, specifically on Ga–Ni and Ge–Ni abundances. The classes were named I–IV, based on decreasing Ga and Ge. Subgroups within these classes were named A, B, and so on. Wasson and Kallemeyn (2002) and Wasson (2011) subsequently found that gold concentrations delineated these classes better than did nickel concentrations. Figure 10.18 illustrates the classification and chemical variation among iron meteorites based on the relationships between gold, cobalt, and gallium concentrations.

The chemical variations within individual subclasses of irons are consistent with those produced by fractional crystallization of metallic liquid. The clear implication is that all irons from an individual subclass come from a single parent body. Perhaps some 75 parent bodies are represented by the suite of analyzed irons. There is a general consensus that iron meteorites, with a few notable exceptions, represent the cores of asteroids or *planetesimals*. Cooling rates, estimated from textures and diffusion profiles, are typically in the range of a few tens of degrees per million years. This slow cooling indicates the irons formed in the interior of bodies with diameters in the range of a few tens to a few hundreds of kilometers. A few classes of irons, most notably the IAB, may represent impact melts rather than segregated cores. Based on the oxygen isotope composition of their silicate

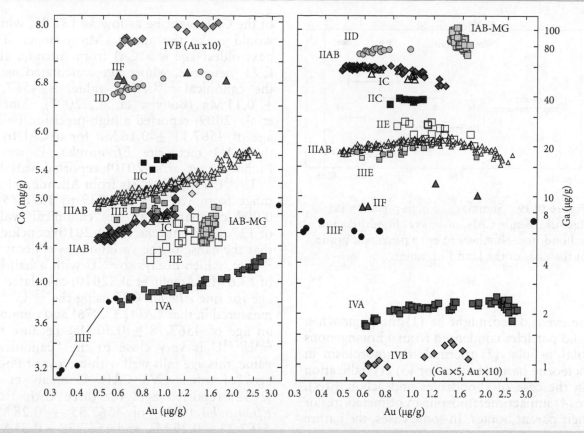

Figure 10.18   Co–Au and Ga–Au plots illustrating the compositional distinctions and variations in groups of iron meteorites. IAB-MG refers to the "main group" of IAB irons, while other subgroups of the IAB irons show more scatter. From J.T. Wasson (unpublished), reproduced by permission.

inclusions, Wasson and Kallemeyn (2002) speculated that the main group of IAB irons represent a single melt body formed by an impact on a carbonaceous chondrite asteroid, and that the elemental trends were formed by crystal separation during downward flow in the asteroid.

10.3.2.3   Stony-irons

The main classes of stony-irons are the *pallasites* and the *mesosiderites*. Pallasites consist of a network of Fe–Ni metal with nodules of olivine. They probably formed at the interface between molten metal and molten silicate bodies, with olivine sinking to the bottom of the silicate magma. Mesosiderites consist of an odd pairing of metal and silicate. The silicate portion is very similar to diogenites – brecciated pyroxene and plagioclase – and a genetic relationship is confirmed by oxygen isotopes (discussed later). The

metal fraction seems closely related to IIIAB irons. It is possible they formed as the result of a collision of two differentiated asteroids, with the liquid core of one asteroid mixing with the regolith of the other.

10.4   TIME AND THE ISOTOPIC COMPOSITION OF THE SOLAR SYSTEM

10.4.1   Meteorite ages

10.4.1.1   Conventional methods

Meteorite ages are generally taken to be the age of solar system. Before we discuss meteorite ages in detail, we need to consider the question of precisely what event is being dated by radiometric chronometers. In Chapter 8, we found that radioactive clocks record the last time the isotope ratio of the daughter element (e.g., $^{87}Sr/^{86}Sr$) was homogenized. This is usually some thermal event. In the context of what we know of early solar system history,

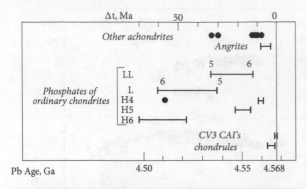

Figure 10.19 Summary of high-precision Pb ages of Allende CAIs, ordinary chondrites, and achondrites. Numbers refer to petrologic grade at the ends of the L and LL range.

the event dated might be (1) the point when solid particles condensed from a homogenous solar nebula, (2) thermal metamorphism in meteorite parent bodies, or (3) crystallization (in the case of chondrules and achondrites), or (4) impact metamorphism of meteorites or their parent bodies. In some cases, the nature of the event being dated is unclear.

The most precise ages of meteorites have been obtained using the U-Pb chronometer (Figure 10.19). Advances in analytical techniques have remarkably improved precision over the last decade or so, to the point that ages with uncertainties of only a few 100,000 years can be obtained. However, some of the issues that traditionally plague geochronology come into focus, including lack of complete initial isotopic homogeneity and deviations from closed system behavior. In addition, new issues arise, including uncertainties in half-lives of the parents and uncertainty in, as well as variation of, the $^{238}U/^{235}U$ ratio (Amelin et al., 2009). Progress is being made in resolving these issues, but further research remains necessary.

The oldest Pb–Pb ages come from CAIs. CAIs are rich in refractory elements like U and depleted in volatile elements like Pb so they are good targets for U–Pb dating. At present, the oldest high-precision date is 4568.2 ± 0.3 Ma for a CAI from the CV3 meteorite *NWA2364* calculated using the "canonical" $^{238}U/^{235}U$ ratio of 137.88 (Bouvier and Wadhwa, 2010). Bouvier and Wadhwa (2010) speculated that the $^{238}U/^{235}U$

of the CAI might be as low as 137.81, which would make the age 0.3 Ma younger. The next oldest age is a CAI from *Allende*, also CV3 meteorite, whose age, calculated using the canonical $^{238}U/^{235}U$ value, is 4567.59 ± 0.11 Ma (Bouvier et al., 2007). Amelin et al. (2009) reported a high-precision Pb-Pb age of 4567.11 ± 0.16 Ma for a CAI from the CV3 meteorite *Efremovka*. However, Brennecka et al.'s (2010) reported variable $^{238}U/^{235}U$ ratios in CAIs from Allende, which range from 137.409 ± 0.039 to 137.885 ± 0.009, compared with the "canonical" value of 137.88. Brennecka et al. (2010) concluded that the cause of the variability was decay of ^{247}Cm, which decays to ^{235}U with a half-life of 13.6 Ma. Amelin et al. (2010) calculated an age for one *Allende* CAI using the $^{238}U/^{235}U$ measured in that CAI (137.876) and obtained an age of 4567.18 ± 0.50 Ma. Because the $^{238}U/^{235}U$ is very close to the "canonical" value, this age falls well within error of Pb-Pb ages of other CV3 CAIs. Connelly et al. (2012) reported ages for three CAIs from *Efremovka* (CV3) of 4567.35 ± 0.28 Ma, 4567.23 ± 0.29 Ma, and 4567.38 ± 0.31 Ma. Bouvier and Wadhwa (2010) speculated that the slightly older age of *NWA2364* inclusions compared with those of *Allende* and *Efremovka* might reflect aqueous alteration of the latter after incorporation into the CV3 parent body.

Recent high precision Pb-Pb studies of chondrules using measured $^{235}U/^{238}U$ ratios reveal a range of ages from 4567.7 ± 0.6 Ma, essentially the same as CAI's, down to 4562.7 ± 0.5 Ma for *Gujba* (CB3) (Connelly et al., 2012, Bollard et al., 2012, 2017), with most falling in the range of 4567 to 4564 Ma.

Phosphates also have high U/Pb ratios and high-precision ages have been obtained for a variety of equilibrated (i.e., petrologic classes 4–6) ordinary chondrites, whose ages, recalculated based on $^{238}U/^{235}U$ = 137.786, range from 4563 to 4498 Ma (Göpel et al., 1994; Binova et al., 2007; Bouvier et al., 2007; Blackburn et al., 2017). The phosphates are thought to have formed during metamorphism; these ages represent the age of metamorphism of these meteorites. The oldest of these meteorites was H4 chondrite *Ste. Marguerite*. The age of CAIs from CV3 meteorites thus seem 3 Ma older than the oldest precise ages obtained on ordinary chondrites. No attempt has been made at high-precision dating of CI chondrites, as they are too fine-grained to separate phases.

Among achondrites, the chronology of the angrites is perhaps best documented. The oldest high-precision Pb-Pb age is 4563.36 ± 0.34 Ma using the measured $^{238}U/^{235}U$ for this meteorite of 137.778 and other high precision ages range downward to 4557.01 ± 0.27 Ma. Thus, differentiation, cooling, and crystallization of the angrite parent body apparently lasted some 6 million years. Wadhwa et al. (2009) reported an age of 4566.5 ± 0.2 Ma for unusual basaltic achondrite, *Asuka 881394*. Bouvier et al. (2011) determined an age of 4562.89 ± 0.59 Ma for another unusual basaltic achondrite, *NWA2976*. *Ibitira*, a unique unbrecciated eucrite, has an age of 4556 ± 6 Ma. These ages are similar to those of chondrites, demonstrating that the parent body of these objects formed, melted, and differentiated, and the outer parts crystallized, within a very short time interval. Not all achondrites are quite so old, however. A few other high-precision ages (those with quoted errors of less than 10 Ma) are available and they range from this value down to 4529 ± 5 Ma for *Nuevo Laredo* and 4510 ± 4 Ma for *Bouvante*. Thus, the total range of high-precision ages in achondrites is about 50 million years, a range that reflects melting, metamorphism and impacts rather than parent body formation.

Iron meteorites appear to be similarly old. Blichert-Toft et al. (2010) determined Pb–Pb ages in troilite (FeS) of 4565.3±0.1 Ma and 4544±7 Ma and for in the IVA iron *Muonionalusta* and *Gibeon*, respectively. Smoliar et al. (1996) reported Re–Os ages of 4558 ± 12 and 4537 ± 8 Ma for IIIA and IIA irons, respectively. Re–Os ages of other irons from 4456 to 4569 Ma but are of lower precision.

K–Ar ages of meteorites are often much younger. This probably reflects Ar outgassing as a result of collisions, and the ages probably date impact metamorphism.

10.4.1.2 Extinct radionuclides

There is abundant and compelling evidence that certain short-lived nuclides once existed in meteorites. This evidence consists of anomalously high abundances of the daughter nuclides in certain meteorites, and fractions of meteorites that correlate with the abundance of the parent element. The first of these to be discovered was the ^{129}I–^{129}Xe decay (Reynolds, 1960). Since then, 18 other *extinct radionuclides* have been discovered. The most significant of these are listed in Table 10.4. These provide evidence of nucleosynthesis occurring shortly before the solar system formed. To understand why, consider the example of ^{129}I. It decays to ^{129}Xe with a half-life of ~16 Ma. Hence 16 Ma after they were created, only 50% of the original atoms of ^{129}I would remain. After two half-lives or 32 Ma, only 25% would remain, after four half-lives or 64 Ma only 6.125% of the original ^{129}I would remain, and so on. After 10 half-lives, or 160 Ma, only $\frac{1}{2}^{10}$ (0.1%) of the original amount would remain. Anomalously high abundance of ^{129}Xe relative to other Xe isotopes that correlate with iodine concentration in a meteorite indicates some ^{129}I was present when the meteorite, or its parent body, formed. From this we can conclude that ^{129}I had been synthesized not more than roughly 10^8 years before the meteorite formed. This time constraint is further reduced by the identification of even shorter-lived radionuclides such as ^{26}Al which decays to ^{26}Mg. That ^{26}Al is the source of the ^{26}Mg is evidenced by the correlation between ^{26}Mg and the Al/Mg ratio. The half-life of ^{26}Al is 0.717 Ma and the production ratio for $^{26}Al/^{27}Al$ in red giants is thought to be around 10^{-3} to 10^{-4}. The $^{26}Al/^{27}Al$ initial ratios in CAIs of 5 × 10^{-5} indicates that nucleosynthesis occurred no more than a few million years before formation of these CAIs.

These short-lived "fossil" radionuclides also provide a means of relative dating of meteorites and other bodies, because the abundance of the extinct radionuclide at the time an object formed can be deduced. Consider Figure 10.20 where $^{53}Cr/^{52}Cr$ in inclusions from *Allende* are plotted as a function of the $^{55}Mn/^{52}Cr$. Provided that (1) all inclusions formed at the same time, (2) all remained closed to Mn and Cr since that time, and (3) ^{53}Mn was present when they formed and has since fully decayed, we can derive

Table 10.4 Short-lived radionuclides in the early solar system.

Radionuclide	Half-life, Ma	Decay	Daughter	Abundance ratio
^{10}Be	1.387	β^-	^{10}B	^{10}Be/^{9}Be ~ 8.8×10^{-4}
^{26}Al	0.717	β^-	^{26}Mg	^{26}Al/^{27}Al = $5.3 \pm 1.2 \times 10^{-5}$
^{36}Cl	0.301	β^-, β^+	^{36}Ar, ^{36}S	^{36}Cl/^{35}Cl ~ 1.8×10^{-5}
^{41}Ca	0.102	e.c.	^{41}K	^{41}Ca/^{40}Ca ~ 4×10^{-9}
^{53}Mn	3.74	e.c.	^{53}Cr	^{53}Mn/^{55}Mn = $6.71 \pm 0.56 \times 10^{-6}$
^{60}Fe	2.62	β^-	^{60}Ni	^{60}Fe/^{56}Fe ~ 1×10^{-8}
^{92}Nb	34.7	β^-	^{92}Zr	^{92}Nb/^{93}Nb ~ 1×10^{-5}
^{107}Pd	6.5	β^-	^{107}Ag	^{107}Pd/^{108}Pd $5.9 \pm 2.2 \times 10^{-5}$
^{129}I	15.7	β^-	^{129}Xe	^{135}Cs/^{133}Cs ~ 4.8×10^{-4}
^{135}Cs	2.3	β^-	^{135}Ba	^{129}I/^{127}I ~ 1×10^{-4}
^{146}Sm	68	α	^{142}Nd	^{146}Sm/^{144}Sm = 0.0094
^{182}Hf	8.9	β^-	^{182}W	^{182}Hf/^{180}Hf = $9.81 \pm 0.41 \times 10^{-5}$
^{244}Pu	80.0	α, SF	Xe	^{244}Pu/^{238}U ~ 7×10^{-3}
^{247}Cm	15.6	α, SF	^{235}U	^{247}Cm/^{235}U ~ 2×10^{-3}

Based on summaries in Dauphas and Chassidon (2011) and Davis and MacKeegan (2014).

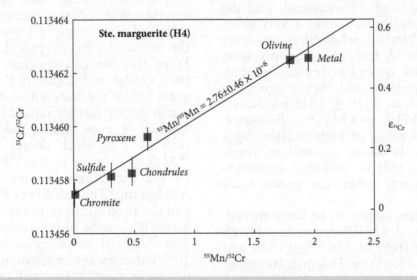

Figure 10.20 Correlation of the ^{53}Cr/^{52}Cr ratio with ^{55}Mn/^{52}Cr ratio in inclusions from the ordinary chondrite *Ste. Marguerite* (H4). Data from Trinquier et al. (2008).

the following equation from the fundamental equation of radioactive decay:

$$\left(\frac{^{53}Cr}{^{52}Cr}\right) = \left(\frac{^{53}Cr}{^{52}Cr}\right)_0 + \left(\frac{^{53}Mn}{^{55}Mn}\right)\left(\frac{^{55}Mn}{^{52}Cr}\right) \quad (10.1)$$

where, as usual, the subscript naught refers to the initial ratio. On a plot of ^{53}Cr/^{52}Cr versus ^{55}Mn/^{52}Cr such as Figure 10.20, this is the equation of a line with a slope equal to the initial ^{53}Mn/^{55}Mn ratio, in this case 2.76×10^{-6} (note that the right-hand scale expresses the ^{53}Cr/^{52}Cr ratio in the epsilon notation introduced in Chapter 8, or deviations in parts per thousand from a terrestrial standard, whose value is 0.1134569).

As time passes, the abundance of the radioactive nuclide will decrease, so that the

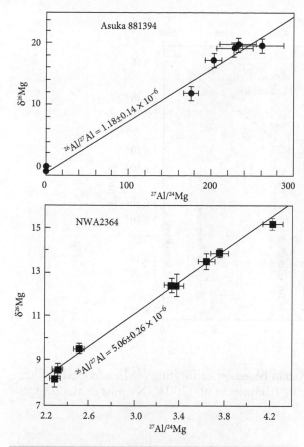

Figure 10.21 Comparison of Al-Mg isotope systematics for two different meteorites. The lower diagram shows minerals separated from a CAI in CV3 *NWA2364*, one of the oldest objects in the solar system (Bouvier and Wadhwa, 2010). The upper diagram shows plagioclase and pyroxene separates from the eucrite *Asuka 881394* (Nyquist et al., 2003). The latter has an initial $^{27}Al/^{26}Al$ more than 40 times lower than *NWA2364*, implying it is some 4 million years younger.

initial ratio determined in diagrams such as Figure 10.20 will be lower for younger objects. This is illustrated in Figure 10.21, which compares ^{26}Al-^{26}Mg systematics in a CAI from the CV3 meteorite *NWA2364* with those in a younger object, the eucrite *Asuka 881394*. The difference in initial $^{26}Al/^{27}Al$ indicates the latter object is some 4 million years younger than the former.

Extinct radionuclides can thus be used to establish a relative time scale of events in the early solar system. A number of factors, however, hinder this process. First, it

is possible that these recently synthesized radionuclides might not have been uniformly distributed through the solar nebula. Indeed, as we shall see in a subsequent section, there is evidence of isotopic heterogeneity in meteorites and their components. Second, isotopic heterogeneity unrelated to decay of extinct radionuclides might be present in the daughter elements. Indeed, there is some evidence for this in the case of Cr, but a correction can be made by measuring an additional Cr isotope, ^{54}Cr. In the case of a light element such as Mg, mass fractionation arising from chemical or physical processes might affect the $^{26}Mg/^{24}Mg$. This too can be corrected by measuring ^{25}Mg, provided the fractionation was mass-dependent. Finally, cosmic ray spallation reactions have been demonstrated in meteorites, and these can affect some of the elements of interest, but again, corrections for these effects can generally be made. In addition, there are, of course, the same issues with conventional radiogenic isotopic geochronology, such as open system behavior.

Provided these can be overcome, an absolute chronology can be established by calibrating relative ages determined from extinct radionuclides with high precision Pb-Pb ages. For the earliest objects, the short-lived nuclides ^{53}Mn and ^{26}Al have proved most useful. Figure 10.22 illustrates such a time scale, anchored on objects dated by both Pb-Pb and ^{26}Al or ^{53}Mn. The chronology begins with the CAI from *NWA2364*. Objects such as *Lewis Hills 86010*, *St. Marguerite*, and *D'Orbigny* provide other anchors. Objects such as *Orgueil*, which has not been dated by conventional radiometric methods, can be placed on the time scale based on their apparent initial $^{26}Al/^{27}Al$ or $^{53}Mn/^{55}Mn$ ratios.

There is debate over the homogeneity of $^{26}Al/^{27}Al$ in the early solar system. In particular, Pb-Pb ages show that chondrule formation began simultaneous with or shortly after CAI formation, yet their $^{26}Al/^{27}Al$ are systematically lower than ~1×10^{-5}. Schiller et al. (2015) obtained a high precision Pb-Pb age on the angrite NWA1670 of 4564.4 ±2 Ma, yet the $^{26}Al/^{27}Al$ initial ratio is 5.9×10^{-7}, which implies an age 4.6 million years rather than 3.3 younger than CAIs assuming an initial $^{26}Al/^{27}Al$ ratio of 5.2×10^{-5}. This and other evidence suggest ^{26}Al has inhomogeneously

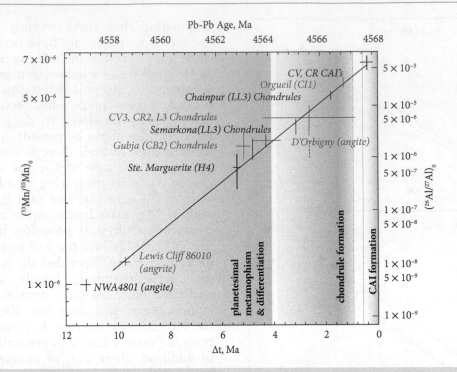

Figure 10.22 Time scale of events in the early solar system based on calibrating ^{53}Mn and ^{26}Al extinct radionuclide chronology to Pb-Pb ages. Based on data in Trinquier et al. (2008), Nyquist et al. (2009), Bouvier and Wadhwa (2010), Katsuyuki et al. (2010).

distributed in the early solar system and that an initial ^{26}Al/^{27}Al ratio of 1.2×10^{-5} is more appropriate for objects other than CAIs. In contrast, Budde et al. (2018) point to the good agreement of ^{26}Al-^{26}Mg, ^{182}Hf/^{182}W, and ^{207}Pb-^{206}Pb ages of CR chondrites and their chondrules and argue for homogeneous distribution of ^{26}Al in the solar system.

Ages, both conventional radiometric ages and those based on extinct radionuclides, can represent different things. In the case of CAIs, their age might either represent the formation or time of aqueous alteration of the parent body. For the LL3 chondrules, this age may well represent the formation age, but some evidence suggests the chondrule event lasted for as much as 4 million years, so all chondrules need not have the same age. In other cases, such as *Ste. Marguerite*, the age may represent the age of metamorphism. In others, such as the HED parent body (Vesta), it represents melting and differentiation. The angrites show a range of ages. *D'Orbigny* is a fine-grained igneous rock that probably formed near the surface of the angite parent

body, which would have cooled quickly. *Lewis Cliff 86010* is a coarse-grained rock from the interior of the angrite parent body that would have cooled more slowly.

As mentioned, the extinct radionuclides also provide evidence of one or more nucleosynthetic events shortly before, or perhaps even during, the formation of the solar system. ^{41}Ca, which decays to ^{41}K with a half-life of 150,000 yrs, provides perhaps the most stringent constraint of the time of nucleosynthesis. Extinct ^{41}Ca has been identified in CAIs from several CV3 and CM2 meteorites, which had an apparent ^{41}Ca/^{40}Ca ratio of 4×10^{-9} when they formed. Interestingly, not all CAIs show evidence of the presence of ^{41}Ca, suggesting the CAI forming event lasted at least several hundred thousand years.

Isotopic variations of Ag resulting from the decay of ^{107}Pd (half-life 6.5 Ma) in iron meteorites indicate core formation in meteorite parent bodies began, and was largely complete, within about 15 Ma of the CAI formation. The ^{182}Hf-^{182}W has provided much more stringent contrasts of core formation in the parent bodies of iron meteorites. Kruijer et al. (2014) showed that the cores segregated in the parent bodies of magmatic iron groups IIAB IID, IVA, and IVB between 0.7 to 3 Ma after CAI formation. However, some nonmagmatic irons of groups IAB and IIE, which may have originated as impact melts, formed later, as much as 28 Ma after core formation. We will see at the end of this chapter that ^{182}Hf-^{182}W has also provided important constraints on timing of formation of the Earth's core.

As we saw earlier in this chapter, heavy elements are synthesized mainly in red giant stars, supernovae, and neutron star mergers. On a galactic scale, these sources will continually inject newly synthesized nuclides into the interstellar medium. Those nuclides that are unstable will steadily decay away. These two competing processes will result in steady-state abundance of these nuclides in the interstellar medium. The abundances of ^{60}Fe, ^{92}Nb, ^{129}I, ^{146}Sm, and ^{182}Hf listed in Table 10.4 roughly match the expected steady-state galactic abundances and hence may not require a specific synthesis event. However, the abundances of ^{10}Be, ^{26}Al, ^{36}Cl, and ^{41}Ca in the early solar system require synthesis of these nuclides at the time, or just before, the solar system formed.

The conventional view is that these nuclides were synthesized in a red giant and/or a supernova in the region where the solar system formed just shortly before its formation. Some of these elements, such as ^{26}Al, ^{36}Cl, ^{41}Ca and ^{107}Pd, are efficiently synthesized by the s-process in AGB red giants; but they can also be synthesized in supernovae. While strong stellar winds of AGB stars can disperse newly synthesized elements, supernovae are more efficient dispersal mechanisms. Furthermore, AGB stars are relatively small (1.5 to 3 solar masses) and would have had main sequence lives of hundreds of millions or billions of years or more before reaching the AGB stage. On the other hand, very massive stars last only tens of millions of years and hence may never leave their nebular nurseries before exploding in supernovae. Indeed, one popular hypothesis is that the formation of the solar system was actually triggered by a supernova shock wave. Boss and Vanhala (2001) provide a good discussion of this view.

Beryllium is not synthesized in stars, hence the presence of ^{10}Be in CAIs and other primitive chondritic components is problematic for the red giant/supernova injection hypothesis. Most workers now agree that ^{10}Be was synthesized entirely by spallation. Spallation may also have been responsible for a small fraction of the ^{26}Al, ^{36}Cl, and ^{41}Ca, and perhaps some of the ^{53}Mn as well. However, it cannot account for the ^{60}Fe and cannot account for all of the ^{26}Al, ^{36}Cl, ^{41}Ca, and ^{53}Mn. Huss et al. (2009) concluded that intermediate-mass *asymptotic giant branch* (AGB) stars (a variety of red giant) and supernovae of stars with precursor masses in the range of about 20–60 solar masses are the most likely sources.

10.4.2   Cosmic ray exposure ages and meteorite parentbodies

As we saw in Chapter 8, cosmic rays colliding with matter in meteorites and planetary bodies produce new nuclides through spallation. The cosmic rays only penetrate to a limited depth (of the order of a meter or less: there is no cutoff, the flux falls off exponentially), so that only small bodies or the surfaces of larger bodies experience significant production of cosmogenic nuclides. The rate of production of nuclides by cosmic ray bombardment can be estimated from experimental physics if the cosmic ray flux is known. Thus, assuming a more or less constant cosmic ray flux, the length of time an object has been exposed to cosmic rays, the *cosmic-ray exposure age* (CRE) can be calculated from the amounts of cosmogenic nuclides. A number of nuclides have proved useful in determining exposure ages, including radioactive ones such those listed in Table 8.5, as well as rare stable ones such as ^{3}He, ^{21}Ne, and ^{83}Kr.

The main finding of these studies can be briefly summarized as follows (Herzog and Caffee, 2014):

- Exposure ages of meteorites are much younger than their formation ages and increase in the order stones < stony irons < irons.
- Exposure ages of stones rarely exceed 100 Ma, the CRE ages of stony irons are typically between 50 and 200 Ma, and the

CRE ages of irons vary with group but commonly exceed 200 Ma.
- Exposure ages are neither uniformly distributed nor tightly clustered.

Determination of exposure ages requires knowledge of production rates. These depend on the flux of primary cosmic ray particles (mainly hydrogen and helium nuclei), their energies, the production rate of secondary particles such as neutrons, pions, and muons, reaction cross sections, and the composition of the meteorite. Considerable research has gone into determining reaction cross sections by artificially irradiating materials, but uncertainties nevertheless remains. Further complications arise as the meteorite components may have experienced several episodes of cosmic ray exposure, such as on the parent body surface or even before accretion into the parent body; this is particularly an issue with brecciated meteorites containing a variety of components (polymict breccias).

Exposure ages for ordinary chondrites are shown in Figure 10.23; exposure ages for irons were shown in Figure 8.34. As can be seen, meteorites became small bodies accessible to cosmic rays comparatively recently. Before that, they must have been stored in larger parent bodies where they would have been protected from cosmic ray bombardment. The H chondrite distribution shows peaks at 6–10, 33, and 24 Ma, the latter peak dominated by H6 meteorites. L chondrites show peaks at 5, 28, and 40 Ma. The more limited data for L chondrites shows peaks around 15 Ma and about 30 Ma (one meteorite has an exposure age of ~0.5 Ma, which is not shown on the graph). Exposure ages of most other chondrite classes show a similar distribution, but the CM, CI, and CR groups are notably younger, with average exposure ages of 2, 3, and 8 Ma, respectively. These distributions suggest occasional collisions among large bodies eject fragments into orbits that ultimately intersect the Earth. The greater exposure age of irons and shorter exposure ages of the CM, CI, and CR groups likely reflects their greater strength and resistance to break-up.

SNC meteorites have particularly young exposure ages, with many shergottites having exposure ages of a few million years and

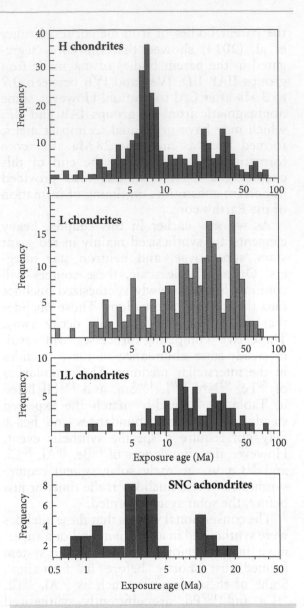

Figure 10.23 Cosmic ray exposure ages of ordinary chondrites and SNC achondrites. Based on compilations of Graf and Marti (1995), Marti and Graf (1992), and Herzog and Caffee (2014).

younger. Nakhlite ages cluster around 10 Ma, perhaps reflecting a single impact.

10.4.3 Asteroids as meteorite parentbodies

The limited variability in composition within meteorite classes and the compositional gaps between different classes suggest that all meteorites of a common class share a close genetic history. Relatively young cosmic ray exposure ages, the clustering of exposure ages of meteorite classes, extensive thermal metamorphism (reaching perhaps 700°C) in chondrites and melting of differentiated meteorites, and slow cooling rates of iron meteorites all indicate that meteorites were once parts of larger *parent* bodies of 10–1000 km. Orbits for some observed falls have been reconstructed, and these reconstructed orbits confirm that many meteorites originate in the main asteroid belt, located between Mars and Jupiter at ~2.1–3.3 AU, which contains ~2 million objects larger than 1 km. For comparison, some 100–150 distinct parent bodies are needed to explain the diversity of meteorite compositions. There are some compositional similarities between different classes that in some cases suggest a genetic relationship between them, and possible derivation from the same parent body. For example, the aubrites and e-chondrites are both highly reduced. Many of the pallasites seem related to the IIIAB irons and may come from the same parent body. Other pallasites seem more closely related to ordinary chondrites and to IAB irons.

Much of what we know about asteroids is based on remote observation. Based on their orbital inclination and eccentricity, many can be grouped into families such that all members of a family appear to have originated by collisional disruption of a single parent body. For example, the Vesta family includes *4 Vesta**, with a diameter of ~500 km and some 15000 other bodies with diameters < 10 km.

Asteroids are also classified by their reflectance spectra, which is the fraction of solar electromagnetic energy reflected from the surface as a function of energy. The reflectance spectrum is in turn a function of chemistry and mineralogy, which allows inferences about the chemistry and mineralogical composition of asteroids. More details can be found in the review of Burbine (2014). Asteroids can then be divided into a dozen or more classes based on their reflectance spectra, the largest of which are the C-, S-, and X-groups. Reflectance spectra of meteorites can be measured in the laboratory and compared with those of asteroids. C-class objects,

which are most common, include 1 *Ceres* and have surface compositions similar to carbonaceous chondrites. S-class objects have surfaces dominated by olivine and pyroxenes and are likely related to ordinary chondrites or primitive achondrites such as ureilites, acapulcoites/lodranites, and winonaites. The X-class includes subclasses E, which may be the source of aubite achondrites and M, thought to consist of metal and are likely the source of iron meteorites. Radar reflections from the asteroid 16 *Psyche* suggest a Ni-Fe composition, similar to iron meteorites. However, based on spectral and density measurements by the European Space Agency *Rosetta* probe, the M-type asteroid 21 *Lutetia* appears to be similar in composition to enstatite chondrites or the CB, CH, or CR carbonaceous chondrites, which are metal-rich. P-type asteroids show spectral similarities to the CI and CM chondrites. The Jupiter "Trojans" (those in the same orbit as Jupiter) consist primarily of D-type asteroids, which appear to be rich in water and organic matter and similar to the unusual primitive carbonaceous chondrite *Tagish Lake*. A smaller but important class are the V-types, which include 4 *Vesta* and many, but not all, of the Vestoid orbital family as well as a number of asteroids in near-Earth orbits. Their spectra match those of the HED achondrites (Figure 10.24), hence *Vesta*, or a smaller asteroid derived from it, have long been assumed to be the source of HED meteorites.

The asteroid belt appears to be compositionally zoned, with X- and S-types predominating in the inner part, C-types being more common in the outer main belt, while P- and D-types dominate beyond the main belt. This zonation corresponds to increased abundance of volatiles such as carbon and water and decreased abundance of silicates and metal with heliocentric distance.

There have now been a number of spacecraft missions that have orbited or landed on asteroids. The first of these was NASA's *NEAR Shoemaker* mission. The spacecraft began orbiting the S-type asteroid *433 Eros*, appropriately enough, on 14 February 2000 (Valentine's Day). Before doing so, it made a

---

* Asteroids are given both names and numbers, the latter in the order in which they were discovered. Vesta was the fourth asteroid discovered, following 1 *Ceres*, 2 *Pallas*, and 3 *Jono*. Although part of the asteroid belt, Ceres has now been classified as a minor planet.

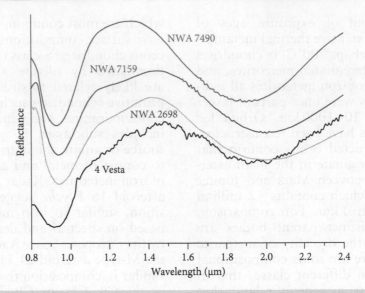

Figure 10.24   Comparison of the laboratory-determined reflectance spectrum of HED meteorites with that of the asteroid *Vesta*. From Ferrari et al. (2017).

close flyby of the C-asteroid *433 Mathilde*, confirming on the basis of density and spectrum a strong similarity to carbonaceous chondrites. Before it was deliberately crashed onto the surface, *NEAR Shoemaker* was able to make not only visible and IR spectral measurements of Eros, but also X-ray and gamma-ray observations and found a close similarity to ordinary chondrites. Despite many problems, Japan Aerospace Exploration Agency's JAXA) *Hayabusa* spacecraft landed on the small (maximum dimension 500 m) asteroid *25153 Itokawa* in 2005 and returned a small (<1 g) sample to Earth five years later. Analysis of the sample confirmed a LL4 to LL6 composition, which was also consistent with spectral observation.

NASA's DAWN spacecraft orbited *4 Vesta* (Figure 10.25), the third largest asteroid by size and second largest by mass, from July 2011 to September 2012. Density (3456 kg/m^3) and gravitational moment of inertia determined by the DAWN mission indicate that Vesta is a differentiated body with an iron core that constitutes about 18% of its mass, a silicate mantle, and a basaltic crust, which has been highly disrupted by impacts (Russell et al., 2012). Spectral measurements confirmed a close similarity to HED achondrites (De Sanctis et al., 2012) and there is a broad (but not unanimous) consensus that Vesta is indeed

the source of these meteorites (Mittlefehlt, 2015). Although individual meteorites give younger ages, a ^{53}Mn-^{53}Cr eucrite–diogenite whole rock isochron suggests it differentiated and melted within 3 million years of CAI formation (Trinquier et al., 2008).

After orbiting *Vesta* for 14 months, DAWN then went onto *1 Ceres* (Figure 10.26), orbiting that body from July 2015 through October 2018. *Ceres*, with a mass of 9.39×10^{20} kg, a mean diameter of 956 km, and a density of 2162 kg/m^3, is the largest asteroid, comprising roughly a third of the mass of the asteroid belt. Spectroscopic observations reveal the surface consists of serpentine, ammoniated clays, salts, Mg-Ca carbonates, water ice, and organic matter (DeSantis et al., Prettyman et al., 2017). Unlike *Vesta*, no specific meteorite class can be matched to *Ceres'* inferred surface composition. Based on gravity and topographic measurements by the DAWN spacecraft, Ceres has a ~430 km thick rocky mantle with a density of ~2400 kg/m^3 and surrounded by an outer ~40 km thick crust with a density of ~1300 kg/m^3 (Fu et al., 2017). The mantle density is similar to that of carbonaceous chondrites, while the crust density is indicative of a composition rich in water ice mixed with denser materials such as hydrated silicates, various salts and possibly clathrates. The subdued topography and absence of large craters in comparison to

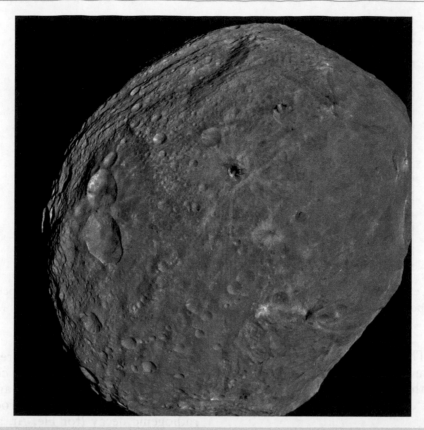

Figure 10.25 Photograph of the asteroid 4 *Vesta* taken by the NASA *DAWN* spacecraft in 2011. Vesta, the second most massive asteroid, has a semi-major orbital axis of 2.4 AU, a mean diameter of 500 km and a mean density of 3420 kg/m^3. It is believed to be the parent body of the HED meteorites. NASA photo.

Vesta suggests the upper mantle is weak and contains interstitial brine, with a viscosity comparable to the Earth's asthenosphere. The brine occasionally rises to the surface depositing salts including Na_2CO_3 that produce the bright spots such as those in *Occator Crater* (DeSanctis et al., 2016). Cryovolcanism (that is, volcanism involving water and ice rather than silicates) has also been identified on Ceres, most notably the relatively young (<250 Ma) cryovolcanic dome known as *Ahuna Mons* (Ruesch et al., 2016). McSween et al. (2018) concluded that *Ceres* has a bulk composition similar to CI/CM carbonaceous chondrites but has undergone differentiation and more extensive hydrothermal alteration than the parent bodies of those meteorites due to its greater size.

Although meteorites represent a wide range of compositions, there is no particular reason to believe that those in collections are representative samples of the compositions of asteroids. The reflectance spectra of most asteroids do not match those of any of meteorite classes – suggesting they have a greater variety of compositions than represented by meteorites. Future space missions should reveal more about asteroids. The Japanese Space Agency's *Hayabusa2* spacecraft arrived at the small (< 1km) C-type asteroid *162173 Ryugu* in 2018 and successfully sampled the surface in 2019 and is scheduled to return samples in late 2020. Spectral analysis from the spacecraft indicates as composition similar to thermal- or shock-metamorphosed carbonaceous chondrites of class CI or CM Katazato et al. (2019). NASA's OSIRIS-Rex mission arrived at the small (~525 m diameter) carbon-rich B-class near-Earth asteroid *101955 Bennu* in 2018. This asteroid appears to be a particularly loosely aggregated rubble pile; spectral imaging indicates it is rich in hydrated silicates and has a close affinity with aqueous altered CM chondrites (Hamilton

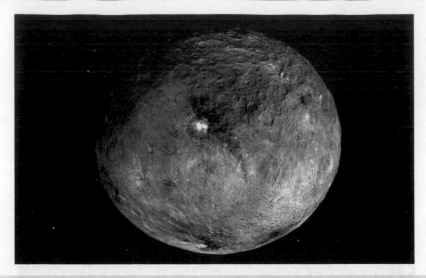

Figure 10.26 Ceres as photographed by NASA's DAWN spacecraft. Bright spots have been identified as salt deposits deposited by hydrothermal fluids rising from depth. NASA photo.

et al., 2019). The spacecraft is scheduled to sample the surface in August, 2020 and return samples to Earth in 2021.

10.4.4 Isotopic anomalies in meteorites

10.4.4.1 Neon alphabet soup and star dust

Since Thomson's discovery that elements could consist of more than one isotope in 1912, scientists have realized that the isotopic composition of the elements might vary in the universe. They also realized that these variations, if found, might provide clues as to how the elements came into being. As the only available extraterrestrial material, meteorites were of obvious interest in this respect. However, isotopic analyses of meteorites, by Harold Urey among others, failed to reveal any differences between meteorites and terrestrial materials. This apparent isotopic homogeneity was raised as an objection to the polygenetic hypothesis of Burbidge, Burbidge, Fowler, and Hoyle (1957), since isotopic variations in space and time were an obvious prediction of this model. Within a few years of its publication, however, John Reynolds, a physicist at the University of California, Berkeley, found isotopic variations in noble gases, particularly neon and xenon (Reynolds, 1960).

Noble gases are present in meteorites at concentrations that are often as low as 1 part in 10^{10}. Though they are fairly readily isolated and analyzed at these concentrations, their isotopic compositions are nonetheless sensitive to change due to processes such as radiogenic decay (for He, Ar, and Xe), spallation and other cosmic ray-induced nuclear processes, and solar wind implantation. In addition, mass fractionation can significantly affect the isotopic compositions of the lighter noble gases (He and Ne). Through the late 1960s, all isotopic variations in meteoritic noble gases were thought to be related to these processes. For example, Ne isotopic variations could be described as mixtures of three components, "Neon A" or planetary (similar in composition to the Earth's atmosphere), "Neon B," or solar, which differed from atmospheric due to mass fractionation, and "Neon S," or spallogenic (cosmogenic) (Figure 10.27). The isotopic variations in Xe discovered by Reynolds were nonetheless significant because they were due to the decay of ^{129}I and ^{244}Pu, which must have been only recently (on a cosmic time scale) synthesized.

In 1969, the picture became more complex when evidence of a ^{22}Ne-rich component, named "Neon E" was found in the high-temperature (900–1100°C) release fractions of six carbonaceous chondrites (Black and Pepin, 1969). However, the carrier of Neon-E proved difficult to identify. It was not until the late 1980s that E. Anders and his colleagues at the University of Chicago (e.g.,

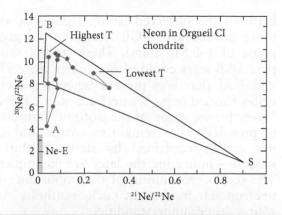

Figure 10.27 Neon isotopic compositions in a step-heating experiment on *Orgueil* CI chondrite, which produced the first evidence of "pre-solar" or exotic Ne. The points connected by the line show the changing Ne isotope ratios with increasing temperature. The shaded area is the original estimate of the composition of the pure Ne-E component. Also shown are the compositions of Ne-A (solar), Ne-B (planetary), and Ne-S (spallogenic). After Black and Pepin (1969).

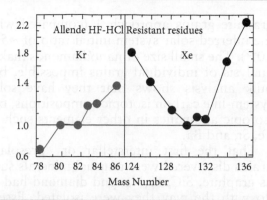

Figure 10.28 The isotopic composition of Kr and Xe of the Xe-HL component in *Allende* matrix. Xe-HL is characteristically enriched in both the light and heavy isotopes, while the lighter noble gases show enrichment only in the heavy isotopes. After Anders (1988).

Tang and Anders, 1988) found that Neon-E is associated with fine-grained (<6 µm) graphite and SiC (silicon carbide) of the matrix. Ne-E actually consists of two isotopically distinct components: Ne-E(L), which was found to reside in graphite, and Ne-E(H) which resides in SiC. The $^{20}Ne/^{22}Ne$ ratio of Ne-E(L) is less than 0.01, while that of Ne-E(H) is less than 0.2.

The other key noble gas in this context is xenon. Having nine isotopes rather than three and with contributions from both ^{129}I decay and fission of Pu and U, isotopic variations in Xe are bound to be more complex than those of Ne. On the other hand, its high mass minimizes mass fractionation effects. Eventually, two isotopically distinct components were identified: Xe-HL, so named because it shows enrichments in both the heaviest and lightest Xe isotopes (Figure 10.28), and the Xe-S component, so named because it is enriched in the s-process-only isotopes, ^{128}Xe and ^{130}Xe. Anders' University of Chicago group eventually identified the carrier of Xe-HL as *nanodiamonds* (typical dimension: ~2.6 nm) and that of Xe-S as SiC.

Once these interstellar grains were isolated, it was possible to study their isotopic

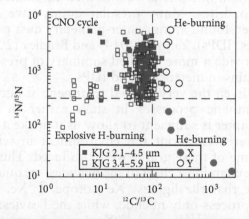

Figure 10.29 Isotopic composition of C and N in SiC from *Murchison* (CM2) meteorite. Dashed lines show the isotopic composition of normal solar system C and N. Populations X and Y, which are anomalous here, are anomalous in other respects as well. After Anders and Zinner (1993).

compositions in detail using ion microprobes. Very large variations in the isotopic composition of carbon and nitrogen were found in silicon carbide (Figure 10.29). The SiC grains do not form a single population but represent a number of populations of grains, each produced in a different astronomical environment. Large isotopic variations occur in a number of other elements, including Mg, Si, Ca, Ti, Sr, Zr, Mo, Ba, Nd, Sm, and Dy. Inferred initial $^{26}Al/^{27}Al$ ratios in some graphite grains approach 1 (compared with the inferred solar system initial ratio of $\sim 5 \times 10^{-5}$). The small size of nanodiamonds makes analysis of individual grains impossible, but bulk analysis shows that they have solar system-like carbon isotopic compositions, but isotopic anomalies in other elements such as Te, Sr, and Ba.

That the first interstellar or "presolar" grains discovered were unusual minerals such as graphite, SiC, Si_3N_4, and diamond had to do with the way they were isolated. Essentially, they were the residues after the rest of the meteorite was dissolved away. Subsequently, however, similar isotopic anomalies have been found in Si_3N_4, spinel, hibonite, a variety of metal carbides, TiO_2, Fe-Ni metal and olivine, with abundances of silicate presolar grains reaching several hundred parts per million in some unequilibrated carbonaceous chondrites. Many presolar grains have also been found among inter-planetary dust particles (IDPs). Zinner (2014) and Bradley (2014) provide a more detailed summary of presolar grains in meteorites and IDPs.

Even the very limited treatment of nucleosynthetic processes in stars earlier in this chapter is sufficient to allow us to make a few inferences about the environment in which some of these grains were produced. Thus, if we examine a chart of the nuclides, we quickly see that the lightest Xe isotope, ^{124}Xe, is a p-process-only nuclide, while the heaviest Xe isotopes, ^{134}Xe and ^{136}Xe, are r-process-only nuclides. The p- and r-processes occur in explosive events including supernovae and neutron star mergers, so Xe-HL was likely produced in an explosive event and captured by nanodiamonds condensing from the outflow. Xe-S is enriched in ^{128}Xe and ^{130}Xe, which are s-process-only isotopes. The s-process, of course, operates mainly in red giants, so we might guess the SiC condensed from the outflow of red giants. Carbon and nitrogen in the SiC are, in most cases, enriched in ^{14}N and ^{13}C relative to normal solar system nitrogen and carbon. As we noted earlier in the chapter, there tends to be some net production of ^{14}N and consumption of ^{12}C in the CNO cycle, which operates in main sequence stars, but also in the H-burning shell of red giants. As it turns out, our guess of red giants as sources of this SiC would be a good one. Theoretical studies show a close match between the observed isotopic patterns and those produced in AGB stars (the red giant phase of $1-3\,M_\odot$ stars). These studies show that AGB stars could also produce the ^{107}Pd and ^{26}Al that was present when the meteorites formed (e.g., Wasserburg et al., 1994). Nevertheless, many of the isotopic variations in presolar grains remain enigmatic and are not readily explained by stellar evolution models – indicating the later are incomplete. The recent recognition of the importance of neutron star mergers in nucleosynthesis may spur a deeper understanding.

10.4.4.2 Isotopic variations in bulk meteorites

Until 1973, O isotope variations in meteorites were thought to be simply the result of mass-dependent fractionation, as they are on Earth. But when R. Clayton of the University of Chicago went to the trouble of measuring ^{17}O (0.037% of O) as well as ^{18}O and ^{16}O, he found that these variations were not consistent with simple mass-dependent fractionation. This is illustrated in Figure 10.30. On a plot of $^{17}O/^{16}O$ versus $^{18}O/^{16}O$, almost all terrestrial materials (atmospheric ozone is a notable exception, as we learned in Chapter 9) plot on a line with a slope of 0.52 – the *terrestrial fractionation line (TFL)*. Lunar samples fall on this same line, but meteorites and meteoritic components do not. In fact, anhydrous minerals from carbonaceous chondrites fall along a line (CCAM: carbonaceous chondrite anhydrous minerals) in Figure 10.30) with a slope of ~ 1. Most CAIs also fall along the same line and extent to extremely ^{16}O enriched compositions ($\delta^{18}O$ as low as -50‰). CI chondrites are a notable exception to the carbonaceous chondrite trend, with extreme ^{16}O depletion and falling close to the TFL, such that most noncarbonaceous materials could be derived from CI materials mainly by mass dependent fractionation.

The initial interpretation was that this reflected mixing between a more or less pure ^{16}O component, such as might be created by helium burning, injected into the solar nebula by a red giant, and a component of "normal" isotopic composition. A decade later, experiments conducted by Thiemens and Heidenreich (1983) suggested a different interpretation. They found that ozone

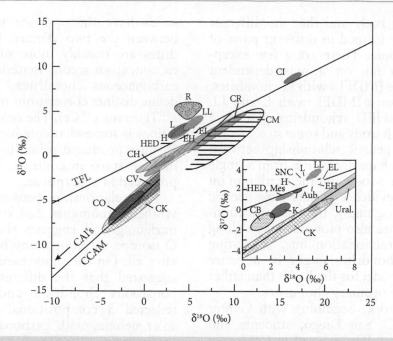

Figure 10.30 Variation of O isotope ratios in meteorites: CO, CK, etc., carbonaceous chondrites; H, L, LL, R, K, chondrites; EL and EH: enstatite chondrites; HED: howardites, eucrites, diogenites; Ural, ureilites; SNC, Shergottiites, Naklites, Chausigny (from Mars). Aub: aubrites, Ang: angrites, Mes: Mesosiderites. The black dot shows the isotopic composition of the Earth and Moon.

produced by a high-frequency electrical discharge experienced mass-independent fractionation, i.e., where the ozone was equally enriched in ^{17}O and ^{18}O relative to ^{16}O. The experiment demonstrates that a slope of 1 on the $\delta^{17}O$—$\delta^{18}O$ diagram could be produced by chemical processes. Thiemens suggested this kind of fractionation arises because nonsymmetric (e.g., $^{16}O^{17}O$ or $^{18}O^{16}O$) molecules have more available energy levels than symmetric (e.g., $^{16}O^{16}O$) molecules (as we saw in Chapter 9, symmetry enters into the calculation of the partition function).

A somewhat different idea was proposed by Clayton (2002). He suggested that the anomalies arose through radiation self-shielding in the solar nebula. In his model, ultraviolet radiation from the early protosun dissociated carbon monoxide, which would have been among the most abundant gases in the solar nebula. Because $C^{16}O$ was far more abundant than either $C^{17}O$ or $C^{18}O$, the radiation of the wavelength necessary to dissociate $C^{16}O$ would have been quickly absorbed as it traveled outward from the Sun. At greater distance from the Sun, radiation of the frequency necessary to dissociate $C^{16}O$ would have been

absent, while that needed to dissociate $C^{17}O$ and $C^{18}O$ would still be available. At those distances, $C^{17}O$ and $C^{18}O$ would have been preferentially dissociated by UV radiation, and equally so, making ^{17}O and ^{18}O preferentially available for reaction to form silicates and other meteorite components. CAIs, the most ^{18}O-rich material, would have formed close to the Sun, and then expelled back out by the X-wind (discussed below). Clayton's model also predicted that the Sun itself should be poor in ^{18}O and ^{17}O compared with meteorites and the Earth – closer in composition to the CAIs. This prediction was confirmed based on analysis of solar wind collected by NASA's Genesis mission. After correcting for fractionation in the collectors, McKesson et al. (2011) estimated the composition of the solar wind as $\delta^{18}O = -103 \pm 3.3‰$ and $\delta^{17}O = -80.8 \pm 5‰$, plotting well to the lower left of Figure 10.30.

While variations *between* classes are mostly mass-independent, variations *within* ordinary chondrite and achondrite classes fall along mass-dependent fractionation lines. This strongly suggests that, for the most part, different groups could not have come from the same parent body and that the different groups probably formed in different parts of the presolar nebula. There are a few exceptions: IIE irons fall on a mass-dependent fractionation line (MDFL) with H-chondrites, IVA irons plot on a MDFL with L and LL chondrites, and HED achondrites plot on a MDFL with IIIAB irons and some stony-irons. This suggests a genetic relationship between these objects, perhaps derivation from a single parent body. The Moon and the Earth plot on a single MDFL, evidence of their close genetic relationship. Intriguingly, the enstatite chondrites and aubrites also plot essentially along the terrestrial fractionation line, suggesting that enstatite chondrites might be a better compositional model for the Earth than either carbonaceous or ordinary chondrites.

Subsequent work, beginning with Gunter Lugmair of U.C. San Diego, students and colleagues of Jerry Wasserburg* at California Institute of Technology and students and colleagues of Claude Allègre at the University of Paris, began to reveal systematic nucleosynthetic anomalies in a variety of elements in bulk samples of meteorites. The earliest discoveries were in Ti and Cr isotopic compositions; advances in analytical instrumentation, most notably multi-collector ion probe and MC-ICP-MS, has greatly accelerated this work and isotopic variations in many elements have now been documented. Some of these variations appear to be correlated, such as Ca and Ti, over all meteorite classes. In contrast, Cr and Ti isotope ratios appear to define two correlations: one among carbonaceous chondrites, another among all other objects.

Warren (2011) pointed out that isotopic compositions reveal a dichotomy between carbonaceous chondrites and all other solar system material and that they correlate, at least in part with oxygen isotope variations. Carbonaceous chondrites have ε^{50}Ti and ε^{48}Ca, and ε^{54}Cr > 0 while other solar system

solids have the opposite with a large gap between the two (Figure 10.31). CI chondrites are notably more similar isotopically to noncarbonaceous material than are other carbonaceous chondrites. The two groups define distinct correlations in some cases (e.g., ε^{50}Ti versus ε^{54}Cr). The origin of these correlations is somewhat enigmatic: both ^{54}Cr and ^{48}Ti are produced primarily by the s-process in AGB stars; in contrast, ^{48}Ca can only be produced in supernovae.

The relationship between these nucleosynthetic anomalies and oxygen isotopes is intriguing and suggests the possibility that O isotope variations may be nucleosynthetic after all. On the other hand, Warren (2011) suggested that the difference between carbonaceous chondrites and other material reflected a compositional zonation in the solar nebula, with carbonaceous chondrites having originally accreted in the outer solar system and other meteorite parent bodies, and the Earth, in the inner solar system. In that case, both mass independent O isotope fractionations and nucleosynthetic variations in other isotopes would both be functions of heliocentric distance and hence correlate. At present, there is no clear consensus on this question.

Nucleosynthetic isotopic variations have also been found in heavier elements such as Nd and siderophile ones such as Mo and Ru and the dichotomy seen in stony meteorites extends to iron meteorites as well (Figure 10.32). Curiously, however, CI chondrites have Mo and Ru isotopic compositions that plot in the noncarbonaceous field. The inverse correlation between ^{92}Mo anomalies and ^{100}Ru suggests these variations result from incomplete mixing of nuclides synthesized in different cosmic environments: ^{92}Mo is a p-process nuclide and hence produced by explosive nucleosynthesis (neutron star mergers and supernovae) while ^{100}Ru is synthesized by the s-process in red giants. To

---

* Gerald Wasserburg (1927–2016) completed his PhD dissertation of K-Ar dating under Harold Urey at the University of Chicago in 1954. He joined the faculty at California Institute of Technology in 1955 and remained there for the rest of his long career. In the 1960s, he established the "Lunatic Asylum" mass spectrometry laboratory in anticipation of the Apollo lunar samples and then rapidly produced high precision lunar ages. He was responsible for a remarkable number of discoveries, including the evidence of ^{26}Al in meteorites, and many analytical advances in mass spectrometry. He won the Crafoord Prize in Geosciences jointly with Claude Allègre in 1982. Nevertheless, one of his proudest achievements was serving as a rifleman in the U.S. Army in Europe during World War II.

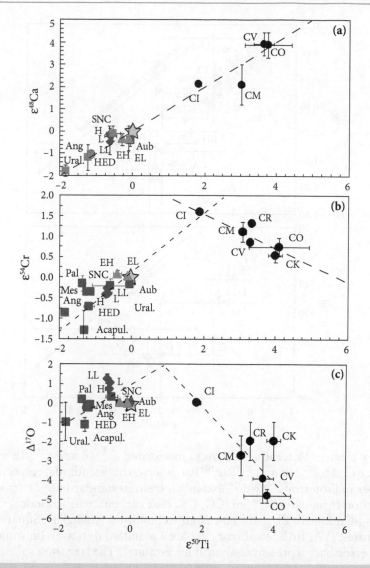

Figure 10.31 Isotopic variations in meteorites. ε^{50}Ti, ε^{54}Cr and ε^{50}Ca are the deviation in parts per 10,000 from the ^{50}Ti/^{47}Ti, ^{54}Cr/^{52}Cr and ^{48}Ca/^{44}Ca ratios, respectively, in terrestrial standards. Δ^{17}O is the displacement of δ^{17}O in per mil from the terrestrial fractionation line. CO, CK, etc.: carbonaceous chondrites (circles); H, L, LL: ordinary chondrites (diamonds); EL and EH: enstatite chondrites (triangles); squares are achondrites and stony irons: HED, Ural. (uralites), SNC, Aub (aubrites) Ang: angrites, Mes (mesosiderites), Pal. (pallasites). Error bars are 1 standard deviation (in some cases, only a single analysis exists, and error bars represent analytical uncertainty). The star shows the isotopic composition of the Earth and Moon.

a first approximation then, carbonaceous chondrites appear to be enriched in debris from neutron star mergers or supernovae relative to noncarbonaceous material, with lesser contributions of this debris in other solar system bodies. The Earth is in contrast relatively enriched in s-process nuclides: Qin and Carlson (2016) point out that all meteorites are deficient in s-process nuclides of

Sr, Zr, Mo, Ba, and Nd. These deficits are quite small, however, ranging from 0.02% to 0.004%. Resolvable s-process deficits have not been identified in heavier elements such as Hf, W, and Os.

In detail, these isotopic variations cannot be explained by simple binary mixing between s- and r/p-process nuclides; at least several nucleosynthetic sources must have been

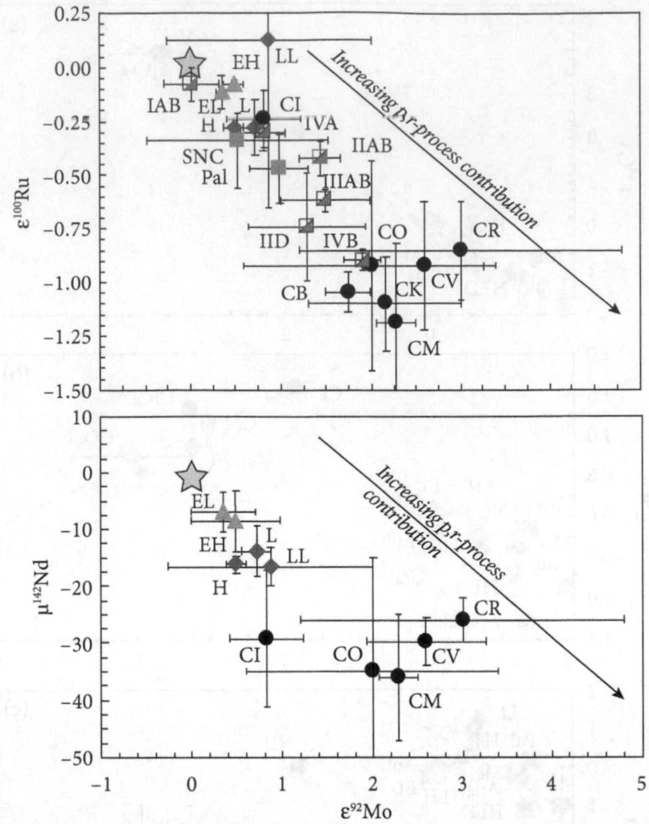

Figure 10.32   Heavy element isotopic variations in meteorites. ε^{92}Mo and ε^{100}Ru are the deviation in parts per 10,000 from ^{92}Mo/^{47}Mo and ^{100}Ru/^{101}Ru in terrestrial standards, respectively. $\mu_{142\text{Nd}}$ is the deviation in parts per million from ^{142}Nd/^{144}Nd in a terrestrial standard after correction for the radiogenic contribution from extinct ^{146}Sm. CO, CK, etc.: carbonaceous chondrites (circles); H, L, LL: ordinary chondrites (diamonds); EL and EH: enstatite chondrites (triangles); Square: SNC; half-filled squares: iron meteorites, IVA, IIAB, etc. Error bars are 1 standard deviation (in some cases, only a single analysis exists, and error bars represent analytical uncertainty). The star shows the isotopic composition of the Earth and Moon.

contributed material that was not completely homogenized in the solar nebula. Because isotopic variations would be quickly homogenized in the solar nebula, Qin and Carlson (2016) as well as other scientists suspect the addition of ejecta from a supernova and/or a neutron star merger as, or shortly before, planetesimals had already begun forming.

10.5   ASTRONOMICAL AND THEORETICAL CONSTRAINTS ON SOLAR SYSTEM FORMATION

As the preceding sections show, meteorites provide some tremendous insights into the formation of our solar system. However, before

we try to use these observations to draw some conclusions about how the solar system and the Earth formed, we need to consider several other sets of observations. The first of these is astronomical observations of star formation occurring elsewhere in the universe. As it turns out, star formation is more or less an everyday event in the universe, and we can watch it happening. The second is theoretical considerations of how solids condense from a hot gas of solar composition. We have seen that meteorites provide evidence for at least transient high temperatures in the early solar system. What can we expect when hot gas cools? We can use thermodynamics to find the answer. Finally, any model of solar

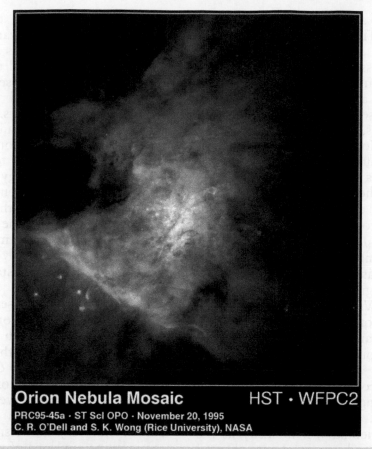

Orion Nebula Mosaic HST · WFPC2
PRC95-45a · ST ScI OPO · November 20, 1995
C. R. O'Dell and S. K. Wong (Rice University), NASA

Figure 10.33 The Great Nebula in Orion, shown in a Hubble Space Telescope photograph, is a cloud of gas and dust within which stars are forming. NASA Photo.

system formation must successfully produce the solar system as it exists. Consequently, we'll briefly review the nature of our solar system.

10.5.1 Evolution of young stellar objects

Astronomical observations have established that stars form when fragments of large molecular clouds collapse, as is occurring in the Great Nebula in Orion (Figure 10.33). Such clouds may have dimensions in excess of 10^6 AU* and masses greater than 10^6 M_O (solar masses). Typically, about 1% of the mass of these clouds consists of submicron-sized dust, about 1% is gaseous molecules heavier than He, and the remainder is H_2 and He gas. Gravity tends to make such clouds collapse upon themselves. Thermal

and rotational motion and internal pressures generated by turbulence tend to expand it. Magnetic fields, which couple to the ionized fraction of the gas, also play a key role in stabilizing these clouds. A careful balance between the forces tending to collapse and forces tending to expand the cloud can result in the cloud being stable indefinitely. Collapse of a part of a nebula can occur through the removal of a supporting force, magnetic fields in particular, or by an increase in an external force, such as a passing shock wave.

In addition to random density perturbations that can result in spontaneously collapse, shock waves could trigger cloud collapse and star formation. One potential source of shock is supernovae. Another is the density waves that manifest themselves as the spiral arms of the galaxies. We can think of the galactic arms

* AU stands for Astronomical Unit, which is the Earth–Sun distance, or 1.49×10^8 km.

as being similar to a traffic jam on the galactic orbital freeway. Though traffic continues to flow through the region of the traffic jam, there is nevertheless a sort of self-perpetuating high concentration of stars in the traffic jam itself. As clouds are pulled into the arms, they are compressed by collisions with other matter in the arms, leading to collapse of the clouds and initiation of star formation.

The Taurus-Auriga cloud complex is a good example of a region in which low-mass stars similar to the Sun are currently forming. The cloud is about 6×10^5 AU across, has a mass of roughly $10^4\, M_\odot$, a density of 10^2–10^3 atoms/cm^3, and a temperature around 10 K. Embedded within the cloud are clumps of gas and dust with densities two orders of magnitude higher than the surrounding cloud. Within some of these clumps are luminous protostars. About 100 stars with mass in the range of 0.2–3 $M_\odot$ have been formed in this cloud in the past few million years.

As the cloud collapses, it will warm adiabatically, resulting in thermal pressure that opposes collapse. Magnetic fields inherited from the larger nebula will intensify as the system contracts and can accelerate charged particles away from the forming star. Further intensification of the magnetic field occurs as an increasing fraction of the material ionizes as temperature increases. Finally, even small amounts of net angular momentum inherited from the larger nebula will cause the system to spin at an increasing rate as it contracts. For a cloud to collapse and create an isolated star, it must rid itself of over 99% of its angular momentum in the process. Otherwise the resulting centrifugal force will break up the star before it can form. Much of what occurs during early stellar evolution reflects the interplay between these factors.

Protostellar evolution of moderate-sized stars (i.e. stars similar to the Sun) can be divided into five phases, labeled –I, 0, I, II, and III, based on the spectra of their electromagnetic emission (Lada and Shu, 1990; Boss, 2005a). The first phase, the –I phase, is the initial collapse of a molecular cloud to form a protostellar core and nebular disk. No astronomical examples of this phase have been found, so understanding this phase depends entirely on theoretical calculations. Model calculations show that once a cloud fragment or clump becomes unstable, supersonic inward motion develops and proceeds rapidly as long as the cloud remains transparent and the energy released by gravitational collapse can be radiated away. Once the cloud becomes optically dense, the collapse slows. At this point, the protostellar core has a radius of about 10 AU and a mass of perhaps only 1% of the final mass of the star. When temperatures in the core reach 2000 K, energy is consumed in the dissociation of hydrogen, allowing further collapse and bringing the radius down to several times that of the eventual star. For a star of about 1 solar mass, the time scale is thought to be roughly 10^6 to 10^7 years. It is longer for smaller stars and shorter for larger ones.

For subsequent phases, observations at wavelengths ranging from radio waves to X-rays provide a wealth of information on protostellar objects and their nebulae. During phase 0, the protostar is deeply embedded in its surrounding cocoon of gas and dust and cannot be directly observed. At the beginning of the phase, the mass of the protostellar core is still very much smaller than that of the infalling envelope of gas and dust. In the meantime, angular momentum progressively flattens the envelope into a rotating disk, the stellar nebula. Material from the surrounding envelope continues to accrete to the disk, but mass is also transferred from the disk to the protostellar core at a slow rate.

The object L1551 IRS5 in the constellation Taurus is considered the prototypical Class I object, and detailed observations of it over the last decades have provided considerable insights into this stage of stellar evolution. Observations at radio wavelengths revealed there are actually two protostars about 45 AU apart with a combined mass of about 1.5 $M_\odot$ (binary stars are more common than individual stars such as the Sun). The protostars are deeply embedded in circumstellar disks that have diameters of about 20 AU, which are, in turn, embedded in a much larger envelope (~200 AU). Mass accretion rates are approximately $10^{-5}\, M_\odot$ yr^{-1} Because they are optically thick, temperatures in the interior of these disks cannot be determined directly. However, surface temperatures of the disks can be determined from the infrared spectra of these objects, 50–400 K at 1 AU, and drop off exponentially with distance. Models that reproduce these surface temperatures have disk interior temperatures of 200–1500 K at 1 AU and decrease exponentially with radial distance. The highest temperatures, which are enough to vaporize silicates, are likely short-lived and persist only for a period of perhaps 10^5 yr during which accretion rates are highest. More moderate temperatures, in the range of 200–700 K, could persist in the inner part of the disk for substantially longer than this.

A very interesting feature of L1551 and other Class I objects is strong "bipolar flows" or jets oriented perpendicular to the circumstellar disks that extend some 1000 AU from the disks. The jets consist of gas moving outwards at velocities of 200–400 km/sec. Within these jets, temperatures may locally reach 100,000 K. As the material in the jets collides with the interstellar medium it creates a shock wave that in turn generates X-rays. The physics that generates these jets is incompletely understood, but magnetic fields undoubtedly play a dominant role. In one theory, the X-wind model (Shu et al., 1997; Shang et al., 2000), the bipolar outflows are the cores of a much broader outflow that emerges from the innermost part of the circumstellar disk as it interacts with the strong magnetic field of the central protostar. As Shang et al. (2000) wrote, "in the X-wind model, the combination of strong magnetic fields and rapid rotation of the young star-disk system acts as an 'eggbeater' to whip out part of the material from the surrounding disk while allowing the rest to sink deeper in the bowl of the gravitational potential well." The jets and associated X-wind remove both mass and angular momentum from the system (Figure 10.34). The latter is particularly important, because as material is accreted from the disk to the star, conservation of angular momentum causes the star to rotate ever faster. Shang et al. (2000) estimated that about a third of the mass accreted to the disk and a larger fraction of the angular momentum is carried away by the X-wind. If this angular momentum were not lost in some way, the resulting centrifugal force would break up the star.

Phase II is represented by so-called *classical T-Tauri stars*, of which the star T-Tauri (now known to be a binary pair) is the type

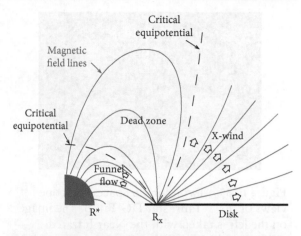

Figure 10.34 Cartoon illustrating the X-wind model. As a result of interaction with the accretion disk, the magnetic field lines of the young star truncate at radius R_x, which typically has a value of a few stellar radii. For disk material in stable orbit at R_x, two equipotential surfaces emerge from the disk, such that gravity dominates inward of the inner one and centrifugal force dominates outward of the outer one. Between these surfaces is a "dead zone" into which matter cannot freely flow. Thus, matter swept off the disk surface by magnetic forces will either be pulled into the star in "funnel flow" or move away from the star as the X-wind. After Shang et al. (2000).

example. During this phase, a visible star begins to emerge from its cocoon of gas and dust, but its brightness is highly variable, and it remains surrounded by its circumstellar disk (Figure 10.35). The star has a cool surface (4000 K) but luminosity several times greater than mature stars of similar mass (Figure 10.1). The luminosity is due entirely to continued accretion and gravitational collapse – fusion has not yet ignited in its interior. A T-Tauri star of one solar mass would still have a diameter several times that of the Sun. X-ray bursts from these stars suggest a more active surface than that of mature stars, likely driven by strong stellar magnetic fields and their interaction with the accretion disk. The surrounding disk is still warm enough to give off measurable IR radiation.

Accretion to the star continues in this phase but has dropped to rates of 10^{-6} to 10^{-8} $M_\odot$ per year. The accretion continues to drive bipolar outflows and associated X-wind

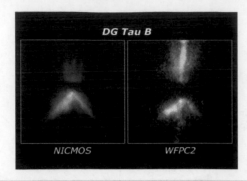

Figure 10.35 Two Hubble Space Telescope views of the T-Tauri star DG Tau B. The image on the left is taken with the Near Infrared Camera and Multi-Object Spectrometer, and the image on the right is taken with the Wide Field Planetary Camera. The accretion disk is a dark horizontal band in both images. The bipolar outflows appear green in the visible image. Infrared interferometry indicates there is a gap of about 0.25 AU between the star and the inner edge of the disk, which extends out about 100 AU. NASA photo.

(Figure 10.34). Typical mass loss rates from the flows and winds are $10^{-8}\,M_\odot$ per year. These winds may be important for a number of reasons. First, the mass loss is significant compared with the inferred accretion and infall rates, meaning there may be little or no net accretion. Second, the outwardly blowing gas has the potential to fling solids outward to the cooler, more distant parts of the disk. In the process, solids are lifted above the disk close to the star where they are exposed to the intense radiation and heated. This may explain the presence of CAIs in carbonaceous chondrites.

Both Class I and II objects can go through occasional *FU Orionis outbursts*. The name FU Orionis derives from a star in the Orion nebula that suddenly brightened by six magnitudes over six months in 1936. Its luminosity has been slowly decaying since then. During such an outburst, the disk outshines the central star by factors of 100–1000, and a powerful wind emerges, producing mass losses of $10^{-6}\,M_\odot$ per year in the case of FU Orionis, some four orders of magnitude greater than in quiescent T-Tauri stars. These outbursts are thought to be the result of greatly enhanced mass accretion rates, perhaps as high as $10^{-4}\,M_\odot$ per year. The cause

is unclear: thermal instability in the disk, changes in the structure of the disk, and the gravitational effects of a binary companion or a giant planet or protoplanet orbiting close to the star have all been suggested as causes of the enhanced mass accretion rate. The outbursts appear to last no more than a century. By the end of Phase II, the star is perhaps 2–6 million years old.

Phase III is represented by *weak-lined T-Tauri stars*, so called because spectral emission and absorption lines are much weaker than in classical T-Tauri stars. In addition, the excess infrared emission that characterizes classical T-Tauri stars is absent in weak-lined T-Tauri stars. The inference from these observations is that the disk has largely dissipated by this stage. Like classical T-Tauri stars, weak-lined T-Tauri stars are cooler yet more luminous than mature main sequence stars of similar mass so that they are closer to the main sequence on the Hertzsprung–Russell diagram than classical T-Tauris.

Weak-lined T-Tauris are particularly luminous in the X-ray part of the spectrum. These X-rays are thought to be produced in hot plasma during magnetic reconnection events above the stellar surface. The same process in the Sun produces bright solar flares, but flares of weak-lined T-Tauris are 100–1000 times more powerful. Outflows and winds subside to those of typical main sequence stars as accretion ends and the star reaches its final mass. During the weak-lined T-Tauri phase, the star contracts to its final radius and density. At the end of this process, fusion ignites in the core and the luminosity and temperature of the star settles onto the main sequence. The entire process from Phase 0 through Phase III consumes perhaps 10 million years.

10.5.2 The condensation sequence

In the preceding section, we saw young stellar objects and their surrounding disks are highly energetic and at least locally very hot environments. We learned earlier that chondrites contain a variety of components, chondrules, CAIs, and AOAs, that provide evidence of at least local and transient high temperatures as our solar system was born. It is thus useful to ask what the chemical effects of temperature are. For example, suppose we heated material of solar or chondritic composition to the point where everything evaporated. What would be the sequence of condensation? The theoretical condensation sequence was been calculated from thermodynamic data, originally by Larimer, Grossman, and Lewis, all of whom worked with E. Anders at the University of Chicago (e.g., Larimer, 1967; Grossman, 1972). The most recent and widely used computations are those of Lodders (2003). The condensation temperature of an element reflects its vapor pressure, its tendency to react with other elements to form compounds in the gas or solid solutions or alloys in the solid, and its abundance in the gas. Let's consider two examples of condensation sequence calculations. First, consider iron, which is a particularly simple case since much of it condenses as Fe metal:

$$Fe_{(g)} \rightarrow Fe_{(s)} \qquad (10.2)$$

Assuming ideality, the partial pressure of iron is simply its mole fraction in the gas times the total pressure (P_T). Since hydrogen and helium are by far the dominant elements in the gas (~98%), the mole fraction can be approximated as the solar $Fe/(^1/_2 H + He)$ ratio (the $^1/$ arises from hydrogen's presence as H_2). Thus, the partial pressure of Fe is written as:

$$p_{Fe} = \frac{[Fe]_\odot}{\frac{1}{2}[H]_\odot + [He]_\odot} P_T \qquad (10.3)$$

where $[Fe]_\odot$, $[H]_\odot$, and $[He]_\odot$ are the solar abundances of Fe, H, and He, respectively, and P_T is total pressure. Once condensation begins, we can express the equilibrium constant for this reaction as the ratio of the partial pressure of Fe in the gas to the concentration in the solid:

$$K = \frac{p_{Fe}}{[Fe]_s} \qquad (10.4)$$

where $[Fe]_s$ is the concentration in the solid and p_{Fe} is the partial pressure in the gas.

Condensation begins when the partial pressure of Fe exceeds the equilibrium vapor pressure of solid Fe. Since:

$$\Delta G = -RT \ln K \qquad (3.86)$$

the equilibrium constant can be written as:

$$\ln K = -\frac{\Delta H_v}{RT} + \frac{\Delta S}{R} \qquad (10.5)$$

where ΔH_V and ΔS_V are the enthalpy and entropy of vaporization. Once condensation of an element begins, its partial pressure drops by $(1 - \alpha)$ where α is the fraction condensed. Hence, eqn. 10.3 becomes:

$$p_{Fe} = \frac{(1 - \alpha)[Fe]_\odot}{\frac{1}{2}[H]_\odot + [He]_\odot} P_T \qquad (10.6)$$

Combining eqns. 10.4, 10.5, and 10.6, the equation describing condensation is:

$$\ln(1 - \alpha) = -\frac{\Delta H_v}{RT} + \frac{\Delta S}{R} - \ln \frac{[Fe]_\odot}{\frac{1}{2}[H]_\odot}$$
$$+ \ln P_T + \ln [Fe]_s \qquad (10.7)$$

Now consider elements such as Mg and Al, which form a variety of compounds in both the gas and solid, complicating the calculation considerably. The reaction for the condensation of spinel can be written as:

$$Mg_{(g)} + 2Al(OH)_{(g)} + O_2 \rightleftharpoons MgAl_2O_{4(s)} + H_2 \qquad (10.8)$$

The equilibrium constant for this reaction is:

$$K = \frac{a_{MgAl_2O_4} p_{H_2}}{p_{Mg} p_{AlOH}^2 p_{O_2}} \qquad (10.9)$$

The first step is to compute partial pressures of the gaseous species. For example, for AlOH and O_2 above, we may write reactions:

$$2H_2O \rightleftharpoons 2H + O_2 \quad \text{and} \quad Al + OH \rightleftharpoons AlOH$$

and calculate equilibrium constants for them from the free energies of these species, for example:

$$K = \frac{p_{AlOH}}{p_{Al} p_{OH}} \qquad (10.10)$$

For each element, an additional constraint is imposed by the total abundance of that element in the gas. Thus, for example:

$$[AlOH]_g + [Al]_g + [AlH] + [Al_2O]$$
$$+ [AlH] + \ldots = [Al]_\odot - [Al]_s \qquad (10.11)$$

where $[Al]_s$ is the total of aluminum in condensed phases. Combining equilibrium constant equations such as 10.10 with mass balance equations such as 10.11 leads to a series of simultaneous equations. These can be solved by successive approximation using a computer in a procedure very similar to the speciation calculations discussed in Chapter 6. Values of equilibrium constants such as eqns. 10.9 and 10.10 can then be computed from ΔH and ΔS using eqn. 10.5, which requires an extensive and accurate thermodynamic database.

Further complications arise when solid solutions form. For example, the spinel term in eqn. 10.8 is 1 if spinel is treated as a pure $MgAl_3O_4$. But other elements can substitute for Mg and Al forming solid solutions that reduced the activity of $MgAl_2O_4$. Values for activity coefficients are difficult to obtain, and ideal solid solution is generally assumed. Thus, the activity of spinel in eqn. 10.9 would be replaced by its mole fraction. Solid solution generally results in the condensation of an element at a higher temperature than if a pure component were the condensed phase.

Some elements, such as Au and Cu, will condense primarily either by reaction with already-condensed Fe metal grains, or by condensing with Fe metal. At lower temperatures (around 670 K for $P_T = 10$ Pa), the Fe metal will react with H_2S gas to form FeS.

Moreover, from the onset of condensation a small but increasing amount of the Fe will react with H_2O gas to form FeO that dissolves in the silicates. A marked increase in the Fe content of silicates occurs around 400–500 K.

Figure 10.36 shows two theoretical sequences calculated by Larimer (1967). In the "fast cooling" sequence, matter does not react with nebular gas after it has condensed – this is effectively fractional condensation, analogous to fractional crystallization. At any given time and temperature, the solid phases are a mixture of material condensed over a range of temperature. In the "slow cooling" or "equilibrium" sequence, condensed material continually reacts and re-equilibrates with the gas as temperature drops, so that at any time, the solid assemblage is equilibrium with the gas at that temperature. Figure 10.37 illustrates the minerals expected at a given temperature somewhat more clearly. The condensation sequence depends critically on total pressure and H pressure; the sequence shown is for relatively low total and H pressure. At relatively high pressure, metallic Fe liquid is the first phase to condense.

In a nutshell, the sequence goes like this (Figure 10.37). The first elements to condense would be Re and the most refractory of the platinoid metals (Os, Ir, Ru), which would condense as metallic phases. Since these are extremely rare elements, they would

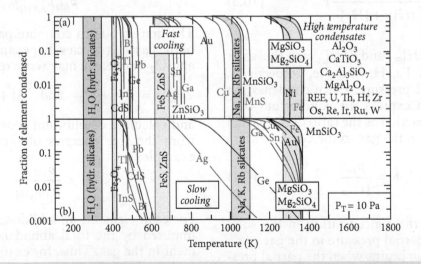

Figure 10.36 Condensation sequence of a gas with solar composition. Fast cooling (a) leads to disequilibrium and condensation of pure elements and compounds. In slow cooling (b), condensed solids are assumed to continually re-equilibrate with the gas. Black lines show condensation of silicates and oxides; red lines show condensation of metals and sulfides. After Larimer (1967).

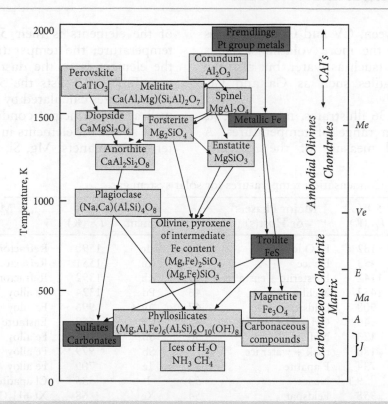

Figure 10.37 Simplified mineralogical condensation sequence. After McSween (1987).

likely form very small metal grains. As we noted earlier, small nuggets of such metal, called *Fremdlinge* (the German word for "strangers"), are found as inclusions in CAIs. Following this would be condensation of oxides and silicates of Ca, Al, and Ti. They should be rich in refractory trace elements such as U, Th, Zr, Ba and the REE. This closely matches the composition of the CAIs, which is the key piece of evidence suggesting that CAIs are high-temperature condensates.

Next in the condensation sequence should come metallic Fe–Ni and compounds richer in the moderately refractory elements such as Mg and Si: olivines and pyroxenes. If the cooling takes place under equilibrium conditions, the high-temperature (CAI) assemblage should react to form anorthite as well, and at lower temperature when Na condenses, plagioclase. These phases are the ones that predominate in chondrules, with the important caveat that chondrules are poorer in metal than the condensation sequence would predict. Since these phases condense at temperatures similar to Fe–Ni metal, some process must

have separated metal from silicates before formation of the chondrules.

The Fe should also largely react out to form more Fe-rich olivine and pyroxene. At lower temperature, S condenses and reacts with Fe to form sulfides. At even lower temperature, the Fe reacts with O to form magnetite and the silicates react with water vapor to form hydrated silicates. Sulfates, carbonates and organic compounds will also form around these temperatures.

If equilibrium conditions prevail, only the last-named compounds would exist when condensation was complete, but all might exist if disequilibrium prevails. The CI chondrites seem to be very similar in composition and petrography to equilibrium condensates (down to 300 K or so), or, more accurately, accretions of equilibrium condensates. The other carbonaceous chondrites approximate, to varying degrees, aggregates of disequilibrium condensates. In particular, the CV and CO chondrites contain both the highest temperature condensates (CAIs) and lowest temperature material (hydrated silicates) and much of everything expected to condense inbetween. CV and CO chondrites are depleted in the more volatile elements and compounds (such as water, but also the moderately volatiles such as Ga and Ge, alkalis).

As Figure 10.36 illustrates, most elements condense over a range of temperatures. A commonly used measure of the *volatility* of the elements is their 50% condensation temperature: the temperature at which half the element is in the dust and 50% in the gas. Table 10.5 lists the 50% condensation temperatures calculated by Lodders (2003) as well as the principal condensing phase. The most abundant elements in the Earth and the terrestrial planets, Mg, Si, and Fe have 50%

Table 10.5 50% Condensation temperatures for solar system gas.

Z	Element	50% T_C (K)	Major Phase(s) or Host(s)	Z	Element	50% T_C (K)	Major Phase(s) or Host(s)
1	H	182	H_2O ice	42	Mo	1590	Refractory alloy
2	He	< 3	He ice	44	Ru	1551	Refractory alloy
3	Li	1142	Forsterite + enstatite	45	Rh	1392	Refractory alloy
4	Be	1452	Melilite	46	Pd	1324	Fe alloy
5	B	908	Feldspar	47	Ag	996	Fe alloy
6	C	40	$CH_4 \cdot 7H_2O + CH_4$ ice	48	Cd	652	Enstatite + troilite
7	N	123	$NH_3 \cdot H_2O$	50	Sn	704	Fe alloy
8	O	180	rock + water ice	51	Sb	979	Fe alloy
9	F	734	F apatite	52	Te	709	Fe alloy
10	Ne	9.1	Ne ice	53	I	535	Cl apatite
11	Na	958	Feldspar	54	Xe	68	$Xe \cdot 6H_2O$
12	Mg	1336	Forsterite	55	Cs	799	feldspar
13	Al	1653	Hibonite	56	Ba	1455	Titanate
14	Si	1310	Forsterite + enstatite	57	La	1578	Hibonite + titanate
15	P	1229	Schreibersite	58	Ce	1478	Hibonite + titanate
16	S	664	Troilite	59	Pr	1582	Hibonite + titanate
17	Cl	948	Sodalite	60	Nd	1602	Hibonite
18	Ar	47	$Ar \cdot 6H_2O$	62	Sm	1590	Hibonite + titanate
19	K	1006	Feldspar	63	Eu	1356	Hibonite + titanate + feldspar
20	Ca	1517	Hibonite + gehlenite	64	Gd	1659	Hibonite
21	Sc	1659	Hibonite	65	Tb	1659	Hibonite
22	Ti	1582	Titanate	66	Dy	1659	Hibonite
23	V	1429	Titanate	67	Ho	1659	Hibonite
24	Cr	1296	Fe alloy	68	Er	1659	Hibonite
25	Mn	1158	Forsterite + enstatite	69	Tm	1659	Hibonite
26	Fe	1334	Fe alloy	70	Yb	1487	Hibonite + titanate
27	Co	1352	Fe alloy	71	Lu	1659	hibonite
28	Ni	1353	Fe alloy	72	Hf	1684	HfO_2
29	Cu	1037	Fe alloy	73	Ta	1573	Hibonite + titanate
30	Zn	726	Forsterite + enstatite	74	W	1789	Refractory alloy
31	Ga	968	Fe alloy + feldspar	75	Re	1821	Refractory alloy
32	Ge	883	Fe alloy	76	Os	1812	Refractory alloy
33	As	1065	Fe alloy	77	Ir	1603	Refractory alloy
34	Se	697	Troilite	78	Pt	1408	Refractory alloy
35	Br	546	Cl apatite	79	Au	1060	Fe alloy
36	Kr	52	$Kr \cdot 6H_2O$	80	Hg	252	Troilite
37	Rb	800	Feldspar	81	Tl	532	Troilite
38	Sr	1464	Titanate	82	Pb	727	Fe alloy
39	Y	1659	Hibonite	83	Bi	746	Fe alloy
40	Zr	1741	ZrO_2	90	Th	1659	Hibonite
41	Nb	1559	Titanate	92	U	1610	Hibonite

condensation temperatures between 1300 and 1350 K. Elements condensing at higher temperatures are considered "refractory". Significantly, the relative concentrations of refractory elements are nearly constant in the various classes of chondrites.

10.5.3 The solar system

The solar system consists of a central star, the Sun, and a variety of bodies that orbit it. These are somewhat arbitrarily divided into planets, asteroids, comets, Kuiper Belt objects, and so on. Table 10.6 lists some data regarding the planets. The planets can be divided into three groups based on their size, density and composition. These are:

- The terrestrial planets:
  Mercury
  Venus
  Earth–Moon
  Mars
  (asteroids)
- The gas giants:
  Jupiter
  Saturn

- The outer icy planets:
  Uranus
  Neptune
- The Kuiper Belt and Oort Cloud.

Seven of the eight planets have orbits that fall on a single plane, ±3°. The same seven have nearly circular orbits, with eccentricities all less than 0.1. Mercury's orbit is inclined some 7° with an eccentricity of 0.2. (Pluto's orbit is inclined 17° and has an eccentricity of 0.25; this highly anomalous orbit is one reason it is no longer considered a planet). Most major satellites of the planets also orbit in nearly the same plane. The Sun's equator is inclined some 7° to this plane. Thus, the angular momentum vectors of major solar system objects are all rather similar, consistent with formation from a single rotating nebula. Rotational vectors of planets are generally inclined to their orbital vectors, some highly so, and Venus and Neptune have retrograde rotations (as does Pluto).

Pluto, once considered the ninth planet, is more than an order of magnitude smaller than Mercury and smaller than some of the major

Table 10.6 Physical data regarding major solar system objects.

	Mass kg	Semi-major axis AU	Radius km	Density (g/cc)	1 atm density (g/cc)	Principal atmospheric components
Sun	1.99×10^{30}			6.7×10^5	1.4	
Mercury	3.35×10^{23}	0.39	2.44×10^3	5.42	5.3	–
Venus	4.87×10^{24}	0.72	6.05×10^3	5.24	3.95	CO_2, N_2, Ar
Earth	5.98×10^{24}	1.0	6.38×10^3	5.52	4.03	N_2, O_2, Ar
Moon	7.35×10^{22}		1.74×10^3	3.3	3.4	–
Mars	6.42×10^{23}	1.6	3.39×10^3	3.93	3.7	CO_2, N_2, Ar
Asteroids	4×10^{21}	2.1 – 3.1				
Ceres	9.4×10^{20}	2.76	473	2.16		–
Vesta	2.6×10^{20}	2.36	267	3.46		–
Jupiter	1.90×10^{27}	5.2	6.99×10^4	1.31		H, He
Io	8.63×10^{22}		1.82×10^3	3.42		–
Europa	4.71×10^{22}		1.55×10^3	3.03		–
Ganymede	1.51×10^{23}		2.63×10^3	1.98		–
Callisto	1.06×10^{23}		2.40×10^3	1.83		–
Saturn	5.69×10^{26}	9.6	5.95×10^4	0.69		H, He
Titan	1.38×10^{23}		2.58×10^3	1.88		N_2, CH_4
Uranus	8.73×10^{25}	19.1	2.54×10^4	1.30		H, He, CH_4
Neptune	1.03×10^{26}	30.8	2.13×10^4	1.76		H, He, CH_4
Triton	2.14×10^{22}		1.35×10^3	2.08		
Pluto	1.30×10^{22}	39.5	1.19×10^3	1.85		

satellites of Jupiter, Saturn, and Neptune, and has a very anomalous orbit. It is more correctly considered a dwarf planet within the *Kuiper Belt*. NASA's New Horizon spacecraft flew by Pluto in 2016. Its surface consists largely of ices of N_2, CO, and water. Its bulk density is, however, greater than these materials, indicating it has a rocky core. The Kuiper Belt, which lies between 30 to 50 AU from the Sun, is a great ring of debris, similar to the asteroid belt but of much lower density material – presumably dominated by hydrocarbons and ices of H_2O, CH_4, and NH_3 with lesser amounts of silicates. This is the source region for short-period, (Jupiter family) comets. There are estimated to be over 70,000 Kuiper Belt objects with a diameter greater than 100 km, with a total mass similar to that of the Earth. In addition to Pluto and Charon, five of these objects have diameters greater than 1000 km, two of which, Makemake and Haumea (named for Polynesian gods), are considered dwarf planets. On January 1, 2019, the New Horizon's space craft flew by the Kuiper Belt object 2014 MU_{69}, informally named *Ultima Thule*, whose semi-major axis is 42 AU. (This object has now been officially named "Arrokoth", the word for sky in the Powhatan language of Native Americans from Virginia and Maryland.) It is a contact binary (two approximately spherical bodies in contact) with a diameter of ~30 km and distinctly red in color, due to conversion of simple organic molecules to complex ones known as tholins by long exposure to ultraviolet radiation (a process known as *space weathering*). Spectral analysis identified water ice and methanol on the surface and an apparent absence of silicates.

There are a number of large objects even further out, so-called *Scattered Disk Objects*, that, as the name implies, are thought to have formed closer in and have been gravitationally flung into distant orbits. *Eris*, discovered in 2003, has a radius just slightly less than Pluto's but is more massive (1.66×10^{22} kg) and like Pluto is considered a minor planet. Spectroscopy indicates a surface of methane ice, but like Pluto, its density requires a rocky core. It has an elliptical orbit with semi-major axis of 68 AU and has a moon of its own: *Dysnomia*. Sedna, also discovered in 2003, is the most distant. It has a radius of roughly 1000 km and a semi-major axis of 507 AU but its orbit is highly elliptical so it can be as

close as 76 AU and as distant as 940 AU. It is one of the reddest objects in the Solar System, again a result of space weathering.

Finally, the Oort Cloud is a region from 50,000 to 100,000 AU where long-period comets are thought to originate. Comets appear to be low-density icy dustballs. Based on spectral analysis of Comet Tempel 1 by NASA's Deep Impact mission and of 67P/Churyumov–Gerasimenko by ESA's Rosetta mission, they consist principally of water ice, CO_2 ice, HCN, a variety of hydrocarbons including polycyclic aromatic hydrocarbons and the amino acid glycine, amorphous carbon, and lesser amounts of silicates, sulfides, and iron alloys. While most comets are quite small, a few km to tens of km in diameter, it is estimated that the total mass of the Oort cloud is 5–100 Earth masses.

The terrestrial planets all consist of silicate mantles surrounding Fe–Ni metal cores and thin atmospheres that are highly depleted in H and He compared with the Sun. The gas giants Jupiter and Saturn are much more similar in composition to the Sun. The atmosphere of Jupiter is 81% H and 18% He by mass (compared with 71% and 28%, respectively, for the Sun), with CH_4 and NH_3 making up much of the rest. Saturn's atmosphere consists of 88% H and 11% He. The He depletion of both these atmospheres relative to the solar composition reflects a concentration of He in the interior. On the whole, the H/He ratio of Jupiter is close to the solar value. Elements heavier than He are about five times enriched in Jupiter compared with the Sun. The H/He ratio of Saturn as a whole is about a factor of 3 lower than the solar ratio and elements heavier than He are roughly 15 times enriched compared with the Sun. The nature of these planets' interiors is not entirely certain. Jupiter probably has a core consisting of liquid or solid metal and silicates with a mass roughly 15 times that of the Earth. Saturn probably has a similar core with a mass 100 times that of the Earth. Surrounding the core are layers of liquid metallic H and ordinary liquid H, both containing dissolved He. The icy planets consist of outer gaseous shells composed of H and He in roughly solar ratio with a few percent CH_4 surrounding mantles consisting of liquid H_2O, CH_4, H_2S, NH_3, H, and He, and finally liquid silicate-metal cores. Elements heavier than He are about 300 times enriched in Neptune and Uranus compared with the Sun.

In a gross way, this compositional pattern is consistent with a radial decrease in nebular temperature: the terrestrial planets are strongly depleted in the highly volatile elements (e.g., H, He, N, C) and somewhat depleted in moderately volatile elements (e.g., K, Pb). From what can be judged from reflectance spectra, the asteroids also fit this pattern: the inner asteroids (sunward of 2.7 AU) are compositionally similar to the achondrites and ordinary chondrites, which are highly depleted in volatile and moderately volatile elements. The outer asteroids (beyond 3.4 AU) are richer in volatile elements and appear to be similar to carbonaceous chondrites.

The planet that we know the most about (other than Earth) is Mars. Observations on Mars and its composition come from (1) spectroscopic measurements by a variety of spacecraft orbiting Mars (as of 2020 there were six active), (2) a variety of landers equipped with instruments capable of analyzing rocks and soil (to date there have been ten successful landers; two currently active), and (3) the SNC meteorites (of which there are 224, the vast majority of which are finds). Most of these are igneous cumulates, although the shergottites include basalts. NWA 7034 and 7533 (thought to be pieces of the same meteoroid) are regolith breccias consisting of clasts of a variety of igneous rocks.

This has enabled the achievement of a vague understanding of the composition and history of Mars. Comparing the volatile inventory of Mars with that of Earth, Mars at first appears depleted in volatile elements. It has a much smaller atmosphere than the Earth (surface pressures are 0.006 atm). Like Venus, the Martian atmosphere is dominated by CO_2 (95%) with N_2 (2.7%) and Ar (1.6%) as the other major components. Principal trace gases include CO (0.8%) and O_2 (0.14%). However, it is now clear from orbiter and surface observations that significant amounts of liquid water existed on the Martian surface during at least its first billion years or so, and there is evidence of some small ephemeral streams at present and enough subsurface soil ice to cover the planet in ~11 m of water. To attain the necessary temperatures, Mars must have had CO_2 pressures at its surface of 5–10 atm. This early atmosphere has been lost, a consequence of several factors, including lower gravity and the lack of a geomagnetic field that serves to prevent erosion of the atmosphere by the solar wind. Thus, the depletion of highly volatile elements on Mars may be a secondary feature.

Like the Earth, Mars has an iron core, but it appears to comprise a smaller fraction of planetary mass than Earth's. The Martian core has a diameter of ~1600–1750 km and is about 22–24% of the mass of the planet (the Earth's core, by comparison is 32% of the mass of the planet). The core may be liquid and of lower density (6200–6400 km/m^3) than the Earth's, indicating it contains up to 18% sulfur. Mars presently has a weak geomagnetic field, but remnant magnetism in some of the more ancient regions on Mars, suggests that it did prior to 3.7 Ga (see review by Mangold et al., 2016). The smaller core translates into a more iron-rich Martian crust and mantle and is likely due to accretion of Mars under more oxidizing conditions than the Earth, resulting in a lower $Fe_{metal}/Fe_{silicate}$ ratio. Both SNC meteorites and lavas analyzed by the NASA rovers are richer in Fe than comparable terrestrial basalts (Figure 10.38).

The surface of Mars consists almost exclusively of basalts (as does the Earth's oceanic crust) and sedimentary rocks derived from them. However, the NASA's Curiosity rover has identified several examples of siliceous/feldspathic rocks in Gale Crater that are more similar to the Earth's continental crust (Figure 10.38). One clast in the meteorite NWA 7533 is also feldspathic. It is possible that basaltic lava flows have buried a more silicic crust, but even in that case the Martian crust, which is anywhere from a few tens to 100 km thick, in certainly more mafic than the terrestrial continental crust.

There is evidence that Mars is richer in moderately volatile elements than the Earth. Analyses of both Martian soil and the composition of SNC meteorites suggest a K/U ratio about 19,000, whereas this ratio is about 13,800 in the Earth. The significance of this ratio is that U is a highly refractory element while K is a moderately volatile one. Sr–Nd isotope systematics of SNC meteorites appear to define an array shifted to higher $^{87}Sr/^{86}Sr$ ratios compared with that of the Earth, implying Mars has a Rb/Sr ratio of about 0.07 compared with 0.03 for the Earth (Taylor, 1992). Pb isotope ratios of these meteorites indicate a $^{238}U/^{204}Pb$ ratio of about 5 for Mars, compared with the

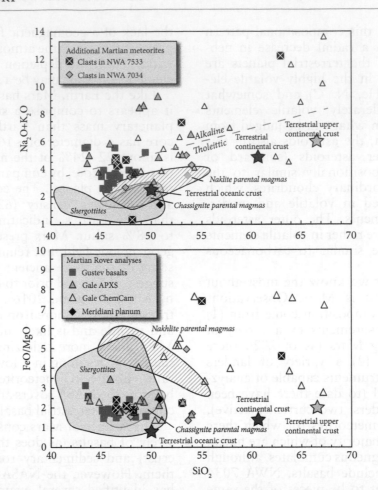

Figure 10.38 Total alkalis and FeO/MgO ratios as a function of SiO$_2$ in Martian rocks. Martian rocks are richer in alkalis and iron than terrestrial rocks. Mars appears to have a lower Al/Si ratio (not shown). Nakhlite and Chassignite parental magmas are calculated based on melt inclusions. Gustev analyses are based on APXS (alpha particle X-ray spectroscopy) analyses of the Spirit rover. The Curiosity rover can perform analyses with a laser spectrometer (ChemCam) as well as with an APXS. The latter are of better quality. Based on data summary in Filberto (2017).

terrestrial value of ~8.5. These ratios are consistent with the higher K/U of Mars, since Rb and Pb are both moderately volatile elements while Sr is a refractory one. Analyses by Martian rovers support this inference as well: as Figure 10.38 shows Martian igneous rocks typically have higher total alkalis than terrestrial rocks of similar SiO$_2$ content. Thus, Martian moderately volatile/refractory element ratios appear to be systematically higher than terrestrial ratios.

The comparison of Venus with the Earth is particularly interesting, although we know considerably less about Venus than Mars. What we do know is based on gamma ray

spectroscopy and analyses of the Venetian atmosphere by spacecraft. Although the two planets are of similar size, the Venetian atmosphere is almost 100 times more massive than that of the Earth. Whereas Venus's atmosphere is dominated by CO$_2$, that of the Earth is dominated by N$_2$ and O$_2$. However, the differences in both atmospheric mass and the abundance of CO$_2$ may reflect the difference in geological and biological processes, on the two planets. Both planets appear to have similar relative abundances of carbon and nitrogen (Prinn and Fegley, 1989). In the case of the Earth, however, most of the surficial carbon is locked up in carbonates and organic carbon in rocks. This is in part due to biological activity, as is the presence of O_2 in the terrestrial atmosphere. There are, however, other differences that are more fundamental. The Venusian atmosphere is richer in noble gases than is the terrestrial one. The $^{40}Ar/^{36}Ar$ ratio of the Venusian atmosphere is about 1.15, whereas that of Earth is about 300. Since ^{40}Ar is produced by radioactive decay, it is related to the planetary K/Ar ratio. The terrestrial $^{40}Ar/^{36}Ar$ is high because the Earth captured relatively little primordial Ar; most of the Ar in the terrestrial atmosphere has been produced by decay of ^{40}K. Apparently, Venus captured much more primordial Ar and has a much lower K/Ar ratio than the Earth.

Although the Earth is depleted in noble gases relative to Venus, it appears to be richer in H_2O (Prinn and Fegley, 1989). Furthermore, the Earth and Venus appear to have similar K/U ratios, implying similar depletions in the moderately volatile elements (Taylor, 1991). Thus, in detail, we find the compositional differences between planets cannot entirely be explained by radially decreasing temperature. The noble gas-rich nature of Venus is just one example.

10.5.4 Other solar systems

Thousands of planets (>4100 in 2020) have been discovered orbiting thousands of stars, with >650 systems with multiple planets. Several thousand more suspects have been identified but not yet confirmed. Multiple planets occur in about 15% of the cases, and as many as eight planets have been found orbiting a single star. Metal-rich stars are more likely to have planets than metal-poor ones; for stars having Fe/H ratios comparable to or greater than that of the Sun, planets have been found in about 15% of stars. Large planets orbiting near their stars are more readily detectable than small ones so that the range of sizes and orbital radii of known planets represents a highly skewed sample of all exoplanets. Despite this bias, several conclusions from the discovery of exoplanets can be drawn. First, they are not rare and terrestrial planets are common. It appears that 1 in 5 sun-like stars has an Earth-sized planet in the habitable zone (the range of orbital distance where liquid water could be present on the planet's surface) (Petagura et al., 2013). Second, unlike our own solar system, large planets can be present quite near the star. Thus, planets and solar systems may be a normal consequence of star formation, but the distribution of planets in our solar system is not necessarily typical. Finally, planets are generally substantially smaller than their central stars; they are not merely undersized companions. This latter observation suggests that planets form in a fundamentally different way than stars and isolated brown dwarfs.

10.6 BUILDING A HABITABLE SOLAR SYSTEM

We have now examined star and solar system formation from several perspectives: observations about the planets of our own solar system, observations about conditions in the early solar system derived from meteorites, observations about stars presently forming elsewhere in the galaxy, and attempts to reproduce processes and conditions in both laboratory or mathematical and computer simulations. Let's first review how these observations constrain the process of formation of *our* solar system 4.56 billion years ago, and then try to deduce what actually happened back then.

10.6.1 Summary of observations

The planets in our solar system show a very strong compositional zonation. The four inner planets are strongly depleted in volatile elements relative to the Sun, with some suggestion of increasing depletion inward. The two giant planets have compositions close to that of the Sun, indicating they formed from a nebula that had experienced little chemical fractionation. The outer icy planets appear slightly depleted in the most volatile elements (H and He) but are much more volatile-rich that the terrestrial planets.

Chondrites can be viewed as mixtures of four principal components: CAIs, chondrules, amoeboid olivine aggregates (AOAs), and matrix. The CAIs consist of highly refractory material, very similar in composition to what we would expect if chondritic dust were heated to the point where 95% of it evaporated, or, conversely, the first 5% of a gas of chondritic composition condensed. In either case, minimum *sustained* temperatures of ~1700 K are required (assuming pressures of around 10 Pa). AOAs appear to have condensed from a gas at only slightly lower temperatures. Chondrules also provide evidence of high temperatures, but in these cases the peak temperatures must have been of short duration. Finally, chondritic groundmass material, particularly in carbonaceous chondrites, includes a variety of materials that condensed at low temperatures as well as presolar grains – ejecta of red giants, supernovae, and kilonovae – that escaped nebular processing. While most such grains (e.g., graphite, SiC, diamond) are inherently refractory and hence would not necessarily be destroyed by high temperatures, these grains retain significant quantities of noble gases, suggesting they never experienced substantial heating.

Variations in the chemical composition of chondrites clearly indicate chemical heterogeneities within the solar nebula. Much of this is clearly related to volatility, implying significant variation in temperature in space and/or time in the nebula. Other variations relate to oxygen fugacity. Since H_2 is the principal reductant and it dominates the gas while O constitutes a significant fraction of condensed matter, variation in oxygen fugacity most likely reflects variation in the ratio of gas to dust. In addition, there must have been significant variations in the metal/silicate ratio within the nebula to explain chondritic variations.

Planetesimals that were the parent bodies of achondrites and irons underwent sufficient heating to melt and differentiate. They formed, melted, and differentiated within a few million years of the CAI formation (taken as the beginning of the solar system).

We can also draw a number of conclusions from the isotopic compositions of meteorites. First, nucleosynthetic-related variations in isotopic composition of a number of elements have now been demonstrated in bulk meteorites. These variations are small, generally a few hundredths of a percent and less, in contrast with the much larger isotopic heterogeneity of presolar grains preserved in some chondrites. Second, these variations reveal a clear distinction between carbonaceous chondrites and other meteoritic and planetary material, suggesting these two groups

formed in different regions of an incompletely homogenized solar nebula. Third, short-lived radionuclides were present when the solar system formed. While some of these do not rise above galactic background levels, some must have been produced and injected into the molecular cloud that ultimately formed the solar system during or very shortly before its collapse. Decay of these radionuclides, as well as conventional dating, suggests the time scale for nebular processing and planet formation was a few million to a few tens of millions of years.

Chondrites are a mix of materials formed under different conditions in different environments. With some understanding of the star-forming process, we can begin to discern what these environments were, as illustrated in Figure 10.39.

10.6.2   Formation of the planets

Nebular disks and possible planets have been imaged in very young stars still surrounded by nebular disks (e.g., Lagrange et al., 2010). These, together with theoretical modeling and observations on meteorites and the present solar system, inform our understanding of how the planets formed. As we saw earlier, observed surface temperatures of nebular disks imply that interior temperatures in the inner parts of these disks are hot enough to vaporize much of the dust. On time scales of perhaps 10^5 years or so, the inner disk cools to the point where solids can recondense. CAIs condensed at temperatures of 1700 K or so and likely formed at this time. Many were subsequently reheated and subjected to partial melting and evaporative loss. The ages of such "processed" CAIs are indistinguishable from apparently "primary" CAIs, suggesting that the period of their formation was short, perhaps 50,000 years. They likely formed close to the protosun of the Sun as it transitioned from a Class I to Class II young stellar object. Having formed close to the Sun, CAIs may have then blown out into cooler, distant parts of the nebula by X-winds. Initial $^{26}Al/^{27}Al$ ratios in at least some amoeboid olivine aggregates suggest they formed around the same time. They condensed from somewhat cooler nebular gas (~1400 K) than the CAIs.

In the inner solar system, the condensed dust comprised only about 0.5% of the mass

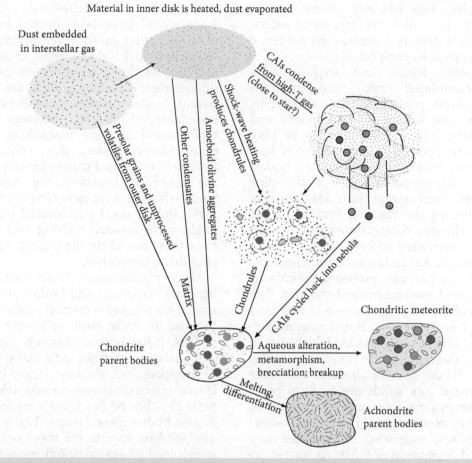

Figure 10.39   The processes involved in the formation of chondrites and their components. Modified from McSween (1987).

of the nebula; in the outer solar system, the dust included ices and comprised 2% of the mass. Gravity caused the dust to settle quickly to the mid-plane of the nebula, so that its concentration there would be much higher and as the dust-to-gas ratio controlled oxygen fugacity, the mid-plane would have had higher oxygen fugacity. Van der Waals and electrostatic forces caused the dust grains to stick together and aggregate into dustballs, much as dust in a house aggregates.

A significant fraction of the dust melted to form chondrules. Pb–Pb and extinct radionuclide dating indicates this mainly occurred 1–2 million years later. Heating experienced by chondrules was transient, lasting minutes to hours. This kind of transient thermal event matches well that expected for shock waves within the solar nebula (Desch and Connolly, 2002). The rapid cooling experienced by chondrules suggests ambient temperatures,

other than during these transient events, were low, perhaps 300 K.

Temperatures in the inner solar system remained warm enough so that ices of water, methane, ammonia, and other volatiles, could not condense. At greater distances, the orbit of Jupiter and outward, the disk remained cool enough so that silicates and iron were never vaporized. More volatile compounds such as water might have been vaporized, but they recondensed long before the nebula dissipated. The radial distance beyond which ice condenses, ~160 K at prevailing pressures, is known as the *snowline*. The snowline would have migrated inward as the nebula cooled. Dust accumulation would have proceeded more rapidly outside the snowline. Nebular evolution models generally place the snowline between the asteroid belt and the orbit of Jupiter. Much of the water and other ices probably condensed on preexisting dust particles. This has two effects: first, it enhances the overall concentration of solids, and second, it greatly increases the tendency of grains to stick to each other.

Electrostatic forces can explain how solids accumulated into millimeter- to centimeter-sized particles. How these particles grew into kilometer-sized blocks and planetesimals, where gravity began to play a role, is less certain. What is clear is that the transition from dust to km-sized bodies must have happened quickly or the dust would have been swept into the growing Sun. Because of the thermal motion of gas molecules, the gas rotates more slowly than the velocity necessary to keep solid bodies in stable orbit: the Keplerian velocity. Thus, gas drag slows the particles embedded within it, and they spiral inward toward the Sun. The effect is negligible for micron-sized dust and bodies of a km or more, but it is quite significant for meter-sized blocks, which would spiral into the Sun from the Earth's orbit in 100 yrs. Thus, growth through this stage must be fast if any solids are to form planets. Further evidence of rapid formation of planetesimals comes from Hf-W ages of some iron meteorites, suggesting they formed only ~1 Ma after formation of CAIs. At least some planetesimals must have accreted, melted, and differentiated into silicate and iron parts very quickly, perhaps during, or even before, the time of chondrule formation.

Two mechanisms for rapid growth of planetesimals have been proposed. In the first, transient high-pressure regions in the disk concentrate meter-sized boulders, and this concentration of solids then affects flow of the gas so as to capture solids drifting inward. Computer simulations suggest this process leads to gravitationally bound "clumps" that collapse to form planetesimals on time scales of a few orbital periods, that is, a few years to a few decades (Johansen et al., 2007). The second mechanism is concentration of particles in low-vorticity (i.e., slowly rotating) eddies within the disk. Simulations show that particles can become sufficiently concentrated in these eddies to become gravitationally bound and eventually contract to form ~100 km size planetesimals (Chambers, 2010). This process could form planetesimals in a few thousand years beyond the snowline but might require a few million years inside it.

Early-formed planetesimals larger than 1 km would have been heated by decay of ^{26}Al such that interior temperatures would have reached 1500 K within ~500,000 years and thus would have undergone metal-silicate differentiation. Such bodies are the likely parent bodies of the achondritic and iron meteorites. Collisions between planetesimals would in some cases lead to growth of ever-larger bodies. But collisions would also have destroyed some planetesimals, with the debris replenishing the nebular dust, from which a new generation of planetesimals could grow. Later-formed planetesimals might have avoided melting and differentiation: these are likely the parent bodies of the chondritic meteorites.

Once planetesimals have formed, gravity becomes important and bodies grow by collision. This process is termed *oligarchic growth* because it ends with relatively few large objects. It has been extensively modeled with computer simulations and is comparatively well understood. Because larger bodies have larger effective cross-sections (their gravity pulls in other bodies from a wider area), the largest bodies grow fastest. The larger bodies also tend to acquire the most regular orbits, which lead to gravitational focusing and further enhancement of growth rates. This leads to very rapid growth in the early stages. As a relatively few large objects become dominant, growth slows. Models suggest that bodies the size of the Moon (0.01 M_E or Earth masses) or Mars (0.1 M_E) could have formed in the inner solar system within 10^5 to 10^7 years. The Hf-W system provides a tighter constraint on this stage. Hf-W data on SNC meteorites suggest that Mars had formed and differentiated into a silicate mantle and iron core about 2 million years after CAI formation (Dauphas and Chaussidon, 2011). Vesta's core also appears to have formed quite early (Touboul et al., 2015).

Once Mars-sized bodies have formed, only a few large bodies are present within any planet-forming zone, and collisions between them become infrequent. Consequently, growth slows dramatically. Indeed, simulations suggest it might have required an additional 10^8 years for bodies the size of the Earth or Venus to form (Chambers and Wetherill, 1998). Another feature of the late stages of accretion is that the collisions involve very large bodies and are consequently catastrophic. The energy released in these collisions leads to extensive melting. As we shall see below, this fits very well with the evidence that we have for the origin of the Earth and the Moon.

There is somewhat more uncertainty as to how the giant planets formed. There are essentially two classes of theories. The first and best established is that of *core accretion*. It proposes that rocky, icy cores of giant planets accreted in a process very similar to that described above, albeit enhanced by the presence of ice beyond the snowline. Once these cores reached a size of 10 Earth masses, they would have had sufficient gravity to capture gas from the solar nebula and eventually become gas giants. Simulations assuming a *minimum mass nebula* (one that has just enough mass to produce the observed planets) take much too long to form giant planets in this way – 10^8 years or more but are quicker if the nebula has more mass. It appears that nebular disks dissipate within 6 Ma (Haisch et al., 2001) – the gas giants must have captured their gas shrouds within this time. The other theory is disk instability (Boss, 1997). It posits that a density perturbation in the disk could cause a clump of gas to become massive enough to be self-gravitating. Once that happens, the clump could collapse into a gas giant planet on time scales of 10^3 to 10^4 yrs.

The observation of gas giant planets orbiting close to their stars has inspired astronomers to rethink planet formation as it is unlikely that they would have formed there. Hydrodynamic models show that gas giants that formed quickly while the nebular disk was still present experience tidal interactions with the disk that causes them to migrate inward. That may explain gas giants close to their stars and has consequences for evolution of our solar system. Models show that with two giant planets such as Jupiter and Saturn, both would migrate inward, but with Saturn migrating more rapidly (Walsh et al., 2011). Eventually they become locked in a 2:3 orbital resonance and the migration reverses. This is known as the Grand Tack model (in reference to the sailing maneuver) and is illustrated in Figure 10.40. This happens after about 100,000 years when Jupiter is orbiting at about 1.5 AU. At this time,

the inner solar system was populated by a large number of planetesimals. In the model, these are shepherded by Jupiter's gravity into a tight distribution around 1 AU. As Saturn and Jupiter begin to migrate outward, many of these planetesimals are scattered outward. At the same time, planetesimals that had formed outward of Jupiter's orbit are scattered inward. On much longer time scales, the inner solar planetesimals accrete to form the terrestrial planets.

Perhaps the most significant success of the Grand Tack model is that it correctly predicts the eventual masses and orbits of the terrestrial planets; other models generally fail to predict the position and size of Mars. It also can explain the zonation in the asteroid belt, with S-types predominant in the inner part and C-types in the outer part. The C-type asteroids are presumably the parents of carbonaceous chondrites. Carbonaceous chondrites are richer in volatiles and hence likely formed at greater heliocentric distance than other chondrites. They are also isotopically distinct (Section 10.4.3.2), also indicating they formed in a different part of the nebula. Scattering during the Grand Tack explains their presence in the inner solar system. Finally, scattering of such volatile-rich objects into the inner solar system may explain the Earth's volatile inventory, particularly its water, as high temperatures in the inner solar system should have prevented their condensation (Morbidelli et al., 2012).

Within less than 10^7 years, the nebula cleared. This resulted from a variety of factors. While the planets were forming, gas and dust were steadily spiraling into the central star. Small bodies that were not swept up by the forming planets would have been flung, through gravitational interaction with the planets, into the central star or out to interstellar space. As the Sun became hotter and more luminous, the nebula would have dissipated through *photoevaporation*. (This term is a bit of a misnomer, since most of the nebula is gas to begin with, it does not evaporate in the conventional sense. But it does absorb radiational energy, which is converted to kinetic energy through collisions. The fast-moving gas molecules then escape to interstellar space.) Finally, the enhanced solar winds of the T-Tauri stage would have helped to drive out remaining gas and dust.

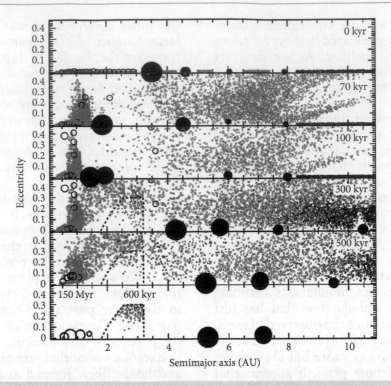

Figure 10.40 The Grand Tack model. Jupiter, Saturn, Uranus and Neptune are represented by large black filled circles; S-type planetesimals are represented by red circles, initially located between 0.3 and 3.0 AU. The C-type planetesimals starting between the giant planets are shown as light blue squares, and the outer-disk planetesimals as dark black squares. Planetary embryos are represented by large open circles. Panels show the inward then outward migration of the giant planets. Once Jupiter reaches its final position at 5 AU at around 500 kyr, inner disk planetary embryos sweep up the remaining planetesimals within 150 Myr leaving four terrestrial planets remaining. From Walsh et al. (2011).

As a result, the nebula clears mostly from the inside out. However, nearby large stars could also erode the nebula from the outside through photoevaporation. This is observed to occur in the Orion nebula. The nebula must have cleared of gas before the inner terrestrial planets were able to accumulate their full share of volatile elements, perhaps because the inner solar system cleared before it cooled enough for these elements to condense. It may have cleared before the ice giants, Uranus and Neptune, were able to capture their full complement of gas.

10.6.3 Chemistry and history of the Moon

Before we consider the formation of the Earth specifically, there is one more set of observations we need to consider: observations about the Moon. The Moon is of interest for several reasons. First, it is the only solar system body, other than the Earth, that we have returned samples from and has been explored by men. From this exploration we have found that much of the earliest history of the Moon is preserved, giving us unique insights into early planetary history. In contrast, no trace remains of the earliest period of Earth history. Second, the Moon is closely associated with the Earth, not only physically, but chemically as well. Lunar oxygen isotope compositions fall on the terrestrial fractionation line (Figure 10.30), implying the Moon and Earth share a similar O isotopic composition. The Moon is nearly unique in this respect: with the exception of E-chondrites, all analyzed meteoritic material, including the SNC meteorites, falls off this line. This strongly implies that the Moon and the Earth are closely related. As we shall see, the Moon holds important clues to the earliest part of Earth's history.

10.6.3.1 Geology and history of the Moon

The Moon can be divided into three geologic provinces: the highlands – mountainous regions apparently consisting largely of anorthosite; the uplands – areas of mild relief covered by a blanket of ejecta from large impacts; and the Mare – the large craters filled with basaltic lavas. Much of the surface of the Moon is covered with fine debris of impacts, called the regolith, consisting of rock and mineral fragments, glass, and some meteorite particles. For the most part, it seems to be locally derived; thus the regolith in the Mare differs from that of the highlands, although large impacts would have showered debris over large regions. Basalt from the Mare encompasses a variety of magma types, including both incompatible-rich and incompatible-poor types, and both quartz-normative and olivine-normative tholeiites. Highland rocks include anorthosite (nearly monomineralic calcic plagioclase), anorthositic gabbro (plagioclase and pyroxene with lesser amounts of olivine), dunite, and K-rich basalts. The highlands are extremely brecciated; most of these rock types have been found only as clasts in breccias. Table 10.7 shows some representative compositions of lunar rocks, and Figure 10.41 shows rare earth patterns of the same rock types. Anorthosite and anorthositic gabbro from the highlands are characterized by LREE enrichment and strong negative Eu anomalies, consistent with crystallization of a plagioclase-rich assemblage from a magma. In contrast, the basalts, including KREEP (an acronym derived from its potassium, REE and phosphorus-rich chemistry), have negative Eu-anomalies. This provides clues to the lunar history described below.

Most of the lunar Mare are thought to have been created by large impacts between 4.2 and 3.8 Ga. Subsequently, the Mare were flooded by basalt to a depth of 5–10 km. These were partial melts generated at 100

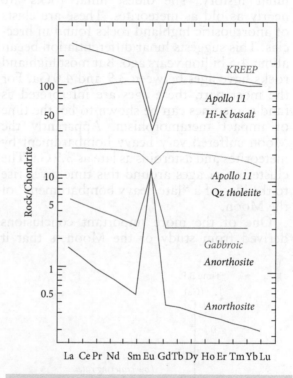

Figure 10.41 Rare earth patterns of representative lunar rocks. Based on Taylor (1975).

Table 10.7 Representative compositions of lunar rocks.

	Highland anorthosite	Highland anorth. gabbro	Mare qz. tholeiite	Low-K Mare basalt	Fra Mauro KREEP
SiO$_2$	44.3	44.5	46.1	40.5	48.0
TiO$_2$	0.06	0.39	3.35	10.5	2.1
Al$_2$O$_3$	35.1	26.0	9.95	10.4	17.6
FeO	0.67	5.77	20.7	18.5	10.9
MnO	–	–	0.28	0.28	–
MgO	0.80	8.05	8.1	7.0	8.7
CaO	18.7	14.9	10.9	11.6	10.7
Na$_2$O	0.8	0.25	0.26	0.41	0.7
K$_2$O	–	–	0.07	0.10	0.54
P$_2$O$_5$	–	–	0.08	0.11	–
Cr$_2$O$_3$	0.02	0.06	0.46	0.25	0.18
Total	100.5	99.9	100.3	99.7	99.4

or so km depth. The flooding occurred over an extensive time: 3.9–3.1 Ga. Mare flooding was the last major lunar geologic event. Subsequent to that time, the only activity has been continual bombardment by meteorites and asteroids, which continued to produce minor disruption of the surface and build-up of the regolith, and rare volcanism.

Figure 10.42 illustrates the highlights of lunar history. The oldest lunar rocks are nearly as old as meteorites. These are clasts of anorthositic highland rocks found in breccias. This suggests lunar differentiation began about 4.5 billion years ago. But most highland rocks have ages between 3.9 and 4.0 Ga. For the most part, these ages are interpreted as (and sometimes can be shown to be) the time of impact metamorphism. Apparently the Moon suffered very heavy bombardment by meteorites and asteroids as late as 3.9 Ga. The clustering of ages around this time gives rise to the idea of a "late heavy bombardment" of the Moon.

One of the more important conclusions derived from study of the Moon is that it underwent very extensive melting just after its formation, perhaps forming a magma ocean 100 km or more deep. The anorthosite of the highlands is thought to have originated by plagioclase flotation in the magma ocean (i.e., anorthosite "icebergs" forming the lunar crust). The lunar crust seems to have been largely in place within 100–200 Ma after the Moon's formation. Fractional crystallization of this magma ocean and flotation of plagioclase accounts for the general Eu depletion (Figure 10.41) observed in basalts derived from the lunar mantle. KREEP are basalts that are particularly enriched in incompatible elements and are thought to be derived from a part of the mantle that was the last part of the magma ocean to crystallize and was consequently strongly enriched in incompatible elements.

10.6.3.2 Composition of the Moon

Table 10.8 compares the composition of the Earth and the Moon. There are several notable differences. First, the Moon is depleted in moderately volatile elements (Na, K) compared with the Earth. Second, although the silicate part of the Moon (mantle + crust) is richer in iron than the silicate Earth, the bulk lunar composition is dramatically poorer in total iron (FeO plus Fe metal) than the Earth and the other terrestrial planets. The latter reflects the small size of the lunar

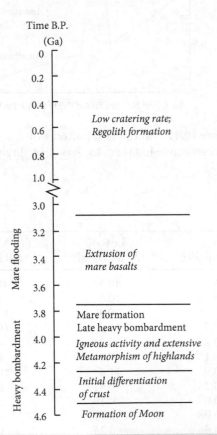

Figure 10.42 Highlights of lunar chronology.

Table 10.8 Comparison of the compositions of the Earth and Moon.

	Bulk Earth	Bulk Moon	Silicate Earth	Silicate Moon
SiO_2	30.3	47.1	45.0	48.2
TiO_2	0.14	0.3	0.17	0.2
Al_2O_3	3.00	3.5	4.6	3.59
FeO	5.43	10.4	8.05	10.6
MgO	25.5	33.0	37.8	33.7
CaO	2.40	2.8	3.55	2.84
Na_2O	0.24	0.05	0.36	0.05
K_2O	0.02	< 0.01	0.029	< 0.01
Fe	28.4	1.0		
ΣFe	32.65	8.33		
Core %	32.5%	1.2%		
Mantle %	67.5%	98.8%		

Terrestrial composition from McDonough and Sun (1995). Lunar composition based on Dauphas et al. (2014).

core, which has a diameter of <380 km, representing <2% of the mass of the planet. Like the Earth's core, it appears to consist of a solid inner core and a liquid outer one. The lunar depletion in Fe compared with the Earth extends to all siderophile and chalcophile elements as well. The Moon is also much more depleted in the volatile elements. Despite these compositional differences, the similarity of terrestrial and lunar oxygen, titanium, and chromium isotope compositions strongly suggests they formed from the same part of the solar nebula.

10.6.4 The giant impact hypothesis and formation of the Earth and the Moon

As we saw, there is evidence that the Moon was extensively melted very early in its history. We saw that numerical simulations of planetary growth within the solar nebula predict collisions between very large bodies in the final stages of accretion. We have also seen that the Moon and the Earth share a nearly identical oxygen chromium and titanium isotopic compositions. In addition, the Earth–Moon system has an anomalously large amount of angular momentum compared with the other planets, and most of this angular momentum resides in the Moon. Finally, the Earth's rotational axis is tilted 23° with respect to its orbital plane. These observations led Hartmann and Davis (1975) to propose that the Earth collided with a large body in the final stages of its accretion. A fraction of the mass of the two bodies was blasted into orbit around the Earth, with much of this debris later coalescing to form the Moon. This idea has become known as the *giant impact hypothesis* and is now widely accepted (even if the details are still debated). The energy released in this impact would have been sufficient to melt the entire mantle of the Earth and to vaporize some of it, though this depends on how efficiently the impact energy is dissipated.

Increasingly detailed numerical simulations of the impact have been produced over the past several decades (see reviews of Canup (2004) and Asphaug (2014)). From these, a "standard model" that best matches observations has emerged. During the latest stages of accretion, the Earth was struck at ~45° angle by a body, called *Theia*, that is 10 to 15% of its mass, roughly the size of Mars. The impact velocity was slightly greater than its escape velocity (11 km/s) and threw material, much of in a molten or gaseous state into low Earth orbit, which eventually spreads as rotating disk around the Earth. Some of it fell back to Earth, but most of the material outside the Earth's Roche limit (the distance beyond which a body can escape tidal disruption by the Earth; about 2.9 Earth radii or ~18,000 km) coalesced into a single body within about a year. Much, perhaps 75% or so, of the material that formed the Moon originated from the impactor, the remainder from the Earth's mantle. Metallic cores are assumed to have already formed in both the Earth and the impactor. Most of the impactor's core merged with that of the Earth, with only a small fraction going into the Moon. The Moon initially forms just outside the Earth's Roche limit and quickly migrates outward over the next tens of thousands of years due to tidal interactions and resonances (the Moon's present orbital radius is 384,000 km).

The simulations lead to a Moon highly depleted in iron and other siderophile elements relative to the Earth. The volatile element depletion of the Moon is a consequence of the evaporative loss of these elements during impact. According to some, the otherwise close chemical and isotopic similarity of the Earth and Moon is a potential problem, given the diversity of compositions exhibited by meteorites. This seems to require that the Earth and Theia formed in the same orbital feeding zone and hence was compositionally quite similar to the proto-Earth (Asphaug, 2014; Dauphas et al., 2014). Consequently, models for the formation of the Moon continue to be investigated and debated.

10.6.5 Tungsten isotopes and the age of the Earth

If the giant impact hypothesis is correct, then the impact marks the time of final segregation of the metal and silicate in the Earth (and the Moon) as well as the completion (or nearly so) of accretion of the Earth. If we can date this event, we can determine the age of the Earth. Looking at Table 10.4, we see that the ^{182}Hf-^{182}W decay pair is just the ticket: Hf is lithophile and should be concentrated in the mantle while W is siderophile and should be concentrated in the core. Furthermore, the half-life of ^{182}Hf is 9 Ma, just right for examining events in the first few tens of millions of years in the solar system. Both are highly refractory elements, which has the advantage that one can reasonably assume that bodies such as the Earth should have an approximately chondritic Hf/W ratio. Over the last two decades this system has provided very interesting insights into the early history of Mars, the Moon, and the Earth.

Because the variations in ^{182}W/^{183}W ratio are quite small, they are generally presented and discussed in the same ε notation used for Nd and Hf isotope ratios (Chapter 8). However, as we noted before, ε_W is the deviation in parts per 10,000 *from a terrestrial tungsten standard*, and $f_{Hf/W}$ is the fractional deviation of the Hf/W ratio from the chondritic value. Assuming that the silicate Earth has a uniform W isotope composition identical to that of the standard then the silicate Earth has ε_W of 0 by definition (we'll see in the next chapter that W isotopes are not quite uniform in the Earth). The basic question can be posed this way: if the ^{182}W/^{183}W ratio in the silicate Earth is higher than in chondrites, it would mean that much of the Earth's tungsten had been sequestered in the Earth's core before ^{182}Hf had entirely decayed. Since the half-life of ^{182}Hf is 9 Ma and using our rule of thumb that a radioactive nuclide is fully decayed in 5 to 10 half-lives, this would mean the core must have formed within 45–90 million years of the time chondritic meteorites formed.

Yin et al. (2002) and Kleine et al. (2002) reported W isotope ratios in carbonaceous chondrites that were 1.9–2.6 epsilon units lower than the terrestrial standard. Furthermore, Kleine et al. (2002) analyzed a variety of terrestrial materials and found they all had W isotopic compositions that are some 2 epsilon units higher than those of the chondrites. There are difficulties in making highly precise isotopic determinations of lunar materials because of cosmic-ray induced nuclear reactions for which a correction must be made. Doing so, Touboul et al. (2007) showed that the Moon has an isotopically homogenous W composition identical to that of the silicate Earth.

Precisely what the high ^{182}W/^{183}W ratio of the silicate Earth implies for the timing of formation of the Earth's core depends on how the core formed. In the simplest model, the core segregated instantaneously from homogenously accreted metal and silicate. The ^{182}W excess of the Earth's mantle of ε_W = +1.9 and a Hf/W ratio of 17 then indicates this happened ~30 Ma after the time of CAI formation. This model is unlikely, however, since we know that metal cores segregated from even relatively small planetesimals within a few million years after the start of the solar system. Large planets such as the Earth formed by collisions between and accretion of already-differentiated planetary embryos. Thus, most studies have considered continuous core formation models. In such models, there is no single "age" for core formation; rather, the reported ages correspond to either the mean time of core formation (formation of 63% of the core). In continuous formation models, W isotopic evolution will depend on the degree to which tungsten in the silicate and metal parts of colliding bodies equilibrate. Tungsten isotope systematics have been variously interpreted as implying core formation ages anywhere from 30 Ma to >100 Ma after CAI formation. However, the identical ^{182}W/^{184}W ratios of the Moon and silicate Earth, despite the two bodies having different Hf/W ratios, indicate that the Moon-forming giant impact occurred after extinction of ^{182}Hf, or more than 60 Ma after CAI formation. The minimum age of the Moon is constrained by Sm-Nd ages of the lunar anorthositic crust. This age is 4.456 ± 0.040 Ga, or about 100 Ma after the beginning of the solar system. Touboul et al. (2007) thus estimated an age for the Moon of 62 +90/−10 Ma after the start of the solar system (Figure 10.43). Kleine et al. (2014) subsequently concluded the silicate Moon has a slightly higher ε_W of around +0.17, but this does not substantially change this interpretation.

The Moon-forming event was likely the last major one in the Earth's formation. However, it is widely thought that an additional half percent or so of the Earth's mass was added subsequently, a so-called *late accretionary veneer*. The evidence for such a veneer comes

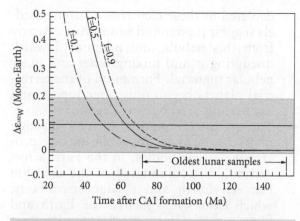

Figure 10.43 Difference between ε_W of Earth and Moon (Δ) versus the age of the giant impact (and presumably final core formation of the Earth). f is the relative difference in Hf/W ratios between the silicate part of the Earth and the silicate part of the Moon. The gray area shows the best value of $\Delta\varepsilon_W$ (0.09 ± 0.10). The Hf/W ratio of the Earth's mantle is estimated at 18 and that of the Moon is 26.5, corresponding to an f of 0.47 (solid line). After Touboul et al. (2007).

from the abundances of siderophile elements, particularly highly siderophile elements, such as the platinum group elements, in the Earth's mantle: although depleted relative to chondrites they abundant than predicted by metal-silicate partitioning. Furthermore, they are present in nearly chondritic abundances whereas experiments predict depletions of platinum group metals that vary by an order of magnitude or more (e.g., Walker, 2016). The solution appears to be to add a half percent or so of chondritic material after final segregation of the core (Chou, 1978). This produces roughly chondritic relative abundances of these highly siderophile elements at appropriate levels in the silicate Earth.

10.7 SUMMARY

The focus of this chapter was the formation of the Earth, but to fully understand that we needed to also understand formation of the elements, star formation, and solar system formation.

- The chemical elements have been synthesized in four cosmic settings. H and He were produced in the Big Bang 13.8

billion years ago. Elements up to iron were, and continue to be, synthesized by fusion in the red giant phase of large stars; some heavier nuclides are also synthesized in later generation red giants by slow neutron capture (the s-process). Heavier elements are synthesized by rapid neutron capture (r-process) and the p-process in supernova explosions and neutron star mergers (kilonova). Li, Be, and B are synthesized by spallation reactions induced by high-energy cosmic rays.

- Meteorites are our most important clue to processes occurring in the early solar system. One class of meteorites, the chondrites, are collections of nebular dust, although they have been variably aqueously or metamorphically altered in their parent bodies. They consist of calcium–aluminum inclusions (CAIs) and amoeboid olivine aggregates, which condensed at highest temperatures, chondrules, once molten droplets, and matrix which contains low temperature condensates as well as minute pre-solar grains of SiC, graphite, diamond, and other minerals with highly anomalous isotopic compositions. Among chondrites, carbonaceous chondrites, which are rich in volatiles as well as refractory elements, are the most primitive and most closely match the composition of the condensable part of the solar nebula. Achondrites are igneous rocks formed during differentiation of asteroidal bodies while most iron meteorites are the cores of disrupted asteroids.

- Meteorites contain the decay products such as ^{26}Mg and ^{53}Cr of short-lived radionuclides such as ^{26}Al and ^{53}Mn that had been synthesized shortly before the solar system formed. They not only provide evidence of continuing nucleosynthesis in the young solar system's galactic neighborhood, together with conventional radiometric chronometers they also provide useful chronometers to establish the timing and sequence of events in solar system formation. This chronology begins 4.568 million years ago with the formation of CAIs from a hot solar nebula and formation of planetesimals and planetary embryos followed within a few million years.

- The solar system formed when part of a vast nebula of gas and dust gravitationally collapsed. Conservation of angular momentum as the collapse occurred resulted in the formation of a spinning nebular disk around the forming Sun. Small differences in isotopic compositions of some elements between carbonaceous chondrites and other meteorites indicate that the nebula was inhomogeneous with compositional differences between the inner and outer regions. The inner part of this nebula was too hot for the more volatile elements to condense and as a consequence the terrestrial planets are depleted in these elements. Current models suggest Jupiter and Saturn formed early from this nebula and migrated inward, disrupting it and mixing outer and inner nebular material. Formation of the terrestrial planets began only after Jupiter and Saturn migrated outward again.

- The terrestrial planets formed through oligarchic growth and collisions of progressive larger bodies. In the Earth's formation, this culminated in the collision between Earth and a Mars-sized body, which extensively melted the Earth and flung material into orbit around it that coalesced to form the Moon.

REFERENCES AND SUGGESTIONS FOR FURTHER READING

Abbott, B.P. et al. 2017, GW170817: Observation of gravitational waves from a binary neutron star inspiral. *Physical Review Letters* 119, 161101. doi: 10.1103/PhysRevLett.119.161101.

Allegre, C.J. 2001. Condensed matter astrophysics: constraints and questions on the early development of the Solar System. *Philosophical Transactions of the Royal Society of London* A359: 2137–55.

Amelin, Y. 2008. U-Pb ages of angrites. *Geochimica et Cosmochimica Acta* 72: 221–32. doi: 10.1016/j.gca.2007.09.034.

Amelin, Y., Connelly, J., Zartman, R.E., et al. 2009. Modern U–Pb chronometry of meteorites: Advancing to higher time resolution reveals new problems. *Geochimica et Cosmochimica Acta* 73: 5212–23. doi: 10.1016/j.gca.2009.01.040.

Amelin, Y., Kaltenbach, A., Iizuka, T., et al. 2010. U–Pb chronology of the Solar System's oldest solids with variable $^{238}U/^{235}U$. *Earth and Planetary Science Letters* 300: 343–50. doi: 10.1016/j.epsl.2010.10.015.

Anders, E. 1988. Circumstellar material in meteorites: noble gases, carbon and nitrogen, in *Meteorites and the Early Solar System* (eds J.F. Kerridge and M.S. Matthews). Tuscon, University of Arizona Press.

Anders, E. and Zinner, E. 1993. Interstellar grains in primitive meteorites: diamond, silicon carbide, and graphite. *Meteoritics* 28: 490–514.

Asphaug, E. 2014. Impact origin of the moon? *Annual Review of Earth and Planetary Sciences* 42: 551–78. doi: 10.1146/annurev-earth-050212-124057.

Blinova, A., Amelin, Y., and Samson, C., 2007, Constraints on the cooling history of the H-chondrite parent body from phosphate and chondrule Pb-isotopic dates from Estacado: *Meteoritics and Planetary Science* 42: 1337–1350. doi: 0.1111/j.1945-5100.2007.tb00578.x.

Bethe, H. and Brown, G.E. 1985. How a supernova explodes. *Scientific American* 252: 60–8.

Black, D.C. and Pepin, R.O. 1969. Trapped neon in meteorites – II. *Earth and Planetary Science Letters* 6: 395–405. doi: 10.1016/0012-821x(69)90190-3.

Blackburn, T., Alexander, C.M.O.D., Carlson, R., and Elkins-Tanton, L.T., 2017, The accretion and impact history of the ordinary chondrite parent bodies: *Geochimica et Cosmochimica Acta* 200:201–217. doi: 10.1016/j.gca.2016.11.038.

Bollard, J., Connelly, J.N., and Bizzarro, M., 2015, Pb-Pb dating of individual chondrules from the CB$_a$ chondrite Gujba: Assessment of the impact plume formation model: *Meteoritics and Planetary Science*. 50:1197–1216. doi: 10.1111/maps.12461.

Bollard, J., Connelly, J.N., Whitehouse, M.J., Pringle, E.A., Bonal, L., Jørgensen, J.K., Nordlund, Å., Moynier, F., Bizzarro, M., 2017. Early formation of planetary building blocks inferred from Pb isotopic ages of chondrules. *Science Advances* 3: 1–9. doi: 10.1126/sciadv.1700407.

Boss, A.P. 1997. Giant planet formation by gravitational instability. *Science* 276: 1836–9.

Boss, A.P. 2005a. The solar nebula, in *Meteorites, Comets and Planets* (ed. A.M. Davis), pp. 63–82. Amsterdam, Elsevier.

Boss, A.P. 2005b. Evolution of the solar nebula. VII. Formation and survival of protoplanets formed by disk instability. *Astrophysical Journal* 629: 535–48.

Boss, A.P., Keiser, S.A., 2012. Supernova-triggered molecular cloud core collapse and the rayleigh-taylor fingers that polluted the solar nebula. *The Astrophysical Journal Letters* 756, L9. doi: 10.1088/2041-8205/756/1/L9.

Bouvier, A., Blichert-Toft, J., Moynier, F., Vervoort, J.D. and Albarede, F. 2007. Pb–Pb dating constraints on the accretion and cooling history of chondrites. *Geochimica et Cosmochimica Acta* 71: 1583–1604. doi: 10.1016/j.gca.2006.12.005.

Bouvier, A., Spivak-Birndorf, L.J., Brennecka, G.A. and Wadhwa, M. 2011. New constraints on early Solar System chronology from Al–Mg and U–Pb isotope systematics in the unique basaltic achondrite Northwest Africa 2976. *Geochimica et Cosmochimica Acta* 75: 5310–23. doi: 10.1016/j.gca.2011.06.033.

Bouvier, A. and Wadhwa, M. 2010. The age of the Solar System redefined by the oldest Pb-Pb age of a meteoritic inclusion. *Nature Geoscience* 3: 637–41. doi: 10.1038/ngeo941.

Bradley, J. P. (2014) 1.8 - Early solar nebula grains – interplanetary dust particles in: Holland, H. D. and Turekian K. K. (eds) *Treatise on Geochemistry* (second edition), vol. Elsevier, Oxford, pp. 287–308.

Brennecka, G.A., Weyer, S., Wadhwa, M., et al. 2010. ^{238}U/^{235}U variations in meteorites: extant ^{247}Cm and implications for Pb-Pb dating. *Science* 327: 449–51. doi: 10.1126/science.1180871.

Budde G., Kruijer T. S, and Kleine T. 2018. Hf-W chronology of CR chondrites: Implications for the timescales of chondrule formation and the distribution of ^{26}Al in the solar nebula. *Geochimica et Cosmochimica Acta* 222: 284–304. doi: 10.1016/j.gca.2017.10.014.

Burbidge, E.M., Burbidge, G.R., Fowler, W.A. and Hoyle, F. 1957. Synthesis of the elements in stars. *Reviews in Modern Physics* 29: 547–650.

Burbine, T.H., 2014. 2.14 - Asteroids A2 - Holland, Heinrich D, in: Turekian, K.K. (Ed.), *Treatise on Geochemistry* (Second Edition). Elsevier, Oxford, pp. 365–415. doi: 10.1016/B978-0-08-095975-7.00129-7.

Canup, R. M. 2004. Dynamics of lunar formation. *Annual Review of Astronomy and Astrophysics* 42: 441–75. doi: 10.1146/annurev.astro.41.082201.113457.

Canup, R. M., Barr, A. C., Crawford, D. A. 2013. Lunar-forming impacts: High-resolution SPH and AMR-CTH simulations. *Icarus* 222: 200–19. doi: 10.1016/j.icarus.2012.10.011.

Chambers, J.E. 2010. Planetesimal formation by turbulent concentration. *Icarus* 208: 505–17. doi: 10.1016/j.icarus.2010.03.004.

Chambers, J.E. and Wetherill, G.W. 1998. Making the terrestrial planets: N-body integrations of planetary embryos in three dimensions. *Icarus* 136: 304–27.

Chen, J.H. and Wasserburg, G.J. 1983. The least radiogenic Pb in iron meteorites. *Fourteenth Lunar and Planetary Science Conference, Abstracts*, Part I, Lunar and Planet Science Institute, Houston, pp. 103–4.

Chou, C.-L., 1978. Fractionation of siderphile elements in the Earth's upper mantle and lunar samples. *Proceedings of the Lunar and Planetary Science Conference 9*, 163–165.

Clayton, R.N. 2002. Self-shielding in the solar nebula. *Nature* 415: 860–61.

Clayton, R.N., Onuma, N. and Mayeda, T.K. 1976. A classification of meteorites based on oxygen isotopes. *Earth and Planetary Science Letters* 30: 10–18.

Connelly, J. N., Bizzarro, M., Krot, A.N., Nordlund, Å., Wielandt, D., and Ivanova, M.A., 2012. The absolute chronology and thermal processing of solids in the solar protoplanetary disk. *Science* 338, 651–55. doi: 10.1126/science.1226919.

Connelly, J. N., Bollard, J., and Bizzarro, M., 2017, Pb–Pb chronometry and the early Solar System: Geochimica et Cosmochimica Acta 201: 345–63. doi: 10.1016/j.gca.2016.10.044.

Cowley, C.R. 1995. *Cosmochemistry*. Cambridge, Cambridge University Press.

Crabb, J. and Schultz, L. 1981. Cosmic-ray exposure ages of the ordinary chondrites and their significance for parent body stratigraphy. *Geochimica et Cosmochimica Acta* 45: 2151–60.

Dauphas, N., Burkhardt, C., Warren, P. H., et al. 2014. Geochemical arguments for an earth-like moon-forming impactor. *Philosophical Transactions of the Royal Society A: Mathematical, Physical and Engineering Sciences* 372. doi: 10.1098/rsta.2013.0244.

Dauphas, N., Chaussidon, M., 2011. A perspective from extinct radionuclides on a young stellar object: the sun and its accretion disk. *Annual Review of Earth and Planetary Sciences* 39, 351–86. doi: 10.1146/annurev-earth-040610-133428.

Davis, A. M., and McKeegan, K. D., 2014, Short-lived radionuclides and early solar system chronology. in Holland, Heinrich D, Turekian, K. K., ed., *Treatise on Geochemistry (Second Edition)*: Oxford, Elsevier, p. 361–95. doi: 10.1016/B978-0-08-095975-7.00113-3.

De Sanctis, M.C., Ammannito, E., Capria, M.T., et al. 2012. Spectroscopic characterization of mineralogy and its diversity across Vesta. *Science* 336: 697–700.

De Sanctis, M. C., Ammannito, E., Raponi, A., et al. 2015. Ammoniated phyllosilicates with a likely outer Solar System origin on (1) Ceres. *Nature* 528:241–245. doi: 10.1038/nature16172.

De Sanctis, M. C., Raponi, A., Ammannito, E., et al. 2016. Bright carbonate deposits as evidence of aqueous alteration on 1 Ceres. *Nature* 536: 54–58. doi: 10.1038/nature18290.

De Sanctis, M. C., Ammannito, E., McSween, H. Y., et al. 2017. Localized aliphatic organic material on the surface of Ceres. *Science* 355: 719–722. doi: 10.1126/science.aaj2305.

Desch, S.J. and Connolly, H.C. 2002. A model of the thermal processing of particles in solar nebula shocks: Application to the cooling rates of chondrules. *Meteoritci and Planetary Science* 37: 183–207.

Ferrari, Marco and Dirri, Fabrizio and Palomba, Ernesto and Stefani, Stefania and Longobardo, Andrea and Rotundi, Alessandra. (2017). FT-IR and μ-IR characterization of HED meteorites in relation to infrared spectra of Vesta-like asteroids. www.copernicus.org.

Freiburghaus, C., Rembges, J. F., Rauscher, T., Kolbe, E., Thielemann, F. K., Kratz, K. L., Pfeiffer, B. and Cowan, J. J., 1999a. The Astrophysical r-Process: A comparison of calculations following adiabatic expansion with classical calculations based on neutron densities and temperatures. *The Astrophysical Journal* 516(1), 381–98.

Freiburghaus, C., Rosswog, S. and Thielemann, F. K., 1999. r-process in neutron star mergers. *The Astrophysical Journal Letters* 525(2), L121–4.

Filiberto, J. 2017. Geochemistry of Martian basalts with constraints on magma genesis. *Chemical Geology* 466: 1–14. doi: 10.1016/j.chemgeo.2017.06.009.

Fu, R. R., Ermakov, A. I., Marchi, S., et al. 2017. The interior structure of Ceres as revealed by surface topography. *Earth and Planetary Science Letters* 476: 153–64. doi 10.1016/j.epsl.2017.07.053.

Graf T., Marti K. 1995. Collisional history of H chondrites. *Journal of Geophysical Research: Planets* 100: 21247–21263. doi: 10.1029/95JE01903.

Goodrich, C.A. and Delaney, J.S. 2000. Fe/Mg–Fe/Mn relations of meteorites and primary heterogeneity of primitive achondrite parent bodies. *Geochimica et Cosmochimica Acta* 64: 149–60. doi: 10.1016/s0016-7037(99)00107-6.

Göpel, C., Manhès, G. and Allègre, C. 1994. U-Pb systematics of phosphates from equilibrated ordinary chondrites. *Earth and Planetary Science Letters* 121: 153–71.

Grossman, L. 1972. Condensation in the primitive solar nebula. *Geochimica et Cosmochimica Acta* 36: 597–619.

Karl E. Haisch, Jr., Elizabeth, A. L., Charles, J. L. 2001. Disk frequencies and lifetimes in young clusters. *The Astrophysical Journal Letters* 553: L153.

Hartmann, W.K. and Davis, D.R. 1975. Satellite-sized planetesimals and lunar origin. *Icarus* 24: 504–15.

Herzog GF, Caffee MW (2014) 1.13 - Cosmic-Ray Exposure Ages of Meteorites A2 - Holland, Heinrich D. In: Turekian KK (ed) *Treatise on Geochemistry (Second Edition)*, Elsevier, Oxford, pp. 419–454. doi: 10.1016/B978-0-08-095975-7.00110-8.

Huss, G.R., Meyer, B.S., Srinivasan, G., Goswami, J.N. and Sahijpal, S. 2009. Stellar sources of the short-lived radionuclides in the early solar system. *Geochimica et Cosmochimica Acta* 73: 4922–45. doi: 10.1016/j.gca.2009.01.039.

Johansen, A., Oishi, J.S., Low, M., et al. 2007. Rapid planetesimal formation in turbulent circumstellar disks. *Nature* 448: 1022–5. doi: 10.1038/nature06086.

Kasen, D., Metzger, B., Barnes, J., Quataert, E. and Ramirez-Ruiz, E. 2017. Origin of the heavy elements in binary neutron-star mergers from a gravitational-wave event. *Nature* 551, 80–84. doi: 10.1038/nature24453.

Katsuyuki Y, Seiji M, Akane Y, Eizo N. 2010. ^{53}Mn-^{53}Cr Chronometry of Cb Chondrite: Evidence for Uniform Distribution of ^{53}Mn in the Early Solar System. The Astrophysical Journal 723:20.

Kleine, T., Münker, C., Mezger, K. and Palme, H. 2002. Rapid accretion and early core formation on asteroids and the terrestrial planets from Hf-W chronometry. *Nature* 418: 952–4.

Lada, C.J. and Shu, F.H. 1990. The formation of Sun-like stars. *Science* 248: 564–72.

Lagrange, A.-M., Bonnefoy, M., Chauvin, G., et al. 2010. A giant planet imaged in the disk of the young star β Pictoris. *Science* 329: 57–9.

Larimer, J.W. 1967. Chemical fractionations in meteorites – I: condensation of the elements. *Geochimica et Cosmochimica Acta* 31: 1215–38.

Lattimer, J. M. and Schramm, D. N., 1974. Black-hole-neutron-star collisions. *The Astrophysical Journal* 192, L145–7.

Li-Xin L, Bohdan P (1998) Transient events from neutron star mergers. *The Astrophysical Journal Letters* 507(1):L59 .LIGO Scientific Collaboration and Virgo Collaboration.

Lodders, K., 2003. Solar system abundances and condensation temperatures of the elements. *The Astrophysical Journal* 591: 1220–47. doi; 10.1086/37549.2.

Lodders K. 2010 Solar System Abundances of the Elements. in: *Principles and Perspectives in Cosmochemistry* Goswami A, Reddy BE (eds.). Berlin, Heidelberg, Springer Berlin Heidelberg, pp. 379–417.

Krot, A.N., Amelin, Y., Cassen, P., et al. 2005. Young chondrules in CB chondrites from a giant impact in the early solar system. *Nature* 436: 989. doi: 10.1038/nature03830.

Krot, A. N., Amelin, Y., Cassen, P., and Meibom, A., 2005, Young chondrules in CB chondrites from a giant impact in the early Solar System: *Nature*. 436: 989–992. doi: 10.1038/nature03830.

Mangold, N., Baratoux, D., Witasse, O., et al. 2016. Mars: A small terrestrial planet. *The Astronomy and Astrophysics Review* 24: 15. doi: 10.1007/s00159-016-0099-5.

Marti K., Graf T. 1992. Cosmic-ray exposure history of ordinary chondrites. *Annual Review of Earth and Planetary Sciences* 20:221–43.

McDonough, W.F. and Sun, S.-S. 1995. The composition of the Earth. *Chemical Geology* 120: 223–53.

McKeegan, K. D., Kallio, A. P. A., Heber, V. S., et al. 2011. The oxygen isotopic composition of the sun inferred from captured solar wind. *Science* 332: 1528–32. doi: 10.1126/science.1204636.

McSween, H.Y., Jr. 1987. *Meteorites and their Planet Bodies*. New York, Cambridge University Press.

McSween, H. Y., Emery, J. P., Rivkin, A. S., et al. 2018. Carbonaceous chondrites as analogs for the composition and alteration of ceres. *Meteoritics & Planetary Science*: doi: 10.1111/maps.12947.

McSween, H.Y. and Huss, G.R. 2010. *Cosmochemistry*. New York, Cambridge University Press.

McSween, H.Y., Taylor, G.J. and Wyatt, M.B. 2009. Elemental composition of the Martian crust. *Science* 324: 736–9. doi: 10.1126/science.1165871.

Meteorit. Planet. Science 42: 1183–95. doi: 10.1111/j.1945-5100.2007.tb00568.x.

Mittlefehldt, D.W., 2015. Asteroid (4) Vesta: I. The howardite-eucrite-diogenite (HED) clan of meteorites. *Chemie der Erde - Geochemistry* 75: 155–183. doi: 10.1016/j.chemer.2014.08.002.

Morbidelli, A., Lunine, J. I., O'Brien, D. P., et al. 2012. Building terrestrial planets. *Annual Review of Earth and Planetary Sciences* 40: 251–275. doi: 10.1146/annurev-earth-042711-105319.

Nyquist, L.E., Kleine, T., Shih, C.Y. and Reese, Y.D. 2009. The distribution of short-lived radioisotopes in the early solar system and the chronology of asteroid accretion, differentiation, and secondary mineralization. *Geochimica et Cosmochimica Acta* 73: 5115–36.

Nyquist, L.E., Reese, Y., Wiesmann, H., Shih, C.Y. and Takeda, H. 2003. Fossil ^{26}Al and ^{53}Mn in the *Asuka* 881394 eucrite: evidence of the earliest crust on asteroid *4 Vesta*. *Earth and Planetary Science Letters* 214: 11–25.

Palme H. 2017. Cosmic Elemental Abundances. in: *Encyclopedia of Geochemistry* (ed. W. M. White). Springer International Publishing, Heidelburg, doi: 10.1007/978-3-319-39193-9_335-1

Petaev, M.I. and Wood, J.A. 1998. The condensation with partial isolation (CWPI) model of condensation in the solar nebula. *Meteorit. Planet. Science* 33: 1123–37. doi: 10.1111/j.1945-5100.1998.tb01717.x.

Petigura, E. A., Howard, A. W., Marcy, G. W. 2013. Prevalence of earth-size planets orbiting sun-like stars. *Proceedings of the National Academy of Sciences* 110: 19273–19278. doi: 10.1073/pnas.1319909110.

Pian, E., D'Avanzo, P., Benetti, S., Branchesi, M., Brocato, E., et al., 2017. Spectroscopic identification of r-process nucleosynthesis in a double neutron-star merger. *Nature* 551, 67–70. doi: 10.1038/nature24298.

Prettyman, T. H., Yamashita, N., Toplis, M. J., *et al.* 2017. Extensive water ice within Ceres' aqueously altered regolith: Evidence from nuclear spectroscopy. *Science* 355: 55–59. doi: 10.1126/science.aah6765.

Prinn, R.G. and Fegley, B. 1989. Solar nebula chemistry: origin of planetary, satellite and cometary volatiles, in *Origin and Evolution of Planetary and Satellite Atmospheres* (eds S.K. Atreya, J.B. Pollack and M.S. Mathews). Tuscon, University of Arizona Press.

Qin, L., Carlson, R. W. 2016. Nucleosynthetic isotope anomalies and their cosmochemical significance. *Geochemical Journal* 50: 43–65. doi: 10.2343/geochemj.2.0401

Rauscher, T., Dauphas, N., Dillmann, I., Fröhlich, C., Zs, F. and Gy, G., 2013. Constraining the astrophysical origin of the p-nuclei through nuclear physics and meteoritic data. *Reports on Progress in Physics* 76(6), 066201 (38p). doi: 10.1088/0034-4885/76/6/066201.

Reynolds, J.R. 1960. Isotopic composition of xenon from enstatite chondrites. *Zeitshrift für Naturforschung* 15a: 1112–14.

Ruesch, O., Platz, T., Schenk, P., et al. 2016. Cryovolcanism on Ceres. *Science*, 353: aaf4286-1–aaaf4286-8. doi:10.1126/science.aaf4286-1.

Russell, C.T., Raymond, C.A., Coradini, A., et al. 2012. DAWN at Vesta: testing the protoplanetary paradigm. *Science* 336: 684–6.

Schulz, T., Upadhyay, D., Münker, C., Mezger, K., 2012. Formation and exposure history of non-magmatic iron meteorites and winonaites: Clues from Sm and W isotopes. *Geochimica et Cosmochimica Acta* 85: 200–212. doi: 10.1016/j.gca.2012.02.012.

Scott E.R.D. and Krot A.N. 2014. 1.2 - Chondrites and Their Components A2 - Holland, Heinrich D. In: Turekian KK (ed) *Treatise on Geochemistry (Second Edition)*, vol. Elsevier, Oxford, pp. 65–137, doi: 10.1016/B978-0-08-095975-7.00104-2.

Shang, H., Shu, F.H., Lee, T. and Glassgold, A.E. 2000. Protostellar winds and chondritic meteorites. *Space Sci. Review* 92: 153–76.

Shu, F.H., Shang, H., Glassgold, A.E. and Lee, T. 1997. X-rays and fluctuating X-winds from protostars. *Science* 277: 1475–9.

Smoliar, M.I., Walker. R.J. and Morgan, J.W. 1996. Re-Os ages of Group IIA, IIIA, IVA, *and IVB iron meteorites*. *Science* 271: 1099–1102. doi: 10.1126/science.271.5252.1099.

Sugiura, N. and Krot, A.N. 2007. ^{26}Al–^{26}Mg systematics of Ca-Al-rich inclusions, amoeboid olivine aggregates, and chondrules from the ungrouped carbonaceous chondrite Acfer 094.

Tang, M. and Anders, E. 1988. Isotopic anomalies of Ne, Xe, and C in meteorites. II. Interstellar diamond and SiC: Carriers of exotic noble gases. *Geochimica et Cosmochimica Acta* 52: 1235–44.

Taylor, S.R. 1975. *Lunar Science: A Post-Apollo View*. New York, Pergamon Press.

Taylor, S.R. 1991. Accretion in the inner nebula: the relationship between terrestrial planetary compositions and meteorites. *Meteoritics* 26: 267–77.

Taylor, S.R. 1992. *Solar System Evolution: A New Perspective*. Cambridge, Cambridge University Press.

Thielemann, F. K., Arcones, A., Käppeli, R., Liebendörfer, M., Rauscher, T., Winteler, C., Fröhlich, C., Dillmann, I., Fischer, T., Martinez-Pinedo, G., Langanke, K., Farouqi, K., Kratz, K. L., Panov, I. and Korneev, I. K., 2011. What are the astrophysical sites for the r-process and the production of heavy elements? *Progress in Particle and Nuclear Physics* 66: 346–353. doi: https://doi.org/10.1016/j.ppnp.2011.01.032.

Thiemens, M.H. and Heidenreich, J.E. 1983. The mass independent fractionation of oxygen – A novel isotopic effect and its cosmochemical implications. *Science* 219: 1073–5.

Touboul, M., Kleine, T., Bourdon, B., Palme, H. and Wieler, R. 2007. Late formation and prolonged differentiation of the Moon inferred from W isotopes in lunar metals. *Nature* 450: 1206–9.

Touboul, M., Kleine, T., Bourdon, B., Palme, H. and Wieler, R. 2009. Tungsten isotopes in ferroan anorthosites: Implications for the age of the Moon and lifetime of its magma ocean. *Icarus* 199: 245–9.

Touboul, M., Sprung, P., Aciego, S. M., et al. 2015. Hf–W chronology of the eucrite parent body. *Geochimica et Cosmochimica Acta* 156: 106–121. doi 10.1016/j.gca.2015.02.018.

Trinquier, A., Birck, J.L., Allègre, C.J., Göpel, C. and Ulfbeck, D. 2008. ^{53}Mn–Cr systematics of the early Solar System revisited. *Geochimica et Cosmochimica Acta* 72: 5146–63.

Van Schmus, W.R. and Wood, J.A. 1967. A chemical petrologic classification for the chondritic meteorites. *Geochimica et Cosmochimica Acta* 31: 747–65.

Wadhwa, M., Amelin, Y., Bogdanovski, O., et al. 2009. Ancient relative and absolute ages for a basaltic meteorite: Implications for timescales of planetesimal accretion and differentiation. *Geochimica et Cosmochimica Acta* 73: 5189–5201. doi: 10.1016/j.gca.2009.04.043.

Walker, R.J., 2016. Siderophile elements in tracing planetary formation and evolution. *Geochemical Perspectives* 5, 1–145. doi: 10.7185/geochempersp.5.1.

Walsh, K. J., Morbidelli, A., Raymond, S. N., et al. 2011. A low mass for Mars from Jupiter's early gas-driven migration. *Nature* 475: 206. doi: 10.1038/nature10201.

Warren, P. H. 2011. Stable-isotopic anomalies and the accretionary assemblage of the Earth and Mars: A subordinate role for carbonaceous chondrites. *Earth and Planetary Science Letters* 311: 93–100. doi: https://doi.org/10.1016/j.epsl.2011.08.047.

Wasserburg, G.J., Busso, M., Gallino, R. and Raiteri, C.M. 1994. Asymptotic giant branch stars as a source of short-lived radioactive nuclei in the solar nebula. *Astrophysical Journal* 424: 412–28.

Wasson, J.T. 1974. *Meteorites*. Berlin, Springer-Verlag.

Wasson, J.T. 1985. *Meteorites: Their Record of Early Solar System History*. New York. W.H. Freeman.

Wasson, J.T. 2011. Relationship between iron-meteorite composition and size: Compositional distribution of irons from North Africa. *Geochimica et Cosmochimica Acta* 75: 1757–72.

Wasson, J.T. and Kallemeyn, G.W. 1988. Compositions of chondrites. *Philosophical Transactions of the Royal Society of London* A325: 535–44.

Wasson, J.T. and Kallemeyn, G.W. 2002. The IAB iron-meteorite complex: A group, five subgroups, numerous grouplets, closely related, mainly formed by crystal segregation in rapidly cooling melts. *Geochimica et Cosmochimica Acta* 66: 2445–73.

Yin, Q., Jacobsen, S.B., Blichert-Toft, J., Télouk, P. and Albarède, F. 2002. A short timescale for terrestrial planet formation from Hf-W chronometry of meteorites. *Nature* 418: 949–51.

Zinner, E. 2014. 1.4 Presolar grains. In: Holland HD, Turekian KK (eds) *Treatise on Geochemistry* (second edition), vol. 1 Elsevier, Oxford, pp. 181–213.

PROBLEMS

1. On the extract of the chart of the nuclides below, identify the mode of origin (S, R, or P process) of the stable isotopes of W, Re, Os, and Ir by writing S, R, or P in the box for each (remember, some nuclides can be created by more than one process). Identify those isotopes you feel should be most abundant and those least abundant. On the chart below, mass numbers are given for only the stable isotopes. As a start, assume the S-process path starts at ^{181}Ta. Assume the unstable isotopes will decay before capturing a neutron during the S-process.

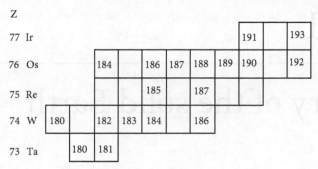

2. One calcium-aluminum inclusion in the Allende meteorite has $\delta^{26}Mg$ values, which implies a $^{26}Al/^{27}Al$ ratio of 0.46×10^{-4} at the time of its formation. A second inclusion apparently formed with a $^{26}Al/^{27}Al$ ratio of 1.1×10^{-4}. The half-life of ^{26}Al is 7.2×10^5 years. Assuming both these inclusions formed from the same cloud of dust and gas and that the $^{26}Al/^{27}Al$ ratio in this cloud was uniform, what is the time interval between formation of the two inclusions?

3. Assuming that the oxygen in CV chondrites with $\delta^{18}O = 0‰$ and $\delta^{17}O = +6$ is a mixture of oxygen having an oxygen isotope composition lying on the terrestrial fractionation line in Figure 10.30 and pure ^{16}O, how much ^{16}O would have to be added to oxygen lying on the terrestrial fractionation line to reproduce their oxygen isotopic composition?

4. Using the partition coefficients in Table 7.5, estimate the fraction of plagioclase that would have to fractionally crystallize from a lunar magma ocean to produce the Eu anomaly of KREEP shown in Figure 10.41. (Hint: Concern yourself only with the Eu/Sm ratio.)

5. Make a plot of the log of the fraction of Os condensed from a gas of "solar" composition as a function of temperature (e.g., a plot similar to Figure 10.34). Assume a total pressure of 10^{-4} atm, ΔH°_V of $738\,kJ/mol$, ΔS°_V of $139\,J/mol$, and the solar system abundances in Table 10.2. Assume the solid is pure Os metal. (Hint: About 50% will be condensed at 1737 K.)

Chapter 11

Geochemistry of the solid Earth

11.1 INTRODUCTION

Having considered the Earth's formation in the last chapter, let's now use the tools of geochemistry we acquired in the first nine chapters to consider how the Earth works. The Earth, unlike many of its neighbors, has evolved over its long history and it remains geologically active. Four and a half billion years later, it is very different place than when it was first formed. Certainly, one of the main objectives of geochemistry is to understand this activity and unravel this history.

The solid Earth consists of three distinct layers: crust, mantle, and core. As Table 11.1 shows, these layers become progressively denser with depth, reflecting a stable density stratification of the planet (one shared by the other terrestrial planets). Unfortunately, only the shallowest of these layers, the crust, is accessible to direct study. The crust may be the most interesting part of the planet, but it represents less than half a percent of its mass. Even most of the crust is out of reach; the deepest boreholes drilled in the continental crust, which is 35 km thick on average, are only 12 km deep; the deepest borehole in oceanic crust, which is on average 7 km thick, is only 1.5 km deep. Thus, study of the solid Earth necessarily relies on indirect approaches and on rare samples of deeper material brought to the surface through geologic processes.

We'll begin by considering the geochemistry of the mantle because it represents the largest fraction of the Earth, both by mass (67%) and volume (88%). Although remote, the mantle is important for a number of reasons. For one, the crust has been formed from it. For another, convection within the Earth's mantle drives most tectonic activity that affects us on the surface, including plate motions, earthquakes, and volcanoes. We'll next consider the core. As we found in the previous chapter, the core formed quite early – more or less simultaneously with formation of the Earth itself. While we do have a few samples of the Earth's mantle, we have none at all from the core, so its composition and history must be inferred entirely indirectly. Lastly, we'll consider the Earth crust – the part of the Earth we inhabit and are most familiar with. The crust, particularly the continental crust, is the most varied part of our planet. We'll find that the continental crust, unlike the core which is as old as the Earth itself, has grown through geologic time through magmatism. The crust, particularly the continental crust, is where most of the geologic history of the planet is preserved.

11.2 THE EARTH'S MANTLE

Parts of the mantle are occasionally thrust to the surface as so-called alpine peridotites; fracture zones in the oceanic crust also

Geochemistry, Second Edition. William M. White.
© 2020 John Wiley & Sons Ltd. Published 2020 by John Wiley & Sons Ltd.
Companion website: www.wiley.com/go/white/geochemistry

Table 11.1 Volumes and masses of the Earth's shells.

	Thickness(km)	Volume10^{12} km^3	Mean densityKg/m^3	Mass10^{24} kg	Mass percent
Atmosphere			0.000005	0.00009	
Hydrosphere	3.80	0.00137	1026	0.00141	0.024
Crust	17	0.0087	2750	0.024	0.4
Mantle	2883	0.899	4476	4.018	67.3
Core	3471	0.177	10915	1.932	32.3
Whole Earth	6371	1.083	5515	5.974	100.00

occasionally expose mantle rocks. And volcanic eruptions sometimes carry small pieces of the mantle to the surface as *xenoliths*. Nevertheless, much of what we know about the mantle has been deduced indirectly. Indirect methods of study include determination of geophysical properties such as heat flow, density, electrical conductivity, and seismic velocity. Another indirect method of study is examination of volcanic rocks produced by partial melting of the mantle. Finally, cosmochemistry provides an important constraint on the composition of the Earth.

The mantle was once viewed as being homogenous, but we now realize that the chemistry of the mantle is heterogeneous on all scales. On a large scale, the mantle appears to consist of a number of reservoirs that have complex histories. The best evidence for this large-scale heterogeneity comes from trace element and isotope ratio studies of volcanic rocks, but there is also evidence that the major element composition of the mantle varies. While trace elements may vary by an order of magnitude or more, the major element variations are much more subtle, just as they are in volcanic rocks and in the crust. Isotope studies have proven tremendously valuable in understanding the mantle for several reasons. First, unlike trace element and major element concentrations, isotope ratios do not change during the magma generation process (except by mixing of the magma with other components such as assimilated crust). Second, radiogenic isotope ratios provide *time-integrated* information about the parent/daughter ratios, and therefore allow inferences about the history of the mantle. Additional evidence of mantle heterogeneity comes from seismic wave velocities, which have revealed large scale structures in the lower mantle.

The crust has been created by extraction of partial melts from the mantle (i.e., volcanism), a process that has continued over the Earth's entire history. For the major elements, crust formation affects the composition of the mantle only slightly because the volume of the crust is so small. For example, if the SiO_2 concentration in the mantle were originally 45%, extraction of continental crust containing an average of 60% from that mantle would reduce the SiO_2 concentration in the mantle only to 44.6%. Of course, extraction of partial melt undoubtedly has changed the major element composition of the mantle *locally*, as we will see in subsequent sections of this chapter. Furthermore, the concentrations of highly incompatible elements in the mantle have changed significantly as a result of formation of the crust. The idea that the core and the mantle segregated very early and the crust formed subsequently gives rise to the notion of a *primitive mantle* composition. This composition is the composition of the mantle before the crust was extracted from it. It is identical to the composition of the crust plus the mantle. Thus, the terms *primitive mantle composition* and *bulk silicate Earth composition* (BSE) are synonymous. One of the objectives of this chapter will be to estimate this composition.

In attempting to assess this composition, we have essentially three kinds of constraints. The first is *geophysical*: geophysical measurements of moment of inertia and seismic wave velocities allow us to constrain the density, compressibility, and rigidity of the mantle, which in turn constrains its composition. The second is *cosmochemical*: the Earth formed from the solar nebula, which has a chondritic composition for condensable elements, and therefore the composition of the Earth should relate in some rational way to the composition of chondrites. Finally, the mantle can be directly sampled locally where it is tectonically exposed or where small pieces, xenoliths, are carried to the surface by volcanic eruptions and indirectly through volcanism. Most lavas erupted on the surface of the Earth are basalts and they are products of partial melting of the mantle. The composition of the mantle thus must be such that it yields basaltic magma upon melting. There is, however, a caveat: only the uppermost mantle is sampled by volcanism and tectonism. Even the most deeply generated magmas, kimberlites, come from the upper few hundred km of the mantle, which is nearly 3000 km deep.

11.2.1 Structure of the mantle and geophysical constraints on mantle composition

Geophysical measurements constrain the physical properties of the Earth and its interior, and therefore constrain its composition. One set of constraints comes from the velocities with which seismic waves travel through the Earth. There are two kinds of seismic body waves: *compressional* or P-waves, in which particle motion is parallel to wave motion, and *shear* or s-waves, in which particle motion is perpendicular to wave motion. P-waves can travel through all media, they are the familiar sound waves in air, but S-waves can only travel through solids. Indeed, the absence of S-wave arrivals in a zone between about $100°$ and $180°$ from an earthquake epicenter, the so-called S-wave shadow zone, provided the original and still most compelling evidence that the outer core is liquid. The velocity with which seismic waves travel depends on the square root of the ratio of the elastic modulus to density. For S-waves this is:

$$v_s = \sqrt{\frac{\mu}{\rho}} \qquad (11.1)$$

where ρ is density and μ is the rigidity modulus, defined as the ratio of shear stress to shear strain. For P-waves the velocity is given by:

$$v_p = \sqrt{\frac{\kappa + \frac{4}{3}\mu}{\rho}} \qquad (11.2)$$

where κ is the bulk modulus. κ is defined as:

$$k = -V\left(\frac{\partial P}{\partial V}\right)_s \qquad (11.3)$$

κ is very nearly the inverse of the compressibility, which we defined in eqn. 2.12. The only difference is that in eqn. 2.12 we specified constant temperature in the differential term and here we are specifying constant entropy. This makes things slightly easier for the present problem because the temperature of the Earth would continually increase downward as a result of adiabatic compression, even if there were no other sources of heat in the Earth's interior. Here, we allow for that adiabatic temperature increase. Density, rigidity modulus, and compressibility all depend on the phases constituting a material rather than just composition. However, assuming equilibrium conditions prevail, only one phase assemblage will be possible for a given composition, pressure, and temperature.

We can independently constrain density, and its variation through the Earth, in another way. The density of the Earth is, of course, simply its mass divided by its volume. The latter has been known since the Greek mathematician Eratosthenes determined the Earth's radius in the third century BC. The mass of the Earth was first estimated from the strength of its gravitational field by Isaac Newton in the seventeenth century and refined by Henry Cavendish in the eighteenth century. Using the most recent determinations of these parameters, we can calculate a mean density for the Earth of $5515\,kg/m^3$. Comparing this value with the value of typical crustal rocks, which are in the range of $2000–3000\,kg/m^3$, immediately leads to the conclusion that density must be greater in the Earth's interior.

Let's now see how we can combine density and seismic wave velocity variations to provide further constraints. First, we know that density is inversely related to volume:

$$\frac{d\rho}{\rho} = -\frac{dV}{V} \qquad (11.4)$$

We can now rewrite eqn. 11.3 in terms of density rather than volume as:

$$\kappa = -\rho\left(\frac{\partial P}{\partial \rho}\right)_s \qquad (11.5)$$

Then we note that pressure will vary with radial distance from the center of the Earth as:

$$\frac{dP}{dr} = -\rho(r)g(r) \qquad (11.6)$$

where $g(r)$ is the gravitational acceleration at radial distance r. We can combine these equations to write:

$$\left(\frac{\kappa}{\rho}\right) = g(r)\rho(r)\frac{dr}{d\rho} \qquad (11.7)$$

Interestingly, it is readily shown that the term on the left can be derived from the seismic wave velocities:

$$\left(\frac{\kappa}{\rho}\right) = V_P^2 - \frac{4}{3}V_s^2 \qquad (11.8)$$

The acceleration of gravity at radial distance r is given by:

$$g(r) = \frac{G}{r^2}\int 4\pi\rho(r)r^2 dr \qquad (11.9)$$

Combining eqns. 11.7 through 11.9, we have:

$$\frac{d\rho(r)}{dr} = \frac{G}{r^2}\frac{\rho(r)}{V_P^2 - \frac{4}{3}V_S^2}\int 4\pi\rho(r)r^2 dr \quad (11.10)$$

Equation 11.10 describes how density changes in a self-compressing, but otherwise uniform sphere and is known as the *Adams–Williamson equation*. It allows us to predict the density of an infinitesimally small layer at distance r_1 from seismic velocities, provided we know the density at an adjacent layer, r_2. Since we can directly measure density at the surface, we can begin there and work our way through the Earth calculating density variation.

Yet another parameter, the Earth's *moment of inertia*, also allows us to constrain how density changes with depth. The moment of inertia for a rotating sphere of uniform density is simply:

$$I = \frac{2}{5}mr^2 \qquad (11.11)$$

where m is mass and r is radius. For a spherical body whose density is a function of depth, the moment of inertia is:

$$I = \int \frac{4}{3}\pi r^4\rho(r)dr \qquad (11.12)$$

The situation for the Earth is slightly more complex since it is not strictly spherical.

Rather, as a consequence of centrifugal force, it is an oblate spheroid with its equatorial radius slightly greater than polar radius. The gravitational interaction between the slightly flattened Earth and the Moon results in a gravitational torque which causes the Earth's rotational vector to itself slowly rotate, or *precess*. From the rate of this precession, about one rotation per 25,000 years, a moment of inertia of 8.07×10^{37} kg/m^2 can be calculated.

Any density distribution we obtain using seismic velocities and the Adams–Williamson equation can also be used to calculate a moment of inertia, which should agree with that measured from precession. The two do not agree if we take the Earth to be a self-compressing sphere, for the simple reason that there are density discontinuities in the Earth that result from compositional changes, such as at the core–mantle boundary, as well as phase changes. The problem then is to produce a density model for the Earth that accounts for both observed seismic wave velocities and moment of inertia. The generally accepted model that accomplishes this is called the *Preliminary Reference Earth Model* or *PREM* (Dziewonski and Anderson, 1981). PREM is based on many decades of seismic records. Density and seismic wave velocities of the PREM model are shown in Figure 11.1. The term *preliminary* clearly implies further refinements are expected, but no revisions have been made since the model was agreed upon in 1981. In part, this is because geophysicists have tended to focus on local deviations from PREM rather than further refine the spherically symmetric model, but it also means that the PREM model is actually reasonably mature. There are indeed quite significant deviations from spherical symmetry, and these have important geochemical implications.

11.2.2 Cosmochemical constraints on mantle composition

In Chapter 10 we considered the composition of that part of our cosmic environment accessible to sampling: meteorites, the solar surface (deduced from optical spectra and solar wind particles), Mars, and the Moon. The meteorite data provide first-order constraints on the formation and composition of the Earth. Since the mantle comprises roughly 99% of the silicate part of the Earth, these

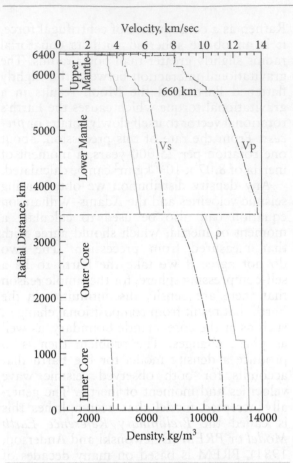

Figure 11.1 The PREM model of seismic velocities and density variation through the Earth.

constraints also apply to the mantle. Two important cosmochemical assumptions guide our thinking about the composition of the Earth: *(1) the entire solar system formed at one time from a nebula of gas and dust, (2) the composition of this nebula was similar to that of CI chondrites for the condensable elements.* The last statement should not be construed to mean that all bodies that formed from this nebula are necessarily of chondritic composition. Indeed, we will see that the Earth's composition differs from chondritic in significant ways.

Thus far there is essentially no evidence to contradict the first assumption (provided we interpret "simultaneous" in a geologic sense; i.e., this process may have lasted 10^8 or more years); the second assumption certainly holds to a first approximation, but clear evidence of isotopic heterogeneity within the solar nebula

revealed in the last decade or so demonstrate that the solar nebula was not as well mixed as we once thought. In addition, processes occurring within the nebula resulted in chemical variation that in turn produced a range of compositions of chondritic parent bodies and in the planets as well.

In summary, we can draw the following conclusions relevant to the formation and composition of the Earth from cosmochemistry:

1. The material from which the solar system formed was of approximately CI chondritic composition (plus gases).
2. Processes within the solar nebula, most particularly related to volatility produced planets and asteroids of variable composition. This is apparent from the composition of chondrites as well as from density variations of the planets.
3. Formation of planets and asteroids occurred within roughly 10^8 years of the beginning of the solar system (which we take to be the formation of its oldest objects, the CAIs found in chondrites).
4. Asteroids and the terrestrial planets underwent differentiation in which iron metal and silicate segregated. Based on the age of differentiated meteorites, and on W isotope ratios of iron meteorites in particular, the process began very early – within a few million years at most of the beginning of the solar system.
5. The final stages of the formation of planetary assembly involved energetic collisions of large bodies. In the case of the Earth–Moon system, this culminated in a collision that released vast amounts of energy. A consequence of this was a magma ocean on the Moon and, most likely, the Earth as well.

11.2.3 Observational constraints on mantle composition

Figure 11.1 implies that the uppermost mantle should consist of rock having a density of around 3400 kg/m³ and P-wave velocities of about 8 km/sec. Rocks having these properties are occasionally exposed at the surface in alpine massifs or collision zones, in fracture zones along mid-ocean ridges, and in sections

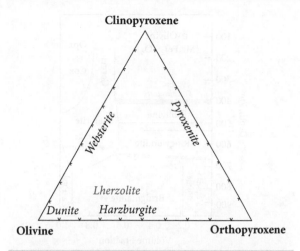

Figure 11.2 Ternary diagram illustrating ultramafic rock nomenclature based on minerals present.

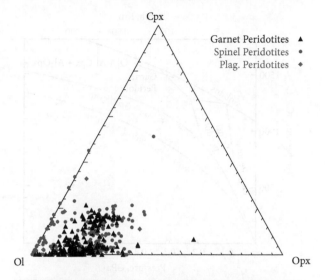

Figure 11.3 Modal mineralogy of peridotite xenoliths in the Deep Lithosphere Database of EarthChem (www.earthchem.org) projected onto the ol-opx-cpx plane.

of obducted oceanic crust called ophiolites. These rocks are invariably Mg- and Fe-rich silicates. In addition, rocks having these properties are sometimes found as xenoliths in lavas; they are also Mg- and Fe-rich silicate rocks that consist dominantly of olivine and pyroxenes. Chondrites, of course, are also Mg- and Fe-rich and consist dominantly of olivine and pyroxenes, so it would seem that such rocks meet both geophysical and cosmochemical constraints for mantle composition. The term *ultramafic* is applied to silicate rocks rich in magnesium and iron, which typically consist of olivine and pyroxenes (but upon hydration at or near the surface transform to minerals such as serpentine or talc). *Peridotite* is a rock dominated by olivine (*peridot* is the gem name for olivine). If the olivine exceeds 90% of the rock, it is termed a *dunite* (Figure 11.2). Rocks with substantial amounts of both pyroxenes as well as olivine are *lherzolites*. This can be prefaced by the name of the Al-bearing phase (e.g., spinel lherzolite), whose nature depends on pressure. Figure 11.3 shows the compositions of ultramafic xenoliths found in the volcanic rocks of Kilbourne Hole, New Mexico. The most common types are lherzolites. This suggests we should be focusing our attention on lherzolite as mantle material. Lherzolite also has appropriate density and seismic velocities to be mantle material. Furthermore, upon

partial melting at high pressures it produces basaltic magma. Ringwood (1975) coined the term *pyrolite,* meaning pyroxene–olivine rock, for the lherzolitic composition matching geophysical and cosmochemical constraints.

11.2.4 Mantle mineralogy and phase transitions

11.2.4.1 Upper mantle phase changes

Figure 11.1 shows that a number of discontinuities exist in the density and seismic velocity profiles of the Earth. Some of these are clearly compositional changes, such as the shallowest discontinuity which is the crust–mantle boundary, and the core–mantle boundary. Others represent phase changes that are isochemical, or nearly so. The most prominent is the inner core–outer core boundary, which is principally a solid–liquid phase change. The discontinuities within the mantle now all appear to be *primarily* phase changes. In the upper 200 km or so of the mantle, the only important phase changes are the nature of the aluminous phase (Figure 11.4), which changes from plagioclase to spinel ($MgAl_2O_4$) and then to garnet with increasing pressure.

The garnet lherzolite assemblage remains stable to depths of about 300 km (although

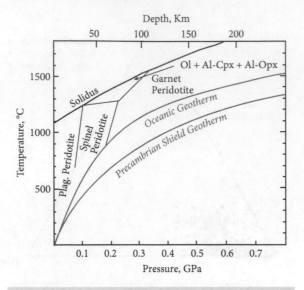

Figure 11.4 Upper mantle phase diagram.

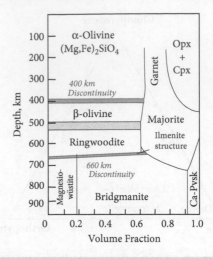

Figure 11.5 Mineral assemblages in the upper 1000 km of the mantle. After Ringwood (1991).

with some changes in the mineral compositions – the basis of thermobarometry we discussed on Chapter 4). At this depth, appreciable amounts of pyroxenes begin to dissolve in garnet, forming a solid solution with the general composition $M_2(M,Si,Al_2)Si_3O_{12}$ where M is Mg, Fe, or Ca. This garnet, called *majorite*, differs from those found at lower pressure in that up to a quarter of the silicon atoms are in octahedral coordination (i.e., surrounded by six oxygens rather than four). The octahedral coordination is favored because anions such as oxygen are more compressible than are cations such as silicon. When compressed, more oxygens can be packed around each silicon atom. This phase change is a gradual one, with complete conversion of pyroxenes to majorite at about 460 km depth (Figure 11.5). The phase change results in a roughly 10% increase in density of the "pyroxene" component.

11.2.4.2 *The transition zone*

Between 400 and 670 km depth, seismic velocities increase more rapidly than elsewhere (Figure 11.1); this depth interval is called the transition zone. At about 400 km, or 14 GPa, olivine undergoes a structural change from the low-pressure or α form, to the β form, also known as *wadsleyite*. In

contrast to the pyroxene-to-majorite phase change, this phase boundary is relatively sharp, with a transition interval of 9–17 km. It results in an 8% increase in density.

At about 500 km depth or so, olivine undergoes a further structural change to the γ form or *ringwoodite*. The structure is similar to that of $MgAl_2O_4$ spinel, and this phase is sometimes, somewhat confusingly, referred to simply as spinel. The change from β to ringwoodite is thought to be more gradual than the α–β transition, occurring over a depth interval of 30 km, and involves only a 2% increase in density. In both the β and γ phases, silicon remains in tetrahedral coordination.

Within the transition zone, some of the Mg and Ca in majorite begins to exsolve to form $CaSiO_3$ in the perovskite structure and $MgSiO_3$ in the ilmenite structure. The proportion of $CaSiO_3$ perovskite increases with depth until majorite disappears at about 720 km. $MgSiO_3$ ilmenite persists only to 660 km.

As Figure 11.1 shows, a sharp and large increase in seismic velocity occurs at around 660 km depth or roughly 23 GPa; this is called the *660 seismic discontinuity* (although the exact depth is still debated, perhaps partly because it varies slightly as described below). This depth marks the beginning of the lower mantle. At this depth γ-olivine reorganizes

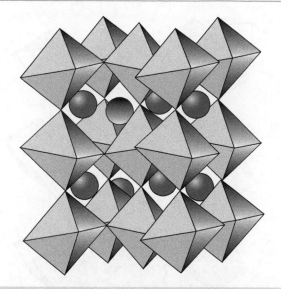

Figure 11.6 The structure of bridgmanite ($MgSiO_3$). The structure consists of corner-sharing SiO_6 octahedra with Mg^{2+} in dodecahedral sites. After Poirier (1991).

to form *bridgmanite** (Mg,Fe)SiO_3 and (Mg,Fe)O *magnesiowüstite* (perhaps more properly called *ferripericlase*), with the Fe going preferentially in the magnesiowüstite. This phase change results in a density increase of about 11%. Bridgmanite has the chemical stoichiometry of pyroxene, but the silicons are in octahedral coordination. The structure, illustrated in Figure 11.6, is similar to that of the "high-temperature" cuprate superconductors discovered in the 1980s.

High-pressure experiments carried out with a diamond anvil show that the transition is quite sharp, occurring within a pressure interval of 0.15 GPa at 1600°C. The transition has a negative Clapeyron slope of approximately P (GPa) = 27.6 − 0.0025 T (K) (e.g., Chopelas et al., 1994), so that it will occur at somewhat shallower depths in hot regions, such as areas of mantle upwelling, and at greater depths in cooler regions, such as subducted lithosphere, though since the Clapeyron slope is shallow, the effect is small. The effect of these differences is to resist motion across

the boundary but is not sufficient to prevent motion across this boundary. Indeed, seismic tomographic images show that some slabs of subducting oceanic crust encounter resistance at 660 km but most ultimately sink through the boundary. Bridgmanite is a much stronger mineral than the tetrahedral silicates of the upper mantle, which helps to explain the roughly 10–30 times greater viscosity of the lower mantle deduced from postglacial isostatic adjustment.

11.2.4.3 *The lower mantle*

The lower mantle, the region between the 660 km seismic discontinuity and the core–mantle boundary at 2890 km, is substantially less accessible to study than is the upper mantle. The principal constraint on its composition is seismic velocities and density. The assumption that the Earth is approximately chondritic forms another constraint. It is generally agreed that the lower mantle is similar in compositional to the upper mantle, composed dominantly of SiO_2, MgO, and FeO with lesser amounts of CaO, Al_2O_3, TiO_2, and so on, but how similar is an open question. In particular, some disagreement remains as to whether it might be slightly richer in FeO or SiO_2 than the upper mantle. For example, Lee et al. (2004) concluded a molar Mg/(Mg+Fe) ratio of about 0.85 for the lower mantle is a better fit to seismic observations than the upper mantle value of 0.9. However, Mattern et al. (2005) have concluded that the density and elastic properties of lower mantle phases are not sufficiently well known to distinguish between these alternatives, as the predicted densities of the pyrolite and chondritic models of the lower mantle differ by less than 0.06 g/cc. Uncertainties about the temperature of the lower mantle and the coefficient of thermal expansion of lower mantle materials compound the problem.

Bridgmanite and magnesiowüstite remain the principal phases to depths of 2600 km or so. For a "pyrolite" mantle, the phase

---

* This phase was long called simply Mg-perovskite because it has the crystal structure of perovskite ($CaTiO_3$) but was not given a mineral name because, although it could be synthesized in high pressure laboratory experiments, it had not been found in nature. Once it was identified in a heavily shocked meteorite, it was given the name bridgmanite in 2014 in honor of Percy Bridgman (1882–1961) of Harvard University, who won the 1946 Nobel Prize in physics for his high-pressure research.

assemblage will be 65–80% bridgmanite, 15–30% magnesiowüstite, and about 7% Ca-perovskite. Bridgmanite is thus the most abundant mineral in the Earth. There had been some debate about how Al_2O_3 is accommodated in lower mantle minerals. Experiments beginning with Frost et al. (2004) showed that it is readily accommodated in bridgmanite through a coupled substitution with ferric iron forming a $MgSiO_3$–$FeAlO_3$ solid solution. However, $Fe^{3+} < Al^{3+}$ in a pyrolite composition, which would appear to limit the amount of Al^{3+} that can be accommodated this way; however, this substitution is sufficiently energetically favorable that iron undergoes disproportionation:

$$3Fe^{2+} \rightarrow 2Fe^{3+} + Fe^0 \qquad (11.13)$$

As a consequence, the lower mantle is expected to contain about 1% metallic Fe. This substitution becomes less favorable with increasing pressure so that the amount of metallic Fe decreases somewhat with depth.

Experiments and *ab initio* simulations show that in the lowermost mantle $MgSiO_3$ transforms from the perovskite-like structure of bridgmanite to one in which the silica octahedral are organized into sheets (Figure 11.7) at about 120–138 GPa in the temperature range of 2500–3500 K (Murakami et al., 2004; Oganov and Ono, 2004). As this mineral has not been found in nature it is referred to simply as post-perovskite phase or pPv. The post-perovskite phase accepts considerably less Fe^{3+} than bridgmanite so that most remaining metallic iron recombines with Fe^{3+} to form Fe^{2+} (Gialampouki et al., 2018).

The pressure of the bridgmanite–post-perovskite transition appears to depend strongly on composition (particularly Fe and Al) as well as temperature, but may occur at roughly 2700 km, just a few hundred kilometers above the base of the mantle (Hirose et al., 2017). This depth corresponds to the top of a region of unusual and highly seismic variable properties known as D″ (pronounced "dee-double prime"). These variations could result from the compositional and temperature dependence of this phase transition, implying compositional and/or temperature variations within D″. With its sheet-like structure, post-perovskite is weaker than

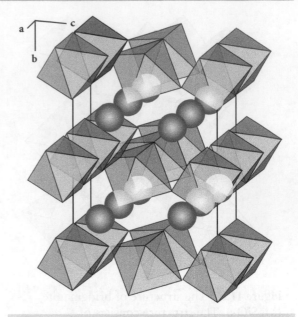

Figure 11.7 Structure of the post-perovskite phase in the lowermost mantle. Metal ions (Mg, Fe) in red and silica octahedra in gray are organized into sheets.

bridgmanite, which would lower the viscosity and enhancing convection.

Further study of the rich variety of reflected and refracted waves passing through this region and modeling of their waveforms ultimately resulted in recognition of small (up to a few hundred km long) and thin (10–40 km) "ultra-low shear-wave velocity zones" (ULVZs) where P- and S-wave velocities are up to 10% and 30%, respectively, lower than in surrounding regions directly above the core–mantle boundary. These seismic velocities are most easily explained as increases in density of up to 10%. The smaller reduction in P-wave velocities led Williams and Garnero (1996) to suggest these were regions of partial melting. In part because iron would preferentially partition into them, such high-pressure melts may be denser than surrounding solid and simply pond near the base of the mantle. Alternatively, the ULVZs may be compositionally distinct layers, possibly due to iron enrichment produced by reactions between core and mantle. While several dozen ULVZs have been detected, much of the core–mantle boundary region is under-sampled by seismic rays so it is difficult to define the size of these features and it is possible there are many more that have not been identified.

Several ULVZs appear to be larger, with lateral dimensions of ~800 km, and are located directly below hot spots such as Iceland and Hawaii (e.g., Yuan and Romanowicz, 2017). Furthermore, French and Romanowicz (2015) found broad quasi-vertical "conduits" of low seismic velocity beneath many hotspots "extending from the ULVZs on the core–mantle boundary to 1,000 km or so below Earth's surface, where some are deflected horizontally, as though entrained into more vigorous upper-mantle circulation." These may well be the mantle plumes postulated by Morgan (1971).

Geochemists and geophysicists had long debated the question as to whether the mantle might be chemically layered. One of the observations suggesting layering was the difference between mid-ocean ridge and oceanic island basalts, particularly in radiogenic isotope ratios, which requires they be derived from reservoirs separated for long periods. Seismologists could clearly resolve the radial seismic velocities variations shown in Figure 11.1, but they could not resolve lateral ones. As described above, however, the seismic discontinuities in the mantle can be explained by isochemical phases changes without invoking compositional changes. But as this debate continued and the earthquake catalogue became more extensive and computers became more powerful, seismological evidence of lateral heterogeneity in the deep mantle gradually came into focus. Tomographic images show subducting oceanic lithosphere descending through the entire mantle, which was inconsistent with compositional layering. Seismic tomography eventually also revealed two regions of anomalously slow S-wave velocity, one beneath Africa and the other beneath the South-Central Pacific, known as *large low shear-wave velocity provinces,* or *LLSVPs* (Figure 11.8). The LLSVPs are also apparent in distortions of the Earth's gravitational field (i.e., geoid anomalies). Although variations in mantle seismic velocity have generally been interpreted in terms of temperature (slow equals hot), a number of other factors suggest that the LLSVPs are also compositionally distinct. These include the sharp boundaries of the LLSVPs, which would be quickly "softened" by thermal diffusion, the very low S-wave velocities combined with moderate decreases in P-wave velocities, and changes in the ratio of S-wave velocity to

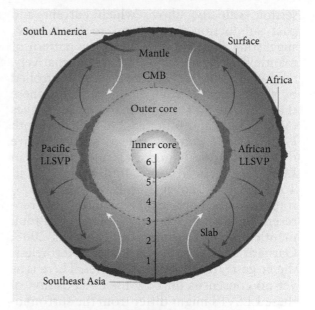

Figure 11.8 Cross-section of the Earth illustrating the two LLSVPs at the base of the mantle. Subduction zones and descending oceanic lithosphere are preferentially located away from these features while mantle plumes preferentially rise from the margins of them. From Romanowicz (2017).

bulk sound speed. Lau et al. (2017) used GPS measurements of solid earth tides to determine that lower portions of the LLSVPs are ~0.5% denser than surrounding mantle. On the other hand, Koelemeijer et al. (2017) found from free oscillations that LLSVPs were less dense than surrounding mantle. Both studies noted the possibility that they were density stratified, with greater density near their bases, which could resolve these apparently inconsistent results.

Remarkably, the LLSVPs are almost antipodal and centered near the equatorial plane, leading Dziewonski et al. (2010) to suggest they have stabilized the Earth's rotational axis close to the present one for geologically long periods, perhaps the entire history of the Earth. Mantle plumes are preferentially located on the margins of these features (Burke et al., 2008; French and Romanowicz, 2015), as are the ULVZs (McNamara et al., 2010). Based on the position of large igneous provinces (LIPS) through time, Burke (2011) argued that these LLSVPs are effectively permanent and stationary. They are surrounded by regions of anomalously fast seismic velocity, above which current and past subduction zones tend to be located, suggesting the LLSVPs control the pattern of mantle flow and plate tectonics. Alternatively, the location of the LLSVPs may be controlled by the descent of cold subducted lithosphere that focuses hot dense material into piles (McNamara and Zhong, 2005). The South Pacific LLSVP extends at least 400 km above the core–mantle boundary while the African LLSVP extends at least 1000 km above it. Together, they cover a substantial (~30%) fraction of the core–mantle boundary but represent only a small fraction (~2.5%) of total mantle volume (Hernlund and Houser, 2008; Garnero et al., 2016). They are, nevertheless, the largest structures in the planet. There is as yet no consensus on how the composition of these LLSVPs might differ from the surrounding mantle or how they formed, but there is no shortage of ideas. We'll consider these in due course.

11.3   ESTIMATING MANTLE AND BULK EARTH COMPOSITION

11.3.1   Major element composition

We can draw several important conclusions from the geophysical properties of the mantle when we combine them with the laboratory experiments of mineral physicists and petrologists. The first of these is that the material that best meets combined cosmochemical, geophysical, and mineral physics has the composition of *pyrolite*. Second, as appears that all seismic discontinuities can be explained by phase changes, there is no compelling geophysical evidence that the mantle is compositionally layered (although we cannot rule it out). On the other hand, there is good evidence of lateral heterogeneity in the lower mantle. While we do not know how the LLSVPs and ULVZs differ compositionally from the rest of the mantle, their combined mass is relatively small. Consequently, we will assume that the composition of the upper mantle is also the composition of the whole mantle. As we noted at the beginning of Section 11.2, this composition will be the same as the *primitive mantle* and bulk silicate Earth compositions for major elements.

Table 11.2 compares several estimates for the major and minor element composition of the bulk silicate Earth with the composition of CI chondrites. The first thing to notice is that there is reasonably broad agreement among the five estimates of the bulk silicate earth composition, despite some differences in details. Clearly, chondrites are much richer in siderophile elements (e.g., Fe, Ni) than all these estimated mantle compositions. The chondritic composition matches the upper mantle composition much better after a sufficient amount of the siderophile elements has been removed to form the Earth's core. However, even after removing the siderophile and highly volatile elements, there are significant differences between the apparent composition of the mantle and chondrites.

First, the mantle is depleted in the alkali elements (e.g., K and Na in Table 11.2). The depletion in alkalis is also apparent by comparing Sr isotope ratios of the mantle and chondrites, as Gast demonstrated in 1960. Some of the Rb depletion of the mantle may be explained by extraction of the Rb into the crust. Indeed, more than half the Earth's Rb may be in the crust. However, the terrestrial Rb/Sr ratio appears to be nearly an order of magnitude lower than chondritic (0.03 vs. 0.25) even when crustal Rb is considered. Independent of Sr isotope considerations, a number of other studies have demonstrated depletion of K, Rb, and Cs in the Earth. This depletion is thought to encompass all the moderately volatile elements. Many of the moderately volatile elements are siderophile or chalcophile, so their depletion in the mantle may also reflect extraction into the core.

The Earth's depletion in moderately volatile elements is not entirely surprising, given that it, along with the other terrestrial planets, is obviously depleted in the atmophile elements. Since the depletion in the highly volatile elements is a feature shared by all the terrestrial planets, it is probably due to temperatures in the inner solar system being too high for these elements to condense completely during the period that the planetesimals that ultimately formed the terrestrial planets were accreting. High temperatures achieved during formation of the Earth (due to release of gravitational energy during collisions), particularly as a result of the giant impact, may have also contributed to volatile loss.

Table 11.2 reflects a general agreement that at least the upper mantle is depleted in silicon

Table 11.2 Comparison of bulk silicate earth major and minor element compositions.

	CI chondrites	CI chondritic mantle[1]	Hart and Zindler[2]	McDonough and Sun[3]	Palme and O'Neill[4]	Lyubetskaya and Korenaga[5]	O'Neill and Palme[6]
SiO_2	22.89	49.77	45.96	45.0	45.4	44.95	45.40
Al_2O_3	1.60	3.48	4.06	4.45	4.49	3.52	4.29
FeO	23.71	6.91	7.54	8.05	8.10	7.97	8.10
MgO	15.94	34.65	37.78	37.8	36.77	39.95	36.77
CaO	1.30	2.83	3.21	3.55	3.65	2.79	3.52
Na_2O	0.671	0.293	0.332	0.36	0.33	0.30	0.281
K_2O	0.067	0.028	0.032	0.029	0.031	0.023	0.019
Cr_2O_3	0.387	0.409	0.468	0.384	0.368	0.385	0.368
MnO	0.250	0.112	0.130	0.135	0.136	0.131	0.136
TiO_2	0.076	0.166	0.181	0.20	0.21	0.158	0.183
NiO	1.371	0.241	0.277	0.25	0.24	0.252	0.237
CoO	0.064	0.012	0.013	0.013	0.013	0.013	0.013
P_2O_5	0.212	0.014	0.019	0.021	0.20	0.15	0.015
Sum	69.79	100.0	100.0	100.2	99.8	100.0	99.3

[1]After removing volatiles and siderophile elements and some oxygen from the mantle to form the core. Hart and Zindler (1986).
[2]Hart and Zindler (1986).
[3]McDonough and Sun (1995).
[4]Palme and O'Neill (2003).
[5]Lyubetskaya and Korenaga (2007).
[6]Calculated from the equations of O'Neill and Palme (2008).

relative to a "chondritic" mantle (whether this is true of the entire mantle is a question we will return to). The silicon depletion can be demonstrated in several ways. For example, Hart and Zindler (1986) showed that Mg/Si and Al/Si ratios of chondritic and terrestrial samples plot along separate arrays that intersect at the low-Si end of the chondritic range (Figure 11.9). Hart and Zindler argued that the "meteorite array" reflects fractionation during processes occurring in the solar nebula or during planet formation (e.g., evaporation, condensation), whereas the "terrestrial array" reflects processes occurring in the Earth's mantle such as partial melting and crystallization.

Table 11.2 also shows that there is broad agreement among the most recent compositional estimates on the FeO concentration in the upper mantle. This reflects the observation that mantle peridotites have uniform concentrations of FeO of about $8 \pm 1\%$.

11.3.2 Trace element composition

Let's consider how other elemental concentrations are estimated in one of the more recent

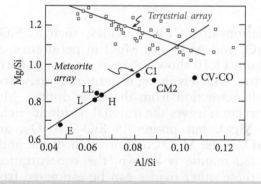

Figure 11.9 Variation of Mg/Si as a function of Al/Si in terrestrial mantle xenoliths and meteorites. The data suggest the Earth is depleted in Mg and Si relative to chondrites. After Hart and Zindler (1986).

studies, that of Palme and O'Neill (2003). They began by adopting an FeO concentration of 8.1%. According to them, the least modified peridotites have a MgO/(MgO+FeO) molar ratio (this ratio is referred to as the Mg-number, often written Mg#) of 0.89. Using that value and FeO of 8.1%, they calculated an MgO concentration for the mantle of 36.77%. They then examined the

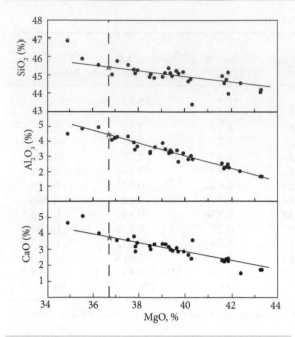

Figure 11.10 Correlation of SiO_2, Al_2O_3, and CaO with MgO in peridotites from the central Dinarides (Balkans). Stars indicate estimated mantle primitive compositions. After Palme and O'Neill (2003).

relationship of other oxides, such as SiO_2, CaO, and Al_2O_3, to MgO in peridotites. As Figure 11.10 shows, there is an inverse correlation, which is thought to result largely from melt extraction from these peridotites. Melt extraction leaves the residual peridotite richer in MgO, but poorer in SiO_2, Al_2O_3, and CaO. If the MgO concentration of the undepleted mantle is known, the concentrations of these other oxides can be estimated from these correlations. Values deduced in this way correspond to Mg/Si and Ca/Al molar ratios of 0.94 and 1.1 respectively. The Mg/Si ratio is quite different from the CI chondrite ratio of 0.83, but in good agreement with the values of 0.95 and 0.94 estimated by Hart and Zindler (1986) and McDonough and Sun (1995), respectively, who used similar approaches.

The Ca/Al ratio agrees well with all other estimates for the Earth as well as with the CI chondrite value. The 50% condensation temperatures of these elements in a gas of solar composition at 10 Pa are: Al, 1650 K; Ca, 1518 K; Mg, 1340 K; and Si, 1311 K (Lodders, 2003). Lithophile elements with 50%

condensation temperatures above that of Mg are generally present in constant proportions in chondritic meteorites. It would seem that temperatures in the solar nebula were never hot enough to fractionate these elements. Consequently, a widely held assumption is that these elements, referred to as *refractory lithophile elements*, should be present in the Earth and other planetary bodies in chondritic proportions. One of these elements is Ti, so using this logic, the Ti concentration can be estimated from the Ti/Al ratio in chondrites and the Al concentration deduced above. The remaining elements in Table 11.2 are either to varying degrees volatile (Na, K, Mn) and/or siderophile (Ni, Co, Cr); the latter likely are concentrated in the Earth's core. Palme and O'Neill (2003) estimated concentrations of some of these elements from the correlations with MgO in peridotites. The K/U and K/La ratios in the mantle and crust show only limited variation. U and La are refractory lithophile elements, so Palme and O'Neill used these ratios to estimate K.

The approach used by Lyubetskaya and Korenaga (2007) was similar to that of earlier studies in that they relied on correlations between elements in peridotites. However, they employed more sophisticated statistical techniques including principal component analysis and Monte Carlo simulations. Principal component analysis is based on correlations among variables (concentrations in this case) and attempts to define secondary variables, the "components", that predict the behavior of the primary variables. They found that a single such component could predict 82% of the compositional variance. They interpreted this component as the effects of melt extraction, an interpretation consistent with earlier studies. As did earlier studies, they adopted the assumption that refractory lithophile elements are present in the Earth in chondritic relative proportions, but rather than adopt strict ratios, they incorporated that assumption into their statistical analysis as a *cost function*. As Table 11.2 shows, their results are broadly similar to earlier studies, but there are a few important differences. Their estimate of MgO is significantly higher than earlier ones and their estimated concentrations of refractory lithophile elements (Al, Ca, Ti) significantly lower. They also found that the Earth had lower concentrations of the incompatible elements Na, K, and P than estimated in earlier studies.

The ^{142}Nd/^{144}Nd ratio of the Earth differs from that of chondrites, which raised the possibility that the Sm/Nd ratio of the Earth, or at least the part accessible to sampling, may not be chondritic (Boyet and Carlson, 2005). (Recall that ^{142}Nd is produced by α-decay of ^{146}Sm, which has a half-life of 103 Ma.) It is now clear that ^{142}Nd/^{144}Nd varies between classes of chondritic meteorites and this is primarily a nucleosynthetic effect resulting from incomplete mixing of s- and r-process nuclides (produced, respectively, in red giants and neutron star mergers) in the solar nebula (Burkhardt et al., 2016). Carbonaceous chondrites are most different from the Earth with ε_{142Nd} of -0.32 ± 0.13 (2σ), ordinary chondrites have of -0.16 ± 0.09, and enstatite chondrites have ^{142}Nd/^{144}Nd most similar to the Earth with ε_{142Nd} of -0.10 ± 0.12 (Boyet et al., 2018). The Earth thus falls at the extreme of enstatite chondrite ε_{142Nd}, leaving the possibility that the Earth's Sm/Nd is slightly different than chondritic, with the terrestrial ^{143}Nd/^{144}Nd being a few epsilon units higher than the chondritic value of 0.

If the Earth's Sm/Nd is nonchondritic, how might this difference come about? It is difficult to see how processes operating in the solar nebula could have fractionated these elements significantly as both are refractory lithophile elements with nearly identical condensation temperatures. Boyet and Carlson (2005) suggested that Earth underwent early differentiation forming an *early-enriched reservoir* (EER), such as a primordial crust that sank into the deep mantle and has not been sampled since where it remains because of its high density. The LLSVPs discussed in the previous section are a possible location for this reservoir. If this is the case, this EER has not participated in mantle or crust evolution as all post-Archean rocks have identical nonchondritic ^{142}Nd/^{144}Nd within 10 ppm. In this scenario, while the Earth has chondritic Sm/Nd and ε_{Nd}, and by extension chondritic relative proportions of other refractory elements, the *observable Earth*, the part that has given rise to the continental and all mantle and mantle-derived materials found at the Earth's surface.

Another possibility, *collisional erosion,* was suggested by Caro et al. (2008) and O'Neill and Palme (2008). As we discussed in Chapter 10, planetary bodies are thought to form through the process of *oligarchic growth*. The initial stages of this process involve aggregations of dust-sized particles to form sand-sized particles, which in turn aggregate to form pebble-sized particles, and so on. The later stages of this process involve infrequent, energetic collisions between large bodies. Sufficient energy is released in these collisions that the growing planet extensively melts. Between collisions, one might reasonably expect a primitive basaltic crust to form through crystallization at the surface. Caro et al. (2008), O'Neill and Palme (2008), and Caro and Bourdon (2010) suggest that a substantial fraction of this crust was blasted away in these collisions, leaving the Earth depleted in elements that were concentrated in that crust: incompatible elements.

Common to both these hypotheses is the idea that planetary melting and consequent differentiation began during, rather than after, planetary accretion. Both hypotheses rely on the idea of formation, through melting and fractional crystallization, of a primitive crust enriched in incompatible elements. Such a crust would have a low Sm/Nd ratio, leaving the remainder of the planet with a higher Sm/Nd ratio than the material from which it accreted. In the Boyet and Carlson (2005) hypothesis, this early crust sinks into the deep mantle, where it remains as an isolated reservoir. In the collisional erosion hypothesis, this early crust is lost from the Earth.

There is some evidence to support the possibility that the terrestrial Sm/Nd is somewhat higher than chondritic and therefore that ε_{Nd} of the Earth is greater than 0:

Oceanic island basalts with the most primordial ^{3}He/^{4}He and other noble gas isotope ratios tend to have positive ε_{Nd}, most commonly in the range of +3 to +5 (e.g., Farley et al., 2002; Class and Goldstein, 2005). Similarly, OIB with Pb isotopes ratios most similar to that expected of the silicate Earth also have ε_{Nd} in this range (Jackson and Carlson, 2011).

Mantle–continental crust mass balance calculations based on Nd isotope ratios suggest that extraction partial melts to form the continental crust has depleted only 25–33% of the mantle in incompatible elements (e.g., DePaolo, 1980), whereas it appears that nearly 50% of the silicate Earth's inventory of the most incompatible elements such as Rb and Th are in the continental crust. We'll consider this further in a subsequent section.

Although the isotopic compositions of a number of elements in enstatite chondrites, most notably oxygen, are more similar to terrestrial values than those of other chondrites, Si isotope ratios of enstatite chondrites are dramatically different from terrestrial ones (section 9.9.5) and this cannot be entirely accounted for by volatile loss and core formation. Dauphas (2017) estimates enstatite chondrite-like material contributed ~70% of the mass of the Earth with the remainder from ordinary and carbonaceous chondrite, implying the Earth formed from material with $\varepsilon_{142Nd} \sim -0.12$. As the observable Earth has $\varepsilon_{142Nd} = 0$, some material must either be hidden or lost.

11.3.3   Composition of the bulk silicate earth

Many previous estimates of bulk Earth composition relied on the assumption that refractory lithophile elements are present in the Earth in chondritic relative proportions. If, however, Earth's $^{142}Nd/^{144}Nd$ ratio is nonchondritic, this assumption must be discarded or modified. O'Neill and Palme (2008) suggested a way to modify the chondritic assumption to account for erosional loss of a primitive crust, and we will follow their approach here. A quite similar approach can be taken to estimate the composition of the *observable Earth* if instead of being lost to space, this early crust was sequestered in the deep mantle as Boyet and Carlson (2005) suggested.

O'Neill and Palme (2008) begin by assuming that the growing proto-earth partially melted to produce a proto-crust of mass fraction f_{pc}^1. The concentration of an element, i, in the proto-crust is given by the batch melting equation (eqn. 7.38):

$$\frac{c_i^{pc}}{c_i^o} = \frac{1}{D_i + f_{pc}^1(1 - D_i)} \quad (11.14)$$

where D_i is the bulk partition coefficient of i. They assume that some of this crust corresponding to a mass fraction f_{pc}^2 is removed by

erosion, along with a fraction of the residue of crust formation, f_{res}^2. The depletion of element i in the bulk silicate Earth is then:

$$\frac{c_i^{BSE}}{c_i^o} = \frac{f_{pc}^1(1 - D_i) + D_i(1 - f_{res}^2) - f_{pc}^2}{[D_i + f_{pc}^1(1 - D_i)](1 - f_{res}^2 - f_{pc}^2)} \quad (11.15)$$

The unknowns in this equation are the three mass fraction terms and the partition coefficients. For the latter, O'Neill and Palme adopt the bulk partition coefficients for formation of the basaltic oceanic crust of Workman and Hart (2005). The difference in $^{142}Nd/^{144}Nd$ between the Earth and chondrites implies a terrestrial Sm/Nd ratio 6% greater than chondritic. Based on this, O'Neill and Palme (2008) concluded that $f_{pc}^1 = 0.026$ and $f_{pc}^2 = 0.014$. In other words, the proto-crust was about 2.6% of the mass of the Earth and about $0.014/0.026 = 54\%$ of this crust was lost. Assuming instead an initial $^{142}Nd/^{144}Nd$ equal to that of enstatite chondrites, which may be a better compositional model for the Earth than ordinary chondrites, implies a terrestrial Sm/Nd ratio only 3% greater than chondritic, which implies an ε_{Nd} of the Earth (or the observable Earth) of +3.6. A value of $\varepsilon_{Nd} = +6$ for the Earth should be considered an upper limit.

Table 11.3 compares bulk silicate Earth compositions estimated by McDonough and Sun (1995) and Lyubetskaya and Korenaga (2007) with a "Depleted Earth" composition computed using the approach described above, assuming a 3% Sm/Nd difference from chondritic, and where the c^o values are taken from Palme and O'Neill (2003). For those elements where Workman and Hart (2005) did not give D values, bulk D values were estimated (e.g., Li, Na), or concentrations were adjusted appropriately based on Palme and O'Neill's original estimate. Other concentrations, such as Cu and Zn, which were derived by Palme and O'Neill (2003) by correlation with MgO are taken unmodified from Palme and O'Neill (2003). K in Table 11.3 (and K_2O in Table 11.2) has been estimated using a revised estimate of the K/U ratio of the Earth of Arevalo et al. (2009) of 13,800, based largely on an upwardly revised K/U in MORB. The Pb concentration is constrained

Table 11.3 Bulk silicate Earth compositions.

	McDonough and Sun (1995)	Lyubetskaya and Korenaga (2007)	"Depleted Earth"		McDonough and Sun (1995)	Lyubetskaya and Korenaga (2007)	"Depleted Earth"
Li	1.60	1.60	1.52	Ag	0.008	0.004	0.004
Be	0.07	0.05	0.06	Cd	0.040	0.050	0.064
B	0.30	0.17	0.23	In	0.011	0.010	0.012
C	120.00	–	100.00	Sn	0.130	0.103	0.125
F	25.00	18.00	22.88	Sb	0.0055	0.0070	0.0089
Na	2670	2220	2590	Te	0.012	0.008	0.008
Mg	228000	234100	221700	I	0.010	0.010	0.001
Al	23500	18700	23500	Cs	0.021	0.016	0.015
Si	210000	210900	212200	Ba	6.60	5.08	5.03
P	90	66	76	La	0.648	0.508	0.555
S	250	230	200	Ce	1.68	1.34	1.53
Cl	17	1.4	8.5	Pr	0.254	0.203	0.235
K	240	190	226	Nd	1.25	0.99	1.16
Ca	25300	20000	25541	Sm	0.406	0.324	0.389
Sc	16	13	16	Eu	0.154	0.123	0.147
Ti	1205	950	1176	Gd	0.544	0.432	0.523
V	82	74	86	Tb	0.099	0.080	0.097
Cr	2625	2645	2520	Dy	0.674	0.540	0.666
Mn	1045	1020	1050	Ho	0.149	0.121	0.149
Fe	62600	62000	63000	Er	0.438	0.346	0.440
Co	105	105	102	Tm	0.068	0.054	0.068
Ni	1960	1985	1860	Yb	0.441	0.346	0.440
Cu	30	25	20	Lu	0.068	0.054	0.068
Zn	55	58	54	Hf	0.283	0.227	0.269
Ga	4.0	4.2	4.4	Ta	0.037	0.030	0.031
Ge	1.1	1.2	1.2	W	0.029	0.012	0.012
As	0.050	0.050	0.057	Re	0.0003	0.0003	0.0003
Se	0.075	0.075	0.079	Os	0.0034	0.0034	0.0034
Br	0.05	0.004	0.022	Ir	0.0032	0.0032	0.0032
Rb	0.60	0.46	0.47	Pt	0.0071	0.0066	0.0066
Sr	19.90	15.80	17.48	Au	0.0010	0.0009	0.0009
Y	4.30	3.37	4.12	Hg	0.0100	0.0060	0.0060
Zr	10.50	8.42	9.64	Tl	0.0035	0.0002	0.0024
Nb	0.66	0.46	0.45	Pb	0.150	0.144	0.120
Mo	0.050	0.030	0.034	Bi	0.0025	0.0040	0.0044
Ru	0.005	0.005	0.005	Th	0.080	0.063	0.063
Rh	0.0009	0.0009	0.0009	U	0.02	0.0173	0.0164
Pd	0.0039	0.0036	0.0033				

All concentrations in ppm. "Eroded Earth" is modified from Palme and O'Neill (2003) as described in the text.

by Pb isotope systematics such that the U/Pb ratio should be 0.133.

The data in Table 11.3 correspond to $^{147}Sm/^{144}Nd$, $^{87}Rb/^{86}Sr$, and $^{176}Lu/^{177}Hf$ ratios of 0.2024, 0.0764, and 0.03565, respectively, which, in turn, correspond to present-day $^{143}Nd/^{144}Nd$, $^{87}Sr/^{86}Sr$, and $^{176}Hf/^{177}Hf$ ratios of 0.51282, 0.70404, and 0.28287, respectively. This implies $\varepsilon_{Hf} = +6.4$ for the bulk silicate Earth.

The compositions listed in Table 11.3 differ mainly in the estimated abundances of highly incompatible element abundances, with McDonough and Sun (1995) (the latter differ only marginally from other estimates such as those of Palme and O'Neill, 2003) estimates being higher for these elements. These differences have implications for the mantle structure and convection. First, the heat-producing elements K, U, and Th are among the most incompatible elements. The latter two estimates are about 20% lower than McDonough and Sun, meaning there would be less energy being generated to power mantle convection. Also, assuming the McDonough and Sun K concentration implies that more than half of the mantle has retained its ^{40}Ar inventory, implying a large "primitive mantle" reservoir, which has been an argument for layered mantle convection. The "depleted Earth" and Lyubetskaya and Korenaga estimates remove geochemical objections to whole mantle convection, which is not supported by seismic imaging. We will explore some of these implications further in a subsequent section.

Figure 11.11 shows the abundances of the elements in the silicate Earth relative to CI chondritic (i.e., solar system) concentrations as a function of their 50% condensation temperatures under conditions relevant to the solar nebula. The refractory lithophile elements are enriched in the Earth by factors ranging from 1.5 to 2.8. Other elements show variable depletions. In general, the volatile lithophile elements are depleted roughly as a

function of their 50% condensation temperatures, consistent with inferences made by Paul Gast and others half a century ago. We also see that many elements are more depleted in the silicate Earth than we would predict from their volatility. We can infer from this that these elements are concentrated in the core, whose composition we consider in the following section.

11.4 THE EARTH'S CORE AND ITS COMPOSITION

11.4.1 Geophysical constraints

That the Earth has an iron core is known even to schoolchildren. But how do we know this, since no one has ever seen or touched the core? Here again we turn to geophysics. Perhaps the first hint came when seventeenth-century physicist William Gilbert observed that the Earth's magnetic field was similar to a bar magnet. The plot thickened in 1634, so to speak, when Henry Gellibrand demonstrated temporal changes in the Earth's magnetic

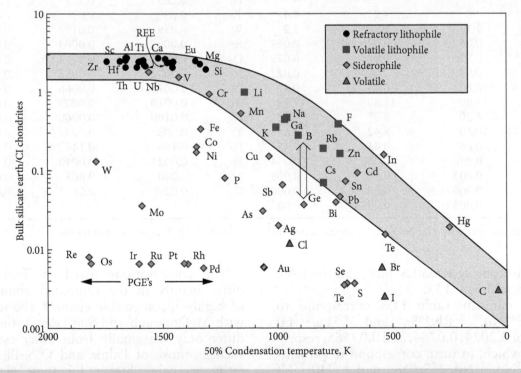

Figure 11.11 Abundances of the elements in the bulk silicate Earth (Table 11.3) relative to CI chondrite abundances as a function of 50% nebular condensation temperature (Lodders, 2003). Primary controlling factors are volatility and the siderophile or chalcophile nature of the element.

field. Further evidence emerged when measurement of the mass of the Earth indicated a high-density interior. Unequivocal evidence of this, however, did not emerge until two centuries later. As seismology advanced in the early twentieth century, work by British seismologists Richard Dixon Oldham and Harold Jefferies demonstrated that the Earth has a liquid core and Danish seismologist Inga Lehmann demonstrated that the inner part of this core was solid. Both the inner and outer core have lower seismic velocities than silicate, suggesting the core has higher density. Comparison of density solutions to the Adams–Williamson equation (eqn. 11.10) with the Earth's moment of inertia provided evidence that the Earth's high-density interior could not be simply a consequence of self-compression: the material in the core must have intrinsically higher density. At this point, we can turn to cosmochemistry to ask what heavy element is abundant enough to make up the core, which makes up 16% of the volume and 32% of the mass of the Earth. Looking at Figure 7.1 and Table 10.2, one answer immediately emerges: iron. Iron has a lower melting point than peridotite, making it easy to explain the transition from solid mantle to liquid iron outer core. Iron is also a conductor, and can, when placed in motion, generate a magnetic field. Looking further at cosmic abundances and at Table 10.2, we see that nickel is a cosmically abundant element, yet depleted in the silicate Earth. It seems likely then that the core contains substantial Ni as well.

Geophysical measurements can take us a bit further in deducing the composition of the core. Any combination of iron and nickel alone would produce a core that is 5–10% denser than the actual core (Birch, 1964; Anderson and Issak, 2002). Consequently, the core must contain a significant concentration of some lighter element or elements; how much depends on which element. If we again glance at cosmic abundances, and also consider which elements are underabundant in the silicate Earth and which could possibly alloy with iron, the main possibilities are H, C, O, Si, and S. At one time or another, a case has been made for each of these. To further constrain the composition of the core, however, we need to look to meteorites.

11.4.2 Cosmochemical constraints

Iron meteorites are now understood to represent pieces of asteroidal cores. In this sense, they may provide a good analogy for the Earth's core. Indeed, Emil Wiechert's 1897 proposal that the Earth had an iron core was inspired in part by the existence of iron meteorites. Iron meteorites are not pure iron, but rather are typically iron–nickel alloy with additional components. These include sulfides and phosphides. Iron meteorites also have high concentrations of siderophile and chalcophile trace elements. However, the analogy between the Earth's core and iron meteorites must be approached with caution because iron meteorites come from relatively small bodies whose core formed under relatively low-pressure conditions. The Earth's core may have formed under much higher-pressure conditions.

Cosmochemistry can nevertheless provide some constraints on core composition. We again make the assumption that the Earth formed within a nebula of chondritic composition. The Earth's composition should therefore be related in some way to that of chondrites. As Figure 11.11 shows, the mantle is depleted in many elements relative to chondrites. Many of these elements are siderophile, and it seems reasonable to think they are concentrated in the core. However, the Earth is also depleted in volatile elements. If an element is both siderophile and volatile, how do we know if their depletion in the mantle reflects concentration in the core because of their siderophile character, or depletion in the Earth as a whole because of their volatile character? The approach taken by McDonough (2018) has been to plot an element's bulk silicate Earth concentration against its 50% condensation temperature as in Figure 11.11. The 50% condensation temperature is the temperature at which 50% of the element condenses in a nebula of solar composition at 10 Pa (10^{-4} atm). When we do this a clear trend emerges (Figure 11.11). Lithophile elements with condensation temperatures above that of Mg (1336 K) are enriched in the silicate Earth. At lower condensation temperatures, a clear trend of decreasing concentration with decreasing condensation temperature emerges. Since the lithophile elements should not be concentrated in the core, this trend should be due to volatility alone.

Most siderophile and chalcophile elements plot below this trend. Let's assume that when we consider the composition of the whole Earth, all elements plot along this volatility trend. That allows us to estimate the concentrations of all elements, and the concentration of elements in the core can be estimated by subtracting the mantle concentration from the whole Earth concentration. This is illustrated for Ge in Figure 11.11. Ge has only about 14% of the mantle concentration we would expect based on its condensation temperature of 883 K. Presumably then, 86% of the Earth's Ge inventory is in the core. From this, we can calculate that the concentration of Ge in the core should be about 20 ppm. In contrast, Ga, which is partly siderophile, is not depleted in the mantle beyond what would be expected from its volatility alone, hence its concentration in the core is probably quite low. Working through the periodic table in this way, McDonough (2018) derived the estimated core compositions listed in Table 11.4.

We should treat the values in Table 11.4 with considerable caution for several reasons. First, they are largely based on the assumption that the Earth differs from a chondritic composition mainly because of volatility. Second, they are based on estimated mantle compositions that are themselves uncertain as discussed in the preceding section. Third, the volatility trend defined by the lithophile elements is a broad one, with the extent of depletion at a given temperature varying by

close to an order of magnitude, and this produces considerable error in the estimate of the core concentration of any particular element. Finally, the most volatile lithophile elements have condensation temperatures around 700 K; beyond that, the trend must be extrapolated. We cannot be sure that the slope of the trend remains constant at low temperatures.

In McDonough's model, the core contains significant concentrations of S and minor concentrations of C and H, but the low density of the core arises mainly from the high concentration of Si and O. If Si is indeed present in the core, this could partly explain the deficiency of the mantle in Si relative to chondrites and, as noted in Chapter 9, this is also consistent with Si isotope ratios. On the other hand, the concentration of Si relative to more refractory elements such as Mg, Al, and Ca does vary even among carbonaceous chondrites, so it is clear that Si was fractionated from these elements within the solar nebula. It is perhaps best to state that cosmochemical considerations allow a significant concentration of Si in the core, but they do not require it.

In contrast, cosmochemical considerations would seem to preclude sulfur being the main light element in the core. Relative to CI chondrites, sulfur is depleted in almost all other classes of chondrites. Ahrens and Jeanloz (1987) found that a core with about 11% S would match the observed seismic properties of the core reasonably well. For the core to contain 11% S would require the Earth to have a S concentration that is 65%

Table 11.4   Estimated composition of the Earth's core.

Element	Conc.	Element	Conc.	Element	Conc.
H	600	Cu	125	Te	0.85
C	2000	Ge	20	I	0.13
N	75	As	5	Cs	0.065
O %	2.0	Se	8	W	0.47
Si %	4.0	Br	0.7	Re	0.23
P	200	Mo	5	Os	2.8
S %	1.9	Ru	4	Ir	2.6
V	150	Rh	0.74	Pt	5.7
Cr %	0.75	Pd	3.1	Au	0.5
Mn	300	Ag	0.15	Hg	0.05
Fe %	85.5	Cd	0.15	Tl	0.03
Co	2490	Sn	0.5	Pb	0.4
Ni %	5.1	Sb	0.13	Bi	0.03

Concentrations in ppm, except % where noted. From McDonough (2003; 2018).

of the CI chondrite concentration. Given the low condensation temperature of S (664 K), this seems improbable.

11.4.3 Experimental constraints

Experiments on the partitioning of elements between iron and silicate phases provide potential constraints both on the composition of the core and on how it formed. The idea that O might be the light element in the core is based on the observation that at high pressure FeO is miscible in Fe liquid, although it is not at low pressure. The eutectic composition in the Fe–FeO system at 16 GPa (still well below core pressures) contains about 10% FeO (e.g., Ringwood and Hibbertson, 1990). Furthermore, the Earth's mantle is depleted in oxygen compared with CI chondrites. The relevance of this last observation, however, is questionable, since the oxidation state of chondrites, and therefore presumably the solar nebula, clearly varied widely. Significant in this respect is the observation that, based on O and Ti isotope ratios, the Earth appears to be more closely related to the highly reduced enstatite chondrites (Figures 10.30 and 10.31) than to other chondrites, particularly CI chondrites. In addition, O'Neill et al. (1998) pointed out that at higher pressures, oxygen solubility in iron liquids decreases with pressure. They argued that the core could contain at most about 2% oxygen.

The problem of oxygen illustrates a fundamental dilemma with partitioning studies: partition behavior depends on temperature, pressure, and composition. This is just as true for metal–silicate partitioning as it is for the silicate mineral–liquid partitioning we considered in Chapter 7. Indeed, the problem of temperature- and pressure-dependence is considerably more severe when we consider planetary-scale differentiation, because the range of temperatures and pressures over which partitioning might have occurred is enormous. Today, the metal–silicate boundary, which is the core–mantle boundary, is at 135 GPa and 3000–4000 K. However, that need not be the environment in which metal and silicate phases equilibrated. Suppose that material was accreted incrementally to the surface of a growing Earth, separated into metal and silicate portions, with subsequent sinking of the metal without further equilibration. In that case, the pressure of equilibration was close to 0.1 MPa and temperatures far lower than the present core–mantle boundary. Yet another scenario is that metal and silicate last equilibrated at the base of a magma ocean at mid-mantle pressures. As we shall see, this may be the most plausible scenario. Experiments have also demonstrated the role of composition in partitioning between metal and silicate melts. In particular, the concentrations of nickel and sulfur in the metallic liquid, the degree of polymerization of the silicate liquid, and the oxygen fugacity are important. Despite this complexity, experimental determination of partition coefficients has elucidated aspects of core formation.

For example, experiments by Righter et al. (1997) demonstrated a strong dependence of the metal–silicate partitioning of Ni, Co, W, and Mo on oxygen fugacity, pressure, temperature, and silicate liquid composition. The effects of oxygen fugacity and pressure are illustrated in Figure 11.12. In the graph, oxygen fugacity is expressed as the difference between actual oxygen fugacity and that of the iron–wüstite buffer (see Chapter 3).

Given the strong T, P, and composition dependence of metal/silicate partition coefficients, it seems useful to consider the thermodynamics of metal–silicate partitioning in more detail. In general, it is assumed that the partitioning occurs between silicate and metal *liquids*. It is, of course, possible that equilibration occurs below the liquidus of either or both phases, implying solids are present as well. However, as long as the solids are in equilibrium with their respective liquids, they will also be in equilibrium with the other liquid. Hence, provided two liquids are present, we need only focus on the liquid–liquid partition coefficients. Righter et al. (1997) considered the following reaction:

$$M^{n+}O_{n/2\,sil\,liq} = M_{met\,liq} + \frac{n}{4}O_{2\,gas} \quad (11.16)$$

where n is the valence of element M. For this equation, we can write the following equilibrium constant expression:

$$-\frac{\Delta G^{\circ}}{RT} = \ln\left(\frac{a_m f_{O_2}^{n/4}}{a_{MO_{n/2}}}\right) \quad (11.17)$$

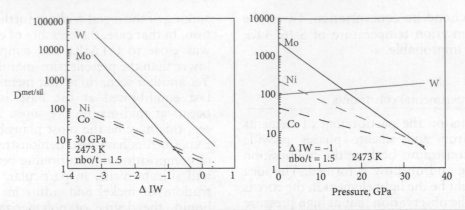

Figure 11.12 Dependence of experimentally determined liquid metal–liquid silicate partition coefficients of Ni, Co, W, and Mo on oxygen fugacity and pressure. ΔIW is the difference, in log units, between oxygen fugacity in the experiment and the oxygen fugacity of the iron–wüstite buffer (Chapter 3). *nbo/t* is the ratio of nonbridging oxygens to tetrahedral cations in the silicate melt (Chapter 7). After Righter et al. (1997).

where, as usual, a denotes activity and f denotes fugacity. Rearranging this expression, we have:

$$\ln\left(\frac{a_M}{a_{MO_{n/2}}}\right) = -\frac{n}{4}\ln(f_{O_2}) - \frac{\Delta G^o}{RT} \quad (11.18)$$

The free energy change will, of course, be temperature- and pressure-dependent. To express that dependence, we can expand the ΔG term to:

$$\Delta G^o = \Delta H^o - T\Delta S^o + P\Delta V^o \quad (11.19)$$

Substituting this into eqn. 11.18, we have:

$$\ln\left(\frac{a_M}{a_{MO_{n/2}}}\right) = -\frac{n}{4}\ln(f_{O_2}) - \frac{\Delta H^o}{RT}$$
$$+ \frac{\Delta S^o}{R} - \frac{P\Delta V^o}{RT} \quad (11.20)$$

It is certainly more convenient to deal with the ratio of concentrations rather than the ratio of activities. The ratio of concentrations is in this case identical to the distribution coefficient. So we can again modify our equation to become:

$$\ln D_M^{met/sil} = -\frac{n}{4}\ln(f_{O_2}) - \frac{\Delta H^o}{RT} + \frac{\Delta S^o}{R}$$
$$- \frac{P\Delta V^o}{RT} - \ln\lambda_M + \ln\lambda_{MO_{n/2}}$$
$$(11.21)$$

where the gamma terms are the activity coefficients, which we expect will be functions of the compositions of the two liquids. We now have an expression for the temperature, pressure, and compositional dependence of the partition coefficient.

On this theoretical basis, Righter et al. (1997) expressed metal/silicate partition coefficients as follows:

$$\ln D_M^{met/sil} = a\ln(f_{O_2}) + \frac{b}{T} + \frac{cP}{T} + d(nbo/t)$$
$$+ e\ln(1 - X_S) + f \quad (11.22)$$

where *nbo/t* is the nonbridging oxygen/ tetrahedral cation ratio (see Chapter 7) and X_S is the mole fraction of sulfur in the metallic liquid. Comparing equations 11.21 and 11.22, we see that the parameter a should relate to the valence of the metal, b to ΔH°, c to ΔV°, d and e to the compositions of the silicate and metal liquids respectively, and f to ΔS°.

Understanding how elements partition between metal and silicate can help us understand the conditions under which the core formed. As Figure 11.11 suggests, the two siderophile elements Ni and Co are about equally depleted in the Earth's mantle. Put another way, the ratio of concentrations of these elements is nearly chondritic, even if their concentrations are not. That suggests the metal–silicate partition coefficients for the two elements should be similar. Early experiments at 1 atmosphere and moderate temperature indicated the partition coefficient of Ni was more than an order of magnitude higher than that of Co. As Figure 11.12 shows, however, the partition coefficients for these elements converge at higher pressure. Using eqn. 11.22, Righter et al. (1997) found that they could explain the concentrations of siderophile elements Ni, Co, Mo, W, and P in the Earth's mantle if metal–silicate equilibration last occurred at 27 GPa and 2200 K with an oxygen fugacity 0.15 log units lower than the iron–wüstite buffer. Interestingly, this pressure corresponds to the top of the upper mantle – the depth where perovskite first becomes stable. Li and Agee (1996) reached a remarkably similar conclusion based on independently determined partition coefficients for Ni and Co. Righter et al. found that the value of X_S necessary to fit the predicted and actual mantle abundances of these elements is about 0.15, which corresponds to a sulfur concentration in the core of about 6%. This is distinctly higher than the S content estimated from cosmochemical considerations.

Ga is apparently not depleted in the mantle beyond what can be explained by its volatility, suggesting it is not concentrated in the core. Since Ga is concentrated in iron meteorites, this is surprising (McDonough, 2003). However, Righter and Drake (2000) demonstrated that the Ga partition coefficient is strongly dependent on T, P, and composition. At low pressure, Ga will partition about equally between metal and silicate, but the partition coefficient decreases with pressure, such that it is less than 0.1 at pressures above 20 GPa. This explains the apparent contradiction. Asteroids are small bodies and their cores segregated at low pressure, hence Ga in the metal is expected. But if the Earth's core segregated at pressures above 20 GPa, we would expect it to contain relatively little Ga.

Whether an element's metal/silicate partition coefficient is greater or less than 1 effectively determines whether that element is siderophile or lithophile. For most elements, the partition coefficient is either greater or less than one under all reasonable conditions. For a few elements, however, the partition coefficient may be greater or less than one depending on T, P, and composition. Thus some elements may be lithophile under some conditions and siderophile (or chalcophile) under others. Let's consider two examples.

The first is the case of Nb, which has traditionally been considered a strictly lithophile element since it is not found in metal phases of meteorites. Nb is also highly refractory; being both refractory and lithophile, we would expect its ratio to other such elements, such as Ta and La, to be chondritic in the bulk silicate Earth. However, Nb/Ta and Nb/La ratios in both the continental crust and upper mantle are distinctly lower than chondritic. Wade and Wood (2001) found that Nb becomes siderophile (i.e., $D^{met/sil} > 1$) at 25 GPa when f_{O_2} is more than about 2.5 log units below the iron–wüstite buffer. Indeed, they found Nb partition coefficients were similar to those of V and Cr. As Figure 11.11 shows, these two elements are depleted in the mantle and hence likely present in the core (Table 11.4). They also found that the partition coefficient for Nb was an order of magnitude or so greater than that of Si. Thus, they argued, if Si is present in the core, Nb should also be present. At f_{O_2} of 2 log units below the iron–wüstite buffer, which is thought to be appropriate for core formation, they found the Nb partition coefficient was about 0.47. This would reduce the silicate Earth Nb concentration by about 20%, or enough to explain the apparent Nb deficit.

A second element of great interest is potassium. Potassium is, of course, radioactive and therefore a source of energy. If it is present in the core, it could supply some or all of the energy needed to drive the geodynamo that sustains the Earth's magnetic field. The dynamo certainly requires energy to be sustained, but how much is unclear. One source of energy is the latent heat of crystallization released as the solid inner core crystallizes from the outer core. In some models of the dynamo, this would provide sufficient energy to drive the dynamo, but not in others. This has raised the question of whether one or more radioactive elements could be present in the core. Attention has focused on K because an iron–potassium–sulfide phase (djerfisherite) has been found in enstatite chondrites. Several experimental partitioning studies have been carried out as a result. In experiments at 1.5 GPa, Chabot and Drake (1999) found that K partitions into the metal phase only when S is present. Gessmann and Wood (2002) found that K partitioning depended strongly on the oxygen concentration (or the activity of FeO) of the metal sulfide liquid. The partition coefficient decreased with increasing pressure, although only modestly. Under the range of conditions and compositions explored, K was never strongly siderophile, but Gessmann and Wood (2002) suggested that the core could contain up 250 ppm K under the very restricted condition where the core is rich in both O and S. At this concentration, K would presently supply about 2 TW of power to the core, which is probably less than is being generated by latent heat of crystallization. K is depleted in the silicate Earth relative to chondrites. As Figure 11.11 shows, most of this depletion is likely due to its volatility. However, K plots on the low side of the lithophile volatility trend in Figure 11.11, which would allow for some K in the core. Indeed, the core could contain up to 400 ppm K and still plot within the gray band of the volatility trend in Figure 11.11. We may say that the cosmochemical and experimental constraints allow, but do not require, up to 250 ppm or so K in the core, but that is considerably less than some geophysicists have argued for. Arguments for the presence of U in the core (e.g., Wohlers et al., 2015) can be similarly dismissed on the basis of the Earth's nearly chondritic Th/U ratio (McDonough, 2018).

Some siderophile concentrations in the silicate Earth are difficult to explain by metal/silicate partitioning alone. Most notably, the noble metals (Ru, Rh, Pd, Re, Os, Ir, Pt, Au) are strongly and uniformly depleted in the silicate Earth compared with chondrites (Figure 11.11). That they are strongly depleted is not surprising: experiments show these elements are highly siderophile with low-pressure metal/silicate partition coefficients in the range of 10^3 to 10^5. Given the large range of partition coefficients, it is surprising that these elements are so uniformly depleted in the mantle. The uniform nature of this depletion is emphasized by $^{187}Os/^{188}Os$ ratios in the mantle. The $^{187}Os/^{188}Os$ ratio of the modern mantle is within the range of chondrites (albeit higher than the average), suggesting the bulk silicate Earth has a Re/Os ratio that is within a few percent of chondritic. This requires that the partition coefficients of Re and Os be nearly identical. While some partition coefficients do tend to converge at high pressure, it seems improbable that the partition coefficients of these elements could be so similar.

As we noted in Chapter 10, one commonly invoked solution to this dilemma is the addition of a "late accretionary veneer" to the Earth after the core had segregated. The idea is that after the core had completely segregated, a small amount of chondritic material continued to accrete to the Earth. To understand the effect of this, imagine that after core segregation, the noble metal concentrations in the mantle reflect metal/silicate equilibrium. This would leave concentrations that are 10^3 to 10^5 times lower than chondritic. Now image that 0.5% of chondritic material is added. This would raise the concentrations of all these elements in the mantle to just over 0.5% of chondritic. This readily explains why the abundances of these elements in the mantle range from 0.006 to 0.008 times chondrites.

In summary, there is very little we know with absolute certainty about the Earth's core. Nevertheless, it is possible to make a variety of inferences about its composition and formation through geophysical, cosmochemical, and experimental approaches. Our knowledge about the core is, however, hardly static. Beyond the fundamental geophysical inferences of the late nineteenth and early twentieth centuries, much of what we have learned about the core has come in the last few decades. Future studies will likely greatly refine our knowledge of the core.

11.5 MANTLE GEOCHEMICAL RESERVOIRS

Up to this point, we have considered the mantle as a uniform body of rock. In addition to seismological evidence for lateral heterogeneity in the deep mantle that cannot be explained by phase changes alone, there is a very considerable body of geochemical evidence that the mantle is quite heterogeneous and that this heterogeneity has mainly developed over geologic time, probably through geologic processes similar to those still operating today. The *prima facie* evidence of this is isotopic variations in oceanic basalts, first discovered in the 1960s (e.g., Gast et al., 1964; Hedge, 1966). The significance of radiogenic isotope ratios is that they are not changed by the processes of magma genesis. Thus isotopic variations in basalts implied isotopic variation in their mantle sources. Those isotopic variations in turn must result from longstanding variations in the ratios of radioactive parent to radiogenic daughter elements. Much of the present research in mantle geochemistry focuses on just what processes might have produced these chemical variations. Before we consider that question, let's first review the evidence for heterogeneity in the mantle. We will begin by considering evidence from oceanic basalts.

11.5.1 Evidence from oceanic basalts

Isotopic variations in basalts, which are partial melts of the mantle, provide clear evidence that the mantle is currently heterogeneous. Oceanic basalts provide better evidence of this than continental basalts because the possibility of the latter being contaminated by the crust through which they pass is much reduced. This is true for three reasons: oceanic crust is much thinner, it has a higher solidus, and it is compositionally similar to melts of the mantle (so that when assimilation does occur, its chemical effects are minimized). Of course, not all continental basalts are necessarily contaminated, nor is there a guarantee that all oceanic basalts have not been (indeed, as we noted in Chapter 9, boron isotopes provide evidence of just such assimilation).

The fundamental isotopic variation in oceanic basalts is between basalts erupted at mid-ocean ridges (mid-ocean ridge basalts or MORB) and those erupted on oceanic island volcanoes (oceanic island basalts or OIB). Table 11.5 and Figures 11.13 through 11.15 illustrate the differences in composition between these two groups. Although the data overlap considerably, MORB on average have lower $^{87}Sr/^{86}Sr$ and $^{206}Pb/^{204}Pb$ and higher ε_{Nd} than OIB. The difference in mean isotopic composition is only part of the story, however. The isotopic compositions of MORB are also considerably more uniform than those of the OIB, as is apparent from the statistics in Table 11.5. The relative variance, as measured by the standard deviations, is about half as great for MORB as for OIB. This implies that MORB are derived from a more uniform source than are OIB. Another interesting feature is that both the MORB and OIB data are skewed towards higher $^{87}Sr/^{86}Sr$, $^{206}Pb/^{204}Pb$, and $^3He/^4He$ and lower ε_{Nd}. Statistical tests (Student's t-test and F-test) demonstrate that the MORB and OIB means and variances of

Table 11.5 Statistical summary of the isotopic composition of oceanic basalts.

		MORB	OIB
$^{87}Sr/^{86}Sr$			
	Mean	0.70287	0.70370
	Mode	0.70280	0.70319
	Median	0.70279	0.70349
	SD	0.00045	0.00089
	SD %	0.06%	0.13%
	Skewness	1.76	1.44
	n	2135	3224
ε_{Nd}			
	Mean	8.67	4.96
	Mode	9.98	6.28
	Median	9.22	5.37
	SD	2.29	2.69
	SD %	26.41%	54.23%
	Skewness	−1.69	−0.7
	n	1678	2580
$^{206}Pb/^{204}Pb$			
	Mean	18.45	18.97
	Mode	18.29	18.87
	Median	18.40	18.89
	SD	0.423	0.75
	SD %	2.29%	3.95%
	Skewness	0.326	1.22
	n	1495	2581
$^3He/^4He$ (R/R_A)			
	Mean	8.81	12.24
	Mode	8.1	8.2
	Median	8.1	11.8
	SD	2.538	4.88
	SD %	28.81%	39.87%
	Skewness	1.053	4.88
	n	573	759

each isotope ratio differ significantly to a high degree of confidence. Consequently, we can conclude that MORB and OIB represent two distinct populations and are derived from two distinct mantle reservoirs.

Some MORB, particularly those from ridge segments near oceanic islands (e.g., the Mid-Atlantic Ridge near the Azores and Iceland; the Galapagos Spreading Center near the Galapagos, etc.), have geochemical characteristics similar to OIB, something first demonstrated by the pioneering work of Jean-Guy Schilling (e.g., Schilling, 1973). If these MORB are excluded, the difference between MORB and OIB is greater. For example, by eliminating the 276 samples in this category, the mean MORB ε_{Nd} increases to 9.23, close to the median of the entire data-set. The data

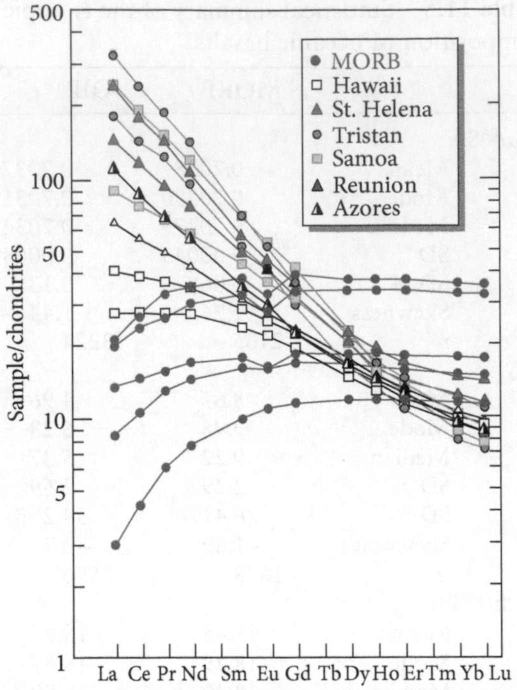

Figure 11.13 Rare earth patterns of mid-ocean ridge basalts and oceanic island basalts.

also show considerably less dispersion, with the standard deviation dropping from 2.3 to 1.7, and are less skewed, the skewness changing from −1.7 to −0.8. This suggests that the mantle supplying melt to oceanic island volcanoes also supplies melt to mid-ocean ridges near these volcanoes, just as Schilling (1973) argued.

While there is broad similarity in isotope ratios of MORB from all regions, but there are systematic, but sometimes subtle, variations in the composition of MORB between ocean basins and four distinct domains can be identified (Meyzen et al., 2007; White and Klein, 2014): Pacific, North Atlantic, South Atlantic, and Indian. The most notable of these are the differences between Indian Ocean MORB and MORB from other regions. Indian Ocean MORB have low $^{206}Pb/^{204}Pb$ and high $^{207}Pb/^{204}Pb$ and $^{208}Pb/^{204}Pb$ for a given $^{206}Pb/^{204}Pb$ and high $^{87}Sr/^{86}Sr$ (Dupré and Allègre, 1983). North Atlantic MORB tend to have the highest ε_{Nd} and ε_{Hf} for a given $^{87}Sr/^{86}Sr$. Boundaries between these domains can be sharp or diffuse. The Pacific–Indian boundary

is located within the Australian–Antarctic discordance and is quite sharp (Klein et al., 1988; Pyle et al., 1992). By contrast, the boundary between the South Atlantic and Indian domains, which occurs west of the Andrew Bain Fracture Zone (the Antarctic–Nubian, Somalian triple junction) located at 30° E on the Southwest Indian Ridge, is gradual (Meyzen et al., 2007). The boundary between the North and South Atlantic provinces, located near 23° S, is also diffuse.

Figure 11.13 illustrates the difference between MORB and OIB in rare earth elements. Relative to chondrites, the light rare earths elements (LREE) in MORB are underabundant compared with the heavier rare earths – MORB are light rare earth-depleted compared with chondrites. The easiest way to produce such depletion in the mantle is to extract a melt from it. The incompatible elements, including the light rare earths, partition preferentially into the melt, leaving a residue depleted in LREE. When the mantle melts again, the basalt produced will inherit this LREE depletion (though the basalt will not be as LREE-depleted as its mantle source). Thus, the mantle source of mid-ocean ridge basalts appears to have suffered melt extraction in the past.

OIB are LREE-enriched to varying degrees. Part of the difference in rare earth patterns of MORB and OIB may be due to the smaller degrees of melting involved in the generation of some, and perhaps most, OIB magmas. Many, though by no means all, OIB are alkaline, whereas almost all MORB are tholeiitic. Alkali basalts are produced by smaller degrees of melting than tholeiites. The relatively large degree of melting that generates MORB probably reflects their generation where lithosphere is thin or nonexistent. This allows the mantle to rise to very shallow depths, allowing high degrees of melting (Chapter 7). On the other hand, there is fairly strong evidence, based ultimately on olivine-melt geothermometry (Chapter 4), that mantle plumes are hotter than the mantle beneath mid-ocean ridges, which, of course, favors higher degrees of melting. Putirka et al. (2007) found the mantle potential temperature beneath mid-ocean ridges is 1454±81°C and that the potential temperature of the Hawaiian, Samoan, and Icelandic plumes to be 268°C to 162°C hotter; flood basalts appear to be produced from

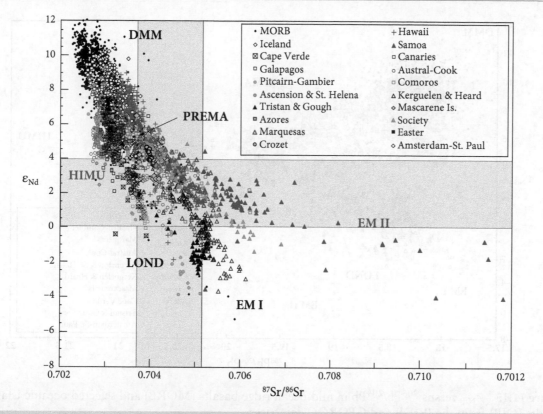

Figure 11.14 ε_{Nd} versus $^{87}Sr/^{87}Sr$ in MORB and selected OIB from the PetDB and GEOROC databases. The gray background shows the range of estimates of bulk silicate Earth composition discussed in Section 11.3.3.

similar mantle temperatures to OIB (Putirka et al., 2018). Herzberg et al. (2007) calculated lower potential temperatures for ambient mantle, in the range of 1280–1400°C, but nevertheless conclude that plumes are typically 200–300°C hotter. Consequently, not all the differences in rare earth patterns between OIB and MORB can be attributed to degree of melting – there must also be differences in the rare earth abundances in the mantle sources of these magmas.

In Figure 11.14, MORB typically plot at higher ε_{Nd} and lower $^{87}Sr/^{86}Sr$ ratios than the estimated composition of bulk silicate Earth, while the opposite is true of OIB. These ratios imply that the mantle that gives rise to MORB magmas, the "MORB source," has low time-integrated Rb/Sr and high Sm/Nd ratios. Since Rb and Nd are more incompatible than Sr and Sm, respectively, these differences are consistent with the differences in rare earth patterns (and the abundances of other incompatible elements) apparent in Figure 11.13. From this we infer that the mantle source of

most MORB has experienced depletion in incompatible elements while the sources of many OIB have experienced enrichment in incompatible elements.

Interestingly, many of the OIB arrays in Figure 11.14 appear to converge near $\varepsilon_{Nd} = +5$ and $^{87}Sr/^{86}Sr = 0.7035$. A similar convergence is apparent in the plot of ε_{Nd} versus $^{206}Pb/^{204}Pb$ in Figure 11.15. A number of papers have noted this convergence of OIB isotopic arrays and have also pointed these OIB have particularly high $^{3}He/^{4}He$. This "common" component has variously been called PREMA (Zindler and Hart, 1986), PHEM (Farley et al., 1992), FOZO (Hart et al., 1992), and "C" (Hanan and Graham, 1996). This convergence falls somewhat to the "depleted" side of the range of estimates of primitive mantle Sr, Nd and Hf isotopic compositions. This and the more "primitive" isotopic compositions of He and other noble gases certainly suggests the possibility that mantle plumes producing these volcanoes

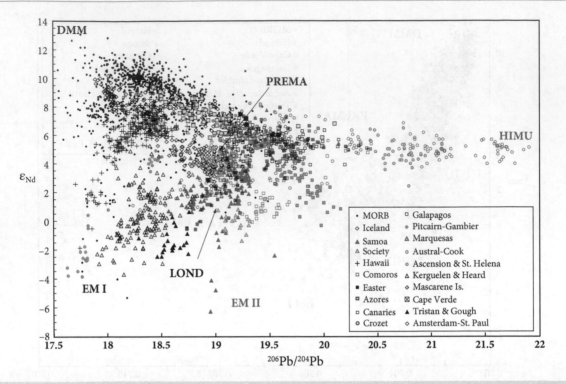

Figure 11.15 ε_{Nd} versus $^{206}Pb/^{204}Pb$ in mid-ocean ridge basalts (MORB) and selected oceanic island basalts (OIB) from the PetDB and GEOROC databases.

contain a primitive component derived from a reservoir unmodified or nearly so over Earth's history. We'll examine this in more detail in a subsequent section.

We conclude that MORB and OIB are derived from distinct mantle reservoirs. Deciphering the physical properties of these reservoirs, where they are located, their size, and how they have evolved is more difficult. At one extreme, these two reservoirs might coexist as different lithologies in the same volume of mantle. As we discuss below, the consensus view, however, is that MORB are derived from the shallow mantle, while oceanic island basalts are derived from deep mantle carried to the surface by rising mantle plumes. If this is the case, the OIB produced by melting of these plumes provide insights into the composition of the deep mantle.

11.5.2 Evolution of the depleted MORB mantle

MORB is easily the most abundant magma type on Earth – this magma forms the entire oceanic crust, produced at a rate of about 3 km³/yr. This suggests the source of MORB,

"depleted MORB mantle" or DMM, is a major reservoir, if not the major reservoir, in the mantle. This observation immediately inspires several questions. Where it is located? How has it evolved? How big is it? We consider these questions in the following paragraphs.

There are some fairly straightforward reasons for thinking MORB come from the shallow mantle and OIB from the deep mantle. MORB are by far the most abundant volcanic rocks on the Earth and they are also compositionally the most uniform. This kind of basalt is erupted, with a few exceptions, wherever plates are moving apart. The pull of subducting lithospheric slabs has been shown to be the primary force acting on lithospheric plates (Forsyth and Uyeda, 1975) and upwelling of mantle beneath mid-ocean ridges appears to be largely a passive response to the plate motion induced by this slab-pull.

The incompatible element depletion that we can infer for the DMM from both the trace element and isotopic composition of MORB is most easily explained by extraction of partial melts from it, which carried away the incompatible elements. Where have these melts gone? They have gone to form the oceanic and continental crusts, both of which are created by partial melting of the mantle and are incompatible element-enriched. Of these, the continental crust is by far the more significant reservoir of incompatible elements. The oceanic crust is less enriched in incompatible elements and furthermore is temporary. It survives less than 100 million years on average before being subducted back into the mantle. We can use isotopic compositions and trace elements to make a simple first-order estimate of the volume of the DMM, if we assume that it is the only incompatible element depleted reservoir and that the continental crust is the only complementary incompatible element enriched reservoir.

This is essentially a mass balance problem among a number of reservoirs, so, following DePaolo (1980), we begin by writing a series of mass balance equations. The first is mass of the reservoirs:

$$\sum_j M_j = 1 \quad (11.23)$$

where M_j is the mass of reservoir j as a fraction of the total mass of the system, in this case the silicate Earth. We can also write a mass balance equation for any element i as:

$$\sum_j M_j C_j^i = C_0^i \quad (11.24)$$

where C_0 is the concentration in the bulk silicate Earth. For an isotope ratio, R, of element i, or for an elemental ratio of which element i is the denominator, the mass balance equation is:

$$\sum_j M_j C_j^i R_j^i = C_0^i R_0^i \quad (11.25)$$

Our problem assumes the existence of three reservoirs: the continental crust, the mantle depleted by crust formation, and the undepleted, or primitive, mantle, which constitutes the remainder of the mantle (we explicitly ignore oceanic island basalt reservoirs and continental lithosphere in this calculation). We also assume the primitive mantle composition is equal to that of bulk silicate earth.

These mass balance equations can be combined to solve for the mass ratio of depleted mantle to continental crust:

$$\frac{M_{DM}}{M_{CC}} = \frac{C_{CC}^i}{C_0^i} \frac{(R_{CC}^i - R_{DM}^i)}{(R_0^i - R_{DM}^i)} - 1 \quad (11.26)$$

where the subscripts DM and CC refer to depleted mantle and continental crust respectively. A number of solutions to the mass balance equations are possible; we want an expression based on values that are relatively well constrained. Isotopic compositions of mantle reservoirs are well constrained because the magmas they produce have the same isotopic composition, but this is not true of elemental concentrations. We do have good estimates of concentrations in the crust because great numbers of samples can be analyzed and an average value computed (section 11.6.2). Once we have solved for the mass of depleted mantle, however, it would be straightforward to solve for the depleted mantle concentration.

The Nd isotope system is perhaps best suited for this question since $^{143}\mathrm{Nd}/^{144}\mathrm{Nd}$ ratios constrain the Nd isotopic composition of the Earth and, being a refractory lithophile element, its concentration in the bulk silicate Earth is also constrained (though not precisely). The Nd concentration and the Sm/Nd ratio of the continental crust are also better constrained than many other elements. The Sm/Nd ratio and $^{143}\mathrm{Nd}/^{144}\mathrm{Nd}$ of the crust are related through isotopic evolution, specifically:

$$\frac{^{143}\mathrm{Nd}}{^{144}\mathrm{Nd}} = \left(\frac{^{143}\mathrm{Nd}}{^{144}\mathrm{Nd}}\right)_0 + \left(\frac{^{147}\mathrm{Sm}}{^{144}\mathrm{Nd}}\right)(e^{\lambda t} - 1)$$
$$(11.27)$$

Because the half-life of $^{147}\mathrm{Sm}$ is long compared with the age of the Earth and because we do not need the level of precision necessary for geochronology, we can linearize this equation as:

$$\frac{^{143}\mathrm{Nd}}{^{144}\mathrm{Nd}} = \left(\frac{^{143}\mathrm{Nd}}{^{144}\mathrm{Nd}}\right)_0 + \left(\frac{^{147}\mathrm{Sm}}{^{144}\mathrm{Nd}}\right)\lambda t \quad (11.28)$$

The continental crust is certainly not of a single age – it was not created at a single time t. However, because it is linear, this equation remains valid for an average crustal age, T. Hence we may write:

$$\left(\frac{^{143}Nd}{^{144}Nd}\right)_{CC} = \left(\frac{^{143}Nd}{^{144}Nd}\right)_{Pm} + \left(\frac{^{147}Sm}{^{144}Nd}\right)_{CC} \lambda T \tag{11.29}$$

where the superscript T denotes the value at time T. The $^{143}Nd/^{144}Nd$ value of the primitive mantle can be calculated at any time from the present-day $^{143}Nd/^{144}Nd$ and $^{147}Sm/^{144}Nd$ of primitive mantle. The $^{143}Nd/^{144}Nd$ of the continental crust calculated in eqn. 11.29 can then be used in eqn. 11.26 to calculate the mass fraction of depleted mantle.

Now let's assign some values to these equations. The mass of the continental crust as a fraction of the mass of the silicate Earth, M_{CC}, is about 0.0055. For the $^{143}Nd/^{144}Nd$ value of the depleted mantle, we'll choose 0.51310 (equal to the median value for MORB listed in Table 11.5, $\varepsilon_{Nd} \approx +9$). Based on assessments of the composition of continental crust (section 11.6.2), a good estimate for its $^{147}Sm/^{144}Nd$ ratio is 0.123. We can use these values to calculate the mass of depleted mantle as a fraction of the silicate Earth as a function of the other parameters in eqn. 11.26, namely the ratio of Nd concentration in the crust to primitive mantle, C_{CC}/C_0, the average age of the continents, T, and the $^{143}Nd/^{144}Nd$ of the silicate Earth (expressed in epsilon units). Figure 11.16 shows the results. A good estimate of C_{CC}/C_{PM} is about 19, but there is considerable uncertainty. A good estimate of T is about 2 Ga, but there is easily 10% uncertainty in this value. If the Sm/Nd ratio is chondritic, Figure 11.16 suggests that the depleted mantle constitutes 40% or less, and more likely only about 25% of the mantle. However, as we have seen, the $^{142}Nd/^{144}Nd$ of terrestrial materials suggests the possibility Sm/Nd ratio of the Earth or the observable part of it is greater than chondritic. If the terrestrial Sm/Nd ratio is 3% greater than chondritic, corresponding to the composition listed in Table 11.3, then the ε_{Nd} value of the Earth is +3.6 and the depleted mantle constitutes 40–60% of the mantle. If the Sm/Nd ratio of the Earth is 6% greater than chondritic, as Caro and Bourdon (2010) argued, then the ε_{Nd} of the Earth is +6.9, and so the mantle depleted by crust extraction constitutes at least 70% of the entire mantle

and possibly all of it. This higher estimates of ε_{Nd} of the Earth assumes that the difference in $^{142}Nd/^{144}Nd$ between the Earth and chondrites is entirely due to differences in Sm/Nd. However, as we have seen, much or all of this difference appears to be due to isotopic heterogeneity in the solar system due to incomplete mixing of nucleosynthetic products. Reasonable estimates of ε_{Nd} for the Earth are thus probably ≤ 6. Choosing a value of 2 would imply that DMM, the MORB source reservoir, constitutes 30–50% of the mantle.

Now let's consider a simple mass balance for a highly incompatible element such as Rb. McDonough and Sun estimate a bulk silicate Earth Rb concentration of 66 ppm. The estimated crustal composition of Rudnick and Gao (2014) in Table 11.11 is 49 ppm with other estimates ranging from 32 to 87 ppm. This requires quantitatively extracting Rb from 45% (with a range from 29–78%) of R the mantle. Of course, Rb has not been quantitatively extracted: there is some left. Salters and Stracke (2005) estimate 0.05 ppm Rb in the depleted mantle; this would increase our mean estimate of the mass of depleted mantle to 52% (slightly less using Workman and Hart's (2005) depleted mantle composition). These estimates barely overlap those we derived based on the Nd isotope mass balance above assuming ε_{Nd} of the Earth is 0. Using the "depleted Earth" composition in Table 11.3, we find Rb must be extracted from 57% (range: 37–91%) to account for the Rb in the continental crust, which increases to 66% assuming the Rb depleted mantle composition of Salters and Stracke (2005). This agrees somewhat better with our Nd isotope mass balance calculation assuming a $\varepsilon_{Nd} > 0$ for the Earth. Clearly, these sorts of mass balance calculations have considerable uncertainty, but we can infer that something like half the mantle has been depleted in incompatible elements to create the continental crust. We should also note that we have not considered the possibility of enriched reservoirs in the mantle, of which oceanic island basalts, to topic to which we now turn, provide evidence. Any enriched reservoirs within the mantle would increase the calculated fraction of DMM. Their mass, and therefore their effect on mass balance, is, however, very uncertain.

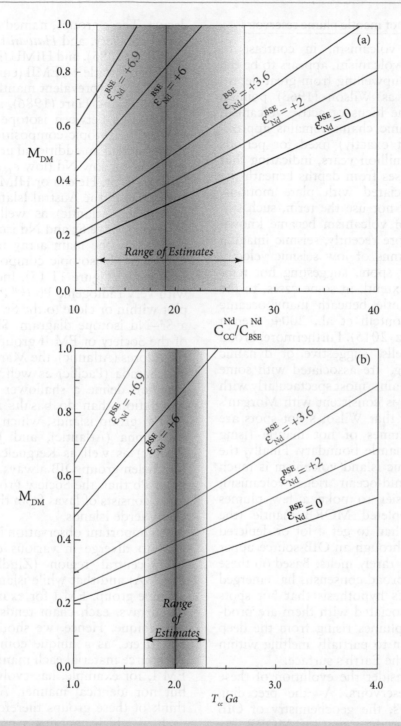

Figure 11.16 Mass fraction of the depleted mantle (DMM) calculated from eqn. 11.26. The top graph shows this for a variety of values of ε_{Nd} for the silicate Earth and a range of C_{CC}/C_{PM}, assuming an average age of the continents of 2 Ga. The lower graph shows the same thing for $C_{CC}/C_{PM} = 19$ and a range for the average age of the continents.

11.5.3 Evolution of mantle plume reservoirs

Oceanic island volcanism, in contrast to mid-ocean ridge volcanism, appears to be the result of mantle upwelling from great depth. To begin with, as Wilson (1963) pointed out long ago, the locus of active volcanism in oceanic volcanic chains remains approximately (but not exactly) fixed for periods as long as 100 million years, indicating that this upwelling rises from depths beneath the convection associated with plate motions. Although he did not use the term, such stationary points of volcanism became known as *hot spots*. More recently, seismic imaging has shown columns of low seismic velocity below these hot spots, suggesting hot temperatures, that extend, in some cases, to the base of the mantle beneath many oceanic islands (e.g., Montelli et al., 2004; French and Romanowicz, 2015). Furthermore, broad topographic swells, suggestive of dynamic mantle upwelling, are associated with some oceanic island chains, most spectacularly with Hawaii. All this is consistent with Morgan's (1971) proposal that Wilson's hot spots are produced by plumes of hot mantle rising from the core–mantle boundary. Finally, the volume of oceanic island volcanism is much smaller than mid-ocean ridge volcanism; it would be easier to poke a few plumes through the depleted MORB mantle (the asthenosphere) than to get a lot of depleted MORB mantle through an OIB-source upper mantle that only rarely melts. Based on these observations, a broad consensus has emerged around Morgan's hypothesis that hot spots and the OIB associated with them are products of mantle plumes rising from the deep mantle and begin to partially melting within 100–200 km of the Earth's surface.

Let's now consider the evolution of these deep mantle reservoirs. As the preceding discussion shows, the geochemistry of OIB is more variable than that of MORB. As discussed above, primitive mantle could be a component of many plumes, but the diversity of isotopic compositions in Figures 11.14 and 11.15 clearly requires a variety of OIB mantle sources with distinct chemical histories. Although there are many mantle plumes and each is to some degree geochemically unique, there appears to be a much smaller number of geochemical reservoirs from which they are drawn. These groups, named *St. Helena, Kerguelen, Society,* and *Hawaii* for a type island by White (1985), and HIMU (high U/Pb), EMI (enriched mantle 1), EMII (enriched mantle 2) and PREMA (prevalent mantle) in the review of Zindler and Hart (1986), show distinctive correlations between isotope ratios or have distinctive isotopic compositions. Hart et al. (1986) added an additional group, LOND, an acronym for low-Nd (low ε_{Nd}). For example, OIB of the St. Helena or HIMU group, which encompasses the Austral Islands (Pacific) and Ascension (Atlantic) as well as St. Helena (Atlantic), have Sr and Nd isotope ratios that plot below the main array in Figure 11.14, and their Pb isotopic compositions are very radiogenic (Figure 11.15). Indeed, all basalts with very radiogenic Pb ($^{206}Pb/^{204}Pb > 20.5$) plot within or close to the St. Helena field on a Sr–Nd isotope diagram. Similarly, islands of the Society or EM II group, which include the Azores (Atlantic), the Marquesas (Pacific), and Samoa (Pacific) as well as the Societies (Pacific), define a shallower Sr–Nd isotope correlation than do basalts from Kerguelen (EM I) group islands, which include Tristan da Cunha (Atlantic), and Juan Fernandez (Pacific) as well as Kerguelen (Indian). The Kerguelen group OIB always have less radiogenic Pb than the Society group. The LOND array consists of lavas from the Comoros and Cape Verde Islands.

An important observation is that the arrays tend to diverge in various directions away from central region (Zindler and Hart's PREMA) and that while island chains within a single group, EM I for example, form similar arrays, each chain tends to be more or less unique. Hence, we should not think of EM I, etc. as a unique component. Rather, the source material each mantle plume within EM I, for example, has evolved in a similar, but not identical manner. A useful way to think of these groups therefore is as *genera*, each of which contains a number of species represented by a single mantle plume.

Figure 11.17 compares the incompatible element abundances in MORB and typical OIB alkali basalt in a *spider diagram*, in which elements are ordered according to their incompatibility and normalized to primitive mantle values. This kind of plot is also sometimes referred to as an *extended rare earth plot*. Melts will be enriched, and residual

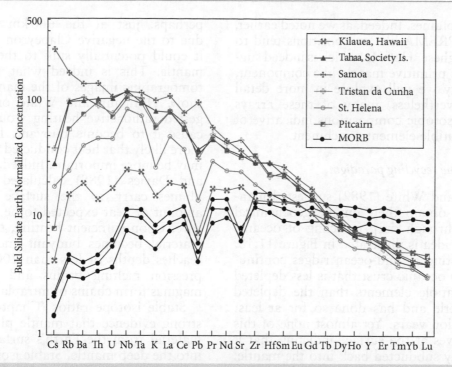

Figure 11.17 "Spider diagram" displaying concentrations of trace elements normalized to the bulk silicate Earth values in Table 11.3 and ordered by increasing compatibility. The plot shows representative patterns for MORB, Hawaii, St. Helena (HIMU), Tristan da Cunha and Pitcairn (EMI), and Samoa and the Society Islands (EMII). Modified from White (2010).

solids depleted, in elements on the left. The spider diagrams suggest complex histories of enrichment and depletion, but, with some exceptions, most notably Pb, an element's enrichment is related to its compatibility. Melting in the lower mantle, because it would involve minerals with very different structures, would create patterns of enrichment and depletion quite different from those in Figure 11.17 (e.g., Corgne et al., 2005; White, 2015). Thus while plumes may come from the deep mantle, they carry chemical signatures created in the upper mantle. This is a very important observation indeed.

Weaver (1991) suggested EMI, EMII, and HIMU basalts could be distinguished based on their trace element geochemistry. However, Willbold and Stracke (2006) found that while HIMU basalts can indeed be distinguished from EM basalts, there are no systematic differences between EMI and EMII. They found that HIMU have lower Rb/Sr, Ba/La, K/La, Th/U, Pb/Ce, and higher U/Pb, Lu/Hf, and Nb/Rb than EM basalts. The distinctions identified by Willbold and Stracke (2006) can be seen to some degree in Figure 11.17.

Most oceanic basalts display negative Pb anomalies (the continental crust and island arc volcanic rocks typically have positive Pb anomalies), but they are smaller for the EM basalts, and the Samoan example actually has a slightly positive Pb anomaly. HIMU basalts, exemplified in Figure 11.16 by the St. Helena sample, are enriched in Nb and Ta relative to K and Rb, have particularly strong depletion in Pb. The HIMU basalts also show a decrease in normalized abundance with increasing incompatibility for elements more incompatible than Nb.

Determining how these distinct geochemical reservoirs have evolved is among the most vexing problems in mantle geochemistry. The principal observation to be explained is that mantle plumes invariably have less depleted isotopic signatures than MORB, and the isotopic compositions of some indicate net enrichment in incompatible elements. As we discussed in the previous section, the convergence of several OIB arrays around isotopic compositions close to those expected of primitive mantle in Figure 11.14 suggests primitive mantle could indeed be a component of at least some plumes. Indeed, as we noted earlier, OIB with PREMA-like compositions tend to have the highest $^3He/^4He$, which is indeed suggestive of a primitive mantle-like component, a possibility we will explore in more detail below. Nevertheless, many of these arrays extend to isotopic compositions indicative of an incompatible element enrichment.

11.5.3.1 The recycling paradigm

Hofmann and White (1982) suggested mantle plumes obtain their unique geochemical signature through deep recycling of oceanic crust. This idea is illustrated in Figure 11.18. Partial melting at mid-ocean ridges continually creates oceanic crust that is less depleted in incompatible elements than the depleted upper mantle and has done so for at least several billion years. Yet almost none of this crust survives at the surface indicting that it is inevitably subducted back into the mantle. The question is, what becomes of it then? Hofmann and White noted that once oceanic crust reaches depths of about 90 km it converts to eclogite, which is denser than peridotite. Because it is rich in Fe and garnet-forming components, it remains denser than peridotite at all depths greater than 90 km (except,

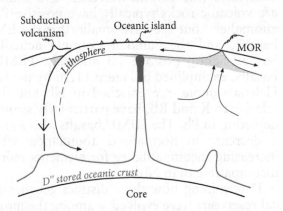

Figure 11.18 Cartoon illustrating the oceanic crustal recycling model of Hofmann and White (1982). Oceanic crust is transformed into eclogite and post-eclogite assemblages upon subduction. It separates from the less dense underlying lithosphere and sinks to the deep mantle where it accumulates. Eventually, it becomes sufficiently hot to form plumes that rise to the surface, producing oceanic island volcanism. After Hofmann and White (1982).

perhaps, just at the 660 km discontinuity due to the negative Clapeyron slope). Thus, it could potentially sink to the base of the mantle. This is indeed what some seismic tomographic images of the mantle appear to show. Hofmann and White originally suggested radioactive heating would ultimately cause it to become buoyant. It now seems more likely that heat conducted from the core may be more important. Indeed, Sleep (1990) and Davies (1988) calculated that mantle plumes carry to the surface roughly the amount of heat expected to be lost from the core. Upon sufficient heating, the subducted material becomes buoyant and rises. As it reaches depths of less than 200 km, decompression melting begins and the resulting magmas form chains of intraplate volcanoes.

Stable isotope ratios (Chapter 9) provide strong evidence that mantle plumes contain material from the Earth's surface subducted into the deep mantle. Stable isotope fractionation factors vary inversely with the square of temperature, therefore, ratios that differ significantly from primordial values "must either have been affected by low temperature processes or must contain a component that was at one time at the surface of the Earth" (Taylor and Sheppard, 1986).

Perhaps the most dramatic evidence is the discovery of mass independently fractionated (MIF) sulfur in olivine sulfide inclusions from the central Pacific volcanoes of Mangaia (Cabral et al., 2013) and Pitcairn Islands (Delavault et al., 2016) (Figure 11.19). These discoveries are particularly important because MIF sulfur was produced only in the Archean O_2- and O_3-free atmosphere and thus unequivocally demonstrate both the presence of recycled oceanic crust in mantle and the antiquity of this recycled material.

Eiler et al. (1997) reported $\delta^{18}O_{SMOW}$ values are as high as 6.1‰ in basalts from the Society Islands and concluded that the Society Islands mantle source contained up to 5% of a sedimentary component, assuming a $\delta^{18}O_{SMOW}$ of +15‰ for the sediment. Subsequently, Workman et al. (2008) reported variations of $\delta^{18}O_{SMOW}$ in olivines in Samoan lavas that correlated positively with $^{87}Sr/^{86}Sr$ and $^{207}Pb/^{204}Pb$ and incompatible element ratios, confirming the presence of recycled material in the source of Samoan lavas. Eiler et al. (1997) also found that some basalts from

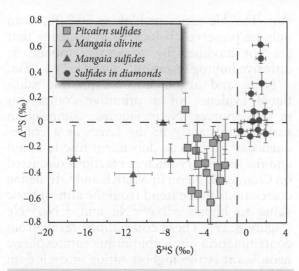

Figure 11.19   Mass independently fractionated sulfur isotopes diamonds from the Orapa kimberlite mine in Botswana and sulfides and olivines from Mangaia and Pitcairn Islands. Dashed lines indicate estimated bulk mantle composition. Mass independent fractionation of sulfur was restricted to the Archean/early Proterozoic atmosphere. This sulfur must have been subducted and stored in the mantle for 2.3 billion years. Data from Farquhar et al. (2002), Cabral et al. (2013), and Delavault et al. (2016).

HIMU island chains had low $\delta^{18}O_{SMOW}$. As we found in Chapter 9, isotope ratios of carbon, sulfur, lithium, calcium, and silicon in oceanic island basalts provide additional evidence that mantle plumes contain material once at the surface of the Earth.

An alternative origin for enriched mantle plume reservoirs was proposed by McKenzie and O'Nions (1983). They noted the common evidence for incompatible element enrichment in the subcontinental lithosphere (which we discuss in the next section) and suggested that under certain circumstances, such as continent–continent collisions, this lithosphere may occasionally detach from the crust above it. Because it is cold, it would also sink to the deep mantle. Workman et al. (2004) suggested a variation on this idea, namely that the enriched component in EMII type OIB was metasomatized oceanic lithosphere. In both the McKenzie and O'Nions and Workman and Hart hypotheses, lithosphere is enriched as small-degree melts percolate into it from below. As in the case

of the Hofmann and White model, it would be stored at the base of the mantle where it is heated by the core, eventually rising to form mantle plumes. Because mantle plumes come in several geochemical varieties, it is possible that both mechanisms operate.

Indeed, other processes may be involved as well. One of these is subduction erosion, which refers to the mechanical removal of the base of the overriding plate in a subduction zone by the subducting one. There is considerable geologic evidence that in sediment-starved subduction zones, such as the Peru–Chile margin, the subducting oceanic crust can abrade and ultimately significantly erode the overriding continental plate from below and some geochemical evidence that this has occurred in Central America. Other evidence includes crystalline basement in the landward trench wall, tilting and subsiding erosional surfaces on the overriding plate (e.g., Scholl and von Huene, 2007), and the paucity of pre-Neoproterozoic blueschist facies rocks (Stern, 2011). The importance of this process is now widely recognized (e.g., Clift et al., 2009; Stern and Scholl, 2010; Stern, 2011). Stern (2011) and Stern and Scholl (2010) argue that the continental mass loss due to subduction erosion exceeds that due to sediment subduction while Clift et al. (2009) estimate that subduction erosion losses are 80% of sediment subduction losses.

Another process is lower crustal floundering, also called, somewhat inaccurately, delamination (Kay and Kay, 1991; Kay et al., 1994). In regions where compression thickens continental crust, there is evidence that the lowermost crust can sink into the mantle. Again, the Peru–Chile margin provides an example where the convergence with the Nazca plate has greatly thickened the South American plate, creating the Andes. The crust here is so thick that the lowermost crust is converted to eclogite, which is denser than mantle peridotite. This eclogitic lower crust can then sink into the mantle, as some seismic imaging appears to show.

11.5.3.2   A primitive component as well

The evidence we have considered so far suggests that the sources of oceanic island basalts are more evolved that the MORB source. Paradoxically, OIB reservoirs appear to be more primitive from the noble gas isotope perspective. As we found in Chapter 8, He continually escapes from the atmosphere to space and while ^{4}He is continually produced by alpha decay of U and Th. ^{3}He/^{4}He ratios are quite low, lower than in the atmosphere, in the continental crust as a consequence of this radiogenic production. The higher atmospheric ratios are maintained by escape of He from the mantle, primarily along mid-ocean ridges. MORB has a relatively uniform ^{3}He/^{4}He with R/R_A of 8.8±2.5 (recall that R/R_A is the ratio relative to atmospheric). As Figure 11.20 shows, while some OIB such as St. Helena and Tristan da Cunha have low ^{3}He/^{4}He as we would expect if they contained a recycled crustal component, most OIB have ^{3}He/^{4}He higher than MORB, with ratios in excess of 30 R/R_A in basalts from Iceland, the Galápagos, and Hawaii; ratios as high as 50 R/RA were reported in early Cenozoic Iceland mantle plume picrites on Baffin Island (Starkey et al., 2009; Stuart et al., 2003). Because He is lost from the mantle during melting and degassing and because ^{4}He is generated by radioactive decay, these high ^{3}He/^{4}He ratios indicate that some OIB sources tap a relatively primordial, less degassed mantle reservoir. Porcelli and Elliott (2008) found

that ^{3}He/^{4}He ratios as high as 50 R/R_A "can only be preserved if located in domains that are not modified by convective mixing or diffusive homogenization" over the last 3 Ga.

Neon and argon isotopes provide additional evidence of a primitive component in OIB sources. Neon isotopic composition varies significantly in the Earth as a consequence of both mass dependent fractionation and the secondary nuclear reactions discussed on Chapter 8. Neon in MORB and OIB define correlations that extend from the atmospheric value to higher ^{20}Ne/^{22}Ne and ^{21}Ne/^{22}Ne (Figure 8.29). These correlations result from contamination by ubiquitous atmospheric neon, so it is the highest ratios in each data set that are of interest. As we pointed out in Chapter 8, neon in MORB and OIB is less fractionated and more solar-like than atmospheric Ne, with ^{20}Ne/^{22}Ne values that in some cases exceed the meteoritic Neon B value. Mantle Ne is also variably enriched in ^{21}Ne compared with solar Ne as a consequence of nuclear reactions initiated by α-decay or fission; the extent of this enrichment is effectively a measure of the (U + Th)/Ne ratio in mantle source. ^{21}Ne/^{22}Ne ratios are higher in MORB than OIB, implying the MORB source has higher (U + Th)/Ne

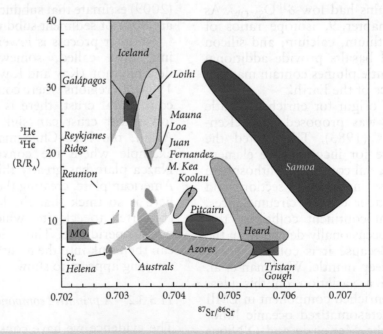

Figure 11.20 ^{3}He/^{4}He vs. ^{87}Sr/^{86}Sr in MORB and OIB. High ^{3}He/^{4}He in many OIB requires a less degassed, more primitive component in their sources. The highest ^{3}He/^{4}He are associated with intermediate ^{87}Sr/^{86}Sr values.

and thus that it is more degassed than OIB sources. Indeed, submarine basalts from the Galapagos and Hawaii and subglacial basalts from Iceland have Ne isotopic compositions that plot close to a mass dependent fractionation line passing through the solar wind composition, indicative of a source with very high Ne/(U + Th), i.e., a relatively undegassed source. Similarly, $^{40}Ar/^{36}Ar$ ratios in basalts, once corrected for atmospheric contamination, are a measure of the time-integrated K/Ar ratio of the mantle source. $^{40}Ar/^{36}Ar$ ratios in MORB can be as high as 40,000, while in OIB they are <10,000, again indicating that OIB reservoirs have experienced less degassing.

Xe, which has nine stable isotopes, offers insights into Earth's early evolution. It is also the rarest and most challenging to analyze of the noble gases, and consequently, high quality data is limited; in addition to data on MORB, useful Xe isotope data are available only from one subglacial Icelandic basalt, submarine basalts from Loihi Seamount, Hawaii, xenoliths from Reunion, and olivines from Samoa. While limited, this dataset is nonetheless highly informative. Xe isotope ratios vary as a consequence of beta decay of the extinct ^{129}I ($t_{1/2}$ = 15.7 Ma) producing ^{129}Xe, fission of ^{238}U and extinct ^{244}Pu ($t_{1/2}$ = 82 Ma) which produce the heavy isotopes, ^{131}Xe, ^{132}Xe, ^{134}Xe, and ^{136}Xe, in somewhat different proportions so with high quality

data it is possible to distinguish plutonium from uranium contributions. As with the other noble gases, samples are variably contaminated by atmospheric Xe, so it is the ratios most offset from the atmospheric value in any particular dataset that are significant (which is not to imply that any particular mantle reservoir is necessarily isotopically homogeneous). Figure 11.21 shows that Xe in MORB is more enriched in both ^{129}Xe and ^{136}Xe than OIB, and both are enriched in these radiogenic isotopes than atmospheric Xe. This indicates that the MORB source has a higher time-integrated I/Xe and (U + Pu)/Xe and is therefore more degassed than the OIB ones. Because of the short half-life of ^{129}I (and ^{244}Pu), this difference must have been established very early in Earth's history and sustained ever since. Indeed, Mukhopadhyay (2012) argues that "Because ^{129}I became extinct about 100 million years after the formation of the Solar System, OIB and MORB mantle sources must have differentiated by 4.45 billion years ago and subsequent mixing must have been limited." The Icelandic subglacial sample was particularly gas-rich, enabling Mukhopadhyay (2012) to identify Xe produced by plutonium fission and he found that, "the Iceland plume source also has a higher proportion of Pu- to U-derived fission Xe, requiring the plume source to be less degassed than MORBs."

Rizo et al. (2016) and Mundl et al. (2017) reported small variations in the $^{182}W/^{184}W$

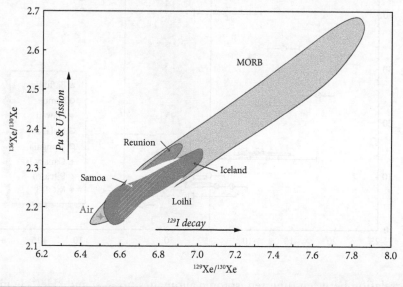

Figure 11.21 $^{129}Xe/^{130}Xe$ vs. $^{136}Xe/^{130}Xe$ in MORB and oceanic island basalts.

ratio in flood basalts of the Ontong-Java Plateau, related to the Louisville Ridge plume, and Baffin Island, related to the Iceland plume, and in some oceanic island basalts. ^{182}W is the decay product of the extinct radionuclide ^{182}Hf (Chapters 8 and 10). The variations are significantly smaller than observed in meteorites and Archean terrestrial rocks; indeed, they are shown in Figure 11.21 in μ_{182W} units, deviations in part per million from the terrestrial standard (a factor of 100 smaller than ε-units). Given the short half-life of ^{182}Hf ($\sim$ 9Ma), variations in ^{182}W must have been produced in the first few tens of millions of years of Earth history. Alternatively, they may have been inherited from the heterogeneity of the proto-planetary bodies which accreted to form the Earth. In either case, the significance is that these primordial heterogeneities have been preserved without being entirely erased by subsequent convective mixing in the mantle. Horan et al. (2018) and Peters et al. (2018) reported variations in $^{142}Nd/^{144}Nd$ at the level of a few parts per million from Samoa and Réunion, suggesting heterogeneity from the Earth's earliest differentiation survives in the mantle. On the other hand, de Leeuw et al. (2017) found no significant deviations from the assumed

bulk Earth $^{142}Nd/^{144}Nd$ value in the Baffin Island basalts that have $^{3}He/^{4}He$ R/R_A as high as 50.

Figure 11.22 shows that there is a weak negative correlation between μ_{182W} and $^{3}He/^{4}He$ such that basalt with the highest $He/^{4}He$ have the most negative μ_{182W}. The Baffin Island sample, which has the highest $^{3}He/^{4}He$ yet has a positive μ_{182W}. Thus, it appears that the reservoir with the least degassed He isotope signature also has the most anomalous μ_{182W}. Some caution is necessary in interpreting these very small variations, however. Kruier and Kleine (2018) also analyzed basalts from the Ontong-Java Plateau and found no excess radiogenic ^{182}W. Instead, they found mass-independent fractionations of W isotopes related to the nuclear field shift effect, which they inferred occurred during chemical purification of samples.

We can conclude that mantle plumes contain both material recycled from the Earth's surface as well relatively undegassed and primitive material that hosts heterogeneity produced very in the Earth's earliest history. We also know that most mantle plumes appear to be spatially associated with the LLSVPs in the deep mantle. There is no

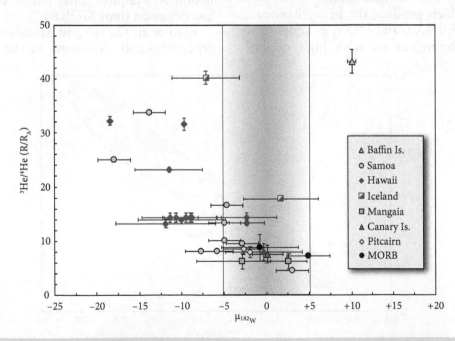

Figure 11.22   Correlation between tungsten isotopic anomalies (μ_{182W}) and $^{3}He/^{4}He$ measured in Samoan, Hawaiian and Baffin Island basalts. Data from Rizo et al. (2016) and Mundl et al. (2017).

consensus as to how these features differ compositionally from surrounding mantle, but one set of ideas involves primordial material residual from the Earth's earliest differentiation. For example, Ballmer et al. (2017) have proposed that the LLSVPs are regions of higher Si/Mg ratios that formed fractionation during magma-ocean crystallization and/or early crust extraction that left the shallow mantle Si-depleted. The higher Si/Mg ratio would result in higher density and bridgmanite/ferripericlase ratio, which, because of the greater strength of bridgmanite, also increases viscosity, both of which would tend to stabilize the LLSVPs. Other proposals include their formation as remnants of an FeO-rich basal magma ocean (Labrosse et al., 2007; Jackson et al., 2017), or accumulation of an early primitive crust (Tolstikhin and Hofmann, 2005; Boyet and Carlson, 2005). Another set of ideas proposes that they are accumulations of subducted oceanic crust that accumulate in piles stabilized by their greater density (e.g., Brandenburg and van Keken, 2007; Tackley, 2011; Mulyukova et al., 2015).

A particularly intriguing possibility that united both sets of ideas was proposed by Li et al. (2014) – namely, that the LLSVPs consist of both dense residues of early mantle differentiation and subducted oceanic crust. In their geodynamic model, subduction and convection sweep an initial, dense primitive layer at the base of the mantle into *thermochemical piles*. Subducted oceanic crust is episodically flushed into the more primitive reservoir and stirred into it through internal convection in the pile. The fraction of oceanic crust in the piles eventually reaches 1% to more than 15%, depending on viscosity and buoyancy contrasts. Mantle plumes form on these piles and consist of a mixture of ambient deep mantle, relatively recently subducted oceanic crust, and primitive material and older oceanic crust from the LLSVPs and the fraction of primitive material in the LLSVPs diminishes slowly over time. Adapting this model to the radiogenic and stable isotopic observations, plumes of the PREMA genus would contain relatively large fractions of material from the LLSVPs while plumes of the other genera would contain only secondary amounts of LLSVP material. We should not, however, think we have solved the riddle of mantle plumes and LLSVPs.

11.5.4   The subcontinental lithospheric mantle

Beneath continents there are regions of mantle through which heat is conducted rather than convected. These regions tend to have fast seismic velocities, suggesting they are cold compared with the convective mantle. Xenoliths derived from these regions (their depth of origin can be established using the thermobarametric techniques covered in Chapter 4) are often harzburgitic (i.e., dominantly olivine–orthopyroxene), which has lower density than cpx-bearing peridotites such as pyrolite. This subcontinental lithosphere is of quite variable thickness: it is only tens of kilometers thick under tectonically active areas such as the Great Basin of the Western US but is more than 200 km thick under the South African craton. In addition to seismic properties, basalts and xenoliths derived from this region inform us of the character and history of the subcontinental lithospheric mantle (SCLM).

Olivines in peridotite xenoliths from Archean cratons, such as the Kaapvaal of Southern Africa, have a mean Mg# (molar ratio: (Mg/(Mg+Fe) × 100) of 92.6 compared with a value of 89.3 expected of a pyrolite composition such as in Table 11.2 (Pearson and Wittig, 2014). By comparison, Mg numbers for abyssal peridotites dredged from fracture zones on mid-ocean ridges are ~90.5 and values for xenoliths in oceanic island basalt are ~91.5. Mg numbers in residual olivines increase during partial melting because Fe partitions into a partial melt more than Mg, so this tells us that a fundamental feature of the SCLM, at least under ancient cratons, is that they have suffered extensive melt extraction, more so than is typical beneath mid-ocean ridges or oceanic islands. Figure 11.23 compares olivine Mg#'s in peridotite xenoliths as a function of depth (determined from geobarometric techniques) under two such cratons, the Slave Province of Canada and the Kaapvaal of Southern Africa, with values predicted for polybaric melting residues for mid-ocean ridges and mantle plumes. Pearson and Wittig (2014) concluded that the Slave Province trend, as well as others not shown, is best approximated by melting under a hot ridge, such as would be expected in a hotter Archean mantle. Melting under such conditions would have created oceanic crust much thicker than modern crust. The

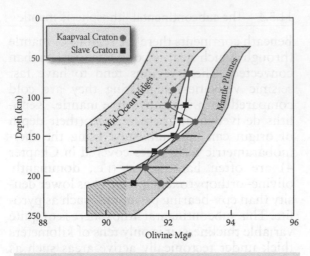

Figure 11.23 Olivine Mg#s in peridotite xenoliths from the Kaapvaal and Slave Cratons as a function of equilibration depth inferred from geobarometry compared with profiles expected from polybaric melt extraction. Modified from Pearson and Wittig (2004).

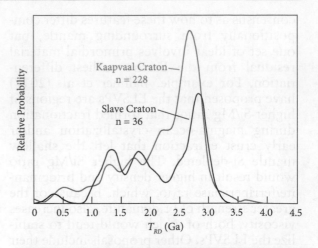

Figure 11.24 Probability functions for Re–Os model rhenium depletion ages measured in peridotites from the Kaapvaal and Slave Provinces. Modified from Pearson and Wittig (2014).

Kaapvaal trend shows a distinct kink to higher temperatures, but it would fit a *hot ridge profile* if it were moved upward by ~75 km. Consequently, Pearson and Wittig explain the distinct kink in the Kaapvaal trend as a result of subduction stacking of two mantle residue columns produced at a hot ridge.

Geochronological studies, particularly Re–Os, show that the timing of these melt depletion events typical match the age of the crust above it. Figure 11.24 shows rhenium-depletion model ages (T_{RD}) (eqn. 8.31) calculated from $^{187}Os/^{188}Os$ ratios in peridotite xenoliths from the Kaapvaal and Slave Provinces. These ages are determined from comparison of $^{87}Os/^{188}Os$ ratios with the primitive mantle evolution line and assume that Re was quantitatively removed at the time (i.e., Re/Os = 0 since that time). Because Re is unlikely to be quantitively removed, these will, if anything, underestimate actual ages somewhat. Although some ages are younger, the mode in both cases is Archean. This is true of other cratons as well: there is long-term coupling of crust with the underlying mantle lithosphere. There is a sharp drop-off in ages at around 3 Ga, even though some of the overlying crustal rocks are older, suggesting the subcontinental lithosphere began to stabilize around this time (Pearson and Wittig, 2014). There are exceptions, however. For example,

in the Archean North China craton, Re–Os T_{RD} ages are often much younger, ranging from Mesoproterozoic to recent, suggesting that the original Archean lithosphere has been removed.

In contrast to the olivine Mg# and Re–Os signatures of melt depletion, many xenoliths show evidence of melt or fluid *metasomatic* enrichment, particularly those from cratons. This is sometimes apparent in the presence of minerals we would not expect in melt-depleted peridotite, such as phlogopite mica and amphibole, and in rock textures. In other cases, the metasomatism is cryptic in that it is discernible only in mineral compositions, particularly in incompatible element abundances such as light rare earths and occasionally stable isotopic compositions. One example is so-called *sinusoidal* REE patterns. Figure 11.25 shows a number of such patterns in garnets in harzburgite xenoliths as well as one "normal" garnet from a lherzolite reported by Stachel et al. (1998), all from the Robert Victor diamond mine in South Africa. There have been a variety of explanations for these patterns and, although the details vary, all involve incompatible element re-enrichment in some form. A model proposed by Viljoen et al. (2015) is illustrated in Figure 11.25. The first step is two episodes of melt extraction from an undepleted lherzolite to produce the depleted harzburgite protolith. This depleted protolith then reacts with a

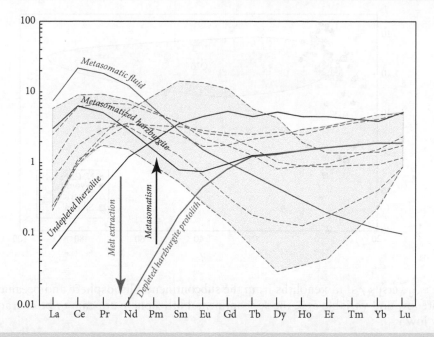

Figure 11.25   REE patterns in garnets from a lherzolite xenolith (solid black line) and harzburgite xenoliths (dashed lines) in kimberlites of the Robert Victor mine. Viljoen et al. (2015) proposed a model in which melt extraction in the garnet stability field is followed by further extraction in the spinel stability field to produce the depleted harzburgite protolith shown. Dehydration of a modestly LREE-depleted eclogite then produces the metasomatic fluid which reacts with the depleted harzburgite garnet to produce the sinusoidal REE pattern.

hydrous fluid that was in equilibrium at a modestly LREE-depleted eclogite at high temperature and pressure. The fluid then reacts with the harzburgite garnet to produce the sinusoidal REE patterns.

Nd, Hf, and Sr isotopic compositions indicate this enrichment is ancient. Figure 11.26 illustrates the point. γ_{Os} values in cratonic xenoliths are predominantly negative, indicative of depletion in Re relative to Os, as would occur through partial melting. In contrast, while some xenoliths have positive (depleted) ε_{Nd} signatures (indeed ε_{Nd} values has high as ~500 and ε_{Hf} values over 1000 have been measured), negative ε_{Nd} values predominate, indicative of incompatible element enrichment. Apparently, neither Re nor Os are transported effectively by metasomatic fluids, reflecting a complex dependence of the Re and Os partition coefficients on fluid composition and oxygen fugacity. The greater isotopic heterogeneity of the subcontinental lithosphere probably reflects its long-term stability: the greater the age of the material, the more variations in parent–daughter ratios will be expressed as variations in radiogenic isotope ratios. Convective mixing in the suboceanic mantle will tend to destroy heterogeneity in the suboceanic mantle.

In some cases, the metasomatism appears to have occurred shortly before eruption of the alkali basalts or kimberlites that carried the xenoliths to the surface. Evidence of this includes zoning in garnet inclusions is diamonds: diffusion would have erased this zoning had the garnets resided in the mantle at high temperature on geologic time scales (Shimizu, 2019). This also implies the diamonds within the xenoliths formed shortly before eruption (which in some cases was billions of years ago); in other cases, it appears to predate eruption by hundreds of millions of years. And there is evidence in some cases of multiple metasomatic events. The highly variable patterns of mineralogy, incompatible element abundances, and radiogenic isotope ratios indicate a variety of fluid compositions have been involved, ranging from H_2O-rich fluids to small-degree silicates and carbonatitic melts.

The sources of these fluids are likely to be diverse as well. Upwelling asthenospheric mantle or mantle plumes may not be able to

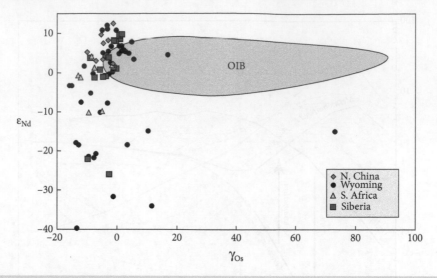

Figure 11.26 ε_{Nd} versus γ_{Os} in xenoliths from the subcontinental lithosphere and oceanic island basalts. Despite low and variable ε_{Nd}, the subcontinental lithosphere appears to be characterized by systematically low γ_{Os}.

rise through the rigid lithosphere but may nonetheless produce very small degree melts that can migrate upward into the lithosphere, reacting with it as they rise. Depending on the mantle composition and melting conditions, this process could alternatively produce carbonatitic melts as well. Yet another possibility is that hydrous fluids released during dehydration of subducting oceanic lithosphere may migrate into the continental lithosphere and react with it (Hawkesworth et al., 1990). Judging from studies of island arc magmas, such fluids appear to be particularly enriched in soluble incompatible elements, such as the alkalis and alkaline earths.

11.6 THE CRUST

We now turn our attention to the crust. Although the crust forms only a small fraction of the mass of the Earth (about 0.5%), it is arguably the most varied and interesting fraction. Further, it is the fraction we can examine directly and therefore know most about. The crust has formed through igneous processes from the mantle over geologic time. There are two fundamental kinds of crust: oceanic and continental. Oceanic crust, created mainly by magmatism at mid-ocean ridges, is basaltic in composition, thin, ephemeral, and relatively uniform. The continental crust is much thicker, essentially permanent, and on average

andesitic in composition. The continental crust is also much more varied. Although it too has formed by magmatism, its evolution is far more complex than that of oceanic crust. Though we have an excellent understanding of how oceanic crust forms, our understanding of the processes that have led to the present continental crust is far from complete. Subduction-related, or "island arc" volcanism appears to play a particularly important role in the formation of the continental crust, so we will pay special attention to processes in island arcs.

11.6.1 The oceanic crust

In a 1962 paper that he called "an essay in geopoetry," Harry Hess (1962) summarized his radical views on seafloor spreading. He speculated that mid-ocean ridges were produced by rising mantle convection currents, and that these convection currents then moved laterally away from the mid-ocean ridges, producing the phenomenon of continental drift. This concept now forms the basis of plate tectonics, the fundamental paradigm of geology. Hess did miss one detail, however. He thought the oceanic crust was hydrated mantle, consisting of "serpentinized peridotite, hydrated by release of water from the mantle over the rising limb of a convection current." However, as we saw in Chapter 7, when hot mantle decompresses as it rises, it melts. This melting generates the basaltic magma that forms the oceanic crust. In some respects, though, Hess's mistake is very minor indeed. Oceanic crust is very ephemeral, and for this reason, it is sometimes better to think of it as part of the mantle reservoir than the crustal one. Nevertheless, igneous processes at mid-ocean ridges have fascinated many geochemists, and much has been learned about them since Hess's paper.

11.6.1.1 Structure of the oceanic crust

Seismic studies show that the oceanic crust has a layered structure (Figure 11.27). Studies of the rare cases where oceanic crust is exposed on land, formations called *ophiolites*, together with drilling into the oceanic crust and examining ocean crustal cross-sections exposed in faulted regions, such as Hess Deep along the East Pacific Rise, have helped in interpreting this seismic structure. The uppermost layer, which is not present at mid-ocean ridges, consists of sediments (seismic layer 1). The sediment is underlain by basaltic lava flows, and below that by the dikes that fed their eruption; together these constitute seismic layer 2. Seismic layer 3 consists of

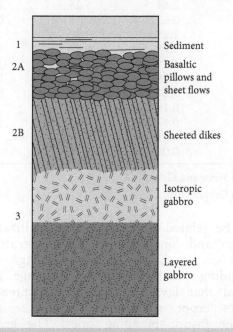

Figure 11.27 Cross-section of typical oceanic crust (not to scale). Numbers on the left side refer to the layers identified in seismic velocity profiles.

gabbros formed by basaltic magmas that crystallized in place (isotropic gabbros) and accumulations of minerals that crystallized from the basaltic magma in crustal magma chambers (layered gabbros). Because of the latter, the gabbros are probably somewhat more mafic on average than are the basalts. On average, the lava flows are about 800 m thick; the sheeted dike complex appears to be of roughly similar thickness. The gabbros are roughly 5 km thick on average and thus compose most of the oceanic crust, which is typically 6–7 km thick.

The structure of the oceanic crust reflects its construction. Melts rise buoyantly from the mantle because they are less dense than surrounding peridotite. Once they reach the crust, however, they are no longer buoyant, and they pond within the crust. Seismic studies show that beneath fast-spreading ridges, such as the East Pacific Rise, a permanent pool of melt exists about 2 km beneath the seafloor. This "melt lens" is only about a kilometer or so wide and 100 m or so deep. Beneath it is a seismically slow area interpreted as a crystal mush. In some cases, pools of melt also appear to exist in the deeper crust. Magma enters the crust at temperatures of 1200°C or so, but conductive heat loss at the seafloor, and more importantly, circulation of seawater (which emerges to form hydrothermal vents and "black smokers") withdraws heat rapidly from the magma. Consequently, it cools and crystallizes. Judging from the thickness of the lower crustal gabbros, most magma crystallizes in place within the lower oceanic crust. As crystallization proceeds, however, the residual magma becomes less dense, so that some of the magma, perhaps 20–25%, will rise toward the seafloor. About half of that will erupt to form lava flows and half will freeze in conduits to form the sheeted dike complex. From this we can see that MORB will not be compositionally representative of the whole oceanic crust; rather, it will be richer in incompatible elements that remain in the melt during fractional crystallization and poorer in compatible elements that partition into the crystallizing minerals.

11.6.1.2   Composition of the oceanic crust

The basaltic lavas, MORB, that constitute the uppermost oceanic crust are readily sampled by dredging, coring, and submersible sampling at mid-ocean ridges. Their average composition is listed in Tables 11.6 and 11.7, along with the average composition of lower crustal gabbros. While the MORB average is based on thousands of samples, the lower crustal average is based on far fewer samples; there is consequently much more uncertainty about the composition of the lower oceanic crust. The bulk composition of the oceanic crust listed in Table 11.6 and 11.7 was computed by White and Klein (2014) based on the assumption that (1) MORB are the product of fractional crystallization within the oceanic crust, and (2) the composition of the magma parental to MORB was in equilibrium with mantle olivine just before it entered the oceanic crust.

Mantle olivine is typically 90–92% forsterite, and according to experimental studies such as that of Roeder and Emslie (1970) (see Chapter 4), a melt in equilibrium with it would have an Mg# of around 72. The average composition listed in Table 11.6 has an Mg# of 59, and therefore could not be in equilibrium with mantle olivine. The bulk composition of the oceanic crust was determined by using the MELTS program (Chapter 4) to correct for fractional crystallization.

The composition of MORB is remarkably uniform throughout the ocean basins and in this respect is unique among igneous rocks. Nevertheless, some systematic regional variations have been identified. Some of this

Table 11.6   Major element composition of the oceanic crust.

	Average MORB[1]	Lower crust[2] average	Bulk crust[1]
SiO_2	50.06	50.6	50.1
TiO_2	1.52	0.78	1.1
Al_2O_3	15.00	16.7	15.7
ΣFeO	10.36	7.50	8.3
MnO	0.19	0.14	0.11
MgO	7.71	9.4	10.3
CaO	11.46	12.5	11.8
Na_2O	2.52	2.35	2.2
K_2O	0.19	0.06	0.11
P_2O_5	0.16	0.02	0.1

[1] From White and Klein (2014).

[2] From Coogan (2014).

Table 11.7   Trace element composition of the oceanic crust (all concentrations in ppm).

	Bulk crust*	Ave. MORB*	Lower crust†
Li	3.52	6.63	
Be	0.31	0.64	
B	0.80	1.80	
K	651	1237	
Sc	36.2	37	37
V	177	299	209
Cr	317	331	308
Co	31.7	44	50
Ni	134	100	138
Cu	43.7	80.8	71
Zn	48.5	86.8	38
Rb	1.74	4.05	
Sr	103	138	115
Y	18.1	32.4	13.9
Zr	44.5	103	28.4
Nb	2.77	6.44	0.93
Cs	0.02	0.05	
Ba	19.4	43.4	
La	2.13	4.87	0.86
Ce	5.81	13.1	2.75
Pr	0.94	2.08	0.52
Nd	4.90	10.4	2.78
Sm	1.70	3.37	1.1
Eu	0.62	1.20	0.58
Gd	2.25	4.42	1.6
Tb	0.43	0.81	0.31
Dy	2.84	5.28	2.09
Ho	0.63	1.14	0.46
Er	1.85	3.30	1.34
Tm	0.28	0.49	
Yb	1.85	3.17	1.27
Lu	0.28	0.48	0.19
Hf	1.21	2.62	
Ta	0.18	0.417	
Pb	0.47	0.657	
Th	0.21	0.491	
U	0.07	0.157	

*From White and Klein (2014)

†From Coogan (2014).

can be related to fractional crystallization. Rubin and Sinton (2007) demonstrated a more or less linear decrease in Mg# with spreading rate through the entire range. They found that lavas at the slowest-spreading ridges erupt roughly 20°C hotter than at the fastest-spreading ones. The greater extent of crystallization and lower eruption temperatures at fast-spreading ridges is likely a direct consequence of the crystallization occurring primarily in the shallow magma chambers beneath fast-spreading ridges.

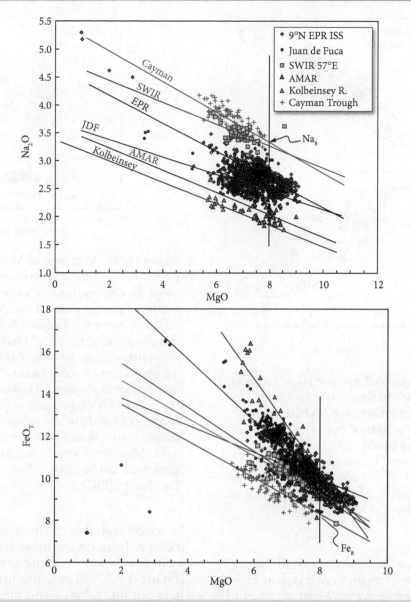

Figure 11.28 Na_2O and FeO vs. MgO in MORB from various regions of the mid-ocean ridge system plot along liquid lines of descent of varying slopes. $Fe_{8.0}$ and $Na_{8.0}$ are the value of Na_2O and FeO, respectively, of the lines at 8.0% MgO. EPR, Integrated Study Site (ISS) at 9°N on the East Pacific Rise; JDF, Juan de Fuca Ridge; SWIR, Southwest Indian Ridge; Kolbeinsey Ridge is north of Iceland, and the Cayman Trough is a small spreading center in the Caribbean. After White and Klein (2014).

The other factors that control the major element composition of MORB are the degree and depth of melting, which are in turn controlled by mantle temperature. The 1987 study of Klein and Langmuir elucidated how these factors affect both the thickness and composition of oceanic crust. However, to assess mantle temperature requires first correcting for the effects of fractional crystallization. Klein and Langmuir (1987) found that basalts in each

region they consider formed coherent trends in plots of oxides versus MgO concentrations (Figure 11.27). Klein and Langmuir assumed these trends represented liquid line of descents produced by fractional crystallization within crustal magma chambers. To compare compositions from different regions, they defined the parameter $Na_{8.0}$ as the value of Na_2O where the trends crossed 8 wt. % MgO (Figure 11.28). $Fe_{8.0}$ was defined in a similar way.

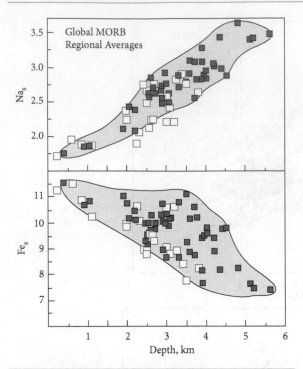

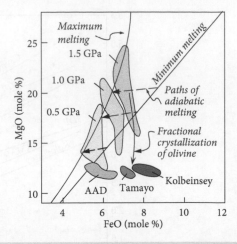

Figure 11.29 Regional average $Fe_{8.0}$ and $Na_{8.0}$ versus axial depth in the mid-ocean ridge system. Solid red squares are MORB from "normal" ridge segments; open squares are from ridges influenced by the Galapagos, Azores, Jan Mayen, Tristan, Iceland, and Bouvet hotspots. The shaded field encompasses normal ridge basalts. After Klein and Langmuir (1987) and White and Klein (2014).

Figure 11.30 Variation of MgO and FeO in partial melts of mantle peridotite. Grayed fields show the compositions of experimentally produced partial melts of peridotite at three different pressures. Colored fields show compositions of high MgO basalts from the Australian–Antarctic Discordance (AAD), the Tamayo Fracture Zone of the East Pacific Rise, and Kolbeinsey regions. Dashed arrows show the path of melt composition produced by melting of adiabatically rising mantle. The curved arrow shows how the composition of a 1.5 GPa melt will evolve due to fractional crystallization of olivine. After Klein and Langmuir (1987).

Corrected for fractional crystallization in this way, major elements correlate in revealing and predictable ways. Most significantly, regions with low mean $Na_{8.0}$ are also characterized by high mean $Fe_{8.0}$. Furthermore, these major element variations correlate with physical characteristics of the ridge axis from which they were recovered. Regional averages of $Na_{8.0}$ and $Fe_{8.0}$, for example, showed a positive and an inverse correlation, respectively, with the average ridge depth from which the lavas were recovered (Figure 11.29). $Na_{8.0}$ correlated inversely with seismically and geologically determined estimates of the crust thickness in each region.

These relationships can be understood as the interplay between the extent of melting and the pressure of melting. Sodium is moderately incompatible during melting of mantle minerals (D~0.02–0.03), and therefore will

be concentrated in the melt at small extents of melting. Iron, on the other hand, varies in the melt as a function of the pressure of melting (Figure 11.30). Thus, the inverse correlation between mean $Na_{8.0}$ and mean $Fe_{8.0}$ suggests that there is a positive correlation between the mean extent of melting and the mean pressure of melting.

Klein and Langmuir concluded that mantle temperature was the key factor in accounting for both depth of the ridge axis and the composition of melts erupted, because mantle temperature affects both degree of melting and the mean depth of melting. Shallow segments of the mid-ocean ridge system overlie relatively hot mantle. The hot mantle intersects the solidus at greater depth and ultimately melts to a greater degree (Figure 11.31). Hotter mantle is less dense and therefore more buoyant, so that ridges overlying hotter mantle will be more elevated. Cooler mantle will not begin to melt until it reaches

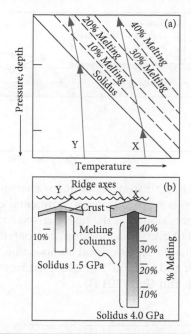

Figure 11.31 (a) Pressure–temperature relationship of adiabatically rising mantle undergoing melting. Hotter mantle (X) intersects the solidus at higher pressure and ultimately melts to a higher degree than cooler mantle (Y). The break in slope occurs because energy is consumed by melting (enthalpy of fusion). (b) Relationship between axial depth, crustal thickness, melting, and mantle temperature. Hotter mantle (X) maintains the ridge at higher elevation because of its buoyancy. It also has a deeper melt column and melts to a greater degree, producing thicker crust than cooler mantle (after Klein and Langmuir, 1989).

shallower depths, and the total extent of melting will be more limited. Klein and Langmuir concluded that a range in degree of melting of 8–20% and in mean pressure of melting of 0.5–1.6 GPa were required to produce the range in compositions observed. The hottest regions of the mantle occur near mantle plumes such as Iceland. Some of the coolest mantle is found at the Australian–Antarctic Discordance, a region where the ridge is particularly deep and isotope studies have suggested is a boundary between mantle convection cells. Overall, the data suggest a range in mantle temperature of some 250°C.

11.6.2 The continental crust

The continental crust is the part of the Earth that is most readily sampled and the part with which we are most familiar. It is, however, likely the most variable part of the Earth in every respect, including compositionally. It is the part of the Earth where geology reveals the planet's history. In this respect, the continents are arguably the most interesting part of the planet. We'll first consider the composition of the continental crust, then we'll see what geochemistry can reveal about its creation and evolution.

The continental crust is extremely heterogeneous, thus the task of estimating its overall composition is a difficult one. Furthermore, only the upper part of the continental crust is exposed to direct sampling: the deepest scientific borehole, drilled by the Russians in the Kola Peninsula, reached only 12 km. The average thickness of the continental crust is about 35 km so we have been able to directly sample only the upper third. Fortunately, however, tectonic and volcanic processes sometimes bring bits of the deep crust to the surface where it can be studied. Nevertheless, geochemists must rely heavily on inferences made from indirect observations to estimate the composition of the continental crust. Beginning with Frank Clarke (1924) and Victor Goldschmidt (1933), a number of such estimates of the composition of the continental crust have been made. These have become increasingly sophisticated with time. Among the most widely cited works are those of Taylor and McLennan (1985, 1995), Weaver and Tarney (1984), Wedepohl (1995) and Rudnick and Fountain (1995). These estimates are not entirely independent. For example, Weaver and Tarney (1984) and Rudnick and Fountain (1995) both rely in part on versions of Taylor and McLennan's studies. Taylor and McLennan in turn rely on the work of Shaw (1967) for many elements, as does Wedepohl (1995). In the following, we will focus particularly on the estimates of Taylor and McLennan (1985, 1995), and Wedepohl (1995), and the work of Rudnick and Gao (2014).

We can divide the problem of estimating crustal composition into two parts. The first is to estimate the composition of the upper, accessible parts of the crust. This is referred to as the *upper crust*. Direct observations provide the most important constraints on the composition of this part of the crust. The second problem is the composition of the deeper, less accessible part of the crust. For this part of the crust, indirect observations, particularly geophysical ones such as seismic velocity and heat flow, provide key constraints on composition. As we shall see, these observations indicate that the continental crust is compositionally stratified, with the lower part being distinctly more mafic – richer in Mg and Fe and poorer in SiO_2 and incompatible elements. Some workers divide the deep crust into a middle and lower crust, while others consider only a single entity that they refer to as the lower crust. The chemical contrasts between the upper and deep crust are much stronger in continents than in the ocean. We should emphasize, however, that any division of the continental crust into layers is done for convenience and is somewhat arbitrary. The continental crust does not appear to have a systematic layered structure that directly results from its creation in the way that the oceanic crust does. Instead, chemical zonation in the continents has evolved over time through a variety of processes.

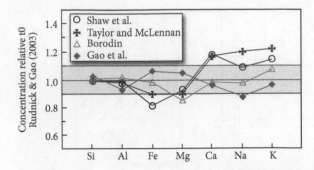

Figure 11.32 Estimates of major element concentrations in the upper continental crust by studies mentioned in the text, derived from analysis of crustal composites, compared with the estimated upper crustal composition of Rudnick and Gao (2014). Modified from Rudnick and Gao (2014).

11.6.2.1 Composition of the upper continental crust

Historically, two approaches to estimating the composition of the upper continental crust have been used. The first approach, pioneered by F.W. Clarke in 1889, is to average analyses of samples taken over a large area. An alternative is to mix sample powders to form *composites* of various rock types and thus reduce the number of analyses to be made (e.g., Shaw et al., 1986; Wedepohl, 1995; Borodin, 1998; Gao et al., 1988). Studies of these kinds consistently produce an average upper crustal composition similar to that of granodiorite, with the concentrations of major oxides agreeing within 20% and most within 10% (Figure 11.32). This is encouraging since granodiorite is the most common igneous rock in the crust. Such estimates also tend to produce relatively similar average concentrations for minor and trace elements. Most modern estimates of upper crustal major element and soluble trace element concentrations are based on this approach.

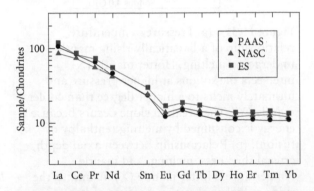

Figure 11.33 Rare earth patterns of Post-Archean Australian Shale (PAAS) composite, the North American Shale Composite (NASC) and the European Shale (EC) composite. After Taylor and McLennan (1985).

The second method is to let the Earth make the composites for us. A couple of kinds of such materials are available. Goldschmidt (1933) suggested the use of glacial clays in meltwater lakes adjacent to the Pleistocene ice front. An alternative but similar approach is to use *loess*, which is fine-grained aeolian material of Pleistocene age. The most readily available of this kind of natural composite, however, is simply sediments. One advantage of sediments over glacial material is that whereas most glacial deposits are of Pleistocene age (but there are various glacial deposits of ages ranging up to 2.3 Ga), sediments of all ages are available so that secular variations in crustal composition can be determined.

The advantages of using geologic composites should be obvious, but there are disadvantages as well. The primary problem is that chemical fractionations are involved in producing sediments from their parents. Weathering of rock typically produces three fractions: sands consisting of resistant minerals, clays, and a solution. These products are transported with varying degrees of efficiency away from the site of production. Since elements tend to be concentrated in one of these three fractions, none of the fractions will have a composition representative of the parent rock. Because it is produced primarily by physical, rather than chemical, action, glacial loess is less susceptible to this kind of chemical fractionation, although some fractionation nevertheless occurs. For example, loess is enriched in SiO_2, Hf, and Zr as a consequence of its enrichment in mechanically and chemically stable minerals, such as quartz and zircon. That, in turn, results from clays being carried further from their site of origin by wind and water. Loess is also depleted in Na, Mg, and Ca, reflecting loss by leaching.

Numerous studies have shown that when rock weathers to produce a sediment, the rare earth pattern of the parent is usually preserved in the sediment. This is because the rare earths are concentrated in the clay fraction, which ultimately forms shales. Other Group 3 elements (Sc and Y), as well as Th, behave similarly to the rare earths during weathering. Furthermore, rare earth patterns are remarkably similar in different shales, suggesting shales are indeed good averages of crustal composition. This is illustrated in Figure 11.33, which compares three shale composites from three continents. Because of these properties of the rare earths, S.R. Taylor and colleagues at the Australian National University used them as a point of departure for estimating the composition of the upper continental crust.

Most recent estimates of crustal composition are based on a combination of both approaches, together with assumptions about the ratios between elements. The most recent comprehensive study of crustal composition is that of Rudnick and Gao (2014). Their estimate of upper crustal composition is compared with earlier ones in Table 11.8. The estimates are broadly similar, and indicate an upper crust of *granodioritic* or *tonalitic* composition (although there are a few elements, such as Cu, where there is considerable disagreement). As we noted above, however, that Rudnick and Gao (2014) relied heavily on earlier work.

11.6.2.2   Composition of the middle and lower continental crust

Rocks from the middle and lower crust, typically in *amphibolite* and *granulite* metamorphic facies, are sometimes exposed at the surface by tectonic processes and hence can provide insights into the nature of this part of crust. Amphibolites are, as their name implies, metamorphic rocks that are relatively rich in amphibole, a mineral that contains water in its structure, but less water than mica-bearing rocks, such as granites and rocks metamorphosed at lower temperature. Granulites, on the other hand, are anhydrous, with pyroxene replacing amphibole and biotite. Middle crustal cross-sections of amphibolite- to granulite-facies rocks contain a wide variety of lithologies, including metasedimentary rocks, but they are dominated by igneous and metamorphic rocks of dioritic, granodioritic, and granitic composition.

However, these *granulite terranes* have often been subjected to retrograde metamorphism (metamorphism occurring while temperatures and pressure decrease), which compromises their value. Furthermore, questions have been raised as to how typical they are of lower continental crust. These questions arise because granulite terranes are generally significantly less mafic than xenoliths from the lower crust. Xenoliths perhaps provide a better direct sample of the lower crust, but they are rare. The point is, any estimate of the composition of the middle and lower crust will have to depend on indirect inference and geophysical constraints as well as analysis of middle and lower crustal samples. There are two principal geophysical constraints:

1.   Heat flow in the continental crust. A portion of the heat flowing out of the crust is produced by radioactive decay of K, U, and Th within the crust (other radioactive elements do not contribute significantly

Table 11.8 Composition of the upper continental crust.

Oxide (wt %)	T & M	Wedepohl	R & G	Normative mineralogy (T & M)	
SiO_2	66.0	64.9	66.6	Quartz	15.7
TiO_2	0.5	0.52	0.64	Orthoclase	20.1
Al_2O_3	15.2	14.6	15.1	Albite	13.6
FeO	4.5	3.97	4.09	Diopside	6.1
MnO	0.07	0.07	0.07	Hypersthene	9.9
MgO	2.2	2.24	2.20	Ilmenite	0.95
CaO	4.2	4.12	4.24		
Na_2O	3.9	3.46	3.56		
K_2O	3.4	4.04	3.19		
P_2O_5	0.20	0.15	0.15		

	T & M	Wedepohl	R & G		T & M	Wedepohl	R & G
Li	20	22	21	Sb	0.2	0.31	0.4
Be	3	3.1	2.1	Te			
B	15	17	17	I		1.4	1.4
C		3240		Cs	3.7	5.8	4.9
N		83	83	Ba	550	668	624
F		611	557	La	30	32.3	31
S		953	621	Ce	64	65.7	63
Cl		640	370	Pr	7.1	6.3	7.1
Sc	11	7	14	Nd	26	25.9	27
Ti	3000	3117		Sm	4.5	4.7	4.7
V	60	53	97	Eu	0.88	0.95	1.0
Cr	35	35	92	Gd	3.8	2.8	4.0
Co	10	11.6	17	Tb	0.64	0.5	0.7
Ni	20	18.6	47	Dy	3.5	2.9	3.9
Cu	25	14.3	28	Ho	0.8	0.62	0.83
Zn	71	52	67	Er	2.3		2.3
Ga	17	14	17.5	Tm	0.33		0.30
Ge	1.6	1.4	1.4	Yb	2.2	2.0	2.0
As	1.5	2	4.8	Lu	0.32	0.27	0.31
Se	0.05	0.083	0.09	Hf	5.8	5.8	5.3
Br		1.6	1.6	Ta	2.2	1.5	0.9
Rb	112	110	82	W	2	1.4	1.9
Sr	350	316	320	Re ppb	0.4		0.198
Y	22	20.7	21	Os ppb	0.05		0.031
Zr	190	237	193	Ir ppb	0.02		0.022
Nb	25	26	12	Pt ppb			0.5
Mo	1.5	1.4	1.1	Au ppb	1.8		1.5
Ru ppb			0.34	Hg ppb		56	50
Pd ppb	0.5		0.52	Tl	0.75	0.75	0.9
Ag ppb	50	55	53	Pb	20	17	17
Cd ppb	98	102	90	Bi	0.12	0.16	0.16
In ppb	50	61	56	Th	10.7	10.3	10.5
Sn	5.5	2.5	2.1	U	2.8	2.5	2.7

Concentrations in ppm except where noted. T & M: Taylor and McLennan (1985, 1995); Wedepohl: Wedepohl (1995); R & G: Rudnick and Gao (2014).

to heat generation because of their long half-lives and low abundances). The concentrations of these elements can be related to rock type, as indicated in Table 11.9.

2. Seismic velocities in the continental crust. Seismic velocities depend on density, compressibility and the shear modulus (eqns. 11.1 and 11.2), which can in turn be related to composition.

Both tell us something of first-order importance about the nature of the continental crust: it is vertically zoned, becoming more mafic (i.e., richer in Fe and Mg and poorer in Si and incompatible elements, including K, U, and Th) with depth. Let's consider them in greater detail.

The average heat flow of the continents is about $60 \, mW/m^2$. This heat has two components: heat conducted out of the mantle, and heat generated by radioactive decay within the continents. The concentrations of K, U, and Th observed at the surface of the crust would produce more heat than is observed to be leaving the continental crust, if these concentrations were uniform through the crust. Thus, the concentrations of these elements must decrease at depth. The problem is complicated, however, by variations in the "mantle" heat flow, which averages about $20 \, mW/m^2$. Heat flow varies significantly with tectonic age. If, as we believe, the continental crust is created by magmatism, it will be initially hot and then cool over time. Subsequent episodes of magmatism may also heat the crust. In addition, variation in mantle heat flow can result from different thicknesses of the lithosphere (Vitorello and Pollack, 1980; Nyblade and Pollack, 1993). The lithosphere is a conductive boundary layer, so that the thicker the lithosphere, the lower the mantle heat flow out of the top of it. Nyblade and Pollack (1993) have argued that regions of old Archean crust are underlain by particularly thick mantle lithosphere, an argument supported by geochronological and thermobarometric studies of mantle xenoliths from these regions. Despite these complexities, heat flow in the crust suggests the middle and lower crust contains significantly less K, U, and Th than the upper crust. This, in turn, suggests the deep continental crust is more mafic than the upper crust.

Seismic velocity profiles vary widely from place to place, as does crustal thickness, but in general P-wave velocities increase with depth in the crust from about 6 km/s in the upper crust to about 7 km/sec in the lower crust. Rudnick and Fountain (1995) examined a global database of seismic cross-sections and found that they can be divided into nine classes, which are illustrated in Figure 11.34. One must next relate seismic velocity to composition by making measurements of seismic velocity in the laboratory on samples of known composition. For example, Figure 11.35 shows the relationship between SiO_2 and seismic velocity in a variety of rock types. Recalling that seismic wave speed is inversely related to density (eqns. 11.1 and 11.2), we might have expected mafic rocks, which are poorer in SiO_2 and richer in Mg and Fe and are denser (Table 11.9), to have slower seismic velocities. However, the lower compressibility of mafic rocks results in seismic waves traveling faster through them. Seismic velocities are thus consistent with the heat flow evidence that the middle and lower crust is more mafic than the upper crust.

To produce an estimate of crustal composition, Rudnick and Fountain assigned

Table 11.9 U, Th, and K concentrations and heat production in various rock types. From Pollack (1982).

Igneous rock type	U (ppm)	Th (ppm)	K (%)	Th/U	K/U	Density g/cm³	Heat production $10^{-6} \, W/m^{-3}$
Granite/rhyolite	3.9	16.0	3.6	4.1	0.9×10^4	2.67	2.5
Granodiorite/dacite	2.3	9.0	2.6	3.9	1.1×10^4	2.72	1.5
Diorite/andesite	1.7	7.0	1.1	4.1	0.7×10^4	2.82	1.1
Gabbro/basalt	0.5	1.6	0.4	3.2	0.8×10^4	2.98	0.3
Peridotite	0.02	0.06	0.006	3.0	0.3×10^4	3.28	0.01
Continental crust	1.25	4.8	1.25	3.8	1.0×10^4	—	0.8

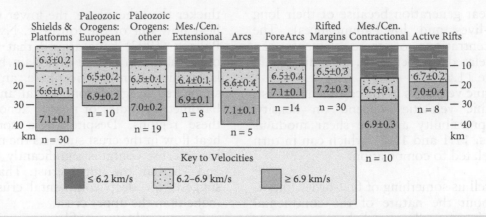

Figure 11.34   Seismic velocity structure of the continental crust, illustrating its three-layered nature. Velocity structure falls into nine types. The number of profiles used to construct each type is shown below each type. After Rudnick and Fountain (1995).

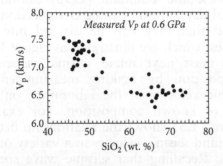

Figure 11.35   Correlation between measured seismic velocity (v_P) and SiO_2 concentration. After Rudnick and Fountain (1995).

an average lithology to the seismic sections shown in Figure 11.34. They then assigned a composition to each lithology using a database of the composition of lower crustal xenoliths. Then by estimating the areal extent of each type of crustal section, and averaging analyses of middle and lower crustal rocks and xenoliths, they produced compositional estimates. These estimates were subsequently updated by Rudnick and Gao (2014) and are listed Table 11.10. This table shows that the composition of the lower crust corresponds to that of tholeiitic basalt; in metamorphic terminology it would be a mafic granulite. The composition of the middle crust corresponds to that of an andesite. At the prevailing pressures and temperatures this rock would be an amphibolite, consisting mainly of amphibole and plagioclase.

Wedepohl (1995) used the European Geo-traverse as a model of the seismic structure of the crust. This seismic cross-section runs from northern Scandinavia to Tunisia and crosses a great variety of tectonic provinces, ranging from the Archean Fennoscandian Shield to the young fold belts of the Alpine orogen. He assigned three lithologies to three ranges of seismic velocities: sediments, granites, and gneisses ($V_P < 6.5$ km/s) corresponding to the upper crust, felsic granulites ($6.5 < V_P < 6.9$ km/s), and mafic granulites ($6.9 < V_P < 7.5$ km/s). He used a database of compositions of felsic and mafic granulites from both xenoliths and exposed terranes to calculate an average composition for each of the latter two. He then computed a lower crustal composition by weighting felsic and mafic granulites in the proportions that their characteristic seismic velocities were observed in the European Geotraverse. His estimate of the composition of the lower crust is also listed in Table 11.10.

Rare earth patterns of upper, middle, and lower crust as estimated by Rudnick and Gao (2014) are compared in Figure 11.36. The negative Eu anomaly in the upper crust and slight positive anomalies in the middle and lower crust (such positive anomalies are typical of many granulites) are an interesting feature of these patterns. Eu is strongly held in plagioclase (Chapter 7). The presence of plagioclase in the melting residue would result in melts having a negative Eu anomaly and the residue having a positive one. Thus,

Table 11.10   Composition of the middle and lower continental crust.

Major oxides, %			
	R & G middle	Wedepohl lower	R & G lower
SiO$_2$	63.5	59.0	53.4
TiO$_2$	069	0.85	0.82
Al$_2$O$_3$	15.0	15.8	16.9
FeO	6.02	7.47	8.57
MnO	0.10	0.12	0.10
MgO	3.59	5.32	7.24
CaO	5.25	6.92	9.59
Na$_2$O	3.39	2.91	2.65
K$_2$O	2.30	1.61	0.61
P$_2$O$_5$	0.15	0.20	0.10

Trace elements in ppm except as noted.							
	R & G middle	Wedepohl lower	R & G lower		R & G middle	Wedepohl lower	R & G lower
Li	12	13	13	Sb	0.28	0.3	0.1
Be	2.29	1.7	1.4	I		0.14	0.1
B	17	5	2	Cs	2.2	0.8	0.3
C		588		Ba	532	568	259
N		34	34	La	24	26.8	8
F	524	429	570	Ce	53	53.1	20
S	20	408	345	Pr	5.8	7.4	2.4
Cl	182	278	250	Nd	25	28.1	11
Sc	19	25.3	31	Sm	4.6	6.0	2.8
V	107	149	196	Eu	1.4	1.6	1.1
Cr	76	228	215	Gd	4.0	5.4	3.1
Co	22	38	38	Tb	0.7	0.81	0.48
Ni	33.5	99	88	Dy	3.8	4.7	3.1
Cu	26	37.4	26	Ho	0.82	0.99	0.68
Zn	69.5	79	78	Er	2.3		1.9
Ga	17.5	17	13	Tm	0.32	0.81	0.24
Ge	1.13	1.4	1.3	Yb	2.2	2.5	1.5
As	3.1	1.3	0.2	Lu	0.4	0.43	0.25
Se	0.064	0.17	0.2	Hf	4.4	4.0	1.9
Br		0.28	0.3	Ta	0.6	0.84	0.6
Rb	65	41	11	W	0.60	0.6	0.6
Sr	282	352	348	Re, ppb			0.18
Y	20	27.2	16	Os, ppb			0.05
Zr	149	165	68	Ir, ppb			0.05
Nb	10	11.3	5	Pt, ppb	0.85		2.7
Mo	0.60	0.6	0.6	Au, ppb	0.66		1.6
Ru, ppb			0.75	Hg, ppb	0.0079	0.021	0.014
Pd ppb	0.76		2.8	Tl	0.27	0.26	0.32
Ag ppb	48	80	65	Pb	15.2	12.5	4
Cd, ppb	0.061	0.101	0.1	Bi	0.17	0.037	0.2
In, ppb		0.052	0.05	Th	6.5	6.6	1.2
Sn	1.3	2.1	1.7	U	1.3	0.93	0.2

R & G, Rudnick and Gao (2003); Wedepohl, Wedepohl (1995).

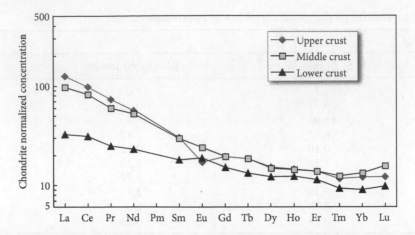

Figure 11.36 Comparison of chondrite-normalized rare earth patterns in upper, middle, and lower crust. Data of Rudnick and Gao (2014).

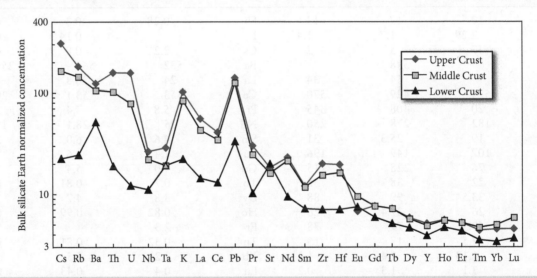

Figure 11.37 Enrichment of incompatible elements relative to bulk silicate Earth (Table 11.3) in the upper, middle, and lower crust. Data from Rudnick and Gao (2014).

these anomalies suggest that the crust has at least partially differentiated to form distinct layers through partial melting, with granitic melts forming the upper crust and granulitic residues of gabbroic composition forming the lower crust. The actual evolution of the deep crust is undoubtedly more complex, however. Intrusion and ponding of mantle-derived magmas of basaltic composition is likely to be another important reason for the mafic composition of the lower crust.

Figure 11.37 is a spider diagram comparing incompatible element abundances in the upper, middle, and lower continental crust.

The lower and middle crusts are less enriched in incompatible elements than the upper crust. This is also consistent with the idea that magmatic processes have been important in creating the compositional layering observed in the crust.

11.6.2.3 Composition of the total continental crust

The approach used by most workers to estimate the composition of the total continental crust is simply to calculate a weighted average of crustal layers considered above. This was done, for example, by Rudnick and Fountain (1995) and Weaver and Tarney (1984), both of whom divided the crust into an upper, lower, and middle section, and relied on Taylor and McLennan's upper crustal estimate. Weaver and Tarney (1984) used average Lewisian* amphibolite as their middle crust composition and average Lewisian granulite as their lower crust composition. Shaw et al. (1986) and Wedepohl (1995) used a similar approach but divided the crust only into upper and lower parts. An important step in this approach is estimating the thickness of the various sections. Here, seismological constraints again come into play.

Taylor and McLennan (1985, 1995) used an entirely different approach to estimating total crustal composition, one based on the "andesite model" of Taylor (1967). Taylor (1967) noted the role played by subduction-related volcanism in creation of the continental crust and assumed that, on average, the crust consisted of island arc andesite. Thus average island arc andesite was used as the estimated composition of the continental crust. This approach was modified in subsequent work, as Taylor concluded that while post-Archean crust was created at subduction zones, Archean crust was not, and it is compositionally different. Taylor and McLennan (1985) essentially modified the Taylor (1967) andesite model for their estimate of Archean crustal composition.

Estimates of the major element composition of the continental crust by Weaver and Tarney (1984), Shaw et al. (1986), Taylor and McLennan (1995), Wedepohl (1995), and Rudnick and Gao (2014) are given in Table 11.11. Also listed are estimates of trace element concentrations by Taylor and McLennan (1995), Rudnick and Gao (2014), and Wedepohl (1995). The ranges of estimates for SiO_2 and Al_2O_3 in Table 11.11 vary by about 10% and 8% respectively; the range in Mg# (52 to 57) is similarly only about 10%. Interestingly, earlier estimates of crustal SiO_2 and Al_2O_3, going back to Goldschmidt (1933), also fall within this range. Thus, we can conclude with some confidence that the continental crust on the whole is similar to that of diorite (or andesite).

The details of the composition of the crust are less certain, however. Ranges for the other oxides are substantially larger: 75% for FeO, 68% for MgO, and 100% for MnO. Of these estimates, the composition of Taylor and McLennan is the most mafic, and that of Weaver and Tarney the least mafic (ranges for FeO and MnO decrease to 30% and 21%, respectively, if the estimates of Taylor and McLennan are excluded). Wedepohl's (1995) estimated crustal composition is significantly more enriched in incompatible elements than that of Taylor and McLennan (1985, 1995); Rudnick and Gao's (2014) estimated concentrations of incompatible elements generally fall between the two.

Figure 11.38 compares the rare earth element patterns in the total continental and oceanic crusts. Both are enriched in all rare earths relative to bulk silicate Earth as well as chondrites, but whereas the continental crust is enriched in the light rare earths relative to the heavy rare earths, the opposite is true of the oceanic crust. The light rare earth enrichment of the continental crust is clear evidence that it, like the oceanic crust, originated as a partial melt of the mantle. The oceanic crust is produced through comparatively large extents of melting of mantle that has already been depleted in incompatible elements through previous episodes of melting. The evolution of the continental crust has been more complex. The strong light rare earth enrichment, however, suggests that relatively small degree melts have been important in its evolution. Figure 11.39 compares the incompatible enrichment of the oceanic and continental crusts. Although the continental crust is enriched in incompatible elements overall, it is significantly less enriched in Nb and Ta and elements of similar incompatibility such as U and K; the oceanic crust exhibits a slight excess enrichment in these elements. It is also much more strongly enriched in Pb than either Ce or Sr, which are similarly incompatible. This negative Nb-Ta "anomaly" and positive Pb anomaly are characteristics of continental crust. They provide important hints as to the mechanism(s) by which continental crust has been created, a topic we consider in the next section.

---

* The Lewisian, which crops out in northwest Scotland, is perhaps the classic exposure of lower crust.

Table 11.11   Composition of the continental crust.

Major oxides, wt. %					
	R & G	T & M	W & T	We	Shaw
SiO_2	60.6	57.3	63.2	61.5	63.2
TiO_2	0.7	0.9	0.6	0.68	0.7
Al_2O_3	15.9	15.9	16.1	15.1	14.8
FeO	6.7	9.1	4.9	5.67	5.60
MnO	0.1	0.18	0.08	0.10	0.09
MgO	4.7	5.3	2.8	3.7	3.15
CaO	6.4	7.4	4.7	5.5	4.66
Na_2O	3.1	3.1	4.2	3.2	3.29
K_2O	1.8	1.1	2.1	2.4	2.34
P_2O_5	0.1		0.19	0.18	0.14

Trace elements (in ppm unless otherwise noted)							
	R & G	T & M	We		R & G	T & M	We
Li	16	13	18	Sb	0.2	0.2	0.3
Be	1.9	1.5	2.4	Te, ppb			5
B	11	10	11	I, ppb	700		800
C			1990	Cs	2	1	3.4
N	56		60	Ba	456	250	584
F	563		525	La	20	16	30
S	404		697	Ce	43	33	60
Cl	244		472	Pr	4.9	3.9	6.7
Sc	21.9	30	16	Nd	20	16	27
V	138	230	98	Sm	3.9	3.5	5.3
Cr	135	185	126	Eu	1.1	1.1	1.3
Co	26.6	29	24	Gd	3.7	3.3	4.0
Ni	59	105	56	Tb	0.6	0.6	.65
Cu	27	75	25	Dy	3.6	3.7	3.8
Zn	72	80	65	Ho	0.77	0.78	0.8
Ga	16	18	15	Er	2.1	2.2	2.1
Ge	1.3	1.6	1.4	Tm	0.28	0.32	0.3
As	2.5	1	1.7	Yb	1.9	2.2	2.0
Se	0.13	0.05	0.12	Lu	0.30	0.3	0.35
Br	0.88		1.0	Hf	3.7	3	4.9
Rb	49	32	78	Ta	0.7	1	1.1
Sr	320	260	333	W	1	1	1.0
Y	19	20	24	Re, ppb	0.188	0.4	0.4
Zr	132	100	203	Os, ppb	0.041	0.005	0.05
Nb	8	11	19	Ir, ppb	0.037	0.1	0.05
Mo	0.8	1	1.1	Pt, ppb	1.5		0.4
Ru, ppb	0.6		0.1	Au, ppb	1.3	3	2.5
Rh, ppb			0.06	Hg, ppb	30		40
Pd, ppb	1.5	1	0.4	Tl, ppb	500	360	520
Ag, ppb	56	80	70	Pb	11	8	14.8
Cd, ppb	80	98	100	Bi	0.18	0.06	0.085
In, ppb	52	50	50	Th	5.6	3.5	8.5
Sn	1.7	2.5	2.3	U	1.3	0.91	1.7

R & G, Rudnick and Gao (2003); T & M, Taylor and McLennan (1985, 1995); We, Wedepohl (1995); Shaw, Shaw et al. (1986); W & T, Weaver and Tarney (1984).

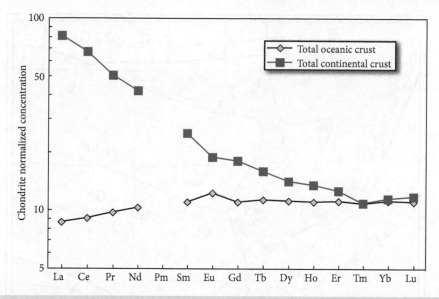

Figure 11.38   Comparison of rare earth patterns of the total oceanic (estimate of White and Klein listed in Table 11.7) and continental crust (estimate of Rudnick and Gao listed in Table 11.11).

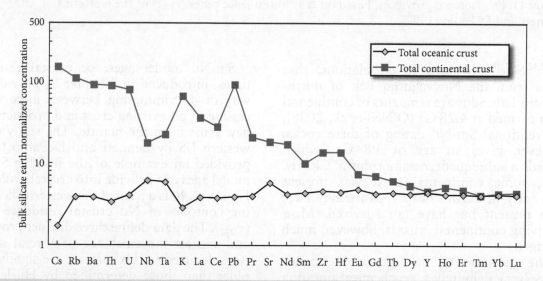

Figure 11.39   Comparison of the incompatible element enrichments of the continental (estimate of Rudnick and Gao listed in Table 11.11) and oceanic crusts (estimate of White and Klein listed in Table 11.7). Negative Nb-Ta and positive Pb anomalies, as well as the overall incompatible element enrichment, are significant characteristics of continental crust.

11.6.3   Growth of the continental crust

The oceanic crust is ephemeral; its mean age is about 60 Ma and, with the exception of possible Permian age crust preserved in the Eastern Mediterranean, it is nowhere older than about 167 Ma. In contrast, the continental crust is much older. The oldest crustal rocks, Acasta gneisses in Canada's Northwest Territory, give Sm-Nd ages of 4.02 Ga.

Individual zircons within these gneisses are as old as 4.2 Ga, and zircons separated from a sandstone metamorphosed about ~3 Ga ago in the Jack Hills of western Australia have U-Pb ages as old as 4.0–4.4 Ga. The chemical and O isotopic compositions of the Jack Hills zircons clearly show they were once part of "continental"-type rocks such as granites or granodiorites. There are also hints, based

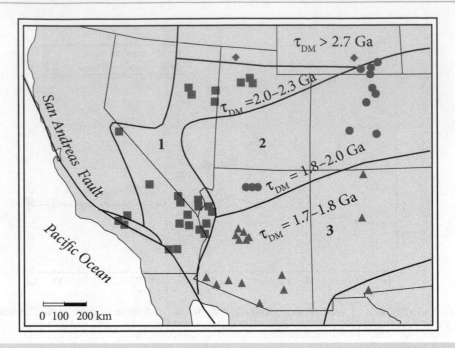

Figure 11.40   Isotopic provinces, based on crustal residence times (τ_{DM}) of the western US. After Bennett and DePaolo (1987).

on $^{142}Nd/^{144}Nd^*$–Sm/Nd correlations, that rocks from the Nuvvagittuq Belt of northwestern Labrador are remnants of continental crust formed at 4.28 Ga (O'Neil et al., 2008). Conventional Sm-Nd dating of these rocks, however, gives an age of 3.84 Ga, which records a subsequent metamorphism. Clearly, then, some continental crust was present very early in Earth's history. More may have been present but have not survived. Most surviving continental crust is, however, much younger.

The question of the age of the continents (very definitely a geochemical question because radiometric dating is a geochemical technique) is a complex one. Because metamorphism, remelting, and so on, can reset radiometric ages, the radiometric age of a rock does not necessarily correspond to the time the material that makes up that rock first became part of the continental crust. Indeed, early efforts to systematically determine the ages of continents and rates of continental growth, such as that of Hurley and Rand (1969), substantially underestimated continental ages because they relied heavily on Rb-Sr ages, which are readily reset during metamorphism.

Sm-Nd model ages, or crustal residence times, introduced in Chapter 8, provide one way of discriminating between mere reprocessing of preexisting crust and production of new crust from the mantle. The study of the western US by Bennett and DePaolo (1987) provided an example of the insights Sm-Nd model ages can provide into crustal evolution. Figure 11.40 is a map of the western US showing contours of Nd crustal residence times (τ_{DM}). The data define three distinct provinces and suggest the existence of several others. Ages deduced in this manner are significantly older than those determined by Hurley and Rand (1969) using Rb-Sr dating. Figure 11.41 shows the initial ε_{Nd} values of granites from the three numbered provinces of Figure 11.40 plotted as a function of their crystallization age. Although the crystallization ages are often much younger, crustal residence times indicate Provinces 1–3 all formed between 1.65 and 1.8 Ga. Only in Province 3 do we find rocks, tholeiitic, and calc-alkaline greenstones, whose Sm-Nd crustal residence age is equal to their crystallization ages. From this we can conclude that only Province 3 was a completely new addition to the crustal

---

* Recall that ^{142}Nd is the daughter of the extinct radionuclide ^{146}Sm, whose half-life is 103 Ma.

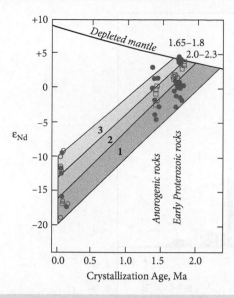

Figure 11.41 Initial ε_{Nd} as a function of crystallization age for igneous rocks of the western US. Groupings 1, 2, and 3 refer to provinces shown in Figure 11.39. After Bennett and DePaolo (1987).

mass at that time. In the other regions, the oldest rocks have initial ε_{Nd} values that plot below the depleted mantle evolution curve. This suggests that mantle melts mixed with preexisting crust, perhaps by assimilating crust (which would have a lower melting temperature) as they ascended through it. Alternatively, preexisting continental crust might have been carried into the mantle by sediment subduction or subduction erosion (topics we consider below) and mixed with mantle before melting.

Throughout the region there were subsequent episodes of magmatism, including one in the Tertiary. However, the initial ε_{Nd} lies along the same growth trajectory as the older rocks. This suggests that these subsequent igneous events involved remelting of existing crust with little or no new material from the mantle.

As we found in Chapter 8, zircons are, in a sense, the ultimate natural clock/time capsule. They readily survive transport in rivers and streams and, using the approach described in Section 8.4.6 it is possible to determine crystallization ages even when they have been partially reset during metamorphism. Thus zircon dating has become the "gold standard" for determining rates of continental growth.

Figure 11.42 shows the distribution of ages of over 410,000 detrital zircons compiled by Puetz et al. (2017). The distribution is clearly nonuniform with peaks at 2.7, 1.9, and 1.2 Ga, 0.6, and 0.25 Ga. Condie (1998) first pointed out the episodicity seen in this plot with far less complete data and pointed out that the peaks coincide with periods of supercontinent assembly.

The peaks in zircon numbers are superimposed on a general decrease with age in the number of zircons, reflects the limited survival of sedimentary sequences and the continental crust itself. As we discussed in Section 11.5.3.1, we know a significant amount of crust has been subducted into the mantle and hence continental crust can be destroyed. These factors result in the bias toward young zircon ages seen in Figure 11.42. The lower graph shows the average initial ε_{Hf} and $\delta^{18}O$ measured in zircons as a function of age (ε_{Hf} and $\delta^{18}O$ are available for only a fraction of the dated zircons). Low ε_{Hf} and high $\delta^{18}O$ indicate the zircon crystallized from a magma containing a large crustal component, while high ε_{Hf} and low $\delta^{18}O$ indicate a large mantle component and hence that the magma was a new addition to continental crust. Although the zircons record variable proportions of crust and mantle in the magmas from which they crystalize, recycled crust is an important component in most of these. Also, zircons are an imperfect recorder of crustal growth, as they are often not present in mafic igneous rocks and mantle melts are almost always mafic in nature.

One of the most enduring debates in geology is the question of how crustal mass has changed over time and what the *net* rate of crustal growth has been. Armstrong (1968, 1981) argued that despite evidence of new additions to crust throughout geologic time, the mass or volume of continents has remained essentially unchanged since 4 Ga because the rate of destruction of continental crust has matched the rate of production. While this perhaps represents an extreme view of crustal volume through time, there is widespread recognition that crust has been destroyed as well as produced through geologic time. Continental crust can be destroyed through weathering, erosion transport, and ultimately subduction of sediment, subduction erosion, and lower crustal foundering. There is good

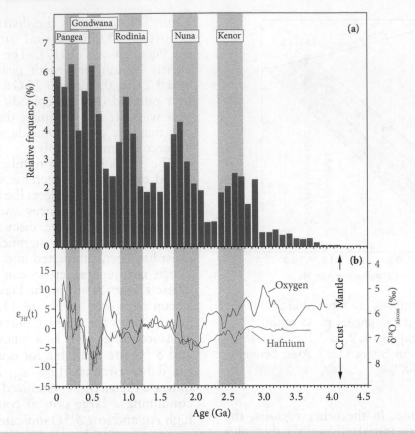

Figure 11.42   (a) Histogram of more than 400,000 U-Pb zircon ages. Data from the compilation of Puetz et al. (2011). (b) Averaged $\delta^{18}O$ and initial ε_{Hf} in zircons. Modified from Hawkesworth (2016).

evidence that all three mechanisms operate, and some have argued that the rates of the first two processes are sufficient to balance rates of new crustal addition and, therefore, produce a steady-state crustal volume, just as Armstrong proposed (e.g., Stern and Scholl, 2010). However, no consensus has formed around these views and the question of the net change in crust mass is still debated.

Because we must take account of how much crust is lost, the zircon age distribution in Figure 11.42 does not directly translate into crustal volumes (e.g., Hawkesworth et al., 2010; Parman, 2015). Based on a smaller (but nevertheless large) set of zircon ages, Condie and Aster (2010) derived the estimate of the fraction of juvenile crust preserved shown in Figure 11.43. They estimate only about 6%, of the surviving crust is older than 3 Ga, that about a third of the extant crust formed in the Archean, roughly 20% in the Proterozoic and only 14% during the last 400 Ma. Declining rates of magmatism and tectonism should, in

any case, come as no surprise since radioactive decay and original heat provide the energy for those processes.

While there is widespread agreement that the uneven distribution of crustal ages is real, there is some debate as to its cause. One interpretation is that they result from peaks of pulses of magmatic activity associated with mantle plumes (Albarède, 1998; Arndt and Davaille, 2013; Condie, 1998). An alternative is that they represent preferential preservation of juvenile crust during supercontinent assembly (e.g., Condie and Aster, 2010; Hawkesworth et al., 2010). Hawkesworth et al. (2016, 2019) point out that although large volumes of magma are generated in subduction zones, the crust generated this way is more likely to be destroyed through erosion and subduction, subduction erosion, and continental foundering. They argue that while less magma is generated in continental collisions, the likelihood of preservation of this crust is higher. In other words, they

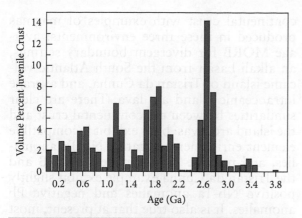

Figure 11.43 Estimated volume of juvenile crustal preserved as a function of geologic age, estimated from a database of 40,000 U-Pb ages of granitoid and detrital zircons. After Condie and Aster (2010).

argue that these peaks do not record crustal production rates, merely preservation during supercontinent assembly. Condie et al. (2016), however, point out that the peaks do not coincide exactly with supercontinent assembly, and instead argue that crust is more likely to be preserved in the accretionary phases preceding supercontinent assembly and that therefore the zircon age peaks do reflect crustal production rates. They agree, however, that a significant fraction of the Earth's early crust has been lost.

Archean continental crust is distinct in many ways from post-Archean crust, including compositionally, as Taylor and McClellan (1985) emphasized. Some aspects of this are most easily documented in the work of Keller and Schoene (2012), who analyzed a petrologic database of 70,000 analyses (Figure 11.44). Post-Archean crust is less mafic, more incompatible element-enriched and appears to have been produced by smaller extents of melting of the mantle, with the transition occurring around 2.5 Ga. Hawkesworth et al. (2016), among many others, also infer that the Archean crust was more mafic but infer the transition from relatively thin, mafic crust to thicker, more silicic crust occurred earlier, around 3 Ga.

These differences, for example the extent of melting (Figure 11.44), are consistent with a hotter Archean mantle (e.g., Herzberg et al., 2010). But does it indicate a change in the way crust was created and/or a more fundamental

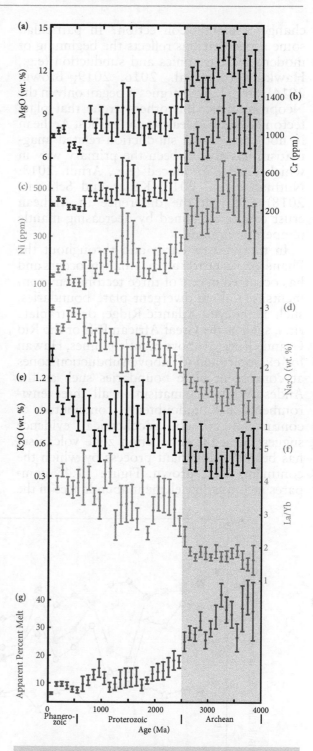

Figure 11.44 Statistical distribution of the chemistry of continental igneous rocks as a function of age binned in 100 Ma increments. Lowest graph shows inferred percent melting from major element chemistry. From Keller and Schoene (2012).

change is mantle convection? In particular, some argue that this reflects the beginning of modern plate tectonics and subduction (e.g., Hawkesworth et al., 2016, 2019; Brown, 2014); Stern (2008) argues it began only in the Neoproterozoic. But others argue that plate tectonics began at least in the earliest Archean if not earlier and subduction-related magmatism has always been the primary way in which crust is produced (e.g., Arndt, 2013; Nutman et al., 2015; Keller and Schoene, 2018) and that the difference in Archean crust can be explained by decreasing mantle temperatures alone.

In the modern Earth and throughout the Phanerozoic, nearly all magmatism occurs and has occurred in one of three tectonic environments: (1) along divergent plate boundaries, such as the Mid-Atlantic Ridge, or intraplate rifts, such as the Great African Rift or the Rio Grande Rift; (2) above mantle plumes, Hawaii for example; and (3) above subduction zones at convergent plate boundaries such as the Andes. While magmatism in all three environments has undoubtedly contributed to continental growth, geochemical evidence suggests that the subduction zone volcanism has been the dominant process by which the continents have grown. Figure 11.45 compares incompatible element abundances in the

continental crust with examples of magmas produced in these three environments: average MORB for divergent boundary settings, an alkali basalt from the South Atlantic volcanic island of Tristan da Cunha, and average intraoceanic island arc lava. There are clear similarities between the continental crust and the island arc lavas: both exhibit incompatible element enrichment, negative Nb-Ta anomalies, and positive Pb anomalies. MORB and the Tristan da Cunha example exhibit slightly positive Nb-Ta anomalies and negative Pb anomalies. It is also true that at present, most new additions to crust occur in subduction zones, for example, the volcanoes of the Andes and Alaskan Peninsula. We consider the geochemistry of subduction zone volcanism in more detail in the following section.

In summary, we can say that continental crust has been produced over much, if not all of the history of the Earth, but crustal production has been very much episodic. The rate of new crust production has been declining over the last 2.5 Ga. Rates of crustal destruction and *net* continental growth remain controversial. Although subduction-related volcanism seems to have been the dominant mode of crustal growth at least through the Proterozoic and Phanerozoic, other mechanisms have played a role. The Wrangalia Terrane in NW

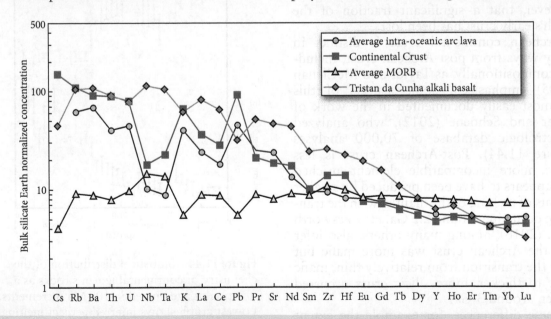

Figure 11.45 Normalized incompatible enrichments of average MORB (Table 11.7), average composition of lavas from eight intra-oceanic island arcs studied by Porter and White (2009), and an alkali basalt from Tristan da Cunha, compared with incompatible element enrichment of the total continental crust (Table 11.11). Only the pattern of island arc lavas matches that of the continental crust.

British Columbia and Alaska is widely considered to consist in part of oceanic plateaus. The plateaus were produced over a mantle plume in Paleozoic times and later accreted to the North American continent by plate tectonic processes. The Coast Ranges of Oregon represent another example of accreted oceanic crust. Mantle plumes surfacing beneath continents also produce magmas that add mass to the continents. The most voluminous eruptions occur in the initial stages of the plume, when the large buoyant plume head approaches the surface. Under these circumstances, enormous volumes of basalt erupt. Examples of such flood basalts include the Siberian Traps, the Karoo of South Africa, the Deccan of India, the Parana of Brazil, and the Columbia River of the northwest US. Gravity anomalies suggest even greater volumes of basaltic magma were trapped at deep crustal levels. This process, whereby dense basaltic magma crystallizes near the base of the crust, is sometimes called *underplating*. This may be an important mechanism of crustal growth, but with few samples from this region, the overall importance of this process is difficult to evaluate.

Continental rifts can also be sites of voluminous eruption of basaltic magma. A well-documented example is the Proterozoic Keweenawan or Mid-Continent Rift of the US, which formed some 1–1.2 Ga ago. Though now mostly covered by Phanerozoic sediments, where it is exposed the rift consists of a trough 150 km wide and 1500 km long filled with up to 15 km of volcanics, primarily basalt, and clastic sediments derived from them. Modern examples of continental rifts include the Rio Grande Rift of New Mexico and the East African Rift. Rift volcanism could also produce significant underplating.

11.6.4 Refining the continental crust

Regardless of tectonic environment, nearly all mantle-derived magmas are mafic in composition: typically they are basaltic, although under certain rare circumstances mantle melting can yield high-magnesian andesites. Even these rare mantle-derived andesites are notably poorer in SiO_2 and generally richer in MgO and FeO than the continental crust compositions listed in Table 11.11. If the continental crust has been produced by partial melting of the mantle, why then is it not basaltic in composition, as is the oceanic crust? Rudnick and Gao (2014) considered four possibilities:

1. Magmas have already evolved to andesitic composition by the time they cross the crust–mantle boundary (the Moho). The complementary mafic cumulates are left behind in the upper mantle. However, peridotite xenoliths from the upper mantle are predominantly restitic peridotite, not cumulates; hence this idea is not supported by observation.

2. Lower crustal floundering, or delamination, may occur when continental crust is thickened in compressional environments, such as convergent plate boundaries. The mafic lower crust is transformed into eclogite, which is denser than the underlying mantle. This process would preferentially remove the mafic part of the crust, leaving a residual crust that consequently becomes more silicic. A related process not considered by Rudnick and Gao (2014) is subduction erosion. Lower crust is more likely to be removed by subduction erosion than upper crust, but not to the extent of crustal floundering.

3. Preferential loss of Mg and Ca from continents by weathering and erosion. Mg is then taken up by the oceanic crust during hydrothermal alteration; Ca is precipitated as carbonate sediment. Both are returned to the mantle by subduction.

4. Under hotter conditions of the Archean, melting of subducting oceanic crust may have been much more common, giving rise to silicic melting, particularly the trondhjemite, tonalite, and granodiorite suites common to the Archean. However, Taylor and McLennan's (1985) estimate of Archean crustal composition is slightly more mafic that their estimate of present composition, which is inconsistent with this idea.

At present, the question of how the continental crust has evolved to obtain its present composition is still unresolved. This is thus clearly a fruitful area for future research.

11.7   SUBDUCTION ZONE PROCESSES

Subduction zones are, of course, the regions where oceanic lithosphere returns to the mantle, but the composition of subducting lithosphere is not the same as the lithosphere created at mid-ocean ridges. The crust, and to some degree the underlying mantle, has chemically exchanged with seawater, giving up some elements, such as Si, and acquiring others, such as Mg. Furthermore, it acquires a veneer of sediment, some of which is subducted as well. Thus subduction zones represent a chemical flux from the crust to the mantle. Magmas erupted in volcanic arcs that form above subduction zones are sourced predominantly from the mantle above the subducting lithosphere. These magmas represent a flux from the mantle to the crust. Subduction zones are thus unique in being areas of two-way chemical exchange between crust and mantle and, as we noted in the previous section, the principal environment in which new continental crust is created.

11.7.1   Major element composition

Island arc volcanics (IAV) (we will use the term *island arc* for all subduction zone magmatism, including continental margin type) are not much different in major element composition from other volcanic rocks. Compared with MORB, the major differences are that IAV magmas tend to be richer in water and that siliceous compositions are more common. Magmas found in island arcs appear to be predominantly andesitic. It is unlikely, however, that andesite is the principal magma produced in arcs, as andesites are rarely produced by partial melting of mantle peridotite. A safer bet is that the primary magmas are predominantly basaltic, although as Grove et al. (2012) point out, primary basaltic magmas produced by hydrous melting, as IAV magmas are, tend to be slightly richer in SiO$_2$ than anhydrous ones. Most island arc andesites are produced by fractional crystallization of such basaltic parents. Nevertheless, while primary andesitic magmas are rare, they do sometimes occur and are thought to be the result of melting at high water content and low pressure or by melting of basalt or eclogite. These can be distinguished from those produced by fractional crystallization by their much higher MgO contents and are appropriately enough called high-magnesium andesites. Boninites are an extreme example of such magmas and can have MgO contents of >20%.

Most IAV are silica-saturated or oversaturated; silica- undersaturated magmas (alkali basalts) are rare, although they do occur. In that sense, we might call them *tholeiitic*. However, in the context of island arc magmas, the term *tholeiite* has a more restrictive meaning. Two principal magma series are recognized, one called *tholeiitic*, the other called *calc-alkaline*. There are two principal differences between these rock series.

First, tholeiites differentiate initially toward higher Fe and tend to maintain higher Fe/Mg than the calc-alkaline lavas (Figure 11.46). The lower Fe content of the calc-alkaline series reflects iron oxide crystallization and suppression of plagioclase crystallization as a result of (1) higher pressure, (2) higher water content of the magma, (3) higher oxygen fugacity (Grove and Baker, 1984). The latter two factors are linked (Brounce et al., 2014) although water itself is not the oxidant. Rather, it appears that soluble oxidized species of sulfur and carbon (i.e., sulfates and carbonates) associated with water are the oxidants (e.g., Rielli et al., 2017). Subduction zone magmas have higher water content than most magmas found in other tectonic

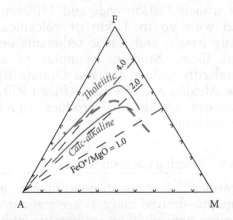

Figure 11.46   AFM (A: K$_2$O + Na$_2$O; F: FeO + MnO; M: MgO) diagram illustrating the difference between the tholeiitic and calc-alkaline lava series of island arcs. Calc-alkaline rocks plot below the heavy dashed dark line, tholeiites above.

settings (e.g., Plank et al., 2013). In the tholeiitic series, the crystallization sequence is typically olivine, plagioclase, followed by clinopyroxene. At higher pressure or higher water content, clinopyroxene supplants plagioclase as the second crystallizing phase, which buffers the iron concentration. Kay et al. (1983) found that, at least for the Aleutians, occurrence of these two magma series could be related to tectonic environment. Tholeiites occur in extensional environments within the arc where magmas can ascend relatively rapidly into the upper crust where they undergo fractional crystallization at low pressure. Calc-alkaline lavas tend to occur in compressional environments within the arc where they cannot so readily ascend to shallow depths, and hence undergo crystallization at greater depth. Crustal thickness also plays a role: thicker crust results in magmas stagnating at greater depth where plagioclase precipitation is suppressed (Miyashiro, 1974); not surprisingly, calc-alkaline magmas are more common where the crust is thick (Turner and Langmuir, 2015).

The second difference is that calc-alkaline magmas are richer in alkalis (K_2O and Na_2O) relative to calcium than tholeiites. Indeed, the calc-alkaline magmas are defined as those that have $Na_2O + K_2O \approx CaO$, whereas tholeiites have $Na_2O + K_2O < CaO$, and alkaline rocks have $Na_2O + K_2O > CaO$.

IAV also tends to be somewhat poorer in Ti than MORB and OIB. Perfit et al. (1980) argued that the difference in Ti content is due to early crystallization of oxides, ilmenite ($FeTiO_3$) in IAV, which buffers the Ti concentration. That, in turn, they argued, reflected higher oxygen fugacities. Kelley and Cottrell (2009) have confirmed higher fugacities in IAV by measuring the Fe^{2+}/Fe^{3+} ratio in glass inclusions in phenocrysts using synchrotron-based near-edge structure (μ-XANES) spectroscopy. The technique makes use of small differences in the X-ray adsorption spectra of Fe^{2+} and Fe^{3+}. A synchrotron produces a sufficiently high-intensity X-ray beam for this analysis to be performed on the micron scale. They found that while MORB had $Fe^{3+}/\Sigma Fe$ of 0.13 to 0.17, IAV had $Fe^{3+}/\Sigma Fe$ of 0.18 to 0.32. They also found that $Fe^{3+}/\Sigma Fe$ correlated directly with water content.

Plank and Langmuir (1988) investigated the factors that control the variation in major

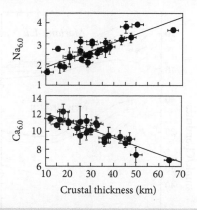

Figure 11.47 Correlation of $Ca_{6.0}$ and $Na_{6.0}$ with crustal thickness in island arc basalts. $Ca_{6.0}$ and $Na_{6.0}$ are the CaO and Na_2O concentrations after correction for fractional crystallization to 6.0% MgO. After Plank and Langmuir (1988).

element composition of island arc basalts. They treated the data in a manner analogous to Klein and Langmuir (1987), correcting regional datasets to a common MgO content, but they used 6% MgO rather than the 8% used by Klein and Langmuir. They found that $Na_{6.0}$ and $Ca_{6.0}$ (i.e., Na_2O and CaO concentrations corrected for fractional crystallization to 6% MgO in a manner analogous to the Klein and Langmuir approach discussed in Section 11.6.1.2) correlated well with crustal thickness (Figure 11.47). Because island arc volcanoes are located ~100±25 km above the subducting lithosphere, they argued that crustal thickness determines the height of the mantle column available for melting such that thicker crust results in smaller melting extents. Turner and Langmuir (2015) confirmed the correlation with crustal thickness but interpret this as the effect of mantle wedge thermal structure, reasoning that thicker crust implies deeper isotherms in the mantle wedge. We explore this further below.

11.7.2 Trace element composition

The differences in trace elements between island arc volcanics and those from other tectonic environments are more significant than the differences in major elements. Rare earths, however, are not particularly distinctive. There is a very considerable range in rare

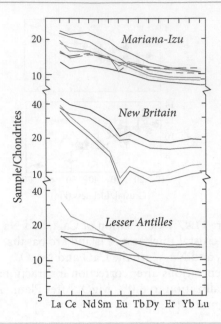

Figure 11.48   Rare earth element patterns of some typical island arc volcanics. After White and Patchett (1984).

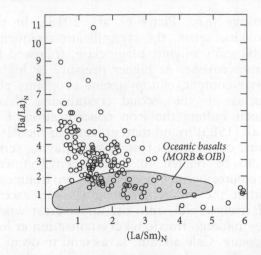

Figure 11.49   Relative alkali–alkaline earth enrichment of IAV illustrated by plotting the $(Ba/La)_N$ ratio vs. the $(La/Sm)_N$ ratio. The subscript N denotes normalization to chondritic values. After Perfit et al. (1980).

11.7.3   Isotopic composition and sediment subduction

earth patterns: from light rare earth (LRE) depleted to LRE enriched (Figure 11.48). IAV are virtually never as LRE depleted as MORB, but absolute REE concentrations are often low, and it is not unusual for the middle and heavy rare earths to be present at lower concentrations than in MORB. One other aspect is of interest. Ce anomalies occur in some IAV, whereas they are never seen in MORB or OIB, though they have been observed in continental carbonatites and kimberlites. Ce anomalies occur in sediment, so there is a suspicion that the anomalies in IAV are inherited from subducted sediment.

An important and ubiquitous feature of island-arc volcanics is a relative depletion in Nb and Ta, which can be seen in Figure 11.45. Island-arc volcanics are also richer in the incompatible alkalis and alkaline earths (K, Rb, Cs, Sr, and Ba) relative to other incompatible elements, when compared with MORB or OIB. This is illustrated in Figure 11.49, using the Ba/La ratio. Though both IAV and oceanic basalts can have a large range in rare earth patterns, as illustrated by the range in La/Sm ratios, the Ba/La ratios of IAV are generally higher.

Island arcs overlie subduction zones, which raises the obvious question of the degree to which subducting lithosphere, including oceanic crust, overlying sediment and under-lying mantle lithosphere, might contribute to island arc magmas. These questions have been most successfully addressed through isotope geochemistry. Sr isotope ratios are generally higher, and Nd isotope ratios generally lower than in MORB, with $^{87}Sr/^{86}Sr$ ratios around 0.7033–0.7037 and ε_{Nd} of +6 to +8 being fairly typical of intraoceanic IAV. This range overlaps considerably with MORB and OIB. Turner and Langmuir (2015) found that Sr and Nd isotope ratios in IAV are often similar to those in volcanics from behind the arc, although this was not the case for Pb isotopes. Reasoning that the back-arc volcanoes sample mantle uninfluenced by subduction, they concluded that much of the Sr and Nd isotopic variation in IAV simply reflects intrinsic heterogeneity in the mantle wedge. There is, however, a tendency for IAV to have slightly higher Sr isotope ratios for a given Nd isotope ratio and hence plot to the right of the oceanic basalt array on a Nd–Sr isotope ratio plot such as Figure 11.14. This shift to higher Sr isotope ratios is thought to reflect a contribution of subducted oceanic crust to IAV magma sources. Weathering and hydrothermal alteration of the oceanic crust shifts Sr isotope ratios of the oceanic crust toward seawater (higher) values but does not affect Nd isotope ratios because of the extremely low concentration of Nd in seawater.

Clear evidence of sediment involvement comes from Pb isotope studies of island arc volcanics, beginning with the work of Armstrong and Cooper (1971). As Figure 11.50 shows, $^{206}Pb/^{204}Pb$ isotope ratios overlap MORB values, but $^{207}Pb/^{204}Pb$ ratios are typically higher in IAV than most oceanic basalts. They tend to form steeper arrays on $^{207}Pb/^{204}Pb-^{206}Pb/^{204}Pb$ plots and overlap the field of marine sediments. For most island arcs, Pb isotope ratios in the volcanic rocks lie between sediment local to the arc and the MORB field (Karig and Kay, 1981). On the whole then, Pb in island arc magmas appears to be a mixture of Pb from local sediment and local upper mantle.

Finally, ^{10}Be provides unequivocal evidence of sediment involvement in IAV magma sources. As we found in Chapter 8, ^{10}Be is a cosmogenic isotope produced in the atmosphere by cosmic ray spallation of ^{14}N. It has a half-life of only 1.4 million years, so we would not expect to find significant amounts of ^{10}Be in the Earth's interior; any present when the Earth formed has long since

decayed away. ^{10}Be created in the atmosphere is purged by rainfall and is strongly absorbed by clays of sediment and soil. Tera et al. (1986) found measurable quantities ($>10^6$ atoms per gram) of ^{10}Be in some arc lavas, while ^{10}Be was completely absent in nonarc lavas. The interpretation is that the ^{10}Be originates from sediment subducted to the magma genesis zone.

Subducted sediment also influences the trace element compositions of arc lavas. Plank and Langmuir (1993) carried out careful study of the composition of volcanic rocks from eight arcs and the sediments being subducted beneath them. By analyzing representative samples from the sediments and considering the proportions of sediment types being carried beneath the arc, they estimated the flux of elements being carried by sediment beneath the arc. They found they could relate the degree of enrichment of most incompatible elements to the sediment flux of that element. For example, the Ba/Na and Th/Na ratios (after correction for fractional crystallization to 6% MgO) correlate strongly with the Ba and Th sediment fluxes (Figure 11.51). Different arcs are enriched to different degrees in these elements: for example, the Lesser Antilles arc has moderate Th/Na ratios but low Ba/Na ratios. The difference appears to be due to the difference in the sediment flux.

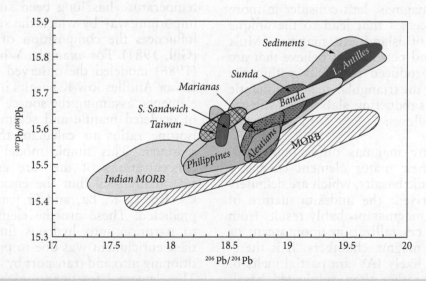

Figure 11.50   Pb isotope ratios in island arc volcanics. Fields for the South Sandwich, Lesser Antilles, Aleutians, Marianas, Philippines, Taiwan, Banda, and Sunda arcs are compared with fields for Atlantic and Pacific MORB (field labeled MORB) and Indian Ocean MORB, and modern marine sediments.

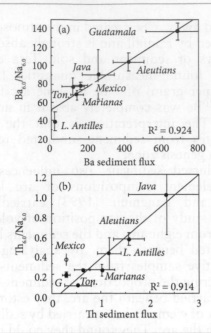

Figure 11.51 Relationship between Ba/Na$_{6.0}$ and Th/Na$_{6.0}$ in volcanic rocks from eight arcs, and Ba and Th sediment flux beneath those arcs. Horizontal bars represent uncertainty in the amount of sediment subducted; vertical bars reflect the variance of the ratio in the arc volcanics. G: Guatemala; Ton: Tonga. From Plank and Langmuir (1993).

11.7.4 Magma genesis in subduction zones

Now that we have an overview of the composition of arc magmas, let's consider in more detail the processes that lead to the unique geochemistry of island arc magmas. Most geochemists and petrologists believe that arc magmas are produced primarily within the mantle wedge, the triangular region of mantle overlying the subducting slab. The evidence for this is as follows:

1. Primary arc magmas differ only modestly in their major element chemistry from oceanic basalts, which are definitely mantle-derived; the andesitic nature of many arc magmas probably results from fractional crystallization upon ascent or in crustal magma chambers. It is therefore most likely IAV are partial melts of peridotite rather than subducted basalt or sediment.
2. Radiogenic isotopic and trace element systematics generally allow only a small

fraction of sediment (generally a few percent or less) to be present in arc magma sources. Relatively high ^{3}He/^{4}He ratios in arc lavas confirm a mantle source.

3. Rare earth patterns of island arc magmas are consistent with these magmas being generated by partial melting of peridotite, and, with rare exceptions, not by partial melting of eclogite, which would be the stable form of subducted basalt beneath island arc volcanoes. Because the heavy rare earths partition strongly into garnet (e.g., Figure 7.18), melts of eclogite should show steep rare earth patterns, with low concentrations of the heavy rare earths. This is not generally the case. Some rare high-magnesium andesites, sometimes called *adakites* (after a well-documented occurrence on Adak Island in the Aleutians) with steep rare earth patterns may represent exceptions to this rule and may indeed be generated by small extents of melting of subducted oceanic crust (Kay, 1978; Defant and Drummond, 1990). It also is possible that such *slab melts* were more common several billion years ago when the Earth was hotter.

How do arc magmas acquire their distinctive geochemical "flavor"? Dehydration and migration of fluids released as hydrous minerals break down under increasing pressure and temperature has long been suspected as an important way by which the subducting slab influences the composition of IAV magmas (Gill, 1981). For example, White and Dupré (1986) modeled the observed enrichment of Lesser Antilles low-K basalts in incompatible elements, assuming the source was a mixture of depleted mantle and sediment, using Nd isotope ratios to calculate the fraction of sediment. This simple model predicted the concentrations of the rare earths and Th reasonably well, but the enrichment of Pb, Cs, Rb, U, K, Ba, and Sr was greater than predicted. These are the elements expected to partition into hydrous fluid, suggesting their enrichment was due to preferential partitioning into and transport by aqueous fluids. The elemental fractionation that occurs during transport from the subducting slab into the mantle wedge is further illustrated in Figure 11.52a. Ba/La ratios in Marianas arc

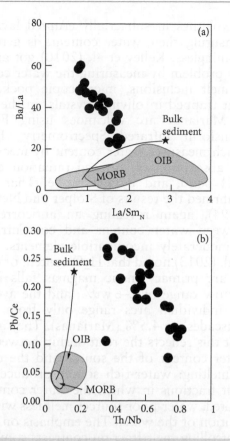

Figure 11.52 (a) Ba/La versus La/Sm$_N$ (subscript N denotes the chondrite-normalized ratio) in Marianas arc lavas compared with those in MORB, OIB, and bulk sediment from ODP Hole 801B, located on the Pacific plate outboard of the Marianas arc. The line connecting MORB and sediment is a mixing curve. Ba enrichment in the arc cannot be explained simply by mixing of MORB and sediment. (b) Pb/Ce versus Th/Nb in Marianas arc lavas. Th is enriched more than can be explained by sediment–MORB mixing alone. Elliott et al. (1997) suggested sediment melting is responsible. Modified from Elliott et al. (1997).

lavas plot systematically above a mixing line between MORB and sediment subducting beneath the arc (sampled in ODP Hole 801) on a plot of Ba/La vs. La/Sm.

Hydrous fluids alone, however, seem incapable of explaining all the distinctive chemical features of IAV. This is illustrated in Figure 11.52b, which shows that Th/Nb are systematically higher than predicted from simple MORB–sediment mixing. Elliott et al.

(1997) concluded from these relationships and $^{230}Th/^{232}Th$ isotope ratios (recall that ^{230}Th is the intermediate daughter of ^{238}U) that both a hydrous fluid and a silicate melt were involved in transport of sediment. They proposed that this melt was a hydrous partial melt of the subducted sediments. Melting appears necessary to account for the fractionation between Th and Nb, neither of which is particularly soluble in aqueous fluids. Furthermore, lavas with the highest Th/Nb also show the greatest light rare earth enrichment.

Early models of the thermal structure of subduction zones indicated that the subducting crust and sediment was too cold to melt. More recent analyses such as that of Syracuse et al. (2010) show that the slab surface easily exceeds its wet solidus, which is at 700–800°C, beneath the Cascadian arc where a young, hot plate is descending at moderate velocity, but melting might not occur beneath Tonga where old lithosphere is descending rapidly. The role of water must be emphasized. While the wet solidus of subducting crust and sediment water is exceeded, temperatures remain well below the dry solidus.

Melting within the mantle wedge is a consequence in large part of water derived from the subducting slab. Both hydrous fluids and water-rich melts will migrate upward into the overlying mantle wedge inducing melting in the mantle wedge. Figure 11.53 is a $P-T$ phase diagram showing that under water-saturated conditions, the peridotite solidus is depressed by hundreds of degrees compared with the dry solidus. At 1.5 GPa (corresponding roughly to 50 km depth in the Earth), peridotite begins to melt at over 400°C cooler temperatures than under "dry" conditions. The effect is even larger at higher pressure. Figure 11.53 also shows that at pressures above about 2 GPa, ilmenite along with chlorite would be stable at and above the solidus. The stability of ilmenite is particularly significant because Nb and Ta strongly partition into it. Thus, the characteristic Nb-Ta depletion of island arc lavas, and indeed the entire continental crust (Figure 11.45), may be due to residual ilmenite being present during the initial states of melting deep within the arc.

Work beginning with Stolper and Newman (1994) has produced evidence directly relating water content to melting in subduction zones.

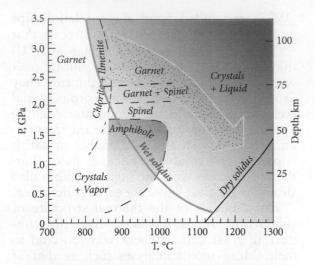

Figure 11.53 Experimentally based phase diagram for the system peridotite + water. The peridotite dry solidus is from Hirschmann (2000). Under "wet" conditions, meaning vapor present, peridotite begins to melt along the curve labeled "Wet solidus" at temperatures hundreds of degrees lower. Curved dashed lines are the chlorite + ilmenite- and amphibole-out curves. Straight dashed lines illustrate the progressive replacement of spinel with garnet. The broad stippled arrow shows the path the melts take in P–T space as they rise through the mantle wedge (Figure 11.55). Modified from Grove et al. (2006).

They showed that water concentrations correlated inversely with the concentrations of moderately incompatible elements, such as Ti, Zr, and Na, in lavas from the Marianas back-arc spreading center and proposed that extent of melting varied inversely with the amount of H_2O-rich component in the source mixture. Water, or H, controls the extent of melting so that smallest extents of melting (about 5%) occur in H_2O-poor sources and give rise to incompatible element-rich basalts, while the highest extents (over 20%) in H_2O-rich sources and incompatible element-poor basalts. But water, or H, also behaves as an incompatible element, so its concentration is high in the small degree melts and lower in high degree melts (eqn. 7.38). Stolper and Newman (1994) focused on back-arc basalts because they erupt at under several km of water, where pressure prevents degassing of the lavas. Degassing typically results in the loss of

all volatiles in subaerially erupted lavas, so measuring their water contents is generally meaningless. Kelley et al. (2010) got around this problem by measuring the water content of melt inclusions, microscopic pockets of melt trapped in olivine crystals, in the lavas of Marianas arc volcanoes using Fourier Transform Infrared Spectrometry (FTIR), which measures water content by measuring the absorption of infrared radiation by the O-H bond, and by ion probe. That study confirmed the results of Stolper and Newman (1994), again revealing an anticorrelation between water content and concentrations of moderately incompatible elements. Plank et al. (2013) noted that H_2O content of nearly all arc primary mafic magmas falls in the narrow range of 2–6 wt%, and the averages for individual arcs range only from 3.2% (Cascades) to 4.5% (Marianas). They argued that this reflects the relationship between the water content of the source and the degree of melting: water-rich source produce large melt fractions in which the water content is diluted, water-poor sources melt less with less dilution of the water. The emphasis on water controlling melt fraction contrasts somewhat with the conclusion of Turner and Langmuir (2015) who argued that mantle wedge temperature controls melting. The likely reality is that melting is controlled by a combination of water content and temperature.

Figure 11.54 summarizes magma generation at subduction zones. Sediment and hydrothermally altered oceanic crust carry water and incompatible elements into the mantle as oceanic lithosphere is subducted. The wedge immediately above the subducting slab has an inverted thermal gradient: temperature decreases with depth because the cold subducting slab acts as a heat sink. Immediately above the slab, the mantle is too cold to melt, even though it is water saturated. However, 10 km above the slab, temperatures approach 1000°C, well above the wet solidus. Consequently, melting begins. These initial melts may contain as much as 28% water (Grove et al., 2006), but as they rise, continued melting progressively dilutes the water content.

An alternative hypothesis to explain subduction zone volcanism proposes that instead of fluids or melts migrating into the mantle wedge, a *mélange* composed of sediment,

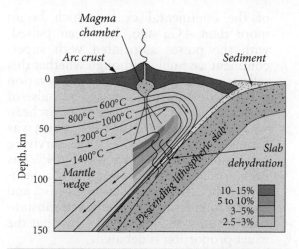

Figure 11.54 Cross-section of a subduction zone illustrating island-arc magma genesis. Arrows show the direction of mantle flow. Red color shading indicates the predicted extent of partial melting in the mantle wedge resulting from the addition of water. Modified from Grove et al. (2006).

basalt and hydrated peridotite can rise as a buoyant diapir into the overlying mantle wedge and melt to produce arc magmas (e.g., Marschall and Schumacher, 2012). Intimate mixtures such as these are found in a number of former subduction zones around the world, such the Catalina Schist and Franciscan Formations of coastal California, the western Alps, Norway, etc. They form as water released from the subducting slab hydrates the overlying mantle, producing notably weak minerals such as serpentine and chlorite. Shearing along the plate interface fault zone then mixes the rocks on both sides. These mélanges are low density and weak, particularly when lubricated by fluids or partial melts and hence can rise into the overlying mantle wedge as sheet-like structures parallel to the trench, ridge-like structures perpendicular to the trench, or flattened wave-like instabilities (e.g., Zhu et al., 2009). Because of the inverted thermal structure of the mantle wedge, these diapirs increasingly melt as they rise. Nielsen and Marschall (2017) argue that melting of such physical mixtures of sediment, basalt, and mantle peridotite better explains trace element and isotopic systematics than the conventional "chemical mixtures" described above.

11.8 SUMMARY

In this chapter we focused on the composition and evolution of the solid Earth.

- We first considered the geophysical (density, seismic velocities, moment of inertia) and cosmochemical constraints on the Earth's structure. A peridotitic mantle and Fe-Ni core best matches these constraints. Seismic discontinuities in the mantle can be explained by phase changes of this material, most importantly a change from tetrahedral to octahedral coordination of Si at 660 km depth.

- For major elements, the mantle composition is the same as the bulk silicate Earth composition. A variety of constraints allows us to estimate this composition, which by all estimates is poorer in Si, Na, and K and richer in Mg, Ca, Al, and Fe (after accounting for Fe in the core) than one composed of CI chondrite material. Refractory lithophile trace elements such as the REE have chondritic relative abundances or nearly so but volatile and siderophile elements are depleted relative to chondrites.

- The Earth's core consists of ~86% Fe and ~5% Ni. Density constraints require it contain one or more light elements; the most likely candidates are Si, O, S, and C. Based on their deficit in the mantle compared with chondrites, the core also contains most of the Earth's inventory of refractory siderophile elements such as the platinum group elements. Although highly depleted, these elements are present in nearly chondritic proportions in the mantle, which is inconsistent with metal-silicate partitioning. The most likely explanation is the addition of a *late accretionary veneer* of 0.5% or so of chondritic material to the mantle after the core formed.

- Isotope ratios and trace element concentrations demonstrate that the mantle is heterogeneous. Mid-ocean ridge basalts are derived from mantle depleted in incompatible elements by previous melt extraction. This depleted reservoir occupies the upper mantle and may constitute ~50% or so of the total mantle. In contrast, oceanic island basalts are produced by melting of mantle plumes rising from the deep mantle that are spatially associated with the periphery of large low S-wave velocity provinces in the lowermost mantle. They derive from four or five distinct reservoirs that contain material that was once at the surface of the Earth and has been subducted into the deep mantle, some quite anciently based on MIF sulfur anomalies. As evidenced by noble gas and tungsten isotope ratios, they also contain primitive material that has been isolated in deep mantle for 4.5 billion years.

- The subcontinental lithosphere beneath cratons is highly melt-depleted and can be as old as the crust above it or nearly so. Subsequent to melt depletion, much of the subcontinental lithosphere has been re-enriched in incompatible elements by silicate, carbonatite, and aqueous fluids percolating into it from below.

- Both the oceanic and continental crust have been created by partial melting of the mantle. While the basaltic oceanic crust is constantly produced and destroyed, the continental crust is semipermanent with the approximate composition of diorite or andesite and enriched in incompatible elements. The lower crust is approximately basaltic or basaltic andesite in composition and substantially more mafic than the granodioritic upper crust. Production

of the continental crust, which began more than 4 Ga ago, has been pulsed, with the pulses associated with supercontinent assembly, although whether this reflects pulsed magmatism or preservation is debated. A particularly strong pulse of crust creation occurred in the late Archean when about a third of the extant crust was formed; only about 6% of the surviving crust is older than 3 Ga and only 14% has been produced over the last 400 Ma. Sediment subduction, subduction erosion, and crustal foundering have removed substantial amounts of continental crust, but the exact proportion is debated.

- At present, most new continental crust is being created in subduction zones. Based on the similarity of trace elements between island arc volcanics and the continental crust, notably the alkali enrichment, Ta-Nb depletion, and Pb enrichment, it appears this has been the principal mechanism of crust creation through time. These distinctive features arise because of the role of water in subduction zone volcanism. Water released from the subduction lithosphere carries a number of elements into the mantle wedge. The water also greatly suppresses the solidus which results in the melting producing these magmas. The water-rich, oxidized conditions also stabilize secondary minerals that retain Ta and Nb.

REFERENCES AND SUGGESTIONS FOR FURTHER READING

Ahrens, T.J. and Jeanloz, R. 1987. Pyrite: shock compression, isentropic release, and composition of the Earth's core. *Journal of Geophysical Research* 92: 10363–75.

Albarède, F. 1995. *Introduction to Geochemical Modeling*. Cambridge, Cambridge University Press.

Albarède, F. 1998. The growth of continental crust. *Tectonophysics* 296: 1–14. doi: 10.1016/S0040-1951(98)00133-4.

Allègre, C.J., Poirier, J.P., Humler, E. and Hofmann, A.W. 1995. The chemical composition of the Earth. *Earth and Planetary Science Letters* 134: 515–26.

Anderson, O.L. and Issak, D.G. 2002. Another look at the core density deficit in the Earth's outer core. *Phys. Earth Planet. Interiors* 131: 19–27.

Arevalo, R., Jr., McDonough, W.F. and Luong, M. 2009. The K/U ratio of the silicate Earth: Insights into mantle composition, structure and thermal evolution. *Earth and Planetary Science Letters* 278: 361–9.

Armstrong, R.L. 1968. A model for the evolution of strontium and lead isotopes in a dynamic Earth. *Reviews in Geophysics* 6: 175–99.

Armstrong, R.L. 1981. Radiogenic isotopes: the case for crustal recycling on a near-steady-state no-continental-growth Earth, in *The Origin and Evolution of the Earth's Continental Crust* (eds S.M. Windley and B.F. Windley), pp. 259–87. London, The Royal Society.

Armstrong, R.L. and Cooper, J.A. 1971. Lead isotopes in island arcs. *Earth and Planetary Science Letters* 35: 27–37.

Arndt, N. T. 2013. The formation and evolution of the continental crust. *Geochemical Perspectives* 2: 405–533.

Bennett, V.C. and DePaolo, D.J. 1987. Proterozoic crustal history of the western United States as determined by neodymium isotope mapping. *Bulletin of the Geological Society of America* 99: 674–85.

Birch, F. 1964. Density and composition of mantle and core. *Journal of Geophysical Research* 69: 4377–88.

Borodin, L.S. 1998. Estimated chemical composition and petrochemical evolution of the upper continental crust. *Geochemistry International* 37: 723–34.

Boyet, M. and Carlson, R.L. 2005. ^{142}Nd evidence for early (>4.3 Ga) global differentiation of the silicate Earth. *Science* 309: 576–81.

Boyet, M., Bouvier, A., Frossard, P., Hammouda, T., Garçon, M., Gannoun, A. 2018. Enstatite chondrites EL3 as building blocks for the Earth: The debate over the ^{146}Sm–^{142}Nd systematics. *Earth and Planetary Science Letters* 488: 68–78. doi: 10.1016/j.epsl.2018.02.004.

Brounce, M. N., Kelley, K. A., Cottrell, E. 2014. Variations in Fe^{3+}/$\sum$Fe of Mariana arc basalts and mantle wedge f_{O_2}. *Journal of Petrology* 55: 2513–36. doi: 10.1093/petrology/egu065.

Brown, M. 2014. The contribution of metamorphic petrology to understanding lithosphere evolution and geodynamics. *Geoscience Frontiers* 5: 553–69. doi: /10.1016/j.gsf.2014.02.005.

Burke, K. 2011. Plate tectonics, the wilson cycle, and mantle plumes: geodynamics from the top. *Annual Review of Earth and Planetary Sciences* 39: 1–29. doi: 10.1146/annurev-earth-040809-152521.

Burke, K., Steinberger, B., Torsvik, T. H., Smethurst, M. A. 2008. Plume Generation Zones at the margins of Large Low Shear Velocity Provinces on the core–mantle boundary. *Earth and Planetary Science Letters* 265: 49–60. doi: 10.1016/j.epsl.2007.09.042.

Cabral, R. A., Jackson, M. G., Rose-Koga, E. F., et al. 2013. Anomalous sulphur isotopes in plume lavas reveal deep mantle storage of Archaean crust. *Nature* 496: 490–3. doi: 10.1038/nature12020.

Caro, G. and Bourdon, B. 2010. Non-chondritic Sm/Nd ratio in the terrestrial planets: Consequences for the geochemical evolution of the mantle crust system. *Geochimica et Cosmochimica Acta* 74: 3333–49.

Caro, G., Bourdon, B., Halliday, A.N. and Quitte, G. 2008. Super-chondritic Sm/Nd ratios in Mars, the Earth, and the Moon. *Nature* 452: 336–9.

Chabot, N. and Drake, M.J. 1999. Potassium solubility in metal: the effects of composition at 15 kbar and 1900°C on partitioning between iron alloys and silicate melts. *Earth and Planetary Science Letters* 172: 323–35.

Chopelas, A., Beohler, R. and Ko, T. 1994. Thermodynamics and behavior of γ-Mg$_2$SiO$_4$ at high pressure: implications for Mg$_2$SiO$_4$ phase equilibrium. *Phys. Chem. Mineral* 21: 351–9.

Clarke, F.W. 1924. *The Data of Geochemistry*, 5th ed. *US Geological Survey Bulletin* 770. Washington, US Government Printing Office.

Class, C. and Goldstein, S., 2005. Evolution of helium isotopes in the Earth's mantle. *Nature* 436, 1107–12. doi: 10.1038/nature0390.

Clift, P. D., Vannucchi, P., Morgan, J. P. 2009. Crustal redistribution, crust–mantle recycling and Phanerozoic evolution of the continental crust. *Earth-Science Reviews* 97: 80–104. doi: 10.1016/j.earscirev.2009.10.003.

Condie, K. C. 1998. Episodic continental growth and supercontinents: a mantle avalanche connection? *Earth and Planetary Science Letters* 163: 97–108. doi: 10.1016/s0012-821x(98)00178-2

Condie, K. C., Arndt, N., Davaille, A., Puetz, S. J. 2017. Zircon age peaks: Production or preservation of continental crust? *Geosphere* 13: 227–34.

Condie, K.C. and Aster, R.C. 2010. Episodic zircon age spectra of orogenic granitoids: The supercontinent connection and continental growth. *Precambrian Research* 180: 227–36. doi: 10.1016/j.precamres.2010.03.008.

Coogan, L. A. (2014) 4.14 - The Lower Oceanic Crust. In: Holland H.D., Turekian K.K. (eds) *Treatise on Geochemistry* (Second Edition), vol. 3. Elsevier, Oxford, pp 497–541. doi: 978-0-08-098300-4.

Corgne, A., Liebske, C., Wood, B.J., Rubie, D.C. and Frost, D.J. 2005. Silicate perovskite-melt partitioning of trace elements and geochemical signature of a deep perovskitic reservoir. *Geochimica et Cosmochimica Acta* 69: 485–96.

Dauphas, N. 2017. The isotopic nature of the Earth's accreting material through time. *Nature* 541: 521–4. doi: 10.1038/nature20830.

Davies, G.F. 1988. Ocean bathymetry and mantle convection, 1, large-scale flow and hotspots. *Journal of Geophysical Research* 93: 10467–80.

de Leeuw, G. A. M., Ellam, R. M., Stuart, F. M. and Carlson, R. W. 2017. ^{142}Nd/^{144}Nd inferences on the nature and origin of the source of high ^{3}He/^{4}He magmas. *Earth and Planetary Science Letters* 472, 62–68. doi: 10.1016/j.epsl.2017.05.005.

Defant, M.J. and Drummond, M.S. 1990. Derivation of some modern arc magmas by melting of young subducted lithosphere. *Nature* 347: 662–5.

Delavault, H., Chauvel, C., Thomassot, E., Devey, C. W., Dazas, B. 2016. Sulfur and lead isotopic evidence of relic Archean sediments in the Pitcairn mantle plume. *Proceedings of the National Academy of Sciences* 113: 12952–6. doi: 10.1073/pnas.1523805113.

DePaolo, D.J. 1980. Crustal growth and mantle evolution: inferences from models of element transport and Nd and Sr isotopes. *Geochimica et Cosmochimica Acta* 44: 1185–96.

Dupré, B. and Allègre, C. J., 1983. Pb-Sr isotope variations in Indian Ocean basalts and mixing phenomena. *Nature* 303, 142–6.

Dziewonski, A. M., Lekic, V., Romanowicz, B. A. 2010. Mantle Anchor Structure: An argument for bottom up tectonics. *Earth and Planetary Science Letters* 299: 69–79. doi: 10.1016/j.epsl.2010.08.013.

Dziewonski, A.M. and Anderson, D.L. 1981. Preliminary Reference Earth Model (PREM). *Phys. Earth Planet. Int.* 25: 297–356.

Eiler, J. M., Farley, K. A., Valley, J. W., et al. 1997. Oxygen isotope variations in oceanic basalt phenocrysts. *Geochimica et Cosmochimica Acta* 61: 2281–93.

Elliott, T., Plank, T., Zindler, A., White, W. and Bourdon, B. 1997. Element transport from subducted slab to juvenile crust at the Mariana arc. *Journal of Geophysical Research* 102: 14,991–15,019. doi: 10.1029/97JB00788.

Farley, K. A., Natland, J. H. and Craig, H., 1992. Binary mixing of enriched and undegassed (primitive?) mantle components (He, Sr, Nd, Pb) in Samoan lavas. *Earth and Planetary Science Letters* 111, 183–199. doi: 10.1016/0012-821X(92)90178-X.

Forsyth, D. and Uyeda, S., 1975. On the relative importance of the driving forces of plate motion. *Geophysical Journal International* 43(1), 163–200. doi: 10.1111/j.1365-246X.1975.tb00631.x.

French, S. W., Romanowicz, B. 2015. Broad plumes rooted at the base of the Earth's mantle beneath major hotspots. *Nature* 525: 95–99. doi: 10.1038/nature14876.

Frost, D. J., Liebske, C., Langenhorst, F., McCammon, C. A., Trønnes, R. G., Rubie, D. C. 2004. Experimental evidence for the existence of iron-rich metal in the Earth's lower mantle. *Nature* 428: 409–12. doi: 10.1038/nature02413.

Gao, S., Luo, T.-C., Zhang, B.-R., et al. 1988. Chemical composition of the continental crust as revealed by studies in East China. *Geochimica et Cosmochimica Acta* 62: 1959–75.

Garnero, E. J., McNamara, A. K., Shim, S.-H. 2016. Continent-sized anomalous zones with low seismic velocity at the base of Earth's mantle. *Nature Geoscience* 9: 481–89. doi: 10.1038/ngeo2733.

Gast, P. W., Tilton, G. R., Hedge, C. 1964. Isotopic composition of lead and strontium from Ascension and Gough Islands. *Science* 145: 1181–5. doi: 10.2307/1714243.

Gessmann, C.K. and Wood, B.J. 2002. Potassium in the Earth's core? *Earth and Planetary Science Letters* 200: 63–78.

Gialampouki, M. A., Xu, S., Morgan, D. 2018. Iron valence and partitioning between post-perovskite and ferropericlase in the Earth's lowermost mantle. *Physics of the Earth and Planetary Interiors* 282: 110–6. doi: 10.1016/j.pepi.2018.06.005.

Gill, J.B. 1981. *Orogenic Andesites and Plate Tectonics*. Berlin, Springer Verlag.

Goldschmidt, V.M. 1933. Grundlagen der quantitativen Geochemie. *Fortschritte der Mineralogie, Kristallographie und Petrographie* 17: 112–56.

Grove, T. L., Till, C. B., Krawczynski, M. J. 2012. The role of H_2O in subduction zone magmatism. *Annual Review of Earth and Planetary Sciences* 40: 413–39. doi: 10.1146/annurev-earth-042711-105310.

Grove, T.L. and Baker, M.B. 1984. Phase equilibrium controls on the tholeiitic versus calc-alkaline differentiation trends. *Journal of Geophysical Research* 89: 3253–74. doi: 10.1029/JB089iB05p03253.

Grove, T.L., Chatterjee, N., Parman, S.W. and Médard, E. 2006. The influence of H_2O on mantle wedge melting. *Earth and Planetary Science Letters* 249: 74–89. doi: 10.1016/j.epsl.2006.06.043.

Hanan, B.B. and Graham, D.W. 1996. Lead and helium isotope evidence from oceanic basalts for a common deep source of mantle plumes. *Science* 272: 991–5.

Hart, S. R., Gerlach, D. C., White, W. M. 1986. A possible new Sr-Nd-Pb mantle array and consequences for mantle mixing. *Geochimica et Cosmochimica Acta* 50: 1551–7. doi: 10.1016/0016-7037(86)90329-7.

Hart, S.R. 1984. The DUPAL anomaly: A large-scale isotopic mantle anomaly in the Southern Hemisphere. *Nature* 309: 753–7.

Hart, S.R. and Zindler, A. 1986. In search of a bulk-earth composition. *Chemical Geology* 57: 247–67.

Hart, S.R., Hauri, E.H., Oschmann, L.A. and Whitehead, J.A. 1992. Mantle plumes and entrainment: isotopic evidence. *Science* 256: 517–20.

Hawkesworth, C. J., Cawood, P. A., Dhuime, B. 2016. Tectonics and crustal evolution. *GSA Today* 26: 4–11. doi: 10.1130/GSATG272A.1.

Hawkesworth, C., Cawood, P. A., Dhuime, B. 2019. Rates of generation and growth of the continental crust. *Geoscience Frontiers* 10: 165–73. doi: 10.1016/j.gsf.2018.02.004.

Hawkesworth, C.J., Dhuime, B., Pietranik, A.B., et al. 2010. The generation and evolution of the continental crust. *Journal of the Geological Society of London* 167: 229–48. doi: 10.1144/0016-76492009-072.

Hawkesworth, C.J., Kempton, P.D., Rogers, N.W., Ellam, R.M. and van Calsteren, P.W. 1990. Continental mantle lithosphere, and shallow level enrichment processes in the Earth's mantle. *Earth and Planetary Science Letters* 96: 256–68.

Hedge, C.E. 1966. Variations in radiogenic strontium found in volcanic rocks. *Journal of Geophysical Research* 71: 6119–26.

Hernlund, J. W., Houser, C. 2008. On the statistical distribution of seismic velocities in Earth's deep mantle. *Earth and Planetary Science Letters* 265: 423–37. doi: 10.1016/j.epsl.2007.10.042.

Herzberg, C., Asimow, P. D., Arndt, N., et al. 2007. Temperatures in ambient mantle and plumes: Constraints from basalts, picrites, and komatiites. *Geochemistry, Geophysics, Geosystems* 8: doi: 10.1029/2006gc001390.

Herzberg, C., Condie, K., Korenaga, J. 2010. Thermal history of the Earth and its petrological expression. *Earth and Planetary Science Letters* 292: 79–88. doi: 10.1016/j.epsl.2010.01.022.

Hess, H.H. 1962. History of ocean basins, in *Petrologic Studies: A Volume In Honor Of A. F. Buddington* (eds A.E.J. Engel, et al.), pp. 599–620. Boulder, CO, Geological Society of America.

Hirschmann, M.M. 2000. Mantle solidus: Experimental constraints and the effects of peridotite composition. *Geochem. Geophys. Geosyst.* 1: doi: 10.1029/2000gc000070.

Hofmann, A.W. and White, W.M. 1982. Mantle plumes from ancient oceanic crust. *Earth and Planetary Science Letters* 57: 421–36.

Horan, M. F., Carlson, R. W., Walker, R. J., Jackson, M., Garçon, M. and Norman, M., 2018. Tracking Hadean processes in modern basalts with 142-Neodymium. *Earth and Planetary Science Letters* 484, 184–91. doi: 10.1016/j.epsl.2017.05.005.

Hurley, P.M. and Rand, J.R. 1969. Pre-drift continental nuclei. *Science* 164: 1229–42.

Jackson, M. G., Carlson, R. W. 2011. An ancient recipe for flood-basalt genesis. *Nature* 476: 316 doi: 10.1038/nature10326.

Jackson, M.G., Hart, S.R., Koppers, A.A.P., et al. 2007. The return of subducted continental crust in Samoan lavas. *Nature* 448: 684–7.

Karig, D.E. and Kay, R.W. 1981. Fate of sediments on the descending plate at convergent margins. *Philosophical Transactions of the Royal Society of London Series A* 301: 233–51.

Kay, R.W. 1978. Aleutian magnesian andesites: melts from subducted Pacific Ocean crust. *Journal of Volcanology and Geothermal Research* 4: 117–32.

Kay, R.W. and Kay, S.M. 1991. Creation and destruction of lower continental crust. *Geologische Rundschau* 80: 259–78.

Kay, S.M., Coira, B. and Viramonte, J. 1994. Young mafic back arc volcanic rocks as indicators of continental lithospheric delamination beneath the Argentine Puna plateau, central Andes. *Journal of Geophysical Research* 99: 24323–39. doi: 10.1029/94jb00896.

Kay, S.M., Kay, R.W. and Citron, G.P. 1983. Tectonic controls on tholeiitic and calc-alkaline magmatism. *Journal of Geophysical Research* 87: 4051–72.

Keller, B., Schoene, B. 2018. Plate tectonics and continental basaltic geochemistry throughout Earth history crustal evolution. *Earth and Planetary Science Letters* 481: 290–304. doi 10.1016/j.epsl.2017.10.031.

Keller, C. B., Schoene, B. 2012. Statistical geochemistry reveals disruption in secular lithospheric evolution about 2.5 Gyr ago. *Nature* 485: 490–3. doi: 10.1038/nature11024.

Kelley, K.A. and Cottrell, E. 2009. Water and the oxidation state of subduction zone magmas. *Science* 325: 605–7. doi: 10.1126/science.1174156.

Kelley, K.A., Plank, T., Newman, S., et al. 2010. Mantle melting as a function of water content beneath the Mariana Arc. *Journal of Petrology* 51: 1711–38. doi: 10.1093/petrology/egq036.

Klein, E.M. and Langmuir, C.H. 1987. Ocean ridge basalt chemistry, axial depth, crustal thickness and temperature variations in the mantle. *Journal of Geophysical Research* 92: 8089–115.

Klein, E.M. and Langmuir, C.H. 1989. Local versus global variations in ocean ridge basalt composition: a reply. *Journal of Geophysical Research* 94: 4241–52.

Lee, K.K.M., O'Neill, B., Panero, W.R., et al. 2004. Equations of state of the high-pressure phases of a natural peridotite and implications for the Earth's lower mantle. *Earth and Planetary Science Letters* 223: 381–93.

Li, J. and Agee, C. 1996. Geochemistry of mantle–core differentiation at high pressure. *Nature* 381: 686–9.

Lodders, K., 2003. Solar system abundances and condensation temperatures of the elements. *Astrophys. Journal* 591: 1220–47.

Lyubetskaya, T. and Korenaga, J. 2007. Chemical composition of Earth's primitive mantle and its variance: 1. Method and results. *Journal of Geophysical Research* 112: B03211. doi: 10.1029/2005jb004223.

Marschall, H. R., Schumacher, J. C. 2012. Arc magmas sourced from mélange diapirs in subduction zones. *Nature Geoscience* 5: 862. doi: 10.1038/ngeo1634.

Mattern, E., Matas, J., Ricard, Y. and Bass, J. 2005. Lower mantle composition and temperature from mineral physics and thermodynamic modelling. *Geophys. J. Int.* 160: 973–90.

McDonough, W. F., 2018. Earth's Core. In: *Encyclopedia of Geochemistry* (ed White, W. M.), pp. 418–429, Springer International Publishing, Cham.

McDonough, W.F. and Sun, S.-S. 1995. The composition of the Earth. *Chemical Geology* 120: 223–53.

McKenzie, D.P. and O'Nions, R.K. 1983. Mantle reservoirs and ocean island basalts. *Nature* 301: 229–31.

McNamara, A. K., Zhong, S. 2005. Thermochemical structures beneath Africa and the Pacific Ocean. *Nature* 437: 1136–1139. doi: 10.1038/nature04066.

Meyzen, C. M., Blichert-Toft, J., Ludden, J. N., Humler, E., Mevel, C. and Albarede, F., 2007. Isotopic portrayal of the Earth's upper mantle flow field. *Nature* 447, 1069–1074. doi: 10.1038/nature05920.

Miyashiro, A. 1974. Volcanic rock series in island arcs and active continental margins. *American Journal of Science* 274: 321–355. doi: 10.2475/ajs.274.4.321.

Montelli, R., Nolet, G., Dahlen, F.A., et al. 2004. Finite-frequency tomography reveals a variety of plumes in the mantle. *Science* 303: 338–43.

Morgan, W.J. 1971. Convection plumes in the lower mantle. *Nature* 230: 42–3.

Mundl, A., Touboul, M., Jackson, M. G., et al. 2017. Tungsten-182 heterogeneity in modern ocean island basalts. *Science* 356: 66–69. doi: 10.1126/science.aal4179.

Murakami, M., Hirose, K., Kawamura, K., Sata, N. and Ohishi, Y. 2004. Post-perovskite phase transition in $MgSiO_3$. *Science* 304: 855–8.

Nielsen, S. G., Marschall, H. R. 2017. Geochemical evidence for mélange melting in global arcs. *Science Advances* 3: e1602402. doi: 10.1126/sciadv.1602402.

Nutman, A. P., Bennett, V. C., Friend, C. R. L. 2015. Proposal for a continent 'Itsaqia' amalgamated at 3.66 Ga and rifted apart from 3.53 Ga: Initiation of a Wilson Cycle near the start of the rock record. *American Journal of Science* 315: 509–36. doi: 10.2475/06.2015.01.

Nyblade, A.A. and Pollack, H.N. 1993. A global analysis of heat flow from Precambrian terrains: implications for the thermal structure of Archean and Proterozoic lithosphere. *Journal of Geophysical Research* 98: 12207–18.

O'Neil, J., Carlson, R.L., Francis, D. and Stevenson, R.K. 2008. Neodymium-142 evidence for Hadean mafic crust. *Science* 321: 1828–31.

O'Neill, H.S., Canil, D. and Rubie, D.C. 1998. Oxide–metal equilibria to 2,500 degrees C and 25 GPa: implications for core formation and the light component in the Earth's core. *Journal of Geophysical Research* 103: 1223–60.

O'Neill, H.S.C. and Palme, H. 2008. Collisional erosion and the non-chondritic composition of the terrestrial planets. *Philosophical Transactions of the Royal Society of London Series A* 366: 4205–38. doi: 10.1098/rsta.2008.0111.

Oganov, A.R. and Ono, S. 2004. Theoretical and experimental evidence for a post-perovskite phase of $MgSiO_3$ in Earth's "D" layer. *Nature* 430: 445–8.

Palme, H. and O'Neill, H.S.C. 2003. Cosmochemical estimates of mantle composition, in *The Mantle and Core. Treatise on Geochemistry*, 2 (ed. R.L. Carlson), pp. 1–38. Amsterdam, Elsevier.

Parman, S. W. 2015. Time-lapse zirconography: imaging punctuated continental evolution. *Geochemical Perspectives Letters* 1: 43–52. doi: 10.7185/geochemlet.1505.

Pearson, D. G., Wittig, N. (2014) 3.6 - The formation and evolution of cratonic mantle lithosphere – evidence from mantle xenoliths. In: Holland H.D. and Turekian, K.K. (eds) *Treatise on Geochemistry (Second Edition)*, vol. 3. Elsevier, Oxford, pp 255–92. doi: 978-0-08-098300-4.

Pearson, D.G., Canil, D. and Shirey, S.B. 2003. Mantle samples included in volcanic rocks: xenoliths and diamonds, in *The Mantle and Core. Treatise on Geochemistry*, 2 (ed. R.L. Carlson), pp. 171–275. Amsterdam, Elsevier.

Perfit, M.R., Gust, D.A., Bence, A.E., Arculus, R.J. and Taylor, S.R. 1980. Chemical characteristics of island-arc basalts: implications for mantle sources. *Chemical Geology* 30: 227–56.

Peters, B. J., Carlson, R. W., Day, J. M. D. and Horan, M. F., 2018. Hadean silicate differentiation preserved by anomalous $^{142}Nd/^{144}Nd$ ratios in the Réunion hotspot source. *Nature* 555, 89. doi: 10.1038/nature25754.

Plank, T. and Langmuir, C.H. 1988. An evaluation of the global variations in the major element chemistry of arc basalts. *Earth and Planetary Science Letters* 90: 349–70.

Plank, T. and Langmuir, C.H. 1993. Tracing trace elements from sediment input to volcanic output at subduction zones. *Nature* 362: 739–42.

Plank, T., Kelley, K. A., Zimmer, M. M., Hauri, E. H., Wallace, P. J. 2013. Why do mafic arc magmas contain ~4 wt% water on average? *Earth and Planetary Science Letters* 364: 168–79. doi: 10.1016/j.epsl.2012.11.044.

Poirier, J.-P. 1991. *Introduction to the Physics of the Earth's Interior*. Cambridge, Cambridge University Press.

Pollack, H.N. 1982. The heat flow from the continents. *Annual Review of Earth and Planetary Sciences* 10: 459–81.

Porter, K.A. and White, W.M. 2009. Deep mantle subduction flux. *Geochem. Geophys. Geosyst.* 10: Q12016. doi: 10.1029/2009gc002656.

Puetz, S. J., Ganade, C. E., Zimmermann, U., Borchardt, G. 2018. Statistical analyses of global U-Pb database 2017. *Geoscience Frontiers* 9: 121–45. doi: 10.1016/j.gsf.2017.06.001.

Putirka, K. D., Perfit, M., J., R. F., Jackson, M. G. 2007. Ambient and excess mantle temperatures, olivine thermometry, and active vs. passive upwelling. *Chemical Geology* 241: 177–206. doi: 10.1016/j.chemgeo.2007.01.014.

Putirka, K., Tao, Y., Hari, K. R., Perfit, M. R., Jackson, M. G., Arevalo, R. 2018. The mantle source of thermal plumes: Trace and minor elements in olivine and major oxides of primitive liquids (and why the olivine compositions don't matter). *American Mineralogist* 103: 1253–70. doi: 10.2138/am-2018-6192.

Rielli, A., Tomkins, A. G., Nebel, O., et al. 2017. Evidence of sub-arc mantle oxidation by sulphur and carbon. *Geochemical Perspectives Letters* 3: 124–32. doi: 10.7185/geochemlet.1713

Righter, K. and Drake, M.J. 2000. Metal/silicate equilibrium in the early Earth – new constraints from the moderately volatile siderophile elements Ga, Cu, P, and Sn. *Geochimica et Cosmochimica Acta* 64: 3581–97.

Righter, K., Drake, M.J. and Yaxley, G. 1997. Prediction of siderophile element metal-silicate partition coefficients to 20 GPa and 2800°C: the effects of pressure, temperature, oxygen fugacity, and silicate and metallic melt compositions. *Physics of the Earth and Planet Interiors* 100: 115–34.

Ringwood, A.E. 1975. *Composition and Petrology of the Earth's Mantle*. New York, McGraw-Hill.

Ringwood, A.E. 1991. Phase transformation and their bearing on the constitution and dynamics of the mantle. *Geochimica et Cosmochimica Acta* 55: 2083–10.

Ringwood, A.E. and Hibbertson, W. 1990. The system Fe–FeO revisited. *Phys. Chem. Mineral.* 17: 313–19.

Rizo, H., Walker, R. J., Carlson, R. W., et al. 2016. Preservation of Earth-forming events in the tungsten isotopic composition of modern flood basalts. *Science* 352: 809–12. doi: 10.1126/science.aad8563.

Roeder, P.L. and Emslie, R.F. 1970. Olivine-liquid equilibrium. *Contributions to Mineralogy and Petrology* 29: 275–89.

Romanowicz, B. 2017. The buoyancy of Earth's deep mantle. *Nature* 551: 308–9. doi: 10.1038/551308a.

Rubin, K.H. and Sinton, J.M. 2007. Inferences on mid-ocean ridge thermal and magmatic structure from MORB compositions. *Earth and Planetary Science Letters* 260: 257–76.

Rudnick, R.L. and Fountain, D.M. 1995. Nature and composition of the continental crust: a lower crustal perspective. *Reviews in Geophysics* 33: 267–309.

Rudnick, R. L., Gao, S. (2014) 4.1 - Composition of the Continental Crust. In: Holland HD, Turekian KK (eds) *Treatise on Geochemistry* (Second Edition), vol. Elsevier, Oxford, pp 1–51. 978-0-08-098300-4.

Salters, V. J. M., Stracke, A. 2004. Composition of the depleted mantle. *Geochemistry Geophysics Geosystems* 5: Q05B07. doi: 10.1029/2003GC000597.

Schilling, J.-G. 1973. Iceland mantle plume: geochemical study of the Reykjanes Ridge. *Nature* 242: 565–71.

Scholl, D. W., von Huene, R. 2007. Crustal recycling at modern subduction zones applied to the past—Issues of growth and preservation of continental basement crust, mantle geochemistry, and supercontinent reconstruction. *Geological Society of America Memoirs* 200: 9–32. doi: 10.1130/2007.1200(02).

Shaw, D.M. 1967. An estimate of the chemical composition of the Canadian Shield. *Canadian Journal of Earth Sciences* 4: 829.

Shaw, D.M., Gramer, J.J., Higgins, M.D. and Truscott, M.G. 1986. Composition of the Canadian Precambrian shield and the continental crust of the earth, in *The Nature of the Lower Continental Crust* (eds J.B. Dawson, J. Hall and K.H. Wedepohl), pp. 275–82. Oxford, Blackwell Scientific Publishers.

Shimizu, N., 2019. Big-Picture Geochemistry from microanalyses – my four-decade odyssey in SIMS. *Geochemical Perspectives* 8(1): 1–104. doi: 10.7185/geochempersp.8.1.

Sleep, N.H. 1990. Hotspots and mantle plumes: some phenomenology. *Journal of Geophysical Research* 95: 6715–36.

Stachel, T., Viljoen, K. S., Brey, G., Harris, J. W. 1998. Metasomatic processes in lherzolitic and harzburgitic domains of diamondiferous lithospheric mantle: REE in garnets from xenoliths and inclusions in diamonds. *Earth and Planetary Science Letters* 159: 1–12. doi: 10.1016/S0012-821X(98)00064-8

Stern, C. R. 2011. Subduction erosion: Rates, mechanisms, and its role in arc magmatism and the evolution of the continental crust and mantle. *Gondwana Research* 20: 284–308. doi: 10.1016/j.gr.2011.03.006.

Stern, R. J. 2008. Modern-style plate tectonics began in Neoproterozoic time: An alternative interpretation of Earth's tectonic history. *Geological Society of America Special Papers* 440: 265–280. doi: 10.1130/2008.2440(13).

Stern, R. J., Scholl, D. W. 2010. Yin and yang of continental crust creation and destruction by plate tectonic processes. *International Geology Review* 52: 1–31. doi: 10.1080/00206810903332322.

Stolper, E. and Newman, S. 1994. The role of water in the petrogenesis of Mariana trough basalts. *Earth and Planetary Science Letters* 121: 293–325.

Sun, N., Wei, W., Han, S., et al. 2018. Phase transition and thermal equations of state of (Fe,Al)-bridgmanite and post-perovskite: Implication for the chemical heterogeneity at the lowermost mantle. *Earth and Planetary Science Letters* 490: 161–69. doi: 10.1016/j.epsl.2018.03.004.

Syracuse, E. M., van Keken, P. E., Abers, G. A. 2010. The global range of subduction zone thermal models. *Physics of the Earth and Planetary Interiors* 183: 73–90. doi: 10.1016/j.pepi.2010.02.004.

Taylor, H. P., Sheppard, S. M. F. (1986) Igenous rocks: I. Processes of isotopic fractionation and isotope systematics. In: Valley JW, Taylor HP, O'Neil JR (eds) *Stable Isotopes in High Temperature Geological Processes* vol 16. Mineral. Soc. Am., Washington, pp 227–71.

Taylor, S.R. 1967. The origin and growth of continents. *Tectonophysics* 4: 17–34.

Taylor, S.R. and McLennan, S.M. 1985. *The Continental Crust: its Composition and Evolution*. Oxford, Blackwell Scientific Publishers.

Taylor, S.R. and McLennan, S.M. 1995. The geochemical evolution of the continental crust. *Reviews in Geophysics* 33: 241–65.

Tera, F., Brown, L., Morris, J., et al. 1986. Sediment incorporation in island-arc magmas: inferences from [10]Be. *Geochimica et Cosmochimica Acta* 50: 535–50.

Turner, S. J., Langmuir, C. H. 2015. What processes control the chemical compositions of arc front stratovolcanoes? *Geochemistry, Geophysics, Geosystems* 16: 1865–93. doi: 10.1002/2014GC005633

Viljoen, K. S., Harris, J. W., Ivanic, T., Richardson, S. H., Gray, K. 2014. Trace element chemistry of peridotitic garnets in diamonds from the Premier (Cullinan) and Finsch kimberlites, South Africa: Contrasting styles of mantle metasomatism. *Lithos* 208–9: 1–15. doi: 10.1016/j.lithos.2014.08.010

Vitorello, I. and Pollack, H.N. 1980. On the variation of continental heat flow with age and the thermal evolution of continents. *Journal of Geophysical Research* 85: 983–95.

von Huene, R. and Scholl, D.W. 1991. Observations at convergent margins concerning sediment subduction, subduction erosion, and the growth of continental crust. *Reviews in Geophysics* 29: 279–316.

Wade, J. and Wood, B.J. 2001. The Earth's 'missing' niobium may be in the core. *Nature* 409: 75–8.

Weaver, B.L. 1991. Trace element evidence for the origin of oceanic basalts. *Geology* 19: 123–6.

Weaver, B.L. and Tarney, J. 1984. Major and trace element composition of the continental lithosphere, in *Structure and Evolution of the Continental Lithosphere, Physics and Chemistry of the Earth vol. 15* (eds. H.N. Pollack and V.R. Murthy), pp. 39–68. Oxford, Pergamon Press.

Wedepohl, K.H. 1995. The composition of the continental crust. *Geochimica et Cosmochimica Acta* 59: 1217–32.

Weichert, E. 1897. Über die Massenverteilung im Innern der Erde. Königliche Gesellschaft der Wissenschaften zu Göttingen Nachrichten, Mathematisch-physikalische Klasse, 3: 221–43.

White, W. M. and Klein, E. M., 2014. 4.13 - Composition of the Oceanic Crust. In: *Treatise on Geochemistry* (Second Edition) (eds. Holland, H. D. and Turekian, K. K.), pp. 457–96, Elsevier, Oxford. doi: 10.1016/B978-0-08-095975-7 .00315-6.

White, W. M. 2015. Probing the Earth's deep interior through geochemistry. *Geochemical Perspectives* 4: 95–251. doi: 10.7185/geochempersp.4.2.

White, W. M., Klein, E. M. (2014) 4.13 - Composition of the oceanic crust. In: Holland HD, Turekian KK (eds) *Treatise on Geochemistry* (Second Edition), vol. 3. Elsevier, Oxford, pp 457–96. doi: 978-0-08-098300-4.

White, W.M. 1985. Sources of oceanic basalts: radiogenic isotope evidence. *Geology* 13: 115–18.

White, W.M. 2010. Oceanic island basalts and mantle plumes: the geochemical perspective. *Annual Review of Earth and Planetary Sciences* 38: 133–60. doi: 10.1146/annurev-earth-040809-152450.

White, W.M. and Dupré, B. 1986. Sediment subduction and magma genesis in the Lesser Antilles: Isotopic and trace element constraints. *Journal of Geophysical Research* 91: 5927–41.

White, W.M. and Patchett, P.J. 1984. Hf-Nd-Sr isotopes and incompatible element abundances in island arcs: implications for magma origins and crust-mantle evolution. *Earth and Planetary Science Letters* 67: 167–85.

Willbold, M. and Stracke, A. 2006. Trace element composition of mantle end-members: Implications for recycling of oceanic and upper and lower continental crust. *Geochem. Geophys. Geosyst.* 7: Q04004. doi: 10.1029/2005gc001005.

Williams, Q., Garnero, E. J. 1996. Seismic evidence for partial melt at the base of earth's mantle. *Science* 273: 1528–30. doi: 10.2307/2891048.

Wilson, J.T. 1963. Evidence from islands on the spreading of the ocean floors. *Nature* 197: 536–8.

Wohlers, A. and Wood, B. J., 2015. A Mercury-like component of early Earth yields uranium in the core and high mantle ^{142}Nd. *Nature* 520(7547), 337–40. doi: 10.1038/nature14350.

Workman, R. and Hart, S.R. 2005. Major and trace element composition of the depleted MORB mantle (DMM). *Earth and Planetary Science Letters* 231: 53–72. doi: 10.1016/j.epsl.2004.12.005.

Workman, R. K., Eiler, J. M., Hart, S. R., Jackson, M. G. 2008. Oxygen isotopes in Samoan lavas: Confirmation of continent recycling. *Geology* 36: 551–54. doi: 10.1130/g24558a.1.

Workman, R., Hart, S.R., Jackson, M., et al. 2004. Recycled metasomatized lithosphere as the origin of the Enriched Mantle II (EM2) end member: Evidence from the Samoan Volcanic Chain. *Geochem. Geophys. Geosyst.* 5, doi: 10.1029/2003GC000623.

Zhu, G., Gerya, T. V., Yuen, D. A., Honda, S., Yoshida, T., Connolly, J. A. D. 2009. Three-dimensional dynamics of hydrous thermal-chemical plumes in oceanic subduction zones. *Geochemistry, Geophysics, Geosystems* 10: doi: 10.1029/2009GC002625.

Zindler, A. and Hart, S.R. 1986. Chemical geodynamics. *Annual Review of Earth and Planetary Science* 14: 493–571.

PROBLEMS

1. Assuming the Earth has a CI chondritic composition (Table 10.2), and using the values for the mass of the core and the mass of the mantle + crust in Table 11.1, calculate what the concentrations of Re, Ir, Mo, and Ag should be in the bulk silicate Earth if the core formed by an equilibrium process, when the silicate/metal partition coefficients are 5×10^{-4}, 5×10^{-5},

8×10^{-4}, and 0.01, respectively. Compare your results with the primitive mantle and core values in Tables 11.3 and 11.4.

2. Calculate new bulk silicate Earth concentrations of Re, Ir, Mo, and Ag by adding 1% CI chondritic material to your results from Problem 1. Again compare these results with primitive mantle values of Table 11.3.

3. Assume that the bulk silicate Earth has an $^{87}Sr/^{86}Sr$ ratio of 0.7045, an initial ratio equal to BABI (0.69897), and an age of 4.45 Ga.

   (a) Calculate the bulk silicate Earth $^{87}Rb/^{86}Sr$ ratio.
   (b) Using your result in (a) and assuming that ^{86}Sr is 9.86% of Sr and ^{87}Rb is 27.83% of Rb (by mass), what is the bulk Earth Rb/Sr ratio?

4. Suppose a rising mantle plume is 220°C hotter than surrounding mantle and that the ringwoodite–bridgmanite transition occurs at 660 km depth in the surrounding mantle. Assume a homogenous mantle density of 3.5 g/cc at 660 km and above and the Clapeyron slope mentioned in section 11.2.4.2. At what depth does the ringwoodite–bridgmanite transition occur within the mantle plume?

5. Whether the ringwoodite–bridgmanite transition is endothermic or exothermic will affect mantle convection (i.e., sinking of lithospheric slabs, rise of mantle plumes). Discuss why this is so, explaining the effects of both endothermic and exothermic phase transitions on rising mantle plumes and sinking lithospheric slabs.

6. Use a mass balance approach and equations 11.22 through 11.28 to calculate the Nd concentration of the mantle depleted by continent formation (DMM). Take the initial and present chondritic $^{143}Nd/^{144}Nd$ ratios to be 0.506677 and 0.512630, the ε_{Nd} of the Earth to be +2, ε_{Nd} of the depleted mantle to be +9, the $^{147}Sm/^{144}Nd$ ratio of the continental crust to be 0.123, and the average age of the continents (T) to be 2.0 Ga. Use the Nd concentrations of the silicate Earth and the continental crust given in Tables 11.4 and 11.11.

7. A peridotite xenolith in a kimberlite from the Kapvaal Craton has $^{187}Re/^{188}Os = 0.096$ and a $^{187}Os/^{188}Os = 0.1086$. The eruption age of the kimberlite is 90 Ma.

   (a) Using the decay constant listed in Table 8.2. calculate the $^{187}Os/^{188}Os$ at the eruption age.
   (b) Calculate the rhenium depletion age (T_{RD}) and Re–Os defined in eqn. 8.31. Use the silicate Earth $^{187}Os/^{188}Os$ of 0.1296 and $^{187}Re/^{188}Os$ of 0.4343 in place of the chondritic value.
   (c) Calculate the Re–Os (T_{MA}) model age (eqn. 8.32), again using silicate Earth $^{187}Os/^{188}Os$ and $^{187}Re/^{188}Os$.

Chapter 12

Organic geochemistry, the carbon cycle, and climate

12.1 INTRODUCTION

Organic compounds are ubiquitous on the surface of the Earth: some form living organisms, some are a result of pollution, some have been leaked or excreted from living cells, but most are breakdown products of dead organisms. Organic substances may be either water-insoluble or water-soluble. Rain typically contains about 1 µg/l dissolved organic carbon (DOC), the ocean around 0.5 mg/l DOC, and soils up to 250 mg/l. In general, insoluble organic compounds in soil are more abundant than water-soluble ones. In addition to their value as fuel (as peat, coal, petroleum, and natural gas), organic substances are also important in controlling the properties of soil, as weathering agents, and as a significant fraction of surficial carbon, which cycles between the ocean, the atmosphere, the biosphere, soils, and rocks as both inorganic carbon in carbonates and organic carbon. Organic carbon reservoirs (biota, soils, coal, petroleum, etc.) exchange with the atmosphere, where carbon (as CO_2 and CH_4) plays an important role in regulating the Earth's surface temperature. Hence, organic carbon plays an indirect role in climate regulation. All of these provide very good reasons for acquiring a better understanding of organic substances and the role they play in geologic processes.

In this chapter, we begin with a brief review of some basic biology and organic chemistry. We then examine how organic compounds are produced and used by organisms. We continue to survey the distribution of organic compounds in water and soil. We then examine their geochemical properties and their roles as complexing agents and adsorbents. We then examine how certain organic molecules preserved in sediment provide insights into past environments and the processes by which sedimentary organic matter evolves into important energy and chemical resources such as coal, gas, and oil. In the final section, we examine how carbon cycles between various organic and inorganic forms at the surface of and within the Earth. We'll see that the abundance of CO_2 and CH_4 in the atmosphere has controlled the Earth's climate throughout its entire history.

Before we do that, however, we need to define the term *organic* in a chemical sense. One definition of an "organic" compound is one produced biologically. However, some compounds generally considered organic can be produced both abiologically* and biologically, so this is not very useful. A common definition of an "organic compound" is one that contains carbon atoms. However, carbonates, carbides, simple carbon oxides (e.g., CO_2, CO), and native carbon (graphite, diamond) are not considered

---

* For example, abiogenic methane is found in hydrothermal fluids on mid-ocean ridges.

Geochemistry, Second Edition. William M. White.
© 2020 John Wiley & Sons Ltd. Published 2020 by John Wiley & Sons Ltd.
Companion website: www.wiley.com/go/white/geochemistry

organic compounds. An alternative, and perhaps preferred definition of an organic compound, and the one we will adopt, is one containing at least one carbon–hydrogen bond. We will refer to molecules synthesized by life as *biomolecules*.

12.2 A BRIEF BIOLOGICAL BACKGROUND

Living organisms are the ultimate source of most, but not all, organic compounds in the environment. While life forms are extremely varied, the basic principles of cell operation are the same in all organisms (excluding viruses). For example, all derive the energy they require through oxidation of organic compounds, all contain DNA (deoxyribonucleic acid), which is the blueprint for synthesis of proteins that carry out various intracellular functions, and all use ATP (adenosine triphosphate) in intracellular energy transactions. This commonality suggests all organisms have evolved from a common ancestor.

Over the years, there have been a variety of schemes for classifying organisms, and one may still find any one of several in biological textbooks. Figure 12.1 presents a simplified version of the Woese system, based on ribosomal RNA sequencing, which has now achieved wide acceptance. It divides organisms into three kingdoms: *archaea* (sometimes called *archaebacteria*), *eubacteria* (or more often simply *bacteria*), and *eukaryotes*. The first two of these kingdoms consist only of simple unicellular organisms; all other organisms, including many unicellular and all multicellular organisms, are eukaryotes. Eubacteria and archaea are collectively called *prokaryotes*. The two groups differ from each other and from eukaryotes in fundamental ways, including metabolic pathways, gene transcription, and the nature of their cell membranes. In contrast to eukaryotes, the DNA of prokaryotes is not contained within a membrane-bound nucleus; rather, it is simply concentrated within one or more regions, called nucleoids, of intracellular fluid, or cytoplasm (a protein solution). In both eukaryotes and prokaryotes, messenger RNA (m-RNA) carries information from DNA, in the form of a complement of a portion of a DNA strand, to *ribosomes*, which consist of RNA and proteins, where proteins are synthesized

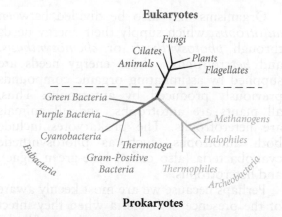

Figure 12.1 Phylogenetic relationships among organisms. The eubacteria and the archeobacteria constitute the prokaryotes; all other organisms are eukaryotes.

by transcription RNA (t-RNA) from amino acids.

An additional contrast between prokaryotes and eukaryotes is that eukaryotes contain a variety of specialized intracellular structures whereas, except for ribosomes, prokaryotes do not. These include *mitochondria*, where energy is generated by oxidation of carbohydrates, *chloroplasts* (in plants), where *photosynthesis* takes place, the Golgi apparatus, which is involved in modifying proteins, and networks of structural proteins that, among other things, participate in changing the shape of the cell so that it can move. Some of these *organelles*, such as the mitochondria and chloroplasts, have their own DNA with affinities to that of bacteria. Gene structure, translation and transcription in eukaryotes is more similar to archaea than bacteria. This and other evidence suggest these organelles may have evolved from bacterial cells living symbiotically within archaeal hosts that ultimately evolved into eukaryotes. Eukaryotic cells typically have dimensions of about 10 μm, and are therefore much larger than prokaryotes, which generally have largest dimensions of 1 μm or less. The eukaryotes may be further divided into a variety of kingdoms that include single-celled organisms (such as diatoms and foraminifera) called *protists,* and multicellular organisms including as *archaeplastids* (plants) and *metazoans* (such as us).

Organisms may also be divided between *autotrophs*, which supply their energy needs through *photosynthesis* or *chemosynthesis*, and *heterotrophs*, whose energy needs are supplied by assimilating organic compounds previously produced by autotrophs. Thus, all plants are autotrophs, and all animals are heterotrophs. The prokaryotes include both autotrophs, such as photosynthetic cyanobacteria (also called blue-green algae), and heterotrophs.

Perhaps because we are most keenly aware of the presence of bacteria when they infect us, we often think of bacteria as "bad." However, bacteria are ubiquitous and play essential roles in every ecosystem; only a small fraction is pathogenic. All chemosynthetic organisms are prokaryotes, as are most organisms capable of living without free oxygen (anaerobes). Archaea may be found in extremely hostile environments such as saline lakes and fumaroles. Prokaryotes play the most important roles in converting the chemical products of life to organic substances found in sediments and soils; thus they are of particular interest in geochemistry. Some prokaryotes reduce sulfate, others oxidize sulfide, some produce methane (methanogens), while others consume it (methanotrophs). Some prokaryotes reduce nitrogen to ammonia, a process called nitrogen fixation, others oxidize ammonia to nitrate (nitrification), and yet others convert nitrate to nitrogen (denitrification). Symbiotic bacteria are essential to the digestive systems of higher animals. *E. coli* in the human gut, for example, synthesizes a number of essential nutrients that are then assimilated through intestinal walls. On the whole, prokaryotes play a much more important role in biogeochemical cycling and geochemical processes than do the metazoans.

12.3 ORGANIC COMPOUNDS AND THEIR NOMENCLATURE

Organic chemistry can be an intimidating subject involving a bewildering array of compounds and their names whose properties depend as much on the details of their structures as on their composition. A complex nomenclature of organic chemistry has evolved because simply referring to an organic compound by its chemical formula is not sufficient to identify the compound.

Compositionally identical compounds can have different structures and different properties. This is, of course, true of inorganic compounds as well, for example, quartz and tridymite, calcite and aragonite, or graphite and diamond, but isomers (i.e., compositionally identical but structurally distinct compounds) are particularly common among organic compounds. Organic nomenclature is in some sense a language unto itself. Like any language, organic nomenclature has a "grammar." Once this grammar is mastered, the exact composition and structure of a compound can be communicated through its name alone. We will need to concern ourselves with only a part of that language.

We will make one simplification in the brief review of organic chemistry that follows: we will ignore, with some exceptions, the phenomenon known as *stereoisomerism*. Organic compounds that are otherwise structurally identical but are mirror images of one another are said to be *stereoisomers*. The difference in physical properties of stereoisomers can be quite small, no more than the direction of rotation of the plane of polarized light, but the difference is often biochemically important. One aspect is important for us: life creates biomolecules in only a single specific version of possible stereoisomers, for example, it creates only left-handed versions of amino acids, but outside cells these can invert to other versions. Indeed, the extent to which some molecules transform to other isomers can be used as a measure of the heating experienced (thermal maturity) of such molecules, as we will find in Section 12.7.5.

12.3.1 Hydrocarbons

Hydrocarbons are the simplest group of organic compounds in that they consist only of carbon and hydrogen. We can think of hydrocarbons, and indeed all organic compounds, as consisting of a basic skeleton of carbon atoms to which various functional groups can be attached to form other sorts of organic compounds. The simplest skeleton is that of the *aliphatic* hydrocarbons, also called *paraffins*, or *alkanes*, which consist of a straight or branched chain or ring of carbon and hydrogen atoms. If they are unbranched, they called *normal* or *n-alkanes*. Individual compounds are given names with a prefix indicating the number of CH_x (x = 2, 3) groups present and an ending "-ane". The simplest such molecule is *methane*, CH_4. Ethane consists of two CH groups, propane 3, and butane 4. Beyond that, the root of the name is based on the Greek word for the number of carbon atoms in the chain (i.e., pentane for five carbons, hexane for 6, heptane for 7, etc.). Radicals formed by removing a hydrogen from a terminal carbon are named by replacing the "-ane" suffix with "-yl", e.g., methyl, butyl. As a group, the radicals formed from alkanes in this way are called *alkyls* (and, thus, the group name also conforms to the naming convention).

Alkanes are not, however, actually straight chains, although they are often represented that way, as they are Figure 12.2a, which shows methane and butane with bonds drawn as right angles. Figure 12.2b provides a "stereochemistry" illustration of methane, showing that the bond angles between hydrogens are actually 109.5°, not 90°. Figures 12.2c and d shows 3-D model illustrations of the methane and n-butane molecules.

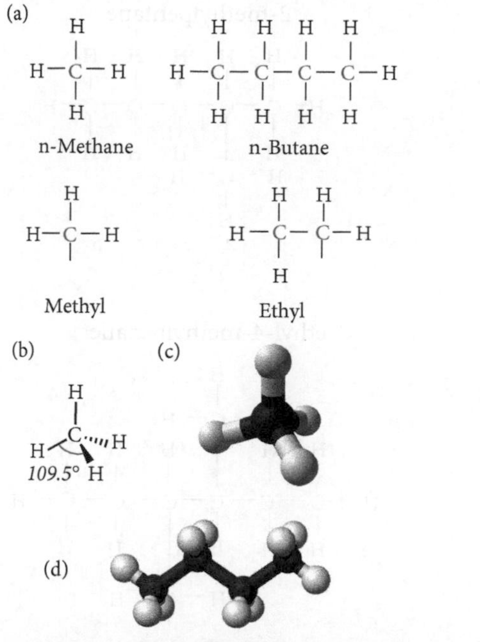

Replacing one of the hydrogens with a carbon atom, to which additional hydrogens are attached, forms branched chains. These are named by prefixing the designations of the side chains to the name of the longest chain in the formula. A number is prefixed indicating the carbon, counting from the nearest end, to which the secondary chain is attached. An example is 2-methylpentane (Figure 12.3). If more than one secondary group is attached, the groups are listed in alphabetical order, for example, 3-ethyl-4-methylhexane shown in Figure 12.3. When several of the same group are attached, a multiplier corresponding to that number precedes the name of the group. An example is 2,4-dimethylhexane, shown in Figure 12.3.

Compounds where all carbon atoms have single bonds to four other atoms are said to be *saturated* (the term *saturated* arises from carbon being bonded to the maximum possible number of hydrogens: i.e., the carbon is hydrogen-saturated). Carbon atoms that are double bonded are termed *olefinic units*. Compounds containing one or more

Figure 12.2 (a) Simple *n*-alkanes or chain hydrocarbons. The suffix "-ane" is used to refer to molecules and the suffix "-yl" is used to refer to equivalent radicals formed by removing a hydrogen. (b) A stereochemistry illustration of the methane molecule, showing the 109.5° angle between H-C-H bonds. The solid wedge indicates a bond coming out of the paper toward the viewer, a hashed wedge is one pointing away from the viewer, and a straight line is a bond in the plane of the paper. (c) 3-D illustration of methane. (d) Stick illustration of n-butane.

pairs of doubly bonded carbons are said to be *unsaturated*. Unsaturated, unbranched acyclic hydrocarbons having one double bond are named by replacing the suffix "-ane" by "-ene," for example, ethylene: $CH_2=CH_2$. A number is used to specify the location (the carbon atom) of the double bond, for example, 2-butene, shown in Figure 12.4. If there are more than two double bonds the ending becomes "-adiene," "-atriene,", and so on. Generic names are *alkene*, *alkadiene*, for example. Triple carbon bonds are also possible, in which case the suffix becomes "yne" for a single triple bond. Acetylene* (Figure 12.4) is an example of a compound

---

* As is the case in most languages, there are words in the lexicon of organic geochemistry that do not conform to the standard grammar. Acetylene, named before the naming conventions were developed, is an example of such an irregular term.

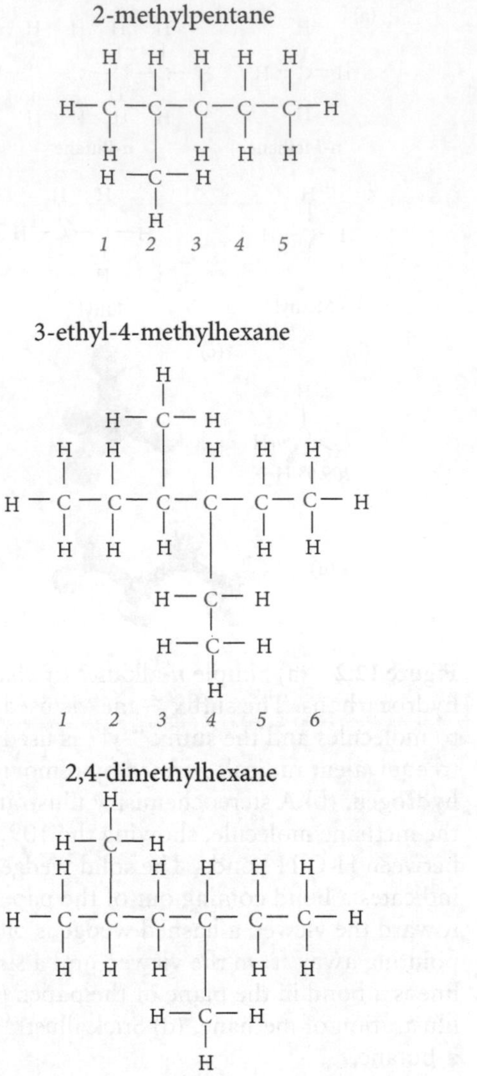

2-methylpentane

3-ethyl-4-methylhexane

2,4-dimethylhexane

Figure 12.3 Some simple and branched hydrocarbon chains.

$$>C=C<$$

Olefinic group

2-Butene

$$H-C\equiv C-H$$

Acetylene

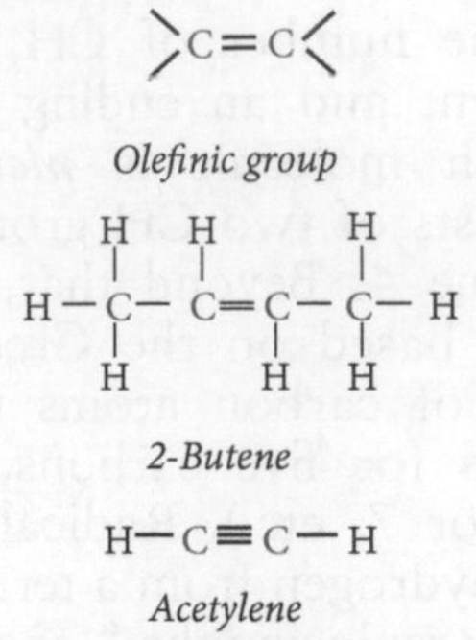

Figure 12.4 Some simple doubly and triply-bonded hydrocarbons.

containing triply bonded carbon. Particularly stable compounds result when carbon bonds are *conjugated*, that is, alternately singly and doubly bonded, for example, –C=C–C=C–C=C–.

Instead of forming chains, the C atoms may form rings; the resulting compounds are called *cyclic hydrocarbons*. Naming conventions for the simple groups are similar to those for chains with the prefix "cyclo-" used to indicate the cyclic nature (e.g., cyclopropane).

A particularly important cyclic structure is the benzene ring, which consists of six conjugately bonded carbon atoms lying in a single plane (Figure 12.5). Compounds based on this structure are particularly stable and are referred to as *aromatic*.* Representation of this structure as alternating single and double bonds is not entirely accurate. The carbon–carbon bond in a saturated alkane such as ethane is 1.54 Å in length; the double bond in ethylene is 1.33 Å in length. All carbon–carbon bonds in the benzene ring are found to be intermediate in length (1.40 Å). Thus, bonding is delocalized, that is, all carbon–carbon bonds in the ring are of approximately equal strength and the double bonds appear to be shared among all carbon atoms in the ring, forming a conjugated π-bond. This delocalization is responsible for the particular stability of this structure. Several rings may be joined to form *polyaromatic units*. Conjugated bonds, which can occur both in aromatic and aliphatic units, absorb visible light, and therefore can give rise to color in organic compounds, as they do chlorophyll and hemoglobin.

12.3.2 Functional groups

From these basic hydrocarbon structures, a great variety of other organic compounds may be formed by replacing hydrogen or carbon atoms with other atoms or functional groups. The basic naming conventions discussed above for hydrocarbons also apply to these other organic molecules. Additional names, prefixes, and suffixes are used to indicate the presence of attached groups replacing hydrogen or other atoms replacing carbon in

---

* Some aromatic compounds, such as benzene, toluene, and a variety of chlorinated phenols, are highly toxic. Because of the stability imparted by the conjugate bonding, they are particularly environmentally hazardous.

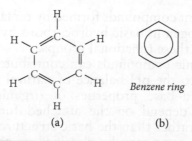

Figure 12.5 Two representations of the benzene ring, the foundation of aromatic hydrocarbons. In (b) the C and H atoms have been removed for clarity. Since all carbon–carbon bonds in the benzene ring are equivalent, (b) is actually a better representation than (a).

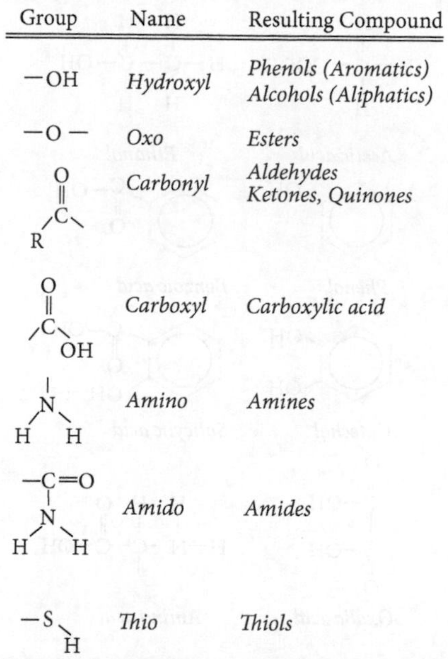

Group	Name	Resulting Compound
—OH	Hydroxyl	*Phenols (Aromatics)* *Alcohols (Aliphatics)*
—O—	Oxo	*Esters*
$\overset{O}{\underset{R}{\parallel\!C}}$	Carbonyl	*Aldehydes* *Ketones, Quinones*
$\overset{O}{\underset{OH}{\parallel\!C}}$	Carboxyl	*Carboxylic acid*
$\underset{H\quad H}{N}$	Amino	*Amines*
$\underset{H\quad H}{\overset{C=O}{N}}$	Amido	*Amides*
$\underset{H}{—S}$	Thio	*Thiols*

Figure 12.6 Important functional groups found in organic compounds.

the basic structure. Attachment of functional groups gives rise to additional nomenclature: the first carbon, which has a functional group attached is designated alpha (α), with succeeding carbons then designated β, γ, etc.

The most important functional groups are summarized in Figure 12.6. The *hydroxyl* (OH) unit may be attached to an aromatic ring to form *phenols* or to aliphatic units to form *alcohols*. The biologically important *carbohydrates* (e.g., sugars and starches) are compounds with the general formula of $(CH_2O)_n$. They are either aliphatic or cyclic hydrocarbons in which some hydrogens have been replaced by OH. The carbonyl group (C=O) forms *aldehydes* when the "R" in Figure 12.6 is H, *ketones* when "R" is either an aliphatic or aromatic group, and *quinones* when the carbon in the group is incorporated into an aromatic ring.

A particularly important functional group from a geologic perspective is the *carboxylic acid* group: COOH. The suffix "-oic acid" is used to designate compounds formed by carboxylic acid groups: for example, methanoic acid is a carboxyl group with an attached H, ethanoic acid is a carboxyl group with an attached CH_3, and benzoic acid is a carboxyl group with an attached benzene ring. Many of the carboxylic acids also have more familiar names, for example, ethanoic acid is more commonly called *acetic acid*, while methanoic acid is also called *formic acid*. Two

carboxyl units bound together form *oxalic acid*. A compound formed by replacement of the dissociable hydrogen in the carboxyl group with some other group is designated by the suffix "-ate."

Carboxyl groups attached to hydrocarbon chains form *fatty acids*, which are important lipids. If one or more of the hydrogens attached to the carbon chain or ring is substituted by a hydroxyl group, the compound is known as a *hydroxy acid* (e.g., salicylic acid illustrated in Figure 12.7). If a doubly bonded oxygen is substituted for two hydrogens attached to the chain (forming a *carbonyl group*), the compound is known as a *keto acid*. Hydroxy acids such as lactic acid, and keto acids such as pyruvic acid, are important in both the Calvin cycle, by which autotrophs synthesize organic compounds, and the Krebs cycle, by which organisms oxidize organic compounds to release energy. We'll discuss those cycles below.

Esters form by combining a carboxylic acid and an alcohol. In the reaction between these two, the OH is removed from the carboxyl group and the H is removed from the alcohol, leaving the two units

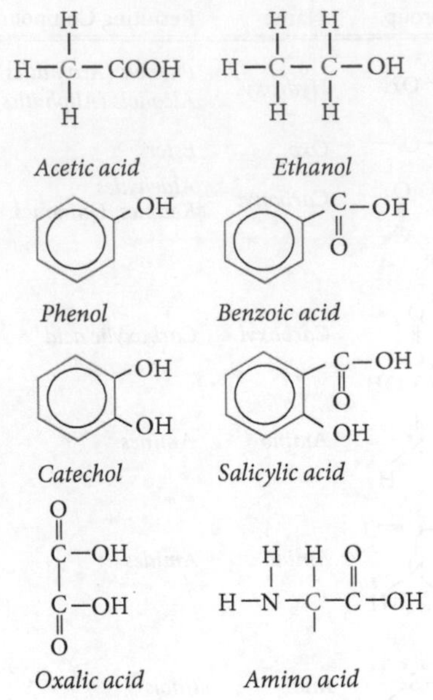

Figure 12.7 Some examples of compounds formed by substituting functional groups for hydrogen in basic hydrocarbon structures.

bound by an oxygen atom.* An example of such a reaction is the formation of pentyl acetate (which gives bananas their familiar odor) from acetic acid and pentyl alcohol:

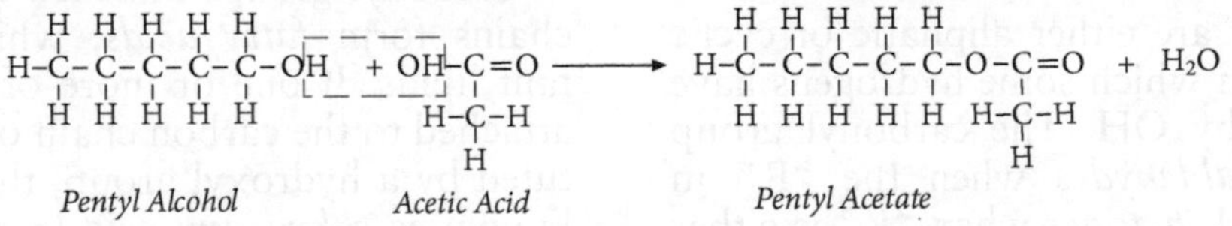

Another geologically and biologically important functional group is the *amino group*, NH_2 (the name being derived from ammonia, NH_3). *Amino acids*, the building blocks of proteins, consist of molecules containing both amino and carboxylic groups. Other important functional groups include the carbonyl group, consisting of a carbon that is double-bonded to oxygen, and thiol groups, where S replaces O in the OH group (i.e., SH). Figure 12.7 illustrates a few of the

important compounds formed by replacement of hydrogen in basic hydrocarbons by one or more of these functional groups.

Organic compounds can contribute significantly to the pH balance of natural waters. The acid–base properties of organic compounds depend on the attached functional groups rather than the basic structure. Functional groups may be classified as acidic or basic depending on their tendency to give up (acidic) or accept (basic) hydrogen ions. Carboxyl groups tend to be strongly acidic, phenols and quinones tend to be mildly acidic. Alcohols, ethers, ketones, and aldehydes are generally classified as neutral. The nitrogen in amines and amides binds free hydrogen ions fairly easily, hence these groups are basic.

The acid–base properties of organic substances are also directly related to their solubility in water. Because water can more easily dissolve ionic substances than neutral ones, strongly basic or strongly acidic compounds (those that have given up or accepted a proton) tend to be more soluble than neutral compounds. Thus, carboxylic acids are very water-soluble, amines somewhat less soluble, and neutral compounds such as esters and ether least soluble.

It is also possible to substitute other elements for carbon in the basic hydrocarbon structure; such atoms are known as *heteroatoms*. Examples are illustrated in Figure 12.8. The pyranyl group is a particularly important one because it forms the basis of many cyclic carbohydrates; the pyridinyl group is an important component of nucleic acids. The pyrrole group is a component of porphyrins, an important class of molecules that include hemes such as

* Such a reaction, in which two molecules combine to form a larger molecule with the elimination of a small molecule (H_2O in this case), is called a condensation reaction or simply *condensation*. The reverse of this reaction is termed *hydrolysis*.

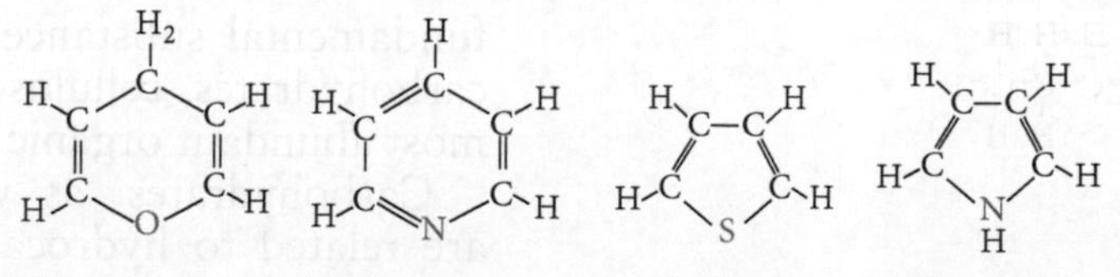

(a) *Pyranyl*   (b) *Pyridinyl*   (c) *Thiophenyl*   (d) *Pyrrole*

Figure 12.8 Examples of functional groups formed by replacing one carbon atom in the cyclic skeleton with atoms of O, N, and S. The names of the resulting compounds are (a) pyranyl, (b) pyridinyl, (c) thiophenyl, and (d) pyrrole.

hemoglobin, chlorins, bacteriochlorins, and chlorophylls.

12.3.3 Short-hand notations of organic molecules

For both aliphatic and cyclic molecules, the number of carbons in the ring or chain is commonly denoted with C followed by a subscript corresponding to the number of carbons, such as C_6 for hexane. It is often convenient to use this notation in condensed structural formulae of long-chained aliphatic compounds. The basic repeating unit of such chains is CH_2. The number of repetitions can be expressed by enclosing the CH_2 in parentheses, followed by a subscript indicating the number or repetitions, for example $(CH_2)_6$. Groups placed on the ends then precede and follow. Thus, for example, stearic acid, a fatty acid consisting of a 17-carbon chain with a hydrogen on one end and a carboxyl group on the other, can be denoted as $CH_3(CH_2)_{16}C(O)OH$. If a double carbon bond occurs, this is designated by $CH=CH$ inserted at the appropriate place. Thus, palmitoleic acid, a common unsaturated fatty acid with the formula $CH_3(CH_2)_5CH=CH(CH_2)_7C(O)OH$ is a 15-carbon aliphatic compound with an olefinic unit between the seventh and eighth carbons, a H on one end, and a carboxyl acid group on the other. If an attached group occurs in the middle part of the chain, then the formula for the group is inserted in parentheses after the appropriate number of repetitions of the CH_2 unit, and a second $(CH_2)_n$ follows, n indicating the number of subsequent repetitions. Thus, the formula for 3-methylhexane would be $CH_3(CH_2)_2(CH_3)(CH_2)_2CH_3$.

There are also several conventions for illustrating the structure of organic molecules. We have already seen one: the hexagon with an enclosed circle to denote the benzene ring (Figure 12.5). As in this illustration, carbons and hydrogens bonded to them are often omitted from representations of organic molecules. We infer a carbon at each bend of the line as well as hydrogens bonded to it. A similar shorthand may be used for aliphatic molecules as well, as illustrated in Figure 12.9. We may summarize these abbreviated illustrations as follows. A carbon is inferred at each change in angle of the line as well as at the ends if nothing else is shown. Double bonds are indicated by double lines. Each carbon may have one or more hydrogens bonded to it. Since carbon always forms four bonds, the number of hydrogens is easily deduced as 4 minus the number of other bonds shown. Solid wedges represent bonds pointing above-the-plane of the drawing, dashed wedge represent bonds pointing below the plane.

12.3.4 Biologically important organic compounds

Life is, of course, based on organic compounds. A remarkable variety of organic compounds can be found in even the simplest cells. Many of these compounds are incredibly complex, commonly having molecular weights exceeding 10,000 Daltons. The most important of these compounds can be divided into a few fundamental classes: carbohydrates, proteins, lipids, nucleotides, and nucleic acids. Essentially, all naturally occurring organic compounds originate from these classes, and most from the first three. Here we briefly review the chemistry of these biologically important compounds.

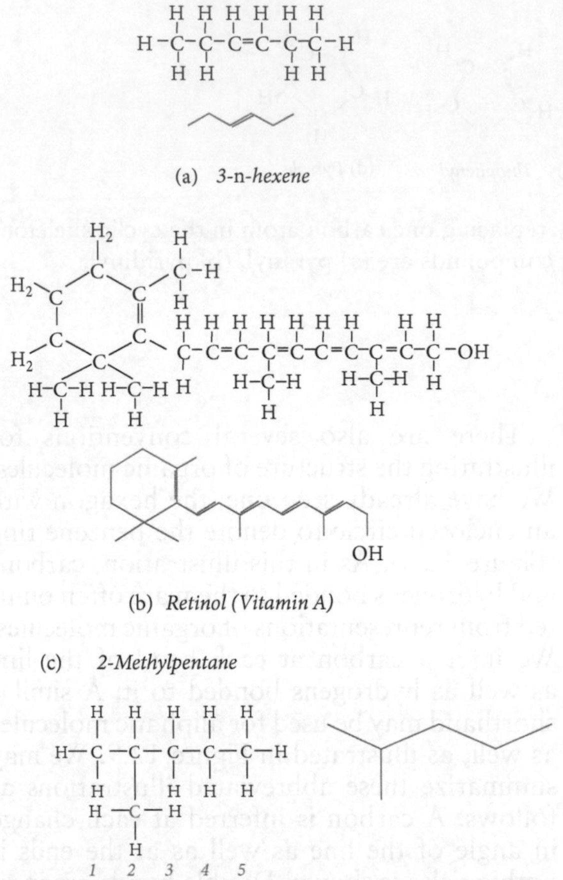

(a) *3-n-hexene*

(b) *Retinol (Vitamin A)*

(c) *2-Methylpentane*

Figure 12.9 Shorthand structural representation of 3-n-hexene and retinol. Carbon and hydrogen atoms are not shown in the short-hand representations. Carbons occur at each joint in the lines as well as at ends of lines. Number of hydrogens bonded to each carbon is equal to 4 minus the number of other bonds shown.

12.3.4.1 Carbohydrates

Carbohydrates could be considered the most fundamental of the biologically important compounds in the sense that they are the direct products of photosynthesis and chemosynthesis). Nearly all life is ultimately dependent on photosynthesis, and virtually all other compounds necessary for life are synthesized in cells from carbohydrates. Some organisms, such as those of hydrothermal vent communities, depend on chemosynthesis rather than photosynthesis, but carbohydrate is also the immediate product of chemosynthesis. Thus, carbohydrates may be regarded as the fundamental substance of life. Furthermore, carbohydrates, cellulose in particular, are the most abundant organic compounds in nature.

Carbohydrates, as we mentioned earlier, are related to hydrocarbons by substitution of hydroxyl groups for hydrogen atoms. Two of the simplest carbohydrates are the sugars glucose and fructose, both of which have the composition $C_6H_{12}O_6$. Both can exist as straight chains or cyclic structures (Figure 12.10), though the cyclic structures predominate. Glucose and fructose are examples of *monosaccharides*, the mono- prefix indicating they consist of single chains or rings. General names for these compounds are formed from the Greek prefix corresponding to the number of carbons and the suffix *-ose*. Thus, fructose and glucose are hexose sugars and ribulose (a building block of nucleic acids) is a pentose sugar. Two monosaccharide units may be linked together by elimination of H_2O to form a disaccharide (another example of a *condensation* reaction). Sucrose, or common table sugar, is the condensation product of glucose and fructose.

Molecules consisting of 10 or more monosaccharide units are called *polysaccharides*. Among the biologically most important polysaccharides are *cellulose* and *starch*. Cellulose, the basic structural material of plants, has the general formula of $(C_6H_{10}O_5)_n$ and consists of long (i.e., $n \geq 10,000$) chains of glucose units cross-linked to each other by hydrogen bonds. Cellulose is an example of a *homogenous polysaccharide*, that is, one that is formed by linkage of a single kind of monosaccharide. *Chitin*, the material forming hard structures in arthropods, mollusks, and some fungi and algae, is also a homogenous polysaccharide. It is related to cellulose by replacing one of the hydroxyl groups with an amino group to form glucosamine, an *amino sugar*. Starch, which serves to store energy in plants, is also a $(C_6H_{10}O_5)_n$ polysaccharide in its simplest form. Starches, however, also include heterogeneous polysaccharides, that is, polysaccharides containing more than one kind of monosaccharide unit. Water-soluble starches consist of relatively short chains ($n \approx 25$); insoluble starches are typically longer, up to 500,000 Daltons. From a geochemical perspective, an important difference between cellulose and starch is that the former is much more stable and less readily metabolized.

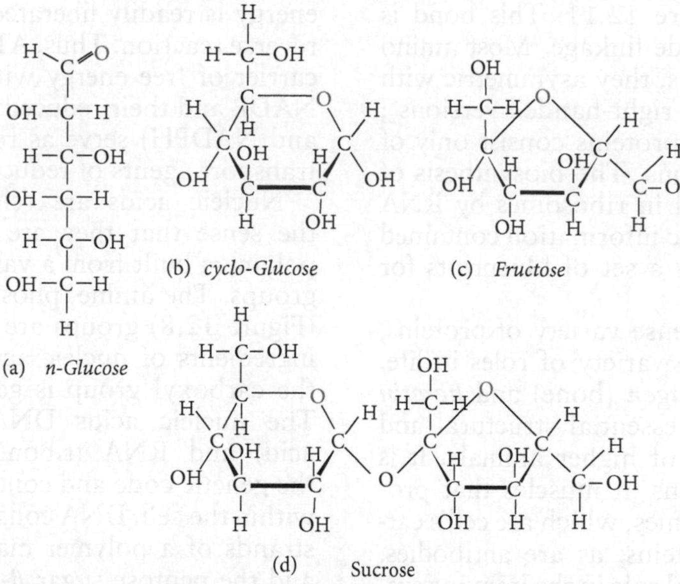

Figure 12.10 Some simple sugars. (a) linear glucose, (b) cyclic glucose, (c) fructose, and (d) sucrose, a disaccharide formed by condensation of glucose and fructose.

Although organisms generally store energy in the form of complex carbohydrates and lipids, these are always first converted back to glucose before oxidation releases this energy.

12.3.4.2 Nitrogen-bearing organic compounds: proteins, nucleotides, and nucleic acids

Amino acids are the basic building blocks of proteins. There are 20 common amino acids, the essential characteristic of which is the presence of both an amine and a carboxylic group (Figure 12.7). The simplest amino acids are *glycine*, which consists of the amino acid group with a hydrogen at the free position (Figure 12.11), and alanine, which has CH_3 at the free position. Amino acids may be characterized as neutral, acidic, or basic. *Acidic* amino acids have an additional carboxylic group, which acts as a proton donor. *Basic* amino acids, such as lysine ($COOHCH((CH_2)_4NH_2) NH_2$), have an additional amine group, which can act as a proton acceptor. Neutral amino acids, such as glycine and alanine, have equal numbers of carboxylic and amine groups.

Proteins are formed by condensation of many amino acid units into polymers called peptides. The simplest proteins consist of 40 amino acid units; the most complex ones consist of more than 8000 units. With 20 basic

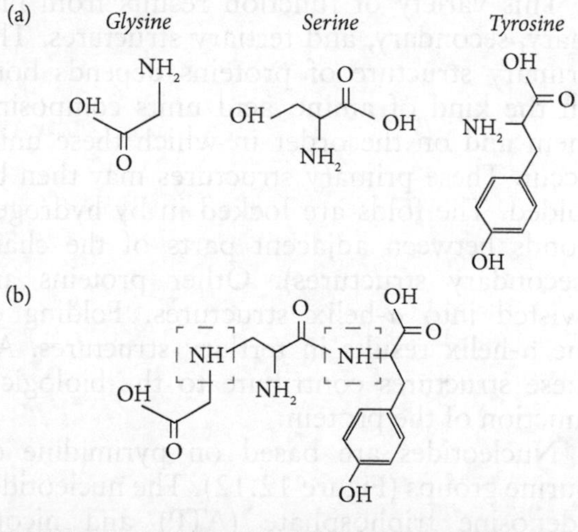

Figure 12.11 (a) Three of the 20 common amino acids that can combine to form proteins. (b) Peptide segment consisting of glycine, serine and tyrosine bound by peptide linkages (dashed boxes). Peptide linkage occurs between amine and carboxyl groups with the elimination of H_2O.

building blocks, the possible combinations are virtually limitless, making the diversity of life possible. The condensation reaction forming peptides consists of linking the carboxylic group of one amino acid to the amine group of another with the elimination of water, as illustrated in Figure 12.11. This bond is referred to as a peptide linkage. Most amino acids are *chiral*, that is, they asymmetric with have left-handed and right-handed versions*; with rare exceptions proteins consist only of the left-handed versions. The biosynthesis of proteins is performed in ribosomes by RNA molecules. The genetic information contained in DNA is essentially a set of blueprints for protein synthesis.

There are an immense variety of proteins, and they play a wide variety of roles in life. Proteins such as *collagen* (bone) and *keratin* (hair, claws) are the essential structural and connective materials of higher animals. It is contraction of proteins in muscles that provide movement. Enzymes, which are cell's catalysts, are often proteins, as are antibodies, which play an essential role in the immune system. Proteins also act to store and transport various elements and compounds; hemoglobin is a good example of such a protein. Hormone proteins serve as messengers and regulators.

This variety of function results from primary, secondary, and tertiary structures. The primary structure of proteins depends both on the kind of amino acid units composing them and on the order in which these units occur. These primary structures may then be folded. The folds are locked in by hydrogen bonds between adjacent parts of the chain (secondary structures). Other proteins are twisted into α-helix structures. Folding of the α-helix results in tertiary structures. All these structures contribute to the biological function of the protein.

Nucleotides are based on pyrimidine or purine groups (Figure 12.12). The nucleotides adenosine triphosphate (ATP) and nicotinamide adenine dinucleotide phosphate (NADP), illustrated in Figure 12.12, play key roles in both the creation and storage of chemical energy (photosynthesis) and its transfer and release (respiration) in organisms. Another nucleotide, nicotinamide adenine dinucleotide (NAD), plays an important role in respiration. ATP can be formed from ADP (adenosine diphosphate) by the addition of an inorganic phosphate ion. This process, called *phosphorylation*, involves a free energy change (ΔG) of about $+40\,kJ/mol$. That

energy is readily liberated on demand by the reverse reaction. Thus, ATP serves as a general carrier of free energy within cells. NAD and NADP and their reduced equivalents (NADH and NADPH) serve as redox couples and as transport agents of reduced hydrogen.

Nucleic acids are related to proteins in the sense that they are nitrogen-containing polymers built from a variety of fundamental groups. The amine, phosphate, and pyridinyl (Figure 12.8) groups are among the essential ingredients of nucleic acids. Unlike proteins, the carboxyl group is generally not present. The nucleic acids DNA (deoxyribonucleic acid) and RNA (ribonucleic acid) contain the genetic code and control protein synthesis within the cell. DNA consists of two backbone strands of a polymer made up of phosphate and the pentose sugar β-D-ribofuranose connected to each other by pairs of the four nucleotides, cytosine, thymine, adenine, and guanine. The genetic information is encoded in the sequence of pairs.

12.3.4.3 Lipids

Unlike carbohydrates and proteins, *lipids* are defined not by their composition and structure but by their behavior: lipids are those organic substances that are water-insoluble but are soluble in organic solvents such as chloroform, toluene, acetone, and ether. Lipids include fats, oils, waxes, steroids, and other compounds. Common fats and oils are generally *triglycerides*, which are esters of three *fatty acids* and *glycerol*, an alcohol (Figure 12.13). Fatty acids are straight-chained (aliphatic) carboxylic acids (i.e., an alkane with a carboxyl group at one end). They typically range in length from C_{12} to C_{36}. Because they are generally formed by successive additions of acetyl (C_2) units, fatty acids have predominantly even numbers of carbon atoms. Unsaturated fats, such as *oleic acid* ($CH_3(CH_2)_7CH=CH(CH_2)_7$ $CH(O)OH$), predominate in plants, whereas saturated fats, such as *stearic acid* ($CH_3(CH_2)_{16}$ $CH(O)OH$), predominate in animals. In *phospholipids*, one of the fatty acids in the triglyceride is replaced by a phosphate unit, which is in turn often linked to a nitrogen base. Many

---

* The chirality of amino acids is generally denoted by L- (left) or D- (right, from the Latin dexter). This, however, is an older convention. A newer one uses S- (left, from the Latin sinister) and R- (right).

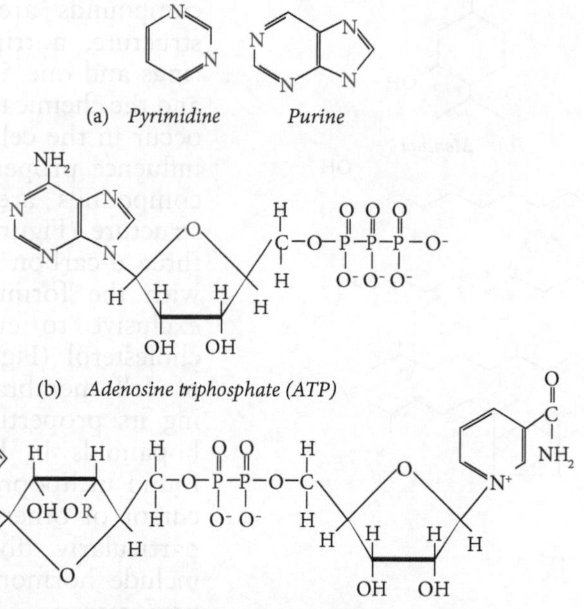

(a) *Pyrimidine*   *Purine*

(b) *Adenosine triphosphate (ATP)*

(c) *Nicotinamide adenine dinucleotide (R = H)*
Nicotinamide adenine dinucleotide phosphate (R = PO₃²⁻)

Figure 12.12   (a) The structure of pyrimidine and purine groups, essential components of nucleotides. (b) Structure of the nucleotide adenosine triphosphate (ATP). (c) Structure of NAD or nicotinamide adenine dinuclotide (when the radical labeled R is H) and NADP or nicotinamide adenine dinucleotide phosphate (when R is phosphate).

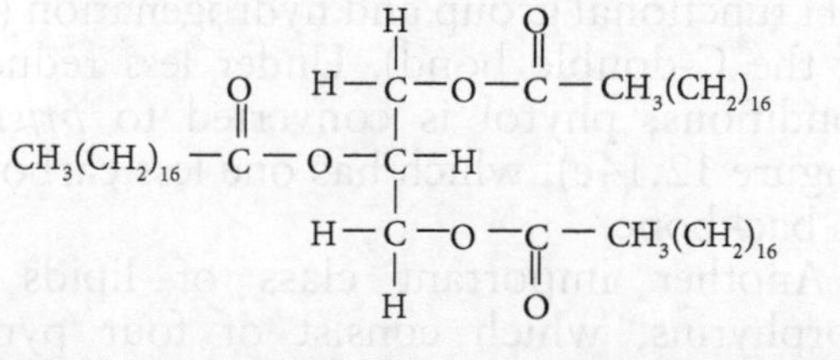

Figure 12.13   A triglyceride fat formed from the alcohol glycerol and three molecules of stearic acid.

acids and fatty alcohols, both of which have chain lengths generally in the range of C_{24} to C_{28}. The fatty alcohols also have predominantly even number of carbon atoms because they are synthesized from fatty acids. Plant waxes also contain long, straight-chained hydrocarbons (C_{23} to C_{33}). These hydrocarbons typically have odd numbers of carbons because they are formed by *decarboxylation** of fatty acids. *Cutin*, which forms protective coatings on plants, is a polymerized hydroxy fatty acid (commonly C_{16} or C_{18}).

Another important class of lipids is the *terpenoids*. Terpenoids display a great diversity of structure, but the basic unit of all terpenoids is the isoprene unit, a branched, five-carbon chain with a methyl group attached to the second carbon atom (Figure 12.14a). Terpenoids are named on the basis of the number of isoprene units present: monoterpenoids have two, sesquiterpenoids have three, diterpenoids have four, triterpenoids have six, tetraterpenoids have

glycolipids, which are combinations of lipids and carbohydrates, are triglycerides in which one of the fatty acids is replaced by a sugar. Ether lipids are glycerides formed from straight-chained alcohols, called n-*alkanols* or fatty alcohols, rather than fatty acids.

Waxes are a mixture of many constituents. Among the most important are wax esters, which are esters of straight-chained fatty

---

* *Decarboxylation* is a process whereby a CO_2 molecule is lost from an organic compound. *Carboxylation* is the addition of a CO_2 molecule to an organic molecule.

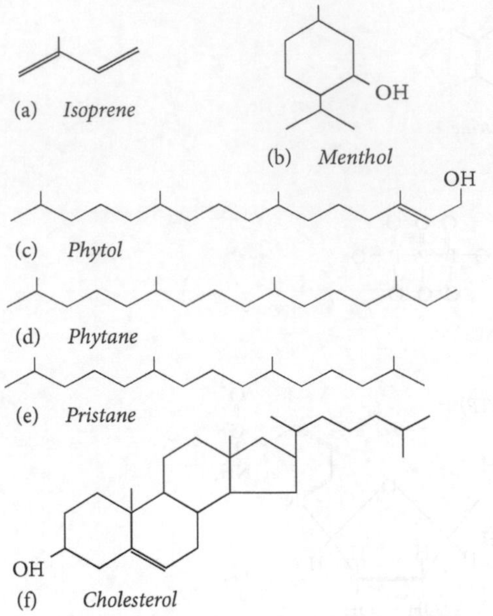

Figure 12.14 Terpenoids. (a) Isoprene, the building block of all terpenoids. (b) Menthol, a simple cyclic monoterpenoid. (c) Phytol, an acyclic diterpene. (d) Phytane, an acyclic diterpane derived from phytol. (e) Pristane, one of the possible products of diagenesis involving phytane. (f) Cholesterol, illustrating the pyrrole ring system (shown in red) shared by all steroids. The carbon atoms at the apices of the phenols and the hydrogens bound to them are not shown.

eight. Molecules consisting of more than eight isoprene units are termed polyterpenoids. Terpenoids may be cyclic (e.g., menthol, cholesterol, Figures 12.14b and 14f) or acyclic, such as phytol, saturated or unsaturated, such as phytol derivatives pristane and phytane. Among other things, terpenoids serve as pheromones (scents, attractants), hormones, antibiotics, resins, and vegetable and animal oils.

Two important classes of terpenoids are hopanoids and steroids, both of which are synthesized from squalene ($C_{30}H_{50}$), a noncyclic triterpenoid (Figure 12.15a). These noncyclic terpenoids, often referred to as acyclic *isoprenoids*, are important from a geochemical perspective because they are common components of sedimentary organic matter. Furthermore, they are also found as biomarkers (Section 12.7.5) in petroleum, having survived diagenesis. Hopanoid

compounds are variations on the hopane structure, a triperpane with four 6-carbon rings and one 5-carbon ring (Figure 12.15b) and the chemical formula $C_{30}H_{52}$. Hopanoids occur in the cell membranes of bacteria and influence properties such as rigidity. Steroid compounds are variations on the sterane structure (Figure 12.15c), which consists of three 6-carbon rings and one 5-carbon one with the formula $C_{17}H_{28}$ and are almost exclusive to eukaryotes. Steroids such as cholesterol (Figure 12.14f) are components of cell membranes in eukaryotes, modifying its properties and a manner analogous hopanoids in bacteria. Cholesterol is also found in lipoproteins and serves as the precursor of other animal steroids, which play particularly diverse roles in animals and include hormones such as testosterone and progesterone.

Isoprenoid derivatives of *phytol*, a diterpenoid that forms part of chlorophyll-a (Figure 12.15d), are among the most common molecules in sedimentary organic matter. Under strongly reducing conditions during diagenesis, phytol (Figure 12.14c) is converted to *phytane* (Figure 12.14d), whose backbone is a 16-carbon chain, through the loss of the OH functional group and hydrogenation (loss of the C-double bond). Under less reducing conditions, phytol is converted to *pristane* (Figure 12.14e), which has one less carbon in its backbone.

Another important class of lipids are porphyrins, which consist of four pyrrole rings with (chlorophylls) or without (hemes, cytochromes) the fifth isocyclic ring. The pyrrole units form a ring structure in which bonding electrons are conjugate, that is delocalized, much like benzene rings. These electrons interact strongly with visible light with the result that porphyrins are generally strongly colored, chlorophyll being a prime example. There are a number of varieties of chlorophyll, but all have four pyrrole rings surrounding a central Mg atom (Figure 12.15d). In heme, a component of the protein hemoglobin, the central atom is Fe (Figure 12.15e), which undergoes redox reactions to transport oxygen to tissues. Porphyrin derivatives of chlorophyll and hemes are important components of sedimentary organic matter, with the central Mg or Fe atom often replaced by another transition

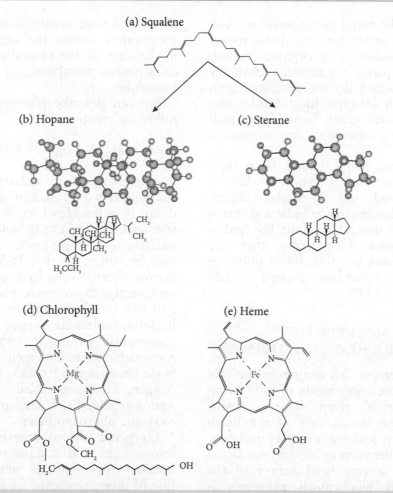

Figure 12.15 (a) Squalene, the precursor to a variety of terpenoids, including hopanoids and steroids, which are variations on the basic framework of (b) hopane and (c) sterane. (d) Structure of the porphyrin chlorophyll-a, with the pyrrole units shown in red. The linear chain is a phytol unit. (e) Structure of heme, another porphyrin, which among other things is the oxygen-carrying unit in the hemoglobin protein.

metal. Melanin is another molecule that interacts strongly with light and gives color to life, although not a porphyrin, also contains pyrrole units.

Like proteins, lipids play a variety of roles in life. Phospholipids are the primary constituents of cell membranes. Per unit weight, lipids release twice as much energy as carbohydrates upon oxidation. Thus, fats and oils serve as efficient stores of energy for both plants and animals. Lipids include pigments that are essential in photosynthesis in plants (e.g., chlorophyll) and vitamin A production in mammals (e.g., carotenoids). Waxes such as cutin form protective barriers. Other lipids act as sex pheromones (i.e., providing scent) or hormones, or assist in digestion. From a

geochemical perspective, lipids are important because many can survive diagenesis and hence play an important role in petroleum formation. Furthermore, some are unique to groups of organisms and are useful in reconstructing past environments; such compounds are called *biomarkers*.

12.3.4.4 Lignin and tannins

Lignin is another important structural material in higher plants. It forms a network around cellulose to provide structural rigidity in wood. It is second only to cellulose as the most abundant organic molecule in the biosphere. It is a rigid, high molecular weight polyphenol. As such, it is quite stable and resistant to bacterial decomposition. As a result, it is a very important contributor to soil and terrestrial sedimentary organic carbon. Because marine plants are almost exclusively algae, most of which do not produce lignin (red algae, which are often multicellular seaweeds, are the exception), lignin is a much less important contributor to marine organic carbon.

Tannins, whose name derives from their use in tanning leather, are another class of compounds found only in higher plants. They occur predominantly in bark and leaves and function to make the plant less palatable to herbivores. Like lignin, they are high molecular weight (500–3000) polyphenol compounds. Functional groups include carboxyl as well as OH.

12.4 THE CHEMISTRY OF LIFE: IMPORTANT BIOCHEMICAL PROCESSES

Our main concern in this chapter is the role played by organic compounds outside living tissue. Nevertheless, given the remarkable impact of life on the surface of the Earth and its chemistry, and the interplay and constant exchange between living and nonliving organic matter, a very brief survey of the more important biochemical processes is appropriate at this point.

As was mentioned earlier, autotrophs produce the energy they need by synthesis of organic compounds from inorganic ones. Most autotrophs are phototrophic: they use light energy to synthesize organic compounds, a process known as photosynthesis. Some bacteria use chemical energy, derived from the oxidation of H_2S or other reduced species, to synthesize organic compounds in a process called chemosynthesis.

12.4.1 Photosynthesis

Like most biochemical processes, photosynthesis is a complex one, involving many steps catalyzed by a variety of compounds. The details of the photosynthetic processes also vary somewhat between photosynthetic bacteria and true plants. In true plants, photosynthesis takes place within specialized intracellular organelles called chloroplasts. In prokaryotic bacteria such specialized intracellular units are absent. In these organisms, the site of photosynthesis may be in internal membranes within the cell protoplasm, as in the case of the cyanobacteria, or on the cell's plasma membrane, as in the case of the halophiles.

We can describe photosynthesis with the following reaction:

$$CO_2 + 2H_2A + light \rightarrow CH_2O + H_2O + 2A$$
$$(12.1)$$

In photosynthetic eukaryotes and cyanobacteria, A is oxygen, and hence the hydrogen donor is water. However, in anaerobic photosynthesis, carried out by some photosynthetic bacteria (e.g., the purple sulfur bacteria), A may be sulfur, so that H_2S is the hydrogen donor; alternatively, hydrogen may be taken up directly. In oxygenic photosynthesis, carried out by plants and most photosynthetic bacteria, molecular oxygen is a byproduct of photosynthesis, and it is this process that is responsible for free oxygen in the atmosphere. Some bacteria can fix CO_2 without liberating oxygen, a process called *anoxygenic photosynthesis*. Our brief description will focus on oxygenic photosynthesis.

Oxygenic photosynthesis can be divided into a light and a dark stage. The former involves two separate photoreactions. The first of these, governed by Photosystem II (or PS II) is the photodissociation of water. This reaction can be described as:

$$2H_2O \rightarrow 4H^+ + 4e^- + O_2 \qquad (12.2)$$

Among the enzymes mediating this process is a Mn-bearing protein, which takes up the liberated electrons by oxidation of Mn. The hydrogen ions and electrons produced by PS II travel along distinct chemical pathways, which in the case of the electrons include Fe- and Cu-bearing proteins, to the site of the dark reactions. The energy from the electrons liberated in PS II is used for phosphorylation of ADP to ATP. In Photosystem I (PS I), $NADP^+$ is reduced to NADPH. In the subsequent dark reactions, this NADPH acts as an electron donor in the reduction of CO_2, and the energy for this reaction is supplied by ATP.

Energy to drive both PS I and PS II is captured by chlorophyll (Figure 12.15c) or some other light-absorbing pigment. Several varieties of chlorophyll are generally present within a given plant. Chlorophyll-a is the principal photosynthetic pigment in plants and cyanobacteria. Some other bacteria, such as the green sulfur bacteria, utilize a closely related substance called bacteriochlorophyll. The halophiles, members of the archaea, use retinol (the same light-sensitive pigment in the human retina) rather than chlorophyll to capture light energy. All chlorophylls strongly absorb light in the red and blue parts of the visible spectrum (the green color of plants results from a lack of absorption of green light).

It is in the dark stage of photosynthesis that carbohydrate is actually synthesized. At this point, there is a divergence in the chemical pathways. In C_3 plants the initial carbohydrate produced (3-phosphoglycerate) is a 3-carbon chain. This process, known as the *Calvin cycle* (illustrated in Figure 12.16), is used by all marine plants and about 90% of terrestrial plants. The first step is *phosphorylation*, or the addition of a phosphate group. In this reaction, ribulose 5-phosphate, a C_5 sugar containing one phosphate, is converted to ribulose 1,5-bisphosphate (Figure 9.14), with the additional phosphate coming from the ATP generated during the light stage. In the next step, an enzyme called *ribulose bisphosphate carboxylase oxygenase* (RUBISCO) catalyzes a reaction in which ribulose 1,5-bisphosphate reacts with one molecule of CO_2 to produce three molecules of 3-phosphoglyceric acid. Then an additional phosphate is added to each of these molecules to form 1,3-bisphosphoglycerate. This phosphate is then replaced by hydrogen supplied by NADPH to form glyceraldehyde 3-phosphate. It is in this step that the new CO_2 in the phosphoglycerate is reduced. The ΔG for reduction of CO_2 to CH_2O is about +480 kJ/mol. Most of this energy is supplied by the oxidation of two molecules of NADPH to $NADP^+$. Most of the resulting triose phosphate is converted back to ribulose bisphosphate for further synthesis, but about 25% is oxidized to form glycolate rather than glycerate, but is then recovered, although at an energy cost, through the photorespiration cycle. Monosaccharides produced in this way, if not immediately metabolized, are eventually converted to more complex carbohydrates or other essential compounds.

The other photosynthetic pathway is the Hatch–Slack cycle, used by the C_4 plants, which include hot-region grasses and related

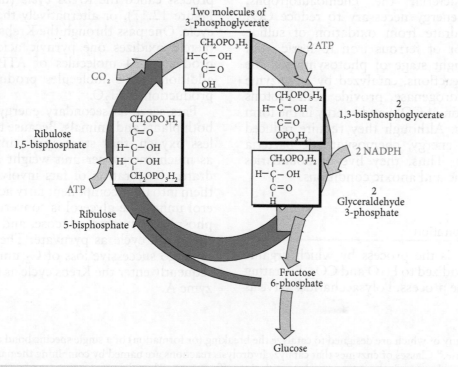

Figure 12.16 The Calvin cycle of dark reactions for the production of glucose in C_3 plants.

crops such as maize and sugarcane. These plants use *phosphoenol pyruvate carboxylase* (PEP) (Figure 9.15) to initially fix the carbon and form oxaloacetate, a compound that contains four carbons. CO_2 is fixed in outer mesophyll cells as oxaloacetate and is then transported, either as malate or asparatate, to inner "bundle sheath" cells, where it is decarboxylated and refixed in the Calvin cycle. C_4 photosynthesis appears to have evolved only recently; the oldest known C_4 plants are of late Miocene Age (though there has been speculation that they may have first evolved in the Cretaceous). Because the C_4 pathway is somewhat more efficient at low ambient concentrations of CO_2, there has been speculation that their appearance may reflect an evolutionary response to low atmospheric CO_2 concentrations of the late Cenozoic.

From a geochemical perspective, the most important aspect of the difference between C_3 and C_4 plants is the difference in carbon isotope fractionation during photosynthesis. The C_3 pathway produces a $\delta^{13}C$ fractionation of from -20 to $-30‰$, whereas the C_4 pathway produces a much smaller fractionation of about $-13‰$. Further aspects of this are discussed in Chapter 9.

Some bacteria, the chemoautotrophs, obtain the energy necessary to reduce CO_2 to carbohydrate from oxidation of sulfide to sulfate or of ferrous iron to ferric iron. As in the light stage of photosynthesis, the oxidation reactions, catalyzed by an enzyme called dehydrogenase, provide both protons (for reduction of CO_2) and energy in the form of electrons. Although they require reduced species for energy, chemosynthetic bacteria are aerobic. Thus, they live at boundaries between oxic and anoxic conditions.

12.4.2   Respiration

Respiration is the process by which organic carbon is oxidized to H_2O and CO_2, liberating energy in the process. Polysaccharides serve as the primary energy storage in both plants and animals. In plants, these generally take the form of starches, while in animals, glycogen serves as the primary energy store. In both cases, single glucose molecules are first liberated from these chains before being oxidized. This is accomplished through the catalytic action of an enzyme.* Glucose molecules liberated from complex carbohydrates in this fashion are then converted to two molecules of pyruvic acid in a multistep process called *glycolysis*. Although ATP is consumed in the initial phosphorylation steps, it is produced in subsequent steps and the entire process results in a net production of two ATP molecules per glucose molecule (and an additional energy gain of 80 kJ/mol).

Glycolysis does not release CO_2. Energy released in the process comes primarily from oxidation of hydrogen in the sugar to hydroxyl and resulting conversion to pyruvic acid. Thus, glycolysis releases only a small fraction of the energy stored in the glucose molecule. Under aerobic conditions, further energy may be obtained by oxidation of the pyruvic acid molecule. This oxidation of pyruvic acid occurs within the mitochondria of eukaryote cells (in contrast, glycolysis occurs within the general cell cytoplasm) in a process called the *Krebs cycle* (illustrated in Figure 12.17), or alternatively the *citric acid cycle*. One pass through the Krebs cycle completely oxidizes one pyruvic acid molecule, producing 34 molecules of ATP, with two additional ATP molecules produced by the production of H_2O.

Fats serve as secondary energy storage in both plants and animals. Because they contain less oxygen, they store approximately twice as much energy per unit weight as carbohydrates. Respiration of fats involves breaking them into their component fatty acid and glycerol units. The glycerol is converted to triose phosphate, much as glucose, and then enters the Krebs cycle as pyruvate. The fatty acids undergo successive loss of C_2 units that subsequently enter the Krebs cycle as acetyl coenzyme A.

---

* Enzymes, many of which are designed to catalyze the breaking (or formation) of a single specific bond are denoted by the ending "-ase." Classes of enzymes that catalyze hydrolysis reactions are named by combining the name of the class of compounds whose hydrolysis they catalyze with the suffix "-ase". Thus, lipases catalyze the hydrolysis of lipids, amylases catalyze the hydrolysis of starches, etc.

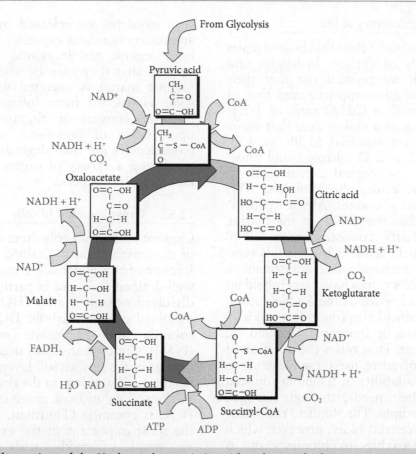

Figure 12.17 Illustration of the Krebs cycle, or citric acid cycle, in which organic matter is oxidized by organisms to produce energy in the form of ATP. New pyruvate enters the cycle at the top and is oxidized to acetyl coenzyme A (CoA) plus CO_2. This acetate is then combined with an oxaloacetate produced in an earlier Krebs cycle to form citrate. Two additional CO_2 molecules as well as additional hydrogens are then successively stripped, ultimately producing NADH and $FADH_2$ (FAD, flavin adenosine dinucleotide) and an oxaloacetate. The hydrogens in $FADH_2$ and NADH are ultimately combined with free oxygen to form H_2O.

When oxygen is absent in cells, hydrogen carried by NADH cannot be oxidized to H_2O and the Krebs cycle cannot operate. Some organisms that normally live under aerobic conditions can switch to an alternative metabolism, called *fermentation*, which does not require oxygen. Such organisms are called *facultative anaerobes*. When such cells are deprived of oxygen, NAD^+ is regenerated from NADH by reduction of pyruvic acid. Lactic acid* ($CH_3CH(OH)COOH$) is one of several possible products of this process.

In alcoholic fermentation, carried out by yeast, CO_2 is removed from pyruvate leaving acetaldehyde (CH_3CHO), which is then reduced to ethanol (CH_3CH_2OH). Many bacteria can live only by fermentation and are poisoned by free oxygen. Such organisms are called *obligate anaerobes*. These bacteria, of which there is a great variety, carry out cellular respiration by using a compound other than oxygen as an electron acceptor. The most common such receptors are nitrate, nitrite, and sulfate, but some bacteria can reduce Fe^{3+} and Mn^{4+}.

* Animal muscle cells have the ability to switch to fermentation when oxygen cannot be supplied fast enough for NAD^+ to be regenerated. Under these circumstances, lactic acid, the formal name for which is 2-hydroxy-propanoic acid, builds up in muscle tissue. Nerve and brain cells, however, cannot respire anaerobically and quickly die when deprived of oxygen.

12.4.3 The stoichiometry of life

Previous sections make clear that biomolecules consist primarily of carbon, hydrogen, and oxygen; indeed, we pointed out that their elemental composition approximates that of simple sugars, with a C:H:O ratio of 1:2:1. Yet the foregoing also makes clear that many other elements are essential to life as well. Following C, H, and O, nitrogen and phosphorus, which are critical constituents of proteins, nucleic acids, phospholipids, ATP, etc., are the most abundant. Alfred Redfield[*] (1934) found that the C:N:P in the marine biomass was fairly consistent with C:N:P ratios of 105:16:1. While it is clear this ratio does vary and perhaps some adjustment is possible, Redfield's values have largely held up over time and is known as the *Redfield ratio*. Interestingly Redfield also found this was also the average ratio of inorganic C, fixed N, and P in seawater. That raises the question of whether life evolved to use these elements in ratio to their availability or the biota adjusted this ratio to their needs, through nitrogen fixation, for example. The Redfield ratio does not hold for terrestrial biota, however, which has far more variable stoichiometry and is substantially poorer in N and P, with mean C:N:P ratio of 970:28:1 (Elser et al., 2000). In freshwater systems, the ratio is more consistent with a mean value of 307:30:1. Much of the rest of the periodic table is essential to some form of life as well, but the abundance ratios of other elements tend to be even more variable.

12.5 ORGANIC MATTER IN NATURAL WATERS AND SOILS

Nonliving organic matter is present in soil and water. Some of this consists of extracellular polymeric substances, most commonly polysaccharides, secreted by microorganisms ranging from bacteria to plankton for a variety of reasons: to protect against desiccation, to extract nutrients or scavenge dissolved organic matter, for and intracellular communication, recognition, and adhesion. In soils, such *exudates* are released by both macro- and microorganisms expressly to break down both organic and inorganic components in soils so that they may be assimilated. Some organic matter is excreted waste products, some is leached from foliage. Most of it, however, consists of biomolecules released upon death of organisms or cells. These biomolecules are then degraded by microbes producing a variety of organic molecules as metabolic biproducts.

12.5.1 Organic matter in soils

Organic matter can constitute a third or more of the mass of poorly drained soils, although fractions from 6–10% are more common in well-drained soils and is partitioned between dissolved organic matter (DOM, also called dissolved organic carbon: DOC) in the soil solution and particulate organic matter (POM). Concentrations of organic matter are highest in the surface soil layers (O and A; see Chapter 13) and low in the deeper layers (C). In the soil solution, a range of 2–30 mg C/L DOC is common (Thurman, 1985). Among the most important of the exudates in soils are simple carboxylic acids such as acetic and oxalic acid (Figure 12.7), formic acid (HCOOH), and citric acid. The hydrogen of phenol OH groups may also dissociate and a variety of phenolic acids also contribute to soil acidity and rock weathering, though less so than the carboxylic acids. These simple organic acids are commonly present in relatively high concentrations around plant roots in soils, though on average their concentration is less than 1 mM in the soil solution (Drever and Vance, 1994). Because of the presence of both these acids many soils are slightly acidic. These acids contribute both directly (through surface complexation reactions) and indirectly (as proton donors, by increasing the solubility of cations through complex formation) to the weathering of rocks, although the overall extent to which weathering is accelerated by plants is unclear. As many simple biomolecules are readily metabolized by microbes, their residence time in the soil

---

[*] Alfred Redfield (1890–1983) received his PhD from Harvard in 1918. He immediately joined the faculty of the Harvard Medical School where he mostly remained until 1931 when he became director of the Marine Biological Laboratory at Woods Hole, where he remained until his retirement in 1956. The A.C. Redfield Award is awarded annually by the American Society of Limnology and Oceanography for lifetime achievement in limnology and oceanography.

is generally quite short, a matter of days or less. Concentrations of these molecules are maintained by continuous production by the biota. In contrast, the residence time of some organic matter in soils can be thousands of years and longer.

Traditionally, soil organic matter has been divided into humin, humic acid, and fluvic acid based on extraction techniques in which the pH is first adjusted to 13 by addition of NaOH, which dissolves the soluble fraction from residual *humin*. The solution containing the soluble fraction, collectively called *humic substances*, is then acidified to a pH of 1. The dark precipitate that forms at this point is known as *humic acid*, the fraction that remains in solution is known as *fulvic acid*. Thus, by definition, solubility decreases from fluvic through humic acid to insoluble humin.

As is summarized in Table 12.1, humic acid is richer in carbon and nitrogen and poorer in oxygen, hydrogen, and sulfur than fulvic acid. Humic acid contains a greater proportion of aromatic groups while the higher content of polar groups such as carboxyl accounts for the higher solubility of fulvic acid. Soil fulvic acids tend to have molecular weights $<2 \times 10^3$, while the molecular weight of soil humic acids can exceed 10^6.

Based on these compositions, the approximate chemical formula for average humic acid is $C_{187}H_{189}O_{89}N_9S$ and that of fulvic acid is $C_{68}H_{91}O_{48}N_3S$. Compared with the composition of living organisms, humic and fulvic acids have substantially lower ratios of hydrogen, oxygen, and nitrogen to carbon. Stevenson and Vance (1989) estimated the average content of functional groups in soil humic and fulvic acids as 7.2 and 10.3 meq/g, respectively. Humin tends to be even richer in carbon and nitrogen and poorer in sulfur than either fulvic or humic acids.

In the traditional view of humic substances, which dates back centuries, fragments of biomolecules progressively assemble into large, dark-colored macromolecules in a process known as humification. The resulting *geomolecules* were thought to be rich in structures particularly resistant to decomposition. Thus while biodegradation progressively broke down biopolymers into smaller fragments, humification would synthesize them into larger ones in progression from fluvic acid through humic acid to humin and ultimately peat and kerogen (e.g., Hedges, 1988). Assembly into higher molecular weight compounds also resulted not only in increased resistance to decay but decreased solubility as humification progressed.

In the last decade or two a new view of humic substances has emerged (e.g., Kleber and Reardon, 2017; Lehmann and Kleber, 2015), in that to the extent that such "geopolymers" exist, they are products only of the harsh chemical extraction process described above rather than natural products. This new view is based on a variety of observations. First, new advanced analytical techniques such as solid state nuclear magnetic resonance spectrometry, X-ray photoelectron

Table 12.1 Composition of soil humic and fulvic acids.

Element wt. %	Humic acid		Fulvic acid	
	Mean	Range	Mean	Range
C	56.	53–59	45.7	40.7–50.6
H	4.6	3.0–6.5	5.4	3.8–7.0
N	3.2	0.8–5.5	2.1	0.9–3.3
O	35.5	32–38.5	44.8	39–50
S	0.8	0.1–1.5	1.9	0.1–3.6
Functional groups (meq/g)				
Total acidic groups		5.6–8.9		6.4–12.2
Carboxyl	3.6	1.5–6.0	8.2	5.2–11.2
Phenolic OH	3.1	2.1–5.7	3.0	0.3–5.7
Alcoholic OH		0.2–4.9		2.6–9.5
Quinoid/keto C=O		0.1–5.6		0.3–3.1
Methyloxy OCH₃		2.1–5.7	3.0	0.3–5.7

After Schnitzer (1978).

spectroscopy, and synchrotron spectro-microscopy have allowed analysis of the chemical characteristics of these substances *in situ* without extraction procedures that modify them. These techniques have failed to find evidence of the hypothesized macromolecules. Instead, these techniques reveal that humic substances are mixtures of plant and microbial constituents in various stages of degradation. Furthermore, there is no thermodynamic basis for humification: ΔG for such synthesis reactions are generally positive. Finally, while some biomolecules are certainly more readily degraded than others, experimental evidence has shown that decomposer organisms have the ability to decompose even presumably recalcitrant and persistent materials more quickly than previously anticipated.

In place of the humification model a new "soil continuum model" has emerged in which soil organic matter as a continuum of progressively decomposing organic compounds spanning the full range from intact plant material to highly oxidized carbon in carboxylic acids (Lehmann and Kleber, 2015). These reactions transform the originally nonpolar plant components into amphiphilic molecules (i.e., consisting of separate hydrophobic (nonpolar) and hydrophilic (polar) parts). These amphiphiles form membrane-like aggregates in which the exterior surfaces are hydrophilic and the interiors hydrophobic that can then undergo hydrophobic absorption on mineral surfaces or form micelle-like aggregates in solution. Thus, rather than being *supermolecules,* humic substances appear to be *supramolecular* aggregates loosely held together by noncovalent bonds and by entropic interactions in which entropy is maximized by dispersing energy throughout the system (Wershaw, 1993). In this model, decomposition is associated with increasing oxidation and hence higher fractions of polar molecules such as carboxylic acids and consequently greater solubility. In this view, the division into fluvic and humic acids is a completely arbitrary one (although perhaps nevertheless sometimes useful).

An additional realization was that sorption onto mineral surfaces protects of organic matter from microbial degradation. Enzyme-catalyzed hydrolysis often requires a precise and unique physical alignment of the enzyme and reactant. The part of the surface of an organic molecule adsorbed on to an inorganic surface will not be accessible to the enzyme. Keil et al. (1994) found that adsorption onto mineral surfaces slowed remineralization by up to five orders of magnitude. Adsorption is also reversible, with a dynamic equilibrium between dissolved and sorbed organic fractions. Thus in the soil continuum model, adsorption on mineral surfaces explains the persistence of organic matter over long periods rather than reorganization into inherently recalcitrant geomolecules. In this model, kerogen, the final residuum of sedimentary organic matter, arises from the combined effects of exclusion of decomposer organisms, anoxia, and heat and pressure associated with burial rather than synthesis of geomolecules (Kleber and Reardon, 2017).

Organic compounds in soils, particularly carboxylic acids such as oxalic acid, play an important role in podzolization – the depletion of Fe and Al in the upper soil horizons and their enrichment in lower horizons (Chapter 13). This occurs as a result of the ability of carboxylic acids to form soluble complexes with Fe and Al. Iron and Al carboxylate complexes form in the upper soil layers, where organic acid concentrations are high, then are carried to deeper levels by water flow. At deeper levels, bacteria oxidize the carboxylate, and the Fe and Al precipitate as hydroxides. We will examine metal–organic complexation further in a subsequent section.

12.5.2 Dissolved organic matter in aquatic and marine environments

Virtually all water at the surface of the Earth contains organic substances. While these can be divided into particulate and dissolved organic matter, there is actually a continuum, with colloids being intermediate between dissolved and particulate forms. Colloids have sizes in the range of 1 nanometer to 0.2 μm, corresponding to molecular weights greater than about 10,000. They may be specifically referred to as colloidal organic matter (COM). "Dissolved" organic matter has been traditionally defined as that which passes through a 0.7 μm but this has been reduced to 0.45 μm in some definitions as filtration technology has improved. 0.2 μm

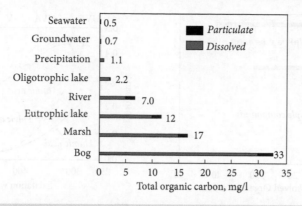

Figure 12.18 Average concentration of dissolved, particulate, and total organic carbon in various natural waters. After Thurman (1985).

filters are also increasingly used; organic matter passing through such filters is sometimes referred to as *ultrafiltered dissolved organic matter* (UDOM). Together, dissolved, colloidal and particulate organic carbon are called *total organic matter* (TOM). Those organic substances that are volatile at ambient temperature constitute volatile organic carbon (VOC).

Figure 12.18 illustrates the range in concentrations of dissolved and particulate organic carbon. Groundwater and seawater typically have the lowest organic carbon concentrations, while wetland waters (bogs, swamps, marshes) have the highest. The interstitial waters of the upper layers of soils often have DOC concentrations in the range of 20–30 mg C/L. Organic carbon concentrations in natural waters generally show strong seasonal variations, as factors such as rainfall, biological productivity, and microbial activity all vary seasonally. Rain and snow also typically contain some DOC. In freshwater systems, DOM ranges from 30–1000 µM but can be higher in wetlands, bogs, and swamps. In all environments dissolved organic matter predominates over particulate.

Some components of dissolved organic matter are readily identified analytically as biomolecules synthesized by organisms to directly support life, including those exuded by cells, and their breakdown products of biomolecules. These include carboxylic acids, phenols, carbohydrates, amino acids and their polymers, nucleic acids, and hydrocarbons. Among carboxylic acids, hydroxy acids such

as glycolate produced by photorespiration are the most common followed by various fatty acids. Most of the dissolved carbohydrates are polysaccharides such as cellulose, but roughly 1% of DOM is monomeric carbohydrates such as glucose and glucosamine. Concentrations of dissolved free amino acids are generally quite low, averaging only about 0.04 µM, but like monomeric carbohydrates they show a large range with maximum concentrations approaching 4 µM in some lakes (Kaplan and Newbold, 2003). Glycine, aspartate, glutamate, alanine, leucine, and serine are usually the most common dissolved amino acids.

Most components are difficult to specifically identify and are classed as fulvic and humic acids (Figure 12.19). The definition of aquatic and marine humic substances differs from the definition in soils: they are defined as polyelectrolytic acids that can be removed from solution through absorption on weak-base ion exchange resins (e.g., the acrylic-ester resin XAD-8) or through some similar procedure. The analytical distinction between humic and fulvic acids is, however, the same as for soils. Modern analytical techniques can now reveal much about the nature of these materials, including molecular weight, the nature of functional groups, C/H, C/O, C/N ratios, cyclic vs. aliphatic, etc.; consequently, the terms fulvic and humic acid are less commonly used. Nevertheless, is still not possible to identify the molecules that make up humic substances. This is because, for example, the nature of an organic molecule depends not just on what functional groups

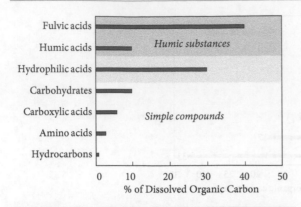

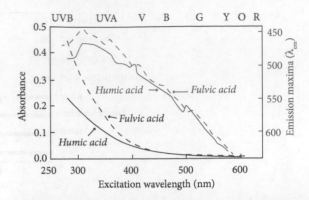

Figure 12.19 Components of dissolved organic carbon in typical river water. After Thurman (1985).

Figure 12.20 Black lines show the absorbance, the fraction of light absorbed (black) per meter path length, as a function of wavelength by Suwannee River humic (soil lines) and fulvic (dashed lines) acids. Red lines show the wavelength of maximum luminescence as a function of the excitation wavelength. For example, light with a wavelength of 400 nm excites emissions at ~500 nm from both humic and fulvic acid. Letters across the top refer, respectively, to UVA, UVB, violet, blue, green, yellow, orange, and red light. Modified from Del Vecchio and Blough (2004).

are attached, but where they are attached. Table 12.2 lists the chemical composition of humic and fulvic acids from a variety of environments.

Another part of the definition of humic substances is that they are colored or *chromophoric*, inspiring the German word for them *Gelbstoff* (yellow substance) and Henry David Thoreau calling it "meadow tea." Absorption spectra of fulvic and humic acids show highest absorbance in the ultraviolet with absorbance smoothly decreasing exponentially with increasing wavelength, as in the examples in Figure 12.20 from the Suwannee River of the southeastern US. Two parameters are often used to characterize the optical

properties of DOM: the first is the *specific UV absorbance* (SUVA$_{254}$), which is the fraction absorbed per meter of path length at wavelength $\lambda = 254$ nm divided by the DOC concentration in mg C/L; the second is the slope, $S = d\ln A/d\lambda$, of the absorbance curve. Terrestrial DOC has higher SUVA$_{254}$ than

Table 12.2 Composition of fulvic and humic acids dissolved in natural waters.

	C	H	O	N	P	S	Ash
Groundwater							
Biscayne aquifer fulvic	55.4	4.2	35.4	1.8			0.04
Biscayne aquifer humic	58.3	3.4	30.1	5.8			10.4
Seawater							
Saragasso Sea fulvic	50.0	6.8	36.4	6.4		0.46	
Lake water							
Lake Celyn humic	50.2	3.1	44.8	1.9			
Lake Celyn fulvic	43.5	2.7	51.6	2.2			
Stream water							
Ogeechee Stream fulvic	54.6	4.97	38.2	0.87	0.62	0.74	0.86
Ogeechee Stream humic	55.9	4.19	36.5	1.27	0.25	0.93	1.13

Data from Aiken et al. (1985). Concentrations in weight percent.

marine DOC. In fresh waters, allochthonous* DOC generally has a higher slope (meaning less absorption at long wavelengths) than autochthonous DOC. Luminescence also decreases with wavelength, although not as smoothly, and is offset to longer wavelength relative to the excitation wavelength. The light absorbing properties of humic substances effects the biota in opposing ways. First, the strong absorbance in the ultraviolet protects aquatic organisms from UV radiation; second, high concentrations of humic substances reduce the quality and quantity of light available for photosynthesis and hence impose limits on primary productivity.

The optical properties of DOM provide insights into its nature. The smooth absorption pattern of humic substances is not what one would expect from the interaction of light with individual parts of molecules (chromophores) or atoms. Del Vecchio and Blough (2004) concluded that it instead arises from intramolecular electron-transfer interactions between hydroxy-aromatic donors and quinone acceptors formed by the partial oxidation of lignin, polyphenol, tannin, and melanin precursors. This is consistent with the emerging view that humic substances are supramolecular aggregations rather than supermolecules.

Most of the DOM in streams and rivers is *allochthonous*, derived from terrestrial plants in the watersheds they drain, and about 10% of net primary productivity[†] within the watershed finds its way into streams. While some enters the stream directly in the form of throughfall, falling leaves, etc., most of this enters the stream via the soil where it has experienced varying degrees of degradation by soil organisms. Soil-derived DOM can include organic molecules adsorbed to soil particles that are desorbed when pH rises. Once dissolved in streams, some organic matter that has persisted in soils can be rapidly mineralized (Kaplan and Cory, 2016), which may relate to desorption from minerals surfaces and hence loss of the protective effect

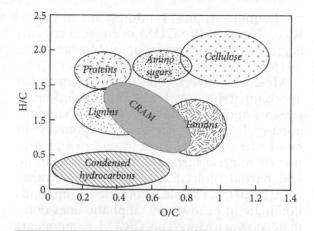

Figure 12.21   Van Krevelen diagram showing composition of various biomolecules and *carboxyl-rich alicyclic molecules* (CRAM) molecules common to stream, river, and deep ocean water.

provided by those surfaces discussed in the previous section.

Most (~70%) of the components making up DOM in first-order (headwater) streams consists of molecules of similar composition across climates and biomes, suggesting similar sources, such as soil-derived organic matter and/or similar diagenetic degradation processes (Jaffe et al., 2012; Mosher et al., 2015). Common to most streams are a group of poorly characterized molecules known as *carboxyl-rich alicyclic[‡] molecules* or CRAM as well as lignin-like compounds. This composition is shown in Figure 12.21 on a van Krevelen diagram, which is a plot of the H/C atomic ratio vs. the O/C atomic ratio and which was originally developed for analysis of kerogen (and which we will use for that purpose in subsequent sections).

On the other hand, Jaffe et al. found that streams in each biome have some distinct molecular patterns. For example, microbial inputs were clearly noticeable in their Kansas grassland stream characterized by open canopy and limestone outcrops and they found high amino acid abundances at

---

* Meaning originating in a place other than where it is found; in this case terrestrial organic matter. Autochthonous is the antonym and refers to organic matter synthesized by aquatic or marine organisms within the body of water.

[†] Net primary productivity is the gross productivity less the amount consumed by cellular respiration of autotrophs and is thus the amount available for growth, consumption by heterotrophs, export, etc.

[‡] alicyclic molecules are ones consisting of both aliphatic and ring structures

the Luquillo tropical rainforest site in Puerto Rico. The diversity DOM in tropical streams is greater than temperate ones (Mosher et al., 2015).

Molecular diversity decreases downstream approximately twofold in second-order streams and threefold by fifth-order streams, apparently as the unique components in first-order streams are lost by selective adsorption on mineral surfaces, photodisintegration, and partial molecular degradation (Mosher et al., 2015). While aromatic compounds dominate in headwaters, aliphatic ones dominate downstream. The CRAM components common to most headwater streams remain present downstream and they are dominant in estuaries and the oceans, reflecting a recalcitrance to bacterial degradation.

Most of the organic matter exported from terrestrial environments to streams is remineralized by bacterial respiration within them, although the exact fraction is uncertain. Estimates of the mount of organic matter exported from the terrestrial landscape to streams range from 1.9×10^{15} g C/yr to 2.7×10^{15} g C/yr with 0.6×10^{15} g C/yr exported to the oceans (Cole et al., 2007; Battin et al., 2009), with the remainder returned to the atmosphere as CO_2 or buried in continental sediments. This export to the oceans represents 33–47% of terrestrial net productivity, significantly higher than earlier estimates. How much of the riverine organic matter actually makes it into the ocean is somewhat unclear. Sholkovitz (1976) showed that much aquatic DOM flocculates and is removed as salinity increases in estuaries (along with inorganic components, as we will see in the Chapter 14). DOM is also removed by photo- and bacterial degradation in estuaries. In a study based on a global database, Massicotte et al. (2017) found a rapid decease in colored terrestrial DOM in a salinity range between 2 and 8‰. Other studies, however, suggest that the recalcitrant CRAM component persists and may contribute significantly to marine DOM (e.g., Cao et al., 2018).

Despite the apparent low concentrations, 35–300 μM, the amount of dissolved organic matter in the oceans, ~55×10^{15} mol C, vastly exceeds the living marine biomass; indeed, it is comparable to the total biomass, terrestrial and marine (Hedges, 2002), although

far smaller than the mass of dissolved inorganic carbon (HCO_3^-). It is, consequently, a major part of the carbon cycle. In contrast to streams and rivers, most DOM in the open ocean is autochthonous, ultimately originating through photosynthesis in the photic zone in the upper 100 m or so. Marine net primary production is estimated at 8 to 11×10^{15} mol C/yr and perhaps half is funneled into DOM through release of exudates, upon cell death, sloppy feeding by zooplankton, or the excretion of their waste products (Dittmar and Stubbins, 2014). DOM then funneled into the *microbial loop*, in which these dissolved organic compounds are consumed by microbes who are in turn fed on by protozoan plankton, who are in turn fed on by larger plankton, up through top predators, the wastes and remains of which again become DOM to be consumed by microbes (Azam et al., 1983).

Figure 12.22a shows the concentration of DOM in hydrocasts in the North Pacific and North Atlantic. Concentrations are highest in the mixed surface layer and decline before reaching approximately constant values at depths of around 1000 m, which corresponds to the base of thermocline (see Chapter 14). Particulate organic matter shows a similar pattern. Most of the organic matter, including most of the DOM, is consumed and converted back to CO_2 within the surface water and a small fraction is photooxidized. Perhaps 10% of gross primary production is exported from surface waters into deeper water as sinking organic particles and as DOM through mixing, water mass downwelling, and absorption on sinking organic and inorganic particles. As Figure 12.22a shows, much of that is consumed and remineralized within the thermocline. A small fraction survives into the deep ocean.

Figure 12.22b shows the ^{14}C activity (see Chapter 8) in DOM in the North Pacific and North Atlantic, expressed as a per-mil deviation of the fractionation-corrected (based on $\delta^{13}C$) activity from the international activity standard (which is the activity of an NBS oxalic acid standard in 1950):

$$\Delta^{14}C = \left(\frac{[^{14}C]_s}{[^{14}C]_{std}} - 1 \right) \times 1000 \qquad (12.3)$$

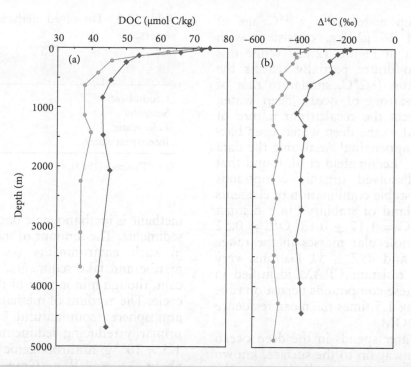

Figure 12.22   (a) Dissolved organic matter as a function of depth from the northeast Pacific Ocean (blue circles) and the northeast Atlantic Ocean (red diamonds). Pacific data from Loh and Bauer (2000), station M. Atlantic data from geotraces.org database (GEOTRACES Station GA03-5; Joaquin E Chaves, NASA Goddard Space Flight Center). (b) $\Delta^{14}C$ from the North Pacific and the Sargasso Sea, Central Atlantic. From Druffel and Bauer (2000).

where $[^{14}C]_s$ is the measured radioactivity of the sample corrected for fractionation and $[^{14}C]_{std}$ is the 1950 activity of the standard (Stuvier and Polach, 1977), and hence is a measure of the time since the carbon was part of the atmospheric reservoir. DOM in both surface and deep water is older in the Pacific than in the Atlantic, which reflects the way water circulates in the global ocean, as we will discuss in Chapter 14. But in both oceans, DOM in deep water is substantially older than in surface water. The "age" of DOM in N. Atlantic deep water is approximately 4000 years while that of Pacific is approximately 6000 years (Druffel and Bauer, 2000); again, while this partly reflects how the ocean circulates, these ages are substantially older than that of dissolved inorganic carbon.

This indicates that DOM that survives into the deep water is quite resistant to microbial attack; furthermore, it appears to become increasingly recalcitrant as it ages. Indeed, it is apparent from Figure 12.22b that some older carbon is present even in the DOM of

surface waters, indicating some old DOM survives deep circulation to be returned to the surface. Flerus et al. (2012) measured molecular mass of DOM components using Fourier transform ion cyclotron resonance mass spectrometry and found that $\Delta^{14}C$ decreased with increasing average molecular masses and decreasing H/C ratios. In detail, some masses correlated positively with $\Delta^{14}C$ while others correlated negatively. Those that correlated positively, such as $C_{13}H_{18}O_7$, had an average mass of 315 Daltons and H/C ratio of 1.45; those that correlated negatively and hence were the older components, such as $C_{21}H_{26}O_{11}$, had an average mass of 420 Daltons and H/C ratio of 1.17.

Lecthenfeld et al. (2014) used these results to calculate first-order rate equation for decay of DOM:

$$DOM(t) \, (\mu M/kg) = 131e^{(-2.23 \times 10^{-4}t)} \quad (12.4)$$

(compare eqn. 5.25). This equation, for example, predicts a DOM concentration for Pacific deep water with a ^{14}C age of 6000 years of 36 μM/kg, consistent with what is shown in Figure 12.22. While this slow decay no doubt partially reflects the low temperatures (~2°C, similar to that of a food refrigerator) of ocean deep water, it mainly reflects the recalcitrant nature of DOM exported to the deep water: even bacteria find it unappetizing! Analyzing the data of Flerus et al., Lecthenfeld et al. found that the marine dissolved organic compounds with the most stable combination of elements formed an "island of stability" in a distinct window of H/C = 1.17 ± 0.13, O/C = 0.52 ± 0.10, and molecular masses in the range of 360 ± 28 and 497 ± 51 Daltons, very similar to the resistant CRAM identified in fresh water. These compounds persist on time scales exceeding 1.5 times the mean residence time of total DOM.

The time water spends in the deep ocean before it returns again to the surface, known as the ventilation time, is quite long, on the order of 10^3 years (Chapter 14). The process by which organic carbon produced by photosynthesis in surface water sinks into the deep ocean to be stored there on millennial times scales is known as the *biological pump*. The deep ocean contains approximately 40×10^{15} mol C and is, consequently, one of the larger pools of carbon in the exogenic carbon cycle; a topic we will explore later in this chapter. Shifts in carbon between the deep ocean and atmosphere were an important feedback mechanism on the Pleistocene glacial cycles (Chapter 9).

12.5.3 Hydrocarbons in natural waters

A variety of hydrocarbons that originate both from natural and anthropogenic sources are also present in natural waters. These may be divided into volatile and nonvolatile hydrocarbons, depending on the vapor pressure and boiling point. Short-chained hydrocarbons tend to be volatile, and this volatility limits their abundance in natural waters. Longer-chained hydrocarbons are not volatile. Their abundance in natural waters is often limited by their generally low solubility. Of the volatile hydrocarbons, methane is by far the most abundant. Some methane can be produced in the digestive tracts of higher animals, but the principal source of natural

Table 12.3 Dissolved methane in natural waters.

	Concentration μg/l
Groundwater	10–10,000
Seawater	10–100
Lake water	10–10,000
Interstitial water	100–10,000

From Thurman (1985).

methane is methanogenic bacteria in reducing sediments. The amount of methane produced in such environments (swamps, reducing marine and lake sediments, etc.) is a significant, though minor, part of the global carbon cycle. The amount of methane released to the atmosphere from natural sources annually, primarily reducing sediments in wetlands, is 1.5×10^{14} g; anthropogenic activities release about 3.6×10^{14} g (Graedel and Crutzen, 1993). The abundance of methane in natural waters is summarized in Table 12.3. The solubility of methane in water is 800 μg/L (at 20°C and 0.1 MPa). At concentrations above this level, methane bubbles will form and this process is undoubtedly important in the transport of methane from sediment interstitial waters of swamps and shallow lakes to the atmosphere. Some of this methane redissolves in the lake water and is oxidized by methanotrophic bacteria.

Among the most common nonvolatile and semi-volatile hydrocarbons in seawater are C_{15} and C_{17} n-alkanes (pentadecane and heptadecane) and isoprenoids, primarily pristane and phytane. C_{15} and C_{17} originate by decarboxylation of C_{16} and C_{18} fatty acids (palmitic and stearic acids), which are derived from zoo- and phytoplankton. *Halogenated hydrocarbons* (i.e., hydrocarbons where one or more hydrogens are replaced by a halogen) are of particular interest because of their toxicity. Most halogenated hydrocarbons in natural waters are anthropogenic, having been directly manufactured and discarded or leaked into natural waters. Others, such as chloroform and trichloromethane, can arise indirectly through chlorination of drinking water. However, some halogenated hydrocarbons do occur naturally in seawater at very low abundances.

12.6 CHEMICAL PROPERTIES OF ORGANIC MOLECULES

12.6.1 Acid–base properties

As we noted above, the carboxyl group can dissociate to give up a hydrogen atom:

$$RCOOH \rightleftharpoons H^+ + RCOO^- \qquad (12.5)$$

(we use R here as a general representation for the remainder of the molecule). Like other acids, organic acids will increasingly dissociate as pH increases. As for other reactions, we can write an equilibrium constant expression:

$$K = \frac{a_{H^+} a_{RCOO^-}}{a_{RCOOH}} \qquad (12.6)$$

The equilibrium constant is commonly reported as pK_a, which, analogous to pH, is the negative of the logarithm of the equilibrium constant. The Henderson–Hasselback equation relates pK_a, activity quotient, and pH:

$$pK_a = -\log K_a = -\frac{a_{RCOO^-}}{a_{RCOOH}} - pH \qquad (12.7)$$

Expressed in this way, the pK_a is the pH where half of the acid molecules are dissociated and half undissociated. Most carboxylic acids have pK_a values of 1–5, which is below the pH of most streams and lakes, and within or lower than the pH range of soils (generally 3–8). Thus carboxylic acids will be dissociated in most environments. As we have seen, carboxylic acids (both simple and as functional groups on humic substances) are important components of dissolved and soil organic matter. The effect of increasing concentrations of organic molecules will thus be to increase the concentration of protons, lowering the pH of natural waters.

The hydrogen in OH groups of phenols can also dissociate and hence contribute to solution or soil acidity. Phenols are, however, much weaker acids than carboxylic acids; whereas the pK_a values of carboxylic acids are typically 3 or so, pK_a for phenols are more typically 8. Thus at pH typical of most natural waters and soils, phenols will be only partially or not dissociated.

In contrast to carboxyl and phenol groups, nitrogen-containing groups, such as amino groups, are basic as they tend to bind free protons. By analogy to pK_a, defined above, we can define a pK_b, which is the pH when half the molecules or groups will be protonated and half unprotonated. Aliphatic amines are the most strongly basic, with typical pK_b values of 10–12. At pH values below this, they will be protonated. Thus, in most natural waters and soils they will bear a positive charge and behave as cations. Because particle surfaces are typically negatively charged, these organic cations are readily bound to particle surfaces and removed from solution. Aromatic amines are typically weaker bases, with pK_b values around 4–6. They will be protonated only in acidic waters and soils.

As we have seen, humic and fulvic acids are generally the most abundant organic substances in natural waters and soils. They often contribute significantly to the acidity of waters and soils. Under some circumstances, such as lowland tropical rivers or swamps, they are the principal negative ions present. They typically contain 10^{-2} eq/g ionizable acid groups per weight of organic carbon. Carboxyl groups are most common, but other functional groups are also present. As a result, humic substances cannot be characterized by a single pK_a. Their titration curves (Figure 12.23) typically have a "smeared out" appearance, a result both of the variety of functional groups present, and electrostatic interactions between these groups.

12.6.2 Complexation

Another important geochemical property of organic molecules is their ability to form complexes with metals, especially transition metals and aluminum. Complexation between metal ions and organic anions is similar, for the most part, to complexation between metals and inorganic anions. An important difference is that many organic compounds have more than one site that can bind to the metal. Compounds having this property are referred to as *multidentate* (Section 6.3.3). Complex formation with multidentate ligands is called *chelation*, and the complexes formed are called *chelates*. A simple example is the oxalate ion, $C_2O_4^{2-}$, which consists of two carboxyl groups (Figure 12.6) and is *bidentate*. Citric acid and glutamic acid (an amino acid) are tridentate ligands (although the

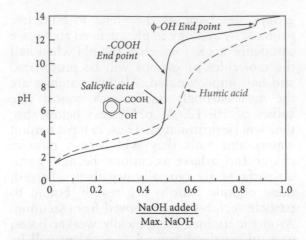

Figure 12.23 Comparison of titration curves of salicylic and humic acid. The salicylic acid shows two endpoints, corresponding to its carboxyl and phenol functional groups. The humic acid shows a smeared-out titration curve corresponding to a continuously changed pK_a. After Morel (1983).

amino group will not dissociate and take on a positive charge, it nevertheless has an electron pair available to share). As organic functional groups are only weakly acidic, metal–organic complexation is strongly pH-dependent.

A metal–oxalate complex results in the formation of a ring (Figure 12.24), with the two oxygens that are singly bonded to carbon each binding to the metal. In the oxalate complex, the ring has five members; a six-member ring would be formed in a metal-malonate; seven-member rings are formed by phthalate or succinate chelates. A metal glycine complex forms two rings on opposing sides of the metal (Figure 12.24). Salicylate is another example of a bidentate ion (Figure 12.25a). In this case, the binding sites are formed by two different functional groups: carboxyl and hydroxyl. Multidentate ions can very strongly bind trace metals. A few natural chelators that are specific for Fe have been characterized. One, enterobactin, is illustrated in Figure 12.25b. Not all organic anions are multidentate, of course. Benzoate, acetate, and phenol are examples of unidentate organic anions. It is also possible for a single metal ion to bind to more than one organic ligand, as illustrated in Figure 12.25c.

Just as for inorganic metal–ligand complexes, we can define stability constants (β)

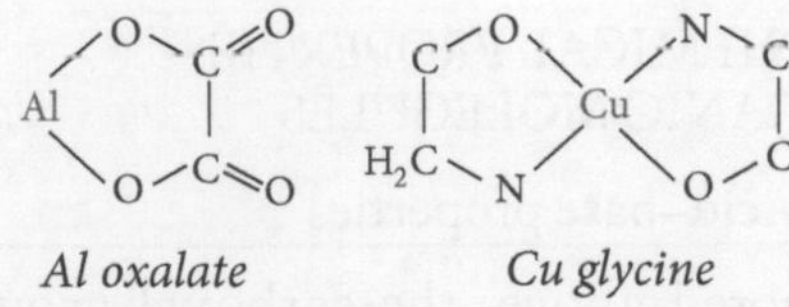

Figure 12.24 Examples of rings formed by chelates.

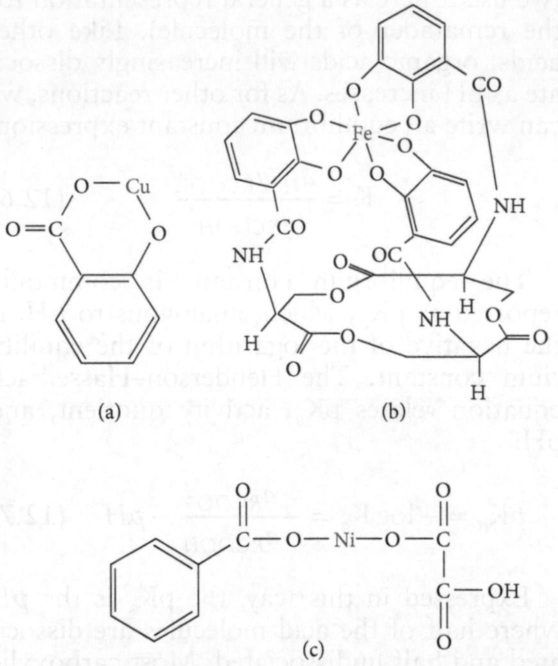

Figure 12.25 (a) Copper salicylate complex. The Cu ion is bound to both the carboxylic and phenol groups. (b) Enterobactin, a natural iron-chelating agent. (c) Ni complexed by separate benzoate and acetate groups.

and apparent stability constants (β^*) for metal–organic ligand complexes. Table 12.4 lists some examples of stability constants for metal–organic complexes that we might expect to find in natural waters. Several generalizations may be made. First, as we saw for inorganic ligands in Chapter 6, the higher the valence state of the metal, the stronger the complex. Thus, in Table 12.4, Fe^{3+} forms stronger complexes with all listed ligands than does Fe^{2+}. Second, there is high degree of correlation between the equilibrium constants of all organic ligands for a given metal. For the divalent metals, stability of metal–organic complexes follow the Irving–Williams series

(Chapter 6), $Pb^{2+} > Cu^{2+} > Ni^{2+} > Co^{2+} > Zn^{2+} > Cd^{2+} > Fe^{2+} > Mn^{2+} > Mg^{2+}$.

It is the functional groups of organic molecules that are primarily responsible for metal ion complexation. Amines, azo-compounds (compounds containing a $-N=N-$ group linking two other groups), ring nitrogen, carboxyl, ether, and ketone are all important in complex formation. Tetrapyrrole pigments, or porphyrins such as chlorophyll and heme, are very strong metal ion complexing agents, particularly for transition metals such as Fe, Zn, and Ni. In the case of dissolved chlorophyll and similar molecules, complex formation occurs through replacement of Mg with a transition metal ion. In this instance, the metal is bound to two nitrogens (see Figure 12.15).

Multidentate complexes are generally more stable than corresponding unidentate ones. We can see this in Table 12.4, where the stability constants for citrate, with three carboxyl-binding sites, are higher than those for acetate with one carboxyl-binding site. Another interesting property of multidentate ligands is that the degree of complexation decreases less strongly with dilution than for monodentate complexes.

As was the case for inorganic complexes, the stability of metal–organic complexes is invariably strongly pH-dependent. The reason for this is simple: hydrogen ions will be major competitors for sites on functional groups. Indeed, we can write the complexation reaction as:

$$m\,M + l\,HL \rightleftharpoons M_mL_l + H^+$$

the equilibrium constant for this reaction is then:

$$K = \frac{[M_mL_l]a_{H^+}}{[M]^m[L]^l} \tag{12.8}$$

or in log form:

$$\log K = \log[M_mL_l] - m\log[M] - l\log[L] - pH \tag{12.9}$$

We should also note that for multidentate ions, mixed hydrogen–metal and hydroxide–ligand complexes are possible. These will become increasingly important at low and high pH respectively. For example, at pH values below about 3, the CuHCitrate complex will be dominant over the simple Cu-citrate complex. In another example, the FeOHGlycolate complex will be more

Table 12.4 Stability constants ($\log \beta$) for metal ion–organic ligand complexation.

	Glycine[1]	Glutamate[2]	Acetate[3]	Citrate[4]	Malonate[5]	Salicylate[6]
H^+	9.78	9.95	4.76	6.4	5.7	13.74
Na^+				1.4	0.7	
K^+				1.3		
Mg^{2+}	2.7	2.8	1.3	4.7	2.9	
Ca^{2+}	1.4	2.1	1.2	4.7	2.4	0.4
Al^{3+}			2.4			12.02
Ba^{2+}	0.8	2.2	1.1	4.1	2.1	0.2
Fe^{3+}	10.8	13.8	4.0	13.5	9.3	17.6
Fe^{2+}	4.3	4.6	1.4	5.7		7.4
Ni^{2+}	6.2	6.5	1.4	6.7	4.1	7.8
Cu^{2+}	8.6	8.8	2.2	7.2	5.7	11.5
Zn^{2+}	5.4	5.8	1.6	6.1	3.8	7.7
Pb^{2+}	5.5		2.7	5.4	4.0	
Hg^{2+}	10.9		6.1	12.2		
Ag^+	3.5		0.7			

[1] $NH_2CH_2(COO)^-$

[2] $(HOOC)(CH_2)_2CHNH_2COO^-$

[3] $C_2H_5COO^-$

[4] $(HOOC)CH_2C(OH)(COOH)CH_2COO^-$ [5] $CH_3CH_2COO^-$ [6] $HOC_6H_4COO^-$

From Morel and Hering (1993).

important at all pH values than the simple Fe-glycolate complex. For clarity, we have omitted stability constants for these mixed complexes from Table 12.4. Nevertheless, as these examples show, these mixed complexes must often be considered in speciation calculations, particularly at high and low pH. A more complete compilation of stability constants for metal–organic complexes may be found in Morel and Hering (1993).

The degree to which dissolved trace metals in natural waters are complexed by organic ligands varies by environment. There is now a good body of observational evidence showing that a large fraction of at least some trace metals (particularly Fe, Cu, and Zn) is complexed by organic ligands in streams, lakes, and ocean surface waters (e.g., Ellwood, 2004; Hoffmann et al., 2007). In some cases, more than 99% of the metal in solution is present as organic complexes. In the ocean surface water, the extent of complexation seems to be somewhat variable; in deep water, far smaller fractions of metal ions appear to be organically complexed (e.g., Baars and Croot, 2011). The organic ligands complexing these metals can have stability constants in excess of 10^{20} (e.g., Witter et al., 2000). Many of these complexing agents, such as trihydroxamate siderophore desferriferrioxamine B, are produced by organisms to acquire essential trace metals such as Fe. They may produce others to defend themselves against the toxicity of others, such as Cu and Pb. It is interesting in this respect that Zn falls into both categories. It is toxic at high concentrations, yet essential for a number of biological processes such as nucleic acid transcription.

Example 12.1 Speciation of organic ligands in fresh water

Using the stability constants in Table 12.4, and the calculated free ion activities for major cations in Example 6.8, calculate the speciation of glycine, citrate, and salicylate. Assume total activities of glycine, citrate, and salicylate of 1.25×10^{-8}, 5×10^{-8}, and 1×10^{-8} M/l, respectively.

Answer: For each ligand, we can write a conservation equation:

$$\Sigma L = L^- + HL + AL + BL + CL + \ldots \tag{12.10}$$

where L^- is the free ligand, HL is the undissociated acid, and AL, BL, CL, etc., are the various metal ligand complexes. For each species we may also write:

$$[ML] = \beta \times [M] \times [L^-] \tag{12.11}$$

where [ML], [M], and [L^-] are the concentrations of the complex, free metal ion or proton, and free ligand, respectively. Substituting eqn. 12.11 into 12.10, we have:

$$\Sigma L = [L^-] + \beta_{HL}[H][L^-] + \beta_{AL}[A][L^-] + \beta_{BL}[B][L^-] + \ldots \tag{12.12}$$

Rearranging, we have:

$$[L^-] = \frac{\Sigma L}{1 + \beta_{AL}[A] + \beta_{BL}[B] + \ldots} \tag{12.13}$$

Since the concentrations of the organic ligands are much lower than those of the major cations, we can assume that organic complexation does not affect activities of the major cations. Equation 12.13 gives us the free ion concentration. From that, we can calculate the concentration of each of the complexes using eqn. 12.11. The result is shown in the adjacent table. We see that at the pH of this example (8), glycine and salicylate are essentially completely undissociated. Citrate is almost completely dissociated, but is 95% complexed by Mg and Ca.

Speciation of organic ligands			
	Glycine	Citrate	Salicylate
H	98.24%	0.11%	100.00%
Na	0.00%	0.03%	0.00%
K	0.00%	0.01%	0.00%
Mg	0.11%	31.77%	0.00%
Ca	0.01%	63.55%	0.00%
Free ligand	1.63%	4.53%	0.00%
Activity of free ligand	2.04×10^{-10}	2.26×10^{-09}	1.82×10^{-14}

Example 12.2 demonstrates that glycine, a common amino acid, and citrate, a common hydroxy-carboxylic acid, and salicylate, a common phenolate, will complex only a small fraction of the total Cu in fresh water with typically low concentrations of these substances. We see that this is due to several factors. First, at this pH, most of the glycine and salicylate are undissociated (Example 12.1), and therefore unavailable to bind Cu, and 95% of the citrate is complexed with Ca and Mg. Second, the greater abundance of inorganic anions such as hydroxyl and carbonate results in their dominating the speciation of Cu. However, one should avoid drawing the conclusion that organic trace metal complexes are inevitably insignificant. We considered only three species in this example, and while they strongly bind copper, all are at fairly low concentrations. Other organic anions, particularly including humates, are often present at sufficient concentration to complex a significant fraction of some trace metals. Problems 8 and 9 at the end of this chapter illustrate that situation.

Example 12.2 Speciation of Cu in fresh water

Use the adjacent stability constants as well as those for glycine, citrate, and salicylate in Table 12.4 to calculate the speciation of Cu in the water sample analysis in Example 6.8, assuming $\Sigma Cu = 10^{-9}$ M. Use the calculated free ion concentrations of anions in Examples 6.8 and 12.1.

Cu stability constants	
	$\log \beta$
$CuOH^+$	6.3
$Cu(OH)_2$	11.8
$CuCl^+$	0.5
$CuCO_3$	6.7
$CuSO_4$	2.4

Answer: In calculating trace element speciation, it is common to assume that complexation with trace metals does not reduce the free ion concentrations of the anions. For this assumption to be valid, the free ion concentrations of the anions should greatly exceed those of the trace metal. This condition is met in this case for the inorganic anions, but not for the organic ones. Nevertheless, we will proceed by making this assumption initially and subsequently examine its validity and make the necessary corrections. We proceed much as we did in Example 12.1, by writing a conservation equation for copper:

$$\Sigma Cu = [Cu^{2+}] + [CuOH^+] + [Cu(OH)_2] + [CuCl^-] +$$

$$[CuCO_3] + [CuSO_4] + [CuGly] + [CuCit] + [CuSal] \qquad (12.14)$$

For each species, we also write a mass action equation, for example:

$$CuCit = \beta_{CuCit} \times [Cu^{2+}] \times [Cit] \qquad (12.15)$$

Substituting the mass action equations into eqn. 12.14 and solving for $[Cu^{2+}]$, we have:

$$[Cu^{2+}] = \frac{\Sigma Cu}{1 + \sum \beta_{CuL_i}[L]} \qquad (12.16)$$

We can then calculate the concentrations of the individual species using eqn. 12.15. The results are shown in the adjacent table. We see that Cu is dominantly complexed by hydroxyl and carbonate. The three organic complexes account for only about 1% of the total copper.

Calculated copper speciation		
	Conc.	**%**
$CuOH^+$	2.12×10^{-10}	21.22%
$Cu(OH)_2$	6.71×10^{-11}	6.71%
$CuCl^+$	7.03×10^{-14}	0.01%
$CuCO_3$	5.97×10^{-10}	59.69%
$CuSO_4$	4.41×10^{-12}	0.44%
CuGly	8.63×10^{-12}	0.86%
CuCit	3.82×10^{-12}	0.38%
CuSal	6.12×10^{-13}	0.06%
Cu^{2+}	1.06×10^{-10}	10.63%

Now let's examine our initial assumption that Cu speciation does not reduce the free ion activities of the anions. With the exception of copper salicylate, the concentration of each species is far less than the free ion concentration of the corresponding anion. In the case of salicylate, however, the concentration exceeds the total free ion concentration of salicylate, a clear indication that our initial assumption was invalid. We could address this problem by performing an iterative calculation such as that used in Example 6.8. However, an examination of the situation reveals a simpler approximate solution. The concentration of free salicylate is far below that of free copper. Furthermore, the stability constant for copper salicylate is very large. In these circumstances, all available salicylate will be complexed with free copper, so we may replace our calculated CuSal concentration with that of the free salicylate concentration we calculated in Example 12.1, 1.82×10^{-14} M. This is a trivial fraction of the total copper. Stream and lake water is likely to contain trace concentrations of other metals that are strongly bound by salicylate, such as Fe. This would further reduce the copper salicylate activity.

Humic substances are certainly capable of complexing metals but the complexation behavior is complex and is difficult to characterized by a single stability constant. This is true for several reasons. First, they do not have unique compositions and contain a variety of functional group with different pK_a. Second, these functional groups are close enough to one another that the electrostatic charge of one site can affect the complexing properties of an adjacent one (recall that this was also true of surfaces: see Chapter 6). A full treatment of this problem is beyond the scope of this book but may be found in Morel and Hering (1993).

12.6.3 Adsorption phenomena

12.6.3.1 The hydrophobic effect and hydrophobic adsorption

Water molecules near large nonpolar molecules such as long chain hydrocarbons cannot orient their polar OH bonds as they normally would to reduce electrostatic repulsions and minimize interaction energy, the presence of a large nonpolar molecule is energetically unfavorable (Figure 12.26). As a result, solution of such substances, called *hydrophobic substances*, in water is associated with a large ΔH_{sol} and large ΔG_{sol}. Thus one characteristic of hydrophobic substances is limited solubility in water. A second characteristic is that when they are present in solution, they are readily absorbed on surfaces (think of grease sticking to your pan), particularly nonpolar ones, such as those of organic solids (such as your hands).

Hydrophobic adsorption differs from other types of adsorption phenomena in that adsorption occurs not as a result of an affinity of the surface for the solute, but because of incompatibility of the hydrophobic compound with water. When a hydrophobic molecule is located on a surface, water molecules are present on one side only, and there is less disruption of water structure than when water molecules are located on both sides. Thus the interaction energy is lower when the substance is located on a surface rather than in solution. Other types of adsorption involve electrostatic or van der Waals interactions or formation of bonds between the surface and the solute. While electrostatic and, particularly, van der Waals interactions generally contribute to hydrophobic adsorption, they are of secondary importance compared with the minimization of interaction energy between the solute and water.

Hydrophobic adsorption can be described by a simple model of partitioning of the hydrophobic species between water and an absorbent. The adsorption partition coefficient, K_P, is defined as:

$$K_P = \frac{\text{moles sorbate/mass solid}}{\text{moles solute/volume solution}}$$

(12.17)

and is typically expressed in units of liters/kilogram. The magnitude of the adsorption partition coefficient for hydrophobic species is related in a simple way to the solubility of the species in water, as illustrated in Figure 12.27a: the least soluble compounds are most strongly adsorbed. The aqueous solubility of such species may be further related to the octanol–water partition coefficient (Figure 12.27b). Octanol is a largely nonpolar molecule, so that there is little structure or ordering of molecules in liquid octanol as there is in water. Thus there is no disruption of solvent molecules when a nonpolar solute is dissolved in octanol. The octanol/water partition coefficient is thus a measure of the "hydrophobicity" of organic molecules. The adsorption coefficient for

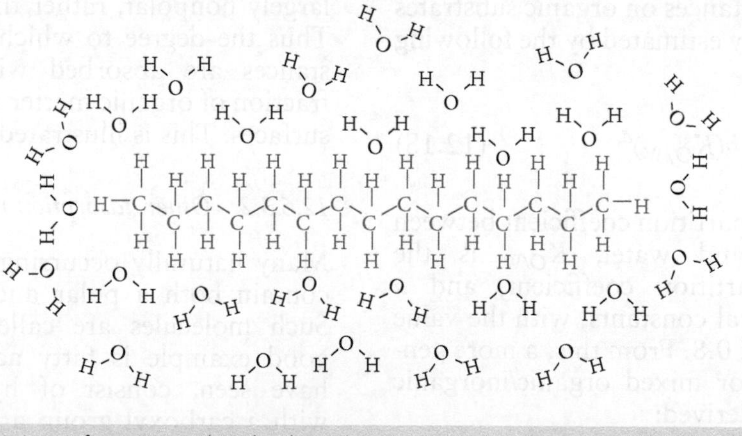

Figure 12.26 Disruption of water molecules by a large nonpolar organic molecule, in this case a C_{15} n-alkane.

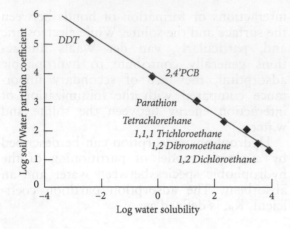

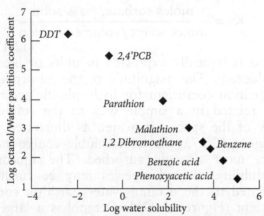

Figure 12.27 (a) Soil/water adsorption partition coefficients for a variety of organic compounds as a function of water solubility, determined by Chiou et al. (1979). (b) Octanol/water partition coefficients as a function of water solubility of organic compounds, determined by Chiou et al. (1977).

hydrophobic substances on organic substrates may be empirically estimated by the following relationship:

$$K_{om} = b(K_{O/W})^a \qquad (12.18)$$

where K_{om} is the partition coefficient between organic solids and water, $K_{O/W}$ is the octanol/water partition coefficient, and a and b are empirical constants, with the value of a being around 0.8. From this, a more general expression for mixed organic/inorganic surfaces may be derived:

$$K_{om} = bf_{OC}(K_{O/W})^a \qquad (12.19)$$

where f_{oc} is the fraction of organic matter in the solid. Comparing eqns. 12.18, 12.19, we see that:

$$K_{om} = K_P/f_{OC} \qquad (12.20)$$

In general, the solubility of organic molecules decreases with increasing molecular weight. This observation, known as *Traube's rule*, is apparent from Figure 12.27. Small polar molecules such as phenoxyacetic acid and benzoic acid have higher solubilities and lower octanol/water partition coefficients than do large nonpolar ones such as DDT and PCBs (polychlorinated biphenols). It is easy to understand why this should be so: the larger the molecule, the greater the volume of water whose structure is disrupted. In addition, the tendency of a molecule to be absorbed and the strength of this adsorption increases with atomic weight. In part, this is true for the same reason solubility decreases: a greater volume of water is disrupted by large molecules. However, as we noted above, van der Waals interactions between the adsorbed substance and the surface also contribute to hydrophobic adsorption. These interactions increase with increasing size of the molecule. Van der Waals interactions contribute a surface binding energy of roughly 2.5 kJ/mol per CH_2 group on the surface. Clearly, the more CH_2 groups involved, the more strongly the substance will be bound to the surface. For this reason, polymers are readily adsorbed to surfaces even if the adsorption free energy per segment is small. Adsorption of large polymers can be virtually irreversible.

Hydrophobic molecules are adsorbed preferentially to organic surfaces, which are largely nonpolar, rather than inorganic ones. Thus the degree to which hydrophobic substances are absorbed will depend on the fraction of organic matter that makes up solid surfaces. This is illustrated in Figure 12.28.

12.6.3.2 Other adsorption mechanisms

Many naturally occurring organic molecules contain both a polar and a nonpolar part. Such molecules are called *amphipathic*. A good example is fatty acids, which, as we have seen, consist of hydrocarbon chains with a carboxyl group attached to one end. The hydrocarbon chain is nonpolar and hydrophobic. The carboxyl group, however,

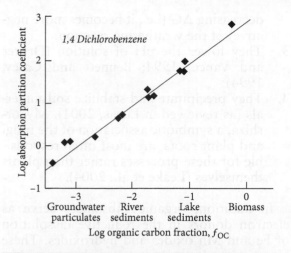

Figure 12.28 Adsorption partition coefficient for 1,4-dichlorobenzene plotted as a function of fraction of organic carbon in the solid absorbent. Other hydrophobic molecules show similar relationships. After Schwarzenbach and Westall (1981).

is quite polar upon dissociation. The carboxyl group itself is readily soluble in water (as demonstrated by the high solubilities of simple carboxylic acids such as formic and acetic acids) and is hence *hydrophilic*. Carboxyl groups are therefore not subject to hydrophobic adsorption except at very low pH, where they are undissociated. They can, however, bind to polar solid surfaces in much the same way as inorganic ions. These include reactions such as *ligand exchange*:

$$S \equiv OH + COOH(CH_2)_n CH_3 \rightleftharpoons$$
$$S \equiv COO(CH_2)_n CH_3 + H_2O \quad (12.21)$$

where the carboxyl group, less its hydrogen, exchanges for an OH group bound to surface site $S\equiv$. Polar function groups or organic anions may also bind to surfaces through *water bridging*, in which complexation with a water molecule solvating an exchangeable cation at a surface occurs:

$$S \equiv M^+(H_2O) + COOHR \rightleftharpoons$$
$$S \equiv M^+(H_2O)-COOHR \quad (12.22)$$

This mechanism is most likely to occur where M is strongly solvated (Mg^{2+} for instance). Where M is not strongly solvated, *cation bridging*, in which there is a direct

bond between the acid functional group and the metal, can occur:

$$S \equiv M^+ + COO^-R \rightleftharpoons S \equiv M^+-COO^-R \quad (12.23)$$

For cationic functional groups, such as quarternized nitrogen, *cation ion exchange* reactions such as:

$$S \equiv M^+ + NH_3^+R \rightleftharpoons S \equiv NH_3^+R + M^+ \quad (12.24)$$

where an organic cation replaces a metal cation at a surface are possible. For anionic functional groups, such as carboxylic acids, anion ion exchange can occur. This is the analogy of 12.21 with the signs reversed, for example, a carboxyl group in anion form replacing a surface OH⁻ group.

All the above reactions may occur at either organic or inorganic surfaces. *Hydrogen bonding* in which a hydrogen is shared between a surface O atom and an O atom in a dissolved organic such as a carboxyl or phenol group, can occur at organic surfaces, for example:

$$S \equiv H^+ + COO^-R \rightleftharpoons S \equiv H^+-COO^-R \quad (12.25)$$

Hydrogen bonding is not restricted to acids. Organic bases, notably those containing nitrogen groups such as amines and pyridines, can also form hydrogen bonds with a hydrogen at a solid surface. Hydrogen bonding between dissolved organics and mineral surfaces is less important because the oxygens of mineral surfaces are not as electronegative as in organic compounds.

Many organic compounds will thus be subject to several types of adsorption: nonpolar parts may be adsorbed to surfaces through hydrophobic bonding, while polar groups may bind through the mechanisms just described.

12.6.3.3 Dependence on pH

Figure 12.29 shows the effect of pH on the adsorption of humic acid on Al_2O_3; the extent of adsorption is greatest at a pH of about 3 and is generally greater at low pH than at high pH. This pH dependence arises because the availability of hydrogen ions in solution will affect the charge on a solid surface in contact with that solution. At pH below the

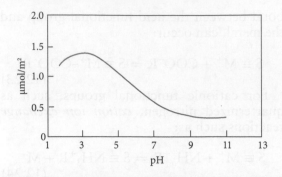

Figure 12.29 Adsorption of humic acid on δ-Al$_2$O$_3$ as a function of pH. After Stumm (1992).

isoelectric point of a mineral, mineral surfaces will be protonated and will carry a positive charge; at higher pH the mineral surface will bear a negative charge. Furthermore, dissociation and protonation of organic functional groups, which will affect the extent of adsorption through the mechanisms discussed above, is pH-dependent.

Clearly, pH will also affect the mechanism of adsorption. Carboxyl acid groups of a humic acid molecule might bind to a surface through cation bridging at high pH where the surface has a net negative charge. At low pH, carboxyl groups will bind to a protonated surface through hydrogen bonding. At a pH close to that of the isoelectric point of a mineral, its surface will be neutral, in which case a humic acid would be subject to hydrophobic adsorption through its nonpolar parts. Thus, the mechanism of adsorption and the strength of the bond formed between adsorbent and adsorbate will be influenced by pH.

12.6.3.4 Role in weathering

Adsorption and the formation of surface complexes play a key role in weathering reactions. Organic acids can play an important role in accelerating weathering reactions in several ways:

1. They form surface complexes, particularly surface chelates that weaken metal–oxygen bonds in the crystal and thus promote removal of metals from the surface.
2. They form complexes with metals in solution, reducing the free ion activities and

decreasing ΔG (i.e., it becomes more negative) of the weathering reaction.
3. They lower the pH of solution (Drever and Vance, 1994; Bennett and Casey, 1994).
4. They precipitate and stabilize soil minerals (as reviewed in Lucas, 2001). Mycorrhiza, a symbiotic association of the fungi and plant roots, are most often responsible for these processes rather than plants themselves (Leake et al., 2004).

In addition, organic substances serve as electron donors in the reductive dissolution of Fe and Mn oxides and hydroxides. These effects have been demonstrated in a variety of laboratory experiments (e.g., Furrer and Stumm, 1986; Zinder et al., 1986) and electron microscopy of minerals exposed to high concentrations of organic acids in both natural and laboratory situations (e.g., Bennett and Casey, 1994).

Furrer and Stumm (1986) investigated the effect of a variety of simple organic acids on dissolution of δ-Al$_2$O$_3$ and demonstrated a first-order dependence of the dissolution rate on the surface concentration of organic complexes:

$$\mathfrak{R} = k\,[\text{S}{\equiv}\text{L}]$$

where $[\text{S}{\equiv}\text{L}]$ is the surface concentration of organic complexes. Bidentate ligands that form mononuclear surface complexes seemed particularly effective in increasing dissolution rate. (There appears to be some evidence that formation of polynuclear surface complexes retards dissolution; Grauer and Stumm, 1982.) Five- and six-member chelate rings were more effective in enhancing dissolution rate than seven-member rings (Figure 12.30). Though monodentate ligands such as benzoate were readily adsorbed on to the surface, they had little effect on dissolution rate. Similarly, Zinder et al. (1986) demonstrated a first-order dependence of the dissolution rate of goethite (FeOOH) on oxalate concentration.

In general, low molecular weight organic acids are more effective in accelerating mineral dissolution than larger molecules, such as humic and fulvic acids. Zhang and Bloom (1999) found that the relative effectiveness of ligands in promoting dissolution of hornblende was oxalic > citric > tannic >

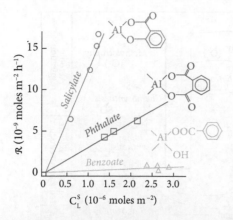

Figure 12.30 Rate of ligand-promoted dissolution of δ-Al_2O_3 as a function of organic ligand surface concentration. Chelates forming five- and six-member rings, such as those formed by salicylate, produced faster dissolution than seven-member rings, such as those formed by phthalate. Unidentate ligands, such as benzoate, have only a small effect on dissolution rate. From Furrer and Stumm (1986).

polygalacturonic > fulvic acid. Similarly, Khademi et al. (2010) showed that the addition of oxalic acid, a dicarboxylic acid (Figure 12.7), to soil significantly increased soluble P concentrations and enhanced plant uptake of P, while addition of citric acid, a tricarboxylic acid, had a lesser effect. Thus similar organic acids can have dissimilar weathering effects.

Field studies show that high concentrations of organic acids, either natural or anthropogenic, clearly accelerate weathering (Bennett and Casey, 1994). However, in most circumstances, the concentrations of organic acids are low, and probably have only a small effect on weathering rates (Drever and Vance, 1994). Organic acids dissolved in formation waters of petroleum-bearing rocks may also enhance porosity by dissolving both carbonates and silicates (Surdam et al., 1984). This enhanced porosity is essential to the migration and recovery of petroleum.

12.7 SEDIMENTARY ORGANIC MATTER

Essentially all bodies of surface water harbor life, and therefore the production of organic carbon in aquatic and marine environments is ubiquitous. Most sedimentary rocks, however, contain rather little organic matter (a fraction of a percent is typical). This is a testament to the efficiency of life: virtually all organic carbon produced by autotrophs is subsequently *remineralized* to CO_2 by respiration. Indeed, most of the organic carbon synthesized in the oceans and deep lakes never reaches the sediment; it is consumed within the water column. Organic carbon that does manage to reach the bottom is subject to consumption by organisms living on and within the sediment. Although macrofauna play a role in remineralization, it is bacteria that are responsible for most of it (in soils, by contrast, fungi are often the dominant consumers of organic matter). Concentrations of bacteria in the surface layers of marine sediments are typically in the range of 10^8 to 10^{10} cells per gram dry weight (Deming and Baross, 1993). The role of bacteria in the cycling of carbon, nitrogen, and sulfur is summarized in Figure 12.31.

These observations raise the question of why any organic matter survives. Why do most sediments contain some organic matter? How does it escape bacterial consumption? And why do some sediments, particularly those that give rise to exploitable petroleum and coal, contain much more organic matter? What special conditions are necessary for this to occur?

Organic matter preserved in ancient sediments, and particularly coal, gas, and oil, differ chemically from living organisms. Since these resources derive from the remains of once-living organisms, we might ask how these chemical differences arise. Are the differences due to chemical transformations of simple organic molecules or selective preservation of more complex ones? Do the differences arise early, during the diagenesis of still young, poorly compacted sediment, or late, under the influence of heat and pressure? We explore these questions in the following sections, where we examine sedimentary organic matter, its diagenesis, and the formation of petroleum, gas, and coal deposits.

12.7.1 Preservation of organic matter

Our previous focus has been on dissolved organic matter, but it is primarily the particulate organic matter that is preserved in marine and many aquatic sediments. Factors

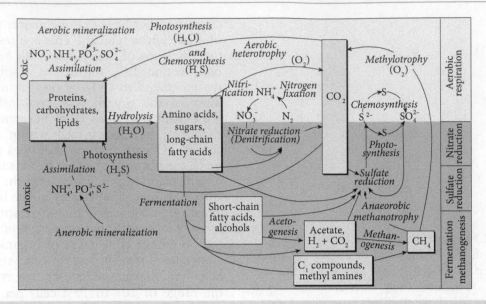

Figure 12.31 The role of bacteria in the cycling of carbon, nitrogen, and sulfur between inorganic and various organic forms. After Killops and Killops (2005).

that affect preservation of these remains include the flux of organic matter to the sediment, bulk sediment accumulation rate, grain size, and availability of oxygen. The flux of organic matter to the sediment depends in turn on its rate of production in surface waters, the flux of allochthonous material, and the depth of the overlying water column. Free-floating single-celled autotrophs (algae and photosynthetic bacteria), collectively called phytoplankton, are responsible for almost all the primary production of organic carbon in marine ecosystems, as well as many fresh water ones. Productivity depends mainly on the availability of nutrients, which in the ocean depends on the proximity to coasts and ocean circulation, a topic we will explore in detail in Chapter 14. Most organic matter, both particulate and dissolved, falling though the water column from the upper photic zone (200 meters at most) is rapidly remineralized by bacteria and animals as it descends through the water column. Hence the greater the water depth, the less organic matter reaches the sediment. Continental shelves are both highly productive, due in part to the supply of nutrients to continents, and shallow depth, making these prime areas for accumulation of organic-rich sediment. In addition, *allochthonous* remains of higher plants delivered by rivers can constitute a significant fraction of the accumulating organic

matter, particularly near major deltas. Indeed, roughly 90% of terrestrial organic matter delivered to the oceans is buried in shelf and slope sediment and as much as a quarter of the total carbon buried in marine sediments may be of terrestrial origin (Sikes, 2017). This terrestrial material has already been degraded in soils, rivers, estuaries, and coastal marine environments and hence is recalcitrant and more likely to be preserved.

Hemingway et al. (2019) examined the relationship between organic carbon bond strengths and age using ramped oxidative pyrolysis followed by ^{14}C analysis of the evolved CO_2 across a range of organic matter. In ramped pyrolysis, the sample is subjected to increasing temperatures (about 150–700°C), in this case in the presence of O_2, over the course of several hours and the amount of CO_2 evolved at each temperature measured. Knowing time and temperature (and assuming a value for the frequency factor) the activation energy, E_A, can be determined from the Arrhenius equation (eqn. 5.25). Figure 12.32 shows the distribution of estimate activation energies for a variety of organic matter. Dissolved organic matter (both riverine and marine) shows the least diversity of bond strengths and ages, being young compared to soil, sediment, and particulate organic matter, all of which show a diversity of bond

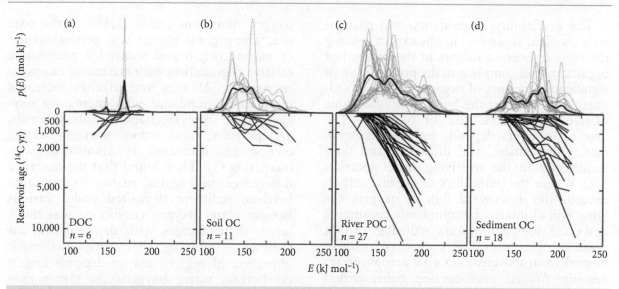

Figure 12.32 Top: probability density functions of activation energies determined by ramped oxidative pyrolysis for dissolved organic matter (average: 163.3 kJ/mol), soil organic matter (average: 153.8 kJ/mol), riverine particulate matter (average: 160.1 kJ/mol), and sediment organic matter (average: 168.9 kJ/mol). Gray lines show patterns for individual samples, black lines are mean values. Bottom: ^{14}C ages as a function of activation energies; each line represents an individual sample. n is the number of samples analyzed. From Hemingway et al. (2019).

strengths. In general bond strength increases with age. This might seem consistent with condensation, but Hemingway et al. point out that condensation this would require DOC to behave similarly, which is not observed. Furthermore, it would not necessarily lead to the formation of stronger bonds. Instead, they argue that the persistence and increasing bond strength with age reflects interactions with mineral surfaces. This interaction must involve more than merely hydrophobic or van der Waals adsorption, which would not explain increasing bond strength, but rather must involve formation of chemical bonds between the organic matter and mineral surfaces. Formation of organometallic complexes with metals derived from mineral breakdown may also contribute to preservation of organic matter.

Organic carbon concentrations are inversely correlated with sediment grain-size for several reasons. First, low-density organic particles can only accumulate where water velocities are low enough to allow finer particles to settle out. Second, a significant fraction of the organic matter in sediments can be present as coatings on mineral grains.

Small grains have higher surface areas per unit mass or volume, and therefore would have higher organic content. And as we have seen, organic matter associated with mineral surfaces is more likely to survive consumption by heterotrophs in the sediment. Third, the permeability of fine-grained sediments is lower than that of coarse-grained ones. Where permeability is low, the flux of oxygen into the sediments will also be low.

Organisms living on and in the sediment respire some of the organic matter, but some is reprocessed into the living biomass and becomes detritus again upon death of the organism. Sedimentary organic matter may experience multiple cycles of such reprocessing and organic matter derived from these organisms, particularly the microbes, can constitute a significant fraction of the organic matter ultimately preserved. The macrofauna play a key role in mixing the sediment, a process known as bioturbation. As we found in Section 5.7, bioturbation produces a biodiffusive flux that, much like molecular diffusion, moves components down concentration gradients.

The availability of oxidants, and particularly oxygen, is, as one might expect, among the most important factors in the survival of organic matter. Simply put, the preservation of significant amounts of organic matter in sediment requires that the burial flux of organic matter exceeds the flux of oxidants. The flux of oxidants depends on sedimentation rate, bioturbation, and diffusion, and their availability in the overlying water (Section 5.7). Where the burial flux of organic carbon exceeds the downward flux of oxygen, the latter will ultimately be completely consumed, and conditions will become reducing. At that point aerobic respiration must cease; with it, bioturbation also ceases as all macrofauna are aerobic. Anoxia may develop either within the sediment, or within the water column itself. Situations where deep water becomes anoxic are rare in the modern ocean (indeed, in most of the deep ocean, conditions do not become anoxic even in the upper few meters of sediment); it occurs only in a few basins where circulation of deep water is restricted, such as the Black Sea. However, anoxicity appears to have been more common at certain times in the geologic past, such as the Cretaceous, when ocean circulation was different. Anoxicity is perhaps more common in lakes, where the abundance of nutrients, and hence biological productivity, is higher than in the open ocean.

Whether preservation of high organic matter concentrations in sediments requires anoxic bottom water is a matter of debate. Calvert and Pederson (1992) argued that the extent of decomposition of marine organic matter is similar under oxic and anoxic conditions, although terrestrial organic matter tends to be degraded less by sulfate reducers. Nierop et al. (2017) examined sedimentary organic matter in sediments at three sites from the northern Arabian Sea where monsoonal upwelling results in high productivity in surface waters and an oxygen minimum zone develops between 150 and 1200 m depth as a consequence of consumption of abundant descending detritus. The bottom at one of the sites was within the oxygen minimum zone with 2 μM dissolved oxygen in the bottom water, a second just below it with 14.3 μM O_2 and a third well below it with 42.8 μM O_2. They found that total sedimentary organic carbon decreased from 6.6% at the oxygen minimum site to 1.1% at the oxic site. The organic matter was predominantly of marine origin and showed a progressive relative degradation with increasing exposure to oxygen. Alkanes and alkenes increased in relative abundance with increasing oxygen while polysaccharides, alkylphenols, alkylpyrroles, and other nitrogen-bearing compounds decreased in abundance with increasing O_2. They found that the observed differences in organic matter composition between sediment deposited under various bottom water oxygen conditions was much larger than changes with depth within the sediment, implying that early diagenetic alteration of organic matter depends largely on bottom water oxygenation rather than subsequent anaerobic degradation within the sediments.

12.7.2   Diagenesis of marine sediments

Diagenesis in the context of organic matter refers to biologically induced changes in organic matter composition that occur in recently deposited sediment. As we noted in previous sections, decomposition begins within the water column and continues once the organic matter reaches the sediment surface. Burial by subsequently accumulating sediment eventually isolates it from the water. Where the organic burial flux is high enough, oxygen is eventually consumed, but life and decomposition persist as respiration continues through fermentation, which narrowly defined refers to reactions in which an internal, rather than external, source of electron acceptors (oxidants) is used. An example familiar to brewers and vintners is the fermentation of glucose to alcohol:

$$C_6H_{12}O_6 \rightarrow 2C_2H_5OH + 2CO_2   (12.26)$$

In this example, part of the glucose molecule is reduced to ethanol and part is oxidized to CO_2. There is a limit to how much of the organic matter can be oxidized in this way, however. Remaining organic matter is subsequently attacked by a series of bacterial communities utilizing a progression of electron receptors (oxidants) at decreasing $p\varepsilon$. We can predict the order of the use of these

Table 12.5   Free energy changes for bacterial reactions.

Reaction	ΔG (kJ/mol CH_2O)
$CH_2O + O_2 \rightarrow CO_{2(aq)} + H_2O$	-493
$5CH_2O + 4NO_3^- \rightarrow 2N_2 + 4HCO_3^- + CO_{2(aq)} + 3H_2O$	-472
$CH_2O + 3CO_{2(aq)} + H_2O + 2MnO_2 \rightarrow 2Mn^{2+} + 4HCO_3^-$	-348
$^\dagger 3CH_2O + 4H^+ + 2N_2 + 3H_2O \rightarrow 3CO_{2(aq)} + 4NH_4^+$	-125
$CH_2O + 7CO_{2(aq)} + 4Fe(OH)_3 \rightarrow 4Fe^{2+} + 8HCO_3^- + 3H_2O$	-103
$2CH_2O + SO_4^{2-} \rightarrow H_2S + 2HCO_3^-$	-99
$2CH_2O \rightarrow CH_4 + CO_{2(aq)}$	-88
$^\dagger 3CH_2O + 2N_2 + 7H_2O \rightarrow 3CO_{2(aq)} + 4NH_4(OH)$	-54

Modified from Berner (1981).

†Because the speciation of ammonia is pH-dependent, the ΔG of the nitrogen fixation reaction depends strongly on pH.

oxidants from the ΔG of the redox reactions involved, shown in Table 12.5. Thus, moving downward in a column of accumulating sediment, we expect to see, following consumption of free oxygen, a series of zones where nitrate, Mn(IV), Fe(III), sulfate, and nitrogen reduction occur. Each of these zones will be colonized by a bacterial flora adapted to conditions in that zone.*

The bulk of the organic matter in sediments exists in solid form, yet only dissolved compounds can cross cell membranes and be a useful source of nutrition to microbes. For this reason, bacteria release exoenzymes that first break insoluble complex organic molecules into smaller soluble ones. Complex organic molecules usually cannot be oxidized completely by a single organism, because no single organism is likely to produce all the necessary enzymes. Instead, consortia of bacteria break down macromolecules. In each step, some energy is released and smaller molecules are produced as waste; these are subsequently attacked by other bacteria. Thus, proteins, carbohydrates, and lipids are broken down into amino acids, simple sugars, and long-chain fatty acids. These smaller molecules can be attacked by fermenting bacteria that produce acetic acid, other short-chained carboxylic acids, alcohols, hydrogen, and CO_2. In the final step, these are converted to methane (CH_4) by methanogenic bacteria. During this process, the remains of bacteria themselves can become a significant part of the sedimentary

organic matter. The stepwise oxidation results in an interdependence between the various bacterial species within each community, as many species are dependent on the "waste" products of other species.

Both the abundance of organic matter and of bacteria decrease with depth in marine sediments, the highest concentrations of both being found in the upper 10 cm. Not surprisingly, decomposition rates decrease when conditions become anoxic (summarized in Henrichs, 1993). Thus most remineralization occurs in the uppermost couple of meters, and the bulk of the organic matter buried beneath this depth is preserved long term (Henrichs, 1993).

What molecules are preserved? Decay resistance increases in the order nucleic acids, proteins, carbohydrates, lipids, and structural macromolecules. Nucleic acids have been found only in sediments younger than 10,000 years (but have been found in older permafrost and preserved in bone for longer periods), and proteins and carbohydrates are rarely found in sediments older than 1 million years (Briggs and Summons, 2014). Cellular structural materials (e.g., components of cell walls), appear to be particularly resistant to bacterial decomposition and form the bulk of the preserved organic matter. Examples of these resistant materials are algaenans, which are found in the cell walls of marine algae, and phlorotannins, found in brown algae such as kelp (de Leeuw and Largeau, 1993). Allochthonous material derived from higher

* Examined at the microscopic level, separation of bacterial species is not quite this simple or complete. For example, within the oxic zone, there are anoxic microenvironments where anaerobic bacteria flourish.

plants may also contribute a number of resistant aromatic-rich compounds (see below) to sediments in marginal marine environments. However, a small fraction of readily metabolized compounds is also preserved. These molecules may survive because they are located in micro-environments such as within resistant structures (e.g., spores, pollen) that shield them from bacterial enzymes. Similarly, we might expect proteinaceous material in carbonate shells to be somewhat protected from bacterial enzymes. Adsorption to inorganic particulates also affords a degree of protection, and most organic matter in marine sediments appears to be adsorbed on-to particle surfaces.

Methane produced by methanogenic bacteria during diagenesis can react with water to form solid ice-like methane clathrate, which consists of a methane molecule locked in a cage of hydrogen-bonded water molecules (Figure 12.33) with the overall composition of approximately $CH_4(H_2O)_{5.75}$. Depending on water temperature, the methane clathrate structure becomes stable at water depths of 300–400 m or more and is favored by increasing pressure and decreasing temperature (Figure 12.34). An enormous mass of methane clathrate, in the range of 500–2500 carbon gigatons (700–3000 Gt methane), appears to exist in continental margin sediments (Milkov, 2004). A somewhat smaller mass of methane clathrate is also present in the deep permafrost of the Arctic tundra. This methane represents both a blessing and a threat. On the positive side, the amount of methane clathrate likely exceeds the total recoverable natural gas in other deposits by a factor of 2−10 and compares with an overall global inventory of fossil fuels of 5000–15,000 Gt carbon. On the negative side, methane clathrate is only marginally stable in the majority of its occurrences. Destabilization of methane clathrate, either through ocean or atmospheric warming or geologic events such as submarine landslides, could release massive quantities of methane to the atmosphere, which, because methane is a powerful greenhouse gas, could have profound climatic consequences.

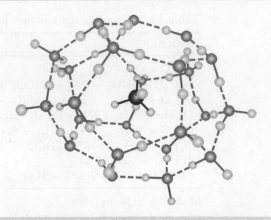

Figure 12.33 Structure of methane clathrate. Red atoms are oxygen, white are hydrogen, and black is carbon. Dashed red lines represent hydrogen bonds.

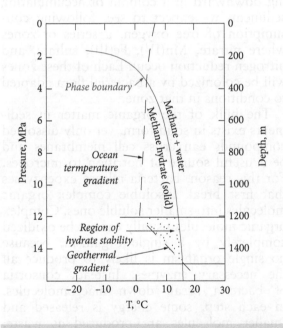

Figure 12.34 Phase diagram for the stability of methane clathrate in marine sediments. In this example, the ocean floor is at 1200 m depth and there is a roughly 200 m deep layer where methane clathrate is stable. After Kvenvolden (1993).

12.7.3 Diagenesis of aquatic sediments

On the whole, diagenesis in freshwater sediment is similar to marine diagenesis. As is also the case in marine sediments, most of the organic detritus in aquatic environments originates from autotrophs, with animals contributing less than 10%. Perhaps the principal difference in diagenesis between large lakes and the ocean is the much lower sulfate concentrations in lakes. Sulfate is important both as an oxidant and because sulfur can be incorporated into organic molecules (primarily lipids) during early diagenesis, a process known as "natural vulcanization." Otherwise, the same sequence of oxidant usage and decomposition occurs, and most of the remineralization occurs near the sediment–water interface.

In small lakes, the bulk of the organic matter reaching the sediment may be allochthonous organic matter derived from terrestrial plants. Higher plants living within the water may also contribute organic matter, and such material is dominant in swamps and marshes. The significance of this is that higher plants contain a greater abundance of aromatic compounds than algae. We found earlier in the chapter that aromatic compounds are often particularly stable. Thus, it is no surprise to find that aromatics such as lignins and tannins produced by higher plants are particularly resistant to bacterial decomposition and, hence, are more easily preserved in sediment.

12.7.4 Summary of diagenetic changes

Changes in sedimentary organic matter occurring as a result of diagenesis can be summarized as follows:

- Functional groups, such as carboxyl, amino, and hydroxyl, are preferentially removed from their parent molecules.
- Loss of these functional groups decreases the oxygen, and to a lesser degree, the hydrogen, content of the organic matter.
- The abundance of readily metabolized organic compounds decreases. Nucleic acids and amino acids and related compounds appear to be the most labile (most readily destroyed), followed by carbohydrates, particularly simple ones and those synthesized for energy storage (e.g., starch) rather than structural (e.g., cellulose) purposes. The simple molecules in these groups (e.g., amino acids, glucose)

are most labile of all. Lipids are the least labile.
- Unsaturated compounds decrease in abundance compared with their saturated equivalents due to hydrogenation of double carbon bonds.
- Aliphatic compounds decrease in abundance compared with aromatic ones. This results partly from aromatization of unsaturated aliphatic compounds and partly from the more resistant nature of aromatics.
- Short-chained molecules (e.g., alkanes, fatty acids), decrease in abundance relative to their long-chain equivalents.
- In high-sulfur environments, such as marine sediments, H_2S (produced by sulfate-reducing bacteria) is incorporated into carbon double bonds in long-chain compounds such as isoprenoids to produce thiol functional groups. These can subsequently form cyclic structures and ultimately aromatic thiophenyls.
- All along, bacterial remains are progressively added to the mixture, and are progressively decomposed along with the organic matter originally deposited.

The principal product of these processes is *kerogen*, the name given to the mixture of complex organic compounds that dominates the organic fraction in sediments.

12.7.5 Biomarkers

Biomarkers, sometimes called *geochemical fossils*, are organic compounds that have a chemical structure known to be present in an organism or class of organisms or be related to such compounds by diagenetic microbial or chemical transformation pathways. The term is used mainly in the context of sedimentary organic matter, including coal and oil, but is sometimes also applied to soil, aquatic, and marine organic matter. These molecules have sometimes lost functional groups such that only the basic hydrocarbon skeleton remains; this skeleton can nonetheless be associated with a specific environment or class of organisms.

Although he did not use the term, the biomarker concept dates to the realization

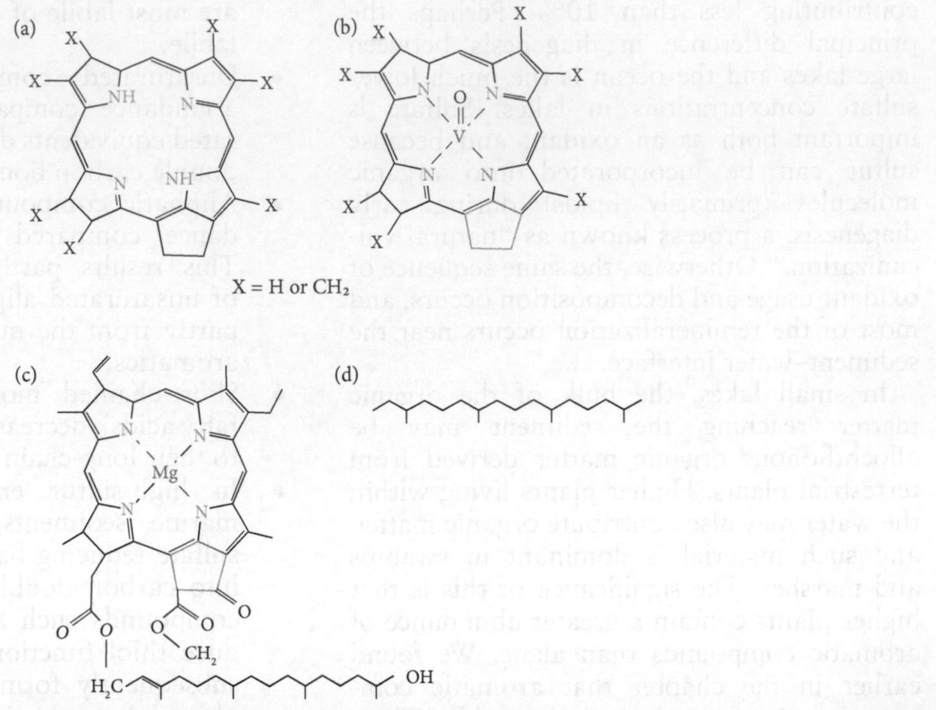

Figure 12.35 (a) Deoxophylloerythroetioporphyrin (DPEP), a derivative of the chlorin unit of chlorophyll, is found in petroleum. (b) In petroleum, the center of the tetrapyrrole ring is often occupied by a transition metal such as vanadium to form a metalloporphyrin. (c) The chlorophyll molecule. (d) Phytane; once the isoprenoid phytol chain is detached from the chlorin unit it can be transformed to pristane or phytane, which are also common biomarkers in petroleum.

by Alfred Treibs* (1934) that alkyl metalloporphyrins in petroleum were derived from chlorophyll (Figure 12.35). This discovery was also the first clear demonstration of the ultimate biological origin of petroleum. Biomarkers present in sediments, petroleum, coal, and extracted from kerogen contain information about biological evolution, ancient ecosystems, biodiversity, paleotemperatures and climate, and sediment depositional environments. They are quite commonly used in petroleum exploration to relate petroleum to its source rock, i.e., the deposit in which it formed.

Most organic molecules that survive on geologic time scales are derived from lipids or structural molecules such as lignins and algaenans. Diagenesis and subsequent catagenesis strips functional groups, heteroatoms, and

most double bonds, so that what remains in petroleum and coal are mainly hydrocarbons, of which n-alkanes are the most abundant. Phytane and pristane (Figure 12.14), components of chlorophyll, are the most commonly recognized. Alkanes with an odd number of carbons are particularly common because oxidation of fatty acids, which are almost entirely even-numbered, yields odd-numbered alkanes. Figure 12.36 illustrates how three important hydrocarbon biomarkers are derived from their biomolecule precursors.

Although phytane and pristane and alkanes derived from fatty acids are not, many biomarkers are taxon-specific and can be used to identify the organic inputs to sedimentary organic matter from various domains, kingdoms, families, and even individual species. This makes them extremely useful in tracing

* Alfred E. Treibs (1899–1983) was a German organic chemist considered to be the founder of organic geochemistry. Treib received his PhD at the Technische Universität München in 1925 on the thermal degradation of cholesterol. He continued working there until 1936 when he was forced to leave for his anti-Nazi sympathies. He returned in 1946 as head of department. The Geochemical Society annually awards the Alfred Treibs Award for major achievements over a period of years in organic geochemistry.

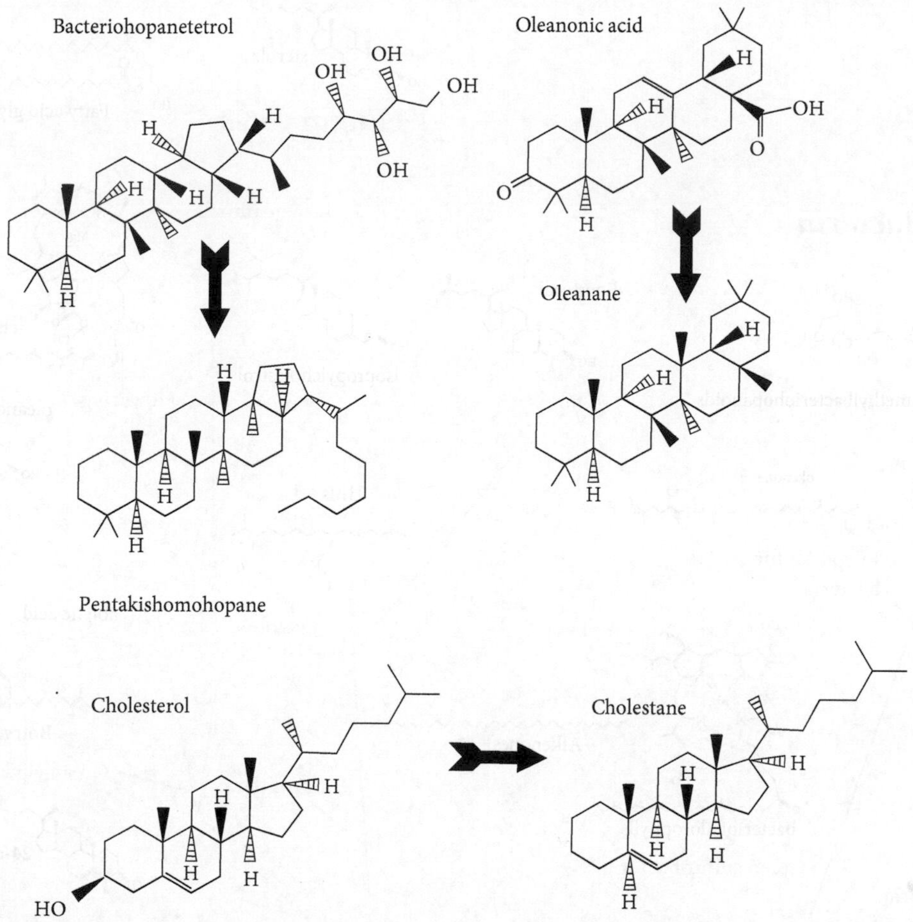

Figure 12.36 Transformation of biomolecules to hydrocarbon biomarkers during diagenesis. Hopanes such as pentakishomohopane, bacterial biomarkers, are derived from bacteriohopanoids; oleanane, a biomarker for flowering plants, is derived from oleanonic acid and other oleanoids; and cholestane, a biomarker for animals, is derived from cholesterol through loss of double bonds, chain shortening, and loss of O and OH.

biological evolution, including in some cases something morphological fossils cannot: tracing evolution of biochemistry and biosynthetic pathways (e.g., Volkman, 2014). Figure 12.37 illustrates how various biomarkers relate to the tree of life. One class of triterpenoids, the hopanoids serve as biomarkers for bacteria while sterols, another class of triterpenoids, are biomarkers for eukaryotes (while bacteria can synthesize some simple sterols, complex sterols with modified side chains are unique to eukaryotes). Many triterpenoids are taxon-specific. For example, dinosteranes are unique to dinoflagellates and highly branched isoprenoids (HBIs) are unique to diatoms, which account for roughly 30% of all primary productivity in marine and

lacustrine environments. On the other hand, botrycocci, a strictly freshwater green alga, produce botryococcenes, a branched triterpene. Oleananes, derived from oleanonic acid, are produced only by flowering plants, whereas certain tetracyclic diterpenoids such as abietic acid and phyllocladanes are produced by gymnosperms (a group that includes conifers, cycads, and ginkos).

In most instances, the biomarker record is correlated with the fossil record. For example, diatoms first appear in the fossil record in the mid-Cretaceous and the first occurrence of HBIs occurs only shortly before the first diatom fossils. Dinoflagellates underwent a major radiation in the Mesozoic and dinosteranes are abundant in the Mesozoic and

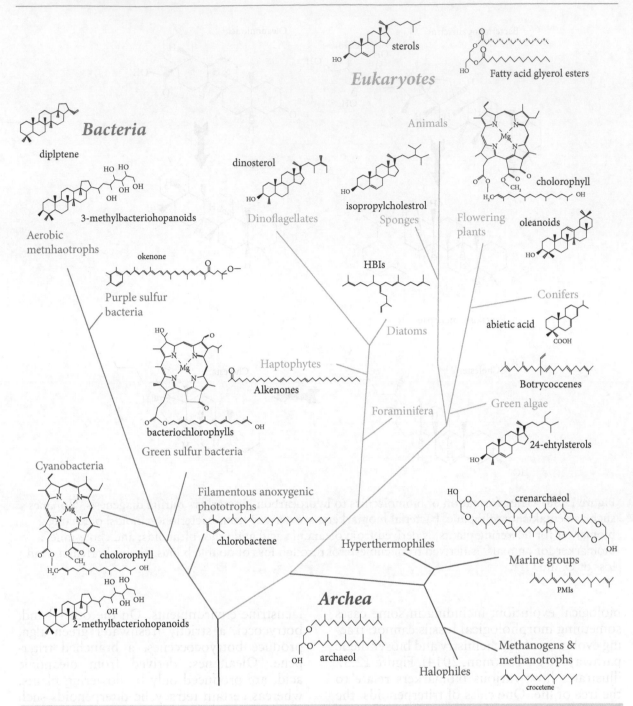

Figure 12.37 The relationship between biomarkers and the tree of life. Based on Briggs and Summons (2014) and Gaines et al. (2009).

younger sediments, although dinoflagellate fossils and dinosteranes do occur rarely in Paleozoic sedimentary rocks. Oleanoids become common in the sedimentary record in the Cretaceous at about the same time that flowering plants appear in the fossil record. In these cases, biomarkers can be used to

identify the age of petroleum source rocks and the environment in which they formed: for example, the presence of dinosteranes generally indicates a Triassic or younger source while highly branched isoprenoids indicate a Cretaceous or younger source (He et al., 2017).

Many organisms, such as most plankton, are not readily fossilized so biomarkers provide an important adjunct to the fossil record. This is particularly true in Precambrian rocks where morphological fossils are rare and even when present their identity is often ambiguous. Although hydrocarbons have been found in some Archean metasediments, it has not been possible to relate any Archean organic matter found so far to specific biomolecules due to thermal alteration, so the biomarker record begins in the Proterozoic. Alleon et al. (2017) used X-ray absorption near edge structure (XANES) to identify a variety of biomolecular signatures sharing strong similarities to those of modern cyanobacteria and modern micro-algae in organic microfossils from the 1.88 Ga Gunflint Chert of Canada, including amide, carbonyl, phenolic, carboxylic, and hydroxyl groups. These are the oldest known biomolecules; however, they cannot be related to specific taxa. Brocks and Schaefer (2008) identified the carotenoid okenane in organic matter of the 1.64 Ga Barney Creek Formation of Australia, which is to date the oldest identified biomarker. Okenane is strictly a biomarker for anoxic and sulfidic conditions in the presence of light, from which Brocks and Schaefer (2008) inferred to presence of photosynthetic purple sulfur bacteria at this time.

Molecular clocks and morphological fossils suggest eukaryotes evolved in the early or middle Proterozoic, but the oldest unambiguous eukaryotic sterane biomarkers are found in 747 Ma Neoproterozoic sedimentary rocks from the Grand Canyon of Arizona, where they are not only present but predominant; they also occur in similar age rocks from Australia. Although they may have evolved earlier, variety of evidence suggests a major diversification of autotrophic eukaryotes in the Neoproterozoic immediately preceding the Cryogenian snowball Earth glaciation and the evolution of metazoans. The earliest evidence of animal life occurs roughly 100 Ma later: Love et al. (2009) reported the presence of 24-isopropylcholestanes, a sponge biomarker, in ~645 Ma sediments from the Huqf Supergroup of Oman, which, in turn, is ~100 Ma older than physical sponge fossils. Nettersheim et al. (2019) have reported that Rhizaria, a group of heterotopic protists that includes radiolarians and foraminifera, also produce 4-isopropylcholestanes, placing the inferred early appearance of sponges in question.

Biomarkers are also useful in reconstructing the paleo-environments. For example, plants produce leaf wax n-alkanes with differing carbon chain lengths. Vascular plant (grasses, trees, and shrubs) waxes contain predominantly long-chain (C29–C31) n-alkanes, whereas sphagnum mosses are predominantly medium-chain length (C23–C25) n-alkanes, hence the ratio of medium- to long-chain n-alkalines can be used to estimate the relative contributions of mosses and aquatic macrophytes vs. higher plants, the former predominating in humid periods. Variations in the chain length distributions of plant wax n-alkanes in lacustrine and peat deposits have been related to differences in growing season temperatures (Bechtel and Püttmann, 2017).

Some biomarkers are used as paleotemperature proxies. By altering the structure of their glycolipid membranes, microbes can raise or lower the melting point of their cell structures and adjust membrane fluidity to adapt to different temperatures. Glycerol dialkyl glycerol tetraethers (GDGTs) are membrane lipids produced by Thaumarchaeota, a phylum of ammonia-oxidizing Archea. They are abundant and well preserved in marine (and lacustrine) sediments and relatively easy to isolate and identify. Schouten et al. (2002) noted the number of cyclopentane rings in the GDGTs in marine surface sediments was related to overlying sea-surface temperatures (SSTs): GDGTs with no rings dominated in polar regions, whereas GDGTs with two or three rings were more abundant in tropical regions. This, as well as observations from culturing and mesocosm experiments showing that the number of cyclopentane rings in GDGTs increases with increasing growth temperature, led to the development of the TEX86 (TetraEther indeX of 86 carbons) paleothermometer, which is the ratio of GDGTs with two or three rings plus crenarchaeol (a GDGT with four cyclopental rings; Figure 12.36) to those with one, two, or three rings plus crenarchaeol. Calibration showed that this ratio relates to temperature as:

$$TEX_{86} = 0.015 \times T + 0.28 \qquad (12.27)$$

where temperature is in Celsius. A number of other slightly different calibrations are reviewed in Tierney (2014). This has been used to reconstruct SSTs from the early Eocene onward.

Biomarkers can also reveal information about the conditions of diagenesis. Pristane is derived from phytol by oxidation and decarboxylation of phytol, while phytane can be derived by dehydration and reduction. Thus, the pristane/phytane ratios indicate the redox state of diagenesis: pristane/phytane ratios <1 indicate anoxic conditions, while ratios >1 indicate oxic conditions. Similarly, a high relative abundance of bacterial C_{35}-homohopanes also indicates deposition under anoxic conditions.

Finally, biomarkers can be used to assess the extent to which kerogen has been converted to petroleum, which requires temperatures above 80°C, that is its *thermal maturity* (discussed below). Many of the most common biomarkers of thermal maturity are based on stereo-isomerization, which involves simply spatial rearrangement of bonds. These can be either *enantiomers* (Figure 12.38a), mirror-image structures having one or more asymmetric carbon atoms that can be related to each other by reflection, or *diastereomers* (Figure 12.38b), which differ at one or more asymmetric centers, but are identical at others and hence not related by reflection. Biosynthesis produces molecules with distinct bond orientations to perform specific functions. These molecules are not necessarily the most thermodynamically stable. In some cases, both isomers might be equally stable, in which case entropy tend to produce equal mixtures of both as temperature increases, a process known as *racemization*.

In other cases, the biosynthesized version of a molecule may be less stable than an alternative, in which case heating drives conversion to the alternative orientation. An example is the hopanoid 17α-trisnorhopane, which upon heating is progressively converted to 18α-trisnorneohopane (Figure 12.37c). In immature sediments the 18α-trisnorneohopane/(18α-trisnorneohopane + 17α-trisnorhopane) value is 0 and is 100% upon full thermal maturation (Peters, 2017).

12.7.6   Kerogen and bitumen

Kerogen is defined as sedimentary organic matter that is insoluble in water, alkali, nonoxidizing acids, and organic solvents (such as benzene/methanol, toluene, methylene chloride). It is usually accompanied by a smaller fraction of carbon disulfide-soluble organic matter, called *bitumen*. Kerogen, an inhomogeneous macromolecular aggregate, constitutes 90% or more of organic matter in sedimentary rocks (much of the remainder being dispersed bitumen). Sedimentary rocks contain something like 15×10^{21} kg C as organic carbon; it is three orders of magnitude more abundant than coal, petroleum, and gas combined, and four orders of magnitude more abundant than the living biomass. Kerogen is thus a major reservoir in the global carbon cycle. Carbon cycles into this reservoir by sedimentary deposition and out of it by erosion and oxidation. The rate at which carbon is added is estimated at 4 to 10×10^{13} kg/yr, from which we can see that this reservoir turns over only slowly (Petsch, 2017). Nevertheless, because of the magnitude of the reservoir and the fluxes, small imbalances in the fluxes can result in large changes in other surficial carbon reservoirs, including the atmosphere, which can consequently have a major impact on climate, as we will describe in Section 12.9. An example of just such an imbalance was the Carboniferous Period when burial of large amounts of carbon in sediments (giving the period its name) triggered a major glaciation.

Kerogen is also the precursor to petroleum, although it varies widely in its petroleum potential. Kerogen that is rich in aliphatic compounds, generally derived from lacustrine and marine algae, has good petroleum potential and is called *sapropelic kerogen. Humic kerogen*, derived principally from the remains of higher plants, is rich in aromatic compounds, but has poor petroleum potential. Kerogen can be broken down into its constituent molecules upon heating in an inert environment, a procedure known as *pyrolysis*. The compositions, although not their structures, of these constituent molecules can be determined by various chromatographic techniques combined with mass spectrometry. Pyrolysis in some ways mimics petroleum

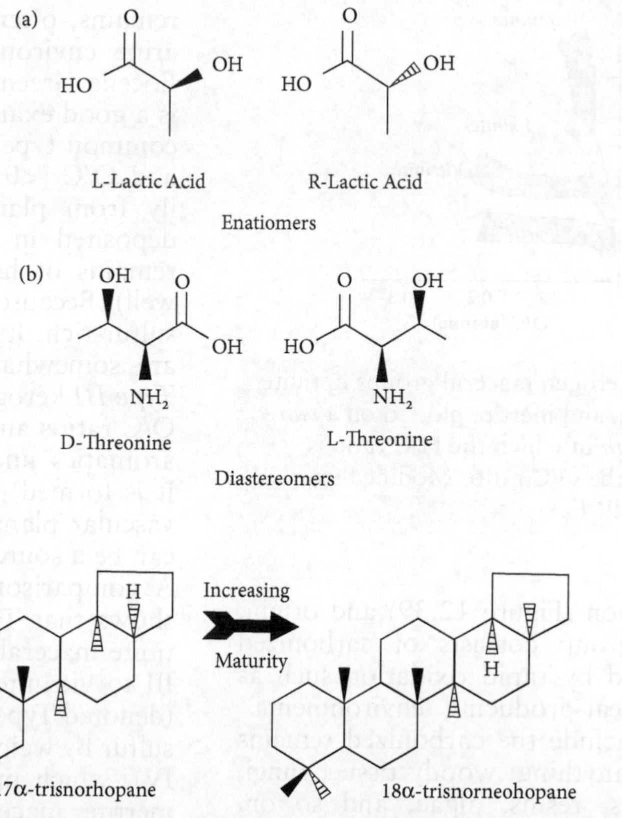

Figure 12.38 (a) Stereoisomers (enantiomers) of lactic acid; the two versions are mirror images. (b) Stereoisomers (diastereomers) of the amino acid threonine. (c) 17α-trisnorhopane transforms to 18α-trisnorneohopane as sediment is buried, which provides a useful measure of thermal maturity.

formation, which involves thermal disaggregation of kerogen, although the heating occurs much more slowly on much longer time scales in the latter case.

Carbon and hydrogen are the main constituents of kerogen. Hydrogen concentrations range from 5–18% (atomic), depending on type and degree of evolution. Oxygen concentrations typically range from 0.25–3%, again depending on type and degree of evolution. Besides C, H, and O, kerogen typically contains 1–3% N and 0.25–1.5% S (though the latter can be higher). A variety of trace metals, notably V and Ni, are also found bound to organic molecules in kerogen.

The structure of kerogen and the manner in which it forms is only partly understood. As with humic substances, it was once thought to form through condensation reactions. It is now thought to form through a combination of selective preservation of certain biopolymers and other materials that are most resistant to degradation and physical protection such as association with mineral surfaces or encapsulation within degradation-resistant biopolymers that prevent or slow degradation (Tegelaar et al., 1989; Petsch, 2017). Cross-linking of peptides, sulfurization of carbohydrates, and formation of nitrogen-bearing heterocyclic components may be important in some settings.

12.7.6.1 Kerogen classification

Microscopic examination reveals that kerogen includes identifiable plant remains, amorphous material, and rare animal remains known as *macerals*. The amorphous material in kerogen may occur as mottled networks, small dense rounded grains, or clumps. Schemes for classifying macerals were first developed to describe coal and later applied to kerogen. Tissot and Welte (1984) divided them into four groups. These groups differ in

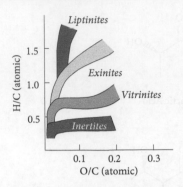

Figure 12.39 Kerogen maceral groups liptinite, exinite, vitrinite, and inertite plotted on a *van Krevelen diagram* in which the H/C ratio is plotted against the O/C ratio. Modified from van Krevelen (1961).

both composition (Figure 12.39) and origin. The *inertite* group consists of carbonized remains formed by rapid oxidation such as wildfires in peat-producing environments.* Inertite may include the carbonized remains of just about anything: woody tissue, fungi, spores, cuticles, resins, algae, and so on. Inertite has low H/C and O/C ratios and, as its name implies, is rather inert. *Vitrinite* is preserved in woody tissue. *Exinite* includes lipid-rich materials derived from leaf cuticle, spores, pollen, algae, plant waxes, resins, fats, and oils. The fourth group, *liptinite*, is similar in many respects to exinites, but whereas exinites have recognizable shapes, liptinites are amorphous bodies and are derived primarily from algal remains and usually have higher H/C ratios than exinites.

Compositionally, kerogen is usually classified into types I through III, based on bulk H/C and O/C ratios, characteristics of which are listed in Table 12.6. Techniques such as solid-state nuclear magnetic resonance and Raman and infrared spectroscopy can elucidate the abundance of different types of functional groups in kerogen but cannot reveal three-dimensional structure. *Type I* kerogen has high H/C and low O/C atomic ratios. It is rich in lipids, especially long-chain aliphatics and amino acids and has high petroleum potential. It consists primarily of liptinites derived from algal and bacterial

remains, often deposited in aquatic or estuarine environments. Kerogen found in the Eocene Green River Shale of the western US is a good example. *Type II* kerogen, the most common type, has intermediate H/C (~1.25) and O/C (<0.2) ratios. It is derived primarily from planktonic and bacterial remains deposited in marine environments (though remains of higher plants can contribute as well). Because of its marine origin, it is often sulfur-rich. Its lipid content and oil potential are somewhat lower than Type I kerogen. *Type III* kerogen has low H/C ratios and high O/C ratios and is similar to coal. It is rich in aromatics and poor in aliphatic structures. It is formed principally from the remains of vascular plants. Its oil potential is poor, but can be a source of gas (particularly methane). A comparison of Figures 12.39 and 12.40 shows that Type I kerogen is related to liptinite macerals, Type II to exinites, and Type III to vitrinites. High-sulfur Type II kerogen (denoted Type II-S) can contain 10% or more sulfur by weight. A fourth kerogen type (Type IV), which more or less corresponds to the inertite maceral group, is sometimes also defined. However, inertite has no petroleum potential, so there is less interest in this type. These types can be further subdivided based on their petroleum-potential "maturity" into subtypes A through D, with A corresponding to immature and D to overmature.

12.7.6.2   Bitumen

The fraction of sedimentary organic matter that is soluble in carbon disulfide is called *bitumen* and includes solids, liquids, and gases. At the end of diagenesis, bitumens generally constitute less than 3–5% of the total organic carbon (the remainder being kerogen), though this figure is occasionally higher. During subsequent thermal evolution; however, the fraction of bitumen increases at the expense of kerogen (Section 12.8). Bitumen consists primarily of two fractions: *asphaltenes* and *maltenes*. These fractionations are defined, like humic substances, by their solubility. *Maltenes* are soluble in light hydrocarbons such as hexane, whereas asphaltenes are not and are more similar

---

* One such modern environment is the Okefenokee Swamp in southern Georgia (US). Wildfires often follow major droughts that occur at ~25 year intervals. These fires may burn the peat to a depth of 30 cm.

Table 12.6   Kerogen composition

	Type I A	Type II A	Type III A
H/C	1.53	1.17	0.87
O/C	0.051	0.097	0.111
N/C	0.029	0.029	0.017
S/C	0.14	0.014	0.002
% aromatic C	29	40	52
average number of C atoms per polyaromatic cluster	16	12	19
% aromatic C with attachments	40	43	35
% O in carboxylic groups	0.8	1.3	2.1
Pyrrolic (mol %N)	27	52	57
Pyridinic (mol %N)	20	17	31
Amino Acid (mol %N)	9	4	0

From Ungerer (2015)

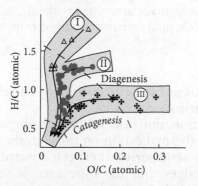

Figure 12.40 H/C and O/C ratios of the three types of kerogen. Open triangles, Type I; closed red circles, Type II; crosses, Type III. Arrows show the direction of compositional evolution during diagenesis and subsequent thermal maturation (catagenesis and metagenesis). Dashed lines show boundaries between regions of diagenesis, catagenesis, and metagenesis. After Tissot and Welte (1984).

to kerogen. Maltenes can be subdivided into *petroleum*, which consists of a variety of hydrocarbons, and *resins*. Resins and asphaltenes, unlike hydrocarbons, are rich in heteroatoms such as N, S, and O. Resins tend to be somewhat richer in hydrogen (H/C atomic ~ 1.4) and poorer in N, S, and O (7–11 wt. %) than asphaltenes (H/C atomic ~ 1.2, N, S, O ~ 8–12%). Components of both have molecular weights greater than 500 and commonly several thousand.

The hydrocarbon fraction consists of both aliphatic and aromatic components. The aliphatic component can further be divided into acyclic alkanes, referred to as *paraffins*, and cycloalkanes, referred to as *naphthenes*.

The lightest hydrocarbons, such as methane and ethane, are gases at room temperature and pressure; heavier hydrocarbons are liquids whose viscosity increases with the number of carbons. The term *oil* refers to the liquid bitumen fraction. *Pyrobitumens* are materials that are not soluble in CS_2 but break down upon heating (pyrolysis) into soluble components.

12.7.7 Isotope composition of sedimentary organic matter

12.7.7.1 Bulk isotopic composition

The isotopic composition of sedimentary organic matter and its derivatives, such as coal and oil, depend on (1) the isotopic composition of the originally deposited organic matter and (2) isotopic fractionations occurring during diagenesis and subsequent thermal evolution. The ultimate source of carbon in sedimentary organic matter is atmospheric CO_2 or marine. As we found in Chapter 9, isotopic fractionation during photosynthesis results in organic carbon being substantially lighter (lower $\delta^{13}C$) that either atmospheric or dissolved CO_2. Terrestrial C_3 plants typically have $\delta^{13}C$ of −25 to −30‰, C_4 plants have $\delta^{13}C$ of −10 to −15‰, while marine autotrophs are somewhat more variable in isotopic composition (−5 to −30‰), though on average they are heavier than terrestrial C_3 plants.

Most living organisms have δD_{SMOW} in the range of −60 to −150‰. Due to hydrogen isotope fractionation in the hydrologic cycle (Chapter 9), terrestrial plants tend to be

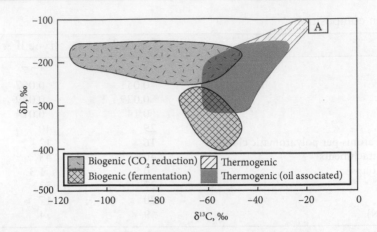

Figure 12.41 Isotopic composition of methane from various sources. *Biogenic* methane is methane produced by methanogens during diagenesis; "oil associated" is methane associated with oil; "A" is the composition of abiogenic methane from mid-ocean ridge hydrothermal systems. Modified from Schoell (1984) and Whiticar (1990).

more deuterium-depleted than marine ones, and terrestrial plants from cold climates are particularly depleted. Lipids are depleted in δD relative to bulk organic matter by 60‰ or more. Most kerogen, coal, and oil show about the same range in δD as do organisms, but lipid-rich kerogen and oil can have substantially lower δD.

For the most part, isotopic fractionation of carbon during diagenesis of organic matter is small. As a result, the $\delta^{13}C$ of sedimentary organic matter is typically within a few per mil of the $\delta^{13}C$ of the biomass from which it is derived. Sedimentary organic matter and humic substances in soil and water tend to be slightly more depleted (by 2 to 3 per mil) in ^{13}C than the organisms from which they are derived, though there are cases where the opposite has been observed. There are several possible causes for this (reviewed in Tissot and Welte, 1984; Macko et al., 1993). Functional groups, such as carboxyl, tend to be relatively ^{13}C rich. Loss of functional groups during diagenesis drives the residual organic carbon to lower $\delta^{13}C$. Preferential remineralization of proteins and carbohydrates leaves a lipid-rich residue, which will be isotopically light. $\delta^{15}N$ generally decreases somewhat during diagenesis due to bacterial utilization of short chain peptides following peptide bond hydrolysis (Macko et al., 1993). Sulfur isotope ratios are little affected by catagenesis and migration,

and thus are sometimes used to correlate oils with their source rocks.

Fractionation of carbon isotope ratios during thermal evolution through the oil generation stage drives kerogen toward higher $\delta^{13}C$ values, but the difference is small. In immature kerogen, bitumens are depleted in $\delta^{13}C$ compared with kerogen, but this difference decreases with increasing maturity (Schoell, 1984). Sofer (1984) found that $\delta^{13}C$ of the oils are within 2‰ of the isotopic composition of their source kerogen. Because the isotopic composition of oil is similar to that of its parent kerogen, isotope ratios are a widely used exploration tool in the petroleum industry.

There can be significant fractionation of carbon and hydrogen isotopes in the generation of methane. As Figure 12.41 shows, methane produced by methanogenic bacteria, called *biogenic methane*, during diagenesis is highly depleted in ^{13}C. Methane produced during catagenesis, termed *thermogenic methane* is depleted in both ^{13}C and deuterium compared with associated oil and kerogen. These fractionations reflect the lower strength of $^{12}C-^{12}C$ bonds compared with $^{13}C-^{12}C$ bonds and, therefore, the greater ease with which the former are broken. As the metagenesis stage is entered, however, the isotope fractionation between methane and residual kerogen decreases and the isotopic composition of methane generated during this stage approaches that of kerogen. This is just what we would expect from both the inverse relationship between the fractionation factor and temperature and the decreasing fractionation as reactions proceed to completion (see Chapter 9).

Methane in mid-ocean ridge hydrothermal vent fluids (Figure 12.40) has systematically higher $\delta^{13}C$ (−15 to −20‰) than biogenic and thermogenic methane derived from sedimentary organic matter (Schanks et al., 1995), demonstrating its abiogenic origin. The isotopic fractionation between methane and CO_2 in these fluids suggests equilibration at temperatures in the range of 600–800°C (Whelan and Craig, 1983).

12.7.7.2 Compound-specific isotopic analysis

An isotope ratio for bulk organic matter is, however, a tool of rather limited usefulness. For example, bulk isotopic analysis cannot be used to discriminate between depositional environments, as the isotopic differences between marine and terrestrial organic matter are neither sufficiently large nor sufficiently systematic. A far more powerful tool emerges through isotopic analysis of specific compounds, particularly biomarkers. Compound specific isotopic analysis has proved useful not only in petroleum exploration, but in environmental and paleontological research. For example, the carbon isotopic distinctions between C_3, C_4, and marine biomass (Figure 9.20) is retained in derivative n-alkanes in sedimentary matter. Based on isotopic analysis of C_{29} n-alkanes derived from leaf waxes in sediment cores from Central American lakes, Huang et al. (2001) showed that over the last 27,000 years, the ratio of C_4 to C_3 plants in the region depended on climate. As the last glacial period ended, conditions in the region became more arid, and the proportion of C_4 plant-derived alkanes increased. Isotopic compositions of specific components of petroleum gases (e.g., methane, ethane, propane) can be related to specific classes of kerogen and correlated to their source rocks (Killops and Killops, 2005).

Isotopic analysis of C_{37} alkadienone (i.e., a 37-carbon chain including a ketone bond and two unsaturated carbons; Figure 12.42) in marine sediments has proved to be useful in reconstructing variations in atmospheric

CO_2. This molecule is a component of cell membranes of a specific group of haptophyte algae (e.g., coccolithophorids such as *Emiliania huxleyi*), and is particularly resistant to diagenetic change (indeed, it survives in petroleum). The idea behind this approach is that lower atmospheric CO_2 levels should result in greater isotopic fractionation between atmospheric CO_2 and organic matter produced by photosynthesis. This is true because isotopic fractionation during photosynthesis, Δ, depends on the extent to which intracellular CO_2 is fixed into organic matter (Section 9.5.1). When CO_2 is abundant, photosynthesis ^{12}C is selectively fixed and the fractionation is large. When CO_2 is less abundant, photosynthesis is less selective, proportionally more ^{13}C is fixed into organic matter, and the fractionation is smaller. Pagani et al. (1999) analyzed $\delta^{13}C$ in C_{37} alkadienone and in carbonate shells of planktonic foraminifera in Tertiary marine sediments, the latter being a measure of dissolved inorganic CO_2. Combining paleotemperature estimates based on $\delta^{18}O$ (discussed in Chapter 9), and $[CO_2]_{aq}$ estimated from the fractionation between C_{37} alkadienone and carbonates in the same sediments, Pagani et al. estimated atmospheric CO_2 through most of the Miocene and late

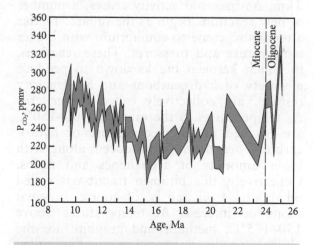

Figure 12.42 Atmospheric CO_2 concentration during the Miocene calculated from the difference between $\delta^{13}C$ in C_{37} diunsaturated alkenones and carbonate in sediments from DSDP site 588 by Pagani et al. (1999). The gray area falls between maximum and minimum values calculated using different assumptions.

Oligocene (Figure 12.42). The results were surprising because they showed that CO_2 was near its preindustrial modern level throughout most of the Miocene. Thus, the cooling that occurred in the late Miocene was not due to decreasing atmospheric CO_2 as was widely suspected. P_{CO_2} does appear to have declined sharply at the Oligocene–Miocene boundary, coinciding with a known glacial event, but otherwise there is little relationship with apparent climate change over this period. There is generally good agreement between this method of estimating atmospheric CO_2 concentrations and estimates based on boron isotopic measurements in foraminiferal shells discussed in Chapter 9.

12.8 PETROLEUM AND COAL FORMATION

12.8.1 Petroleum

12.8.1.1 Catagenesis and metagenesis

As sedimentary organic matter is buried, it experiences progressively higher temperatures and pressures. Although most bacterial decomposition occurs quickly, in the upper meter or so, it may continue at a much slower pace almost indefinitely. Indeed, bacteria have been found in subsurface rocks at temperatures of up to 75°C and depths of nearly 3 km. As bacterial activity ceases, a number of new reactions begin as the organic matter attempts to come to equilibrium with higher temperature and pressures. These reactions, in which kerogen breaks down to produce a variety of hydrocarbons and a refractory residue, are collectively called *catagenesis*. As temperatures in the range of 100–150°C are reached a complex mixture of hydrocarbons, *petroleum*, is produced, along with lesser amounts of asphaltenes and resins. Collectively, this bitumen fraction is called oil or *crude oil* and is, of course, of great economic interest. At temperatures above 150–175°C, methane and graphite are the ultimate products, created in a process called *metagenesis*.

One of the principal effects of diagenesis is the assembly of randomly structured supramolecular aggregates. During catagenesis, this process is reversed as kerogen breaks down into comparatively simple hydrogen-rich molecules (hydrocarbons) and a more structured hydrogen-depleted carbon residue. The hydrogen-rich phase is mobile and will migrate out of the source rock if a migration pathway exists. The refractory carbon-rich residue is immobile and remains in place.

Whereas diagenesis is a result of microbial metabolic activity, catagenesis is a thermodynamic and kinetic response to increasing temperature and pressure in which kerogen undergoes rearrangement to take on a more ordered and compact structure. As this occurs, the alignment of molecular nuclei, each composed of two or more aromatic sheets, becomes increasingly parallel, the number of sheets per nucleus increases, and the space between them decreases. Aliphatic units that are peripheral to the aromatic nuclei as well as those that bridge nuclei are progressively eliminated to become part of the petroleum, with longer chains eliminated preferentially. Since most of the remaining functional groups in kerogen are attached to these aliphatic units, these are also eliminated. Heteroatoms (N, S, and O) are also eliminated in this process. Aromatic units increase in abundance relative to aliphatic units. This results from aromatization of cyclic aliphatic structures as well as elimination of aliphatic structures. As unsaturated n-alkanes have two or more hydrogens per carbon atom whereas aromatic units have one or fewer hydrogens per carbon, the compositional effect of catagenesis on kerogen is a decrease in the H/C ratio, as well as the O/C ratio. This compositional evolution is illustrated by the arrows in Figure 12.40.

The degree of thermal maturation of kerogen can be monitored from its H/C and O/C ratios. In the *oil window*, the point where maximum hydrocarbon generation occurs, the H/C ratio is less than 1 and the O/C ratio less than 0.1. Kerogen with H/C ratios lower than 0.5 is over-mature, that is, it has already entered the metagenesis stage where methane is the principal hydrocarbon product. Kerogen maturity can also be monitored by measuring *vitrinite reflectance*. Kerogen in the diagenetic stage reflects light only weakly, but as its structure becomes denser and more ordered during catagenesis, more incident light is reflected. Vitrinite reflectance increases from about 0.2% in recent sedimentary organic matter to 4% or more in

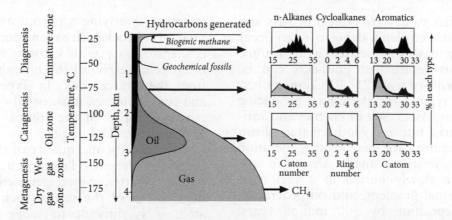

Figure 12.43   General scheme for hydrocarbon generation as a function of depth and temperature. Composition of the hydrocarbons generated is shown in the graphs to the right. Temperature and depth scales assume a geothermal gradient of 40°C per km. After Tissot and Welte (1984).

over-mature kerogen. In the oil-generating stage of catagenesis, vitrinite reflectance is typically in the range of 0.6–1.3%.

Figure 12.43 summarizes the generation of oil and gas as a function of temperature. During catagenesis, heteroatomic bonds are the first to be broken as they are generally weaker than carbon–carbon bonds. Hydrocarbons released during this stage are those attached to the kerogen structure with heteroatoms or merely trapped within it; often these are only slightly modified from their biomolecular form. Thus, the hydrocarbon fraction of bitumen in immature kerogen is dominated by biomarkers.

As temperature increases, carbon–carbon bonds are also broken, in a process called *cracking*. Carbon–carbon bonds in the centers of chains are slightly weaker than those on the ends. As these begin to break, hydrocarbon fragments are released that progressively dilute biomarkers. Also because of this effect, the size of hydrocarbons evolved decreases with increasing maturity. The first hydrocarbons to evolve in the oil window have on average relatively high molecular weight, $C_{35}H_{54}$. This decreases to less than $C_{10}H_{18}$ at the peak of the oil window and continues to decrease at higher temperatures.

As temperatures approach and exceed 150°C, smaller hydrocarbons ($\leq C_5$) become dominant. These are gases at surface temperature and pressure. Dissolved in them, however, are lesser amounts of longer chains ($\geq C_6$). These condense to liquids upon reaching the surface and hence are called *condensates*. Hydrocarbons that are gas-dominated yet contain a significant amount of longer hydrocarbons are called *gas condensates*, and this stage of catagenesis, corresponding roughly to 150–180°C, is called the *wet gas zone*. At higher temperatures, the liquid hydrocarbons are completely eliminated by C-C bond breakage. The upper temperature limit of oil stability has been revised upward as examples of n-alkanes in reservoirs at 200°C have been reported (e.g., Vandenbroucke et al., 1999), and reaction modeling suggests light oil could persist to 250°C (Dominé et al., 2002). Eventually, all C-C hydrocarbon bonds are broken, leaving methane as the sole hydrocarbon, accompanied by a nearly pure carbon residue. This stage of evolution is referred to as metagenesis or the *dry gas zone*.

Rates of reactions involved in catagenesis show an exponential temperature dependence, as we might expect (Chapter 5). Reaction rates roughly double for every 5–10°C increase in temperature. Because of this, catagenesis depends not just on temperature, but on time as well, or more specifically, on the heating rate. Heating rate in turn depends on (1) the burial rate and (2) the geothermal gradient. The burial rate depends primarily on the rate at which the sedimentary basin subsides. The geothermal gradient at the surface of the Earth varies widely, from 10°C/km to 80°C/km or even higher in geothermal areas. Values at the low end of this range are typical of old continental shields; higher values are typical of rifts and oceanic crust. Petroleum deposits often occur in subsiding basins associated with tectonic activity, thus geothermal gradients can be high. Gradients of from 25–50°C/km are perhaps most typical for petroleum-producing environments. As a result of kinetics and variations in burial rate and geothermal gradient, the time required for petroleum generation will vary. In western Canada, Devonian sediments were slowly buried in a region of low geothermal gradient, and oil generation followed deposition by 300 million years. In contrast, 10-million-year-old sediments in the Los Angeles Basin are already generating petroleum and much younger sediments are generating petroleum in regions of very high geothermal gradient in the Bransfield Strait of Antarctica. The temperature required for the onset of petroleum generation varies inversely with time. For example, this threshold is about 60°C in Lower Jurassic sediments of the Paris Basin, but is 115°C in Mio-Pliocene sediments of the Los Angeles Basin. The temperature and depth scales in Figure 12.43 correspond to a relatively high geothermal gradient (40°C/km). Reaction rates also depend on the type of kerogen involved. Labile reactive kerogen (Type I) reacts at relatively low temperatures; refractory Type III can require substantially higher temperatures for petroleum generation (as high as 250°C). Since long aliphatic chains are unstable at these temperatures, the principal product of Type III kerogen is methane.

12.8.1.2 Migration and post-generation compositional evolution

Most petroleum source rocks are fine-grained. Subjected to the pressure of burial, their porosities are typically quite low. Liquid and gaseous hydrocarbons are expelled once the source rock becomes saturated. Being less dense than both rock and water, petroleum and gas tend to migrate upward. The mechanisms of migration of hydrocarbons are not fully understood, but probably involve both passage through microfractures and diffusion through the kerogen matrix. Migration will continue until the petroleum reaches an impermeable barrier, a "trap," or the surface. From the standpoint of economic recovery, the ideal situation is a trap, such as a clay-rich

sediment, overlying a porous and permeable "reservoir" rock such as sandstone. Expulsion efficiencies vary with kerogen type. In Type I kerogen, nearly all the oil can be expelled from the source rock. In Type III kerogen and coal, however, most or all of the oil may remain trapped in the source rock and be ultimately cracked to gas.

The quantity and quality of the petroleum generated depends largely on the type of organic matter. Since petroleum tends to migrate out of the source rock as it is created, it is difficult to judge the amount of petroleum generated from field studies. However, both mass balance calculations on natural depth sequences and laboratory pyrolysis experiments on immature kerogen give some indication of the petroleum generation potential (Tissot and Welte, 1984; Rullkötter, 1993). Type I kerogen yields up to 80% light hydrocarbons upon pyrolysis. Mass balance studies of Type II kerogen indicate a hydrocarbon generation potential of up to 60%. Type III kerogens yield much less hydrocarbon upon pyrolysis (<15%).

Chemical changes may occur in several ways during and after migration. Fractionation during migration can occur as a result of the differing diffusivity and viscosity of hydrocarbons: light hydrocarbons are more diffusive and less viscous than heavy ones. As a result, they will migrate more readily, and the hydrocarbons in a reservoir are often enriched in the light fraction compared with the source rock. Asphaltenes are insoluble in light hydrocarbons and may precipitate as a consequence of this process. Polar constituents in oil, asphaltenes, and resins, may be absorbed by mineral surfaces and are less readily expelled from the source rock, resulting in a depletion in these components in oil in reservoir rocks compared with source rock bitumen. The more water-soluble components of petroleum may dissolve in water either flowing through a reservoir or encountered by migrating petroleum. This process, called *water washing*, will deplete the petroleum in these water-soluble components. Aerobic bacteria encountered by petroleum can metabolize petroleum components, in a process called *biodegradation*. Long, unbranched alkyl chains are preferentially attacked, followed by branched chains, cycloalkanes, and acyclic isoprenoids. Aromatic steroids are the least affected. Finally, further thermal evolution can occur after migration, resulting in an increase in methane and aromatic components at the expense of aliphatic chains.

12.8.1.3 Composition of crude oils

Figure 12.44 summarizes the compositions of crude oils. Average "producible" crude oils contain 57% aliphatic hydrocarbons (with a slight dominance of acyclic over cyclic), 29% aromatic hydrocarbons, and 14% resins and asphaltenes. On an elemental basis, it consists approximately of 82–87% C, 12–15% H, 0.1–5% each of S

and O, and 0.1–1.5% N by weight. On an atomic basis, this corresponds to a C:H:S:O:N ratio of 100:190:2.5:4:1.5. The distribution of n-alkanes differs widely between various types of crudes, as shown in Figure 12.45. Among cycloalkanes, those with two to four rings generally predominate. Alkylated compounds dominate the aromatic fraction; those with one to three additional carbons are most common. Aromatics decrease in abundance with increasing number of rings, so that benzene derivatives (one ring) are most common, followed by naphthalenes (two rings), and so on. Molecules containing both saturated and unsaturated rings (naphthnoaromatics) are also present, typically in an abundance

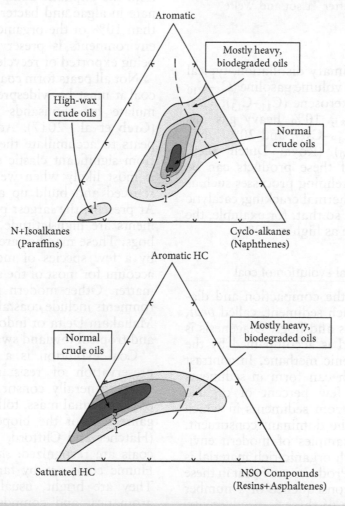

Figure 12.44 Ternary diagrams representing the composition of crude oils. (a) Isofrequency contours of hydrocarbons boiling above 210°C in 541 crude oils divided between aromatics, cyclo-alkanes, and n- and isoalkanes. (b) Isofrequency contours of saturated hydrocarbons, aromatic hydrocarbons, and NSO compounds (wt. percent in the fraction boiling above 210°C) in 636 crude oils. After Tissot and Welte (1984).

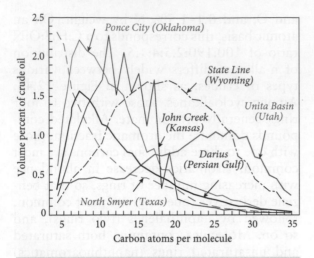

Figure 12.45   Distribution of n-alkanes in different crude oils. After Tissot and Welte (1984).

of 5 wt.%. Upon primary distillation, typical crude oil yields 27% volume gasoline (C_4–C_{10} compounds), 13% kerosene (C_{11}–C_{13}), 12% diesel fuel (C_{14}–C_{18}), 10% heavy gas oils (e.g., heating oil) (C_{19}–C_{25}), and 20% lubricating oil (C_{26}–C_{40}) (Royal Dutch Shell, 1983). The ratio of these products can be changed by further refining processes such as solvent extraction, thermal cracking, catalytic cracking, and so on, so that, for example, the gasoline yield can be as high as 50%.

12.8.2   Compositional evolution of coal

Coal is formed by the compaction and diagenesis of organic-rich sediment, called *peat*, deposited in swamps and bogs. Diagenesis is mostly anerobic and is accompanied by the production of biogenic methane. In contrast to petroleum, which can form in sediments containing only a few percent of organic matter, coal forms from sediments in which organic content is the dominant constituent. There are many examples of modern environments where such organic-rich material is now accumulating. Production of peat in these environments is a consequence of a number of factors. The first of these is productivity. Wetlands are generally characterized by high biological productivity; hence there is a high flux of organic matter to the sediment. The second factor is hydrology. Peat formation occurs where there is an excess of inflow and

precipitation over outflow and evaporation. This maintains a waterlogged soil as peat accumulates. Waterlogged conditions restrict the flux of oxygen into the sediment, resulting in conditions becoming anoxic immediately below the sediment–water interface. The third factor is the abundance of dissolved organic acids, some resulting from fermentation, others exuded by mosses and bacteria. These acids lower pH and inhibit the activity of decomposing bacteria. Finally, the primary producers in such environments are bryophytes (mosses) and vascular plants. As we noted above, these contain relatively high concentrations of aromatic compounds, which are more resistant to decomposition than the aliphatic compounds that predominate in algae and bacteria. Nevertheless, less than 10% of the organic production in these environments is preserved as peat, the rest being exported or recycled.

Not all peats form coal. For peat to become coal it must be widespread, thick, and accumulate for thousands of years (or more) (Greb et al., 2017). Additionally, for thick peats to accumulate they must be protected from significant clastic sediment input. This is most likely when wetlands are rainwater sourced and build up above the landscape. At present, the largest peat-forming environments are high-latitude (>45°) marshes and bogs. These marshes are typically dominated by a few species of moss (*Sphagnum*) that account for most of the accumulating organic matter. Other modern peat-producing environments include coastal swamps, such as the Mahakam Delta of Indonesia, and temperate and tropical lowland swamps.

Coal evolution is a process of selective preservation of resistant organic remains, which generally constitute a minor fraction of the original mass, followed by minor reorganization of the biopolymers that survive (Hatcher and Clifford, 1997). Two types of coals are recognized: *sapropelic* and *humic*. Humic coals are by far the most common. They are bright, usually stratified, rich in aromatics, and composed primarily of the remains of higher plants. Less common sapropelic coals are dull, rarely stratified, and derived from lipid-rich organic matter such as the remains of algae (boghead coals or torbanites) or spores (cannel coals). The primary

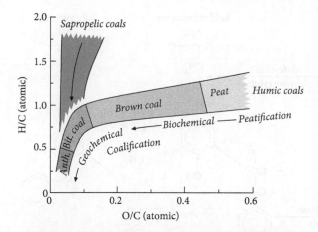

Figure 12.46 Chemical evolution of coals on a van Krevelen diagram. After Killops and Killops (2005).

maceral group of humic coals is vitrinite, while that of sapropelic coals is exinite.

The evolution of coal, illustrated in Figures 12.46 and 12.47, is generally broken down into two phases: *peatification* and *coalification*. Coalification is subdivided into *biochemical* and *geochemical* stages. Together, peatification and the biochemical stage of coalification are equivalent to diagenesis, while the geochemical stage of coalification is comparable to catagenesis. During peatification, bacterial and fungal attack results in depolymerization and removal of functional groups from the original biomolecules (Figure 12.47). This process is begun by aerobic organisms and continued by anaerobic bacteria once conditions become reducing. This is accompanied by the evolution of various gases, primarily CO_2 and H_2O with lesser amounts of NH_3, N_2, and CH_4, and condensation of the degradation products into humic substances. As in diagenesis, the concentrations of the most labile components decrease, while those of more refractory ones increase. The latter include lignins and tannins, and lipids derived from leaves, spores, pollen, fruit, and resin. Another important process during peatification is compaction and expulsion of water.

During biochemical coalification, continued loss of functional groups drives the O/C ratio to lower values with only a slight decrease in H/C ratio. Remaining labile components are metabolized and

refractory material continues to condense to aromatic-dominated structures (Figure 12.47). The final product of the diagenetic phase is *lignite*, also called *brown coal,* which contains 50–60% C and 5–7% H. The carbon structure of lignites is dominated by catechol-like carbon rings (benzene rings with two hydroxyl groups attached, Figure 12.47) This material may be accompanied by a small bitumen fraction, derived primarily from lipid components and moisture content may be as high as 75%. It typically has an energy content of 10–20 MJ/kg.

Temperature and pressure increase with burial and this initiates the geochemical stage of coalification. Coal at this stage contains 1–2% N and generally less than 1% S. Continued compaction results in a continued decrease in the water present. Loss of functional groups such as methoxyl (–OCH$_3$), hydroxyl (–OH), carboxyl (COOH–), and carbonyl (C=O) further reduces moisture content and the O/C ratio with only a minor decrease in the H/C ratio. By the time the O/C ratio reaches 0.1, most of the functional groups have been lost. Color changes from brown to black. The resulting material is now called *bituminous coal*, which requires temperatures in the range of 40–100°C, comparable to the "oil window" in petroleum genesis. Bituminous coal has a fairly bright appearance and contains 75% or more C and a water content less than 10% and an energy content of 25–35 MJ/kg. At this point, vitrinite reflectance reaches 0.5%, and 70% or more the carbon is in aromatic. Volatile hydrocarbons are progressively lost in the bituminous stage as temperatures reach and surpass ~150°C, comparable to the metagenesis stage of petroleum evolution.

With further heating, aromatization of cycloalkyl structures becomes the dominant process, releasing methane. Aromatization and loss of methane reduces the H/C ratio, which decreases rapidly upon further heating. In the temperature range of 200–250°C, *anthracite* is formed as the H/C ratio decreases below 0.5%. Anthracite is characterized by vitrinite reflectance of >2.5% and a carbon content of greater than 90%. Of this carbon, 90% or more is in aromatic structures. As in kerogen, these aromatic structures initially take the form of randomly ordered nuclei. During the geochemical stage of coalification,

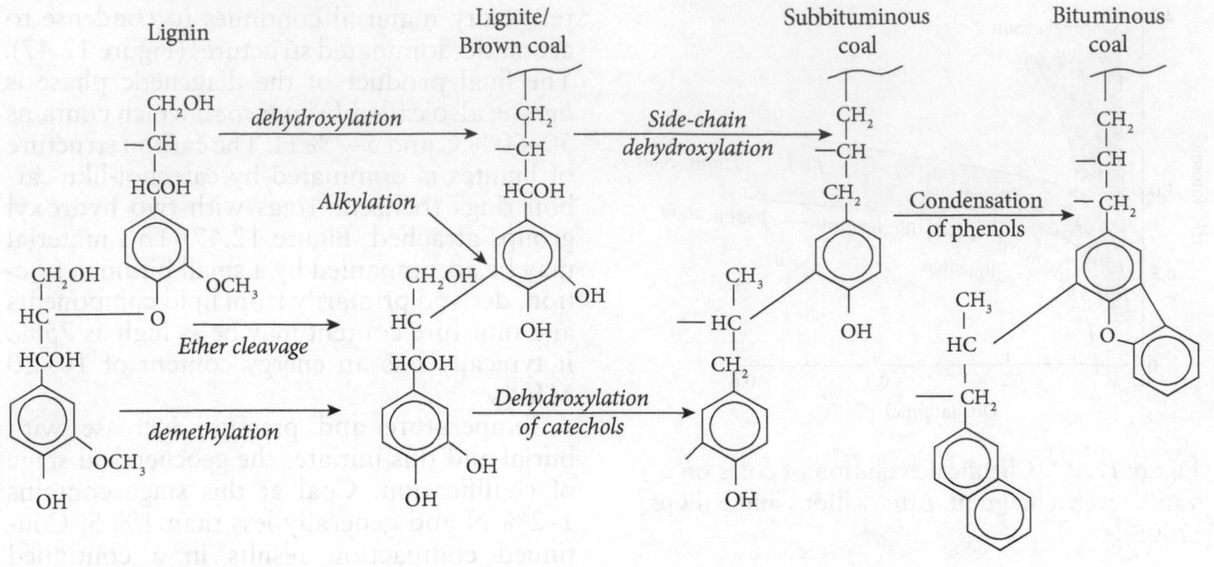

Figure 12.47   Chemical evolution of lignite to coal. After Hatcher and Clifford (1997).

these nuclei become increasingly ordered, so that by the anthracite stage, they are arranged as approximately parallel sheets, progressing toward the arrangement in graphite.

12.9   THE CARBON CYCLE AND CLIMATE

In 1896, building on the 1824 work of Joseph Fourier and the 1860 work of Irish physicist John Tyndall, Svante Arrhenius (whom we met in Chapter 5 in connection with his contribution to kinetic theory) published a paper titled "*On the influence of carbonic acid in the air upon the temperature of the ground*" in which he suggested that the concentration of atmospheric CO_2 might be increasing as a result of the extensive burning of coal that began with the Industrial Revolution. Taking note of the way in which CO_2 absorbs infrared radiation, he supposed that increasing atmospheric CO_2 variations would result in warming of the Earth's surface temperature. Arrhenius thus provided the first warning that burning of fossil fuels would result in greenhouse-driven climate change, as well as the key to understanding how climate has evolved over the entirety of Earth's history. In this section, we will briefly review greenhouse climate theory and carbon geochemistry.

12.9.1   Greenhouse energy balance

A greenhouse remains warm because visible light from the Sun is readily transmitted through glass. The radiative energy is adsorbed by the ground and objects within the greenhouse and is converted to thermal motion of atoms and molecules: heat. As a consequence of that heat, the atoms radiate electromagnetic radiation. The wavelength emitted by those atoms is in the infrared in accordance with Wein's Law, which states that the wavelength of maximum spectral emittance of black body radiation is inversely related to temperature. Glass, however, absorbs much of that infrared radiation rather than transmitting it. The glass then re-emits radiation, about half of which is directed downward back into the greenhouse. This effectively traps energy in the greenhouse and it warms. As it does, it emits radiation more intensely (according to Stephan's law, the intensity of black-body radiation is proportional to the fourth power of temperature) until an equilibrium sets in such that as much radiative energy escapes the greenhouse as arrives from Sun.

The Earth's atmosphere works in much the same way. Visible light from the Sun is largely passed through the atmosphere and is adsorbed by the Earth's surface, which then radiates in the infrared. That radiation is adsorbed when its frequency matches a resonance of the vibrational frequency of the bonds in a particular molecule. The principal gases in the modern Earth's atmosphere, N_2, O_2, and Ar, are monatomic or symmetric diatomic molecules that do not absorb in the infrared part of the spectrum (nor, for that matter, in the visible: if they did, we would not be able to see the Sun, or each other). However, certain trace gases in the atmosphere, notably H_2O, CO_2, CH_4, and N_2O, strongly absorb certain wavelengths of infrared radiation. CO_2, CH_4, and N_2O absorb at different frequencies and thus each independently affects the atmospheric energy balance. However, the absorption bands of H_2O and CO_2 do overlap somewhat. Because relatively small amounts of these gases can absorb a large fraction of the radiation at specific frequencies, the effect of these gases on atmospheric energy balance does not scale linearly with their concentrations, but rather with the log of their concentrations. Thus, for example, small changes in the abundance of CH_4 have a much greater effect on the energy balance than do small changes in more abundant CO_2, even though CO_2 absorbs at frequencies close to the Earth's maximum spectral emittance and is thus inherently a more effective greenhouse gas than CH_4, which adsorbs on the edge of the Earth's spectrum.

The combined effect of these gases is to absorb much of the infrared radiated by the Earth's surface and to raise the average temperature of the Earth's surface from 254 K ($-19°C$) to 286 K ($+13°C$). H_2O is the most powerful of the greenhouse gases, because it absorbs over a relatively wide range of frequencies and because its concentration is relatively high (its atmospheric concentration can approach 4% on a very hot, humid day). However, the residence time of water in the atmosphere is quite short, so that its effect alone can only be limited. Its concentration in the atmosphere is strongly related to temperature and at 254 K, the atmosphere would contain very little water indeed. Water thus merely amplifies the effect of the primary greenhouse gases, CO_2, CH_4, and N_2O: warmer temperatures lead to more evaporation and more humid air and a stronger greenhouse effect, which leads to warmer temperatures, and so on. Variations in atmospheric H_2O are important in short-term, local variations in temperature (this is why, for example, nights in humid regions are warm and nights in arid ones are cold). On long time scales, however, it is variations in the concentration of CO_2 and the other trace greenhouse gases that control climate. Variations in atmospheric greenhouse gas concentrations are a result of how carbon is cycled between the atmosphere and other reservoirs and how the Earth and life have evolved over the last 4.5 Ga. We consider these in the following sections.

12.9.2 The exogenous carbon cycle

We will adopt (and anglify) the French word *exogène* to refer to the Earth's surface, including the atmosphere, biosphere, hydrosphere, cryosphere, soil, weathering crystalline rock, and reactive, unlithified sediments. Carbon in the exogene cycles between a variety of forms, of which organic carbon is one. This carbon cycle is illustrated in Figure 12.48. Roughly 120 gigatons (Gt) of carbon are fixed into organic carbon by photosynthesis of terrestrial plants every year. About half of this, around 60 Gt, is quickly returned to the atmosphere through respiration of plants or animals feeding on them, while the other half flows into a reservoir of nonliving organic carbon that includes leaf litter, dead trees, peat, and soil organic matter. The mass of this reservoir is approximately steady-state, so that roughly this same amount is oxidized to CO_2 every year. This nonliving organic carbon reservoir contains more than twice as much carbon as the terrestrial biota (more than 99% of which is plants), and more carbon than is present in the atmosphere and biota combined. Atmospheric CO_2 also readily exchanges with dissolved forms of carbon in the ocean, with roughly 90 Gt of CO_2 dissolving in the ocean and a similar amount exsolving out every year.

The marine dissolved carbon reservoir, about 98% of which is HCO_3^-, contains roughly 50 times as much carbon as does the atmosphere. Carbon is cycled far more rapidly between organic and inorganic forms in the marine environment: net primary production (photosynthesis minus respiration) of marine phytoplankton is roughly 40 Gt, about 80% of terrestrial net primary production, even though the marine biosphere

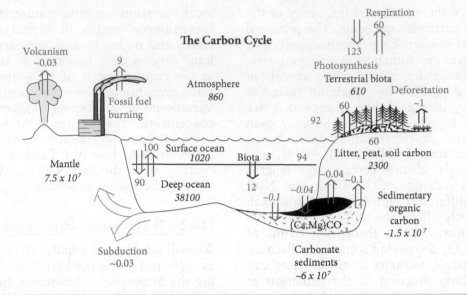

Figure 12.48 The carbon cycle. Numbers in italic show the amount of carbon (in 10^{15} grams or gigatons, Gt) in the atmosphere, oceans, terrestrial biosphere, and soil (including litter, debris, and so on). Fluxes (red) between these reservoirs (arrows) are in Gt/yr. Magnitudes of reservoirs and fluxes are principally from Siegenthaler and Sarmiento (1993) and Falkowski et al. (2000). Anthropogenic fluxes based on data from the Intergovernmental Panel on Climate Change (www.IPCC.ch).

is 200 times smaller than the terrestrial one. Only the surface layer of the ocean, roughly the upper 200 m, exchanges readily with the atmosphere. The deep ocean, which contains the bulk of the dissolved carbon, is isolated from the atmosphere. Carbon flows into that reservoir either through downwelling of surface waters, which occurs mostly near the poles, or through falling organic remains such as dead organisms and fecal pellets. Nearly all of that falling organic matter is remineralized to HCO_3^- before reaching the sediment. The process by which carbon moves from the atmosphere into the marine biota via photosynthesis and from there into the deep ocean through sinking organic particles and then into dissolved CO_2 is known as the *biologic pump*.

In addition, carbonate shells of planktonic organisms, most notably coccolithophorids and foraminifera, also fall through the water column to the deep water. Those shells falling below about 4000 m largely redissolve because water at this depth becomes strongly corrosive to calcite. One reason it is corrosive is that its pH is lower (Chapter 14). As a consequence, dissolved CO_2 concentrations are significantly higher in the deep ocean

than in surface waters. This deep-water CO_2 eventually returns to the surface through upwelling and mixing, but the process is slow. The average ventilation time of the ocean (i.e., the average time deep water spends out of contact with the atmosphere) is about 1000 years based on ^{14}C analysis. Thus the biologic pump acts to sequester CO_2 from the atmosphere and maintains atmospheric CO_2 levels some 150–200 ppm lower than it would be otherwise (Falkowski et al., 2000).

On short geologic time scales, 100,000 years and less, atmospheric CO_2 levels are controlled by the balance of carbon fluxes into and out of the oceans and the terrestrial biosphere and soils. Over the last million years or so, these fluxes have varied in response to glacial cycles driven by Milankovitch forcing (Chapter 9), resulting in variation in atmospheric CO_2 concentrations. As Figure 12.49 shows, CO_2 varied from around 190 ppm in glacial episodes to around 280 ppm in interglacial episodes.

Glacial cycles affect CO_2 fluxes in a number of ways. The first of these is volume and temperature of the oceans. The smaller the volume of the ocean, the less CO_2 it can hold; ocean volume over the past million years or

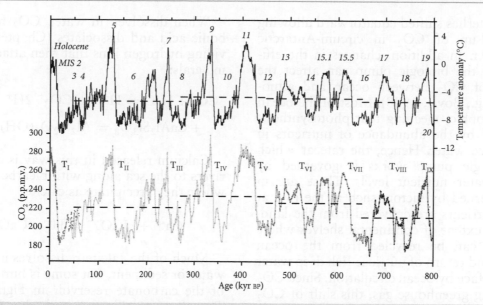

Figure 12.49 Comparison of CO_2 in bubbles (gray shows analytical uncertainties) in the EPICA ice core with temperatures calculated from δD. From Luthi et al. (2008). Numbers on the temperature plot are marine isotope stages. T_I, T_{II}, etc., on the CO_2 plot are terminations of glaciations.

so has been controlled by waxing and waning of ice sheets. CO_2 is more soluble in water at lower temperature, so low temperatures favor a net flux of CO_2 from the atmosphere to the oceans. The effect of glacial cycles on these two factors is thus opposite. Glacial cycles also affect the terrestrial biota, but, again, with opposing effects. During glacial times, sea-level drops and the area available for terrestrial vegetation expands, but expansion of glaciers also reduces this area. Precipitation patterns also change from glacial to interglacial times and also affect the biota as the total area of arid regions changes. Variations in climate and the terrestrial biota in turn drive variations in the mass of carbon stored as dead organic matter in soils, forest litter, and peat. Interestingly, there is no simple relationship between the mass of living carbon in a biome and the mass of dead carbon stored in soils. For example, two biomes, tundra and grassland, account for only 4% of the terrestrial biomass yet account for more than a quarter of organic matter stored in soils. This is because organic matter decays slowly in these environments, whereas it decays quite quickly in tropical and temperate forests, which together account for over 60% of the terrestrial biomass. Thus, the interaction between climate, atmospheric CO_2, and the

terrestrial biomass is complex and it has not yet been fully quantified.

The most important changes in CO_2 fluxes into and out of the atmosphere in glacial–interglacial cycles appear to be those into and out of the oceans. These fluxes change in response to climate-driven changes in ocean circulation, which change the ventilation time of the ocean and hence the storage of CO_2 in the deep ocean. As Figure 12.48 shows, the deep ocean contains far more CO_2 (mainly as bicarbonate) than the atmosphere, so only a small fraction of CO_2 would have to shift between the atmosphere and the deep ocean to produce the changes in atmospheric concentration seen in Figure 12.49 (to find out what fraction, work Problem 12.12). Toggweiler et al. (2006) suggested that the key ocean circulation changes result from a climate-driven migration of the westerly winds in the Southern Ocean. In the present interglacial climate, the most intense westerly winds are located south of the Antarctic polar front. As a result of a phenomenon called Ekman transport, these winds drive water away from Antarctica, and as a result, water rises, or "upwells" from depth, allowing CO_2 build-up in the deep ocean to vent to the atmosphere, keeping atmospheric CO_2 concentrations high. During glacial times, these westerlies shifted equatorward allowing for build-up of CO_2 in circum-Antarctic deep water. In addition, changes in the efficiency of the biologic pump can affect the balance of CO_2 between ocean and atmosphere (e.g., Boyle, 1988). Most of the ocean is oligotrophic, meaning that photosynthesis is limited by the abundance of nutrients in the surface water. Hence, the rate at which the biologic pump works is governed by surface-water nutrient levels. These are in turn governed by factors such as the rate at which nutrients are delivered from the land, the areal extent of continental shelves where nutrients can be recycled from the ocean bottom and return of nutrient-rich deep water to the surface by ocean circulation. Since CO_2 is a strong greenhouse gas, this shift of CO_2 between the atmosphere and the ocean serves to amplify the climate changes that produce them in the first place. In this way, the quite small changes in geographic and temporal distribution of insolation were strengthened sufficiently to produce the remarkable glacial–interglacial cycles of the Pleistocene.

12.9.3   The deep carbon cycle

Carbon at the Earth's surface is also part of a deeper, slower cycle. A small fraction of the organic carbon fixed every year is buried in sediments (Figure 12.48). Much of what is initially buried in sediment is remineralized during diagenesis, but a fraction is sequestered from the exogene for long geologic times. Some fraction of this sedimentary organic carbon eventually returns to the exogene through weathering, and some fraction is subducted into the mantle. Some of the subducted organic carbon is returned relatively quickly to the atmosphere through subduction-related volcanism or decarbonation during metamorphism, while some continues into the deep mantle. Diamonds in a Brazilian kimberlite containing isotopically light carbon indicative of an ultimate biological origin (Chapter 9) together with silicate inclusions indicative of a lower mantle origin demonstrate that the deep carbon cycle extends into the lower mantle (Walter et al., 2011). That, too, can eventually return to the atmosphere through mid-ocean ridge and mantle plume-related volcanism.

When dissolved in water, CO_2 forms carbonic acid and dissociates (Chapter 6), providing hydrogen ions that then attack silicate minerals:

$$CO_2 + H_2O \rightleftharpoons H^+ + HCO_3^-\ 2H^+$$

$$+\ CaAl_2Si_3O_8 \rightleftharpoons Al_2Si_2O_5(OH)_4 + Ca^{2+}$$

Calcium released in this way is carried by rivers to the sea along with bicarbonate ions, where they precipitate as calcite:

$$Ca^{2+} + HCO_3^- \rightleftharpoons H^+ + CaCO_3$$

Much of the calcite redissolves in the deep water or sediment, but some is buried as part of the carbonate reservoir in Figure 12.48. The effect of silicate weathering is thus to remove CO_2 from the atmosphere. Weathering of limestone has no effect on atmospheric CO_2, however, as carbonate ions produced in the process are also precipitated as calcite in the ocean and thus carbonate is recycled back into the sedimentary carbonate reservoir. As is the case with sedimentary organic carbon, sedimentary carbonate can eventually return to the atmosphere through metamorphism or volcanism.

Together, the relative rates of volcanism and metamorphism, weathering of sedimentary organic matter, weathering of silicates (and resulting burial of sedimentary carbonate), and burial of sedimentary organic matter control the amount of carbon in the oceans, atmosphere, and biosphere on long geologic time scales (more than about 1 Ma), as first noted by J. J. Ebelmen (1847) and T. C. Chamberlin (1899). For the last few hundred million years, the masses of carbon in the various reservoirs illustrated in Figure 12.48 have been approximately constant and the fluxes into and out of the exogene have been more or less at steady state. The modifier "more or less" is important. Indeed, the GEOCARB model developed by R.A. Berner and his colleagues at Yale University over the years (e.g., Berner et al., 1983; Berner, 2006) begins with an assumption of steady-state and then uses carbon isotope variations in marine sediments as primary input to model variations in atmospheric CO_2 that result from imbalances in fluxes into and out of the exogene (i.e., deviations from steady-state). The model includes various feedbacks between climate and geochemical processes. Small changes in these fluxes likely account, at least in part, for the climatic extremes that have occurred in the Phanerozoic. The Pleistocene glaciations are only the most recent example; glaciations also occurred at the end of the Ordovician and from the late Carboniferous into the Permian. In between these times, the Earth experienced warm periods, such as the Cretaceous, when the poles were ice-free and the oceans circulated in a far different manner than they do today.

The fluxes into and out of the exogene do not operate independently but are coupled in complex ways that result in both positive and negative feedbacks, just as is the case in the exogenous carbon cycle. These are illustrated in Figure 12.50. Just as in the exogenous cycle, these feedbacks are linked to and affect climate. In the exogenic carbon cycle, the Milankovitch cycles (Chapter 9) drove the Pleistocene glacial cycles. In the deep carbon cycle, the principal external factor is tectonic activity, including continental position, uplift, volcanism, and metamorphism. Continent position and uplift directly affect global climate, but in opposing ways. Continent position and climate can both also affect

ocean circulation, albeit in complex ways, which affects nutrient distribution; that, in turn, affects the marine biota and the biologic pump. Volcanism, which is closely linked to tectonic activity, affects the terrestrial biota directly by supplying soil nutrients and emitting CO_2 and reduced gases to the atmosphere (the latter consuming O_2). Uplift increases erosion and weathering, which supplies nutrients to both the terrestrial and marine biota. Increasing terrestrial and marine bioproductivity results in more burial of organic matter in sediments, which draws down atmospheric CO_2 and increases atmospheric O_2. Increasing atmospheric O_2 increases weathering of sedimentary organic carbon, while increasing CO_2, along with organic acids produced by the terrestrial biota, increases silicate weathering. O_2 affects the biota through wildfires: fires are more likely and burn more extensively at higher O_2 levels. Increased atmospheric CO_2 increases temperature, which increases silicate weathering rates, but increased silicate weathering decreases atmospheric CO_2, which decreases global temperature, which decreases weathering, resulting in a natural thermostat (a topic we'll examine further in the following chapter). Climate also affects the terrestrial biota, albeit in complex ways.

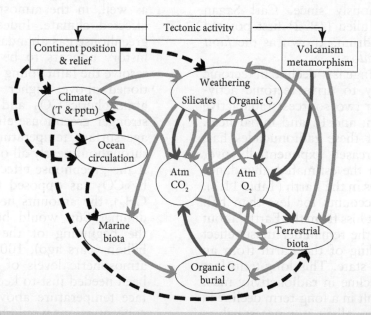

Figure 12.50   Feedback loops in the long-term carbon cycle. Dark arrows represent positive feedbacks, red arrows represent negative feedbacks, and dashed arrows are complex feedbacks (can be positive or negative). After Berner (1999) and Killops and Killops (2005).

These complex interactions have led to variations in climate through the Phanerozoic, such as the Ordovician, Permo-Carboniferous, and Pleistocene glaciations.

12.9.4 Evolutionary changes affecting the carbon cycle

Superimposed on the interactions illustrated in Figure 12.50 have been three long-term unidirectional changes that have affected the Earth, the carbon cycle, and climate.

- The first of these has been a steady increase in brightness of the Sun and, consequently, insolation. Stars grow progressively brighter over their main sequence (Figure 10.1) lifetimes. The Sun is now about 30% brighter than it was 4.5 billion years ago when it first became a main sequence star. This increase in insolation would result in a surface temperature increase of nearly 22°C, all other factors being equal. Interestingly, the present mean global surface temperature is now about 13°C. This implies a mean surface temperature on the young Earth of −9°C, well below freezing, all other things being equal. Yet, the existence of sedimentary rocks as old as 3.8 Ga demonstrates liquid water has been present on Earth nearly continuously since. Carl Sagan and George Mullen (1972) first pointed out this conundrum, known as the *faint young Sun paradox*.

- The second change is a decline in tectonic activity. Energy to drive tectonic activity comes from two sources: radioactive decay of U, Th, and K, and initial heat. The activity of these radionuclides has, of course, decreased exponentially over time. Based on the estimated abundance of radionuclides in the Earth (Table 11.3), radioactivity accounts for less than half the present heat loss from the Earth (about 45 terawatts); the remainder must reflect long-term cooling of the Earth from an initially hotter state. This long-term loss of heat and decline in radioactivity must necessarily result in a long-term decline in tectonic activity, albeit not necessarily a steady one.

- The third change has been the evolution of life, which has had a profound effect on the nature of the atmosphere, and, as a result, on climate. The Earth's atmosphere in Hadean and early Archean times would have certainly been much different from the present one. Oxygen would have been absent; instead CO_2 would likely have been the dominant component, as it is in the atmospheres of Mars and Venus. It was likely modestly reducing, with some CH_4 present. Life was established on Earth by at least 3.5 Ga and probably by 3.8 Ga (Chapter 9), but how much earlier than the Great Oxidation Event at 2.3–2.4 Ga oxygenic photosynthesis arose is debated. Oxygenic photosynthesis, of course, converts atmospheric CO_2 to organic matter, producing O_2 as a byproduct. Most of the organic matter produced by photosynthesis has simply recycled back into CO_2 through respiration, consuming the O_2 originally produced. However, some of that organic matter has been buried in sediment, leading to a drawdown of atmospheric CO_2 and build-up of O_2. Indeed, there is more than three times as much organic carbon in sediments as needed to account for the O_2 in the atmosphere (this implies large amounts of O_2 have been consumed by oxidizing sulfur and iron). The drawdown of CO_2, and probably methane as well, in the atmosphere has in turn affected climate. Indeed, a decrease in greenhouse gas abundances over geologic history appears to be the only way to resolve the faint young Sun paradox mentioned above: higher concentrations of atmospheric CO_2 and CH_4 provided a stronger greenhouse effect, keeping average surface temperatures above freezing throughout nearly all of the Precambrian. If the greenhouse effect is provided only by CO_2 (as opposed to both CO_2 and CH_4), the amounts needed to offset the dimmer Sun would be considerable: at the beginning of the Proterozoic (2.5 billion years ago), 100 times the present atmospheric levels of CO_2 would have been needed just to keep the average surface temperature above freezing. Lower concentrations would have been needed if substantial methane were present. At the beginning of the Phanerozoic when the Sun was only 5% dimmer than at present, a CO_2 concentration some 10 times the present one would have been needed to maintain an average surface temperature similar to the modern one (assuming CH_4 concentrations similar to modern ones). While photosynthetic life was present in the oceans in the Proterozoic, it was not until the mid-Paleozoic that land plants evolved. The evolution of terrestrial flora provided a new environment in which organic matter could be buried, namely bogs, swamps, and mires; much of this was ultimately converted to coal. Vast deposits of coal produced during the aptly named Carboniferous period testify to the effect of the new terrestrial biota on the carbon cycle. In addition, the invasion of the land by plants accelerated silicate weathering by releasing organic acids from roots. Finally, the evolution of terrestrial flora would have changed climate directly by changing the Earth's albedo (reflectivity). Deserts, sand, and soil, presumably the Precambrian land cover, reflect 25–30% of solar radiation back into space. Forests reflect only about 10% of that radiation. Thus the decrease in albedo in the Paleozoic that would have resulted from the evolution of terrestrial flora would have effectively increased insolation and had a warming effect on climate.

12.9.5   The carbon cycle and climate through time

Figure 12.51 shows Berner's (2006) modeled atmospheric CO_2 concentrations over the Phanerozoic. The model is based on the relationships illustrated in Figure 12.50 and the record of carbon isotope ratios in marine carbonate sediments. The latter change in response to shifts in the fluxes shown in Figure 12.50. For example, because organic carbon has low $\delta^{13}C$, an increased burial of organic carbon drives the isotopic composition of carbon in the exogene toward more positive values. Increased weathering of organic carbon would have the opposite effect. These isotopic shifts are reflected in the isotopic composition of carbonates precipitated from the oceans. The overall picture suggested by this model is one of declining

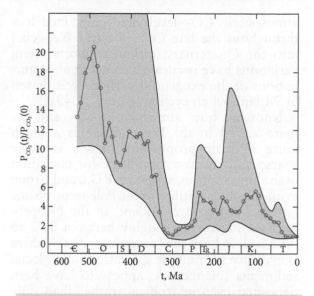

Figure 12.51   Concentration of atmospheric carbon dioxide over the Phanerozoic. Open red circles show the modeled values of Berner (2006). Errors on the model are shown as gray and are taken from Berner and Kothavala (2001).

atmospheric CO_2, but if the model is correct, the decline has not been steady. A decline through the Ordovician led to a glacial epoch in the Late Ordovician–Early Silurian. The Berner model suggests this was due to weathering of silicate rock (specifically volcanics). Continental position likely also played a role, as most evidence of glaciation comes from areas positioned near the South Pole at the time. CO_2 recovered in the Silurian and Devonian, but declined again in the Carboniferous, leading to the Permo-Carboniferous glaciation. This time the cause appears to be burial of vast amounts of organic carbon in bogs, swamps, and mires that was ultimately transformed into coal. Atmospheric CO_2 recovered in the Mesozoic, but not to levels seen in the Paleozoic (remember, however, that progressively less CO_2 needed to maintain the same temperature). After reaching concentrations perhaps five times greater than present ones, atmospheric CO_2 declined in the late Cretaceous and early Tertiary periods. An essentially similar decline over this period is seen in the model of Hansen and Wallmann (2003). Judging from CO_2 levels deduced from diunsaturated alkenones (Figure 12.49) and boron isotopes (Figure 9.47), long-term atmospheric CO_2 levels have remained low throughout the late Cenozoic (the Neogene) into the Quaternary, although shorter term variations have occurred as a result of perturbations of the exogenous carbon cycle driven by Milankovitch cycling (Figure 12.49).

Knowing how atmospheric CO_2 and climate varied in the Precambrian is a much more difficult proposition. There is some sparse and equivocal evidence for glaciation in the late Archean, around 2.8 Ga, and strong evidence of glaciations in the Paleoproterozoic (around 2.4 to 2.2 Ga) and in the Neoproterozoic (occurring roughly between 0.72 to 0.63 Ga). In both the Proterozoic events, there is evidence of multiple glaciations and glacial sediments (diamictites) appear to have been deposited even at tropical, rather than only polar, latitudes, based on paleomagnetic inclinations and other geologic indicators. Other evidence, such as the presence of marine sediments above and below them, indicate that they were deposited at low elevation. Examples of glacial sediments from the Neoproterozoic are found on every continent except Antarctica. These observations suggest that these glaciations were far more severe than Phanerozoic glaciations that followed (in the Pleistocene glaciations, for example, ice sheets reached no further south than about 40°N, although mountain glaciers were, and still are, present at lower latitudes). This has given rise to the "snowball Earth" hypothesis: glacial events so severe that oceans were entirely frozen over or nearly so (Kirschvink, 1992; Hoffman et al., 2017). The record of the Paleoproterozoic glaciation is more equivocal, but low-latitude paleomagnetic inclinations in sequences in North America, the Baltic, and South Africa suggest that this too was a severe snowball-Earth-like glaciation that extended to the tropics.

The Paleoproterozoic diamictites in the Huronian Formation in Canada occur stratigraphically above an older conglomerate containing abundant detrital pyrite (FeS_2) and uraninite (UO_2). That reduced minerals could survive erosion, transport, and deposition indicates they were deposited in an oxygen-free atmosphere. Other examples of detrital pyrite and uraninite in Archean-age sediments are relatively common, but with rare exceptions they do not occur in sediments younger than 2.2 Ga. The Huronian diamictites are overlain by redbeds: sandstones consisting of quartz coated with hematite. The hematite is indicative of an oxidizing atmosphere. Redbeds are absent in sediments older than 2.4 Ga, but occur more or less throughout the subsequent geologic record. Detrital pyrite and uraninite and redbeds are part of a suite of evidence, including mass-independent sulfur isotope fractionation discussed in Chapter 9, that significant amounts of atmospheric O_2 first appeared between 2.4 and 2.2 Ga. This period is known as the Great Oxidation Event (GOE). Detailed evidence suggests this event was prolonged over 100–200 million years with oscillations in atmospheric oxygen and climate (Gumsley et al., 2017; Philippot et al., 2018). Oxygen levels in the atmosphere before the GOE were likely a factor of 10^{-5} lower than present and rose to perhaps 10% or so of present atmospheric levels during the GOE (e.g., Canfield, 2005; Holland, 2006). The close correspondence in time of the Paleoproterozoic Huronian glaciation with the appearance of atmospheric oxygen is perhaps not a coincidence. Kasting and Ono (2006) suggested that the rise in atmospheric oxygen caused a drop in atmospheric CH_4. A drop in concentration of the latter from 10^{-3} to 10^{-5} would have produced a temperature decrease of 20°C, enough to trigger glaciation unless the early Proterozoic climate was much warmer than today. Alternative explanations involve pulses of mantle plume volcanism or rift-related volcanism (e.g., Melezhik, 2006; Gumsley et al., 2017) that resulted in an increase in silicate weathering and drawdown of atmospheric CO_2.

Why did the atmosphere remain oxygen-free throughout the Archean and then rapidly increase in the Paleoproterozoic? It's possible that this was the time when oxygenic photosynthesis first evolved. Others argue, however, that it evolved earlier. One reason for a delay is that iron and sulfur in the exogene would have been in reduced states. Before oxygen could build up in the atmosphere, the stock of reduced iron and sulfur would have to be substantially exhausted. Indeed, the observation that post-GOE Proterozoic paleosols, like modern ones, are iron-rich (because Fe^{3+} is insoluble and cannot be leached from them) while Archean paleosols are iron-poor (because Fe^{2+} was leached) is evidence that considerable oxygen was consumed in the process of oxidizing soil Fe. Similarly, gypsum-bearing evaporites first appearing in the geologic record around the time of the GOE is evidence that oxygen was consumed in oxidizing sulfur to sulfate. Only after the stock of reduced Fe and S were oxidized could atmospheric O_2 rise. This would have taken time; just how long would have depended on the rate of oxygen production as well as the supply of reduced components to the surface by volcanism. Large variations in $\delta^{13}C$ in marine carbonates deposited around this time, including excursions to values as negative as −12‰, suggest disturbances to the exogenic carbon cycle which in turn support the possibility of changes in atmospheric greenhouse gas levels. Alternatively, Kump et al. (2001) and Holland (2009), among others, have suggested that a change in the composition of volcanic gases would account for the timing of the GEO. Although they differ in detail, both models conjecture that the mantle, and consequently volcanic gases, became more oxidizing through time as a result of subduction. Condie et al. (2009) argue that there is evidence for a near absence of volcanism beginning around 2.45 Ga and lasting for 200–250 Ma. The resulting reduction in the flux of reductants would have allowed oxygen to build up in the atmosphere. Cox et al. (2018) emphasize the role of phosphorus as a limiting nutrient and point out that it would have become more available as the mantle cooled and new additions to crust became more siliceous (Figure 11.44). Others have argued that the GEO corresponds to the beginning of modern subduction-driven plate tectonics. Thus in some hypotheses both the GEO and the associated Paleoproterozoic glaciations resulted from processes and changes occurring deep within the Earth as well as changes occurring at the surface.

The geologic record of the Neoproterozoic glaciations is more complete, and there is compelling evidence that at least two of the glaciations, the Sturtian from about 717–659 Ma and the Marinoan from around 650–635 Ma, were severe and extended into the tropics (Hoffman et al., 2017). Although there is not yet complete agreement on this point, a consensus appears to be forming that not only were the continents largely ice-covered, but most or all of the ocean was frozen over by a thick layer of ice: a snowball Earth. Indeed, the evidence of extensive glaciation is compelling enough that this period of the Neoproterozoic has been named the *Cryogenian*. In both cases, glacial sediments are overlain by thick deposits of marine carbonates, termed *cap carbonates* and are preceded by large negative $\delta^{13}C$ anomalies that return to positive values just before the glaciation. A possible earlier Kaigas glaciation at around 740 Ma and a later Gaskers glaciation at around 582 Ma appear to have been less severe.

There is no consensus as to the cause of the Cryogenian glaciations but essentially all hypotheses involve a disturbance of the carbon cycle, either through silicate weathering and burial of carbonate or burial of organic carbon (Godderis et al., 2011). But like the Paleoproterozoic Huronian glaciation, the Neoproterozoic glaciations are associated with dramatic changes in atmospheric oxygen levels. There is now an evolving consensus based on a variety of geochemical evidence preserved in Proterozoic sedimentary rocks that while oxygen levels may have been 10% or more of present levels in the first few hundred million years following the GOE, atmospheric O_2 fell back to 1% or perhaps even 0.1% of present levels and persisted for the next billion and a half years into the Cryogenian (Lyons et al., 2014). During this time oxygen was present in the surface ocean, but the deep ocean would have been reducing and sulfidic (much like the modern Black Sea). As the Cryogenian ended, oxygen levels began to rise again dramatically, although perhaps not reaching present levels until the Devonian when forests evolved. As we noted earlier, the oldest steroid biomarkers of eukaryotes occur at 717 Ma and there is microfossil evidence of a major radiation of green algae around 750–800 Ma. This raises the intriguing possibility that evolution might have led to an increase in photosynthesis and decreased atmospheric CO_2 and increased O_2 (e.g., Tziperman et al., 2011). Remains of eukaryotes would be more likely to sink quickly through the water column and be buried, reducing deep water oxygen demand, allowing oxygen to penetrate to ocean deep water (Lenton et al., 2014). That in turn would have allowed the evolution and expansion of bottom-dwelling animals.

The termination of the Marinoan glaciation marks the beginning of the last period of the Proterozoic: the Ediacaran. Although one more, less severe, glaciation occurred in the Ediacaran, climate appears to have again stabilized. It was during the Ediacaran period that multicellular animals appeared (although sponge-like fossils from the Cryogenian have been reported). The appearance of metazoans (animals) at the time that atmospheric O_2 was rising toward present levels is undoubtedly not coincidental. It appears to provide a resolution to Darwin's dilemma:

> There is another . . . difficulty, which is much more serious. I allude to the manner in which species belonging to several* of the main divisions of the animal kingdom suddenly appear in the lowest known fossiliferous rocks. If the theory be true, it is indisputable that before the lowest Cambrian stratum was deposited, long periods elapsed . . . and that during these vast periods, the world swarmed with living creatures. But to the question why we do not find rich fossiliferous deposits belonging to these assumed earliest periods before the Cambrian system, I can give no satisfactory answer. The case at present must remain inexplicable; and may be truly urged as a valid argument against the views here entertained. (Darwin, 1859)

This has long puzzled subsequent generations of paleontologists as well. Geochemistry has provided the answer even if the details remain debated: oxygen levels were too low to support multicellular life until the latest Proterozoic.

12.9.6   Fossil fuels and anthropogenic climate change

About 250 years ago humans began to replace traditional energy sources – muscle power, animal power, wind, and water – with a new energy source, coal. The coal-fired steam engine, first developed by James Watt to pump water from (ironically enough) a coal mine, was soon adapted to saw wood and cut stone, mill flour, spin yarn and weave fabric, provide transport, dig canals, and so on. Beginning in the mid-nineteenth century, petroleum and natural gas began to

supplement and partly replace coal and coal gas. The primary combustion product of all of these is, of course, CO_2. In 1896, Svante Arrhenius wrote the paper mentioned earlier, predicting that burning of fossil fuels should lead to an increase in atmospheric CO_2 concentration, which should in turn enhance the greenhouse effect and increase global temperatures. Arrhenius thought this was a good thing because it might prevent the Earth from entering another ice age and it should enhance agricultural productivity and help to feed a growing global population. At the time, there were no systematic measurements of atmospheric CO_2 concentration and global temperature recordkeeping was just beginning, so the theory could not be tested.

Arrhenius's theory was mostly ignored (but not entirely, see Chamberlin, 1899) by the scientific community until the 1950s, when Roger Revelle and Hans Suess (1955) took notice of it. Suess was a nuclear chemist at the US Geological Survey working on ^{14}C dating and noticed an apparent decrease in the specific activity* of ^{14}C in the atmosphere. He suspected that increasing atmospheric CO_2 derived from fossil fuel, which would have a specific activity of 0, might be the answer. When he moved to University of California–San Diego to join oceanographer Roger Revelle at the Scripps Institution of Oceanography, they recruited Charles Keeling to begin measuring atmospheric CO_2 on a regular basis. Keeling soon set up a monitoring station atop Mauna Loa in Hawaii and, subsequently, one at the South Pole. Monthly measurements of atmospheric CO_2 have been made at these stations ever since (eventually, similar stations would be set up all over the world); the data from Mauna Loa are shown in Figure 12.52. Superimposed on the wiggles caused by the seasonal cycle of photosynthesis in the terrestrial biosphere is a clear increase in atmospheric CO_2 from a seasonally adjusted value of 316 ppm in 1960 to 410 ppm in 2018, corresponding to an average annual increase of 1.6 ppm. The rate of rise has been increasing, however, and in the first decades of the twenty-first century approached 2 ppm per year. The increase in rate of rise is consistent with the increasing rate of emissions: the

---

* Recall from Chapter 8 that the specific activity is defined as the activity (decays per unit time) of 14C divided by the amount of C in grams.

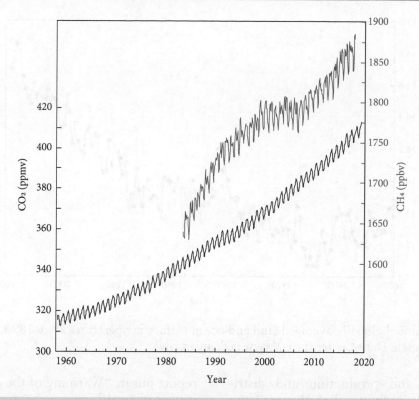

Figure 12.52   Concentrations of atmospheric CO_2 and CH_4 measured at the Scripps Institution of Oceanography Mauna Loa observatory over the last half-century. Sources: Scripps CO_2 program (http://scrippsco2.ucsd.edu) and NOAA Earth System Research Laboratory (https://www.esrl.noaa.gov/gmd/ccgg/trends/full.html).

Intergovernmental Panel on Climate Change (IPCC) estimates that the carbon emitted by fossil-fuel burning increased from an average of 6.4 ± 0.4 gigatons of carbon (GtC) per year in the 1990s to 9.5 ± 0.6 GtC per year in 2011 (Stocker et al., 2013). In addition to fossil-fuel burning, the IPCC estimates that an additional 0.9 GtC per year is being added to the atmosphere through "land use change" (primarily cutting of tropic forests).

From the total emissions of around 10 GtC/yr, we would predict atmospheric CO_2 should be increasing by about 4 ppm per year, more than twice the actual rate. Put another way, the actual increase in atmospheric CO_2 is only around 4–5 GtC/yr. This difference reflects carbon transfer into other exogenous reservoirs shown in Figure 12.48. Various studies suggest the ocean is taking up about 2 GtC/yr. One consequence of this is ocean acidification: ocean surface pH has declined from an estimated preindustrial value of 8.17 to a present value of 8.07, corresponding to a 26% increase in H^+.

The remaining 2–3 GtC/yr being released by fossil-fuel burning and tropical deforestation is apparently being taken up by the Northern Hemisphere biosphere. There could be several reasons for the expansion of the Northern Hemisphere biosphere. First, as agriculture became more efficient in the twentieth century, land cleared for agriculture has been abandoned and is returning to forest. Second, emissions from fossil-fuel burning, including both CO_2 and nitrates, may be stimulating plant growth. Third, warming is enabling expansion of boreal forests into regions previously covered by tundra.

Figure 12.52 also shows the change in atmospheric methane concentrations. These concentrations exceed the natural range of the last 650,000 years (320–790 ppb) as determined from ice cores. The principal anthropogenic sources are farm animals (ruminants such as cows and sheep) and rice farming. Other sources include landfills, sewage treatment facilities, biomass burning (incomplete burning produces methane as

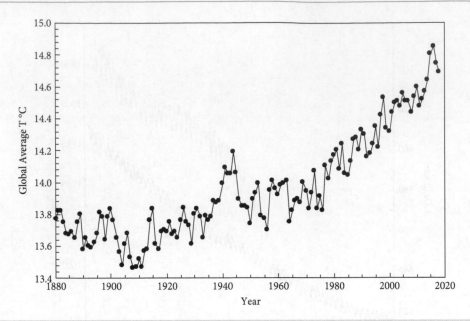

Figure 12.53   Annual globally averaged land and ocean surface temperatures since 1890. Data source: US National Climate Data Center (http://www.ncdc.noaa.gov).

well as CO_2) and production and distribution of hydrocarbons (i.e., losses from oil and gas wells and distribution lines). In addition, warming may be enhancing release of methane from permafrost and methane clathrates on Arctic continental shelves.

Figure 12.53 shows annual globally averaged land and ocean surface temperatures from 1880–2018. It is apparent that surface temperatures have increased over the period, albeit irregularly. The net increase is about 1°C. Most of that change has occurred since 1960 (the same year CO_2 measurements began); indeed, the rate of increase over the last 50 years (0.16°C per decade) has been more than double the average rate of increase over the last 120 years. There are numerous other indicators of changing climate as well: the average temperature of the oceans has increased to depths of at least 3000 m (the ocean has been absorbing more than 80% of the heat added to the climate system); mountain glaciers have receded and snow and ice cover has declined, as has Arctic Sea ice (the latter dramatically); the average atmospheric water vapor content has increased (in a way more or less consistent with the extra water vapor that warmer air can hold) and precipitation patterns have changed; and sea level has been rising at a rate of 1.8 mm/yr over the last 50 years. As the 2013 IPCC

report put it, "Warming of the climate system is unequivocal."

In the previous sections, we saw how climate has changed over Earth's history and that these changes occurred as a result of changes to the carbon cycle and atmospheric greenhouse gas concentrations. Given what Earth's history teaches us, the observed recent climate change (Figure 12.53), the observed increase in greenhouse gas concentrations (Figure 12.52), and the known amounts of fossil fuels that have been burned, can we really question whether fossil-fuel burning is leading to climate change? The 2013 IPCC report (Stocker et al., 2013) states: "It is extremely likely that human influence has been the dominant cause of the observed warming since the mid-20th century." Climate change does have some "upside" effects, such as those noted by Arrhenius, but it also has negative effects, and the latter are likely to outweigh the former. Furthermore, our understanding of the climate system remains quite limited. There are reasons to think that temperature increases in the future might be more rapid than the present moderate rate (as an example, CO_2 solubility decreases with temperature, so the ocean may not take up as much CO_2 in the future). We have seen that in the Pleistocene, climate swung rapidly between cold and warm states.

These observations should give us pause about the world's current equivocal, even cavalier, attitude toward climate change, and motivate us to move beyond fossil fuels as our principal energy source. Our fossil fuel reserves, while ultimately limited, remain vast, and we could continue to rely on them through much of this century and perhaps beyond. The Saudi oil minister once observed that the Stone Age did not end because people ran out of stones. Many alternatives to fossil fuels are either now cost-effective or are approaching that point. Parts of the solution are relatively simple, other parts may require ingenuity and some modest short-term sacrifice in return for much greater long-term benefit, but we must begin to earnestly strive to reach the goal of a carbon-free economy within the next few decades if we are to avoid severe disruption to our world.

12.10 SUMMARY

In this chapter we reviewed organic geochemistry, the origin of fossil fuel resources, and the carbon cycle and its control on climate.

- We began by reviewing the basic architecture of organic molecules, the simplest of which are hydrocarbons, which consist of chains and rings of carbon and hydrogen. Various biomolecules derive from this architecture by the addition of other elements, most notably O, N, S, and P, that form a variety of functional groups that modify that architecture. In the marine environment, C, N, and P are present in organisms in the Redfield ratio of 105:16:1.
- Biologically important organic molecules can be divided into a few fundamental classes: carbohydrates, proteins, lipids, nucleotides, and nucleic acids. Of these, lipids and polyphenols such as lignins and tannins are most likely to survive in soils and sediment.
- Organic matter in soils consists of the breakdown products of plant and microbe as well as molecules exuded by them and typically makes up 5% and more of the soil mass. Soil humic substances consist of supramolecular aggregates of everything from intact plant material to highly oxidized carbon in carboxylic acids loosely held together by noncovalent bonds. Adsorption on mineral surfaces explains the persistence of organic matter over long periods.
- Dissolved organic matter (DOM) in streams and lakes varies depending on environment, particularly in headwater streams, but a predominant component in most are carboxyl-rich alicyclic molecules or CRAM with H:C ratios between 0.5 and 1.5 and O:C ratios between 0.4 and 0.8. Streams export roughly a third to half of net terrestrial productivity to the oceans, and oceans CRAM also dominates the composition of marine DOM.
- In the marine environment, primary productivity in ocean surface waters is quickly recycled through the microbial loop that sustains both microbes and protozoans, which then form the basis of the marine food web. Roughly 10% of surface water productivity is exported to deep water as sinking organic particles where it remains millennial time scales. This *biological pump* effectively sequesters carbon in deep ocean water, isolating it from exchange with the atmosphere. Shifts in carbon between the deep ocean and atmosphere were an important feedback mechanism on the Pleistocene climate cycles.
- Organic molecules readily complex transition metals and aluminum and a large fraction of at least some trace metals dissolved in a natural water is complexed by organic ligands. This complexation plays an important role in the evolution of soil profiles in soil fertility.
- Organic matter in sediments undergoes a series of microbially mediated reactions that remove functional groups, decreases oxygen and hydrogen, decreases short-chained molecules relative to long-chain equivalents, decreases the abundance of unsaturated compounds, and increases the abundance of aromatic ones relative to aliphatic ones.
- The product of these diagenetic processes is *kerogen*, which can be divided into three types: Type I consists of macerals of exinite (charred material) and vitrinite (woody material), Type II consists of exinite (pollen plant waxes, resins, and oils), and Type III consists of algal remains.

- Despite the diagenetic processing, the basic structure of some molecules is retained and can be related to specific classes of organisms. These *biomarkers* have proved useful in reconstructing paleo-environments, including temperatures and CO_2 levels, in constraining biological evolution, and assessing the thermal maturity of kerogen.
- Under heat and pressure, marine and aquatic kerogen undergoes *catagenesis*, that at temperatures in the range of 75–150°C produces a mixture of liquid hydrocarbons known as petroleum. Further temperature increases convert these to short-chained hydrocarbon gases.
- Accumulation of plant remains in bogs and mires produces peat, which upon heating increases aromatic components and decreases H/C and O/C ratios to produce coal.
- Organic matter is an important part of the carbon cycle and burial of organic matter through time has produced the free oxygen present in the atmosphere. Oxygen levels increased in several distinct steps, the first of these around 2.3 Ga in the Great Oxidation Event, the second around 600 Ma, simultaneous with the first appearance of animals. Subsequent increases occurred with the evolution of terrestrial flora in the Paleozoic.
- Atmospheric CO_2 levels have been a major control on climate throughout Earth's history. Far higher atmospheric CO_2 in the early Earth maintained a habitable climate despite the much lower solar radiative flux. Draw-down of greenhouse gases associated with increases in O_2 resulting in severe glaciations in the Proterozoic. A less-severe glaciation resulted from the sequestration of organic carbon in the Carboniferous. While Milankovitch variations were the pacemaker of Pleistocene glaciations, the principal cause was a shift of CO_2 from the atmosphere to the deep ocean.
- Burning of fossil fuels over the last two centuries, and particularly in the last 50 years, has increased atmospheric CO_2 to levels well above any occurring in the Quaternary, which is demonstrably increasing global temperatures.

REFERENCES AND SUGGESTIONS FOR FURTHER READING

Aiken, G.R., McKnight, D.M., Wershaw, R.L. and MacCarthy, P. (eds). 1985. *Humic Substances in Soil, Sediment, and Water*. New York, Wiley Interscience.

Alleon, J., Bernard, S., Le Guillou, C., et al. 2016. Molecular preservation of 1.88 Ga Gunflint organic microfossils as a function of temperature and mineralogy. *Nature Communications* 7: 11977. doi: 10.1038/ncomms11977.

Arrhenius, S. 1896. Über den Einfluss des Atmosphärischen Kohlensäurengehalts auf die Temperatur der Erdoberfläche. *Proceedings of the Royal Swedish Academy of Science* 22: 1–101.

Azam, F., Fenchel, T., Field, J. G., Gray, J. S., Meyer-Reil, L. A. and Thingstad, F. 1983. The ecological role of water-column microbes. *Marine Ecology Progress Series* 10: 257–63.

Baars, O. and Croot, P.L. 2011. The speciation of dissolved zinc in the Atlantic sector of the Southern Ocean. *Deep Sea Research II* 58: 2720–32.

Battin, T. J., Luyssaert, S., Kaplan, L. A., Aufdenkampe, A. K., Richter, A. and Tranvik, L. J. 2009. The boundless carbon cycle. *Nature Geoscience* 2: 598–600. doi: 10.1038/ngeo618.

Bechtel, A. and Püttmann, W. 2017. Biomarkers: Coal. In: White WM (ed.) *Encyclopedia of Geochemistry*. Springer International Publishing, Cham, pp 1–14. 978-3-319-39193-9.

Bennett, P.C. and Casey, W. 1994. Chemistry and mechanisms of low-temperature dissolution of silicates by organic acids, in *Organic Acids in Geological Processes* (eds E.D. Pitman and M.D. Lewan), pp. 162–200. Berlin, Springer Verlag.

Bergmann, W. 1978. Zur Struckturaufklärung von Huminsäuren aus Abwasser. PhD thesis, University of Tübingen.

Berner, R.A. 1981. *Early Diagenesis, A Theoretical Approach*. Princeton, Princeton University Press.

Berner, R.A. 1999. A new look at the long-term carbon cycle. *GSA Today* 9(11): 1–6.

Berner, R.A. 2006. GEOCARBSULF: A combined model for Phanerozoic atmospheric O_2 and CO_2. *Geochimica et Cosmochimica Acta* 70: 5653–64.

Berner, R.A. and Kothavala, Z. 2001. Geocarb III: a revised model of atmospheric CO_2 over Phanerozoic time. *American Journal of Science* 301: 182–204. doi: 10.2475/ajs.301.2.182.

Berner, R.A., Lasaga, A.C. and Garrells, R.M. 1983. The carbonate–silicate geochemical cycle and its effect on atmospheric carbon dioxide over the past 100 million years. *American Journal of Science* 283: 641–83.

Boyle, E.A. 1988. The role of vertical chemical fractionation in controlling late Quaternary atmospheric carbon dioxide. *Journal of Geophysical Research* 93: 701–15.

Brocks, J. J. and Schaeffer, P. 2008. Okenane, a biomarker for purple sulfur bacteria (Chromatiaceae), and other new carotenoid derivatives from the 1640Ma Barney Creek Formation. *Geochimica et Cosmochimica Acta* 72: 1396–414. doi: 10.1016/j.gca.2007.12.006.

Calvert, S.E. and Pederson, T.F. 1992. Organic carbon accumulation and preservation: how important is anoxia, in *Organic Matter: Productivity, Accumulation, and Preservation in Recent and Ancient Sediments* (eds J.K. Whelan and J.W. Farmington), pp. 231–63. New York, Columbia University Press.

Canfield, D.E. 2005. The early history of atmospheric oxygen: homage to Robert M. Garrels. *Annual Review of Earth and Planetary Science* 33: 1–36. doi: 10.1146/annurev.earth.33.092203.122711.

Cao, X., Aiken, G. R., Butler, K. D., et al. 2018. Evidence for major input of riverine organic matter into the ocean. *Organic Geochemistry* 116: 62–76. doi: 10.1016/j.orggeochem.2017.11.001.

Chamberlin, T. C. 1899. An attempt to frame a working hypothesis of the cause of glacial periods on an atmospheric basis. *Journal of Geology* 7: 545–84. doi: 10.1086/608449.

Chiou, C.T., Freed, V.H., Schmedding, D.W. and Kohnert, R.L. 1977. Partition coefficient and bioaccumulation of selected organic chemicals. *Environmental Science and Technology* 11: 475–8.

Chiou, C.T., Peters, L.J. and Freed, V.H. 1979. A physical concept of soil–water equilibria for nonionic organic compounds. *Science* 206: 831–2.

Cole, J. J., Prairie, Y. T., Caraco, N. F., et al. 2007. Plumbing the Global Carbon Cycle: Integrating Inland Waters into the Terrestrial Carbon Budget. *Ecosystems* 10: 172–85. doi: 10.1007/s10021-006-9013-8.

Darwin, C. 1859. *On the Origin of Species*. Routledge, London.

de Leeuw, J.W. and Largeau, C. 1993. A review of the macromolecular organic compounds that comprise living organisms and their role in kerogen, coal, and petroleum formation, in *Organic Geochemistry: Principles and Applications* (eds M.H. Engel and S.A. Macko), pp. 23–72. New York, Plenum.

Del Vecchio, R. and Blough, N. V. 2004. On the origin of the optical properties of humic substances. *Environmental Science and Technology* 38(14), 3885–91. doi: 10.1021/es049912h.

Deming, J.W. and Baross, J.A. 1993. The early diagenesis of organic matter: bacterial activity, in *Organic Geochemistry: Principles and Applications* (eds M.H. Engel and S.A. Macko), pp. 119–44. New York, Plenum.

Dittmar, T., Stubbins, A. 2014. 12.6 – Dissolved organic matter in aquatic systems. In: Holland HD, Turekian KK (eds) *Treatise on Geochemistry* (Second Edition), vol. Elsevier, Oxford, pp 125–156 978-0-08-098300-4.

Dominé, F., Bounaceur, R., Scacchi, G., et al. 2002. Up to what temperature is petroleum stable? New insights from a 5200 free radical reactions model. *Organic Geochemistry* 33: 1487–99. doi: 10.1016/s0146-6380(02)00108-0.

Drever, J.I. and Vance, G.F. 1994. Role of soil organic acids in mineral weathering processes, in *Organic Acids in Geological Processes* (eds E.D. Pitman and M.D. Lewan), pp. 138–61. Berlin, Springer Verlag.

Druffel, E. R. M., Bauer, J. E. 2000. Radiocarbon distributions in Southern Ocean dissolved and particulate organic matter. *Geophysical Research Letters* 27: 1495–8. doi: 10.1029/1999gl002398.

Ebelman, J. J., 1847. Sur la décomposition des roches. *Annales des Mines* 4, 627–54.

Ellwood, M.J. 2004. Zinc and cadmium speciation in subantarctic waters east of New Zealand. *Marine Chemistry* 87: 37–58. doi: 10.1016/j.marchem.2004.01.005.

Engel, M. and Macko, S.A. 1993. *Organic Geochemistry: Principles and Applications*. New York, Plenum Press.

Falkowski, P., Scholes, R.J., Boyle, E., et al. 2000. The global carbon cycle: A test of our knowledge of Earth as a system. *Science* 290: 291–6.

Flerus, R., Lechtenfeld, O. J., Koch, B. P., et al. 2012. A molecular perspective on the ageing of marine dissolved organic matter. *Biogeosciences* 9: 1935–55. doi: 10.5194/bg-9-1935-2012.

Fogel, M.L. and Cifuentes, M.L. 1993. Isotope fractionation during primary production, in *Organic Geochemistry: Principles and Applications* (eds M.H. Engel and S.A. Macko), pp. 73–98. New York, Plenum.

Fourier, J.-B.J. 1824. Remarques générales sur les températures du globe terrestre et des espaces planétaires. *Annales de Chimie et de Physique* 27: 136–67.

Furrer, G. and Stumm, W. 1986. The coordination chemistry of weathering I. Dissolution kinetics of δ-Al_2O_3 and BeO. *Geochimica et Cosmochimica Acta* 50: 1847–60.

Gaines, S. M., Eglinton, G., Rullkötter, J. 2009. *Echos of Life: What Fossil Molecules Reveal About Earth History*. Oxford University Press, New York

Godderis, Y., Le Hir, G. and Donnadieu, Y. 2011. Modelling the snowball Earth. *Geological Society London Memoirs* 36: 151–61. doi: 10.1144/m36.10, 2011.

Graedel, T.E. and Crutzen, P.J. 1993. *Atmospheric Change: An Earth System Perspective*. New York, W.H. Freeman.

Grauer, R. and Stumm, W. 1982. Die Koordinationschemie oxidisher Grenzflächen und ihre Auswirkung auf die Auflösingskinetic oxidisher Festphasen in wässrigen Lösungen. *Colloid. Polymer Science* 260: 959–70.

Greb, S. F., Eble, C. F., Hower, J. C. 2017. Coal. In: White WM (ed) *Encyclopedia of Geochemistry*, Springer International Publishing, Cham, pp 1–16 978-3-319-39193-9.

Gumsley, A. P., Chamberlain, K. R., Bleeker, W., Söderlund, U., de Kock, M. O., Larsson, E. R. and Bekker, A., 2017. Timing and tempo of the Great Oxidation Event. *Proceedings of the National Academy of Sciences* 114(8): 1811–16. doi: 10.1073/pnas.1608824114.

Hansen, K.W. and Wallmann, K. 2003. Cretaceous and Cenozoic evolution of seawater composition, atmospheric O_2 and CO_2: A model perspective. *American Journal of Science* 303: 94–148. doi: 10.2475/ajs.303.2.94.

Hatcher, P.G. and Clifford, D.J. 1997. The organic geochemistry of coal: from plant materials to coal. *Organic Geochemistry* 27: 251–74. doi: 10.1016/s0146-6380(97)00051-x.

He, M., Moldowan, M. J., Peters, K. E. (2017) Biomarkers: Petroleum. In: White WM (ed) *Encyclopedia of Geochemistry*. Springer International Publishing, Cham, pp 1–13. 978-3-319-39193-9.

Hedges, J. I. 2002. Why dissolved organics matter. In: Hansell DA, Carlson CA (eds) *Biogeochemistry of marine dissolved organic matter*. Elsevier, London, pp 1–33.

Hemingway, J. D., Rothman, D. H., Grant, K. E., et al. 2019. Mineral protection regulates long-term global preservation of natural organic carbon. *Nature* 570: 228–31. doi: 10.1038/s41586-019-1280-6.

Henrichs, S.M. 1993. Early diagenesis of organic matter: the dynamics (rates) of cycling of organic compounds, in *Organic Geochemistry: Principles and Applications* (eds M.H. Engel and S.A. Macko), pp. 101–17. New York, Plenum.

Hoffman, P. F., Abbot, D. S., Ashkenazy, Y., et al. 2017. Snowball Earth climate dynamics and Cryogenian geology–geobiology. *Science Advances* 3: e1600983. doi: 10.1126/sciadv.1600983.

Hoffmann, S.R., Shafer, M.M. and Armstrong, D.E. 2007. Strong colloidal and dissolved organic ligands binding copper and zinc in rivers. *Environmental Science and Technology* 41: 6996–7002. doi: 10.1021/es070958v.

Holland, H.D. 2006. The oxygenation of the atmosphere and oceans. *Philosophical Transactions of the Royal Society of London B* 361: 903–15. doi: 10.1098/rstb.2006.1838.

Holland, H.D. 2009. Why the atmosphere became oxygenated: a proposal. *Geochimica et Cosmochimica Acta* 73: 5241–55. doi: 10.1016/j.gca.2009.05.070.

Huang, Y., Street-Perrott, F.A., Metcalfe, S.E., et al. 2001. Climate change as the dominant control on glacial–interglacial variations in C_3 and C_4 plant abundance. *Science* 293: 1647–51. doi: 10.1126/science.1060143.

Jaffé, R., Yamashita, Y., Maie, N., et al. 2012. Dissolved Organic Matter in Headwater Streams: Compositional Variability across Climatic Regions of North America. *Geochimica et Cosmochimica Acta* 94: 95–108. doi: 10.1016/j.gca.2012.06.031.

Kaplan, L. A., Newbold, J. D. (2003) 4 – The Role of monomers in stream ecosystem metabolism. In: Findlay SEG, Sinsabaugh RL (eds) *Aquatic Ecosystems* Academic Press, Burlington, pp 97–119 978-0-12-256371-3.

Kasting, J.F. and Ono, S. 2006. Palaeoclimates: the first two billion years. *Philosophical Transactions of the Royal Society of London B* 361: 917–29. doi: 10.1098/rstb.2006.1839.

Keil, R. G., Montluçon, D. B., Prahl, F. G., Hedges, J. I. 1994. Sorptive preservation of labile organic matter in marine sediments. *Nature* 370: 549–52. doi: 10.1038/370549a0.

Khademi, Z., Jones, D., Malakouti, M. and Asadi, F. 2010. Organic acids differ in enhancing phosphorus uptake by *Triticum aestivum* L. – effects of rhizosphere concentration and counterion. *Plant and Soil* 334: 151–9. doi: 10.1007/s11104-009-0215-7.

Killops, S.D. and Killops, V.J. 2005. *Organic Geochemistry* (2nd edn). Malden, MA, Blackwell.

Kirschvink, J.L. 1992. Late Proterozoic low-latitude global glaciation: the snowball Earth, in *The Proterozoic Biosphere* (eds J.W. Schopf and C. Klein), pp. 51–2. New York, Cambridge University Press.

Kleber, M., Reardon, P. 2017. Biopolymers and Macromolecules. In: White WM (ed.) *Encyclopedia of Geochemistry* vol. l. Springer International Publishing, Cham, pp 1–5. 978-3-319-39193-9.

Kump, L.R., Kasting, J.F. and Barley, M.E. 2001. Rise of atmospheric oxygen and the 'upside-down' Archean mantle. *Geochem. Geophys. Geosyst.* 2:. doi: 10.1029/2000gc000114.

Kvenvolden, K.A. 1993. *A Primer on Gas Hydrates. USGS Professional Paper 1570*. Washington, DC, US Government Printing Office.

Leake, J., Johnson, D., Donnelly, D., et al. 2004. Networks of power and influence: the role of mycorrhizal mycelium in controlling plant communities and agroecosystem functioning. *Canadian Journal of Botany* 82: 1016–45.

Lechtenfeld, O. J., Kattner, G., Flerus, R., McCallister, S. L., Schmitt-Kopplin, P., Koch, B. P. 2014. Molecular transformation and degradation of refractory dissolved organic matter in the Atlantic and Southern Ocean. *Geochimica et Cosmochimica Acta* 126: 321–37. doi: 10.1016/j.gca.2013.11.009.

Lehmann, J., Kleber, M. 2015. The contentious nature of soil organic matter. *Nature* 528: 60. doi: 10.1038/nature16069.

Lenton, T. M., Boyle, R. A., Poulton, S. W., Shields-Zhou, G. A. and Butterfield, N. J., 2014. Co-evolution of eukaryotes and ocean oxygenation in the Neoproterozoic era. *Nature Geoscience* 7: 257–65. doi: 10.1038/ngeo2108.

Loh, A. N., Bauer, J. E. 2000. Distribution, partitioning and fluxes of dissolved and particulate organic C, N and P in the eastern North Pacific and Southern Oceans. *Deep Sea Research Part I: Oceanographic Research Papers* 47: 2287–316. doi: 10.1016/S0967-0637(00)00027-3.

Lucas, Y., 2001. The role of plants in controlling rates and products of weathering: importance of biological pumping. *Annual Review of Earth and Planetary Science* 29: 135–63. doi: 10.1146/annurev.earth.29.1.135.

Luthi, D., Le Floch, M., Bereiter, B., et al. 2008. High-resolution carbon dioxide concentration record 650,000-800,000 years before present. *Nature* 453: 379–82.

Macko, S.A., Engel, M.H. and Parker, P.L. 1993. Early diagenesis of organic matter in sediments: assessment of mechanisms and preservation by the use of isotopic molecular approaches, in *Organic Geochemistry: Principles and Applications* (eds M.H. Engel and S.A. Macko), pp. 211–23. New York, Plenum.

Mason, S.F. 1991. *Chemical Evolution: Origin of the Elements, Molecules, and Living Systems*. Oxford, Oxford University Press.

Massicotte, P., Asmala, E., Stedmon, C., Markager, S. 2017. Global distribution of dissolved organic matter along the aquatic continuum: Across rivers, lakes and oceans. *Science of the Total Environment* 609: 180–91. doi: 10.1016/j.scitotenv.2017.07.076.

McKnight, D.M., Pereira, W.E., Ceazan, M.L. and Wissmar, R.C. 1982. Characterization of dissolved organic materials in surface waters within the blast zone of Mount St. Helens, Washington. *Organic Geochemistry* 4: 85–92.

Melezhik, V.A. 2006. Multiple causes of Earth's earliest global glaciation. *Terra Nova* 18: 130–7. doi: 10.1111/j.1365-3121.2006.00672.x.

Milkov, A.V. 2004. Global estimates of hydrate-bound gas in marine sediments: how much is really out there? *Earth Science Reviews* 66: 183–97.

Morel, F.M.M. 1983. *Principles of Aquatic Chemistry*. New York, Wiley Interscience.

Morel, F.M.M. and Hering, J.G. 1993. *Principles and Applications of Aquatic Chemistry*. New York, John Wiley and Sons, Ltd.

Mosher, J. J., Kaplan, L. A., Podgorski, D. C., McKenna, A. M., Marshall, A. G. 2015. Longitudinal shifts in dissolved organic matter chemogeography and chemodiversity within headwater streams: a river continuum reprise. *Biogeochemistry* 124: 371–385. doi: 10.1007/s10533-015-0103-6

Nettersheim, B. J., Brocks, J. J., Schwelm, A., et al. 2019. Putative sponge biomarkers in unicellular Rhizaria question an early rise of animals. *Nature Ecology and Evolution* 3: 577–81. doi: 10.1038/s41559-019-0806-5.

Pagani, M., Arthur, M.A. and Freeman, K.H. 1999. Miocene evolution of atmospheric carbon dioxide. *Paleoceanography* 14: 273–92. doi: 10.1029/1999pa900006.

Peters, K. E., Michael Moldowan, J. 2017. Biomarkers: Assessment of thermal maturity. In: White WM (ed) *Encyclopedia of Geochemistry*. Springer International Publishing, Cham, pp 1–8. 978-3-319-39193-9.

Petsch, S. 2017. Kerogen. In: White WM (ed.) *Encyclopedia of Geochemistry*. Springer International Publishing, Cham, pp 1–5. 978-3-319-39193-9.

Philippot, P., Ávila, J. N., Killingsworth, B. A., Tessalina, S., Baton, F., Caquineau, T., Muller, E., Pecoits, E., Cartigny, P., Lalonde, S. V., Ireland, T. R., Thomazo, C., van Kranendonk, M. J. and Busigny, V. 2018. Globally asynchronous sulphur isotope signals require re-definition of the Great Oxidation Event. *Nature Communications* 9: 1–10. doi: 10.1038/s41467-018-04621-x.

Revelle, R., Suess, H. E. 1957. Carbon dioxide exchange between atmosphere and ocean and the question of an increase of atmospheric CO_2 during the past decades. *Tellus* 9: 18–27.

Royal Dutch Shell. 1983. *The Petroleum Handbook*. Amsterdam, Elsevier.

Rullkötter, J. 1993. The thermal alteration of kerogen and the formation of oil, in *Organic Geochemistry: Principles and Applications* (eds M.H. Engel and S.A. Macko), pp. 101–17. New York, Plenum.

Sagan, C., Mullen, G. 1972. Earth and Mars: Evolution of atmospheres and surface temperatures. *Science* 177: 52–56. doi: 10.1126/science.177.4043.52.

Schanks, W.C., Böhlke, J.K. and Seal, R.R. 1995. Stable isotopes in mid-ocean ridge hydrothermal systems: interactions between fluids, minerals, and organisms, in *Seafloor Hydrothermal Systems, Geophysical Monograph Vol. 91* (eds S.E. Humphris, R.A. Zierenberg, L S. Mullineaux and R.E. Thomson), pp. 194–221. Washington, American Geophysical Union.

Schnitzer, M. 1978. Humic substances: chemistry and reactions, in *Soil Organic Matter* (eds M. Schnitzer and S.H. Khan). Amsterdam, Elsevier.

Schoell, M. 1984. Stable isotopes in petroleum research, in *Advances in Petroleum Geochemistry, 1* (eds. J. Brooks and D. Welte), pp. 215–45. London, Academic Press.

Schouten, S., Hopmans, E. C., Schefuß, E., Sinninghe Damsté, J. S. 2002. Distributional variations in marine crenarchaeotal membrane lipids: a new tool for reconstructing ancient sea water temperatures? *Earth and Planetary Science Letters* 204: 265–74. doi: 10.1016/S0012-821X(02)00979-2.

Schwarzenbach, R.P. and Westall, J. 1981. Transport of nonpolar organic compounds from surface waters to groundwater. Laboratory sorption studies. *Environmental Science And Technology.* 15: 1360–67.

Sholkovitz, E. R. 1976. Flocculation of dissolved organic and inorganic matter during the mixing of river water and seawater. *Geochimica et Cosmochimica Acta* 40: 831–45. doi: 10.1016/0016-7037(76)90035-1.

Siegenthaler, U. and Sarmiento, J.L. 1993. Atmospheric carbon dioxide and the ocean. *Nature* 365: 119–25.

Sofer, Z. 1984. Stable carbon isotope compositions of crude oils: application to source depositional environments and petroleum alteration. *American Association of Petroleum Geologists Bulletin* 68: 31–49.

Sposito, G. 1989. *The Chemistry of Soils*. New York, Oxford University Press.

Stevenson, F.J. and Vance, G.F. 1989. Naturally occurring aluminum–organic complexes, in *The Environmental Chemistry of Aluminum* (ed. G. Sposito), pp. 117–45. Boca Raton, FL, CRC Press.

Stocker, T. F., Qin, D., Plattner, G.-K., et al. 2013. IPCC, 2013: Climate change 2013: The physical science basis. *Contribution of Working Group I to the Fifth Assessment Report of the Intergovernmental Panel on Climate Change*. Cambridge, 1535 pp.

Stuiver, M., Polach, H. A. 1977. Discussion: reporting of ^{14}C data. *Radiocarbon* 19: 355–63. doi: 10.1017/S0033822200003672.

Stumm, W. 1992. *Chemistry of the Solid–Water Interface*. New York, Wiley Interscience.

Stumm, W. and Morgan, J.J. 1996. *Aquatic Chemistry*. New York, Wiley Interscience.

Surdam, R.C., Boese, S.W. and Crossey, L.J. 1984. The geochemistry of secondary porosity, in *Clastic Diagenesis, American Association of Petroleum Geologists Memoir 37* (eds. R.A. McDonald and R.C. Surdam), pp. 127–49. Tulsa, American Association of Petroleum Geologists.

Tegelaar, E.W., Derenne, S., Largeau, C. and de Leeuw, J.W. 1989. A reappraisal of kerogen formation. *Geochimica et Cosmochimica Acta* 53: 3103–7.

Thurman, E.M. 1985. *Organic Geochemistry of Natural Waters*. Dordrecht, Martinus Nijhoff/Dr W. Junk Publishers.

Tierney, J. E. (2014) 12.14 – Biomarker-based inferences of past climate: The TEX86 paleotemperature proxy. In: Holland HD, Turekian KK (eds.) *Treatise on Geochemistry* (2nd ed.), vol. Elsevier, Oxford, pp 379–93. 978-0-08-098300-4.

Tissot, B.P. and Welte, D.H. 1984. *Petroleum Formation and Occurrence*. Berlin, Springer Verlag.

Toggweiler, J.R., Russell, J.L. and Carson, S.R. 2006. Midlatitude westerlies, atmospheric CO_2, and climate change during the ice ages. *Paleoceanography* 21: PA2005,10.1029/2005pa001154.

Treibs, A. 1934. Chlorophyll- und Häminderivate in bituminösen Gesteinen, Erdölen, Erdwachsen und Asphalten. Ein Beitrag zur Entstehung des Erdöls. *Justus Liebigs Annalen der Chemie* 510: 42–62. doi: 10.1002/jlac.19345100103.

Tziperman, E., Halevy, I., Johnston, D. T., Knoll, A. H. and Schrag, D. P. 2011. Biologically induced initiation of Neoproterozoic snowball-Earth events. *Proceedings of the National Academy of Sciences* 108(37): 15091–6. doi: 10.1073/pnas.1016361108.

Ungerer, P., Collell, J., Yiannourakou, M. 2015. Molecular modeling of the volumetric and thermodynamic properties of kerogen: influence of organic type and maturity. *Energy and Fuels* 29: 91–105. doi: 10.1021/ef502154k.

van Krevelen, D.W. 1961. *Coal: Typology, Chemistry, Physics and Constitution*. Amsterdam, Elsevier.

Vandenbroucke, M., Béhar, F. and Rudkiewicz, J.L. 1999. Kinetic modelling of petroleum formation and cracking: implications from the high pressure/high temperature Elgin Field (UK, North Sea). *Organic Geochemistry* 30: 1105–25.

Walter, M. J., Kohn, S. C., Araujo, D., et al. 2011. Deep mantle cycling of oceanic crust: evidence from diamonds and their mineral inclusions. *Science* 334: 54–57. doi: 10.1126/science.1209300.

Wershaw, R. 1993. Model for Humus in Soils and Sediments. *Environmental Science and Technology* 27: 814–16.3/20/2020 6:08:13 PM doi: 10.1021/es00042a603.

Whelan, J.A. and Craig, H. 1983. Methane, hydrogen and helium in hydrothermal fluids at 21°N on the East Pacific Rise, in *Hydrothermal Processes at Seafloor Spreading Centers* (eds P. Rona, K. Boström, L. Laubier and K.L. Smith), pp. 391–409. New York, Plenum.

Whiticar, M.J. 1990. A geochemial perspective of natural gas and atmospheric methane. *Organic Geochemistry* 16: 531–47. doi: 10.1016/0146-6380(90)90068-b.

Witter, A.E., Hutchins, D.A., Butler, A., and Luther, I.G.W. 2000. Determination of conditional stability constants and kinetic constants for strong model Fe-binding ligands in seawater. *Marine Chemistry* 69: 1–17.

Zhang, H. and Bloom, P.R. 1999. Dissolution kinetics of hornblende in organic acid solutions. *Soil Science Society of America Journal* 63: 815–22. doi: 10.2136/sssaj1999.634815x.

Zinder, B., Furrer, G. and Stumm, W. 1986. The coordination chemistry of weathering: II. Dissolution of Fe(III) oxides. *Geochimica et Cosmochimica Acta* 50: 1861–9.

PROBLEMS

1. Sketch the structure of the following:

   (a) Citric acid: $HOC(CH_2CO_2H)_2CO_2H$

   (b) Tartaric acid: $HO_2CCH(OH)CH(OH)CO_2H$ (2,3,-dihydroxybutanedioic acid)

2. Write the chemical formula and sketch the structure of 2-hydroxypropanoic acid (lactic acid).

3. Suppose you could follow the pathway of individual atoms during photosynthesis. While this is not possible, something similar can be done by isotopic labeling of water and CO_2. If ^{18}O-labeled water is added to a suspension of photosynthesizing chloroplasts, which of the following compounds will first show enrichment in ^{18}O: ATP, NADPH, O_2, or 3-phosphoglycerate? If you repeat the experiment with ^2H-labeled water and ^{13}C-labeled CO_2, which of these molecules will first show enrichment in these isotopes?

4. The first and second acidity constants of oxalic acid $((COOH)_2)$ are $pK_{a1} = 1.23$ and $pK_{a2} = 4.19$. What is the pH of a solution formed by dissolving 1 mole of oxalic acid in 1 kg of water?

5. If the 1 M oxalic acid solution of Problem 4 is titrated with 1 M NaOH, how will pH change as a function of the amount of base added? Make a plot of pH versus amount of base added.

6. The rate of bond cleavage during the thermal maturation of kerogen approximately doubles for every 10°C rise in temperature. Thermal maturation reaches a peak at ~100°C. Based on this and assuming that these reaction rates show an Arrhenius temperature dependence (eqn. 5.25), estimate the activation energy for these reactions.

7. Astrophysicist Thomas Gold suggested that most petroleum deposits are formed by abiogenic organic carbon (mainly in the form of methane) diffusing out of the mantle. There are few, if any, geochemists that agree. Describe at least three geochemical observations that support the "conventional" theory that petroleum is formed from sedimentary kerogen, which in turn is derived from the remains of once-living organisms.

8. Bartschat et al. (1992) modeled the metal-complexing behavior of humic acid as that of two ligands: a bidentate carboxylic ligand (e.g., malonate) and a bidentate phenol one (e.g., catechol), and that the effective concentrations of these are 10^{-3} mol/g humate and 5×10^{-4} mol/g humate, respectively. Using the following apparent stability constants, calculate the fraction of copper complexed if the humate concentration is 10 mg/l, the pH 8, and the total copper concentration is 10^{-8} M. Assume that copper and humate are the only species present.

   Apparent stability constants:

"Malonate":	H_2L	$\beta_2 = 8.7$
	HL	$\beta_1 = 5.7$
	CuL	$\beta_{Cu} = 5.7$
"Catechol":	H_2L	$\beta_2 = 9.1$
	HL	$\beta_1 = 12.4$
	CuL	$\beta_{Cu} = 13.4$

9. Repeat the calculation in Problem 8 above, but for pH 5.5.

10. The adjacent table lists organic solid/water (K_{OM}) and octanol/water (K_{OC}) partition coefficients for some nonpolar compounds. Are these data consistent with eqn. 12.15? What values do you determine for constants *a* and *b*? (*HINT:* Use linear regression.)

	K_{OW}	K_{OM}
Acetophenone	38.90	42.66
Benzene	128.82	83.18
Tetrachloroethylene	398.11	208.93
Napthalene	2290.87	1288.25
Parathion	6456.54	1148.15
Pyrene	151356.12	83176.38
Chlorobenzene	512.86	389.05
DDT	1548816.62	138038.43
2,4,5,2',4',5'-PCB	5248074.60	218776.16

11. Sediment from a highly eutropic lake was found to have an organic carbon fraction of 5.8%. Using the adsorption partition coefficient for DDT listed in Problem 10, predict the concentration of DDT in the sediment if the lake water has a DDT concentration of 3 $\mu g/l$ and the sediment contains 5.8% organic matter.

12. During Pleistocene glacial periods atmospheric CO_2 concentration was 180 ppm. During interglacial periods, The atmospheric concentration of CO_2 rose 280 ppmv, which is equal to a total of 600×10^{15} g of carbon as CO_2. The deep ocean contained $38,000 \times 10^{15}$ g carbon (as bicarbonate) during interglacial periods. What fraction of deep ocean carbon was been transferred to the atmosphere to account for the increased atmospheric CO_2?

Chapter 13

The land surface: weathering, soils, and streams

13.1 INTRODUCTION

The Earth is unique, at least in our Solar System, in having liquid water continually present on the surface. Indeed, oceans cover two-thirds of the Earth's surface: we live on a watery planet. On land, water continually transforms the surface both physically and chemically through weathering and erosion. We saw in Chapter 11 that the upper continental crust has the approximate average composition of granodiorite, and that the oceanic crust consists of basalt. But a random sample of rock from the crust is unlikely to be either; indeed, it may not be an igneous rock at all. At the very surface of the Earth, sediments and soils predominate. Both are ultimately produced by the interaction of water with "crystalline rock" (by which we mean igneous and metamorphic rocks). Clearly, to fully understand the evolution of the Earth, we need to understand the role of geochemical processes involving water. Beyond that, water is, of course, essential to life, another apparently unique feature of our planet, again at least in this solar system. Finally, water is central to human activity and survival. Human activity has had a broad effect on water chemistry, and with 7 billion people now living on the planet, finding clean water has become a major concern. Here, however, we will use the tools developed in the first part of the book to understand the natural chemistry of water and its interaction with rock to produce soil and sediment.

We can broadly distinguish two kinds of aqueous solutions: continental waters and seawater. Continental waters by this definition include groundwater, fresh surface waters (river, stream and lake waters), and saline lake waters. The compositions of these fluids are quite diverse. Seawater, on the other hand, is reasonably uniform, and it is by far the dominant fluid on the Earth's surface. Hydrothermal fluids, which may be derived from ordinary groundwater, formation pore waters, seawater, or magmatic water are three classes of water produced when water is heated and undergoes accelerated interactions with rock and often carry a much higher concentration of dissolved constituents. Precipitation from these hydrothermal solutions have formed many of the ore deposits that allowed us to move beyond the Stone Age and remain critical to modern civilization. This topic too we will examine in Chapter 15.

The part of the planet we will discuss in this chapter is known as the *Critical Zone*, defined as the region from the top of vegetation to the bottom of circulating groundwater (Ashley, 1998; Richardson, 2018). It is so called because all terrestrial life (apart from extremophile communities living at greater depth) lives within and depends on processes occurring within this zone. One can broaden this statement further and argue that essentially all life depends on the Critical Zone because weathering and erosion within it provide the nutrients upon which marine life depends. Within this zone air, water, rock,

Geochemistry, Second Edition. William M. White.
© 2020 John Wiley & Sons Ltd. Published 2020 by John Wiley & Sons Ltd.
Companion website: www.wiley.com/go/white/geochemistry

and life interact in manifold complex ways to produce Earth's unique outer skin: soil rather than the regoliths of other bodies such as the Moon and Mars. Most importantly, human life depends on it. Not only do we live within the Critical Zone, we depend on it for water, most of our food, and many other resources, ranging from wood to the many kinds of mineral deposits that form within it and which we'll explore in Chapter 15. It is also within this zone that human activity has had its greatest impact, some of which threatens our health, our food and water, and indeed the sustainability of human culture. A realization of the importance of this zone has led to an international effort to establish a number of critical zone observatories in different settings around the world to monitor surface and subsurface water, meteorological conditions, soil properties, and vegetation in order to understand processes and changes occurring in the Critical Zone. We will examine some of the human-induced changes and threats in Chapter 15. In this chapter we will focus on the natural geochemical processes occurring within the Critical Zone.

13.2   REDOX IN NATURAL WATERS

The surface of the Earth represents a boundary between regions of very different redox state. The atmosphere contains free oxygen and therefore is highly oxidizing. In the Earth's interior, however, conditions are reducing, there is no free oxygen, and Fe is primarily in the 2+ valence state. Natural waters exist in this boundary region and their redox state is highly variable. Biological activity is the principal cause of this variability. Autotrophs, which range from bacteria through algae to plants, use solar energy and *photosynthesis* to drive thermodynamically unfavorable reduction reactions that produce free O_2, the ultimate oxidant, on the one hand and organic matter, the dominant reductant, on the other. Indeed, as we found in the previous chapter, it is photosynthesis that is responsible for the oxidizing nature of the atmosphere and the redox imbalance between the Earth's exterior and interior. Both plants and animals (heterotrophs) liberate stored chemical energy

by catalyzing the oxidation of organic matter in a process called *respiration*. The redox state of solutions and solids at the Earth's surface is largely governed by the balance between photosynthesis and respiration. By this we mean that most waters are in an oxidized state because of photosynthesis and exchange with the atmosphere. When they become reducing, it is most often because respiration exceeds photosynthesis and they have been isolated from the atmosphere. Water may also become reducing as a result of reaction with sediments deposited in ancient reduced environments, but the reducing nature of those ancient environments usually also resulted from biological activity. Oxidation of reduced primary igneous rocks also consumes oxygen, and this process governs the redox state of some systems, mid-ocean ridge hydrothermal solutions for example. On a global scale, however, these processes are of secondary importance for the redox state of natural waters.

The predominant participants in redox cycles are C, O, N, S, Fe, and Mn. There are several other elements, for example, Cr, V, As, Eu, and Ce, that have variable redox states; these elements, however, are always present in trace quantities and their valence states reflect, rather than control, the redox state of the system. Although phosphorus has only one valence state (5+) under natural conditions, its concentration in solution is closely linked to redox state because the biological reactions that control redox state also control phosphorus concentration as it is adsorbed on Fe oxide surfaces.

Water in equilibrium with atmospheric oxygen has a $p\varepsilon$ of +13.6 (at pH = 7). At this $p\varepsilon$, thermodynamics tells us that all carbon should be present as CO_2 (or related carbonate species), all nitrogen as NO_3^-, all S as SO_4^{2-}, all Fe as Fe^{3+}, and all Mn as Mn^{4+}. This is clearly not the case and this disequilibrium reflects the kinetic sluggishness of many, though not all, redox reactions*. Given the disequilibrium we observe, the applicability of thermodynamics to redox systems would appear to be limited. Thermodynamics may nevertheless be used to develop partial equilibrium models. In such models, we can make

---

* While this may make life difficult for geochemists, it is also what makes it possible in the first place. We, like all other organisms, consist of a collection of reduced organic species that manage to persist in an oxidizing environment!

use of *redox couples* that might reasonably be at equilibrium to describe the redox state of the system. In Chapter 3, we introduced the tools needed to deal with redox reactions: E_H, the hydrogen scale potential (the potential developed in a standard hydrogen electrode cell) and $p\varepsilon$, or electron activity. We found that both may in turn be related to the Gibbs free energy of reaction through the Nernst equation (eqn. 3.109). These are all the tools we need; in this section, we will see how we can apply them to understanding redox in aqueous systems.

Table 13.1 lists the $p\varepsilon°$ of the most important redox half reactions in aqueous systems. Also listed are $p\varepsilon_W$ values. $p\varepsilon_W$ is the $p\varepsilon°$ when the concentration of H^+ is set to 10^{-7} (pH = 7). The relation between $p\varepsilon°$ and $p\varepsilon_W$ is simply:

$$p\varepsilon_W = p\varepsilon° + \log [H^+]^\nu = p\varepsilon° - \nu \times 7$$

where ν is the stoichiometric coefficient of the hydrogen ion in the reaction. Reactions are ordered by decreasing $p\varepsilon_W$ from strong oxidants at the top to strong reductants at the bottom. In this order, each reactant can oxidize any product below it in the list, but not above it. Thus, sulfate can oxidize methane to CO_2, but not ferrous iron to ferric iron. Redox

reactions in aqueous systems are often biologically mediated. In the following section, we briefly explore the role of the biota in controlling the redox state of aqueous systems.

13.2.1 Biogeochemical redox reactions

As we noted above, photosynthesis and atmospheric exchange maintain a high $p\varepsilon$ in surface waters. Water does not transmit light well, so there is an exponential decrease in light intensity with depth. As a result, photosynthesis is not possible below depths of 100–200 m even in the clearest waters. In murky waters, photosynthesis can be restricted to the upper few meters or less. Below this *photic zone*, biologic activity and respiration continue, sustained by falling organic matter from the photic zone. In the deep waters of lakes and seas where the rate of respiration exceeds downward advection of oxygenated surface water, respiration will consume all available oxygen. Once oxygen is consumed, a variety of specialized bacteria continue to consume organic matter and respire utilizing oxidants other than oxygen. Thus, $p\varepsilon$ will continue to decrease.

Since bacteria exploit first the most energetically favorable reactions, Table 13.1 provides a guide to the sequence in which oxidants

Table 13.1 $p\varepsilon$ of principal aquatic redox couples.

	Reaction	$p\varepsilon°$	$p\varepsilon_W$
1	$\frac{1}{4}O_{2(g)} + H^+ \rightleftharpoons \frac{1}{2}H_2O$	+20.75	+13.75
2	$\frac{1}{5}NO_3^- + \frac{6}{5}H^+ + e^- \rightleftharpoons \frac{1}{10}N_2 + \frac{3}{5}H_2O$	+21.05	+12.65
3	$\frac{1}{2}MnO_{2(s)} + 2H^+ + 2e^- \rightleftharpoons \frac{1}{2}Mn^{2+} + H_2O$	+20.8	+9.8[†]
4	$\frac{1}{9}NO_3^- + \frac{10}{9}H^+ + e^- \rightleftharpoons \frac{1}{9}NH_4^- + \frac{1}{3}H_2O$	+14.9	+6.15
5	$Fe(OH)_{3(s)} + 3H^+ + 3e^- \rightleftharpoons Fe^{2+} + 3H_2O$	+16.0	+1.0[†]
6	$CH_2O^* + H^+ + 3e^- \rightleftharpoons \frac{1}{2}CH_3OH$	+4.01	−3.01
7	$\frac{1}{8}SO_4^{2-} + \frac{5}{4}H^+ + e^- \rightleftharpoons \frac{1}{8}H_2S + \frac{1}{2}H_2O$	+5.25	−3.5
8	$\frac{1}{8}SO_4^{2-} + \frac{9}{8}H^+ + e^- \rightleftharpoons \frac{1}{8}HS^- + \frac{1}{2}H_2O$	+4.25	−3.6
9	$\frac{1}{8}CO_{2(g)} + H^+ + e^- \rightleftharpoons \frac{1}{8}CH_4 + \frac{1}{4}H_2O$	+2.9	−4.1
10	$\frac{1}{6}N_{2(g)} + \frac{4}{3}H^+ + e^- \rightleftharpoons \frac{1}{3}NH_4^+$	+4.65	−4.7
11	$\frac{1}{4}CO_{2(g)} + H^+ + e^- \rightleftharpoons \frac{1}{4}CH_2O^* + \frac{1}{4}H_2O$	−0.2	−7.2

*The concentration of Mn^{2+} and Fe^{2+} are set to 1 μM.

[†]We are using "CH_2O", which is formally formaldehyde, as an abbreviation for organic matter generally (for example, glucose is $C_6H_{12}O_6$).

are consumed as pε decreases. From it, we can infer that once all molecular oxygen is consumed, reduction of nitrate to molecular nitrogen will occur (reaction 2). This process, known as *denitrification*, is carried out by bacteria, which use the oxygen liberated to oxidize organic matter and the net energy liberated to sustain themselves. At lower levels of pε, other bacteria reduce nitrate to ammonia (reaction 4), a process called *nitrate reduction*, again using this reaction as an electron sink to organic matter. At about this pε level, Mn^{4+} will be reduced to Mn^{2+}. At lower pε, ferric iron is reduced to ferrous iron. The reduction of both Mn and Fe may also be biologically mediated in whole or in part. These conditions are commonly achieved in water-logged soils and bogs and in ground-waters moving through reduced bedrock such as black shales.

From Table 13.1, we can expect that *fermentation* (reaction 6 in Table 13.1) will follow reduction of Fe. Fermentation can involve any of a number of reactions, only one of which, reduction of organic matter (carbohydrate) to methanol, is represented in Table 13.1. In fermentation reactions, further reduction of some of the organic carbon provides a sink of electrons, allowing oxidation of the remaining organic carbon; for example, in glucose, which has six carbons, some carbons are oxidized to CO_2 while others are reduced to alcohol or acetic acid. While these kinds of reactions can be carried out by many organisms, it is bacterial-mediated fermentation that is of geochemical interest.

At lower pε, sulfate is used as the oxidant by sulfate-reducing bacteria to oxidize organic matter. Sulfide-bearing waters can occur in *meromictic* lakes, ponds and seas. In such water bodies, respiration produces reducing conditions and a density gradient maintained by chemical or thermal stratification prevents deep water from coming in contact with the atmosphere. The largest such example is the Black Sea, where H_2S is present below about 100 to 200m. At even lower pε, nitrogen is reduced to ammonia (reaction 9), a process known as *dissimilatory nitrogen reduction*, with the nitrogen serving as the electron acceptor for the oxidization of organic matter. This reaction is distinct from *nitrogen fixation*, which is carried out primarily under aerobic conditions and catalyzed by molybdenum-, vanadium, or iron-containing *nitrogenase* enzymes. The ability to catalyze this reaction is limited to a few specific classes of microbes, called diazotrophs, that include cyanobacteria, green sulfur bacteria, a few classes of Archaea, and rhizobia bacteria that live symbiotically in the root nodules of some plants, most notably legumes. All other life ultimately depends on the nitrogen fixed by these microbes (although lightning also produces some nitrogen oxides from N_2).

To summarize, in a water or soil where the availability of oxygen is less than that of organic matter, we would expect to see oxygen consumed first, followed by reduction of nitrate, manganese, iron, sulfur, and finally nitrogen. This sequence is illustrated on a pε-pH diagram in Figure 13.1. We would expect to see a similar sequence with depth in a column of sediment where the supply of organic matter exceeds the supply of oxygen and other oxidants.

13.2.2 Eutrophication

The extent to which the redox sequence described above proceeds in a body of water depends on several factors. The first of these is temperature structure, because this governs the advection of oxygen to deep waters. As mentioned above, light (and other forms of electromagnetic energy) is not transmitted

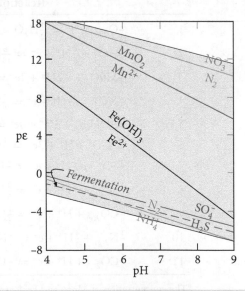

Figure 13.1 Important biogeochemical redox couples in natural waters.

well by water. Thus, only surface waters are heated by the Sun. Above 4°C water density decreases as the temperature of surface water rises (fresh water reaches its greatest density at 4°C). These warmer surface waters, known in lakes as the *epilimnion*, generally overlie a zone where temperature decreases rapidly, known as the *thermocline* or *metalimnion*, and a deeper zone of cooler water, known as the *hypolimnion*. This temperature stratification produces a stable density stratification which limits vertical advection of water and dissolved constituents, including oxygen and nutrients. In tropical lakes and seas, this stratification can be permanent. In temperate regions, however, there is an annual cycle in which stratification develops in the spring and summer. As the surface water cools in the fall and winter, its density decreases below that of the deep water and vertical mixing occurs. The second important factor governing the extent to which reduction in deep water occurs is nutrient levels. Nutrient levels limit the amount of production of organic carbon by photosynthesizers (in lakes, phosphorus concentrations are usually limiting; in the oceans, nitrate and micronutrients such as iron appear to be limiting). The availability of organic carbon in turn controls *biological oxygen demand* (BOD). In water with high nutrient levels there is a high flux of organic carbon to deep waters and hence higher BOD.

In lakes with high nutrient levels, the temperature stratification described above can lead to a situation where dissolved oxygen is present in the epilimnion and absent in the hypolimnion. Regions where dissolved oxygen is present are termed *oxic*, those where sulfide or methane are present are called *anoxic*. Regions of intermediate pϵ are called *suboxic*. Conditions where oxygen levels are insufficient to sustain life are called *hypoxic* (or dysoxic): dissolved oxygen below 2 mg/l are lethal to fish and most fish avoid areas where dissolved oxygen is below 4 mg/l, although invertebrates may survive in lower dissolved oxygen conditions. Lakes are where suboxic or anoxic conditions exist as a result of high biological productivity are said to be *eutrophic*. This occurs naturally in many bodies of water, particularly in the tropics where stratification is permanent. Although not a lake, the Black Sea is the largest body of water where anoxic conditions prevail at depth and it is entirely a natural phenomenon. It can also occur, however, as a result of the addition of pollutants such as sewage or phosphate- or nitrate-rich agricultural runoff. Addition of the nutrients enhances productivity and availability of organic carbon, and ultimately BOD. Where this occurs naturally, ecosystems have adapted to this circumstance and only anaerobic bacteria are found in the hypolimnion. When it results from pollution, it can be catastrophic for macrofauna such as fish. We will consider one example of eutrophication and hypoxia, Lake Erie, its causes, and remediation efforts in Chapter 15.

13.2.3 Redox buffers and transition metal chemistry

The behavior of transition metals in aqueous solutions and solids in equilibrium with them is particularly dependent on redox state. Many transition metals have more than one valence state within the range of pϵ of water. In a number of cases, the metal is much more soluble in one valence state than in others. The best examples of this behavior are provided by iron and manganese, both of which are much more soluble in their reduced (Fe^{2+}, Mn^{2+}) than in oxidized (Fe^{3+}, Mn^{4+}) forms. Redox conditions are thus a strong control on the concentrations of these elements in natural waters.

Because of the low solubility of their oxidized forms, the concentrations of Fe, Mn, and similar metals are quite low under "normal" conditions, i.e., high pϵ and near-neutral pH. There are two common circumstances where higher Fe and Mn concentrations in water occur. The first is when sulfide ores are exposed by mining and oxidized to sulfate, e.g.:

$$2FeS_{2(s)} + 2H_2O + 7O_2 \rightleftharpoons 4H^+ + 4SO_4^{2-}$$

This can dramatically lower the pH of streams draining such areas. The lower pH, in turn, allows higher concentrations of dissolved metals (e.g., Figure 13.2), even under oxidizing conditions; we'll examine the example of the Rio Tinto in Chapter 15. The second circumstance where higher Fe and Mn concentrations occur is under suboxic or anoxic conditions that may occur in deep waters of lakes and seas as well as sediment

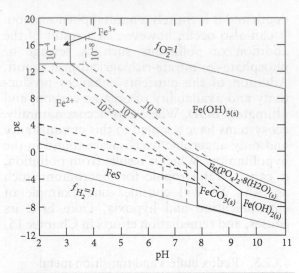

Figure 13.2 Contours of dissolved Fe activity as a function of $p\varepsilon$ and pH computed for P_{CO_2} = 10^{-2}, $\Sigma S = 10^{-4}$, $\Sigma PO_4 = 10^{-4}$, and $\Sigma Fe =$ 10^{-6} (solid line; dashed lines show other Fe concentrations), conditions that often occur in soils and natural waters, although sulfur concentrations can be higher in acid mine drainage. Although pyrite (FeS_2) is generally the thermodynamically stable sulfide (FeS_2), mackinawite (FeS) is kinetically favored and forms metastably in place of pyrite. Stability fields for ferric hydroxide ($Fe(OH)_3$ (which is also unstable and is replaced by goethite, FeO(OH), or ferrihydrite, $Fe_{10}O_{14}(OH)_2$), siderite ($FeCO_3$), vivianite $Fe(PO_4)_2 \cdot 8(H_2O)$, and ferrous hydroxide are also shown. The latter is unstable and reacts to form "green rust" as a result of incorporation of foreign anions and water molecules between brucite-like layers.

pore waters. Under these circumstances Fe and Mn are reduced to their soluble forms, allowing much higher concentrations.

In cases where precipitation or dissolution involves a change in valence or oxidation

state, the solubility product must include $p\varepsilon$ or some other redox couple, e.g.:

$$Fe_2O_3 + 6H^+ + 2e^- \leftrightharpoons 2Fe^{2+}$$

$$+ 3H_2O \quad K = \frac{a_{Fe^{2+}}^2}{a_{H^+}^6 a_{e^-}^2} \quad (13.1)$$

In Chapter 3, we noted that $p\varepsilon$ is often difficult to determine. One approach to the problem is to assume the redox state in the solution is controlled by a specific reaction. The controlling redox reactions will be those involving the most abundant species; very often, this is sulfate reduction:

$$\frac{1}{8}SO_4^{2-} + \frac{5}{4}H^+ + e^- \leftrightharpoons \frac{1}{8}H_2S + \frac{1}{2}H_2O$$
$$(13.2)$$

in which case $p\varepsilon$ is given by:

$$pe = pK - \frac{1}{8}\log\left(\frac{a_{H_2S}}{a_{SO_4^{2-}}}\right) - \frac{5}{4}pH \quad (13.3)$$

Under the assumption that this reaction controls the redox state of the solution, electrons may be eliminated from other redox reactions by substituting the above expression. For example, iron redox equilibrium may be written as:

$$\frac{1}{8}SO_4^{2-} + \frac{5}{4}H^+ + Fe^{2+}$$

$$\leftrightharpoons \frac{1}{8}H_2S + \frac{1}{2}H_2O + Fe^{3+} \quad (13.4)$$

In this sense, the $p\varepsilon$ of most natural waters will be controlled by a *redox buffer*, a concept we considered in Chapter 3. Example 13.1 illustrates this approach.

Example 13.1 Redox state of lake water

Consider water of the hypolimnion of a lake in which all oxygen has been consumed and with the following initial composition: $SO_4^{2-} = 2 \times 10^{-4}$ M, $\Sigma Fe^{3+} = 10^{-6}$ M, Alk = 4×10^{-4} eq/L, $\Sigma CO_2 =$ 1.0×10^{-3} M, $\Sigma CH_2O = 2 \times 10^{-4}$ M (here again we use CH_2O as shorthand for organic matter), pH = 6.3. Determine the pH, $p\varepsilon$, and speciation of sulfur and iron when all organic matter is consumed and redox equilibrium is achieved. The first dissociation constant of H_2S is 10^{-7}.

Answer: Let's first consider the redox reactions involved. The species involved in redox reactions will be those of carbon, sulfur, and iron. The concentration of iron is small, so its oxidation state will reflect, rather than control, that of the solution. Thus, oxidation of organic matter will occur though reduction of sulfate. We can express this by combining reactions 7 and 11 in Table 13.1 (we see from the dissociation constant that H_2S will be the dominant sulfate species at pH below 7, so we chose reaction 7 rather than reaction 8):

$$\frac{1}{2}SO_4^{2-} + H^+ + CH_2O \rightleftharpoons \frac{1}{2}H_2S + \frac{1}{2}H_2O + CO_2 \quad K = 10^{10.54} \tag{13.5}$$

From the magnitude of the equilibrium constant (obtained from the $p\varepsilon°$ values in Table 13.1), we can see that right side of this reaction is strongly favored. Since sulfate is present in excess of organic matter, this means sulfate will be reduced until all organic matter is consumed, which will leave equimolar concentrations of sulfate and sulfide (10^{-4} M each). Therefore, the redox state of the system will be governed by that of sulfur. The redox state of iron can then be related to that of sulfur using reaction 13.4, for which we calculate an equilibrium constant of $10^{-10.75}$ from Table 13.1.

The next problem we face is that of choosing components. As usual, we chose H^+ as one component (and implicitly H_2O as another). We will also want to choose a sulfur, carbon, and iron species as a component, but which ones? We could choose the electron as a component, but consistent with our conclusion above that the redox state of the system is governed by that of sulfur, a better choice is to choose both sulfate and sulfide, specifically H_2S, as components. We can also see from Table 13.1 that Fe should be largely reduced, so we chose Fe^{2+} as the iron species. pH will be largely controlled by carbonate species, since these are more than an order of magnitude more abundant than sulfate species; oxidation of organic matter will increase the concentration of CO_2, which will lower pH slightly. In Figure 6.1, we can see that at pH below 6.4, H_2CO_3 will be the dominant carbonate species, so we chose CO_2 as our component and made H_2CO_3 as the sum of CO_2 and H_2O (the latter, along with H^+ are implicit components; see Chapter 6). Our components are therefore H+, SO_4^-, CO_2, Fe^{2+}, and H_2S. The species of interest will include H^+, OH^-, SO_4^{2-}, H_2S, HS^-, H_2CO_3, HCO_3^-, as well as the various species of Fe (Fe^{2+}, Fe^{3+}, $Fe(OH)^{2+}$, $Fe(OH)_2^+$, $Fe(OH)_3$ (we assume that CO_3^{2-}, HSO_4^- and S^{2-} concentrations are negligible at this pH and neglect them throughout).

Our next step is to determine pH. For *TOTH* we have:

$$TOTH = [H^+] - [OH^-] - [HCO_3^-] - [HS^-] + \frac{5}{4}\Sigma Fe^{3+} \tag{13.6}$$

The presence of the Fe^{3+} term may at first be confusing. To understand why it occurs, we can use equation 13.4 to express Fe^{3+} as the algebraic sum of our components:

$$Fe^{3+} = \frac{1}{8}SO_4^{2-} + \frac{5}{4}H^+ + Fe^{2+} - \frac{1}{8}H_2S - \frac{1}{2}H_2O \tag{13.7}$$

The first four terms on the right-hand side of equation 13.6 are simply alkalinity plus additional CO_2 produced by oxidation of organic matter, so 13.6 may be rewritten as:

$$TOTH \cong \frac{5}{4}\Sigma Fe^{3+} - Alk - [CH_2O] \tag{13.8}$$

Inspecting equation 13.6, we see that HCO_3^- is by far the largest term. Furthermore, the Fe term in equation 13.8 is negligible, so we have:

$$TOTH \cong [HCO_3^-] = 6 \times 10^{-4} \tag{13.9}$$

The conservation equation for carbonate is:

$$\Sigma H_2CO_3 = [H_2CO_3] + [HCO_3^-] = \Sigma CO_2 + \Sigma CH_2O = 1.3 \times 10^{-3}$$

Hence:

$$H_2CO_3 = \Sigma H_2CO_3 - HCO_3^- = (1.3 - 0.6) \times 10^{-3}$$

We can use this to calculate pH since:

$$K = \frac{[HCO_3^-][H^+]}{[H_2CO_3]} = 10^{6.25}$$

Solving for $[H^+]$ and substituting values, we find that pH = 6.28.

For the conservation equation for sulfate, we will have to include terms for both Fe^{3+} (equation 13.4) and organic matter. Writing organic matter as the algebraic sum of our components we have:

$$CH_2O = \frac{1}{2}H_2S + H_2O + CO_2 - \frac{1}{2}SO_4^{2-} - H^+$$

The amount of sulfate present will be that originally present less that used to oxidize organic matter. The only other oxidant present in the system is ferric iron, so the amount of sulfide used to oxidize organic matter will be the total organic matter less the amount of ferric iron initially present. The sulfate conservation equation is then:

$$\Sigma SO_4 = [SO_4^-] - \frac{1}{2}\Sigma CH_2O + \frac{1}{8}\Sigma Fe^{3+} \cong 1.0 \times 10^{-4}M$$

(the Fe term is again negligible). The amount of sulfide present will be the amount created by oxidation of organic matter, less the amount of organic matter oxidized by iron, so the sulfide conservation equation is:

$$\Sigma H_2S = H_2S + HS^- = \frac{1}{2}\Sigma CH_2O - \frac{1}{8}\Sigma Fe^{3+} = 1.0 \times 10^{-4}M \qquad (13.10)$$

We now want to calculate the speciation of sulfide. We have

$$\Sigma H_2S = H_2S + HS^- = 1.0 \times 10^{-4}M$$

and

$$K_{1-H_2S} = \frac{[H^+][HS^-]}{[H_2S]} = 8.9 \times 10^{-8}$$

Solving these two equations, we have:

$$[HS^-] = \frac{10^{-4}}{10^{7.05}10^{-6.28}} = 1.45 \times 10^{-5}$$

The concentration of H_2S is then easily calculated as 8.55×10^{-5} M. We can now calculate the $p\varepsilon$ of the solution by substituting the above values and the $p\varepsilon°$ for reaction 7 in Table 13.1 into equation 13.3. Doing so, we find $p\varepsilon$ is –2.59.

Calculating the Fe^{3+}/Fe^{2+} ratio and the concentration of Fe^{3+} is left to the reader as problem 13.8.

The development of anoxic conditions leads to an interesting cycling of iron and manganese within the water column. Below the oxic–anoxic boundary, Mn and Fe in particulates are reduced and dissolved. The metals then diffuse upward to the oxic–anoxic boundary where they again are oxidized and precipitate. The particulates then migrate downward, are reduced, and the cycle begins again, a process sometimes referred to as an *iron (or manganese) shuttle*.

A related phenomenon can occur within sediments. Even where anoxic conditions are not achieved within the water column, they can be achieved within the underlying sediment. Indeed, this will occur where the burial rate of organic matter is high enough to exceed the supply of oxygen. Figure 13.3 shows an example, namely a sediment core from southern Lake Michigan studied by Robbins and Callender (1975). The sediment contains about 2% organic carbon in the upper few centimeters, which decreases by a factor of 3 down core. The concentration of acid-extractable Mn in the solid phase (Figure 13.3a), presumably surface-bound Mn and Mn oxides, is constant at about 540 ppm in the upper 6 cm, but decreases rapidly to about 400 ppm by 12 cm. The concentration of dissolved Mn in the pore water increases from about 0.5 ppm to a maximum of 1.35 ppm at 5 cm and then subsequently decreases (Figure 13.3) to a constant value of about 0.6 ppm in the bottom half of the core.

Because the Lake Michigan region is heavily populated, it is tempting to interpret the data in Figure 13.3, particularly the increase in acid-extractable Mn near the core top, as being a result of recent pollution. However, Robbins and Callender (1975) demonstrated that the data could be explained with a simple steady-state diagenetic model involving Mn reduction, diffusion, and reprecipitation (as $MnCO_3$). In Chapter 5, we derived the diagenetic equation:

$$\left(\frac{\partial C_i}{\partial t}\right) = -\left(\frac{\partial F_i}{\partial x}\right)_t + \sum R_i \qquad (5.173)$$

The first term on the right is the change in total vertical flux with depth, the second is the sum of rates of all reactions occurring. There are two potential flux terms in this case, pore water advection (due to compaction) and diffusion. There are also several reactions occurring: dissolution or desorption associated with reduction and a precipitation reaction. If the system is at steady state, then $\partial c/\partial t = 0$. Assuming steady-state, Robbins and Callender (1975) derived the following version of the diagenetic equation:

$$\phi D \frac{d^2 c}{dz^2} - v \frac{dc}{dz} - \phi k_1 (c - c_f) + \phi k_0(z) = 0$$

$$(13.11)$$

where ϕ is porosity (assumed to be 0.8), D is the diffusion coefficient, v is the advective

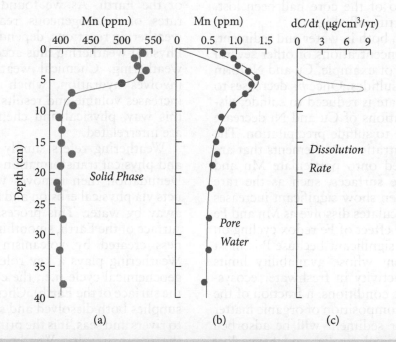

(a) (b) (c)

Figure 13.3 (a) Concentration of acid-leachable Mn in Lake Michigan sediment as a function of depth. (b) Dissolved Mn in pore water from the same sediment core. Solid line shows the dissolution–diffusion–reprecipitation model of Robbins and Calender (1975) constrained to pass through 0 concentration at 0 depth. Dashed line shows the model when this constraint is removed. (c) Dissolution rate of solid Mn calculated from rate of change of concentration of acid-leachable Mn and used to produce the model in (b). From Robbins and Calender (1975).

velocity (−0.2 cm/yr), k_1 is the rate constant for reprecipitation reactions, and k_0 is the dissolution rate (expressed as a function of depth). The first term is the diffusive term, the second the advective, the third the rate of reprecipitation, and the fourth is the dissolution rate. The dissolution rate is related to the change in concentration of acid-extractable Mn. The last term may then be written as:

$$\phi k_0(z) = \phi \frac{R}{\phi} \frac{\partial c_s}{\partial x} \qquad (13.12)$$

where R is the rate of addition and/or removal of Mn to or from the pore water (g/cm^2/yr) and c_s is the concentration in the solid. Using least squares regression, Robbins and Callender found that the parameters that best fit the data were $D = 9 \times 10^{-7}$ cm^2/sec (30 cm^2/yr), $k_1 = 1$ yr^{-1}, and $c_f = 0.5$ ppm. The solid line in Figure 13.3b represents the prediction of equation 13.11 using these values and assuming c_0 (porewater concentration at the surface) is 0. The dashed line in Figure 13.3b assumes $c_0 = 0.6$ ppm. The latter is too high, as c_0 should be the same concentration as lake water. Robbins and Callender speculated that the top cm or so of the core had been lost, resulting in an artificially high c_0.

Redox cycling, both in water and sediment, can affect the concentrations of other several other elements. For example, Cu and Ni form highly insoluble sulfides. Once $p\varepsilon$ decreases to levels where sulfate is reduced to sulfide, dissolved concentrations of Cu and Ni decrease dramatically due to sulfide precipitation. The dissolved concentrations of elements that are strongly absorbed onto particulate Mn and Fe oxyhydroxide surfaces, such as the rare earths and P, often show significant increases when these particulates dissolve as Mn and Fe are reduced. The effect of Fe redox cycling on P is particularly significant because P is most often the nutrient whose availability limits biological productivity in freshwater ecosystems. Under oxic conditions, a fraction of the P released by decomposition of organic matter in deep water or sediment will be adsorbed by particles (particularly Fe) and hence lost from the ecosystem to sediment. If conditions become anoxic, iron dissolves and adsorbed P is released into solution, where it can again become available to the biota. As a result, lakes that become eutrophic due to P pollution can remain so long after the pollution ceases because P is simply internally recycled under the prevailing anoxic conditions. Worse yet, once conditions become anoxic, nonanthropogenic P can be released from the sediment, leading to higher biological production and more severe anoxia.

13.3 WEATHERING, SOILS, AND BIOGEOCHEMICAL CYCLING

Weathering is the process by which rock is broken down into relatively fine solids (soil or sediment particles) and dissolved components and ultimately removed from the surface. Physical weathering involves fracturing of rock through processes such as frost and root weathering, thermal cycling, pressure release, abrasion, and expansion associated with hydration. An important aspect of physical weathering is that it increases surface area. Chemical weathering, which will be our focus, could be more precisely described as the process by which rocks originally formed at higher temperatures come to equilibrium with water at the temperature, pressure, and oxygen fugacity prevailing at the surface of the Earth. As we found in Section 6.5, rates of heterogeneous reactions, such as weathering reactions, depend on surface area. Physical weathering thus accelerates chemical weathering. Chemical weathering generally involves hydration, which weakens it and increases volume and results in fracturing. In this way, physical and chemical weathering are interrelated.

Weathering refers simply to the chemical and physical transformation of rock in place. Denudation then removes weathering products via physical erosion and as solutes carried away by water. This process transforms the surface of the Earth, smoothing out the roughness created by volcanism and tectonism. Weathering plays a key role in the *exogenic* geochemical cycle (i.e., the cycle operating at the surface of the Earth). Chemical weathering supplies both dissolved and suspended matter to rivers and seas. It is the principal reason that the ocean is salty. Weathering also supplies nutrients to the biota in the form of dissolved components in the soil solution; without weathering terrestrial life would be far different and far more limited. Weathering can be an important source of ores. Bauxite, which is the primary ore of Al, is the product of extreme weathering that leaves a soil residue containing very high concentrations of aluminum oxides and hydroxides. Laterites developed on peridotites can also be important ores for Ni and Co; indeed, laterites are the source of about 50% of world Ni production and 80% of world heavy rare earth resources. We'll address these topics in Chapter 15.

In previous chapters, we examined many important weathering reactions from thermodynamic and kinetic perspectives. In this chapter, we will step back to look at weathering on a broader scale and examine weathering in nature and the interrelationships among chemical weathering, biological processes, and soil formation. We then discuss in some detail the question of what controls weathering rates. Finally, we look at the composition of rivers and streams.

Most rock at the surface of the Earth is overlain by a thin veneer of a fine material known as a regolith. It consists of a mixture of weathering products and organic matter; the biologically active part of the regolith is referred to as soil. Most weathering reactions occur within the soil, or at the interface between the soil and bedrock, so it is worth briefly considering soil and its development. Undisturbed, mature soils are layered, forming a soil profile (Figure 13.4). The nature of these layers varies, depending on climate (temperature, amount of precipitation, etc.), vegetation (which, in turn, depends largely on climate), time, and the nature of the underlying rock. Consequently, no two soil profiles will be identical. What follows is a general description of an idealized soil profile. Real profiles are likely to differ in some respects from this.

13.3.1   Soil profiles

The uppermost soil layer, referred to as the O *horizon*, consists entirely, or nearly so, of organic material whose state of decomposition increases downward and is generally no more than a few cm thick. This layer is best developed in forested regions or waterlogged soils depleted in O_2 where decomposition is slow. In other regions it may be incompletely developed or entirely absent.

Below this organic layer lies the upper mineral soil, designated as the *A horizon* or the zone of removal, which ranges in thickness

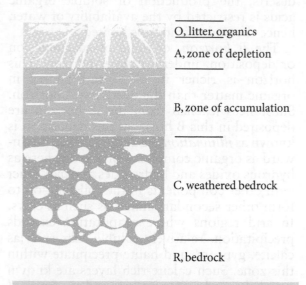

Figure 13.4   Soil profile, illustrating the O, A, B, and C horizons described in the text. Not all soils conform to this pattern.

from several centimeters to a meter or more. In addition to a variety of minerals, this layer contains a substantial organic fraction, which is dominated by an amorphous mixture of organic substances collectively called *humus* (Chapter 12). Weathering reactions in this layer produce a soil solution rich in silica and alkali and alkaline earth cations that percolates downward into the underlying layer. Such movements of components with soil are known as *translocations*; removal of components from soil horizons is known as *eluviation*. In temperate forested regions, where rainfall is high and organic decomposition slow, Fe and Al released by weathering reactions can be complexed by organic acids and carried downward into the underlying layer. The downward transport of Fe and Al is known as *podzolization*. In some tropical regions, organic decomposition can be sufficiently rapid and complete that there is little available soluble organic acid to complex and transport Fe and Al, so podzolization may not occur. As a result, Fe and Al accumulate in the A horizon as hydrous oxides and hydroxides. Where leaching is particularly strong, the A horizon may be underlain by a thin whitish highly leached, or *eluviated*, layer known as the *E horizon*, enriched in highly resistant minerals such as quartz. In grasslands and deserts, the production of soluble organic acids is restricted by the availability of water, hence the podzolization is limited.

The *B horizon*, or zone of accumulation or deposition, underlies the A horizon. This horizon is richer in clays and poorer in organic matter than the overlying A horizon. Substances leached from the A horizon are deposited in this B horizon; this deposition is known as *illluviation*. Fe and Al carried downward as organic complexes precipitate here as hydrous oxides and hydroxides and may react with other components in the soil solution to form other secondary minerals such as clays. In arid regions where evaporation exceeds precipitation, relatively soluble salts such as calcite, gypsum, and halite precipitate within this zone. Such calcite-rich layers are known as *caliche*, and are typically found at a depth of 30 to 70 cm. The clay-rich nature of the B horizon, particularly when cemented by precipitated calcite or Fe oxides, can greatly restrict permeability of this layer. Such impermeable layers are sometimes referred to as *hardpan*.

In tropical regions, where weathering is intense and has continued for millions of years in the absence of disturbances such as glaciation or tectonics, the soil profile may be up to 100 m thick. In the absence of podzolization, there is often little distinction between the A and B horizons. Base cations are nearly completely leached in tropical soils, leaving a soil dominated by minerals such as kaolinite, gibbsite, and Fe oxides and hydroxides. Such soils are called *laterites*. In the most extreme cases, SiO_2 may be nearly completely leached as well, leaving a soil dominated by gibbsite. The ratio of SiO_2 to $Al_2O_3 + Fe_2O_3$ (the latter collectively called the sesquioxides) is a useful index of the intensity of weathering within the soil as well as to the extent of podzolization. Typical values for the A and B horizons of several climatic regions are summarized in Table 13.2. Often, separate layers can be recognized within the B and the other horizons and these are designated B_1, B_2, etc. downward.

The *C horizon* underlies the B horizon and directly overlies the bedrock. It consists of partly weathered rock, often only coarsely fragmented, and its direct weathering

Table 13.2 $SiO_2/(Al_2O_3 + Fe_2O_3)$ ratios of soils.

	A horizon	B horizon
Boreal	9.3	6.7
Cool-temperate	4.07	2.28
Warm-temperate	3.77	3.15
Tropical	1.47	1.61

From Schlesinger (1991).

products. In soils that develop directly from local materials, it is mineralogically related to the underlying bedrock and has little organic matter. When readily weatherable silicates have been largely or wholly replaced *in situ* by clays and oxides but textures and structures are sufficiently well preserved that the nature of the original parent can be recognized, this material is called *saprolite*. Alternatively, soil may develop on volcanic ash, glacial till, or material transported by wind (loess) or water (alluvium). Loess up to 100 m thick was deposited in some northern regions during the Pleistocene glaciations. Soils in deserts surrounded by mountains and in river floodplains generally develop from alluvium rather than underlying bedrock. In such cases, the C layer will be unrelated to underlying bedrock or may be underlain by fossil soils (paleosols). While organic acids dominate weathering reactions in the upper part of the soil profile, carbonic acid, produced by respiration within the soil and transported downward by percolating water, is largely responsible for weathering in the C horizon.

The full development of an "equilibrium" soil profile requires time, thousands to hundreds of thousands of years. How much time depends largely on climate: soil development is most rapid in areas of warm temperature and high rainfall. Regions that have been disturbed in the recent geologic past may show incomplete soil formation, or soils uncharacteristic of the present climate. For example, many desert soils contain clay layers developed during wetter Pleistocene times. Soils in recently glaciated areas of North America and Northern Europe remain thin and immature, even though 12,000 years have passed since the last glacial retreat. Floods and landslides are other natural processes that can interrupt soil development. Soil profiles are also be disturbed by agriculture and the resulting increase in erosion and other anthropogenic effects such as acid rain. Thus, actual soil profiles often deviate from that illustrated in Figure 13.4.

13.3.2 Chemical cycling in soils

An example of chemical variations in a soil profile is shown in Figure 13.5. This soil,

developed on a beach terrace in Mendocino County, California, was studied by Brimhall and Dietrich (1987) and illustrates processes occurring in a podzol (a soil that has experienced podzolization). The parent material is uplifted Pleistocene beach sand and is poor in most elements other than Si compared with common rocks. The soil profile consists

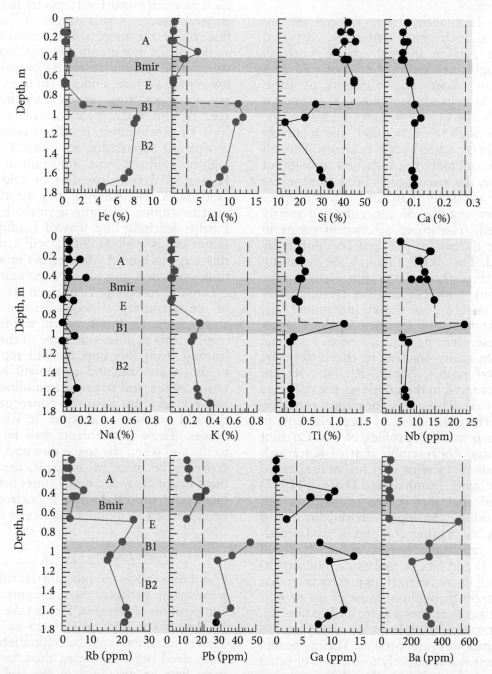

Figure 13.5 Concentration profiles in a podzol developed on a Pleistocene beach terrace in Mendocino, California. Dashed line shows the concentration in the parent beach sand. Soil horizons are simplified from those identified by Brimhall and Dietrich (1987). Data from Brimhall and Dietrich (1987).

of an A horizon, the upper part of which is densely rooted and rich in organic matter, underlain by a transitional layer that Brimhall and Dietrich (1987) labeled Bmir. This is, in turn, underlain by a white and red banded E horizon, which is, in turn, underlain by the B horizon, which Brimhall and Dietrich (1987) divided into B1 and B2 layers based on appearance.

Figure 13.5 shows that Fe and Al are slightly, though not uniformly, depleted from the surface through the E horizon, and strongly enriched in the B horizon. This reflects the downward transport of these elements as organic complexes. Si is present in roughly the same concentrations as in the parent through the E horizon, and relatively depleted in the underlying B horizons. Though Si has undoubtedly been leached throughout the profile, the leaching has been more severe for other elements than for Si in the upper profile, leaving the Si concentration nearly unchanged. The lower Si concentrations in the lower horizon in part reflects dilution by Fe and Al. The alkalis and alkaline earths are depleted throughout the profile relative to the parent, but while Na and Ca are uniformly depleted (as is Sr, not shown in Figure 13.5), K, Rb, and Ba are enriched in the B horizon. These elements are all readily soluble and hence easily leached in the upper part of the soil profile, but K, Rb, and Ba are readily accepted in the interlayer sites of clays (Section 6.5). Thus, clay formation probably explains the enrichment of these elements in the B horizon. The profiles of Pb, Cu (not shown), and Ga resemble that of K (though they are also somewhat enriched at the base of the A horizon). Brimhall and Dietrich (1987) speculated that this is due to oxidation of metal sulfides in the upper horizon, downward transport as organic complexes, and reduction and reprecipitation as sulfides in the B horizon. Ti and Nb, as well as Zr (not shown) are present in concentrations greater than that of the parent throughout most of the profile. These elements are not soluble and do not usually form soluble complexes hence they are virtually immobile in soil profiles. Their apparent enrichment is due entirely to the loss of other elements. Indeed, Brimhall and Dietrich use Nb and Ti concentrations to calculate that there has been a mass loss, through leaching, of 60% in the upper part of the soil profile.

From the preceding discussion, we can infer that considerable chemical cycling occurs within soils. The initial stages of weathering occur within the C horizon, where the most unstable primary minerals (e.g., feldspars, sulfides, carbonates, ferromagnesian silicates such as olivine, pyroxenes, amphiboles) undergo reaction. More resistant minerals such as quartz and Fe–Ti oxides may remain largely intact at this stage. A substantial fraction of the more soluble elements, such as the alkalis and alkaline earths, may be lost to solution at this stage. As erosion slowly lowers the surface, minerals in the C horizon are, in effect, transported into the B horizon. At this stage, new secondary minerals form by precipitation from and reaction with downward percolating solutions. In areas of high rainfall, the most important of these are Fe and Al oxyhydroxides; in arid regions, carbonates and sulfates may be the dominant new minerals forming in the B horizon. Further leaching and loss of readily soluble elements, as well as SiO_2, will continue in this stage in humid climates. As erosion continues to lower the surface, the material will eventually reach the A horizon. Here, many of the secondary minerals formed in the C and B horizons break down, releasing their components to solution. Some of the material leached from this horizon will reprecipitate in the B horizon and some will be lost to groundwater and streams. In addition, plants actively take up dissolved constituents from the soil solution and store it within their tissues. These constituents will be returned to the soil when the tissue dies and is broken down by bacteria. In principle, then, an element might be cycled many times between the biota, and the O, A, and B horizons before being carried away in groundwater flow.

13.3.3   Biogeochemical cycling

The biota plays a substantial role in the weathering process and in controlling the composition of streams. Plants take up a host of elements in inorganic form as nutrients. These nutrients may be ultimately derived from dead organic matter, dust, or rock but must first be dissolved in the soil solution. P_{CO_2} is substantially higher in soil than in the atmosphere; this is a direct result of respiration of organisms in the soil. This higher P_{CO_2}

lowers pH and hence accelerates weathering reactions. Organic acids produced by plants and bacteria have the same effect. In addition, many weathering reactions, including, but not limited to, redox reactions, may be directly catalyzed by soil bacteria. The biota will thus influence both soil and water chemistry within a watershed and can be a significant reservoir for some elements.

Living organisms consist of a bewildering variety of organic compounds (some of these were discuss is Chapter 12). From a geochemical perspective, it is often satisfactory to approximate the composition of the biomass as CH_2O (e.g., the composition of glucose, a simple sugar, is $C_6H_{12}O_6$). A better approximation for the composition of land plants would be $C_{1200}H_{1900}O_{900}N_{25}P_2S_1$ (Berner and Berner, 1996). A great many other nutrients*, however, are essential for life (e.g., Mg and Fe are essential photosynthesis; Mo is essential for nitrogen fixation) and are taken up by plants and microbes in smaller amounts. These can be divided up into *macronutrients*, which occur in plants at concentrations in excess of 500 ppm and include N, P, K, Ca, Mg, and S, and *micronutrients*, which occur at lower concentrations and include, among others, B, Fe, Mn, Cu, Zn, Mo, Co, and Cl. In all, life utilizes a substantial fraction of the periodic table. Other elements are also taken up by plants and stored in tissue, even though they play no biochemical role (as far as we know), simply because plants cannot discriminate sufficiently against them. Table 13.3 lists the concentrations of the elements in average dried plant matter. The actual composition of plants varies widely, however; grasses, for example, can contain over 1% SiO_2 (>4600 ppm Si).

There are four sources of nutrients in an ecosystem: the atmosphere, dead organic matter, water, and rock. The atmosphere is the obvious direct source of CO_2 and O_2 and the indirect source of N and H_2O in terrestrial ecosystems. However, it may also be the direct or indirect source of a number of other nutrients, which arrive either in atmospheric dust or dissolved in rain. Along coasts, rain and fog is particularly enriched in aerosols (sea salt finds its way into the atmosphere when waves

Table 13.3 Elemental composition of dried plant matter.

Element	Percent	Element	ppm
C	49.65	V	1
N	0.92	Mn	400
O	43.2	Cr	2.4
Total	93.77	Fe	500
Element	ppm	Co	0.4
Li	0.1	Ni	3
B	5	Cu	9
Na	200	Zn	70
Mg	700	Se	0.1
Al	20	Rb	2
Si	1500	Sr	20
P	700	Mo	0.65
S	500	Ag	0.05
K	3000	Ba	30
Ca	5000	U	0.05
Ti	2		

From Brooks (1972).

break throwing up fine droplets that subsequently evaporate) and this can be a significant source of those components. Plants can take up some of these atmosphere-delivered nutrients directly through foliage; most, however, cycle through the soil solution and are taken up by roots, which is the primary source of nutrients. Because equilibrium between surface adsorbed and dissolved species is achieved relatively quickly, elements adsorbed on the surfaces of oxides, clays, and organic solids represent an intermediate reservoir of nutrients. For example, phosphorus, often the growth-limiting nutrient, is readily adsorbed on the surface of iron oxides and hydroxides. For this reason, the surface properties of soil particles are an important influence on soil fertility. Even in relatively fertile soils the concentration of key nutrients such as phosphorus may be effectively zero in the soil immediately adjacent roots, and the rate of delivery to plant may be limited by diffusion.

In most ecosystems, particularly mature ones, detritus, that is dead organic matter, is the most important source of nutrients. In the Hubbard Brook Experimental Forest in New Hampshire, USA, for example, this recycling supplies over 80% of the required P, K, Ca, and Mg (Schlesinger, 1991). For the most part,

---

* A nutrient is an element or compound essential to life that cannot be synthesized by the organism and therefore must be obtained from an external source.

this recycling occurs as leaf tissue dies, falls to the ground, and decomposes in an annual cycle. Additionally, some fraction of nutrients is recycled more directly. Nutrients may be leached from leaves by precipitation, another *translocation*. Plants also recycle nutrients internally by withdrawing them from leaves and stems before the annual loss of this material and storing them for use in the following season. For this reason, the concentration of nutrients in litterfall is lower than in living tissue. Not surprisingly, the fraction of nutrients recycled in this way, and overall nutrient use efficiency, is higher in plants living on nutrient-poor soils (Schlesinger, 1991).

Rainwater passing through the vegetation canopy will carry not only nutrients leached from foliage, but also compounds dissolved from dust and aerosols deposited on leaves, known as *dry deposition*. Fog and mist will also deposit solutes on plant leaves. The term *occult* deposition refers to both dry deposition and deposition from mist and fog. The total flux of solutes dissolved from leaf surfaces, including both the occult deposition and translocation fluxes, and carried to the soil by precipitation is called *throughfall*, and can be quite significant in regions where there is a high aerosol flux. Such regions may be either those downwind from heavily populated areas, where the atmosphere contains high levels of nitrate and sulfate from fossilfuel burning, or arid regions, where there is abundant dust in the atmosphere. Table 13.4 compares the concentration of nutrients measured in throughfall and bulk precipitation and demonstrates the importance of translocation and occult deposition in the Vosage Mountains of France.

Dry deposition can be a particularly important source of nutrients in some cases. An extreme example has been documented in Hawaiian soils by Chadwick et al. (1999) and Kurtz et al. (2001). Marine aerosols are the dominant source of nutrients such as K, Ca, Si (an important nutrient for grasses), and P in the oldest, most intensely weathered soils. Sr and Nd isotope ratios in these soils are indistinguishable from average continental crust and with much higher $^{87}Sr/^{86}Sr$ and lower ε_{Nd} than Hawaiian lavas upon which the soils developed (Chapter 11). Furthermore, quartz and mica, which are not present in Hawaiian basalts, become the dominant nonclay minerals in these older soils.

The biota affects the composition of soil and stream water in another way as well. Some fraction of soil water taken up by roots is ultimately lost from the plant to the atmosphere through leaf stomata (the openings allowing CO_2 into the leaf). This loss of water is called *transpiration*. Water lost through transpiration and that lost by direct evaporation from the ground surface are often collectively called *evapotranspiration*. As one might expect, transpiration varies seasonally: transpiration is high in spring and summer and minimal in winter and also depends on climate. Evapotranspiration concentrates dissolved solids in soil and stream water.

The rate at which *litter*, the dead vegetation lying above the O horizon, decomposes, and hence "turns over" depends strongly on climate. Table 13.5 lists the mean residence times of bulk organic matter and nutrients in the surface litter of forest ecosystems, which ranges from hundreds of years in boreal forests to a year or less in tropical

Table 13.4 Concentrations in bulk precipitation and throughfall in the Vosage, France.

	NH$_4$	Na	K	Mg	Ca	H$^+$	Cl	NO$_3$	SO$_4$
Concentration (µeq/L)									
Bulk precipitation	19.1	10.0	2.8	4.5	11.9	33.9	12.5	24.1	41.5
Throughfall	36.9	46.4	52.7	17.8	65.5	114.8	63.4	48.3	185.0
Fluxes (moles/ha/y)									
Bulk precipitation	270	142	39	32	84	480	177	340	290
Throughfall	385	484	550	93	642	1197	661	817	966
Difference	115	342	511	61	558	717	484	477	676
Occult precipitation	115	342	102	31	206	1282	484	477	676
Translocation	0	0	409	30	352	−565	0	0	0

Data from Probst et al. (1990).

Table 13.5 Residence times of organic matter and nutrients in forest litter (years).

Region	Organic matter	N	P	K	Ca	Mg
Boreal forest	353	230	324	94	149	455
Temperate forest						
coniferous	17	17.9	15.3	2.2	5.9	12.9
deciduous	4	5.5	5.8	1.3	3	3.4
Mediterranean	3.8	4.2	3.6	1.4	5	2.8
Tropical rainforest	0.4	2	1.6	0.7	1.5	1.1

From Schlesinger (1991).

rainforests. K is recycled more rapidly than bulk organic matter, but recycling times for other nutrients are generally comparable to that of bulk organic matter. Although animals, particularly those living in the soil such as termites and worms, play a role in organic decomposition, most of it is carried out by soil fungi and bacteria. These soil microbes can comprise up to 5% of the organic carbon in soils, with fungi dominating over bacteria in well-drained soils. Decomposition of organic matter is accomplished by extracellular enzymes released by these organisms. Because soil microbes concentrate them, a particularly high fraction of organically bound N and P in soils is contained in the microbial biomass.

The cation exchange capacity of humic substances in soil is important in providing a reservoir of nutrients to plants. Organic compounds can also complex metals in the soil and this can result in transport of Al and Fe, which are otherwise immobile. In most cases, the mass of soil humus exceeds the combined mass or living vegetation and litter. The residence time of humus in the soil may exceed that of litter by several orders of magnitude, with measured mean ^{14}C ages ranging upward to thousands of years (Figure 12.32). Plants take up nutrients in their roots, which in some cases can extend deep into the soil. These nutrients are then incorporated into wood, leaves, etc. above the soil. When these tissues die, nutrients in them are eventually released at top of the soil profile. This process is known as *biolifting*.

Provided a system is stable on time scales which are long relative to biochemical cycling, this together with abiological processes will reach steady-steady state such that soil profiles do not change with time.

13.4 WEATHERING RATES

Silicate weathering, and the subsequent precipitation of carbonate in the ocean, consumes CO_2 in reactions such as:

$$CaAl_2Si_2O_8 + H_2CO_3 + 3H_2O$$
$$\rightleftharpoons CaCO_3 + Al_2Si_2O_5(OH)_4$$

In this way, weathering is an important control on the concentration of atmospheric CO_2, which is in turn an important control on global temperature and climate.* Hence whatever factors control the weathering rate also influence climate. What are these factors? Walker et al. (1981) proposed there is a negative feedback in the climate system such that increases in atmosphere CO_2 increases temperature, which increases silicate weathering rates, which in draws down atmospheric CO_2, decreasing temperature. In a series of papers, this idea was further developed in by Robert Berner and colleagues in the BLAG (Berner, Lasaga and Garrels) model (Berner et al., 1983) and succeeding GEOCARB models (Berner, 1996). Others, including Edmond et al. (1995), have argued that despite the obvious temperature effect on reaction rates, global temperature exerts little control on weathering rates in nature, and that tectonic uplift and exposure of fresh rock has a much stronger influence on weathering rate, and ultimately on global climate.

There are also anthropogenic influences on weathering rates. Agriculture and harvesting of forests have had a clear impact on erosion, and probably chemical weathering rates as well. Combustion products of fossil fuels

* As we noted in Chapter 6, this is often called the Urey reaction after H. C. Urey. However, both Jacques-Joseph Ebelman (1814–1852) and Thomas C. Chamberlin (1843–1928), a predecessor of Urey's at the University of Chicago, had previously pointed out the influence of silicate weathering on atmospheric CO_2 concentrations. Chamberlin (1899), citing earlier work of Tyndall and Arrhenius, proposed silicate weathering and draw-down of atmospheric CO_2 was responsible for the ice ages. While silicate weathering ultimately proved not be the cause, Chamberlin's recognition of the influence of silicate weathering on climate was certainly a remarkably early insight.

released to the atmosphere include nitrates and sulfates that acidify precipitation (*acid rain*). The resulting decrease in pH affects chemical weathering rates (and had a clear adverse effect on the biota in some localities). On the other hand, weathering consumes H^+ and thereby increases pH; in some localities this buffering effect of weathering reactions is sufficient to entirely overcome the effects of acid rain. Understanding the impact of acid rain and the degree to which its effects are mitigated by weathering is an important goal of many weathering studies. We'll pursue this further in Chapter 15.

In Chapter 5, we examined the kinetics of weathering reactions mainly from the perspective of laboratory experiments. We noted there that laboratory rates appear to be orders of magnitude faster than rates observed in nature, which emphasizes the importance of studying weathering outside the laboratory. Here we explore attempts to investigate the weathering process as it occurs in nature. Two approaches to this have been developed. The first is to examine the weathering products that remain where they were produced, primarily within soils and regoliths and it includes both solids and the soil solution, which we will call the *in situ* approach. The second approach, which can be called the *watershed* approach, is to examine the weathering products being carried away by flowing surface or groundwater.

13.4.1   The *in situ* approach

13.4.1.1   *Weathering profiles*

While soils may develop on material deposited on top of bedrock, for example flood deposits, loess, glacial till, volcanic ash, etc., most soils are derived directly from underlying bedrock from weathering reactions occurring in the soil at the soil–bedrock interface. This provides an opportunity to estimate reaction rates and relate the products of weathering reactions to the reactants and, in some cases, allows reaction mechanisms to be deduced. Weathering rates depend on great variety of factors including temperature, atmospheric inputs (dry deposition as well as precipitation), biota, erosion rates, nature of the bedrock, etc. so rates and mechanisms will vary with locality. However, by studying

weathering in a variety of environments and overall picture can be developed on the importance of these individual factors assessed.

Weathering rates can be defined as:

$$\Re = \frac{\Delta M}{S \Delta t} \qquad (13.13)$$

where ΔM is the change in moles over time Δt and S is the mineral surface area (A. F. White and Buss, 2014). Establishing weathering rates is done by sampling solid and aqueous compositions with depth in a *weathering profile*. Figure 13.6 shows idealized fully developed weathering profiles overlying parental bedrock at depth z. The vertical shallow segment reflects the complete depletion of a weathering mineral while the vertical segment at the greatest depths corresponds to the unweathered bedrock (the protolith). The gradient at intermediate depths defines the *reaction zone* where weathering is occurring. Weather rates for a component are calculated as:

$$\Re_{solute} = \frac{q_h}{\nu S_v} \frac{dc_{solute}}{dz} \qquad (13.14)$$

$$\Re_{solid} = \frac{\omega}{\nu S_v} \frac{dc_{solid}}{dz} \qquad (13.15)$$

where ν is the stoichiometric coefficient of the component, q_h is the hydrologic flux (m/s) (equivalent to the solute velocity), ω is the weathering velocity (rate at which reaction front progresses downward in m/s), and S_V is the specific surface area per unit volume. Here we see the importance of surface area in reaction rates, and hence the role of physical weathering. Determination of weathering rates requires measurement of specific surface area, which is most commonly done using the BET method discussed in Chapter 5. The solute gradient is stationary under constant hydrologic conditions (and may change seasonally as hydrologic conditions vary) and thus provides a "snapshot" of contemporary weathering rates. In contrast, the solid gradient develops over much longer time scales, thousands to millions of years, and reflects average weathering rates over those time scales. Absolute mass changes with depth per unit volume are much greater in the solid than the solute profile so that $dc_{solid}/dz > dc_{solute}/dz$.

The thickness of the reaction zone depends on the ratio of the weathering reaction rate to

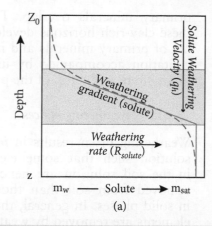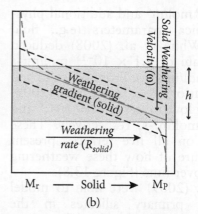

(a) (b)

Figure 13.6 Idealized weathering profiles. The soil surface is z_0. (a) show the mass of a component (generally an ion or a species such as H_4SiO_4) in the soil solution: m_w corresponds to the net solute component derived from weathering and m_p is the solute composition in equilibrium with the protolith. On short time scales under constant hydrologic conditions, the solute profile is constant. (b) illustrates the mass of a component (which may be a mineral such as biotite or an element such as Mg) decreases from the residual regolith at the top (M_R) to the protolith composition (M_p) at the base. The reaction zone (shaded) consists of the gradient at intermediate depths. The parallel dashed lines illustrate the progressive change in the solid-state weathering profile on long time scales. From A. F. White and Buss (2014).

the weathering velocity and in turn on whether weathering is rate-limited or transport-limited (Lichtner, 1988). In the latter case, where fluid advection is slow and transport occurs via diffusion the thickness of the reaction zone, h, is

$$h = \left(\frac{\phi D}{kA}\right)^{1/2} \quad (13.16)$$

where ϕ is the porosity, D the diffusion coefficient, A the surface area and k the reaction rate constant. Where components are transported by advection, the thickness is:

$$h = \frac{v}{kA} \quad (13.17)$$

where v is the advective velocity.

A series of uplifted Pleistocene marine terraces in Santa Cruz, California, have provided a fairly straightforward example of how weathering rates can be assessed from field studies. The protoliths are beach sands derived locally from granite and are relatively mineralogically simple (quartz, plagioclase, alkali feldspar). [10]Be cosmic-ray exposure dating (Chapter 8) shows the five terraces range from 65,000 to 226,000 years and have experienced little erosion. Figure 13.7

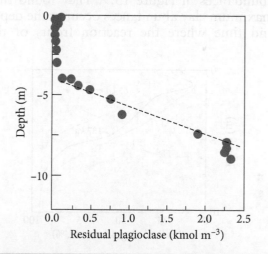

Figure 13.7 Plagioclase abundances in the soil of a 194 ka beach terrace near Santa Cruz, California. Dashed line shows the inferred gradient through the reaction zone. Plagioclase is nearly completely removed in the upper 3 meters and near protolith abundances below 8 m. From A. F. White and Buss (2014).

shows the change in plagioclase abundance with depth in the 194 ka terrace. Plagioclase is essentially completely removed in the upper 3 meters and attains nearly protolith abundances at 8 meters. From the intervening gradient of 0.48 mol/m⁴ and additional physical and chemical parameters (e.g., fluid fluxes), A. F. White et al. (2008) deduced a weathering rate of 1.1×10^{-15} mol/m²-s ($= 3.5 \times 10^{-5}$ mol/m²-ka). From Na solid and solute concentrations A. F. White et al. (2009) deduced slightly lower rates, 5.86×10^{-16} and 8.82×10^{-16} mol/m²-s, respectively. These same analyses on all five terraces present provides a picture of how these weathering profiles evolve over time (Figure 13.8).

Maher et al. (2009) were able to model weathering of primary silicates in the 226 ka Santa Cruz Terrace using laboratory-determined rate laws and rate constants, including the parallel rate equation of Hellmann and Tisserand (2006) discussed in Chapter 5 (eqn. 5.160). They used a reactive transport model to predict dissolution rates as well as downward aqueous flow, precipitation of secondary minerals such as kaolinite, and cation exchange by clays (Section 6.4.4). The predictions of the model are compared with measured plagioclase and kaolinite abundances in Figure 13.9. They found that maximum clay abundances occur at the depth and time where the reaction fronts of the

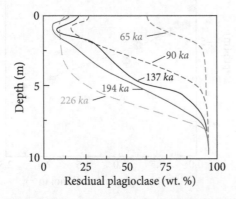

Figure 13.8 Comparison of plagioclase abundances in the weathering profiles of five uplifted beach terraces near Santa Cruz, California. ranging in age from 65 ka to 226 ka. Significant plagioclase remains near the surface of the youngest profile, but the profile thickens as weathering penetrates deeper with age. The oldest three soils have fully developed weathering profiles that can be modeled with a constant weathering velocity, ω. The abundance of plagioclase in the upper part of the older profile is thought to be due to aeolian input. Based on data in A. F. White et al. (2009).

primary minerals overlap. They concluded these clay-rich horizons develop by weathering of primary minerals and *in situ* clay precipitation accompanied by under-saturation of kaolinite at the top of the profile.

13.4.1.2 Weathering indices

Weathering often results in *incongruent* dissolution such that some elements dissolve in the soil solution are hence more readily removed from soil than those that remain in solid phases. In general, the rate at which elements are removed by weathering depends on their solubility. In the case of A-type metals (Section 6.3.3), which are the primary constituents of silicate rocks, solubility generally decreases with valence state. Thus, the alkalis are generally the most soluble and, among major elements, Al the least, along with Fe under oxidizing conditions. Potassium, however, can be an exception. While it is easily removed from primary minerals, it is also readily incorporated in the clay products of weathering. Based on this contrasting mobility, geochemists have developed a number of weathering indices by ratioing immobile and mobile elements. One such ratio is the *Chemical Index of Alteration* (CIA): $100 \times Al_2O_3/(Al_2O_3 + CaO^* + Na_2O + K_2O)$, where CAO* is CaO in the silicate fraction and excludes CaO in carbonate and apatite. The *Chemical Index of Weathering* (CIW): $100 \times Al_2O_3/(Al_2O_3 + CaO + Na_2O)$ is similar but excludes potassium for the reason mentioned above. Both indices increase with increasing weathering. An alternative is the *Weathering Index of Parker* (WIP): $100 \times [2(Na_2O/0.35) + MgO/0.9 + 2(K_2O/0.25)] + CaO/0.7$, which is based on bond strengths with oxygen and decreases with weathering. This avoids the assumption that sesquioxides such as Al_2O_3 and Fe_2O_3 are immobile. As noted above, these can become mobile when complexed by organic compounds and iron is mobile when reduced to the Fe^{2+} form.

Figure 13.10 compares density profiles with these three weathering indices in soil developed on Hawi basalts of Kohala Volcano, Hawaii, and the Occoquan granite of Virginia. Assuming weathering is isovolumetric, the weathering index profile should match the density profile. A. F. White (2008) concluded all three of these indices are effective

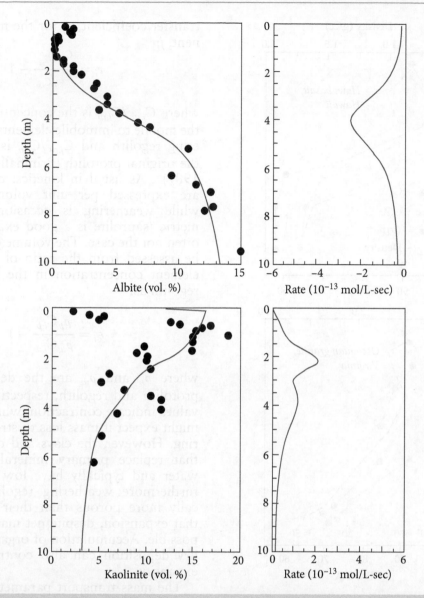

Figure 13.9 Left frames compared the observed (solid points) abundances of albite and kaolinite in the 226 ka Santa Cruz terrace with the abundances (red lines) predicted by the model of Maher et al. (2009). Right frames show the modelled reaction rates. From Maher et al. (2009).

indicators of weathering, although the match seems somewhat better for the granite than the basalt, because basalt weathering is often not isovolumetric. Price and Vieble (2003) found that for heterogeneous protoliths such as metamorphic rocks, the WIP was the most appropriate.

Many trace elements are even more immobile Al or Fe^{3+}, such as the rare earths, which have a valence of 3+ and the high-field-strength elements (Chapter 7) with valences of 4+ (Zr, Hf, Th) and 5+ (Nb, Ta). Thus, these elements and ratios to them are

also used as weathering indices. Kurtz et al. (2000) compared the relative mobilities of Al, Zr, Nb, Hf, Ta, and Th in weathering profiles developed on six Hawaiian basalts ranging in age from 0.3 ka (Kilauea) to 4100 ka (Kauai). The study included soils developed on the 150 ka Hawi basalts of Kohala Volcano mentioned above at different elevations and consequently climates. They found that while Nb/Ta ratios were identical in soils to those of the protoliths, ratios of Al, Hf and Th to Nb varied, particularly in deeper soils. Zr/Nb ratios were constant in dry climate Kohala

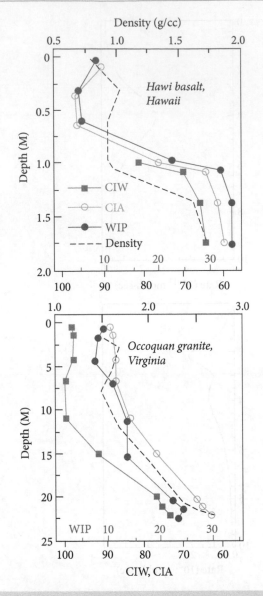

Figure 13.10 CIA, CIW, and WIP weathering indices and density in soils developed on basalt and granite as a function of depth. Adapted from A. F. White (2008).

soils but were otherwise variable as well. This suggests ratios to Ta or Nb may be the best weathering index.

Once an element, i, is shown to be immobile, the ratio of concentration of mobile elements (or components such as SiO_2) to the immobile one can be used to determine a mass

transfer coefficient, τ_j for the mobile component, j:

$$\tau_i = \frac{C_{j,w}/C_{i,w}}{C_{j,p}/C_{i,p}} - 1 \qquad (13.18)$$

where $C_{j,w}/C_{i,w}$ is the concentration ratio of the mobile to immobile elements in the weathering regolith and $C_{j,p}/C_{i,p}$ is that ratio in the original protolith (Brimhall and Dietrich, 1987)*. As usual in kinetics, concentrations are expressed per unit volume. However, while weathering is occasionally isovolumetric (saprolite is a good example), this is often not the case. The volume change, ε, can be assessed from the ratio of the immobile element concentration in the protolith and regolith as:

$$\varepsilon_i = \frac{\rho_p C_{i,p}}{\rho_w C_{i,w}} - 1 \qquad (13.19)$$

where ρ_p and ρ_w are the densities of the protolith and regolith, respectively. Negative values indicate contraction, which is what we might expect if mass loss to streams is occurring. However, the clays and oxyhydroxides that replace primary minerals incorporate water and typically have low densities and furthermore weathering regoliths are typically more porous than their protoliths so that expansion, despite net mass loss, is also possible. Accumulation of organic debris and dry deposition can also contribute to soils expansion.

The mass transport parameter, τ_j, for any particular element or component will not be constant through the regolith but by measuring concentrations through the weathering profile, the net loss of the component j from the regolith by integrating:

$$\Delta M_j = \left(\rho_p \frac{C_{j,p}}{m_j}\right) \int_0^z -\tau_j dz \qquad (13.20)$$

over the depth of the profile, z. Here ΔM is in units of moles and m_j is the molecular weight of component j.

* Quantifying weathering rates in this manner was first done by Jacques Joseph Ebelmen (1814–1852), a French engineer and professor at the Ecole des Mines in Paris. Ebelmen was also a pioneer in elucidating the carbon cycle (which he called the "carbon rotation." His name is one of the 72 names of French scientists, engineers, and mathematicians inscribed on the Eiffel Tower.

The distribution of components within weathering profiles are, of course, affected by some of the processes mentioned in Section 13.3. Components can be added to the soil by dry and wet deposition. Even relatively immobile elements can be leached from the upper parts of weathering profiles and reprecipitated at depth when complexed by organic ligands. Plants remove components from the soil and incorporate them into biomass. Biolifting can result in elements being redistributed upward in the weathering profile. In addition, in many environments there are seasonal variations in precipitation that consequently result in variable solute profiles. These factors must be considered in assessing weathering rates.

13.4.1.3 Erosion

Physical erosion affects the weathering profile and the total weathering flux. The Santa Cruz terrace profiles (Figure 13.9) illustrate the case where physical erosion is negligible and the weathering profile moves downward through time. In the absence of erosion, the reaction front can eventually reach a depth where transport of weathering products becomes negligible and the profile approximately steady state. This characterizes some tropical soils in flat terrain where soil depths can reach 100 m. More often, physical erosion also occurs, and we can imagine the reaction zone moving upward through time. The depth of the reaction zone will then depend on the difference between the weathering velocity and the erosion rate. Steadystate is reached when these are equal.

Where physical erosion occurs, the rate of removal in moles/m^2 of an immobile component, i, is:

$$DC_{i,p} = EC_{i,w} \qquad (13.21)$$

where D is a *denudation rate* in m/s, E is the erosion rate in m/s and, as before, $C_{i,p}$ and $C_{i,w}$ are the concentrations per unit volume (m^3). For a mobile element that is also lost through solution due to chemical weathering, the rate is:

$$DC_{j,p} = EC_{j,w} + W_j \qquad (13.22)$$

where Wj is the chemical weathering flux of component j in m/s. These equations can be combined to give weathering flux W_j of component j normalized to the total denudation rate for component j, which is equal to the negative of the mass transfer τ_j defined in eqn. 13.18:

$$\frac{W_j}{DC_{j,p}} = \left[1 - \frac{C_{j,w}/C_{i,w}}{C_{j,p}/C_{i,p}} \right] = -\tau_i \qquad (13.23)$$

The term on the right is also known as the *chemical depletion factor* (CDF). Where erosion rates are rapid relative to weathering rates, physical removal of components dominates, and the CDF is small. Where erosion rates are slow, chemical removal can dominate and the CDF will be high.

13.4.2 The watershed approach

A second approach to determining weathering rates is to assess the rates at which weathering products are carried away by flowing water on the scale of a watershed. Soluble products of chemical weathering will generally first be incorporated into the soil solution and then groundwater before entering streams and ultimately rivers. The term *watershed* is effectively synonymous with *drainage basin*. It is the area over which precipitation collects and exits through a single stream. Watersheds can vary in scale from the smallest stream to the largest of rivers. The weathering flux J_w of component i in a watershed is calculated as:

$$J_{i,w} = J_{i,total} - J_{i,atm} \pm J_{i,bio} \pm J_{i,exch} \pm J_{i,anthr} \qquad (13.24)$$

where J_{atm} is the atmospheric input, J_{bio} is the biological flux into or out of the stream water, J_{exch} represent adsorption–desorption in the soils and rocks, J_{total} is the anthropogenic flux, and J_{total} is the total flux out of the watershed. J_{total} is measured from:

$$J_{i,total} = \frac{c_i q_h}{A_w} \qquad (13.25)$$

where c_i is the concentration of i in stream water, q_h is the hydrologic flux (volume of water flowing out of the watershed per unit time) and A_w is the land surface area of the watershed. The anthropogenic flux can include aerosols, fertilizers, removal of agricultural or lumbering products, wastes. etc. Since the interest here is natural processes, areas impacted by anthropogenic processes are avoided. One approach is to focus on small, remote watersheds because avoiding anthropogenic impacts in large watersheds is nearly impossible. This approach is illustrated in Figure 13.11. We begin with the example of the study of small watersheds in the Southern Appalachians of the United States.

13.4.2.1 Watersheds in the Coweeta Basin, Southern Appalachians

The US Forest Service's Coweeta Hydrologic Laboratory comprises an area of 1625 hectares located in Appalachian Mountains of southwestern-most South Carolina. The area has been relatively undisturbed by human activity for nearly a century. It is characterized by steep slopes with 900 m of relief with elevations up to 1585 m and high average annual rainfall. Bedrock consists of a variety of metasediments and metavolcanics consisting of quartz, muscovite, biotite, plagioclase (oligoclase) and almandine garnet along with a variety of accessory minerals. This is overlain by soil averaging about 6 m in depth; 95% of this thickness is saprolite (C horizon). The area is divided into several watersheds, each drained by a different stream.

As biomass uptake, stream water composition and hydrologic flux data has been

compiled continuously over decades by the Forest Service, M. A. Velbel and his colleagues have been studying weathering processes since the 1980s. They found from petrographic, electron microprobe, and X-ray diffraction analyses that the primary weathering reactions were the breakdown of biotite, garnet, and feldspar. On the other hand, muscovite and quartz were not appreciably weathered, and abundances of other minerals were too small to affect the mass balance (Velbel, 1985a). Biotite weathers initially to form vermiculite. The primary lattice structure is partially preserved in this process, which involves loss of K (and some Mg), oxidation of Fe, and uptake of dissolved Fe and Al, and some Ca. The reaction may be written as:

$$K_{0.85}Na_{0.02}(Mg_{1.2}Fe^{2+}_{1.2}Al_{0.45})(Al_{1.2}Si_{2.8})$$
$$\times O_{10}(OH)_2 + 0.19O_2 + 0.078H^+$$
$$+ 0.31H_2O + 0.016Ca^{2+} + 0.04Na^+$$
$$+ 0.35Al + 0.3Fe(OH)^+_2$$
$$\rightleftharpoons K_{0.25}Na_{0.06}Ca_{0.016}(Mg_{1.1}Fe^{2+}_{0.5}Fe^{3+}_{1.1})$$
$$\times (Al_{1.2}Si_{2.8})O_{10}(OH)_2 \cdot 0.133Al_6(OH)_{15}$$
$$+ 0.6K^+ + Mg^{2+} \tag{13.26}$$

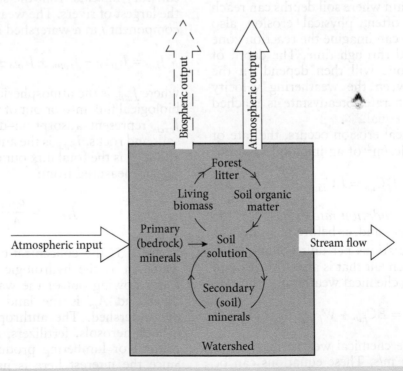

Figure 13.11 Watershed mass balance. Illustration of the mass balance approach to weathering on a watershed scale.

Almandine garnet weathers congruently. Within the C horizon, local reprecipitation of the iron as goethite and some of the aluminum as gibbsite produces a protective surface layer on almandine and weathering reactions are limited by the rate of transport of reactants and products across the layer. In higher soil horizons, organic chelating agents remove the iron and aluminum and weathering is limited only by the rate of surface reactions.

Plagioclase weathers by selective attack at defects in the lattice (Section 5.6). In early stages of weathering, components are removed in solution and reprecipitated elsewhere. The weathering reaction for plagioclase composition of An32 may be described as:

$$Ca_{0.32}Na_{0.68}(Al_{1.32}Si_{2.68})O_8 + 2.64H^+$$
$$+ 5.36H_2O \rightleftharpoons 0.32Ca^{2+} + 0.68Na^+$$
$$+ 2.68H_4SiO_{4(aq)} + 1.32\,Al(OH)_{2(aq)}^+$$
$$(13.27)$$

In the upper part of the soil profile, rapid flushing keeps dissolved silica concentrations low so that kaolinite stability is not attained, and aluminum released by plagioclase weathering precipitates as gibbsite, or is consumed in the production of vermiculite and chlorite from biotite. Deeper in the soil profile, water is in prolonged contact with rock and acquires enough aluminum and silica to reach kaolinite saturation. Because the bedrock is highly impermeable, most water is eventually shunted laterally downslope and does not penetrate the bedrock. What little water does penetrate forms smectite in voids and fractures. These reactions, however, have little effect on stream chemistry.

The concentrations of major cations in stream water, well water from the base of the saprolite (saprolite–bedrock interface), and soil water (sampled at 25 cm depth) are shown in Table 13.6. It is apparent from the table that well water is nearly identical in composition to stream water. This indicates that stream water chemistry is determined entirely by reactions occurring as it percolates through the saprolite, with the exception that, after leaving the subsurface, the water equilibrates with the atmosphere, resulting

Table 13.6 Composition of water from the Coweeta watershed (ppm).

	K	Na	Ca	Mg	pH
Soil	0.92	0.27	2.48	1.27	6.12
Well	0.60	1.08	1.12	0.65	5.10
Stream	0.59	1.08	1.06	0.64	6.64

From Velbel (1985b).

in a loss of CO_2 and increase in pH. On a plot of log (a_K+/a_H+) vs. log $(a_{H_2SiO_4})$, the composition of stream waters plot in the kaolinite stability field (Figure 6.21). Other aspects of stream chemistry indicate that both gibbsite and kaolinite form, consistent with the observation that both minerals occur as weathering products.

Weathering rates can be determined using a system of simultaneous mass balance equations (Velbel 1985a; Velbel and Price, 2007). For each component or element, i, the net flux of the element out of the watershed (steam output minus rain input) can be expressed as the sum of its production or consumption in each weathering reaction as well as by the biomass, i.e.:

$$\Delta m_i = \sum_j^\phi v_{i,j}\alpha_j \qquad (13.28)$$

where Δm_i is the net flux of element i out of the watershed, $v_{i,j}$ is the stoichiometric coefficient of element i in the weathering reaction of phase j, and α_j is the weathering rate of phase j, which are the unknowns. The stoichiometric coefficients are determined from mineral analyses and Δms (in units of kg/ha) are determined from the product of stream compositions and hydrologic flux out of the watershed. Where there is net biomass uptake, the biomass becomes a "fictive" phase whose stoichiometric coefficients and mass flux are determined by biomass composition and net primary production. The stoichiometric coefficient of an element is negative if the element is taken up and consumed in the weathering of a phase. For example, weathering of biotite in reaction 13.27 takes up Ca and Na from the soil solution, consequently the stoichiometric coefficients of Ca and Na in biotite are negative.

This equation can be written in matrix form as:

$$\Delta \mathbf{M} = \mathbf{NA} \qquad (13.28a)$$

where $\Delta \mathbf{M}$ is the flux vector, $\mathbf{N}$ is the matrix of stoichiometric coefficients, and $\mathbf{A}$ is the weathering rate vector. The equation requires that $\mathbf{N}$ be a square matrix and consequently the number of components or elements must match the number of phases. For example, for the elements K, Na, Ca, and Mg we have four components with which we can write four equations and solve for four unknowns, the rates of plagioclase, biotite, and garnet weathering and net biomass uptake. The procedure in explained in Example 13.2.

If the modal abundances and surface areas of minerals in the protolith are known, rates calculated in this way can be converted to moles per unit surface area per time for comparison with laboratory-determined rates. For example, Price et al. (2014) determined the weathering rate of biotite to be 7.2×10^{-14} mol m^{-2} s^{-1}, which is at least an order

Example 13.2 Determining weathering fluxes

For a system of n components and ϕ minerals, equation 13.28 can be written in matrix form as:

$$
\begin{bmatrix} \Delta m_1 \\ \Delta m_2 \\ \vdots \\ \Delta m_n \end{bmatrix}
=
\begin{bmatrix}
v_{1,1} & v_{1,2} & \cdots & v_{1,n} \\
v_{2,1} & v_{2,2} & \cdots & v_{2,n} \\
\vdots & \vdots & \cdots & \vdots \\
v_{\phi,1} & v_{\phi,2} & \cdots & v_{\phi,n}
\end{bmatrix}
\begin{bmatrix} \alpha_1 \\ \alpha_2 \\ \vdots \\ \alpha_\phi \end{bmatrix}
\qquad (13.29)
$$

Velbel and Price (2007) provided the data in the adjacent table for the stoichiometric coefficients and weathering fluxes.

	Plag	Garn	Biot	Biomass	Δm (mol, ha^{-1} yr^{-1})
Ca	0.32	0.2	−0.16	0.144	**45.4**
Mg	0	0.5	0.01	0.056	94.2
K	0	0	0.6	0.150	74.2
Na	0.68	0	−0.04	0.008	359

Answer: The matrix form of this equation suggests that solving for weathering rates is most readily addressed using MATLAB. A simple MATLAB script to solve this equation is shown below:

```
%nu values: stoichiometric coefficients
% Ca, Mg, K, Na
%Plag, Garn, Biot,Biomass
Nu = [0.32,0.2,-0.016,0.144;0,0.5,0.1,0.056;0,0,0.6,0.150;0.68,0,-0.04,0.008]
%mass fluxes vector
dmass = [45.5;94.2;74.2;359]

alpha = Nu\dmass
alpha =
1.0e+03 *
0.56780
0.24000
0.43147
-1.2312
```

Thus, the weathering rates are plagioclase = 568 mol/ha, garnet = 240 mol/ha, and biotite = 431 mol/ha with a biomass nutrient uptake of 1231 mol/ha.

of magnitude lower than laboratory determined rates, which range from 1.2×10^{-10} to 5×10^{-13}. When normalized to estimated mineral surface areas, the reactions rates of garnet and plagioclase calculated by Velbel (1985a) were a factor of 8 and 2, respectively, lower than measured laboratory rates. As we pointed out in Chapter 6, this is often the case. There are several possible explanations, including aging and formation of protective surface layers on natural surfaces, the possible inhibitory effect of dissolved components, and local approach to equilibrium in natural systems, which reduces the reaction affinity and slows the rate. In addition, water flow is heterogeneous and not all mineral surfaces are in contact with pore fluid and participate in reactions.

Velbel calculated that 40,000 years were required to weather all the garnet, 140,000 years to react all the biotite, and 160,000 years to weather all the plagioclase in the soil horizon. The calculated *saprolitization rate* was 3.8 to 15 cm/1000 yrs, the lower rate applying to complete destruction of primary minerals. This rate is nearly equal to the long-term denudation, or erosion, rate (rate at which rock is removed from the surface) for the southern Appalachians. This agreement suggests the system is in steady-state, i.e., weathering penetrates bedrock at the same rate at which weathering products are removed by erosion, maintaining a constant thickness weathering profile. However, based on sediment export rates, the short-term denudation rates for the region were much slower, by as much as a factor of 20. This indicates that up to 96% of erosion occurs not by steady-state removal of soil, but by infrequent catastrophic events such as landslides and severe storms. This is entirely consistent with convention geologic views as well as other studies of erosion in the Southern Appalachians.

Equation 13.28 can also be used to calculate the rate of production of secondary minerals. Considering additional phases requires that we include additional elements in our mass balance. For example, Velbel and Price (2005) added Si, Al, and the rare earths La and Dy to the four elements used in 13.2 in order to solve for additional phases allanite (all), vermiculite (verm), kaolinite (kaol), gibbsite (gibbs), and biomass (biom) to write the following equation for the Coweeta W2 watershed:

$$
\begin{array}{l}
\begin{array}{r}
Si \\ Al \\ Mg \\ Ca \\ Na \\ K \\ Dy \\ La
\end{array}
\left[
\begin{array}{cccccccc}
3.16 & 2.72 & 3.0 & 2.75 & 2.75 & 2.0 & 0.0 & 0.0 \\
2.2 & 1.28 & 2.0 & 1.57 & 1.81 & 2.0 & 1.0 & 0.0 \\
0.06 & 0.0 & 0.36 & 1.42 & 1.54 & 0.0 & 0.0 & 0.56 \\
1.58 & 0.28 & 0.41 & 0.001 & 0.026 & 0.0 & 0.0 & 0.144 \\
0.16 & 0.72 & 0.0 & 0.041 & 0.032 & 0.0 & 0.0 & 0.008 \\
0.0 & 0.0 & 0.0 & 0.88 & 0.48 & 0.0 & 0.0 & 0.15 \\
0.000731 & 0.0 & 0.000565 & 0.0 & 0.0000734 & 0.0 & 0.0 & 0.0 \\
0.00533 & 0.0 & 0.0 & 0.0 & 0.000506 & 0.0 & 0.0 & 0.0
\end{array}
\right]
\end{array}
$$

$$
\times
\begin{bmatrix}
\alpha_{all} \\ \alpha_{plag} \\ \alpha_{garn} \\ \alpha_{biot} \\ \alpha_{verm} \\ \alpha_{kaol} \\ \alpha_{gibb} \\ \alpha_{biom}
\end{bmatrix}
=
\begin{bmatrix}
1277 \\ 0 \\ 94.2 \\ 45.4 \\ 359 \\ 74.2 \\ 0.005 \\ 0.03
\end{bmatrix}
\tag{13.30}
$$

Solving this for the weathering rates is left to the reader as Problem 13.9.

13.4.2.2 *Thermodynamic and kinetic assessment of stream compositions*

Thermodynamics can provide insights as to what reactions are occurring in the weathering process. Let's consider the thermodynamics and kinetics of weathering in more detail, using the Coweeta study as an example. Figure 13.12 shows the stability diagram for the system $K_2O-Al_2O_3-SiO_2-H_2O$ with the expected paths a solution would take in weathering of K-feldspar. Stream composition data from the Coweeta Watershed (Velbel, 1985b) are plotted on the diagram as crosses. The data plot within the kaolinite stability field, consistent with Velbel's observation that kaolinite is forming within the saprolite. The rainwater, however, is presumably much more dilute. Let's arbitrarily assume it plots at point A in Figure 13.12. This point plots within the gibbsite stability field. Thus, from thermodynamics, we expect the initial weathering of feldspar will produce gibbsite. The reaction (considering only the K-component in the alkali feldspar solid solution) is:

$$
\begin{aligned}
KAlSi_3O_8 + 7H_2O + H^+ \\
\rightleftharpoons K^+ + 3H_4SiO_{4(aq)} + Al(OH)_{3(s)}
\end{aligned}
\tag{13.31}
$$

This reaction produces dissolved K^+ and H_2SiO_4 and consumes H^+, so the composition

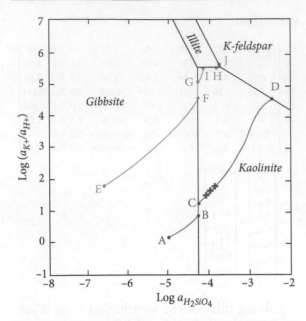

Figure 13.12 Stability diagram for the system H_2O, K^+, H^+, and H_2SiO_4 showing the solution path as weathering proceeds. (Illite is a clay mineral that is compositionally and structurally similar to the muscovite, $KAl_3Si_3O_{10}(OH)_2$, but is more compositionally and structurally variable.) Crosses show measured compositions of Coweeta stream water (Velbel, 1985b). Rainwater is presumably much more dilute, for example, the composition marked by point A. As weathering proceeds, the composition proceeds toward point C and then to point D. Path E–J is a hypothetical path of another possible solution. See text for discussion.

of the water will evolve up to the right toward point B on Figure 13.12 (the exact path depends on solution alkalinity because species such as H_2CO_3 can dissociate to partially replace the H^+ consumed in reaction 13.31.

When point B is reached, both kaolinite and gibbsite are stable, and any additional H_4SiO_4 produced by weathering of feldspar reacts with gibbsite to produce kaolinite:

$$KAlSi_3O_8 + 3Al(OH)_{3(s)} + H_4SiO_{4(aq)} + H^+$$
$$\rightleftharpoons K^+ + 2Al_2Si_2O_5(OH)_4 + 3H_2O \quad (13.32)$$

The path is thus vertical along B–C until all gibbsite is consumed. Once it is consumed,

further weathering produces kaolinite and dissolved K and H_2SiO_4 by the reaction:

$$KAlSi_3O_8 + 3H_2O,$$
$$\rightleftharpoons K^+ + 2H_4SiO_{4(aq)} + \frac{1}{2}Al_2Si_2O_5(OH)_4 \quad (13.33)$$

The path (C–D) is steeper because less H_2SiO_4 is produced in weathering to kaolinite than to gibbsite. Eventually, the K^+ and H_2SiO_4 concentrations reach the point (D) where feldspar is stable, at which point no further weathering occurs because the solution is in equilibrium with K-feldspar.

Depending on the initial solution composition, other reaction paths are also possible. For example, a solution starting at point E in the gibbsite field would initially evolve in a similar manner to one starting at point A: producing first gibbsite then kaolinite. However, the solution starting at point E would eventually reach the illite stability field at point H. At this point, kaolinite is converted to illite through the reaction:

$$KAlSi_3O_8 + Al_2Si_2O_5(OH)_4 + 3H_2O$$
$$\rightleftharpoons KAl_3Si_3O_{10}(OH)_2 + 2H_4SiO_{4(aq)} \quad (13.34)$$

The K^+/H^+ ratio is unaffected in this reaction, but the H_2SiO_4 concentration continues to increase. Once all kaolinite is consumed, further weathering of K-feldspar produces additional illite and the concentrations of H^+ decreases and K^+ and H_2SiO_4 increase again in the reaction:

$$3KAlSi_3O_8 + 12H_2O + 2H^+$$
$$\rightleftharpoons 2K^+ + KAl_3Si_3O_{10}(OH)_2 + 6H_4SiO_{4(aq)} \quad (13.35)$$

This continues until the stability field of K-feldspar is reached. Paths beginning at higher K^+/H^+ ratios may miss the kaolinite stability fields altogether.

This purely thermodynamic analysis does not predict the phases actually found in the weathering profile as both kaolinite and gibbsite occur together, while the water composition plots well within the kaolinite-only stability field. The problem arises because we have ignored kinetics. In essence, we have assumed that the dissolution of K-feldspar is slow, but that the solution quickly comes to equilibrium with secondary minerals such as gibbsite and kaolinite, which is not the case.

Let's consider the progress of the reaction along path A–D from a kinetic perspective with a reaction progress variable, ξ, which we define as the number of moles of feldspar consumed. We assume that the reaction rate of each reaction, i, depends on the extent of disequilibrium and can be described by the equation:

$$\mathfrak{R}_i = k_i \frac{\Delta G}{RT} \qquad (13.36)$$

where $\mathfrak{R}$ is the reaction rate, k is the rate constant that includes a mineral surface area per mass term, and ΔG is the affinity of the reaction (in essence, we are assuming these reactions are elementary). The activity–activity and reaction progress diagrams computed by Lasaga et al. (1994) assuming the value of k is 10 times as large for the precipitation of gibbsite and kaolinite as for feldspar dissolution in Figure 13.12. The activity–activity diagram is similar to that in Figure 13.13, although there is almost no vertical path along the gibbsite–kaolinite boundary. The reaction progress diagram, however, is quite different. We see that gibbsite and kaolinite now coexist over a wide region. This is a simple consequence of assuming finite rates for the precipitation reactions. Thus, it is not surprising that Velbel (1985) found that gibbsite and kaolinite coexisted in the weathering profile even though the stream compositions plot within the kaolinite-only field.

13.4.3 Factors controlling weathering rates

Lasaga et al. (1994) proposed the following general form for the net rate law for weathering reactions:

$$\mathfrak{R} = k_0 A_{min} e^{-E_A/RT} a_{H+}^n \prod_i a_i^m f(\Delta G_r) \qquad (13.37)$$

where A_{min} is the mineral surface area, $ke^{-E_A/RT}$ expresses the usual dependence of the rate on temperature and activation energy (E_A), a_{H_+} expresses the dependence on pH to some power n, the terms a_i represent possible catalytic or inhibitory effects of other ions with an exponential effect m, and $f(\Delta G_r)$

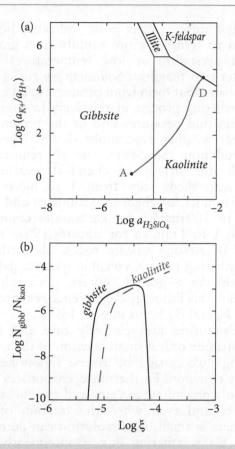

Figure 13.13 (a) Stability diagram as in Figure 13.12. (b) Reaction progress diagram computed assuming the rate constants for precipitation of gibbsite and kaolinite are 10 times faster than that for the dissolution of feldspar. From Lasaga et al. (1994).

expresses the dependence of the reaction rate on reaction affinity, i.e., the extent of disequilibrium.

The terms in equation 13.37 relate to lithology, climate and hydrology, topography, and mechanical erosion (i.e., mass wasting), and biota. We consider each of these in turn.

13.4.3.1 Lithology

The composition of the protolith undergoing weathering, particularly the solubility of the phases constituting the rocks or extent to which they are out of thermodynamic equilibrium under prevailing conditions, exerts a strong influence on weathering rates. Chemical sediments such as evaporites and limestones are highly soluble and weather very rapidly. Basalts, which crystallize from high temperature, water-poor silicate liquids, tend to weather more rapidly than granites that crystallize at low temperatures from water-rich magmas. Sedimentary rocks such as shales that form from products of an earlier weathering process at relatively low temperatures and pressures and in the presence of water weather even more slowly. Chemical denudation rates (the rate of removal of rock due strictly to chemical weathering) of watersheds vary from 1 to more than 600 mm/ka for chemical sediments and from 14 to 100 mm/ka for carbonates compared with 1 to 4 mm/ka for silicates (Viers et al., 2014). Among silicate rocks, the order of weathering rates is volcanic rocks > gabbro, sandstone > granite, gneiss, mica schist > shale. This lithology dependence is express by the k_0 and E_A terms in eqn. 13.37.

Evaporites are relatively rare and hence contribute only a small fraction of the weathering flux carried by rivers; limestones are very common. Furthermore, carbonates commonly precipitate as veins and vugs of silicate rocks and even when the fraction of carbonate is small, its dissolution can dominate the Ca weathering flux. Consequently, dissolved Ca/Na ratios in watersheds underlain by silicate rocks can have Ca/Na ratios up to an order of magnitude higher than that of the protolith. This is particularly true in watersheds with wetter climate, rapid tectonic uplift, and high rates of mechanical erosion. Carbonate constitutes only 1% of Himalayan glacial tills yet contribute 90% of the weathering-derived Ca; overall, carbonate weathering provides 87% of the Ca to Himalayan rivers and 80% of the Ca Loch Vale granite watershed of Colorado. Under stable, steady state conditions, Ca/Na ratios in streams tend to decline as carbonates are weathered out. Although they account for a significant fraction of the Ca weathering flux, carbonate weathering has no effect on atmospheric CO_2 as the weathering products simply reprecipitate as carbonate.

Permeability and porosity are also important, both because they control flow rate of water through the protolith and determine the surface area (A_{min} in eqn. 13.38) across which weathering reactions can occur. Although the specific surface area in fine-grained rocks exceeds those of coarse-grained ones, the former tend to have lower porosity and permeability unless extensively fractured. Basaltic lavas that fracture extensively during rapid cooling hence have relatively high permeability that also favors their rapid weathering. Basaltic weathering is also important because basalts are rich in Ca and Mg; basalts typically contain 10–12% CaO and 6–10% MgO, compared with 4% and 2% in average upper continental crust, and a large fraction of the Ca released by weathering ultimately precipitates as carbonate and a significant fraction of Mg does as well. As first pointed out by Chamberlin (1899) and Urey (1952) and as we discussed in Chapter 12, the process of silicate weathering and subsequent precipitation of carbonate exerts a strong control on atmospheric CO_2 and consequently climate. Because of this, and their higher weatherability, basalt weathering accounts for ~30% of global CO_2 consumption despite occupying only 3–5% of the land surface.

13.4.3.2 Climate: temperature, precipitation, and hydrology

Rates of weathering reactions have an Arrhenius-like temperature dependence and increase with temperature. Equilibration with water is the driving force of almost all weathering reactions; in the absence water high temperature minerals can persist metastably indefinitely. Water also carries away weathering products, both solids and dissolved, helping to expose fresh bedrock. When flow of water is slow through the weathering regolith, reaction rates become diffusion limited, even when temperatures and surface area of fresh protolith are high. Thus, we expect weathering rates to depend on climate.

The temperature sensitivity of weathering rates is of particular interest because the CO_2-driven climate system should have a built-in thermostat-like negative feedback as proposed by Walker et al. (1981). Higher global temperatures should increase weathering rates and draw down atmospheric CO_2 thus reducing global temperatures. Lower temperatures should have the opposite effect. The temperature dependence of weathering reactions is well demonstrated in laboratory experiments, but does this hold in nature where other variables are involved?

The $e^{-E_A/RT}$ term in equation 13.37 expresses the usual Arrhenius exponential temperature dependence of reaction rates. We thus expect weathering rates to be higher in warm climates than in cold ones, and this is indeed observed. The exact degree of temperature dependence will in turn depend on the activation energy. Various studies of the dependence of weathering rates on climate suggest the activation energy for chemical weathering is in the range of 40 to 80 kJ/mol (Lasaga et al., 1994). This is consistent with the average of activation energies of weathering reactions determined in laboratory studies (e.g., Table 5.4). This activation energy means that an increase in temperature of 8°C would result in a doubling of the weathering rate if all other factors remain constant.

Li et al. (2016) examined the temperature dependence of basaltic weathering in basaltic watersheds globally and found a strong correlation ($r^2 = 0.75$) of bicarbonate ion (which they used as a proxy for weathering rate) with mean annual temperature among inactive volcanic areas (active volcanic areas typically had higher weathering rates but were excluded due to complicating factors such as hydrothermal activity). From this, they calculated an activation energy, E_A, of 41.6 kJ/mol. This is at the low end of the range cited above, consistent with the more rapid weathering rates of basalt. These results are consistent with the thermostat-like effect proposed by Walker et al. (1981). They speculate that weathering of the anomalously large volumes of mafic volcanics erupted in the Cretaceous could account for the approximately 8°C cooling over the Cenozoic.

For other rocks types and other components such as Si and Na, however, the relationship between weathering rates and temperature is less clear, with modest temperature dependence in some cases and none in others. This clearly implies other factors must play a role as well.

Precipitation is obviously also important. The surface area term in eqn. 13.37 is the surface area across which reaction occurs, which effectively is the mineral surface area in contact with water. Where rainfall is insufficient to continually wet all grain surfaces, the rate of chemical weathering will be lower and consequently chemical weathering is slow in arid regions. In some cases, this also results in low overall erosion rates as fresh rock is stronger than weathered rock. In others,

where glaciation or high relief leads to high mass wasting and high erosion, much of the material being removed will be fresh rock rather than weathering products.

In addition to the wetting mineral surface area, which depends on the volume of water in the regolith, the rates at which water flows through the regolith will also be important. Maher (2010) found that weathering reactions rates declined exponentially with fluid flow rates. We found in Chapter 6 that when a system is far from equilibrium and reaction affinity, ΔG_r, is above a critical value, reaction rates are independent of ΔG_r; at lower values, they depend on ΔG_r and are much slower. When the flow rate is slow, reaction rates approach saturation in weathering products, slowing reaction rates (this is expressed by the ΔG term in eqn. 13.37).

Consequently, the relationship between precipitation and weathering is not simple. Bluth and Kump (1994) examined chemical weathering rates (using concentrations of bicarbonate and SiO_2 in streams as proxies for these rates) and found that the concentrations of SiO_2 and bicarbonate in streams from a given region remain constant over a large range of runoff; since the fluxes of these components are the product of concentration and runoff, this indicates that fluxes and consequently weathering rates increase with increasing precipitation. However, they found that bicarbonate and SiO_2 concentrations can level off and even drop when runoff exceed 100 cm/yr, indicating that additional precipitation is acting to merely dilute weathering products rather than increase weathering rates. Li et al. (2016) also found that bicarbonate concentrations in streams in basaltic watersheds were constant over a large range of runoff, implying the weathering flux increases with runoff, but the correlation was weaker and they were "not able to distinguish whether or not there is a significant runoff control on basalt weathering rates."

13.4.3.3 Topography and mechanical erosion

Chemical weathering operates most rapidly when primary mineral surfaces are exposed to attack by water. Mechanical erosion processes such as landslides remove weathering products and expose fresh rock to water, consequently enhancing chemical weathering.

In the absence of mechanical erosion, low rates of chemical weathering can occur even in humid tropical areas, such as the Orinoco, Amazon, and Congo basins. This results in is what Stallard and Edmond (1983) called a "transport-limited" regime, where thick lateritic soil (up to 100 m) insulates the underlying bedrock from chemical attack. Edmond et al. (1995) pointed out that such transport-limited regimes can occur even in areas of high elevation, such the Guiana Shield in South America: "elevation *per se* is not the determining variable, but rather the mechanism by which it is produced." In tectonically active areas, such as the Andes, faulting generates high relief and exposes fresh rock to chemical weathering. The Himalayas represent a contrasting regime where mechanical erosion is so rapid that rock is removed before it can be weathered. Chemical weathering accounts for only 10% of the total denudation rate in the Ganges watershed compared with 45% for the Congo River (White and Buss, 2014). In such cases, chemical weathering is said to be *kinetic-limited*. Not all tropical areas necessarily have low rates of chemical erosion, however, the Rio Icacos watershed in the tropical Luquillo Mountains of Puerto Rico, the site of a Critical Zone Observatory, exhibits particularly high rates of chemical weathering. Despite relatively thick regolith, active tectonics results in relatively frequent landslides exposed fresh rock to chemical attack.

Glacial erosion can also exert a strong influence on chemical weathering. Not only do glaciers expose fresh rock, but upon receding they leave a layer of finally ground fresh rock known as glacial till that is ripe for chemical attack. Chemical weathering rates are consequently higher in formerly glaciated areas than regions of comparably cool climates, but nevertheless lower than warm climate regions.

In summary, tectonics exert a strong control on global chemical weathering rates and, ultimately, atmospheric CO_2 levels and climate, as Edmond et al. (1995) emphasized.

13.4.3.4 Role of biota

The role of the biota on chemical weathering is complex, but on balance the effect of the biota appears to be to increase weathering rates. Roots stabilize the soil against mechanical erosion, which inhibits the exposure of fresh rock to weathering. On the other hand, plants can contribute to mechanical weathering by root wedging, increasing the surface area of minerals exposed to water. Roots also generate porosity that increases water circulation of water and contact between the soil solution and minerals. Plants take up a variety of nutrients from the soil solution, and thus increase the disequilibrium between primary minerals and the soil solution. In mature ecosystems, however, growth and decomposition approach a steady-state balance so that there is little net loss of nutrients to the biota. Soil microbes can catalyze the oxidation and reduction of redox sensitive elements such as Fe and S. Transpiration also returns water to the atmosphere, and hence increases rainfall.

Respiration within the soil by plant roots and more importantly soil organisms and microbes that survive on their detritus increases soil P_{CO_2} and thereby decrease soil pH through respiration. This effect is important as P_{CO_2} in soils can exceed that of the atmosphere by up to two orders of magnitude in some soils. As we noted earlier, formation and dissolution of carbonic acid is the principal source of H^+ in most natural systems. Decomposition products include organic acids, which also reduce soil pH. This dependence is expressed by the a_{H+} term in eqn. 13.37. Organic substances can also have an effect on weathering rates beyond merely decreasing pH in two ways. First, organic acids can form surface complexes that directly promote weathering. Second, they form soluble complexes with otherwise highly insoluble cations such as Al^{3+} and Fe^{3+}, enabling their transport within and out of the regolith in solution.

Several studies, summarized by Viers et al. (2014), have demonstrated a relationship between dissolved organic carbon (DOC) and dissolved cations in rivers and streams suggesting DOC enhances weathering rates by up to a factor of 3–4, yet other studies have found no relationship. Drever (1994) noted that the rates of many weathering reactions are independent of pH in the pH range of 4.5–8 (he acknowledged that weathering rates of many ferromagnesian silicates, such as pyroxenes and amphiboles, do increase with decreasing pH in this range). Most soil solutions have pH above 4.5, so pH decreases due to the biota have only a small effect on weathering rate. He also argued that while there is evidence that organic acids, such as oxalic acid, accelerate weathering reactions, the effect is typically only a factor of 2 at concentrations in the mM range. In nature, the organic acid concentrations are much lower and the overall effect of organic acids is likely to be small.

Drever (1994) also pointed out that the effect of plants will be different depending on whether weathering is rate-limited or transport-limited. The Alps, where physical weathering and transport of weathering products is rapid, are a good example of the former. Here, plants should cause an increase in weathering through increasing the contact time between water rock, hence their effect is to increase chemical weathering rates. The Amazon Basin, where weathering rates are extremely low, is an example of the latter. The combination of subdued topography and dense vegetation limits the rate at which weathering products can be transported, leading to the accumulation of extremely deep (100 m) soils. The thick soil cover isolates the bedrock from incoming precipitation. Thus, the effect of plants in the transport-limited environment is to reduce chemical weathering rates. Just how much terrestrial life has accelerated weathering is a matter of debate. Schwartzman and Volk (1989), for example, concluded that the existence of land plants has increased weathering rates by 2–3 orders of magnitude. On the other hand, Drever (1994) argued that the direct chemical effect of land plants on weathering rates is probably no more than a factor of 2.

13.5   THE COMPOSITION OF RIVERS

Rivers return precipitation falling on land to the oceans (although a few, such as the Jordan and Volga Rivers, empty into inland saline lakes and seas) and in the process also accumulate dissolved solids produced by weathering. As Table 13.7 shows, even rainwater is not pure. Sea salts enter the atmosphere via water droplets produced by breaking waves in the ocean. Since this is a physical process, there is little associated chemical fractionation and hence these seawater-derived salts are present in rain in the same concentrations as in seawater; these are known as

cyclic salts. The cyclic contribution is quite significant for Na^+, Cl^-, and SO_4^-, but small for most other components. Rain can also dissolve aerosols from other sources, such as continent-derived dust, compounds transpired by trees, volcanic ash and gases, pollutants, etc., so that total ionic concentrations in rain are not always in the same proportion as in seawater. In addition to cyclic salts and solutes derived from weathering, rivers also contain a variety of organic compounds derived from biological activity as we found in Chapter 12, as well as suspended solids, including both organic matter and mineral grains derived from erosion.

The dissolved organic carbon (DOC), defined as the fraction of organic matter that passes through a 0.45 μm filter, consists mainly of humic and fluvic acids, as discussed in Chapter 12. They are derived primarily from humic substances in soils, although they tend to be somewhat more oxidized, and play an important role in forming soluble complexes of Al^{3+}, Fe^{3+}, as well as trace metals including Cu, Ag, Hg, and Pb. They are also a significant source of acidity in rivers draining organic-rich soils. Rivers on average also contain 330–400 μmol/kg of particulate organic matter (Perdue and Ritchie, 2014).

Many rivers have also been significantly impacted by anthropogenic activity: agriculture, logging, mining, industrial and household wastes, road deicing, etc. Table 13.7 consequently lists two averages for river water composition. One, from Gaillardet et al. (1999), is based on 62 of the largest rivers that drain 54% of the land area and constitute 53% of the discharge to the ocean. The other average labeled *pristine* is from Maybeck (2003) and represents a discharge-weighted average of some 1200 rivers and streams that have been minimally impacted by anthropogenic activity. This data set includes some large rivers such as the Mackenzie, Lena, and Mekong, but is mainly based on smaller tributaries and has poor coverage in some areas such as Europe, east and south Asia, and Australia. Because they are based on different data sets, the difference cannot entirely be attributed to anthropogenic activity.

As we might expect, the major cations in rivers are the alkalis and alkaline earths, reflecting their relatively high solubility. Dissolved silica (usually reported as SiO_2

Table 13.7 Average composition of dissolved loads of rivers.

	Ave upper crust mg/g	Ave rain mg/kg	Major rivers dissolved mg/kg	Pristine rivers dissolved mg/kg	Percent cyclic	River dissolved/ crust × 10⁴	Anthropogenic flux to oceans 10⁶ t/yr
Na	25.4	0.9	6.8	5.5	11%	2.66	78
K	28	0.23	1.5	1.7	1%	0.64	5
Ca	30	1	15.9	11.9	0.1%	53.71	47
Mg	14.9	0.15	4.1	2.9	2%	2.73	10.5
Cl	0.07	1.13	8.4	8.3	18%	120.0	93
SO_4	6.05	2.02	11.3	8.4	11%	1.87	124
HCO_3	–	–	60.3	48.7		–	100
SiO_2	666	–	7.6	8.7	~0%	0.01	
TDS	–	5.43	115.9	96.1			
DOC			~440 µmol/kg				

Average rain and percent cyclic are from Berner and Berner (1996). Average upper crust from Table 11.8, average major river dissolved load is the discharge weighted average of data from Gaillardet et al. (1999), average pristine river dissolved load and anthropogenic flux are from Meybeck (2003). DOC is dissolved organic matter and is from Perdue and Ritchie (2014).

although the dominate dissolved species is H_4SiO_4) is also among the major dissolved species in rivers, though nevertheless depleted compared with its concentration in the crust. This contrasts with seawater, where biological utilization results in very low SiO_2 concentrations. Ca^{2+} is the dominant cation in rivers and HCO_3^- is the dominant anion. Again, this contrasts with seawater, where Na^+ and Cl^- are the dominant ions. We'll call this the *river water paradox* and we will resolve it in the next chapter. To a significant extent, enrichment of rivers in Ca reflects the importance

of carbonate rock weathering in controlling river composition, as we discuss below.

The composition of river water varies widely from the averages given in Table 13.7, as may be seen in Table 13.8, which lists the compositions of a number of the large rivers in the Gaillardet data set, selected to illustrate the range observed. River water composition depends on a number of factors, which we discuss below.

Stallard and Edmond (1983) proposed a classification of rivers of the Amazon basin based on total dissolved cations (ΣZ^+) and

Table 13.8 Compositions of representative major rivers.

	Amazon	Congo	Mackenzie	Changjiang (Yangtze)	Mississippi	Lena	Rhone	Murray–Darling
Discharge	6590	1200	308	928	580	525	54	23.6
Na	80	96	330	222	478	196	491	4391
K	21	43	26	36	72	18	54	154
Ca	135	56	890	973	850	428	1770	525
Mg	37	59	337	292	366	210	267	700
Cl	61	37	226	151	294	343	640	4886
SO_4	47	15	369	164	266	142	479	396
HCO_3	344	258	1803	2311	1902	870	2885	1541
SiO_2	115	157	50	127	127	97	67	83
TDS (mg/L)	44	35	209	221	216	112	500	1271
Alk (µeq)	382	179	1879	2044	2164	851	1959	–
ΣZ^+ (µeq)	445	369	2810	2788	2982	1490	9085	6995
$SiO_2/\Sigma Z^+$	0.26	0.43	0.02	0.04	0.04	0.07	0.01	0.01
Na/Na+Ca	0.28	0.63	0.27	0.19	0.36	0.31	0.22	0.89

Concentrations in µmol/l. Concentrations and discharge from Gaillardet et al. (1999); alkalinity from Moon et al. (2014).

argued that the principal factors controlling water chemistry are rock type and the intensity of weathering. Berner and Berner (1996) showed that this classification can be extended to all rivers. The classification is as follows.

Transport-Limited Silicate Terranes. Stallard and Edmond's (1983) first category was rivers with $\Sigma Z^+ < 200$ µeq/l. They found that such rivers drain intensely weathered materials in a transport-limited regime (e.g., Rio Negro). Rivers in this category are enriched in SiO_2, Al, Fe, and organic anions and have a low pH. Perhaps the most significant feature of rivers in this category is their high $SiO_2/\Sigma Z^+$. The $SiO_2/\Sigma Z^+$ of the average upper crust is 2.4; silicate sedimentary rocks, derived from rocks that have been partially weathered, have a somewhat higher ratio (e.g., the ratio in riverine suspended matter is 3.2). Taken to its extreme, weathering leaves a residue of Fe and Al oxides, quantitatively stripping the alkali and alkaline earth cations as well as silica. Thus, the most intense weathering would produce water with $SiO_2/\Sigma Z^+$ of 2.4 or above, depending on the nature of the bedrock. Where weathering is somewhat less intense, leaving a kaolinite residue for example, the expected ratio would be closer to 1. Whereas in most rivers the bicarbonate concentration approximately equals the total alkalinity, indicating carbonate species are the principal nonconservative ions, bicarbonate concentrations are much lower than alkalinity in rivers in this category. This is a result of high concentrations of organic anions, which account for much of the alkalinity total anion concentration. None of the world's major rivers fall in this category, but many of the Amazon tributaries studied by Stallard and Edmond (1983) do.

Weathering-Limited Silicate Terranes. Stallard and Edmond (1983) proposed that rivers with ΣZ^+ between 200 and 450 µeq/l drained "weathering-limited siliceous terranes" (e.g., Congo, Amazon). In such regions the rate of erosion exceeds the rate of chemical weathering, and cations are leached from minerals in preference to SiO_2, leading to lower $SiO_2/\Sigma Z^+$ ratios of the water, while the availability of fresh rock results in higher ΣZ^+ and TDS than in transport-limited regimes. Rivers in this category have $SiO_2/\Sigma Z^+$ of between 0.1 and 0.5.

Carbonate Terranes. Stallard and Edmond's third category was those rivers with ΣZ^+ between 450 and 3000 µeq/l. Such rivers have low $Na^+/(Na^+ + Ca^{2+})$ and high Ca, Mg, alkalinity, and SO_4^{2-} (from oxidation of pyrite in reduced shales). Also, such rivers tend to have 1:1 ratios of Na to Cl and (Mg+Ca) to $(HCO_3^- + SO_4^{2-})$. These features indicate that these ions are derived from weathering of carbonates and evaporite minerals such as halite and gypsum. Rivers with these properties tend to drain areas underlain by marine sediments containing carbonates, reduced shales, and minor evaporites. Most of the world's major rivers fall in this category (Berner and Berner, 1996).

Evaporite Terranes. The fourth category of Stallard and Edmond was those rivers with $\Sigma Z^+ > 3000$ µeq/l. These rivers also tend to have 1:1 ratios of Na to Cl and (Mg+Ca) to $(HCO_3^- + SO_4^{2-})$. Such high ionic strength rivers drain terranes with abundant evaporites. Rivers in this and the third category have $SiO_2/\Sigma Z^+$ less than 0.1.

Stallard and Edmond (1983) used a ternary plot of carbonate alkalinity (HCO_3^-), SiO_2, and $Cl^- + SO_4^{2-}$ (all in µeq/l) to illustrate the differences between these categories of rivers. A similar plot is shown in Figure 13.14, showing data from river compositions listed in Table 13.8. Some caution is needed in comparing this with Stallard and Edmonds original work, however, because these data are not corrected for cyclic salts as were Stallard and Edmond's. In addition, the rivers they considered in the Amazon Basin will largely minimally impacted by anthropogenic activity. The combination of cyclic salts and anthropogenic activity can be significant. Several European rivers, including the Rhine, Weser, and Seine, whose high salinity places them in "evaporite" region of the plot do so only because of pollution.

Although it is clear that rock type and weathering regime (transport-limited vs. weathering limited) are the most important factors controlling river chemistry, we should emphasize that both precipitation and evapotranspiration do play some role, albeit a lesser one. The concentration of cyclic salts in rain decrease with distance from the ocean. For example, concentrations of Cl^- in rain in the Amazon Basin measured by Stallard and

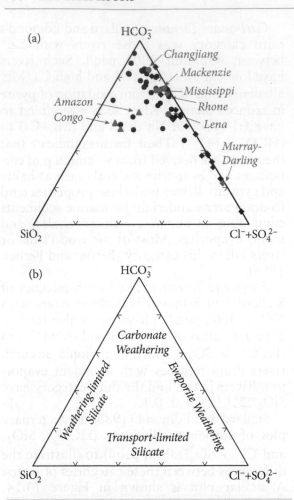

Figure 13.14   Ternary plot of SiO_2, $Cl + SO_4$, and HCO_3 used by Stallard and Edmond (1983) to illustrate how rock type and weathering intensity control river composition. (a) shows representative major rivers. Symbols are triangles: $\Sigma Z^+ < 450$ µmol/l; circles: $450 < \Sigma Z^+ < 3000$ µeq; diamonds: $\Sigma Z^+ > 3000$ µmol/l. Red symbols are rivers listed in Table 13.8. (b) shows where Stallard and Edmond's (1983) four categories plot.

Edmond (1981) were as high as 200 mg/l near the mouth of the Amazon and as low as 2 mg/l at a station on the western edge of the basin 2800 km from the Atlantic coast. For this reason, coastal rivers in areas of high rainfall, such as the southeastern US, are the ones most likely to be truly precipitation-dominated (Berner and Berner, 1996). Evaporative concentration, and precipitation of calcite in soils, does indeed influence the composition of rivers draining arid regions such as the Colorado and Rio Grande in the Southwestern

US. This is also true of the Murray-Darling River of southeastern Australia, which plots in the evaporite-weathering quadrant of Figure 13.14. However, there are no evaporites in the Murray-Darling watershed; rather the enrichment is due principally to high rates of evaporation and evapotranspiration in the arid watershed which it drains. Agricultural use of water increases evaporation, as two-thirds or more of the water used in irrigation evaporates. Thus, this activity serves to further increase the level of dissolved solids in such rivers.

Figure 13.14 nevertheless emphasizes the dominance of carbonate weathering in global river compositions. Crystalline rocks account for 34% of the global surface rock outcrop, but only 12% of the global riverine dissolved load. In contrast, evaporites constitute only 1.25% of the outcrop, but 17% of the dissolved load. Carbonate rocks, which constitute 16% of the outcrop, account for half of the global riverine dissolved load. This is consistent with the observation that most major rivers fall in Stallard and Edmond's carbonate category, even though carbonate rocks are rarer than silicate ones. At a more fundamental level, it is the weathering rates of minerals that are important in controlling river chemistry: silicates release dissolved components slowly; carbonates and evaporites dissolve rapidly.

13.6   CONTINENTAL SALINE WATERS

Saline waters result from evaporative concentration of fresh water. Saline lakes are common in arid regions, such as parts of China and Africa, the high plateaus of the Andes, and the western US. These can be valuable sources of resources, including of course various salts, but also lithium and boron; we'll discuss this in Chapter 15. Table 13.9 lists the principal components in a number of saline lakes and seas. As we might expect, there is a fair variation in concentrations with total salinity varying from about 1/7 of seawater to nearly 10 times that of seawater, Close inspection reveals something we might not expect: the relative concentrations of these elements also vary greatly. Some are carbonate brines, some are chlorine-rich brines. Some have high sulfate concentrations, and some do not. Na^+ is always a major cation,

Table 13.9   Saline lake brines.

	Qinghai China	Dead Sea	Mono Lake, Calif.	Lake Tyrell, Australia	Lake Borogia Kenya	Abert Lake, Oregon	Caspian Sea	Great Salt Lake	Nyer Co China	Pyramid Lake, Nevada
SiO_2	—		14		–	645	–	48	–	1
Na	46,800	39,330	21,500	77160	24,600	119,000	2280	81782	28000	1,630
K	16,540	6500	1170	660	435	3,890	70	2642	20440	134
Ca	610	17,750	4.5	514	12	–	270	208	920	10
Mg	15,320	40,450	34	7100	<1	–	540	4623	76080	113
Li		~20	10				0.3	21	23	
SO_4	56550	760	7380	15450	330	9,230	2100	9136	16900	264
Cl	88440	212600	13500	131000	5680	115,000	3850	81782	272970	1,960
ΣCO_3	4760	290	18680	290	30000	60,300	200	509	~0	1,390
Total	186570	317680	56600	21990	61060	309,000	9320	147211	415310	5,510
pH			9.6			9.8		7.4		

Concentrations in mg/l. From Deocampo and Jones (2014) and Eugster and Hardie (1978).

but relative proportions of Ca^{2+} and Mg^{2+} vary greatly. What leads to this diversity in composition?

We found in the previous section that rivers are quite diverse in their compositions, reflecting the lithology and weathering processes in the watersheds they drain. The lithology of the watersheds within which saline lakes occur is similarly the key factor in their subsequent chemical evolution. Most important, perhaps, is the role of crystallization in magnifying relatively small differences in the composition of source waters. This is very similar to the role played by crystallization in producing compositional diversity in igneous rocks. Indeed, it is possible to describe saline lake compositions in terms of *normative compositions*, that is the equilibrium mineral assemblage that would result if they completely crystallize, in a way very much analogous to the C.I.P.W. norms of igneous petrology. Let's briefly consider what happens when dissolved solids in water are evaporatively concentrated.

In almost all natural waters, the first mineral to precipitate will be calcite or aragonite. If the molar concentrations of Ca^{2+} and CO_3^{2-} are equal, the precipitation of calcite does not change the relative concentrations of these two ions. If, however, the concentration of CO_3^{2-} exceeds that of Ca^{2+}, even by a small amount, then crystallization of calcite leads to an increase in the relative concentration of CO_3^{2-} and a decrease in the concentration of Ca^{2+}. As long as evaporation continues,

calcite will continue to crystallize and Ca concentrations will continue to decrease, and those of CO_3^{2-} to increase. This process leads to a Ca-poor, CO_3^{2-}-rich brine. If the opposite is true, namely $Ca^{2+} > CO_3^{2-}$, then Ca^{2+} increases and CO_3^{2-} decreases, leading to a CO_3^{2-} poor brine. Although dolomite $(Mg,Ca)(CO_3)_2$ might be thermodynamically predicted to precipitate as well as this stage, dolomite reactions are so sluggish that this rarely occurs. This illustrates the concept of *chemical divide*, which is shown in Figure 13.15. Depending on initial composition, an evaporating solution will come to forks in the compositional evolution paths that lead to very different compositions.

Depending on the path taken at the calcite divide, the next divide is either precipitation of gypsum or precipitation of a Mg mineral, either magnesite or sepiolite $(Mg_4Si_6O_{15}(OH)_2 \cdot 6H_2O$, a mixed-layer, 2:1 clay) if silica concentrations are sufficient. Gypsum precipitation leads to either a sulfate-depleted or calcium-depleted brine. Both dolomite and sepiolite crystallization lead to either magnesium- or carbonate-poor brines, because sepiolite precipitation also consumes carbonate:

$$Mg^{2+} + 2HCO_3^- + 3H_4SiO_4$$
$$\rightleftharpoons MgSi_3O_6(OH)_2 + 2CO_2 + 6H_2O$$

If sulfate remains after gypsum precipitation exhausts dissolved Ca^{2+} or some Mg^{2+}

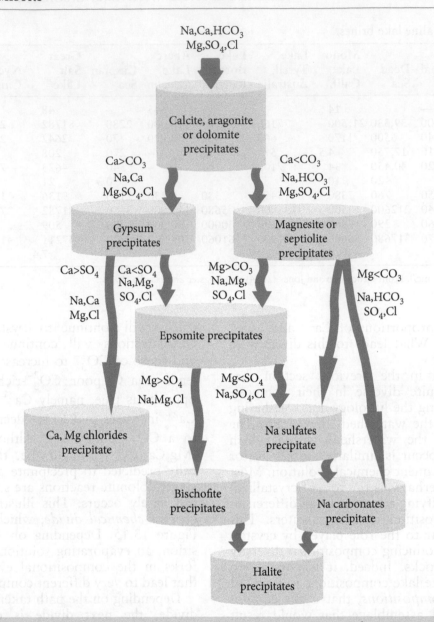

Figure 13.15 Chemical evolution of saline lakes and seas as a consequence of evaporative concentration and mineral precipitation. Based on Eugster and Hardie (1988), Drever (1988), and Deocampo and Jones (2014).

remains after sepiolite precipitation, magnesium sulfates such as epsomite, $(Mg(SO_4) \cdot 7(H_2O))$, will precipitate; if sulfate remains following this, sodium-bearing sulfates such as mirabilite $(Na_2SO_4 \cdot 10H_2O)$ or mixed sulfates such as leonite $(K_2SO_4 \cdot MgSO_4 \cdot 4H_2O)$ can subsequently precipitate. Carbonate remaining after magnesite precipitation may precipitate as sodium carbonates or bicarbonates such as trona $(Na_3H(CO_3)_2 \cdot 2H_2O)$ or nahcolite $(NaHCO_3)$ or a variety of mixed sulfate-carbonates containing Ca, Mg, and

K as well as Na. Chlorides are usually the most soluble salts of the cations present in saline lakes, and consequently they precipitate only when evaporative concentration reaches extreme stages. Chlorides might include not only halite but syvite (KCl) and bischofite $(MgCl_2 \cdot 6H_2O)$.

Several other factors play a role in determining the composition of saline waters. One is whether sulfate reduction occurs. The solubility of oxygen decreases with increasing salinity, so reducing conditions are more likely in saline than in fresh waters. Sulfate reduction obviously depletes sulfate (by converting it to insoluble sulfides), but it also increases carbonate alkalinity by production of carbonate because sulfate reduction occurs through the bacterial oxidation of organic matter. Another factor is ion exchange and absorption. Many saline lakes are fed by subsurface flow, providing the opportunity for ion exchange with clays and other minerals. This accounts for the low K concentrations of most saline waters.

Finally, cyclic wetting and drying can lead to significant changes in saline lake chemistry. These will occur not only on an annual basis, but as a consequence of longer cycles of drought and wetter climate. It can also lead to some interesting effects resulting from kinetics. In dry periods, evaporite minerals will precipitate subsurface. Rains in many dry areas come as occasional or rare cloud bursts and may wet the soil long enough to dissolve highly soluble salts such as sodium chloride, but not long enough to achieve equilibrium with less soluble salts such as gypsum. This can lead to sulfate concentrations lower than that expected from evaporative concentration of rainwater.

If inflow is primarily from low-density stream and river water, strong vertical density stratification can also result. In addition, horizontal circulation is often limited, particularly where marshy shallow regions of the lake exist. Thus, at any given time or place, the composition of saline lakes and seas may differ significantly from those listed in Table 13.9.

13.7   SUMMARY

In this chapter we focused on the *Critical Zone*, defined as the region from the top of vegetation to the bottom of circulating groundwater and comprising the biota, surface and groundwaters, regolith and soil. Nearly all life, and certainly all human life, depends on the geochemistry of this zone.

- We began by considering the oxidation–reduction chemistry of this zone, which is the interface between the oxidized atmosphere and the more reducing conditions of the Earth's interior. The primary reductant is organic matter and hence redox state is ultimately controlled by the balance between photosynthesis and respiration. A single dominant redox buffer, such as sulfate–sulfide, often controls the redox state of a system, and consequently the redox state of other redox-sensitive elements, such as Fe and Mn, can be assessed through that redox buffer.

- Soils develop through weathering of rock. Movement of components, *translocations*, in the weathering zone lead to the development of distinct soil horizons. These begin with the surface A horizon, consisting of organic matter and thoroughly weathered material. Components are leached from the A horizon and are carried downward to accumulate in the B horizon. Primary weathering reactions occur principally in the underlying C horizon, which consists of partially weathered rock. The biota contributes to soil development through the addition of organic matter including organic acids capable of complexing insoluble metals, and biolifting of nutrients through the soil. In some soils, dry and wet deposition from the atmosphere can be an important influence on soil chemistry.

- We then returned to the topic of weathering rates. In addition to supplying nutrients to the biota, weathering is important because silicate weathering consumes atmospheric CO_2 and hence influence climate on long time scales. Two approaches have been used to assess weathering rates. The *in situ* method examines concentration gradients within the weathering profile and the rate at which the weathering reaction zone progresses downward through time. We introduced several indices to assess the extent of weathering of parent rock: these are based on the relative concentrations of elements of differing mobility; Al and Fe are the least mobile and the alkalis and alkaline earths the most mobile. The second approach is the watershed approach, which is based on the fluxes of elements carried out of watersheds by streams. Both approaches produce weathering rates lower than those measured in the laboratory. Lithology, climate and hydrology, topography, and the biota all exert important controls on weathering rates.

- Streams and rivers carry away weathering products to the ocean or inland seas. Most rivers are an approximate dilute calcium bicarbonate solution but also contain *cyclic salts,* those present in precipitation and derived from seawater. The compositions of rivers depend on the weathering of the lithology of the areas they drain. They can be classified as transport-limited silicate terranes, weathering limited silicate terranes, carbonate terranes, and evaporite terranes. Nearly half the rivers fall in the carbonate terrane category even though carbonate rocks comprise only 16% of rock outcrop, reflecting the more rapid weathering of carbonates relative to silicates.

- Saline lakes show a much greater compositional diversity than rivers. This results from the existence of compositional divides as crystallization proceeds, which magnifies compositional differences in the input to these lakes.

REFERENCES AND SUGGESTIONS FOR FURTHER READING

Ashley, G. M., 1998. Where are we headed? "Soft" rock research into the new millennium. In: *Geological Society of America Annual Meeting Program with Abstracts* p. A-148, Geological Society of America.

Berner, R. A., A. C. Lasaga, and R. M. Garrells. 1983. The carbonate-silicate geochemical cycle and its effect on atmospheric carbon dioxide over the past 100 million years. *American Journal of Science* 283: 641–83.

Berner, E. K. and R. A. Berner, 1996. *Global Environment: Water, Air, and Geochemical Cycles.* Upper Saddle River, NJ, Prentice Hall.

Berner, R. A. 2006. GEOCARBSULF: A combined model for phanerozoic atmospheric O_2 and CO_2. *Geochimica et Cosmochimica Acta* 70: 5653–64. doi: 10.1016/j.gca.2005.11.032.

Bluth, G. J. S. and L. R. Kump. 1994. Lithologic and climatological controls of river chemistry. *Geochimica et Cosmochimica Acta* 58: 2341–60.

Brantley, S. L. and Y. Chen.1995. Chemical weathering rates of pyroxenes and amphiboles. in *Chemical Weathering Rates of Silicate Minerals, Reviews in Mineralogy 31*, ed. A. F. White and S. L. Brantley. 119–72. Washington: Min. Soc. Amer.

Brimhall, G. H. and W. E. Dietrich. 1987. Constitutive mass balance relations between chemical composition, volume, density, porosity, and strain in metasomatic hydrochemical systems: results on weathering and pedogenesis. *Geochimica et Cosmochimica Acta* 51: 567–88.

Brooks, R. R. 1972. *Geobotany and Biogeochemistry in Mineral Exploration.* New York: Harper and Row.

Chamberlin, T. C. 1899. An attempt to frame a working hypothesis of the cause of glacial periods on an atmospheric basis. *Journal of Geology* 7: 545–84. doi: 10.1086/608449.

Deocampo, D. M., Jones, B. F. (2014) Geochemistry of saline lakes. In: Holland H. D., Turekian K. K. (eds) *Treatise on Geochemistry (second edition),*. Elsevier, Oxford, pp. 437–69. doi: 10.1016/B978-0-08-095975-7.00515-5.

Drever, J. I., 1988. *The Geochemistry of Natural Waters,* Prentice Hall, Englewood Cliffs, 437 p.

Drever, J. I. 1994. The effect of land plants on weathering rates of silicate minerals. *Geochim. Cosmochim. Acta* 58: 2325–32.

Edmond, J. M., M. R. Palmer, C. I. Measures, B. Grant and R. F. Stalland. 1995. The fluvial geochemistry and denudation rate of the Guayana Shield in Venezuela, Columbia, and Brazil. *Geochimica et Cosmochimica Acta* 59: 3301–26.

Eugster, H. P. and L. A. Hardie, 1978, Saline Lakes, in *Lakes—Chemistry, Geology, Physics,* A. Lerman (ed.), Springer-Verlag, New York, pp. 237–93.

Kurtz, A. C., Derry, L. A., Chadwick, O. A., Alfano, M. J. 2000. Refractory element mobility in volcanic soils. *Geology* 28: 683–86. doi:10.1130/0091-7613(2000.)

Lasaga, A. C., Soler, J. M., Ganor, J., Burch, T. E., Nagy, K. L. 1994. Chemical weathering rate laws and global geochemical cycles. *Geochimica et Cosmochimica Acta* 58: 2361–86. doi: 10.1016/0016-7037(94)90016-7.

Lichtner, P. C. 1988. The quasi-stationary state approximation to coupled mass transport and fluid-rock interaction in a porous medium. *Geochimica et Cosmochimica Acta* 52: 143–65. doi: 10.1016/0016-7037(88)90063-4.

Maher, K. 2010. The dependence of chemical weathering rates on fluid residence time. *Earth and Planetary Science Letters* 294: 101–10. doi: 10.1016/j.epsl.2010.03.010.

Maher, K., Steefel, C. I., White, A. F., Stonestrom, D. A. 2009. The role of reaction affinity and secondary minerals in regulating chemical weathering rates at the Santa Cruz soil chronosequence, California. *Geochimica et Cosmochimica Acta* 73: 2804–31. doi: 10.1016/j.gca.2009.01.030.

Meybeck, M. 2003. Global occurrence of major elements in rivers. In: Holland HD, Turekian KK (eds) *Trestise on Geochemistry* vol 5. pp. 207–23.

Morel, F. M. M. and J. G. Hering. 1993. *Principles and Applications of Aquatic Chemistry*. New York: John Wiley and Sons.

Price, J. R., Velbel, M. A. 2003. Chemical weathering indices applied to weathering profiles developed on heterogeneous felsic metamorphic parent rocks. *Chemical Geology* 202: 397–416. doi: 10.1016/j.chemgeo.2002.11.001.

Price, J. R., Velbel, M. A. 2014. Rates of biotite weathering, and clay mineral transformation and neoformation, determined from watershed geochemical mass-balance methods for the Coweeta Hydrologic Laboratory, Southern Blue Ridge Mountains, North Carolina, USA. *Journal of Aquatic Geochemistry* 20: 203–24. doi: 10.1007/s10498-013-9190-y.

Probst, A., E. Dambrine, D. Viville and B. Fritz. 1990. Influence of acid atmospheric inputs on surface water chemistry and mineral fluxes in a declining spruce stand within a small granitic catchment (Vosage Massif, France). *Journal of Hydrology* 116: 101–24.

Perdue, E. M., Ritchie, J. D. (2014) Dissolved organic matter in freshwaters. In: Holland, H. D. and Turekian KK (ed) *Treatise on Geochemistry* (second edition) vol. 7 Elsevier, Oxford, pp. 237–272. doi: 10.1016/B978-0-08-095975-7.00509-X.

Richardson, J. B., 2017. Critical Zone. In: *Encyclopedia of Geochemistry* (ed White, W. M.), pp. 326–331, Springer International Publishing, Cham. doi: 10.1007/978-3-319-39312-4_355

Richardson, S. M. and H. Y. McSween. 1988. *Geochemistry: Pathways and Processes,* New York: Prentice Hall.

Robbins, J. A. and E. Callender. 1965. Diagenesis of manganese in Lake Michigan sediments. *American Journal of Science* 275: 512–33.

Schlesinger, W. H. 1991. *Biogeochemistry*. San Diego: Academic Press.

Schwartzman, D. W. and T. Volk. 1989. Biotic enhancement of weathering and the habitability of the Earth. *Nature* 340: 457–60.

Sposito, G. 1989. *The Chemistry of Soils*. New York: Oxford University Press.

Stallard, R. F. and J. M. Edmond. 1981. Geochemistry of the Amazon 1. Precipitation chemistry and the marine contribution to the dissolved load at the time of peak discharge. *Journal of Geophysical Research* 86: 9844–58.

Stallard, R. F. and J. M. Edmond. 1983. Geochemistry of the Amazon 2. The influence of geology and weathering environment on the dissolved load. *Journal of Geophysical Research* 88: 9671–88.

Stumm, W. and J. J. Morgan. 1995. *Aquatic Chemistry,* New York: Wiley and Sons.

Urey, H. C. 1952. On the early chemical history of the earth and the origin of life. *Proceedings of the National Academy of Sciences* 38: 351–63.

Velbel, M. A. 1985a. Geochemical mass balances and weathering rates in forested watersheds of the southern blue ridge. *American Journal of Science* 285: 904–30. doi: 10.2475/ajs.285.10.904.

Velbel, M. A. 1985b. Hydrogeochemical constraints on mass balances in forested watersheds of the Southern Appalachians. in *The Chemistry of Weathering*, ed. J. I. Drever. 231–47. Dordrecht: D. Reidel Publ. Co.

Velbel, M. A., Price, J. R. 2007. Solute geochemical mass-balances and mineral weathering rates in small watersheds: Methodology, recent advances, and future directions. *Applied Geochemistry* 22: 1682–700. doi: 10.1016/j.apgeochem.2007.03.029.

Walker, J. C. G., Hays, P. B., Kasting, J. F. 1981. A negative feedback mechanism for the long-term stabilization of earth's surface temperature. *Journal of Geophysical Research: Oceans* 86: 9776–82. doi: 10.1029/JC086iC10p09776.

White, A. F. 2008. Quantitative approaches to characterizing natural chemical weathering rates. In: Brantley SL, Kubicki JD, White AF (eds/) *Kinetics of Water-Rock Interaction*, vol. Springer, New York, pp. 469–543.

White, A. F. and A. E. Blum. 1995. Effects of climate on chemical weathering in watersheds. *Geochim. Cosmochim. Acta* 59: 1729–47.

White, A. F., Buss, H. L. 2014. Natural weathering rates of silicate minerals In: Holland, H. and Turekian K.K. (eds) *Treatise on Geochemistry* (second edition) vol. 7. Elsevier, Oxford, pp. 115–55.

White, A. F., Schulz, M. S., Stonestrom, D. A., et al. 2009. Chemical weathering of a marine terrace chronosequence, Santa Cruz, California. Part II: Solute profiles, gradients and the comparisons of contemporary and long-term weathering rates. *Geochimica et Cosmochimica Acta* 73: 2769–803. doi: 10.1016/j.gca.2009.01.029.

White, A. F., Schulz, M. S., Vivit, D. V., Blum, A. E., Stonestrom, D. A., Anderson, S. P. 2008. Chemical weathering of a marine terrace chronosequence, Santa Cruz, California I: Interpreting rates and controls based on soil concentration–depth profiles. *Geochimica et Cosmochimica Acta* 72: 36–68. doi: 10.1016/j.gca.2007.08.029.

PROBLEMS

1. Consider the weathering of An33 plagioclase (i.e., 33% anorthosite; 67% albite) to kaolinite.

   (a) Write a balanced dissolution reaction similar to reaction 13.27 for this plagioclase such that kaolinite and dissolved Na^+, Ca^{2+}, and H_4SiO_4 are the products.

   (b) Assuming that plagioclase is an ideal solution, use the data below to calculate its standard state free energy.

(c)   Write an equilibrium constant expression for this reaction with the usual assumptions that the solids and the water are pure substances.

Component	$\Delta G°_f$ (kJ/mol)
Kaolinite	−3797.8
Albite	−3708.3
Anorthite	−4215.6
$H_4SiO_{4(aq)}$	−1310.1
H_2O	−238.1
H^+	0
$Na+$	−261.9
Ca^{2+}	−553.5

(d)   Using the data above, calculate the standard state free energy change (ΔG_r) for your reaction.

(e)   Calculate the equilibrium constant at standard state conditions (25°C and 0.1 MPa) from the ΔG_r you calculated above.

(f)   Assuming the composition of the soil solution in Table 13.6 is in equilibrium with this plagioclase and kaolinite, predict the concentration of H_2SiO_4 from your reaction.

2. The dissolution of *talc* may be described by the reaction:

$$Mg_3Si_4O_{10}(OH)_2 + 6H^+ + 4H_2O \rightleftharpoons 3Mg^{2+} + 4H_4SiO_4 \qquad (1)$$

that of *brucite* as:

$$Mg(OH)_2 + 2H^+ \rightleftharpoons Mg^{2+} + 2H_2O \qquad (2)$$

and that of *serpentine* as:

$$Mg_3Si_2O_5(OH)_4 + 6H^+ \rightleftharpoons 3Mg^{2+} + 2H_4SiO_4 + H_2O \qquad (3)$$

The solubility of quartz is 10^{-4}. The ΔG for these reactions (at 25° C) have been estimated as −114.09 kJ, −96.98 kJ, and −193.96 kJ respectively. On a plot of log ($[Mg^{2+}]/[H^+]^2$) vs. log $[H_4SiO_4]$ show the stability fields for talc, brucite, and serpentine.

3. For this question, assume that the dissolution rate of brucite (reaction 6.57) shows the dependence on Q/K given in equation 5.68, that K= $10^{-11.6}$, the value of $\mathfrak{R}_+$ is 10^{-6} moles/sec, a temperature of 25°C. and no other ions are present in solution. If all other factors remain constant, how will the dissolution rate change as equilibrium between brucite and solution is approached? Make a plot showing the relative reaction rate as a function of log ($[Mg^{2+}]/[H+]^2$).

4. The following is an analysis of the Congo River in Africa (units are mg/l).

pH	6.87
Ca^{2+}	2.37
Mg^{2+}	1.38
Na^+	1.99
K^+	1.40
Cl^-	1.40
SO_4	1.17
HCO_3	13.43
SiO_2	10.36
TDS	54

(a) Calculate the *alkalinity* of this water. How does alkalinity compare with the bicarbonate concentration (*HINT:* be sure to use molar units in your comparison).

(b) Calculate the $Na^+/(Na^+ + Ca^{2+})$, ΣZ^+ (µeq/l) and $SiO_2/\Sigma Z^+$ molar ratios of this river.

(c) Into which of Stallard and Edmond's categories would this river fall?

5. Referring to Figure 13.12, would you expect the Congo River water of Problem 4 to be in equilibrium with gibbsite, kaolinite, illite, or K-feldspar?

6. Assuming a dissolved aluminum concentration of 0.05 mg/l in Congo River water, calculate the concentration of the various aluminum species at the pH given in problem 4 and the equilibrium constants in equations 6.79 through 6.82.

7. Use the analysis of Congo River water in Problem 4 for these questions.

(a) What is the ionic strength of this solution?

(b) Calculate the Debye–Hückel activity coefficients for this solution.

(c) The bicarbonate concentration reported is actually total carbonate (ΣCO_2). Calculate the actual concentrations of each of the three carbonate species.

8. Complete the problem posed in Example 13.1 by calculating the Fe^{3+}/Fe^{2+} ratio and the concentration of Fe^{3+}.

9. Compute the solution to Eqn. 13.30 to find the weathering fluxes for allanite (all), plagioclase, garnet, biotite, vermiculite (verm), kaolinite (kaol), gibbsite (gibbs), and biom. This is most easily done in MATLAB, but Excel has matrix functions that can be used to solve this as well.

Chapter 14

The ocean as a chemical system

14.1 INTRODUCTION

Antoine Lavoisier* called seawater "the rinsings of the Earth." Given the tenuous understanding of geological processes existing at the time (the late eighteenth century), this is a remarkably insightful observation. Most of the salts in the oceans are derived from weathering of the continents and delivered to the oceans by rivers. Nevertheless, the composition of seawater is quite unlike what you would obtain by evaporative concentration of river water. Also unlike rivers, the concentrations of the major ions in seawater are remarkably uniform, being present in constant proportions and varying only with total salt content: salinity.

In contrast, the trace and minor element composition of seawater varies substantially, and elements can cycle between different forms within the oceans including both organic and inorganic solids as well as various dissolved species. This internal cycling is intimately tied to the various physical, geological, and biological processes occurring within the ocean. The biota plays a particularly crucial role both in internal chemical cycling and in controlling the overall composition of seawater.

Lavoisier's statement also reminds us that the oceans are part of a grander geochemical system. Sediments deposited in the ocean provide a record of that system. On human time scales at least, the ocean appears to be very nearly in steady state. It is tempting to apply Lyell's principle of uniformitarianism and assume that the composition of the seawater has also been constant on geologic time scales. Some aspects of seawater composition do change over time, as we found in Chapters 8 and 9. Precisely because these variations are related to changes in other geological processes, such as plate tectonics, climate, life, and atmospheric chemistry, they can tell us much about the Earth's history and the workings of the planet. Interpreting these past changes begins with an understanding of how the modern ocean works and the controls on its composition. This understanding is our main goal for this chapter.

14.2 SOME BACKGROUND OCEANOGRAPHIC CONCEPTS

Understanding ocean chemistry requires some understanding of the physical, biological, and geological processes occurring within the ocean. The concentrations of dissolved elements vary both vertically and horizontally in the ocean as a consequence of these processes. In this section, we will define some oceanographic terms and then briefly consider the circulation within the oceans.

---

* Antoine Lavoisier, born in France in 1743, is often called the father of modern chemistry. Unfortunately, his day job was as tax collector and he died at the guillotine in 1794.

Geochemistry, Second Edition. William M. White.
© 2020 John Wiley & Sons Ltd. Published 2020 by John Wiley & Sons Ltd.
Companion website: www.wiley.com/go/white/geochemistry

14.2.1 Salinity, chlorinity, temperature, and density

Salinity is effectively, but not exactly, the total dissolved solids in seawater. It is defined as the weight in grams of the dissolved inorganic matter in one kilogram of water after all the bromide and iodide have been replaced by the equivalent amount of chloride and all carbonate converted to oxide (CO_2 driven off). This unfortunate definition is due to the difficulty early scientists, including Robert Boyle*, had in measuring salinity by drying to constant weight. Salinity is now determined by measuring electrical conductivity, which increases in direct relation to the concentrations of ions in water, and hence with salinity. Salinity, denoted S, is defined as having units of parts per thousand (‰). Because it is defined this way, it is unnecessary to state the units, although they sometimes are.

Another useful definition is chlorinity, which is the halide concentration in grams per kilogram measured by titration with silver and calculated as if all the halide were chloride (total halides are actually 0.043% greater than chlorinity). Chlorinity, which also has defined units of part per thousand, can also be measured by conductivity. As we shall see, Cl is always present in seawater as a constant proportion of total salt, and therefore there is a direct relationship between chlorinity and salinity. By definition:

$$S = 1.80655 \, Cl \qquad (14.1)$$

Standard seawater, which is slightly more concentrated than average seawater, has a salinity of 35.000 and a chlorinity of 19.374 parts per thousand. Open ocean water rarely has a salinity greater than 38 or less than 33.

Temperature, along with salinity, determines the density of seawater. Since density differences drive much of the flow of ocean water, these are key oceanographic parameters. Temperature in the oceans can be reported as either *potential* or *in situ* temperature, but the former is the most commonly used. *In situ* temperature is the actual temperature of a parcel of water at depth. Potential temperature, denoted θ, is the temperature the water would have if brought to the surface. The difference between the two is thus the temperature difference due to adiabatic expansion. Since water cools when it expands, potential temperature is always less than *in situ* temperature (except for surface water). The difference is small, on the order of 0.1°C. While this difference is important to physical oceanographers, it is generally negligible for our purposes.

Temperature and salinity, and therefore also density, can be changed only at the surface. Temperature is changed only through solar heating of the upper few meters or loss of heat to the atmosphere at the surface, any heat lost or gained from the ocean floor being negligible. Salinity too is only changed at the surface through dilution by rain or river water or concentration by evaporation or freezing. Once a water mass leaves the ocean surface, these properties can only change through mixing with other water masses. Thus, oceanographers refer to temperature and salinity as *conservative* properties of seawater.

The density of seawater is 2–3% greater than that of pure water. Average seawater, with a salinity of ~35 and a temperature of 20°C, has a density of 1.0247 g/cc. Density is usually reported as the parameter σ, which is the *per mil* deviation from the density of pure water (1 g/cc). Thus if density is 1.0247 g/cc, σ is 24.7. Again, one can distinguish between *in situ* and potential density, potential density being the density water would have if brought to the surface and is always lower than or *in situ* density (surface water again being an exception). The difference is small and generally negligible for our purposes.

14.2.2 Circulation of the ocean and the structure of ocean water

Ocean circulation, like that of the atmosphere, is ultimately driven by differential heating of the Earth: more solar energy is gained at low latitudes than at high latitudes. Because the mechanisms of surface and vertical circulation in the oceans are somewhat different, it is convenient to treat them separately.

---

* Robert Boyle (1627–1691) was another of the founders of modern chemistry. He defined the chemical element as the practical limit of chemical analysis, and deduced the inverse relationship between the pressure and volume of gas, a version of the ideal gas law.

14.2.2.1 Surface circulation

Surface circulation of the ocean is driven primarily by winds; hence the surface circulation is sometimes also called the *wind-driven circulation*. The important features are as follows:

- Both north and south of the climatic equator, known as the Inter-Tropical Convergence, or ITCZ, water moves from east to west, driven by the trade winds. These currents are known as the North and South Equatorial Currents. Between these two currents, a narrow and generally weak Equatorial Counter Current can run from west to east.

- Two large gyres operate in both the Atlantic and Pacific Oceans, one in the Northern and one in the Southern Hemisphere. Rotation is clockwise in the Northern Hemisphere and counterclockwise in the Southern Hemisphere. The Coriolis Force, an apparent force that results from the Earth's rotation, is largely responsible for this circular current pattern. These currents are most intense along the western boundaries of ocean basins, a phenomenon also due to the Earth's rotation. Examples of intense western boundary currents are the Gulf Stream and Kuroshio Current.

- The circulation in the Indian Ocean is somewhat similar but undergoes radical seasonal changes in response to the monsoons. In Northern Hemisphere summer, the North Equatorial Current reverses and joins the equatorial countercurrent to become the Southwest Monsoon Current. The Somali Current, which flows to the southwest along the African Coast in Northern Hemisphere winter, reverses direction to flow northeastward in Northern Hemisphere summer.

- Water moves from west to east in Southern Ocean (the globe-encircling belt of ocean south of Africa and America). This is called the Antarctic Circumpolar Current or West Wind Drift. Directly adjacent the Antarctic coast, a counter current, called the Polar Current, runs east to west. The Antarctic Convergence or Antarctic Polar Front occurs at the interface between these currents.

14.2.2.2 Density structure and deep circulation

The deep circulation of the oceans is driven by density differences. Seawater density is controlled by temperature and salinity, so this circulation is also called the *thermohaline circulation* Most of the ocean is stably stratified; that is, each layer of water is denser than the layer above and less dense than the layer below. Where this is not the case, a water mass will move up or down until it reaches a level of equilibrium density. *Upwelling* of deeper water typically occurs where winds or currents create a *divergence* of surface water. *Downwelling* occurs where winds or currents produce a *convergence* of surface water. Wind and current-driven upwelling and downwelling link the surface and deep circulation of the ocean.

In the modern ocean, temperature differences dominate density variations and are principally responsible for deep circulation. This may not have always been the case, however. During warmer periods in the past, such as the Cretaceous, deep circulation may have been driven principally by salinity differences.

Figure 14.1 shows how temperature, salinity, and density typically vary with depth in temperate and tropical regions. The upper hundred meters or so is nearly uniform in temperature and salinity due to mixing by waves (the actual depth of the mixed layer varied both seasonally and geographically, depending largely on wave height). Below the upper mixed layer is a region called the *thermocline*, where temperature decreases rapidly. Salinity may also change rapidly in this region; a region where salinity changes rapidly is called a *halocline*. The temperature changes cause a rapid increase in density with depth, and this region of the water column is called the *pycnocline*. Below the pycnocline, temperature and salinity vary less with depth. In polar regions, water may be essentially isothermal throughout the water column.

The pycnocline represents a strong boundary to vertical mixing of water and effectively isolates surface water from deep water. This leads to a useful chemical simplification: the two-box model of the ocean. In this model, the ocean is divided into a box representing the surface water above the pycnocline, and one representing the deep water below it (Figure 14.2). Fluxes between these boxes

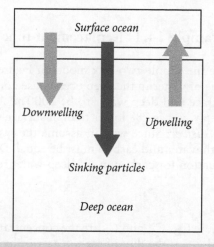

Figure 14.2 A two-box model of the ocean and the fluxes between them.

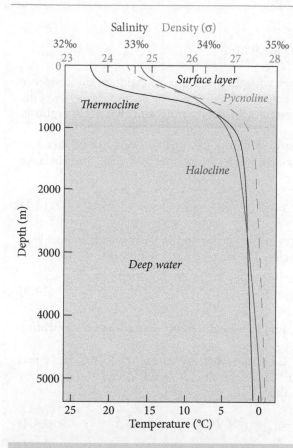

Figure 14.1 Typical temperate region temperature, salinity, and density variations with depth. Although a thermocline is present in most regions of the ocean, its exact nature and depth varies with location and season. Polar regions exhibit little change in temperature with depth. Salinity profiles are most more variable. Since density is primarily a function of temperature, the pycnocline usually mimics the thermocline.

can occur both because of advection of water (upwelling and downwelling) and because of falling particles, both organic and inorganic. The upper box exchanges with the atmosphere and receives all the riverine input. All photosynthetic activity occurs in the upper box because light does not penetrate below 100 m (only 0.5% of the incident sunlight penetrates to a depth of 100 m, even in the clearest water). On the other hand, the flux out of the ocean (to sediment) of both particles and dissolved solids occurs through the lower box. Since the depth of the surface layer varies in the ocean and the density boundary is gradational rather than sharp, any definition of the size of the boxes is rather arbitrary. The depth of the boundary between

the surface and deep layer may be variably defined, depending on the particular problem at hand. In Example 14.1, for instance, we define it as 1000 m.

Water flows across the pycnocline only a few limited regions; we can divide these into regions of "intermediate water" formation and "deep water" formation (*formation* refers to a water mass acquiring its temperature and salinity characteristics at the surface before sinking through the pycnocline). Intermediate waters do not usually penetrate below depths of 1500 m; deep water may penetrate to the bottom. There are four principal regions of intermediate water production. The first is in the Mediterranean where evaporation increases salinity of surface water to 37–38‰ and that water is cooled in winter causing it to sink. It flows out of the Strait of Gibraltar and sinks and spreads out in the Atlantic at a depth of about 1000 m. This water is known as Mediterranean Intermediate Water (MIW). Another intermediate water, known as Antarctic Intermediate Water (AAIW), is produced at the convergence at the Antarctic Polar Front at about 50°S. North Pacific Intermediate Water and North Atlantic Intermediate Water are produced at the Arctic Polar Front at 50–60° N. Of these water masses, AAIW is the densest and most voluminous. Because of this limited exchange of water across the thermocline, the ventilation time of the oceans, the time between when water sinks through the thermocline and reemerges at the surface, is long (see Example 14.1).

Example 14.1 Replacement time of deep ocean water

Use the simple two-box model in Figure 14.2 together with the following to estimate the residence time of water in the deep ocean based on the specific activity of ^{14}C. Take the boundary between the surface and deep water to be 1000 m. Assume the system is at steady state and that $^{14}C/C$ ratio in deep water is 7% lower than in surface water.

Answer: Since we can assume the system is at steady state, the upward and downward fluxes of both water and carbon must be equal. Denoting the water flux by J_W we may write the mass balance equation for carbon in the deep-water reservoir as:

$$J_W C_D = J_W C_S + J_{CP} \tag{14.2}$$

where C_D, C_S, and C_P are the concentrations of carbon in deep water and shallow water, respectively, and F_{CP} is the flux of carbon carried by sinking particles. The sinking particle flux is thus just:

$$J_{CP} = J_W(C_D - C_S) \tag{14.3}$$

In other words, the sinking particle flux must account for the difference in carbon concentration between the surface and deep water.

We may now write a mass balance equation for ^{14}C in deep water by setting the loss of ^{14}C equal to the gain of ^{14}C. ^{14}C is lost through the upward flux of water and radioactive decay and gained by the downward flux of water and the sinking particle flux.

$$J_W C_D (^{14}C/C)_D - \lambda V_D C_D (^{14}C/C)_D = J_W C_S (^{14}C/C)_S + J_{CP} (^{14}C/C)_S \tag{14.4}$$

where $(^{14}C/C)_D$, and $(^{14}C/C)_S$ are the $^{14}C/C$ ratios in deep and shallow water, respectively, V_D is the volume of deep water, and λ is the decay constant of ^{14}C. We have implicitly assumed that sinking particles have the same ^{14}C activity as surface water. Substituting 14.3 into 14.4, we have

$$J_W C_D (^{14}C/C)_D + \lambda V_D C_D (^{14}C/C)_D = J_W C_S (^{14}C/C)_S + J_W(C_D - C_S)(^{14}C/C)_S \tag{14.5}$$

Rearranging and eliminating terms we have

$$\lambda V_D (^{14}C/C)_D = J_W (^{14}C/C)_S - J_W (^{14}C/C)_D \tag{14.6}$$

Another rearrangement and we arrive at:

$$V_D/J_W = [1 - (^{14}C/C)_D/(^{14}C/C)_S]/[\lambda(^{14}C/C)_D/(^{14}C/C)_S] \tag{14.7}$$

As we shall see later in this chapter, we define steady-state residence time as the amount in a reservoir divided by the flux into it or out of it. Thus the above equation gives the residence time of water in the deep ocean (notice it has units of time). Substituting 0.1209×10^{-3} yr^{-1} for λ (Table 8.5) and 0.93 for $(^{14}C/C)_D/(^{14}C/C)_S$, we calculate a residence time of 623 years.

We can also use this equation to calculate the average upward velocity of water. Rearranging 14.7, we have:

$$J_W = V_D\lambda(^{14}C/C)_D/(^{14}C/C)_S/[1 - (^{14}C/C)_D/(^{14}C/C)_S] \tag{14.8}$$

If we express the volume of the deep ocean as the average depth, d, times area, A, we have:

$$J_W = Ad\lambda(^{14}C/C)_D/(^{14}C/C)_S/[1 - (^{14}C/C)_D/(^{14}C/C)_S] \tag{14.9}$$

Dividing both sides by A, we have:

$$J_W/A = d\lambda(^{14}C/C)_D/(^{14}C/C)_S/[1 - (^{14}C/C)_D/(^{14}C/C)_S] \qquad (14.10)$$

F_W/A is the velocity. Taking d as 3000 m, we calculate F_W/A as 4.81 m/yr. (This calculation follows a similar one in Broeker and Peng, 1983). Based on some 15000 ^{14}C analyses of ocean water throughout the world, Matsumoto (2007) determined ages of deep water ranging from 290 years for the Atlantic and Southern Oceans to 900 years for the Pacific. The oldest water, in excess of 1000 years, is found in the North Pacific and northern Indian Oceans.

There are only two regions of deep-water production, both at high latitudes. Antarctic Bottom Water (AABW), which is the densest and most voluminous deep water in the ocean, is produced primarily in the Weddell Sea. Cold winds blowing from Antarctica cool it, while freezing of sea ice increases its salinity. The other deep-water mass, North Atlantic Bottom Water (NADW), is produced around Iceland in winter when winds cause upwelling and cooling of saline MIW. NADW then sinks and flows southward along the western boundary of the Atlantic. It upwells off the coast of Antarctic where it cools and sinks to becomes part of the AABW.

Water returns to the surface through a slow, diffuse upward advection. Final return from the thermocline to the surface occurs in localized zones of wind-driven upwelling, including in the along the equator, where the trade winds create a divergence of surface water, along the west coasts of continents, where winds blowing along the coast drive the water offshore (this is a process known as Ekman transport and is related to the Coriolis force), and at the Antarctic divergence south of the Antarctic Polar Front.

Figure 14.3 shows a somewhat more complete box model of ocean circulation. The great mixing zone is the Southern Ocean. It is here that NADW mixes with AABW and flows into all the world's oceans. Both the Indian and Pacific import deep water and export surface water, which also passes through the Southern Ocean. In contrast, the Atlantic exports deep water in the form of NADW and imports surface water. This has consequences for biological productivity because, as we will see, surface water is nutrient-poor while deep water is nutrient rich.

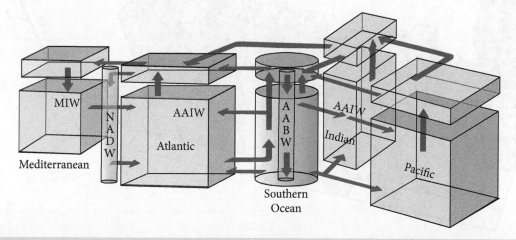

Figure 14.3 Box model of ocean circulation through the major ocean basins, all of which are connected through the Southern Ocean. Red arrows show major flows. MIW: Mediterranean Intermediate Water, NADW: North Atlantic Deep Water, AAIW: Antarctic Intermediate Water, AABW: Antarctic Bottom Water. Boxes are not to scale. Adapted from de Baar (2018).

14.3 COMPOSITION OF SEAWATER

Table 14.1 lists the concentrations and chemical form of the elements in seawater. Concentrations range over 12 orders of magnitude (16 if H and O are included). From Figure 14.4 we see that the most abundant elements in seawater are those on the "wings" of the periodic table, the alkalis, the alkaline earths, and the halogens. In the terminology we introduced in Chapter 6, these elements form "hard" ions that have inert gas electronic structures. Bonding of these elements is predominantly ionic; they have relatively small electrostatic energy and large radii (low z/r ratio), so that in solution they are present mainly as free ions rather than complexes. Elements in the interior of the periodic table are generally present at lower concentrations. These elements have higher z/r ratios, form bonds of a more covalent character, and are strongly hydrolyzed. The latter tendency leads to their rapid removal by adsorption on particle surfaces. A few elements are exceptions to this pattern. These are elements, such as S, Mo, Tl, and U, that form highly soluble oxyanion complexes, for example, SO_4^{2-}, MoO_4^{2-}, UO_4^{2-} or soluble simple ions (e.g., Tl^+).

Although solubility provides a guide to elemental concentrations in seawater, the composition of seawater is not controlled by solubility. Rather, the composition of seawater is controlled by a variety of processes, from tectonism on the planetary scale to surface adsorption/desorption reactions at the atomic scale.

Many of the same processes that remove the elements from seawater, and thus play a role in controlling its composition, also impose vertical, and to a lesser degree horizontal, concentration gradients in the ocean. Table 14.1 also assigns each element to one of four categories based on their vertical distribution in the water column. C: conservative; CG: conservative gas; N: biologically

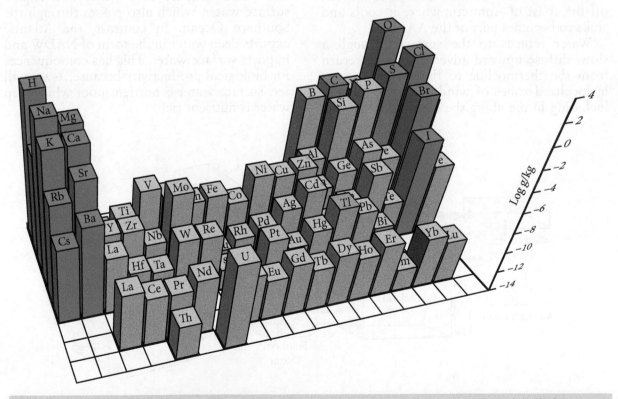

Figure 14.4 Histogram of elemental abundances in seawater. In addition to H and O, the alkalis, alkaline earths, and halogens plus C and S (in red), which form oxide anion radicals, are the most abundant elements in seawater. These abundances are strikingly different than terrestrial abundances shown in Figure 7.1.

Table 14.1 Concentrations of elements dissolved in seawater and river water.

Element	Average seawater concentration (μg/kg)	Average seawater concentration (μmol/kg)	Open ocean range	Principal dissolved Species	Distribution	River water concentration (μg/liter)
H	1.1×10^8	54.2×10^6		H_2O	C	1.1×10^8
He	7.2×10^{-3}	0.0018		He	NG	—
Li	179	25.9		Li^+	C	1.84
Be	0.0021	2.3×10^{-5}	4–30 pM	$BeOH^+$, $Be(OH)_2$	S/N	0.0089
B	4.16×10^3	416		$B(OH)_3$, $B(OH)_4^-$	C	10.2
C	2.7×10^4	2250	2.05–2.35 mM	HCO_3^-, CO_3^{2-}	N	—
C(org.)	1×10^2	8.3		various	—	—
N_2	8260	590		N_2	CG	—
NO_3	186	30	<1–45 μM	NO_3^-	N	—
O	8.9×10^8	55.6×10^6		H_2O	C	—
O	2800	175	1–350 μM	O_2	inverse N	—
F	1292	68		F^-	C	5.3
Ne	0.164	0.0081		Ne	CG	—
Na	1.076×10^7	4.68×10^5		Na^+	C	5,300
Mg	1.288×10^6	5.30×10^4		Mg^{2+}, $MgSO_4$	C	3,100
Al	0.054	0.002	0.3–40 nM	$Al(OH)_4^-$	S	50
Si	2800	100	0.5–180 μM	H_4SiO_4	N	5,000
P	71	2.3	<0.1–3.5 μM	HPO_4^-	N	–
S	8.99×10^5	2.80×10^5		SO_4^{2-}, $NaSO_4^{2-}$	C	2840
Cl	1.936×10^7	5.46×10^5		Cl^-	C	4700
Ar	636	15.9		Ar	CG	—
K	3.97×10^5	1.02×10^4		K^+	C	1450
Ca	4.05×10^5	1.01×10^4		Ca^{+2}, $CaSO_4^{2-}$	~C	14,500
Sc	0.00072	1.6×10^{-6}	8–20 pM	$Sc(OH)_3$	S/N	1.2
Ti	0.0070	1.5×10^{-4}	6–250 pM	$Ti(OH)_4$, $Ti(OH)_2$	S/N	0.49
V	1.78	0.035	30–37 nM	HVO_4^{2-}	~C	0.71
Cr	0.208	0.004	3–5 nM	CrO_4^-	S/N	0.7
Mn	0.016	3.0×10^{-4}	0.08–5 nM	Mn^{+2}	S	34
Fe	0.028	5.0×10^{-4}	0.03–3 nM	$Fe(OH)^{2+}$	S/N	66
Co	0.002	4×10^{-5}	3–300 pM	Co^{2+}	S	0.15
Ni	0.45	8×10^{-3}	2–12 nM	Ni^{2+}	N	0.8
Cu	0.19	3×10^{-3}	0.5–5 nM	$CuCO_3$	S/N	1.48
Zn	0.39	0.006	0.05–10 nM	Zn^{+2}	N	0.6
Ga	0.0014	2×10^{-4}	5–60 pM	$Ga(OH)_4^-$	S/N	0.03
Ge	0.03	4.7×10^{-4}	0.4–0.5 nM	$CH_3Ge(OH)_3$	N	6.8×10^{-3}
As	1.72	0.023	17–25 nM	$HAs(OH)_4^{2-}$	S/N	0.62
Se	0.14	0.0018	0.5–2.3 nM	SeO_4^{2-}	N	0.07
Br	6.7×10^4	840		Br^-	C	20
Kr	0.32	0.0038		Kr	CG	—
Rb	128	1.4		Rb^+	C	1.63
Sr	7884	90		Sr^{+2}	~C	60
Y	0.018	2×10^{-4}	60–300 pM	$Y(CO_3)^-$	S/N	0.04
Zr	0.015	1.6×10^{-4}	9–300 pM	$Zr(OH)_5^-$	S/N	0.036
Nb	0.00027	3×10^{-6}	1–4 pM	$Nb(OH)_6^-$	S?	1.7×10^{-3}

(Continued)

Table 14.1   *(continued)*

Element	Average seawater concentration (μg/kg)	Average seawater concentration (μmol/kg)	Open ocean range	Principal dissolved Species	Distribution	River water concentration (μg/liter)
Mo	10	0.104	–	MoO_4^{2-}	C	0.42
Ru	2×10^{-8}	$\sim2\times10^{-8}$	–	$Ru(OH)_n^{4-n}$	?	—
Rh	8×10^{-5}	8×10^{-7}	0.4–1 pM	$Rh(OH)_n^{3-n}$	N	—
Pd	7.5×10^{-5}	7×10^{-7}	0.2–1 pM	$PdCl_4^-$	N	0.028
Ag	0.0022	2×10^{-5}	1–35 pM	$AgCl_2^-, AgCl_3^{2-}$	N	0.3
Cd	0.067	6×10^{-4}	1–1050 pM	$CdCl^0$	N	0.08
In	1.2×10^{-5}	1×10^{-7}	0.04–2 pM	$In(OH)_3$	S	—
Sn	4.7×10^{-4}	4×10^{-6}	1–20 pM	$SnO(OH)_3^-$	S	—
Sb	0.19	0.0016	–	$Sb(OH)_6^-$	C	0.07
Te	7.6×10^{-5}	6×10^{-7}	0.5–1.5 pM	$Te(OH)_6$	S	—
I	57.1	0.450	0.35–0.46 μM	IO_3^-	$\sim$C	0.05
Xe	0.065	5×10^{-4}	–	Xe	CG	—
Cs	0.3	2.2×10^{-3}	–	Cs^+	C	0.011
Ba	15	0.11	30–150 nM	$Ba^{+2}, BaCl^+$	N	23
La	3.89×10^{-3}	2.6×10^{-5}	8–40 pM	$La(CO_3)^+$	S/N	0.12
Ce	8.55×10^{-4}	6.1×10^{-6}	1.5–8 pM	$Ce(CO_3)^+$	S/N	0.26
Pr	5.91×10^{-4}	4.2×10^{-6}	1–10 pM	$Pr(CO_3)^+$	S/N	0.04
Nd	2.55×10^{-3}	1.8×10^{-5}	4–50 pM	$Nd(CO_3)^+$	S/N	0.152
Sm	5.08×10^{-4}	3.4×10^{-6}	1–10 pM	$Sm(CO_3)^+$	S/N	0.036
Eu	1.44×10^{-4}	9.7×10^{-7}	0.2–2.5 pM	$Eu(CO_3)^+$	S/N	0.001
Gd	7.27×10^{-3}	4.6×10^{-6}	1.5–12 pM	$Gd(CO_3)^+$	S/N	0.04
Tb	1.17×10^{-4}	7.4×10^{-7}	0.5–2 pM	$Tb(CO_3)^+$	S/N	0.0098
Dy	9.15×10^{-4}	5.6×10^{-6}	2–14 pM	$Dy(CO_3)^+$	S/N	5.5×10^{-3}
Ho	2.47×10^{-4}	1.5×10^{-6}	0.8–2.2 pM	$Ho(CO_3)^+$	S/N	0.0071
Er	8.48×10^{-4}	5.1×10^{-6}	1.5–12 pM	$Er(CO_3)^+$	S/N	0.02
Tm	1.29×10^{-4}	7.7×10^{-7}	0.3–2 pM	$Tm(CO_3)^+$	S/N	3.3×10^{-3}
Yb	8.78×10^{-4}	5.1×10^{-6}	1–12 pM	$Yb(CO_3)^+$	S/N	0.04
Lu	1.51×10^{-4}	8.6×10^{-7}	0.2–2 pM	$Lu(CO_3)^+$	S/N	2.4×10^{-3}
Hf	5.35×10^{-5}	9×10^{-7}	0.06–1 pM	$Hf(OH)_4$	S/N	5.9×10^{-3}
Ta	1.81×10^{-5}	1×10^{-7}	0.01–0.3 pM	$Ta(OH)_5$	S?	1.1×10^{-3}
W	0.0101	5.5×10^{-5}	–	WO_4^{2-}	C	0.1
Re	0.00745	4×10^{-5}	–	ReO_4^-	C	0.0004
Os	1.14×10^{-6}	6×10^{-9}	3–8 fM	—	?	9×10^{-6}
Ir	2×10^{-7}	$\sim8\times10^{-10}$	0.5–1 fM	$Ir(OH)_3$?	S?	—
Pt	1.6×10^{-4}	8×10^{-7}	0.2–1.5 pm	$PtCl_4^{2-}$ (?)	$\sim$C	—
Au	9.8×10^{-6}	5×10^{-8}	10–100 fM	$AuCl_2^-, Au(OH)_3$	S?	0.0001
Hg	2×10^{-4}	1×10^{-6}	0.2–10 pM	$HgCl_4^{2-}$	S/N	–
Tl	1.4×10^{-2}	7×10^{-5}	60–75 pM	$Tl^+, TlCl$	$\sim$C	3.5×10^{-4}
Pb	2.1×10^{-3}	1×10^{-5}	4–150 pM	$PbCO_3$	S	0.08
Bi	2.1×10^{-5}	1×10^{-7}	10–500 fM	$BiO^+, Bi(OH)_2^+$	S	—
Ra	1.3×10^{-7}	5.8×10^{-10}		Ra^{+2}	N	2.4×10^{-8}
Th	7×10^{-5}	3×10^{-7}	0.3–0.6 pM	$Th(OH)_4$	S	0.041
U	3.2	0.0135	–	$UO_2(CO_3)_3^{4-}$	C	0.37

Category: C: Conservative, N: Nutrient/Biologically Controlled, S: Scavenged CG: conservative gas; NG: nonconservative gas. Sources: Bruland (2014), Broecker and Peng (1982) de Barr et al. (2018) and Gaillardet et al. (2014); REE abundances based on literature compilation of recent high-quality analyses.

controlled, "nutrient-type" distribution); S: scavenged. In the following sections, we will examine the behavior of each of these groups and the processes responsible for these gradients.

14.3.1 Speciation in seawater

Table 14.1 lists the principal species present for each element. The major ions in seawater, Na^{2+}, K^+, Mg^{2+}, Ca^{2+}, Cl^-, SO_4^{2-}, and HCO_3^- are predominantly present (>95%) as free ions.where n is the stoichiometric

coefficient and is 2 for Ni(OH)2 and 1 for all others. Many of the trace metals, however, are present primarily as complexes. The wide variety of elements and the relatively high concentrations of ligands in seawater lead to the formation of a variety of complexes. The fraction of each element present as a given species may be calculated if the stability constants are known (Chapter 6). Calculation of major ion speciation requires an iterative procedure, similar to that in Example 6.7. Calculation of trace element speciation is fairly straightforward, as demonstrated in Example 14.2.

Example 14.2 Inorganic complexation of Ni in seawater

Using the stability constants (β^0) for Ni complexes and the free ligand concentrations in the table below, calculate the fraction of total dissolved Ni in each form. Assume a temperature of 25°C. Use the following apparent stability constants from Millero and Schreiber (1982) and ligand concentrations in the table below.

Complex	Log β^*	Log [Ligand]
NiOH$^+$	4.2	−6
Ni(OH)$_2$	7.46	−6
NiCl$^+$	2	−2.6
NiCO$_3$	0.8	−4.5
Ni(SO$_4$)	4.4	−1.6

Answer: The concentration of each complex is given by:

$$[NiL_n] = \beta^*[Ni^{2+}][L]^n \qquad (14.11)$$

where n is the stoichiometric coefficient and is 2 for Ni(OH)$_2$ and 1 for all others. The conservation equation for Ni is:

$$\Sigma Ni = [Ni^{2+}] + [Ni(OH)^-] + [Ni(OH)_2] + [NiCl^-] + [NiSO_4] + [NiCO_3]$$

We can rewrite this as:

$$\Sigma Ni = [Ni^{2+}](1 + \beta^*[Ni^{2+}][OH^+] + \beta^*[Ni^{2+}][OH^+]^2 + \dots)$$

The fraction of Ni present as each species is then $[NiL_n]/\Sigma Ni$ where NiL_n is calculated in equation 14.11. Ni is present predominantly as free ion and carbonate, with minor amounts of the sulfate and as chloride.

Principal Ni Complexes in Seawater		
	Log [NiL]/[Ni]	% form
Ni^{2+}	0	45.1%
$NiOH^+$	−1.80	0.7%
$Ni(OH)_2$	−4.54	0%
$NiCl^+$	−0.6	11.5%
$NiSO_4$	−0.80	8.0%
$NiCO_3$	−0.10	33.7%

Table 14.2 Major ions in seawater.

Ion	g/kg (ppt) at S = 35‰	Percent of dissolved solids
Cl^-	19.354	55.05
SO_4^{2-}	2.712	7.68
HCO_3^-	0.140	0.41
$B(OH)_4^-$	0.0323	0.07
Br^-	0.0673	0.19
F^-	0.0013	0.00
Na^+	10.77	30.61
Mg^{2+}	1.290	3.69
Ca^{2+}	0.412	1.16
K^+	0.399	1.10
Sr^{2+}	0.008	0.03

14.3.2 Conservative elements

The conservative elements share the property of being always found in constant proportions to one another and to salinity in the open sea, even though salinity varies. All the major ions in seawater, except for bicarbonate, are included in this group. Their concentrations are listed in Table 14.2. This constancy of the major ion composition of seawater, which is typically expressed as a ratio to Cl, is sometimes called the *law of constant proportions* and has been known for nearly two centuries. For most purposes, we may state that concentrations of these elements vary in the ocean only through dilution or concentration of dissolved salts by addition or loss of pure water. While chemical and biological processes occur within the ocean do change seawater chemistry, they have an insignificant effect on the concentrations of conservative elements.

The major ions do vary in certain unusual situations, namely (1) in estuaries, (2) in anoxic basins (where sulfate is reduced), (3) when freezing occurs (sea ice retains more sulfate than chloride), (4) in isolated basins where evaporation proceeds to the point where salts begin to precipitate, and (5) as a result of hydrothermal inputs to restricted basins (e.g., red sea brines). Ca and Sr are slight exceptions to the rule in that they are inhomogeneously distributed even in the open ocean, though only slightly. The concentrations of these elements, as well as that of HCO_3^-, vary as a result of biological production of organic carbon, calcium carbonate, and strontium sulfate* in the surface water and sinking of the remains of organisms into deep water. Most of these biologically produced particles break down in deep water, releasing these species into solution, resulting in a flux of these elements from surface waters to deep waters (the *biological pump* we discussed in Chapter 12; we explore this further below). As a result, deep water is about 15% enriched in bicarbonate, 1% enriched in Sr, and 0.5% enriched in Ca relative to surface water. As we shall see, these biological processes also create much larger vertical variations in the concentrations of many minor constituents.

Some minor and trace elements are also present in constant proportions; these include Rb, Mo, Cs, Re, Tl, and U. Vanadium is nearly conservative, with a total range of only about ±15%. These elements share the properties that they are not extensively utilized by the biota and form soluble ions or radicals that are not surface reactive.

14.3.3 Dissolved gases

The concentrations of dissolved gases in the oceans are maintained primarily by exchange with the atmosphere. These gases may be

---

* A class of protozoans called *Acantharia* build shells of $SrSO_4$.

divided into conservative and nonconservative ones. The noble gases, except He, and nitrogen constitute the conservative gases. As their name implies, their concentrations are not affected by internal processes in the ocean. Concentrations of these gases are governed entirely by exchange with the atmosphere. Since they are all minor constituents of seawater, we can use Henry's law to describe the equilibrium solubility of atmospheric gases in the ocean:

$$C = kp \qquad (14.12)$$

where C is the concentration in seawater, k is the Henry's law constant, or Bunsen absorption coefficient, and p is the partial pressure of the gas in the atmosphere. The light noble gases are the least soluble; the heavy noble gases and CO_2 are the most soluble. The conservative gases are not uniformly distributed in the ocean. This is because of the temperature dependence of gas solubility: they are more soluble at lower temperature. Over a temperature range of 0° to 30° C, this produces a variation in dissolved concentration of about a factor of two for several gases. The temperature dependence is strongest for the heavy noble gases and CO_2, and weakest for the light noble gases. Thus, the concentration of conservative gases in seawater depends on the temperature at which atmosphere–ocean equilibration occurred. An interesting aspect is that gas solubilities are nonlinear functions of temperature, and consequently, mixing of water masses formed at different temperatures can lead to concentrations above the solubility curves.

O_2 and CO_2 are the principal nonconservative gases. They vary because of photosynthesis and respiration. Nitrogen is also biologically utilized. However, only a small fraction of the dissolved N_2 is present as "fixed" nitrogen (as NH_4, NO_3^-, and N_2O); hence, N_2 behaves effectively as a conservative element.

Helium is another nonconservative gas because of the input of He to the ocean by hydrothermal activity at mid-ocean ridges. Elevated He concentrations and high $^3He/^4He$ ratios found at mid-depth, particularly in the Pacific, reflecting this injection of mantle He by mid-ocean ridge hydrothermal systems. In the following sections, we examine the variation of O_2 and CO_2 in the ocean in greater detail.

14.3.3.1 O_2 variation in the ocean

Biological activity occurs throughout the oceans but is concentrated in the surface water because only there is there sufficient light for photosynthesis. This part of the water column is called the *photic zone*. Both respiration and photosynthesis occur in the surface water, but the rate of photosynthesis exceeds that of respiration in the surface ocean, so there is a net O_2 production at the surface layer. Most organic matter produced in the surface ocean is also consumed there, but a fraction sinks into the deep water. This sinking organic matter is consumed by bacteria and scavenging deep water organisms and results in a net consumption of oxygen.

Within the deep water, two factors govern the distribution of oxygen. The first is the "age" of the water, the time since it last exchanged with the atmosphere. The longer deep water has been away from the surface, the more depleted in oxygen it becomes. The second factor is the abundance of organic matter. The abundance of food decreases with depth, so that respiration, and hence oxygen consumption, is highest just below the photic zone and lowest in the deepest water. Thus, the vertical distribution of oxygen is often characterized by an oxygen minimum that typically occurs within the thermocline. Primary production (i.e., photosynthesis) in the surface waters varies geographically (for reasons we will subsequently discuss), and oxygen is more depleted in deep water underlying high biological productivity regions that lie beneath regions of low productivity.

Figure 14.5 compares the distribution of O_2 with temperature and salinity in a north–south profile of the Atlantic Ocean. Highest O_2 concentrations are found in surface waters at high latitudes, where the water is cold and the solubility of O_2 is highest. The minimum O_2 concentrations are found in mid-latitudes at depths of 500–1000 m. Oxygen minima at these depths characterize most of the world ocean at temperate and tropical latitudes. At greater depth, as well as at higher latitudes, the concentration of O_2 is higher because the water there is generally younger, i.e., it has more recently exchanged at the surface. Because deep water in the Pacific and Indian Oceans is older than deep water in the Atlantic, oxygen concentrations are generally

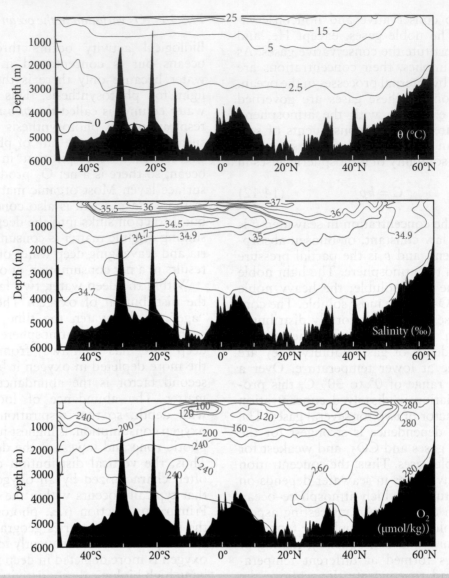

Figure 14.5 Variation of potential temperature, salinity, and oxygen concentrations in the Atlantic Ocean along the GEOTRACES transect GA02. Based on visualization in the eGEOTRACES website (http://www.egeotraces.org) by Reiner Schlitzer.

lower. This is apparent from a comparison of Figures 14.5 and 14.6; the latter shows data from a hydrographic station in the tropical south-central Pacific where deep water O_2 levels are generally <150 µmol/kg compared with generally >240 µmol/kg in the Atlantic.

A particularly strong O_2 depletion occurs beneath high-productivity regions of the eastern equatorial Pacific and conditions are locally suboxic (i.e., no free O_2). Anoxic conditions develop in deep water in basins where the connection to the open ocean is restricted. The best example is the Black Sea.

The Black Sea is a 2000 m deep basin whose only connection with the rest of the world ocean is through the shallow Bosporus Strait. As a result, water becomes anoxic at a depth of about 100 m. Anoxia is also present in the Curacao Trench, off the northern coast of South America. Here, anoxia is a result both of restricted circulation and high productivity in the overlying surface water. Anoxic conditions also develop in some deep fjords.

Hypoxic conditions, in which oxygen is not entirely consumed but fall below the level necessary to support animal life, occur locally

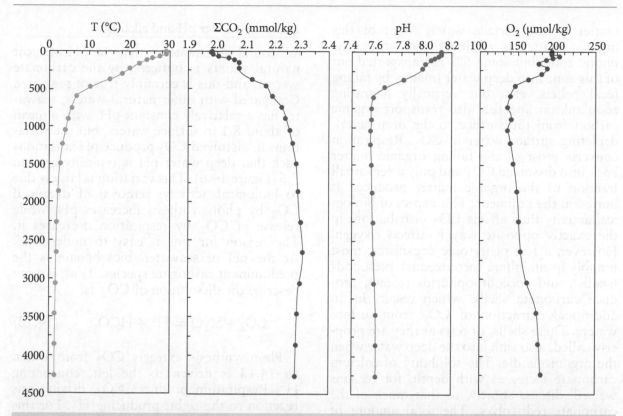

Figure 14.6 Variation of temperature, ΣCO_2, pH, and O_2 with depth in the south-central tropical Pacific at 150°W, 16.5°S; R/V Ron Brown Cruise P16C, Station 1, 4 Apr. 2015. ΣCO_2 is low and O_2 is high in the photic zone as a result of photosynthesis then increasing and decreasing, respectively through the thermocline. pH varies inversely with ΣCO_2. Data retrieved from: NOAA CLIVAR and Carbon Hydrographic Data Office, La Jolla, CA, USA; DOI: 10.7942/C2WC7C https://cchdo.ucsd.edu.

in several regions in the oceans. In some of these areas, hypoxia is human-caused. One example in the northwestern Gulf of Mexico, where the Mississippi carries excess fertilizer-derived nitrate and phosphate into the ocean, resulting in excessive phytoplankton productivity in spring. When these die, they are decomposed by bacteria below the photic zone, which drives O_2 to critically low levels.

14.3.3.2 Distribution of CO_2 in the ocean

As we found in previous chapters, most CO_2 dissolved in water will be present as carbonate or bicarbonate ion. Nevertheless, we will often refer to all species of carbonate and CO_2 simply as CO_2 or ΣCO_2. CO_2 cannot be treated strictly as a dissolved gas, as there are sinks and sources of CO_2 other than the atmosphere. For example, some of the dissolved CO_2 is delivered to the ocean by rivers

as bicarbonate ion. These properties make the distribution of CO_2 more complicated than that of other gases.

Like oxygen, CO_2 concentrations are affected by biological activity, and its solubility is affected by temperature. These factors result in a significant geographic variation in CO_2 in surface water. Surface water is often supersaturated with respect to the atmosphere in equatorial regions as upwelling brings CO_2-riched deeper water to the surface and warming decreases it solubility. The greatest degree of undersaturation occurs in polar regions, where photosynthesis decreases CO_2 and cooling increases its solubility. There is a net flux of CO_2 from the ocean to the atmosphere in low latitudes and a net flux from the atmosphere to the ocean in high latitudes.

Biological activity is responsible for vertical variation in CO_2 in the ocean (Figure 14.6). Photosynthesis converts CO_2 to organic matter in the surface water. Most of this organic matter is remineralized within the photic zone, but some 5% is transported out of this zone into deep water (mainly by falling fecal pellets, etc.; but vertically migrating zooplankton and fish also transport organic carbon from the surface to the deep layer), depleting surface water in CO_2. Respiration converts most of the falling organic matter back into dissolved CO_2 and only a very small fraction of the organic matter produced is buried in the sediment. This aspect of biological activity thus affects CO_2 distributions in the exactly opposite way it affects oxygen. However, a few planktonic organisms, most notably foraminifera (protozoans), pteropods (snails), and coccolithophorids (algae), produce carbonate shells, which results in an additional extraction of CO_2 from surface waters. These shells, or tests as they are properly called, also sink into the deep water when the organisms die. The solubility of calcium carbonate increases with depth, for reasons we will discuss shortly, so that much of the carbonate redissolves. The total amount of dissolved CO_2 converted to carbonate is small compared with that converted to organic carbon. However, a much large fraction of biogenic carbonate sinks out of the photic zone, so that the downward flux of carbon in carbonate represents about 6% of the total downward flux of carbon. A larger fraction of carbonate produced is also buried, so that the flux of carbon out of the ocean is due primarily to carbonate sedimentation rather than organic matter sedimentation.

The transport of CO_2 from surface to deep water as organic matter and biogenic carbonate is called *the biological pump*. As we might expect, the biological pump produces an enrichment of CO_2 in the deep ocean over the shallow ocean (Figure 14.6).

Biological activity also produces a variation in the isotopic composition of carbon in seawater, as we found in Chapter 9, and imposes a gradient in $\delta^{13}C$ on the water column (Figure 9.17). The extent of depletion of ^{12}C in surface water will depend on biological activity: $\delta^{13}C$ will be higher in productive waters than in unproductive waters. The extent of enrichment of ^{12}C in deep water depends on the age of the deep water, as does CO_2. "Old" deep water will have lower $\delta^{13}C$ than "young" deep water.

14.3.4 Seawater pH and alkalinity

We found in Chapter 6 that the pH of most natural waters is buffered by the carbonate system, and this is certainly true of seawater. Compared with other natural waters, seawater has a relatively constant pH, with a mean of about 8.1 in surface waters, but the variations in dissolved CO_2 produce pH variations such that deep water pH is typically around 7.6 (Figure 14.6). This variation is largely due to biological activity: removal of dissolved CO_2 by photosynthesis increases pH, while release of CO_2 by respiration decreases it. The reason for this is easy to understand. At the pH of seawater, bicarbonate is the predominant carbonate species. Thus, we can describe the dissolution of CO_2 as:

$$CO_2 + H_2O \rightleftharpoons H^+ + HCO_3^- \qquad (14.13)$$

Photosynthesis extracts CO_2 from water, so 14.13 is driven to the left, consuming H^+. Respiration produces CO_2, driving this reaction to the right, producing H^+. For this reason, the pH of the ocean decreases with depth. In Figure 14.6 we see pH and O_2 declining and CO_2 increasing rapidly below the photic zone. In this region, there are relatively high amounts of organic matter falling from the photic zone and hence high net respiration.

pH is also affected by precipitation and dissolution of calcium carbonate. Since bicarbonate is the most abundant carbonate species, the precipitation reaction is effectively:

$$Ca^{2+} + HCO_3^- \rightleftharpoons H^+ + CaCO_3 \qquad (14.14)$$

Here is it easy to see that precipitation of calcium carbonate decreases pH while dissolution increases it. Thus, production of biogenic carbonate in surface water and its dissolution in deep water acts to reduce the vertical pH variations produced by photosynthesis and respiration.

Another important parameter used to describe ocean chemistry, and one closely related to pH, is alkalinity. In Chapter 6 we defined alkalinity as the sum of the concentration (in equivalents) of bases that are titratable with strong acid. It is a measure of acid-neutralizing capacity of a solution. An operational definition of total alkalinity for seawater is (in anoxic environments, we would need to include the HS^- ion):

$$Alk = [HCO_3^-] + [CO_3^{2-}] + [B(OH)_4^-]$$
$$+ 2[HPO_4^{2-}] + [NO_3^-] + [OH^-] - [H^+]$$
$$(14.15)$$

Often, particularly in surface water, the phosphate and nitrate terms are negligible and becomes nearly identical to *carbonate alkalinity*, which is:

$$CAlk = [HCO_3^-] + 2[CO_3^{2-}] + [OH^-] - [H^+]$$
$$(14.16)$$

(which is identical to 6.33). One of the reasons that alkalinity is important is that it can be readily determined by titration.

In Chapter 6, we stated that alkalinity is "conservative," meaning that it cannot be changed except by the addition or removal of components. It is important to understand that alkalinity is not conservative in an oceanographic sense, as is, for example, salinity. In an oceanographic sense, we define a *conservative* property to be one that changes only at the surface by concentration or dilution. Concentration and dilution affect alkalinity; indeed, these processes are the principal cause of variation in alkalinity (alkalinity is strongly correlated with salinity). However, precipitation and dissolution in the ocean do significantly affect alkalinity (whereas the effect on salinity is negligible), so alkalinity is not conservative. Indeed, alkalinity typically varies systematically with depth, being greater in deep water than in the surface water.

What causes this depth variation? It might be tempting to guess that photosynthesis and respiration are responsible. However, these processes have no direct effect on alkalinity. When CO_2 dissolves in water, it dissociates to produce a proton and a bicarbonate ion. In the alkalinity equation, these exactly balance, so there is no effect on alkalinity. Production and oxidation of organic matter do affect alkalinity through the uptake and release of phosphate and nitrate, but the concentration of these nutrients is generally small. The main cause of the systematic variation of alkalinity in the water column is carbonate precipitation and dissolution. For every mole of calcium carbonate precipitated, a mole of carbonate is removed, and alkalinity increases by two equivalents, and vice versa, so the effect is quite significant.

14.3.5 Carbonate dissolution and precipitation

From the preceding sections, we can see that precipitation of calcium carbonate in surface waters and its dissolution at depth is an important oceanographic phenomenon. Carbonate sedimentation is also an important geological process in other respects, including its role in the global carbon cycle. In the previous two chapters we found that silicate weathering consumes atmospheric CO_2 and exerts a strong control on climate on geological time scales. CO_2 consumed in weathering is carried to the oceans in the form of bicarbonate ion. The final step is precipitation of carbonate and burial in sediment. However, much of the carbonate precipitated in the surface water redissolves in the deep water. This dissolution produces alkalinity that consumes H^+ when the carbonate ion is converted to bicarbonate.

Let's examine carbonate precipitation and dissolution in a little more detail. Two forms of calcium carbonate precipitate from seawater. Most carbonate shell-forming organisms, including the planktonic foraminifera and coccolithophorids that account for most carbonate precipitated, precipitate calcite. Pteropods and many corals, however, precipitate aragonite, even though aragonite, the high-pressure form of calcium carbonate, is not thermodynamically stable anywhere in the ocean. The surface ocean is everywhere supersaturated with respect to both calcite and aragonite, usually to depths of 1000 m or more.* Nevertheless, except in some rather rare and unusual situations, carbonate precipitation occurs only when biologically mediated. There are two interesting geochemical questions here. First, why does the

---

* You might ask how aragonite can be supersaturated if it is not thermodynamically stable. It is supersaturated because aragonite has a lower Gibbs free energy than dissolved calcium carbonate in surface seawater, but aragonite has a higher Gibbs free energy than calcite, so it is unstable with respect to calcite.

ocean go from supersaturated at the surface to undersaturated at depth, and second, why doesn't calcium carbonate precipitate without biological intervention?

There are three reasons why the oceans become undersaturated with respect to calcium carbonate at depth. First, increasing P_{CO_2} of deep water drives pH to lower levels, increasing solubility. This might seem counterintuitive, as one might think that that increasing P_{CO_2} should produce an increase in the carbonate ion concentration and therefore drive the reaction toward precipitation. However, increases in P_{CO_2} and ΣCO_2 with depth produce a *decrease* in CO_3^{2-} concentration. This is most easily understood if we express the carbonate ion concentration as a function of total CO_2 using the solubility and dissociation constants for the carbonate system (eqns. 6.18 through 6.20):

$$[CO_3^{2-}] = \{(\Sigma CO_2)/(10^{-2\,pH}/K_1K_2)\}$$
$$+ 10^{-pH}/K_2 + 1)\} \qquad (14.17)$$

This equation shows that while carbonate is proportional to ΣCO_2, it is inversely proportional to the square of $[H^+]$. The pH drop resulting from production of CO_2 by respiration is thus dominant. Carbonate ion concentrations drop by over a factor of three from the surface waters to the waters with the highest dissolved CO_2.

The second reason is that the solubility of calcium carbonate increases with increasing pressure. This results from the positive ΔV of the precipitation reaction. Calcite and aragonite are about twice as soluble at 5000 m (corresponding to a pressure of 500 atm or 50 MPa) than at 1 atmosphere. The solubility of $CaCO_3$ is also temperature-dependent (see Example 14.3) and salinity-dependent (due to the effect of ionic strength on the activity coefficients), but both these effects are rather small in the open ocean. Finally, the dissociation constants of carbonic acid also vary with temperature and pressure such that in deep cold water there is less CO_3^{2-} for a given amount of ΣCO_3.

Example 14.3 Calcite solubility in the oceans

Calculate the calcite saturation state, Ω, of Pacific Ocean water shown in Figure 14.6 at the surface water and at the bottom at 4272 m depth. Temperature, total CO_2, salinity, and pH were measured at 29.12°C, 1982.6 μmol/kg, 35.66, and 8.12, respectively, in the surface water and 1.44°C, 2283.7 μmol/kg, 34.70, and 7.60, respectively, in the bottom water corresponding to a pressure of 42.7 MPa. Assume a Ca^{2+} concentration of 10.28 mmol/kg at both depths. The volume change for ΔV_r for calcite precipitation is 31.8 cc/mol. Lueker et al. (2000) parameterized the apparent first and second carbonic acid dissociation constants in seawater as:

$$pK_1 = 3633.86/T - 61.2172 + 9.6777\ln(T) - 0.011555S + 0.0001152S^2 \qquad (14.18)$$

$$pK_2 = 471.78/T - 25.929 + 3.16967\ln(T) - 0.01781S + 0.0001122S^2 \qquad (14.19)$$

The calcite solubility constant as a function of temperature and salinity (Mucci, 1983) is:

$$pK_{cal}^{app} = -191.945 - 0.77993T + 2903.293/T + 711.595\log(T)$$
$$+(-0.77712 + 2.8426T + 178.34)S^{0.5} - 0.07711S + 4.129S^{1.5} \qquad (14.20)$$

Answer: the saturation state, Ω, is the ratio of the ion activity product, Q, to the equilibrium constant, $[Ca^{2+}][CO_3^{2-}]/K^{app}$. We can begin by calculating the carbonate ion concentration from ΣCO_2 and pH from eqn. 14.17 using the calculated equilibrium constants in eqns. 14.18, 14.19 above for given temperatures and salinities. Doing so we calculate the carbonate concentrations of 0.283 and 0.0370 mmol/kg for the surface and deep water, respectively.

The ion activity products are calculated (units in mol/kg) as:

$$Q = [CO_3^{2-}][Ca^{2+}] \tag{14.21}$$

The results are 2.83×10^{-6} and 3.80×10^{-7} for the surface and deep water, respectively. Next, we need to correct for the effects of pressure.

We calculate the log solubility products using eqn. 14.20 as -6.19 and -6.18 for the surface and deep water, respectively (not much difference!). We calculate saturation state for the surface water, Ω, of 4.49. However, we also need to correct the deep-water K_{cal} value for the effects of pressure. Assuming both calcite and seawater are incompressible, we can use eqn. 3.97a to do this. Rewritten in $\log_{10}$ rather than natural log, this equation becomes:

$$\log K_{T,P} = \log K_{T,0.1} - \frac{\Delta V_r (P - 0.1)}{2.303 RT}$$

Using this equation, we calculate pK_{cal} as -5.92 and calculate $\Omega = 0.31$, indicating the bottom water is undersaturated with respect to calcite. (Using a slightly more sophisticated approach which accounts for the dependence of ΔV_r on T and P, the value of Ω is a little lower, 0.1).

The kinetics of carbonate precipitation are still not fully understood, in spite of several decades of research. Quite a bit is known, however, particularly about the calcite precipitation and dissolution. A number of laboratory studies (e.g., Chou et al., 1989; Zuddas and Mucci, 1994) have concluded that the principal reaction mechanism of calcite precipitation in seawater is:

$$Ca^{2+} + CO_3^{2-} \rightleftharpoons CaCO_3 \tag{14.22}$$

In other words, this simple stoichiometric expression best represents what actually occurs on the molecular level (however, other mechanisms appear to predominate at lower pH). In Chapter 5, we found that the net rate of reaction can be expressed as:

$$\mathfrak{R}_{net} = \mathfrak{R}_+ - \mathfrak{R}_- \tag{5.57}$$

If 14.22 is the elementary reaction describing precipitation, then:

$$\mathfrak{R}_+ = k_+ [Ca^{2+}][CO_3^{2-}] \tag{14.23}$$

and

$$\mathfrak{R}_- = k_- [CaCO_3] \tag{14.24}$$

where k_+ and k_- are the forward and reverse rate constants, respectively. Taking the concentration of $CaCO_3$ in the solid is 1, the net reaction rate should be:

$$\mathfrak{R}_{net} = k_+ [Ca^{2+}][CO_3^{2-}] - k_- \tag{14.25}$$

However, Zhong and Mucci (1993) and Zuddas and Mucci (1994) found that under conditions of constant $[Ca^{2+}]$, the overall rate equation is:

$$\mathfrak{R}_{net} = k_f [CO_3^{2-}]^3 - k_r \tag{14.26}$$

where k_f is an *apparent* precipitation rate constant and k_r is the apparent dissolution rate constant (incorporating both the rate constant and Ca^{2+} concentration). In other words, the reaction is third order with respect to the carbonate ion, rather than first order as expected, and the rate accelerates exponentially with increasing oversaturation. This indicates that other processes must be involved and strongly influence the reaction rate. Many such factors have been identified and include ionic strength (i.e., salinity), Mg^{2+}, and SO_4^{2-}, the calcium to carbonate ratios, humic acids and other organic substances, temperature, and pressure. Some of these factors are discussed in Lopez et al. (2009).

In contrast to the precipitation reaction, dissolution of carbonate appears to begin close to the depth where undersaturation is reached. However, dissolution is not instantaneous. If we could remove the water from the ocean basins, we would find that mountain peaks in the ocean are covered with white carbonate sediment, but that carbonate is absent from the deeper plains and valleys. This picture is very much reminiscent of what we often see in winter: mountain peaks covered with snow while the valleys are bare. Although the snow line, the elevation where we first find snow, depends on temperature, it does not necessarily correspond to the 0°C isotherm. Indeed, it will generally be somewhat lower than this. The snow line is that elevation where the rate at which snow falls just matches the rate at which it melts. The *carbonate compensation depth* (CCD) is the depth at which the rate at which carbonate accumulates just equals the rate at which it dissolves and hence is not preserved in the sediment. By the analogy of melting snow discussed above, this is sometimes called the *marine snowline*. The depth of the CCD varies

as a consequence of varying calcite saturation state, which in turn reflects the variation in ΣCO_2 and pH. It is deepest in the Atlantic, where it as deep as 5500 m, and shallowest in the North Pacific, where it is as shallow as 3500 m. It averages about 4500 m.

To investigate the kinetics of calcite dissolution, Peterson (1966) hung carefully weighed spheres of calcite at various depths in the ocean. He recovered the spheres 265 days later and reweighed them to determine the rate of dissolution as a function of depth. The results showed a rapid increase in dissolution at a depth of about 3500 m (Figure 14.7), a region known as the *lysocline*. Since then, this experiment has been duplicated several times with increasing sophistication.

A large number of calcite dissolution laboratory studies have also been carried out showing that at the pH and ionic strength of seawater, the dominant dissolution

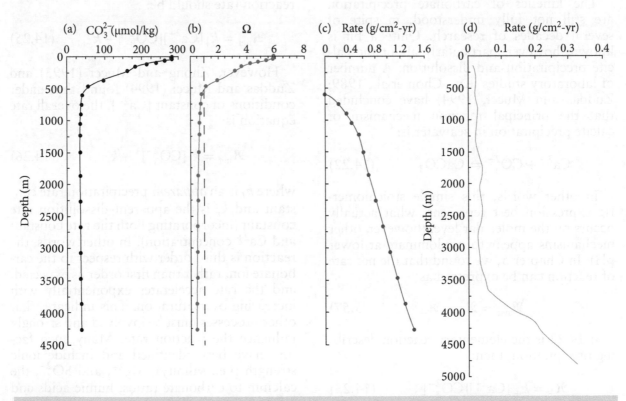

Figure 14.7 (a) CO_3^{2-} measured in the same water column as Figure 14.6 (R/V Ron Brown Cruise P16C, Station 1 at 150°W, 16.5°S). (b) Ω calculated from CO_3^{2-} shown in a. using carbonic acid dissociation constants of Lueker et al. (2000), T, S, and pH shown in Figure 14.6 and assuming $[Ca^{2+}]$ = 10.28 mmol/kg. (c) Dissolution rate calculated from eqn. 14.28 and Ω shown in (b). with k based on values of Sudhas et al. (2015): $k = 3.32 \times 10^{-7} exp(E_A/RT)$ for $\Omega \geq 0.7$ and $k = 1.40 \times 10^{-6} exp(E_A/RT)$ for $\Omega < 0.7$ with $E_A = 20$ kJ/mol-K. (d) Dissolution rate of calcite as a function of depth in the Pacific at 19° N 156° W as determined by the Peterson spheres experiment.

mechanism appears to be the reverse of 14.22, i.e.:

$$CaCO_3 \rightleftharpoons Ca^{2+} + CO_3^{2-} \qquad (14.27)$$

Like precipitation rates, these studies have shown that dissolution rates depend on a number of factors, including the Mg content of the calcite, pressure, the nature of the mineral surface area, and the presence of organic molecules, particularly carbonic anhydrase, an enzyme that catalyzed calcium carbonate dissolution. Nevertheless, Subhas et al. (2015) found that most calcite dissolution rates could be fit to the following equation:

$$\mathfrak{R} = k(1 - \Omega)^{3.9} \qquad (14.28)$$

where Ω is the saturation index, Q/K (compare eqn. 5.68a). Thus, we expect dissolution to accelerate exponentially the more undersaturated the solution becomes. Furthermore, a number of workers have found the reaction rate is a discontinuous function of Ω, being higher far from equilibrium than near equilibrium, with the change occurring at $\Omega \approx 0.7$. This suggests a change in dissolution mechanism from defect-driven near equilibrium to etch pit driven, analogous to that observed in silicates discussed in Section 5.6. Figure 14.7 compares CO_3^{2-}, Ω, and dissolution rates calculated with eqn. 14.28 for the same hydrocast, as shown in Figure 14.6 with the results of the Peterson spheres experiment. The two show comparable dissolution rates that accelerate with depth, although the pattern is somewhat different. This may be partly due to the hydrocast and spheres experiments being in different locations in the central tropical Pacific.

14.3.6 Nutrient elements

Biology plays an enormous role in the abundance and distribution of elements in the ocean. The reverse is also true, the oceans are *oligotrophic*: that is the rate of primary productivity is limited by the abundance of nutrients. Hence ocean chemistry, in particular the availability of nutrients, exerts a primary control on the rate of photosynthesis, hence the abundance of life. The areas of highest primary productivity include coasts, where runoff from continents provides

nutrients and shallow depth allows them to be recycled into the surface water and regions where nutrient-rich water upwells from depth. Upwelling is particularly common along west coasts of continents where prevailing winds drive water away from the coast (in a process known as Ekman transport) and deep water upwells to replace it. Upwelling also occurs along the equator and in other areas where winds create a *divergence* in surface water motion and deep water upwells to replace it. Consequently, primary production in the oceans is concentrated in a relatively small fraction of the ocean's surface area. The centers of the great ocean gyres, such as the Sargasso Sea in the North Atlantic, in the oceans are particularly nutrient-poor and have particularly low levels of primary production.

The two primary nutrients that limit primary production in the ocean are phosphorus and nitrogen. As we found in Chapter 12, nitrogen is a component of proteins and nucleic acids, and phosphorus, is a component of nucleotides including nucleic acids and molecules such as ATP. For most primary producers, nitrogen must be "fixed" before it can be utilized. Thus, NO_3^-, N_2O, and NH_4^+ are nutrients, but N_2 is not (in the following discussion we shall use N to refer only to fixed inorganic nitrogen). Photosynthetic cyanobacteria are the exception: they have the ability to convert N_2 to NH_3. Biological uptake of these nutrients typically leads to their severe depletion in surface waters. Indeed, nitrate levels can fall below detection limits. The uptake of nutrients from surface waters and their release into deep water from falling organic particles imposes a vertical gradient on the concentration on nutrients in the ocean (Figure 14.8). As we noted in Chapter 12, Redfield (1958) found that C, N, and P were present in living tissue in nearly constant proportions of 106:16:1. These are called the *Redfield ratios* (more recent work suggests that marine organic matter is actually richer in carbon and just slightly poorer in nitrogen, with the best current estimate of the Redfield ratios being about 126:16:1). Thus we would expect the concentration of phosphate and nitrate to be highly correlated in seawater, and this is indeed the case.

Several planktonic organisms, most notably diatoms, which are among the most important

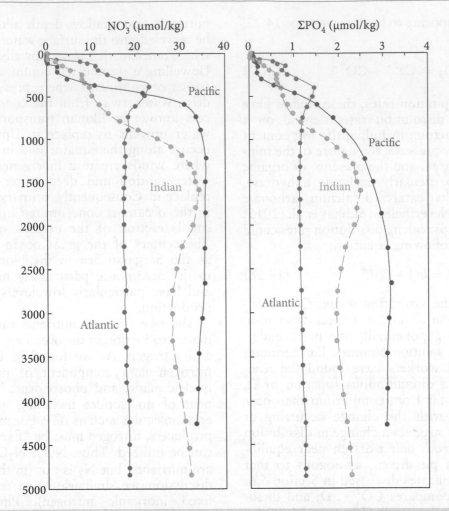

Figure 14.8 Typical depth profiles of two primary nutrients, nitrate and phosphate, in the Atlantic, Pacific, and Indian Oceans. Nutrient concentrations approach 0 in the surface water. Notice also the lower concentrations in the Atlantic. Data from NOAA Ocean Carbon Data System. Pacific profile at 16.5°S, 150°W (RV Ron Brown, station P16N–1), Atlantic profile at 41.3°N, 51.8°W (RV Atlantis II, Station A20–12), Indian Profile at 33°S, 36.95°W (RV Roger Revelle, Station IO5:2009–26).

photosynthesizers in the ocean, build tests of SiO_2. For such organisms, Si is as important a nutrient as N and P and it is also strongly depleted in surface water (Figure 14.9). It appears that in many regions of the oceans, the availability of one or all of these three nutrients limits biological productivity. For this reason, these elements are known as *biolimiting*.

Maxima in nutrient profiles sometimes occur in the depth range of 500–1500 m, reflecting relatively high rates of respiration at these depths. Since nutrients accumulate in deep water, we would expect that the oldest waters would have the highest nutrient concentrations. This is indeed the

case. Atlantic deep water shows considerably less enrichment in nutrients than the Pacific because it is younger. The flux of deep water from the Atlantic to in the Pacific results in a flux of nutrient elements from the Atlantic to the Pacific and an enrichment the Pacific in nutrient elements compared with the Atlantic.

Many other elements are also essential for life. B, Na, Mg, S, Cl, K, Ca, and Mo are widely or universally required, and F, Br and Sr are required by some species. However, these elements are sufficiently abundant in seawater that biological activity produces no, or very little, variation in their concentration in the ocean. For this reason, these elements are sometimes referred to as *biounlimited*.

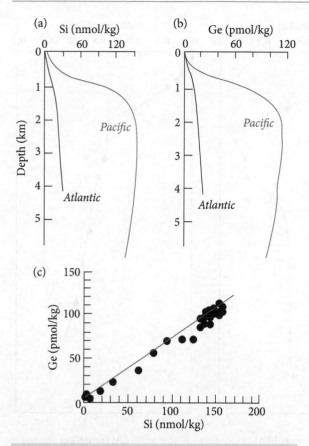

Figure 14.9 Depth profiles of Si (a) and (b) Ge in the Pacific (red lines) and the Atlantic (black lines). (c) shows the correlation between Si and Ge in the Pacific profile. After Froelich et al. (1985).

In addition to these elements, a number of transition metals are also required for life in the order Fe>Zn>Mn>Cu>Ni≫Co. Other elements are widely or universally required as well, including V, Se, Cd, and I. These elements are in sufficiently low concentrations that biological activity imposes vertical and horizontal gradients in their concentration in the ocean (Figure 14.10). Some of these elements, such as Fe and Zn, also show severe depletion in the surface water and are also classed as *biolimiting* elements. Indeed, Martin and Gordon (1988) argued that the availability of Fe limits phytoplankton growth in high productivity regions such as the Subarctic, Antarctic, and equatorial Pacific, where major nutrients are not severely depleted. In these areas, Fe concentrations are as low as 0.1 nmol/l. That Fe can indeed limit

phytoplankton productivity was confirmed in an experiment carried out in the eastern equatorial Pacific in 1993. Fe (as $FeSO_4$) was added to a $100\,km^2$ patch of ocean where Fe concentrations were very low. Within days of adding Fe, phytoplankton productivity increased significantly (Martin et al., 1994).

For other elements, such as V, Cr, and Se, only slight surface water depletion occurs. Elements such as these are known as *biointermediate*. The distributions of several of these biolimiting and biointermediate elements are also significantly affected by nonbiological processes in the ocean and as a result, they can have vertical concentration profiles that differ from the classic nutrient profile seen in Figures 14.7 and 14.10. This can occur as a result of oxidation and reduction, adsorption onto particle surfaces, a process called scavenging, and by other inputs, such as input from hydrothermal vent sediments of the ocean bottom, and wind-blown dust dissolving in the surface water. We will consider the vertical distribution of these elements, and the processes creating them, in more detail in the next section.

A number of other elements show nutrient type distribution patterns, that is depleted in surface water and enriched in deep water, even though they have no known biological function. These include As (which show only slight surface depletion), Sc, Ge, Pd, Ag, Cd, Ba, the rare earths, Pt, and Ra. Ba appears to be precipitated by biological mediated reactions (as $BaSO_4$), but the details are not yet understood. Other elements are inadvertently taken up; that is, organisms appear unable to discriminate between them and needed elements. Ge provides a well-documented example. It is chemically similar to Si and apparently is taken up by diatoms in place of Si. Like Si, its concentration approaches 0 in surface water and increases with depth (Figure 14.9b). Overall, its concentration is strongly correlated with that of Si (Figure 14.9c). Clearly, biological activity controls the distribution of Ge in the oceans.

Comparing the phosphate and silicate profiles in Figures 14.8 and 14.9, we see that concentrations of nitrate increase more rapidly than that of SiO_2. Essentially, all release of nitrate occurs within the upper 1500 m, and most within the upper 500 m.

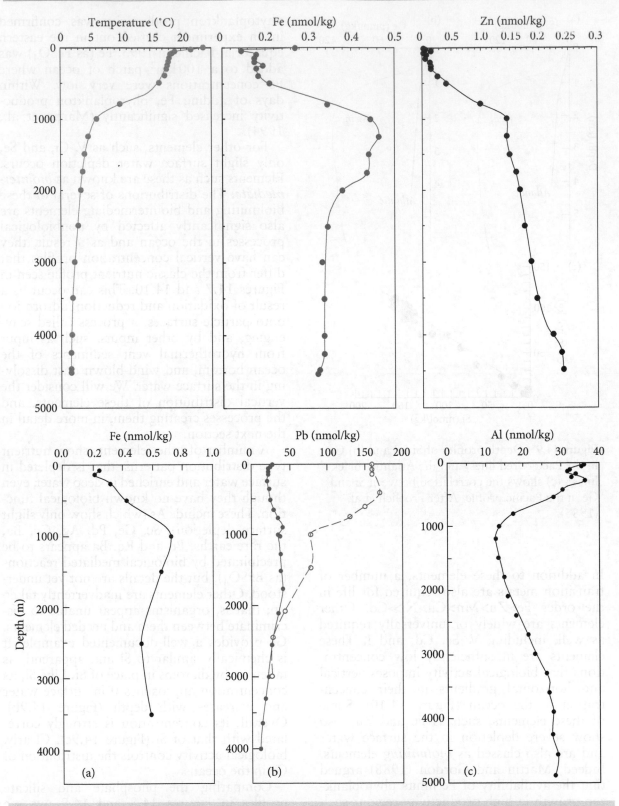

Figure 14.10   Depth profiles of micronutrients Fe and Zn in the North Atlantic at the GEOTRACES Bermuda Atlantic Times Series station at (31°46′N, 64°05′W). In addition to being taken up by phytoplankton, Fe is scavenged by particle adsorption. This profile shows an elevated surface concentration, likely due to deposition of Sahara dust. Data from Middag et al. (2015).

Maximum Si concentrations are only reached at depths of 1500 to 2000 m. The reason for these different distributions is straightforward. Nitrate is released by biological mediated decomposition of soft tissues of organisms. Si is incorporated only in the inorganic tests, which undergo dissolution without biological mediation (although, passage of these tests through the gut of higher organisms undoubtedly speeds dissolution). This abiologic dissolution is slower. Elements that are rapidly released are referred to as *labile* elements; those released more slowly, such as Si, are referred to as *refractory*. Labile nutrients include nitrate, Mn, Cd, and Ni. Since Ge concentrations correlated with those of Si, we include it with the refractory elements. Some refractory elements are present only in soft tissues of plankton, but nevertheless undergo only slow release including Cu, Zn, and Fe.

It is interesting to compare the distribution of Si and Ca, both utilized by organisms to build tests. In both cases, the tests have some tendency to dissolve after the organism dies, and thus there is some recycling of these elements back to seawater. But because the flux of Si to the oceans is low, the concentration is relatively low everywhere and the plankton utilizes essentially all of it in surface waters. Plenty of Ca is supplied to the oceans and its concentration is high; organisms utilize only a portion of that available. The relative variation in Ca concentration relative to salinity is only about 0.5%. Both Si and Ca provide good demonstrations of the lack of control of equilibrium thermodynamics on ocean chemistry, at least on a large scale. Calcium carbonate is oversaturated in surface water yet precipitates only in biologically mediated reactions. In contrast, the ocean is everywhere undersaturated with respect to opal (though is locally oversaturated with respect to quartz), yet it is biologically precipitated and redissolved only upon death of the organism. The oceans are certainly an example of a kinetically controlled, rather than thermodynamically controlled, system.

14.3.7 Particle-reactive elements

The vertical and horizontal distributions of many trace elements in seawater, including Al, Ga, Sn, Te, Hg, Pb, Bi, and Th are controlled by abiological reactions with particles. The distribution of Be, Cu, Ti, Cr, Mn, Fe, Zr, the REE, Hf, and Pt are controlled by both biological processes and abiological reactions with particles. A common characteristic of these elements is that they are strongly hydrolyzed in seawater – that is, at seawater pH they react with water to form hydroxo-complexes (e.g., $Al(OH)_3$, $Al(OH)_4^-$). Most particles in the ocean have large numbers of O-donor surface groups. Cations that form hydroxo-complexes in solution can form surface complexes and consequently are readily adsorbed to these surfaces, hence the term *particle reactive*. Oxidative and reductive dissolution and precipitation are also important for redox sensitive elements such as Mn, Fe, and Ce. (Even here, however, the biota plays some role, in the production of particles, production of organic molecules that coat particles and affect their surface properties, and in catalyzing oxidation and reduction reactions.)

Particle-reactive elements show a variety of vertical concentration profiles. Al, Ga, Te, Hg, Pb, Bi, and Th are typically enriched in surface water and depleted at depth. Figure 14.11 provides examples of Pb and Al. In the case of Al, Ga, Sn, Te, and Pb, the surface enrichment results in wind-borne particulates to the surface water followed by partial dissolution. A significant part of the aeolian flux of Sn, Te, Hg, and Pb may be anthropogenic, hence their distributions and surface maxima may not be steady-state features. Pb concentrations have been particularly affected by emissions of tetraethyl Pb in gasoline (a topic we will examine in the following chapter). Concentrations of dissolved Pb in North Atlantic surface water have declined by a factor of three, however, since Pb was eliminated from gasoline beginning in the 1970s and as the gasoline-derived Pb has been progressively scavenged by particles (Bruland, 2014). Mercury in the atmosphere and land and sea surfaces is also anthropogenic, but although emissions have been substantially reduced, large amounts of anthropogenic Hg remain present due to its peculiar chemistry (we'll discuss this in the following chapter).

Mid-depth maxima may occur because of hydrothermal input. Most hydrothermal activity in the oceans occurs on mid-ocean ridges, whose depth is typically in the range of 2500–3500 m. Hydrothermal fluids mix with

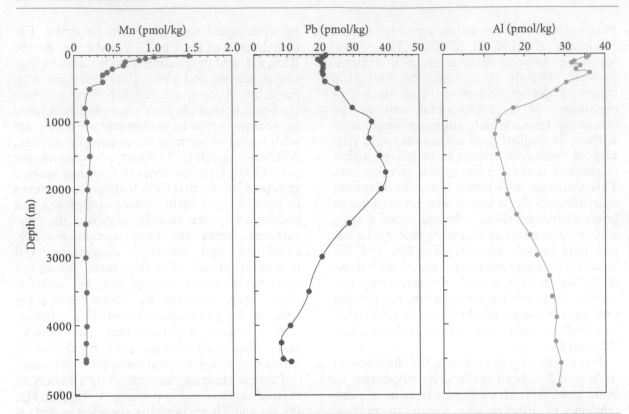

Figure 14.11   Depth profiles of particle-reactive metals Mn, Pb, and Al in the North Atlantic. Much of the marine inventory of these elements is derived from partial dissolution of wind-blown dust and other aerosols. Once dissolved, all three elements are scavenged by particle adsorption; in addition, Mn is taken up by phytoplankton. Hydrothermal fluids and fluxes from sediment pore water are additional sources that can increase abundances of these elements in deep water. Mn and Pb data are from the same GEOTRACES Bermuda Atlantic Times Series station as in Figure 14.10 (Middag et al., 2015); Al is from the GEOTRACES database, RV Knorr cruise GA03 station 8 at 35.42°N, 66.54°W.

surrounding water and form slightly warm, and therefore slightly buoyant, plumes, which rise hundreds of meters above the ridge crest and are then transported laterally by ocean currents and then progressively scavenged by particles. Profiles of Mn, and to a lesser degree Fe, can also show maxima associated with the O minimum. These maxima occur because particle-bound Mn and Fe are reduced in O-poor water and the reduced species, Fe^{2+} and Mn^{2+}, are much more soluble than the oxidized species. In reducing conditions then, Fe and Mn can be released from particulates rather than scavenged.

14.3.8   One-dimensional advection–diffusion model

Let's examine these concentration–depth profiles in a bit more detail. Concentration

profiles such as these can be readily modeled using a *one-dimensional advection–diffusion model* (Craig, 1974). The essential assumption of such a model is that the profile observed is a steady-state feature; that is that the variation with depth is the same today as it was, say, 1000 years ago. Let's begin by considering the simple case of the vertical variation of conservative property of ocean water, such as salinity, between fixed values of salinity at the top and bottom of the water column. Salinity will vary only because of transport of water (by our definition of conservative, chemical and biological processes have no effect). Two kinds of transport are of interest: turbulent transport and vertical velocity of the water. Turbulent transport is also known as *eddy diffusion*. Its is mathematically analogous to chemical diffusion and may be described by the equation:

$$\frac{\partial c}{\partial t} = K\frac{\partial^2 c}{\partial z^2} \qquad (14.29)$$

where c is a concentration (such as salinity), K is the eddy diffusion coefficient and has units of m^2/yr, and z is depth. Notice that eqn. 14.29 is identical to Fick's second law (eqn. 5.75), except that we have replaced the chemical diffusion coefficient, D, with K (and x with z; here we define z as being positive upward). Adding a term for vertical velocity, we have:

$$\frac{\partial c}{\partial t} = K\frac{\partial^2 c}{\partial z^2} - \omega\frac{\partial c}{\partial z} \qquad (14.30)$$

Notice that the velocity term in eqn. 14.30 is exactly analogous to the one in eqn. 5.165, which we derived from sediment diagenesis. At steady-state $\partial c/\partial t = 0$, so:

$$K\frac{\partial c^2}{\partial z^2} = \omega\frac{\partial c}{\partial z} \qquad (14.31)$$

This is a second-order differential equation with respect to c, the solution depends on the boundary conditions. These are that c is fixed at $c = c_0$ at the bottom of the mixing zone ($z = 0$) and $c = c_Z$ at the top of the mixing zone ($z = Z$) of total height Z. The solution to this equation is:

$$c(z) = (c_Z - c_0)f(z) + c_0 \qquad (14.32)$$

where

$$f(z) = \frac{e^{z\omega/K} - 1}{e^{Z\omega/K} - 1} \qquad (14.33)$$

Since $c(z)$ is a linear function of $f(z)$, eqn. 14.32 can be used to test the appropriateness of the one-dimensional model. The key parameter is the ratio ω/K, which has units of inverse distance and is the *scale height*. It can be estimated by fitting eqn. 14.32 to a conservative parameter such as temperature or salinity over the interval where T–S is linear. If a truly conservative parameter is plotted against $f(z)$, a straight line should result. Any deviation from linearity would indicate there is significant horizontal advection and that the one-dimensional model is not appropriate. Provided horizontal advection is not

occurring, we can use eqn. 14.32 to determine whether a particular species is conservative or not: any deviation from linearity on a plot of c versus $f(z)$ would indicate nonconservative behavior.

Now let's consider a nonconservative species that is actively scavenged from seawater through surface adsorption on particles. We assume that the adsorption rate is proportional to the concentration, i.e., first-order kinetics. The change of concentration with time can then be described as:

$$\frac{\partial c}{\partial t} = K\frac{\partial^2 c}{\partial z^2} - \omega\frac{\partial c}{\partial z} - \psi c \qquad (14.34)$$

where ψ is the scavenging rate constant, which we assume is constant with depth. The sign of ψ is such that positive ψ corresponds to removal from seawater, i.e., adsorption, and negative for addition to seawater, i.e., decay or desorption. ψ has units of inverse time; the inverse of ψ is known as the *scavenging residence time* and is denoted τ_ψ.

At steady-state:

$$K\frac{\partial c^2}{\partial z^2} = \omega\frac{\partial c}{\partial z} + \psi c \qquad (14.35)$$

The solution to this equation, which again depends on the boundary conditions, is:

$$c(z) = c_Z F(\psi, z) + c_0 G(\psi, z) \qquad (14.36)$$

where:

$$F(\psi, z) = \frac{e^{-[Z-z]\omega/2K}\sinh\left(Az\omega/2K\right)}{\sinh\left(AZ\omega/2K\right)} \qquad (14.36a)$$

$$G(\psi, z) = \frac{e^{z\omega/2K}\sinh\left(A[Z-z]\omega/2K\right)}{\sinh\left(AZ\omega/2K\right)} \qquad (14.36b)$$

and

$$A = (1 + {}^{4K\psi}/\omega^2)^{1/2} \qquad (14.36c)$$

Finally, suppose that in addition to these processes, there is a steady production of the dissolved species through biological decomposition of organic matter. The production in that case would be independent of concentration. The rate of change of concentration is then:

$$\frac{\partial c}{\partial t} = K\frac{\partial^2 c}{\partial z^2} - \omega\frac{\partial c}{\partial z} - \psi c + J \qquad (14.37)$$

where J is the rate of production. For simplicity, we assume that it is independent of depth. At steady state:

$$K\frac{\partial c^2}{\partial z^2} + J = \omega\frac{\partial c}{\partial z} + \psi c \qquad (14.38)$$

J can also be used to represent *zero-order removal* – zero-order implying that the rate of removal is independent of concentration. The solution to eqn. 14.38 is given by replacing all concentration terms in 14.35 by $(c - J/\psi)$. For $\psi = 0$, i.e., for no scavenging, this solution is clearly undefined. The solution in that case is:

$$c(z) = (c_Z - c_0)f(z) + c_0 + J/\omega[z - Zf(z)] \qquad (14.39)$$

The value of the vertical velocity, ω, can be determined from ^{14}C analyses. Typical values are in the range of 3–4 m/yr. K is typically in the range of 2000–3000 m^2/yr. Example 14.4 illustrates an application of this model.

Example 14.4 Advection–diffusion model

Given the data of Middag et al. (2015) from the Bermuda Atlantic Times Series station in the table below, fit the advection–diffusion model to trace metal concentrations through the water column. Estimate values of ψ, the absorption rate for each element.

Trace metal and salinity data of Middag et al. (2015) for the Bermuda Atlantic time series station.

Depth (m)	Salinity	Zn (pmol/kg)	Pb (pmol/kg)	La (pmol/kg)
1001	35.07	1.44	35.54	22.60
1249	35.01	1.44	34.71	23.02
1500	34.99	1.49	37.88	23.16
1749	34.97	1.61	39.78	22.14
2000	34.97	1.68	38.48	21.59
2501	34.96	1.75	28.59	22.75
3001	34.93	1.86	20.38	25.50
3500	34.90	1.96	16.32	30.58
4002	34.89	2.23	10.65	34.62
4249	34.88	2.38	8.03	36.89
4474	34.88	2.41	8.63	38.00

Answer: The first step is to determine whether the conservative parameters can be fit to an advection–diffusion model by testing for linearity between temperature and salinity. Doing so, we find that they are linear only beneath the thermocline (Figure 14.12a). Next, we want to determine K and ω by fitting the model to salinity. The best way to do this is to plot salinity vs. depth using Excel or MATLAB, then plot the model based on eqn. 14.31 and adjust K and ω until the model curve matches the data. Guesses in the range of K between 2000 and 3000 m^2/yr and ω between 3 and 4 m/yr are a good place to start. Figure 14.12b shows that $K = 2500$ m^2/yr and $\omega = 4$ m/yr provide a reasonable, albeit not perfect, match to the data.

Next, we want to plot the trace metals as a function of depth. With K/ω now fixed, we can compute functions F and G and the model concentrations from eqn. 14.36. As we did to fit K/ω, we can make successive guesses for ψ until our model curve matches the data. Figure 14.12c shows that we can fit the Zn data reasonably well with a value of ψ of 0.00075 yr^{-1}, giving an estimated scavenging residence time of ~1300 yrs. However, Pb cannot be fit to this model: even with ψ set to 0 (Figure 14.12d), Pb concentrations plot above the model curve, indicating dissolved Pb is being added to the water. The most likely explanation is that Pb absorbed on to the surfaces of particles in the surface water is being released in the deep water. As we noted earlier, gasoline-derived Pb had greatly increased Pb concentrations in the surface water in the mid- to late-twentieth century and that Pb has subsequently been removed by adsorption. These Pb-rich particles are now falling through the water column and releasing Pb. Organic particles would be particularly likely to release adsorbed Pb as they are broken down by bacteria. Thus, dissolved Pb is being added to the deep water, particularly in the region around 2000 m. A simple advection–diffusion–scavenging model cannot be fit to the Pb data. The data can be fit approximately with zero-order removal expressed by eqn. 14.38. Finding the solution for Pb as well as finding a fit to the La is left to the reader in the problems at the end of this chapter.

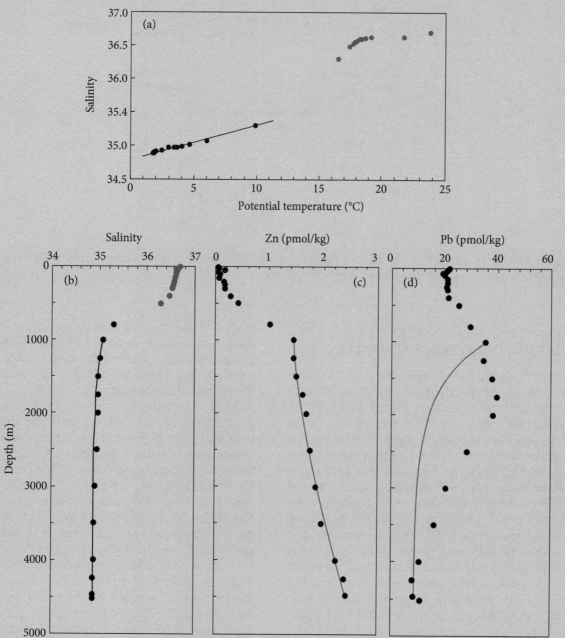

Figure 14.12   (a) A plot of salinity against potential temperature is linear only below the thermocline (black symbols), indicating an advection–diffusion model is only applicable to this part of the water column. (b) Salinity data from this station fit to the model with $K = 2500 \, m^2/yr$ and $\omega = 4 \, m/yr$ (black line). (c) The advection–diffusion model (red line) fit to the Zn data for $\psi = 0.00075 \, yr^{-1}$. (d) The model does not fit Pb; instead the Pb curve suggests production (i.e., a positive value of J in eqn. 14.39).

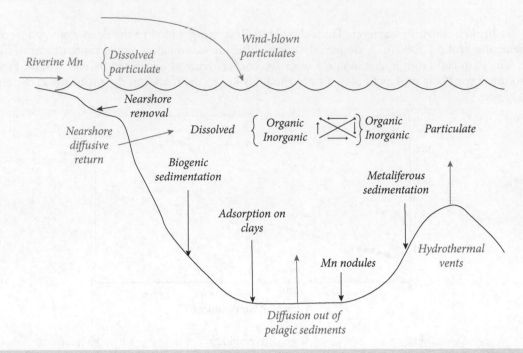

Figure 14.13   Marine geochemistry of Mn, illustrating the range of possible sources and sinks, as well as internal processing, of dissolved material in seawater.

14.4   SOURCES AND SINKS OF DISSOLVED MATTER IN SEAWATER

In 1899, John Jolly* attempted to estimate the age of the Earth from the mass of salts in the sea and the mass added annually by rivers. His result, 90 million years, is a factor of 50 less than the actual age of the Earth. His mistake was to not recognize that the ocean is a dynamic, open system, and it is ultimately the balance between addition and removal of an element that dictates salt concentration in the ocean. This was recognized by Georg Forschhammer in 1865 when he wrote: "The quantity of different elements in seawater is not proportional to the quantity of elements which river water pours into the sea, but is inversely proportional to the facility with which the elements are made insoluble by general chemical or organo-chemical actions in the sea." One of our objectives in this chapter will be to examine the budget of dissolved matter in the oceans; that is, to determine the sources and sinks and the rates at which salts are added and removed from the oceans. As we shall see, rivers are not the only source of dissolved salts in the sea.

Many elements are present in seawater at concentrations far below their equilibrium solubility. Ca is present in surface water at concentrations above solubility. Clearly then, the composition of seawater is not controlled by thermodynamic solubility (only the concentrations of the conservative gases are controlled by equilibrium solubility). Rather, the composition of seawater is kinetically controlled; specifically, it is controlled by the rates at which dissolved matter is added to and removed from seawater. Rivers represent the principal "source" of dissolved solids in seawater and sediments represent the principal *sink*. There are, however, a variety of other sources and sinks. These are shown in Figure 14.13, which illustrates the marine

* John Jolly (1857–1933) was an Irish geologist and professor at the Trinity College, Dublin. Although estimate using this method was far too young, it was comparable to other estimates of the time. Shortly after its publication, Jolly became interested in using radioactivity to determine geologic ages. Working with Ernest Rutherford, in 1913 he estimated the age of the Devonian period as 400 Ma – in excellent agreement with the modern geologic time scale. He later pioneered radiotherapy treatments for cancer.

geochemical cycle of Mn. For any given element, one of these sources or sinks might be dominant, and some or most of the sources might be negligible. In the following sections, we will discuss each of these sources and sinks. Before we do, however, we introduce a useful concept in marine geochemistry, that of residence time.

14.4.1 Residence time

An important concept in the chemistry of sea-water is that of *residence time* (Barth, 1952; Goldberg and Arrhenius, 1958). Residence time, τ, is defined as the ratio of the mass of an element in the ocean divided by the flux to the ocean, i.e.:

$$\tau = \frac{A}{dA/dt} \qquad (14.40)$$

where A is the mass of the element of interest in the oceans, and dA/dt is the flux to seawater. Implicit in the residence time concept is the assumption that *the oceans are in steady state*, that is, the composition does not change with time, so that the flux of an element into seawater equals the flux out of seawater. Thus, it does not matter whether we use the flux into or out of the oceans in eqn. 14.40. If we do this calculation for water and assume that river water is the only flux, the volume of the oceans is 1.35×10^{18} m^3 and the riverine flux is 3.74×10^{13} m^3/yr, we find the residence time of water in the oceans to be 36,100 years.

If river water is the principal source of the element, eqn. 14.40 can be re-expressed as:

$$\tau = \frac{C_{SW}}{C_{RW}} 3.61 \times 10^4 yr \qquad (14.41)$$

where C_{SW} and C_{RW} are the concentrations in seawater and river water, respectively. For example, rivers are the principal source of Na in seawater. The concentration of sodium is 5.3 mg/kg in river water and 10.77 g/kg in seawater, so we calculate a residence time of 74 Ma for Na (only a little different from Jolly's result). We have ignored the groundwater flux, which is poorly constrained; inclusion of the groundwater flux would decrease the residence time. On the other hand, about half the sodium in river water is derived from

cyclic salts; i.e., it has simply been cycled through the hydrologic system. If we don't count this cycling in the residence time, then sodium has an ocean residence time of about 150 Ma. Many elements have several sources, and the fluxes from these are poorly known; thus, residence times are not well known for all elements. Nevertheless, residence times of elements in seawater clearly vary greatly: from 65 Ma for Cl (including the cyclic flux) to a few tens of years for Th.

Table 14.3 lists the estimate residence times for some elements in seawater. Over the last two decades, remarkable progress has been made in defining the composition of seawater. However, the fluxes of many elements to (and from) the ocean remain poorly known. As a result, the residence times of most elements are poorly constrained and may be significantly revised in the future. Nevertheless, we can see residence times correlate with abundance: the major ions have long residence times, the least abundant have short ones. The short residence time of Ca also resolves

Table 14.3 Residence times of selected elements in seawater.

Element	Mean conc. (μmol/l)	Residence time, years
Cl	5.46×10^5	6.5×10^7
Na	4.68×10^5	4.5×10^7
Mg	4.3×10^4	1.01×10^7
B	416	1.19×10^7
S	2.8×10^5	9.42×10^6
K	1.02×10^4	8.45×10^6
Sr	90	3.51×10^6
Li	25.9	1.79×10^6
Ca	1.01×10^4	8.16×10^5
F	68	5×10^5
U	0.0135	4×10^5
P	2.3	4.88×10^4
Si	100	1.03×10^4
Yb	6×10^{-6}	2200
Hf	9×10^{-7}	1300
La	3.5×10^{-5}	650–1600
Nd	3×10^{-5}	400–950
Al	0.002	200
Fe	5×10^{-5}	200–500
Ti	1.5×10^{-4}	150
Th	3×10^{-7}	45

Concentrations from Table 14.1. Residence times from Lecuyer (2016) and MBARI periodic table of the elements (https://www.mbari.org/science/upper-ocean-systems/chemical-sensor-group/periodic-table-of-elements-in-the-ocean/).

the *river water paradox* that we mentioned in Chapter 13. As we found in that chapter, Ca is the most abundant cation in river water, yet it is not in seawater. Although the Ca flux to the ocean is large, the flux out of the ocean is also large, resulting in a much shorter residence time than the other major ions in seawater.

14.4.2 River and groundwater flux to the oceans

Overall, rivers are the main source of dissolved salts in the ocean, though they are not the main source for all elements. Current estimates of average concentrations in river water are given in Table 14.1. We discussed the factors that controlled river water composition in Chapter 13. These were the chemical composition of the source rock, climate, topography, and intensity of weathering in the catchment basin. In addition to river flow, groundwater can discharge directly into the ocean, and this is another important source of continent-derived constituents to the oceans.

As we found in Chapter 13, rivers vary widely in composition. Since there are many rivers and each is in some respect unique, the task of estimating the average composition of rivers and their combined flux to the ocean is not an easy one. The largest 20–25 rivers carry only 15% of the flux to the oceans. Relatively few rivers have been subjected to thorough geochemical investigations; for some trace elements there are few data for any rivers. Furthermore, the composition of many rivers has been disturbed by mining, agricultural, industrial, and other activities, so that their modern compositions may not be representative of the composition in the past. Finally, the composition of most rivers varies with river flow rate. Thus, characterizing the composition of a river requires many measurements made over the course of a year or more. For all these reasons, there is considerable uncertainty in the concentrations listed in Table 14.1 and Gaillardet et al. (2014) point out these values can only be considered "as a first order approximation."

14.4.2.1 Estuaries

Even if the compositions of rivers were well known, the task of determining the riverine flux to the sea would still be a very difficult one. The reason for this is that estuaries act as flow-through chemical reactors, in which dissolved components are added and removed, and what flows out is not the same as what flows in. This is a consequence of changes in solution chemistry that occur when river and seawater mix.

An estuary is that portion of a river into which coastal seawater is carried by tidal forces; estuaries are thus zones of mixing of freshwater and seawater. They may be lagoons behind barrier islands (e.g., Pamlico Sound, North Carolina), river deltas (e.g., the Rhine Delta), drowned river valleys (e.g., the Gironde Estuary, France, Chesapeake Bay, US), tectonic depressions (e.g., San Francisco Bay), or fjords (Sannich Inlet, British Columbia). In the typical estuary, there is a downstream flow of fresh water at the surface and an upstream flow of seawater at depth; the fresh water overlies the saltwater because of its lower density. Depending on the geometry of the estuary, the strength of the tides, and the strength of the river flow, these two layers will mix to varying degrees. Low river flow and strong tides produce a "well-mixed" estuary in which there is little vertical gradient in salinity; strong river flow and weak tides lead to a "salt wedge" estuary in which a strong pycnocline develops at the interface between the layers.

The principal chemical changes that can occur in estuaries are as follows:

- Changes in ionic strength
- Changes in the concentrations of major cations, which affects speciation of minor components and surface charge on particles
- Changes in pH
- Changes in the concentration and nature of suspended matter
- Changes in the redox state within the water and sediment.

These changes can result in either release of ions to solution, or, more frequently, scavenging of them, resulting in estuaries often acting as a filter to remove both dissolved and suspended components before river water reaches the ocean.

Whether a dissolved component is added to or removed from solution in an estuary can be readily determined by plotting its concentration against salinity. The concentration of any *conservative* species in an estuary will be a function only of the proportions in which sea and river water are mixed. Salinity is always conservative in estuaries because addition or removal of major species does not occur to a significant extent. The concentration of any other conservative species should therefore be a simple linear function of salinity. As illustrated in Figure 14.14, any deviation from linearity indicates removal or addition of that species.

Figure 14.15 shows concentration–salinity diagrams for metal ions in the Gironde estuary, France. Fe concentrations plot along a curve below the conservative mixing line, consistent with its removal by colloid flocculation. Ni plots along a curve above the mixing line, indicating addition. Kraepiel et al. (1997) concluded that this results from desorption of Ni from particle surfaces, which, in turn, is due to a decrease in the free Ni ion concentration due to complexation by inorganic ligands. Essentially, all the Ni addition occurs at salinities lower than 10‰; at high salinities Ni behave nearly conservatively. Pb experiences only modest removal. Pb is strongly particle-reactive, so conservative behavior would be surprising. Its apparent conservative behavior could be due to a combination of addition by desorption from solid surfaces and removal by coprecipitation with colloidal Fe hydroxides and humates.

As seawater mixes with river water, the resulting increase in ionic strength and change in pH and solute composition compresses the surface double layer of suspended material and induces the flocculation, as we discussed in Section 6.6. This can dramatically affect trace metal chemistry. As a practical matter, aquatic and marine chemists often consider "particulates" to be the material retained on 0.2 μm filters (earlier studies used 0.4 μm filters) and anything passing through such a filter to be "dissolved."

Unfortunately, nature is not so neat: in reality there is a continuum between truly dissolved substances and readily recognizable particles. Materials between these are termed *colloids*. Colloids are often defined as having sizes in the range of 10^{-9} to 10^{-6} m and may be separated by special techniques such as ultrafiltration and membrane filtration. Entities at the small end of this range would consist of $\sim 10^3$ or fewer atoms and approach the size of larger humic acids, while particles at the large end of the range would be retained on filter paper. Colloids are a surprisingly important component of natural waters.

Much of the nominally dissolved Fe in rivers is likely bound to organic colloids or present as Fe oxide or clay colloids (e.g., Boyle et al., 1977). A significant fraction of humic acids is probably also colloidal (Sholkovitz,

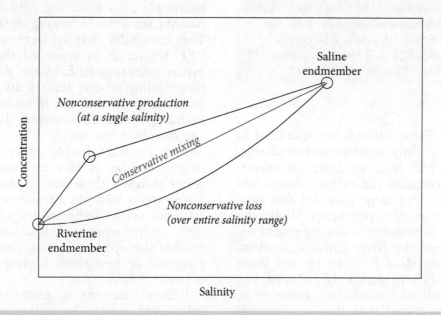

Figure 14.14 Use of concentration–salinity diagrams in estuaries.

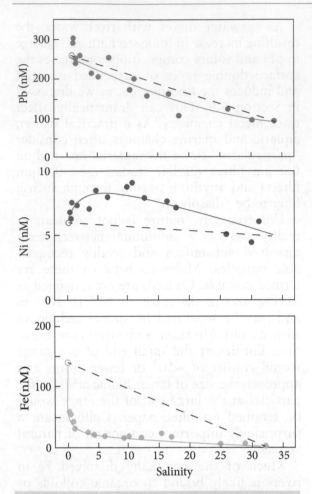

Figure 14.15 Concentrations of dissolved Pb, Ni, and Fe measured in the Gironde estuary, Bordeaux, France. Open symbols are the average concentration of the river water. Dashed line is the conservative mixing line, solid lines are a visual best fit to the data. The curves indicate removal of Fe and Pb and addition of Ni. Modified from Kraepiel et al. (1997).

et al., 1978). These colloids are stabilized in solution by a slightly negative surface charge. Upon mixing with high ionic strength seawater, cations neutralize the surface charge and they coagulate into large particles that flocculate. In a series of experiments, Sholkovitz (1976, 1978) demonstrated that mixing of sea and water from the River Luce of Scotland results in removal of 95% of Fe and lesser amounts of humic acids, Cu, Ni, Mn, Al, Co, and Cu by colloid flocculation. Boyle et al. (1977) demonstrated that this is a general phenomenon occurring in many estuaries and

estimated that 90% of the riverine flux of Fe is removed in this way. However, some flocculant in estuaries can be remobilized within the estuary and the Fe desorbed, particularly at the river month some of which may be exported to the ocean (Zhu et al., 2018). Additionally, removal of Fe in some glacial and other boreal estuaries appears to be less effective (e.g., Kritzberg, 2014).

Subsequent studies have revealed that other elements, such as the rare earths, are also removed by this process. For example, Hoyle et al. (1984) found that 95% of the heavy rare earths and 65% of the light rare earths dissolved in river water were removed by flocculating colloids in the estuary of the River Luce and removal was correlated with Fe removal. REE removal has been documented in other estuaries, including in the Amazon, where Rousseau et al. (2015) found that 95% of both the light and heavy rare earths were removed (Figure 14.16a). Most of this removal occurs at salinities below 10, at which point Rousseau et al. found that 73% of the "dissolved" Nd was in the fine colloidal fraction. Rousseau et al. made two other interesting observations. First, there is a slight increase in dissolved Nd between salinities of 17 and 28 (Figure 14.16b). Second, the ε_{Nd} of the particulate fraction was lower (–10.7) than that of the dissolved fraction (–8.9). As salinity increases, the ε_{Nd} of the dissolved Nd decreases more rapidly than predicted by simple binary mixing between river and Atlantic seawater in this region ($\varepsilon_{Nd} \approx -12.1$). They concluded that the increase in dissolved REE results from some of the particulate matter releasing REE to the dissolved fraction during mixing with seawater between salinities of 17 and 28; in addition, isotopic exchange occurred between the particulate and dissolved fraction.

The electric double layer surrounding larger particles is also compressed as river water mixes with seawater, causing them to flocculate as well. As a result of this process, estuaries act as sediment traps. As much as 90% of the suspended load of rivers never reaches the open sea, being deposited in the estuarine or near-shore environment instead (Chester, 1990).

Redox reactions in estuaries also affect solution chemistry. Increasing pH speeds the oxidation of ferrous to ferric iron, while the

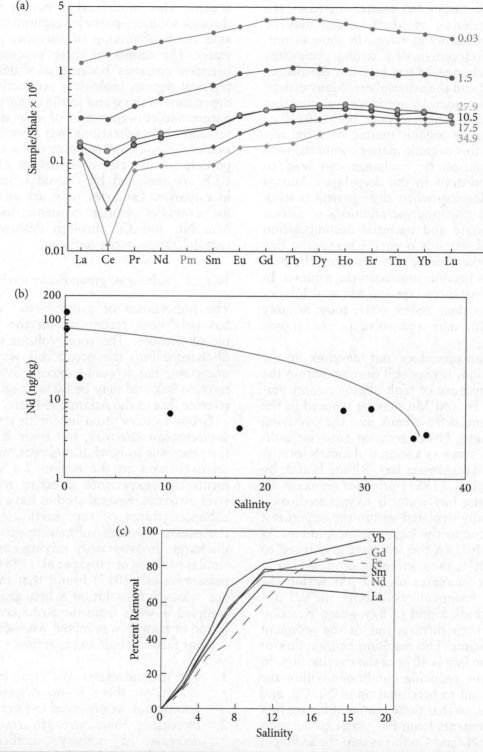

Figure 14.16   (a) Shale-normalized are earth patterns (Section 7.2.2.4) in the dissolved fraction of waters in the Amazon River estuary. Removal apparently occurs by coagulation and precipitation of colloids. Ce is removed even more rapidly than other rare earths due to its IV valence state. (b) Dissolved concentration of Nd as a function of salinity. Red line shows the expected distribution for mixing between river and seawater. Most removal occurs at salinities below 10. There is an increase in dissolved Nd between salinities of 17 and 28 due to dissolution of particles in the suspended load. (c) Percent removal of REE as a function of salinity. Data from Rousseau et al. (2015).

presence of suspended matter increases the rate of oxidation of Mn^{2+}, which may be present metastably in rivers. In some estuaries, the development of a strong pycnocline inhibits exchange of gas between the deeper saline layer and the atmosphere. Many estuaries are surrounded by wetlands (salt marshes or wetlands), which export both dissolved and particulate organic matter. Bacterial respiration of this organic matter combined with limited atmospheric exchange can lead to anoxic conditions in the deep layer. Anoxia may also develop when algal growth is stimulated by anthropogenic additions of nitrate and phosphate and bacterial decomposition of the algal remains occurs. Chesapeake Bay is a good example of this situation; the bottom waters become anoxic in the summer. In such circumstances, species whose solubility depends on their redox state, most notably Fe and Mn, may redissolve in the anoxic water.

Where anoxia does not develop in the water column, it may still develop within the sediment because of high organic sedimentation rates. Fe and Mn may be reduced in the sediment and diffuse back into the overlying water column. Hence reduced estuarine sediments may serve as a source of metals ions. A study of Narragansett Bay, Rhode Island, by Elderfield et al. (1981) provides an example. Although the bay water is oxygenated, oxygen is rapidly depleted within the uppermost sediment due to the high organic content. As a result, Mn^{4+} in the sediment is reduced to soluble Mn^{2+}, the concentration of which in pore water increases to 2–7 μM within the top cm of sediment, well above the 0.5 μM concentrations found in Bay water. Remobilized Mn then diffuses out of the sediment into Bay water. The resulting benthic flux of Mn into the Bay is 40% of the riverine flux. In contrast, the reducing conditions within the sediment lead to precipitation of Ni, Cd, and Cu sulfides, so that there is a net diffusive flux of these elements from Bay water to the sediment. The Ni and Cu fluxes into the sediment were 60% and 30% of the riverine fluxes, respectively. In Buzzards Bay, Massachusetts, Elderfield and Sholkovitz (1987) concluded that benthic flux of REE to the bay was of a magnitude similar to the riverine flux.

In summary, chemical processes in estuaries significantly modify the riverine flux to the oceans. This modification occurs primarily through solution–particle reactions that occur as a result of mixing of seawater and river water. The results of these processes differ between estuaries because of differences in physical regime, biological productivity, residence time of the water in the estuary, and the nature and concentration of suspended matter. Some generalizations may nevertheless be made. A significant fraction of Fe and other particle-reactive elements such as Al and the REE are removed by colloidal flocculation in estuaries. Estuaries may act as a source for a number of other elements, such as Ba, Mn, Ni, and Cd through desorption and remobilization in the sediment.

14.4.2.2 Submarine groundwater discharge

The importance of groundwater discharge has only been recognized in the last couple of decades. The total volume of waters discharged into the ocean this way is very uncertain, but it could exceed 30% of the riverine flux and may be 80% or more of the riverine flux in the Atlantic (Moore, 2010).

Groundwaters often mix with seawater in *subterranean estuaries*, but there is evidence the process is somewhat different, with larger relative fluxes to the oceans for some elements that experience effective removal in river estuaries. Several studies have suggested enhanced fluxes of rare earth elements to the oceans through submarine groundwater discharge. In laboratory mixing experiments similar to those of Hoyle et al. (1984), Johannesson et al. (2017) found that rare earths are released to solution when groundwater sampled in wells from the Kona coast of the island of Hawaii was mixed with seawater. Four factors could explain this:

1. The groundwaters are poor in organic matter, so there is no organic matter flocculation as observed in rivers.
2. Increasing ionic strength results in a decrease in activity coefficients, as would be predicted, for example, by the Debye–Hückel equation (eqn. 3.74). This decreases the *effective* concentration of dissolved species and thus drives precipitation–dissolution and adsorption–desorption reactions toward dissolution and desorption.

3. Mixing of groundwater and seawater also results increased concentration of Mg^{2+} and Ca^{2+}, which then displace trace metals on cation binding sites of colloids and solids.

4. Increased carbonate concentrations results in rare earths forming relatively soluble dicarbonato complexes such as $La(CO_3)_2^-$.

Kona coastal seawater is also notably enriched in REE compared with North Pacific surface seawater, which is consistent with a significant REE groundwater discharge flux. Similar results elsewhere, including Rhode Island and Florida, suggest enhanced REE fluxes in groundwater discharge, even when those elements experience partial removal when mixing with seawater and this may be a main source of REE in seawater (Kim and Kim, 2014).

Submarine groundwater discharge may be an important source for other elements as well, including fixed nitrogen, Si, Ra, Ba, Sr, V, Cr, and a number of other trace metals. Windom et al. (2006) found greatly elevated Fe concentrations in the 240 km long Patos Lagoon in Brazil, which is separated from the ocean by a barrier island, as well as more moderately elevated Fe in the adjacent coastal surf zone. Although some loss of Fe occurs during mixing, they concluded that there was a net groundwater flux that was 10% of the aerosol Fe flux to the South Atlantic Ocean. Other studies, however, have found that Fe and some other redox-sensitive elements are removed in subterranean estuaries resulting in a low net flux to the ocean.

14.4.3 The hydrothermal flux

One of the most exciting developments in geochemistry in the past 40 years has been the discovery of hydrothermal vents at mid-ocean ridges and other submarine volcanoes. Simply the sight of 350°C water, black with precipitate, jetting out of the ocean bottom, surrounded by a vibrant if bizarre community of organisms living in total darkness at depths of 2500 m or more is exciting (Figure 14.17). But these phenomena were exciting for other reasons as well. Hydrothermal systems are sites of active ore deposition, so scientists were able to directly analyze the kinds of

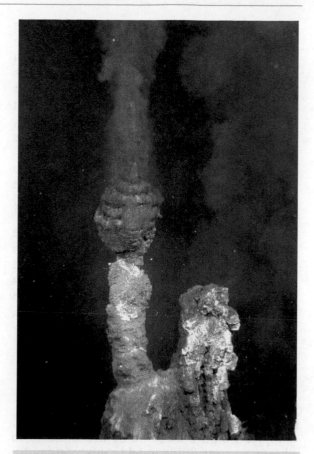

Figure 14.17 An active hydrothermal chimney (black smoker) in the caldera of Brothers Volcano in the Kermadec island arc. Photo from NOAA Vents Program.

fluids that produce volcanogenic massive sulfide ores (we'll pursue this topic further in the next chapter). Hydrothermal activity is also an important source for some elements in the oceans, and an important sink for others and has a profound effect on the composition of the oceanic crust. Thus, the discovery of hydrothermal vents has provided geochemists with the opportunity to put into place a major piece of the great geochemical puzzle.

The fluids emanating at hydrothermal vents consist of seawater that has undergone extensive reaction at a variety of temperatures with the oceanic crust. These fluids are compositionally diverse, but are typically reduced (sulfide-bearing), acidic, metal-rich, and can be as hot as 400°C. Upon mixing with cold, oxidized, alkaline seawater, the metals precipitate to build sulfide "chimneys" that emit jets of water from which metals precipitate as sulfides and Fe-Mn oxyhydroxides, hence the name *black smokers*. The vent fluids can then continue to mix with surrounding seawater rising hundreds of meters or more as buoyant *hydrothermal plumes*, in which the original vent fluid has been diluted by a factor of 10^4, but which can still be identified by anomalous high levels of metals and ^{3}He in them.

Based on available heat in the upper oceanic crust at mid-ocean ridges and Tl mass balance in the oceans, the high temperature (>350°C) water flux is in the range of 0.2 to 6×10^{13} kg/yr. As the oceanic crust and lithosphere continues to cool as it moves away from ridges, lower temperature circulation, and basalt–seawater interaction continues. This could produce an additional flux of 0.2 to 1×10^{17} kg/yr, although the chemical effects are likely small at low temperatures. For comparison, the riverine flux is $\sim 4 \times 10^{16}$ kg/yr.

14.4.3.1 *The composition of hydrothermal fluids*

Samples of pure vent fluids are difficult to obtain, as vent fluids quickly mix with ambient seawater, although great progress has been made in developing samplers that can be inserted directly into the vent. The pure vent fluid end-member of the sampled mixture can nevertheless be calculated. This is straightforward provided the concentration of at least one property of the vent fluid is known. Since the temperature of the vent fluid can be determined, this provides the key to calculating the vent fluid end-member composition. The first vents discovered, on the Galapagos Spreading Center, were diffuse, low temperature vents (<13°C). A strong inverse correlation between Mg and temperature was observed, and Edmond et al. (1979) concluded that the pure hydrothermal fluid had a temperature of 350°C and a Mg concentration of 0. Subsequent studies have shown that in most, but not all cases, Mg is quantitatively extracted from seawater in high-temperature hydrothermal systems. The concentrations of all other species in the vent fluid can be obtained from the intercept (Mg = 0) of a plot of the concentration of the species of interest. Figure 14.18 illustrates this for two different fluids at the TAG hydrothermal site

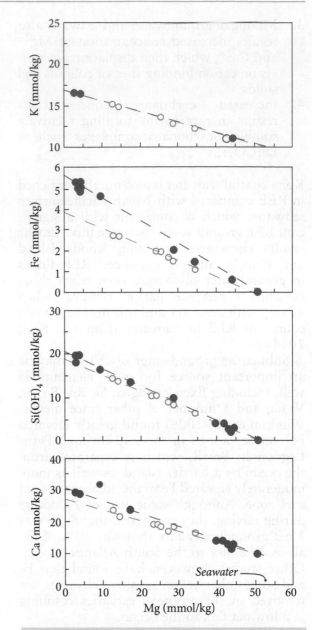

Figure 14.18 (a) Ca, Si(OH)$_4$, Fe, and K vs. Mg in hydrothermal fluids from TAG area of the Mid-Atlantic Ridge at 26°N. Closed red symbols are samples from a black smoker (360–366°C), open symbols for a white smoker (273–301°C). From Edmond (1995).

on the Mid-Atlantic Ridge: fluids venting from hotter black smoker and those from somewhat cooler white smokers. The white smoker fluids are poorer in Ca, Fe, Cu, and Si and also depleted in H$_2$S but enriched in Zn. The white smoker fluid is thought to result from subsurface mixing of 15–20%

seawater (Edmond et al., 1995). This results in precipitation of anhydrite, pyrrhotite (FeS), and silica, which removes Ca, Fe, H_2S, and Si and dissolution of sphalerite, which enriches them in Zn.

Table 14.4 lists compositions of several representative hydrothermal vent fluids. The composition of hydrothermal fluids depends on a number of factors: the nature of the rock through which water has circulated, depth of circulation (and therefore temperature and pressure of reaction, the water/rock ratio, subsurface mixing with seawater (as in the example of the TAG fluids above) phase separation, magmatic degassing, and, for fluid temperatures below 115°C, biological activity. We'll now consider these processes in more detail.

14.4.3.2 Evolution of hydrothermal fluids

In a simplistic fashion, hydrothermal systems can be divided into three zones (Alt, 1995). The principal processes and reactions involved are shown schematically in Figure 14.19. The first is the *recharge zone*, where seawater enters the oceanic crust and is heated as it penetrates downward. In this zone, which may be several kilometers or more off the axis of the ridge, flow is diffuse, so water/rock ratios are low as are temperatures (<200°C). The second is the *reaction zone*. This is thought to be often located near the base of the sheeted dikes and in the upper gabbros (see Chapter 11) at a depth of 1500–2000 m. In this zone, water reacts with hot rock at temperatures ≥350°C and the primary chemical characteristics of the fluid are determined. Because of reduced

Table 14.4 Composition of representative hydrothermal vent fluids.

	21°N	South cleft plume	Axial virgin mound	Lost City BH vent	TAG black	TAG white	Seawater
	EPR	JDF	JDF	MAR	MAR	MAR	
T °C	273–355	224	299	90	335–350	270–301	—
pH	3.3–3.8	3.2	4.4	10.5	3–3.8	3	7.6
Li µmol/kg	891–1322	1718		46	367–411	352	25.9
B µmol/kg	500–548	496	450	38	520		416
CO_2 mmol/kg	5.7	3.7–4.5	285				~2250
CH_4	28–1380			1.26	–		
Na mmol/kg	432–513	796	148	495	564	549	468
Al µmol/kg	4.0–5.2				5		0–0.04
Si mmol/kg	15.6–19.5	23.3	13.5	0.073	18.3		0–0.18
H_2S mmol/kg	6.6–8.4	3.5	18	0.07	5.9	0.5–3	26.9*
$\delta^{34}S$	+1.4–3.4	+5.7	+7.3				+21
Cl mmol/kg	489–579	1087	176	541	559	636	546
K mmol/kg	32.5–49.2	51.6	6.98	10.5	17–20	17–20	10.2
Ca mmol/kg	11.7–20.8	96.4	10.2	26.9	29	27	10.1
Mn mmol/kg	0.67–1.0	3.59	142		830	756	<0.005
Fe mmol/kg	.75–2.43	18.7	12	0.0006	3600	3835	<0.003
Cu µmol/kg	0.02–44	1.5	0.4		115	3	<0.005
Zn µmol/kg	40–106	780	2.2		42	350	<0.01
Ge nmol/kg	130–170	150–260					<0.0005
Br µmol/kg	802–929	1832	250	851	847		840
Rb µmol/kg	27–33	37		2.8	10.7	9.4	1.4
Sr µmol/kg	65–97	312	46	102	100		90
$^{87}Sr/^{86}Sr$	0.7032			0.7064	0.7038	0.7046	0.70918
Cs nmol/kg	202			16	179	113	2.2
Ba µmol/kg	8–16						<0.15
Tl nmol/kg		110					<0.0001
Pb nmol/kg	183–359	1630					<0.00015

Compiled from the literature. EPR: East Pacific Rise, MAR: Mid-Atlantic Ridge. JDF: Juan de Fuca Ridge.

*Concentration of *sulfate* in seawater.

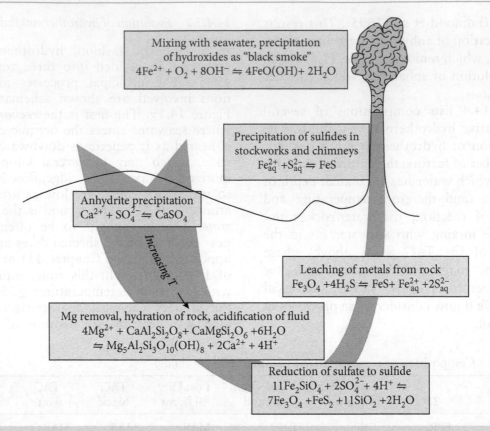

Figure 14.19 Some of the important reactions occurring in mid-ocean ridge hydrothermal systems.

permeability, water/rock ratios are lower here than in the overlying recharge zone. Phase separation, in which the fluid separates into a high-density brine and lower-density fluid, may occur in this zone. The third zone, in which the final composition of the fluid is achieved, is the *upflow zone*. Here water rises rapidly and cools somewhat as it does so. Precipitation of some sulfides may occur. Boiling may also occur if pressure is sufficiently low and temperature sufficiently high. Since upflow appears to be quite focused, water/rock ratios are low. Mixing with low temperature seawater may also occur, which can lead to extensive sulfide and anhydrite precipitation.

- Recharge zone

As water enters the oceanic crust, low-temperature reactions occur in which volcanic glass and minerals are transformed to clays such as celadonite, nontronite, and smectite, and oxyhydroxides. In this process, the rock may up alkalis Li, K, Rb, Cs, B, and U from seawater. As seawater warms to temperatures around 130° C, anhydrite ($CaSO_4$) begins to precipitate. By 200°C, essentially all the Ca^{2+} and two-thirds of the SO_4^{2-} (as well as a significant fraction of the Sr) are lost in this way. Addition of Ca^{2+} to the fluid by reaction with basalt can result in further removal of sulfate. Thus the fluid entering the reaction zone is severely depleted in calcium and sulfate. Anhydrite is rare in altered oceanic crust as much of the anhydrite precipitated in this way later dissolves when the crust cools.

The third major reaction in the recharge zone is loss of Mg^{2+} from seawater to the oceanic crust. This occurs through replacement of primary igneous minerals and glass by clay minerals such as sapolite and smectites, for example, by replacing plagioclase:

$$2Mg^{2+} + 4H_2O + 2SiO_2 + 2CaAl_2Si_2O_8$$
$$\rightleftharpoons Mg_2Ca_2Al_4Si_6O_{20}(OH)_4 + 4H^+$$
(14.42)

- Reaction zone

At higher temperatures, chlorite and tremolite may form, for example, by replacing plagioclase and pyroxene:

$$2Mg^{2+} + 8H_2O + CaMgSi_2O_6$$
$$+ CaAl_2Si_2O_8 \rightleftharpoons Mg_5Al_2Si_3O_{10}(OH)_8$$
$$+ Ca^{2+} + H_4SiO_4 + 2H^+ \qquad (14.43)$$

$$Mg^{2+} + 4CaMgSi_2O_6 + H_4SiO_4$$
$$\rightleftharpoons Ca_2Mg_5Si_8O_{22}(OH)_2 + 2Ca^{2+}$$
$$+ 2H_2O \qquad (14.44)$$

The significance of these reactions is not only loss of Mg from the solution, but also the production of H^+ (or, equivalently, the consumption of OH^-). *It is these reactions that account, in part, for the low pH of hydrothermal vent solutions.* This greatly increases the fluid's capacity to leach and transport metals.

Von Damm and Bischoff (1987) used measured SiO_2 concentrations in Juan de Fuca vent fluids together with thermodynamic data to estimate that the fluids equilibrated with quartz at pressures of 46–48 MPa and temperatures of 390–410°C. Geothermometry performed on minerals in altered oceanic crust indicate temperatures as high as 400–500°C. By comparing a thermodynamic model of hydrothermal interactions and assuming fluids are in equilibrium with the assemblage anhydrite-plagioclase-epidote-pyrite-magnetite, Seyfried and Ding (1995) estimated temperatures of 370–385°C and 30–40 MPa for equilibration of fluids from the 21°N on the EPR and the MARK area of the Mid-Atlantic Ridge.

Under these conditions, reactions would include the formation of amphiboles, talc, actinolite, and other hydrous silicates from reactions involving ferromagnesian silicates (olivines and pyroxenes) and the formation of epidote from plagioclase:

$$Ca^{2+} + 2H_2O + 3CaAl_2Si_2O_8$$
$$\rightleftharpoons 2Ca_2Al_3Si_3O_{12}(OH) + 2H^+ \qquad (14.45)$$

as well as the exchange of Na^+ for Ca^{2+} in plagioclase, a process termed *albitization*, with precipitation of quartz:

$$2Na^+ + 2CaAl_2Si_2O_8 \rightleftharpoons 2NaAlSi_3O_8 + SiO_2$$
$$+ 2Ca^{2+} \qquad (14.46)$$

The evidence for albitization comes not only from the identification of albitized plagioclase in hydrothermally altered rocks, but also the inverse correlation between Na/Cl and Ca/Cl in hydrothermal fluids.

In addition, the fluid will be reduced by oxidation of ferrous iron in the rock, e.g.:

$$2SO_4^{2-} + 4H^+ + 11Fe_2SiO_4$$
$$\rightleftharpoons FeS_2 + 7Fe_3O_4 + 11SiO_2 + 2H_2O \qquad (14.47)$$

The solubility of transition metals and S increase substantially at temperatures above 350°C, so sulfides in the rock are dissolved, e.g.:

$$Cl^{2-} + 2H^+ + FeS \rightleftharpoons H_2S_{aq} + FeCl^0_{aq} \qquad (14.48)$$

The metals released will be essentially completely complexed by chloride, which is by far the dominant anion in the solution, as most sulfate has been removed or reduced and sulfide and CO_2 will be largely protonated at the prevailing pH. Isotopic compositions of H_2S in vent fluids indicates most sulfur is derived from dissolution of sulfides in the rock, with a smaller contribution from reduction of seawater sulfate.

Whereas the alkalis Li, K, Rb, Cs, B, and U are taken up by the rock at low temperature, they are released at high temperature. Loss of K, Rb, and Cs begins around 150°C, but loss of Li probably does not begin until higher temperatures are reached (Na, though, is actively taken up even at high temperatures by albitization).

Fluids make their closest approach to the magma chamber in the reaction zone, and magmatic volatiles may be added to fluids within this zone. Hydrothermal vent fluids with seawater chlorinities have CO_2 concentrations as high as 18 mmol/kg, which is substantially more than seawater (~2 mmol/kg). The isotopic composition of this carbon ($\delta^{13}C \approx -4‰$ to $-10‰$) is similar to that of mantle carbon (see Chapter 9) and distinct from that of seawater bicarbonate ($\delta^{13}C \approx 0$), implying that the excess CO_2 is probably of magmatic origin. Other magmatic volatiles present in the fluid may include He, H_2, H_2O, and CH_4 (which is not present as such in the magma, but forms by reaction between CO_2 and H_2 at temperatures below ~500°C). In most cases, any contribution of magmatic H_2O will be insignificant compared with seawater-derived H_2O. However, fluids from vents at 9–10°N on the EPR have negative δD values, which suggest a small but significant contribution of magmatic water. These vents developed and were sampled shortly after an eruption in 1991. Shanks et al. (1995) calculated that the observed δD values could be explained by addition of 3% magmatic water and that this water could be supplied by degassing of a dike 20 km long, 1.5 km deep and 1 m wide. The magmatic water would be exhausted in about 3 years.

The reaction zone is also the region where phase separation is most likely to occur. Below its critical point, at 29.8 MPa and 407°C, seawater boils to produce a low-salinity vapor phase and a liquid whose salinity initially approximates that of the original liquid. In the case of hydrothermal fluids, the vapor produced would be strongly enriched in H_2S and CO_2 as well as other volatiles. As boiling continues, the liquid becomes increasingly saline. Above its critical point, seawater separates into a dense brine and a less saline fluid that is enriched in H_2S, CO_2, CH_4, and H_2. As phase separation continues, the fluid becomes increasingly dilute while the brine becomes more concentrated. The phase diagram (P-X) for the system H_2O-NaCl shown in Figure 14.20 illustrates this. Seawater behaves approximately as 3.5% NaCl solution, which corresponds to a NaCl mole fraction of 1.1%. Imagine a parcel of water heated to 400°C in the deep crust at a pressure of 30 MPa; these temperatures and pressures, seawater will be a supercritical fluid. Now imagine that it begins to rise. At about 28 MPa it will begin to separate into low salinity fluid and one more saline than seawater. As it continues to rise, the low salinity fluid becomes increasingly less saline and the high salinity one more saline. By the time it rises to 22 MPa, the low salinity fluid would

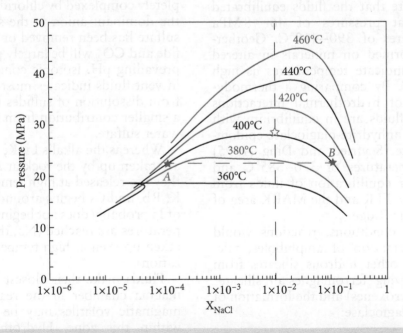

Figure 14.20 Pressure-composition phase diagram for the system H_2O—NaCl. A seawater-like NaCl solution has an NaCl mole fraction of 0.011 (vertical dashed line) and will plot just above the two-phase region at 400°C, and 29 MPa (open star). As the water rises and pressure drops to 22 MPa, it will have separated into a low salinity fluid of composition A with $X_{NaCl} \approx 0.00011$ that coexists (horizontal dashed red line) a brine of composition B with $X_{NaCl} \approx 0.12$. Phase relationships based on the model of Driesner and Heinrich (2007).

have a NaCl mole fraction of ~0.011% and the high salinity fluid would contain a NaCl mole fraction of ~12%.

Phase separation and mixing between the fluids produced by it provides the best explanation for the large variations in Cl content observed in hydrothermal fluids, which otherwise behaves nearly conservatively. Brines produced by phase separation may be too dense to rise to the seafloor. Instead, they may reside at depth for prolonged periods of time, slowly mixing with less saline fluids. The high chlorinity of the North Cleft vent fluid (Table 14.4) results from mixing of a fluid with seawater chlorinity and a high salinity brine (Von Damm, 1988). The low-chlorinity Virgin Mound fluid is an example of a fluid whose chlorinity has been reduced and CO_2 elevated by mixing with a low-salinity vapor phase produced by boiling. The shallow depth of this vent would mean that hydrothermal fluids would reach the critical point nearly 1500 m beneath the seafloor, providing ample opportunity for subcritical phase separation during ascent.

The chlorinity of fluids also influences other compositional factors. The solubility of H_2S decreases with increasing chlorinity. However, concentrations of metals such as Fe and Cu increase with increasing chlorinity because of the formation of metal chloride complexes. Thus, a fluid produced by mixing will likely be out of equilibrium with the rock and thus undergo further reaction after mixing (Seyfried and Ding, 1995).

Two other important controls on solution chemistry are pH and f_{O2}. The Fe/Cu ratio of the fluid is sensitive to both, high f_{O2} and low pH favoring a high Fe/Cu ratio. Fe/Cu ratios of hydrothermal fluids indicate pH values in the reaction zones of 4.8 to 5.2 and f_{O2} buffered by the assemblage anhydrite-magnetite-pyrite (Seyfried and Ding, 1995) at temperatures of ~400°C and ~40 MPa.

The composition of fluids venting in Lost City hydrothermal system at 30°N on the Mid-Atlantic Ridge is the most distinctive of those listed in Table 14.4. The system is developed on peridotite exposed in detachment fault some 15 km from the plate boundary. The fluids are alkaline rather than acidic, Si and metal poor, and rich in H_2, CH_4, and other abiologically synthesized dissolved

organic compounds. Rather than being driven by magmatic heat, the heat source for this system is the exothermic nature of reactions such as the conversion of olivine to serpentine and brucite:

$$2\,Mg_2(SiO_4) + 3H_2O \rightleftharpoons Mg_3Si_2O_5(OH)_4 \\ + Mg(OH)_2 \quad (14.49)$$

The alkaline nature of these fluids is due to the Ca- and Si-poor nature of peridotite and the relatively low temperatures. The absence of a stable Ca-bearing phase such as tremolite in the metamorphic assemblage such as chlorite and tremolite, results in the consumption rather than production of acidity (as in reactions 14.43 through 14.45 in basalt). For example, diopside is converted to serpentine in basalt through reactions such as

$$2Mg(OH)_2 + 2CaMgSi_2O_6 + 4H^+ \\ \rightleftharpoons 2Mg_3Si_2O_5(OH)_4 + 2Ca^{2+} \quad (14.50)$$

The Rainbow hydrothermal field at 36°C on the Mid-Atlantic Ridge is also developed on peridotite, but the fluids are hot (360°C) and acidic (pH ~3). Here temperatures are sufficiently high to stabilize Ca-bearing minerals such as tremolite, which accounts for the difference in pH. The Rainbow vent fluid chemistry is strongly influenced by phase separation that produces Cl-rich brines and have dramatically higher transition metals (particularly Fe and Mn) and rare earth concentrations than in other MAR fluids due to Cl-complexation.

Another interesting aspect of the Lost City and particularly the Rainbow fluids are the high H_2 and CH_4 concentrations (these are present in lower concentrations in other vent fluids as well). These form from reactions such as

$$15(Mg_{1.8}Fe_{0.2})SiO_4 + 20.5H_2O \\ \rightleftharpoons 7.5Mg_3Si_2O_5(OH)_4 + 4.5Mg(OH)_2 \\ + Fe_3O_4 + H_2 \quad (14.51)$$

Upon mixing with seawater, methane and a variety of simple organic molecules such as carboxylic acids and alcohols can form by reaction between H_2 and CO_2 (Shock and Schulte, 1998).

- Upflow zone

The density decrease caused by heating eventually forces the hydrothermal fluid to rise to the seafloor. The concentrated flow out of vents indicates that the upwelling zone can be quite narrow and flow is strongly focused through fractures. Upflow zones in exposed sections of oceanic crust (ophiolites) are altered to the assemblage epidote-quartz-titanite or actinolite-albite-titanite-chlorite assemblages. This is consistent with thermodynamic calculations for hydrothermal fluids. The fluid experiences decompression as it rises and may experience phase separation at this point.

Most fluids appear to have undergone some conductive cooling during ascent. Cooling of fluids induces precipitation of sulfides and quartz. In ophiolites, upflow zones are marked by mineralized alteration pipes, or stockworks. The solubility of Cu shows the strongest temperature dependence, followed by Fe, thus the concentrations of these two elements may drop significantly during upflow as a consequence of precipitation of chalcopyrite:

$$Cu^+ + Fe_3O_4 + FeS_2 + 0,5\,H_2O$$
$$\rightleftharpoons CuFeS_2 + 1.5Fe_2O_3 + H^+$$

In the Guaymas Basin of the Gulf of California and the Escanaba Trough of the southern Gorda Ridge, vent fluids exit through sediment cover. Hydrothermal fluids must traverse up to 500 m of sediment before exiting to seawater. Vent temperatures are somewhat cooler, ~220°C for the Gorda Ridge system. The compositions of these fluids are distinct, they are mildly acidic (pH 5.5 to 5.9). Guaymas Basin fluids are somewhat richer in K, Rb, Cs, Ca, and Sr, due to dissolution of carbonate and leaching of sediment. Both the Guaymas and Escanaba fluids are transition metal-poor, as a result of sulfide precipitation in the sediment. The Guaymas fluid is also rich in hydrocarbons, which are produced by thermal degradation of organic matter in the sediment.

Finally, hydrothermal fluids eventually mix with seawater, either in the shallow subsurface or as they exit the seafloor. This induces additional cooling and precipitation. Along with sulfide precipitation, mixing causes the seawater-derived sulfate to precipitate as anhydrite. Precipitation of anhydrite accounts for the white "smoke" of white smokers. Precipitation at the sea surface quickly builds chimneys, which can reach more than 10 meters above the seafloor. Chimneys consist primarily of Fe and Cu sulfides such as pyrite (FeS_2), marcasite (FeS_2), pyrrhotite (FeS), chalcopyrite ($CuFeS_2$), bornite (Cu_5FeS_4), cubanite ($CuFe_2S_3$) with lesser amounts of sphalerite (ZnS), wurtzite (ZnS), galena (PbS), silica, silicates, anhydrite, and barite ($BaSO_4$). In contrast to the chimneys of high temperature metal-rich fluids, the Lost City chimneys consist of calcite, aragonite, and brucite. For the most part, chimneys are rather fragile structures subject to weathering in which anhydrite redissolves and sulfides oxidize to oxyhydroxides once venting terminates.

- Hydrothermal plumes

As vent fluid is diluted with seawater, a hydrothermal "plume" is created, which can rise hundreds of meters above the vent site because of its slightly warmer temperature and therefore lower density than surrounding water. Precipitation of sulfide "smoke" immediately above the vent removes up to half the dissolved Fe. Much of the remainder is oxidized to Fe^{3+} and precipitated as oxyhydroxides in the plume. The half-life for Fe^{2+} oxidation in seawater is anywhere from a few minutes to a day or more, depending on O_2 concentrations and pH. During Fe precipitation, a number of elements may be coprecipitated, including Mn, P, V, Cr, and As. The kinetics of Mn oxidation are considerably slower, and Mn precipitation is generally delayed until the plume reaches neutral buoyancy and begins to spread out horizontally. Mn oxidation appears to be bacterially mediated. The Fe-Mn particles produced within the plume strongly scavenge particle-reactive elements, such as Th, Be, and the rare earths, from seawater.

14.4.3.3　Hydrothermal fluxes

The importance of mid-ocean ridge hydrothermal systems in controlling the composition of seawater was immediately realized upon the discovery of the first hydrothermal vents.

Initially, it appeared that estimating the flux of elements into and out of the oceanic crust was straightforward (e.g., Edmond et al., 1989). Unfortunately, the problem has proven to be not so simple as hydrothermal vent fluids are more variable in both flow and composition than initially thought. A variety of different approaches can be taken to estimate the total flux. One is to ratio average elemental concentrations in vent fluids to the heat flux, and another is to ratio them to the ^{3}He flux, which is a conservative tracer in the ocean. Yet another approach is to consider the change in composition of hydrothermally altered oceanic crust in drill holes or ophiolites with pristine oceanic crust. Other approaches are based on mass balance of isotope ratios.

Table 14.5 lists the estimated fluxes of a variety of elements published in the last 20 years or so. Most fluxes are uncertain by an order of magnitude or more. It is nevertheless interesting to observe that many of these estimated fluxes are comparable to the riverine fluxes, emphasizing the importance of hydrothermal activity in controlling seawater composition, even if we do not understand that control yet. Hydrothermal activity appears to be an important source of Li, K, Mn, Fe, Zn, Ba, and Tl and an important sink for Mg, U, and perhaps S.

There are, however, a number of caveats to the fluxes listed in Table 14.5. First, it includes only high-temperature hydrothermal systems at mid-ocean ridges. It excludes hydrothermal activity associated with subduction zone and mid-plate volcanism, although these are likely to be less than 10% of ridge crest activity. However, it also excludes low-temperature activity. Based on Tl isotopes, Nielsen et al. (2006) estimated water flux of the latter at 0.2 to 5.4×10^{17} kg/yr, which greatly exceeds their estimate of the high-temperature flux of 7.2×10^{12} kg/yr. Because low-temperature systems are likely to chemically less reactive, fluxes per kg are likely far lower, but the integrated total may be larger. Furthermore, the fluxes may oppose the high-temperature ones, which appears to be the case for elements such as Si, Ca, K, Rb and Tl, and many transition metals dissolved in vent fluids quickly precipitate when those fluids mix with seawater, making the net flux uncertain. As a consequence, low temperature hydrothermal activity may be a net sink for some elements present in high-temperature vent fluids. The hydrothermal flux nevertheless appears to be significant for some of these elements. For example, Tagliabue et al. (2010) estimated the global hydrothermal dissolved Fe flux at 9×10^8 mol/yr. Although this is only ~1–5% of the flux listed in Table 14.5, it is nonetheless a major source of Fe in the Southern Ocean where the riverine flux is negligible. In addition, the riverine fluxes listed do not take account of the processes in estuaries discussed in section 14.4.2.1 that alter (in most instances significantly reduce) the fluxes of particle reactive and redox sensitive elements.

Table 14.5 Global fluxes to seawater from high-temperature ridge crest hydrothermal activity.

	Axial hydrothermal mol/yr	Riversmol/yr
Li	5×10^9 to 4×10^{10}	1.0×10^{10}
Mg	-3×10^{12}	5.2×10^{12}
Si	-3×10^{11} to 1×10^{12}	6.5×10^{12}
S	-1.5×10^{11} to -2.3×10^{11}	3.2×10^{12}
K	1×10^{11} to 6.9×10^{11}	1.2×10^{12}
Ca	3×10^{12}	1.3×10^{13}
Mn	1×10^{10} to 9×10^{10}	2.3×10^{10}
Fe	2×10^{10} to 9×10^{10}	4.4×10^{10}
Cu	3×10^8 to 8×10^9	8.7×10^8
Zn	1×10^8 to 3×10^9	3.5×10^8
Rb	1×10^8 to 1×10^9	7.1×10^8
Sr	0 to 1×10^9	2.6×10^{10}
Ba	2×10^8 to 1×10^9	6.3×10^9
Tl	1×10^3 to 3×10^5	1.3×10^3
Pb	1×10^4 to 2.8×10^5	1.5×10^7
U	-1×10^5 to -1.6×10^7	5.9×10^7

Hydrothermal fluxes from Elderfield and Schultz (1996), Nielson et al. (2006) Coogon et al. (2016) Coogan et al. (2017) and Staudigel (2014); Riverine fluxes from Gaillardet et al. (2014) and Maybeck (2003). The S flux listed is the net of sulfate and sulfide fluxes.

14.4.4 The atmospheric source

The atmosphere is, of course, the principal source and sink of dissolved gases in the ocean, but it is also a surprisingly important source of other dissolved constituents, as well as particulate matter, in the oceans. These other constituents are derived from particles in the atmosphere called *aerosols*. Aerosols have several sources: sea spray, mineral dust derived from soils and desert sands, volcanic eruptions, condensation reactions in the atmosphere, the biosphere, particularly fires, and anthropogenic activity such as combustion of fossil fuels, mining and mineral processing, agriculture, and the production and consumption of various chemicals. Of these sources, sea spray is the most important; however, sea spray does not represent a true flux to the oceans, as it is derived directly from them, and we will not consider it further. Excluding sea spray, dust from arid regions is the dominant source of aerosols, but this varies by element. Desert dust, shown in a NASA Goddard Space Flight Center computer simulation in Figure 14.21, is the dominant source of Al, Si, Mn, Fe, and rare earths, but anthropogenic activity and fire sources also contribute to those elements and may be especially important for Cu, Zn, Pb, and volcanoes may be the dominant source of Cd (Mahowald et al., 2018).

The material flux from the atmosphere to the ocean may occur through *dry deposition*, which includes both settling of particles from the atmosphere and gas adsorption, and *wet deposition*. Wet deposition includes all matter, both particulate and gaseous, first scavenged from the atmosphere by precipitation (i.e.,

rain and snow) before being delivered to the oceans.

Though marine chemists agree that the atmosphere is an important source for a variety of species in the ocean, quantifying the atmosphere-to-ocean flux is difficult. Relative to the surface area of the oceans, the number of observations is comparatively small. Furthermore, deposition rates are very heterogeneous in both space and time, due to variations in climate, wind patterns, and aerosol source distribution, such that more than half the total annual aerosol deposition may occur over a small number of days (Mahowald et al., 2018). Charles Darwin (1846) summarized observations of dust falling on ships in the Central Atlantic and concluded, "I think there can be no doubt that the dust which falls in the Atlantic does come from Africa." Modern observations show that particle concentrations can briefly reach extremely high levels, up to 700 µg per cubic meter of air, over the Atlantic between 30° and 5°N, where the Northeast Trade Winds carry mineral dust from Saharan dust storms. There are also significant fluxes of dust from the Asian and Australian deserts to the North Pacific, though it is smaller than

Figure 14.21 Global distribution of dust, sea salt, and carbon and sulfates in a global atmospheric modeling run on the Discover supercomputer at the NASA Center for Climate Simulation at Goddard Space Flight Center. Image credit: William Putman, NASA/Goddard.

the Saharan dust plume. In contrast, particle concentrations are less than 0.01 µg/m³ over remote areas of the South Pacific. Aerosol composition also varies widely, due both to difference in sources and fractionation that occurs as the aerosol is transported from the source area. This variability, together with the vastness of the ocean, makes it difficult to derive global fluxes.

The next question we must ask is, what fraction of the particulate matter deposited on the ocean surface dissolves? In experiments with natural aerosols collected over the Red Sea, Mackey et al. (2015) found that solubilities in a seven-day period ranged from 74% of Cd and 65% of Zn to 2.5% of Al and 0.04% of Fe. Solubilities, however, vary depending on the aerosol source and the extent of atmospheric processing. Iron solubility is inversely proportional to the total aerosol loadings and increases with distance from source. These observations reflect several factors (Sholkovitz et al., 2012; Mahowald et al., 2018). First, Fe derived from combustion and anthropogenic sources is substantially more soluble (up to 50% or more) than Fe in mineral dust, but is variably diluted by insoluble Fe in desert-derived dust storms. Second, the combustion/ anthropogenic Fe is present in smaller particles, which are inherently more reactive (see Chapter 5) and also settle out of the atmosphere more slowly than larger particles. Third, in the atmosphere, reaction with sulfate and organic acids in droplets can increase solubility and longer transit allows more time for such processing.

Table 14.6 lists estimates of the atmospheric flux of dissolved matter from the oceans and compares them with the riverine flux. While the atmospheric flux is insignificant for major ions in seawater such as Mg, Ca, and K, it can be significant for Al, Si, transition metals, and REE. The fluxes of some of these elements, including Cu, Zn, and Pb, include a significant anthropogenic component and hence do not necessarily represent the steady state for these elements. Indeed, anthropogenic sources likely constitute 75% of the aerosol Pb flux and in the recent past it may have contributed >95% of the Pb flux to the North Atlantic. This flux has decreased dramatically since the elimination of tetraethyl Pb gasoline in most countries and now may be below 50% (Bridgestock

Table 14.6 Atmospheric flux to the oceans.

Element	Atmospheric dissolved flux	Riverine flux	Atmosphere/ River
Al	6.8×10^{10}	4.4×10^{10}	1.52
Mg	2.8×10^{11}	5.5×10^{12}	0.05
Si	1×10^{12}	6.5×10^{12}	0.12
P	3.8×10^{8}	6.5×10^{10}	0.006
K	4.2×10^{9}	1.2×10^{12}	0.003
Ca	57×10^{10}	1.3×10^{13}	0.005
Mn	2.6×10^{8}	2.3×10^{10}	0.011
Fe	1.1×10^{9}	4.4×10^{10}	0.025
Ni	9.6×10^{7}	5.1×10^{8}	0.19
Cu	1.5×10^{8}	8.7×10^{8}	0.21
Zn	1.0×10^{9}	3.5×10^{8}	4.43
Cd	2.1×10^{6}	2.7×10^{7}	0.08
La	4.1×10^{6}	3.2×10^{7}	0.13
Nd	4.4×10^{6}	3.9×10^{7}	0.11
Yb	3.2×10^{5}	3.5×10^{6}	0.09
Pb	1.4×10^{8}	1.5×10^{7}	95

All fluxes are in moles/yr. Atmospheric fluxes from Duce et al. (1991), Greaves et al. (1993), and Zhang et al. (2015). Fluxes of Duce et al. modified based on revised solubilities given by Mackey et al. (2015). Riverine fluxes from Gaillardet et al. (2014), Berner and Berner (1996), and Seitzinger et al. (2015).

et al., 2016); we'll discuss Pb pollution in more detail in the following chapter.

The riverine fluxes listed in Table 14.6 are the fluxes delivered to estuaries. As discussed in Section 14.4.2.1, 90–95% of Fe and substantial fractions of other particle-reactive elements may be removed in estuaries and never reach the ocean. Reducing the riverine Fe flux by 95% would produce an atmospheric to riverine flux ratio of 0.5. Furthermore, what Fe does make it to the ocean is removed by uptake by phytoplankton or adsorption on particle surfaces in coastal regions. Aerosols are consequently a primary source of Fe to the remote open ocean. While smaller fractions of other transition metals may be removed in this way, it is nonetheless clear that aerosols represent a major flux of these elements to the remote areas of the ocean as well. As we found in section 14.3.5, Fe and several other transition metals can be biolimiting. Thus, aerosols are an important aspect in maintaining open ocean photosynthesis and thus in the biological pump.

14.4.5 Sedimentary sinks and sources

Sediments are, in one way or another, the major sink for dissolved matter in the oceans. There are several ways in which elements dissolved in seawater find their way into sediments: (1) biologic uptake, (2) scavenging by organic particles, (3) scavenging or adsorption by or reaction with clay and other particles, (4) precipitation of, coprecipitation with, or adsorption by hydroxides and oxides, and (5) precipitation as evaporite salts. In addition, diffusion of dissolved species into or out of sediments may occur. In the latter case, sediments may serve as a source, rather than a sink, for an element.

14.4.5.1 Biogenic sediments

As we have seen, biological activity controls the distribution of not only the major nutrients but also the micronutrients in the ocean. Biogenic particles incorporated in sediment are an important sink for such elements. Over 50% of the ocean floor is covered with biogenic siliceous and calcareous oozes. In addition to primary productivity, which is controlled by the abundance of nutrients in the surface water, bathymetric and oceanographic factors also control the distribution of biogenic sediments. A variety of organisms build shells and tests of calcium carbonate. Mollusks and reef-building corals are obvious examples, but largely restricted to coasts. The shallow water carbonate production is estimated to be 2 to 3×10^{13} moles/yr, primarily from corals (Anderson, 2014). In the pelagic (i.e., open ocean photic zone) realm, coccolithophorid algae account for most of the pelagic calcium carbonate production ($\sim$60%) with smaller significant fractions of production from planktonic foraminifera (heterotrophic protists) and pteropods (pelagic snails), which precipitate aragonite rather than calcite. The total open ocean production is estimated to be 0.6 to 1.5×10^{14} moles per year.

As we found in Section 14.3.5, deep water is undersaturated in calcium carbonate such that much of the production of calcium carbonate in the surface water redissolves in the deep water and calcareous sediment is absent below the carbonate compensation depth at $\sim$4500 m. Even in shallow water, diagenetic processes within the sediment result in some dissolution. Consequently, the next export of calcium from the oceans to sediments is smaller, about 0.9 to 1.4×10^{13} moles per year in shallow water and $\sim$8 $\times 10^{12}$ moles per year in the open ocean (Andersson, 2014).

This total export of 1.8 to 2.3×10^{13} moles/year exceeds the sum of the riverine plus hydrothermal fluxes listed in Table 14.5 of $\sim$1.5 $\times 10^{13}$ moles/year. While there is considerable uncertainty in these fluxes, it is likely that sources and sinks of Ca in the oceans are out of balance. That misbalance is due to a post-glacial increase in carbonate precipitation on continental margins that has resulted from the $\sim$130 m sea-level rise since the last glacial maximum, increasing the surface area of the shallow ocean. It is just such areas that reef-building corals inhabit, so carbonate precipitation and accumulation increased as sea-level rose. Kleypas (1997) estimates that the carbonate production on coral reefs increased from 2.6×10^{12} moles/yr at the last glacial maximum to a maximum rate of 1×10^{13} moles/yr 2000 years ago. That was somewhat offset by a decrease in carbonate accumulation in the deep ocean. As more carbonate is lost to shallow sediments, the saturation state in the ocean decreases, the carbonate compensation depth increases and less carbonate is exported to the deep ocean.

Counterintuitively, this increased export of $CaCO_3$ from the ocean during sea-level rise would have resulted in an increase in atmospheric CO_2. The reason for this is that since bicarbonate is the overwhelming predominant carbonate form in the ocean, the overall precipitation reaction can be written as:

$$Ca^{2+} + 2HCO_3^- \rightleftharpoons CaCO_3 + H_2CO_3$$

increasing the saturation state of ocean water with respect to atmospheric CO_2:

$$H_2CO_3 \rightleftharpoons H_2O + CO_2$$

The process may have accounted for an increase in atmospheric CO_2 by as much as 20 ppm, or $\sim$20% of the post-glacial rise, contributing to post-glacial warming.

Marine carbonates contain variable amounts of Mg, and the Mg/Ca ratio is a function of temperature and can therefore be used as a paleothermometer (Herzberg and

Schmidt, 2018). Pelagic carbonate is generally Mg-poor, with coccolithophores generally having less than 4 mol% Mg and planktonic foraminifera an order of magnitude less. Shallow water carbonates often contain more Mg, up to ~20 mol% in the case of coralline red algae. However, the solubility of calcite increases with Mg contents above about 2%, so high-Mg calcite is less likely to survive. Assuming a molar Mg content of 2% for marine carbonates, and the carbonate export rate given above, gives a flux of roughly 4.5×10^{11} moles/yr of Mg or about 8% of the riverine flux.

Although dolomite $(Mg,Ca(CO_3)_2)$ is common in the geologic record, it rarely precipitates directly from seawater; this paradox is known as the *dolomite problem*. It does, however, form during diagenesis of carbonate platforms such as the Bahamas. As dolomite formation results in significant Mg isotope fractionation whereas uptake in high temperature hydrothermal systems does not, Tipper et al. (2006) estimated that dolomite formation accounts for at least ~10% of the Mg flux out of the modern ocean.

The major sink for Mg is clearly ridge crest hydrothermal activity; this flux (Table 14.4) plus the biogenic carbonate and dolomite fluxes are roughly ~75% of the riverine flux. Given the uncertainties, it's possible these fluxes balance. It is also possible there are other important sinks as well. Formation of authigenic clays in the near-shore environment may be one such sink (Rude and Aller, 1989).

The primary sources and sinks of bicarbonate (as distinct from CO_2) are the same as those of Ca: river water and carbonate precipitation. In addition, however, bacterial sulfate reduction in sediment provides an additional source:

$$2CH_2O + SO_4^{2-} \rightleftharpoons H_2S + 2HCO_3^-$$

According to Berner and Berner (2016), this flux is 2.4×10^{12} moles, which is about 7% of the riverine flux.

A preindustrial balanced flux of CO_2 between atmosphere and oceans of about 75×10^{15} mol/yr is well established from ^{14}C studies. The riverine flux dissolved organic carbon is far smaller, about 3×10^{13} mol/yr, most of which is remineralized in the ocean.

Photosynthesis in the ocean produces about 4×10^{15} mol/yr, of which 25% sinks out of the photic zone. However, almost all of this is remineralized in deep water, with a net export to sediments of only $\sim 1.7 \times 10^{12}$ mol/yr, or roughly 0.02% of primary production. Burial of significant fractions (>0.5 wt%) of organic matter in sediments is unusual and restricted to highly productive upwelling areas where the accumulation of organic matter exceeds the oxygen supply.

Diatoms account for most of the silica uptake in the ocean, but silicoflagellates (algae), radiolarians (protists) and sponges also utilize silica. In contrast to calcium carbonate, which is supersaturated in the surface water and increasingly undersaturated with depth, silica is everywhere undersaturated in the ocean and most undersaturated in the surface water due to biological uptake (Figure 14.9). Siliceous oozes are found beneath the Pacific equatorial high-productivity belt and beneath the productive waters at high latitudes, particularly in the Southern Ocean and cover roughly 8% of the ocean floor. In addition to productivity, water temperature appears to control the distribution of silicic oozes. Dutkiewicz et al. (2015) found that diatomaceous oozes are present only where seawater surface temperatures are below ~6°C. Gnanadesikan (1999) proposed an Arrhenius temperature dependence (eqn. 5.25) with the frequency factor $A = 1.32 \times 10^{16}$/day and the barrier energy $E_B = 95.45$ kJ. This gives a fractional dissolution rates of 0.25 per day at 25°C and 0.01 per day at 2°C. Much of the silica is packaged into fecal pellets, which sink at rates of ~50 m/day. Except in cold polar waters, most of the silica will redissolve in the surface water; silica that survives sinking beneath the thermocline is more likely to survive in sediment.

Tréguer and De La Rocha (2013) estimate the total biogenic silica production in the photic zone to be 240×10^{12} mol/yr, of which over a half occurs in the coastal zone and a third in the Southern Ocean. Of this, some 6.3×10^{12} mol/yr is buried in sediment. This approximately balances the riverine input, but an additional sink is necessary to balance hydrothermal and aerosol fluxes (Tables 14.5 and 14.6) as well as fluxes from submarine groundwater discharge (6×10^{11}) and dissolution of sediments and basalt on the seafloor

($\sim 2 \times 10^{12}$). They argue uptake by sponges in shallow water represents an additional sink of 3.6×10^{12} mol/yr.

14.4.5.2 Evaporites

For the remainder of the major ions in seawater, Cl^-, SO_4^{2-}, Na^{2+}, and K^+, sources appear to greatly exceed sinks, hence the steady-state assumption appears to fail. For Na, cation exchange can account for only about 10% of the riverine flux. For K, authigenic clay formation and cation exchange represent major fluxes, but the magnitudes of them are uncertain. For SO_4^{2-}, biogenic reduction may account for about 40% of the riverine flux. For chloride, the only significant sink is burial in pore waters. Yet, evidence suggests that the salinity of seawater has been approximately constant over the Phanerozoic (e.g., Lowenstein et al., 2014), indicating that the steady-state hypothesis holds, at least in the approximate, on such long time-scales.

Evaporites represent the missing flux for these elements. Evaporites, as their name implies, form when seawater is evaporatively concentrated to the point where salts precipitate. This process begins when salinity reaches ~ 50 with precipitation of aragonite, which is followed by precipitation of gypsum at a salinity of ~ 145 (about forty times concentration of the original seawater), followed by halite precipitation at salinities of ~ 350 (Warren, 2018). Subsequent evaporation causes precipitation of various K and Mg salts such as carnallite ($KMgCl_3 \cdot 6H_2O$), sylvite (KCl), and epsomite ($MgSO_4 \cdot 7H_2O$).

At present, however, the evaporite flux from seawater is negligible. Rather than being steady-state, evaporite formation is episodic. The last major evaporite event was the Messinian, when tectonics cyclically closed the connection between the Atlantic and Mediterranean through the Strait of Gibraltar between 5.9 and 5.3 million years ago. Because evaporation exceeds precipitation and river runoff in the Mediterranean Sea, once the connection to the ocean closes, it would dry up rather quickly (how quickly is left to the reader as Problem 11). As a result, 4.86×10^{18} kg of salt, mainly gypsum and halite, was precipitated in the Mediterranean (Ryan, 2008), representing 5% of the salt

in the oceans and resulting in a corresponding decrease in global ocean salinity. The geologic record shows that over the previous 150 million years smaller evaporite events occurred sporadically with a frequency of every 10–20 million years. Over the longer term, even larger evaporite events occurred, such as in the Gulf of Mexico 150 million years ago, Eastern Europe (270 and 370 million years ago), etc. Over geologic time, evaporites are a major sink for major ions. It is estimated that evaporite deposits contain about as much Cl as do the oceans.

In detail, the major ion composition of seawater has not been steady-state over the Phanerozoic. Studies of fluid inclusions in evaporite minerals have revealed comparatively large variations in the relative abundances of SO_4^{2-}, Mg^{2+}, and Ca^{2+} (Figure 14.22). The mineralogy of carbonate produced by marine calcifiers has varied as well such that during times of high Mg/Ca ratios in seawater, aragonite, and high-Mg calcite predominates, while low-Mg calcite predominates when Mg/Ca is low; this is often referred to as a variation between *aragonite seas* and *calcite seas* (Lowenstein et al., 2014). The cause of this variation is debated. Some researchers favor a model in which the ratio of the Mg-rich hydrothermal flux to the Ca-rich riverine flux is the controlling factor, which in turn reflects relative rates of

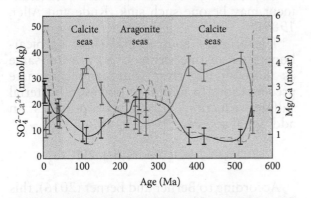

Figure 14.22 Variation of Ca^{2+} (black line) and SO_4^{2-} (red line) concentrations and the Mg/Ca ratio (dashed line) in seawater deduced from fluid inclusions in evaporites over Phanerozoic time. Error bars show the actual Ca^{2+} and SO_4^{2-} data. Based on Lowenstein et al. (2003) and Reis (2010).

seafloor spreading and continental weathering. Others favor a control by the extent of dolomitization. That in turn reflects the ratio of carbonate precipitated in shallow to deep environments, as dolomitization is largely restricted to shallow water carbonates.

14.4.5.3 Red clays, metalliferous sediments, and Mn nodules

Roughly 40% of the ocean floor is covered by fine clays. They consist of the finest fraction of riverine sediment, atmospheric dust, and volcanic ash. In remote regions of the Pacific, they consist almost entirely of the latter two. These lithogenous particles scavenge particle-reactive elements from seawater as they fall through the water column and also as they lie on the bottom before burial when sedimentation rates are low. The iron-rich scavenged, or *hydrogenous*, component gives them a red color and they are known as *red clays*.

In the Nares Abyssal Plain (northeast of Puerto Rico) of the Atlantic, both red and gray clays occur: the gray clays are fine-grained distal ends of turbidity currents originating on the North American shelf, while red clays are restricted to areas of low sedimentation rate. Table 14.7 lists the transition metal and REE concentrations (Figure 14.23) and $^{87}Sr/^{86}Sr$ of the two clays. Thomson et al. (1984) found, however, that after leaching the $^{87}Sr/^{86}Sr$ of the two were identical, implying the clays had the same source and the difference in composition reflected the hydrogenous component. By comparing the elemental fluxes with the total sediment flux, they were able to determine the *authigenic flux* that was independent of the flux of sediment to the bottom. When sedimentation rates are high, the authigenic component is simply highly diluted. Lack of this dilution at low sedimentation rates results in high Fe and Mn concentrations.

Another means of removal of elements from seawater is precipitation of oxides and hydroxides, principally of Mn and Fe, and coprecipitation or adsorption of particle-reactive elements by them. This occurs in two principal ways. The first is in hydrothermal plumes. As we found above, hydrothermal fluids are enriched in Fe^{2+} and Mn^{2+} and Fe quickly oxidizes and precipitates while oxidation and precipitation of Mn is

Table 14.7 Concentrations and fluxes in clays of the Nares Abyssal Plain.

	Red clay	Grey clay	Authigenic flux
	ppm	ppm	$\mu mol/cm^2-10^3$ yr
V	174.1	154.7	0.12
Cr	97.2	94.1	–
Mn	3020	769	23
Fe	55633	51727	49
Co	47.1	24.1	0.22
Ni	93.9	55.8	0.29
Cu	102.2	45.1	0.41
Zn	128.8	117.1	0.11
Rb	153.9	162.5	0
Sr	147.2	157.7	0
Nb	19.5	16.7	0
La	47.9	41.3	19×10^{-3}
Ce	123.0	89.1	106×10^{-3}
Nd	42.8	38.8	13×10^{-3}
Sm	8.2	7.3	2.9×10^{-3}
Eu	1.7	1.5	0.5×10^{-3}
Gd	6.9	6.4	1.6×10^{-3}
Dy	5.9	5.4	1.4×10^{-3}
Er	3.2	2.9	0.62×10^{-3}
Yb	3.0	2.6	0.64×10^{-3}
$^{87}Sr/^{86}Sr$	0.7228	0.7284	0.70919

Data from Thomason et al. (1984).

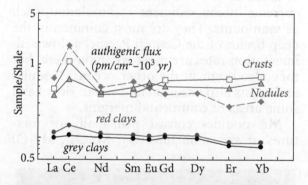

Figure 14.23 Shale-normalized average Mn nodule and crust rare earth patterns (see Chapter 7) of gray and red clays of the Nares Abyssal Plain, Atlantic Ocean, and the authigenic scavenging flux calculated by Thomson et al. (1984). Clay and flux data from Thomson et al. (1984); nodule and crust data from compilation of Li and Schoonmaker (2014).

delayed until the plume becomes neutrally buoyant and spreads laterally. As precipitated particles settle out of the water column, they then scavenge other particle-reactive elements from seawater as they sink. When these hydrothermal particles are abundant, they produce so-called *metalliferous sediment*. The highest concentrations of metalliferous sediments are found near the mid-ocean ridges, particularly adjacent fast spreading regions of the EPR, but the influence of hydrothermal plumes can still be seen thousands of kilometers from the ridge crest.

Mn and Fe oxides and hydroxides may also precipitate directly on the seafloor, with a previously existing surface acting as a nucleation site. In sediment-covered areas, shards of volcanic glass, shark's teeth and other such particles may serve as a nucleation site, with the Mn-Fe precipitates eventually forming a coating of up to 10 or more centimeters diameter. Typically, they form flattened spheres with botryoidal, smooth, or rough surfaces. These are known as *manganese nodules* (Figure 14.24). Solid surfaces, such as the surface of a lava flow, may also provide a nucleation site. In this case, the Mn-Fe precipitates will form a coating on the surface up to several cm thick. Such coatings are known as *manganese crusts*. Nodules and crusts grow extremely slowly and occur only in areas of low sedimentation rate or where currents inhibit sediment accumulation, such as seamounts. They are most common in the deep basins of the Central Pacific, as low sedimentation rates are most common there, but they also occur in the other oceans. Nodules and crusts also occur on mid-ocean ridges and some areas of continental margins.

Mn nodules consist principally of mixtures of vernadite (δMnO_2: $MnO_{2-x} \cdot xH_2O$)

Figure 14.24 Photograph of a manganese nodule recovered from the Pacific Ocean. Source Wikipedia (www.wikipedia.org). Reproduced under Creative Commons License.

birnessite: $((Na_{0.3}Ca_{0.1}K_{0.1})(Mn^{4+},Mn^{3+})_2 O_4 \cdot 1.5H_2O$; also called 7Å manganite)), todorokite: $((Na,Ca,K,Ba,Sr)_{1-x}(Mn,Mg,Al)_6 O_{12} \cdot 3-4H_2O$; also called 10Å manganite), buserite ($Na_4Mn_{14}O_{27} \cdot 21H_2O$), and various iron oxy-hydroxides (i.e., $FeOOH \cdot nH_2O$). Older layers of thick crusts contain secondary carbonate fluorapatite ($Ca_5(PO_4,CO_3)_3(F,Cl)$), which forms by replacement and precipitation in pore spaces (Mizell and Hein, 2018). Both nodules and crust can contain quartz and other detrital minerals that have sunk through the water column.

Average composition is given in Table 14.8. Compositions, however, vary with location and water depth. Mineralogy also influences composition. The structure of todorokite allows the incorporation of large amounts of diagenetic elements such as Ni and Cu, and whereas vernadite has a layered structure and tends to accumulate more Co. Mn oxides have a strong negative surface charge at seawater pH, which results in cation adsorption. On the other hand, negative and neutrally charged complexes are adsorbed onto the slightly positively charged surface of Fe oxyhydroxides. After adsorption, redox sensitive metals can be oxidized to an immobile state on the surface. For example, Co, Ce, and Tl^+ oxidize to Co^{3+}, Ce^{4+}, and Tl^{13+}, respectively, on the surface of the Mn oxide while Te^{4+} and Pt^{2+} may oxidize to Te^{6+} and Pt^{4+} on the surface of the Fe oxyhydroxide (Mizell and Hein, 2018).

Nodules may grow both by precipitation from seawater and by precipitation from metals diffusing upward though sediment pore waters. Nodules growing from seawater are called *hydrogenetic*; those growing from sediment pore waters and are called *diagenetic*. Hydrogenetic nodules tend to be dominantly composed of δMnO_2, while toderokite is more common in diagenetic nodules. Diagenetic growth may be oxic or suboxic. Suboxic diagenetic nodules occur beneath high productivity regions where sufficient organic matter reaches the sediment becomes reducing at depth, thereby mobilizing redox-sensitive metals such as Mn Fe, Ni and Cu. Nodules forming by oxic diagenesis tend to be the most enriched in Ni, Cu, and Zn. Hydrogenetic nodules are the richest in Fe, Co, Te, platinum group metals, and

Table 14.8 Average composition of Mn nodules and crusts.

	Nodule (%)	Crust (%)		Nodule (ppm)	Crust (ppm)
Na	1.7	1.5	Sr	830	1200
Mg	1.6	0.88	Te	10	–
Al	2.7	0.41	Ba	2300	1000
Si	7.7	2.2	La	157	190
P	0.15	0.25	Ce	530	900
Ca	2.3	2.2	Nd	158	150
V	0.05	0.05	Sm	36	30
Cr	0.004	0.001	Eu	9	8.1
Mn	18.6	20.2	Gd	32	39
Fe	12.5	14.9	Er	18	24
Co	0.27	0.59	Yb	20	24
Ni	0.66	0.37	Pt	0.002	0.0004
Cu	0.45	0.009	Au	0.002	0.25
Zn	0.12	0.054	Pb	900	1400

From Li and Schoonmaker (2014) and Mizell and Hein (2018).

rare earth elements. Most nodules are of mixed diagenetic and hydrogenetic origin. This produces a compositional difference between the tops and bottoms of nodules. In contrast to nodules, crusts, which form on impermeable surfaces, grow by accretion from seawater and are thus strictly hydrogenetic. Near mid-ocean ridges and other submarine volcanos, crusts may grow primarily from metals provided by hydrothermal vents. Such crusts are termed *hydrothermal*.

Crusts and hydrogenetic nodules grow at rates as low as 1–5 mm/Ma. Growth rates of suboxic diagenetic nodules may be substantially higher, up to 250 mm/Ma, but nodules of mixed origin, which are the majority, grow at rates of a few cm per million years. The slow growth explains why nodules are found only in areas of low sedimentation rate: in other areas, the nuclei are buried before having a chance to grow. Nevertheless, many nodules appear to grow at rates lower than the local sedimentation rate. Even more puzzling, they are concentrated at the sediment surface: there are typically twice as many nodules at the sediment surface as buried in the upper meter of sediment. The reason for this is debated; one hypothesis is that the action of burrowing organisms keeps the nodules at the sediment surface.

14.4.5.4 Porewater fluxes into and out of sediments

Sediments can serve as a sink for dissolved matter in yet another way: through diffusion of dissolved components into sediments. Dissolved components may also diffuse out of sediment poor waters into seawater, or pore water may be expelled by compaction. In these latter cases, sediments serve as a source of dissolved matter in seawater. As we found in Chapter 5, diffusion occurs only when a compositional gradient exists. Sediment pore water originates simply as seawater trapped between sedimentary particles, so its composition is initally identical to seawater. Diagenetic reactions occurring within the sediment, however, produce changes in pore water composition, establishing chemical gradients that drive diffusion into or out of the sedimentary column (Section 5.7). Furthermore, as sediment is buried beneath subsequently accumulating material, it is compacted, driving pore water back into the overlying seawater, producing an advective flux of porewater-enriched components to seawater. In the last two decades or so, there has been increasing recognition of the importance of these sedimentary fluxes, particularly for trace elements such as iron and the rare earths.

Iron is, of course, a redox-sensitive element and it is dramatically more soluble in the ferrous form compared with the ferric one. The oxidation state of seawater makes the insoluble ferric form the predominant one nearly everywhere in the oceans. Where sufficient organic matter is present in sediments, respiration consumes oxygen (and produces CO_2) and Fe can be reduced though dissimilatory iron reduction (DIR) in which bacteria use iron as an electron receptor to oxidize organic matter. Based on measured Fe fluxes from sediments on the California margin, Elrod et al. (2004) estimated that the flux of Fe from continental shelf sediment undergoing DIR to the ocean photic zone was equal to or greater than that of the aerosol flux. A study of Fe isotopes by Homoky et al. (2009) on sediments

from the northern California margin adjacent the Eel River at a water depth of 120 m confirmed the importance of DIR. The sediments contain roughly 1% organic matter and respiration results in conditions becoming sufficiently reducing within the upper 4 cm for most of the nitrate to be reduced to ammonia (Figure 14.25). At the same depth, dissolved Fe concentrations in the pore

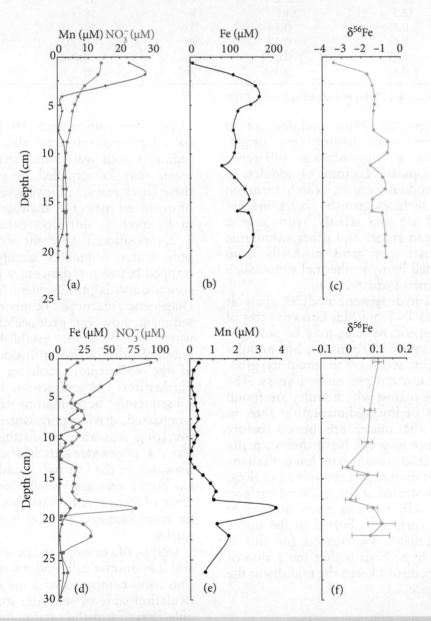

Figure 14.25   Concentrations of dissolved nitrate, manganese (a), and iron (b) and δ^{56}Fe (c) in pore water in sediments from the northern California margin near the Eel River. Concentrations of dissolved nitrate, iron (d), and manganese (e) and δ^{56}Fe (f) in pore water in sediments from the Crozet Plateau in the southwestern Indian Ocean. Data from Homosky et al. (2009).

water reach >150 μM, over a factor of 5000 greater than typical seawater concentrations. Although it can occasionally be more negative, seawater typically has $\delta^{56}Fe$ values between about −0.5‰ and +0.5‰. DIR reduction results in a large negative Fe isotope fractionation with $\Delta\delta^{56}Fe$ typically in the range of −1.5 to −3 (Section 9.9.6). The more negative $\delta^{56}Fe$ in pore water in the "Eel River" site indicate that Fe is being mobilized through DIR. The dissolved iron can then diffuse into seawater. Manganese too appears to be mobilized in this process, although not as dramatically as Fe.

Homoky et al. (2009) also studied sediment pore waters at a site on Crozet Plateau in the southwestern Indian Ocean at a water depth of 3200 m. Here organic matter contents are lower, generally <0.5%. Nitrate contents indicate conditions become suboxic at depth. Dissolved Fe concentrations in this pore waters are generally an order of magnitude lower than at the Eel River site (Figure 14.25d), but nonetheless more than 500 times typical seawater concentrations, as are Mn pore water concentrations. In this case, however, $\delta^{56}Fe$, which average about +0.05‰ (Figure 14.25f), are essentially unfractionated from $\delta^{56}Fe$ in those of igneous rocks. This indicates Fe is being released to solution primarily by nonreductive dissolution of volcanoclastic sediment, rather than DIR.

Diagenetic reactions within sediments can result in a flux of other elements to seawater as well. Li is readily incorporated into interlayer sites of clays at low temperature, with some clays' mineral containing as much as 500 ppm Li. Because of its high hydration energy, however, Li is easily removed from these sites and as sediments are subjected to increasing temperatures during burial, Li is lost from solid phases to pore waters. This sets up a concentration gradient that drives diffusion of Li out of the sediment. Perhaps more importantly, Li-rich pore waters are expelled from the sediment as it compacts, producing an advective flux of dissolved Li to seawater. Li concentrations in pore waters and coexisting sediments from ODP Site 688 on the Peru margin analyzed by Martin et al. (1991) shown in Figure 14.26 provide an example. Concentrations increase from near seawater values (25.6 μM) at the top of the

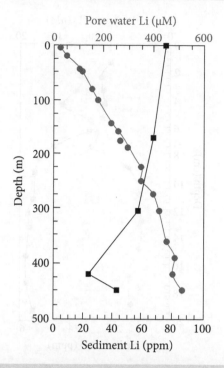

Figure 14.26 Li concentrations in pore waters (red circles) and sediments (black squares) from ODP Site 688 on the Peru continental shelf. Data from Martin et al. (1991).

core to over 500 μM at a depth of 450 m. At the same time, Li concentrations in the sediment decrease. This decrease in Li concentration in sediments does not seem sufficient to account for the increase in pore water Li concentrations and Martin et al. (1991) speculated that additional Li was being released from underlying basement rocks.

Martin et al. (1991) estimated that expulsion of pore waters of continental margin sediments supplied 1 to 3×10^{10} moles/yr of Li to the ocean. This flux may exceed both the riverine flux ($1.1–1.7 \times 10^{10}$ moles/yr) and the hydrothermal flux ($0.5–4 \times 10^{10}$ moles/yr). Thus sediment pore water may be the dominant source of dissolved Li in the oceans. Even the minimum value represents a substantial flux, supplying 20–25% of the marine Li.

Uranium is present in seawater in the VI state, generally as the soluble uranyl tricarbonate species ($UO_2(CO_3)_3^{-4}$). However, the reduced species, U^{4+}, is relatively insoluble. Figure 14.27 shows an example of the U profiles determined by Klinkhammer and Palmer (1991) in cores taken from the California

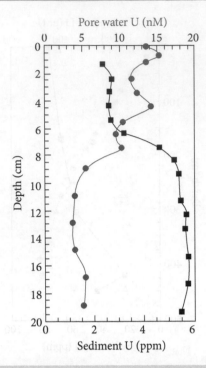

Pore water U (nM)

Sediment U (ppm)

Figure 14.27 U concentration in sediment (black squares) and pore waters (red circles) from the California shelf sediments. Data from Klinkhammer and Palmer (1991).

continental shelf just south of the Monterey Fan. Between 1 and 2 cm depth, U pore water concentrations slightly exceed the seawater concentration (13.4 nmol/l). This results from release of U from labile organic phases in the upper part of the core. Below 4 cm, however, consumption of the 2.5% organic carbon in this core leads to suboxic conditions and reduction of U^{6+} to U^{4+} and dissolved U pore water concentrations decrease to 4–5 nmol/l and concentrations in the coexisting solid phase increase sharply from 2.7 to 5.7 ppm at depth. Once reduced, U is immobilized in the solid phase. This produces a concentration gradient that causes U to diffuse downward from seawater into the sediment. Klinkhammer and Palmer (1991) estimated that suboxic diagenesis results in a U flux out of seawater of 0.67×10^{10} g/yr, making it the single largest sink for dissolved U in the oceans.

Jeandel and Oekers (2015) argue that dissolution of riverine particulate matter, including in estuaries, in the water column, and on continental margin sediments is an important source of transition metals and REE represent an important source of these

elements in seawater, accounting for as much as 95% of the Nd flux. Even for relatively soluble elements, dissolution of riverine particulate matter appears to be an important source to seawater. For example, within the Mediterranean, the Si flux from dissolution of riverine particulates appears to be comparable to the dissolved riverine flux and globally it may account for up to 45% of the Sr flux.

14.5　SUMMARY

In the chapter we examine the chemistry of the oceans, which account for ~98% of surface water on the planet and covers two-thirds of its surface.

- *Salinity* and *temperature* can be changed only at the ocean surface and thus are said to be *conservative* properties. Together, temperature and salinity control seawater density; density differences drive the deep circulation of the oceans. The *thermocline* from roughly 200 m to 1000 m limits exchange of surface and deep water. Broadly, water is cooled and sinks at high latitudes and returns to the surface through mixing and upwelling on a time scale of ~1000 years. The surface circulation of the oceans is driven by winds.
- Seawater is saline and alkaline because it contains the weathering products of the continents. The principal ions in seawater are Na^+, Mg^{2+}, Ca^{2+}, K^+, Cl^-, and SO_4^{2-}. These are always present in constant proportions and vary only in proportion to salinity (Ca^{2+} varies slightly). Hence, they too are considered conservative.
- Minor and trace elements fall into several categories based on their vertical distribution in the water column. A few elements, such as Rb, Sr, F, Mo, Tl, and U, are also present in constant proportions and hence are conservative. Gases including N and noble gases are also conservative within the ocean, but their concentrations nevertheless vary because of the temperature dependence of gas solubility at the surface. A few elements, notably Al and Pb, are enriched in the surface water and depleted in deep water because they are mainly delivered to the oceans by winds and are rapidly absorbed onto particle

surfaces (*scavenging*). Most minor and trace elements show a *nutrient* type distribution of depletion in surface water and enrichment in deep water. This reflects uptake by phytoplankton in the surface water and release in deep water as organic remains decay. The distribution of some elements, however, is controlled by a combination of scavenging and biological uptake and decay.

- The process of biological uptake of bicarbonate by photosynthesis in the surface water and decay in the deep water, where it is sequestered on long time scales, is known as the *biological pump*. It plays an important role on atmospheric CO_2 concentrations and hence on climate. Increased concentrations of dissolved CO_2 decrease deep water pH (from 8.1 to 7.6). As a result of this decrease in pH and the increase in pressure, calcium carbonate is undersaturated in deep water, whereas it is oversaturated in surface water. Shells produced by plankton in the surface water thus tend to dissolve in the deep water, which further enhances the biological pump. The depth at which no carbonate shells survive, ~4500 m, is known as the carbonate compensation depth.

- Concentrations of elements in the seawater are controlled both by the rate at which they are added and the rate at which they are removed. Assuming that the oceans are in steady-state, we can define a residence time, τ, for an element as:

$$\tau = \frac{A}{dA/dt} \qquad (14.40)$$

where A is the mass of the element of interest in the oceans, and dA/dt is the flux to or from seawater. Residence times vary from ~65 million years for Cl to a few tens of years for particle reactive elements such as Th.

- Rivers are the principal source of many elements in seawater. However, changes in ionic strength and pH due to mixing between river and seawater results in either release of ions to solution and, more frequently, scavenging of them by flocculating particles and colloids.

- Ridge crest hydrothermal activity is an important source for Li, K, Mn, Fe, Zn, Ba, and Tl and an important sink for Mg, U, and perhaps S. This interaction between seawater and the hot, young oceanic crust is also important in producing a class of ore deposits known as volcanogenic massive sulfides discussed in the next chapter.

- The atmosphere is an important source of some transition metals such as Zn and Cu and is the dominant source, of course, of dissolved gases but also of highly particle reactive elements such as Al and Pb.

- Sediments are the primary sink for most elements in seawater. Over 50% of the ocean floor is covered with biogenic calcareous and siliceous oozes. Precipitation and accumulation of calcium carbonate sediments on the ocean floor is the final step in weathering cycle's control on atmospheric CO_2 and hence long-term climate. The high removal rates of calcium carbonate by biological activity in the oceans also resolved the *river water paradox* that river water is approximately a calcium bicarbonate solution while seawater is an approximately sodium chloride solution. Diffusion out of sediments, however, can also be a source of elements in seawater. Evaporites are the most important sink for some elements, most notably Na^+ and Cl^-, but they form discontinuously through time. The most recent event, the Messinian salinity crisis occurred when the Strait of Gibraltar cyclically closed between 5.96 and 5.33 Ma, and enough salt was precipitated in the Mediterranean basin to reduce global ocean salinity by 5%.

REFERENCES

Alt, J. C. 1995. Subseafloor processes in mid-ocean ridge hydrothermal systems. In *Seafloor Hydrothermal Systems: Physical, Chemical, Biological and Geological Interactions, Geophysical Monograph Vol. 91*, S. E. Humphris, R. A. Zierenberg, L. S. Mullineaux and R. E. Thomson. ed., pp. 85–114. Washington: AGU.

Andersson, A. J. (2014) 8.19 – The oceanic $CaCO_3$ cycle. In: Holland HD, Turekian KK (eds) *Treatise on Geochemistry* (second edition), vol. Elsevier, Oxford, pp 519–42.

Barth, T. W. 1952. *Theoretical Petrology*. New York: John Wiley.

Berner, E. K. and Berner, R. A. 2012. *Global Environment*. Princeton University Press, Princeton, NJ.

Boyle, E. A., J. M. Edmond and E. R. Sholkovitz. 1977. The mechanism of iron removal in estuaries. *Geochimica et Cosmochimica Acta* 41: 1313–24.

Broecker, W. S. and T.-H. Peng. 1982. *Tracers in the Sea*. New York: Eldigio Press.

Bridgestock, L., van de Flierdt, T., Rehkämper, M., et al. 2016. Return of naturally sourced Pb to Atlantic surface waters. *Nature Communications* 7: 12921. doi: 10.1038/ncomms12921.

Bruland, K. W., Middag, R. and Lohan, M. C. 2014. 8.2 – Controls of trace metals in seawater. In: *Treatise on Geochemistry* (Second Edition) (ed. Holland, Heinrich D and Turekian, K. K.), pp. 19–51, Elsevier, Oxford. doi: 10.1016/B978-0-08-095975-7.00602-1.

Chester, R. 1990. *Marine Geochemistry*, Unwin Hyman, London, 698.

Chou, L., Garrels, R. M. and R. Wolast. 1989. Comparative study of the kinetics and mechanisms of dissolution of carbonate minerals. *Chemical Geology* 78: 269–82.

Craig, H. 1974. A scavenging model for trace elements in the deep sea. *Earth and Planetary Science Letters* 23: 149–59.

Darwin, C. 1846. An account of the fine dust which often falls on vessels in the Atlantic Ocean. *Quarterly Journal of the Geological Society* 2: 26–30.

de Baar, H. J. W., van Heuven, S. M. A. C. and Middag, R. 2018. Ocean salinity, major elements, and thermohaline circulation. In: *Encyclopedia of Geochemistry* (ed White, W. M.), pp. 1042–8, Springer International Publishing, Cham.

Driesner, T., Heinrich, C. A. 2007. The system H_2O–NaCl. Part I: Correlation formulae for phase relations in temperature–pressure–composition space from 0 to 1000°C, 0 to 5000 bar, and 0 to 1 X_{NaCl}. *Geochimica et Cosmochimica Acta* 71: 4880–901. doi: 10.1016/j.gca.2006.01.033.

Duce, R. A., P. S. Liss, J. T. Merrill, E. L. Atlas, P. Buat-Menard, B. B. Hicks, J. M. Miller, et al. 1991. The atmospheric input of trace species to the world ocean. *Global Biogeochemical Cycles* 5: 193–259.

Dutkiewicz, A., Müller, R. D., O'Callaghan, S., Jónasson, H. 2015. Census of seafloor sediments in the world's ocean. *Geology* 43: 795–8.

Edmond, J. M., Campbell, A. C., Palmer, M. R., et al. 1995. Time series studies of vent fluids from the TAG and MARK sites (1986, 1990) Mid-Atlantic Ridge: A new solution chemistry model and a mechanism for Cu/Zn zonation in massive sulphide ore bodies. In: Parson LM, Walker CL, Dixon DR (eds) *Hydrothermal Vents and Processes*. Geological Society of London, London, pp. 77–86.

Elderfield, H., N. Luedtke, R. J. McCaffrey and M. Bender. 1981. Benthic flux studies in Narragansett Bay. *American Journal of Science* 281: 768–87.

Elderfield, H. and A. Schultz. 1996. Mid-ocean ridge hydrothermal fluxes and the chemical composition of the ocean. *Annual Reviews of Earth and Planetary Science* 24: 191–224.

Elderfield, H. and E. R. Sholkovitz 1987. Rare earth elements in the pore waters of reducing nearshore sediments. *Earth and Planetary Science Letters* 82: 280–8.

Froelich, P. N., G. A. Hambrick, M. O. Andreae, R. A. Mortlock and J. M. Edmond. 1985. The geochemistry of inorganic germanium in natural waters. *Journal of Geophysical Research* 90: 1131–41.

Gaillardet, J., Viers, J. and Dupré, B. 2014. 7.7 – Trace Elements in River Waters. In: *Treatise on Geochemistry* (2nd ed.) (eds. Holland, H. D. and Turekian, K. K.), pp. 195–235. Elsevier, Oxford.

Gnanadesikan, A. 1999. A global model of silicon cycling: Sensitivity to eddy parameterization and dissolution. *Global Biogeochemical Cycles* 13: 199–220. doi: 10.1029/1998GB900013.

Goldberg, E. D. and G. O. S. Arrhenius. 1958. Chemistry of Pacific pelagic sediments. *Geochimica et Cosmochimica Acta* 13: 153–212.

Greaves, M. J., P. J. Statham and H. Elderfield. 1994. Rare earth element mobilization from marine atmospheric dust into seawater. *Marine Chemistry* 46: 255–60.

Homoky, W. B., Severmann, S., Mills, R. A., Statham, P. J., Fones, G. R. 2009. Pore-fluid Fe isotopes reflect the extent of benthic Fe redox recycling: Evidence from continental shelf and deep-sea sediments. *Geology* 37: 751–4 doi: 10.1130/G25731A.1.

Hoyle, J., H. Elderfield, A. Gledhill and M. Greives, The behavior of the rare earth elements during mixing of river and seawaters, *Geochimica et Cosmochimica Acta* 48, 143–9, 1984.

Jeandel, C., Oelkers, E. H. 2015. The influence of terrigenous particulate material dissolution on ocean chemistry and global element cycles. *Chemical Geology* 395: 50–66. doi: 10.1016/j.chemgeo.2014.12.001.

Kim, I. and Kim, G. 2014. Submarine groundwater discharge as a main source of rare earth elements in coastal waters. *Marine Chemistry* 160, 11–17. doi: 0.1016/j.marchem.2014.01.003.

Klinkhammer, G. P. and M. R. Palmer. 1991. Uranium in the oceans: where it goes and why. *Geochimica et Cosmochimica Acta* 55: 1799–896.

Kleypas, J. A. 1997. Modeled estimates of global reef habitat and carbonate production since the last glacial maximum. *Paleoceanography* 12: 533–45. doi: 10.1029/97PA01134.

Kraepiel, A. M. L., J.-F. Chiffoleau, J.-M. Martin and F. M. M. Morel. 1997. Geochemistry of trace metals in the Gironde estuary. *Geochimica et Cosmochimica Acta* 61: 1421–1436.

Kritzberg, E. S., Bedmar Villanueva, A., Jung, M. and Reader, H. E. 2014. Importance of Boreal Rivers in Providing Iron to Marine Waters. *PLOS One* 9(9), e107500. doi: 10.1371/journal.pone.0107500.

Lécuyer, C. 2016. Seawater residence times of some elements of geochemical interest and the salinity of the oceans. *Bulletin de la Société Géologique de France* 187(6), 245–60. doi: 10.2113/gssgfbull.187.6.245.

Li, Y. H., Schoonmaker, J. E. 2014. 9.1 – Chemical composition and mineralogy of marine sediments In: Turekian KK, Holland HD (eds) *Treatise on Geochemistry* (2nd ed.), vol. 9 Elsevier, Oxford, pp. 1–32. doi: 10.1016/B978-0-08-095975-7.00701-4.

Lowenstein, T. K., Hardie, L. A., Timofeeff, M. N. and Demicco, R. V. 2003. Secular variation in seawater chemistry and the origin of calcium chloride basinal brines. *Geology* 31(10), 857–60. doi: 10.1130/G19728R.1.

Lowenstein, T. K., Kendall, B. and Anbar, A. D. 2014. 8.21 – The geologic History of seawater. In: *Treatise on Geochemistry* (2nd ed.) (eds. Holland, H. D. and Turekian, K. K.), pp. 569–622, Elsevier, Oxford.

Lueker, T. J., Dickson, A. G., Keeling, C. D. 2000. Ocean P_{CO2} calculated from dissolved inorganic carbon, alkalinity, and equations for K_1 and K_2: Validation based on laboratory measurements of CO_2 in gas and seawater at equilibrium. *Marine Chemistry* 70: 105–19. doi 10.1016/S0304-4203(00)00022-0.

Lopez, O., Zuddas, P., and Faivre, D. 2009. The influence of temperature and seawater composition on calcite crystal growth mechanisms and kinetics: Implications for Mg incorporation in calcite lattice. *Geochimica et Cosmochimica Acta* 73, 337–47. doi: 10.1016/j.gca.2008.10.022.

Mackey, K. R. M., Chien, C.-T., Post, A. F., Saito, M. A., Paytan, A. 2015. Rapid and gradual modes of aerosol trace metal dissolution in seawater. *Frontiers in Microbiology.* 5: 794. doi: 10.3389/fmicb.2014.00794.

Mahowald, N. M., Hamilton, D. S., Mackey, K. R. M., et al. 2018. Aerosol trace metal leaching and impacts on marine microorganisms. *Nature Communications* 9: 2614. doi: 10.1038/s41467-018-04970-7.

Martin, J. B., M. Kastner and H. Elderfield. 1991. Lithium: sources in pore fluids of Peru slope sediments and implications for oceanic fluxes. *Marine Geology* 102: 281–92.

Martin, J. H. and R. M. Gordon. 1988. Northeast Pacific iron distributions in relation to phytoplankton productivity. *Deep Sea Research* 35: 177–96.

Martin, J. H., K. H. Coale, K. S. Johnson, S. E. Fitzwater, R. M. Gordon, S. J. Tanner and e. al. 1994. Testing the iron hypothesis in ecosystems of the equatorial Pacific. *Nature* 371: 123–29.

Matsumoto, K. 2007. Radiocarbon-based circulation age of the world oceans. *Journal of Geophysical Research* 112. doi: 10.1029/2007JC004095.

Middag, R., Séférian, R., Conway, T. M., John, S. G., Bruland, K. W. and de Baar, H. J. W. 2015. Intercomparison of dissolved trace elements at the Bermuda Atlantic Time Series Station. *Marine Chemistry* 177, 476–89. doi: 10.1016/j.marchem.2015.06.014.

Millero, F. J. and Screiber, D. R. 1982. Use of the ion pairing model to estimate activity coefficients of the ionic components of natural waters. *American Journal of Science* 282, 1508–1540. doi: 10.2475/ajs.282.9.1508.

Mizell, K., Hein, J. R. 2018. Ferromanganese crusts and nodules: Rocks that grow. In: White WM (ed) *Encyclopedia of Geochemistry*, Springer International Publishing, Cham, pp. 477–83.

Moore, W. S. 2010. The effect of submarine groundwater discharge on the ocean. *Annual Review of Marine Science* 2, 59–88. doi: 10.1146/annurev-marine-120308-081019.

Nielsen, S. G., Rehkämper, M., Teagle, D. A. H., Butterfield, D. A., Alt, J. C., Halliday, A. N. 2006. Hydrothermal fluid fluxes calculated from the isotopic mass balance of thallium in the ocean crust. *Earth and Planetary Science Letters* 251: 120–33. doi: 10.1016/j.epsl.2006.09.002.

Peterson, M. N. A. 1966. Calcite: rates of dissolution in a vertical profile in the central Pacific. *Science.* 154: 1542–1544.

Redfield, A. C. 1958. The biological control of chemical factors in the environment. *American Journal of Science* 46: 205–21.

Ries, J. B. 2010. Review: geological and experimental evidence for secular variation in seawater Mg/Ca (calcite-aragonite seas) and its effects on marine biological calcification. *Biogeosciences* 7(9), 2795–849. doi: 10.5194/bg-7-2795-2010.

Ryan, W. B. 2008. Modeling the magnitude and timing of evaporative drawdown during the Messinian salinity crisis. *Stratigraphy* 5(1), 227–43.

Rude, P. D., Aller, R. C. 1989. Early diagenetic alteration of lateritic particle coatings in Amazon continental shelf sediment. *Journal of Sedimentary Research* 59: 704–16.

Seitzinger, S. P., Mayorga, E., Bouwman, A. F., et al. 2010. Global river nutrient export: A scenario analysis of past and future trends. *Global Biogeochemical Cycles* 24: doi: 10.1029/2009GB003587.

Seyfried, W. E. J. and K. Ding. 1995. Phase equilibria in subseafloor hydrothermal systems: a review of the role of redox, temperature, pH and dissolved Cl on the chemistry of hot spring fluids at mid-ocean ridges. In *Seafloor Hydrothermal Systems: Physical, Chemical, Biological and Geological Interactions, Geophysical Monograph Vol. 91*, S. E. Humphris, R. A. Zierenberg, L. S. Mullineaux and R. E. Thomson. ed., pp. 248–72. Washington: AGU.

Shanks, W. C. I., J. K. Böhlke and R. R. Seal II. 1995. Stable isotopes in mid-ocean ridge hydrothermal systems: interactions between fluids, minerals, and organisms. In *Seafloor Hydrothermal Systems: Physical, Chemical, Biological and Geological Interactions, Geophysical Monograph Vol. 91*, S. E. Humphris, R. A. Zierenberg, L. S. Mullineaux and R. E. Thomson. ed., pp. 194–221. Washington: AGU.

Shock, E. L., Schulte, M. D. 1998. Organic synthesis during fluid mixing in hydrothermal systems. *Journal of Geophysical Research* 103: 28513–27. doi: 10.1029/98JE02142.

Sholkovitz, E. R. 1976. Flocculation of dissolved organic and inorganic matter during the mixing of river water and seawater. *Geochimica et Cosmochimica Acta* 40: 831–45.

Sholkovitz, E. R. 1978. The flocculation of dissolved Fe, Mn, Al, Cu, Ni, Co and Cd during estuarine mixing. *Earth and Planetary Science Letters* 41: 77–86.

Sholkovitz, E. R., Sedwick, P. N., Church, T. M., Baker, A. R., Powell, C. F. 2012. Fractional solubility of aerosol iron: Synthesis of a global-scale data set. *Geochimica et Cosmochimica Acta* 89: 173–89. doi: 10.1016/j.gca.2012.04.022.

Stanley, J. K. and R. H. Byrne. 1990. Inorganic speciation of zinc(II) in seawater. *Geochimica et Cosmochimica Acta* 54: 753–60.

Staudigel, H. 2014. 4.16 – Chemical fluxes from hydrothermal alteration of the oceanic crust. In: Turekian KKH, Heinrich D. (ed) *Treatise on Geochemistry* (second edition), vol. 4. Elsevier, Oxford, pp. 583–606. doi: 10.1016/B978-0-08-095975-7.00318-1.

Subhas, A. V., Rollins, N. E., Berelson, W. M., Dong, S., Erez, J., Adkins, J. F. 2015. A novel determination of calcite dissolution kinetics in seawater. *Geochimica et Cosmochimica Acta* 170: 51–68. doi: 10.1016/j.gca.2015.08.011.

Tipper, E. T., Galy, A., Gaillardet, J., Bickle, M. J., Elderfield, H., Carder, E. A. 2006. The magnesium isotope budget of the modern ocean: Constraints from riverine magnesium isotope ratios. *Earth and Planetary Science Letters* 250: 241–53. doi: 10.1016/j.epsl.2006.07.037.

Thomson J., M. S. N. Carpenter, S. Colley, T. R. S. Wilson, H. Elderfield, and H. Kennedy. 1984. Metal accumulation rates in northwest Atlantic pelagic sediments. *Geochimica et Cosmochimica Acta* 48: 1935–48.

Tréguer, P. J., Rocha, C. L. D. L. 2013. The world ocean silica cycle. *Annual Review of Marine Science* 5: 477–501 doi: 10.1146/annurev-marine-121211-172346.

Von Damm, K. L. 1988. Systematics of and postulated controls on submarine hydrothermal solution chemistry. *Journal of Geophysical Research* 93: 4551–62.

Von Damm, K. L. and J. L. Bischoff. 1987. Chemistry of hydrothermal solutions from the southern Juan de Fuca ridge. *Journal of Geophysical Research* 92: 11334-11346.

Warren, J. 2018. Evaporites. In: White WM (ed) *Encyclopedia of Geochemistry*, Springer International Publishing, Cham, pp. 464–71. 978-3-319-39312-4.

Zhang, Y., Mahowald, N., Scanza, R. A., et al. 2015. Modeling the global emission, transport and deposition of trace elements associated with mineral dust. *Biogeosciences* 12: 5771–92. doi: 10.5194/bg-12-5771-2015.

Zhong, S. and A. Mucci. 1993. Calcite precipitation in seawater using a constant addition technique: a new overall reaction kinetic expression. *Geochimica et Cosmochimica Acta* 57: 1409–18.

Zhu, X., Zhang, R., Wu, Y., Zhu, J., Bao, D. and Zhang, J. 2018. The Remobilization and Removal of Fe in Estuary—A Case Study in the Changjiang Estuary, China. *Journal of Geophysical Research: Oceans* 123(4), 2539–53. doi: 10.1002/2017JC013671.

Zuddas, P. and A. Mucci. 1994. Kinetics of calcite precipitation from seawater: I. a classical chemical description for strong electrolyte solutions. *Geochimica et Cosmochimica Acta* 58: 4353–2.

PROBLEMS

1. Assume a simple two-box model of the ocean in which the upper 1000 m has a temperature of 20°C in the tropics and 3°C at high latitude and the lower 3000 m has a temperature of 2°C in both cases. Assuming the temperature dependence of dissolution activation factor and barrier energy and sinking rates of fecal pellets given in Section 14.4.5.1, how much silica survives to reach the bottom at 4000 m in the two cases.

2. In the tropical ocean assume temperature in upper 100 m is 25°C and below that temperature declines linearly with depth to 2°C at 1000 m and is a constant 2°C below that. For the high-latitude ocean assume the upper 100 m is 5°C and below that declines linearly with depth to 2°C at 1000 m and is constant below that. Assuming the temperature dependence of dissolution rates and sinking rates of fecal pellets given in Section 14.4.5.1, what fraction of biogenic silica survives to reach the bottom at a depth of 4000 m in these two cases?

3. A number of reaction mechanisms have been proposed for the precipitation of calcite from seawater. These include

$$Ca^{2+} + HCO_3^- \rightleftharpoons H^+ + CaCO_3$$

$$Ca^{2+} + CO_3^{2-} \rightleftharpoons CaCO_3$$

$$Ca^{2+} + 2HCO_3^- \rightleftharpoons H_2CO_3 + CaCO_3$$

$$Ca^{2+} + HCO_3^- + OH^- \rightleftharpoons H_2O + CaCO_3$$

Suggest a series of laboratory experiments that you could performed that would enable you to distinguish which of these mechanisms actually occurs. Assume you have a well-equipped laboratory in which you can measure all macroscopic properties (i.e., concentrations, partial pressure, pH, carbonate alkalinity). Describe what properties you would measure and how you would use the data you obtained to discriminate between these mechanisms.

4. Use eqn. 14.28 and rate constants given in the caption to Figure 14.7 to calculate the calcite dissolution rates for the surface and deep water saturation states you calculated in Problem 3.

5. Preindustrial atmospheric CO_2 levels were only about 70% of present levels and ocean surface pH was about 0.1 higher than present. Let's assume total CO_2 in the surface water was also only 70% of present and that pH was 8.2. Where not stated, use the values given in Example 14.3 in this problem.
   (a) Use these values to calculate the carbonate ion concentration, the calcite ion activity coefficient (Q), and the calcite saturation state (Ω) for the surface water concentration in Example 14.3.
   (b) Anthropogenic CO_2 has mostly not penetrated deep water yet. If the total carbonate in the deep water in Example 14.3 were to increase by 40% in the future and pH were to drop by 0.1, what would the carbonate ion concentration, the calcite ion activity coefficient (Q), and the calcite saturation state (Ω).

6. Composition of seawater:
   (a) Calculate the molar concentrations of the major ions in seawater listed in Table 14.2.
   (b) Calculate the ionic strength of this solution.
   (c) Calculate the carbonate and total alkalinity of this solution using equilibrium constants in Table 6.1 assuming a pH of 8.1.

7. Use the *one-dimensional advection–diffusion model* in the depth interval of 595–4875 m and the chemical data from the North Pacific in the table below to answer the following questions. For this locality, the K was determined to be 2300 and ω to be 4. Is salinity conservative and the one-dimensional model applicable? Make a plot of S vs. $f(z)$ (eqn. 15.32). Are Cu, Ni, and Al conservative? Are they being produced or scavenged? For each of these elements, find a value of ψ or J to fit the one-dimensional advection–diffusion model to the data.

Chemical data from the North Pacific

Depth	Salinity	Cu	Ni	Al
0		0.54	2.49	90
75	33.98	0.69	2.9	88
185	33.92	0.91	3.79	84
375	34.05	1.45	5.26	70
595	34	1.9	7.49	60
780	34.19	2.15	9.07	52
985	34.37	2.38	9.64	48
1505	34.55	2.8	9.79	45
2025	34.61	3.18	10.6	47
2570	34.65	3.46	10.8	50
3055	34.66	3.9	10.9	54
3533	34.66	4.26	10.7	63
4000	34.67	4.57	10.8	66
4635	34.68	5.03	10.3	74
4875	34.68	5.34	10.4	79

Concentrations in nmol/kg; salinity in ppt.

8. Stanley and Byrne (1990) give the following stability constants for Zn complexes in seawater:

$$Zn^{2+} + Cl^- \rightleftharpoons ZnCl^- \qquad \beta^* = -0.40$$

$$Zn^{2+} + HCO_3^- \rightleftharpoons ZnHCO_3 \qquad \beta^* = 0.83$$

$$Zn^{2+} + CO_3^{2-} \rightleftharpoons ZnCO_3 \qquad \beta^* = 2.87$$

$$Zn^{2+} + 2CO_3^{2-} \rightleftharpoons Zn(CO_3)_2^{2-} \qquad \beta^* = 4.41$$

$$Zn^{2+} + H_2O \rightleftharpoons Zn(OH)^+ + H^+ \quad \beta^* = -9.25$$

$$Zn^{2+} + SO_4^{2-} \rightleftharpoons ZnSO_4 \qquad \beta^* = 0.90$$

Using these stability constants, a pH of 8.1, the ligand concentrations given in Table 14.2 (and the equilibrium concentration of carbonate ion calculated with the equilibrium constant in Table 6.1 for a temperature of 25°C), calculate the fraction of Zn present as each of these species plus free Zn^{2+}.

9. Using the flows of water through mid-ocean ridge crest and flank hydrothermal systems and the mass of the oceans given in Appendix I, how long does it take to cycle the entire ocean through these systems?

10. In the San Clemente Basin, off the southern California coast, Barnes and Cochran (1990) found that U concentrations in sediment pore waters decreased to 3.35 nM/l in the top 7 cm of sediment. Assuming an effective diffusion coefficient (corrected for porosity and tortuosity in the sediment) of 68.1 cm²/yr, calculate the flux (in nM/cm²) of U from seawater to sediment in this locality.

11. Today, annual evaporation of water from the Mediterranean Sea is 3600 km³. Precipitation delivers ~1000 km³/yr and river runoff an additional 500 km³/yr. This is balanced by net influxes of 2100 km³ from through the Strait of Gibraltar and the Dardanelles. The volume of the Mediterranean is 3.75×10^6 km³ and has an average salinity of 38. If the influx from the Atlantic and Black Seas were to cease, how long would it take before calcite begins to precipitate? How long before anhydrite precipitates? How long before halite precipitates?

Chapter 15

Applied geochemistry

15.1 INTRODUCTION

The previous fourteen chapters have focused on developing chemical principles and using those tools to understand how the Earth formed and how it works. In this chapter we turn our attention to using that understanding to meet the needs of society, in particular, meeting the demand for mineral resources and keeping the planet habitable. These two issues are closely related. Exploitation of resources often directly degrades the environment and disposal of wastes degrades it further. As human populations have exploded, the need to continually supply society with earth-derived resources and minimize and remediate our deleterious effects on the environment have become more pressing. Both of these issues involve chemistry.

Ore deposits are localities where elements have been concentrated to levels vastly exceeding typical ones as a consequence of geochemical processes. We will see that virtually every aspect of what we have learned in the previous 14 chapters can be put to good use in understanding how ore deposits form and how we can find future ones. On the other hand, high levels of harmful chemicals, what we refer to as *pollution*, are among the most pressing of environmental issues. Here again, meeting the challenges posed by pollution requires an understanding of the behavior of elements in the natural environment.

15.2 MINERAL RESOURCES

Human history is in some respects the story of the development of increasingly sophisticated tools. The earliest known stone tools date to 3.3 Ma (Harmand et al., 2015) and predate the emergence of our species and even our genus. The early hominin tool kit no doubt included artifacts made from bone, wood, and other animal and plant tissues but it is the stone tools which have survived, giving rise to the term *Stone Age*. All stone is not suitable for making stone tools, so even then our ancestors had to hunt for tool-making resources such as fine-grained volcanic rock, obsidian, and flint. The Stone Age ended when people learned to smelt copper around 7000 years ago. Copper tools are far superior to stone ones, but copper is a relatively soft metal. Superior tools and weapons could be made by alloying copper with tin to make bronze. However, tin and copper ores usually develop in different geological circumstances, as we will see. Copper smelting first began in the Balkans and Middle East, which lack significant tin deposits; tin isotope studies indicate that Cornwall, England, was the tin source of some of the earliest European bronze objects, so bronze required long-distance trade networks. The Bronze Age was succeeded by the Iron Age, which arguably continues to the present: world production of iron ore was 2.5 billion tons in 2019 according to the

US Geological Survey Minerals Commodity Summary of 2019.

While *per capita* use of metals has stabilized in developed countries and use per unit of economic productivity such as gross domestic product (GDP) has dropped, demand for metals and other mineral resources remains high. For example, our use of copper is more extensive than ever: the world produced 20 million tons of copper ore in 2019 according to the USGS. In this respect, the Copper Age continues as well. Furthermore, modern technologies differ from ancient ones in the variety of metals and other materials we use. Nothing illustrates this more than smartphones. At least 80 different elements are incorporated in smartphones or used in their production. These include metals used by ancient societies such as Cu, Sn, Pb, and Fe, as well as many other transition metals such as Ni, Co, and Cr, but also nontraditional ones such as rare earth elements, Li, Ta, As, Br, K, P, etc. In summary, while we have learned to use resources far more efficiently and while recycling materials fills part of the demand*, we will need to continue to extract mineral resources into the distant future. To extract them, we need to first find them and to do that we need to understand how they form and read the clues to their existence. That is the subject to which we now turn.

15.2.1 Ore deposits: definitions and classification

15.2.1.1 Definitions

An *ore deposit* is rock that has an element, compound, or mineral at sufficiently high concentration and in sufficient amount that it can be mined, refined, and sold at a profit. A broad definition would include materials that are extracted in large quantities and used with little further refinement, such as sand, gravel, building stone, limestone, gypsum, salt, etc. Such materials are generally said to be quarried rather than mined. These materials are no less essential to modern society than metals, but space requires we restrict our attention to metal ores. Even so, we have space here to consider only a limited number of types of ore deposits, chosen to illustrate the variety

of geochemical processes involved in their formation.

The concentration necessary for a mineral deposit to be economic varies widely. Iron is by far the metal produced in greatest quantity and it is typically mined at concentrations of around 25% or more, and enrichment factor of only about 6 over average crustal concentrations. Aluminum is economic at similar concentrations, an enrichment factor of only 3 to 4. Copper and neodymium are economic at levels about 0.5 to 1%, representing enrichment factors of 200 to 300%, respectively. At the extreme, gold and platinum are economic at concentrations of 5 to 6 ppm, which are enrichment factors of 4000 and 10,000, respectively. Prices, of course, vary with scarcity: from ~0.1 US\$/kg for Al to ~10,000 US\$/kg for Pt.

We should also point out that some important resources are not "mined" but rather extracted from seawater or brines. Seawater is the principal source of both Mg and Br while sulfur and CO_2 are recovered from oil field brines. Helium is extracted from gas wells. We will discuss one example of metal extracted from brine.

15.2.1.2 Classification of ore deposits

Ore deposits processes can be divided into four broad categories: magmatic, hydrothermal, sedimentary, and weathering, although there is considerable variation within each of these categories and some overlap between them. We'll briefly consider only an example or two of each type to highlight the confluence of geological and chemical processes that create them.

Magmatic ores can be subdivided into orthomagmatic and hydromagmatic deposits. The former are those where the minerals of interest precipitated directly from magma while the latter precipitated from magmatic hydrous fluids. Examples of the former include large layered mafic intrusions in which magma chamber processes can produce bands of minerals such as magnetite and chromite; the Bushveld complex of South Africa is an example. Exsolution of sulfide magma from a silicate one, which can also produce

---

* Recycled metals supplied 49% of overall demand in the US in 2015, the last year for which data are available according to the USGS Commodities Summaries.

ore-grade enrichments of chalcophile and siderophile elements such as Ni and PGEs, is another example of a magmatic ore deposit. The Sudbury Complex of Canada is a prime example. Pegmatite veins, which precipitate at relatively low temperatures ($\sim$500°C) from silica- and H_2O-rich magmatic fluids, and porphyry copper deposits are examples of hydromagmatic ores. Magmatism also plays a role in formation of other deposits where the element of interest is further enriched by secondary processes.

Hydrothermal ores, as the name implies, are those precipitated from hot aqueous solution. These are also the most varied. Magmatism can be the source of heat and metals in these deposits even when it is not the source of the fluid. This is the case in volcanogenic massive sulfides (VMS), which are deposits precipitated at around 350°C and below in situations analogous to mid-ocean ridge hydrothermal systems. In other cases, such as Mississippi Valley type Pb-Zn deposits and sediment-hosted stratiform Cu-Ag deposits, ore precipitated from formation brines at temperatures ranging down to $\sim$100°C.

Sedimentary ores are those precipitated directly from seawater or brines. Banded iron formations (BIFs) are an example. As their name implies, they consist of alternative bands or iron-rich minerals such as magnetite with quartz or calcite. Most formed when reduced Fe dissolved in deep ocean water was oxidized. BIFs are the largest source of Fe globally; examples include the Hamersley Formation of Australia and the Iron Ranges of the Lake Superior area in North America. Salars (salt flats) and their brines, which are important sources of lithium, are another example of sedimentary ores (although the ore in this case may be a solution). Lithium and other elements derived from weathering in volcanic regions are carried by streams to isolated basins and evaporatively concentrated.

Weathering produces some ores, such as laterites, by extensively tropical leaching of soils leaving behind only the most insoluble elements. Bauxites produced in this way are the principal source of Al. Laterites developed on ultramafic igneous rocks are a major source of Ni. Extensively leaching of alkaline, incompatible element-enriched igneous rocks can further enrich such rocks in rare earths, making them important ores

of these elements. *Supergene* deposits form when oxidative weathering of sulfides mobilizes metal elements in the resulting acidic solution. The metals then reprecipitate when pH changes upon contact with ground water. In some cases, this can produce ores of even higher grade than the ore from which they are derived, particularly near the base of the weathering zone where conditions become reducing enough to stabilize sulfide minerals.

Ores can also be classified by when they formed relative to the rocks that host them. *Syngenetic* ores are those that formed simultaneously with their host rock. Orthomagmatic ores are essentially syngenetic by definition, so this term is generally used with respect to other types of ores. *Epigenetic* ores are those that formed after the rocks that host them. Some of these deposits are also referred to as *epithermal*, a term that refers to deposits formed at relatively low temperatures, <300°C, and relatively near the surface in association with volcanic and related intrusive rocks. *Hypogene* ores are those that form at depth within the Earth, specifically below the zone of weathering such as copper porphyry deposits. *Supergene* ores form near or at the surface of the Earth, generally within the weathering zone. This includes not only ore produced primarily by weathering such as bauxites, but also those produced by weathering of primary hypogene ores.

15.2.2 Orthomagmatic ore deposits

Mafic layered intrusions hosting platinum group element (PGE) ores are predominantly of late Archean and early Proterozoic age. They tended to form in periods of supercontinent amalgamation (Section 11.6.3) and crust formation (Maier et al., 2013), occur in stable cratons, and are suspected of having been produced by mantle plumes (Arndt et al., 2017). The 2.7 Ga Stillwater Complex in the Beartooth Mountains of Montana is one example. Within it, the J-M reef horizon has PGE concentrations reaching 30 ppm as disseminated sulfides hosted by gabbro. Perhaps the type example and the largest of these is the enormous 2.05 Ga Bushveld Complex of South Africa: it outcrops over an area of 60,000 km^2 (an area the size of Ireland), is >6 km thick, and has an estimated volume of 0.7 to 1.0 $\times$ 10^6 km^3

(Maier et al., 2013). By comparison, the early Tertiary Skaergaard intrusion of Greenland, one of the largest comparable Phanerozoic layered intrusions and PGE prospects, has an estimated volume of only 280 km³. The Bushveld hosts the world's largest reserves of PGEs, Cr, and V and is a significant source of other metals such as Cu, Ni, and Au. Such a large body of magma may have produced enormous flood basalts as well, but these would have been eroded during the 1.5 Ga unconformity before the overlying Paleozoic Karoo supergroup was deposited. Zeh et al. (2015) obtained high precision zircon ages of 2.055.91±25 Ma on these fine-grained margin rocks and 2.054.87±37 on cumulus rocks from the center of the pluton, implying the Bushveld was intruded and cooled below 600°C within 1 million years.

Stratigraphically, the Bushveld is divided into a Lower Zone (~0.8 km), Lower and Upper Critical Zones (~1.3 km), a Middle Zone (~2.3 km), and an Upper Zone (~1.9 km) (Figure 15.1). Three distinct parental magmas have been identified from analysis of fine-grained margins (which would have chilled rapidly, freezing in the original compositions): a high magnesian andesite termed B1 and two tholeiitic basalts (B2 and B3). The B1 magma is parental to the Lower and Lower Critical Zone, while the Upper Critical Zone can be modeled as crystallizing from mixtures of B1 and B2 magmas; the B3 magma with admixtures of the B2 magma appear to parental to the Main and Upper Zones. Initial $^{87}Sr/^{86}Sr$ ratios for the parental magmas (0.7034 to 0.7077) are well above those expected of 2 Ga mantle. This, as well as initial ε_{Nd} values in the range of −5 to −8 throughout the complex, testifies to extensive crustal assimilation. Initial Os isotope ratios on the other hand are ~0.07 γ_{Os}, indicating that the platinum group elements are exclusively mantle derived or nearly so. Maier et al. (2013) suggest that metasomatized subcontinental lithospheric mantle (see Section 11.5.4) may be a contributor to the PGEs in the Bushveld parental magmas which may also account for some of their incompatible element enrichment. Most sulfur isotope ratios are close to mantle values as well, although locally they provide evidence of a crustal S contribution.

Overall, composition varies upward in a manner consistent with fractional crystallization of these parents, with the dunites and harzburgites cumulates at the base grading to norite (i.e., orthopyroxene-plagioclase gabbro), anorthosite layers, and ultimately diorite in the uppermost sequence. In detail, however, the stratigraphy and mineralogy are far more complex, with frequent layering, some of which is rhythmic, compositional reversals, etc., which is reflected in the highly variable MgO concentrations in Figure 15.1. La/Sm$_N$ ratios are above expected mantle values even in the Lower Zone, but show little systematic variation with height, which is inconsistent with closed system fractional crystallization. Sr isotope ratios also increase upward to values exceeding ratios of the parental magmas or mixtures of them, indicating that fractional crystallization was accompanied by additional crustal assimilation. Given the enormous heat content of this intrusion, this is hardly surprising.

None of the parental magmas are abnormally enriched in PGEs, Cr, V, or S (Figure 15.2). Consequently, we can conclude that local enrichment in these elements is a result of processes that occurred within the intrusion. The PGE ores are hosted in localized "reefs," of which there are several types in the Bushveld:

- Contact reefs along the sidewalls and base of the intrusion, in which PGEs are hosted in disseminated sulfides within mafic and ultramafic sequences, including chromite layers (Figure 15.3) of thicknesses ranging from a few mm to about 1.5 m within mafic and ultramafic sequences, such as LG5 (Figures 15.1 and 15.2), particularly in the lower and critical zones.
- Silicate-hosted reefs in ultramafic sequences containing 1–3% sulfides and up to 10 ppm PGEs, of which the Merensky Reef near the top of the Critical Zone (Figures 15.1 and 15.2) is the type example.
- Reefs of disseminated sulfides in the upper zones hosted by magnetite gabbros or norites such as horizon labeled BV in Figures 15.1 and 15.2.
- Transgressive iron-rich ultramafic "pipes" precipitated from upward or downward percolating melts. The PGEs are present as Fe-Ni sulfides, PGE sulfides such as laurite, and in some cases as PGE alloys.

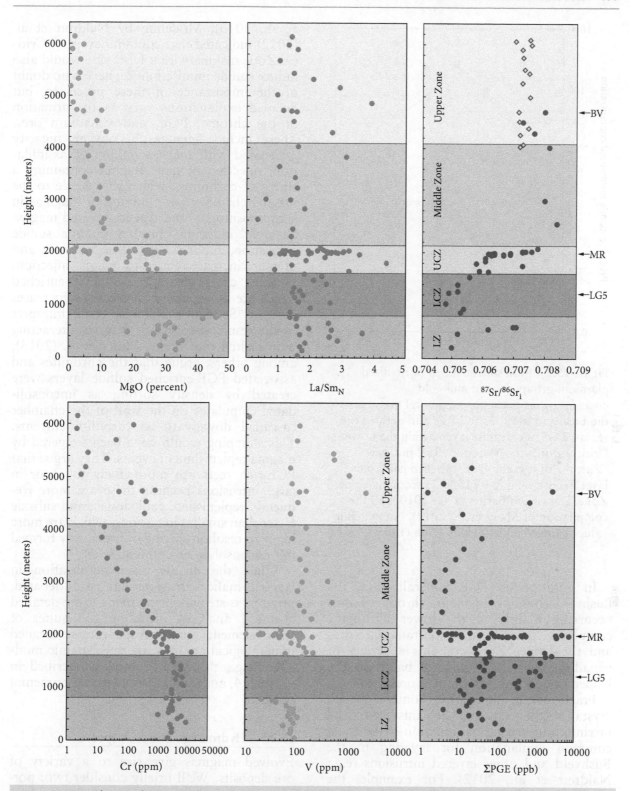

Figure 15.1   Chemical stratigraphy of the Bushveld Intrusion. Data from compilation of Maier et al. (2013) with additional Sr isotope data from Kruger et al. (1987). LZ, LCZ, and UCZ are the Lower Zone, Lower Critical Zone, and Upper Critical Zone, respectively. La/Sm$_N$ is the chondrite-normalized ratio. LG5, MR, and BV show the locations of the noble metal patterns shown in Figure 15.2.

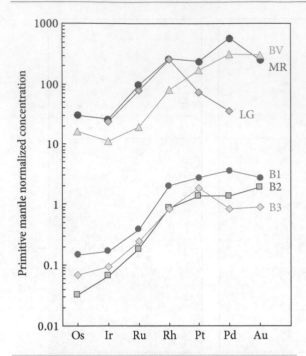

Figure 15.2 Bulk silicate Earth-normalized platinum group element and gold concentrations in the three parental magmas of the Bushveld (B1, B2, and B3), and in three ore seams: LG5, a chromite layer within the Lower Critical Zone, the Merensky Reef in Upper Critical Zone, and a PGE-enriched gabbroic layer in core sample BV1520.33 from the Upper Zone. Data from Barnes et al. (2010) and the compilation of Maier et al. (2013). Normalizing values from McDonough and Sun (1995).

In addition to PGE mineralization, the Bushveld contains 14 major chromite bands occurring mainly in the lower ultramafic cumulates that form important Cr ores and the Upper Zone contains a variety of vanadium-enriched magnetite bands in the Upper Zone form important V ores.

Fractional crystallization combined with crystal settling, magma replenishment and mixing, and crustal assimilation is the most common explanation for layering in the Bushveld and other layered intrusions (e.g., Naldrett et al., 2012). For example, the primary magmas of the Bushveld are all undersaturated with respect to sulfide, but 15% fractional crystallization of the high magnesian andesite (B1) would raise the sulfide concentration to ~600 ppm, enough to exsolve a separate sulfide liquid (Maier et al., 2013). Modeling by Naldrett et al. (2012) indicated that contamination of Critical Zone magma with a felsic one would also induce sulfide immiscibility. There is no doubt of the importance of these processes, but detailed explanations vary as to formation of the chrome, PGE, and vanadium ores. These include changes in oxygen fugacity associated with magma mixing or assimilation, injection of new magma containing a slurry of chromites, which then settle to the magma chamber floor, mixing of magmas in a compositional- and density-layered magma chamber inducing chromite and/or sulfide saturation, mixing with felsic magma, and pressure increase related to magma injection. In some cases, chromite and PGE-enriched bands are associated with dramatic increases in $^{87}Sr/^{86}Sr$, which Naldrett et al. interpret as the result of fresh hot magma interacting with melted roof rocks. Maier et al. (2013), among others, argue that the chromites and associated PGE-enriched sulfide layers were created by density sorting as unconsolidated cumulates on the wall of the chamber cascaded downward as turbidity currents. This slumping would have been triggered by magma replenishment events. They argue that PGE-rich reefs are more likely to occur in large intrusions because these are more frequently replenished, cool slower, and subside faster than smaller intrusions, which are more likely to result in sorting of previously formed but unconsolidated cumulates.

While the details of mineralization in layered mafic intrusions are still debated, progress is steadily being made using detailed chemical analysis, including full suites of trace elements and isotope ratios, detailed mineralogical analysis, thermodynamic modeling (e.g., the MELTS model described in Chapter 4) and comparison with experimental results.

15.2.3   Hydromagmatic ore deposits

Evolved magmas give rise to a variety of ore deposits. We'll briefly consider two: porphyry copper deposits and granite-related tin deposits. In both cases, the ores are deposited from magmatic aqueous fluids at temperatures near the granite solidus. As we noted, Cu and Sn were two of the earliest metals that humans learned to smelt. Alloying copper

Figure 15.3 Chromite bands of Upper Group 3 (UG3) alternating with anorthosite in the Upper Critical Zone of the Bushveld. These layers contain >1% chrome as well as ~4 ppm PGE. Photo by Kevin Walsh & Wikipedia, reproduced under Creative Commons.

with 10% or so of tin produces bronze and it was bronze tools and weapons that powered the great ancient civilizations of the Middle East and Eastern Mediterranean, such as the Minoans, Egyptians, and Babylonians, the Zhou Dynasty in China, and Indus Valley Civilization. Although ores of both copper and tin are both associated with evolved, generally felsic intrusive rocks, the geochemistry of the two elements is quite different and their ores only occasionally occur together. Copper is a highly chalcophile element and primary copper ores are invariably sulfides. Tin, although siderophile, behaves as a lithophile element in the mantle and crust and tin oxide, cassiterite (SnO_2), is the primary ore mineral.

15.2.3.1 Metal solubility in ore forming fluids

The solubility of metals in aqueous fluids is a critical factor in the origin of both hydromagmatic and hydrothermal ore deposits, so before we go further we will briefly review metal solubility in hydrothermal fluids. Transition metals have limited solubility in aqueous solutions under surface conditions of low temperature, high oxidation potential

(i.e. high, $p\varepsilon$ or E_H), near-neutral pH, and dilute solutions. After all, if they did not, our tools, automobiles, infrastructure, etc. would simply dissolve away. Hydrothermal solutions in contrast are, in addition to being hot, more saline, reducing, and acidic than the surface waters we discussed in Chapters 6, 13, and 14. Under these conditions, transition metals become much more soluble.

Chloride is the dominant anion in most, but not all, hydrothermal solutions regardless of their origin and consequently, it is the dominant metal-complexing ligand. Fluorides and bromides can also form soluble metal complexes, but F and Br are generally less abundant than Cl and important only in some circumstances. Increasing temperature decreases the dielectric constant of water that favors increasing ion association (Chapter 4) and, in most cases, increasing numbers of ligands coordinating metal anions in aqueous solution. Thus, in Figure 15.4 for example, Zn^{2+} (or more properly the aquo-complex $Zn^{2+}(H_2O)_6$) is the dominant Zn ion at low temperature even in moderately saline solutions, but as temperature increases $ZnCl^+$ becomes dominant. As salinity increases,

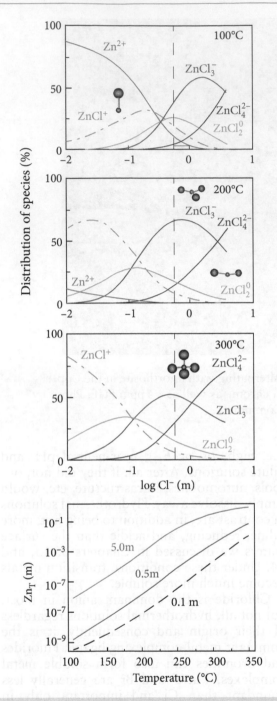

Figure 15.4 Experimentally determined zinc speciation and solubility in NaCl solutions from 100° to 300°C at pH = 6. Upper panels show the speciation of Zn(I) in aqueous NaCl solutions at varying temperatures. Dashed vertical line shows the comparable salinity of seawater. Bottom panel shows total Zn solubility as a function of temperature and NaCl concentration. Modified from Bourcier and Barnes (1987). Molecular stick models based on Brugger et al. (2016); coordinating water molecules are not shown for clarity.

multi-ligand complexes become increasingly important. Just as the reactions between dissolved CO_2 in water greatly increases it solubility over other gases, the formation of multiple species greatly increases the solubility of metal ions in solution. Figure 15.5 shows a similar phenomenon occurs for cobalt complexes. At low to moderate salinities and low temperature, Co^{2+} is the dominant species, but as temperature and salinity increase, multi-ligand complexes become increasingly important such that, for example, in a hydrothermal vent fluid with seawater salinity (~0.56 m) and a temperature of 350°C, $CoCl_4^{2-}$ is the dominant species.

Both copper and tin exist in two valence states in the Earth: Cu(II) and Sn(IV) under surface, oxidizing conditions and Cu(I) and Sn(II) under more reducing conditions such as hydrothermal solution (due to the limited solubility of Sn(IV) in the case of the latter). For both Cu and Sn, various chloride complexes are often the dominant form of these metals in hydrothermal solutions. Unlike Zn and Co, however, increasing temperature does not necessarily result in an increase in the number of coordinating chloride ions. This is apparent in Figure 15.6 where $CuCl_3^{2-}$ is the dominant species at seawater salinity at 50°C but $CuCl_2^-$ dominates at higher temperatures. Tin shows a similar pattern: In low temperature, low salinity fluids, Sn^{2+} dominates with $SnCl^+$, $SnCl_2^0$, $SnCl_3^-$, and $SnCl_4^{2-}$ becoming increasingly important as salinity increases. However, above roughly 200°C, $SnCl_4^{2-}$ becomes unstable and $SnCl_3^-$ is the dominant species in high-temperature saline fluids (Müller and Seward, 2001). Tin solubility also decreases rapidly with increasing pH.

The dominant form of metals in solution depends on a number of other factors as well, including pH, redox state, and availability of ligands. The concentration of sulfur in ore forming fluids increases exponentially with temperature as Fe sulfides dissolve. While copper sulfides have low solubility at low temperature, hydrosulfide complexes, such as $Cu(HS)_2^-$, can become the dominant form of copper in such solution at higher temperatures. Mixed ligand complexes such as $CuCl(HS)^-$ may also be important. Additional mixed complexes such as $KCu(HS)$ and $NaClCuHS$ are also possible at high temperatures and $CoHS^+$ can be an important Co transporting species at

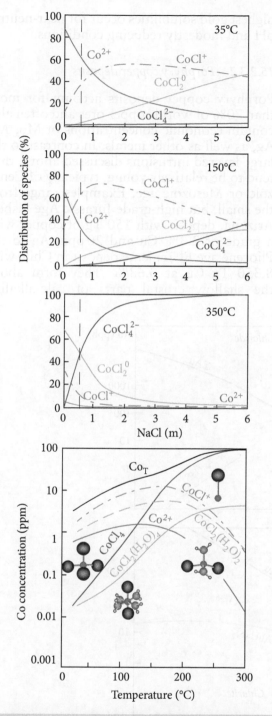

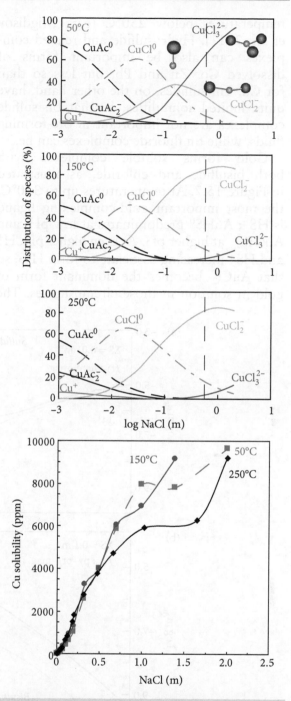

Figure 15.5 Experimentally determined cobalt speciation and solubility in hydrothermal fluids at 60 MPa. Upper panels show the speciation of Co^{2+} in aqueous NaCl solutions at varying temperatures. Dashed vertical line shows the comparable salinity of seawater. Bottom panel shows total Cu solubility in 3m NaCl solution as a function of temperature. Modified from Liu et al. (2011). Molecular stick models based on Brugger et al. (2016); coordinating water molecules are not shown for clarity.

Figure 15.6 Experimentally determined copper speciation and solubility in hydrothermal fluids. Upper panels show the speciation of Cu(I) in aqueous NaCl solutions at varying temperatures (modified from Liu et al. 2001; molecular stick models based on Brügger et al., 2016). Acetate complexes may be important in sedimentary-derived fluids. Dashed vertical line shows the comparable salinity of seawater. Bottom panel shows experimentally determined total Cu solubility in NaCl solutions at pH ~4.5 to 5 as a function of temperature based on data in Liu et al. (2001).

temperatures below 250°C (e.g., Migdisov et al., 2011). Hydrosulfide and related complexes can also be important forms of dissolved Co, Zn and Pb, but less so than for Cu. Tin sulfides, on the other hand, have quite limited solubilities, so that tin sulfide complexes are not important in ore-forming fluids, while tin fluoride complexes can be.

Gold forms soluble complexes with both bisulfide and chloride, as illustrated in Figure 15.7. At temperatures up to 350°C, the most important gold-complexing ligand is HS^-; $AuHS^0$ predominates at low pH, and $Au(HS)_2^-$ at higher pH. At even higher pH, H^+ and HS^- increasing associate to form H_2S, so that $AuCl_2^-$ becomes the dominant form of gold in solution in those circumstances. The highest gold solubilities occur for near-neutral pH and modestly reducing conditions.

15.2.3.2 Porphyry copper deposits

Porphyry copper deposits account for more than 70% of world copper ores and often also contain economic concentrations of Mo, Au, Ag, as well as other metals. In contrast to the large layered intrusions discussed above, they tend to be relatively young, typically of Cenozoic or Mesozoic age. Examples range from the small, but high-grade Jurassic Age Bisbee, Arizona, deposit with 150 Tg of copper with a grade of 2.3% Cu and 20 ppm Ag to the Pliocene age El Teniente deposit in Chile with 8,350 Tg Cu at 0.68%. They form above the shallow crustal parts of calc-alkaline

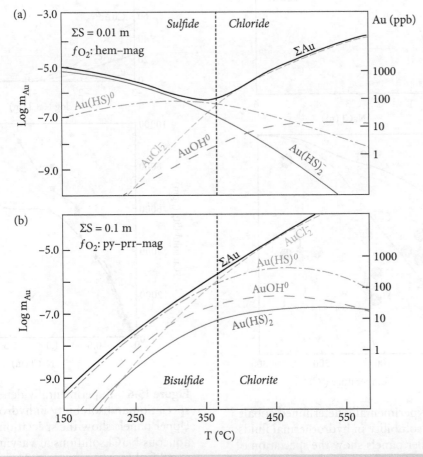

Figure 15.7   Gold solubility and speciation at 100 MPa as a function of temperature for in 1.5 m NaCl and 0.5 m KCl aqueous solutions with pH buffered by the assemblage K-feldspar–muscovite–quartz. (a) $\Sigma S = 0.01$ m and f_{O_2} buffered by the assemblage hematite–magnetite. (b) ΣS and f_{O_2} are buffered by the assemblage pyrite–pyrrhotite–magnetite; the maximum value of ΣS is 0.1 m. The black dashed line (ΣAu) indicates the total solubility of gold. Vertical dotted line divides sulfide and bisulfide dominated fields from chloride dominated fields. From Williams-Jones et al. (2009).

subduction-related plutons that range in composition from dioritic to granitic. Mo-rich porphyry Cu deposits tend to be associated with the more felsic intrusions, while Au-rich ones are associated with the more mafic ones (Stilltoe, 2010). Radiogenic isotope ratios demonstrate that the magmas are primarily of mantle derivation; for example, initial $^{87}Sr/^{86}Sr$ ratios of the Chilean batholiths giving rise to some of the world's largest Cu porphyry deposits are typically 0.703–0.706. In other cases, however, such as the Arizona deposits, magmas have a significant crustal component. Porphyry copper deposits tend to form in mature subduction zone volcanic systems. Among other factors, older volcanic arcs are likely to develop thicker crust in which magma will stagnate and develop large crustal magma chambers. These may persist for millions of years with intermittent recharge of hot basaltic magma that evolves to silicic compositions through fractional crystallization and crustal assimilation. Ore formation generally developed late in the life cycle of individual volcanoes when most or all eruptive activity has ceased.

Ore metals in these deposits do not appear to be derived from subducting oceanic crust and lithosphere; for example, Richards (2015) points out that Cu concentrations in island arc volcanics (IAV) are no higher than in mid-ocean ridge basalts. However, the subducting slab does appear to supply two things that give rise to porphyry copper deposits. The first is the oxidized sulfur. Subduction-related magmas are notably richer in sulfur than mantle-derived ones in other tectonic environments and elevated $\delta^{34}S$ in some of these suggest some of this sulfur is ultimately derived from seawater sulfate (Richards, 2015). Sulfate also carries oxygen from the slab to the mantle wedge and helps to explain why IAV are more oxidized than magmas from other environments (see Section 11.7). The more oxidized nature of these magmas is important because sulfur is present as SO_2 at oxygen fugacities more than 1 log unit above the fayalite-quartz-magnetite buffer (FMQ; Chapter 3) and is soluble as such in silicate magma whereas sulfide can exsolve as a separate sulfide liquid, carrying with it siderophile elements such as Cu, Mo, and Au. Indeed, the abundance of anhydrite in some porphyry deposits reflects predominance of SO_2 in the magmatic fluids. The second thing the subducting slab provides is water. IAV are also water-rich compared with magmas in other environments (see Section 11.7) and porphyry copper ore formation is triggered when the magma becomes saturated in H_2O and exsolves as a separate fluid phase that invades the overlying rock.

Mineralization is focused within a cylindrical volume, typically of only a few km^3 cross-section (but larger in the giant deposits) at depths of 2–4 km within the system of stocks and dikes that formed as magma rose to the surface from 5–15 km deep mid-crustal plutons (Figure 15.8). Hydrothermal alteration in this volume grades concentrically outward from *potassic* (quartz, K-feldspar, biotite, muscovite) to *propylitic* (epidote–chlorite–albite) and *argillic* (quartz-kaolinite-chlorite) at shallow levels, reflecting aqueous fluid–rock reaction at progressively lower temperatures. They typically contain several generations of veining. Early sinuous, ductility deformed veins, so-called *A-veins*, formed from highest temperature fluids ($\gtrsim 600°C$) and are associated with intense potassic alteration but tend to be relatively barren. These are crosscut by linear, parallel, quartz-anhydrite-chalcopyrite ($CuFeS_2$)-molybdenite (MoS_2) "B-veins" deposited from lower temperature fluids (350–550°C). These veins and surrounding zones of disseminated chalcopyrite and bornite (Cu_3FeS_4) have the highest ore concentrations (Figure 15.9). The B veins are, in turn, cut by *pyritic D-veins* deposited from even lower temperature fluids (350–250°C). The picture is thus one of hydrothermal alteration at temperatures decreasing both radially and with time (Stilltoe, 2010; Richards, 2018).

Hydrous alteration minerals precipitated early in the system (e.g., biotite) typically have magmatic O and H isotopic compositions. $\delta^{34}S$ ratios of sulfide minerals in the main ores zones of most deposits tend to be only slightly different from mantle values (Figure 9.36). They do vary somewhat, due to sulfate–sulfide fractionation in magma and fluids, crustal assimilation, and contributions from the subducting lithosphere as noted earlier. This and other evidence establish that the hydrothermal fluids producing these ores are magmatic. $\delta^{18}O$ and δD in later minerals precipitated in

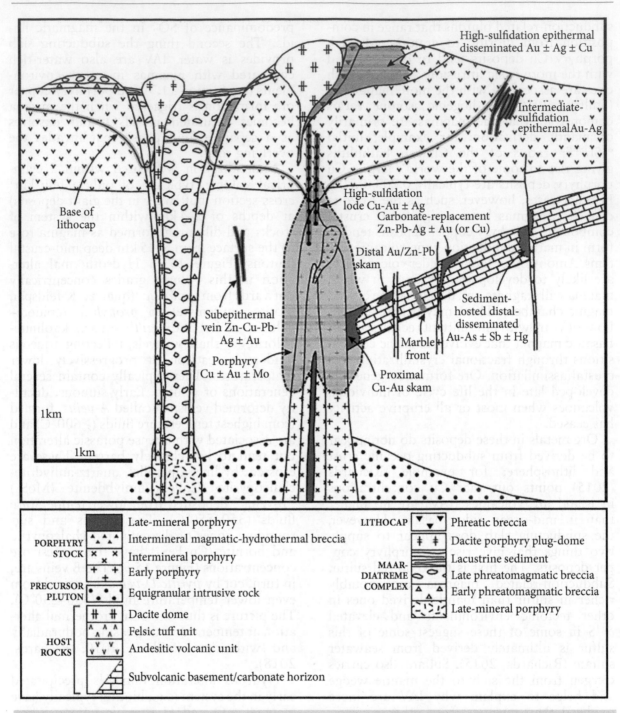

High-sulfidation epithermal
disseminated Au ± Ag ± Cu

Intermediate-
sulfidation
epithermalAu-Ag

Base of
lithocap

High-sulfidation
lode Cu-Au ± Ag
Carbonate-replacement
Zn-Pb-Ag ± Au (or Cu)

Distal Au/Zn-Pb
skarn

Sediment-
hosted distal-
disseminated
Au-As ± Sb ± Hg

Marble
front

Subepithermal
vein Zn-Cu-Pb-
Ag ± Au

Porphyry
Cu ± Au ± Mo

Proximal
Cu-Au skarn

1km

1km

PORPHYRY STOCK
Late-mineral porphyry
Intermineral magmatic-hydrothermal breccia
Intermineral porphyry
Early porphyry

PRECURSOR PLUTON
Equigranular intrusive rock

HOST ROCKS
Dacite dome
Felsic tuff unit
Andesitic volcanic unit
Subvolcanic basement/carbonate horizon

LITHOCAP
Phreatic breccia
Dacite porphyry plug-dome

MAAR-DIATREME COMPLEX
Lacustrine sediment
Late phreatomagmatic breccia
Early phreatomagmatic breccia
Late-mineral porphyry

Figure 15.8 Cross-section of a typical porphyry copper deposit illustrating the various types of ores that can be associated with them. Individual deposits rarely contain all ores and features shown. From Stilltoe (2010).

the lower temperature alteration zones show a paleolatitude dependence (Figure 9.32) indicative of meteoric water ingress, in contrast to $\delta^{18}O$ and δD in the earlier ore-forming fluids, which lack such a dependence.

As we found in Chapter 11, primary mafic arc magmas typically contain ~4 wt% H_2O, and this will increase with fractional crystallization. Such magmas will become water-saturated and begin to evolve a

Figure 15.9 Chalcopyrite, sphalerite, and galena mineralization associated with quartz and pyrite veins cutting dacite porphyry in the Kabba (perseverance) porphyry copper prospect in Arizona. Photo courtesy of Bell Copper.

separate fluid phase at pressures of ~200 MPa (2 kbars), corresponding to the upper parts of crustal magma chambers underlying porphyry copper deposits. Although the exsolved fluid may simply slowly migrate out of the top of the magma chamber, the frequent presence of breccias in porphyry copper deposits provides evidence that initiation of hydrothermal mineralization events can be sudden and catastrophic. Large caldera forming eruptions might be one possible trigger, but mineralization generally post-dates volcanism in any given center. Richards (2018) speculates that other possible triggers are mega-earthquakes that, on geologic time scales, frequently occur in subduction zones, leading to coalescence of bubbles in the magma chamber (much as a shaken soda can) and hydrofracturing, or volcanic edifice collapse, leading to unloading and a sudden decrease in pressure. Regardless of the trigger mechanism, the time scale of the subsequent ore forming event appears to be geologically short. Indeed, vein halo systematics in the Butte, Montana, and Bajo de la Alumbrera, Argentina, deposits suggest the ore forming events may have been nearly explosive and of only ~1000-year duration even for very large deposits (Cathles and Shannon, 2007). Ore may accumulate from a number of such events.

Once an aqueous fluid phase exsolves, components such as Na, K, Ca, CO_2, SO_2, H_2S, etc. will partition into it and the salinity of the fluid can reach tens of percent. Transition metals, particularly those concentrated in porphyry copper deposits such as Cu, Au, and Mo, will partition into this hot, saline fluid where they form the variety of stable soluble complexes discussed above. Partition coefficients between fluid and magma depend on a number of factors, including the mole fraction of CO_2 and salinity (as well as, of course, T and P), but can reach values in excess of 100 for Cu and 1000 for Au (e.g., Frank et al., 2011). This is consistent with the composition of fluid inclusions in mineral grains analyzed from these deposits, which have salinities of tens of percent and Cu concentrations of thousands of ppm or more (Seward, 2014).

CO_2 and NaCl both greatly expand the two-phase region of water (e.g., Schmidt and Bodnar, 2000), so even a fluid initially exsolved as a supercritical one can undergo boiling upon cooling and decompression and separate into a low-density vapor and a smaller amount of high-density saline liquid. The vapor phase nonetheless may contain several percent of NaCl. Experiments show that most metals, including Cu, Mo, and Au, preferentially partition into the brine with partition coefficients depending on the sulfur and CO_2 concentrations as well as P and T (e.g., Zajacz et al., 2017). Nevertheless, significant Cu and Au remain in the vapor and concentrations in the vapor greatly exceed those in the magma. Both types of fluid are observed as fluid inclusions and Audétat et al. (2008)

report Cu concentrations as high as 2% in vapor phase fluid inclusions. There is a broad consensus that metal precipitates from the vapor phase in porphyry Cu–Au deposits, but Audétat et al. (2008) suggests that brine condensates play a central role in the formation of Sn, W, Mo, and REE deposits.

As the fluids migrate upward, they cool and expand, mix with groundwater, and react with the surrounding rock, which results in changes in fluid chemistry, including changes in pH and oxygen fugacity. Whether these changes alone are sufficient to explain the precipitation of copper sulfides is debated and several alternative mechanisms have recently been suggested:

Henley et al. (2015) suggested that sulfide necessary for metal deposition is produced by rapid chemisorption reaction of SO_2 in the vapor phase on plagioclase, which reacts to produce anhydrite, a common mineral in these deposits, and H_2S:

$$3CaAl_2SiO_8 + 4SO_2 + H_2O \rightleftharpoons 3CaSO_4$$
$$+ 3Al_2Si_2O_5 + 3SiO_2 + H_2S \qquad (15.1)$$

This reaction then drives sulfide precipitation from the vapor:

$$CuCl_{2(g)} + FeCl_{2(g)} + 2H_2S_{(g)}$$
$$\rightleftharpoons CuFeS_2 + 4HCl_{(g)} \qquad (15.2)$$

Blundy et al. (2015) point out the paradox that copper enrichment and transport are favored by oxidized, chlorine-rich fluids, yet the ores consist of reduced sulfides. Also, oxidized arc magmas become enriched in copper and chlorine as they evolve, but sulfur can be lost through degassing during this evolution. They argue that such evolved magmas will be too sulfur-poor to produce the observed ores. As they point out, both brine and vapor fluid inclusions are common in porphyry coppers. They suggest that dense metalliferous brines will exsolve slowly from the evolved magma and assemble in the shallow crustal reservoirs. In their model, sulfide ore deposition is triggered when bursts of sulfur-rich vapor exsolve from volatile-rich mafic magma in deeper reservoirs and mix with the metal-rich brines:

$$11H_2S_{(g)} + SO_{2(g)} + 6CuCl_{2(brine)}^-$$
$$+ 6FeCl_{2(brine)} \rightleftharpoons 6CuFeS_2 + 6Cl^-$$
$$+ 18HCl_{(g)} + 2H_2O_{(g)} \qquad (15.3)$$

Finally, Mungall et al. (2015) point out that sulfur emitted during quiescent periods of arc volcanoes far exceeds what can plausibly be supplied by the felsic magmas within the shallow volcanic feeders and must instead be supplied by deeper mafic magma reservoirs. They also point out that sulfide melt inclusions occur along with vapor and brine inclusions in a number of ore deposits and that Cu/Au ratios in associated ore deposits match those of sulfide-silicate partitioning rather than silicate-vapor partitioning. Their experiments show that when a fluid phase exsolves from the magma, tiny exsolved sulfide droplets can adhere to bubbles. Because the mass of the sulfide droplet is small, the compound bubbles remain buoyant and rise through the magma. As they rise and decompress, the sulfide liquid and the metals dissolve into the fluid phase, which then rises into the overlying rock where sulfide ores are precipitated.

15.2.3.3 Tin alkali granite deposits

Let's now turn to tin, the key metal in the transition from the Copper to the Bronze Age. Like porphyry copper deposits, hydromagmatic tin ores precipitate from fluids exsolved from evolved magmatic intrusions. There are important differences, however, in the magmas that give rise to these ores. Whereas porphyry Cu deposits develop from oxidized mantle-derived, subduction-related magmas, Sn deposits develop from reduced granitic magmas formed by crustal melting in continental collisions with little or no mantle contribution. Deposits of W, Ta, Nb, Li, and other elements are sometimes also associated with Sn ores.

Tin is in the 2+ valence state in the mantle and as such behaves as a moderately incompatible element. Indeed Sn/Sm ratios in the mantle appear to be approximately constant at ~0.32 (Jochum et al., 1993), although the ratio is somewhat higher (~0.40) in the crust. The crossover from Sn^{2+} to Sn^{4+} occur at oxygen fugacities near the fayalite-magnetite-quartz (FMQ) buffer (Durasova et al., 1986); consequently, it will be in that state in oxidized magmas such as subduction-related ones as well as systems in contact with the atmosphere. As we learned in Chapter 6, tetravalent elements tend to be highly insoluble in aqueous solution and hence Sn^{4+} tends to be highly immobile in weathering, as are W, Nb, and Ta. Another important aspect of these elements is that they form soluble complexes with Cl and/or F in hydrothermal solutions (e.g., Bhalla et al., 2005; Lecumberri-Sanchez and Bodnar, 2018). This combination of the geochemical properties is the key to formation of these types of ore.

Phanerozoic hydromagmatic tin and tungsten deposits and related lithium-rich pegmatites define discontinuous, several thousand kilometers-long belts, including along the Andes, in Southeast Asia, and in the Paleozoic Acadian, Variscan, and Alleghenian orogenic belts on both sides of the North Atlantic. According to Romer and Kroner (2016), the origin of these deposits involves a series of precursor events to produce a Sn-rich parental granite from which Sn ores can form:

- Intense chemical weathering of sedimentary rocks on a stable continent. As we found in Chapter 13, this results in leaching of soluble elements such as Na, Ca, Mg, and Sr, leaving the residual material enriched in immobile elements and those with a high affinity for clays, including Al, Sn, W, Ta, Nb, B, Li, K, and Rb.
- Tectonic accumulation of these weathered sediments on continent margins. In the North Atlantic region this occurred as a consequence of compressional tectonics during the closure of the Rheic Ocean preceding the formation of Pangea.
- Heating and partial melting of the sedimentary protoliths producing peraluminous alkaline granites enriched in Sn and/or W. In the North Atlantic region this occurred during the continental collision that formation of Pangea and during post-orogenic crustal extension and mantle upwelling. Wolf et al. (2018) argue that production of Sn-rich granites requires either high-temperature, large-degree melting, or multistage melting. At minimal hydrous sediment melting temperatures of 650–750°C, Sn, which will be in the tetravalent state, and related elements behave compatibly, partitioning into solid phases. For example, in their study of migmatites from the Iberian Massif of northwest Spain, tin is sequestered in muscovite, biotite, titanite, magnetite, rutile, and ilmenite and consequently

these minimum melts will be depleted in Sn. At higher temperatures, ~800°C, these phases break down and Sn and related elements enter the melt. Wolf et al. suggest that the most favorable scenario for producing Sn-rich parental granite melts is multistage melting where a low temperature melt is extracted enriching the residue in Sn and related elements, which, upon subsequent melting, produces Sn-rich granitic magma.

- Fractional crystallization of granitic magmas producing further Sn enrichment. In contrast to the oxidized magmas giving rise to porphyry copper deposits, development of tin ores requires reducing conditions at or below the FMQ buffer. Under oxidizing conditions, in which it is in the Sn^{4+} state, tin will partition into oxide phases such as ilmenite, but in more reducing conditions where it is in the Sn^{2+} state, it behaves as an incompatible element in granitic magmas, particularly in peraluminous compositions, and melts rich in Cl and F (Bhalla et al., 2005). The latter not only act as charge-balancing cations but also act as a network-modifying ion (Chapter 4) to breakup the melt structure producing sites for larger cations. Under these circumstances, extensive fractional crystallization further enriches Sn and associated elements in the magma. Williams-Jones et al. (2010) estimate that the Cornubian granites of the Cornwall Peninsula of southwestern England, which host tin deposits mined since the Bronze Age, experienced 10–50% fractional crystallization to achieve the observed Sn concentrations.
- Exsolution of a hydrous fluid from the magma and precipitation of ores in veins, pegmatites and replacement reactions in the solidified granite and the surrounding country rock. As fractional crystallization proceeds, water will eventually reach saturation and an aqueous fluid will exsolve. Under conditions of low pH and high halogen content Sn, W, Nb, and Ta will partition into acidic, halogen-rich saline aqueous fluids. Fluid inclusions in the Cornubian cassiterite and wolframite ores indicate precipitation occurring primarily in the 300–450°C range, but there are indications that B-rich fluids evolved

at near the granite solidus of ~650°C (Jackson et al., 1989). As in the case of porphyry copper deposits, some of the fluid evolution was apparently explosive, as evidenced by breccia pipes and dikes, but in other areas, mineralization is in swarms of quartz veins containing tourmaline $((Li,Na,Ca,Mg,Fe,Al)_4Al_6Si_6O_{18}(BO_3)_3(OH)_3(OH,F))$, mica, and feldspar in addition to ore minerals, pegmatites, and broad alteration zones known as *greisen* consisting of quartz and muscovite ± topaz $(Al_2SiO_4(F,OH)_2)$, fluorite (CaF_2), and tourmaline within the granite. The pegmatites typically consist of quartz, alkali feldspars, tourmaline, micas, chlorite, apatite, fluorite, and sulfides, in addition to either cassiterite or wolframite and likely formed at 400–500°C. Mineralization occurs mainly within the top of the pluton, but also occurs up to 1 km into the surrounding country rocks. In most instances, O and H isotopes indicate the fluid contained significant meteoric water. Precipitation of ores may have occurred due to cooling or mixing of primary magmatic fluids with meteoric water, which results in oxidation or dilution of Cl or F concentrations.

15.2.4   Hydrothermal ore deposits

15.2.4.1   *Volcanogenic massive sulfide deposits*

Volcanogenic massive sulfide (VMS) deposits form in submarine environments from circulating hydrothermal fluids heated by volcanic activity. Ore deposition occurs when upwelling, hot, 350–400°C, hydrothermal fluids cool and mix with cold seawater at or below the seafloor. Most are hosted within volcanics and shallow intrusives, but they can be hosted by overlying sedimentary successions as well. They occur throughout the geologic record with the oldest known deposits in the Pilbara of Western Australia and the Barberton of South Africa dating from 3.45 Ga. More recent examples include the Paleozoic Rio Tinto deposit, which contains more than 500 million tons of ore and which has been mined for 5000 years (with severe environmental consequences that we will consider in Section 15.3.2). Another example is the Mesozoic Troodos deposit on Cyprus,

which was an important source of copper for early Eastern Mediterranean civilizations. The distribution of these deposits is episodic, corresponding with the assembly of supercontinents (Section 11.6.3), including Nuna, Gondwana, and Pangea (Huston et al., 2010). That no doubt reflects a preservation bias, as active ore formation is occurring today on the ocean floor. Indeed, Cathles (2011) estimates that base metal resources on the modern ocean floor contain >600 times the total known VMS reserves on land.

VMS deposits are important sources of base metals, primarily Cu, Pb, Zn, which are so called because of their low value compared with precious metals. In some cases, however, they also contain economic concentrations of Ag, Au, and chalcophile metals such as Co In, Cd, Tl, Ga, Se, Sb, Bi, and in some cases, Sn. Typical grades for VMS ores are 1–2% Cu, 1–4% Zn, 0–1% Pb, ~1 ppm Au, and 10–50 ppm Ag, although some deposits are enriched in some elements by an order of magnitude of more over these values. The deposits are classified by the nature of the host into mafic, bimodal-mafic, mafic-siliciclastic, bimodal-felsic, and bimodal-siliciclastic (Barrie and Hannington, 1999). Mafic volcanic-hosted deposits, which are the least common preserved deposits, tend to be smallest but richest in Cu and Au. The largest deposits are bimodal-siliciclastic, and these are richest in Pb, Zn, and Ag (Hannington, 2014). The metals are derived mainly from the igneous rocks that also provide heat for the system, but Pb isotopes indicate that sediments can contribute some of the Pb and other metals in sediment-hosted deposits.

As the name implies, mineralization is primarily sulfides, including pyrite ± pyrrhotite with associated chalcopyrite $(CuFeS_2)$, sphalerite (ZnS), and galena (PbS), which occur in veins, breccias, and as replacement of original minerals, often in association with chloritic, sericitic, or silicic alteration zones. Most individual deposits tend to small, at least in comparison to be some of the deposits we have already considered. Mid-ocean ridge hydrothermal systems (Chapter 14) exemplify the way in which volcanogenic massive sulfide (VMS) ores are deposited but they can form as in association with any submarine volcanism, including island arcs, back-arc basins, and intracontinental rifts such as the Red Sea.

Unlike the porphyry copper deposits that we considered earlier, the source of hydrothermal fluids is primarily seawater, although in the case of subduction-related volcanism magmatic water might constitute as much as 25%. This inference is supported by the compositions of modern vent fluids and ancient fluid inclusions and by O and H isotope ratios. Modern black smoker fluids have $\delta^{34}S$ values of +1–7‰, indicating that sulfur is mainly igneous-derived with a smaller contribution from seawater. This is consistent with our understanding of the evolution of these fluids, namely that seawater sulfate is removed by anhydrite precipitation before leaching of metals occurs (Chapter 14). Sediment-hosted VMS can have $\delta^{34}S$ that varies significantly from mantle values, indicating a sediment-derived sulfur contribution in some cases (Figure 9.36).

VMS deposits can take a variety of forms. Hydrothermal fluids venting into open, oxic oceanic environments typically form mound and chimney complexes, of which the TAG hydrothermal field on the Mid-Atlantic Ridge is one of several modern examples. In these cases, much of the preserved mineralization occurs in the shallow subsurface. When fluids vent to anoxic bottom waters, large stratiform sheet-like deposits can form. Anoxic marine conditions prevailed prior to the Great Oxidation Event (GOE) at 2.4–2.3 Ga, but periods of basin-wide deep-water anoxia have occurred periodically since then. Local anoxic conditions can also occur when reduced, saline hydrothermal fluids vent into bathymetric depressions forming a pool of dense brine. The Atlantis II deep in the Red Sea is a modern example. Deposits formed in these environments can be significantly larger than those in oxic settings. Finally, sulfides may precipitate or replace existing minerals when hydrothermal fluids rise through overlying sedimentary or volcanic/volcaniclastic rocks. These are sometimes known as stratabound deposits. Sedimented segments of the mid-ocean ridge system such as the Guaymas Basin in the Gulf of California and Middle Valley on the Juan de Fuca Ridge are actively forming modern examples of such deposits.

While mid-ocean ridge hydrothermal "black smokers" may be the most common setting for hydrothermal sulfide deposition, most preserved VMS ore deposits likely formed in different environments for several reasons. First, black smoker vents lose much of their contained metal to a diffuse hydrothermal plume. Second, sulfides deposited on or near the seafloor in oxic environments can quickly oxidize and redissolve. Third, nearly all oceanic crust is subducted into the mantle. Deposits formed in convergent settings, perhaps particularly rifted island arcs (the Marianas is an example), are more likely to be preserved and account for a much larger fraction of VMS deposits in the geologic record. Finally, VMS deposits are frequently hosted by felsic volcanics or bimodal felsic-mafic sequences, whereas felsic rocks are rare on mid-ocean ridges.

The exhaustive studies of modern mid-ocean ridge hydrothermal systems have nevertheless been enormously informative with respect to VMS formation. We reviewed the chemistry of these systems in Section 14.4.3. Here we will briefly reconsider them with respect to ore formation. As we found in the previous section, many transition metals, including Cu, form strong chloride complexes in hot saline solutions, and some may also be transported as hydrosulfide complexes, such as $Au(HS)_2^-$ (Figure 15.7). Consequently, once temperatures reach 350°C and the solution has become acidic and reducing (reactions 14.44 through 14.47), base metals can be quantitatively leached from rock and transported in solution. Base metal solubility depends on a number of factors, including redox state, sulfide content, pH, and salinity, but can be quite high at temperatures above 375°C. The solubility of Cu, for example, in experimental hydrothermal fluids can reach 50 mM (Seyfried and Ding, 1995). Measured concentrations in hydrothermal vent fluids (Table 14.4), however, are often much lower, suggesting that metal deposition occurred at depth or that the fluids in the reaction zone were undersaturated with these metals (Hannington, 2014).

Of the principal base metals found in these deposits, copper is the least soluble, so that as temperatures decrease and pH increases as fluids rise, chalcopyrite, together with pyrite or pyrrhotite, will be the first sulfides to precipitate (Figure 15.10). As the fluids cool further and pH increases due to mixing with seawater, sphalerite and galena precipitate. Consequently, VMS deposits are often zoned, with

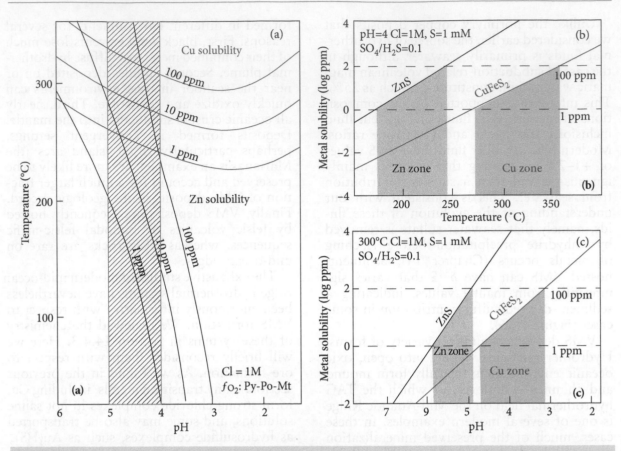

Figure 15.10 Solubility of Cu and Zn sulfides. (a) Solubilities in fluids at the pyrite–pyrrhotite–magnetite fO_2 buffer as a function of pH and temperature. Hashed area shows the typical pH range of hydrothermal fluids. Adapted from Hannington (2014). (b) Chalcopyrite and sphalerite solubility as a function of temperature. Adapted from Hannington (2014). (c) Chalcopyrite and sphalerite solubility as a function of temperature. Adapted from Large (1992). During fluid evolution, temperatures decrease and pH increases, resulting in chalcopyrite precipitated first, followed by sphalerite.

Cu-rich zones occurring at the deepest levels and near the center of upflow regions, whereas sphalerite and galena are found at higher levels and at greater distance from the upflow zone (Figure 15.11). There is also evidence of zone refining occurring, where Zn and Pb sulfides initially precipitated in young systems are redissolved and reprecipitated at high levels as the system heats up. Gold and silver enrichments are usually associated with high concentrations of Zn and Pb.

A modern example of this zonation was found when the Ocean Drilling Program drilled into an active hydrothermal field in the Middle Valley segment of the Juan de Fuca Ridge, where 265°C hydrothermal fluids are venting from the seafloor (Zierenburg et al., 1998). Here the spreading center has been covered by turbiditic sediment. Instead of forming extensive lava flows on the seafloor, basaltic magmas have intruded the overlying sediment in a series of sills. Massive sulfide deposits as thick as 100 m produced by seafloor venting were encountered at shallow depth. These consisted of Fe, Zn, and Cu sulfides, with as much as 51% Zn and 0.62% Cu that showed some textural evidence of dissolution. Below the mound deposits, drilling intersected the feeder zone within the turbidites, which was locally intensely mineralized with up to 50% sulfide minerals. Mineralization consists dominantly of isocubanite ($CuFe_3S_2$) with lesser amounts of pyrrhotite (FeS) and only traces of other sulfides occurring as impregnations and replacement of the host sediments and

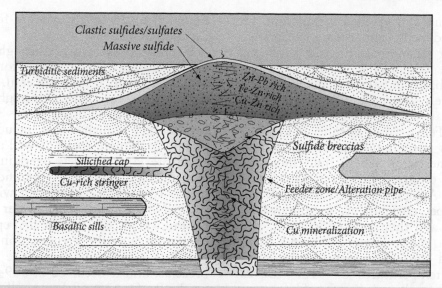

Figure 15.11 Hypothetical illustration of a mafic-siliciclastic volcanogenic massive sulfide deposit such as some Rio Tinto deposits and the actively forming deposit in Middle Valley of the Juan de Fuca Ridge. Red shading illustrates Cu enrichment.

controlled by variation in the original sedimentary texture. At 200 m depth, Cu grade reaches 16%.

15.2.4.2 Stratiform copper deposits

Not all hydrothermal mineral deposits are directly related to volcanism. Stratiform copper deposits are one example. Although these deposits tend to be focused in thin layers that occur over wide regions, the term *stratiform* is, however, somewhat misleading as mineralization often crosses stratigraphic boundaries. Examples include the sediment-hosted stratiform copper deposits such as the Permian Kupferschiefer of Germany and Poland, which has been mined for Cu and Ag since the Middle Ages, and the Neoproterozoic Katanga Copperbelt in Zambia and Democratic Republic of Congo, which also hosts ~35% of the world's Co production. These are products of basin-scale fluid-flow systems in which metals are leached from underlying rock by saline brines under relatively oxidizing conditions. The brines then rise into overlying sediments where sulfides are precipitated. These types of deposit ranges in age from the Paleoproterozoic into the Cenozoic. They share a common origin that includes a source of metals, a source of transporting saline fluids, fluid pathways and thermal or hydraulic pressure to drive the fluids, and chemical and physical "traps" that result in sulfide precipitation.

The Kupferschiefer is perhaps the type example of stratiform copper deposits, having been studied for hundreds of years. The Kupferschiefer is pyrite-, and carbonate-bearing, organic-rich shale, typically only ~0.5m thick, that extends over much of Northern Europe. It was deposited in an intracontinental rift basin and is underlain by the *Rotliegend* red beds, consisting of oxidized terrestrial sediments and heavily albitized and hematized bimodal volcanic rocks. A transgression then deposited the thin (~1-m) laminated, black, coaly *Kupferschiefer* shale in a coal swamp or lagoonal setting and subsequently the *Zechstein* sediments in a series of carbonate-anhydrite-salt deposition cycles. These exhibit the same zonation observed in VMS deposits with Zn- and Pb-rich zones overlying a basal Cu-rich zone. In these areas, ore grades average about 2% Cu and 50 ppm Ag; Pb, Zn, Au, Ni, Mo, and PGEs are extracted as well in some cases. Based on Re-Os ages (245 to 264) Ma, mineralization began shortly after deposition but may have continued for as much as 100 Ma (Alderton et al., 2016).

Dense brines expelled from overlying evaporites through compaction and basin subsidence sank and leached metal from the underlying red beds, and then rose, preferentially along basin margins and basement high into the Kupferschiefer. Only in some areas does base metal enrichment reach economic levels. The ore consists variably of chalcocite (CuS_2), covellite (CuS), chalcopyrite ($CuFeS_2$), and bornite (Cu_5FeS_4), together with lesser amounts of galena and sphalerite. The areas where fluid flux was greatest were oxidized to barren *Rote Faule* (red fooling). Adjacent to the Rote Faule are areas of copper enrichment, surrounded by areas of lead and then zinc enrichment, and finally barren shale. Enrichment and zonation occurred through mineral replacement reactions at temperatures of $\sim$120–150°C. Pyrite in the unaltered Kupferschiefer was replaced by sphalerite, which was, in turn, replaced by galena, and then by chalcopyrite. Replacement proceeded in this fashion through bornite, chalcocite, and finally, in the Rote Faule, hematite. Precipitation has generally been assumed to occur when fluids were reduced by bacterial sulfide reduction or reaction with organic matter, but Wagner et al. (2010) argue from highly variable $\delta^{34}S$ (−44.5 to −3.9‰) that mixing with reduced, alkaline groundwater was important in sulfide precipitation. To account for the volume of copper deposited, brines must have been contributed from a substantial portion of the entire Zechstein basin (Cathles and Adams, 2005).

The Katanga Copperbelt is particularly interesting not only because it is one of the largest known stratiform deposits but also because it is unusually rich in cobalt, a metal critical to modern technologies, including rechargeable Li ion batteries, jet engines, superalloys, magnets, catalysts, etc., not to mention its long history of use as the striking blue pigment "cobalt blue" ($CoAl_2O_4$). Cobalt, like its neighboring elements Fe and Ni, has a mixture of siderophile, lithophile, and chalcophile properties. It occurs in other Cu sulfide deposits, but only rarely at economic levels. Cobalt sulfide solubility is strongly dependent on temperature and salinity, decreasing, for example, from 142 to 0.6 ppm as NaCl decreases from 5 to 0.1 m at 300°C (Figure 15.7). The high salinity and temperatures of the hydrothermal fluids the Katanga ore-bearing fluids compared with similar deposits partly account for their high Co content.

Katanga ores formed through a complex history closely linked to the tectonic evolution of the area, which Saintilan et al. (2018) have attempted to reconstruct through Re-Os dating. The ores are hosted by a $\sim$880−727 Ma sedimentary sequence that includes siliciclastic and carbonate rocks as well as four thick evaporite layers, and is overlain by Cryogenian glacial diamictites. Ore sulfide $\delta^{34}S$ range from −13 to +5‰, consistent with an origin from Neoproterozoic seawater sulfate (El Desouky et al., 2010). The first episode of ore formation was the precipitation of carrolite ($CuCo_2S_4$) replacing evaporite minerals in breccias, and it occurred at 609±5 Ma based on the Re-Os ages of the carrollites (Saintilan et al., 2018). Initial $^{187}Os/^{188}Os$ of 3.2 as well as initial $^{87}Sr/^{86}Sr > 0.708$ indicate a crustal source for the metals, inferred to be Mesoproterozoic basement. These first-phase mineralizing fluids were hot, 270−320°C, and highly saline containing 35−40% NaCl. Saintilan et al. speculate that melt water produced at the termination of the Marinoan ice glaciation penetrated into the underlying sedimentary sequences to depths of ca. 1000 m, dissolving the evaporites. The initiation of the Luftian orogeny involving the collision between the Congo and Kalahari craton then drove fluid circulation into the underlying basement, leaching metals from them before convectively rising and precipitating the ores.

The main episode of Co-Cu mineralization occurred between 540 and 490 Ma when compressional tectonics resulting from the collision between the Congo and Kalahari cratons drive fluid flow. Temperatures were lower (115−220°C) as were salinities, although still very high (12−20% NaCl), than the first phase. Initial $^{187}Os/^{188}Os$ of 3.7 are similar to those of first phase mineralization, and it is possible that dissolution of first phase sulfides were the source of second phase ones. A final late stage of mineralization in which bornite (Cu_5FeS_4) can be seen replacing carrollite occurred around 473±3 Ma based on Re-Os dating of bornite. The bornite has distinctly lower $^{187}Os/^{188}Os$ of 0.4, indicating a contribution of metals from a juvenile igneous source; intrusion of the Hook Batholith may have provided both metals and heat to drive fluid flow.

15.2.4.3   Mississippi Valley-type Zn-Pb deposits

The second type of sediment-hosted hydrothermal deposit we will briefly consider is Mississippi Valley type (MTV) Zn-Pb deposits. Although a few Proterozoic examples are known, they are almost entirely of Phanerozoic age, with a peak in the mid- to late-Paleozoic. This restriction in time may be related to the need for oxidizing conditions necessary to transport metals in relatively cool saline solutions over long distances. The type locality, as the name implies, is the US Mississippi Valley, where deposits occur in a discontinuous belt from Wisconsin and Illinois to Tennessee, Southeast Missouri, and Arkansas. MTV-type mineralization occurs primarily as sphalerite and galena in platform carbonates, and is epigenetic, with mineralization occurring tens to hundreds of millions of years after host-rock deposition. Ore bodies are variable and may be stratabound, pipes, tabular zones, veins, or dissolution breccias, often as replacement of carbonate minerals (Wilkinson, 2014). They typically form in the foreland of passive continental margins undergoing continental collision (Leach et al., 2010). In the Mississippi Valley, deposition was associated with a series of Paleozoic collisions that culminated with the assembly of Pangea. The few available radiometric ages range from 380 Ma, corresponding to the Arcadian orogeny, to 250 Ma, corresponding to the Alleghenian/Ouachita orogeny (Leach et al., 2010).

Individual deposits tend to occur where faults intersecting regional aquifers provide pathways for fluid flow. Fluid inclusions show the mineralizing fluids were saline, with a mode of around 23% NaCl equivalent and relatively cool, with temperatures typically ranging from 75 to 175°C (Bodnar et al., 2014), although exceptionally as cool as 50°C and as hot as 300°C. Some fluid inclusions contain up to 400 ppm Pb, although mean concentrations lower, ~75 ppm (Stoffel et al., 2008). Given the limited solubilities of Zn and Pb sulfides at these low temperatures, the fluid must have been acidic and have high f_{O2} and low f_{H2S}, which Wilkerson (2014) suggests could have formed through sulfide oxidation to sulfuric acid.

Pb isotope ratios and $\delta^{34}S$ in deposits of the Mississippi Valley are heterogeneous, even within individual deposits and these variations are often uncorrelated. This suggests that the source of metals and sulfur differed. $\delta^{34}S$ are mostly positive and broadly consistent with thermal reduction of sulfate that was ultimately derived from Paleozoic seawater, although in some cases negative $\delta^{34}S$ point to bacterial sulfate reduction, which produces larger fractionations (up to –45‰ compared with ~–15‰ for thermal sulfate reduction).

The longstanding model for generation of MTV-type deposits invokes basin-scale fluid induced by continental collision orogenic belts, the Appalachians and/or the Ouachita Mountains in the case of the Mississippi Valley. Alternatively, fluids may have originated more locally from compaction of the Paleozoic sediment (Wilkerson, 2014). Explaining why these fluids were both hot and saline remains problematic. Topographical flow driven by recharge in a distant uplifted orogenic zone tends to cool a basin as it warms its discharge side and tends to flush brines from the basin before the margins are warmed. In addition, a variety of evidence suggests that the total heating time when the deposits formed must have been short, $\lesssim$200,000 yrs. Cathles and Adams (2005) suggest fluid flow was pulsed, driven by sea-level rise in interglacial episodes of the Permo-Carboniferous glaciation.

Regardless of the driving mechanism and source of the water, Na/Br and Cl/Br ratios suggest that the mineralizing fluids acquired their high salinity by incorporating evaporite bitterns and/or dissolution of halite encountered along the flow path. This warm, saline fluid was then able to leach Zn and Pb from rocks along the flow path. Temperatures involved, however, were too low to dissolve significant amounts of Cu, explaining why, unlike other sediment-hosted deposits, Cu-mineralization is absent in MTV-type deposits. In some models, sulfide precipitation was triggered when sulfate in the fluids was reduced through any one of several mechanisms such as reaction with sedimentary organic matter. Alternatively, precipitation may have been triggered when the metal-bearing fluid mixed with a second, sulfide-bearing fluid. That would be consistent with the lack of correlation between Pb and S isotope ratios noted above. Indeed, several recent studies document

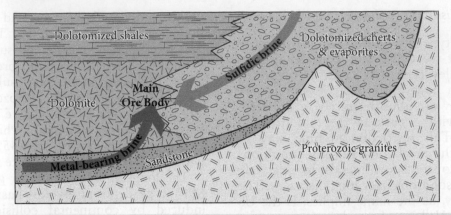

Figure 15.12 Model for the formation of Pb-Zn deposits in the Viburnum District of SE Missouri. From Shelton et al. (2009).

two types of fluid preserved in fluid inclusions: a metal-rich bittern-like fluid and a metal-poor one that acquired its salinity from halite dissolution. Shelton et al. (2009) proposed that metal deposition in Viburnum District of Southeast Missouri was triggered by mixing of a downward-infiltrating, sulfur-bearing, halite-dissolution brine with an upward-migrating, metal-rich seawater-evaporite brine (Figure 15.12).

15.2.5 Sedimentary ore deposits

Sedimentary processes give rise to a number of important ore deposits. One of the most important may be phosphates, an important component of fertilizers that make high-efficiency farming possible and enable a small fraction of the population to grow the food that feeds the rest. We could also include ferromanganese nodules and crusts on the seafloor, which are, in addition to Mn, enriched in Co and Ni and other important transition metals. So far, these have yet to be exploited, although trials are underway. Another is *placer deposits*, which are metals and minerals eroded from a primary ore and deposited elsewhere. The Archean Witwatersr and conglomerate in South Africa hosts one of the largest gold deposits on the planet as well as significant placer diamond deposits. Here, however, we will focus on just two quite different types of deposits, banded iron formations and lithium-rich brines.

15.2.5.1 Banded iron formations

The Bronze Age was succeeded in Europe and the Middle East by the Iron Age. Iron

tools and weapons enabled Roman conquest of the Mediterranean region and construction of massive infrastructure projects, many of which survive today. Iron powered the Qin Dynasty unification of China and the Bantu expansion in Sub-Saharan Africa. Iron remains our most widely used metal. Iron ores form in a variety of environments, but today banded iron formations (BIFs) account for more than 60% of iron ore reserves. They consist of bands of Fe minerals that can include hematite, magnetite, grenolite ($(Fe^{2+},Fe^{3+})_{2\text{-}3}Si_2O_5OH_4$), stilpnomelane ($K(Fe^{2+},Mg,Fe^{3+})_8(Si,Al)_{12}(O, OH)_{27}\cdot n(H_2O)$), and Fe-rich carbonates such as siderite and ankerite alternating with chert (often metamorphosed to quartzite) (Figure 15.13). The original mineralogy, however, is uncertain, having been obscured by diagenesis and metamorphism. It may have consisted of ferruginous clays and ferri-oxihydroxides, although hematite may have been a primary mineral in some cases. Unaltered BIFs contain 20–30% or more Fe, but many BIFs have been enriched through subsequent processes to much higher grades. In some cases, removal of silica and local remobilization of the Fe by basinal fluids or weathering increased the Fe content of remaining rock up to ~65% Fe (Bekker et al., 2010; Arndt et al., 2017). Within the 2.45 Ga Hamersley, Australia, ore body, one of the largest in the world, hematite-carbonate alteration and ore enrichment occurred at temperatures of ~250°C under high fluid pressures.

The largest BIFs, known as *Superior*-type, were deposited in near-shore, continental shelf

Figure 15.13 Archean (2.7 Ga) Algoma-type banded iron formation from the Temagami, Ontario, Canada.

environments between about 2.6 and 2.4 Ga, shortly before and during the Great Oxidation Event (GOE) around 2.3 to 2.4 Ga. A second brief pulse of large deposits occurred around 1.88 Ga. Although they lack direct connection to volcanism, deposition of giant deposits was episodic and coeval with eruption of large igneous provinces (LIPs) produced by mantle plumes (Bekker et al., 2010).

Texturally, they can be divided into banded iron formations, which dominated in the Archean to earliest Paleoproterozoic and granular iron formations (GIFs), which are more common in later Paleoproterozoic deposits. The former were deposited in relatively deep water, below the wave base, likely as density current or turbidites. Indeed, some show graded bedding typical of such deposits and density sorting may in some cases account for the layering. GIFs, on the other hand, show sedimentary features indictive of being deposited on shallow continental shelves above the wave base.

Algoma-type BIFs are stratigraphically associated with submarine volcanic rocks in greenstone belts and, in some cases, with volcanogenic massive sulfide (VMS) deposits (Bekker et al., 2010). They are smaller, more common, and more widely distributed through time, particularly in the Archean. Whereas Superior type deposits first occur around 2.6 Ga, Algoma-type deposits occur

as early as 3.7 to 3.8 Ga in Isua supracrustals (i.e., rocks formed on the Earth's surface) and as late as the Neoproterozoic in association with the *Snowball Earth* glaciations. They were generally deposited in deep water. Ferruginous sediments deposited in brine pools fed by mid-ocean ridge hydrothermal activity within the Red Sea, such the Atlantis Deep, may represent a modern analogy of such environments, although the abundance of sulfides in Atlantis Deep deposits differs from BIFs.

The depositional mechanism of BIFs remains a matter of debate. Until the Great Oxidization GOE, the oceans and atmosphere were reducing. Sulfur, which is insoluble in its reduced form, would have been absent in Archean. Iron is soluble under reducing, S-free conditions and the oceans consequently likely contained considerable levels of dissolved iron supplied by submarine hydrothermal activity. That the peak in BIF deposition at 2.4–2.6 Ga was around the time of the GOE is unlikely to be coincidental. Yet BIFs were deposited both before and after this time, so how dissolved Fe^{2+} was oxidized and precipitated to insoluble Fe^{3+} is unclear. A number of mechanisms have been proposed (Bekker et al., 2010):

- *Abiotic oxidation of Fe^{2+} in O_2-bearing surface waters.* In this model, coastal upwelling of deep water enriched in Fe^{2+}

by submarine hydrothermal springs would mix with surface water oxygenated by photosynthetic cyanobacteria. The central question with respect to this model is, when did oxygenic photosynthesis first evolve? The answer is uncertain. Some argue that it first evolved only shortly before the GEO, others argue it evolved well before then (Section 12.9.5).

- *Metabolic bacterial ferrous iron oxidation.* Some microaerobic bacteria utilize reduced iron to fix CO_2 into organic carbon through reactions such as:

$$6Fe^{2+} + 0.5O_2 + CO_2 + 16H_2O$$

$$\rightleftharpoons [CH_2O] + 6Fe(OH)_3 + 12H^+ \quad (15.4)$$

Such bacteria are abundant at the Loihi seamount iron-rich hydrothermal vents and under low oxygen condition can dominate the iron cycle. This presupposes, however, some oxygen being present and might not be relevant to the earliest BIFs.

- *Anoxygenic photosynthesis.* Various purple and green bacteria can use Fe^{2+} rather than hydrogen as an electron donor in photosynthetic carbon dioxide fixation:

$$4Fe^{2+} + 11H_2O + CO_2$$

$$\rightleftharpoons n[CH2O] + 4Fe(OH)^3 + 8H^+ \quad (15.5)$$

It is certainly possible and perhaps probable that evolution of anoxygenic photosynthesis preceded oxygenic photosynthesis in the iron-rich oceans of the Archean.

- *UV photooxidation.* Prior to oxygenation of the atmosphere and the associated development of an ozone layer, the UV flux at the ocean's surface would have been much greater than in the modern world. UV photooxidation of Fe^{2+} has been demonstrated in the laboratory in simple systems but not in complex solutions such as seawater.

Of course, these possibilities are not mutually exclusive, and it's possible that the mechanism changed with time. Indeed, there is evidence that it did. Rare earth patterns of Archean and earliest Proterozoic BIFs lack the positive Ce anomalies that characterize modern iron-rich marine sediments such

as those of the Atlantis II deep in the Red Sea (Figure 15.14), whereas those deposited around 1.88 Ga have Ce anomalies similar to modern ones (Planavsky et al., 2010). As we learned in Chapter 7, cerium, unlike other rare earths, can be oxidized to Ce^{4+}. This, however, requires higher E_H (or pε) than oxidation of Fe^{2+} to Fe^{3+} or Mn^{2+} to Mn^{4+} (Table 3.3). Planavsky et al. argue that the Paleoproterozoic BIFs record formed in an environment similar to modern redox-stratified basins in which precipitation of Mn^{4+} oxides in oxic shallow water carries metals and Ce oxides downward across a redoxcline to deeper anoxic waters. Oxide dissolution in the anoxic deep water then raises the light-to-heavy REE ratio and increases the concentration of Ce. They argue that this indicates a change from anoxygenic bacterial Fe oxidation prior to the GOE to abiotic oxidation in oxygen-bearing surface waters in the later Paleoproterozoic deposits.

Iron isotope ratios also suggest significant changes in ocean chemistry and marine iron cycle over time. Submarine hydrothermal activity is considered to be the primary source of Fe to Archean and Proterozoic oceans. Primary modern hydrothermal fluids have $\delta^{56}Fe$ of -0.5 to 0‰. As we found in Section 9.9.6, the fractionation in the oxidation of ferrous to ferric iron can be as large as 3‰, although Beard et al. (2010) showed experimentally that the equilibrium fractionation in the geologically relevant precipitation of goethite ($FeO(OH)$) from aqueous ferrous iron was somewhat smaller, about $+1.05$‰. Neo- and Meso-Archean BIF's have positive $\delta^{56}Fe$ with values up to ~$+1$‰ and cherts from this period can have $\delta^{56}Fe$ up to $+2.5$‰, which would be consistent with oxidation of dissolved ferrous iron with $\delta^{56}Fe$ of ~0‰ (Figure 15.15). The absence of light values suggests oxidative removal of iron was limited during this time. Beginning around 2.9 Ga, $\delta^{56}Fe$ become much more variable, perhaps because bacterial dissimilatory iron reduction with fractionation as large as -3‰ was occurring (Johnson et al., 2008). Variation in $\delta^{56}Fe$ in BIFs and other marine sediments became extreme in the period between about 2.7 and 2.3 Ga, with a total range of >5‰ and $\delta^{56}Fe$ as low as -2‰ at 2.3 Ga. Among other factors, this suggests that oxidative removal of Fe from the oceans in large Superior-type BIFs

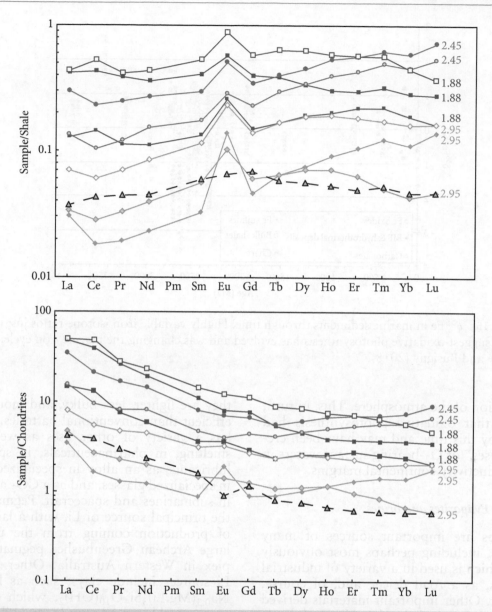

Figure 15.14 Shale- and chondrite-normalized rare earth patterns of Archean and Proterozoic banded iron formations. Red lines are from 1.88 Ga Gunflint (filled squares) and Biwabik (open squares) Iron Formations of North America. Filled circles are from the 2.45 Ga Hamersley Iron Formation of Australia; diamonds are from the ca. 2.7 Ga Manjeri Iron Formation of Zimbabwe (dark filled diamonds) and the 2.95 Ga Pongola Supergroup Iron Formation of South Africa (open diamonds). Dashed black line and triangles are modern metalliferous sediment from the Atlantis II Deep in the Red Sea. The 1.88 Ga BIFs display Ce anomalies similar to modern deposits; earlier BIFs do not, although some are anomalously enriched in La. Numbers indicate the age (in Ga) of the deposit. BIF data from Planavsky et al. (2010) (Gd values for the BIFs are interpolated). Atlantis Deep data are from Laurila et al. (2014).

was a very significant factor in the marine Fe budget. Around 1.8 Ga δ^{56}Fe in BIFs once again became predominantly positive and Fe isotopic variations remained limited through the rest of the Proterozoic and Phanerozoic.

The abundance of large Superior-type BIF deposits just prior to the GOE at 2.3–2.4 Ga, together with the extreme variability in δ^{56}Fe beginning about 2.7 Ga, suggests the marine Fe cycle was already changing prior

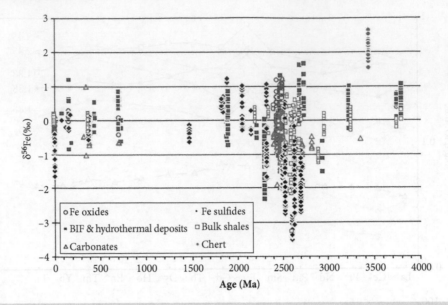

Figure 15.15 δ^{56}Fe in marine sediments through time. Highly variable iron isotope ratios just prior to the GOE suggest oxidative photosynthesis has evolved and was changing the marine iron cycle. From Planavsky and Busigny (2018).

to oxidation of the atmosphere. This, in turn, suggests that oxidative photosynthesis had evolved by this time and may have been creating "oases" of O_2-bearing surface waters in high productivity continental margins.

15.2.5.2   Evaporites and brines

Evaporites are important sources of many materials, including perhaps most obviously halite, which is used in a variety of industrial processes, as a road deicer, and, of course, table salt. Other important materials derived from evaporites or their related brines include gypsum, used in wallboard and plaster, sylvite (KCl) a key ingredient in fertilizers, and boron, magnesium, and bromine. The mineral composition of evaporites depends firstly on whether they form from seawater or saline lakes; as we found in section 13.6, the latter can vary considerably in composition, based on the composition of rocks in the watershed in which they lie and the course of fractional crystallization. Here we will briefly consider another evaporite-derived resource that is critical to a carbon-free energy future: lithium.

Demand for lithium is rapidly accelerating in response to the development of Li-ion batteries in automobiles and electronics because they are lighter, less bulky, and more energy efficient than conventional batteries. Lithium has a variety of other uses as well: in Al smelting, in pharmaceuticals, in specialized lubricants, as an alloy in specialized metals, in specialized glasses, and as a CO_2 absorbent in submarines and spacecraft. Pegmatites are the principal source of Li, with a large share of production coming from the unusually large Archean Greenbushes pegmatite complex in Western Australia. Other potential Li sources include clays, such as hectorite, $Na_{0.3}(Mg,Li)_3Si_4O_{10}(OH)_2$, which can contain several hundred ppm Li, and oil field and geothermal brines. However, brines associated with continental evaporites and saline lakes are a much larger future resource with an estimated total of 21.6 Mt compared with 3.9 Mt in pegmatites. Furthermore, individual pegmatite deposits tend to be considerably smaller than brine deposits (Kesler et al., 2012), so resource development focuses on brines.

Continental saline lakes, which we discussed in Chapter 13, and associated salt flats, or *salars*, form in arid region enclosed basins, thus tectonics and climate control their locations. Li is recovered not from the salt, but rather from the interstitial bittern present beneath the solid crust in actively forming salars. However, most salars do not contain economic Li deposits. That raises the question of what combination of geologic and chemical factors gives rise to such deposits?

Although Li is produced from brine in China and the western United States, the Puna and Altiplano plateaus of the Andes are responsible for most current brine production and hold the majority of future resources. The *Uyuni* salar in Bolivia, with an estimated resource of 10.2 Mt of Li, represents nearly half of the global total; the *Atacama* in Chile holds nearly 30%. Let's consider a somewhat smaller one, the *Salar del Hombre Muerto* in the Puna Plateau of Argentina, which contains several producing wells. A study by Godfrey et al. (2013) provides insights as to why these Andean salars are such rich Li resources. The following summary is based on their work.

The Salar del Hombre Muerto (Figure 15.16) occupies 600 km^2 at an elevation of 4000 m and consists primarily of halite, with occasionally interlayered gypsum, and borax. Lithium concentrations vary, but can exceed 900 ppm in the shallow lake, "the lagoon," covering part of the salar and ranges from 450 to 1100 ppm in the pore fluid within the halite. The watershed consists mainly of young volcanic rocks, most notably the enormous 2 Ma Cerro Galán ignimbrite, but Paleozoic and Cenozoic sediments and Paleozoic metamorphic rocks are present as well. With the exception of one outcrop of fine-grained ash that contains 106 ppm Li, most of these rocks have Li concentrations close to the upper crustal average (~25 ppm). The Rió los Platos, which is the largest source of surface water to the salar, flows northward from Cerro Galán about 150 km and drains almost 80% of the watershed and contains 3.3 ppm Li.

As we found in Section 9.9.2, ^{6}Li partitions into clays during weathering, leaving water enriched in ^{7}Li. Surprisingly, the streams in the area, including the Rió los Platos, have δ^7Li values are comparable to those of the rocks outcropping in the watershed, which range mostly between +3 and +7‰; the del Hombre Muerto lagoon has δ^7Li ≈ +7‰. By comparison, the global riverine average is ~+23‰. This suggests Li in the streams was derived from reactions at temperatures at least modestly elevated above surface ones with minimal clay formation and Li isotope fractionation. Godfrey et al. (2013) suggested that the streams obtain their high Li concentrations from geothermal springs, which typically have 30–40 ppm Li. δ^7Li >10‰

Figure 15.16   The Salar del Hombre Muerto. Photo courtesy of Terry Jordan.

of most of these springs are in the range of 2–5‰, similar to the bedrock.

Additionally, while stream waters in the area and the lagoon have $\delta^{18}O$ and δD values that plot along an evaporative trend (Figure 9.12), the subsurface brine plots close to the meteoric water line, indicating that much of the flux into the salar is subsurface. Once this shallow groundwater enters the salar, it dissolves salt and rises to the surface through capillary action where it evaporates and halite precipitates. Experiments conducted by Godfrey et al. showed that the halite-brine distribution coefficient, $D = (Li/Na)_{halite}/(Li/Na)_{brine}$, was 0.048 at the initial halite-saturation solution and decreased to 0.005 by 90% evaporation. Thus, Li is excluded from crystallizing halite and consequently becomes progressively enriched in the brine.

Godfrey et al. (2013) found that aragonite travertine and geyser deposits formed in a wetter climate around 40,000 years ago were not as enriched in Li and further that they precipitated from solutions with higher δ^7Li than present stream and geothermal springs. The higher δ^7Li suggest more active clay formation in the past. Thus, the present dry climate is an important factor in formation of the Li-rich brines of the Salar del Hombre Muerto. In summary, key factors in formation of Li-rich brines appear to include the presence of glass-rich silicic volcanic ash from which Li can be readily leached, geothermal activity, which also enhances Li leaching, and an arid climate in which clay formation, which would otherwise remove Li from solution, is minimal.

Several other factors make the Salar del Hombre Muerto brine particularly suitable for Li mining. First, the Mg/Li ratio is low, which in turn likely reflects the silicic volcanics in the area. This is important because Li is extracted from brines by fractional crystallization in a series of evaporation ponds. Mg tends to concentrate in the residual brine along with Li in the process, which makes the final lithium carbonate product more difficult to purify. The second factor is the relatively cool climate at this high elevation. This also makes extraction easier because Li remains in solution longer at lower temperatures (lithium carbonate solubility decreases with increasing temperature).

15.2.6 Weathering-related ore deposits

Weathering can give rise to a number of important mineral deposits. As we noted earlier, weathering can produce supergene ores that are enriched relative to the original hypogene ones as bacterially mediated oxidation of sulfides produces sulfuric acid solutions that can leach metals, carry them downward, and reprecipitate them, enriching metals already present. Here we'll focus on just a few examples of primary ores produced by weathering. In these cases, it is the preferential removal of most other metals that results in enrichment of some.

15.2.6.1 Bauxites and laterites

Bauxite is the primary ore of aluminum, the wonder metal of the twentieth century. Until the development of large-scale refinement methods in late nineteenth century, aluminum was considered a precious metal despite its abundance in the crust because of the difficulty in reducing it to metal. Refining aluminum remains a very energy-intensive process and bauxite is often shipped long distances to countries such as Canada and Iceland where cheap hydropower and geothermal energy is available. Today, aluminum ore production (64 million tons in 2019) is second only to iron in world production and because of its light weight, corrosion resistance, strength, and low cost, it has found use in a very wide range of applications.

Bauxite by definition is a rock that contains >45.5% Al_2O_3 and <20% Fe. In detail it is usually a mixture of fine-grained aluminum oxyhydroxides including gibbsite ($Al(OH)_3$), diaspore ($AlO(OH)$), and boehmite ($AlO(OH)$) (Figure 3.2). It develops from extremely weathered soils or paleosols known broadly as *laterites*. The term *laterite*, however, usually implies an Fe-enriched paleosol; these are a major source of Ni as well as Nb, Au, rare earths (REE), and a variety of other metals and in the past have been an important source of Fe, although they are little exploited as such today. We'll use *lateritic* to describe both laterite- and bauxite-related soils and reserve the term *laterite* to describe just the Fe-rich, Al-poor ones.

We found in Chapter 13 that as rock weathers and soil develops on it, the soluble components are progressively dissolved and carried away by groundwater. $SiO_2/(Al_2O_3 + Fe_2O_3)$ ratios decrease and ratios such as the chemical index of weathering ($Al_2O_3/(Al_2O_3 + CaO + Na_2O + K_2O)$) increase reflecting this loss of soluble components. Among the major components in rocks and soils, Al_2O_3 and Fe_2O_3 are the least soluble and their concentrations increase as weathering proceeds, along with those of the high field strength (HFS) elements (Nb, Ta, Zr, and Hf) and, to a lesser extent, others elements such as the REE and Th. Elements, such as Cr, As, Ga, Sc, V and, possibly, Au, can also accumulate in Fe oxides. At the extreme, the soil consists almost entirely of Fe and/or Al oxides and oxyhydroxides.

Most soils, however, never develop to this extreme because erosion removes the soil from the top as the weathering front progresses downward (Section 13.4.1). Development of bauxite and laterite ores can occur only where rate of the downward progression of the weathering front exceeds the erosion rate. Weathering reactions are temperature dependent and also require contact between minerals and water, so extreme weathering is favored by warm, wet climates, although lateritic soils can develop in other climates as well. Most bauxites and laterites have developed between latitudes 35° N and 35° S. Even then, this development is a slow progress and requires millions of years (Butt et al., 2000). This in turn requires long periods of tectonic stability with sufficient relief to allow groundwater containing leaching products to drain away but not so much that erosion dominates over weathering. Parts of Africa, India, South America, Southeast Asia, and Australia have been stable for over 100 Ma, allowing weathering profiles to extend as deep as 150 m. Major bauxite producers include Guinea, Australia, Brazil, and Jamaica, which have tropical climates (Arndt et al., 2015).

If extreme weathering leads to enrichment in Al and Fe, how do we explain the existence of two distinct sets of ores: Al-rich bauxites and Fe-rich laterites? Certainly, the protolith will have something to do with this: soils developed on Fe-rich mafic igneous rocks are more likely to evolve to laterites while those on Al-rich felsic igneous rocks and pelitic sediments are more likely to evolve to bauxites. But Petersen (1971) pointed out another factor, and one we can understand by reconsidering what we learned of the

chemistry of these two elements in previous chapters. We see from Figures 6.18 and 6.19 that as a consequence of progressive formation of hydroxide complexes, aluminum is minimally soluble at a pH of around 6 while ferric iron is minimally soluble at a pH around 9. Al is more soluble than Fe over most of the range of natural water, but particularly at high and low pH, suggesting Al-poor laterites could be generated when the soil solution is slightly acidic or slightly alkaline, particularly the latter. Now let's consider Figures 3.19 and 13.2. We see that at sufficiently low pϵ (or E_H), iron will be predominantly in the soluble ferrous form. We can superimpose Al solubility from Figure 6.18 on Figure 13.2 to produce a composite Pourbaix diagram showing the pϵ-pH regions where bauxites and laterites are most likely to develop (Figure 15.17). Bauxites will develop from soils when conditions become reducing, while laterites will develop under oxidizing conditions, particularly at high or low pH. Peterson argued that bauxites develop in heavily vegetated regions from organic-rich soils that become anoxic with depth while

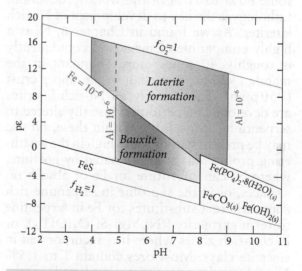

Figure 15.17 Simplified Pourbaix or pϵ-pH diagram showing the conditions in which soil weathering will lead to laterites and bauxites. Red lines show where Al and Fe solubility is 10^{-6}. The intensity of red shading indicates the extent to which Al is more soluble than Fe^{3+} and hence that Fe-rich laterites are likely to form. Blue shading indicates bauxite formation is most likely in reducing conditions in the pH range of 5 to 7.

laterites are associated with more sparsely vegetated areas and pointed out that was indeed consistent with observations. As we noted in Chapter 13, organic acids can complex and transport Fe and Al within soils, but in tropical soils organic decomposition is rapid, leaving little available soluble organic acids.

Tole (1987) carried out a number of experiments to simulate weathering of nepheline ($NaAlSiO_6$) and bauxite development. He found that formation of kaolinite was favored at low pH (<3.8) while gibbsite was favored at higher pH. At even higher pH, he found that muscovite, paragonite, and kaolinite were favored over gibbsite. He concluded that the optimum pH range for bauxite formation was 5 to 7. He also found that equilibrium was slow to be achieved $25°C$, with the solution remaining supersaturated in Al, while this was not the case with higher temperatures and therefore concluded that bauxite formation was more likely in warm climates.

While laterites are now rarely mined for Fe, over the last two decades laterites have replaced magmatic sulfide ores as the dominant source of Ni and now account for some 60% to 70% of the world production. Lithology is the key to development of Ni-rich laterites. As we found in Chapter 6, Ni is a highly compatible element and consequently is roughly 40 times more abundant in the mantle ($\sim$2000 ppm) than in the upper crust ($\sim$50 ppm). Thus, nearly all Ni-rich laterites are developed on peridotite (usually altered to serpentinite) protoliths. Within these, Ni ore may be present within zones high in the weathering profile that are dominated by goethite, where Ni can substitute for Fe or absorb on surfaces, or in the saprolite in serpentine-rich zones where it substitutes for Fe in serpentine to form garnierite ($(Ni,Mg)_3Si_2O_5(OH)_4$), or in clay-rich zones where it is incorporated in smectite clays. Most ores contain 1 to 1.5% Ni. Most deposits have developed in the so-called "wet and dry topics" of highly seasonal rainfall. Major deposits occur in New Caledonia, Cuba, and Indonesia, but deposits have also been found in Oregon, Australia, and the Urals of Russia and Kazakhstan.

15.2.7   Rare earth ore deposits

We consider rare earth ores in a separate section not because they do not fall into any of the above categories, but rather because they fall into many of them. This emphasizes the point that while one type of deposit can account for most of the world's supply of a particular element, in most cases several types of ores are exploited for that element. Another reason to do so is that over the last several decades the rare earths have gone from obscurity to critical resources.

The lanthanide rare earths and their isotopes have proved themselves very useful in understanding a wide range of geochemical phenomena ranging from crust-mantle evolution to ocean circulation, as we have found in the preceding chapters, but many practical uses for them have been found as well over the last half century. Indeed, technology has grown heavily dependent on them; uses include catalysts, which account for 60% of use, and in specialized metals alloys. Twenty-first-century technology is particularly dependent on them. Rare earths are used in fiber optics, phosphors in computer and smart phone displays, CFL and LED lights, lasers, and in a variety of high-tech and medical imagining equipment. Cerium, the most abundant rare earth, is used as a diesel fuel additive and in mechanico-chemical polishing of high-precision optics and semiconductors. Lanthanum is used in nickel-metal hydride rechargeable batteries, which have largely replaced NiCd ones. Rare earths have high magnetic moments because their orbital electron structure contains many unpaired electrons whose spins can align with the magnet field and not cancel out. Neodymium–iron magnets ($Nd_2Fe_{14}B$) have the highest magnetic energy density of any magnet; your headphones or earbuds probably use Nd magnets. Samarium–cobalt magnets ($SmCo_5$), which have somewhat lower energy density but higher Curie point of $\sim$800°C, are used in higher-temperatures applications. Terdnol-D, a terbium–dysprosium alloy ($(Tb_{0.3}Dy_{0.7})Fe_2$), has the highest magnetostriction of any alloy, meaning its volume changes in response to an applied magnetic field, and is used in sonar, ultrasonic devices, and loudspeakers.

In most of these applications, rare earths are used in small amounts, but larger quantities are used in batteries and high-performance magnets in electric motors and turbines. Your smart phone contains only about half a gram of rare earths (usually including 8 or more REE), but each Tesla Model 3 contains more than 6 kg of Nd and wind turbines can contain hundreds of kg.

While rare earths are more abundant in the Earth's crust than many other valuable metals, the mining and extraction of rare earths pose particular challenges. To begin with, few processes concentrate them to ore grade. Furthermore, rare earths can be hosted in a variety of minerals in ore deposits, requiring a variety of extraction methods. They are often associated with high Th concentrations, so that radioactivity can be an issue. Finally, once the rare earths are extracted, they must be separated from each other, which is a difficult process given their chemical similarity to each other (as isotope geochemists well know). Rare earths are more readily extracted from some deposits than others, which is an important factor in deciding whether the deposit is economic. Total world production of rare earth oxides was a mere 210,000 tons in 2019 compared with 2.5 billion tons of iron ore according to the USGS Commodities Summaries. However, demand is increasing nearly 14% per year and they are important to our modern digital, connected society and key to a future fossil fuel-free energy economy (Ganguli and Cook, 2018). At present, China accounts for 70% of global production but trade disputes including export restrictions and tariffs in the last decade have made finding and developing new sources a priority for much of the world. Here we'll briefly review a couple of examples of REE mineral deposits.

Some rare earth deposits are orthomagmatic ores, albeit unusual ones: carbonatites, syenites, and related alkaline magmatic rocks. This type of ore is exemplified by the Mountain Pass deposit of California, a carbonatite which has an average grade of ~9% rare earth oxide. For much of the second half of the twentieth century, this mine was the world's largest producer of rare earths. However, operations ceased in 2002 due primarily to environmental concerns (we'll consider those in later sections) but reopened in 2018. Carbonatites are rare and unusual magmas, consisting primarily of carbonate rather than silicate minerals. The Sulfide Queen carbonatite hosting the Mountain Pass ore deposit is unusual, even by carbonatite standards, in its extreme enrichment in Ba and REE (Figure 15.18). The carbonatite, which forms a tabular intrusion, is generally porphyritic with centimeter-size barite ($BaSO_4$) phenocrysts (25%) in a matrix of calcite and/or dolomite (60–65%), and 10–15% bastnäsite ($[LREE]CO_3F$). Minor REE phases include synchysite ($Ca[LREE](CO_3)_2F$), parisite ($Ca[LREE]_2(CO_3)_3F_2$), and monazite ($(REE,Th)PO_4$ (Castor et al., 2008). Carbonatite dikes also occur in the area and are associated with and intrude Proterozoic ultrapotassic intrusive rocks that occur in a narrow north-trending belt in southeastern California (Castor et al., 2008; Poletti et al., 2016). The ultrapotassic rocks, which range from mafic "shonkinites" to felsic "syenites" are themselves highly enriched in REE (Figure 15.18) and other normally incompatible elements. Poletti et al. (2016) obtained U–Pb ages of the ultrapotassic rocks that ranged from $1429=\pm10$ to 1385 ± 18 Ma; the age carbonatite ore body, 1371 ± 10 Ma, overlaps within error with the young end of that age range. They reported that initial ε_{Nd} of the carbonatite is –3.0, which falls within the range of the ultrapotassic rocks, –3.5 to –12. Despite the negative ε_{Nd} values, a number of factors indicate that both the ultrapotassic and carbonatite magmas were mantle-derived, with little crustal contribution. Castor (2008) proposed that the carbonatites formed by exsolution from a mafic alkaline silicate magma as it evolved, but Poletti et al. (2016) found that such an origin is incompatible with carbonatite–silicate magma partition coefficients. They instead proposed it formed, along with the ultrapotassic magmas, by small extents of melting of previously incompatible element-enriched subcontinental lithosphere.

The Bayan Obo in Northern China is an example of a hydrothermal or hydromagmatic ore, which one is a matter of debate. It is the world's largest known REE deposit and supplanted Mountain Pass as the largest producer of REE in the late 1990s. The ore is hosted within a Proterozoic dolomite; the richest REE ores, with grades of ~6% rare earth oxide, are associated with banded massive magnetite and hematite, which was mined beginning in the 1930s for Fe and Nb. Recovery of REE began in 1957 but Bayan Obo only became a major global REE producer in the 1990s. The evolution of this deposit has been exceedingly complex, with multiple episodes of fluid infiltration, metasomatism,

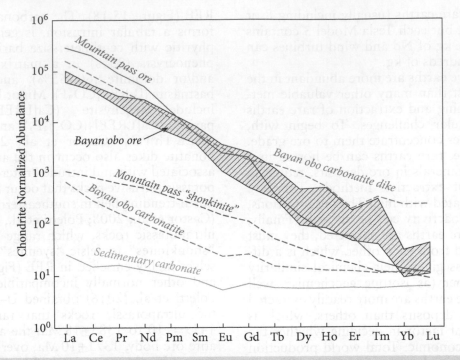

Figure 15.18   Rare earth patterns of the Mountain Pass and Bayan Obo ore deposits. Also shown are carbonatite dike in the Bayan Obo region and a mafic ultrapotassic "shonkinite" from the Mountain Pass area (dashed lines). The pattern of a typical sedimentary carbonate is shown for comparison. Mountain Pass data are from Poletti et al. (2016); Bayan Obo data and sedimentary carbonate are from Yang et al. (2009).

metamorphism, and deformation occurring over a period of a billion years. This evolution has been so complex that whether the host dolomite originated as sedimentary, carbonatite, or sedimentary carbonate intruded by carbonatite is debated. While multiple lines of evidence point to a sedimentary origin of the dolomite, REE abundances in the ore body (Figure 15.18) are far more similar to carbonatites than to sedimentary carbonates. A consensus has emerged, however, that regardless of the origin of the host, the ore formed through several episodes of hydrothermal activity (Yang et al., 2009; Kynicky et al., 2013; Smith et al., 2016; Liu et al., 2018). Smith et al. (2016) argued that ore formation is best explained by infiltration of carbonatite or carbonatite-derived fluids into a sedimentary dolomite limestone. Oxygen and carbon isotope ratios in the ore and the host can be modelled as mixtures of sedimentary carbonate ($\delta^{18}O \sim +20$ to $+25‰$, $\delta^{13}C \sim 0‰$) and carbonatite ($\delta^{18}O \sim +6‰$, $\delta^{13}C \sim -5‰$). Fluid inclusions range from saline (7–10% NaCl) to low salinity (1–6% NaCl), high

CO_2 (54%) fluids. Fluid inclusions in fluorite record temperatures of up to 280°C, but these may only be recording the latest stages of mineralization. Experiments by Migdisov et al. (2009) showed that REE form stable soluble complexes with both fluoride and chloride at these temperatures so such saline fluids could have readily transported REE.

The earliest REE mineralization was the precipitation of monazite and bastnäsite, which took place at around 1.26 Ga based on Sm-Nd mineral isochrons and zircon Th-Pb ages (Smith et al., 2016; Liu et al., 2018). This event is associated with alkaline and carbonatitic magmatism, with the latter being the likely source of the mineralizing fluids. Initial ε_{Nd} values are mainly in the range of -3 to $+0.9$, suggesting an incompatible-enriched subcontinental lithospheric source for the magmatism. Several subsequent stages of metamorphism and deformation occurred, culminating around 381–367 Ma based on zircon U-Pb and Th-Pb zircon ages (Liu et al., 2018). This coincides with the Caledonian orogeny and subduction of the Mongolian oceanic plate beneath the North China craton and resulted in shear foliation and banding of the ores. The REE were extensively remobilized during this event, which resulted in formation of Ca–REE fluorocarbonates such as synchysite, which in turn were replaced by barite and Ba–REE fluorocarbonates such as huanghoite ($BaREE(CO_3)_2F$). This occurred mainly through remobilization of REE within the deposit but a new flux of crustal-derived REE may have also occurred.

Both the Mountain Pass and Bayan Obo deposits are strongly enriched in light rare earths with little if any enrichment in heavy REE (HREE) over typical crustal rocks. Yet as cataloged above, there is strong demand for HREE as well. An entirely different type of deposit, laterites, such as the Zudong deposit in Jiangxi province in South China, account for ~35% of China's REE production and represent over 80% of global HREE resources (Li et al., 2017). Although these ores are of lower grade, 0.05 to 0.2% rare earth oxides, the rare earths are easily extracted, which together with relatively high HREE abundances makes these deposits economic. Let's briefly consider the Zudong deposits in more detail.

The Zudong deposit is developed on the 168 Ma Xinxiu peraluminous granite composed of plagioclase, potassium feldspar, quartz and muscovite. These minerals account for only a fraction of the REE inventory, with 72% of the REE hosted in secondary minerals, synchysite, gadolinite (($REE,Y)FeBeSi_2O_{10}$), and zircon, as well as minor amounts of monazite, fergusonite ($REENbO_4$), and xenotime (($REE,Y)(P,As)O_4$) (Li et al., 2017). As Figure 15.19 shows, this granite is unusually HREE-rich, with La/Yb of ~1, compared with, for example, average continental crust (Figure 11.35). Li et al. (2017) argue that the unusual HREE enrichment reflects a combination of extreme fractional crystallization involving LREE-rich minerals such as monazite as well as further subsolidus hydrothermal *autometasomatism* by fluids exsolved from the cooling intrusion, for which there is a variety of evidence. Regardless of the cause, the HREE enrichment of the laterite clearly reflects the HREE enrichment of the parent.

The area is located within the humid subtropics with over 1500 mm of annual rainfall and consists of low hills with <100 m relief. Consequently, weathering rates are high and erosion rates low allowing development weathering profiles typically ranging from 15–35 m but locally as deep as 60 m (Bao and Zhao, 2008). The humic-rich A horizon is commonly 1–2 m thick and consists mainly of quartz and clays (see Section 13.3 for an explanation of soil horizons). The B horizon above ~6 or 7 m is extensively weathered horizon and consists of <15% K-feldspar

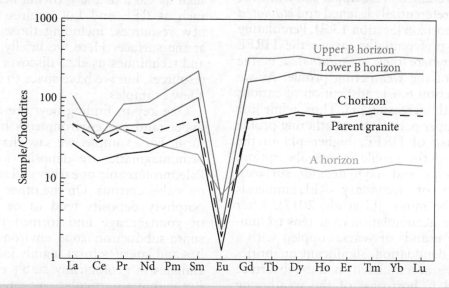

Figure 15.19 Rare earth patterns in the soil profile of the Zudong laterite REE deposit in Jiangxi Province, China. The pattern of the partially weathered C horizon directly above the bedrock is similar to that of the parent material. The A horizon (shallowest) is depleted in most REE. Mining is focused on the REE-rich B horizons.

and around 40% clay minerals and Fe oxyhydroxides; the lower 10–20 m consists of 15–25% K-feldspar and about 30% clays, which include montmorillonite, kaolinite, and gibbsite. The C horizon, which is 3–5 m thick, is saprolitic, retaining the texture of the parent, and contains <20% clay minerals. There is a stepwise transformation of muscovite to smectite, kaolinite, and eventually to gibbsite, with increasing extent of weathering. As is typical of the humid subtropics, the area is heavily vegetated so that soil pH is 4.4–5.2 in the near-surface due to the abundance of organic acids but increases with depth to ~6.3 in the C horizon. Synchysite and gadolinite in the parent rock breakdown fairly readily during weathering, while the remaining minerals are more resistant to weathering and can occasionally be found in the weathering profile, although their contribution to the REE inventory of the laterite is small.

Rare earth patterns of this weathering profile and the parent granite are shown in Figure 15.19. The A horizon is slightly enriched in the La, Ce, and Eu and strongly depleted in Nd, Sm and HREE relative to the parent rock. This reflects a progressively decreasing affinity with atomic number of the REE for clays and oxyhydroxides with atomic number and an increasing tendency to form soluble complexes. The result is that the lightest REE are retained in the upper soil while the HREE are preferentially leached and *eluviated* from the A horizon (Section 13.3). Percolating water thus preferentially carries the HREE downward where they are *illuviated* in the lower part of the weathering profile. As we found in Section 6.6.1, adsorption of cations increases with increasing pH. Thus, while low pH in the upper part of the weathering profile favors release of HREE, higher pH in the lower part of the profile favors absorption, mainly on clay and oxyhydroxide surfaces. Precipitation of secondary REE minerals appears to be minor (Li et al., 2017). After leaching and accumulation over tens to hundreds of thousands of years, coupled with a low rate of denudation, significant quantities of REE have accumulated in the subsurface part (B and C horizons) of the weathering profiles. Because the REE are mainly absorbed on surfaces, they are readily recovered by ion exchange with ammonium sulfate, making this type of deposit economic at low grade.

Figure 15.19 also shows that the REE pattern in the A horizon displays a slight positive Ce anomaly in contrast to the negative anomalies of the parent and deeper horizons. As we found in previous chapters, Ce is in the Ce^{4+} state in oxidizing environments and is much more insoluble in this form and is preferentially retained in the A horizon rather than transported into deeper horizons.

15.2.8 Geochemical exploration: finding future resources

Some ore deposits, such as bauxites and HREE-rich laterites discussed in the previous section, form at the surface and are relatively easily found, but most ores form at depth. Tectonics and subsequent erosion can eventually bring these to the surface. Most of those ore deposits exposed at the surface have long since been discovered; indeed, the Rio Tinto VMS deposit has been mined since prehistory. Many have been largely mined out, such as the rich gold deposits in the western Sierra Nevada that prompted the 1849 California Gold Rush. Furthermore, political instability, which plagues areas like Central Africa and trade disputes, such as those that have threatened global REE supply, mean that we may not always be able to rely on existing supplies of resources. That, together with the continuing need for traditional metals such as Cu and the growing need for others such as REE and Li, requires that we find new resources, including those not exposed at the surface. Here we briefly review tools and techniques used to discover new mineral resources, but we have space to consider only a few examples.

One key to finding new ore deposits is to understand the environments in which they form. For example, we saw that most major orthomagmatic ore deposits formed in the Paleoproterozoic or earlier and are now found on stable cratons. On the other hand, copper porphyry deposits tend to be of Mesozoic or younger age and formed principally in supra-subduction zone environments. Bauxites and laterites form mainly in warm humid climates in geologically stable environments. But within a particular environment, how do we find mineral deposits that are not exposed at the surface?

Geologic mapping is perhaps the first step in mineral exploration. Certainly, in most regions this has already been done in at least low levels of detail, so, for example, we can identify regions where subduction zone volcanism has occurred in the past. The exploration process often begins with more detailed geologic mapping as well as with geophysical methods such as aerial gravity and magnetic surveys. Once an area of interest has been identified, geochemical survey methods come into play, and it is those that we will focus on. Before we get to that, let's consider how elements disperse from ore deposits in the surface environment.

15.2.8.1 Primary dispersion

Formation of ore deposits often involves processes that affect rock well beyond the ore itself, particularly those involving fluid flow. These wider effects of the ore-forming process are known as *primary dispersion* (Kelly et al., 2006; Kyser et al., 2013; Cohen, 2014) (Figure 15.20). Let's consider the example of VMS deposits. In addition to leaching and deposition of metals, fluid flow in VMS systems results in metamorphism of the original igneous rocks. The late Archean

Noranda VMS deposit of Quebec, Canada, is associated with greenschist assemblages of albite, iron-rich chlorite, and course-grained Fe-poor epidote (clinozoisite), and amphibole (ferroactinolite), and this assemblage persists for 10 km or more from the ore deposit (Harrington et al., 2003). This assemblage differs from those produced by the subsequent regional greenschist metamorphism and in local hydrothermal "haloes" associated with barren, lower-temperature hydrothermal systems characterized by Fe-rich epidote, Fe-poor chlorite, and rare actinolite. Water–rock interaction also affects oxygen isotope ratios, and this extends well beyond the area of mineralization (Figure 9.34 shows one example of an oxygen isotope halo). As we found in Chapter 9, $\delta^{18}O$ in water–rock reactions depends on temperature and the water–rock ratio. Low-temperature water–rock interaction, in which the fractionation factor is high, decreases the $\delta^{18}O$ of the water and increases that of the rock, while high-temperature interaction does the opposite. Thus, we expect to find low $\delta^{18}O$ in areas of highest temperature and most intense hydrothermal flow; these are just the areas where we expect ore deposition

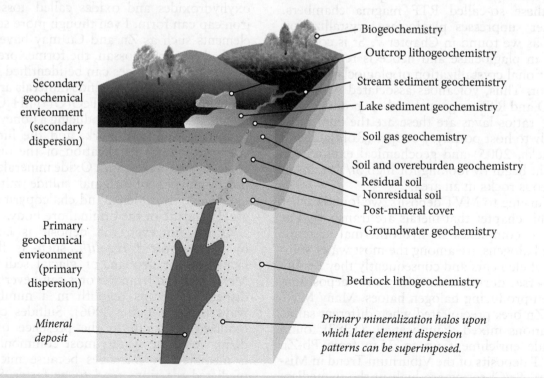

Secondary geochemical envieonment (secondary dispersion)

Primary geochemical envieonment (primary dispersion)

Mineral deposit

Biogeochemistry

Outcrop lithogeochemistry

Stream sediment geochemistry

Lake sediment geochemistry

Soil gas geochemistry

Soil and overeburden geochemistry

Residual soil

Nonresidual soil

Post-mineral cover

Groundwater geochemistry

Bedriock lithogeochemistry

Primary mineralization halos upon which later element dispersion patterns can be superimposed.

Figure 15.20 Primary and secondary dispersion around ore deposits and targets for geochemical exploration. From Kyser et al. (2013).

in VMS systems. In the Noranda district, Cathles (1993) documented a large-scale pattern of low $\delta^{18}O$ that form a 2–3 wide annulus around the Flavrian pluton where $\delta^{18}O$ decreased to as low as 0.6‰. Six fingers of low $\delta^{18}O$ radiated from the annulus and five of the six points toward the most massive sulfide ore deposits. The annulus is, in turn, surrounded by a ~10 km wide band of elevated O isotope ratios where $\delta^{18}O$ is as high as 14‰. Cathles interpreted these as recharge zones: regions where cold seawater was infiltrating downward and interaction with rock at low temperatures or ^{18}O enrichment in an arrested migrating thermal front.

Now let's consider porphyry copper deposits. These are generally products of subduction zone volcanism, but not all subduction zone volcanoes produce porphyry coppers. As we found earlier, porphyry copper deposits form through exsolution of an aqueous fluid from the magma; consequently, they are more likely to form from water-rich magmas. Furthermore, they are more likely to develop in long-lived systems that undergo repeated cycles of replenishment, tapping, and crystallization. As we found in Chapter 7, this leads to high levels of incompatible elements in these so-called RTF magma chambers. Water suppresses plagioclase crystallization and as we found in Chapter 7, Sr is compatible in plagioclase and hence is removed by fractional crystallization of plagioclase, but Y is not. Thus, volcanoes associated with high H_2O and RTF magma chambers produce high Sr/Y ratios lavas are these are the ones most likely to host porphyry copper (Rohrlach and Loucks, 2005) and geochemical exploration might begin by looking for high Sr/Y ratios in igneous rocks in an area.

Turning to MVT deposits, we found earlier in this chapter that metals are transported as halide complexes in hydrothermal systems. The halogens are among the most water soluble of elements and consequently they readily disperse during and after ore deposition, often producing halogen haloes. Many MVT Pb-Zn ores precipitated when different saline solutions mixed. Panno et al. (1983) found halide enrichment up to 75 m from Pb-Zn MVT deposits of the Viburnum Trend in Missouri. Br/Cl ratios are high in the immediate vicinity of the ore and decreased asymptotically away from the ore, reflecting the mixing between the metal-bearing, Br-rich evaporite bittern-derived solution with a halite-dissolution derived one.

15.2.8.2 Secondary dispersion

Elements concentrated in ore deposits represent a low entropy situation and as we found in Chapter 2 entropy tends to increase over time. As a practical matter, this means that elements tend to disperse into the surrounding environment through both physical and chemical processes. This is known as *secondary dispersion*. One example of the former is erosion, which, among other things, leads to placer deposits. Gold placers in the Sacramento River that initiated the California Gold Rush eventually led miners to the primary deposits ('the Mother Lode") in the Sierra Nevada. We will focus here, however, on chemical dispersion.

Many ore deposits contain elements in reduced form and are out of equilibrium with the oxidizing environment at the surface. Although it is rarely the target metal, Fe is the most common one in sulfide deposits. As the oxidized weathering zone cuts down into the sulfide body, a thick layer of ferric oxyhydroxides and oxides called gossan or iron cap can form. Even though more soluble elements such as Zn and Cu may have been leached from the gossan, the former presence of Zn and Cu sulfides can be identified in the gossan texture. When sulfide minerals are oxidized, residual limonite ($FeO(OH)\cdot nH_2O$) can remain in the cavities, producing a honeycomb pattern, called box work, and the limonite color provides an indication of the original chalcopyrite/pyrite ratio. Oxide minerals form pseudomorphs of original sulfide minerals, like sphalerite, galena, and chalcopyrite, that are diagnostic of the original ore body.

Often, however, the deposit is covered by a *transported regolith*, such as fluvial, lacustrine, or aeolian, or glacial sediments. Metals derived from the ores can nevertheless disperse into this regolith in a number of ways (Kelley et al., 2006). Sulfides can be oxidized at depth in the presence of oxidizing meteoric water; most commonly this is mediated by microbes because microbial mediated reactions tend to be faster. Sulfide oxidation also results in decreasing pH, which increases metal solubility. Metals can then be complexed by inorganic or organic ligands that are metabolic byproducts of this process. Below the water table, these ions can diffuse or be advected upward. However, in the unsaturated vadose zone above it, transport is more difficult as diffusion will be very slow. Some metals, e.g., Hg, Pb, can form volatile complexes that can then migrate to the surface in that form. Others can be transported in rising bubbles of microbial-produced gases such as CO_2 and CH_4. Other gases such as COS and CS_2 in soil gas can be indicators of buried sulfide deposits. Earthquakes can occasionally force water to the surface and repeat episodes over long periods can result in build-up of metals at the surface. In the surface zone, metals will absorb onto clay and oxyhydroxide surfaces. Many plants in arid zones have deep root systems that can extend for 10 meters. Uptake of water by these roots can then effectively transport metals to the surface.

Another mechanism is electrochemical dispersion. Reducing (electron-rich) conditions at depth and oxidizing (electron-poor) ones at the surface can result in the establishment of an electrochemical potential between the two. If a conductor such as a steeply dipping sulfide deposit is present, electrons can migrate along it to the oxidized surface environment, which effectively forms a cathode. Cations dissolved in groundwater, such as positively charged metal ions, will migrate toward the cathode at the top of the buried conductor and anions away from it. Even in the absence of a conductor, voltages as low as a few hundred millivolts can result in a voltaic cell with the charges carried by anions rather than by electrons (Hamilton, 1998). This self-potential can be detected directly, but also gives rise to electrochemical diffusion. Reduced species, such as HS^- and Fe^{2+}, migrate upward along this gradient and reactions between oxidized species moving in the opposite direction dissipate charge away from the sulfide. However, as oxidizing agents above the water table are consumed, a "reduced column" forms between the body and the water table. Near the surface, this induces the migration of reduced species upward and outward and oxidized species, including H^+, inward and downward (Figure 15.21). This can result in both metal enrichments and depletions that can occur either in a single geochemical anomaly over the deposit or a

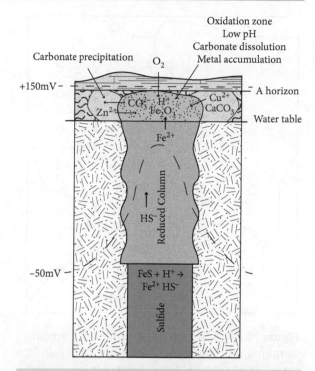

Figure 15.21 Electrochemical dispersion above a buried sulfide ore deposit. The contrast between the reducing environment of the sulfide and the oxidizing environment of the surface produces a small electrical self-potential that enhances diffusion of reduced species to the surface.

double "rabbit-ear" anomaly surrounding it. Low pH results in carbonate dissolution over the column and precipitation around it. This electrochemical diffusion mechanism explains the development of geochemical anomalies that have developed in glacial till deposited only in the last ~10,000 years in regions such as Canada, where simple chemical diffusion is too slow to explain them (Cameron et al., 2004).

15.2.8.3 Exploration process

From regolith and soils, metals can make their way into streams, lakes, and the biosphere (Figure 15.20). Geochemical surveys often begin with sampling stream and pond water and sediment within a drainage basin. This can point to the area of a watershed where a mineral deposit is most likely to be located. This phase can be followed up by soil samples taken on a grid (Figure 15.22) in the focus area. In some cases, this will involve a series

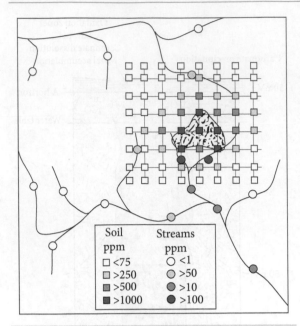

Soil
ppm
☐ <75
▢ >250
▨ >500
■ >1000

Streams
ppm
○ <1
◔ >50
◑ >10
● >100

Figure 15.22 Cartoon illustrating how stream sampling followed by soil sampling on a grid can close in on a hidden ore deposit.

of finer grids. These surveys often also involve sampling the biota. One example of this is that trees record the chemical history of the region in their wood. If the region has been subjected to recent pollution that might misleadingly suggest an ore deposit, older growth rings provide sample material that predates industrial contamination.

Some analyses can be easily done in the field by measuring pH, alkalinity, dissolved oxygen, etc. in water and soil samples. Chemical and mineralogical compositions of sediment and rock samples can also be done in the field using portable X-ray fluorescence (XRF) and portable X-ray diffraction (XRD) units. Other portable instruments now used include portable short-wave infrared (SWIR), Laser Induced Breakdown Spectroscopy (LIBS), and Raman mineral analyzers (Kyser, 2015). Follow-up laboratory studies include petrographic analysis and more complete chemical analysis, including inductively coupled plasma-mass spectroscopy (ICP-MS), trace element analysis and stable isotope ratio analysis. Ore-derived metals in the soil often are mainly absorbed on surfaces, so soil chemical analysis often involves analysis of just the finest particle fractions or leachates. The final step in locating the ore deposit involves

drilling and taking a series of core samples to delineate the area of mineralization and determine whether mining is economically justified. Kyser et al. (2015), Cameron et al. (2004), and Eppinger et al. (2013) provide more examples of the exploration process.

15.3 ENVIRONMENTAL GEOCHEMISTRY

Humans have long had an impact on the environment, but this impact has grown exponentially with population and technology to the point where it threatens the sustainability of civilizations. This threat comes in various forms:

- Land-use changes include conversion of forested areas to agricultural ones, pavement, and buildings in areas of high population density that restrict precipitation infiltration and therefore increase runoff and flooding probability, infilling of wetlands that trap sediment, filter runoff, and are key habitats for many species, including commercially valuable ones.
- Pathogen spread is associated with animal husbandry, butchering operations, and human sewage.
- Toxic organic compounds—some are natural, such as petroleum, and some are man-made, such as pesticides and herbicides—are introduced into the surface environment.
- Increasing concentrations of greenhouse gases are changing global climate.
- The redox state of natural waters changes through, for example, agricultural runoff.
- Toxic metals are introduced into soil, water, and the atmosphere.
- The pH of soil and natural waters changes.

Humanity has been slow to realize these long-developing problems. It was not until the mid-twentieth century when alarm bells began to sound, perhaps most loudly with Rachel Carson's publication of *Silent Spring* in 1962. Geochemistry concerns itself, after all, with the behavior of chemicals in the natural environment, so geochemists have multiple roles to play in understanding and mitigating environmental degradation. Environmental geochemistry is an enormous topic and entire textbooks can and have been written on it. Here we will be able to only briefly touch on a small part of this. Because our space is limited, we will focus only on a few examples where geochemistry can be applied to understand the problems and devise solutions. This certainly includes anthropogenic climate change, which may well be our biggest environmental threat, but we considered that in Chapter 12. Consequently, we will focus on the last three topics.

15.3.1   Eutrophication redux

We discussed the basic limnology and chemistry leading to suboxic or anoxic conditions in deep portions of density-stratified lakes in Section 13.2.2 and noted this can be a natural phenomenon. Here we will briefly consider how this occurs as a result of anthropogenic nutrients driving high levels of productivity, often by cyanobacteria, in the surface waters or *epilimnion* and decay of their remains with consequent oxygen depletion in the deep water or *hypolimnion* of thermally stratified lakes. In temperate zone lakes, stratification and anoxic conditions are seasonal, but can have nonetheless catastrophic consequences for the lake and stream fauna and human health. This is a global problem, nor is it restricted to lakes; other examples including the Caspian and Baltic Seas and coastal areas around the world, including "dead zone" at depth in the Gulf of Mexico south of the Mississippi delta. We will consider one well-studied example: Lake Erie.

Lake Erie is one of the Great Lakes, which form much of the boundary between the United States and Canada. It is the second smallest in area and smallest in volume and depth, with an average depth of 19 m and maximum of 64 m. Water entering through the Detroit River from Lake Huron and the upper Great Lakes represents 95% of inflows; water flows out through the Niagara River through Lake Ontario to the St. Lawrence River. The remainder of inflows come from a 78,000 km^2 watershed that includes industrial centers, the homes of 12 million people, and a large track of intense agriculture, consisting mainly of rotations of maize, soybeans, and wheat.

Dramatic population and industrial growth in the region in the first half of the twentieth century led to such dramatic inflows of sewage and industrial wastes that by the 1960s Lake Erie was called the "Dead Sea of North America." Fires erupted on the Cuyahoga River, which flows through Cleveland, Ohio, more than a dozen times between 1868 and 1969 because of solvents and other flammables on the surface. Concentrations of soluble phosphate in Lake Erie increased from 7.5 µg/L in 1948 to 36 µg/L in 1962. As a result, late summer/fall cyanobacteria blooms and eutrophication were endemic and led to dramatic declines of fish stocks and the consequent collapse of commercial and sport fishing industries. The 1969 river fire proved to be a catalysis for change. Public outrage led to the Great Lakes Water Quality Agreement between Canada and the United States in 1972, with strict limits on releases from point sources such as sewage treatment and industrial effluents, including bans on phosphorus in detergents. Efforts were also made to change agricultural practices to reduce phosphate inputs, including reduced fertilizer application, reduced tillage, grassed waterways, and filter strips. The result was swift and dramatic. The total phosphorus flux into dropped from 28 kt/year in 1968 to the target level of 11 kt/yr in the 1980s, most of it a result of decline from point sources (Scavia et al., 2014). Eutrophication eventually ceased, fish stocks recovered, and public beaches reopened for swimming.

Since the mid-1990s, however, cyanobacteria blooms have increased in frequency, mainly in the western part of the lake (Figure 15.23), and extensive seasonal hypoxia events have returned to Lake Erie's central basin. In addition to causing hypoxia, cyanobacteria release microcystin toxins, which cause liver damage in humans. In August 2014, Toledo, Ohio, at the southwest end of the lake, issued a *Do Not Consume* order for the municipal water supply over several days when hazardous levels of microcystins were detected in the water. This has occurred even though point sources of phosphorus to Lake Erie have remained at or below targets and, although it has varied with annual runoff, total phosphorus has not shown a systematic increase.

Dissolved phosphate concentrations in soil are usually quite low as a consequence of uptake by the biota and adsorption on ferri-oxyhydroxides. Consequently, the 1972 agreement focused on total phosphorus, particularly point sources and particulate phosphate from soil. Renewed eutrophication

Figure 15.23 Algae bloom in western Lake Erie photographed by Moderate Resolution Imaging Spectroradiometer (MODIS) on the NASA's Aqua satellite on October 9, 2011. Such blooms have once again become common in the shallow western part of Lake Erie in the late summer and fall over the last two decades. NASA Image.

appears to be related to dissolved reactive phosphorus (effectively PO_4), which has increased as a fraction of total phosphate (Scavia et al., 2014). Ironically, some of this may be due to adoption of no-tillage and conservation tillage practices meant to reduce the particulate P flux. No-till practices can result in P accumulation in the soil surface layer because crop residue and applied artificial and organic (manure and poultry litter) fertilizers are not mixed into the soil column, where P can be absorbed on particular surfaces and stored. Instead, this highly soluble surface P is readily carried away in storm events (Dolağu et al., 2012). Another issue has been an increasing tendency to fertilize in the fall rather than spring; the runoff potential of dissolved P is higher in fall because the ground is bare and plant roots are not available for nutrient uptake. A switch to organic fertilizers may not be the solution as dissolved P runoff from fields fertilized with manure were three times those fertilized with inorganic fertilizers, although this may be because farmers were targeting nitrogen rather than phosphorus levels (King et al., 2018).

An additional factor in the Lake Erie story is climate change. Both the amount of precipitation and frequency of extreme rainfall events, particularly in late winter and early spring, have increased in the Lake Erie watershed and 60–75% of P inputs are delivered during precipitation-driven high river discharge events (Dolağu et al., 2012; Scavia et al., 2014). This is an issue not just for Lake Erie, but globally and for nitrogen as well, particularly since precipitation and extreme events will increase as climate warms (Sinha et al., 2017). To mitigate the problems of Lake Erie, there is a consensus among researchers on a number of new agricultural practice guidelines, summarized by King et al. (2018):

- Organic fertilizers should be applied to target phosphorus needs rather than nitrogen; residual P leftover from previous applications should be taken into account.
- Soil should be tested at least once per crop rotation and the fertilizer applied limited to necessary amounts based on the test results.
- Fertilizer application prior to large precipitation events should be avoided and fertilizer should be applied as close as possible to planting in the spring or during the

Figure 15.24   The Río Tinto in southern Spain is discolored by the dissolved iron produced by sulfide oxidation as a result of millennia of mining operations. Photo by Carol Stoker, NASA Ames Research Center.

period between June and October (i.e., dry period of the year).

- Fertilizer should be placed in the subsurface at the time of application using injection, banding, or tillage; broadcast fertilizer applications to the soil surface should be avoided.

15.3.2   Toxic metals in the environment

Metallurgy was a key advancement that further spurred the development of civilization. Nevertheless, this had some deleterious effects on the environment and in many cases on human health. This has long been true; the Greek physician Nikander described the symptoms of lead poisoning in lead workers in the second century B.C. (Needleman, 2004). Pb toxicity was restricted and local in classical times, but anthropogenic Pb was already beginning to spread around the world, as evidenced by increased Pb concentrations in layers of Greenland ice cores dating from that period (Hong et al., 2004). Environmental degradation has been greatly amplified

by the burgeoning human population and the enormous power of our technology; after a decline in Medieval times, Pb in late twentieth-century Greenland ice reached levels 25 times higher than in Roman times. Other volatile metals, such as mercury, have also dispersed globally. Many metals, as well as other toxins, bioaccumulate, which can result in dangerously high levels for animals at high trophic levels, including, of course, humans.

As a general rule, "B" type or soft metals are more toxic than "A" type or hard sphere ones (Chapter 6). Table 15.1 lists 13 trace metals and metalloids that are priority pollutants (Adriano, 2001) along with their US Environmental Protection Agency drinking water maximum contaminant levels. The latter provides only a rough guide to toxicity as metals can enter the body in various ways, and their toxicity is affected by that. Ironically, some of these elements are essential to some form of life, including humans in the case of Cr, Cu, Ni, Se, and Sn. Some, such as Cr, are mainly toxic in specific forms: Cr is a

Table 15.1 US EPA maximum contaminant level in drinking water.

	EPA maximum contaminant level mg/L
Ag	0.1
As	0.01
Be	0.004
Cd	0.005
Cr	0.1
Cu	1.3
Hg	0.002
Ni	–
Pb	0.015
Sb	0.006
Se	0.05
Tl	0.002
Zn	5.0

Source: US EPA 2018 Edition of the Drinking Water Standards and Health Advisories Tables.

carcinogen and highly toxic in the Cr^{6+} form, in which it rarely occurs in nature but not in other valence states; Hg is more toxic in the form of methylmercury than in inorganic forms. Interestingly, part of Hg toxicity relates to its interference with essential enzymes that incorporate another element on this list: selenium. Finally, the list is not exclusive; other metals can be toxic at sufficiently high exposures. An interesting example is Al: over the pH range of most natural waters, Al is too insoluble to be present at toxic levels; but in low pH waters, such as those affected by mine drainage or acid rain, Al can reach levels that are toxic to plants and fish. Toxic metals enter soil, water, food, and air in a variety of ways; we have space only to consider a few.

15.3.2.1 Mine wastes

We devoted the first half of this chapter to ore deposits, so let's now consider the environmental consequences of exploiting them. Typically, 95–99% of rock removed in mining operations is waste and in the case of low-grade gold ores that fraction can be even higher. The amount of mine waste generated annually vastly exceeds commercial and household wastes and is comparable to the amount of solids moved by natural geologic processes (Fyfe, 1981). This, in itself, can blight the landscape; further problems arise

from chemical reactions that occur when ore is exposed at the surface and from the toxicity of the metals involved. The problem is particularly acute when the ores consists of sulfide minerals, which is often the case, as we found earlier in this chapter. When exposed to oxygenated water, sulfides quickly undergo oxic weathering, for example, pyrite will react to form ferri-oxyhydroxide and sulfuric acid. The overall reaction may be written as:

$$FeS_2 + \frac{15}{4}O_2 + \frac{7}{2}H_2O \rightleftharpoons Fe(OH)_3 + 2H_2SO_4 \tag{15.4}$$

As we noted earlier, the reaction is usually microbially mediated and occurs in several steps. Sulfur oxidizes before iron, so the initial result is an acidic solution rich in ferrous iron:

$$2FeS_2 + 7O_2 + 2H_2O$$
$$\rightleftharpoons 2Fe^{2+} + 4SO_4^{2-} + 4H^+ \tag{15.5}$$

Oxidation of iron consumes some of this acidity:

$$4Fe^{2+} + 2O_2 + 4H^+ \rightleftharpoons 4Fe^{3+} + 2H_2O \tag{15.6}$$

but eventual precipitation of ferric iron will produce additional acidity:

$$Fe^{3+} + 3H_2O \rightleftharpoons Fe(OH)_3 + 3H^+ \tag{15.7}$$

Acid produced in this way is neutralized in some cases by reaction with carbonates and silicates in surrounding rock. Acid mine drainage forms when the acidifying potential of reactions such as (15.4) are greater than the neutralizing potential of reactions with carbonates and silicates. In general, metals in Table 15.1 are more soluble at lower pH; even in circum-neutral waters, streams can contain elevated levels of more soluble melts and metalloids such as Cu, Ni, Zn, Se, and As.

The consequences of this is what is known as *acid mine drainage* and is illustrated by the Río Tinto (*dyed river*) in Andalusia in southern Spain, which drains the VMS deposits of the same name (a name shared as well by the giant international mining conglomerate Rio Tinto). Some 5000 years of mining of copper, silver, gold, and other metals as well as natural weathering have given the river a reddish hue (Figure 15.24) due to the presence of dissolved iron. The banks are lined with effervescent* sulfate minerals. These minerals precipitate during periods of low flow, storing acidity, and dissolve during periods of high flow. Toxic levels of Cu, Zn, As, and Pb are found in sediments along the entire 95 km length of the river below the mines (Davis, et al., 2000). This chemistry makes the river nearly barren of life except for extremophile microbes and a few rare and unusual eukaryotes. Even within the river's estuary near the town of Palos de la Frontera (from whence Columbus set sail beginning in 1492 on his voyages to America), pH is typically 2.0–2.5 during low tides. Sediment cores from near the mouth of the river reveal Cu concentrations of nearly 3500 ppm in a horizon with a ^{14}C age of 3640 BP, demonstrating the long history of pollution (Davis et al., 2000). Olías et al. (2006) estimated the Zn and Cu fluxes from the Rio Tinto and Rio Odiel, which empties into the same estuary and also drains the mining district, to be 3.5 kt/yr and 1.7 kt/yr, respectively. Based on the data of Gaillardet et al. (2014) this zinc flux would represent 15% of the total riverine flux to the oceans while the rivers constitute only 0.0058% of the riverine water flux. As Table 15.1 shows, the toxicities of Cu and Zn are low, but the rivers also transport 36 t/yr of As, 11 t/yr of Cd, and 27 t/yr of Pb, all of which are highly toxic. However, measurement of trace element concentrations in the Gulf of Cadiz suggest most of this is likely trapped in the estuary (see Chapter 14) and or removed in the near-shore environment (González-Ortegón et al., 2019).

To minimize acid mine drainage, mine tailings, which are the waste rock and fluids generated in producing the ore concentrate, are often stored in containment ponds, but since 1970 over 70 major tailings containment failures have occurred. For example, a tailing containment failure in the Philippines in 1996 released 2–3 million cubic meters of toxic material into the Makulapnit-Boac River in less than a week; a similar failure in Romania killed 1200 tons of fish in the Danube tributaries, resulting in loss of income for thousands of fishermen as well

as contaminating the water supply for towns and farms (Nordstrom, 2011).

Acid mine drainage often worsens when a mine is closed and abandoned because pumps used to keep the water out of open-pit and/or underground mines are turned off, allowing rise of groundwater, which can then react with exposed sulfides in wall rocks. Some 33,000 abandoned mines in the 12 western US states and Alaska have contaminated surface and groundwater and the US government has spent billions of dollars on clean-up, but there is far more to be done. The estimated costs for total worldwide liability associated with the current and future remediation of acid drainage are approximately US$100 billion. Acid mine drainage is not restricted to mines targeting metals; coal mines can be a problem as well. Because coal forms from sediments deposited in highly reducing conditions, pyrite is often present in coal.

Laws and regulations have been in place since the 1970s to reduce or eliminate acid mine drainage and the mine permitting process generally requires a prevention plan. Park et al. (2019) summarized some of these strategies:

- *Oxygen barrier*. Restricting access of oxygen to mine wastes can prevent reaction 4 and the generation of acidity. In some cases, tailings are placed under water, in other cases covered by other materials (Figure 15.25). These can include oxygen-consuming materials such as wood waste, straw mulch, or other organic residues or acid-neutralizing materials such as limestone. In one version of this strategy, "covers with capillary barrier effects" (CCBE), a moisture-saturated layer limits the diffusion of oxygen through the cover. This is done by placing a layer of fine-grained material over a coarser one. The fine-grained material has smaller interstitial pore space, which results in water retention by surface tension (the capillary effect), while the coarser layers drain (Bussière et al., 2004).
- *Blending*. Mixing of mine tailings with acid-consuming material such as limestone can control the oxidation of pyrite

---

* A term derived from the Latin *efflorescere* referring to their tendency to grow in cauliflower-like textures.

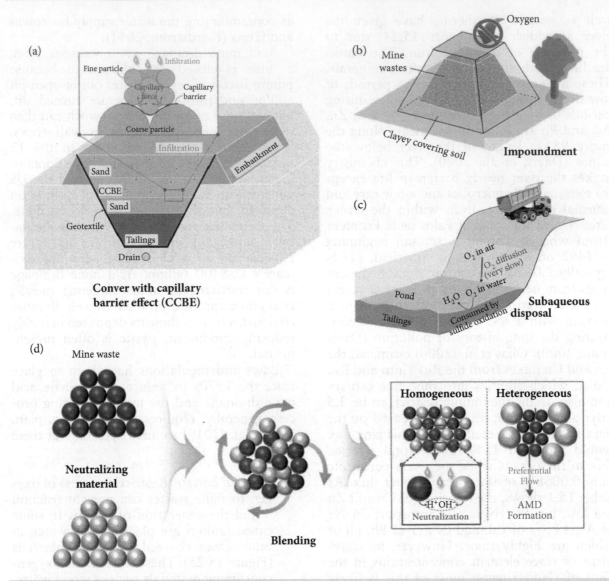

Figure 15.25 Illustration of some containment strategies to prevent acid mine drainage. From Park (2019).

in four ways: precipitation of Fe^{3+} that limits the supply of oxidants, increase in pH that impairs the activity of microbes that catalyze pyrite oxidation, formation of coatings on sulfide surfaces that reduces reactive surface area, and formation of a low permeability layer (hardpan) consisting of ferri-oxyhydroxide and gypsum that limits the diffusion of O and infiltration of water.

- *Bactericides.* Eliminating acidophilic iron- and sulfur-oxidizing bacteria (e.g., *Thiobacillus ferrooxidans* and *Thiobacillus thiooxidans*) that accelerate pyrite

oxidation with bactericides such as sodium lauryl sulfate and sodium dodecylbenzene sulfonate. However, these compounds are water-soluble are easily rinsed away by rain, so the treatment must be continually repeated.

- *Coatings and encapsulation.* The approach here is to produce a surface layer on sulfides that make the mineral surface hydrophobic (minimize water–mineral interaction) and/or reduces the exposed sulfide surface area. For example, an organic molecule such as sodium oleate (CH_3 $(CH_2)_7CH=CH(CH_2)_7COONa$), which is commonly used in flotation separation, forms an inert hydrophobic film on pyrite. The long-term stability of such coatings is unclear, however. Another approach is to induce the formation of ferri-oxyhydroxides, phosphates, silicates, alkoxysilanes, and aluminum-oxyhydroxides on sulfide surfaces by treatment with an oxidizing agent together with a base such as $NaHCO_3$.

• *Recycling.* Promising experiments have been performed incorporating mine wastes into construction materials such as cement and mortar and as a replacement for natural aggregates in road asphalt. Another approach being tested is the creation of "geopolymer bricks" with an amorphous polymeric structure composed of repeating units of –Si–O–Al–O–Si– functional groups. Metals such as Cu, Fe, and Zn can be immobilized within these structures and the bricks then used as a replacement for conventional bricks.

Rare earth mining presents an entirely different hazard: radioactivity – not from the naturally radioactive rare earths, La, Sm, and Lu, but from the relatively high concentrations of U and particularly Th found in REE minerals. Although Th, with a half-life of 14 billion years, is only weakly radioactive, in REE deposits it is in radioactive equilibrium (Chapter 8) with its decay chain, which consists of six alpha and two beta decays. Tailings can consequently be sufficiently radioactive so as to present a radiation hazard. In addition, processing the ore requires large amounts of acid that must be neutralized with alkali, creating additional hazards. Several problems emerged in the Mountain Pass operation in California, including percolation into groundwater from containment ponds and wastewater pipeline failures (EPA, 2012). The remediation and clean-up costs of these problems, together with competition from China, are what caused the mine to shut down in 2002. Similar problems have occurred on an even larger scale at the Bayan Obo mine

in China. In addition, Th-bearing airborne particulates have exposed workers as well as residents in the area to sufficient levels of radioactivity to pose a long-term risk (Wang, et al., 2016).

15.3.2.2 Atmospheric lead

There is perhaps no substance with which humans have so pervasively polluted the planet as lead. Humans began mining and smelting lead in prehistory. Although rarer in the crust than copper, it occurs in a wide variety of ore deposits and is easily smelted and worked. Ancient societies used it for ornamental objects, as a sweetener for wine, in cosmetics, in glass, in solder, in glazes on pottery, sinkers for fishing nets, etc. Many of these uses continue into the modern era. Large-scale use and consequent large-scale lead production pollution began with the Romans who used lead pipe in plumbing (indeed, the English word for plumbing derives from the Latin name for lead: *plumbum*). Although the Romans such as Vitruvius, the Roman engineer who designed Rome's water system, were aware of Pb toxicity at high levels, only in the twentieth century did recognition of its toxicity at trace levels develop. The development of that awareness and the subsequent vast reduction in lead in the environment is one of the great success stories of the scientific environmental regulation over the last 50 years; it also illustrates the role geochemists can play in environmental protection.

Lead production declined after the fall of the Roman Empire, but picked up again in the eleventh and twelfth centuries. The introduction of tetraethyl lead as an *anti-knock* (i.e., ensuring smooth ignition) additive gasoline in the mid-1920s vastly increased the amount of Pb pollution by direct emission of particulate Pb to the atmosphere; it was widely dispersed and virtually unavoidable. Claire Patterson was an isotope geochemist at Cal Tech best known for having established the age of the Earth as 4.55 Ga* (Patterson, 1956). In trying to measure Pb isotopes in meteorites and

---

* Claire Patterson (1922–1995) was awarded the Goldschmidt Prize by the Geochemical Society in 1980. In 1998, the Geochemical Society established the Patterson Award "recognizing an innovative breakthrough of fundamental significance in environmental geochemistry". Patterson put the age of the Earth at 4.55 ± 0.07 Ma. Although we now know the age of the solar system with exquisite precision, we don't know the age of the Earth much better than Patterson's value.

terrestrial materials, Patterson noticed that Pb contamination was everywhere and he had to go to extreme lengths to eliminate it (one immediate outcome of his finding was that by the late 1960s to early 1970s, isotope geochemists followed Patterson by doing their chemistry in clean laboratories with ultra-filtered air and ultra-clean reagents).

Patterson and his colleagues documented the dramatic increase in Pb concentration in Greenland snow over time, particularly after 1940 (Figure 15.26). While the pervasiveness of Pb pollution was clear to policy makers, the question remained as to whether it was harmful, and its pervasiveness complicated the question because almost everyone in the developed world was exposed to it. Patterson (1965) argued that Pb was a health issue.

Pb toxicity at high levels was well known, but its effects at lower levels were not. Patterson (1965) took a geochemical approach to argue that what had become "normal" levels of Pb were not "natural" and were very likely toxic. He pointed out that the ratio of other toxic heavy metals to calcium (e.g., Hg/Ca) in tissue were far lower than crustal ratios and argued that this demonstrated that life strongly discriminated against these toxins. By analogy to these toxic metals, Patterson argued that the "natural" Pb concentration in humans should be about 0.002 ppm in blood, whereas various surveys found an average of about 0.25 ppm in blood – two orders

of magnitude higher. At the time, 0.5 ppm in blood was considered the lower limit for toxicity. Patterson argued that a factor of only two between the 0.5 ppm considered toxic and average at 0.25 ppm made little sense and that Pb likely had deleterious effects at concentrations well below those definitely associated with Pb poisoning. He pointed out fallout from particulate Pb worked its way into water and soil and into the food supply. He also pointed out that Pb in plumbing, "in paints, glazes, enamels, solders, brasses, plastics, and glasses in kitchenware, tableware, and potable liquid dispensing machines," and in solder used to seal food cans also contributed to human Pb intake.

He testified before a US Senate committee in 1966 arguing for the removal of Pb from gasoline. Not surprisingly, his paper and testimony were not well received by the tetraethyl Pb industry. Patterson lost research funding and pressure was applied, unsuccessfully, to have him fired from Cal Tech. Nevertheless, he had successfully raised awareness. In 1973 the US Environmental Protection Agency introduced regulations to phase out Pb in gasoline and between 1976 and 1980, tetraethyl Pb emissions dropped 50% and the blood level in average Americans dropped 37% and levels continued to fall thereafter (Figure 15.27). In 1995, the EPA issued a total ban on Pb in gasoline in automobiles. Regulations in Europe followed those in the United States, although with a time lag, and eventually, most of the world followed as well. Today, tetraethyl Pb is used only in Algeria, Yemen, and Iraq.

Other sources of Pb were eventually also addressed. From 1978 to 1986, Pb was increasingly banned in the US in paint, ceramic glazes, and plumbing. Strick limits were placed on industrial emissions, such smelters, which dropped by more than half between 1970 and 1975 and are now down by more than a factor of 100. Emissions from coal-fired power plants have dropped by an order of magnitude in the US. Tetraethyl lead still is used in fuel for propeller aircraft with internal combustion engines and this remains the major source of emissions to the atmosphere.

In the meantime, much has been learned about the toxicity of Pb and its effects. It reduces fertility and causes gastrointestinal, renal, and muscular problems. Many of lead's toxic properties are due to its ability to mimic or compete with calcium, particularly in neuronal signaling, the brain, and brain

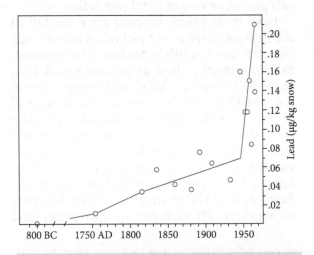

Figure 15.26 Concentration of lead in ice from the Camp Century ice core from Greenland. Concentrations increase from <0.001µg/kg at 800 BC to >0.20 µg/kg in 1960. From Murozumi et al. (1969).

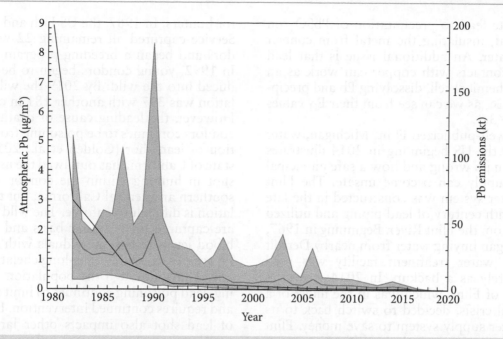

Figure 15.27   Atmospheric concentration of Pb in air over the United States since 1980 (red line shows the mean; shaded area shows 90th percentile variations), and US emissions (black line). Data from US EPA.

development. The effects are worse on young children and fetuses than adults. Clear correlations have been established between Pb levels in children's blood, test scores, reasoning and decision making, educational success, and juvenile delinquency (e.g., Needleman, 2004; Alzer et al., 2018). Indeed, it has been argued that the decline in Pb blood levels in children which began in the mid-1970s is responsible for the decline in violent crime in the United States that began 20 years later.

The legacy of the extensive use of lead is still with us, however. Although Pb is no longer used in new plumbing, Pb plumbing installed before the mid-twentieth century is still in widespread use, particularly in Europe. This might sound scary, but it is *usually* not a problem. The reason is that *scale* typically forms on the inside of piping through precipitation of calcium and magnesium carbonates and, in some cases, iron and manganese hydroxides. This prevents contact of the water with the Pb and therefore inhibits dissolution of Pb. Ancient Rome used Pb piping in its municipal water system, but their water supply, delivered through a famous network of long aqueducts and tunnels, was largely drawn from regions of carbonate bedrock. Consequently, the water had high alkalinity

(hard water) and high pH, conditions under which the solubility of Pb is low and scale developed rapidly on pipes.

Pipe corrosion can be a problem with water of low alkalinity and low pH. One parameter monitored in water supply systems to guard against corrosion is the Langelier Saturation Index (LSI), which describes calcium carbonate solubility:

$$LSI = pH$$
$$+ \log\left(\frac{K_2 \gamma_{Ca^{2+}}[Ca^{2+}]\gamma_{HCO_3^-}[HCO_3^-]}{\gamma_{H^+}K_{sp}}\right)$$
$$(15.8)$$

(see Chapter 6). An LSI value less than zero indicates calcite undersaturation and dissolution is favored. In this circumstance the water could be corrosive to Pb pipes. Measures can be taken to control corrosion, including the addition of corrosion inhibitors such as phosphate to the water supply. Another indicator of corrosion is the chloride-to-sulfate mass ratio (CSMR: Cl^-/SO_4^{2-}). CSMR values less than 0.58 inhibit corrosion while greater ratios tend to favor it, reflecting the greater solubility of $PbCl_2$ than $PbSO_4$. High levels of sulfate lead to precipitation of $PbSO_4$ on Pb metal, insulating the metal from contact with water. An additional issue is that lead solder contacts with copper can work as an electrochemical cell, dissolving Pb and precipitating Cu, as we can see from their E_H values in Table 3.3.

The well-publicized Flint, Michigan, water crisis in the US beginning in 2014 illustrates what can go wrong and how a safe municipal water supply can become unsafe. The Flint city water system was constructed in the late nineteenth century of lead piping and utilized water from the Flint River. Beginning in 1967, Flint began buying water from nearby Detroit and its water treatment facility was used only rarely as a backup. In 2014, however, the city of Flint, which was in the midst of a financial crisis, decided to switch back to its old water supply system to save money. Flint River water is particularly difficult to treat, and the old treatment plant lacked modern monitoring equipment and the treatments involved increased chloride concentrations and decreased alkalinity and pH, both of which made the water more corrosive (Masten et al., 2016). Most critical, perhaps, was a failure to add corrosion inhibitors. Residents soon began complaining about color, taste, and odor of the water, but response was slow. Pb levels were variable in the main system but often exceeded EPA maximum contaminant level of 0.015 mg/l (Table 15.1) and occasionally exceeded 0.02 mg/l. Water tested in one home was 0.1 mg/l. Average blood levels increased by a factor of about 2.5, and levels of over 0.5 ppb (toxic even by pre-1970 standards) were measured in over 500 children. In late 2015, water was switched back to the Detroit system. Damage had been done, however; fertility rates decreased by 12%, fetal death rates increased by 58% and the long-term effects on children have yet to be measured.

Anthropogenic Pb has also had an impact on wildlife, as illustrated by the saga of the California condor, the largest bird in North America, with a wingspan of 3 m and weight of 12 kg. The condor population declined in the twentieth century from a number of causes, perhaps primarily DDT, which thins eggshells, but also ingestion of lead shot, a particular hazard for condors as they feed exclusively on carrion, primarily large mammals such as deer, which are also a favorite target

for hunters. In 1987, the US Fish and Wildlife Service captured all remaining 22 wild condors and began a breeding program in zoos. In 1992, young condors began to be reintroduced into the wild. By 2019 the wild population was 337 with another 181 in captivity. However, the leading cause of death in wild condors continues to be poisoning from ingestion of lead shot (Golden et al., 2016). The state of California has outlawed the use of lead shot in hunting within the condor range in southern and central California, but the regulation is difficult to enforce. The wild condors are captured on a regular basis and their Pb blood levels tested; individuals with elevated Pb are treated with weeks-long chelation therapy. Thus, although the population is growing, lead poisoning continues to limit recovery and requires continued intervention. Ingestion of lead shot also impacts other large birds such as the bald eagle, the golden eagle and certain species of hawks (Golden et al., 2016).

15.3.2.3 Mercury

We'll next consider mercury because this demonstrates how understanding geochemical behavior is necessary to understand the challenges that toxic metals in the environment present. Ice core records show that significant anthropogenic Hg was being released even prior to widespread industrialization in the late nineteenth century, primarily from mining and gold refining, and this Hg continues to cycle through the environment (Schuster et al., 2002). Mercury is among the most toxic of metals, as the US EPA drinking water limits in Table 15.1 suggest. Like Pb, it is a neurotoxin but affects other parts of the body as well, including muscular, renal, and gastrointestinal systems and is more toxic to children and fetuses than adults. Like Pb, anthropogenic Hg has dispersed widely in the environment. Historically, mercury poisoning occurred in connection with mining and industrial uses (e.g., mad hatter's disease); the greatest present concern is consumption of seafood. Like Pb, Hg bioaccumulates and is present in greater quantities in shellfish and fish at the top of the food chain. The threat of Hg in seafood first became apparent in Minamata, Japan, in the mid-twentieth century, where industrial Hg wastes discharged to the Minamata Bay led, ultimately, to more than 1700 deaths and many more severe illnesses (Minamata disease) as a consequence of consumption of locally harvested seafood. Today, organizations such as the US Food and Drug Administration, the European Food Safety Authority, and World Health Organization warn young children, pregnant women, and women of child-bearing age to limit consumption of fish. Typical concentrations range nearly two orders of magnitude, however, from 0.01 ppm in sardines and 0.02 ppm in salmon to ~1 ppm in swordfish and shark, so the main concern is with a limited number of species.

As with Pb, environmental regulation has limited anthropogenic emissions; the UN Minamata Convention on Mercury of 2013 obligates its 126 signatory nations to decrease emissions. In the US, emissions decreased by a factor of 5 between 1993 and 2011 according to US EPA emissions summary. Unlike Pb, however, the amount of anthropogenic Hg in the atmosphere has not dropped precipitously. Indeed, over the last 20 years, the Hg concentration in air over Europe has remained nearly constant at around 1.5 ng/m^3 (UN Environment, 2019). There are several reasons for this: first, there was a single principal source of Pb pollution, tetraethyl Pb in gasoline, which could be readily addressed with regulation; second, while emissions in the developed world have dropped, emissions from developing countries, particularly in Asia, have increased. Third, the geochemistry of Hg differs significantly from that of Pb. As we described in Section 9.10.6, mercury undergoes a variety of mass dependent and mass independent isotopic fractionations (Figure 9.55).

Let's briefly consider the geochemistry of Hg. Like Pb, it is a B-type metal, chalcophile, and incompatible, being enriched in the crust by roughly an order of magnitude over the bulk silicate Earth (Alpers, 2017). Like, Pb, it is divalent within the silicate Earth but unlike, Pb it can occur naturally in the metallic form. The electrochemical potential for reduction of Hg^{2+} to Hg^0 is similar to that for reduction of Fe^{3+} to Fe^{2+} (Table 3.3), consequently redox cycling occurs in the range of common environmental conditions. In the metallic state, Hg is highly volatile, with a melting point of $-38.8°C$ and a vapor pressure of 0.26 pascals at 25°C. Mercury is also more reactive than Pb, particularly with organic matter. It forms highly volatile methyl mercury (CH_3Hg) and dimethyl mercury ($(CH_3)_2Hg$) through both microbially mediated and abiotic reactions; both forms are considerably more toxic than inorganic Hg. As a consequence, Hg naturally enters the atmosphere in gaseous form where it is present primarily (~95%) as Hg^0 and has an estimated atmospheric residence time of ~6 months. Within the atmosphere, it undergoes a variety of redox reactions, including oxidation by halogen radicals and photoreduction and photo-oxidation. Both reduced and oxidized forms are scavenged from the atmosphere by particles or cloud droplets and removed from the atmosphere; Hg in rain is primarily in the oxidized form, but both reduced and oxidized Hg occur in dry deposition. Plants also remove Hg^0 by direct uptake through their stomata.

Primary anthropogenic emissions are currently estimated at ~3100 tons, the vast majority of which is emitted directly to the atmosphere as either Hg^0, volatile organic Hg compounds, or particulate Hg^{2+} compounds. This is five times greater than primary natural emissions of ~600 tons, mainly from volcanic eruptions, fumaroles and hydrothermal activity, most of which is also emitted directly to the atmosphere. Globally, small-scale artisanal mining is the largest source (Hg is used to refine gold in these operations but not in large scale ones), contributing 38% of emissions; industrial emissions, which include large scale mining, constitute 28%, and fossil fuel burning (mainly coal) constitutes 24%. This varies by region, however; whereas artisanal mining is the overwhelming dominant source (>70%) in South America and Sub-Saharan Africa, coal burning is the major source (over 60%) in North America and Europe (UN Environment, 2019).

Figure 15.28 illustrates the present global mercury cycle. An important aspect is that mercury removed from the atmosphere to the land and ocean surface can then be re-emitted to the atmosphere. Hg in terrestrial reservoirs is transferred to the atmosphere by respiration of organic carbon, photoreduction, and biomass burning; and to the surface ocean by river runoff. Mercury in the surface ocean can evaporate to the atmosphere as Hg^0 or methylated species.

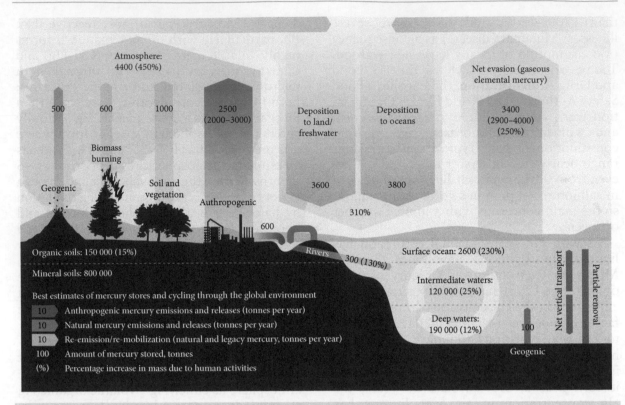

Atmosphere:
4400 (450%)

Net evasion (gaseous
elemental mercury)

500   600   1000   2500
(2000–3000)

Deposition
to land/
freshwater

Deposition
to oceans

3400
(2900–4000)
(250%)

Biomass
burning

Geogenic

Soil and
vegetation

Anthropogenic

3600   3800

310%

600

Organic soils: 150 000 (15%)   Surface ocean: 2600 (230%)

Rivers

300 (130%)

Mineral soils: 800 000

Intermediate waters:
120 000 (25%)

Best estimates of mercury stores and cycling through the global environment

10	Anthropogenic mercury emissions and releases (tonnes per year)
10	Natural mercury emissions and releases (tonnes per year)
10	Re-emission/re-mobilization (natural and legacy mercury, tonnes per year)
100	Amount of mercury stored, tonnes
(%)	Percentage increase in mass due to human activities

Deep waters:
190 000 (12%)   100

Geogenic

Net vertical transport

Particle removal

Figure 15.28   Global mercury cycle. From UN Environment (2019).

Amos et al. (2013) concluded that "legacy" anthropogenic Hg re-emitted from surface reservoirs accounts for 60% of Hg presently in the atmosphere and that ~85% of the Hg in the surface ocean is anthropogenic, with more than half of that from pre-1950 emissions. Mercury in the surface ocean can be adsorbed or converted to particulate form and then sink into the deeper ocean. Deposition of this particulate form in sediments is the ultimate sink in the system, but the rate of removal to sediments is slower than anthropogenic input rates.

Mercury isotope studies are helping to elucidate this complex cycle and can be used to identify and quantify natural and anthropogenic fluxes (Berqquist, 2018). Coal burning releases considerable amounts of Hg, both as gaseous Hg^0 and in particulate matter, and can be identified from its isotopic signature (e.g., Huang et al., 2016). In water, mercury is quickly absorbed onto particle surfaces, where it retains its isotopic signature and can be used to identify anthropogenic Hg sources such as gold mining. The Hg isotopic composition of

fish tissue matches that observed experimentally in aqueous photochemical demethylation reactions, indicating it is generally taken up in the food web as demethylated Hg. Hg isotope ratios in fish vary considerably, for example $\Delta^{201}Hg$ ratios tend to be higher in pelagic and marine fish than in those from near-shore and estuarine environments. Detailed studies can determine where Hg is methylated, the species of Hg accumulated, trophic level, the depth at which it is foraged, the residence time of methyl Hg, and the source of Hg for methylation (Bergquist, 2018).

Because of its complex cycling, mercury is removed from the environment only slowly. As a result, the 2018 UN Global Mercury Assessment concludes, "In the shorter term, mercury in seawater and marine food webs is likely to increase even at current levels of anthropogenic emissions and releases, simply because some of the legacy mercury from soils will continue to be carried by rivers to the sea and to be re-volatilized into the air."

15.3.3 Acid deposition

We now turn to acid deposition as our final environmental issue because it is also an unfolding, albeit incomplete, environmental success story and because it illustrates how the geochemistry we have covered in previous chapters can be put to practical use in identifying and solving an environmental issue. Acid deposition refers primarily to deposition from the atmosphere to the Earth's surface of H_2SO_4 and HNO_3, which are produced by oxidation of SO_2 and NO and NO_2, the latter two collectively referred to as NO_x. Sulfur and nitrogen are common components of coal and petroleum (although concentrations of both vary widely) and are oxidized during combustion. In addition, combustion can produce NO_x from atmospheric N_2. Acid deposition is often referred to as "acid rain" but these pollutants can be removed from the atmosphere by absorption on particulates and subsequent dry deposition as well as by rain (wet deposition). Both occur in the atmosphere naturally, of course, from sources such as microbial activity, lightning, and volcanic eruptions, but anthropogenic sources presently far outweigh natural ones. The primary anthropogenic sources are combustion of fossil fuels, although in the past smelters have been a major source of SO_2 and agriculture is an important source of NH_3, which oxidizes to NO_x, from both fertilizers and animal wastes. Atmospheric SO_2 and NO_x are direct health hazard to humans as they can irritate airways and aggravate respiratory diseases such as asthma and both contribute to formation of smog.

Perhaps somewhat ironically, however, both emissions have beneficial effects by mitigating greenhouse gas climate change. Sulfate formed by photooxidation of SO_2 combines with water to form H_2SO_4, which has a very low vapor pressure and forms small droplets in the atmosphere that reflect sunlight, counteracting warming from anthropogenic CO_2 emissions. Indeed, the eruption of Mt. Pinatubo in 1991 injected enough SO_2 into the stratosphere that global temperatures were reduced by 0.4°C for a year. Nitrate deposition effectively fertilizes forests, which remove CO_2 through photosynthesis. Atmospheric sulfate and nitrate is a complex topic, but we will restrict attention to the resulting acidification of soils, streams, and lakes.

An awareness of the deleterious effects of local sulfur deposition associated with smelters dates to at least the eighteenth century. The term *acid rain* was first used in the nineteenth century and the effects of acidic precipitation on plant life and fish began to become clear in the early twentieth century (Cowling, 1982). Widespread sampling of atmospheric deposition began in Europe in the 1950s and demonstrated that atmospheric processes could disperse pollutants on the scale of 1000 km or more and therefore that this was a continent- or global-scale problem, not just a local one. Many of the early studies was done in Scandinavia, as soils there are particularly prone to acidification, and in Canada, which also has glacial soils prone to acidification and is also the site of the Sudbury smelter, which at one time was the single largest point source of SO_2.

The pH of rainfall in the most polluted areas can fall below 4; this compares with a pH of 5.6 expected in pollution-free rain simply from dissolution of CO_2 (see Chapter 6). However, it is not primarily the direct effect of H^+ ions on organisms that degrades ecosystems, although this can adversely affect aquatic organisms. Rather, nitrate and sulfate deposition affects soil chemistry, and consequently stream and lake chemistry, and it is that which impacts organisms. The primary effect come from decreased base cation (Ca, Mg, Na, K) concentrations and increased Al concentrations in soil and stream and lake water. The extent to which acid deposition affects the biota depends very much on the nature of the soil and bedrock/regolith it developed on. Indeed, differences in soil chemistry mean that there is only a weak correlation between rainfall pH and ecosystem impact. Let's use what we've learned in previous chapters to understand this.

As we found in previous chapters, weathering reactions (e.g., eqns. 5.1 through 5.7) within the soil profile consume acidity. The pH of the soil solution depends on the ratio of the rate of addition of strong acid anions such as NO_3^- and SO_4^{2-} to the rate of acid-consuming weathering reactions. We found in Section 6.2.4 that alkalinity is equivalent to *acid neutralizing capacity* (ANC). As eqn. 6.37 shows, alkalinity depends directly on the concentrations of conservative cations, principally Ca, Mg, Na, and K, all of which are important nutrients. For readily weatherable materials, release of these base cations can maintain high alkalinity and therefore neutralize acidity in the soil solution. Carbonate rocks as well as mafic igneous and metamorphic rocks weather at high rates and produce soils rich in base cations, particularly Ca and Mg. Where soils are developed on such rocks the effects of acid deposition are generally minimal. On the other hand, siliceous igneous and metamorphic rocks and sedimentary rocks lacking significant carbonate content weather more slowly and are poor in Ca and Mg and consequently have limited acid neutralizing capacity. Such soils are the most impacted by acid deposition. Thus, the impact of acid deposition depends on the nature of the regolith or bedrock as well as the acid deposition rate.

Within soils, the nutrients that are readily available to plants are absorbed on mineral surfaces (Section 6.6.1) or in exchangeable sites in clays (Section 6.5.2). When acidity increases, H^+ displaces these nutrients, which include the base cations mentioned above, from surfaces or exchangeable sites in clays into the soil solution from whence they can be carried away by streams. If the rate of loss of these exchangeable cations exceeds their production by weathering of primary minerals, they become depleted in soils, with adverse effects on flora (Norton et al., 2014).

Now let's look at Al solubility as a function of pH (Figure 6.18) once again. The minimum solubility of Al occurs around pH 6. Decreasing pH to 5 results in more than an order of magnitude increase in Al solubility. Decreasing pH to 4 results in a nearly four orders of magnitude increase in solubility over the minimum value. Although aluminum has very limited toxicity to mammals and only at high levels, high concentrations of dissolved inorganic aluminum adversely affects many tree species, including Norway spruce, red spruce, and sugar maples, in several ways, including inhibiting uptake of calcium, the fifth most abundant element in trees, by roots (e.g., Shortle and Smith, 1988). The combination of Ca depletion in soils and high dissolved inorganic Al (organically complexed Al is not toxic to plants, although it can be to mammals and birds) thus combines to produce a

"double whammy" on forests. In addition, acid rain and mist can leach Ca from foliage, leaving trees vulnerable to winter freezing and injury. Beginning in the second half of the twentieth century, these factors led to mass die-offs of trees, particularly spruce, in Scandinavia, at higher elevations in Germany and the Czech Republic, and in eastern Canada and the northeastern US in areas where soils have limited acid neutralizing capacity (Lawrence, 2018). Increased mortality of sugar maples, an economically important species, also became apparent in eastern Canada and the northeastern United States.

In its initial stages, soil acidification can be expected to increase the flux of base cations to streams and lakes as they are leached from soils. As soils become depleted, however, concentrations in aquatic ecosystems decline. *Daphnia* are small crustaceans that are often the primary herbivore in many aquatic systems and require high levels of dissolved Ca for growth and reproduction. Jeziorski et al. (2008) demonstrated declines in daphnia abundances that were correlated with declining Ca concentrations in Canadian Shield lakes. This then impacts fish populations as daphnia are an important prey for juvenile fish and a keystone in the food chain leading to large adult fish. Soil acidification also leads to increases in dissolved inorganic Al in streams and lakes. Aluminum is toxic to gill-breathing animals such as fish and invertebrates by reducing the activities of certain gill enzymes (e.g., Schofield and Trojnar, 1980). Particularly during spring snow melt, there can be a pulse of acidity and dissolved Al that has led to fish kills observed in Scandinavia, Canada, the Adirondack Mountains in the Northeast US, and elsewhere (Norton, 2014).

Recognizing the effects of acid deposition, 32 European countries signed the UN Convention on Long-range Transboundary Air Pollution, which came into force in 1983 to limit SO_2 and NO_X emissions; subsequent conventions imposed stricter limits. Reductions in acid deposition have been achieved through installation of scrubbers on smelters and coal-fired power plants, replacing the latter with other means of electrical power generation, and installation of catalytic converters in cars and trucks, which convert NO_x in exhaust to N_2. Figure 15.29 shows the resulting decrease in sulfate in precipitation in

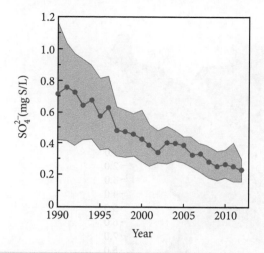

Figure 15.29 Average concentration of sulfate dissolved in rain across Europe (red symbols); shaded area shows 25th and 75th quantiles. Source: European Monitoring and Evaluation Programme Chemical Coordinating Center (https://www.emep.int).

Europe since 1990. In the US, emissions began to be limited with the passage of the Clean Air Act of 1970 and the establishment of the Environmental Protection Agency; more severe limitations came into effect with the Clean Air Act of 1990. These regulations have led to significant reductions in atmospheric sulfate and nitrate emissions; for example, SO_2 emissions in the US fell from 17.26 Mt in 1980 to 0.11 Mt in 2016. Ambient air sulfate concentrations decreased by 40 to 50% in eastern US from 1989–1991 to 2007–2009. Nitrate reductions have been less dramatic (Figure 15.30), in part because of the contribution of agricultural ammonia and the difficulty in completely removing NO_x from automobile exhaust. Reductions have nevertheless been significant: between 1989–1991 and 2007–2009, wet inorganic nitrogen deposition decreases in the eastern United States ranged from 16% in the Midwest to 27% in the mid-Atlantic region and the Northeast (generally nitrate and sulfate emissions and deposition in the western US has been far lower than in the eastern part of the country). In contrast, acid deposition is increasing in East Asia as a result of more coal-fired power plants and more automobiles.

Acid deposition and sulfate and nitrate concentrations in soils, streams, and lakes are decreasing in Europe and North American and soil, stream, and lake chemistry is improving (e.g., Johnson et al., 2018), but ecosystem recovery lags. The Adirondack region of northern New York state in the US illustrates this. The Adirondack Mountains are an area of uplifted Proterozoic igneous and metamorphic rocks covered in most areas by a veneer of Pleistocene glacial till. Soils are thin and have low acid neutralizing capacity, Consequently, Adirondack forests and aquatic ecosystems were particularly badly impacted by acid rain in the second half of the twentieth century; for example, a 1984–1987 survey found that 24% of lakes in region were fishless and numerous species had been lost from those where fish remained (Baldigo et al., 2016). In contrast, the impact of acid deposition on the lowland areas of New York state underlain by Paleozoic sediments with high acid neutralizing capacity has been minimal. Driscroll et al. (2016) found that sulfate concentrations in 48 monitored lakes in the Adirondacks declined at a rate of –2.35 µeq/L/yr between 1982 and 2015. Nitrate is also declining in monitored lakes but at a lower rate of –0.38 µeq/L/yr. In response, base cation fluxes are decreasing, ANC (i.e., alkalinity) and pH are increasing, and inorganic Al concentrations are decreasing. However, recent surveys show that fish populations have generally not recovered, although some positive trends were identified: species richness improved in a third of the lakes and the fraction of lakes with no fish declined (Baldigo et al., 2016).

Figure 15.31 shows the data for one of the monitored lakes, Big Moose Lake. The lake is located at an elevation of 550 m and has a surface area of ~5 km² and an average depth of 7 m; its watershed consists of soils developed on thin till. Sulfate concentrations have declined significantly; nitrate concentrations have declined as well, although to a lesser extent. As a consequence, acid normalizing capacity has increased and there has been a small but statistically significant decrease in total inorganic Al concentrations at a rate of about 4.4 mg/L/yr. The amplitude of seasonal variations, particularly the pulses of high Al concentrations associated with spring snow melt, have become less severe. For example, spring Al concentrations regularly exceeded 400 mg/L in the 1990s, but have usually been

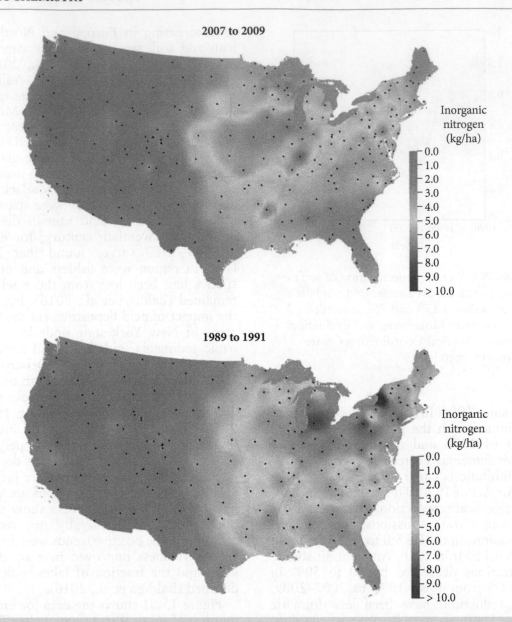

Figure 15.30 Change in nitrate wet deposition over the eastern US between 1989–1991 and 2007–2009; dots are monitoring stations. Source: US National Acid Precipitation Assessment Program Report to Congress 2011.

less than 275 mg/L since 2011. Nevertheless, these values compare with the commonly accepted threshold of about 54 mg/L for brook trout survival in lakes of the northeastern US (Baldigo et al., 2016); even in summer low concentrations have not yet reached these levels in Big Moose Lake. Baldigo et al.

conclude that "additional time may simply be needed for biological recovery to progress, or else more proactive efforts may be necessary to restore natural fish assemblages in Adirondack lakes." This statement could be applied to nearly all other regions severely impacted by acid deposition.

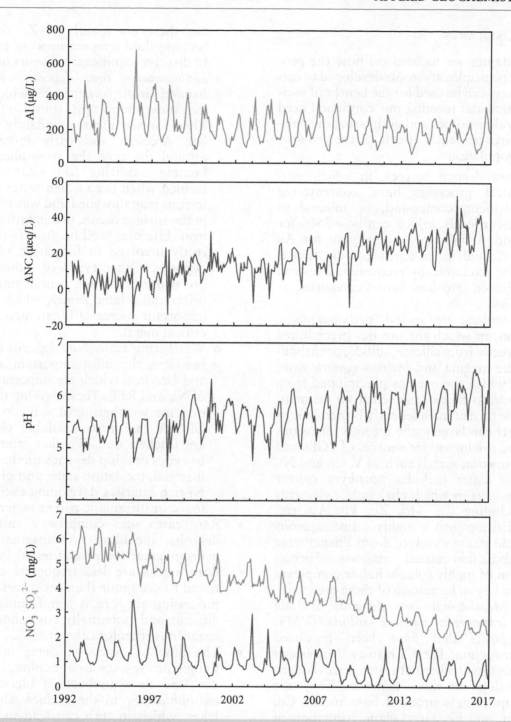

Figure 15.31 Monthly measurements of dissolved nitrate, phosphate, pH, ANC (acid neutralizing capacity = alkalinity) and total Al concentrations in Big Moose Lake in the Adirondack Mtns. between 1992 and 2017. Data from Adirondacks Lakes Survey Corporation (http://www.adirondacklakessurvey .org/index.shtml).

15.4   SUMMARY

In this chapter we focused on how the geochemical principles and tools developed in earlier chapters can be used for the benefit of society, in particular meeting our continued need for mineral resources and addressing the deleterious environmental effects of the growing human population.

- An *ore deposit* is rock in which geochemical processes have concentrated an element, compound, or mineral to sufficient levels, which can be ~25% for Fe and Al but only a few ppm for Au and Pt, and in sufficient amount that it can be economically recovered. Ores can be divided into four broad categories, as follows:
  - *Magmatic ores* include *orthomagmatic ores*, in which the ore has precipitated directly from silicate, sulfide, or carbonatite magma and *hydromagmatic ores*, in which the ore has precipitated from hydrous fluids exsolved from magmas. The former includes early Proterozoic or Archean large mafic igneous intrusions and are important sources of PGEs and transition metals such as V, Cr, and Ni. The latter includes *porphyry copper deposits* in which chalcophile elements including Cu, Mo, Zn, Pb, Au, etc. partition into a highly saline aqueous fluid phase exsolved from Phanerozoic subduction-related magmas. Formation of highly soluble halide complexes is a key to formation of these ores.
  - *Hydrothermal ore deposits* include *volcanogenic massive sulfide* (VMS) deposits that have been produced throughout Earth's history. Mid-ocean ridge hydrothermal systems are modern analogs of these deposits. They are important sources of base metals Cu, Zn, and Pb. Here again, formation of soluble chloride complexes at elevated temperatures and low pH is the key to metal transport. In contrast, *stratiform ore deposits* are unassociated with magmatism and form when warm (~120–250°C), saline fluids that have dissolved base metals from source rocks are reduced or cooled. *Mississippi Valley type* (MTV) also form from precipitation from saline fluids, but these are mainly Pb-Zn deposits because fluid temperatures are too low to dissolve significant amounts of Cu.
  - *Sedimentary ore deposits* include *banded iron formations*, which are the most important source of iron. They formed most abundantly in the late Archean and early Proterozoic around the time the atmosphere first became oxidizing (the GOE). Most formed when deep ocean water rich in ferrous iron upwelled and was oxidized in the surface ocean, precipitating ferric iron. Life may well be directly or indirectly involved in formation of these deposits. Other types of sedimentary ore deposits include continental *evaporites* and related *brines*, which are an important source of Li, an increasingly critical metal.
  - *Weathering-related ore deposits* include *bauxites*, the most important Al ore, and *laterites*, which are important ores of Ni, and REE. These develop through extreme weathering of soils, in which all but the most insoluble elements are leached out. Whether laterites or bauxites develop depends on the parent material, oxidation state, and pH, with Ni-rich laterites developing exclusively mafic or ultramafic parent material.
- Rare earth ores comprise a variety of deposits, including orthomagmatic and hydromagmatic carbonatites to laterites. Rare earths are less frequently concentrated to ore grade than other metals and processing the ore is significantly more difficult and potentially environmentally hazardous than for other metals.
- The flow of excess nutrients in everything from sewage to agricultural runoff can lead to cyanobacteria blooms and *eutrophication* in the surface waters of lakes, which in turn can lead to anoxic conditions in deep water when their remains decay. Restrictions on sewage and industrial wastes as well as particulate phosphate from farms were successful in eliminating anoxia and restoring the health of Lake Erie. However, seasonal anoxia has occasionally returned, making necessary further revisions to agricultural practices to limit runoff of dissolved phosphorus.

- Mining of sulfide ores such as VMS deposits can lead to acidification of soils and streams as sulfide is oxidized to sulfuric acid, a phenomenon known as *acid mine drainage*. This acidity results in high concentrations of toxic metals in streams as they are significantly more soluble at low pH. A number of mitigation strategies are possible, involving limiting contact of sulfides with oxygen or controlling bacteria that catalyze sulfur oxidation.

- *Toxic metals* find their way into the environment in other ways as well. Eliminating Pb from gasoline, smelter effluents, and common products such as paint and plumbing is an example of how societies can control toxic metal pollution through regulation. Regulations have also been successful in reducing primary anthropogenic mercury emissions, but additional limits are needed in the developed world on coal burning and in less-developed countries on use of Hg in small-scale gold refining. However, Hg levels in the atmosphere remain high because of its complex chemistry in which it readily shuttles between the land surface, the atmosphere, and the ocean surface; consequently, removal from the environment is slow.

- Finally, we considered another partial success story, that of *acid rain*. Dry and wet deposition of sulfate and nitrate ultimate derived mostly from burning fossil fuels have acidified soils and streams where soils have developed on bedrock or regolith with poor acid-neutralizing capacity. This acidity results in loss of base cations such as Ca, Mg, Na, and K and high concentrations of dissolved Al, all of which have quite deleterious effects on the flora and streams and lake fauna. While regulations have greatly reduced emissions, affected ecosystems are only slowly recovering.

REFERENCES

Adriano, D. C. 2001. *Trace Elements in the Terrestrial Environment*. Springer-Verlag, New York ISBN

Aizer, A. and Currie, J. 2018. Lead and juvenile delinquency: new evidence from linked birth, school and juvenile detention records. *The Review of Economics and Statistics* 10: 307–41 null. doi: 10.1162/rest_a_00814.

Alderton, D. H. M., Selby, D., Kucha, H., Blundell, D. J. 2016. A multistage origin for Kupferschiefer mineralization. *Ore Geology Reviews* 79: 535–543. doi: 10.1016/j.oregeorev.2016.05.007.

Alpers, C. N. (2017) Mercury. In: White WM (ed) *Encyclopedia of Geochemistry*. Springer International Publishing, Cham, pp 895–900. ISBN 1-6 978-3-319-39193-9.

Amos, H. M., Jacob, D. J., Streets, D. G., Sunderland, E. M. 2013. Legacy impacts of all-time anthropogenic emissions on the global mercury cycle. *Global Biogeochemical Cycles* 27: 410–21. doi:10.1002/gbc.20040.

Arndt, N. T., Fontboté, L., Hedenquist, J. W., Kesler, S. E., Thompson, J. F. H., Wood, D. G. 2017. Future global mineral resources. *Geochemical Perspectives* 6: 1–71.

Arndt, N., Kesler, S. and Ganino, C., 2015. Deposits formed by sedimentary and surficial processes. In: *Metals and Society: An Introduction to Economic Geology* pp. 137–72, Springer International Publishing, Cham.

Audétat, A., Pettke, T., Heinrich, C. A., Bodnar, R. J. 2008. Special paper: the composition of magmatic-hydrothermal fluids in barren and mineralized intrusions. *Economic Geology* 103: 877–908. doi:10.2113/gsecongeo.103.5.877.

Baldigo, B. P., Roy, K. M. and Driscoll, C. T., 2016. Response of fish assemblages to declining acidic deposition in Adirondack Mountain lakes, 1984–2012. *Atmospheric Environment* 146, 223–35. doi: https://doi.org/10.1016/j.atmosenv.2016.06.049.

Bao, Z. and Zhao, Z., 2008. Geochemistry of mineralization with exchangeable REY in the weathering crusts of granitic rocks in South China. *Ore Geology Reviews* 33, 519–35. doi: 10.1016/j.oregeorev.2007.03.005.

Barnes, S.-J., Maier, W. D., Curl, E. A. 2010. Composition of the marginal rocks and sills of the Rustenburg Layered Suite, Bushveld Complex, South Africa: Implications for the formation of the platinum-group element deposits. *Economic Geology* 105: 1491–511. doi: 10.2113/econgeo.105.8.1491.

Barrie, C., Hannington, M. J. R. i. E. G. 1999. Classification of volcanic-associated massive sulfide deposits based on host-rock composition. *Reviews in Economic Geology* 8: 1–11.

Bekker, A., Slack, J. F., Planavsky, N., et al. 2010. Iron formation: the sedimentary product of a complex interplay among mantle, tectonic, oceanic, and biospheric processes. *Economic Geology* 105: 467–508. doi: 10.2113/gsecongeo.105.3.467.

Bergquist, B. A., 2018. Mercury Isotopes. In: Encyclopedia of Geochemistry *(ed White, W. M.)*, pp. 900–6, Springer International Publishing, Cham. doi: 10.1007/978-3-319-39312-4_122

Bhalla, P., Holtz, F., Linnen, R. L., Behrens, H. 2005. Solubility of cassiterite in evolved granitic melts: effect of T, f_{O2}, and additional volatiles. *Lithos* 80: 387–400. doi: 10.1016/j.lithos.2004.06.014.

Blundy, J., Mavrogenes, J., Tattitch, B., Sparks, S., Gilmer, A. 2015. Generation of porphyry copper deposits by gas–brine reaction in volcanic arcs. *Nature Geoscience* 8: 235. doi:10.1038/ngeo2351.

Bourcier, W. L., Barnes, H. L. 1987. Ore solution chemistry; VII, Stabilities of chloride and bisulfide complexes of zinc to 350 degrees C. *Economic Geology* 82: 1839–1863. doi: 10.2113/gsecongeo.82.7.1839.

Brugger, J., Liu, W., Etschmann, B., Mei, Y., Sherman, D. M., Testemale, D. 2016. A review of the coordination chemistry of hydrothermal systems, or do coordination changes make ore deposits? *Chemical Geology* 447: 219–253. doi: 10.1016/j.chemgeo.2016.10.021

Bussière, B., Benzaazoua, M., Aubertin, M., Mbonimpa, M. 2004. A laboratory study of covers made of low-sulphide tailings to prevent acid mine drainage. *Environmental Geology* 45: 609–22. doi:10.1007/s00254-003-0919-6.

Butt, C. R. M., Lintern, M. J., Anand, R. R. 2000. Evolution of regoliths and landscapes in deeply weathered terrain — implications for geochemical exploration. *Ore Geology Reviews* 16: 167–183. doi 10.1016/S0169-1368(99)00029-3.

Cameron, E. M., Hamilton, S. M., Leybourne, M. I., Hall, G. E. M. and McClenaghan, M. B., 2004. Finding deeply buried deposits using geochemistry. *Geochemistry: Exploration, Environment, Analysis* 4(1), 7–32. doi: 10.1144/1467-7873/03-019.

Castor, S. B. 2008. The Mountain Pass rare-earth carbonatite and associated ultrapotassic rocks, California. *The Canadian Mineralogist* 46: 779–806. doi: 10.3749/canmin.46.4.779.

Cathles, L. M. 1993. Oxygen isotope alteration in the Noranda mining district, Abitibi greenstone belt, Quebec. *Economic Geology* 88: 1483–511. doi: 10.2113/gsecongeo.88.6.1483

Cathles, L. M. 2011. What processes at mid-ocean ridges tell us about volcanogenic massive sulfide deposits. *Mineralium Deposita* 46: 639–57. doi: 10.1007/s00126-010-0292-9.

Cathles, L. M., Shannon, R. 2007. How potassium silicate alteration suggests the formation of porphyry ore deposits begins with the nearly explosive but barren expulsion of large volumes of magmatic water. *Earth and Planetary Science Letters* 262: 92–108. doi: 10.1016/j.epsl.2007.07.029.

Cathles, L., Adams, J. J. 2005. Fluid flow and petroleum and mineral resources in the upper (< 20 km) continental crust. *Economic Geology* 100: 77–110.

Cohen, D. R. and Bowell, R. J. 2014. 13.24 - Exploration Geochemistry. In: Treatise on Geochemistry *(Second Edition)* *(eds Holland, H. D. and Turekian, K. K.)*, pp. 623–650, Elsevier, Oxford.

Cowling, E. B. 1982. Acid precipitation in historical perspective. *Environmental Science and Technology* 16(2), 110A–23A. doi: 10.1021/es00096a725.

Daloğlu, I., Cho, K. H., Scavia, D. 2012. Evaluating causes of trends in long-term dissolved reactive phosphorus loads to lake erie. *Environmental Science and Technology* 46: 10660–6. doi: 10.1021/es302315d.

Driscoll, C. T., Driscoll, K. M., Fakhraei, H. and Civerolo, K., 2016. Long-term temporal trends and spatial patterns in the acid-base chemistry of lakes in the Adirondack region of New York in response to decreases in acidic deposition. *Atmospheric Environment* 146, 5–14. doi: 10.1016/j.atmosenv.2016.08.034.

Durasova, N. A., Ryabchikov, I. D., Barsukov, V. L. 1986. The redox potential and the behavior of tin in magmatic systems. *International Geology Review* 28: 305–311. doi:10.1080/00206818609466274

El Desouky, H. A., Muchez, P., Boyce, A. J., Schneider, J., Cailteux, J. L. H., Dewaele, S. and von Quadt, A. 2010. Genesis of sediment-hosted stratiform copper–cobalt mineralization at Luiswishi and Kamoto, Katanga Copperbelt (Democratic Republic of Congo). *Mineralium Deposita* 45(8), 735–63. doi: 10.1007/s00126-010-0298-3.

Eppinger, R. G., Fey, D. L., Giles, S. A., Grunsky, E. C., Kelley, K. D., Minsley, B. J., Munk, L. and Smith, S. M. 2013. Summary of exploration geochemical and mineralogical studies at the giant pebble porphyry Cu-Au-Mo deposit, alaska: implications for exploration under cover. *Economic Geology* 108(3), 495–527. doi: 10.2113/econgeo.108.3.495.

Frank, M. R., Simon, A. C., Pettke, T., Candela, P. A., Piccoli, P. M. 2011. Gold and copper partitioning in magmatic-hydrothermal systems at 800°C and 100MPa. *Geochimica et Cosmochimica Acta* 75: 2470–2482. doi: doi.org/10.1016/j.gca.2011.02.012.

Fyfe, W. 1981. The environmental crisis: quantifying geosphere interactions. *Science* 213: 105–10.

Ganguli, R., Cook, D. R. 2018. Rare earths: A review of the landscape. *MRS Energy and Sustainability* 5: E9. doi:10.1557/mre.2018.7.

Godfrey, L. V., Chan, L. H., Alonso, R. N., Lowenstein, T. K., McDonough, W. F., Houston, J., Li, J., Bobst, A. and Jordan, T. E. 2013. The role of climate in the accumulation of lithium-rich brine in the Central Andes. *Applied Geochemistry* 38, 92–102. doi: 10.1016/j.apgeochem.2013.09.002.

Golden, N. H., Warner, S. E. and Coffey, M. J. 2016. A review and assessment of spent lead ammunition and its exposure and effects to scavenging birds in the United States. In: *Reviews of Environmental Contamination and Toxicology* Volume 237 (ed de Voogt, W. P.), pp. 123–91, Springer International Publishing, Cham.

González-Ortegón, E., Laiz, I., Sánchez-Quiles, D., Cobelo-Garcia, A., Tovar-Sánchez, A. 2019. Trace metal characterization and fluxes from the Guadiana, Tinto-Odiel and Guadalquivir estuaries to the Gulf of Cadiz. *Science of The Total Environment* 650: 2454–66. doi: 10.1016/j.scitotenv.2018.09.290.

Hamilton, S. M., 1998. Electrochemical mass-transport in overburden: a new model to account for the formation of selective leach geochemical anomalies in glacial terrain. *Journal of Geochemical Exploration* 63(3), 155–72. doi: 10.1016/S0375-6742(98)00052-1.

Hannington, M. D. (2014) 13.18 – Volcanogenic massive sulfide deposits. In: Holland HD, Turekian KK (eds) *Treatise on Geochemistry* (Second Edition), vol. Elsevier, Oxford, pp 463–88 978-0-08-098300-4.

Hannington, M. D., Santaguida, F., Kjarsgaard, I. M., Cathles, L. M. 2003. Regional-scale hydrothermal alteration in the Central Blake River Group, western Abitibi subprovince, Canada: implications for VMS prospectivity. *Mineralium Deposita* 38: 393–422. doi: 10.1007/s00126-002-0298-z.

Harmand, S., Lewis, J. E., Feibel, C. S., et al. 2015. 3.3-million-year-old stone tools from Lomekwi 3, West Turkana, Kenya. *Nature* 521: 310. doi:10.1038/nature14464.

Henley, R. W., King, P. L., Wykes, J. L., et al. 2015. Porphyry copper deposit formation by sub-volcanic sulphur dioxide flux and chemisorption. *Nature Geoscience* 8: 210. doi:10.1038/ngeo2367.

Hong, S., Candelone, J.-P., Patterson, C. C., Boutron, C. F. 1994. Greenland ice evidence of hemispheric lead pollution two millennia ago by greek and roman civilizations. *Science* 265: 1841–3. doi: 10.1126/science.265.5180.1841.

Huang, Q., Chen, J., Huang, W., et al. 2016. Isotopic composition for source identification of mercury in atmospheric fine particles. *Atmospheric Chemistry and Physics* 16: 11773–6. doi: 10.5194/acp-16-11773-2016.

Huston, D. L., Pehrsson, S., Eglington, B. M., Zaw, K. 2010. The geology and metallogeny of volcanic-hosted massive sulfide deposits: variations through geologic time and with tectonic setting. *Economic Geology* 105: 571–91. doi: 10.2113/gsecongeo.105.3.571.

Jackson, N. J., Willis-Richards, J., Manning, D. A. C., Sams, M. S. 1989. Evolution of the Cornubian ore field, Southwest England; Part II, Mineral deposits and ore-forming processes. *Economic Geology* 84: 1101–33. doi: 10.2113/gsecongeo.84.5.1101.

Jeziorski, A., Yan, N. D., Paterson, A. M., DeSellas, A. M., Turner, M. A., Jeffries, D. S., Keller, B., Weeber, R. C., McNicol, D. K., Palmer, M. E., McIver, K., Arseneau, K., Ginn, B. K., Cumming, B. F. and Smol, J. P., 2008. The widespread threat of calcium decline in fresh waters. *Science* 322(5906), 1374–7. doi: 10.1126/science.1164949.

Jochum, K. P., Hofmann, A. W., Seufert, H. M. 1993. Tin in mantle-derived rocks: Constraints on Earth evolution. *Geochimica et Cosmochimica Acta* 57: 3585–95. doi: 10.1016/0016-7037(93)90141-I.

Johnson, C. M., Beard, B. L., Roden, E. E. 2008. The iron isotope fingerprints of redox and biogeochemical cycling in modern and ancient earth. *Annual Review of Earth and Planetary Sciences* 36: 457–93. doi: 10.1146/annurev.earth.36.031207.124139.

Johnson, J., Graf Pannatier, E., Carnicelli, S., Cecchini, G., Clarke, N., Cools, N., Hansen, K., Meesenburg, H., Nieminen, T. M., Pihl-Karlsson, G., Titeux, H., Vanguelova, E., Verstraeten, A., Vesterdal, L., Waldner, P. and Jonard, M., 2018. The response of soil solution chemistry in European forests to decreasing acid deposition. *Global Change Biology* 24(8), 3603–19. doi: 10.1111/gcb.14156.

Kelley, D. L., Kelley, K. D., Coker, W. B., Caughlin, B. and Doherty, M. E. 2006. Beyond the obvious limits of ore deposits: The use of mineralogical, geochemical, and biological features for the remote detection of mineralization. *Economic Geology* 101(4), 729–52. doi: 10.2113/gsecongeo.101.4.729.

Kesler, S. E., Gruber, P. W., Medina, P. A., Keoleian, G. A., Everson, M. P. and Wallington, T. J. 2012. Global lithium resources: Relative importance of pegmatite, brine and other deposits. *Ore Geology Reviews* 48, 55–69. doi: 10.1016/j.oregeorev.2012.05.006.

King, K. W., Williams, M. R., LaBarge, G. A., et al. 2018. Addressing agricultural phosphorus loss in artificially drained landscapes with 4R nutrient management practices. *Journal of Soil and Water Conservation* 73: 35–47. doi:10.2489/jswc.73.1.35.

Kruger, F. J., Cawthorn, R. G., Walsh, K. L. 1987. Strontium isotopic evidence against magma addition in the Upper Zone of the Bushveld Complex. *Earth and Planetary Science Letters* 84: 51–58. doi:https://doi.org/10.1016/0012-821X(87)90175-0.

Kynicky, J., Smith, M. P. and Xu, C. 2012. Diversity of rare earth deposits: the key example of China. *Elements* 8, 361–7. doi: 10.2113/gselements.8.5.361.

Kyser, K., Barr, J. and Ihlenfeld, C., 2015. Applied geochemistry in mineral exploration and mining. *Elements* 11(4), 241–6. doi: 10.2113/gselements.11.4.241.

Large, R. R. 1992. Australian volcanic-hosted massive sulfide deposits; features, styles, and genetic models. *Economic Geology* 87: 471–510. doi: 10.2113/gsecongeo.87.3.471.

Laurila, T. E., Hannington, M. D., Petersen, S., Garbe-Schönberg, D. 2014. Early depositional history of metalliferous sediments in the Atlantis II Deep of the Red Sea: Evidence from rare earth element geochemistry. *Geochimica et Cosmochimica Acta* 126: 146–68. doi: 10.1016/j.gca.2013.11.001.

Lawrence, G. B. 2018. Acid deposition. In: Encyclopedia of Geochemistry *(ed White, W. M.)*, pp. 12–15, Springer International Publishing, Cham. doi: 10.1007/978-3-319-39312-4_168.

Leach, D. L., Bradley, D. C., Huston, D., Pisarevsky, S. A., Taylor, R. D. and Gardoll, S. J., 2010. Sediment-hosted lead-zinc deposits in Earth history. *Economic Geology* 105(3), 593–625. doi: 10.2113/gsecongeo.105.3.593.

Lecumberri-Sanchez, P., Bodnar, R. J. 2018. Halogen geochemistry of ore deposits: contributions towards understanding sources and processes. In: Harlov DE, Aranovich L (eds.) *The Role of Halogens in Terrestrial and Extraterrestrial Geochemical Processes: Surface, Crust, and Mantle*, Springer International Publishing, Cham, pp 261–305.

Li, Y. H. M., Zhao, W. W. and Zhou, M.-F. 2017. Nature of parent rocks, mineralization styles and ore genesis of regolith-hosted REE deposits in South China: An integrated genetic model. *Journal of Asian Earth Sciences* 148, 65–95. doi: 10.1016/j.jseaes.2017.08.004.

Liu, W., Borg, S. J., Testemale, D., Etschmann, B., Hazemann, J.-L., Brugger, J. 2011. Speciation and thermodynamic properties for cobalt chloride complexes in hydrothermal fluids at 35–440°C and 600bar: An in-situ XAS study. *Geochimica et Cosmochimica Acta* 75: 1227–48. doi: 10.1016/j.gca.2010.12.002.

Liu, W., McPhail, D. C., Brugger, J. 2001. An experimental study of copper(I)-chloride and copper(I)-acetate complexing in hydrothermal solutions between 50°C and 250°C and vapor-saturated pressure. *Geochimica et Cosmochimica Acta* 65: 2937–48. doi: 10.1016/S0016-7037(01)00631-7.

Liu, Y.-L., Ling, M.-X., Williams, I. S., Yang, X.-Y., Wang, C. Y. and Sun, W. 2018. The formation of the giant Bayan Obo REE-Nb-Fe deposit, North China, Mesoproterozoic carbonatite and overprinted Paleozoic dolomitization. *Ore Geology Reviews* 92, 73–83. doi: 10.1016/j.oregeorev.2017.11.011.

Maier, W. D., Barnes, S.-J., Groves, D. I. 2013. The Bushveld Complex, South Africa: formation of platinum–palladium, chrome- and vanadium-rich layers via hydrodynamic sorting of a mobilized cumulate slurry in a large, relatively slowly cooling, subsiding magma chamber. *Mineralium Deposita* 48: 1–56. doi:10.1007/s00126-012-0436-1.

Masten, S. J., Davies, S. H. and Mcelmurry, S. P. 2016. Flint water crisis: what happened and why? *Journal of the American Water Works Association* 108(12), 22–34. doi: 10.5942/jawwa.2016.108.0195.

McDonough, W. F. and Sun, S. S. 1995. The composition of the Earth. *Chemical Geology* 120, 223–53. doi: 10.1016/0009-2541(94)00140-4.

Migdisov, A. A., Williams-Jones, A. E. and Wagner, T. 2009. An experimental study of the solubility and speciation of the Rare earth elements (III) in fluoride- and chloride-bearing aqueous solutions at temperatures up to 300°C. *Geochimica et Cosmochimica Acta* 73, 7087–109. doi: 10.1016/j.gca.2009.08.023.

Migdisov, A. A., Zezin, D., Williams-Jones, A. E. 2011. An experimental study of cobalt (II) complexation in Cl– and H_2S-bearing hydrothermal solutions. *Geochimica et Cosmochimica Acta* 75: 4065–79. doi: 10.1016/j.gca.2011.05.003.

Mungall, J. E., Brenan, J. M., Godel, B., Barnes, S. J., Gaillard, F. 2015. Transport of metals and sulphur in magmas by flotation of sulphide melt on vapour bubbles. *Nature Geoscience* 8: 216. doi:10.1038/ngeo2373.

Murozumi, M., Chow, T. J. and Patterson, C. 1969. Chemical concentrations of pollutant lead aerosols, terrestrial dusts and sea salts in Greenland and Antarctic snow strata. *Geochimica et Cosmochimica Acta* 33(10), 1247–94. doi: 10.1016/0016-7037(69)90045-3.

Naldrett, A. J., Wilson, A., Kinnaird, J., Yudovskaya, M., Chunnett, G. 2012. The origin of chromitites and related PGE mineralization in the Bushveld Complex: new mineralogical and petrological constraints. *Mineralium Deposita* 47: 209–32. doi: 10.1007/s00126-011-0366-3.

Needleman, H. 2004. Lead poisoning. *Annual Review of Medicine* 55: 209–22. doi:10.1146/annurev.med.55.091902.103653.

Nordstrom, D. K. 2011. Mine Waters: Acidic to Circumneutral. *Elements* 7: 393–8. doi:10.2113/gselements.7.6.393.

Norton, S. A., Kopáček, J. and Fernandez, I. J., 2014. 11.10 – Acid Rain – Acidification and recovery. In: Treatise on Geochemistry *(Second Edition) (eds Holland, H. D. and Turekian, K. K.)*, pp. 379–414, Elsevier, Oxford.

Olías, M., Cánovas, C. R., Nieto, J. M., Sarmiento, A. M. 2006. Evaluation of the dissolved contaminant load transported by the Tinto and Odiel rivers (South West Spain). *Applied Geochemistry* 21: 1733–49. doi 10.1016/j.apgeochem.2006.05.009.

Park, I., Tabelin, C. B., Jeon, S. et al. 2019. A review of recent strategies for acid mine drainage prevention and mine tailings recycling. *Chemosphere* 219: 588–606. doi: 10.1016/j.chemosphere.2018.11.053.

Patterson, C. C. 1965. Contaminated and natural lead environments of man. *Archives of Environmental Health: An International Journal* 11(3), 344–60. doi: 10.1080/00039896.1965.10664229.

Patterson, C. 1956. Age of meteorites and the earth. *Geochimica et Cosmochimica Acta* 16, 230–237.

Petersen, U. 1971. Laterite and bauxite formation. *Economic Geology* 66: 1070–1071. doi:10.2113/gsecongeo.66.7.1070.

Planavsky, N. J., Busigny, V. 2018. Iron isotopes. In: White WM (ed) *Encyclopedia of Geochemistry*, Springer International Publishing, Cham, pp. 756-762 978-3-319-39312-4.

Planavsky, N., Bekker, A., Rouxel, O. J., et al. 2010. Rare earth element and yttrium compositions of Archean and Paleoproterozoic Fe formations revisited: New perspectives on the significance and mechanisms of deposition. *Geochimica et Cosmochimica Acta* 74: 6387–405. doi: 10.1016/j.gca.2010.07.021.

Poletti, J. E., Cottle, J. M., Hagen-Peter, G. A., Lackey, J. S. 2016. Petrochronological constraints on the origin of the Mountain Pass ultrapotassic and carbonatite intrusive suite, California. *Journal of Petrology* 57: 1555–98. doi: 10.1093/petrology/egw050.

Richards, J. P. 2015. The oxidation state, and sulfur and Cu contents of arc magmas: implications for metallogeny. *Lithos* 233: 27–45. doi: 10.1016/j.lithos.2014.12.011.

Richards, J. P. 2018. A Shake-Up in the Porphyry World? *Economic Geology* 113: 1225–33. doi: 10.5382/econgeo.2018.4589.

Rohrlach, B. D. and Loucks, R. R. 2005. Multi-million-year cyclic ramp-up of volatiles in a lower crustal magma reservoir trapped below the Tampakan copper-gold deposit by Mio-Pliocene crustal compression in the southern Philippines. In: Super porphyry copper and gold deposits: A global perspective *(ed Porter, T.)*, pp. 369–407, Adelaide, PGC Publishing.

Romer, R. L., Kroner, U. 2014. Sediment and weathering control on the distribution of Paleozoic magmatic tin–tungsten mineralization. *Mineralium Deposita* 50: 327–338. doi: 10.1007/s00126-014-0540-5

Romer, R. L., Kroner, U. 2016. Phanerozoic tin and tungsten mineralization—Tectonic controls on the distribution of enriched protoliths and heat sources for crustal melting. *Gondwana Research* 31: 60–95. doi: 10.1016/j.gr.2015.11.002.

Saintilan, N. J., Selby, D., Creaser, R. A., Dewaele, S. 2018. Sulphide Re-Os geochronology links orogenesis, salt and Cu-Co ores in the Central African Copperbelt. *Scientific Reports* 8: 14946. doi:10.1038/s41598-018-33399-7.

Scavia, D., David Allan, J., Arend, K. K., et al. 2014. Assessing and addressing the re-eutrophication of Lake Erie: Central basin hypoxia. *Journal of Great Lakes Research* 40: 226–46. doi10.1016/j.jglr.2014.02.004.

Schmidt, C., Bodnar, R. J. 2000. Synthetic fluid inclusions: XVI. PVTX properties in the system H_2O-NaCl-CO_2 at elevated temperatures, pressures, and salinities. *Geochimica et Cosmochimica Acta* 64: 3853–69. doi: 10.1016/S0016-7037(00)00471-3.

Schofield, C. L. and Trojnar, J. R. 1980. Aluminum toxicity to brook trout (*Savelinus fotinalis*) in acidified waters. In: Polluted rain *(eds Toribara, T., Miller, M. and Morrow, P.)*, pp. 341–62, New York, Plenum Press.

Schuster, P. F., Krabbenhoft, D. P., Naftz, D. L., et al. 2002. Atmospheric mercury deposition during the last 270 years: a glacial ice core record of natural and anthropogenic sources. *Environmental Science and Technology* 36: 2303–10. doi:10.1021/es0157503.

Seward, T. M., Williams-Jones, A. E., Migdisov, A. A. 2014. 13.2 – The Chemistry of Metal Transport and Deposition by Ore-Forming Hydrothermal Fluids. In: Holland HD, Turekian KK (eds) *Treatise on Geochemistry* (Second Edition), vol. Elsevier, Oxford, pp 29–57 978-0-08-098300-4.

Seyfried, W. E. J., Ding, K. 1995. Phase equilibria in subseafloor hydrothermal systems: a review of the role of redox, temperature, pH and dissolved Cl on the chemistry of hot spring fluids at mid-ocean ridges. In: Humphris SE, Zierenberg RA, Mullineaux LS, Thomson RE (eds) *Seafloor Hydrothermal Systems: Physical, Chemical, Biological and Geological Interactions*, vol 91. Washington, AGU, pp 248–72.

Shelton, K. L., Gregg, J. M. and Johnson, A. W. 2009. Replacement dolomites and ore sulfides as recorders of multiple fluids and fluid sources in the Southeast Missouri Mississippi Valley-Type District: Halogen-[87]Sr/[86]Sr-δ^{18}O-δ^{34}S Systematics in the Bonneterre Dolomite. *Economic Geology* 104(5), 733–48. doi: 10.2113/gsecongeo.104.5.733.

Sillitoe, R. H. 2010. Porphyry Copper Systems. *Economic Geology* 105: 3–41. doi: 10.2113/gsecongeo.105.1.3.

Sinha, E., Michalak, A. M., Balaji, V. 2017. Eutrophication will increase during the 21st century as a result of precipitation changes. *Science* 357: 405–8. doi: 10.1126/science.aan2409.

Smith, M. P., Campbell, L. S., Kynicky, J. 2015. A review of the genesis of the world class Bayan Obo Fe–REE–Nb deposits, Inner Mongolia, China: Multistage processes and outstanding questions. *Ore Geology Reviews* 64: 459–76. doi 10.1016/j.oregeorev.2014.03.007.

Stoffell, B., Appold, M. S., Wilkinson, J. J., McClean, N. A. and Jeffries, T. E., 2008. Geochemistry and Evolution of Mississippi Valley-Type Mineralizing Brines from the Tri-State and Northern Arkansas Districts Determined by LA-ICP-MS Microanalysis of Fluid Inclusions. *Economic Geology* 103(7), 1411–35. doi: 10.2113/gsecongeo.103.7.1411.

Tole, M. P. 1987. Thermodynamic and kinetic aspects of formation of bauxites. *Chemical Geology* 60: 95–100. doi: 10.1016/0009-2541(87)90114-8.

UN Environment. 2019. *Global Mercury Assessment 2018*. UN Environment Programme, Chemicals and Health Branch, Geneva.

Wagner, T., Okrusch, M., Weyer, S., et al. 2010. The role of the Kupferschiefer in the formation of hydrothermal base metal mineralization in the Spessart ore district, Germany: insight from detailed sulfur isotope studies. *Mineralium Deposita* 45: 217–39. doi: 10.1007/s00126-009-0270-2.

Wilkinson, J. J., 2014. 13.9 – Sediment-hosted zinc–lead mineralization: processes and perspectives. In: *Treatise on Geochemistry (Second Edition)* (eds. Holland, H. D. and Turekian, K. K.), pp. 219–49, Elsevier, Oxford.

Williams-Jones, A. E., Bowell, R. J., Migdisov, A. A. 2009. Gold in solution. *Elements* 5: 281–87.

Wolf, M., Romer, R. L., Franz, L., López-Moro, F. J. 2018. Tin in granitic melts: The role of melting temperature and protolith composition. *Lithos* 310–11: 20–30. doi: 10.1016/j.lithos.2018.04.004.

Yang, X.-Y., Sun, W.-D., Zhang, Y.-X. and Zheng, Y.-F., 2009. Geochemical constraints on the genesis of the Bayan Obo Fe–Nb–REE deposit in Inner Mongolia, China. *Geochimica et Cosmochimica Acta* 73(5), 1417–35. doi: 10.1016/j.gca.2008.12.003.

Zajacz, Z., Candela, P. A., Piccoli, P. M. 2017. The partitioning of Cu, Au and Mo between liquid and vapor at magmatic temperatures and its implications for the genesis of magmatic-hydrothermal ore deposits. *Geochimica et Cosmochimica Acta* 207: 81–101. doi: 10.1016/j.gca.2017.03.015.

Zeh, A., Ovtcharova, M., Wilson, A. H., Schaltegger, U. 2015. The Bushveld Complex was emplaced and cooled in less than one million years – results of zirconology, and geotectonic implications. *Earth and Planetary Science Letters* 418: 103–14. doi: 10.1016/j.epsl.2015.02.035.

Zierenberg, R. A., Fouquet, Y., Miller, D. J., et al. 1998. The deep structure of a sea-floor hydrothermal deposit. *Nature* 392: 485. doi: 10.1038/33126.

PROBLEMS

1. Using the stability constants, β_n, for chlorotin complexes from Müller and Seward (2001) listed in the adjacent table, calculate the *activity ratio* of $SnCl^-$, $SnCl_2$, $SnCl_3^-$, and $SnCl_4^{2-}$ to Sn^{2+} for chloride *activities* of 0.01, 0.1, 0.5, and 1 at 100°C and 250°C. *Note:* β_1 refers to the reaction $Sn^{2+} + Cl^- \rightleftharpoons SnCl^+$, β_2 to the reaction $Sn^{2+} + 2Cl^- \rightleftharpoons SnCl_2$, etc. as explained in Section 6.3.1.

Chlorotin Stability Constants		
	100°C	**250°C**
β_1	1.43	1.65
β_2	2.25	3.06
β_3	2.39	3.96
β_4	1.95	0

2. With the results from problem 1, calculate the percentage of the total tin activity of each species for each case. *Hint:* See Section 6.3.

3. The solubility of chalcolite, Cu_2S, can be expressed as:

$$\tfrac{1}{2}Cu_2S + \tfrac{3}{2}HS^- + \tfrac{1}{2}H^+ \rightleftharpoons Cu(HS)_2^-$$

At 300°C, the equilibrium constant for this reaction is $10^{2.9}$. Considering two other speciation reactions:

$$Cu(HS)_2^- \rightleftharpoons Cu(HS)^0 + HS^- \quad \log K = -3.5$$

and

$$Cu(HS)_2^- \rightleftharpoons Cu^+ + 2HS^- \quad \log K = -12.1$$

Calculate the relative activities of Cu^+, $Cu(HS)^0$, and $Cu(HS)_2^-$ for a solution with a HS^- activity of 0.005 m and a pH of 6, 7, and 8.

4. Using the equilibrium constants in eqns. 6.79 through 6.83, calculate the concentration of total dissolved aluminum in equilibrium with gibbsite at pH 6, 5, and 4.

5. Calculate the acid-neutralizing capacity, which is equivalent to alkalinity (in µeq/L), of the Big Moose Lake water samples shown in the adjacent table using eqn. 6.37.

Big Moose lake water chemistry

	M.W.	Feb, '17	Feb, '93
SO_4^{2-}	96.1	2.807	6.520
NO_3^-	62	0.780	1.787
Cl^-	35.5	0.353	0.300
F^-	19	0.054	0.084
Ca^{2+}	40.08	1.280	2.020
Mg^{2+}	24.3	0.224	0.320
Na^+	23	0.675	0.590
K^+	39.1	0.193	0.320

Concentrations are in mg/L.

Appendix

Constants, units and conversions

PHYSICAL AND CHEMICAL CONSTANTS

Speed of Light (c)	2.998×10^8 m/s
Planck's constant (h)	6.626×10^{-34} J/Hz
Boltzmann's constant (k)	1.380×10^{-23} J/K
Gravitational constant (G)	6.672×10^{-11} N-m^2/kg
Avogadro number (N_A)	6.022×10^{23} mol^{-1}
Gas constant (R)	8.314 J/mol-K
	(1.987 cal/mol-K)
Electron charge (e)	1.602×10^{-19} coulombs (C)
Faraday constant (F)	96.485 C/mole = kJ/V-eq.
Permittivity in vacuum (ε)	8.85×10^{-12} C^2/J-m
Dielectric constant of water	78.54

THE EARTH

Mass of the Earth ($M_\oplus$)	5.97×10^{24} kg
Mantle	4.0×10^{24} kg
Core	1.94×10^{24} kg
Continental crust	2.2×10^{22} kg
Oceans	1.4×10^{21} kg
Atmosphere	5.1×10^{18} kg
Mean radius	6.37×10^6 m
Radius of core	3.47×10^6 m
Radius of orbit	1.49×10^{11} m
The Sun	
Mass ($M_\odot$)	1.99×10^{30} kg
Radius	6.96×10^8 m

Geochemistry, Second Edition. William M. White.
© 2020 John Wiley & Sons Ltd. Published 2020 by John Wiley & Sons Ltd.
Companion website: www.wiley.com/go/white/geochemistry

SI UNITS AND CONVERSIONS

In the SI system, fundamental units are the *kilogram, meter,* and *second*. Consequently, preferred units of volume, pressure, and energy are m^3, pascals, and joules, respectively.

Mass — Kilogram (kg)
- Pound
  - 1 lb = 0.4535 kg
  - (1 kg = 2.205 lb)
- u (unified atomic mass unit)/Dalton
  - 1 u ≡ 1/12 mass of ^{12}C atom
  - 1 u = 1.66×10^{-27} kg
  - 1 u = 931.49 MeV/c^2

Distance — Meter (m)
- Inch
  - 1 in = 0.0254 m
- Ångstrom
  - 1 Å ≡ 10^{-10} m
- Mile (US)
  - 1 mi = 1609 m
- Astronomical unit (AU)
  - 1 AU ≡ 1.49×10^{11} m
- Parsec
  - 1 parsec = 3.084×10^{16} m
  - = 2.07×10^5 AU
  - = 3.26 light-years
- Light-year
  - 1 ly = 6.35×10^4 AU = 9.46×10^{15} m

Force — Newton (N)
- 1 N ≡ 1 kg-m/s^2
- 1 dyne = 10^5 N
- 1 dyne ≡ 1 gm-cm/s^2

Energy — Joule (J)
- 1 J ≡ 1 kg-m^2/s^2
- Erg
  - 1 erg = 10^{-7} J
  - 1 erg = 1 gm-cm^2/sec^2
- Calorie
  - 1 calorie = 4.184 J
- Liter-atmosphere
  - 1 l-atm = 101.29 J
- Liter-Pascal
  - 1 l-Pa = 99.98×10^{-5} J
- Electron volt
  - 1 eV = 1.602×10^{-19} J
- u/Dalton
  - 1 u = 9.315×10^2 MeV
- Volt
  - 1 Volt-coulomb = 1 J
- Kilowatt-hour
  - 1 kWh = 3.6×10^6 J

Pressure — Pascal (Pa)
- Pascal
  - 1 Pa ≡ 1 N/m^2 = 1 kg/m-s^2
- Bar
  - 1 bar = 10^5 Pa (= 0.1 MPa)
- Atmosphere
  - 1 atm = 1.013×10^5 Pa

Volume — Liter (l)
- 1 l ≡ 10^3 cm
- 1 l = 10^{-6} m^3
- US gallon
  - 1 gal = 3.785 l
- Imperial gallon
  - 1 gal = 4.549 l

Concentration
- Molarity — moles/l (M)
- Molality — moles/kg (m)
  - 1 µM (micromole) = 10^{-6} M
  - 1 nM (nanomole) = 10^{-9} M
  - 1 pM (picomole) = 10^{-12} M
  - 1 fM (femtomole) = 10^{-15} M

Radioactivity

Curie (Ci)

Time and Frequency

Area

Acre

Becquerel (Bq)
  1 Bq ≡ 1 decay per second
3.7 × 10¹⁰ Bq
(1 curie is the activity of 1 g of ^{226}Ra)
1 year = 31,557,600 s
1 day = 86,400 s
1 Hertz = 1 cycle/s

Hectare (ha)
  1 ha ≡ 10⁴ m²
  1 ha = 10⁻² km²
  1 acre = 0.4047 ha

SI PREFIXES

peta, P	10^{15}
tera, T	10^{12}
giga, G	10^{9}
mega, M	10^{6}
kilo, k	10^{3}
hector, h	10^{2}
deca, da	10
deci, d	10^{-1}
centi, c	10^{-2}
milli, m	10^{-3}
micro, μ	10^{-6}
nano, n	10^{-9}
pico, p	10^{-12}
femto, f	10^{-15}

Index

Note: Page numbers in *italic* refer to figures; those in **bold** to tables; those followed by 'n' relae to footnotes.

Geochemistry, Second Edition. William M. White.
© 2020 John Wiley & Sons Ltd. Published 2020 by John Wiley & Sons Ltd.
Companion website: www.wiley.com/go/white/geochemistry

LIST OF EXAMPLES